Introduction to Data Communications and Networking

Wayne Tomasi

DeVry University
Phoenix, Arizona

PEARSON
Prentice
Hall

Upper Saddle River, New Jersey
Columbus, Ohio

Editor in Chief: Stephen Helba
Assistant Vice President and Publisher: Charles Stewart
Assistant Editor: Mayda Bosco
Production Editor: Alexandrina Benedicto Wolf
Design Coordinator: Diane Ernsberger
Cover Designer: Thomas Borah
Cover Art: Digital Vision
Production Manager: Matt Ottenweller
Marketing Manager: Ben Leonard

This book was set in Times Roman by Carlisle Communications, Ltd. It was printed and bound by Courier Kendallville, Inc. The cover was printed by The Lehigh Press, Inc.

Pearson Education Ltd.
Pearson Education Singapore Pte. Ltd.
Pearson Education Canada, Ltd.
Pearson Education—Japan

Pearson Education Australia PTY. Limited
Pearson Education North Asia Ltd.
Pearson Educación de Mexico, S.A. de C. V.
Pearson Education Malaysia Pte. Ltd.

10 9 8 7 6 5 4 3 2
ISBN: 0–13–013828–2

To my mother, Mrs. Elizabeth Tomasi Ward, whose direction and gentle suggestions guided me through the early years of my life and helped make me the person I am today.

Preface

This book introduces the fundamental concepts of electronic communications systems, data communications, and networks. Topics covered include wireless and wireline telecommunications systems, basic data communications networks and systems, local area networks, internetworks, and the Internet, including the TCP/IP protocol suite. The text describes how the networks themselves work; it does not describe how the software applications that utilize the network work.

Readers with previous knowledge in basic electronics, including fundamental digital concepts and binary number systems, will have little trouble understanding the topics presented. Each chapter contains numerous examples that emphasize the most important concepts presented. When appropriate, questions and problems are included at the end of each chapter, and answers to selected problems are provided at the end of the book. Chapters and topics within chapters do not need to be covered in the same sequence as they are presented in the book.

ORGANIZATION OF THE TEXT

Chapter 1 gives a brief history of data communications and introduces the fundamental concepts of data communications and networking. Topics covered include definitions of network architectures, topologies, classifications, and data communications protocols and standards. Several primary standards organizations are described along with a brief description of their purpose, including a description of the International Organization for Standardization Open System Interconnect protocol hierarchy. The fundamental components of a data communications circuit, including explanations of data communications circuit arrangements, are also described. This chapter also introduces fundamental network components (e.g., clients, servers, local and network operating systems, network models) and network classifications such as local and metropolitan area networks.

Chapter 2 gives a brief explanation of signal analysis for single-frequency sinusoids and complex repetitive waveforms. Electrical noise and interference are described along with the concept and significance of signal-to-noise ratio. This chapter also gives a brief explanation of conventional analog modulation systems, such as AM and FM. More common digital modulation techniques, such as frequency-shift keying and phase-shift keying, and how digital modulation relates to information capacity, bit rate, and baud are also discussed.

Chapter 3 presents the fundamental concepts of metallic transmission media, including the principles and characteristics of transverse electromagnetic wave propagation. The characteristics of the most common types of metallic transmission media are described, and a detailed explanation is given for the classifications used with standard shielded and unshielded twisted-pair cable. This chapter also includes descriptions of the losses typical to metallic cables.

Chapter 4 compares the advantages and disadvantages between metallic transmission media and optical fiber transmission media. The fundamental concepts of wave propagation over optical fiber cables are given, including explanations for velocity of propagation, refraction, reflection, and modes of propagation commonly used for propagating signals over optical fibers. This chapter also describes the losses typical to optical fiber cables, and optical sources and detectors.

Chapter 5 introduces the advantages and disadvantages of digital (pulse) transmission. The fundamental principles of pulse code modulation, including the concepts of quantization and digital companding, and the steps necessary to produce a basic DS–0 digital signal, are explained.

Chapter 6 describes how signals from multiple sources (both analog and digital) can be multiplexed to improve the capacity of a transmission medium. A detailed explanation is given on how DS–0 signals are combined into T-carrier systems capable of carrying information from thousands of sources simultaneously. This chapter also covers the basic principles of frequency- and wavelength-division multiplexing.

Chapter 7 describes the fundamental concepts of wireless communications systems, including free-space electromagnetic wave propagation. It also describes the basic principles of microwave and satellite radio communications systems.

Chapter 8 introduces the basic concepts of telephone instruments and signals. The operation of a basic telephone set is described along with descriptions of the call progress tones and signals associated with completing a basic telephone call.

Chapter 9 gives a detailed description of a typical telephone circuit, including local loops and trunk circuits. This chapter also describes the transmission parameters for private-line voice and data circuits, including descriptions of typical telephone circuit impairments.

Chapter 10 gives a detailed description of the public telephone network, including descriptions of central office switches and how they are interconnected in tandem to complete local and long-distance telephone calls. The pre- and post-divesture North American Telephone Switching Hierarchies as well as the Signaling System Number 7 (SS7) switching system are also discussed.

Chapter 11 describes the evolution and the fundamental concepts of cellular telephone systems, including descriptions of the basic concepts of cell splitting, frequency reuse, interference, roaming, and cellular telephone call processing. Chapter 12 expands the basic cellular concepts presented in Chapter 11, including descriptions of first- and second-generation personal communications systems (PCS). The basic concepts of analog and digital cellular telephone systems, including descriptions of AMPS, DAMPS, NAMPS EIA IS–54, IS–136, IS–95, and TDMA and CDMA multiple-accessing technologies are also described.

Chapter 13 covers the fundamental concepts of data communications codes, asynchronous and synchronous data formats, and error control. Several means of providing error detection and correction, such as VRC, LCR, CRC, ARQ, and the Hamming code are described.

Chapter 14 introduces fundamental data communications hardware components, including descriptions of data terminal equipment and data communications equipment. The fundamental concepts of UARTs and USRTs are covered along with an introduction of serial interfaces and a detailed description of the RS–232 interface.

Chapter 15 expands the coverage of data terminal and data communications equipment. Data communications modems are explained with detailed descriptions for Bell System–compatible modem standards and current ITU-T modem recommendations. The AT command set is also described.

Chapter 16 gives a detailed description of the data-link protocols used on private data communications networks. Character- and bit-oriented protocols are defined and described along with their applications. Protocols described in detail include XMODEM, YMODEM, KERMIT, Bisync, SDLC, and HDLC.

Chapter 17 gives a detailed description of networking and internetworking fundamentals, including transmission formats, LAN topologies, and collision and broadcast domains. Detailed explanations of the common connectivity devices, such as repeaters, hubs, bridges, switches, routers, and gateways, how they affect collision and broadcast domains, and how they are interconnected to create networks and subnetworks are given.

Chapter 18 gives a detailed description of local area networks, including a comprehensive coverage of the IEEE 802 project. Topics include the various access control methodologies, such as multiple access, carrier sense multiple access, carrier sense multiple access with collision detect, and carrier sense multiple access with collision avoidance. This chapter also includes detailed descriptions of the frame formats and operation of IEEE 802.3 Ethernet and Ethernet II, as well as current Ethernet technologies, such as switched Ethernet, 100 Mbps fast Ethernet, and 1000 Mbps Ethernet.

Chapter 19 introduces the TCP/IP protocol suite and Internet Protocol (IP) addressing. It includes detailed descriptions of the classes of IP addresses available on the global Internet and explains the differences between global, reserved, private, and broadcast addresses.

Chapter 20 gives a detailed description of how classful and classless networks, subnetworks, and supernetworks are formed, including a comprehensive coverage of subnet masks and how they are used to separate a network into subnetworks. Numerous examples are given to illustrate how subnet masks are used in modern IP-based networks. This chapter also describes supernetting and classless IP addressing.

Chapter 21 covers layer 3 networking protocols and gives detailed explanations of address resolution and the address resolution protocol (ARP). It also describes the format for an IP datagram, the functions of each field within the header, and IP options.

Chapter 22 gives a detailed description of Internet control management protocol (ICMP) and shows how ICMP messages are encapsulated within an IP datagram and used to carry diagnostic and error-reporting messages. This chapter emphasizes how ICMP is used to improve the performance of the Internet Protocol.

Chapter 23 introduces port numbers, sockets addresses, and transport-layer protocols. A detailed description is given for the user datagram protocol (UDP) and transmission control protocol (TCP), including the functions of the fields that make up each of their headers. This chapter gives a comparison between connectionless, unreliable and connection-oriented, reliable protocols.

Chapter 24 introduces IP version 6 (IPv6) and compares IPv6 to IPv4. Descriptions are given for hexadecimal colon notation, addressing types, and address allocation. IPv6 header fields, extensions, and extensions headers are described. ICMP version 6 is also described and compared to the current version, ICMPv4.

Chapter 25 describes configuration and domain name protocols, including dynamic host configuration protocol and domain name system. Detailed descriptions are given for the format for domain names and the process of domain resolution.

Chapter 26 introduces several application-layer protocols and the processes they perform. Protocols covered include Telnet, file transfer protocol, trivial file transfer protocol, simple mail transfer protocol, post office protocol, and hypertext transfer protocol.

Chapter 27 discusses integrated services data networks and broadband access technologies. Detailed descriptions are given for public-switched data networks, ISDN, and DSL. Value-added networks, switching networks, ATM, and the CCITT X.1 and X.25 protocols are also described.

ACKNOWLEDGMENTS

I would like to thank the following individuals for reviewing this book and providing constructive criticism and valuable feedback: Costas Vassiliadis, Ohio University; K. R. Kirkendall, Montana State University; Phillip Davis, Delmar College; William Hessmiller, Editors and Training Associates; Jeff Rankinen, Pennsylvania College of Technology; Robert Borns, Purdue University; Joseph Bumblis, Purdue University; and Robert Akl, University of Northern Texas. I would also like to thank my project editor, Ms. Kelli Jauron, for taking time from her family and very busy life to help produce this book. I would also like to thank Kelli for being my friend for the past seven years. I also thank Ms. Holly Henjum for taking over this project halfway through production and completing the job seamlessly. I would like to thank my assistant editor, Mayda Bosco, for all the work she put into this project. Without the contributions from these people, this book would not have been possible.

Wayne Tomasi

Brief Contents

Contents

C H A P T E R 1

Introduction to Data Communications and Networking

CHAPTER OUTLINE

OBJECTIVES

- Define the following terms: *data, data communications, data communications circuit,* and *data communications network*
- Give a brief description of the evolution of data communications
- Define *data communications network architecture*
- Describe data communications protocols
- Describe the basic concepts of connection-oriented and connectionless protocols
- Describe syntax and semantics and how they relate to data communications
- Define *data communications standards* and explain why they are necessary
- Describe the following standards organizations: ISO, ITU-T, IEEE, ANSI, EIA, TIA, IAB, ETF, and IRTF
- Define *open systems interconnection*
- Name and explain the functions of each of the layers of the seven-layer OSI model
- Define *station* and *node*
- Describe the fundamental block diagram of a two-station data communications circuit and explain how the following terms relate to it: *source, transmitter, transmission medium, receiver,* and *destination*
- Describe serial and parallel data transmission and explain the advantages and disadvantages of both types of transmissions

- Define *data communications circuit arrangements*
- Describe the following transmission modes: simplex, half duplex, full duplex, and full/full duplex
- Define *data communications network*
- Describe the following network components, functions, and features: servers, clients, transmission media, shared data, shared printers, and network interface card
- Define *local operating system*
- Define *network operating system*
- Describe peer-to-peer client/server and dedicated client/server networks
- Define *network topology* and describe the following: star, bus, ring, mesh, and hybrid
- Describe the following classifications of networks: LAN, MAN, WAN, GAN, building backbone, campus backbone, and enterprise network
- Briefly describe the TCP/IP hierarchical model
- Briefly describe the Cisco three-layer hierarchical model

1-1 INTRODUCTION

Since the early 1970s, technological advances around the world have occurred at a phenomenal rate, transforming the *telecommunications industry* into a highly sophisticated and extremely dynamic field. Where previously telecommunications systems had only voice to accommodate, the advent of very large-scale integration chips and the accompanying low-cost microprocessors, computers, and peripheral equipment has dramatically increased the need for the exchange of digital information. This, of course, necessitated the development and implementation of higher-capacity and much faster means of communicating.

In the data communications world, *data* generally are defined as information that is stored in digital form. The word *data* is plural; a single unit of data is a *datum*. Data communications is the process of transferring digital information (usually in binary form) between two or more points. *Information* is defined as knowledge or intelligence. Information that has been processed, organized, and stored is called data.

The fundamental purpose of a *data communications circuit* is to transfer digital information from one place to another. Thus, *data communications* can be summarized as the transmission, reception, and processing of digital information. The original source information can be in analog form, such as the human voice or music, or in digital form, such as binary-coded numbers or alphanumeric codes. If the source information is in analog form, it must be converted to digital form at the source and then converted back to analog form at the destination.

A *network* is a set of *devices* (sometimes called *nodes* or *stations*) interconnected by media links. *Data communications networks* are systems of interrelated computers and computer equipment and can be as simple as a personal computer connected to a printer or two personal computers connected together through the *public telephone network*. On the other hand, a data communications network can be a complex communications system comprised of one or more mainframe computers and hundreds, thousands, or even millions of remote terminals, personal computers, and workstations. In essence, there is virtually no limit to the capacity or size of a data communications network.

Years ago, a single computer serviced virtually every computing need. Today, the single-computer concept has been replaced by the networking concept, where a large number of separate but interconnected computers share their resources. Data communications networks and systems of networks are used to interconnect virtually all kinds of digital computing equipment, from *automatic teller machines* (ATMs) to bank computers; personal computers to information highways, such as the *Internet;* and workstations to mainframe computers. Data communications networks can also be used for airline and hotel reservation systems, mass media and news networks, and electronic mail delivery systems. The list of applications for data communications networks is virtually endless.

1-2 HISTORY OF DATA COMMUNICATIONS

It is highly likely that data communications began long before recorded time in the form of smoke signals or tom-tom drums, although they surely did not involve electricity or an electronic apparatus, and it is highly unlikely that they were binary coded. One of the earliest means of communicating electrically coded information occurred in 1753, when a proposal submitted to a Scottish magazine suggested running a communications line between villages comprised of 26 parallel wires, each wire for one letter of the alphabet. A Swiss inventor constructed a prototype of the 26-wire system, but current wire-making technology proved the idea impractical.

In 1833, Carl Friedrich Gauss developed an unusual system based on a five-by-five matrix representing 25 letters (I and J were combined). The idea was to send messages over a single wire by deflecting a needle to the right or left between one and five times. The initial set of deflections indicated a row, and the second set indicated a column. Consequently, it could take as many as 10 deflections to convey a single character through the system.

If we limit the scope of data communications to methods that use *binary-coded* electrical signals to transmit information, then the first successful (and practical) data communications system was invented by Samuel F. B. Morse in 1832 and called the *telegraph.* Morse also developed the first practical *data communications code,* which he called the *Morse code.* With telegraph, dots and dashes (analogous to logic 1s and 0s) are transmitted across a wire using electromechanical induction. Various combinations of dots, dashes, and pauses represented binary codes for letters, numbers, and punctuation marks. Because all codes did not contain the same number of dots and dashes, Morse's system combined human intelligence with electronics, as decoding was dependent on the hearing and reasoning ability of the person receiving the message. (Sir Charles Wheatstone and Sir William Cooke allegedly invented the first telegraph in England, but their contraption required six different wires for a single telegraph line.)

In 1840, Morse secured an American patent for the telegraph, and in 1844 the first telegraph line was established between Baltimore and Washington, D.C., with the first message conveyed over this system being "What hath God wrought!" In 1849, the first slow-speed telegraph printer was invented, but it was not until 1860 that high-speed (15-bps) printers were available. In 1850, Western Union Telegraph Company was formed in Rochester, New York, for the purpose of carrying coded messages from one person to another.

In 1874, Emile Baudot invented a telegraph *multiplexer,* which allowed signals from up to six different telegraph machines to be transmitted simultaneously over a single wire. The *telephone* was invented in 1875 by Alexander Graham Bell, and, unfortunately, very little new evolved in telegraph until 1899, when Guglielmo Marconi succeeded in sending radio (wireless) telegraph messages. Telegraph was the only means of sending information across large spans of water until 1920, when the first commercial radio stations carrying voice information were installed.

It is unclear exactly when the first electrical computer was developed. Konrad Zuis, a German engineer, demonstrated a computing machine sometime in the late 1930s; however, at the time, Hitler was preoccupied trying to conquer the rest of the world, so the project fizzled out. Bell Telephone Laboratories is given credit for developing the first special-purpose computer in 1940 using electromechanical relays for performing logical operations. However, J. Presper Eckert and John Mauchley at the University of Pennsylvania are given credit by some for beginning modern-day computing when they developed the ENIAC computer on February 14, 1946.

In 1949, the U.S. National Bureau of Standards developed the first all-electronic diode-based computer capable of executing stored programs. The U.S. Census Bureau installed the machine, which is considered the first commercially produced American computer. In the 1950s, computers used punch cards for inputting information, printers for outputting information, and magnetic tape reels for permanently storing information. These early computers could process only one job at a time using a technique called *batch processing*.

The first general-purpose computer was an automatic sequence-controlled calculator developed jointly by Harvard University and International Business Machines (IBM) Corporation. The UNIVAC computer, built in 1951 by Remington Rand Corporation, was the first mass-produced electronic computer.

In the 1960s, batch-processing systems were replaced by on-line processing systems with terminals connected directly to the computer through serial or parallel communications lines. The 1970s introduced microprocessor-controlled microcomputers, and by the 1980s personal computers became an essential item in the home and workplace. Since then, the number of mainframe computers, small business computers, personal computers, and computer terminals has increased exponentially, creating a situation where more and more people have the need (or at least think they have the need) to exchange digital information with each other. Consequently, the need for data communications circuits, networks, and systems has also increased exponentially.

Soon after the invention of the telephone, the American Telephone and Telegraph Company (AT&T) emerged, providing both long-distance and local telephone service and data communications service throughout the United States. The vast AT&T system was referred to by some as the "Bell System" and by others as "Ma Bell." During this time, Western Union Corporation provided telegraph service. Until 1968, the AT&T operating tariff allowed only equipment furnished by AT&T to be connected to AT&T lines. In 1968, a landmark Supreme Court decision, the *Carterfone* decision, allowed non-Bell companies to interconnect to the vast AT&T communications network. This decision started the interconnect industry, which has led to competitive data communications offerings by a large number of independent companies. In 1983, as a direct result of an antitrust suit filed by the federal government, AT&T agreed in a court settlement to divest itself of operating companies that provide basic local telephone service to the various geographic regions of the United States. Since the divestiture, the complexity of the public telephone system in the United States has grown even more involved and complicated.

Recent developments in data communications networking, such as the *Internet, intranets,* and the *World Wide Web* (WWW), have created a virtual explosion in the data communications industry. A seemingly infinite number of people, from homemaker to chief executive officer, now feel a need to communicate over a finite number of facilities. Thus, the demand for higher-capacity and higher-speed data communications systems is increasing daily with no end in sight.

The *Internet* is a public data communications network used by millions of people all over the world to exchange business and personal information. The Internet began to evolve in 1969 at the *Advanced Research Projects Agency* (ARPA). ARPANET was formed in the late 1970s to connect sites around the United States. From the mid-1980s to April 30, 1995, the *National Science Foundation* (NSF) funded a high-speed backbone called NSFNET.

Intranets are private data communications networks used by many companies to exchange information among employees and resources. Intranets normally are used for security reasons or to satisfy specific connectivity requirements. Company intranets are generally connected to the public Internet through a *firewall,* which converts the intranet addressing system to the public Internet addressing system and provides security functionality by filtering incoming and outgoing traffic based on addressing and protocols.

The *World Wide Web* (WWW) is a server-based application that allows subscribers to access the services offered by the Web. Browsers, such as Netscape Communicator and Microsoft Internet Explorer, are commonly used for accessing data over the WWW.

1-3 DATA COMMUNICATIONS NETWORK ARCHITECTURE, PROTOCOLS, AND STANDARDS

1-3-1 Data Communications Network Architecture

A *data communications network* is any system of computers, computer terminals, or computer peripheral equipment used to transmit and/or receive information between two or more locations. *Network architectures* outline the products and services necessary for the individual components within a data communications network to operate together.

In essence, network architecture is a set of equipment, transmission media, and procedures that ensures that a specific sequence of events occurs in a network in the proper order to produce the intended results. Network architecture must include sufficient information to allow a program or a piece of hardware to perform its intended function. The primary goal of network architecture is to give the users of the network the tools necessary for setting up the network and performing data flow control. A network architecture outlines the way in which a data communications network is arranged or structured and generally includes the concept of *levels* or *layers* of functional responsibility within the architecture. The *functional responsibilities* include electrical specifications, hardware arrangements, and software procedures.

Networks and network protocols fall into three general classifications: *current, legacy,* and *legendary.* Current networks include the most modern and sophisticated networks and protocols available. If a network or protocol becomes a legacy, no one really wants to use it, but for some reason it just will not go away. When an antiquated network or protocol finally disappears, it becomes legendary.

In general terms, computer networks can be classified in two different ways: *broadcast* and *point to point.* With broadcast networks, all stations and devices on the network share a single communications channel. Data are propagated through the network in relatively short messages sometimes called *frames, blocks,* or *packets.* Many or all subscribers of the network receive transmitted messages, and each message contains an address that identifies specifically which subscriber (or subscribers) is intended to receive the message. When messages are intended for all subscribers on the network, it is called *broadcasting,* and when messages are intended for a specific group of subscribers, it is called *multicasting.*

Point-to-point networks have only two stations. Therefore, no addresses are needed. All transmissions from one station are intended for and received by the other station. With point-to-point networks, data are often transmitted in long, continuous messages, sometimes requiring several hours to send.

In more specific terms, point-to-point and broadcast networks can be subdivided into many categories in which one type of network is often included as a subnetwork of another.

1-3-2 Data Communications Protocols

Computer networks communicate using *protocols,* which define the procedures that the systems involved in the communications process will use. Numerous protocols are used today to provide networking capabilities, such as how much data can be sent, how it will be sent, how it will be addressed, and what procedure will be used to ensure that there are no undetected errors.

Protocols are arrangements between people or processes. In essence, a protocol is a set of customs, rules, or regulations dealing with formality or precedence, such as diplomatic or military protocol. Each functional layer of a network is responsible for providing a specific service to the data being transported through the network by providing a set of rules, called protocols, that perform a specific function (or functions) within the network. *Data communications protocols* are sets of rules governing the orderly exchange of data within the network or a portion of the network, whereas network architecture is a set of layers and protocols that govern the operation of the network. The list of protocols used by a system is called a *protocol stack,* which generally includes only one protocol per layer. *Layered network architectures* consist of two or more independent levels. Each level has a specific set of responsibilities and functions, including data transfer, flow control, data segmentation and reassembly, sequence control, error detection and correction, and notification.

1-3-2-1 Connection-oriented and connectionless protocols. Protocols can be generally classified as either *connection oriented* or *connectionless.* With a connection-oriented protocol, a logical connection is established between the endpoints (e.g., a *virtual circuit*) prior to the transmission of data. Connection-oriented protocols operate in a manner similar to making a standard telephone call where there is a sequence of actions and acknowledgments, such as setting up the call, establishing the connection, and then disconnecting. The actions and acknowledgments include dial tone, Touch-Tone signaling, ringing and ring-back signals, and busy signals.

Connection-oriented protocols are designed to provide a high degree of reliability for data moving through the network. This is accomplished by using a rigid set of procedures for establishing the connection, transferring the data, acknowledging the data, and then clearing the connection. In a connection-oriented system, each packet of data is assigned a unique sequence number and an associated acknowledgment number to track the data as they travel through a network. If data are lost or damaged, the destination station requests that they be resent. A connection-oriented protocol is depicted in Figure 1-1a. Characteristics of connection-oriented protocols include the following:

1. A connection process called a *handshake* occurs between two stations before any data are actually transmitted. Connections are sometimes referred to as *sessions, virtual circuits,* or *logical connections.*
2. Most connection-oriented protocols require some means of acknowledging the data as they are being transmitted. Protocols that use acknowledgment procedures provide a high level of network reliability.
3. Connection-oriented protocols often provide some means of error control (i.e., error detection and error correction). Whenever data are found to be in error, the receiving station requests a retransmission.
4. When a connection is no longer needed, a specific handshake drops the connection.

Connectionless protocols are protocols where data are exchanged in an unplanned fashion without prior coordination between endpoints (e.g., a datagram). Connectionless protocols do not provide the same high degree of reliability as connection-oriented protocols; however, connectionless protocols offer a significant advantage in transmission speed. Connectionless protocols operate in a manner similar to the U.S. Postal Service, where information is formatted, placed in an envelope with source and destination addresses, and then mailed. You can only hope the letter arrives at its destination. A connectionless protocol is depicted in Figure 1-1b. Characteristics of connectionless protocols are as follow:

1. Connectionless protocols send data with a source and destination address without a handshake to ensure that the destination is ready to receive the data.
2. Connectionless protocols usually do not support error control or acknowledgment procedures, making them a relatively unreliable method of data transmission.
3. Connectionless protocols are used because they are often more efficient, as the data being transmitted usually do not justify the extra overhead required by connection-oriented protocols.

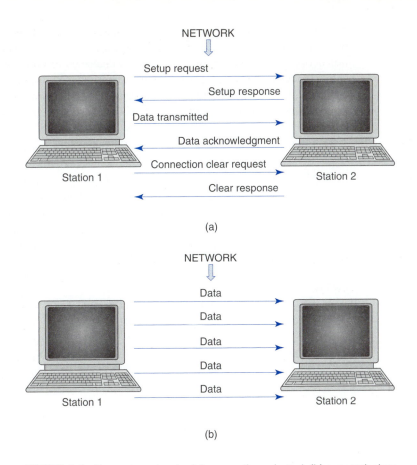

FIGURE 1-1 Network protocols: (a) connection oriented; (b) connectionless

1-3-2-2 Syntax and semantics. Protocols include the concepts of *syntax* and *semantics*. Syntax refers to the structure or format of the data within the message, which includes the sequence in which the data are sent. For example, the first byte of a message might be the address of the source and the second byte the address of the destination. Semantics refers to the meaning of each section of data. For example, does a destination address identify only the location of the final destination, or does it also identify the route the data take between the sending and receiving locations?

1-3-3 Data Communications Standards

During the past several decades, the data communications industry has grown at an astronomical rate. Consequently, the need to provide communications between dissimilar computer equipment and systems has also increased. A major issue facing the data communications industry today is worldwide compatibility. Major areas of interest are software and programming language, electrical and cable interface, transmission media, communications signal, and format compatibility. Thus, to ensure an orderly transfer of information, it has been necessary to establish standard means of governing the physical, electrical, and procedural arrangements of a data communications system.

A standard is an object or procedure considered by an authority or by general consent as a basis of comparison. Standards are authoritative principles or rules that imply a model or pattern for guidance by comparison. *Data communications standards* are guidelines that have been generally accepted by the data communications industry. The guidelines outline procedures and equipment configurations that help ensure an orderly transfer of information between two or more pieces of data communications equipment or two or more data communications networks. Data communications standards are not laws; rather, they are simply suggested ways of implementing procedures and accomplishing results. If everyone complies with the standards, everyone's equipment, procedures, and processes will be compatible with everyone else's, and there will be little difficulty communicating information through the system. Today, most companies make their products to comply with standards.

There are two basic types of standards: *proprietary* (closed) system and *open* system. Proprietary standards are generally manufactured and controlled by one company. Other companies are not allowed to manufacture equipment or write software using this standard. An example of a proprietary standard is Apple Macintosh computers. Advantages of proprietary standards are tighter control, easier consensus, and a monopoly. Disadvantages include lack of choice for the customers, higher financial investment, overpricing, and reduced customer protection against the manufacturer going out of business.

With open system standards, any company can produce compatible equipment or software; however, often a royalty must be paid to the original company. An example of an open system standard is IBM's personal computer. Advantages of open system standards are customer choice, compatibility between vendors, and competition by smaller companies. Disadvantages include less product control and increased difficulty acquiring agreement between vendors for changes or updates. In addition, standard items are not always as compatible as we would like them to be.

1-4 STANDARDS ORGANIZATIONS FOR DATA COMMUNICATIONS

A consortium of organizations, governments, manufacturers, and users meet on a regular basis to ensure an orderly flow of information within data communications networks and systems by establishing guidelines and standards. The intent is that all data communications equipment manufacturers and users comply with these standards. Standards organizations generate, control, and administer standards. Often, competing companies will form a joint committee to create a compromised standard that is acceptable to everyone. The most prominent organizations relied on in North America to publish standards and make recommendations for the data, telecommunications, and networking industries are shown in Figure 1-2.

1-4-1 International Standards Organization (ISO)

Created in 1946, the *International Standards Organization* (ISO) is the international organization for standardization on a wide range of subjects. The ISO is a voluntary, nontreaty organization whose membership is comprised mainly of members from the standards committees of various governments throughout the world. The ISO creates the sets of rules and standards for graphics and document exchange and provides models for equipment and system compatibility, quality enhancement, improved productivity, and reduced costs. The ISO is responsible for endorsing and coordinating the work of the other standards organizations. The member body of the ISO from the United States is the American National Standards Institute (ANSI).

1-4-2 International Telecommunications Union— Telecommunications Sector

The *International Telecommunications Union—Telecommunications Sector* (ITU-T), formerly the Comité Consultatif Internationale de Télégraphie et Téléphonie (CCITT), is one of four permanent parts of the International Telecommunications Union based in Geneva, Switzerland.

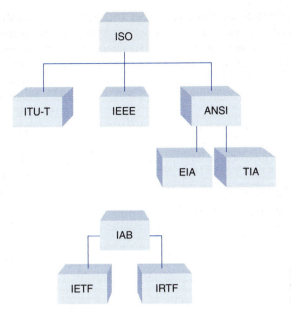

FIGURE 1-2 Standards organizations for data and network communications

Membership in the ITU-T consists of government authorities and representatives from many countries. The ITU-T is now the standards organization for the United Nations and develops the recommended sets of rules and standards for telephone and data communications. The ITU-T has developed three sets of specifications: the V series for modem interfacing and data transmission over telephone lines; the X series for data transmission over public digital networks, e-mail, and directory services; and the I and Q series for Integrated Services Digital Network (ISDN) and its extension Broadband ISDN (sometimes called the Information Superhighway).

The ITU-T is separated into 14 study groups that prepare recommendations on the following topics:

Network and service operation
Tariff and accounting principles
Telecommunications management network and network maintenance
Protection against electromagnetic environment effects
Outside plant
Data networks and open system communications
Characteristics of telematic systems
Television and sound transmission
Language and general software aspects for telecommunications systems
Signaling requirements and protocols
End-to-end transmission performance of networks and terminals
General network aspects
Transport networks, systems, and equipment
Multimedia services and systems

1-4-3 Institute of Electrical and Electronics Engineers

The *Institute of Electrical and Electronics Engineers* (IEEE) is an international professional organization founded in the United States and is comprised of electronics, computer, and communications engineers. The IEEE is currently the world's largest professional society with over 200,000 members. The IEEE works closely with ANSI to develop communications and information processing standards with the underlying goal of advancing theory, creativity, and product quality in any field associated with electrical engineering.

1-4-4 American National Standards Institute

The *American National Standards Institute* (ANSI) is the official standards agency for the United States and is the U.S. voting representative for the ISO. However, ANSI is a completely private, nonprofit organization comprised of equipment manufacturers and users of data processing equipment and services. Although ANSI has no affiliations with the federal government of the United States, it serves as the national coordinating institution for voluntary standardization in the United States. ANSI membership is comprised of people from professional societies, industry associations, governmental and regulatory bodies, and consumer groups.

1-4-5 Electronics Industry Association

The *Electronics Industry Associations* (EIA) is a nonprofit U.S. trade association that establishes and recommends industrial standards. EIA activities include standards development, increasing public awareness, and lobbying. The EIA is responsible for developing the RS (recommended standard) series of standards for data and telecommunications.

1-4-6 Telecommunications Industry Association

The *Telecommunications Industry Association* (TIA) is the leading trade association in the communications and information technology industry. The TIA facilitates business development opportunities and a competitive marketplace through market development, trade promotion, trade shows, domestic and international advocacy, and standards development. The TIA represents manufacturers of communications and information technology products and services providers for the global marketplace through its core competencies. The TIA also facilitates the convergence of new communications networks while working for a competitive and innovative market environment.

1-4-7 Internet Architecture Board

In 1957, the Advanced Research Projects Agency (ARPA), the research arm of the Department of Defense, was created in response to the Soviet Union's launching of *Sputnik*. The original purpose of ARPA was to accelerate the advancement of technologies that could possibly be useful to the U.S. military. When ARPANET was initiated in the late 1970s, ARPA formed a committee to oversee it. In 1983, the name of the committee was changed to the *Internet Activities Board* (IAB). The meaning of the acronym was later changed to the *Internet Architecture Board*.

Today the IAB is a technical advisory group of the Internet Society with the following responsibilities:

1. Oversees the architecture protocols and procedures used by the Internet
2. Manages the processes used to create Internet standards and serves as an appeal board for complaints of improper execution of the standardization processes
3. Is responsible for the administration of the various Internet assigned numbers
4. Acts as representative for Internet Society interests in liaison relationships with other organizations concerned with standards and other technical and organizational issues relevant to the worldwide Internet
5. Acts as a source of advice and guidance to the board of trustees and officers of the Internet Society concerning technical, architectural, procedural, and policy matters pertaining to the Internet and its enabling technologies

1-4-8 Internet Engineering Task Force

The *Internet Engineering Task Force* (IETF) is a large international community of network designers, operators, vendors, and researchers concerned with the evolution of the Internet architecture and the smooth operation of the Internet.

1-4-9 Internet Research Task Force

The *Internet Research Task Force* (IRTF) promotes research of importance to the evolution of the future Internet by creating focused, long-term and small research groups working on topics related to Internet protocols, applications, architecture, and technology.

1-5 LAYERED NETWORK ARCHITECTURE

The basic concept of *layering* network responsibilities is that each layer adds value to services provided by sets of lower layers. In this way, the highest level is offered the full set of services needed to run a distributed data application. There are several advantages to using a *layered architecture*. A layered architecture facilitates *peer-to-peer communications protocols* where a given layer in one system can logically communicate with its corresponding layer in another system. This allows different computers to communicate at different levels. Figure 1-3 shows a layered architecture where layer N at the source logically (but not necessarily physically) communicates with layer N at the destination and layer N of any intermediate nodes.

1-5-1 Protocol Data Unit

When technological advances occur in a layered architecture, it is easier to modify one layer's protocol without having to modify all the other layers. Each layer is essentially independent of every other layer. Therefore, many of the functions found in lower layers have been removed entirely from software tasks and replaced with hardware. The primary disadvantage of layered architectures is the tremendous amount of overhead required. With layered architectures, communications between two corresponding layers requires a unit of data called a *protocol data unit* (PDU). As shown in Figure 1-4, a PDU can be a *header* added at the beginning of a message or a *trailer* appended to the end of a message. In a layered architecture, communications occurs between similar layers; however, data must flow through the other layers. Data flows downward through the layers in the source system and upward through the layers in the destination system. In intermediate systems, data flows upward first and then downward. As data passes from one layer into another, headers and trailers are added and removed from the PDU. The process of adding or removing PDU information is called *encapsulation/decapsulation* because it appears as though the PDU from the upper layer is encapsulated in the PDU from the lower layer during the downward

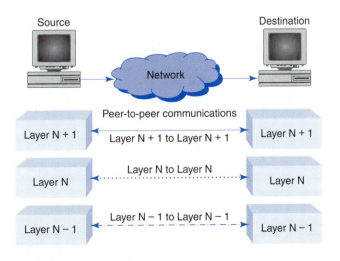

FIGURE 1-3   Peer-to-peer data communications

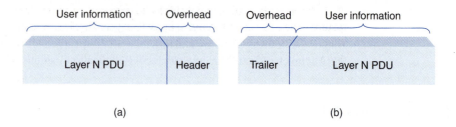

(a)   (b)

FIGURE 1-4 Protocol data unit: (a) header; (b) trailer

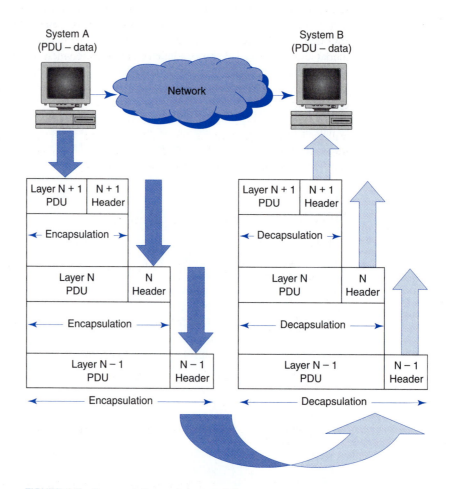

FIGURE 1-5 Encapsulation and decapsulation

movement and decapsulated during the upward movement. *Encapsulate* means to place in a capsule or other protected environment, and *decapsulate* means to remove from a capsule or other protected environment. Figure 1-5 illustrates the concepts of encapsulation and decapsulation.

In a layered protocol such as the one shown in Figure 1-3, layer N receive services from the layer immediately below it (N − 1) and provides services to the layer directly above it (N + 1). Layer N can provide service to more than one entity in layer N + 1 by using a *service access point* (SAP) *address* to define for which entity the service is intended.

Information and network information passes from one layer of a multilayered architecture to another layer through a layer-to-layer *interface*. A layer-to-layer interface defines what information and services the lower layer must provide to the upper layer. A well-defined layer and layer-to-layer interface provide modularity to a network.

1-6 OPEN SYSTEMS INTERCONNECTION

Open systems interconnection (OSI) is the name for a set of standards for communicating among computers. The primary purpose of OSI standards is to serve as a structural guideline for exchanging information between computers, workstations, and networks. The OSI is endorsed by both the ISO and the ITU-T, which have worked together to establish a set of ISO standards and ITU-T recommendations that are essentially identical. In 1983, the ISO and ITU-T (CCITT) adopted a seven-layer communications architecture reference model. Each layer consists of specific protocols for communicating.

The ISO seven-layer open systems interconnection model is shown in Figure 1-6. This hierarchy was developed to facilitate the intercommunications of data processing equipment by separating network responsibilities into seven distinct layers. As with any layered architecture, overhead information is added to a PDU in the form of headers and trailers. In fact, if all seven levels of the OSI model are addressed, as little as 15% of the transmitted message is actually source information, and the rest is overhead. The result of adding headers to each layer is illustrated in Figure 1-7.

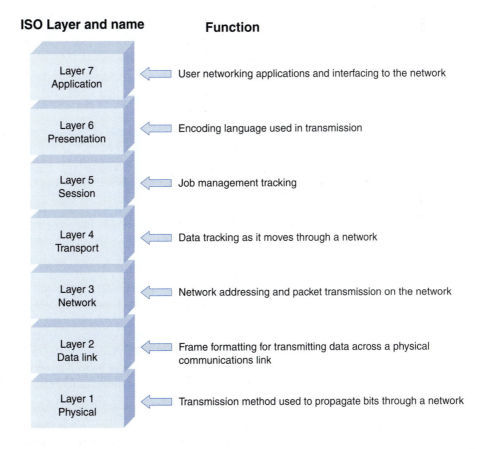

FIGURE 1-6 OSI seven-layer protocol hierarchy

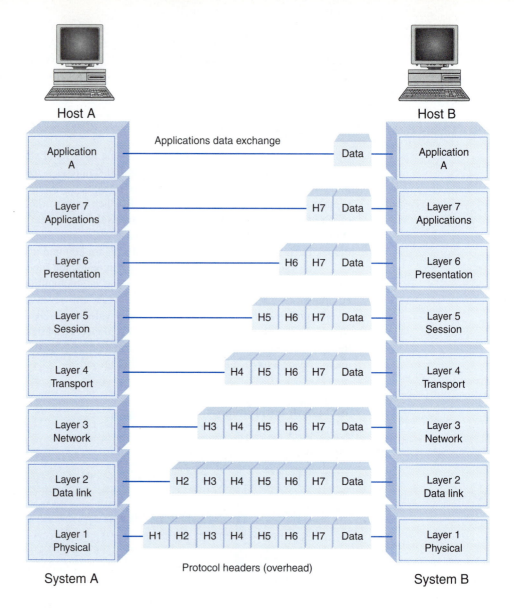

FIGURE 1-7 OSI seven-layer international protocol hierarchy. H7—applications header, H6—presentation header, H5—session header, H4—transport header, H3—network header, H2—data-link header, H1—physical header

In recent years, the OSI seven-layer model has become more academic than standard, as the hierarchy does not coincide with the Internet's four-layer protocol model. However, the basic functions of the layers are still performed, so the seven-layer model continues to serve as a reference model when describing network functions.

Levels 4 to 7 address the applications aspects of the network that allow for two host computers to communicate directly. The three bottom layers are concerned with the actual mechanics of moving data (at the bit level) from one machine to another. The basic services provided by each layer are discussed in a later chapter of this book. A brief summary of the services provided by each layer is given here.

1. *Physical layer.* The physical layer is the lowest level of the OSI hierarchy and is responsible for the actual propagation of unstructured data bits (1s and 0s) through a transmis-

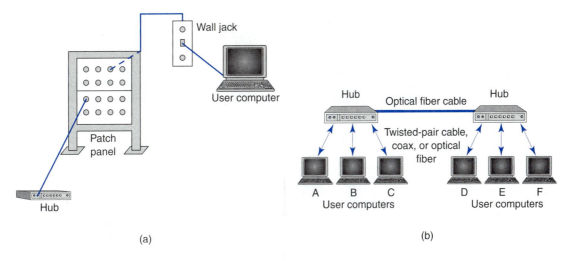

FIGURE 1-8 OSI layer 1—physical: (a) computer to hub; (b) connectivity devices

sion medium, which includes how bits are represented, the bit rate, and how bit synchronization is achieved. The physical layer specifies the type of transmission medium and the transmission mode (simplex, half duplex, or full duplex) and the physical, electrical, functional, and procedural standards for accessing data communications networks. Definitions such as connections, pin assignments, interface parameters, timing, maximum and minimum voltage levels, and circuit impedances are made at the physical level. Transmission media defined by the physical layer include metallic cable, optical fiber cable, or wireless radio-wave propagation. The physical layer for a cable connection is depicted in Figure 1-8a.

Connectivity devices connect devices on cabled networks. An example of a connectivity device is a hub. A hub is a transparent device that samples the incoming bit stream and simply repeats it to the other devices connected to the hub. The hub does not examine the data to determine what the destination is; therefore, it is classified as a layer 1 component. Physical layer connectivity for a cabled network is shown in Figure 1-8b.

The physical layer also includes the *carrier system* used to propagate the data signals between points in the network. Carrier systems are simply communications systems that carry data through a system using either metallic or optical fiber cables or wireless arrangements, such as microwave, satellites, and cellular radio systems. The carrier can use analog or digital signals that are somehow converted to a different form (encoded or modulated) by the data and then propagated through the system.

2. *Data-link layer.* The data-link layer is responsible for providing error-free communications across the physical link connecting primary and secondary stations (nodes) within a network (sometimes referred to as *hop-to-hop* delivery). The data-link layer packages data from the physical layer into groups called blocks, frames, or packets and provides a means to activate, maintain, and deactivate the data communications link between nodes. The data-link layer provides the final framing of the information signal, provides synchronization, facilitates the orderly flow of data between nodes, outlines procedures for error detection and correction, and provides the physical addressing information. A block diagram of a network showing data transferred between two computers (A and E) at the data-link level is illustrated in Figure 1-9. Note that the hubs are transparent but that the switch passes the transmission on to only the hub serving the intended destination.

3. *Network layer.* The network layer provides details that enable data to be routed between devices in an environment using multiple networks, subnetworks, or both. Networking components that operate at the network layer include routers and their software. The

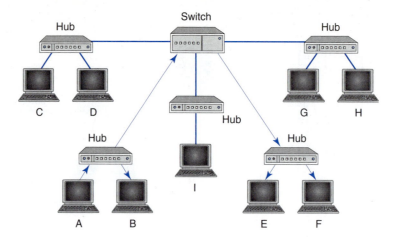

FIGURE 1-9 OSI layer 2—data link

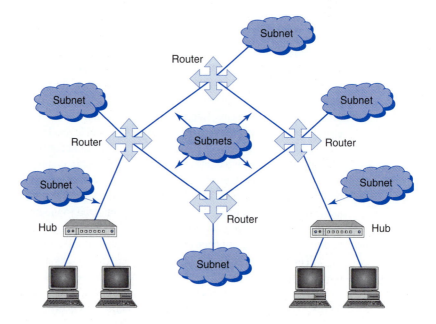

FIGURE 1-10 OSI layer 3—network

network layer determines which network configuration is most appropriate for the function provided by the network and addresses and routes data within networks by establishing, maintaining, and terminating connections between them. The network layer provides the upper layers of the hierarchy with independence from the data transmission and switching technologies used to interconnect systems. It accomplishes this by defining the mechanism in which messages are broken into smaller data packets and routed from a sending node to a receiving node within a data communications network. The network layer also typically provides the source and destination network addresses (logical addresses), subnet information, and source and destination node addresses. Figure 1-10 illustrates the network layer of the OSI protocol hierarchy. Note that the network is subdivided into subnetworks that are separated by routers.

4. *Transport layer.* The transport layer controls and ensures the end-to-end integrity of the data message propagated through the network between two devices, providing the re-

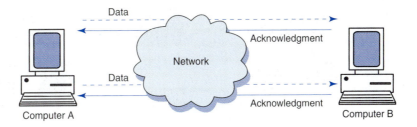

FIGURE 1-11 OSI layer 4—transport

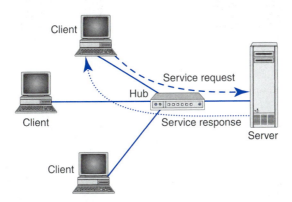

FIGURE 1-12 OSI layer 5—session

liable, transparent transfer of data between two endpoints. Transport layer responsibilities includes message routing, segmenting, error recovery, and two types of basic services to an upper-layer protocol: connectionless oriented and connectionless. The transport layer is the highest layer in the OSI hierarchy in terms of communications and may provide data tracking, connection flow control, sequencing of data, error checking, and application addressing and identification. Figure 1-11 depicts data transmission at the transport layer.

5. *Session layer.* The session layer is responsible for network availability (i.e., data storage and processor capacity). Session layer protocols provide the logical connection entities at the application layer. These applications include file transfer protocols and sending e-mail. Session responsibilities include network log-on and log-off procedures and user authentication. A session is a temporary condition that exists when data are actually in the process of being transferred and does not include procedures such as call establishment, setup, or disconnect. The session layer determines the type of dialogue available (i.e., simplex, half duplex, or full duplex). Session layer characteristics include virtual connections between applications entities, synchronization of data flow for recovery purposes, creation of dialogue units and activity units, connection parameter negotiation, and partitioning services into functional groups. Figure 1-12 illustrates the establishment of a session on a data network.

6. *Presentation layer.* The presentation layer provides independence to the application processes by addressing any code or syntax conversion necessary to present the data to the network in a common communications format. The presentation layer specifies how end-user applications should format the data. This layer provides for translation between local representations of data and the representation of data that will be used for transfer between end users. The results of encryption, data compression, and virtual terminals are examples of the translation service.

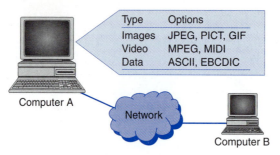

FIGURE 1-13 OSI layer 6—presentation

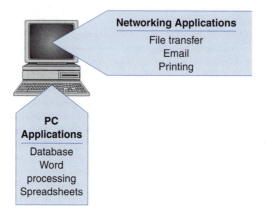

FIGURE 1-14 OSI layer 7—applications

The presentation layer translates between different data formats and protocols. Presentation functions include data file formatting, encoding, encryption and decryption of data messages, dialogue procedures, data compression algorithms, synchronization, interruption, and termination. The presentation layer performs code and character set translation (including ASCII and EBCDIC) and formatting information and determines the display mechanism for messages. Figure 1-13 shows an illustration of the presentation layer.

7. *Application layer.* The application layer is the highest layer in the hierarchy and is analogous to the general manager of the network by providing access to the OSI environment. The applications layer provides distributed information services and controls the sequence of activities within an application and also the sequence of events between the computer application and the user of another application. The application layer (shown in Figure 1-14) communicates directly with the user's application program.

User application processes require application layer service elements to access the networking environment. There are two types of service elements: CASEs (*common application service elements*), which are generally useful to a variety of application processes, and SASEs (*specific application service elements*), which generally satisfy particular needs of application processes. CASE examples include association control that establishes, maintains, and terminates connections with a peer application entity and commitment, concurrence, and recovery that ensure the integrity of distributed transactions. SASE examples involve the TCP/IP protocol stack and include FTP (*file transfer protocol*), SNMP (*simple network management protocol*), Telnet (*virtual terminal protocol*), and SMTP (*simple mail transfer protocol*).

1-7 DATA COMMUNICATIONS CIRCUITS

The underlying purpose of a data communications circuit is to provide a transmission path between locations and to transfer digital information from one station to another using electronic circuits. A *station* is simply an endpoint where subscribers gain access to the circuit. A station is sometimes called a *node*, which is the location of computers, computer terminals, workstations, and other digital computing equipment. There are almost as many types of data communications circuits as there are types of data communications equipment.

Data communications circuits utilize electronic communications equipment and *facilities* to interconnect digital computer equipment. Communications facilities are physical means of interconnecting stations within a data communications system and can include virtually any type of physical transmission media or wireless radio system in existence. Communications facilities are provided to data communications users through public telephone networks (PTN), public data networks (PDN), and a multitude of private data communications systems.

Figure 1-15 shows a simplified block diagram of a two-station data communications circuit. The fundamental components of the circuit are source of digital information, transmitter, transmission medium, receiver, and destination for the digital information. Although the figure shows transmission in only one direction, bidirectional transmission is possible by providing a duplicate set of circuit components in the opposite direction.

Source. The information source generates data and could be a mainframe computer, personal computer, workstation, or virtually any other piece of digital equipment. The source equipment provides a means for humans to enter data into the system.

Transmitter. Source data is seldom in a form suitable to propagate through the transmission medium. For example, digital signals (pulses) cannot be propagated through a wireless radio system without being converted to analog first. The transmitter encodes the source information and converts it to a different form, allowing it to be more efficiently propagated through the transmission medium. In essence, the transmitter acts as an interface between the source equipment and the transmission medium.

Transmission medium. The transmission medium carries the encoded signals from the transmitter to the receiver. There are many different types of transmission media, such as free-space radio transmission (including all forms of wireless transmission, such as terrestrial microwave, satellite radio, and cellular telephone) and physical facilities, such as metallic and optical fiber cables. Very often, the transmission path is comprised of several different types of transmission facilities.

Receiver. The receiver converts the encoded signals received from the transmission medium back to their original form (i.e., decodes them) or whatever form is used in the destination equipment. The receiver acts as an interface between the transmission medium and the destination equipment.

Destination. Like the source, the destination could be a mainframe computer, personal computer, workstation, or virtually any other piece of digital equipment.

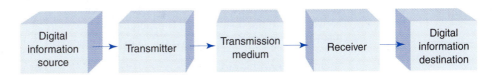

FIGURE 1-15 Simplified block diagram of a two-station data communications circuit

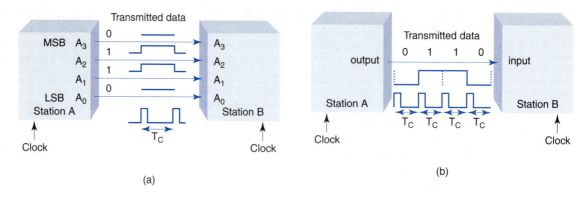

FIGURE 1-16 Data transmission: (a) parallel; (b) serial

1-8 SERIAL AND PARALLEL DATA TRANSMISSION

Binary information can be transmitted either in parallel or serially. Figure 1-16a shows how the binary code 0110 is transmitted from station A to station B in parallel. As the figure shows, each bit position (A_0 to A_3) has its own transmission line. Consequently, all four bits can be transmitted simultaneously during the time of a single clock pulse (T_C). This type of transmission is called *parallel by bit* or *serial by character.*

Figure 1-16b shows the same binary code transmitted serially. As the figure shows, there is a single transmission line, and thus only one bit can be transmitted at a time. Consequently, it requires four clock pulses ($4T_C$) to transmit the entire four-bit code. This type of transmission is called *serial by bit.*

Obviously, the principal trade-off between parallel and serial data transmission is speed versus simplicity. Data transmission can be accomplished much more rapidly using parallel transmission; however, parallel transmission requires more data lines. As a general rule, parallel transmission is used for short-distance data communications and within a computer, and serial transmission is used for long-distance data communications.

1-9 DATA COMMUNICATIONS CIRCUIT ARRANGEMENTS

Data communications circuits can be configured in a multitude of arrangements depending on the specifics of the circuit, such as how many stations are on the circuit, type of transmission facility, distance between stations, and how many users are at each station. A data communications circuit can be described in terms of circuit configuration and transmission mode.

1-9-1 Circuit Configurations

Data communications networks can be generally categorized as either two point or multipoint. A *two-point* configuration involves only two locations or stations, whereas a *multipoint* configuration involves three or more stations. Regardless of the configuration, each station can have one or more computers, computer terminals, or workstations. A two-point circuit involves the transfer of digital information between a mainframe computer and a personal computer, two mainframe computers, two personal computers, or two data communications networks. A multipoint network is generally used to interconnect a single mainframe computer (host) to many personal computers or to interconnect many personal computers.

1-9-2 Transmission Modes

Essentially, there are four modes of transmission for data communications circuits: *simplex, half duplex, full duplex,* and *full/full duplex.*

1-9-2-1 Simplex.   In the simplex (SX) mode, data transmission is unidirectional; information can be sent in only one direction. Simplex lines are also called *receive-only, transmit-only,* or *one-way-only* lines. Commercial radio broadcasting is an example of simplex transmission, as information is propagated in only one direction—from the broadcasting station to the listener.

1-9-2-2 Half duplex.   In the half-duplex (HDX) mode, data transmission is possible in both directions but not at the same time. Half-duplex communications lines are also called *two-way-alternate* or *either-way* lines. Citizens band (CB) radio is an example of half-duplex transmission because to send a message, the *push-to-talk* (PTT) switch must be depressed, which turns on the transmitter and shuts off the receiver. To receive a message, the PTT switch must be off, which shuts off the transmitter and turns on the receiver.

1-9-2-3 Full duplex.   In the full-duplex (FDX) mode, transmissions are possible in both directions simultaneously, but they must be between the same two stations. Full-duplex lines are also called *two-way simultaneous, duplex,* or *both-way* lines. A local telephone call is an example of full-duplex transmission. Although it is unlikely that both parties would be talking at the same time, they could if they wanted to.

1-9-2-4 Full/full duplex.   In the full/full duplex (F/FDX) mode, transmission is possible in both directions at the same time but not between the same two stations (i.e., one station is transmitting to a second station and receiving from a third station at the same time). Full/full duplex is possible only on multipoint circuits. The U.S. postal system is an example of full/full duplex transmission because a person can send a letter to one address and receive a letter from another address at the same time.

1-10   DATA COMMUNICATIONS NETWORKS

Any group of computers connected together can be called a *data communications network,* and the process of sharing resources between computers over a data communications network is called *networking.* In its simplest form, networking is two or more computers connected together through a common transmission medium for the purpose of sharing data. The concept of networking began when someone determined that there was a need to share software and data resources and that there was a better way to do it than storing data on a disk and literally running from one computer to another. By the way, this manual technique of moving data on disks is sometimes referred to as *sneaker net.* The most important considerations of a data communications network are performance, transmission rate, reliability, and security.

Applications running on modern computer networks vary greatly from company to company. A network must be designed with the intended application in mind. A general categorization of networking applications is listed in Table 1-1. The specific application affects how well a network will perform. Each network has a finite capacity. Therefore, network

Table 1-1   Networking Applications

Application	Examples
Standard office applications	E-mail, file transfers, and printing
High-end office applications	Video imaging, computer-aided drafting, computer-aided design, and software development
Manufacturing automation	Process and numerical control
Mainframe connectivity	Personal computers, workstations, and terminal support
Multimedia applications	Live interactive video

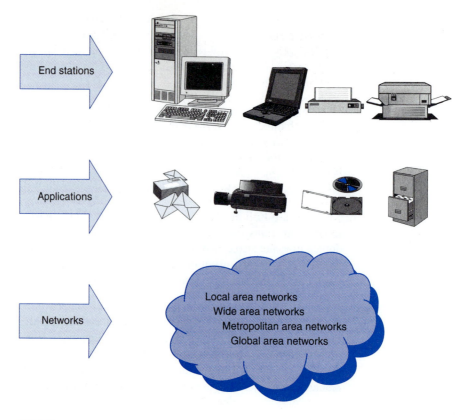

FIGURE 1-17 Basic network components

designers and engineers must be aware of the type and frequency of information traffic on the network.

There are many factors involved when designing a computer network, including the following:

1. Network goals as defined by organizational management
2. Network security
3. Network uptime requirements
4. Network response-time requirements
5. Network and resource costs

The primary balancing act in computer networking is speed versus reliability. Too often, network performance is severely degraded by using error checking procedures, data encryption, and handshaking (acknowledgments). However, these features are often required and are incorporated into protocols.

Some networking protocols are very reliable but require a significant amount of overhead to provide the desired high level of service. These protocols are examples of connection-oriented protocols. Other protocols are designed with speed as the primary parameter and, therefore, forgo some of the reliability features of the connection-oriented protocols. These *quick protocols* are examples of connectionless protocols.

1-10-1 Network Components, Functions, and Features

Computer networks are like snowflakes—no two are the same. The basic components of computer networks are shown in Figure 1-17. All computer networks include some combination of the following: end stations, applications, and a network that will support the data

File request

Copy of requested file

User computer   File server   **FIGURE 1-18** File server operation

traffic between the end stations. A computer network designed three years ago to support the basic networking applications of the time may have a difficult time supporting recently developed high-end applications, such as medical imaging and live video teleconferencing. Network designers, administrators, and managers must understand and monitor the most recent types and frequency of networked applications.

Computer networks all share common devices, functions, and features, including servers, clients, transmission media, shared data, shared printers and other peripherals, hardware and software resources, network interface card (NIC), local operating system (LOS), and the network operating system (NOS).

1-10-1-1 Servers. *Servers* are computers that hold shared files, programs, and the network operating system. Servers provide access to network resources to all the users of the network. There are many different kinds of servers, and one server can provide several functions. For example, there are file servers, print servers, mail servers, communications servers, database servers, directory/security servers, fax servers, and Web servers, to name a few.

Figure 1-18 shows the operation of a *file server*. A user (client) requests a file from the file server. The file server sends a copy of the file to the requesting user. File servers allow users to access and manipulate disk resources stored on other computers. An example of a file server application is when two or more users edit a shared spreadsheet file that is stored on a server. File servers have the following characteristics:

1. File servers are loaded with files, accounts, and a record of the access rights of users or groups of users on the network.
2. The server provides a shareable virtual disk to the users (clients).
3. File mapping schemes are implemented to provide the virtualness of the files (i.e., the files are made to look like they are on the user's computer).
4. Security systems are installed and configured to provide the server with the required security and protection for the files.
5. Redirector or shell software programs located on the users' computers transparently activate the client's software on the file server.

1-10-1-2 Clients. *Clients* are computers that access and use the network and shared network resources. Client computers are basically the customers (users) of the network, as they request and receive services from the servers.

1-10-1-3 Transmission media. *Transmission media* are the facilities used to interconnect computers in a network, such as twisted-pair wire, coaxial cable, and optical fiber cable. Transmission media are sometimes called channels, links, or lines.

1-10-1-4 Shared data. *Shared data* are data that file servers provide to clients, such as data files, printer access programs, and e-mail.

1-10-1-5 Shared printers and other peripherals. *Shared printers* and *peripherals* are hardware resources provided to the users of the network by servers. Resources provided include data files, printers, software, or any other items used by clients on the network.

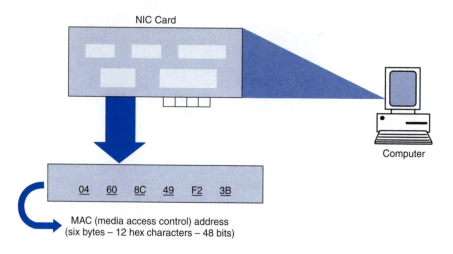

FIGURE 1-19 Network interface card (NIC)

1-10-1-6 Network interface card. Each computer in a network has a special expansion card called a *network interface card* (NIC). The NIC prepares (formats) and sends data, receives data, and controls data flow between the computer and the network. On the transmit side, the NIC passes frames of data on to the physical layer, which transmits the data to the physical link. On the receive side, the NIC processes bits received from the physical layer and processes the message based on its contents. A network interface card is shown in Figure 1-19. Characteristics of NICs include the following:

1. The NIC constructs, transmits, receives, and processes data to and from a PC and the connected network.
2. Each device connected to a network must have a NIC installed.
3. A NIC is generally installed in a computer as a daughterboard, although some computer manufacturers incorporate the NIC into the motherboard during manufacturing.
4. Each NIC has a unique six-byte media access control (MAC) address, which is typically permanently burned into the NIC when it is manufactured. The MAC address is sometimes called the physical, hardware, node, Ethernet, or LAN address.
5. The NIC must be compatible with the network (i.e., Ethernet—10baseT or token ring) to operate properly.
6. NICs manufactured by different vendors vary in speed, complexity, manageability, and cost.
7. The NIC requires drivers to operate on the network.

1-10-1-7 Local operating system. A *local operating system* (LOS) allows personal computers to access files, print to a local printer, and have and use one or more disk and CD drives that are located on the computer. Examples of LOSs are MS-DOS, PC-DOS, Unix, Macintosh, OS/2, Windows 3.11, Windows 95, Windows 98, Windows 2000, and Linux. Figure 1-20 illustrates the relationship between a personal computer and its LOS.

1-10-1-8 Network operating system. The *network operating system* (NOS) is a program that runs on computers and servers that allows the computers to communicate over a network. The NOS provides services to clients such as log-in features, password authenti-

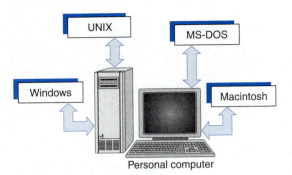

FIGURE 1-20 Local operating system (LOS)

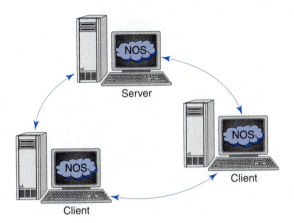

FIGURE 1-21 Network operating system (NOS)

cation, printer access, network administration functions, and data file sharing. Some of the more popular network operating systems are Unix, Novell NetWare, AppleShare, Macintosh System 7, IBM LAN Server, Compaq Open VMS, and Microsoft Windows NT Server. The NOS is software that makes communications over a network more manageable. The relationship between clients, servers, and the NOS is shown in Figure 1-21, and the layout of a local network operating system is depicted in Figure 1-22. Characteristics of NOSs include the following:

1. A NOS allows users of a network to interface with the network transparently.
2. A NOS commonly offers the following services: file service, print service, mail service, communications service, database service, and directory and security services.
3. The NOS determines whether data are intended for the user's computer or whether the data need to be redirected out onto the network.
4. The NOS implements client software for the user, which allows them to access servers on the network.

1-10-2 Network Models

Computer networks can be represented with two basic network models: *peer-to-peer client/server* and *dedicated client/server.* The client/server method specifies the way in which two computers can communicate with software over a network. Although clients and servers are generally shown as separate units, they are often active in a single computer but not at the same time. With the client/server concept, a computer acting as a client initiates a software request from another computer acting as a server. The server computer responds and attempts

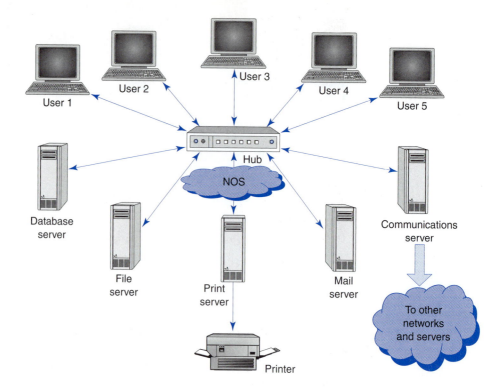

FIGURE 1-22 Network layout using a network operating system (NOS)

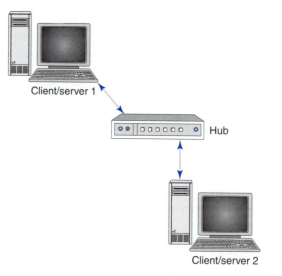

FIGURE 1-23 Client/server concept

to satisfy the request from the client. The server computer might then act as a client and request services from another computer. The client/server concept is illustrated in Figure 1-23.

1-10-2-1 Peer-to-peer client/server network. A *peer-to-peer client/server network* is one in which all computers share their resources, such as hard drives, printers, and so on, with all the other computers on the network. Therefore, the peer-to-peer operating system divides its time between servicing the computer on which it is loaded and servicing

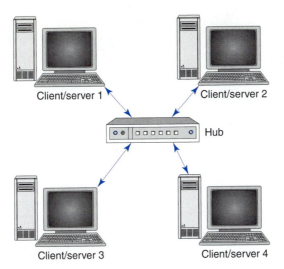

FIGURE 1-24 Peer-to-peer client/server network

Client/server 1

Client/server 2

Hub

Client/server 3

Client/server 4

requests from other computers. In a peer-to-peer network (sometimes called a *workgroup*), there are no dedicated servers or hierarchy among the computers.

Figure 1-24 shows a peer-to-peer client/server network with four clients/servers (users) connected together through a hub. All computers are equal, hence the name *peer*. Each computer in the network can function as a client and/or a server, and no single computer holds the network operating system or shared files. Also, no one computer is assigned network administrative tasks. The users at each computer determine which data on their computer are shared with the other computers on the network. Individual users are also responsible for installing and upgrading the software on their computer.

Because there is no central controlling computer, a peer-to-peer network is an appropriate choice when there are fewer than 10 users on the network, when all computers are located in the same general area, when security is not an issue, or when there is limited growth projected for the network in the immediate future. Peer-to-peer computer networks should be small for the following reasons:

1. When operating in the server role, the operating system is not optimized to efficiently handle multiple simultaneous requests.
2. The end user's performance as a client would be degraded.
3. Administrative issues such as security, data backups, and data ownership may be compromised in a large peer-to-peer network.

1-10-2-2 Dedicated client/server network. In a *dedicated client/server network,* one computer is designated the server, and the rest of the computers are clients. As the network grows, additional computers can be designated servers. Generally, the designated servers function only as servers and are not used as a client or workstation. The servers store all the network's shared files and applications programs, such as word processor documents, compilers, database applications, spreadsheets, and the network operating system. Client computers can access the servers and have shared files transferred to them over the transmission medium.

Figure 1-25 shows a dedicated client/server-based network with three servers and three clients (users). Each client can access the resources on any of the servers and also the resources on other client computers. The dedicated client/server-based network is probably

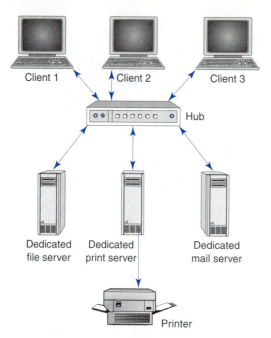

FIGURE 1-25 Dedicated client/server network

the most commonly used computer networking model. There can be a separate dedicated server for each function (i.e., file server, print server, mail server, etc.) or one single general-purpose server responsible for all services.

In some client/server networks, client computers submit jobs to one of the servers. The server runs the software and completes the job and then sends the results back to the client computer. In this type of client/server network, less information propagates through the network than with the file server configuration because only data and not applications programs are transferred between computers.

In general, the dedicated client/server model is preferable to the peer-to-peer client/server model for general-purpose data networks. The peer-to-peer model client/server model is usually preferable for special purposes, such as a small group of users sharing resources.

1-10-3 Network Topologies

Network topology describes the layout or appearance of a network—that is, how the computers, cables, and other components within a data communications network are interconnected, both physically and logically. The *physical topology* describes how the network is actually laid out, and the *logical topology* describes how data actually flow through the network.

In a data communications network, two or more stations connect to a link, and one or more links form a topology. Topology is a major consideration for capacity, cost, and reliability when designing a data communications network. The most basic topologies are *point to point* and *multipoint.* A point-to-point topology is used in data communications networks that transfer high-speed digital information between only two stations. Very often, point-to-point data circuits involve communications between a mainframe computer and another mainframe computer or some other type of high-capacity digital device. A two-point circuit is shown in Figure 1-26a.

A multipoint topology connects three or more stations through a single transmission medium. Examples of multipoint topologies are *star, bus, ring, mesh,* and *hybrid.*

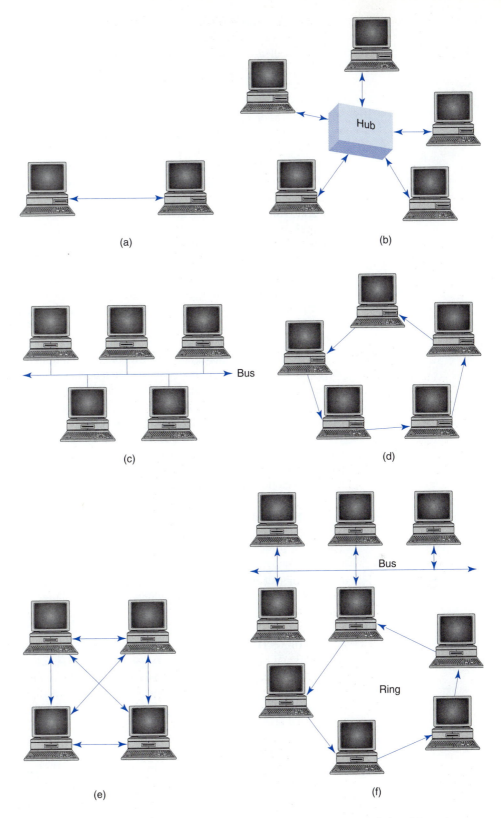

FIGURE 1-26 Network topologies: (a) point to point; (b) star; (c) bus; (d) ring; (e) mesh; (f) hybrid

1-10-3-1 Star topology. A *star topology* is a multipoint data communications network where remote stations are connected by cable segments directly to a centrally located computer called a *hub,* which acts like a multipoint connector (see Figure 1-26b). In essence, a star topology is simply a multipoint circuit comprised of many two-point circuits where each remote station communicates directly with a centrally located computer. With a star topology, remote stations cannot communicate directly with one another, so they must relay information through the hub. Hubs also have store-and-forward capabilities, enabling them to handle more than one message at a time.

1-10-3-2 Bus topology. A *bus topology* is a multipoint data communications circuit that makes it relatively simple to control data flow between and among the computers because this configuration allows all stations to receive every transmission over the network. With a bus topology, all the remote stations are physically or logically connected to a single transmission line called a *bus.* The bus topology is the simplest and most common method of interconnecting computers. The two ends of the transmission line never touch to form a complete loop. A bus topology is sometimes called *multidrop* or *linear bus,* and all stations share a common transmission medium. Data networks using the bus topology generally involve one centrally located host computer that controls data flow to and from the other stations. The bus topology is sometimes called a *horizontal bus* and is shown in Figure 1-26c.

1-10-3-3 Ring topology. A *ring topology* is a multipoint data communications network where all stations are interconnected in tandem (series) to form a closed loop or circle. A ring topology is sometimes called a *loop.* Each station in the loop is joined by point-to-point links to two other stations (the transmitter of one and the receiver of the other) (see Figure 1-26d). Transmissions are unidirectional and must propagate through all the stations in the loop. Each computer acts like a repeater in that it receives signals from down-line computers and then retransmits them to up-line computers. The ring topology is similar to the bus and star topologies, as it generally involves one centrally located host computer that controls data flow to and from the other stations.

1-10-3-4 Mesh topology. In a *mesh topology,* every station has a direct two-point communications link to every other station on the circuit as shown in Figure 1-26e. The mesh topology is sometimes called *fully connected.* A disadvantage of a mesh topology is a fully connected circuit requires $n(n - 1)/2$ physical transmission paths to interconnect n stations and each station must have $n - 1$ input/output ports. Advantages of a mesh topology are reduced traffic problems, increased reliability, and enhanced security.

1-10-3-5 Hybrid topology. A *hybrid topology* is simply combining two or more of the traditional topologies to form a larger, more complex topology. Hybrid topologies are sometimes called *mixed topologies.* An example of a hybrid topology is the *bus star* topology shown in Figure 1-26f. Other hybrid configurations include the *star ring, bus ring,* and virtually every other combination you can think of.

1-10-4 Network Classifications

Networks are generally classified by size, which includes geographic area, distance between stations, number of computers, transmission speed (bps), transmission media, and the network's physical architecture. The four primary classifications of networks are *local area networks* (LANs), *metropolitan area networks* (MANs), *wide*

Table 1-2 Primary Network Types

Network Type	Characteristics
LAN (local area network)	Interconnects computer users within a department, company, or group
MAN (metropolitan area network)	Interconnects computers in and around a large city
WAN (wide area network)	Interconnects computers in and around an entire country
GAN (global area network)	Interconnects computers from around the entire globe
Building backbone	Interconnects LANs within a building
Campus backbone	Interconnects building LANs
Enterprise network	Interconnects many or all of the above
PAN (personal area network)	Interconnects memory cards carried by people and in computers that are in close proximity to each other
PAN (power line area network, sometimes called PLAN)	Virtually no limit on how many computers it can interconnect and covers an area limited only by the availability of power distribution lines

area networks (WANs), and *global area networks* (GANs). In addition, there are three primary types of interconnecting networks: *building backbone, campus backbone,* and *enterprise network.* Two promising computer networks of the future share the same acronym: the PAN (*personal area network*) and PAN (*power line area network,* sometimes called PLAN). The idea behind a personal area network is to allow people to transfer data through the human body simply by touching each other. Power line area networks use existing ac distribution networks to carry data wherever power lines go, which is virtually everywhere.

When two or more networks are connected together, they constitute an *internetwork* or *internet.* An internet (lowercase *i*) is sometimes confused with the *Internet* (uppercase *I*). The term *internet* is a generic term that simply means to interconnect two or more networks, whereas *Internet* is the name of a specific worldwide data communications network. Table 1-2 summarizes the characteristics of the primary types of networks, and Figure 1-27 illustrates the geographic relationship among computers and the different types of networks.

1-10-4-1 Local area network. *Local area networks* (LANs) are typically privately owned data communications networks in which 10 to 100 computer users typically share data resources with one or more file servers. LANs use a network operating system to provide two-way communications at bit rates typically in the range of 10 Mbps to 100 Mbps and higher between a large variety of data communications equipment within a relatively small geographical area, such as in the same room, building, or building complex (see Figure 1-28). A LAN can be as simple as two personal computers and a printer or could contain dozens of computers, workstations, and peripheral devices. Most LANs link equipment that are within a few miles of each other or closer. Because the size of most LANs is limited, the longest (or worst-case) transmission time is bounded and known by everyone using the network. Therefore, LANs can utilize configurations that otherwise would not be possible.

LANs were designed for sharing resources between a wide range of digital equipment, including personal computers, workstations, and printers. The resources shared can be software as well as hardware. Most LANs are owned by the company or organization

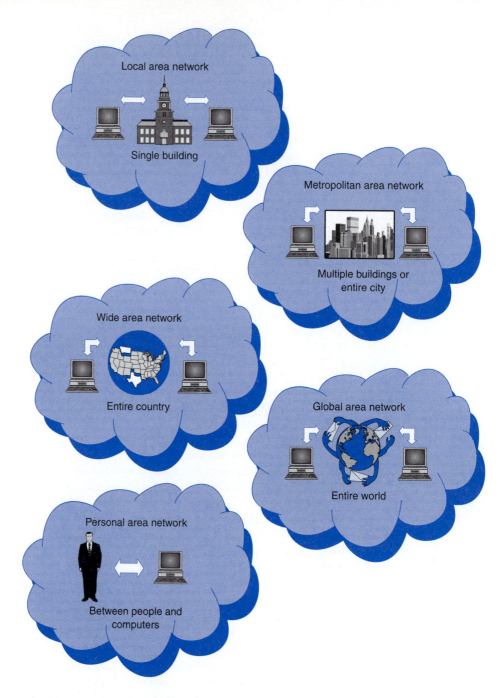

FIGURE 1-27 Computer network types

that uses it and have a connection to a building backbone for access to other departmental LANs, MANs, WANs, and GANs.

1-10-4-2 Metropolitan area network. A *metropolitan area network* (MAN) is a high-speed network similar to a LAN except MANs are designed to encompass larger areas, usually that of an entire city (see Figure 1-29). Most MANs support the transmission of both data and voice and in some cases video. MANs typically operate at

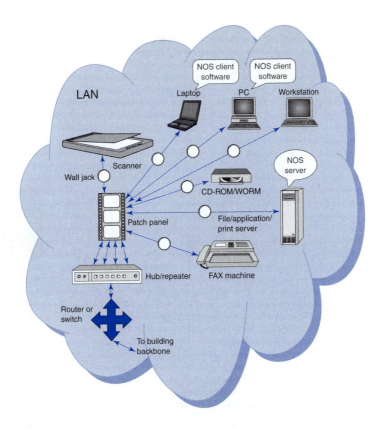

FIGURE 1-28 Local area network (LAN) layout

speeds of 1.5 Mbps to 10 Mbps and range from five miles to a few hundred miles in length. A MAN generally uses only one or two transmission cables and requires no switches. A MAN could be a single network, such as a cable television distribution network, or it could be a means of interconnecting two or more LANs into a single, larger network, enabling data resources to be shared LAN to LAN as well as from station to station or computer to computer. Large companies often use MANs to interconnect all their LANs.

A MAN can be owned and operated entirely by a single, private company, or it could lease services and facilities on a monthly basis from the local cable or telephone company. Switched Multimegabit Data Services (SMDS) is an example of a service offered by local telephone companies for handling high-speed data communications for MANs. Other examples of MANs are FDDI (fiber distributed data interface) and ATM (asynchronous transfer mode).

1-10-4-3 Wide area network. *Wide area networks* (WANs) are the oldest type of data communications network that provide relatively slow-speed, long-distance transmission of data, voice, and video information over relatively large and widely dispersed geographical areas, such as a country or an entire continent (see Figure 1-30). WANs typically interconnect cities and states. WANs typically operate at bit rates from 1.5 Mbps to 2.4 Gbps and cover a distance of 100 to 1000 miles.

WANs may utilize both public and private communications systems to provide service over an area that is virtually unlimited; however, WANs are generally obtained through service providers and normally come in the form of leased-line or circuit-switching technology. Often WANs interconnect routers in different locations. Examples of WANs are

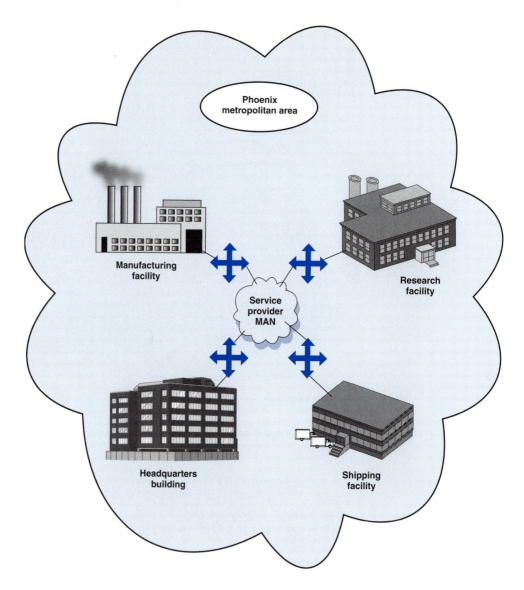

FIGURE 1-29 Metropolitan area network (MAN)

ISDN (integrated services digital network), T1 and T3 digital carrier systems, frame relay, X.25, ATM, and using data modems over standard telephone lines.

1-10-4-4 Global area network. *Global area networks* (GANs) provide connects between countries around the entire globe (see Figure 1-31). The Internet is a good example of a GAN, as it is essentially a network comprised of other networks that interconnects virtually every country in the world. GANs operate from 1.5 Mbps to 100 Gbps and cover thousands of miles.

1-10-4-5 Building backbone. A *building backbone* is a network connection that normally carries traffic between departmental LANs within a single company. A building backbone generally consists of a switch or a router (see Figure 1-32) that can provide connectivity to other networks, such as campus backbones, enterprise backbones, MANs, WANs, or GANs.

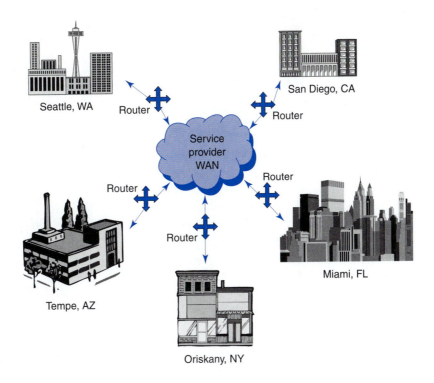

FIGURE 1-30 Wide area network (WAN)

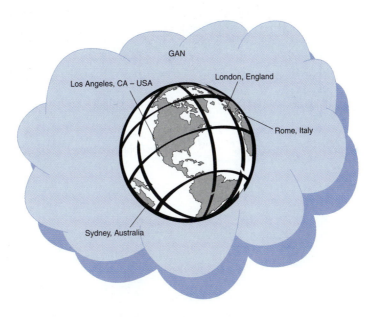

FIGURE 1-31 Global area network (GAN)

1-10-4-6 Campus backbone. A *campus backbone* is a network connection used to carry traffic to and from LANs located in various buildings on campus (see Figure 1-33). A campus backbone is designed for sites that have a group of buildings at a single location, such as corporate headquarters, universities, airports, and research parks.

A campus backbone normally uses optical fiber cables for the transmission media between buildings. The optical fiber cable is used to connect interconnecting devices, such as

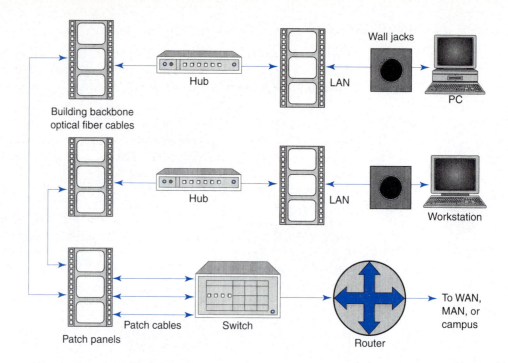

FIGURE 1-32 Building backbone

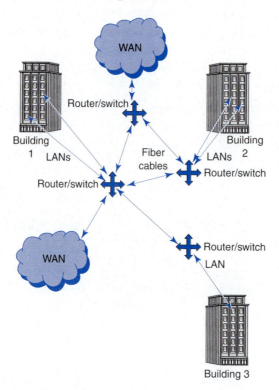

FIGURE 1-33 Campus backbone

bridges, routers, and switches. Campus backbones must operate at relatively high transmission rates to handle the large volumes of traffic between sites.

1-10-4-7 Enterprise networks. An *enterprise network* includes some or all of the previously mentioned networks and components connected in a cohesive and manageable fashion.

The functional layers of the OSI seven-layer protocol hierarchy do not line up well with certain data communications applications, such as the Internet. Because of this, there are several other protocols that see widespread use, such as TCP/IP and the Cisco three-layer hierarchical model.

1-11-1   TCP/IP Protocol Suite

The *TCP/IP protocol suite* (*transmission control protocol/Internet protocol*) was actually developed by the Department of Defense before the inception of the seven-layer OSI model. TCP/IP is comprised of several interactive modules that provide specific functionality without necessarily operating independent of one another. The OSI seven-layer model specifies exactly which function each layer performs, whereas TCP/IP is comprised of several relatively independent protocols that can be combined in many ways, depending on system needs. The term *hierarchical* simply means that the upper-level protocols are supported by one or more lower-level protocols. Depending on whose definition you use, TCP/IP is a hierarchical protocol comprised of either three or four layers.

The three-layer version of TCP/IP contains the *network, transport,* and *application* layers that reside above two lower-layer protocols that are not specified by TCP/IP (the physical and data link layers). The network layer of TCP/IP provides internetworking functions similar to those provided by the network layer of the OSI network model. The network layer is sometimes called the *internetwork layer* or *internet layer.*

The transport layer of TCP/IP contains two protocols: TCP (transmission control protocol) and UDP (user datagram protocol). TCP functions go beyond those specified by the transport layer of the OSI model, as they define several tasks defined for the session layer. In essence, TCP allows two application layers to communicate with each other.

The applications layer of TCP/IP contains several other protocols that users and programs utilize to perform the functions of the three uppermost layers of the OSI hierarchy (i.e., the applications, presentation, and session layers).

The four-layer version of TCP/IP specifies the network access, Internet, host-to-host, and process layers:

> *Network access layer.* Provides a means of physically delivering data packets using frames or cells
>
> *Internet layer.* Contains information that pertains to how data can be routed through the *network*
>
> *Host-to-host layer.* Services the process and Internet layers to handle the reliability and session aspects of data transmission
>
> *Process layer.* Provides applications support

TCP/IP is probably the dominant communications protocol in use today. It provides a common denominator, allowing many different types of devices to communicate over a network or system of networks while supporting a wide variety of applications.

1-11-2   Cisco Three-Layer Model

Cisco defines a three-layer logical hierarchy that specifies where things belong, how they fit together, and what functions go where. The three layers are the core, distribution, and access:

> *Core layer.* The core layer is literally the core of the network, as it resides at the top of the hierarchy and is responsible for transporting large amounts of data traffic reliably and quickly. The only purpose of the core layer is to switch traffic as quickly as possible.

Distribution layer. The distribution layer is sometimes called the *workgroup layer.* The distribution layer is the communications point between the access and the core layers that provides routing, filtering, WAN access, and how many data packets are allowed to access the core layer. The distribution layer determines the fastest way to handle service requests; for example, the fastest way to forward a file request to a server. Several functions are performed at the distribution level:

1. Implementation of tools such as access lists, packet filtering, and queuing
2. Implementation of security and network policies, including firewalls and address translation
3. Redistribution between routing protocols
4. Routing between virtual LANs and other workgroup support functions
5. Definition of broadcast and multicast domains

Access layer. The access layer controls workgroup and individual user access to internetworking resources, most of which are available locally. The access layer is sometimes called the *desktop layer.* Several functions are performed at the access layer level:

1. Access control
2. Creation of separate collision domains (segmentation)
3. Workgroup connectivity into the distribution layer

QUESTIONS

1-1. Define the following terms: *data, information,* and *data communications network.*

1-2. What was the first data communications system that used binary-coded electrical signals?

1-3. Discuss the relationship between network architecture and protocol.

1-4. Briefly describe broadcast and point-to-point computer networks.

1-5. Define the following terms: *protocol, connection-oriented protocols, connectionless protocols,* and *protocol stacks.*

1-6. What is the difference between syntax and semantics?

1-7. What are data communications standards, and why are they needed?

1-8. Name and briefly describe the differences between the two kinds of data communications standards.

1-9. List and describe the eight primary standards organizations for data communications.

1-10. Define the open systems interconnection.

1-11. Briefly describe the seven layers of the OSI protocol hierarchy.

1-12. List and briefly describe the basic functions of the five components of a data communications circuit.

1-13. Briefly describe the differences between serial and parallel data transmission.

1-14. What are the two basic kinds of data communications circuit configurations?

1-15. List and briefly describe the four transmission modes.

1-16. List and describe the functions of the most common components of a computer network.

1-17. What are the differences between servers and clients on a data communications network?

1-18. Describe a peer-to-peer data communications network.

1-19. What are the differences between peer-to-peer client/server networks and dedicated client/server networks?

1-20. What is a data communications network topology?

1-21. List and briefly describe the five basic data communications network topologies.

1-22. List and briefly describe the major network classifications.

1-23. Briefly describe the TCP/IP protocol model.

1-24. Briefly describe the Cisco three-layer protocol model.

C H A P T E R   2

Signals, Noise, Modulation, and Demodulation

CHAPTER OUTLINE

OBJECTIVES

- Define *analog* and *digital signals* and describe the differences between them
- Describe the differences between digital modulation and digital transmission
- Define *signal analysis*
- Describe the parameters amplitude, frequency, and phase
- Describe what is meant by a periodic wave
- Describe the differences between the time domain and the frequency domain
- Define *complex wave*
- Describe the Fourier series
- Define *wave symmetry* and describe what is meant by even, odd, and half-wave symmetry
- Define *frequency spectrum* and *bandwidth*
- Define *electrical noise* and describe the most common types
- Explain signal-to-noise ratio
- Define *modulation*
- Describe the following analog modulation techniques: amplitude modulation, frequency modulation, and phase modulation
- Define the following terms: *information capacity, bits, bit rate, baud,* and M-*ary encoding*
- Describe the following digital modulation techniques: *digital amplitude modulation, frequency-shift keying, phase-shift keying,* and *quadrature amplitude modulation*
- Define *bandwidth efficiency*
- Describe Trellis code modulation
- Describe probability of error and bit error rate

Electrical signals can be in *analog* or *digital* form. With analog signals, the amplitude changes continuously with respect to time with no breaks or discontinuities. Figure 2-1a shows a sine wave, which is the most basic analog signal. As the figure shows, the amplitude of the sine wave varies continuously with time between its maximum value, V_{max}, and its minimum value, V_{min}. The time a sine wave is above its average amplitude equals the time it is below its average value, and the time of the positive and negative half cycles are equal (t).

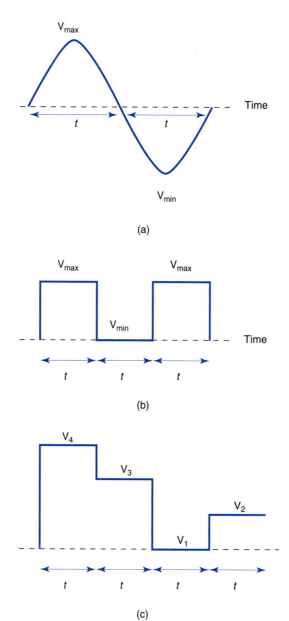

FIGURE 2-1   Electrical signals:
(a) sine wave; (b) binary digital signal;
(c) quaternary digital signal

Digital signals are described as discrete; their amplitude maintains a constant level for a prescribed period of time, and then it changes to another level. If there are only two levels possible, it is called a *binary signal*. All binary signals are digital, but all digital signals are not necessarily binary. Figure 2-1b shows a binary digital signal (called a *pulse*). As the figure shows, the amplitude of the waveform is at its maximum value, V_{max}, for t seconds and then changes to its minimum value, V_{min}, for t seconds. The only time the signal is not at either V_{max} or V_{min} is when it is transitioning between the maximum and minimum values. However, the voltage is stable (constant) only when it is at V_{max} or V_{min}. Figure 2-1c shows a four-level digital signal. Because there are four levels (V_1, V_2, V_3, and V_4), the signal is called a *quaternary* digital signal. Again, the voltage is constant for t seconds, then it changes to one of the three other values.

Although we generally think of data communications systems as being digital, many of them actually utilize both analog and digital signals. Because it is often impractical and sometimes impossible to propagate information signals over metallic or optical fiber cables or through the earth's atmosphere, it is often necessary to change the form of the source information. Converting information signals to a different form is called *modulation*, and the reverse process is called *demodulation*. *Modulate* simply means to change; therefore, when an analog signal is being modulated, some property of it is changing proportional to the *modulating signal*. The modulating signal is the information, and the signal being modulated is called the *carrier*.

The two basic types of electronic communications systems are *analog* and *digital*. An *analog communications system* is a communications system in which energy is transmitted and received in analog form (continuously varying signals, such as sine or cosine waves). With analog communications systems, the information signals are analog (such as voice or music), and they are propagated through the system in analog form as shown in Figure 2-2a.

The term *digital communications,* however, covers a broad range of communications techniques, including *digital transmission* and *digital modulation*. Digital transmission is a true digital system where pulses (discrete levels, such as +5 V and ground) are transferred directly between two or more points in a communications system. Digital transmission systems require a physical facility between the transmitter and receiver, such as a metallic wire or an optical fiber cable. With digital transmission systems, the original source information may be in analog or digital form; however, if it is in analog form, it must be converted to digital prior to transmission. Figure 2-2b shows how digital transmission is accomplished when the original source information is analog or digital. Digital transmission is described in detail in Chapter 3.

Digital modulation is the transmittal of digitally modulated analog signals between two or more points in a communications system. With digital modulation, the modulating and demodulated signals are digital pulses; however, they are carried through the system on an analog signal (the carrier signal). Also, the original source information with digital modulation may be in digital or analog form; however, if it is analog, it must be converted to digital before modulating the carrier. Digital modulation is shown in Figure 2-2c. With digital modulation, the transmission medium may be a physical cable facility or free space (i.e., the earth's atmosphere).

2-2 SIGNAL ANALYSIS

When designing electronic communications systems, it is often necessary to analyze and predict the performance of the circuit on the basis of the voltage distribution and frequency composition of the information signal. This is done with mathematical signal analysis. Although all signals in electronic communications systems are not single-frequency sine or cosine waves, many of them are, and the signals that are not can often be represented by combinations of sine and cosine waves.

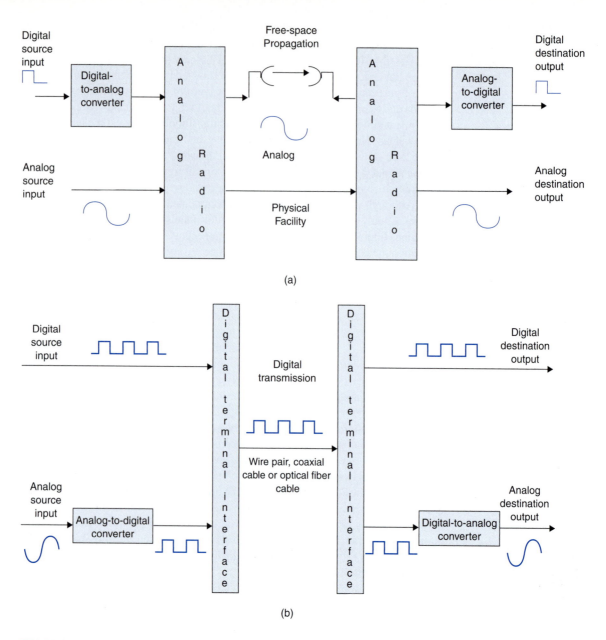

FIGURE 2-2 Analog and digital communications systems: (a) analog communications system; (b) digital transmission

2-2-1 Amplitude, Frequency, and Phase

Figure 2-3 shows three cycles of a sine wave. A *cycle* is one complete variation in the signal, and the *period* is the time the waveform takes to complete one cycle (*T*). One cycle constitutes 360 degrees (or 2π radians). Sine waves can be described in terms of three parameters: *amplitude, frequency,* and *phase.*

2-2-1-1 Amplitude. Amplitude is analogous to magnitude or displacement. The amplitude of a signal is the magnitude of the signal at any point on the waveform. Signal amplitude is generally represented on the vertical axis of a waveform graph as shown in

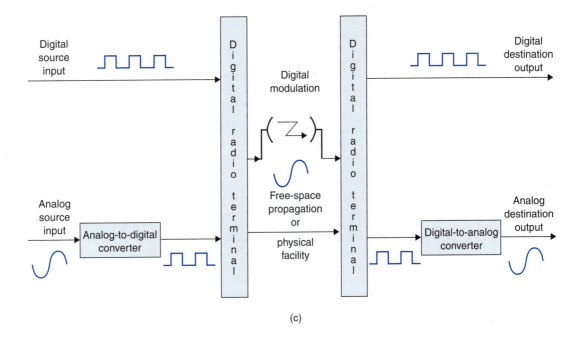

FIGURE 2-2 (c) digital modulation

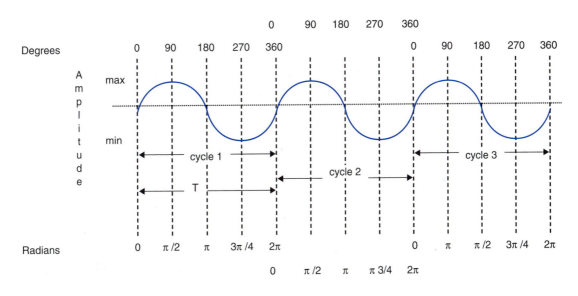

FIGURE 2-3 Three sine waves showing amplitude, frequency, and phase

Figure 2-3. Thus, the amplitude is the vertical displacement or the distance from the waveform to the horizontal axis. The amplitude of electrical signals is generally measured in voltage. The maximum voltage of a signal in respect to its average value (i.e., the vertical center of the waveform) is called its *peak amplitude* or *peak voltage* (V).

2-2-1-2 Frequency. The time of one cycle of a waveform is its period, which is measured in seconds (see Figure 2-3). Time is generally represented on the horizontal axis of a waveform graph. The reciprocal of the time of one cycle of a waveform is frequency (f). Mathematically, the relationship between frequency and period is

$$f = \frac{1}{T} \tag{2-1}$$

where f = cycles per second (hertz)
         T = time of one cycle (seconds)

and $T = \frac{1}{f}$ \hfill (2-2)

Frequency is technically measured in cycles per second but is expressed in hertz, after the German physicist Heinrich Rudolf Hertz. Frequency is measured in hertz (Hz), kilohertz (kHz = 10^3 Hz), megahertz (MHz = 10^6 Hz), gigahertz (GHz = 10^9 Hz), terahertz (THz = 10^{12} Hz), petahertz (PHz = 10^{15} Hz), and exahertz (Ehz = 10^{18} Hz). Common units for time are nanoseconds (10^{-9} seconds), microseconds (10^{-6} seconds), and milliseconds (10^{-3} seconds).

2-2-1-3 Phase.   The phase of a signal is measured in degrees or radians (360 degrees = 2π radians) with respect to a reference point. A phase of 45 degrees means the waveform is shifted one-eighth of a cycle from the reference. A phase shift of 180 degrees corresponds to a shift of half a cycle (i.e., a complete inversion of a sine wave), and a phase shift of 360 degrees corresponds to a shift of one complete cycle. Phase, like time, is shown on the horizontal axis of a signal graph. The cycle begins at 0 degrees, the maximum amplitude occurs at 90 degrees, the waveform crosses the horizontal axis at 0 degrees and again at 180 degrees, the minimum amplitude occurs at 270 degrees, and the end of the waveform is at 360 degrees. For repetitive waveforms, 360 degrees on one cycle is 0 degrees on the following cycle.

Figure 2-4 shows two sine waves (A and B) at the same frequency. However, sine wave A is half the amplitude (5 volts) of sine wave B (10 volts), and sine wave B is shifted 90 degrees in respect to sine wave A.

Example 2-1

A sine wave has a frequency of 1000 Hz (1 kHz). Determine its period.

Solution   From Equation 2-2,

$$T = \frac{1}{1000} = 1 \text{ ms}$$

Example 2-2

A sine wave has a period of 2 ms. Determine its frequency.

Solution   From Equation 2-1,

$$f = \frac{1}{0.002} = 500 \text{ Hz}$$

2-2-2   Periodic Signals

In essence, *signal analysis* is the mathematical analysis of the amplitude, frequency, and phase of a signal. Electrical signals are voltage- or current-time variations. Mathematically, a single-frequency voltage waveform is

$$v(t) = V \sin(2\pi f t + \theta) \tag{2-3}$$

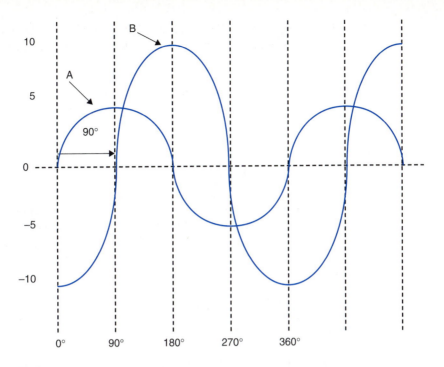

FIGURE 2-4 Comparison of two sine waves of different amplitudes and phases

where $v(t)$ = time-varying voltage sine wave
  V = peak amplitude (volts)
  f = frequency (hertz)
  t = time (seconds)
  θ = phase (degrees or radians)

Equation 2-3 is for a single-frequency, repetitive waveform. Such a waveform is called a *periodic wave* because it repeats at a uniform rate (i.e., each successive cycle of the signal takes exactly the same length of time and has exactly the same amplitude variations as every other cycle). A series of sine, cosine, or square waves constitute an example of periodic waves. Periodic waves can be analyzed in either the *time domain* or the *frequency domain*.

Example 2-3

For the following voltage waveform, determine the frequency (f), peak voltage amplitude (V), and phase (θ):

$$v(t) = 6 \sin(2\pi 10{,}000t + 30)$$

Solution   From Equation 2-3,

$$f = 10{,}000 \text{ Hz or 10 kHz}$$
$$V = 6 \text{ volts peak}$$
$$\theta = 30 \text{ degrees}\quad\text{or}\quad \pi/6 \text{ radians}$$

2-2-2-1 Time domain.   A standard oscilloscope is a time-domain instrument. The display on the cathode ray tube (CRT) is an amplitude-versus-time representation of the

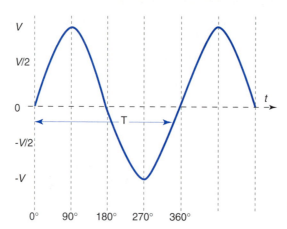

FIGURE 2-5   Time domain representation of a single-frequency sine wave

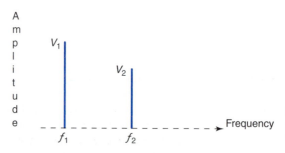

FIGURE 2-6   Frequency spectrum
(frequency domain representation) of
two sine waves

signal and is commonly called a *signal waveform*. Essentially, a signal waveform shows
the shape and instantaneous magnitude of the signal with respect to time but does not directly indicate its frequency content. With an oscilloscope, the vertical deflection is proportional to the amplitude of the input signal, and the horizontal deflection is a function of
time (sweep rate). Figure 2-5 shows the signal waveform for a single-frequency sinusoidal
signal with a peak amplitude of V volts and a frequency of $f = 1/T$ hertz.

   2-2-2-2 Frequency domain.   A spectrum analyzer is a frequency-domain instrument. Essentially, no waveform is displayed on the CRT. Instead, an amplitude-versus-
frequency plot is shown (this is called a *frequency spectrum*). With a spectrum analyzer, the
horizontal axis represents frequency and the vertical axis amplitude. Therefore, there is a
vertical deflection for each frequency present in the waveform. The vertical deflection
(height) of each line is proportional to the amplitude of the frequency that it represents.
   Figure 2-6 shows the frequency spectrum for two sinusoidal signals with peak amplitudes of V_1 and V_2 volts and frequencies of f_1 and f_2 hertz, respectively.

2-2-3   Complex Signals

Essentially, any repetitive waveform that is comprised of more than one harmonically related sine or cosine wave is called a *nonsinusoidal, complex* wave. Thus, a complex wave
is any periodic (repetitive) waveform that is not a sinusoid, such as square waves, rectan-
gular waves, and triangular waves. To analyze a complex periodic wave, it is necessary to
use a mathematical series developed in 1826 by the French physicist and mathematician
Baron Jean Fourier. This series is appropriately called the *Fourier series*.

   2-2-3-1 The Fourier series.   The Fourier series is used in signal analysis to represent the sinusoidal components of nonsinusoidal periodic waveforms (i.e., to change a time-

domain signal to a frequency-domain signal). In general, a Fourier series can be written for any periodic function as a series of terms that include trigonometric functions with the following mathematical expression:

$$f(t) = A_0 + A_1 \cos \alpha + A_2 \cos 2\alpha + A_3 \cos 3\alpha + \dots A_n \cos n\alpha$$

$$+ A_0 + B_1 \sin \beta_1 + B_2 \sin 2\beta + B_3 \sin 3\beta + \dots B_n \sin n\beta \qquad \text{(2-4)}$$

where $\alpha = \beta$

Equation 2-4 states that the waveform $f(t)$ comprises an average (dc) value (A_0) and either sine or cosine functions in which each successive term has a frequency that is an integer multiple of the frequency of the first term in the series. There are no restrictions on the values or relative values of the amplitudes for the sine or cosine terms. Equation 2-4 is expressed in words as follows: any *periodic waveform* is comprised of an average dc component and a series of harmonically related sine or cosine waves. A *harmonic* is an integral multiple of the *fundamental frequency*. The fundamental frequency is the first harmonic and is equal to the frequency (*repetition rate*) of the waveform. The second multiple of the fundamental frequency is called the *second harmonic,* the third multiple is called the *third harmonic,* and so forth. The fundamental frequency is the minimum frequency necessary to represent a waveform. Therefore, Equation 2-4 can be rewritten as

$$f(t) = \text{dc} + \text{fundamental} + \text{2nd harmonic} + \text{3rd harmonic} + \dots n\text{th harmonic} \qquad \text{(2-5)}$$

2-2-3-2 Wave symmetry.   Simply stated, wave symmetry describes the symmetry of a waveform in the time domain, that is, its relative position with respect to the horizontal (time) and vertical (amplitude) axes.

2-2-3-3 Even symmetry.   If a periodic voltage waveform is *symmetric* about the vertical axis, it is said to have *axes,* or *mirror, symmetry* and is called an *even function.* For all even functions, the β coefficients in Equation 2-4 are zero. Therefore, the signal simply contains a dc component and the cosine terms (note that a cosine wave is itself an even function). Even functions satisfy the condition

$$f(t) = f(-t) \qquad \text{(2-6)}$$

Equation 2-6 states that the magnitude and polarity of the function at $+t$ is equal to the magnitude and polarity at $-t$. A waveform that contains only the even functions is shown in Figure 2-7a.

2-2-3-4 Odd symmetry.   If a periodic voltage waveform is symmetric about a line midway between the vertical axis and the negative horizontal axis (i.e., the axes in the second and fourth quadrants) and passing through the coordinate origin, it is said to have *point,* or *skew, symmetry* and is called an *odd function.* For all odd functions, the α coefficients in Equation 2-4 are zero. Therefore, the signal simply contains a dc component and the sine terms (note that a sine wave is itself an odd function). Odd functions must be mirrored first in the Y-axis and then in the X-axis for superposition. Thus,

$$f(t) = -f(-t) \qquad \text{(2-7)}$$

Equation 2-7 states that the magnitude of the function at $+t$ is equal to the negative of the magnitude at $-t$ (i.e., equal in magnitude but opposite in sign). A periodic waveform that contains only the odd functions is shown in Figure 2-7b.

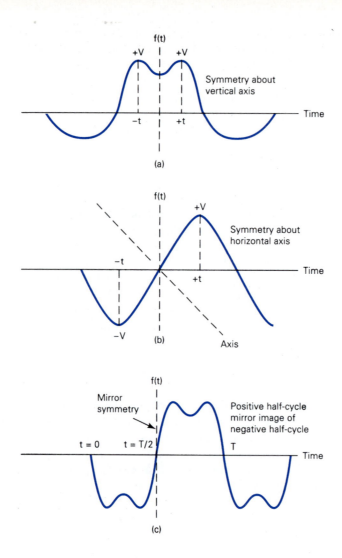

FIGURE 2-7   Wave symmetries: (a) even symmetry; (b) odd symmetry; (c) half-wave symmetry

2-2-3-5 Half-wave symmetry.   If a periodic voltage waveform is such that the waveform for the first half cycle ($t = 0$ to $t = T/2$) repeats itself except with the opposite sign for the second half cycle ($t = T/2$ to $t = T$), it is said to have *half-wave symmetry*. For all waveforms with half-wave symmetry, the even harmonics in the series for both the sine and cosine terms are zero. Therefore, half-wave functions satisfy the condition

$$f(t) = \frac{-f(T + t)}{2}$$

(2-8)

A periodic waveform that exhibits half-wave symmetry is shown in Figure 2-7c. It should be noted that a waveform could have half-wave as well as either odd or even symmetry at the same time.

Table 2-1 summarizes the Fourier series for several of the more common nonsinusoidal waveforms.

Table 2-1 Fourier Series Summary

Waveform	Fourier Series
odd 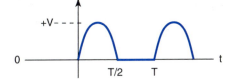	$v(t) = \dfrac{V}{\pi} + \dfrac{V}{2}\sin \omega t - \dfrac{2V}{3\pi}\cos 2\omega t - \dfrac{2V}{15\pi}\cos 4\omega t + \cdots$ $v(t) = \dfrac{V}{\pi} + \dfrac{V}{2}\sin \omega t + \displaystyle\sum_{N=2}^{\infty} \dfrac{V[1 + (-1)^{N}]}{\pi(1 - N^2)}\cos N\omega t$
even 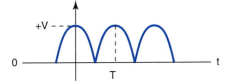	$v(t) = \dfrac{2V}{\pi} + \dfrac{4V}{3\pi}\cos \omega t - \dfrac{4V}{15\pi}\cos 2\omega t + \cdots$ $v(t) = \dfrac{2V}{\pi} + \displaystyle\sum_{N=1}^{\infty} \dfrac{4V(-1)^{N}}{\pi[1 - (2N)^2]}\cos N\omega t$
odd 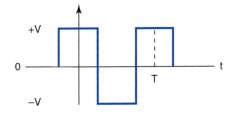	$v(t) = \dfrac{4V}{\pi}\sin \omega t + \dfrac{4V}{3\pi}\sin 3\omega t + \cdots$ $v(t) = \displaystyle\sum_{N=\text{odd}}^{\infty} \dfrac{4V}{N\pi}\sin N\omega t$
even 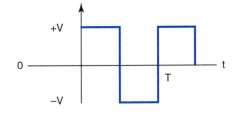	$v(t) = \dfrac{4V}{\pi}\cos \omega t - \dfrac{4V}{3\pi}\cos 3\omega t + \dfrac{4V}{5\pi}\cos 5\omega t + \cdots$ $v(t) = \displaystyle\sum_{N=\text{odd}}^{\infty} \dfrac{V\sin N\pi/2}{N\pi/2}\cos N\omega t$
even 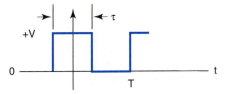	$v(t) = \dfrac{v\tau}{T} + \displaystyle\sum_{N=1}^{\infty} \left(\dfrac{2V\tau}{T} \dfrac{\sin N\omega t/T}{N\pi t/T} \right)\cos N\pi t$
even 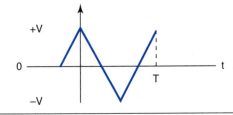	$v(t) = \dfrac{8V}{\pi^2}\cos \omega t + \dfrac{8V}{(3\pi)^2}\cos 3\omega t + \dfrac{8V}{(5\pi)^2}\cos 5\omega t + \cdots$ $v(t) = \displaystyle\sum_{N=\text{odd}}^{\infty} \dfrac{8V}{(N\pi)^2}\cos N\omega t$

Example 2-4

For the square waves shown in Figure 2-8, determine the peak amplitudes and frequencies for the first five odd harmonics.

Solution From inspection of the waveform in Figure 2-8, it can be seen that it is a square wave, the average dc voltage is 0 V, and the waveform has both odd and half-wave symmetry. From examining the waveforms in Table 2-1, it can be determined that the following Fourier series for a square wave with odd symmetry describes the waveform:

$$v(t) = V_0 + \frac{4V}{\pi} \left[\sin(\omega t) + \frac{1}{3} \sin(3\omega t) + \frac{1}{5} \sin(5\omega t) + \frac{1}{7} \sin(7\omega t) + \ldots \right] \qquad \textbf{(2-9)}$$

where $v(t)$ = time-varying voltage function
V_0 = average dc voltage (volts)
V = peak amplitude of the square wave (volts)
$\omega = 2\pi f$ (radians per second)
T = period of the square wave (seconds)
f = fundamental frequency of the square wave ($1/T$ hertz)

The fundamental frequency of the square wave is

$$f = \frac{1}{T} = \frac{1}{0.001} = 1000 \text{ Hz}$$

From Equation 2-9, it can be seen that the frequency and amplitude of the nth odd harmonic can be determined from the following expressions:

$$f_n = n \times f \qquad \textbf{(2-10a)}$$

$$V_n = \frac{4V}{n\pi} \qquad n = \text{odd positive integer value} \qquad \textbf{(2-10b)}$$

where n = nth harmonic (odd harmonics only for a square wave)
f = fundamental frequency of the square wave ($1/T$) (hertz)
V_n = peak amplitude of the nth harmonic (volts)
f_n = frequency of the nth harmonic (hertz)
V = peak amplitude of the square wave (volts)

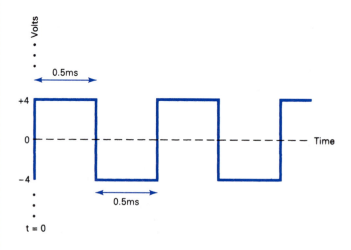

FIGURE 2-8 Waveform for Example 2–1

Substituting $n = 1, 3, 5, 7$, and 9 into Equations 2-10a and 2-10b gives the following frequencies and voltages:

n	Harmonic	Frequency (Hz)	Peak Voltage (V)
1	First	1000	5.09
3	Third	3000	1.69
5	Fifth	5000	1.02
7	Seventh	7000	0.73
9	Ninth	9000	0.57

2-2-4 Frequency Spectrum and Bandwidth

The *frequency spectrum* of a waveform consists of all the frequencies contained in the waveform and their respective amplitudes plotted in the frequency domain. The frequency spectrum for Example 2-4 is shown in Figure 2-9. Frequency spectrums can show absolute values of frequency versus voltage or frequency versus power level, or they can plot frequency versus some relative unit of measurement, such as decibels (dB).

The term *bandwidth* can be used in several ways. The bandwidth of a frequency spectrum is the range of frequencies contained in the spectrum. The bandwidth is calculated by subtracting the lowest frequency from the highest. The bandwidth of the frequency spectrum shown in Figure 2-9 is 8000 Hz (9000 − 1000).

The bandwidth of an information signal is simply the difference between the highest and lowest frequencies contained in the information, and the bandwidth of a communications channel is the difference between the highest and lowest frequencies that the channel will allow to pass through it (i.e., its *passband*). The bandwidth of a communications channel must be sufficiently large (wide) to pass all significant information frequencies. In other words, the bandwidth of a communications channel must be equal to or greater than the bandwidth of the information signal. Speech contains frequency components ranging from approximately 100 Hz to 8 kHz, although most of the energy is distributed in the 400-Hz to 600-Hz band with the fundamental frequency of typical human voice about 500 Hz. However, standard telephone circuits have a passband between 300 Hz and 3000 Hz, as shown in Figure 2-10, which equates to a bandwidth of 2700 Hz (3000 − 300). Twenty-seven hundred hertz is well beyond what is necessary to convey typical speech information. If a cable television transmission system has a passband from 500 kHz to 5000 kHz, it has a bandwidth of 4500 kHz (4.5 MHz). As a general rule, a communications channel cannot propagate a signal through it that is changing at a rate that exceeds the bandwidth of the channel.

In general, the more complex the information signal, the more bandwidth required to transport it through a communications system in a given period of time. Approximately

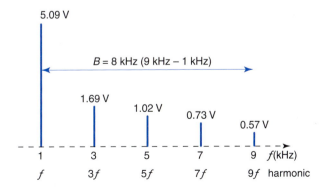

FIGURE 2-9 Frequency spectrum for Example 2–1

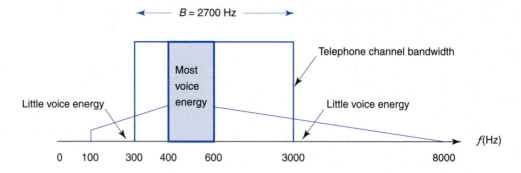

FIGURE 2-10 Voice-frequency spectrum and telephone circuit bandwidth

3 kHz of bandwidth is required to propagate one voice-quality analog telephone conversation. In contrast, it takes approximately 32 kHz of bandwidth to propagate one voice-quality digital telephone conversation. Commercial FM broadcasting stations require 200 kHz of bandwidth to propagate high-fidelity music signals, and almost 6 MHz of bandwidth is required for broadcast-quality television signals.

2-3 ELECTRICAL NOISE AND SIGNAL-TO-NOISE RATIO

2-3-1 Electrical Noise

Electrical noise is defined as any undesirable electrical energy that falls within the passband of the signal. For example, in audio recording, any unwanted electrical signals that fall within the audio frequency band of 0 Hz to 15 kHz will interfere with the music and, therefore, are considered noise. Figure 2-11 shows the effect that noise has on an electrical signal. Figure 2-11a shows a sine wave without noise, and Figure 2-11b shows the same signal except in the presence of noise. The grassy-looking squiggles superimposed on the sine wave in Figure 2-11b are electrical noise, which contains a multitude of frequencies and amplitudes that can interfere with the quality of the signal.

Although there are many types and classifications of electrical noise, this discussion is limited to the types of noise that are the most prevalent and the most interfering to data communications signals: *man-made noise, thermal noise, correlated noise*, and *impulse noise*.

2-3-1-1 Man-made noise. *Man-made noise* is simply noise that is produced by mankind. The predominant sources of man-made noise are spark-producing mechanisms, such as commutators in electric motors, automobile ignition systems, ac power–generating and switching equipment, and fluorescent lights. Man-made noise is impulsive in nature and contains a wide range of frequencies that are propagated through space in the same manner as radio waves. Man-made noise is most intense in the more densely populated metropolitan and industrial areas and is, therefore, sometimes called *industrial noise*.

2-3-1-2 Thermal noise. *Thermal noise* is associated with the rapid and random movement of electrons within a conductor due to thermal agitation and is present in all electronic components and communications systems. Because thermal noise is uniformly distributed across the entire electromagnetic frequency spectrum, it is often referred to as *white noise* (analogous to the color white containing all colors [frequencies] of light). Thermal noise is a form of additive noise, meaning that it cannot be eliminated and that it increases in intensity with the number of devices in a circuit and with circuit length. Therefore, thermal noise sets the upper bound on the performance of a communications system.

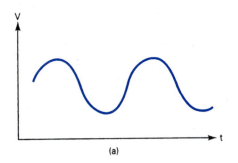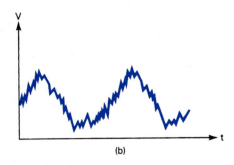

FIGURE 2-11   Effects of noise on a signal: (a) signal without noise; (b) signal with noise

Noise power density is the thermal noise power present in a 1-Hz bandwidth. Mathematically, noise power density is

$$N_o = KT \qquad (2\text{-}11)$$

where N_o = noise power density (watts per hertz [W/Hz])
K = Boltzmann's proportionality constant (1.38×10^{-23} joules per degree Kelvin)
T = absolute temperature (Kelvin) (0 K = $-273°$C, and room temperature = $17°$C or 290 K)

Thermal noise is independent of frequency; thus, the thermal noise present in any bandwidth is

$$N = KTB \qquad (2\text{-}12)$$

where N = thermal noise power (watts)
B = bandwidth (hertz)

Noise power is generally expressed in dBm as

$$N_{(dBm)} = 10 \log\left(\frac{KTB}{0.001}\right) \qquad (2\text{-}13)$$

Rewriting Equation 2-13 gives

$$N_{dBm} = 10 \log\left(\frac{KT}{0.001}\right) + 10 \log B \qquad (2\text{-}14)$$

At room temperature Equation 2-14 can be expressed as

$$N_{dBm} = -174 \text{ dBm} + 10 \log B \qquad (2\text{-}15)$$

Example 2-5

For an electronic device operating at 17°C with a bandwidth of 10 kHz, determine the thermal noise power in watts and dBm.

Solution   Thermal noise power in watts is determined by substituting into Equation 2-12:

$$N = KTB$$
$$= (1.38 \times 10^{-23})(290)(10,000)$$
$$= 4 \times 10^{-17} \text{ watts}$$

Thermal noise in dBm is calculated using Equation 2-13:

$$N_{\text{dBm}} = 10 \log\left(\frac{4 \times 10^{-17}}{0.001}\right)$$

$$= -134 \text{ dBm}$$

or by using Equation 2-15:

$$N_{\text{dBm}} = -174 \text{ dBm} + 10 \log(10{,}000)$$

$$= -134 \text{ dBm}$$

2-3-1-3 Correlated noise. *Correlated noise* is noise that is correlated (mutually related) to the signal and cannot be present in a circuit unless there is a signal—simply stated, *no signal, no noise.* Correlated noise is produced by *nonlinear amplification* and includes *harmonic distortion* and *intermodulation distortion,* both of which are forms of *nonlinear distortion.* All circuits are nonlinear to some extent; therefore, all data communications circuits produce nonlinear distortion. Nonlinear distortion creates unwanted frequencies that interfere with the signal and degrade performance.

Harmonic distortion occurs when unwanted *harmonics* of a signal are produced through nonlinear amplification (*nonlinear mixing*). As previously stated, harmonics are integer multiples of the original signal. The original signal is the first harmonic and is called the *fundamental frequency.* Two times the original signal frequency is the second harmonic, three times is the third harmonic, and so forth. *Amplitude distortion* is another name for harmonic distortion.

There are various degrees of harmonic distortion. Second-order harmonic distortion is the ratio of the rms amplitude of the second harmonic to the rms amplitude of the fundamental frequency. Third-order harmonic distortion is the ratio of the rms amplitude of the third harmonic to the rms value of the fundamental. A more meaningful measurement is total harmonic distortion (TDH), which is the ratio of the quadratic sum of the rms values of all the higher harmonics to the rms value of the fundamental frequency.

Figure 2-12a shows the input and output frequency spectrums for a nonlinear device with a single input frequency (f_1). As the figure shows, the output spectrum contains the original input frequency plus several harmonics ($2f_1, 3f_1, 4f_1$) that were not part of the original signal.

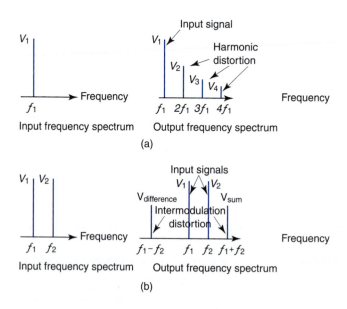

FIGURE 2-12 Correlated noise: (a) harmonic distortion; (b) intermodulation distortion

Intermodulation distortion is the generation of unwanted *sum* and *difference* frequencies produced when two or more signals are amplified in a nonlinear device. The sum and difference frequencies are called *cross products*. The emphasis here is on the word *unwanted* because in communications circuits it is often desirable to produce harmonics or to mix two or more signals to produce sum and difference frequencies. Cross-product frequencies can interfere with the information signals in a circuit or with the information signals in other circuits.

Figure 2-12b shows the input and output frequency spectrums for a nonlinear device with two input frequencies (f_1 and f_2). As the figure shows, the output spectrum contains the two original frequencies plus their sum and difference frequencies ($f_1 - f_2$ and $f_1 + f_2$). In actuality, the output spectrum would also contain the harmonics of the two input frequencies ($2f_1$, $3f_1$, $2f_2$, and $3f_2$), as the same nonlinearities that caused the intermodulation distortion would also cause harmonic distortion. The harmonics have been eliminated from the diagram for simplicity.

2-3-1-4 Impulse noise. *Impulse noise* is characterized by high-amplitude peaks of short duration in the total noise spectrum. As the name implies, impulse noise consists of sudden bursts of irregularly shaped pulses that generally last between a few microseconds and several milliseconds, depending on their amplitude and origin. The significance of impulse noise (hits) on voice communications is often more annoying than inhibitive, as impulse hits produce a sharp popping or crackling sound. On data circuits, however, impulse noise can be devastating.

Common sources of impulse noise include transients produced from electromechanical switches (such as relays and solenoids), electric motors, appliances, electric lights (especially fluorescent), power lines, automotive ignition systems, poor-quality solder joints, and lightning.

2-3-1-5 Signal-to-noise power ratio. *Signal-to-noise power ratio* (S/N) is the ratio of the signal power level to the thermal noise power level. Mathematically, signal-to-noise power ratio is expressed as

$$\frac{S}{N} = \frac{P_S}{P_N} \qquad (2\text{-}16a)$$

where S/N = signal-to-noise power ratio (unitless)
           P_S = signal power (watts)
           P_N = noise power (watts)

The signal-to-noise power ratio is often expressed in dB as

$$\frac{S}{N}(\text{dBm}) = 10 \log \frac{P_S}{P_N} \qquad (2\text{-}16b)$$

Example 2-6

For a circuit with a signal power of 10 W and a thermal noise power of 0.01 W, determine the signal-to-noise power ratio.

Solution   The signal-to-noise power ratio is found by substituting into Equations 2-16a and 2-16b:

$$\frac{S}{N} = \frac{10}{0.01} = 1000$$

$$\frac{S}{N}(\text{dBm}) = 10 \log \frac{10}{0.01} = 30 \text{ dB}$$

As previously stated, Equation 2-3 is the general expression for a *time-varying* sine wave of voltage, such as a high-frequency carrier signal. If the information signal is analog and the amplitude (V) of the carrier is varied proportional to the information signal, *amplitude modulation* (AM) is produced. If the frequency (f) is varied proportional to the information signal, *frequency modulation* (FM) is produced, and if the phase (θ) is varied proportional to the information signal, *phase modulation* (PM) is produced. Frequency and phase modulation are very similar and often combined and are simply called *angle modulation*.

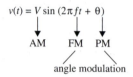

Information signals are transported between transmitters and receivers over some form of transmission medium. However, the original information signals are seldom in a form suitable for transmission. Therefore, they must be transformed from their original form into a form that is more suitable for transmission. The process of impressing relatively low-frequency information signals onto a high-frequency *carrier signal* is called *modulation*. *Demodulation* is the reverse process where the received signals are transformed back to their original form.

Analog modulation is used for the transmission of conventional analog signals, such as voice, music, and video, and it not particularly useful for data communications systems, so only a brief description is given here.

2-4-1 Amplitude Modulation

Amplitude modulation (AM) is the process of changing the amplitude of a relatively high-frequency carrier signal in proportion to the instantaneous value of the modulating signal (information). Amplitude modulation is a relatively inexpensive, low-quality form of modulation that is used for commercial broadcasting of both audio and video signals. Amplitude modulation is also used for two-way mobile radio communications, such as *citizens band* (CB) radio.

AM modulators are two-input devices. One input is a single, relatively high-frequency carrier signal of constant amplitude, and the second input is comprised of relatively low-frequency information signals that may be a single frequency or a complex waveform made up of many frequencies. In the modulator, the information signals act on or modulate the RF carrier, producing a *modulated wave*.

Figure 2-13 illustrates the relationship among the carrier ($V_c \sin[2\pi f_c t]$), the modulating signal ($V_m \sin[2\pi f_m t]$), and the modulated wave ($V_{am}[t]$). The figure shows how an AM waveform is produced when a single-frequency modulating signal acts on a high-frequency carrier signal. Since the output waveform contains all the frequencies that make up the AM signal and it is used to transport the information through the system, the shape of the modulated wave is called the *AM envelope*. Note that when there is no modulating signal, the output waveform is simply the carrier signal. However, when a modulating signal is applied, the amplitude of the output wave varies in accordance with the modulating signal.

2-4-2 Angle Modulation

Angle modulation results whenever the phase angle (θ) of a sinusoidal signal is varied with respect to time. Angle modulation includes both frequency (FM) and phase (PM) modulation. In essence, the difference between frequency and phase modulation lies in which prop-

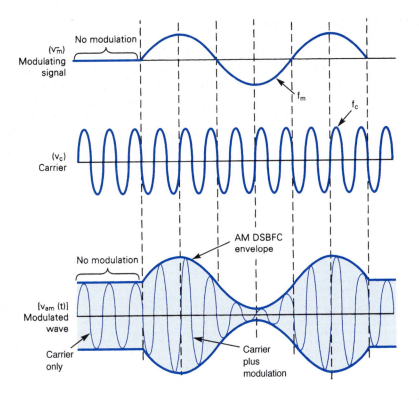

FIGURE 2-13 AM generation

erty of the carrier (frequency or phase) is directly varied by the modulating signal and which property is indirectly varied. Whenever the frequency of a carrier signal is varied, the phase is also varied and vice versa. Therefore, both FM and PM must occur whenever either form of modulation is performed. If the frequency of the carrier is varied directly in accordance with the information (modulating) signal, FM results. If the phase of the carrier is varied directly in accordance with the information signal, PM results.

Figure 2-14 shows an angle-modulated signal ($m[t]$) in the frequency domain. The figure shows how the carrier frequency (f_c) is changed when acted on by a modulating signal ($V_m[t]$). The magnitude and direction of the frequency shift (Δf) is proportional to the amplitude and polarity of the modulating signal (V_m).

Figure 2-15 shows a sinusoidal carrier in which the frequency (f) is changed (deviated) over a period of time. The fat portion of the waveform corresponds to the change in the period of the carrier (ΔT). The minimum period (T_{min}) corresponds to the maximum frequency ($f_{max} = f_c + \Delta f$), and the maximum period (T_{max}) corresponds to the minimum frequency ($f_{min} = f_c - \Delta f$).

Figure 2-16 illustrates both frequency and phase modulation of a sinusoidal carrier by a single-frequency modulating signal. It can be seen that the FM and PM waveforms are identical except for their time relationship (phase). Thus, it is impossible to distinguish an FM waveform from a PM waveform without knowing the dynamic characteristics of the modulating signal. With FM, the maximum frequency deviation (positive and negative change in the carrier frequency) occurs during the maximum positive and negative peaks of the modulating signal (i.e., the frequency deviation is proportional to the amplitude of the modulating signal). With PM, the maximum frequency deviation occurs during the zero crossings in the modulating signal (i.e., the frequency deviation is proportional to the slope of the modulating signal).

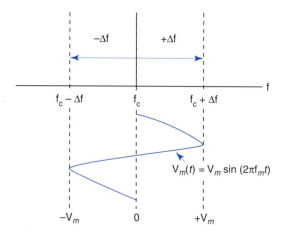

FIGURE 2-14   Angle-modulated wave in the frequency domain

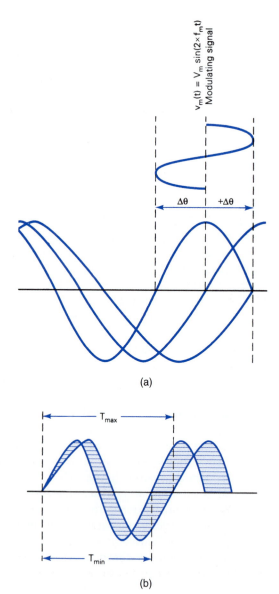

(a)

(b)

FIGURE 2-15   Angle modulation in the time domain: (a) phase changing with time; (b) frequency changing with time

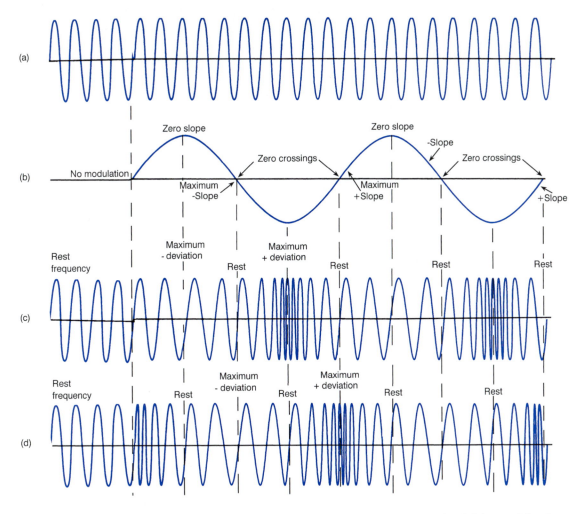

FIGURE 2-16 Phase and frequency modulation of a sine-wave carrier by a sine-wave signal: (a) unmodulated carrier; (b) modulating signal; (c) frequency-modulated wave; (d) phase-modulated wave

2-5 INFORMATION CAPACITY, BITS, BIT RATE, BAUD, AND *M*-ARY ENCODING

2-5-1 Information Capacity, Bits, and Bit Rate

Information theory is a highly theoretical study of the efficient use of bandwidth to propagate information through electronic communications systems. Information theory can be used to determine the *information capacity* of a data communications system. Information capacity is a measure of how much information can be propagated through a communications system and a function of bandwidth and transmission time.

Information capacity represents the number of independent symbols that can be carried through a system in a given unit of time. The most basic digital symbol used to represent information is the *binary digit,* or *bit*. Therefore, it is often convenient to express the information capacity of a system as a *bit rate*. Bit rate is simply the number of bits transmitted during 1 second and is expressed in *bits per second* (bps).

In 1928, R. Hartley of Bell Telephone Laboratories developed a useful relationship among bandwidth, transmission time, and information capacity. Simply stated, Hartley's law is

$$I \propto B \times t \qquad \text{(2-17)}$$

where I = information capacity (bits per second)
 B = bandwidth (hertz)
 t = transmission time (seconds)

From Equation 2-17 it can be seen that information capacity is a linear function of bandwidth and transmission time and is directly proportional to both. If either the bandwidth or the transmission time changes, a directly proportional change occurs in the information capacity.

In 1948, mathematician Claude E. Shannon (also of Bell Telephone Laboratories) published a paper in the *Bell System Technical Journal* relating the information capacity of a communications channel to bandwidth and *signal-to-noise ratio*. The higher the signal-to-noise ratio, the better the performance and the higher the information capacity. Mathematically stated, *the Shannon limit for information capacity* is

$$I = B \log_2\left(1 + \frac{S}{N}\right) \qquad \text{(2-18a)}$$

or

$$I = 3.32 \, B \log_{10}\left(1 + \frac{S}{N}\right) \qquad \text{(2-18b)}$$

where I = information capacity (bps)
 B = bandwidth (hertz)
 $\dfrac{S}{N}$ = signal-to-noise power ratio (unitless)

For a standard telephone circuit with a signal-to-noise power ratio of 1000 (30 dB) and a bandwidth of 2.7 kHz, the Shannon limit for information capacity is

$$I = 2700 \log_2(1 + 1000)$$
$$= 26.9 \text{ kbps}$$

Shannon's formula is often misunderstood. The results of the preceding example indicate that 26.9 kbps can be propagated through a 2.7-kHz communications channel. This may be true, but it cannot be done with a binary system. To achieve an information transmission rate of 26.9 kbps through a 2.7-kHz channel, each symbol transmitted must contain more than one bit.

2-5-2 *M*-ary Encoding

M-*ary* is a term derived from the word *binary*. *M* simply represents a digit that corresponds to the number of conditions, levels, or combinations possible for a given number of binary variables. It is often advantageous to encode at a level higher than binary (sometimes referred to as *beyond-binary encoding* or *higher-than-binary encoding*) where there are more than two conditions possible. For example, a digital signal with four possible conditions (voltage levels, frequencies, phases, and so on) is an *M*-ary system where $M = 4$. If there are eight possible conditions, $M = 8$ and so forth. The number of bits necessary to produce a given number of conditions is expressed mathematically as

$$N = \log_2 M \qquad \text{(2-19)}$$

where N = number of bits necessary
         M = number of conditions, levels, or combinations possible with N bits

Equation 2-19 can be simplified and rearranged to express the number of conditions possible with N bits as

$$2^N = M \qquad \text{(2-20)}$$

For example, with one bit, only 2^1, or two, conditions are possible. With two bits, 2^2, or four, conditions are possible. With three bits, 2^3, or eight, conditions are possible.

2-5-3   Baud and Minimum Bandwidth

Baud is a term that is often misunderstood and commonly confused with bit rate (bps). Baud, like bit rate, is a rate of change; however, baud refers to the rate of change of the signal on the transmission medium after encoding and modulation have occurred. Bit rate refers to the rate of change of a digital information signal, which is usually binary. Baud is the reciprocal of the time of one output *signaling element,* and a signaling element may represent several information bits. A signaling element is sometimes called a *symbol* and could be encoded as a change in the amplitude, frequency, or phase. For example, binary signals are generally encoded and transmitted one bit at a time in the form of discrete voltage levels representing logic 1s (highs) and logic 0s (lows). A baud is also transmitted one at a time; however, a baud may represent more than one information bit. Thus, the baud of a data communications system may be considerably less than the bit rate.

According to H. Nyquist, binary digital signals can be propagated through an ideal noiseless transmission medium at a rate equal to two times the bandwidth of the medium. The minimum theoretical bandwidth necessary to propagate a signal is called the *minimum Nyquist bandwidth* or sometime the *minimum Nyquist frequency.* Thus, $f_b = 2B$, where f_b is the bit rate in bps and B is the *ideal Nyquist bandwidth.* The actual bandwidth necessary to propagate a given bit rate depends on several factors, including the type of encoding and modulation used, the types of filters used, system noise, and desired error performance. The ideal bandwidth is generally used for comparison purposes only.

The relationship between bandwidth and bit rate also applies to the opposite situation. For a given bandwidth (B), the highest bit rate is $2B$. For example, a standard telephone circuit has a bandwidth of approximately 2700 Hz, which has the capacity to propagate 5400 bps through it. However, if more than two levels are used for signaling (higher-than-binary encoding), more than one bit may be transmitted at a time, and it is possible to propagate a bit rate that exceeds $2B$. Using multilevel signaling, the Nyquist formulation for channel capacity is

$$f_b = B \log_2 M \qquad \text{(2-21a)}$$

where f_b = channel capacity (bps)
         B = minimum Nyquist bandwidth (hertz)
         M = number of discrete signal or voltage levels

Equation 2-21a can be rearranged to solve for the minimum bandwidth necessary to pass M-ary digitally modulated carriers:

$$B = \left(\frac{f_b}{\log_2 M} \right) \qquad \text{(2-21b)}$$

If N is substituted for $\log_2 M$, Equation 2-21b reduces to

$$B = \frac{f_b}{N} \qquad \text{(2-21c)}$$

where N is the number of bits encoded into each signaling element.

If information bits are encoded (grouped) and then converted to signals with more than two levels, transmission rates in excess of 2*B* are possible, as will be seen in subsequent sections of this chapter. In addition, since baud is the encoded rate of change, it also equals the bit rate divided by the number of bits encoded into one signaling element. Thus,

$$\text{baud} = \frac{f_b}{N} \tag{2-22}$$

By comparing Equation 2-21c with Equation 2-22, it can be seen that with digital modulation, the baud and the ideal minimum Nyquist bandwidth have the same value and are equal to the bit rate divided by the number of bits encoded.

2-6   DIGITAL MODULATION

Digital modulation is the transmittal of digitally modulated analog signals (carriers) between two or more points in a communications system. Digital modulation is sometimes called *digital radio* because digitally modulated signals can be propagated through earth's atmosphere and used in wireless communications systems. The property that distinguishes a digital modulation system from conventional analog modulation systems (AM, FM, or PM) is the nature of the modulating signal. Both analog and digital modulation systems use analog carriers to transport the information through the system. However, with analog modulation systems, the information signal is also analog, whereas with digital modulation, the information signal is digital.

Referring again to Equation 2-3, if the information signal is digital and the amplitude (*V*) of the carrier is varied proportional to the information signal, a digitally modulated signal called *amplitude-shift keying* (ASK) is produced. If the frequency (*f*) is varied proportional to the information signal, *frequency-shift keying* (FSK) is produced, and if the phase of the carrier (θ) is varied proportional to the information signal, *phase-shift keying* (PSK) is produced. If both the amplitude and the phase are varied proportional to the information signal, *quadrature amplitude modulation* (QAM) results. ASK, FSK, PSK, and QAM are all forms of digital modulation.

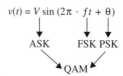

Digital modulation is ideally suited to a multitude of communications applications, including both cable and wireless systems. Applications include the following: (1) relatively low-speed voice-band data communications modems, such as those found in most personal computers; (2) high-speed data transmission systems, such as broadband *digital subscriber lines* (DSL); (3) digital satellite communications systems; and (4) *personal communications systems* (PCS), such as *cellular* telephone systems.

Figure 2-17 shows a simplified block diagram for a digital modulation system. In the transmitter, the precoder performs level conversion and then encodes the incoming data into groups of bits that modulate the analog carrier. The modulated carrier is shaped (filtered), amplified, and then transmitted through the transmission medium to the receiver. The transmission medium can be a metallic cable, optical fiber cable, earth's atmosphere, or a combination of two or more of the transmission systems. In the receiver, the incoming signals are filtered, amplified, and then applied to the demodulator and decoder circuits, which reproduce the original source information. The clock and carrier recovery circuits recover the

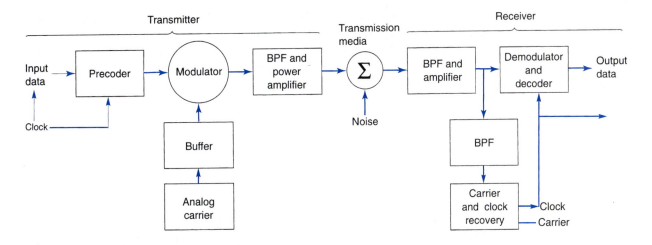

FIGURE 2-17 Simplified block diagram of a digital radio system

analog carrier and digital timing (clock) signals from the incoming modulated wave since they are necessary to perform the demodulation process.

2-6-1 Amplitude-Shift Keying

The simplest digital modulation technique is *amplitude-shift keying* (ASK), where a binary information signal directly modulates the amplitude of an analog carrier. ASK is similar to standard amplitude modulation except there are only two output amplitudes possible. Amplitude-shift keying is sometimes called *digital amplitude modulation* (DAM). Mathematically, amplitude shift keying is

$$v_{ask}(t) = [1 + v_m(t)]\left[\frac{A}{2}\cos(\omega_c t)\right] \tag{2-23}$$

where $v_{ask}(t)$ = amplitude-shift keying wave
          $v_m(t)$ = digital information (modulating) signal (volts)
          $A/2$ = unmodulated carrier amplitude (volts)
          ω_c = analog carrier radian frequency (radians per second, $2\pi f_c t$)

In Equation 2-23, the modulating signal, $v_m(t)$, is a normalized binary waveform, where $+1$ V = logic 1 and -1 V = logic 0. Therefore, for a logic 1 input $v_m(t) = +1$ V, Equation 2-23 reduces to

$$v_{ask}(t) = [1 + 1]\left[\frac{A}{2}\cos(\omega_c t)\right]$$
$$= A\cos(\omega_c t)$$

and for a logic 0 input, $v_m(t) = -1$ V, Equation 2-23 reduces to

$$v_{ask}(t) = [1 - 1]\left[\frac{A}{2}\cos(\omega_c t)\right]$$
$$= 0$$

Thus, the modulated wave, $v_{ask}(t)$, is either $A\cos(\omega_c t)$ or 0. Hence, the carrier is either "on" or "off," which is why amplitude-shift keying is sometimes referred to as *on-off keying* (OOK).

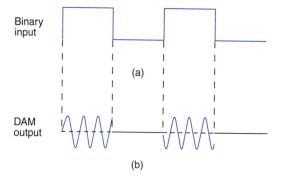

FIGURE 2-18 Digital amplitude modulation: (a) input binary; (b) output DAM waveform

Figure 2-18 shows the input and output waveforms from an ASK modulator. From the figure, it can be seen that for every change in the input binary data stream, there is one change in the ASK waveform, and the time of one bit (t_b) equals the time of one analog-signaling element (t_s). It is also important to note that for the entire time the binary input is high, the output is a constant-amplitude, constant-frequency signal and that for the entire time the binary input is low, the carrier is off. The bit time is the reciprocal of the bit rate, and the time of one signaling element is the reciprocal of the baud. Therefore, the rate of change of the ASK waveform (baud) is the same as the rate of change of the binary input (bps); thus, the bit rate equals the baud. With ASK, the bit rate is also equal to the minimum Nyquist bandwidth. This can be verified by substituting into Equations 2-21c and 2-22 by setting N to 1:

$$B = \frac{f_b}{1} = f_b$$

$$\text{baud} = \frac{f_b}{1} = f_b$$

Example 2-7

Determine the baud and minimum bandwidth necessary to pass a 10-kbps binary signal using amplitude-shift keying.

Solution For ASK, $N = 1$, and the baud and minimum bandwidth are determined from Equations 2-21c and 2-22, respectively:

$$B = \frac{10,000}{1} = 10,000$$

$$\text{baud} = \frac{10,000}{1} = 10,000$$

The use of amplitude-modulated analog carriers to transport digital information is a relatively low-quality, low-cost type of digital modulation and, therefore, is seldom used except for very low-speed telemetry circuits.

2-6-2 Frequency Shift Keying

Frequency shift keying (FSK) is another relatively simple, low-performance type of digital modulation. FSK is a form of constant-amplitude angle modulation similar to standard frequency modulation (FM) except the modulating signal is a binary signal that varies between two discrete voltage levels rather than a continuously changing analog waveform. Consequently, FSK is sometimes called *binary FSK* (BFSK). The general expression for FSK is

$$v_{\text{fsk}}(t) = V_c \cos\{2\pi[f_c + v_m(t)\Delta f]t\} \qquad (2\text{-}24)$$

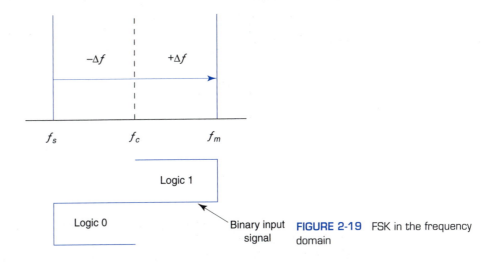

FIGURE 2-19 FSK in the frequency domain

where $v_{fsk}(t)$ = binary FSK waveform
         V_c = peak analog carrier amplitude (volts)
         f_c = analog carrier center frequency (hertz)
         Δf = peak change (shift) in the analog carrier frequency (hertz)
         $v_m(t)$ = binary input (modulating) signal (volts)

From Equation 2-24, it can be seen that the peak shift in the carrier frequency (Δf) is proportional to the amplitude of the binary input signal ($v_m[t]$) and that the direction of the shift is determined by the polarity. The modulating signal is a normalized binary waveform where a logic 1 = +1 V and a logic 0 = −1 V. Thus, for a logic 1 input, $v_m(t)$ = +1, and Equation 2-24 can be rewritten as

$$v_{fsk}(t) = V_c \cos[2\pi(f_c + \Delta f)t]$$

For a logic 0 input, $v_m(t)$ = −1, and Equation 2–24 becomes

$$v_{fsk}(t) = V_c \cos[2\pi(f_c - \Delta f)t]$$

With binary FSK, the carrier center frequency (f_c) is shifted (deviated) up and down in the frequency domain by the binary input signal as shown in Figure 2-19. As the binary input signal changes from a logic 0 to a logic 1 and vice versa, the output frequency shifts between two frequencies: a mark, or logic 1, frequency (f_m) and a space, or logic 0, frequency (f_s). The mark and space frequencies are separated from the carrier frequency by the peak frequency deviation (Δf) and from each other by 2 Δf.

With FSK, frequency deviation is defined as the difference between either the mark or the space frequency and the center frequency or half the difference between the mark and space frequencies. Frequency deviation is illustrated in Figure 2-19 and expressed mathematically as

$$\Delta f = \frac{|f_m - f_s|}{2} \qquad (2\text{-}25)$$

where Δf = frequency deviation (hertz)
     $|f_m - f_s|$ = absolute difference between the mark and space frequencies (hertz)

Figure 2-20a shows in the time domain the binary input to an FSK modulator and the corresponding FSK output. As the figure shows, when the binary input (f_b) changes from a logic 1 to a logic 0 and vice versa, the FSK output frequency shifts from a mark (f_m) to a

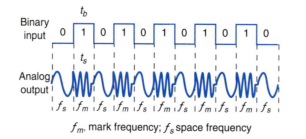

Binary input	Frequency output
0	space (f_s)
1	mark (f_m)

f_m, mark frequency; f_s space frequency

(a)

(b)

FIGURE 2-20a   FSK in the time domain: [a] waveform; [b] truth table

space (f_s) frequency and vice versa. In Figure 2-20a, the mark frequency is the higher frequency ($f_c + \Delta f$), and the space frequency is the lower frequency ($f_c - \Delta f$), although this relationship could be just the opposite. Note also in Figure 2-20a that the time of one bit (t_b) is the same as the time the FSK output is a mark of space frequency (t_s). Thus, the bit time equals the time of an FSK signaling element, and the bit rate equals the baud. Figure 2-20b shows the truth table for a binary FSK modulator. The truth table shows the input and output possibilities for a given digital modulation scheme.

2-6-2-1 FSK baud and bandwidth.

The baud for binary FSK is determined by making $N = 1$ in Equation 2-22:

$$\text{baud} = \frac{f_b}{1} = f_b$$

FSK is the exception to the rule for digital modulation, as the minimum bandwidth is not determined from Equation 2-21c. The minimum bandwidth for FSK is determined from the following formula:

$$B = 2(\Delta f + f_b) \tag{2-26}$$

where B = minimum Nyquist bandwidth (hertz)
Δf = frequency deviation ($|f_m - f_s|$) (hertz)
f_b = input bit rate (bps)

Example 2-8

Determine
a. The peak frequency deviation.
b. Minimum bandwidth.
c. Baud for a binary FSK signal with a mark frequency of 49 kHz, a space frequency of 51 kHz, and an input bit rate of 2 kbps.

Solution

a. The peak frequency deviation is determined from Equation 2-25:

$$\Delta f = \frac{|49 \text{ kHz} - 51 \text{ kHz}|}{2}$$

$$= 1 \text{ kHz}$$

b. The minimum bandwidth is determined from Equation 2-26:

$$B = 2(1000 + 2000)$$

$$= 6 \text{ kHz}$$

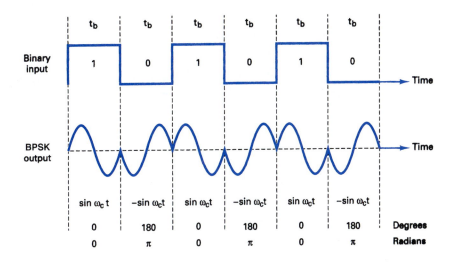

FIGURE 2-21   Output phase–versus–time relationship for a BPSK modulator

c. For FSK, $N = 1$, and the baud is determined from Equation 2-22 as

$$\text{baud} = \frac{2000}{1} = 2000$$

2-6-3   Phase-Shift Keying

Phase-shift keying (PSK) is another form of *angle-modulated, constant-amplitude* digital modulation. PSK is an *M*-ary digital modulation scheme similar to conventional phase modulation except with PSK the input is a binary digital signal and there are a limited number of output phases possible. The input binary information is encoded into groups of bits prior to modulating the carrier. The number of bits in a group ranges from 1 to 12 or more. The number of output phases is defined by *M* as described in Equation 2-20 and determined by the number of bits in the group (*N*).

2-6-3-1 Binary phase-shift keying.
The simplest form of PSK is *binary phase-shift keying* (BPSK), where $N = 1$ and $M = 2$. Therefore, with BPSK, two phases ($2^1 = 2$) are possible for the carrier. One phase represents a logic 1, and the other phase represents a logic 0. As the input digital signal changes state (i.e., from a 1 to a 0 or from a 0 to a 1), the phase of the output carrier shifts between two angles that are separated by 180 degrees. Hence, other names for BPSK are *phase-reversal keying* (PRK) and *biphase modulation*. BPSK is a form of square-wave modulation of a *continuous wave* (CW) signal.

Figure 2-21 shows the output phase–versus–time relationship for a BPSK waveform. As the figure shows, a logic 1 input produces an analog output signal with a 0-degree phase angle, and a logic 0 input produces an analog output signal with a 180-degree phase angle. As the binary input shifts between a logic 1 and a logic 0 condition, and vice versa the phase of the BPSK waveform shifts between 0 degrees and 180 degrees, respectively. For simplicity, only one cycle of the analog carrier is shown in each signaling element, although there may be anywhere between a fraction of a cycle to several thousand cycles, depending on the relationship between the input bit rate and the analog carrier frequency. It can also be seen that the time of one BPSK signaling element (t_s) is equal to the time of one information bit (t_b), which indicates that the bit rate equals the baud.

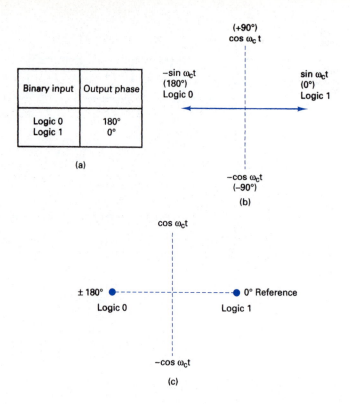

Binary input	Output phase
Logic 0	180°
Logic 1	0°

(a)

(+90°)
cos ω_c t

−sin ω_ct
(180°)
Logic 0

sin ω_ct
(0°)
Logic 1

−cos ω_ct
(−90°)

(b)

cos ω_ct

± 180°   0° Reference

Logic 0   Logic 1

−cos ω_ct

(c)

FIGURE 2-22   BPSK modulator: (a) truth table; (b) phasor diagram; (c) constellation diagram

Figure 2-22 shows the *truth table, phasor diagram,* and *constellation diagram* for a BPSK signal. A constellation diagram, which is sometimes called a *signal state-space diagram,* is similar to a phasor diagram except that the entire phasor is not drawn. In a constellation diagram, only the relative positions of the peaks of the phasors are shown.

2-6-3-2 Higher levels of PSK.   When two or more information bits are combined prior to performing PSK modulation, more than two output phases are possible. This is evident from Equation 2-20.

Quaternary phase-shift keying (QPSK), or *quadrature PSK* as it is sometimes called, is an *M*-ary encoding scheme where $N = 2$ and $M = 4$ (hence the name "quaternary," meaning "4"). With QPSK, four output phases are possible for a single carrier frequency. Because there are four output phases, there must be four different input conditions. The digital input to a QPSK modulator is a binary (base 2) signal, which requires two input bits to produce four input combinations. With two bits, there are four possible conditions: 00, 01, 10, and 11. Therefore, with QPSK, the binary input data are combined into groups of two bits, called *dibits*. In the modulator, each dibit code generates one of the four possible output phases (+45, +135, −45, and −135 degrees). Therefore, for each two-bit dibit clocked into the modulator, a single output change occurs, and the rate of change at the output (baud) is equal to one-half the input bit rate (i.e., two input bits produce one output phase change). Figure 2-23 shows the output phase–versus–time relationship, truth table, and constellation diagram for QPSK.

With 8-PSK, three bits are encoded, forming tribits and producing eight different output phases. With 8-PSK, $N = 3$, $M = 8$, and the minimum bandwidth and baud equal one-

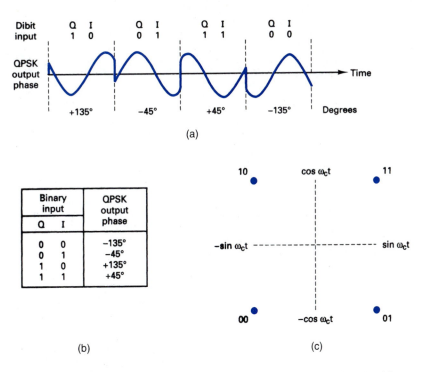

Binary input		QPSK output phase
Q	I	
0	0	−135°
0	1	−45°
1	0	+135°
1	1	+45°

(b)

(c)

FIGURE 2-23 QPSK: (a) output phase–versus–time relationship; (b) truth table; (c) constellation

third the bit rate $(f_b/3)$. Figure 2-24 shows the output phase–versus–time relationship, truth table, and constellation diagram for 8-PSK. As the figure shows, all eight points on the constellation diagram are the same distance from the intersection of the X and Y axes, and each point is separated from its two adjacent phases by 45 degrees.

With 16-PSK, four bits (called *quadbits*) are combined, producing 16 different output phases. With 16-PSK, $N = 4$, $M = 16$, and the minimum bandwidth and baud equal one-fourth the bit rate $(f_b/4)$. Figure 2-25 shows the truth table and constellation diagram for 16-PSK. Comparing Figures 2-23, 2-24, and 2-25 shows that as the level of encoding increases (i.e., the value of N and M increases), the more output phases are possible and the closer each point on the constellation diagram is to adjacent points. For an M-ary PSK system with 64 output phases $(N = 6)$, the angular separation between adjacent phases is only 5.6 degrees. This is an obvious limitation in the level of encoding (and bit rates) possible with PSK, as a point is eventually reached where receivers cannot discern the phase of the received signaling element. In addition, phase impairments inherent on communications lines have a tendency to shift the phase of the PSK signal, destroying its integrity and producing errors.

Table 2-2 summarizes the relationship between number of bits encoded, number of output conditions possible, minimum bandwidth, and baud for ASK, FSK, and PSK. Note that with the three binary modulation schemes (ASK, FSK, and BPSK), $N = 1$, $M = 2$, only two output conditions are possible, and the baud is equal to the bit rate. However, for values of N above 1, the number of output conditions increases, and the minimum bandwidth and baud decrease. Therefore, digital modulation schemes where $N > 1$ achieve *bandwidth compression* (i.e., less bandwidth is required to propagate a given bit rate). When data compression is performed, higher data transmission rates are possible for a given bandwidth.

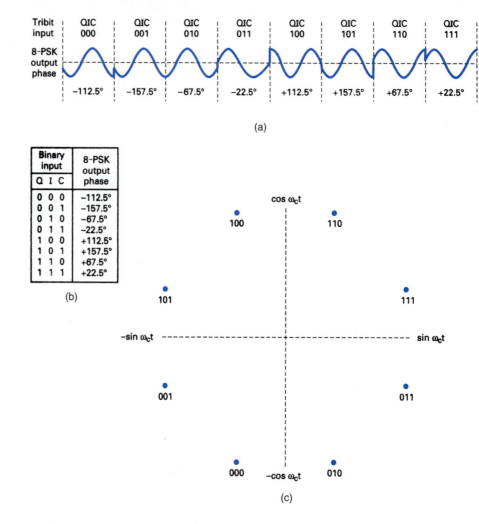

FIGURE 2-24   8-PSK: (a) output phase–versus–time relationship; (b) truth table; (c) constellation diagram

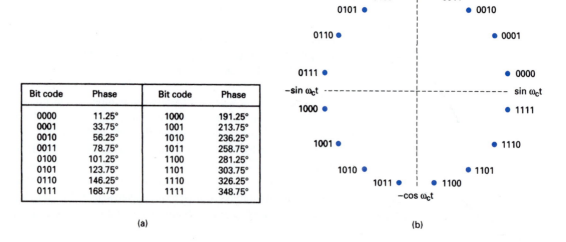

FIGURE 2-25   16-PSK: (a) truth table; (b) constellation diagram

Table 2–2 ASK, FSK, and PSK Summary

Modulation	Encoding Scheme	Outputs Possible	Minimum Bandwidth	Baud
ASK	Single bit	2	f_b [a]	f_b
FSK	Single bit	2	$>f_b$	f_b
BPSK	Single bit	2	f_b	f_b
QPSK	Dibits	4	$f_b/2$	$f_b/2$
8-PSK	Tribits	8	$f_b/3$	$f_b/3$
16-PSK	Quadbits	16	$f_b/4$	$f_b/4$

[a] f_b indicates a magnitude equal to the input bit rate.

2-6-4 Quadrature Amplitude Modulation

Quadrature amplitude modulation (QAM) is a form of digital modulation similar to PSK except the digital information is contained in both the amplitude and the phase of the transmitted carrier. With QAM, amplitude- and phase-shift keying are combined in such a way that the positions of the signaling elements on the constellation diagrams are optimized to achieve the greatest distance between elements, thus reducing the likelihood of one element being misinterpreted as another element. Obviously, this reduces the likelihood of errors occurring.

Equations 2-21c and 2-22 are also valid with QAM for determining the minimum bandwidth and baud. In addition, the degree of bandwidth compression accomplished with QAM is the same as it is with PSK (i.e., B = baud = f_b/N). The primary advantage of QAM over PSK is immunity to transmission impairments, especially phase impairments that are inherent in all communications systems.

Figure 2-26 shows the output phase–versus–time relationship, truth table, and constellation diagram for 8-QAM. As seen in Figure 2-26, with 8-QAM, there are four phases and two amplitudes that are combined to produce eight different output conditions. With 8-QAM (as with 8-PSK), three bits are encoded, forming tribits and producing eight different output conditions. With 8-QAM, $N = 3$, $M = 8$, and the minimum bandwidth and baud equal one-third the bit rate ($f_b/3$).

Figure 2-27 shows the truth table and constellation diagram for 16-QAM. With 16-QAM, there are 12 phases and three amplitudes that are combined to produce 16 different output conditions. With QAM, there are always more phases possible than amplitude. With 16-QAM, four bits (quadbits) are combined, producing 16 different output conditons. With 16-QAM, $N = 4$, $M = 16$, and the minimum bandwidth and baud equal one-fourth the bit rate ($f_b/4$).

Example 2-9

For the following modulation schemes, construct a table showing the number of bits encoded, number of output conditions, minimum bandwidth, and baud for an information data rate of 12 kbps: QPSK, 8-PSK, 8-QAM, 16-PSK, and 16-QAM.

Solution

Modulation	N	M	B (Hz)	Baud
QPSK	2	4	6000	6000
8-PSK	3	8	4000	4000
8-QAM	3	8	4000	4000
16-PSK	4	16	3000	3000
16-QAM	4	16	3000	3000

From Example 2-9, it can be seen that a 12-kbps data stream can be propagated through a narrower bandwidth using either 16-PSK or 16-QAM than with the lower levels of encoding.

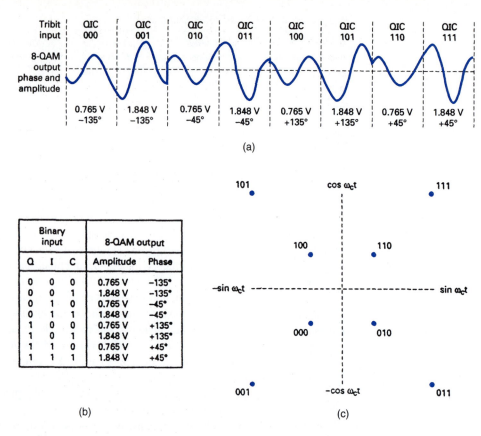

(a)

Binary input			8-QAM output	
Q	I	C	Amplitude	Phase
0	0	0	0.765 V	−135°
0	0	1	1.848 V	−135°
0	1	0	0.765 V	−45°
0	1	1	1.848 V	−45°
1	0	0	0.765 V	+135°
1	0	1	1.848 V	+135°
1	1	0	0.765 V	+45°
1	1	1	1.848 V	+45°

(b)

(c)

FIGURE 2-26 8-QAM: (a) output phase–versus-time relationship; (b) truth table; (c) constellation diagram

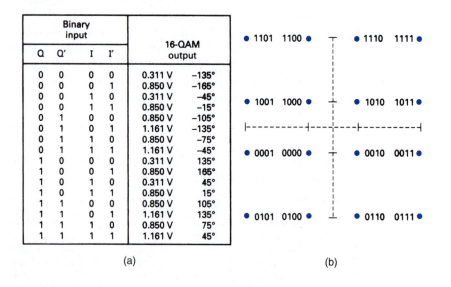

Binary input				16-QAM output	
Q	Q′	I	I′		
0	0	0	0	0.311 V	−135°
0	0	0	1	0.850 V	−165°
0	0	1	0	0.311 V	−45°
0	0	1	1	0.850 V	−15°
0	1	0	0	0.850 V	−105°
0	1	0	1	1.161 V	−135°
0	1	1	0	0.850 V	−75°
0	1	1	1	1.161 V	−45°
1	0	0	0	0.311 V	135°
1	0	0	1	0.850 V	165°
1	0	1	0	0.311 V	45°
1	0	1	1	0.850 V	15°
1	1	0	0	0.850 V	105°
1	1	0	1	1.161 V	135°
1	1	1	0	0.850 V	75°
1	1	1	1	1.161 V	45°

(a)

(b)

FIGURE 2-27 16-QAM modulator: (a) truth table; (b) constellation diagram

2-6-5 Bandwidth Efficiency

Bandwidth efficiency (sometimes called *information density* or *spectral efficiency*) is often used to compare the performance of one digital modulation technique to another. In essence, bandwidth efficiency is the ratio of the transmission bit rate to the minimum bandwidth required for a particular modulation scheme. Bandwidth efficiency is generally normalized to a 1-Hz bandwidth and, thus, indicates the number of bits that can be propagated through a transmission medium for each hertz of bandwidth. Mathematically, bandwidth efficiency is

$$B\eta = \frac{\text{transmission bit rate (bps)}}{\text{minimum bandwidth (Hz)}} \tag{2-27}$$

$$= \frac{\text{bits/second}}{\text{hertz}} = \frac{\text{bits/second}}{\text{cycles/second}} = \frac{\text{bits}}{\text{cycle}}$$

where $B\eta$ = bandwidth efficiency

Bandwidth efficiency can also be given as a percentage by simply multiplying $B\eta$ by 100.

Example 2-10

For an 8-PSK system operating at an information bit rate of 24 kbps, determine:

a. Baud.
b. Minimum bandwidth.
c. Bandwidth efficiency.

Solution

a. Baud is determined by substituting into Equation 2-22:

$$\text{baud} = \frac{24{,}000}{3} = 8000$$

b. Bandwidth is determined by substituting into Equation 2-21c:

$$B = \frac{24{,}000}{3} = 8000$$

c. Bandwidth efficiency is calculated from Equation 2-27:

$$B\eta = \frac{24{,}000 \ \text{bps}}{8000 \ \text{Hz}}$$

$$= 3 \ \text{bits per second per cycle of bandwidth}$$

Example 2-11

For 16-PSK and a transmission system with a 10-kHz bandwidth, determine the maximum bit rate.

Solution   The bandwidth efficiency for 16-PSK is 4, which means that four bits can be propagated through the system for each hertz of bandwidth. Therefore, the maximum bit rate is simply the product of the bandwidth and the bandwidth efficiency, or

$$\text{bit rate} = 4 \times 10000$$

$$= 40{,}000 \ \text{bps}$$

2-6-6 Digital Modulation Summary

The properties of several digital modulation schemes are summarized in Table 2-3.

Table 2–3 ASK, FSK, PSK, and QAM Summary

Modulation	Encoding Scheme	Outputs Possible	Minimum Bandwidth	Baud	Bη
ASK	Single bit	2	f_b^a	f_b	1
FSK	Single bit	2	f_b	f_b	1
BPSK	Single bit	2	f_b	f_b	1
QPSK	Dibits	4	$f_b/2$	$f_b/2$	2
8-PSK	Tribits	8	$f_b/3$	$f_b/3$	3
8-QAM	Tribits	8	$f_b/3$	$f_b/3$	3
16-PSK	Quadbits	16	$f_b/4$	$f_b/4$	4
16-QAM	Quadbits	16	$f_b/4$	$f_b/4$	4
32-PSK	Five bits	32	$f_b/5$	$f_b/5$	5
64-QAM	Six bits	64	$f_b/6$	$f_b/6$	6

[a] f_b indicates a magnitude equal to the input bit rate.

2-6-7 Trellis Code Modulation

Achieving data transmission rates in excess of 9600 bps over standard telephone lines with approximately a 3-kHz bandwidth obviously requires an encoding scheme well beyond the quadbits used with 16-PSK or 16-QAM (i.e., M must be significantly greater than 16). As might be expected, higher encoding schemes require higher signal-to-noise ratios. Using the Shannon limit for information capacity (Equation 2-18b), a data transmission rate of 28.8 kbps through a 3200-Hz bandwidth requires a signal-to-noise ratio of 27 dB. Transmission rates of 56 kbps require a signal-to-noise ratio of 53 dB, which is virtually impossible to achieve over a standard telephone circuit.

Data transmission rates in excess of 56 kbps can be achieved, however, over standard telephone circuits using an encoding technique called *trellis code modulation* (TCM). Dr. Ungerboeck at IBM Zuerich Research Laboratory developed TCM, which involves using *convolutional* (*tree*) codes, which combine encoding and modulation to reduce the probability of error, thus improving the bit error performance. The fundamental idea behind TCM is introducing controlled redundancy in the bit stream with a convolutional code, thus reducing the likelihood of transmission errors. What sets TCM apart from standard encoding schemes is the introduction of redundancy by doubling the number of signal points in a given PSK or QAM constellation.

Trellis code modulation is sometimes thought of as a magical method of increasing transmission bit rates over communications systems using QAM or PSK with fixed bandwidths. Few people fully understand this concept, as modem manufacturers do not seem willing to share information on TCM. Therefore, the following explanation is intended not to fully describe the process of TCM but rather to introduce the topic and give the reader a basic understanding of how TCM works and the advantage it has over conventional digital modulation techniques.

Trellis coding defines the manner in which signal-state transitions are allowed to occur, and transitions that do not follow this pattern are interpreted as transmission errors. Therefore, TCM can improve error performance by restricting the manner in which signals are allowed to transition.

TCM is thought of as a coding scheme that improves on standard QAM by increasing the distance between symbols on the constellation (known as the *Euclidean distance*). The first TCM system used a five-bit code, which included four QAM bits (a quadbit) and a fifth bit used to help decode the quadbit. Transmitting five bits within a single signaling element requires producing 32 discernible signals. A 3200-baud signal using nine-bit TCM encoding produces 512 different codes. The nine data bits plus a redundant bit for TCM requires a 960-point constellation.

2-6-8   Probability of Error and Bit Error Rate

Probability of error (P[e]) and *bit error rate* (BER) are often used interchangeably, although in practice they do have slightly different meanings. P(e) is a theoretical (mathematical) expectation of the bit error rate for a given system. BER is an empirical (historical) record of a system's actual bit error performance. For example, if a system has a P(e) of 10^{-5}, this means that mathematically you can expect one bit error in every 100,000 bits transmitted ($1/10^{-5} = 1/100,000$). If a system has a BER of 10^{-5}, this means that in past performance there was one bit error for every 100,000 bits transmitted. A bit error rate is measured and then compared with the expected probability of error to evaluate a system's performance.

QUESTIONS

2-1. Briefly describe the differences between analog and digital signals.

2-2. Briefly describe the differences between analog and digital communications systems.

2-3. Define *digital transmission* and *digital modulation* and describe the differences between them.

2-4. Define the following terms as they relate to electrical signals: *amplitude, frequency, period,* and *phase.*

2-5. What is a periodic signal?

2-6. Describe the differences between the time domain and the frequency domain.

2-7. Describe a complex signal.

2-8. What is the significance of the Fourier series?

2-9. Describe the following wave symmetries: even, odd, and half-wave.

2-10. Define *frequency spectrum* and *bandwidth.*

2-11. Define *electrical noise.*

2-12. Give a brief description of the following forms of electrical noise: man-made, thermal, correlated, and impulse.

2-13. Describe the difference between harmonic and intermodulation noise.

2-14. Define *analog modulation* and describe the following types: amplitude, frequency, and phase.

2-15. Describe the following terms: *information capacity, bit, bit rate,* and *baud.*

2-16. Briefly describe the significance of the Shannon limit for information capacity.

2-17. What is meant by the term M-*ary* encoding?

2-18. What is the basic difference between bit rate and baud?

2-19. Define *digital modulation.*

2-20. Give a brief description of amplitude-shift keying, frequency-shift keying, phase-shift keying, and quadrature amplitude modulation.

2-21. Describe what is meant by the term *beyond-binary* or *higher-than-binary encoding.*

2-22. Describe a constellation diagram and tell what it is used for.

2-23. Describe the relationship between bit rate, bandwidth, and baud for ASK.

2-24. Describe the relationship between bit rate, bandwidth, and baud for FSK.

2-25. Describe the relationship between bit rate, bandwidth, and baud for BPSK.

2-26. Describe the relationship between bit rate, bandwidth, and baud for QPSK.

2-27. Describe the relationship between bit rate, bandwidth, and baud for 8-PSK.

2-28. Describe the relationship between bit rate, bandwidth, and baud for 16-PSK.

2-29. Describe the relationship between bit rate, bandwidth, and baud for 8-QAM.

2-30. Describe the relationship between bit rate, bandwidth, and baud for 16-QAM.

2-31. Describe bandwidth efficiency and give its significance for digital modulation.

2-32. Briefly describe Trellis code modulation.

2-33. Describe probability of error and bit error rate.

PROBLEMS

2-1. For sine waves with the following frequencies, determine the periods:
  a. 10 kHz
  b. 100 kHz
  c. 1 MHz
  d. 1 GHz

2-2. For sine waves with the following periods, determine the frequencies:
  a. 0.001 s
  b. 0.1 ms
  c. 0.5 μs
  d. 0.2 ns

2-3. For the following voltage waveforms, determine the frequency, period, peak amplitude, and phase:
  a. $v(t) = 4 \sin(2\pi 20,000t + 90)$
  b. $v(t) = 0.5 \sin(2\pi 5000t + -45)$
  c. $v(t) = 12 \sin(2\pi 1,000,000t + 60)$
  d. $v(t) = 8 \sin(2\pi 50,000t - 60)$

2-4. Determine the fundamental frequency for the following square wave:

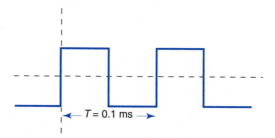

2-5. What kind of symmetry(ies) does the following waveform have?

2-6. Determine the first three harmonics of the following waveform:

2-7. Determine the bandwidth for the following frequency spectrum:

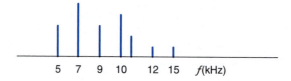

2-8. Determine the thermal noise in watts and dBm for the following conditions:
  a. An electronic device operating at 17°C with a bandwidth of 100 kHz.
  b. An electronic device operating at 17°C with a bandwidth of 10 Hz.
  c. An electronic device operating at 17°C with a bandwidth of 100 MHz.

2-9. For a circuit with a signal power of 100 W and a thermal noise power of 0.002 mW, determine the signal-to-noise power ratio in absolute and dB values.

2-10. Determine the information capacity in bps for a circuit with a 100-kHz bandwidth and a signal-to-noise ratio of 40 dB (10,000).

2-11. Determine the number of conditions possible for a binary code containing the following number of bits:

    a. 3
    b. 5
    c. 7
    d. 12

2-12. Determine the highest bit rate possible for a circuit propagating a four-bit binary code with a bandwidth of 10,000 Hz.

2-13. Determine the minimum bandwidth, baud, and bandwidth efficiency for the following bit rates using amplitude-shift keying (ASK):

    a. $f_b = 200$ bps
    b. $f_b = 1$ kbps
    c. $f_b = 10,000$ bps

2-14. Determine the minimum bandwidth, baud, and bandwidth efficiency for the following bit rates and mark and space frequencies with frequency-shift keying (FSK):

    a. $f_b = 300$ bps, $f_m = 1800$ Hz, $f_s = 1600$ Hz
    b. $f_b = 1200$ bps, $f_m = 2200$ Hz, $f_s = 1200$ Hz
    c. $f_b = 600$ bps, $f_m = 1070$ Hz, $f_s = 1270$ Hz

2-15. Determine the minimum bandwidth, baud, and bandwidth efficiency for the following bit rates and modulation schemes—BPSK, QPSK, 8-PSK, and 16-PSK:

    a. $f_b = 2400$ bps
    b. $f_b = 4800$ bps
    c. $f_b = 9600$ bps

2-16. Determine the minimum bandwidth, baud, and bandwidth efficiency for the following bit rates and modulation schemes—8-QAM and 16-QAM:

    a. $f_b = 2400$ bps
    b. $f_b = 4800$ bps
    c. $f_b = 9600$ bps

2-17. A probability of error of 10-6 means it is probable that there will be _____ bit error for every _____ bits transmitted.

C H A P T E R 3

Metallic Cable Transmission Media

OBJECTIVES

- Define the general categories of metallic transmission lines
- Define *guided* and *unguided transmission lines*
- Define *metallic transmission lines*
- Explain the difference between transverse and longitudinal waves
- Describe the following characteristics of electromagnetic waves: wave velocity, frequency, and wavelength
- Describe balanced and unbalanced transmission lines
- Describe the following types of metallic transmission lines: open-wire, twin-lead, twisted-pair, unshielded twisted pair, shielded twisted pair, and coaxial transmission lines
- Describe the EIA/TIA 568 standard classifications for twisted-pair transmission lines
- Describe plenum cable
- Describe the equivalent circuit for a metallic transmission line
- Define and describe *characteristic impedance* and *propagation constant*
- Explain how waves are propagated down a metallic transmission line
- List and describe the types of losses associated with metallic transmission lines

3-1 INTRODUCTION

The *transmission medium* is included in the lowest layer of the OSI protocol hierarchy — the physical layer. In the most basic terms, a transmission medium is simply the path between a transmitter and a receiver in a communications system. Transmission media should be transparent to data with the sole purpose of transporting bits (1s and 0s) among computers, computer networks, and other computer equipment.

Transmission media can be generally categorized as either *unguided* or *guided*. Guided transmission media are those media with some form of conductor that provides a conduit in which electromagnetic signals are contained. In essence, the conductor directs the signal that is propagating down it. Only devices physically connected to the medium can receive signals propagating down a guided transmission medium. Examples of guided transmission media are copper wire and optical fiber. Copper wires transport signals using electrical current, whereas optical fibers transport signals by propagating electromagnetic waves through a non-conductive material. Unguided transmission media are wireless systems (i.e., those without a physical conductor). Unguided signals are emitted and then radiated through air or a vacuum (or sometimes water). The direction of propagation in an unguided transmission medium depends on the direction in which the signal was emitted and any obstacles the signal may encounter while propagating. Signals propagating down an unguided transmission medium are available to anyone who has a device capable of receiving them. Examples of unguided transmission media are air (earth's atmosphere) and free space (a vacuum).

A *cable transmission medium* is a guided transmission medium and can be any physical facility used to propagate electromagnetic signals between two locations in a communications system. A physical facility is one that occupies space and has weight (i.e., one that you can touch and feel as opposed to a wireless transmission medium, such as earth's atmosphere or a vacuum). Physical transmission media include metallic cables (transmission lines) and optical cables (fibers). *Metallic transmission lines* include *open-wire*, *twin-lead*, and *twisted-pair* copper wire as well as *coaxial cable,* and optical fibers include plastic- and glass-core fibers encapsulated in a wide assortment of cladding materials. *Cable transmission systems* are the most common means of interconnecting devices in local area networks because cable transmission systems are the only transmission medium suitable for the transmission of digital signals. Cable transmission systems are also the only acceptable transmission media for digital carrier systems such as T-carriers.

3-2 METALLIC TRANSMISSION LINES

A *transmission line* is a *metallic conductor system* used to transfer electrical energy from one point to another using electrical current flow. More specifically, a transmission line is two or more electrical conductors separated by a nonconductive insulator (dielectric), such as a pair of wires or a system of wire pairs. A transmission line can be as short as a few inches, or it can span several thousand miles. Transmission lines can be used to propagate dc or low-frequency ac (such as 60-cycle electrical power and audio signals), or they can also be used to propagate very high frequencies (such as microwave radio-frequency signals). When propagating low-frequency signals, transmission line behavior is rather simple and quite predictable. However, when propagating high-frequency signals, the characteristics of transmission lines become more involved, and their behavior is somewhat peculiar to a student of lumped-constant circuits and systems.

3-3 TRANSVERSE ELECTROMAGNETIC WAVES

Basically, there are two kinds of waves: *longitudinal* and *transverse*. With longitudinal waves, the displacement (amplitude) is in the direction of propagation. A surface wave of

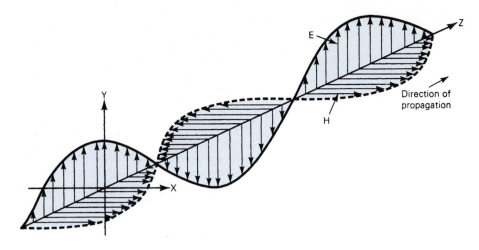

FIGURE 3-1   Transverse electromagnetic wave

water is a longitudinal wave. Sound waves are also longitudinal. With transverse waves, the direction of displacement is perpendicular to the direction of propagation. Electromagnetic waves are transverse waves.

Propagation of electrical power along a transmission line occurs in the form of *transverse electromagnetic* (TEM) *waves*. A wave is an *oscillatory motion*. The vibration of a particle excites similar vibrations in nearby particles. A TEM wave propagates primarily in the nonconductor (dielectric) that separates the two conductors of the transmission line. Therefore, a wave travels or propagates itself through a medium. Electromagnetic waves are produced by the acceleration of an electric charge. In conductors, current and voltage are always accompanied by an electric (E) field and a magnetic (H) field in the adjoining region of space. Figure 3-1 shows the spatial relationships between the E and H fields of an electromagnetic wave. From the figure, it can be seen that the E and H fields are perpendicular to each other (at 90° angles) at all points. This is referred to as *space quadrature*. Electromagnetic waves that travel along a transmission line from the source to the load are called *incident waves,* and those that travel from the load back toward the source are called *reflected waves*.

3-4   CHARACTERISTICS OF ELECTROMAGNETIC WAVES

Three of the primary characteristics of electromagnetic waves are *wave velocity, frequency,* and *wavelength*.

3-4-1   Wave Velocity

Waves travel at various speeds, depending on the type of wave and the characteristics of the propagation medium. Sound waves travel at approximately 1100 feet per second in the normal atmosphere. Electromagnetic waves travel much faster. In free space (a vacuum), TEM waves travel at the speed of light, $c = 186,283$ statute miles per second, or 299,793,000 meters per second, rounded off to 186,000 miles per second and 3×10^8 meters per second, respectively. However, in air (such as earth's atmosphere), TEM waves travel slightly slower, and along a transmission line, electromagnetic waves travel considerably slower.

3-4-2   Frequency and Wavelength

The oscillations of an electromagnetic wave are periodic and repetitive. Therefore, they are characterized by a frequency. The rate at which the periodic wave repeats is its *frequency*. The distance of one cycle occurring in space is called the *wavelength* and is determined from the following equation:

$$\text{distance} = \text{velocity} \times \text{time} \tag{3-1a}$$

If the time for one cycle is substituted into Equation 3-1a, we get the length of one cycle, which is called the wavelength and is represented by the Greek lowercase lambda (λ):

$$\lambda = \text{velocity} \times \text{period}$$

$$= v \times T \tag{3-1b}$$

where λ = wavelength
$\quad\quad\quad v$ = velocity
$\quad\quad\quad T$ = period

And, because $T = 1/f$,

$$\lambda = \frac{v}{f} \tag{3-2}$$

For free-space propagation, $v = c$; therefore, the length of one cycle is

$$\lambda = \frac{c}{f} = \frac{3 \times 10^8_{m/s}}{f_{cycles/s}} = \frac{\text{meters}}{\text{cycle}} \tag{3-3a}$$

To solve for wavelength in feet or inches, Equation 3-3a can be rewritten as

$$\lambda = \frac{1.18 \times 10^9_{in/x}}{f_{cycles/s}} = \frac{\text{inches}}{\text{cycle}} \tag{3-3b}$$

$$\lambda = \frac{9.83 \times 10^8_{ft/s}}{f_{cycles/s}} = \frac{\text{feet}}{\text{cycle}} \tag{3-3c}$$

Figure 3-2 shows a graph of the displacement and direction of propagation of a transverse electromagnetic wave as it travels along a transmission line from a source to a load. The horizontal (X) axis represents distance and the vertical (Y) axis displacement (voltage). One wavelength is the distance covered by one cycle of the wave. It can be seen that the wave moves to the right or propagates down the line with time. If a voltmeter were placed at any stationary point on the line, the voltage measured would fluctuate from zero to maximum positive, back to zero, to maximum negative, back to zero again, and then the cycle repeats.

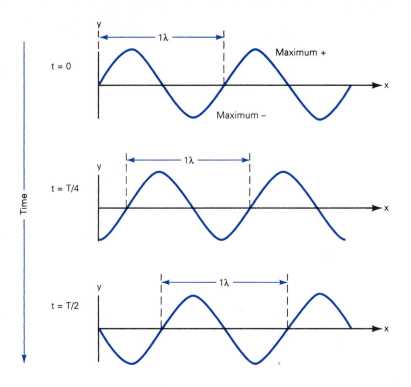

FIGURE 3-2 Displacement and velocity of a transverse wave as it propagates down a transmission line

3-5 TRANSMISSION LINE CLASSIFICATIONS

Transmission lines can be generally classified as *balanced* or *unbalanced*.

3-5-1 Balanced Transmission Lines

With two-wire balanced lines, both conductors carry current; however, one conductor carries the signal, and the other conductor is the return path. This type of transmission is called *differential* or *balanced* signal transmission. The signal propagating down the wire is measured as the potential difference between the two wires. Figure 3-3 shows a balanced transmission line system. Both conductors in a balanced line carry signal currents. The two currents are equal in magnitude with respect to electrical ground but travel in opposite

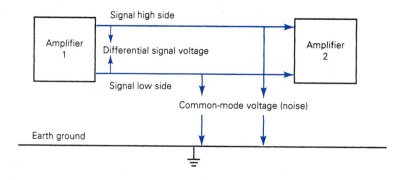

FIGURE 3-3 Differential, or balanced, transmission system

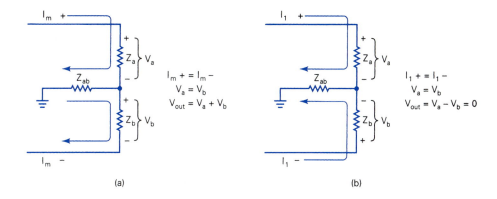

$I_m + = I_m -$
$V_a = V_b$
$V_{out} = V_a + V_b$

$I_1 + = I_1 -$
$V_a = V_b$
$V_{out} = V_a - V_b = 0$

(a)   (b)

FIGURE 3-4 Results of metallic and longitudinal currents on a balanced transmission line: (a) metallic currents due to signal voltages; (b) longitudinal currents due to noise voltages

directions. Currents that flow in opposite directions in a balanced wire pair are called *metallic circuit currents*. Currents that flow in the same direction are called *longitudinal currents*. A balanced wire pair has the advantage that most *noise interference* (sometimes called *common-mode interference*) is induced equally in both wires, producing longitudinal currents that cancel in the load. The cancellation of common mode signals is called *common-mode rejection* (CMRR). Common-mode rejection ratios of 40 dB to 70 dB are common in balanced transmission lines. Any pair of wires can operate in the balanced mode, provided that neither wire is at ground potential.

Figure 3-4 shows the result of metallic and longitudinal currents on a two-wire balanced transmission line. From the figure, notice that the longitudinal currents (often produced by static interference) cancel in the load.

3-5-2 Unbalanced Transmission Lines

With an unbalanced transmission line, one wire is at ground potential, whereas the other wire is at signal potential. This type of transmission line is called *single-ended* or *unbalanced* signal transmission. With unbalanced signal transmission, the ground wire may also be the reference for other signal-carrying wires. If this is the case, the ground wire must go wherever any of the signal wires go. Sometimes this creates a problem because a length of wire has resistance, inductance, and capacitance, and therefore a small potential difference may exist between any two points on the ground wire. Consequently, the ground wire is not a perfect reference point and is capable of having noise induced into it.

Unbalanced transmission lines have the advantage of requiring only one wire for each signal, and only one ground line is required no matter how many signals are grouped into one conductor. The primary disadvantage of unbalanced transmission lines is reduced immunity to common-mode signals, such as noise and other interference.

Figure 3-5 shows two unbalanced transmission systems. The potential difference on each signal wire is measured from that wire to a common ground reference. Balanced transmission lines can be connected to unbalanced lines and vice versa with special transformers called *balums*.

3-6 METALLIC TRANSMISSION LINE TYPES

All data communications systems and computer networks are interconnected to some degree with cables, which are all or part of the transmission medium transporting signals between computers. Although there is an enormous variety of cables manufactured today,

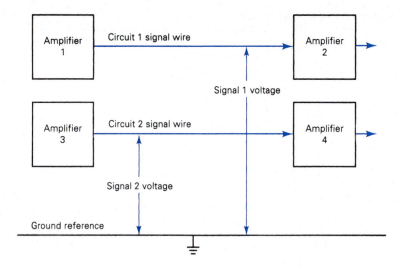

FIGURE 3-5 Single-ended, or unbalanced, transmission system

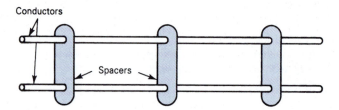

FIGURE 3-6 Open-wire transmission line

only a handful of them are commonly used for data communications circuits and computer networks. Belden, which is a leading cable manufacturer, lists more than 2000 different types of cables in its catalog. The most common metallic cables used to interconnect data communications systems and computer networks today are *parallel-conductor transmission lines* and *coaxial transmission lines*.

3-6-1 Parallel-Conductor Transmission Lines

Parallel-wire transmission lines are comprised of two or more metallic conductors (usually copper) separated by a nonconductive insulating material called a *dielectric*. Common dielectric materials include air, rubber, polyethylene, paper, mica, glass, and Teflon.

The most common parallel-conductor transmission lines are *open wire, twin lead,* and *twisted pair,* including *unshielded twisted pair* (UTP) and *shielded twisted pair* (STP).

3-6-2 Open-Wire Transmission Lines

Open-wire transmission lines are two-wire parallel conductors (see Figure 3-6). Open-wire transmission lines consist simply of two parallel wires, closely spaced and separated by air. Nonconductive spacers are placed at periodic intervals not only for support but also to keep the distance between the conductors constant. The distance between the two conductors is generally between 2 inches and 6 inches. The dielectric is simply the air between and around the two conductors in which the TEM wave propagates. The only real advantage of this type of transmission line is its simple construction. Because there is no shielding, radi-

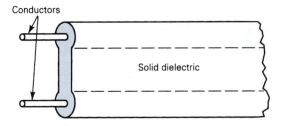

FIGURE 3-7 Twin-lead two-wire transmission line

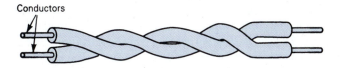

FIGURE 3-8 Twisted-pair two-wire transmission line

ation losses are high, and the cable is susceptible to picking up signals through mutual induction, which produces crosstalk. Crosstalk occurs when a signal on one cable interferes with a signal on an adjacent cable. The primary use of open-wire transmission lines is in standard voice-grade telephone applications.

3-6-2-1 Twin lead. *Twin lead* is another form of two-wire parallel-conductor transmission line and is shown in Figure 3-7. Twin-lead is essentially the same as open-wire transmission line except that the spacers between the two conductors are replaced with a continuous solid dielectric that ensures uniform spacing along the entire cable. Uniform spacing is a desirable characteristic for reasons that are explained later in this chapter. Twin-lead transmission line is the flat, brown cable typically used to connect televisions to rooftop antennas. Common dielectric materials used with twin-lead cable are Teflon and polyethylene.

3-6-2-2 Twisted-pair transmission lines. A *twisted-pair* (TP) transmission line (shown in Figure 3-8) is formed by twisting two insulated conductors around each other. Twisted pairs are often stranded in *units,* and the units are then cabled into *cores* containing up to 3000 pairs of wire. The cores are then covered with various types of *sheaths* forming cables. Neighboring pairs are sometimes twisted with different pitches (twist length) to reduce the effects of *electromagnetic interference* (EMI) and *radio-frequency interference* (RFI) from external sources (usually man-made), such as fluorescent lights, power cables, motors, relays, and transformers. Twisting the wires also reduces crosstalk between cable pairs.

The size of twisted-pair wire varies from 16 gauge (16 AWG [American Wire Gauge]) to 26 gauge. The higher the wire gauge, the smaller the diameter and the higher the resistance. Twisted-pair cable is used for both analog and digital signals and is the most commonly used transmission medium for telephone networks and building cabling systems. Twisted-pair transmission lines are also the transmission medium of choice for most local area networks because twisted-pair cable is simple to install and relatively inexpensive when compared to coaxial and optical fiber cables.

There are two basic types of twisted-pair transmission lines specified by the EIA/TIA 568 Commercial Building Telecommunications Cabling Standard for local area networks: 100-ohm *unshielded twisted pair* (UTP) and 150-ohm *shielded twisted pair*

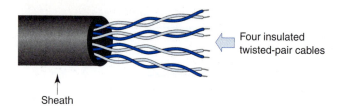

Four insulated
twisted-pair cables

Sheath

FIGURE 3-9   Unshielded twisted-pair (UTP) cable

(STP). A typical network utilizes a variety of cabling technologies, depending on the network's size, its topology, and what protocol is used. The 568 standard provides guidelines for interconnecting various cabling technologies by dividing network-wiring systems into six subsystems: *horizontal cabling, backbone cabling, work area, telecommunications closet, equipment room,* and *building entrance.* The six subsystems specified in the 568 cabling standard are described in more detail in a later chapter.

3-6-2-3 Unshielded twisted-pair.   An *unshielded twisted-pair* (UTP) cable consists of two copper wires where each wire is separately encapsulated in PVC (*polyvinyl chloride*) insulation (see Figure 3-9). Because a wire can act like an antenna, the wires are twisted two or more times at varying lengths to reduce crosstalk and interference. By carefully controlling the number of twists per foot and the manner in which multiple pairs are twisted around each other, manufacturers can improve the bandwidth (i.e., bit rate) of the cable pair significantly. The minimum number of twists for UTP cable is two per foot.

Most telephone systems use UTP cable, and the majority of new buildings are prewired with UTP cable. Generally, more cable is installed than is initially needed, providing room for orderly growth. This is one of the primary reasons why UTP cable is so popular. UTP cable is inexpensive, flexible, and easy to install. UTP cable is the least expensive transmission medium, but it is also the most susceptible to external electromagnetic interference.

To meet the operational requirements for local area networks, the EIA/TIA 568 standard classifies UTP twisted-pair cables into levels and categories that certify maximum data rates and recommended transmission distances for both UTP and STP cables (see Table 3-1). Standard UTP cable for local area networks is comprised of four pairs of 22-gauge or 24-gauge copper wire where each pair of wires is twisted around each other.

There are six primary unshielded twisted-pair cables classified by the EIA/TIA 568 standard: level 1, level 2, category 3, category 4, category 5, enhanced category 5, and category 6.

Level 1. Level 1 cable (sometimes called category 1) is ordinary thin-copper, voice-grade telephone wire typically installed before the establishment of the 568 standard. Many of these cables are insulated with paper, cord, or rubber and are, therefore, highly susceptible to interference caused by insulation breakdown. Level 1 cable is suitable only for voice-grade telephone signals and very low-speed data applications (typically under 2400 bps).

Level 2. Level 2 cable (sometimes called category 2) is only marginally better than level 1 cable but well below the standard's minimum level of acceptance. Level 2 cables are also typically old, leftover voice-grade telephone wires installed prior to the establishment of the 568 standard. Level 2 cables comply with IBM's Type 3 specification GA27-3773-1, which was developed for IEEE 802.5 Token Ring local area networks operating at transmission rates of 4 Mbps.

Category 3. Category 3 (CAT-3) cable has more stringent requirements than level 1 or level 2 cables and must have at least three turns per inch, and no two pairs within

Table 3-1 EIA/TIA 568 UTP and STP Levels and Categories

Cable Type	Intended Use	Data Rate	Distance
Level 1 (UTP)	Standard voice and low-speed data	2400 bps	18,000 feet
Level 2 (UTP)	Standard voice and low-speed data	4 Mbps	18,000 feet
Category 3 (UTP/STP)	Low-speed local area networks	16 Mbps and all level 2 applications	100 meters
Category 4 (UTP/STP)	Low-speed local area networks	20 Mbps and all category 3 applications	100 meters
Category 5 (UTP/STP)	High-speed local area networks	100 Mbps	100 meters
Enhanced category 5 (UTP/STP)	High-speed local area networks and asynchronous transfer mode (ATM)	350 Mbps	100 meters or more
Proposed New Categories			
Category 6 (UTP/STP)	Very high-speed local area networks and asynchronous transfer mode (ATM)	550 Mbps	100 meters or more
Category 7 shielded screen twisted pair (STP)	Ultra-high-speed local area networks and asynchronous transfer mode (ATM)	1 Gbps	100 meters or more
Foil twisted pair (STP)	Ultra-high-speed local area networks and asynchronous transfer mode (ATM); designed to minimize EMI susceptibility and maximize EMI immunity	>1 Gbps	?
Shielded foil twisted pair (STP)	Ultra-high-speed local area networks and asynchronous transfer mode (ATM); designed to minimize EMI susceptibility and maximize EMI immunity	>1 Gbps	?

the same cable can have the same number of turns per inch. This specification provides the cable more immunity to crosstalk. CAT-3 cable was designed to accommodate the requirements for two local area networks: IEEE 802.5 Token Ring (16 Mbps) and IEEE 802.3 10Base-T Ethernet (10 Mbps). In essence, CAT-3 cable are used for virtually any voice or data transmission rate up to 16 Mbps and, if four wire pairs are used, can accommodate transmission rates up to 100 Mbps.

Category 4. Category 4 (CAT-4) cable is little more than an upgraded version of CAT-3 cable designed to meet tighter constraints for attenuation (loss) and crosstalk. CAT-4 cable was designed for data transmission rates up to 20 Mbps. CAT-4 cables can also handle transmission rates up to 100 Mbps using cables containing four pairs of wires.

Category 5. Category 5 (CAT-5) cable is manufactured with more stringent design specifications than either CAT-3 or CAT-4 cables, including cable uniformity, insulation type, and number of turns per inch (12 turns per inch for CAT-5). Consequently, CAT-5 cable has better attenuation and crosstalk characteristics than the lower cable classifications. Attenuation in simple terms is simply the reduction of signal strength with distance, and crosstalk is the coupling of signals from one pair of wires to another pair. Near-end crosstalk refers to coupling that takes place when a transmitted signal is coupled into the receive signal at the same end of the cable.

CAT-5 cable is the cable of choice for most modern-day local area networks. CAT-5 cable was designed for data transmission rates up to 100 Mbps; however, data rates in excess of 500 Mbps are sometimes achieved. CAT-5 cable is unshielded twisted-pair cable comprised of four pairs of wires, although only two (pairs 2 and 3) were intended to be used for connectivity. The other two wire pairs are reserved spares. The following standard color code is specified by the EIA for CAT-5 cable:

Pair 1—blue/white stripe and blue
Pair 2—orange/white stripe and orange
Pair 3—green/white stripe and green
Pair 4—brown/white stripe and brown

Each wire in a CAT-5 cable can be a single conductor or a bundle of stranded wires referred to as *CAT-5 solid* or *CAT-5 flex,* respectively. When both cable types are used in the same application, the solid cable is used for backbones and whenever the cable passes through walls or ceilings. The stranded cable is typically used for patch cables between hubs and patch panels and for drop cables that are connected directly between hubs and computers.

Enhanced Category 5. Enhanced category 5 (CAT-5e) cables are tested more stringently than CAT-5 cables and include several additional measurements. CAT-5e cable is intended for data transmission rates up to 350 Mbps.

Category 6. Category 6 (CAT-6) cable is a recently proposed cable type comprised of four pairs of wire capable of operating at transmission data rates up to 400 Mbps. CAT-6 cable is very similar to CAT-5 cable except CAT-6 cable is designed and fabricated with closer tolerances and uses more advanced connectors.

3-6-2-4 Shielded twisted pair.   *Shielded twisted-pair* (STP) cable is a parallel two-wire transmission line consisting of two copper conductors separated by a solid dielectric material. The wires and dielectric are enclosed in a conductive-metal sleeve called a *foil.* If the sleeve is woven into a mesh, it is called a *braid.* The sleeve is connected to ground and acts as a shield, preventing signals from radiating beyond their boundaries (see Figure 3-10). The sleeve also keeps electromagnetic noise and radio interference produced in external sources from reaching the signal conductors. STP cable is thicker and less flexible than UTP cable, making it more difficult and expensive to install. In addition, STP cable requires an additional grounding connector and is more expensive to manufacture. However, STP cable offers greater security and greater immunity to interference.

There are seven primary STP cables classified by the EIA/TIA 568 standard: category 3, category 4, category 5, enhanced category 5, category 7, foil twisted pair, and shielded-foil twisted pair. Categories 3 through 5e STP have essentially the same parameters as their UTP counterparts, except the added shielding provides greater immunity to interference.

Category 5e. Category 5e STP feature individually shielded pairs of twisted wire.

Category 7. Category 7 *shielded-screen twisted-pair* cable (SSTP) is also called PiMF (*pairs in metal foil*) cable. SSTP cable is comprised of four pairs of 22-AWG

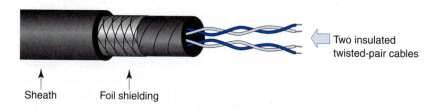

Sheath   Foil shielding

Two insulated twisted-pair cables

FIGURE 3-10   Shielded twisted-pair (STP) cable

Table 3-2 Attenuation and Crosstalk Characteristics of Twisted-Pair Cable

Frequency (MHz)	CAT-3 UTP	CAT-5 UTP	150-ohm STP
Attenuation (dB per 100 meters)			
1	2.6	2.0	1.1
4	5.6	4.1	2.2
16	13.1	8.2	4.4
25	—	10.4	6.2
100	—	22.0	12.3
300	—	—	21.4
Near-End Crosstalk (dB)			
1	41	62	58
4	32	53	58
16	23	44	50.4
25	—	41	47.5
100	—	32	38.5
300	—	—	31.3

or 23-AWG copper wire surrounded by a common metallic foil shield, followed by a braided metallic shield.

Foil twisted pair. Foil twisted-pair cable is comprised of four pairs of 24-AWG copper wire encapsulated in a common metallic-foil shield with a PVC outer sheath. Foil twisted-pair cable has been deliberately designed to minimize EMI susceptibility while maximizing EMI immunity.

Shielded-foil twisted pair. Shielded-foil twisted-pair cable is comprised of four pairs of 24-AWG copper wires surrounded by a common metallic foil shield encapsulated in a braided metallic shield. Shielded-foil twisted-pair cables offer superior EMI protection.

3-6-2-5 Attenuation and crosstalk comparison. Table 3-2 shows a comparison of the attenuation and near-end crosstalk characteristics of three of the most popular types of twisted-pair cable. Attenuation is given in dB of loss per 100 meters of cable with respect to frequency. Lower dB values indicate a higher-quality cable, and the smaller the differences in the dB value for the various frequencies, the better the frequency response. Crosstalk is given in dB of attenuation between the transmit signal and the signal returned due to crosstalk, with higher dB values indicating less crosstalk.

3-6-3 Plenum Cable

Plenum is the name given to the area between the ceiling and the roof in a single-story building or between the ceiling and the floor of the next-higher level in a multistory building. Metal rectangular-shaped air ducts were traditionally placed in the plenum to control airflow in the building (both for heating and for cooling). In more modern buildings, the ceiling itself is used to control airflow because the plenum goes virtually everywhere in the building. This presents an interesting situation if a fire should occur in the plenum because the airflow would not be contained in a fire-resistant duct system. For ease of installation, networking cables are typically distributed throughout the building in the plenum. Traditional (*nonplenum*) cables use standard PVC sheathing for the outside insulator, which is highly toxic when ignited. Therefore, if a fire should occur, the toxic chemicals produced from the burning PVC would propagate through the plenum, possibly contaminating the entire building.

The National Electric Code (NEC) requires plenum cable to have special fire-resistant insulation. Plenum cables are coated with Teflon, which does not emit noxious chemicals when ignited, or special fire-resistant PVC, which is called *plenum-grade PVC*. Therefore, plenum cables are used to route signals throughout the building in the air ducts. Because plenum cables are considerably more expensive than nonplenum cables, traditional PVC coated cables are used everywhere else.

3-6-4 Coaxial (Concentric) Transmission Lines

In the recent past, parallel-conductor transmission lines were suited only for low data transmission rates. At higher transmission rates, their radiation and dielectric losses as well as their susceptibility to external interference were excessive. Therefore, *coaxial* cables were often used for high data transmission rates to reduce losses and isolate transmission paths. However, modern UTP and STP twisted-pair cables operate at bit rates in excess of 1 Gbps at much lower costs than coaxial cable. Twisted-pair cables are also cheaper, lighter, and easier to work with than coaxial cables. In addition, many extremely high-speed computer networks prefer optical fiber cables to coaxial cables. Therefore, coaxial cable is seeing less and less use in computer networks, although it is a very popular transmission line for analog systems, such as cable television distribution networks.

The basic coaxial cable consists of a center conductor surrounded by a dielectric material (insulation), then a *concentric* (uniform distance from the center) shielding, and finally a rubber environmental protection outer jacket. *Shielding* refers to the woven or stranded mesh (or braid) that surrounds some types of coaxial cables. A coaxial cable with one layer of foil insulation and one layer of braided shielding is referred to as *dual shielded*. Environments that are subject to exceptionally high interference use *quad shielding,* which consists of two layers of foil insulation and two layers of braided metal shielding.

The center conductor of a coaxial cable is the signal wire, and the braider outer conductor is the signal return (ground). The center conductor is separated from the shield by a solid dielectric material or insulated spacers when air is used for the dielectric. For relatively high bit rates, the braided outer conductor provides excellent shielding against external interference. However, at lower bit rates, the use of shielding is usually not cost effective.

Essentially, there are two basic types of coaxial cables: *rigid air filled* and *solid flexible*. Figure 3-11a shows a rigid air coaxial line. It can be seen that a tubular outer conductor surrounds the center conductor coaxially and that the insulating material is air. The outer conductor is physically isolated and separated from the center conductor by space, which is generally filled with Pyrex, polystyrene, or some other nonconductive material. Figure 3-11b shows a solid flexible coaxial cable. The outer conductor is braided, flexible, and coaxial to the center conductor. The insulating material is a solid nonconductive polyethylene material that provides both support and electrical isolation between the inner and outer conductors. The inner conductor is a flexible copper wire that can be either solid or hollow (cellular).

Rigid air-filled coaxial cables are relatively expensive to manufacture, and to minimize losses, the air insulator must be relatively free of moisture. Solid coaxial cables have lower losses than hollow cables and are easier to construct, install, and maintain. Both types of coaxial cables are relatively immune to external radiation, radiate little themselves, and are capable of operating at higher bit rates than their parallel-wire counterparts. For these reasons, coaxial cable is more secure than twisted-pair cable. Coaxial cables can also be used over longer distances and support more stations on a shared-media network than twisted-pair cable. The primary disadvantage of coaxial transmission lines is their poor cost-to-performance ratio, low reliability, and high maintenance.

The RG numbering system typically used with coaxial cables refers to cables approved by the U.S. Department of Defense (DoD). The DoD numbering system is used by

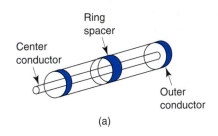

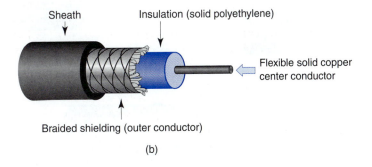

FIGURE 3-11 Coaxial or concentric transmission line: (a) rigid air filled; (b) solid flexible

Table 3-3 Coaxial Cable Characteristics

DoD Reference No.	Characteristic Impedance (ohms)	Velocity Factor	10-MHz Attenuation (dB/p 100 ft)	Conductor Size	Capacitance (pF/ft)
RG-8/A-AU	52	0.66	0.585	12	29.5
RG-8/U foam	50	0.80	0.405	12	25.4
RG-58/A-AU	53	0.66	1.250	20	28.5
RG-58 foam	50	0.79	1.000	20	25.4
RG-59/A-AU	73	0.84	0.800	20	16.5
RG-59 foam	75	0.79	0.880	20	16.9

most cable manufacturers for generic names; however, most manufacturers have developed several variations of each cable using their own product designations. For example, the Belden product number 1426A cross-references to one of several variations of the RG 58/U solid copper-core coaxial cable manufactured by Belden.

Table 3-3 lists several common coaxial cables and several of their parameters. Keep in mind that these values may vary slightly from manufacturer to manufacturer and for different variations of the same cable manufactured by the same company.

3-6-4-1 Coaxial cable connectors.   There are essentially two types of coaxial cable connectors: standard *BNC connectors* and *type-N connectors*. BNC connectors are sometimes referred to as *bayonet mount*, as they can be easily twisted on or off. N-type connectors are threaded and must be screwed on and off. Several BNC and N-type connectors are shown in Figure 3-12.

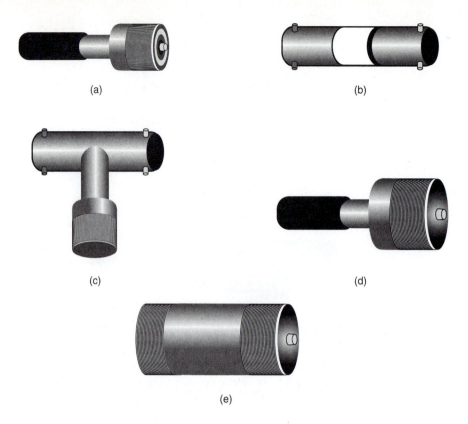

(a)

(b)

(c)

(d)

(e)

FIGURE 3-12 Coaxial cable connectors: (a) BNC connector; (b) BNC barrel; (c) BNC T; (d) Type-N; (e) Type-N barrel

3-7 METALLIC TRANSMISSION LINE EQUIVALENT CIRCUIT

The characteristics of a transmission line are determined by its electrical properties, such as wire *conductivity* and insulator *dielectric constant,* and its physical properties, such as wire diameter and conductor spacing. These properties, in turn, determine the *primary electrical constants*: *series dc resistance* (R), *series inductance* (L), *shunt capacitance* (C), and *shunt conductance* (G). Resistance and inductance occur along the line, whereas capacitance and conductance occur between the conductors. The primary constants are uniformly distributed throughout the length of the line and, therefore, are commonly called *distributed parameters*. To simplify analysis, distributed parameters are commonly given for a unit length of cable to form an artificial electrical model of the line. The combined parameters are called *lumped parameters*. For example, series resistance is generally given in ohms per unit length (i.e., ohms per meter or ohms per foot).

Figure 3-13 shows the electrical equivalent circuit for a metallic two-wire transmission line showing the relative placement of the various lumped parameters. For convenience, the conductance between the two wires is shown in reciprocal form and given as a shunt *leakage resistance* (R_s).

3-7-1 Characteristic Impedance

For maximum power transfer from the source to the load, a transmission line must be terminated in a purely resistive load equal to the *characteristic impedance* of the transmission line. The characteristic impedance (Z_o) of a transmission line is a complex quantity that is expressed in ohms, is ideally independent of line length, and cannot be directly measured. Characteristic impedance (sometimes called *surge impedance*) is defined as the impedance

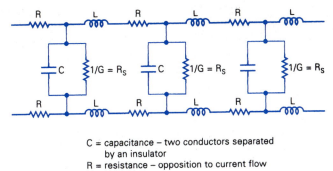

C = capacitance – two conductors separated
   by an insulator
R = resistance – opposition to current flow
L = self inductance
1/G = leakage resistance of dielectric
R_s = shunt leakage resistance

FIGURE 3-13 Two-wire parallel transmission line, electrical equivalent circuit

seen looking into an infinitely long line or the impedance seen looking into a finite length of line that is terminated in a purely resistive load with a resistance equal to the characteristic impedance of the line. A transmission line stores energy in its distributed inductance and capacitance. If a transmission line is infinitely long, it can store energy indefinitely; energy from the source enters the line, and none of it is returned. Therefore, the line acts as a resistor that dissipates all the energy. An infinitely long line can be simulated if a finite line is terminated in a purely resistive load equal to Z_o; all the energy that enters the line from the source is dissipated in the load (this assumes a totally lossless line).

The characteristic impedance of a transmission cannot be measured directly; however, it can be calculated using Ohm's law. When a source is connected to an infinitely long line and a voltage is applied, a current flows. Even though the load is open, the circuit is complete through the distributed constants of the line. The characteristic impedance is simply the ratio of the source voltage (E_o) to the line current (I_o). Mathematically, Z_o is

$$Z_o = \frac{E_o}{I_o} \qquad (3\text{-}4)$$

where Z_o = characteristic impedance (ohms)
         E_o = source voltage (volts)
         I_o = transmission line current (amps)

The characteristic impedance of a two-wire parallel transmission line with an air dielectric can be determined from its physical dimensions (see Figure 3-14a) and the formula

$$Z_o = 276 \log\frac{D}{r} \qquad (3\text{-}5)$$

where Z_o = characteristic impedance (ohms)
         D = distance between the centers of the two conductors (inches)
         r = radius of the conductor (inches)

Example 3-1

Determine the characteristic impedance for an air dielectric two-wire parallel transmission line with a D/r ratio of 12.22.

Solution Substituting into Equation 3-5, we obtain

$$Z_O = 276 \log(12.22) = 300 \text{ ohms}$$

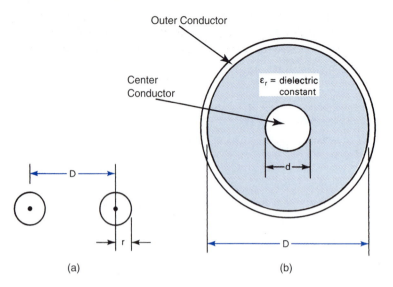

Outer Conductor

Center
Conductor

ε_r = dielectric
constant

(a)

(b)

FIGURE 3-14 Physical dimensions of transmission lines: (a) two-wire parallel transmission line; (b) coaxial-cable transmission line

The characteristic impedance of a concentric coaxial cable can also be determined from its physical dimensions (see Figure 3-14b) and the formula

$$Z_o = \frac{138}{\sqrt{\epsilon_r}} \left(\log\frac{D}{d} \right) \qquad (3\text{-}6)$$

where Z_o = characteristic impedance (ohms)
D = inside diameter of the outer conductor (inches)
ϵ_r = relative dielectric constant of the insulating material (unitless)

Example 3-2

Determine the characteristic impedance for an RG-59A coaxial cable with the following specifications: d = 0.025 in., D = 0.15 in., and ϵ_r = 2.23.

Solution   Substituting into Equation 3-6, we obtain

$$Z_O = \frac{138}{\sqrt{2.23}} \left(\log\frac{0.15 \text{ in}}{0.25 \text{ in}} \right) = 71.9 \text{ ohms}$$

For extremely high frequencies, characteristic impedance can be determined from the inductance and capacitance of the cable using the following formula:

$$Z_O = \sqrt{\frac{L}{C}} \qquad (3\text{-}7)$$

3-8   WAVE PROPAGATION ON METALLIC TRANSMISSION LINES

Electromagnetic waves travel at the speed of light when propagating through a vacuum and nearly at the speed of light when propagating through air. However, in metallic transmission lines where the conductor is generally copper and the dielectric materials vary considerably with cable type, an electromagnetic wave travels much more slowly.

3-8-1 Velocity Factor and Dielectric Constant

Velocity factor (sometimes called *velocity constant*) is defined simply as the ratio of the actual velocity of propagation of an electromagnetic wave through a given medium to the velocity of propagation through a vacuum (free space). Mathematically, velocity factor is

$$V_f = \frac{V_p}{c} \tag{3-8}$$

where V_f = velocity factor (unitless)
   V_p = actual velocity of propagation (meters per second)
   c = velocity of propagation through a vacuum (3×10^8 m/s)

and rearranging Equation 3-8 gives

$$V_f \times c = V_p \tag{3-9}$$

The velocity at which an electromagnetic wave travels through a transmission line depends on the dielectric constant of the insulating material separating the two conductors. The velocity factory is closely approximated with the formula

$$V_p = \frac{1}{\sqrt{\epsilon_r}} \tag{3-10}$$

where ϵ_r is the dielectric constant of a given material (the permittivity of the material relative to the permittivity of a vacuum—the ratio ϵ/ϵ_o, where ϵ is the permittivity of the dielectric and ϵ_o is the permittivity of air).

Velocity factor is sometimes given as a percentage, which is simply the absolute velocity factor multiplied by 100. For example, an absolute velocity factor $V_f = 0.62$ may be stated as 62%.

Dielectric constant is simply the relative permittivity of a material. The relative dielectric constant of air is 1.0006. However, the dielectric constant of materials commonly used in transmission lines ranges from 1.4872 to 7.5, giving velocity factors from 0.3651 to 0.8200. The velocity factors and relative dielectric constants of several insulating materials are listed in Table 3-4.

Dielectric constant depends on the type of insulating material used. Inductors store magnetic energy, and capacitors store electric energy. It takes a finite amount of time for an inductor or a capacitor to take on or give up energy. Therefore, the velocity at which an electromagnetic wave propagates along a transmission line varies with the inductance and capacitance of the cable. It can be shown that time $T = \sqrt{LC}$.

Table 3-4   Velocity Factor and Dielectric Constant

Material	Velocity Factor (V_f)	Relative Dielectric Constant (ϵ_r)
Vacuum	1.0000	1.0000
Air	0.9997	1.0006
Teflon foam	0.8200	1.4872
Teflon	0.6901	2.1000
Polyethylene	0.6637	2.2700
Paper, paraffined	0.6325	2.5000
Polystyrene	0.6325	2.5000
Polyvinyl chloride	0.5505	3.3000
Rubber	0.5774	3.0000
Mica	0.4472	5.0000
Glass	0.3651	7.5000

Therefore, inductance, capacitance, and velocity of propagation are mathematically related by the formula

$$\text{velocity} \times \text{time} = \text{distance}$$

Therefore,

$$V_p = \frac{\text{distance}}{\text{time}} = \frac{D}{T} \tag{3-11a}$$

Substituting $\sqrt{LC}$ for time yields

$$V_p = \frac{D}{\sqrt{LC}} \tag{3-11b}$$

If distance is normalized to 1 meter, the velocity of propagation for a lossless transmission line is

$$V_p = \frac{1\text{m}}{\sqrt{LC_s}} = \frac{1}{\sqrt{LC}}\frac{\text{meters}}{\text{second}} \tag{3-11c}$$

where V_p = velocity of propagation (meters per second)
$\sqrt{LC}$ = seconds

Example 3-3

For a given length of RG 8A/U coaxial cable with a distributed capacitance $C = 96.6$ pF/m, a distributed inductance $L = 241.56$ nH/m, and a relative dielectric constant $\epsilon_r = 2.3$, determine the velocity of propagation and the velocity factor.

Solution   From Equation 3-11c

$$V_p = \frac{1}{\sqrt{(96.6 \times 10^{-12})(241.56 \times 10^{-9})}} = 2.07 \times 10^8 \text{ m/s}$$

From Equation 3-8,

$$V_f = \frac{2.07 \times 10^8 \text{ m/s}}{3 \times 10^8 \text{ m/s}} = 0.69$$

From Equation 3-10,

$$V_f = \frac{1}{\sqrt{2.3}} \approx 0.66$$

3-9   METALLIC TRANSMISSION LINE LOSSES

For analysis purposes, metallic transmission lines are often considered to be totally lossless. In reality, however, there are several ways in which signal power is lost in a transmission line. They include *conductor loss, radiation loss, dielectric heating loss, coupling loss,* and *corona.* Cable manufacturers generally lump all cable losses together and specify them as attenuation loss in decibels per unit length (e.g., dB/m, dB/ft, and so on).

3-9-1   Conductor Losses

Because electrical current flows through a metallic transmission line and the line has a finite resistance, there is an inherent and unavoidable power loss. This is sometimes called *conductor loss* or *conductor heating loss* and is simply an I^2R power loss. Because resistance is distributed throughout a transmission line, conductor loss is directly proportional

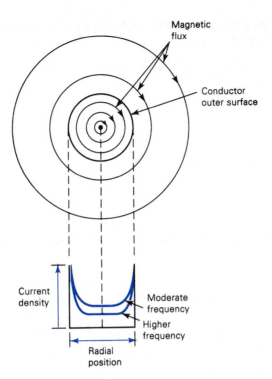

FIGURE 3-15 Isolated round conductor showing magnetic lines of flux, current distributions, and the skin effect

to the square of the line length. To reduce conductor loss, simply shorten the transmission line or use a larger-diameter wire (i.e., one with less resistance).

Conductor loss depends somewhat on frequency because of a phenomena called the *skin effect*. When current flows through an isolated round wire, the magnetic flux associated with it is in the form of concentric circles surrounding the wire core. This is shown in Figure 3-15. From the figure, it can be seen that the flux density near the center of the conductor is greater than it is near the surface. Consequently, the lines of flux near the center of the conductor encircle the current and reduce the mobility of the encircled electrons. This is a form of self-inductance and causes the inductance near the center of the conductor to be greater than at the surface. Therefore, at high frequencies, most of the current flows along the surface (outer skin) of the conductor rather than near its center. This is equivalent to reducing the cross-sectional area of the conductor and increasing the opposition to current flow (i.e., resistance). The additional opposition has a 0-degree phase angle and is, therefore, a resistance and not a reactance.

Conductor loss in metallic transmission lines varies from as low as a fraction of a decibel per 100 meters for rigid air dielectric coaxial cable to as high as 200 dB per 100 meters for a solid dielectric flexible coaxial cable. Because both I^2R losses and dielectric losses are proportional to length, they are generally lumped together and expressed in decibels of loss per unit length (i.e., dB/m).

3-9-2 Dielectric Heating Losses

A difference of potential between two conductors of a metallic transmission line causes *dielectric heating*. Heat is a form of energy and must be taken from the energy propagating down the line. For air dielectric transmission lines, the heating loss is negligible. However, for solid-core transmission lines, dielectric heating loss increases with frequency.

3-9-3 Radiation Losses

If the separation between conductors in a metallic transmission line is an appreciable fraction of a wavelength, the electrostatic and electromagnetic fields that surround the conductor cause the line to act as if it were an antenna and transfer energy to any nearby conductive material. The energy radiated is called *radiation loss* and depends on dielectric material, conductor spacing, and length of the transmission line. Radiation losses are reduced by properly shielding the cable. Therefore, shielded cables (such as STP and coaxial cable) have less radiation loss than unshielded cables (such as twin lead, open wire, and UTP). Radiation loss is also directly proportional to frequency.

3-9-4 Coupling Losses

Coupling loss occurs whenever a connection is made to or from a transmission line or when two sections of transmission line are connected together. Mechanical connections are discontinuities, which are locations where dissimilar materials meet. Discontinuities tend to heat up, radiate energy, and dissipate power.

3-9-5 Corona

Corona is a luminous discharge that occurs between the two conductors of a transmission line when the difference of potential between them exceeds the breakdown voltage of the dielectric insulator. Generally, when corona occurs, the transmission line is destroyed.

QUESTIONS

3-1. In which layer of the OSI protocol hierarchy is the transmission medium found?

3-2. What are the two general categories of transmission media?

3-3. Define *transmission line*.

3-4. Describe a transverse electromagnetic wave.

3-5. Define *wave velocity*.

3-6. Define *frequency* and *wavelength*.

3-7. Describe balanced and unbalanced transmission lines.

3-8. Describe parallel-conductor transmission lines.

3-9. Describe open-wire transmission lines.

3-10. Describe twin-lead transmission lines.

3-11. Describe twisted-pair transmission lines.

3-12. Explain the difference between shielded twisted-pair (STP) and unshielded twisted-pair (UTP) transmission lines.

3-13. Describe level 1, level 2, category 3, category 4, category 5, and category 6 twisted-pair transmission lines.

3-14. Describe enhanced category 5, category 7, foil twisted-pair, and shielded-foil twisted-pair transmission lines.

3-15. Define cable *attenuation* and *crosstalk*.

3-16. Describe the plenum and plenum cable.

3-17. Describe coaxial transmission lines.

3-18. Describe the electrical and physical properties of transmission lines.

3-19. List and describe the four primary constants of a transmission line.

3-20. Define *characteristic impedance* for a transmission line.

3-21. What properties of a transmission line determine its characteristic impedance?

3-22. Define *velocity factor* and *dielectric constant* and tell how they affect the performance of a transmission line.

3-23. What properties of a transmission line determine its velocity factor?

3-24. What properties of a transmission line determine its dielectric constant?

3-25. List and describe the five types of transmission line losses.

PROBLEMS

3-1. Determine the wavelengths for electromagnetic waves in free space with the following frequencies: 1 kHz, 100 kHz, 1 MHz, and 1 GHz.

3-2. Determine the frequencies for electromagnetic waves in free space with the following wavelengths: 1 cm, 1 m, 10 m, 100 m, and 1000 m.

3-3. Determine the characteristic impedance for a two-wire parallel transmission line with an air dielectric and a D/r ratio of 8.8.

3-4. Determine the characteristic impedance for an air-filled coaxial cable transmission line with a D/d ratio of 4.

3-5. Determine the characteristic impedance for a coaxial cable with inductance $L = 0.2\ \mu H/ft$ and capacitance $C = 16$ pF/ft.

3-6. For a given length of coaxial cable with distributed capacitance $C = 48.3$ pF/m and distributed inductance $L = 241.56$ nH/m, determine the velocity factor and velocity of propagation.

C H A P T E R 4

Optical Fiber Transmission Media

CHAPTER OUTLINE

OBJECTIVES

- Define *optical communications*
- Compare the advantages and disadvantages of optical fibers compared to metallic cables
- Describe the electromagnetic frequency spectrum and wavelength spectrum
- Describe several types of optical fiber construction
- Describe the block diagram of an optical fiber communications system
- Explain the physics of light and the following terms: *velocity of propagation, refraction, refractive index, critical angle, acceptance angle, acceptance cone,* and *numerical aperture*
- Describe the units of optical power
- Explain Snell's law
- Define *modes of propagation* and *index profile*
- Describe the three types of optical fiber configurations: single-mode step index, multimode step index, and multimode graded index
- List and describe the losses associated with optical fibers
- Compare the advantages and disadvantages of the various light sources
- Describe the characteristics of light detectors
- Describe the characteristics of LASER diodes

4-1 INTRODUCTION

Optical fiber cables are the newest and probably the most promising type of guided transmission medium for virtually all forms of digital and data communications applications, including local, metropolitan, and wide area networks. With optical fibers, electromagnetic waves are guided through a media composed of a transparent material without using electrical current flow. With optical fibers, electromagnetic light waves propagate through the media in much the same way that radio signals propagate through earth's atmosphere.

In essence, an *optical communications system* is one that uses light as the carrier of information. Propagating light waves through earth's atmosphere is difficult and often impractical. Consequently, optical fiber communications systems use glass or plastic fiber cables to "*contain*" the light waves and guide them in a manner similar to the way electromagnetic waves are guided through a metallic transmission medium.

The *information-carrying capacity* of any electronic communications system is directly proportional to bandwidth. Optical fiber cables have, for all practical purposes, an infinite bandwidth. Therefore, they have the capacity to carry much more information than their metallic counterparts or, for that matter, even the most sophisticated wireless communications systems.

4-2 ADVANTAGES OF OPTICAL FIBER CABLES

Communications using glass or plastic optical fiber cables has several overwhelming advantages over conventional metallic transmission media for both telecommunication and computer networking applications. The advantages of using optical fibers include the following:

1. *Wider bandwidth and greater information capacity.* Optical fibers have greater information capacity than metallic cables because of the inherently wider bandwidths available with optical frequencies. Optical fibers are available with bandwidths up to several thousand gigahertz. The *primary electrical constants* (resistance, inductance, and capacitance) in metallic cables cause them to act like low-pass filters, which limit their transmission frequencies, bandwidth, bit rate, and information-carrying capacity. Modern optical fiber communications systems are capable of transmitting several gigabits per second over hundreds of miles, allowing literally millions of individual voice and data channels to be combined and propagated over one optical fiber cable.

2. *Immunity to crosstalk.* Optical fiber cables are immune to crosstalk because glass and plastic fibers are nonconductors of electrical current. Therefore, fiber cables are not surrounded by a changing magnetic field, which is the primary cause of crosstalk between metallic conductors located physically close to each other.

3. *Immunity to static interference.* Because optical fiber cables are nonconductors of electrical current, they are immune to static noise due to electromagnetic interference (EMI) caused by lightning, electric motors, relays, fluorescent lights, and other electrical noise sources (most of which are man-made). For the same reason, fiber cables do not radiate electromagnetic energy.

4. *Environmental immunity.* Optical fiber cables are more resistant to environmental extremes (including weather variations) than metallic cables. Optical cables also operate over a wider temperature range and are less affected by corrosive liquids and gases.

5. *Safety and convenience.* Optical fiber cables are safer and easier to install and maintain than metallic cables. Because glass and plastic fibers are nonconductors, there are no electrical currents or voltages associated with them. Optical fibers can be used around volatile liquids and gasses without worrying about their causing explosions or fires. Optical fibers are also smaller and much more lightweight and compact than metallic cables. Consequently, they are more flexible, are easier to work with, require less storage space, are cheaper to transport, and are easier to install and maintain.

6. *Lower transmission loss.* Optical fibers have considerably less signal loss than their metallic counterparts. Optical fibers are currently being manufactured with as little as a few tenths of a dB loss per kilometer. Consequently, optical regenerators and amplifiers can be spaced considerably farther apart than with metallic transmission lines.

7. *Security.* Optical fiber cables are more secure than metallic cables. It is virtually impossible to tap into a fiber cable without the user's knowledge, and optical cables cannot be detected with metal detectors unless they are reinforced with steel for strength.

8. *Durability and reliability.* Optical fiber cables last longer and are more reliable than metallic facilities because fiber cables have a higher tolerance to changes in environmental conditions and are immune to corrosive materials.

9. *Economics.* The cost of optical fiber cables is approximately the same as metallic cables. Fiber cables have less loss and require fewer repeaters, which equates to lower installation and overall system costs and improved reliability.

4-3 DISADVANTAGES OF OPTICAL FIBER CABLES

Although the advantages of optical fiber cables far exceed the disadvantages, it is important to know the limitations of the fiber. The disadvantages of optical fibers include the following:

1. *Interfacing costs.* Optical fiber cable systems are virtually useless by themselves. To be practical and useful, they must be connected to standard electronic facilities, which often require expensive interfaces.

2. *Strength.* Optical fibers by themselves have a significantly lower tensile strength than coaxial cable. This can be improved by coating the fiber with standard *Kevlar* and a protective jacket of PVC. In addition, glass fiber is much more fragile than copper wire, making fiber less attractive where hardware portability is required.

3. *Remote electrical power.* Occasionally, it is necessary to provide electrical power to remote interface or regenerating equipment. This cannot be accomplished with the optical cable, so additional metallic cables must be included in the cable assembly.

4. *Losses through bending.* Optical fiber cables are more susceptible to losses introduced by bending the cable. Electromagnetic waves propagate through an optical cable by either refraction or reflection. Therefore, bending the cable causes irregularities in the cable dimensions, resulting in a loss of signal power. Optical fibers are also more prone to manufacturing defects as even the most minor defect can cause excessive loss of signal power.

5. *Specialized tools, equipment, and training.* Optical fiber cables require special tools to splice and repair cables and special test equipment to make routine measurements. Not only is repairing fiber cables difficult and expensive, but technicians working on optical cables also require special skills and training. In addition, sometimes it is difficult to locate faults in optical cables because there is no electrical continuity.

4-4 ELECTROMAGNETIC SPECTRUM

The total electromagnetic frequency spectrum is shown in Figure 4-1. From the figure, it can be seen that the frequency spectrum extends from the subsonic frequencies (a few hertz) to cosmic rays (10^{22} Hz). The light frequency spectrum can be divided into three general bands:

1. *Infrared.* The band of light frequencies that is too high to be seen by the human eye with wavelengths ranging between 770 nm and 10^6 nm. Optical fiber systems generally operate in the infrared band.

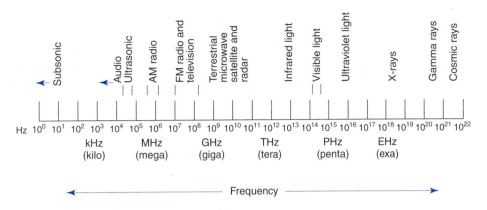

FIGURE 4-1 Electromagnetic frequency spectrum

2. *Visible.* The band of light frequencies to which the human eye will respond with wavelengths ranging between 390 nm and 770 nm. This band is visible to the human eye.

3. *Ultraviolet.* The band of light frequencies that are too low to be seen by the human eye with wavelengths ranging between 10 nm and 390 nm.

When dealing with ultra-high-frequency electromagnetic waves, such as light, it is common to use units of wavelength rather than frequency. Wavelength is the length that one cycle of an electromagnetic wave occupies in space. The length of a wavelength depends on the frequency of the wave and the velocity of light. Mathematically, wavelength is

$$\lambda = \frac{c}{f} \tag{4-1}$$

where λ = wavelength (meters per cycle)
  c = velocity of light (300,000,000 meters per second)
  f = frequency (hertz)

With light frequencies, wavelength is often stated in microns, where 1 micron = 10^{-6} meters (1 μm), or in nanometers (nm), where 1 nm = 10^{-9} meters. However, when describing the optical spectrum, the unit angstrom is sometimes used to express wavelength, where 1 angstrom = 10^{-10} meters or 0.0001 of a micron. Figure 4-2 shows the total electromagnetic wavelength spectrum.

4-5 OPTICAL FIBER COMMUNICATIONS SYSTEM BLOCK DIAGRAM

Figure 4-3 shows a simplified block diagram of a simplex optical fiber communications link. The three primary building blocks are the transmitter, the receiver, and the optical fiber cable. The transmitter is comprised of a voltage-to-current converter, a light source, and a source-to-fiber interface (light coupler). The fiber guide is the transmission medium, which is either an ultrapure glass or a plastic cable. It may be necessary to add one or more regenerators to the transmission medium, depending on the distance between the transmitter and receiver. Functionally, the regenerator performs light amplification. However, in reality the signal is not actually amplified; it is reconstructed. The receiver includes a fiber-to-interface (light coupler), a photodetector, and a current-to-voltage converter.

In the transmitter, the light source can be modulated by a digital or an analog signal. The voltage-to-current converter serves as an electrical interface between the input circuitry

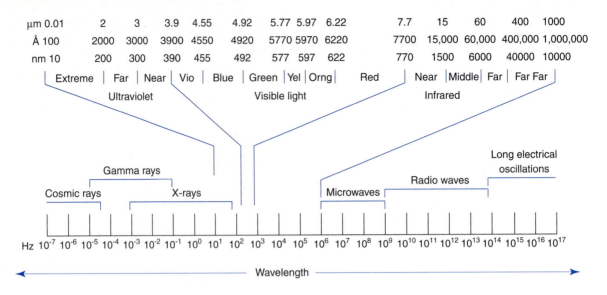

FIGURE 4-2 Electromagnetic wavelength spectrum

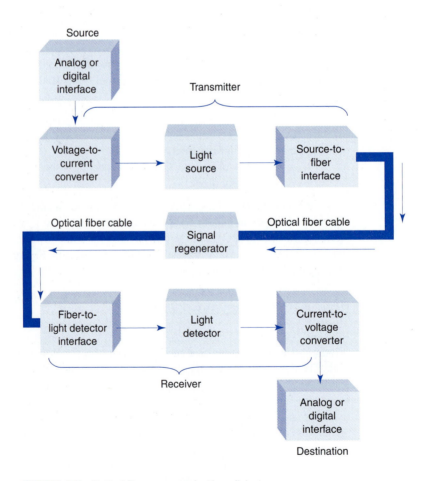

FIGURE 4-3 Optical fiber communications link

and the light source. The light source is either an infrared light-emitting diode (LED) or an injection laser diode (ILD). The amount of light emitted by either an LED or an ILD is proportional to the amount of drive current. Thus, the voltage-to-current converter converts an input signal voltage to a current that is used to drive the light source. The light outputted by the light source is directly proportional to the magnitude of the input voltage. In essence, the light intensity is modulated by the input signal.

The source-to-fiber coupler (such as an optical lens) is a mechanical interface. Its function is to couple light emitted by the light source into the optical fiber cable. The optical fiber consists of a glass or plastic fiber core surrounded by a cladding and then encapsulated in a protective jacket. The fiber-to-light detector-coupling device is also a mechanical coupler. Its function is to couple as much light as possible from the fiber cable into the light detector.

The light detector is generally a PIN (*p*-type-*intrinsic*-*n*-type) diode, an APD (*avalanche photodiode*), or a *phototransistor*. All three of these devices convert light energy to current. Consequently, a current-to-voltage converter is required to produce an output voltage proportional to the original source information. The current-to-voltage converter transforms changes in detector current to changes in voltage.

4-6 OPTICAL FIBER CONSTRUCTION

The actual fiber portion of an optical cable is generally considered to include both the fiber *core* and its *cladding* (see Figure 4-4). A special lacquer, silicone, or acrylate coating is generally applied to the outside of the cladding to seal and preserve the fiber's strength, which helps maintain the cable's attenuation characteristics. The coating also helps protect the fiber from moisture, which reduces the possibility of the occurrence of a detrimental phenomenon called *stress corrosion* (sometimes called *static fatigue*) caused by high humidity. Moisture causes silicon dioxide crystals to interact, causing bonds to break down, which causes spontaneous fractures over a prolonged period of time. The protective coating is surrounded by a *buffer jacket,* which provides the cable with additional protection against abrasion and shock. Materials commonly used for the buffer jacket include steel, fiberglass, plastic, flame-retardant polyvinyl chloride (FR-PVC), Kevlar yarn, and paper. The buffer jacket is encapsulated in a *strength member,* which increases the tensile strength of the overall cable assembly. Finally, the entire cable assembly is contained in an outer polyurethane jacket.

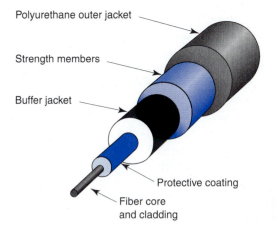

Polyurethane outer jacket

Strength members

Buffer jacket

Protective coating

Fiber core
and cladding

FIGURE 4-4 Optical fiber cable construction

Essentially, there are three varieties of optical fiber cables available today. All three of the cable varieties are constructed of glass, plastic, or a combination of glass and plastic. The three varieties are as follows:

1. Plastic core and cladding
2. Glass core with plastic cladding (often called PCS fiber: *plastic-clad silica*)
3. Glass core and glass cladding (often called SCS: *silica-clad silica*)

Plastic fibers are more flexible and, consequently, more rugged than glass. Therefore, plastic cables are easier to install, can withstand stress better, are less expensive, and weigh approximately 60% less than glass. However, plastic fibers have higher attenuation characteristics and do not propagate light as efficiently as glass. Therefore, plastic fibers are limited to relatively short cable runs, such as within a single building.

Fibers with glass cores have less attenuation than plastic fibers, with PCS being slightly better than SCS. PCS fibers are also less affected by radiation and, therefore, are more immune to external interference. SCS fibers have the best propagation characteristics and are easier to terminate than PCS fibers. Unfortunately, SCS fibers are the least rugged, and they are more susceptible to increases in attenuation when exposed to radiation.

4-6-1 Cable Configurations

There are many different cable designs available today. Figure 4-5 shows examples of several optical fiber cable configurations. With loose tube construction (Figure 4-5a), each fiber is contained in a protective tube. Inside the tube, a polyurethane compound encapsules the fiber and prevents the intrusion of water.

Figure 4-5b shows the construction of a constrained optical fiber cable. Surrounding the fiber are a primary and a secondary buffer comprised of Kevlar yarn, which increases the tensile strength of the cable. Figure 4-5c shows a *multiple-strand* cable configuration, which includes a steel central member and a layer of Mylar tape wrap to increase the cable's tensile strength. Figure 4-5d shows a ribbon configuration for a telephone cable, and Figure 4-5e shows both the end and side views of a plastic-clad silica cable.

4-7 THE PHYSICS OF LIGHT

Although the performance of optical fibers can be analyzed completely by application of Maxwell's equations, this is necessarily complex. For most practical applications, geometric wave tracing may be used instead.

In 1860, James Clerk Maxwell theorized that electromagnetic radiation contained a series of oscillating waves comprised of an electric and a magnetic field in quadrature (at 90-degree angles). However, in 1905, Albert Einstein and Max Planck showed that when light is emitted or absorbed, it behaves like an electromagnetic wave and also like a particle called a *photon,* which possesses energy proportional to its frequency. This theory is known as *Planck's law*. Planck's law describes the photoelectric effect, which states that "when visible light or high-frequency electromagnetic radiation illuminates a metallic surface, electrons are emitted." The emitted electrons produce an electric current. Planck's law is expressed mathematically as

$$E_p = hf \qquad (4\text{-}2)$$

where E_p = energy of the photon (joules)
h = Planck's constant = 6.625×10^{-34} Joules per second
f = frequency of light (photon) emitted (hertz)

Photon energy may also be expressed in terms of wavelength. Substituting Equation 4-1 into Equation 4-2 yields

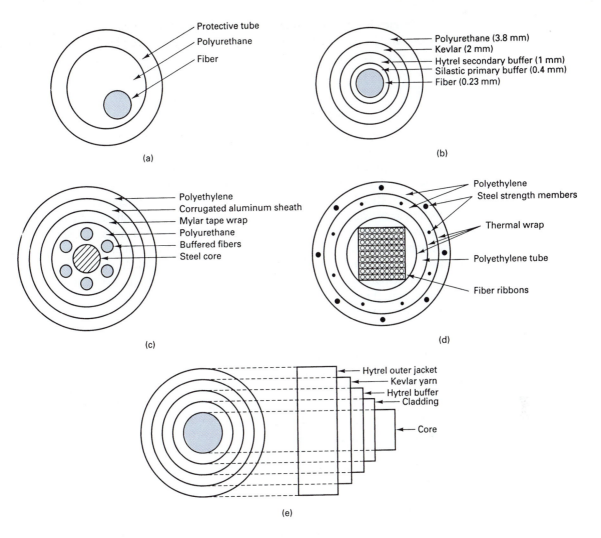

FIGURE 4-5 Fiber optic cable configurations: (a) loose tube construction; (b) constrained fiber; (c) multiple strands; (d) telephone cable; (e) plastic-silica cable

$$E_p = hf \qquad \textbf{(4-3a)}$$

or
$$E_p = \frac{hc}{\lambda} \qquad \textbf{(4-3b)}$$

An atom has several energy levels or states, the lowest of which is the ground state. Any energy level above the ground state is called an *excited state*. If an atom in one energy level decays to a lower energy level, the loss of energy (in electron volts) is emitted as a photon of light. The energy of the photon is equal to the difference between the energy of the two energy levels. The process of decaying from one energy level to another energy level is called *spontaneous decay* or *spontaneous emission*.

Atoms can be irradiated by a light source whose energy is equal to the difference between ground level and an energy level. This can cause an electron to change from one energy level to another by absorbing light energy. The process of moving from one energy level to another is called *absorption*. When making the transition from one energy level to another, the atom absorbs a packet of energy called a *photon*. This process is similar to that of emission.

The energy absorbed or emitted (photon) is equal to the difference between the two energy levels. Mathematically,

$$E_p = E_2 - E_1 \qquad (4\text{-}4)$$

where E_p is the energy of the photon (joules).

4-7-1 Optical Power

Light intensity is a rather complex concept that can be expressed in either *photometric* or *radiometric* terms. *Photometry* is the science of measuring only light waves that are visible to the human eye. Radiometry, on the other hand, measures light throughout the entire electromagnetic spectrum. In photometric terms, light intensity is generally described in terms of luminous *flux density* and measured in lumens per unit area. Radiometric terms, however, are often more useful to engineers and technologists. In radiometric terms, *optical power* measures the rate at which electromagnetic waves transfer light energy. In simple terms, optical power is described as the flow of light energy past a given point in a specified time. Optical power is expressed mathematically as

$$P = \frac{d(\text{energy})}{d(\text{time})} \qquad (4\text{-}5a)$$

or

$$= \frac{dQ}{dt} \qquad (4\text{-}5b)$$

where P = optical power (watts)
dQ = instantaneous charge (joules)
dt = instantaneous change in time (seconds)

Optical power is sometimes called *radiant flux* (ϕ), which is equivalent to joules per second and is the same power that is measured electrically or thermally in watts. Radiometric terms are generally used with light sources with output powers ranging from tens of microwatts to more than 100 milliwatts. Optical power is generally stated in decibels relative to a defined power level such as 1 mW (dBm) or 1 μW (dBμ). Mathematically stated,

$$\text{dBm} = 10 \log\left[\frac{P(\text{watts})}{0.001(\text{watt})}\right] \qquad (4\text{-}6)$$

and

$$\text{dbμ} = 10 \log\left[\frac{P(\text{watts})}{0.000001(\text{watts})}\right] \qquad (4\text{-}7)$$

Example 4-4

Determine the optical power in dBm and dBμ for power levels of (a) 10 mW and (b) 20 μW.

Solution

a. Substituting into Equations 4–6 and 4–7 gives

$$\text{dBm} = 10 \log\frac{10 \text{ mW}}{1 \text{ mW}} = 10 \text{ dBm}$$

$$\text{dBμ} = 10 \log\frac{10 \text{ mW}}{1\text{μW}} = 40 \text{ dBμ}$$

b. Substituting into Equations 4–6 and 4–7 gives

$$\text{dBm} = 10 \log \frac{20 \ \mu\text{W}}{1 \ \text{mW}} = -17 \ \text{dBm}$$

$$\text{dB}\mu = 10 \log \frac{20 \ \mu\text{W}}{1 \ \mu\text{W}} = 13 \ \text{dB}\mu$$

4-8 VELOCITY OF PROPAGATION

In free space (a vacuum), electromagnetic energy, such as light waves, travels at approximately 300,000,000 meters per second (186,000 miles per second). Also, in free space the velocity of propagation is the same for all light frequencies. However, it has been demonstrated that electromagnetic waves travel slower in materials more dense than free space and that all light frequencies do not propagate at the same velocity. When the velocity of an electromagnetic wave is reduced as it passes from one medium to another medium of denser material, the light ray changes direction or refracts (bends) toward the normal. When an electromagnetic wave passes from a more dense material into a less dense material, the light ray is refracted away from the normal. The *normal* is simply an imaginary line drawn perpendicular to the interface of the two materials at the point of incidence.

4-8-1 Refraction

For light-wave frequencies, electromagnetic waves travel through earth's atmosphere (air) at approximately the same velocity as through a vacuum (i.e., the speed of light). Figure 4-6a shows how a light ray is refracted (bent) as it passes from a less dense material into a more dense material. (Actually, the light ray is not bent but, rather, changes direction at the interface.) Figure 4-6b shows how sunlight, which contains all light frequencies (*white light*), is affected as it passes through a material that is more dense than air. Refraction occurs at both air/glass interfaces. The violet wavelengths are refracted the most, whereas the red wavelengths are refracted the least. The spectral separation of white light in this manner is called *prismatic refraction*. It is this phenomenon that causes rainbows, where water droplets in the atmosphere act as small prisms that split the white sunlight into the various wavelengths, creating a visible spectrum of color.

4-8-1-1 Refractive index.
The amount of bending or refraction that occurs at the interface of two materials of different densities is quite predictable and depends on the *refractive indexes* of the two materials. Refractive index is simply the ratio of the velocity of propagation of a light ray in free space to the velocity of propagation of a light ray in a given material. Mathematically, refractive index is

$$n = \frac{c}{v} \qquad \text{(4-8)}$$

where n = refractive index (unitless)
c = speed of light in free space (3×10^8 meters per second)
v = speed of light in a given material (meters per second)

Although the refractive index is also a function of frequency, the variation in most light wave applications is insignificant and, thus, omitted from this discussion. The indexes of refraction of several common materials are given in Table 4-1.

4-8-1-2 Snell's law.
How a light ray reacts when it meets the interface of two transmissive materials that have different indexes of refraction can be explained with Snell's law. A refractive index model for Snell's law is shown in Figure 4-7. The *angle of incidence* is the angle at which the propagating ray strikes the interface with respect to the normal, and the *angle of refraction* is the angle formed between the propagating ray and the normal

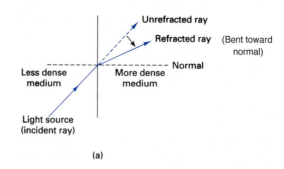

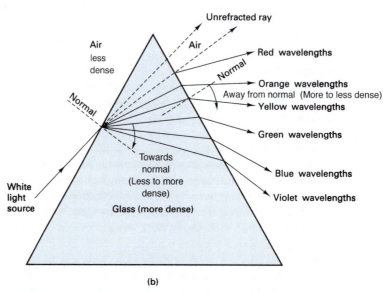

FIGURE 4-6   Refraction of light: (a) light refraction; (b) prismatic refraction

Table 4-1   Typical Indexes of Refraction

Material	Index of Refraction[a]
Vacuum	1.0
Air	1.0003 ($\approx$1)
Water	1.33
Ethyl alcohol	1.36
Fused quartz	1.46
Glass fiber	1.5–1.9
Diamond	2.0–2.42
Silicon	3.4
Gallium-arsenide	2.6

[a]Index of refraction is based on a wavelength of light emitted from a sodium flame (589 nm).

after the ray has entered the second medium. At the interface of medium 1 and medium 2, the incident ray may be refracted toward the normal or away from it, depending on whether n_1 is greater than or less than n_2. Hence, the angle of refraction can be larger or smaller than the angle of incidence, depending on the refractive indexes of the two materials. Snell's law stated mathematically is

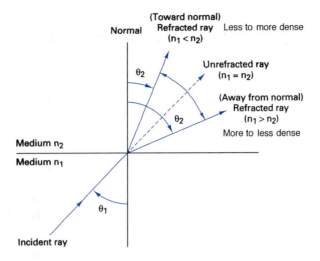

FIGURE 4-7 Refractive model for Snell's law

$$n_1 \sin \theta_1 = n_2 \sin \theta_2 \qquad (4\text{-}9)$$

where n_1 = refractive index of material 1 (unitless)
          n_2 = refractive index of material 2 (unitless)
          θ_1 = angle of incidence (degrees)
          θ_2 = angle of refraction (degrees)

4-8-1-3 Critical angle. Figure 4-8 shows a condition in which an incident ray is striking the glass/cladding interface at an angle (θ_1) such that the angle of refraction (θ_2) is 90 degrees and the refracted ray is along the interface. This angle of incidence is called the *critical angle* (θ_c), which is defined as the minimum angle of incidence at which a light ray may strike the interface of two media and result in an angle of refraction of 90 degrees or greater. It is important to note that the light ray must be traveling from a medium of higher refractive index to a medium with a lower refractive index (i.e., glass into cladding). If the

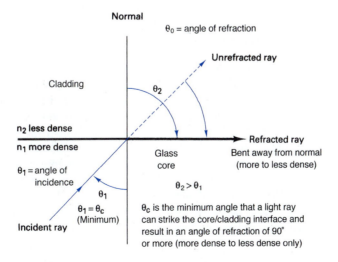

FIGURE 4-8 Critical angle refraction

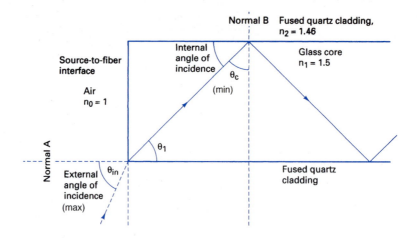

FIGURE 4-9 Ray propagation into and down an optical fiber cable

angle of refraction is 90 degrees or greater, the light ray is not allowed to penetrate the less dense material. Consequently, total reflection takes place at the interface, and the angle of reflection is equal to the angle of incidence. Critical angle can be represented mathematically by rearranging Equation 4-9 as

$$\sin \theta_1 = \frac{n_2}{n_1} \sin \theta_2$$

With $\theta_2 = 90$ degrees, θ_1 becomes the critical angle (θ_c) and

$$\sin \theta_c = \frac{n_2}{n_1} (1) = \sin \theta_c = \frac{n_2}{n_1}$$

and

$$\theta_c = \sin^{-1} \frac{n_2}{n_1} \qquad (4\text{-}10)$$

where θ_c is the critical angle.

From Equation 4-10, it can be seen that the critical angle is dependent on the ratio of the refractive indexes of the core and cladding. For example, a ratio $n_2/n_1 = 0.77$ produces a critical angle of 50.4 degrees, whereas a ratio $n_2/n_1 = 0.625$ yields a critical angle of 38.7 degrees.

4-8-1-4 Acceptance angle, acceptance cone, and numerical aperture. Figure 4-9 shows the source end of a fiber cable and a light ray propagating into and then down the fiber. When light rays enter the core of the fiber, they strike the air/glass interface at normal A. The refractive index of air is approximately 1, and the refractive index of the glass core is 1.5. Consequently, the light enters the cable traveling from a less dense to a more dense medium, causing the ray to refract toward the normal. This causes the light rays to change direction and propagate diagonally down the core at an angle that is less than the external angle of incidence (θ_{in}). For a ray of light to propagate down the cable, it must strike the internal core/cladding interface at an angle that is greater than the critical angle (θ_c). Using Snell's law, it can be shown that the maximum angle that an external light ray may strike the air/glass interface and still enter the core and propagate down the fiber is

$$\theta_{in\,(max)} = \sin^{-1} \frac{\sqrt{n_1^2 - n_2^2}}{n_0} \qquad (4\text{-}11a)$$

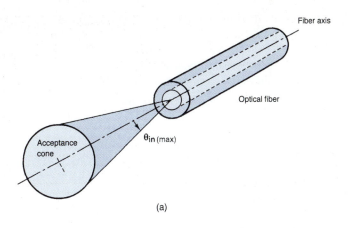

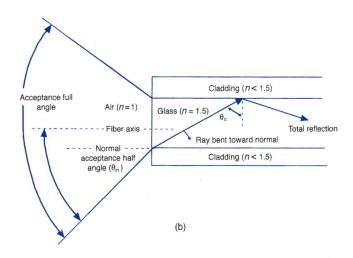

FIGURE 4-10 (a) Acceptance angle; (b) acceptance cone

where $\theta_{in(max)}$ = acceptance angle (degrees)
  n_0 = refractive index of air (1)
  n_1 = refractive index of glass fiber core (1.5)
  n_2 = refractive index of quartz fiber cladding (1.46)

Since the refractive index of air is 1, Equation 4-11a reduces to

$$\theta_{in\,(max)} = \sin^{-1}\sqrt{n_1^2 - n_2^2} \qquad\qquad \textbf{(4-11b)}$$

$\theta_{in(max)}$ is called the *acceptance angle* or *acceptance cone half angle*. $\theta_{in(max)}$ defines the maximum angle in which external light rays may strike the air/glass interface and still propagate down the fiber. Rotating the acceptance angle around the fiber core axis describes the acceptance cone of the fiber input. An acceptance cone is shown in Figure 4-10a, and the relationship between acceptance angle and critical angle is shown in Figure 4-10b. Note that the critical angle is defined as a minimum value and the acceptance angle as a maximum value. Light rays striking the air/glass interface at an angle greater than the acceptance angle will enter the cladding and, therefore, will not propagate down the cable.

Numerical aperture (NA) is closely related to acceptance angle and is the figure of merit commonly used to measure the magnitude of the acceptance angle. In essence, numerical aperture is used to describe the light-gathering or light-collecting ability of an optical fiber (i.e., the ability to couple light into the cable from an external source). The larger the magnitude of the numerical aperture, the greater the amount of external light the fiber will accept. The numerical aperture for light entering the glass fiber from an air medium is described mathematically as

$$NA = \sin \theta_{in} \qquad \text{(4-12a)}$$

and

$$NA = \sqrt{n_1^2 - n_2^2} \qquad \text{(4-12b)}$$

Therefore,

$$\theta_{in} = \sin^{-1} NA \qquad \text{(4-12c)}$$

where θ_{in} = acceptance angle (degrees)
NA = numerical aperture (unitless)
n_1 = refractive index of glass fiber core (unitless)
n_2 = refractive index of quartz fiber cladding (unitless)

A larger-diameter core does not necessarily produce a larger numerical aperture, although in practice larger-core fibers tend to have larger numerical apertures. Numerical aperture can be calculated using Equation 4-12a or 4-12b, but in practice it is generally measured by looking at the output of a fiber because the light-guiding properties of a fiber cable are symmetrical. Therefore, light leaves a cable and spreads out over an angle equal to the acceptance angle.

4-9 PROPAGATION OF LIGHT THROUGH AN OPTICAL FIBER CABLE

Light can be propagated down an optical fiber cable using either reflection or refraction. How the light propagates depends on the *mode of propagation* and the *index profile* of the fiber.

4-9-1 Modes of Propagation

In fiber optics terminology, the word *mode* simply means "path." If there is only one path for light rays to take down a cable, it is called *single mode*. If there is more than one path, it is called *multimode*. Figure 4-11 shows single and multimode propagation of light rays down an optical fiber. As shown in Figure 4-11a, with single-mode propagation, there is only one path for light rays to take, which is directly down the center of the cable. However, as Figure 4-11b shows, with multimode propagation there are many higher-order modes possible, and light rays propagate down the cable in a zigzagging fashion following several paths.

The number of paths (modes) possible for a multimode fiber cable depends on the frequency (wavelength) of the light signal, the refractive indexes of the core and cladding, and the core diameter. Mathematically, the number of modes possible for a given cable can be approximated by the following formula:

$$N \approx \left[\frac{\pi d}{\lambda} \sqrt{n_1^2 - n_2^2} \right]^2 \qquad \text{(4-13)}$$

where N = number of propagating modes
d = core diameter (meters)
λ = wavelength (meters)
n_1 = refractive index of core
n_2 = refractive index of cladding

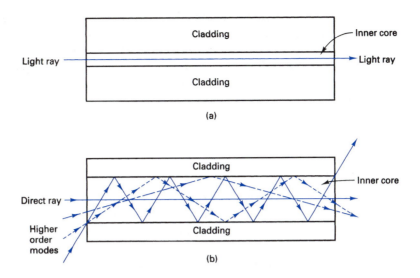

FIGURE 4-11 Modes of propagation: (a) single mode; (b) multimode

A multimode step-index fiber with a core diameter of 50 µm, a core refractive index of 1.6, a cladding refractive index of 1.584, and a wavelength of 1300 nm has approximately 372 possible modes.

4-9-2 Index Profile

The index profile of an optical fiber is a graphical representation of the magnitude of the refractive index across the fiber. The refractive index is plotted on the horizontal axis, and the radial distance from the core axis is plotted on the vertical axis. Figure 4-12 shows the core index profiles for the three types of optical fiber cables.

There are two basic types of index profiles: step and graded. A *step-index* fiber has a central core with a uniform refractive index (i.e., constant density throughout). An outside cladding that also has a uniform refractive index surrounds the core; however, the refractive index of the cladding is less than that of the central core. From Figures 4-12a and b, it can be seen that in step-index fibers there is an abrupt change in the refractive index at the core/cladding interface. This is true for both single and multimode step-index fibers.

In the *graded-index* fiber, shown in Figure 4-12c, it can be see that there is no cladding, and the refractive index of the core is nonuniform; it is highest in the center of the core and decreases gradually with distance toward the outer edge. The index profile shows a core density that is maximum in the center and decreases symmetrically with distance from the center.

4-10   OPTICAL FIBER MODES AND CLASSIFICATIONS

Propagation modes can be categorized as either multimode or single mode, and then multimode can be further subdivided into step index or graded index. Although there are a wide variety of combinations of modes and indexes, there are only three practical types of optical fiber configurations: *single-mode step index, multimode step index,* and *multimode graded index.*

4-10-1   Single-Mode Step-Index Optical Fiber

Single-mode fibers are the dominant fibers used in today's telecommunications and data networking industries. A single-mode step-index fiber has a central core that is significantly smaller in diameter than any of the multimode cables. In fact, the diameter is sufficiently

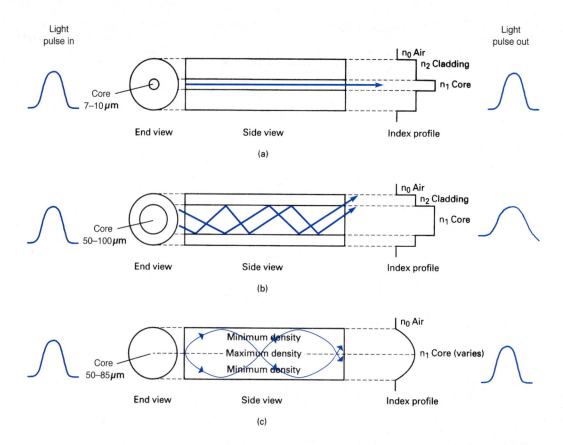

FIGURE 4-12 Core index profiles: (a) single-mode step index; (b) multimode step index; (c) multimode graded index

small that there is essentially only one path that light may take as it propagates down the cable. This type of fiber is shown in Figure 4-13a. In the simplest form of single-mode step-index fiber, the outside cladding is simply air. The refractive index of the glass core (n_1) is approximately 1.5, and the refractive index of the air cladding (n_2) is 1. The large difference in the refractive indexes results in a small critical angle (approximately 42 degrees) at the glass/air interface. Consequently, a single-mode step-index fiber has a wide external acceptance angle, which makes it relatively easy to couple light into the cable from an external source. However, this type of fiber is very weak and difficult to splice or terminate.

A more practical type of single-mode step-index fiber is one that has a cladding other than air, such as the cable shown in Figure 4-13b. The refractive index of the cladding (n_2) is slightly less than that of the central core (n_1) and is uniform throughout the cladding. This type of cable is physically stronger than the air-clad fiber, but the critical angle is also much higher (approximately 77 degrees). This results in a small acceptance angle and a narrow source-to-fiber aperture, making it much more difficult to couple light into the fiber from a light source.

With both types of single-mode step-index fibers, light is propagated down the fiber through reflection. Light rays that enter the fiber either propagate straight down the core or, perhaps, are reflected only a few times. Consequently, all light rays follow approximately the same path down the cable and take approximately the same amount of time to travel the length of the cable. This is one overwhelming advantage of single-mode step-index fibers, as explained in more detail in a later section of this chapter.

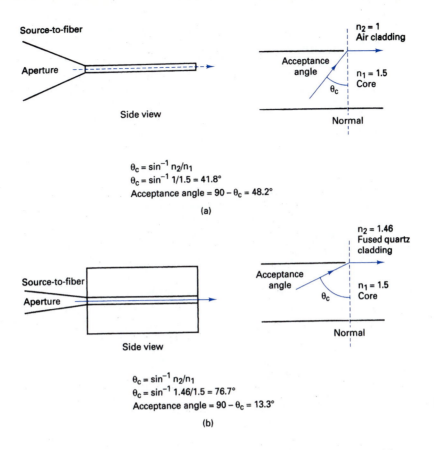

$$\theta_c = \sin^{-1} n_2/n_1$$
$$\theta_c = \sin^{-1} 1/1.5 = 41.8°$$
Acceptance angle $= 90 - \theta_c = 48.2°$

(a)

$$\theta_c = \sin^{-1} n_2/n_1$$
$$\theta_c = \sin^{-1} 1.46/1.5 = 76.7°$$
Acceptance angle $= 90 - \theta_c = 13.3°$

(b)

FIGURE 4-13 Single-mode step-index fibers: (a) air cladding; (b) glass cladding

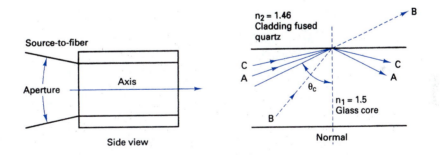

FIGURE 4-14 Multimode step-index fiber

4-10-2 Multimode Step-Index Optical Fiber

A multimode step-index optical fiber is shown in Figure 4-14. Multimode step-index fibers are similar to the single-mode step-index fibers except the center core is much larger with the multimode configuration. This type of fiber has a large light-to-fiber aperture and, consequently, allows more external light to enter the cable. The light rays that strike the core/cladding interface at an angle greater than the critical angle (ray A) are propagated down the core in a zigzag fashion, continuously reflecting off the interface boundary. Light rays that strike the core/cladding interface at an angle less than the critical angle (ray B) enter the cladding and are lost. It can be seen that there are many paths that a light ray may

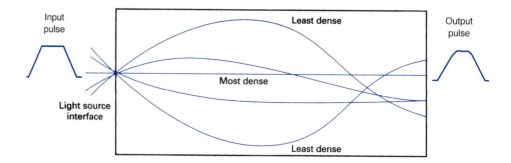

FIGURE 4-15 Multimode graded-index fiber

follow as it propagates down the fiber. As a result, all light rays do not follow the same path and, consequently, do not take the same amount of time to travel the length of the cable.

4-10-3   Multimode Graded-Index Optical Fiber

A multimode graded-index optical fiber is shown in Figure 4-15. Graded-index fibers are characterized by a central core with a nonuniform refractive index. Thus, the cable's density is maximum at the center and decreases gradually toward the outer edge. Light rays propagate down this type of fiber through refraction rather than reflection. As a light ray propagates diagonally across the core toward the center, it is continually intersecting a less dense to more dense interface. Consequently, the light rays are constantly being refracted, which results in a continuous bending of the light rays. Light enters the fiber at many different angles. As the light rays propagate down the fiber, the rays traveling in the outermost area of the fiber travel a greater distance than the rays traveling near the center. Because the refractive index decreases with distance from the center and the velocity is inversely proportional to refractive index, the light rays traveling farthest from the center propagate at a higher velocity. Consequently, they take approximately the same amount of time to travel the length of the fiber.

4-11   OPTICAL FIBER COMPARISON

Single-Mode Step-Index Fiber
Advantages:
1. Minimum dispersion means that all rays propagating down the fiber take approximately the same path; thus, they take approximately the same length of time to travel down the cable. Consequently, a pulse of light entering the cable can be reproduced at the receiving end very accurately.
2. Because of the high accuracy in reproducing transmitted pulses at the receive end, wider bandwidths and higher information transmission rates (bps) are possible with single-mode step index fibers than with the other types of fibers.

Disadvantages:
1. Because the central core is very small, it is difficult to couple light into and out of this type of fiber. The source-to-fiber aperture is the smallest of all the fiber types.
2. Again, because of the small central core, a highly directive light source, such as a laser, is required to couple light into a single-mode step-index fiber.
3. Single-mode step-index fibers are expensive and difficult to manufacture.

Multimode Step-Index Fiber
Advantages:
1. Multimode step-index fibers are relatively inexpensive and simple to manufacture.
2. It is easier to couple light into and out of multimode step-index fibers because they have a relatively large source-to-fiber aperture.

Disadvantages:

1. Light rays take many different paths down the fiber, which results in large differences in propagation times. Because of this, rays traveling down this type of fiber have a tendency to spread out. Consequently, a pulse of light propagating down a multimode step-index fiber is distorted more than with the other types of fibers.
2. The bandwidths and rate of information transfer rates possible with this type of cable are less than with the other types of fiber cables.

Multimode Graded-Index Fiber

1. Essentially, there are no outstanding advantages or disadvantages of this type of fiber. Multimode graded-index fibers are easier to couple light into and out of than single-mode step-index fibers but are more difficult than multimode step-index fibers. Distortion due to multiple propagation paths is greater than in single-mode step-index fibers but less than in multimode step-index fibers. This multimode graded-index fiber is considered an intermediate fiber compared to the other fiber types.

4-12   LOSSES IN OPTICAL FIBER CABLES

Power loss in an optical fiber cable is probably the most important characteristic of the cable. Power loss is often called *attenuation* and results in a reduction in the power of the light wave as it travels down the cable. Attenuation has several adverse effects on performance, including reducing the system's bandwidth, information transmission rate, efficiency, and overall system capacity.

The standard formula for expressing the total power loss in an optical fiber cable is

$$A_{(dB)} = 10 \log\left(\frac{P_{out}}{P_{in}}\right) \tag{4-14}$$

where $A_{(dB)}$ = total reduction in power level, attenuation (unitless)
  P_{out} = cable output power (watts)
  P_{in} = cable input power (watts)

In general, multimode fibers tend to have more attenuation than single-mode cables, primarily because of the increased scattering of the light wave produced from the dopants in the glass. Table 4-2 shows output power as a percentage of input power for an optical fiber cable with several values of decibel loss. A 4-dB cable loss reduces the output power to 50% of the input power.

Attenuation of light propagating through glass depends on wavelength. The three wavelength bands typically used for optical fiber communications systems are centered around

Table 4-2   % Output Power versus Loss in dB

Loss (dB)	Output Power (%)
1	79
3	50
6	25
9	12.5
10	10
13	5
20	1
30	0.1
40	0.01
50	0.001

Optical Fiber Transmission Media

Table 4-3 Fiber Cable Attenuation

Cable Type	Core Diameter (μm)	Cladding Diameter (μm)	NA (unitless)	Attenuation (dB/km)
Single mode	8	125	—	0.5 at 1300 nm
	5	125	—	0.4 at 1300 nm
Graded index	50	125	0.2	4 at 850 nm
	100	140	0.3	5 at 850 nm
Step index	200	380	0.27	6 at 850 nm
	300	440	0.27	6 at 850 nm
PCS	200	350	0.3	10 at 790 nm
	400	550	0.3	10 at 790 nm
Plastic	—	750	0.5	400 at 650 nm
	—	1000	0.5	400 at 650 nm

0.85 microns, 1.30 microns, and 1.55 microns. For the kind of glass typically used for optical communications systems, the 1.30-micron and 1.55-micron bands have less than 5% loss per kilometer, while the 0.85-micron band experiences almost 20% loss per kilometer.

Although total power loss is of primary importance in an optical fiber cable, attenuation is generally expressed in decibels of loss per unit length. Attenuation is expressed as a positive dB value because by definition it is a loss. Table 4-3 lists attenuation in dB/km for several types of optical fiber cables.

The optical power in watts measured at a given distance from a power source can be determined mathematically as

$$P = P_t \times 10^{-Al/10} \tag{4-15}$$

where
P = measured power level (watts)
P_t = transmitted power level (watts)
A = cable power loss (dB/km)
l = cable length (km)

Likewise, the optical power in decibel units is

$$P(\text{dBm}) = P_{in}(\text{dBm}) - A(\text{dB}) \tag{4-16}$$

where
P = measured power level (dBm)
P_{in} = transmit power (dBm)
A = cable power loss, attenuation (dB)

Example 4-5

For a single-mode optical cable with 0.25 dB/km loss, determine the optical power 100 km from a 0.1-mW light source.

Solution Substituting into Equation 4-15 gives

$$P = 0.1 \text{ mW} \times 10^{-\{[(0.25)(100)]/(10)\}}$$

$$= 1 \times 10^{-4} \times 10^{\{[(0.25)(100)]/(10)\}}$$

$$= (1 \times 10^{-4})(1 \times 10^{-2.5})$$

$$= 0.316 \text{ μW}$$

and

$$P(\text{dBm}) = 10 \log\left(\frac{0.316 \text{ μW}}{0.001}\right)$$

$$= -35 \text{ dBm}$$

or by substituting into Equation 4-16,

$$P(\text{dBm}) = 10 \log\left(\frac{0.1 \text{ mW}}{0.001 \text{ W}}\right) - [(100 \text{ km})(0.25 \text{ dB/km})]$$

$$= -10 \text{ dBm} - 25 \text{ dB}$$

$$= -35 \text{ dBm}$$

Transmission losses in optical fiber cables are one of the most important characteristics of the fibers. Losses in the fiber result in a reduction in the light power, thus reducing the system bandwidth, information transmission rate, efficiency, and overall system capacity. The predominant losses in optical fiber cables are the following:

1. Absorption loss
2. Material or Rayleigh scattering losses
3. Chromatic or wavelength dispersion
4. Radiation losses
5. Modal dispersion
6. Coupling losses

4-12-1 Absorption Losses

Absorption loss in optical fibers is analogous to power dissipation in copper cables; impurities in the fiber absorb the light and convert it to heat. The ultrapure glass used to manufacture optical fibers is approximately 99.9999% pure. Still, absorption losses between 1 dB/km and 1000 dB/km are typical. Essentially, there are three factors that contribute to the absorption losses in optical fibers: *ultraviolet absorption, infrared absorption,* and *ion resonance absorption.*

4-12-1-1 Ultraviolet absorption. Ultraviolet absorption is caused by valence electrons in the silica material from which fibers are manufactured. Light ionizes the valence electrons into conduction. The ionization is equivalent to a loss in the total light field and, consequently, contributes to the transmission losses of the fiber.

4-12-1-2 Infrared absorption. Infrared absorption is a result of photons of light that are absorbed by the atoms of the glass core molecules. The absorbed photons are converted to random mechanical vibrations typical of heating.

4-12-1-3 Ion resonance absorption. Ion resonance absorption is caused by OH^- ions in the material. The source of the OH^- ions is water molecules that have been trapped in the glass during the manufacturing process. Iron, copper, and chromium molecules also cause ion absorption.

Figure 4-16 shows typical losses in optical fiber cables due to ultraviolet, infrared, and ion resonance absorption.

4-12-2 Material or Rayleigh Scattering Losses

During manufacturing, glass is drawn into long fibers of very small diameter. During this process, the glass is in a plastic state (not liquid and not solid). The tension applied to the glass causes the cooling glass to develop permanent submicroscopic irregularities. When light rays propagating down a fiber strike one of these impurities, they are diffracted. Diffraction causes the light to disperse, or spread out, in many directions. Some of the diffracted light continues down the fiber, and some of it escapes through the cladding. The light rays that escape represent a loss in light power. This is called *Rayleigh scattering loss.* Figure 4-17 graphically shows the relationship between wavelength and Rayleigh scattering loss.

4-12-3 Chromatic Distortion or Wavelength Dispersion

Light-emitting diodes (LEDs) emit light containing many wavelengths. Each wavelength within the composite light signal travels at a different velocity when propagating through

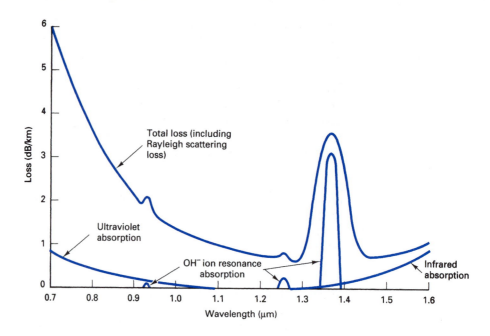

FIGURE 4-16 Absorption losses in optical fibers

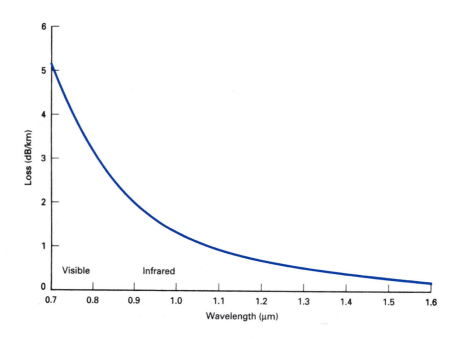

FIGURE 4-17 Rayleigh scattering loss as a function of wavelength

glass. Consequently, light rays that are simultaneously emitted from an LED and propagated down an optical fiber do not arrive at the far end of the fiber at the same time, which results in an impairment called *chromatic distortion* (sometimes called *wavelength dispersion*). Chromatic distortion can be eliminated by using a monochromatic light source such as an injection laser diode (ILD). Chromatic distortion occurs only in fibers with a single mode of transmission.

4-12-4 Radiation Losses

Radiation losses are caused predominantly by small bends and kinks in the fiber. Essentially, there are two types of bends: microbends and constant-radius bends. *Microbending* occurs as a result of differences in the thermal contraction rates between the core and the cladding material. A microbend is a miniature bend or geometric imperfection along the axis of the fiber and represents a discontinuity in the fiber where Rayleigh scattering can occur. Microbending losses generally contribute less than 20% of the total attenuation in a fiber. *Constant-radius bends* are caused by excessive pressure and tension and generally occur when fibers are bent during handling or installation.

4-12-5 Modal Dispersion

Modal dispersion (sometimes called *pulse spreading*) is caused by the difference in the propagation times of light rays that take different paths down a fiber. Obviously, modal dispersion can occur only in multimode fibers. It can be reduced considerably by using graded-index fibers and almost entirely eliminated by using single-mode step-index fibers.

Modal dispersion can cause a pulse of light energy to spread out in time as it propagates down a fiber. If the pulse spreading is sufficiently severe, one pulse may interfere with another. In multimode step-index fibers, a light ray propagating straight down the axis of the fiber takes the least amount of time to travel the length of the fiber. A light ray that strikes the core/cladding interface at the critical angle will undergo the largest number of internal reflections and, consequently, take the longest time to travel the length of the cable.

For multimode propagation, dispersion is often expressed as a *bandwidth length product* (BLP) or *bandwidth distance product* (BDP). BLP indicates what signal frequencies can be propagated through a given distance of fiber cable and is expressed mathematically as the product of distance and bandwidth (sometimes called *linewidth*). Bandwidth length products are often expressed in MHz/km units. As the length of an optical cable increases, the bandwidth (and thus the bit rate) decreases in proportion.

Example 4-6

For a 300-meter optical fiber cable with a BLP of 600 MHz/km, determine the bandwidth.

Solution
$$B = \frac{600 \text{ MHz/km}}{0.3 \text{ km}}$$
$$B = 2 \text{ GHz}$$

Figure 4-18 shows three light rays propagating down a multimode step-index optical fiber. The lowest-order mode (ray 1) travels in a path parallel to the axis of the fiber. The middle-order mode (ray 2) bounces several times at the interface before traveling the length of the fiber. The highest-order mode (ray 3) makes many trips back and forth across the

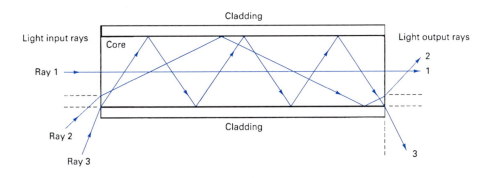

FIGURE 4-18 Light propagation down a multimode step-index fiber

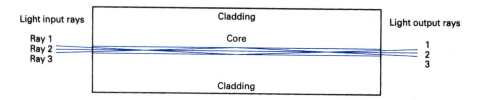

FIGURE 4-19 Light propagation down a single-mode step-index fiber

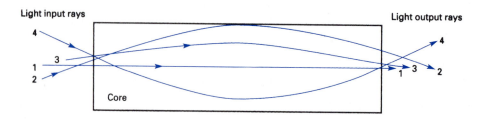

FIGURE 4-20 Light propagation down a multimode graded-index fiber

fiber as it propagates the entire length. It can be seen that ray 3 travels a considerably longer distance than ray 1 over the length of the cable. Consequently, if the three rays of light were emitted into the fiber at the same time, each ray would reach the far end at a different time, resulting in a spreading out of the light energy with respect to time. This is called *modal dispersion* and results in a stretched pulse that is also reduced in amplitude at the output of the fiber.

Figure 4-19 shows light rays propagating down a single-mode step-index cable. Because the radial dimension of the fiber is sufficiently small, there is only a single transmission path that all rays must follow as they propagate down the length of the fiber. Consequently, each ray of light travels the same distance in a given period of time, and modal dispersion is virtually eliminated.

Figure 4-20 shows light propagating down a multimode graded-index fiber. Three rays are shown traveling in three different modes. Although the three rays travel different paths, they all take approximately the same amount of time to propagate the length of the fiber. This is because the refractive index decreases with distance from the center, and the velocity at which a ray travels is inversely proportional to the refractive index. Consequently, the farther rays 2 and 3 travel from the center of the cable, the faster they propagate.

Figure 4-21 shows the relative time/energy relationship of a pulse of light as it propagates down an optical fiber cable. From the figure, it can be seen that as the pulse propagates down the cable, the light rays that make up the pulse spread out in time, which causes a corresponding reduction in the pulse amplitude and stretching of the pulse width. This is called *pulse spreading* or *pulse width dispersion* and causes errors in digital transmission.

4-12-6 Coupling Losses

Coupling losses are caused by imperfect physical connections. In fiber cables coupling losses can occur at any of the following three types of optical junctions: light source-to-fiber connections, fiber-to-fiber connections, and fiber-to-photodetector connections. Junction losses are most often caused by one of the following alignment problems: lateral misalignment, gap misalignment, angular misalignment, and imperfect surface finishes.

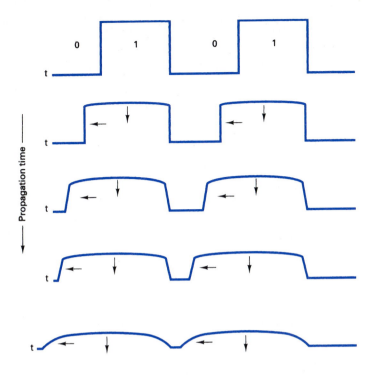

FIGURE 4-21 Pulse width dispersion in an optical fiber cable

4-12-6-1 Lateral displacement. *Lateral displacement (misalignment)* is shown in Figure 4-22a and is the lateral or axial displacement between two pieces of adjoining fiber cables. The amount of loss can be from a couple tenths of a decibel to several decibels. This loss is generally negligible if the fiber axes are aligned to within 5% of the smaller fiber's diameter.

4-12-6-2 Gap displacement (misalignment). *Gap displacement (misalignment)* is shown in Figure 4-22b and is sometimes called *end separation*. When splices are made in optical fibers, the fibers should actually touch. The farther apart the fibers are, the greater the loss of light. If two fibers are joined with a connector, the ends should not touch because the two ends rubbing against each other in the connector could cause damage to either or both fibers.

4-12-6-3 Angular displacement (misalignment). *Angular displacement (misalignment)* is shown in Figure 4-22c and is sometimes called *angular displacement*. If the angular displacement is less than 2 degrees, the loss will typically be less than 0.5 dB.

4-12-6-4 Imperfect surface finish. *Imperfect surface finish* is shown in Figure 4-22d. The ends of the two adjoining fibers should be highly polished and fit together squarely. If the fiber ends are less than 3 degrees off from perpendicular, the losses will typically be less than 0.5 dB.

4-13 LIGHT SOURCES

The range of light frequencies detectable by the human eye occupies a very narrow segment of the total electromagnetic frequency spectrum. For example, blue light occupies the higher frequencies (shorter wavelengths) of visible light, and red hues occupy the lower

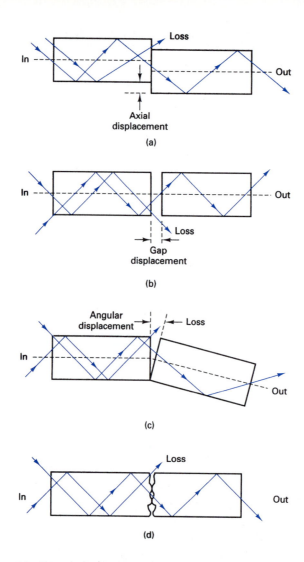

FIGURE 4-22 Fiber alignment impairments: (a) lateral misalignments; (b) gap displacement; (c) angular misalignment; (d) surface finish

frequencies (longer wavelengths). Figure 4-23 shows the light wavelength distribution produced from a tungsten lamp and the range of wavelengths perceivable by the human eye. As the figure shows, the human eye can detect only those light waves between approximately 380 nm and 780 nm. Furthermore, light consists of many shades of colors that are directly related to the heat of the energy being radiated. Figure 4-23 also shows that more visible light is produced as the temperature of the lamp is increased.

Light sources used for optical fiber systems must be at wavelengths efficiently propagated by the optical fiber. In addition, the range of wavelengths must be considered because the wider the range, the more likely the chance that chromatic dispersion will occur. Light sources must also produce sufficient power to allow the light to propagate through the fiber without causing distortion in the cable itself or in the receiver. Finally, light sources must be constructed so that their outputs can be efficiently coupled into and out of the optical cable.

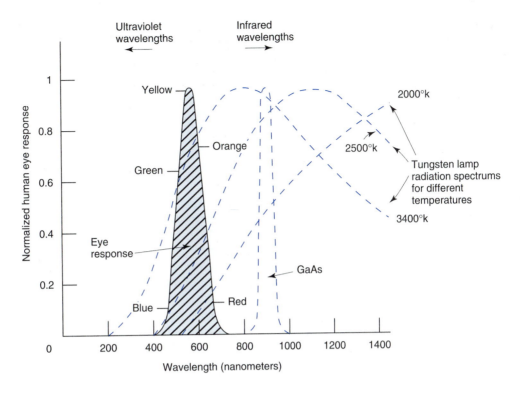

FIGURE 4-23 Tungsten lamp radiation and human eye response

4-13-1 Optical Sources

There are essentially only two types of practical light sources used to generate light for optical fiber communications systems: *light-emitting diodes* (LEDs) and *injection laser diodes* (ILDs). Both devices are constructed from semiconductor materials and have advantages and disadvantages. Standard LEDs have spectral widths of 30 nm to 50 nm, while injection lasers have spectral widths of only 1 nm to 3 nm (1 nm corresponds to a frequency of about 178 GHz). Therefore, a 1320-nm light source with a spectral linewidth of 0.0056 nm has a frequency bandwidth of approximately 1 GHz. Linewidth is the wavelength equivalent of bandwidth.

4-13-1-1 Light-emitting diodes. A light emitting diode (LED) is a p-n junction diode, usually made from a semiconductor material such as aluminum-gallium-arsenide (AlGaAs) or gallium-arsenide-phosphide (GaAsP). LEDs emit light by spontaneous emission—light is emitted as a result of the recombination of electrons and holes. Table 4-4 lists some of the common semiconductor materials used in LED construction and their respective output wavelengths.

Figure 4-24 shows the typical electrical characteristics for a low-cost infrared light-emitting diode. Figure 4-24a shows the output power versus forward current. From the figure, it can be seen that the output power varies linearly over a wide range of input current. Figure 4-24b shows output power versus temperature. It can be seen that the output power varies inversely with temperature between a temperature range of –40°C to 80°C. Figure 4-24c shows relative output power in respect to output wavelength. For this particular example, the maximum output power is achieved at an output wavelength of 825 nm.

Table 4-4 Semiconductor Material Wavelengths

Material	Wavelength (nm)
AlGaInP	630–680
GaInP	670
GaAlAs	620–895
GaAs	904
InGaAs	980
InGaAsP	1100–1650
InGaAsSb	1700–4400

4-13-1-2 Injection laser diodes.   The injection laser diode (ILD) is similar to the LED. In fact, below a certain threshold current, an ILD acts similarly to an LED. Above the threshold current, an ILD oscillates, and lasing occurs. As current passes through a forward-biased p-n junction diode, light is emitted by spontaneous emission at a frequency determined by the energy gap of the semiconductor material.

The radiant output power of a typical ILD is more directive than that of an LED. However, very little output power is realized until a threshold current is reached and lasing occurs. After lasing begins, the optical power increases dramatically, with small increases in drive current. It can also be seen that the magnitude of the optical output power of the ILD is more dependent on operating temperature than is the LED.

Injection laser diodes have several advantages over LEDs and some disadvantages.

Advantages:
1. ILDs emit coherent (orderly) light, whereas LEDs emit incoherent (disorderly) light. Therefore, ILDs have a more direct radian pattern, making it easier to couple light emitted by the ILD into an optical fiber cable. This reduces the coupling losses and allows smaller fibers to be used.
2. The radiant output power from an ILD is greater than that for an LED. A typical output power for an ILD is 5 mW (7 dBm) and only 0.5 mW (–3 dBm) for LEDs. This allows ILDs to provide a higher drive power and to be used for systems that operate over longer distances.
3. ILDs can be used at higher bit rates than LEDs.
4. ILDs generate monochromatic light, which reduces chromatic or wavelength dispersion.

Disadvantages:
1. ILDs are typically 10 times more expensive than LEDs.
2. Because ILDs operate at higher powers, they typically have a much shorter lifetime than LEDs.
3. ILDs are more temperature dependent than LEDs.

4-14 LIGHT DETECTORS

4-14-1 Optical Detectors
There are two devices commonly used to detect light energy in optical fiber communications receivers: PIN (*p*-type-*i*ntrinsic-*n*-type) diodes and APD (*a*valanche *photodi*odes). The PIN diode is probably the most common device used as a light detector in optical fiber communications systems. Essentially, a PIN photodiode operates just the opposite of an LED. APDs are more sensitive than pin diodes and require less additional amplification. The disadvantages of APDs are relatively long transmit times and additional internally generated noise due to the avalanche multiplication factor.

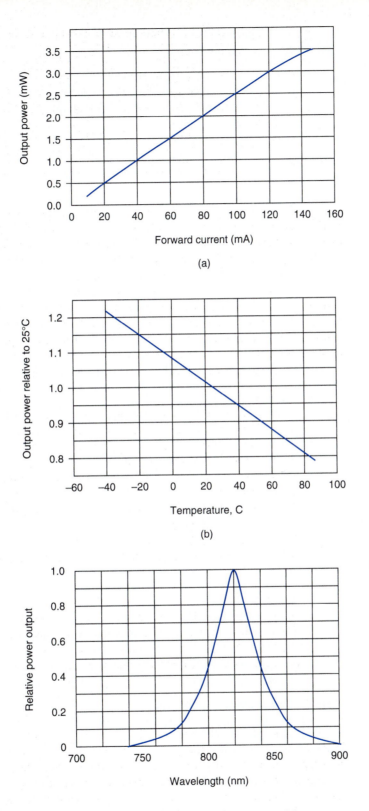

(a)

(b)

FIGURE 4-24 Typical LED electrical characteristics: (a) output power versus forward current; (b) output power versus temperature; (c) output power versus output wavelength

4-14-1-1 Characteristics of light detectors. The most important characteristics of light detectors are the following:

1. *Responsivity.* A measure of the conversion efficiency of a photodetector. It is the ratio of the output current of a photodiode to the input optical power and has the unit of amperes/watt. Responsivity is generally given for a particular wavelength or frequency.
2. *Dark current.* The leakage current that flows through a photodiode with no light input. Thermally generated carriers in the diode cause dark current.
3. *Transit time.* The time it takes a light-induced carrier to travel across the depletion region of a semiconductor. This parameter determines the maximum bit rate possible with a particular photodiode.
4. *Spectral response.* The range of wavelength values that a given photodiode will respond to. Generally, relative spectral response is graphed as a function of wavelength or frequency.
5. *Light sensitivity.* The minimum optical power a light detector can receive and still produce a usable electrical output signal. Light sensitivity is generally given for a particular wavelength in either dBm or dBμ.

4-15 LASERS

Laser is an acronym for *light amplification stimulated by the emission of radiation.* Laser technology deals with the concentration of light into a very small, powerful beam. The acronym was chosen when technology shifted from microwaves to light waves. Basically, there are four types of lasers: gas, liquid, solid, and semiconductor.

1. *Gas lasers.* Gas lasers use a mixture of helium and neon enclosed in a glass tube. A flow of coherent (one frequency) light waves is emitted through the output coupler when an electric current is discharged into the gas. The continuous light-wave output is monochromatic (one color).
2. *Liquid lasers.* Liquid lasers use organic dyes enclosed in a glass tube for an active medium. Dye is circulated into the tube with a pump. A powerful pulse of light excites the organic dye.
3. *Solid lasers.* Solid lasers use a solid, cylindrical crystal, such as ruby, for the active medium. Each end of the ruby is polished and parallel. The ruby is excited by a tungsten lamp tied to an alternating-current power supply. The output from the laser is a continuous wave.
4. *Semiconductor lasers.* Semiconductor lasers are made from semiconductor *p-n* junctions and are commonly called injection laser diodes (ILDs). The excitation mechanism is a direct-current power supply that controls the amount of current to the active medium. The output light from an ILD is easily modulated, making it very useful in many electronic communications applications.

4-15-1 Laser Characteristics

All types of lasers have several common characteristics. They all use (1) an active material to convert energy into laser light, (2) a pumping source to provide power or energy, (3) optics to direct the beam through the active material to be amplified, (4) optics to direct the beam into a narrow powerful cone of divergence, (5) a feedback mechanism to provide continuous operation, and (6) an output coupler to transmit power out of the laser.

The radiation of a laser is extremely intense and directional. When focused into a fine hairlike beam, it can concentrate all its power into the narrow beam. If the beam of light were allowed to diverge, it would lose most of its power.

QUESTIONS

4-1. Define an *optical fiber* transmission system.

4-2. What is the relationship between *bandwidth* and *information capacity*?

4-3. Contrast the *advantages* and *disadvantages* of optical fiber cables compared to metallic transmission lines.

4-4. Briefly describe the *construction* of an optical fiber cable.

4-5. Define how the following terms relate to optical fiber cables: *velocity of propagation, refraction,* and *refractive index.*

4-6. State *Snell's law* for refraction and outline its significance for optical fiber cables.

4-7. Define *critical angle*.

4-8. Describe what is meant by *mode of operation* and *index profile*.

4-9. Describe *step-index* and *graded-index* optical cables.

4-10. Contrast the *advantages* and *disadvantages* of step-index, graded-index, single-mode propagation, and multimode propagation.

4-11. Why is *single-mode propagation* impossible with graded-index optical fibers?

4-12. Describe *source-to fiber aperture*.

4-13. Describe *acceptance angle* and *acceptance cone*.

4-14. Define *numerical aperture*.

4-15. List and briefly describe the *losses* associated with optical fibers.

4-16. List and describe the various types of *coupling losses* associated with optical fibers.

4-17. Describe the two primary types of *light sources* used with optical fibers.

PROBLEMS

4-1. Determine the wavelengths in nanometers, angstroms, and micrometers for the following light frequencies: 3.45×10^{14} Hz, 3.62×10^{14} Hz, and 3.21×10^{14} Hz.

4-2. Determine the light frequency for the following wavelengths: 670 nm, 7800 angstroms, and 710 nm.

4-3. For a glass ($n = 1.5$)/quarts ($n = 1.38$) interface and an angle of incidence of 35 degrees, determine the angle of refraction.

4-4. Determine the critical angle for the fiber described in Problem 4–3.

4-5. Determine the acceptance angle for the fiber described in Problem 4–3.

4-6. Determine the numerical for the fiber described in Problem 4–3.

C H A P T E R   5

Digital Transmission

CHAPTER OUTLINE

OBJECTIVES

- Define *digital transmission*
- List and describe the advantages and disadvantages of digital transmission
- Describe pulse width modulation, pulse position modulation, and pulse amplitude modulation
- Define and describe *pulse code modulation*
- Explain flat-top and natural sampling
- Describe the Nyquist sampling theorem
- Define *quantization* and describe a folded binary code
- Define and explain *dynamic range*
- Explain signal voltage-to-quantization noise ratio
- Explain the difference between linear and nonlinear PCM codes
- Define *companding* and describe analog and digital companding
- Describe digital compression
- Explain how to determine PCM line speed
- Describe delta modulation PCM
- Describe differential PCM

5-1 INTRODUCTION

As stated in Chapter 2, *digital transmission* is the transmittal of digital signals between two or more points in a communications system. The signals can be binary or any other form of discrete-level digital pulses. The original source information may be in digital form, or it could be analog signals that have been converted to digital pulses prior to transmission and converted back to analog signals in the receiver. With digital transmission systems, a physical facility, such as a pair of wires, a coaxial cable, or an optical fiber cable, is required to interconnect the various points within the system. The pulses are contained in and propagate down the cable. Digital pulses cannot be propagated through a wireless transmission system such as earth's atmosphere or free space (vacuum).

Alex H. Reeves developed the first digital transmission system in 1937 at the Paris Laboratories of AT&T for the purpose of carrying digitally encoded analog signals, such as the human voice, over metallic wire cables between telephone offices. Today, digital transmission systems are used not only to carry digitally encoded voice and video signals but also to carry digital source information directly between computers and computer networks. Digital transmission systems use both metallic and optical fiber cables for their transmission medium.

5-1-1 Advantages of Digital Transmission

The primary advantage of digital transmission over analog transmission is noise immunity. Digital signals are inherently less susceptible than analog signals to interference caused by noise because with digital signals it is not necessary to evaluate the precise amplitude, frequency, or phase to ascertain its logic condition. Instead, pulses are evaluated during a precise time interval, and a simple determination is made whether the pulse is above or below a prescribed reference level.

Digital signals are also better suited than analog signals for processing and combining using a technique called multiplexing. Digital signal processing (DSP) is the processing of analog signals using digital methods and includes band-limiting the signal with filters, amplitude equalization, and phase shifting. It is much simpler to store digital signals than analog signals, and the transmission rate of digital signals can be easily changed to adapt to different environments and to interface with different types of equipment.

In addition, digital transmission systems are more resistant to analog systems to additive noise because they use signal regeneration rather than signal amplification. Noise produced in electronic circuits is additive (i.e., it accumulates); therefore, the signal-to-noise ratio deteriorates each time an analog signal is amplified. Consequently, the number of circuits the signal must pass through limits the total distance analog signals can be transported. However, digital regenerators sample noisy signals and then reproduce an entirely new digital signal with the same signal-to-noise ratio as the original transmitted signal. Therefore, digital signals can be transported greater distances than analog signals.

Finally, digital signals are simpler to measure and evaluate than analog signals. Therefore, it is easier to compare the error performance of one digital system to another digital system. Also, with digital signals, transmission errors can be detected and corrected more easily and more accurately than is possible with analog signals.

5-1-2 Disadvantages of Digital Transmission

The transmission of digitally encoded analog signals requires significantly more bandwidth than simply transmitting the original analog signal. Bandwidth is one of the most important aspects of any communications system because it is costly and limited.

Analog signals must also be converted to digital pulses prior to transmission and converted back to their original analog form at the receiver, thus necessitating additional encoding and decoding circuitry. In addition, digital transmission requires precise time synchronization between the clocks in the transmitters and receivers. Finally, digital transmission systems are incompatible with older analog transmission systems.

5-2   PULSE MODULATION

Pulse modulation consists essentially of sampling analog information signals and then converting those samples into discrete pulses and then transporting the pulses from a source to a destination over a physical transmission medium. The four predominant methods of pulse modulation include *pulse width modulation* (PWM), *pulse position modulation* (PPM), *pulse amplitude modulation* (PAM), and *pulse code modulation* (PCM).

1. *PWM.* This method of pulse modulation is sometimes called *pulse duration modulation* (PDM) or *pulse length modulation* (PLM), as the width (active portion of the duty cycle) of a constant amplitude pulse is varied proportional to the amplitude of the analog signal at the time the signal is sampled. PWM is shown in Figure 5-1c. As the figure shows, the amplitude of sample 1 is lower than the amplitude of sample 2. Thus, pulse 1 is narrower than pulse 2. The maximum analog signal amplitude produces the widest pulse, and the minimum analog signal amplitude produces the narrowest pulse. Note, however, that all pulses have the same amplitude.

2. *PPM.* With PPM, the position of a constant-width pulse within a prescribed time slot is varied according to the amplitude of the sample of the analog signal. PPM is shown in Figure 5-1d. As the figure shows, the higher the amplitude of the sample, the farther to the right the pulse is positioned within the prescribed time slot. The highest amplitude sample produces a pulse to the far right, and the lowest amplitude sample produces a pulse to the far left.

3. *PAM.* With PAM, the amplitude of a constant-width, constant-position pulse is varied according to the amplitude of the sample of the analog signal. PAM is shown in Figure 5-1e, where it can be seen that the amplitude of a pulse coincides with the amplitude of the analog signal. PAM waveforms resemble the original analog signal more than the waveforms for PWM or PPM.

4. *PCM.* With PCM, the analog signal is sampled and then converted to a serial n-bit binary code for transmission. Each code has the same number of bits and requires the same length of time for transmission. PCM is shown in Figure 5-1f.

PAM is used as an intermediate form of modulation with PSK, QAM, and PCM, although it is seldom used by itself. PWM and PPM are used in special-purpose communications systems mainly for the military but are seldom used for commercial digital transmission systems. PCM is by far the most prevalent form of pulse modulation and, consequently, will be discussed in more detail in subsequent sections of this chapter.

5-3   PULSE CODE MODULATION

Alex H. Reeves is credited with inventing pulse code modulation (PCM) in 1937 while working for AT&T in Paris. Although the merits of PCM were recognized early in its development, it was not until the mid-1960s with the advent of solid-state electronics that PCM became prevalent. In the United States today, PCM is the preferred method of communications within the public switched telephone network because with PCM it is easy to combine digitized voice and digital data into a single, high-speed digital signal and propagate it over either metallic or optical fiber cables.

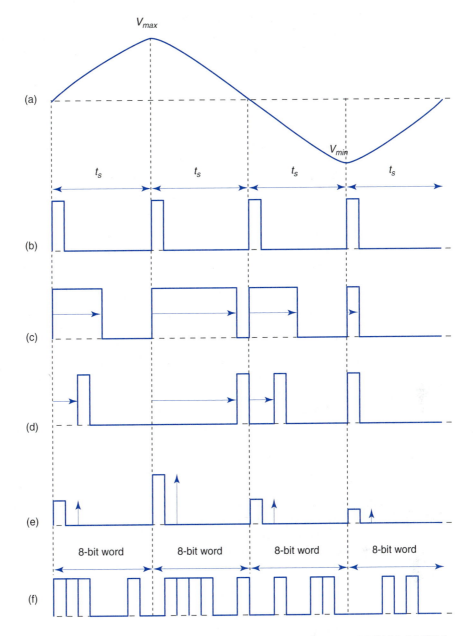

FIGURE 5-1 Pulse modulation: (a) analog signal; (b) sample pulse; (c) PWM; (d) PPM; (e) PAM; (f) PCM

Pulse code modulation is the only one of the digitally encoded modulation techniques shown in Figure 5-1 that is commonly used for digital transmission. The term *pulse code modulation* is somewhat of a misnomer, as it is not really a type of modulation but rather a form of digitally coding analog signals. With PCM, the pulses are of fixed length and fixed amplitude. PCM is a binary system where a pulse or lack of a pulse within a prescribed time slot represents either a logic 1 or a logic 0 condition. PWM, PPM, and PAM are digital but seldom binary, as a pulse does not represent a single binary digit (bit).

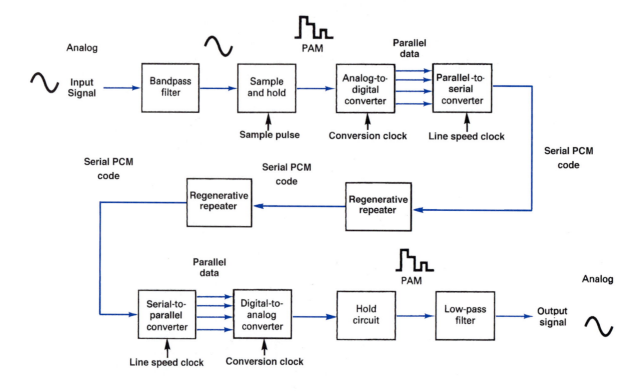

Analog

PAM

Parallel data

Input Signal → Bandpass filter → Sample and hold → Analog-to-digital converter → Parallel-to-serial converter

Sample pulse

Conversion clock

Line speed clock

Serial PCM code

Serial PCM code

Serial PCM code

Regenerative repeater ← Regenerative repeater

Parallel data

PAM

Analog

Serial-to-parallel converter → Digital-to-analog converter → Hold circuit → Low-pass filter → Output signal

Line speed clock

Conversion clock

PCM Receiver

FIGURE 5-2 Simplified block diagram of a single-channel, simplex PCM transmission system

Figure 5-2 shows a simplified block diagram of a single-channel, simplex (one-way only) PCM system. The bandpass filter limits the frequency of the analog input signal to the standard voice-band frequency range of 300 Hz to 3000 Hz. The *sample-and-hold* circuit periodically samples the analog input signal and converts those samples to a multilevel PAM signal. The *analog-to-digital converter* (ADC) converts the PAM samples to parallel PCM codes, which are converted to serial binary data in the *parallel-to-serial converter* and then outputted onto the transmission line as serial digital pulses. The transmission line *repeaters* are placed at prescribed distances to regenerate the digital pulses.

In the receiver, the *serial-to-parallel converter* converts serial pulses received from the transmission line to parallel PCM codes. The *digital-to-analog converter* (DAC) converts the parallel PCM codes to multilevel PAM signals. The hold circuit is basically a low-pass filter that converts the PAM signals back to its original analog form.

Figure 5-2 also shows several clock signals and sample pulses that will be explained in later sections of this chapter. An integrated circuit that performs the PCM encoding and decoding functions is called a codec (coder/decoder), which is also described in a later section of this chapter.

5-3-1 PCM Sampling

The function of a sampling circuit in a PCM transmitter is to periodically sample the continually changing analog input voltage and convert those samples to a series of constant-amplitude pulses that can more easily be converted to binary PCM code. For the ADC to ac-

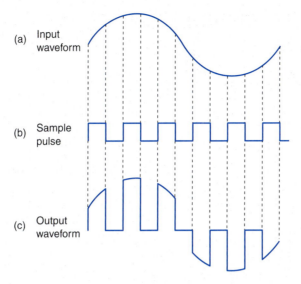

FIGURE 5-3 Natural sampling: (a) input analog signal; (b) sample pulse; (c) sampled output

curately convert a voltage to a binary code, the voltage must be relatively constant so that the ADC can complete the conversion before the voltage level changes. If not, the ADC would be continually attempting to follow the changes and may never stabilize on any PCM code.

Essentially, there are two basic techniques used to perform the sampling function: natural sampling and flat-top sampling. *Natural sampling* is shown in Figure 5-3. Natural sampling is when tops of the sample pulses retain their natural shape during the sample interval, making it difficult for an ADC to convert the sample to a PCM code. With natural sampling, the frequency spectrum of the sampled output is different from that of an ideal sample. The amplitude of the frequency components produced from narrow, finite-width sample pulses decreases for the higher harmonics in a (sin *x*)/*x* manner. This alters the information frequency spectrum, requiring the use of frequency equalizers (compensation filters) before recovery by a low-pass filter.

The most common method used for sampling voice signals in PCM systems is *flat-top sampling,* which is accomplished in a *sample-and-hold circuit.* The purpose of a sample-and-hold circuit is to periodically sample the continually changing analog input voltage and convert those samples to a series of constant-amplitude PAM voltage levels. With flat-top sampling, the input voltage is sampled with a narrow pulse and then held relatively constant until the next sample is taken. Figure 5-4 shows flat-top sampling. As the figure shows, the sampling process introduces an error called *aperture error*, which is when the amplitude of the sampled signal changes during the sample pulse time. This prevents the recovery circuit in the PCM receiver from exactly reproducing the original analog signal voltage. The magnitude of error depends on how much the analog signal voltage changes while the sample is being taken and the width (duration) of the sample pulse. Flat-top sampling, however, introduces less aperture distortion than natural sampling and can operate with a slower analog-to-digital converter.

5-3-2 Sampling Rate

The *Nyquist sampling theorem* establishes the *minimum sampling rate* (f_s) that can be used for a given PCM system. For a sample to be reproduced accurately in a PCM receiver, each cycle of the analog input signal (f_a) must be sampled at least twice. Consequently, the minimum sampling rate is equal to twice the highest audio input frequency. If f_s is less than two

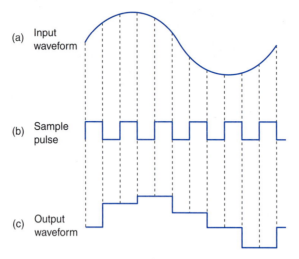

FIGURE 5-4 Flat-top sampling: (a) input analog signal;
(b) sample pulse; (c) sampled output

times f_a, an impairment called *alias* or *foldover distortion* occurs. Mathematically, the minimum Nyquist sampling rate is

$$f_s \geq 2f_a \qquad (5\text{-}1)$$

where f_s = minimum Nyquist sample rate (hertz)
f_a = maximum analog input frequency (hertz)

A sample-and-hold circuit is a nonlinear device with two inputs: the sampling pulse and the analog input signal. Consequently, nonlinear mixing occurs between these two signals. As long as the sampling pulse is at least twice the highest analog input signal, foldover distortion does not occur, and the samples contain all the information from the original signal. However, if the analog input frequency exceeds twice the sampling rate, information is lost, and foldover distortion occurs.

5-3-3   Quantization and the Folded Binary Code

Quantization is the process of converting an infinite number of possibilities to a finite number of conditions. Analog signals contain an infinite number of amplitude possibilities. Thus, converting an analog signal to a PCM code with a limited number of combinations requires quantization. In essence, quantization is the process of rounding off the amplitudes of flat-top samples to a manageable number of levels. For example, a sine wave with a peak amplitude of 5 V varies between +5 V and −5 V passing through every possible amplitude in between. A PCM code could have only eight bits, which equates to only 2^8, or 256, combinations. Obviously, converting samples of a sine wave to PCM requires some rounding off.

With quantization, the total voltage range is subdivided into a smaller number of subranges, as shown in Table 5-1. The PCM code shown in Table 5-1 is a three-bit sign-magnitude code with eight possible combinations (four positive and four negative). The leftmost bit is the sign bit (1 = + and 0 = −), and the two rightmost bits represent magnitude. This type of code is called a folded binary code because the codes on the bottom half of the table are a mirror image of the codes on the top half, except for the sign bit. If the negative codes were folded over on top of the positive codes, they would match perfectly. Also, with a folded binary code there are two codes assigned to zero volts: 100 (+0) and

Table 5-1 Three-Bit PCM Code

Sign	Magnitude		Decimal value	Quantization range
1	1	1	+3	+2.5 V to +3.5 V
1	1	0	+2	+1.5 V to +2.5 V
1	0	1	+1	+0.5 V to +1.5 V
1	0	0	+0	0 V to +0.5 V
0	0	0	−0	0 V to −0.5 V
0	0	1	−1	−0.5 V to −1.5 V
0	1	0	−2	−1.5 V to −2.5 V
0	1	1	−3	−2.5 V to −3.5 V

8 Sub ranges

000 (−0). The magnitude difference between adjacent steps is called the *quantization interval* or *quantum*. For the code shown in Table 5-1, the quantization interval is 1 V. Therefore, for this code, the maximum signal magnitude that can be encoded is +3 V (111) or −3 V (011) and the minimum signal magnitude is +1 V (101) or −1 V (001). If the magnitude of the sample exceeds the highest quantization interval, *overload distortion* (also called *peak limiting*) occurs.

Assigning PCM codes to absolute magnitudes is called quantizing. The magnitude of a quantum is also called the *resolution*. The resolution is equal to the voltage of the *least significant bit* (V_{lsb}) of the PCM code, which is the minimum voltage other than 0 V that can be decoded by the digital-to-analog converter in the receiver. The resolution for the PCM code shown in Table 5-1 is 1 V. The smaller the magnitude of a quantum, the better (smaller) the resolution and the more accurately the quantized signal will resemble the original analog sample.

In Table 5-1, each three-bit code has a range of input voltages that will be converted to that code. For example, any voltage between +0.5 and +1.5 will be converted to the code 101 (+1 V). Each code has a *quantization range* equal to + or − one-half the magnitude of a quantum except the codes for +0 and −0. The 0-V codes each have an input range equal to only one-half a quantum.

Figure 5-5 shows an analog input signal, the sampling pulse, the corresponding quantized signal (PAM), and the PCM code for each sample. The likelihood of a sample voltage being equal to one of the eight quantization levels is remote. Therefore, as shown in the figure, each sample voltage is rounded off (quantized) to the closest available level and then converted to its corresponding PCM code. The PAM signal in the transmitter is essentially the same PAM signal produced in the receiver. Therefore, any round-off errors in the transmitted signal are reproduced when the code is converted back to analog by the DAC in the receiver. This error is called the *quantization error* (Q_e). The quantization error is equivalent to additive white noise as it alters the signal amplitude. Consequently, quantization error is also called *quantization noise* (Q_n). The maximum magnitude for the quantization error is equal to one-half a quantum (±0.5 V for the code shown in Table 5-1).

The quantized signal shown in Figure 5-5c at best only roughly resembles the original analog input signal. This is because with a three-bit PCM code, the resolution is rather poor and also because there are only three samples taken of the analog signal. The quality of the PAM signal can be improved by using a PCM code with more bits, reducing the magnitude of a quantum and improving the resolution. The quality can also be improved by sampling the analog signal at a faster rate. Figure 5-6 shows the same analog input signal shown in Figure 5-5 except the signal is being sampled at a much higher rate. As the figure shows, the PAM signal resembles the analog input signal rather closely.

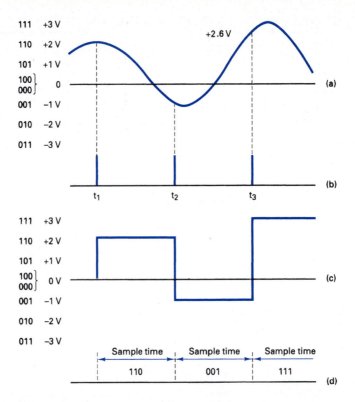

FIGURE 5-5 (a) Analog input signal; (b) sample pulse; (c) PAM signal; (d) PCM code

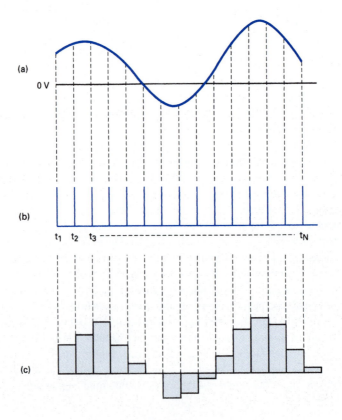

FIGURE 5-6 PAM: (a) input signal; (b) sample pulse; (c) PAM signal

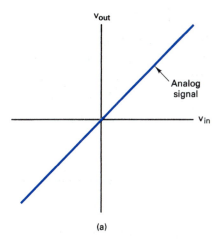

(a)

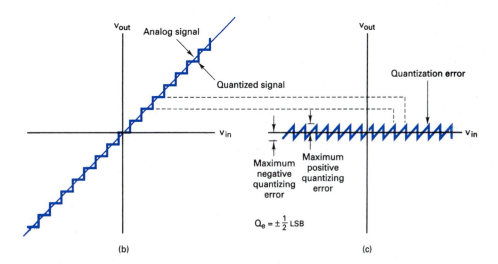

(b)   (c)

FIGURE 5-7   Linear input-versus-output transfer curve: (a) linear transfer function; (b) quantization; (c) Q_e

Figure 5-7 shows the input-versus-output transfer function for a linear analog-to-
digital converter (sometimes called a linear quantizer). As the figure shows for a linear analog input signal (i.e., a ramp), the quantized signal is a staircase function. Thus, as shown in
Figure 5-7c, the maximum quantization error is the same for any magnitude input signal.

Example 5-1

For the PCM coding scheme shown in Figure 5-5, determine the quantized voltage, quantization error (Q_e), and PCM code for the analog sample voltage of $+1.07$ V.

Solution   a. To determine the quantized level, simply divide the sample voltage by resolution and
then round the answer off to the nearest quantization level:

$$\frac{+1.07 \text{ V}}{1 \text{ V}} = 1.07 \Rightarrow 1$$

The quantization error is the difference between the original sample voltage and the quantized level, or

$$Q_e = 1.07 - 1 = 0.07$$

From Table 5-1, the PCM code for $+1$ is 101.

5-4 DYNAMIC RANGE

The number of PCM bits transmitted per sample is determined by several variables, which include maximum allowable input amplitude, resolution, and dynamic range. *Dynamic range* (DR) is the ratio of the largest possible magnitude to the smallest possible magnitude (other than 0 V) that can be decoded by the digital-to-analog converter in the receiver. Mathematically, dynamic range is

$$DR = \frac{V_{max}}{V_{min}} \tag{5-2}$$

where DR = dynamic range (unitless ratio)
V_{min} = the quantum value (resolution)
V_{max} = the maximum voltage magnitude that can be discerned by the DACs in the receiver

Equation 5-2 can be rewritten as

$$DR = \frac{V_{max}}{\text{resolution}} \tag{5-3}$$

For the system shown in Table 5-1,

$$DR = \frac{3 \text{ V}}{1 \text{ V}} = 3$$

However, it is common to express dynamic range as a dB value; therefore,

$$DR = 20 \log \frac{V_{max}}{V_{min}} \tag{5-4}$$

For the system shown in Table 5-1,

$$DR = 20 \log 3 = 9.54 \text{ dB}$$

The number of bits used for a PCM code depends on the dynamic range. The relationship between dynamic range and the number of bits in a PCM code is

$$2^n - 1 \geq DR \tag{5-5a}$$

and for a minimum number of bits

$$2^n - 1 = DR \tag{5-5b}$$

where n = number of bits in a PCM code, excluding the sign bit
DR = absolute value of dynamic range

Why $2^n - 1$? One positive and one negative PCM code is used for 0 V, which is not considered for dynamic range. Therefore,

$$2^n = DR + 1$$

To solve for the number of bits (n) necessary to produce a dynamic range of 3, convert to logs:

$$\log 2^n = \log(DR + 1)$$

$$n\log 2 = \log(DR + 1)$$

$$n = \frac{\log(3 + 1)}{\log 2} = \frac{0.602}{0.301} = 2$$

Table 5-2 Dynamic Range versus Number of PCM Magnitude Bits

Number of Bits in PCM Code (n)	Number of Levels Possible (M = 2^n)	Dynamic Range (dB)
1	2	6.02
2	4	12
3	8	18.1
4	16	24.1
5	32	30.1
6	64	36.1
7	128	42.1
8	256	48.2
9	512	54.2
10	1024	60.2
11	2048	66.2
12	4096	72.2
13	8192	78.3
14	16,384	84.3
15	32,768	90.3
16	65,536	96.3

For a dynamic range of 3, a PCM code with two bits is required. Dynamic range can be expressed in decibels as

$$DR_{(dB)} = 20 \log\left(\frac{V_{max}}{V_{min}}\right)$$

or

$$DR_{(dB)} = 20 \log(2^n - 1) \tag{5-6}$$

where n is the number of PCM bits.

For values of $n > 4$, dynamic range is approximated as

$$DR_{(dB)} \approx 20 \log(2^n)$$

$$\approx 20n \log(2)$$

$$\approx 6n \tag{5-7}$$

Equation 5-7 indicates that there is approximately 6 dB dynamic range for each magnitude bit in a linear PCM code. Table 5-2 summarizes dynamic range for PCM codes with n bits for values of n up to 16.

Example 5-2

For a PCM system with the following parameters, determine (a) minimum sample rate, (b) minimum number of bits used in the PCM code, (c) resolution, and (d) quantization error.

> Maximum analog input frequency = 4 kHz
> Maximum decoded voltage at the receiver = ±2.55 V
> Minimum dynamic range = 46 dB

Solution

a. Substituting into Equation 5-1, the minimum sample rate is

$$f_s = 2f_a = 2(4 \text{ kHz}) = 8 \text{ kHz}$$

b. To determine the absolute value for dynamic range, substitute into Equation 5-4:

$$46 \text{ dB} = 20 \log \frac{V_{max}}{V_{min}}$$

$$2.3 = \log \frac{V_{max}}{V_{min}}$$

$$10^{2.3} = \frac{V_{max}}{V_{min}}$$

$$199.5 = DR$$

The minimum number of bits is determined by rearranging Equation 5-5b and solving for *n*:

$$n = \frac{\log(199.5 + 1)}{\log 2} = 7.63$$

The closest whole number greater than 7.63 is 8; therefore, eight bits must be used for the magnitude. Because the input amplitude range is ±2.55, one additional bit, the sign bit, is required. Therefore, the total number of PCM bits is nine, and the total number of PCM codes is $2^9 = 512$. (There are 255 positive codes, 255 negative codes, and 2 zero codes.)
To determine the actual dynamic range, substitute into Equation 5-6:

$$
\begin{aligned}
DR_{(dB)} &= 20 \log(2^n - 1) \\
&= 20(\log 256 - 1) \\
&= 48.13 \text{ dB}
\end{aligned}
$$

c. The resolution is determined by dividing the maximum positive or maximum negative voltage by the number of positive or negative nonzero PCM codes:

$$\text{resolution} = \frac{V_{max}}{2^n - 1} = \frac{2.55}{2^8 - 1} = \frac{2.55}{256 - 1} = 0.01 \text{ V}$$

The maximum quantization error is

$$Q_e = \frac{\text{resolution}}{2} = \frac{0.01 \text{ V}}{2} = 0.005 \text{ V}$$

5-5 SIGNAL VOLTAGE-TO-QUANTIZATION NOISE VOLTAGE RATIO

The three-bit PCM coding scheme shown in Figures 5–5 and 5–6 are *linear codes*, which means that the magnitude change between any two successive codes is the same. Consequently, the magnitude of their quantization error is also the same. The maximum quantization noise is half the resolution (quantum value). Therefore, the worst possible *signal voltage–to–quantization noise voltage ratio* (SQR) occurs when the input signal is at its minimum amplitude (101 or 001). Mathematically, the worst-case voltage SQR is

$$SQR = \frac{\text{resolution}}{Q_e} = \frac{V_{lsb}}{V_{lsb}/2} = 2$$

For the PCM code shown in Figure 5-5, the worst-case (minimum) SQR occurs for the lowest magnitude quantization voltage (±1 V). Therefore, the minimum SQR is

$$SQR_{(min)} = \frac{1}{0.5} = 2$$

or, in dB,

$$
\begin{aligned}
&= 20 \log(2) \\
&= 6 \text{ dB}
\end{aligned}
$$

For a maximum amplitude input signal of 3 V (either 111 or 011), the maximum quantization noise is also equal to the resolution divided by 2. Therefore, the SQR for a maximum input signal is

$$SQR_{(max)} = \frac{V_{max}}{Q_e} = \frac{3}{0.5/2} = 6$$

or, in dB,
$$= 20 \log 6$$
$$= 15.6 \text{ dB}$$

From the preceding example, it can be seen that even though the magnitude of the quantization error remains constant throughout the entire PCM code, the percentage error does not; it decreases as the magnitude of the sample amplitude increases.

The preceding expression for SQR is for voltage and presumes the maximum quantization error; therefore, it is of little practical use and is shown only for comparison purposes and to illustrate that the SQR is not constant throughout the entire range of sample amplitudes. In reality and as shown in Figure 5-6, the difference between the PAM waveform and the analog input waveform varies in magnitude; therefore, the SQR is not constant. Generally, the quantization error or distortion caused by digitizing an analog sample is expressed as an average signal power–to–average noise power ratio. For linear PCM codes (all quantization intervals have equal magnitudes), the *signal power–to–quantizing noise power ratio* (also called *signal-to-distortion ratio* or *signal-to-noise ratio*) is determined by the following formula:

$$SQR_{(dB)} = 10 \log \frac{v^2/R}{(q^2/12)/R} \qquad \text{(5-8a)}$$

where
$$R = \text{resistance (ohms)}$$
$$v = \text{rms signal voltage (volts)}$$
$$q = \text{quantization interval (volts)}$$
$$v^2/R = \text{average signal power (watts)}$$
$$(q^2/12)/R = \text{average quantization noise power (watts)}$$

If the resistances are assumed to be equal, Equation 5-8a reduces to

$$SQR = 10 \log\left[\frac{v^2}{q^2/12}\right]$$

$$= 10.8 = 20 \log \frac{v}{q} \qquad \text{(5-8b)}$$

5-6 LINEAR VERSUS NONLINEAR PCM CODES

Early PCM systems used *linear codes* (i.e., the magnitude change between any two successive steps is uniform). With linear coding, the accuracy (resolution) for the higher-amplitude analog signals is the same as for the lower-amplitude signals, and the SQR for the lower-amplitude signals is less than for the higher-amplitude signals. With voice transmission, low-amplitude signals are more likely to occur than large-amplitude signals. Therefore, if there were more codes for the lower amplitudes, it would increase the accuracy where the accuracy is needed. As a result, there would be fewer codes available for the higher amplitudes, which would increase the quantization error for the larger-amplitude signals (thus decreasing the SQR). Such a coding technique is called *nonlinear* or *nonuniform encoding*. With nonlinear encoding, the step size increases with the amplitude of the input signal.

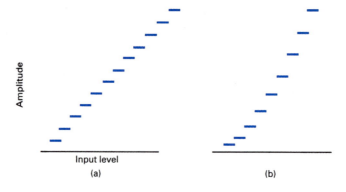

FIGURE 5-8 (a) Linear versus (b) nonlinear encoding

Figure 5-8 shows the step outputs from a linear and a nonlinear analog-to-digital converter. Note that with nonlinear encoding there are more codes at the bottom of the scale than there are at the top, thus increasing the accuracy for the smaller-amplitude signals. Also note that the distance between successive codes is greater for the higher-amplitude signals, thus increasing the quantization error and reducing the SQR. Also, because the ratio of V_{max} to V_{min} is increased with nonlinear encoding, the dynamic range is larger than with a uniform linear code. It is evident that nonlinear encoding is a compromise; SQR is sacrificed for the higher-amplitude signals to achieve more accuracy for the lower-amplitude signals and to achieve a larger dynamic range. It is difficult to fabricate nonlinear analog-to-digital converters; consequently, alternative methods of achieving the same results have been devised.

5-7 COMPANDING

Companding is the process of *compressing* and then *expanding*. With companded systems, the higher-amplitude analog signals are compressed (amplified less than the lower-amplitude signals) prior to transmission and then expanded (amplified more than the lower-amplitude signals) in the receiver. Companding is a means of improving the dynamic range of a communications system.

Figure 5-9 illustrates the process of companding. An analog input signal with a dynamic range of 50 dB is compressed to 25 dB prior to transmission and then in the receiver expanded back to its original dynamic range of 50 dB. With PCM, companding may be accomplished using analog or digital techniques. Early PCM systems used analog companding, whereas more modern systems use digital companding.

5-7-1 Analog Companding

Historically, analog compression was implemented using specially designed diodes inserted in the analog signal path in a PCM transmitter prior to the sample-and-hold circuit. Analog expansion was also implemented with diodes that were placed just after the low-pass filter in the PCM receiver.

Figure 5-10 shows the basic process of analog companding. In the transmitter, the dynamic range of the analog signal is compressed, sampled, and then converted to a linear PCM code. In the receiver, the PCM code is converted to a PAM signal, filtered, and then expanded back to its original dynamic range.

Different signal distributions require different companding characteristics. For instance, voice-quality telephone signals require a relatively constant SQR performance over a wide dynamic range, which means that the distortion must be proportional to signal am-

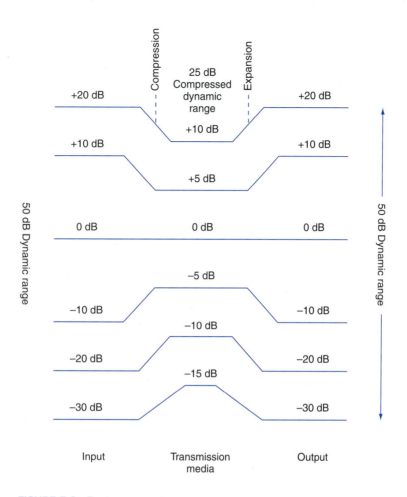

FIGURE 5-9 Basic companding process

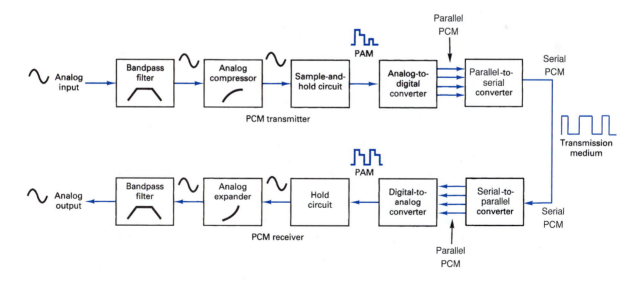

FIGURE 5-10 PCM system with analog companding

plitude for all input signal levels. This requires a logarithmic compression ratio, which requires an infinite dynamic range and an infinite number of PCM codes. Of course, this is impossible to achieve. However, there are two methods of analog companding currently being used that closely approximate a logarithmic function and are often called log-PCM codes. The two methods are μ-*law* and the *A-law* companding.

5-7-1-1 μ-law companding. In the United States and Japan, μ-*law* companding is used. The compression characteristics for μ-*law* is

$$V_{out} = \frac{V_{max} \ln(1 + \mu V_{in}/V_{max})}{\ln(1 + \mu)} \tag{5-9}$$

where V_{max} = maximum uncompressed analog input amplitude (volts)
            V_{in} = amplitude of the input signal at a particular instant of time (volts)
            μ = parameter used to define the amount of compression (unitless)
          V_{out} = compressed output amplitude (volts)

Figure 5-11 shows the compression curves for several values of μ. Note that the higher the μ, the more compression. Also note that for μ = 0, the curve is linear (no compression).

The parameter μ determines the range of signal power in which the SQR is relatively constant. Voice transmission requires a minimum dynamic range of 40 dB and a seven-bit PCM code. For a relatively constant SQR and a 40-dB dynamic range, a μ ≥ 100 is required. The early Bell System PCM systems used a seven-bit code with a μ = 100. However, the most recent PCM systems use an eight-bit code and a μ = 255.

Example 5-3

For a compressor with a μ = 255, determine

a. The voltage gain for the following relative values of V_{in}: V_{max}, 0.75 V_{max}, 0.5 V_{max}, and 0.25 V_{max}.
b. The compressed output voltage for a maximum input voltage of 4 V.
c. Input and output dynamic ranges and compression.

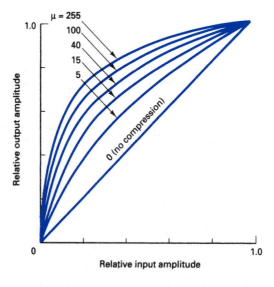

FIGURE 5-11 μ-law compression characteristics

Solution **a.** Substituting into Equation 5-9, the following voltage gains are achieved for the given input magnitudes:

V_{in}	Compressed voltage gain
V_{max}	1.00
$0.75\ V_{max}$	1.26
$0.5\ V_{max}$	1.75
$0.25\ V_{max}$	3.00

b. Using the compressed voltage gains determined in step a, the output voltage is simply the input voltage times the compression gain:

V_{in}	V_{out}
$V_{max} = 4\ V$	4.00 V
$0.75\ V_{max} = 3\ V$	3.78 V
$0.50\ V_{max} = 2\ V$	3.50 V
$0.25\ V_{max} = 1\ V$	3.00 V

c. Dynamic range is calculated by substituting into Equation 5-2:

$$\text{input dynamic range} = 20 \log \frac{4}{1} = 12\ dB$$

$$\text{output dynamic range} = 20 \log \frac{4}{3} = 2.5\ dB$$

$$\text{compression} = \text{input dynamic range minus output dynamic range}$$
$$= 12\ dB - 2.5\ dB = 9.5\ dB$$

To restore the signals to their original proportions in the receiver, the compressed voltages are expanded by passing them through an amplifier with gain characteristics that are the complement of those in the compressor. For the values given in Example 5-3, the voltage gains in the receiver are the following:

V_{in}	Expanded voltage gain
V_{max}	1.00
$0.75\ V_{max}$	0.79
$0.5\ V_{max}$	0.57
$0.25\ V_{max}$	0.33

The overall circuit gain is simply the product of the compression and expansion factors, which equals 1 for all input voltage levels. For the values given in Example 5-3,

$$V_{in} = V_{max} \qquad 1 \times 1 = 1$$
$$V_{in} = 0.75\ V_{max} \qquad 1.26 \times 0.79 \cong 1$$
$$V_{in} = 0.5\ V_{max} \qquad 1.75 \times 0.57 \cong 1$$
$$V_{in} = 0.25\ V_{max} \qquad 3 \times 0.33 \cong 1$$

5-7-1-2 A-law companding. In Europe, the ITU-T has established *A-law* companding to be used to approximate true logarithmic companding. For an intended dynamic range, *A-law* companding has a slightly flatter SQR than μ-law. *A-law* companding, however, is inferior to μ-law in terms of small-signal quality (idle channel noise). The compression characteristic for *A-law* companding is

$$V_{out} = V_{max} \frac{A V_{in}/V_{max}}{1 + \ln A} \qquad\qquad 0 \le \frac{V_{in}}{V_{max}} \le \frac{1}{A} \qquad\qquad \textbf{(5-10a)}$$

$$= \frac{1 + \ln(A V_{in}/V_{max})}{1 + \ln A} \qquad\qquad \frac{1}{A} \le \frac{V_{in}}{V_{max}} \le 1 \qquad\qquad \textbf{(5-10b)}$$

5-7-2 Digital Companding

Digital companding involves compression in the transmitter after the input sample has been converted to a linear PCM code and then expansion in the receiver prior to PCM decoding. Figure 5-12 shows the block diagram for a digitally companded PCM system.

With digital companding, the analog signal is first sampled and converted to a linear PCM code, and then the linear code is digitally compressed. In the receiver, the compressed PCM code is expanded and then decoded (i.e., converted back to analog). The most recent digitally compressed PCM systems use a 12-bit linear PCM code and an eight-bit compressed PCM code. The compression and expansion curves closely resemble the analog μ-law curves with a μ = 255 by approximating the curve with a set of eight straight-line segments (segments 0 through 7). The slope of each successive segment is exactly one-half that of the previous segment.

Figure 5-13 shows the 12-to-eight-bit digital compression curve for positive values only. The curve for negative values is identical except the inverse. Although there are 16 segments (eight positive and eight negative), this scheme is often called *13-segment compression* because the curve for segments +0, +1, −0, and −1 is a straight line with a constant slope and is considered as one segment.

The digital companding algorithm for a 12-bit linear to eight-bit compressed code is actually quite simple. The eight-bit compressed code consists of a sign bit, a three-bit segment identifier, and a five-bit magnitude code that specifies the quantization interval within the specified segment (see Figure 5-14a).

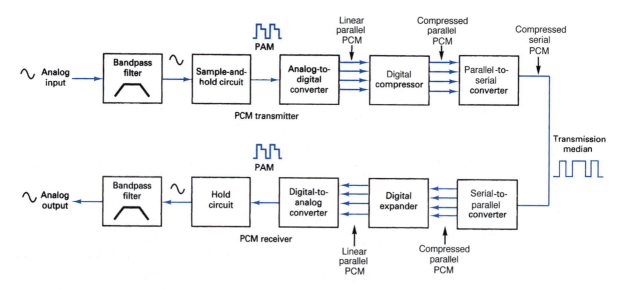

FIGURE 5-12 Digitally companded PCM system

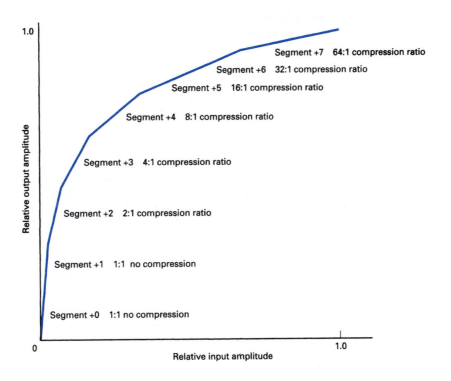

FIGURE 5-13 μ255 compression characteristics (positive values only)

Sign bit 1 = + 0 = −	3-Bit segment identifier 000 to 111	4-Bit quantization interval A B C D 0000 to 1111

(a)

Transmission
media

Transmit   Receive

Encoded
PCM

Decoded
PCM

Segment	12-Bit linear code	8-Bit compressed code
0	s0000000ABCD	s000ABCD
1	s0000001ABCD	s001ABCD
2	s000001ABCDX	s010ABCD
3	s00000ABCDXX	s011ABCD
4	s0001ABCDXXX	s100ABCD
5	s001ABCDXXXX	s101ABCD
6	s01ABCDXXXXX	s110ABCD
7	s1ABCDXXXXXX	s111ABCD

(b)

8-Bit compressed code	12-Bit recovered code	Segment
s000ABCD	s0000000ABCD	0
s001ABCD	s0000001ABCD	1
s010ABCD	s000001ABCD1	2
s011ABCD	s00000ABCD10	3
s100ABCD	s0001ABCD100	4
s101ABCD	s001ABCD1000	5
s110ABCD	s01ABCD10000	6
s111ABCD	s1ABCD100000	7

(c)

FIGURE 5-14 12-bit-to-eight-bit digital companding: (a) eight-bit μ255 compressed code format; (b) μ255 encoding table; (c) μ255 decoding table

In the μ 255-encoding table shown in Figure 5-14b, the bit positions designated with an X are truncated during compression and subsequently lost. Bits designated A, B, C, and D are transmitted as is. The sign bit is also transmitted as is. Note that for segments 0 and 1, the encoded 12-bit PCM code is duplicated exactly at the output of the decoder (compare Figures 5–14b and 5–14c), whereas for segment 7, only the most significant six bits are duplicated. With 11 magnitude bits, there are 2048 possible codes; however, they are not equally distributed among the eight segments. There are 16 codes in segment 0 and 16 codes in segment 1. In each subsequent segment, the number of codes doubles (i.e., segment 2 has 32 codes; segment 3 has 64 codes, and so on). However, in each of the eight segments, only 16 12-bit codes can be produced. Consequently, in segments 0 and 1, there is no compression (of the 16 possible codes, all 16 can be decoded). In segment 2, there is a compression ratio of 2:1 (of the 32 possible codes, only 16 can be decoded). In segment 3, there is a 4:1 compression ratio (64 codes to 16 codes). The compression ratio doubles with each successive segment. The compression ratio in segment 7 is 1024/16, or 64:1.

The compression process is as follows. The analog signal is sampled and converted to a linear 12-bit sign-magnitude code. The sign bit is transferred directly to an eight-bit compressed code. The segment number in the eight-bit code is determined by counting the number of leading 0s in the 11-bit magnitude portion of the linear code beginning with the most-significant bit. Subtract the number of leading 0s (not to exceed 7) from 7. The result is the segment number, which is converted to a three-bit binary number and inserted into the eight-bit compressed code as the segment identifier. The four magnitude bits (A, B, C, and D) represent the quantization interval (i.e., subsegments) and are substituted into the least-significant four bits of the eight-bit compressed code.

Essentially, segments 2 through 7 are subdivided into smaller subsegments. Each segment consists of 16 subsegments, which correspond to the 16 conditions possible for bits A, B, C, and D (0000–1111). In segment 2, there are two codes per subsegment. In segment 3, there are four. The number of codes per subsegment doubles with each subsequent segment. Consequently, in segment 7, each subsegment has 64 codes. In the decoder, the most significant of the truncated bits is reinserted as a logic 1. The remaining truncated bits are reinserted as 0s. This ensures that the maximum magnitude of error introduced by the compression and expansion process is minimized. Essentially, the decoder guesses what the truncated bits were prior to encoding. The most logical guess is halfway between the minimum- and maximum-magnitude codes. For example, in segment 6, the five least-significant bits are truncated during compression; therefore, in the receiver, the decoder must try to determine what those bits were. The possibilities include any code between 00000 and 11111. The logical guess is 10000, approximately half the maximum magnitude. Consequently, the maximum compression error is slightly more than one-half the maximum magnitude for that segment.

Example 5-4

Determine the 12-bit linear code, the eight-bit compressed code, the decoded 12-bit code, the quantization error, and the compression error for a resolution of 0.01 V and analog sample voltages of **(a)** +0.053 V, **(b)** −0.318 V, and **(c)** +10.234 V.

Solution

a. To determine the 12-bit linear code, simply divide the sample voltage by the resolution, round off the quotient, and then convert the result to a 12-bit sign-magnitude code:

$$\frac{+0.053 \text{ V}}{+0.01 \text{ V}} = +5.3, \text{ which is rounded off to 5, producing a quantization error}$$
$$Q_e = 0.3(0.01 \text{ V}) = 0.003 \text{ V}$$

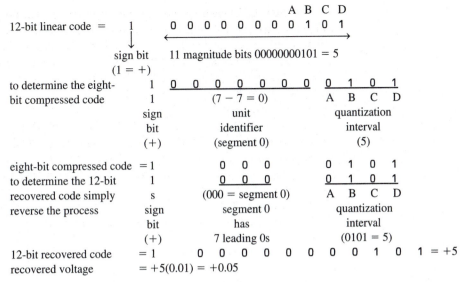

```
                                        A B C D
12-bit linear code  =    1        0 0 0 0 0 0 0 0 1 0 1
                         ↓         ←──────────────────────→
                      sign bit     11 magnitude bits 00000000101 = 5
                      (1 = +)
```

```
to determine the eight-      1    0  0  0  0  0  0  0   0  1  0  1
bit compressed code          1       (7 − 7 = 0)        A  B  C  D
                           sign         unit            quantization
                            bit       identifier           interval
                            (+)      (segment 0)             (5)
```

```
eight-bit compressed code  = 1      0  0  0      0  1  0  1
to determine the 12-bit      1      0  0  0      0  1  0  1
recovered code simply        s   (000 = segment 0)  A  B  C  D
reverse the process        sign     segment 0    quantization
                            bit        has          interval
                            (+)     7 leading 0s   (0101 = 5)
```

```
12-bit recovered code  = 1   0  0  0  0  0  0  0  0  1  0  1 = +5
recovered voltage      = +5(0.01) = +0.05
```

As the example shows, the recovered 12-bit code (+5) is exactly the same as the original 12-bit linear code (+5). Therefore, the decoded voltage (+0.05 V) is the same as the original encoded voltage (+0.5). This is true for all codes in segments 0 and 1. Thus, there is no compression error in segments 0 and 1, and the only error produced is from the quantizing process (for this example, the quantization error $Q_e = 0.003$ V).

b. To determine the 12-bit linear code,

$$\frac{-0.318 \text{ V}}{+0.01 \text{ V}} = -31.8, \text{ which is rounded off to } -32, \text{ producing a}$$
quantization error $Q_e = -0.2(0.01 \text{ V}) = -0.002$ V

```
                                        A B C D
12-bit linear code  =    0        0 0 0 0 0 1 0 0 0 0 0
                         ↓         ←──── 11 magnitude bits ────→
                      sign bit
                      (0 = −)
```

```
to determine the eight-      0    0  0  0  0  0  1   0  0  0  0   0
bit compressed code          0       (7 − 5 = 2)     A  B  C  D   X
                           sign         unit         quantization truncated
                            bit       identifier        interval
                            (−)      (segment 2)          (0)
```

```
eight-bit compressed code  = 0      0  1  0      0  0  0  0
again, to determine          0      0  1  0      0  0  0  0
the 12-bit recovered              (7 − 2 = 5)    A  B  C  D
code, simply reverse       sign     segment 5    quantization
the process                 bit       has 5        interval
                            (−)     leading 0s    (0000 = 0)
                                                 A  B  C  D
```

```
12-bit recovered code  = 0   0  0  0  0  0  1  0  0  0  0  1 = −33
                         ↑                  ↑              ↑
                         s               inserted       inserted
```

decoded voltage = −33(0.1) = −0.33 V

Note the two inserted 1s in the recovered 12-bit code. The least-significant bit is determined from the decoding table shown in Figure 5-14c. As the figure shows, in the receiver the most significant of the truncated bits is always set (1), and all other truncated bits are cleared (0s). For segment 2 codes, there is only one truncated bit; thus, it is set in the receiver. The inserted 1 in bit position 6 was dropped during the 12-bit-to-eight-bit conversion process, as transmission of this bit is redundant because if it were not a 1, the sample would not be in that segment. Consequently, for all segments except segments 0 and 1, a 1 is automatically inserted between the reinserted 0s and the ABCD bits.

For this example, there are two errors: the quantization error and the compression error. The quantization error is due to rounding off the sample voltage in the encoder to the closest PCM code, and the compression error is caused by forcing the truncated bit to be a 1 in the receiver. Keep in mind that the two errors are not always additive, as they could cause errors in the opposite direction and actually cancel each another. The worst-case scenario would be when the two errors were in the same direction and at their maximum values. For this example, the combined error was 0.33 V − 0.318 V = 0.012 V. The worst possible error in segments 0 and 1 is the maximum quantization error, or half the magnitude of the resolution. In segments 2 through 7, the worst possible error is the sum of the maximum quantization error plus the magnitude of the most significant of the truncated bits.

c. To determine the 12-bit linear code,

$$\frac{+10.234 \text{ V}}{+0.01 \text{ V}} = +1023.4, \text{ which is rounded off to 1023, producing a}$$

quantization error $Q_e = -0.4(0.01 \text{ V}) = -0.004 \text{ V}$

```
                                          A   B   C   D
12-bit linear code =      1    0   1   1   1   1   1   1   1   1   1
                          ↓        ←────────  11 magnitude bits  ────────→
                        sign bit
                        (1 = +)
to determine the eight-    1    0   1   1   1   1   1   1   1   1   1
bit compressed code        1                A   B   C   D   X   X   X   X   X
                                                              truncated
```

```
eight-bit compressed code  =    1    1   1   0   1   1   1   1
to determine the 12-bit         1    1   1   0   1   1   1   1
recovered code                  S    segment 6   A   B   C   D
12-bit recovered code      =    1  0   1   1   1   1   1   1   0   0   0   0  = +1008
                                   ↑        A   B   C   D   ↑
                                S    inserted            inserted
decoded voltage            = +1008(0.01) = +10.08 V
```

The difference between the original 12-bit code and the decoded 12-bit code is

$$10.23 - 10.08 = 0.15$$

or

```
1011 1111 1111
1011 1111 0000
─────────────
     1111   = 15(0.01) = 0.15 V
```

For this example, there are again two errors: a quantization error of 0.004 V and a compression error of 0.15 V. The combined error is 10.234 V − 10.08 V = 0.154 V.

5-7-3   Digital Compression Error

As seen in Example 5-4, the magnitude of the compression error is not the same for all samples. However, the maximum percentage error is the same in each segment (other than segments 0 and 1, where there is no compression error). For comparison purposes, the following formula is used for computing the percentage error introduced by digital compression:

$$\% \text{ error} = \frac{\text{12-bit encoded voltage} - \text{12-bit decoded voltage}}{\text{12-bit decoded voltage}} \times 100 \qquad (5\text{-}11)$$

Example 5-5

The maximum percentage error will occur for the smallest number in the lowest subsegment within any given segment. Because there is no compression error in segments 0 and 1, for segment 3 the maximum % error is computed as follows:

$$\text{Transmit 12-bit code} \quad \text{s} \; 0 \; 0 \; 0 \; 0 \; 1 \; 0 \; 0 \; 0 \; 0 \; 0 \; 0$$
$$\underline{\text{Receive 12-bit code} \quad \text{s} \; 0 \; 0 \; 0 \; 0 \; 1 \; 0 \; 0 \; 0 \; 0 \; 1 \; 0}$$
$$0 \; 0 \; 0 \; 0 \; 0 \; 0 \; 0 \; 0 \; 0 \; 0 \; 1 \; 0$$

$$\% \text{ error} = \frac{|1000000 - 1000010|}{1000010} \times 100$$

$$= \frac{|64 - 66|}{66} \times 100 = 3.03\%$$

and for segment 7

$$\text{Transmit 12-bit code} \quad \text{s} \; 1 \; 0 \; 0 \; 0 \; 0 \; 0 \; 0 \; 0 \; 0 \; 0 \; 0$$
$$\underline{\text{Receive 12-bit code} \quad \text{s} \; 1 \; 0 \; 0 \; 0 \; 0 \; 1 \; 0 \; 0 \; 0 \; 0 \; 0}$$
$$0 \; 0 \; 0 \; 0 \; 0 \; 1 \; 0 \; 0 \; 0 \; 0 \; 0$$

$$\% \text{ error} = \frac{|10000000000 - 10000100000|}{10000100000} \times 100$$

$$= \frac{|1024 - 1056|}{1056} \times 100 = 3.03\%$$

As Example 5-5 shows, the maximum magnitude of error is higher for segment 7; however, the maximum percentage error is the same for segments 2 through 7. Consequently, the maximum SQR degradation is the same for each segment.

Although there are several ways in which the 12-bit-to-eight-bit compression and eight-bit-to-12-bit expansion can be accomplished with hardware, the simplest and most economical method is with a lookup table in ROM (read-only memory).

Essentially every function performed by a PCM encoder and decoder is now accomplished with a single integrated-circuit chip called a *codec*. Most of the more recently developed codecs are called combo chips, as they include an antialiasing (bandpass) filter, a sample-and-hold circuit, and an analog-to-digital converter in the transmit section and a digital-to-analog converter, a hold circuit, and a bandpass filter in the receive section.

5-8 PCM LINE SPEED

Line speed is simply the data rate at which serial PCM bits are clocked out of the PCM encoder onto the transmission line. Line speed is dependent on the sample rate and the number of bits in the compressed PCM code. Mathematically, line speed is

$$\text{line speed} = \frac{\text{samples}}{\text{second}} \times \frac{\text{bits}}{\text{sample}} \tag{5-12}$$

where   line speed = the transmission rate in bits per second (bps)
samples/second = sample rate (f_s)
bits/sample = number of bits in the compressed PCM code

Example 5-6

For a single-channel PCM system with a sample rate $f_s = 6000$ samples per second and a seven-bit compressed PCM code, determine the line speed:

Solution $$\text{line speed} = \frac{6000 \text{ samples}}{\text{second}} \times \frac{7 \text{ bits}}{\text{sample}}$$

$$= 42{,}000 \text{ bps}$$

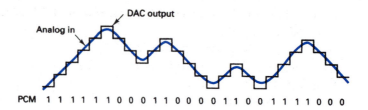

FIGURE 5-15 Ideal operation of a delta modulation encoder

5-9 DELTA MODULATION PCM AND DIFFERENTIAL PCM

Delta modulation PCM uses a single-bit PCM code to achieve digital transmission of analog signals. With conventional PCM, each code is a binary representation of both the sign and the magnitude of a sample of an analog signal. Therefore, multiple-bit codes are required to represent the many amplitudes that a sample can have. With delta modulation PCM, rather than encode and transmit a representation of the sample, only a single bit is transmitted, which simply indicates whether the present sample is larger or smaller in magnitude than the previous sample. The algorithm for delta modulation PCM is quite simple. If the current sample is larger in magnitude than the previous sample, a logic 1 is transmitted, and if the current sample is smaller in magnitude than the previous sample, a logic 0 is transmitted. Figure 5-15 shows the ideal operation of a delta modulation encoder.

In a typical PCM-encoded speech waveform, there are often successive samples taken in which there is little difference between the amplitudes of the two samples. This necessitates transmitting several identical PCM codes, which is redundant. *Differential pulse code modulation* (DPCM) is designed specifically to take advantage of the sample-to-sample redundancies in typical speech waveforms. With DPCM, a binary code proportional to the difference in the amplitude of two successive samples is transmitted rather than a binary code of an actual sample. Because the range of sample differences is typically less than the range of individual sample amplitudes, fewer bits are required for DPCM than for conventional PCM.

QUESTIONS

5-1. Contrast the advantages and disadvantages of *digital transmission.*

5-2. List and briefly describe the four most common methods of *pulse transmission.*

5-3. Which method listed in question 5–2 is the only form of pulse modulation that is used in digital transmission systems? Explain.

5-4. List and describe the primary components of a single-channel PCM system.

5-5. What is the purpose of a *sample-and-hold circuit?*

5-6. What is the difference between *natural* and *flat-top sampling?*

5-7. What is the *Nyquist sampling rate?*

5-8. Define and state the causes of *foldover distortion.* Give an alternate name for it.

5-9. Define *quantization* and *quantization interval.*

5-10. Describe the difference between *magnitude-only* PCM codes and *sign-magnitude* PCM codes.

5-11. Explain *overload distortion.* What is another name for it?

5-12. Describe *quantization noise.* What is another name for it?

5-13. Define *dynamic range.*

5-14. Explain the relationship between *dynamic range, resolution,* and the number of bits in a PCM code.

5-15. Define *signal voltage–to–quantization noise ratio* and give its relationship to *resolution, dynamic range,* and the number of bits in a PCM code.

5-16. Explain the differences between *linear* and *nonlinear* PCM codes.

5-17. Define *companding*.

5-18. What does the parameter μ determine?

5-19. Briefly explain the process of *digital companding*.

5-20. Explain *digital compression error*.

5-21. Contrast *delta modulation* PCM and standard PCM.

5-22. Contrast *differential* PCM and *standard* PCM.

PROBLEMS

5-1. Determine the minimum Nyquist rate for a maximum analog input frequency of the following:

    a. 4 kHz

    b. 10 kHz

    c. 7 kHz

5-2. For the following sample rates, determine the maximum analog input frequency:

    a. 20 kHz

    b. 8 kHz

    c. 6 kHz

5-3. Determine the alias frequency for a 15-kHz sample rate and an analog input frequency of 8 kHz.

5-4. Determine the dynamic range for PCM codes with the following number of magnitude bits:

    a. 10

    b. 13

    c. 8

5-5. Determine the minimum number of PCM bits (including the sign bit) for a dynamic range of 80 dB.

5-6. For a resolution of 0.04 V, determine the voltages for the following linear seven-bit PCM codes:

    a. 0110101

    b. 0000011

    c. 1000001

    d. 0111111

    e. 1000000

5-7. A 12-bit linear PCM code is digitally compressed into eight bits. The resolution = 0.03 V. Determine the following for an analog input voltage of 1.465 V:

    a. 12-bit linear PCM code

    b. 8-bit compressed code

    c. Decoded 12-bit PCM code

    d. Decoded voltage

    e. Percentage error

5-8. For a 12-bit linear PCM code with a resolution of 0.02 V, determine the voltage range that would be converted to the following PCM codes:

    a. 100000000001

    b. 000000000000

    c. 110000000000

    d. 010000000000

    e. 100100000001

    f. 101010101010

5-9. For each of the following 12-bit linear PCM codes, determine the eight-bit compressed code to which they would be converted:

    a. 100000001000

    b. 100000001001

    c. 100000010000

    d. 000000100000

    e. 010000000000

    f. 010000100000

CHAPTER 6

Multiplexing and T Carriers

CHAPTER OUTLINE

OBJECTIVES

- Define *multiplexing*
- Describe time-division multiplexing
- Describe the DS-0 channel
- Describe the frame format and operation for the T1 digital carrier system
- Describe the superframe and extended superframe TDM formats
- Describe the operation of fractional T carriers
- Describe the North American Digital Hierarchy
- Define *line encoding*
- Define the following terms and describe how they affect line encoding: *duty cycle, bandwidth, clock recovery, error detection,* and *detecting* and *decoding*
- Describe the basic T carrier formats
- Describe the European T carrier system
- Explain statistical time-division multiplexing
- List and describe the various frame synchronization techniques
- Describe frequency-division multiplexing
- Describe the North American FDM Hierarchy
- Describe wavelength-division multiplexing
- Describe the synchronous optical network

6-1 INTRODUCTION

Multiplexing is the transmission of information (in any form) from more than one source to more than one destination over the same transmission medium (facility). Although transmissions occur on the same facility, they do not necessarily occur at the same time. The transmission medium may be a metallic wire pair, a coaxial cable, a PCS mobile telephone, a terrestrial microwave radio system, a satellite microwave system, or an optical fiber cable.

There are several domains in which multiplexing can be accomplished, including space, phase, time, frequency, and wavelength. The most predominant methods of multiplexing, however, are *time-division multiplexing* (TDM), *frequency-division multiplexing* (FDM), and *wavelength-division multiplexing* (WDM).

6-2 TIME-DIVISION MULTIPLEXING

With *time-division multiplexing* (TDM), transmissions from multiple sources occur on the same facility but not at the same time. Transmissions from various sources are *interleaved* in the time domain. PCM is the most prevalent encoding technique used for TDM digital signals. With a PCM-TDM system, two or more voice channels are sampled, converted to PCM codes, and then time-division multiplexed onto a single metallic or optical fiber cable. The fundamental building block for most TDM systems in the United States begins with a DS-0 channel (digital signal level 0).

Example 6-1

Figure 6-1 shows the simplified block diagram for a DS-0 single-channel PCM system. For an 8-kHz sample rate and an eight-bit PCM code, determine the line speed.

Solution

$$\text{line speed} = \frac{8000 \text{ samples}}{\text{second}} \times \frac{8 \text{ bits}}{\text{sample}}$$

$$= 64,000 \text{ bps}$$

Figure 6-2a shows the simplified block diagram for a PCM carrier system comprised of two DS-0 channels that have been time-division multiplexed. Each channel's input is alternately sampled at an 8-kHz rate and converted to an eight-bit PCM code. While the PCM code for channel 1 is being transmitted, channel 2 is sampled and converted to PCM code. While the PCM code from channel 2 is being transmitted, the next sample is taken from channel 1 and converted to PCM code. This process continues, and samples are taken alternately from each

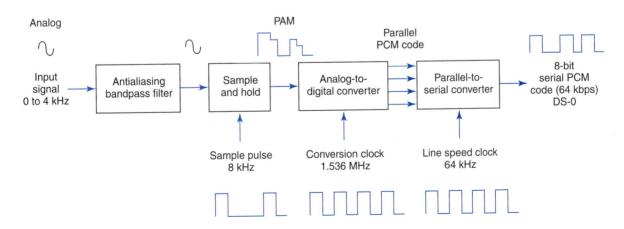

FIGURE 6-1 Single-channel (DS-0-level) PCM transmission system

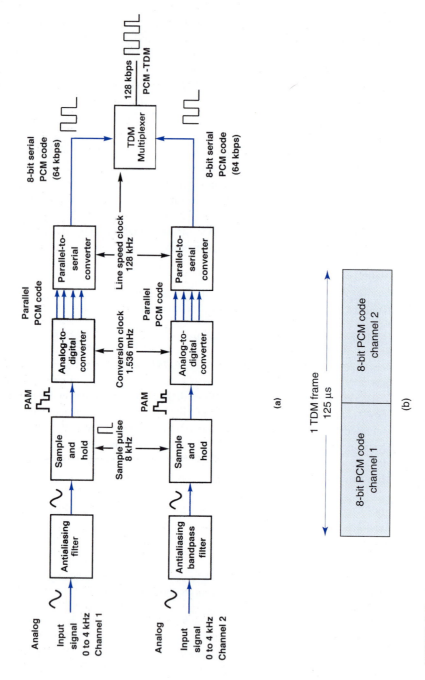

(a)

1 TDM frame
125 μs

8-bit PCM code channel 1	8-bit PCM code channel 2

(b)

FIGURE 6-2 Two-channel PCM-TDM system: (a) block diagram; (b) TDM frame

channel, converted to PCM code, and transmitted. The multiplexer is simply an electronically controlled digital switch with two inputs and one output. Channel 1 and channel 2 are alternately selected and connected to the transmission line through the multiplexer. One eight-bit PCM code from each channel (16 total bits) is called a TDM *frame*, and the time it takes to transmit one TDM frame is called the *frame time*. The frame time is equal to the reciprocal of the sample rate ($1/f_s$, or $1/8000 = 125$ µs). Figure 6-2b shows the TDM frame allocation for a two-channel PCM system with an 8-kHz sample rate. The PCM code for each channel occupies a fixed time slot (epoch) within the total TDM frame.

Example 6-2

For a two-channel PCM system with an 8-kHz sample rate and an eight-bit PCM code, determine the line speed.

Solution A sample is taken from each channel during each frame; therefore, the time allocated to transmit the PCM bits from each channel is equal to one-half the total frame time. Therefore, eight bits from each channel must be transmitted during each frame (a total of 16 PCM bits per frame). Thus, the line speed at the output of the multiplexer is

$$\frac{2 \text{ channels}}{\text{frame}} \times \frac{8000 \text{ frames}}{\text{second}} \times \frac{8 \text{ bits}}{\text{channel}} = 128 \text{ kbps}$$

Although each channel is producing and transmitting only 64 kbps, the bits must be clocked out onto the line at a 128-kHz rate to allow eight bits from each channel to be transmitted in a 126-µs time slot.

6-3 TI DIGITAL CARRIER SYSTEM

A digital carrier system is a communications system that uses digital pulse rather than analog signals to encode information. Figure 6-3a shows the block diagram for the Bell System T1 digital carrier system, which is the North American telephone standard and recognized by the ITU-T as Recommendation G.733. A T1 carrier system time-division multiplexes PCM-encoded samples from 24 voice-band channels for transmission over a single metallic wire pair or optical fiber transmission line. Each voice-band channel has a bandwidth of approximately 300 Hz to 3000 Hz. Again, the multiplexer is simply a digital switch with 24 independent inputs and one time-division multiplexed output. The PCM output signals from the 24 voice-band channels are sequentially selected and connected through the multiplexer to the transmission line.

Simply time-division multiplexing 24 voice-band channels does not in itself constitute a T1 carrier system. At this point, the output of the multiplexer is simply a multiplexed first-level digital signal (DS level 1). The system does not become a T1 carrier until it is line encoded and placed on special conditioned cables called *T1 lines*. Line encoding is described later in this chapter.

With a T1 carrier system, D-type (digital) channel banks perform the sampling, encoding, and multiplexing of 24 voice-band channels. Each channel contains an eight-bit PCM code and is sampled 8000 times a second. Each channel is sampled at the same rate but not necessarily at the same time. Figure 6-3b shows the channel sampling sequence for a 24-channel T1 digital carrier system. As the figure shows, each channel is sampled once each frame but not at the same time. Each channel's sample is offset from the previous channel's sample by 1/24th of the total frame time. Therefore, one 64-kbps PCM-encoded sample is transmitted for each voice-band channel during each frame (a frame time of $1/8000 = 125$ µs). The line speed is calculated as follows:

$$\frac{24 \text{ channels}}{\text{frame}} \times \frac{8 \text{ bits}}{\text{channel}} = 192 \text{ bits per frame}$$

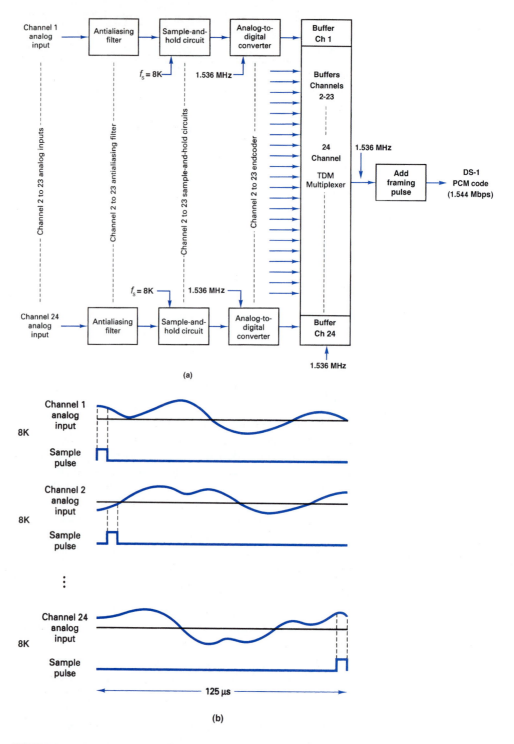

FIGURE 6-3 Bell system T1 digital carrier system: (a) block diagram; (b) sampling sequence

thus $\quad \dfrac{192 \text{ bits}}{\text{frame}} \times \dfrac{8000 \text{ frames}}{\text{second}} = 1.536 \text{ Mbps}$

Later, an additional bit (called the framing bit) is added to each frame. The framing bit occurs once per frame (8000-bps rate) and is recovered in the receiver where it is used to maintain frame and sample synchronization between the TDM transmitter and receiver. As a result, each frame contains 193 bits, and the line speed for a T1 digital carrier system is

$$\dfrac{193 \text{ bits}}{\text{frame}} \times \dfrac{8000 \text{ frames}}{\text{second}} = 1.544 \text{ Mbps}$$

Example 6-3

For a 20-channel PCM/TDM system with an 8-kHz sample rate, 10 bits per sample, and one framing bit per frame, determine the line speed.

Solution The total number of PCM bits per frame is

$$\dfrac{20 \text{ channels}}{\text{frame}} \times \dfrac{10 \text{ bits}}{\text{channel}} = 200 \text{ bits per frame}$$

The total number of bits per frame counting the framing bit is

$$200 \text{ PCM bits} + 1 \text{ framing bit} = 201 \text{ bits per frame}$$

Therefore, the line speed is

$$\dfrac{201 \text{ bits}}{\text{frame}} \times \dfrac{8000 \text{ frames}}{\text{second}} = 1.608 \text{ Mbps}$$

6-3-1 D-Type Channel Banks

Early T1 carrier systems used D1 *digital channel banks* (PCM encoders) with a seven-bit magnitude-only PCM code, analog companding, and a $\mu = 100$. Later-version digital channel banks (D2 and D3) added an eighth bit (the signaling bit) to each PCM code for the purpose of interoffice *signaling* (supervision between telephone offices, such as on-hook, off-hook, dial pulsing, and so forth). Since a signaling bit was added to each sample in every frame, the signaling rate was 8 kbps. Over the years, the T1 carrier system generically progressed, and modern versions use digitally companded, eight-bit sign magnitude–compressed PCM codes with a $\mu = 255$. Also, with the early digital channel banks, the framing bit sequence was simply an alternating 1/0 pattern.

6-3-2 Superframe TDM Format

The 8-kbps signaling rate used with the early digital channel banks was excessive for signaling on standard telephone voice circuits. Therefore, with modern channel banks, a signaling bit is substituted only into the least-significant bit (LSB) of every sixth frame. Hence, five of every six frames have eight-bit resolution, while one in every six frames (the signaling frame) has only seven-bit resolution. Consequently, the signaling rate on each channel is only 1.333 kbps (8000 bps/6), and the effective number of bits per sample is actually $7^{5/6}$ bits.

Because only every sixth frame includes a signaling bit, it is necessary that all the frames be numbered so that the receiver knows when to extract the signaling bit. Also, because the signaling is accomplished with a two-bit binary word, it is necessary to identify the most- and least-significant bits (MSB and LSB, respectively) of the signaling word. Consequently, the superframe format shown in Figure 6-4 was devised. Within each superframe are 12 consecutively numbered frames (1–12). The signaling bits are substituted in frames 6 and 12, the MSB into frame 6, and the LSB into frame 12. Frames 1 to 6 are called the A highway, with frame 6 designated the A channel signaling frame. Frames 7 to 12 are called the B highway, with frame 12 designated the B channel signaling frame. Therefore,

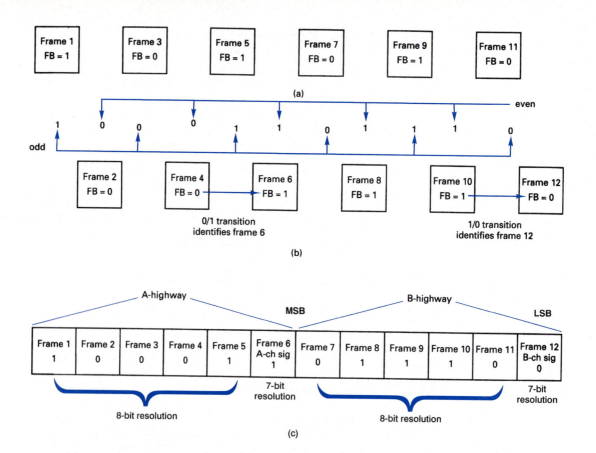

FIGURE 6-4 Framing bit sequence for the T1 superframe format using D2 and D3 channel banks: (a) frame synchronizing bits (odd-numbered frames); (b) signaling frame alignment bits (even-numbered frames); (c) composite frame alignment

in addition to identifying the signaling frames, the sixth and twelfth frames must also be positively identified.

To identify frames 6 and 12, a different framing bit sequence is used for the odd- and even-numbered frames. The odd frames (frames 1, 3, 5, 7, 9, and 11) have an alternating 1/0 pattern, and the even frames (frames 2, 4, 6, 8, 10, and 12) have a 0 0 1 1 1 0 repetitive pattern. As a result, the combined framing bit pattern is 1 0 0 0 1 1 0 1 1 1 0 0. The odd-numbered frames are used for frame and sample synchronization, and the even-numbered frames are used to identify the A and B channel signaling frames (frames 6 and 12). Frame 6 is identified by a 0/1 transition in the framing bit between frames 4 and 6. Frame 12 is identified by a 1/0 transition in the framing bit between frames 10 and 12.

Figure 6-5a shows the framing bit circuitry for the 24-channel T1 carrier system. Note that the bit rate at the output of the TDM multiplexer is 1.536 Mbps and that the bit rate at the output of the 193-bit shift register is 1.544 Mbps. The difference (8 kbps) is due to the addition of the framing bit.

D4 channel banks time-division multiplex 48 voice-band telephone channels and operate at a transmission rate of 3.152 Mbps. This is slightly more than twice the line speed for 24-channel D1, D2, or D3 channel banks because with D4 channel banks, rather than transmitting a single framing bit with each frame, a 10-bit frame synchronization pattern is used.

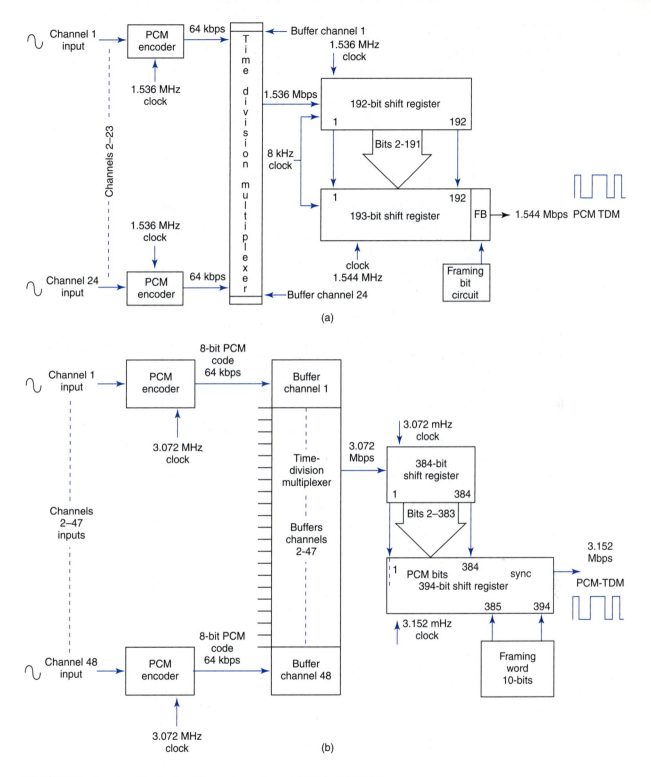

FIGURE 6-5 Framing bit circuitry T1 carrier system: (a) DS-1; (b) DS-1C

Example 6-4

Determine

a. The total number of bits in a D4 (DS-1C) TDM frame
b. The line speed

Solution

a. $\dfrac{8 \text{ bits}}{\text{channel}} \times \dfrac{48 \text{ channels}}{\text{frame}} = \dfrac{384 \text{ bits}}{\text{frame}} + \dfrac{10 \text{ framing bits}}{\text{frame}} = \dfrac{394 \text{ bits}}{\text{frame}}$

and the line speed for DS-1C systems is

b. $\text{line speed} = \dfrac{394 \text{ bits}}{\text{frame}} \times \dfrac{8000 \text{ frames}}{\text{second}} = 3.152 \text{ Mbps}$

The framing for DS-1 (T1) PCM-TDM system or the framing pattern for the DS-1C (T1C) time-division multiplexed carrier system is added to the multiplexed digital signal at the output of the multiplexer. The framing bit circuitry used for the 48-channel DS-1C is shown in Figure 6-5b.

6-3-3   Extended Superframe Format

Another framing format recently developed for new designs of T1 carrier systems is the *extended superframe format*. The extended superframe format consists of 24, 193-bit frames, totaling 4632 bits, of which 24 are framing bits. One extended superframe occupies 3 ms:

$$\left(\dfrac{1}{1.544 \text{ Mbits/sec}} \right) \left(\dfrac{193 \text{ bits}}{\text{frame}} \right) (24 \text{ frames}) = 3 \text{ ms}$$

A framing bit occurs once every 193 bits; however, only 6 of the 24 framing bits are used for frame synchronization. Frame synchronization bits occur in frames 4, 8, 12, 16, 20, and 24 and have a bit sequence of 0 0 1 0 1 1. Six additional framing bits in frames 1, 5, 9, 13, 17, and 21 are used for an error-detection code called CRC-6 (*cyclic redundancy checking*). The 12 remaining framing bits provide for a management channel called the *facilities data link* (FDL). FDL bits occur in frames 2, 3, 6, 7, 10, 11, 14, 15, 18, 19, 22, and 23.

The extended superframe format supports a four-bit signaling word with signaling bits provided in the second least-significant bit of each channel during every sixth frame. The signaling bit in frame 6 is called the A bit, the signaling bit in frame 12 is called the B bit, the signaling bit in frame 18 is called the C bit, and the signaling bit in frame 24 is called the D bit. These signaling bit streams are sometimes called the A, B, C, and D *signaling channels* (or *signaling highways*). The extended superframe framing bit pattern is summarized in Table 6-1.

Table 6-1   Extended Superframe Format

Frame Number	Framing Bit	Frame Number	Framing Bit
1	C	13	C
2	F	14	F
3	F	15	F
4	S = 0	16	S = 0
5	C	17	C
6	F	18	F
7	F	19	F
8	S = 0	20	S = 1
9	C	21	C
10	F	22	F
11	F	23	F
12	S = 1	24	S = 1

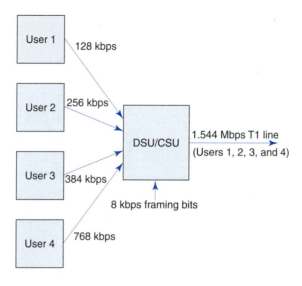

FIGURE 6-6 Fractional T1 carrier service

6-3-4 Fractional T Carrier Service

Fractional T carrier emerged because standard T1 carriers provide a higher capacity (i.e., higher bit rate) than most users require. Fractional T1 systems distribute the channels (i.e., bits) in a standard T1 system among more than one user, allowing several subscribers to share one T1 line. For example, several small businesses located in the same building can share one T1 line (both its capacity and its cost).

Bit rates offered with fractional T1 carrier systems are 64 kbps (1 channel), 128 kbps (2 channels), 256 kbps (4 channels), 384 kbps (6 channels), 512 kbps (8 channels), and 768 kbps (12 channels), with 384 kbps (1/4 T1) and 768 kbps (1/2 T1) being the most common. The minimum data rate necessary to propagate video information is 384 kbps. Fractional T3 is essentially the same as fractional T1 except with higher channel capacities, higher bit rates, and more customer options.

Figure 6-6 shows four subscribers combining their transmissions in a special unit called a *data service unit/channel service unit* (DSU/CSU). A DSU/CSU is a digital interface that provides the physical connection to a digital carrier network. User 1 is allocated 128 kbps, user 2,256 kbps, user 3,384 kbps, and user 4,768 kbps for a total of 1.536 kbps (8 kbps is reserved for the framing bit).

6-4 NORTH AMERICAN DIGITAL MULTIPLEXING HIERARCHY

Multiplexing signals in digital form lends itself easily to interconnecting digital transmission facilities with different transmission bit rates. Figure 6-7 shows the American Telephone and Telegraph Company's (AT&T's) North American Digital Hierarchy for multiplexing digital signals into a single higher-speed pulse stream suitable for transmission on the next-higher level of the hierarchy. To upgrade from one level in the hierarchy to the next-higher level, a special device called *muldem* (*mul*tiplexers/*dem*ultiplexer) is required. Muldems can handle bit-rate conversions in both directions. The muldem designations (M112, M23, and so on) identify the input and output digital signals associated with that muldem. For instance, an M12 muldem interfaces DS-1 and DS-2 digital signals. An M23 muldem interfaces DS-2 and DS-3 digital signals. As the figure shows, DS-1 signals may be further multiplexed or line encoded and placed on specially conditioned cables called

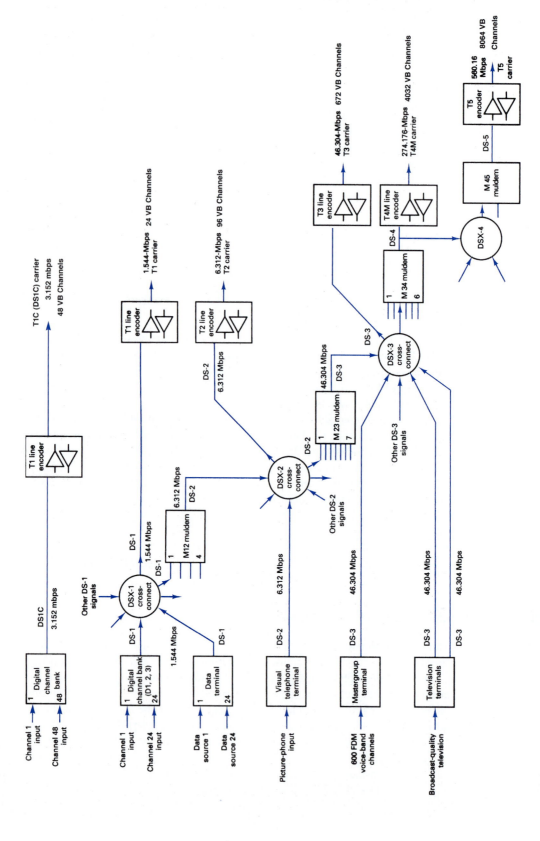

FIGURE 6-7 North American Digital Hierarchy

Table 6-2 North American Digital Hierarchy Summary

Line Type	Digital Signal	Bit Rate	Channel Capacities	Services Offered
T1	DS-1	1.544 Mbps	24	Voice-band telephone or data
Fractional T1	DS-1	64 kbps to 1.536 Mbps	24	Voice-band telephone or data
T1C	DS-1C	3.152 Mbps	48	Voice-band telephone or data
T2	DS-2	6.312 Mbps	96	Voice-band telephone, data, or picture phone
T3	DS-3	44.736 Mbps	672	Voice-band telephone, data, picture phone, and broadcast-quality television
Fractional T3	DS-3	64 kbps to 23.152 Mbps	672	Voice-band telephone, data, picture phone, and broadcast-quality television
T4M	DS-4	274.176 Mbps	4032	Same as T3 except more capacity
T5	DS-5	560.160 Mbps	8064	Same as T3 except more capacity

T1 lines. DS-2, DS-3, DS-4, and DS-5 digital signals may be placed on T2, T3, T4M, or T5 lines, respectively.

Digital signals are routed at central locations called *digital cross-connects*. A digital cross-connect (DSX) provides a convenient place to make patchable interconnects and to perform routine maintenance and troubleshooting. Each type of digital signal (DS-1, DS-2, and so on) has its own digital switch (DSX–1, DSX–2, and so on). The output from a digital switch may be upgraded to the next-higher level of multiplexing or line encoded and placed on its respective T lines (T1, T2, and so on).

Table 6-2 lists the digital signals, their bit rates, channel capacities, and services offered for the line types included in the North American Digital Hierarchy.

6-5 DIGITAL LINE ENCODING

Digital line encoding involves converting standard logic levels (TTL, CMOS, and the like) to a form more suitable to telephone line transmission. Essentially, six primary factors must be considered when selecting a line-encoding format:

1. Transmission voltages and dc component
2. Duty cycle
3. Bandwidth considerations
4. Clock and framing bit recovery
5. Error detection
6. Ease of detection and decoding

6-5-1 Transmission Voltages and DC Component

Transmission voltages or levels can be categorized as being either *unipolar* (UP) or *bipolar* (BP). Unipolar transmission of binary data involves the transmission of only a single nonzero voltage level (e.g., either a positive or a negative voltage for a logic 1 and 0 V [ground] for a logic 0). In bipolar transmission, two nonzero voltages are involved (e.g., a positive voltage for a logic 1 and an equal-magnitude negative voltage for a logic 0 or vice

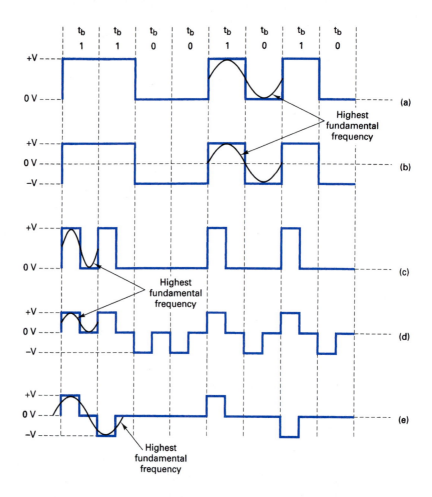

FIGURE 6-8 Line-encoding formats: (a) UPNRZ; (b) BPNRZ; (c) UPRZ; (d) BPRZ; (e) BPRZ-AMI

versa). Assuming an equal probability of the occurrence of a logic 1 or a logic 0, symmetrical bipolar voltages reduce the average power by 50%.

6-5-2 Duty Cycle

The *duty cycle* of a binary pulse can be used to categorize the type of transmission. If the binary pulse is maintained for the entire bit time, this is called *nonreturn to zero* (NRZ). If the active time of the binary pulse is less than 100% of the bit time, this is called *return to zero* (RZ).

Unipolar and bipolar transmission voltages can be combined with either return-to-zero or nonreturn to zero in several ways to achieve a particular line-encoding scheme. Figure 6-8 shows five line-encoding possibilities.

In Figure 6-8a, there is only one nonzero voltage level ($+V$ = logic 1); a zero voltage indicates a logic 0. Also, each logic 1 condition maintains the positive voltage for the entire bit time (100% duty cycle). Consequently, Figure 6-8a represents a unipolar nonreturn-to-zero signal (UPNRZ). Assuming an equal number of 1s and 0s, the average dc voltage of a UPNRZ waveform is equal to half the nonzero voltage ($V/2$).

In Figure 6-8b, there are two nonzero voltages ($+V$ = logic 1 and $-V$ = logic 0), and a 100% duty cycle is used. Therefore, Figure 6-8b represents a bipolar nonreturn-to-zero signal (BPNRZ). When equal-magnitude voltages are used for logic 1s and logic 0s

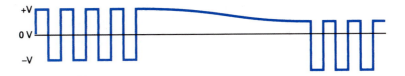

FIGURE 6-9 DC wandering

and assuming an equal probability of logic 1s and logic 0s occurring, the average dc voltage of a BPNRZ waveform is 0 V.

In Figure 6-8c, only one nonzero voltage is used, but each pulse is active for only 50% of a bit time ($t_b/2$). Consequently, the waveform shown in Figure 6-8c represents a unipolar return-to-zero signal (UPRZ). Assuming an equal probability of 1s and 0s occurring, the average dc voltage of a UPRZ waveform is one-fourth the nonzero voltage (V/4).

Figure 6-8d shows a waveform where there are two nonzero voltages (+V = logic 1 and −V = logic 0). Also, each pulse is active only 50% of a bit time. Consequently, the waveform shown in Figure 6-8d represents a bipolar return-to-zero (BPRZ) signal. Assuming equal-magnitude voltages for logic 1s and logic 0s and an equal probability of 1s and 0s occurring, the average dc voltage of a BPRZ waveform is 0 V.

In Figure 6-8e, there are again two nonzero voltage levels (−V and +V), but now both polarities represent logic 1s, and 0 V represents a logic 0. This method of line encoding is called *alternate mark inversion* (AMI). With AMI transmissions, successive logic 1s are inverted in polarity from the previous logic 1. Because return to zero is used, the encoding technique is called *bipolar-return-to-zero alternate mark inversion* (BPRZ-AMI). The average dc voltage of a BPRZ AMI waveform is approximately 0 V regardless of the bit sequence.

With NRZ encoding, a long string of either logic 1s or logic 0s produces a condition in which a receive may lose its amplitude reference for optimum discrimination between received 1s and 0s. This condition is called *dc wandering*. The problem may also arise when there is a significant imbalance in the number of 1s and 0s transmitted. Figure 6-9 shows how dc wandering is produced from a long string of successive logic 1s. It can be seen that after a long string of 1s, 1-to-0 errors are more likely than 0-to-1 errors. Similarly, long strings of logic 0s increase the probability of 0-to-1 errors.

The method of line encoding used determines the minimum bandwidth required for transmission, how easily a clock may be extracted from it, how easily it may be decoded, the average dc voltage level, and whether it offers a convenient means of detecting errors.

6-5-3 Bandwidth Requirements

To determine the minimum bandwidth required to propagate a line-encoded digital signal, you must determine the highest fundamental frequency associated with the signal (see Figure 6-8). The highest fundamental frequency is determined from the worst-case (fastest transition) binary bit sequence. With UPNRZ, the worst-case condition is an alternating 1/0 sequence; the period of the highest fundamental frequency takes the time of two bits and, therefore, is equal to one-half the bit rate ($f_b/2$). With BPNRZ, again the worst-case condition is an alternating 1/0 sequence, and the highest fundamental frequency is one-half the bit rate ($f_b/2$). With UPRZ, the worst-case condition occurs when two successive logic 1s occur. Therefore, the minimum bandwidth is equal to the bit rate (f_b). With BPRZ encoding, the worst-case condition occurs for successive logic 1s or successive logic 0s, and the minimum bandwidth is again equal to the bit rate (f_b). With BPRZ-AMI, the worst-case condition is two or more consecutive logic 1s, and the minimum bandwidth is equal to one-half the bit rate ($f_b/2$).

6-5-4 Clock and Framing Bit Recovery

To recover and maintain clock and framing bit synchronization from the received data, there must be sufficient transitions in the data waveform. With UPNRZ and BPNRZ encoding, a long string of 1s or 0s generates a data signal void of transitions and, therefore, is inadequate for clock recovery. With UPRZ and BPRZ-AMI encoding, a long string of 0s also generates a data signal void of transitions. With BPRZ, a transition occurs in each bit position regardless of whether the bit is a 1 or a 0. Thus, BPRZ is the best encoding scheme for clock recovery. If long sequences of 0s are prevented from occurring, BPRZ-AMI encoding provides sufficient transitions to ensure clock synchronization.

6-5-5 Error Detection

With UPNRZ, BPNRZ, UPRZ, and BPRZ encoding, there is no way to determine if the received data have errors. However, with BPRZ-AMI encoding, an error in any bit will cause a bipolar violation (BPV—the reception of two or more consecutive logic 1s with the same polarity). Therefore, BPRZ-AMI has a built-in error-detection mechanism. T carriers use BPRZ-AMI, with +3 V and −3 V representing a logic 1 and 0 V representing a logic 0.

Table 6-3 summarizes the bandwidth, average voltage, clock recovery, and error-detection capabilities of the line-encoding formats shown in Figure 6-8. From Table 6-3, it can be seen that BPRZ-AMI encoding has the best overall characteristics and is, therefore, the most commonly used encoding format.

6-5-6 Digital Biphase

Digital biphase (sometimes called the *Manchester code* or *diphase*) is a popular type of line encoding that produces a strong timing component for clock recovery and does not cause dc wandering. Biphase is a form of BPRZ encoding that uses one cycle of a square wave at 00 phase to represent a logic 1 and one cycle of a square wave at 180° phase to represent a logic 0. Digital biphase encoding is shown in Figure 6-10. Note that a transition occurs in the center of every signaling element regardless of its logic condition or phase. Thus, biphase produces a strong timing component for clock recovery. In addition, assuming an equal probability of 1s and 0s, the average dc voltage is 0 V, and there is no dc wandering. A disadvantage of biphase is that it contains no means of error detection.

Table 6-3 Line-Encoding Summary

Encoding Format	Minimum BW	Average DC	Clock Recovery	Error Detection
UPNRZ	$f_b/2^*$	+V/2	Poor	No
BPNRZ	$f_b/2^*$	$0 V^*$	Poor	No
UPRZ	f_b	+V/4	Good	No
BPRZ	f_b	$0 V^*$	Best*	No
BPRZ-AMI	$f_b/2^*$	$0 V^*$	Good	Yes*

*Denotes best performance or quality.

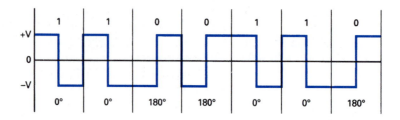

FIGURE 6-10 Digital biphase

T carriers are used for the transmission of PCM-encoded time-division multiplexed digital signals. In addition, T carriers utilize special line-encoded signals and metallic cables that have been conditioned to meet the relatively high bandwidths required for high-speed digital transmission. Digital signals deteriorate as they propagate along a cable because of power losses in the metallic conductors and the low-pass filtering inherent in parallel-wire transmission lines. Consequently, *regenerative repeaters* must be placed at periodic intervals. The distance between repeaters depends on the transmission bit rate and the line-encoding technique used.

Figure 6-11 shows the block diagram for a regenerative repeater. Essentially, there are three functional blocks: an *amplifier/equalizer*, a *timing clock recovery circuit*, and the *regenerator* itself. The amplifier/equalizer filters and shapes the incoming digital signal and raises its power level so that the regenerator circuit can make a pulse–no pulse decision. The timing clock recovery circuit reproduces the clocking information from the received data and provides the proper timing information to the regenerator so that samples can be made at the optimum time, minimizing the chance of an error occurring. A regenerative repeater is simply a threshold detector that compares the sampled voltage received to a reference level and determines whether the bit is a logic 1 or a logic 0.

Spacing of repeaters is designed to maintain an adequate signal-to-noise ratio for error-free performance. The signal-to-noise ratio at the output of a regenerative repeater is exactly what it was at the output of the transmit terminal or at the output of the previous regenerator (i.e., the signal-to-noise ratio does not deteriorate as a digital signal propagates through a regenerator; in fact, a regenerator reconstructs the original pulses with the original signal-to-noise ratio).

6-6-1 T1 Carrier System

T1 carrier systems were designed to combine PCM and TDM techniques for the transmission of 24 64-kbps channels with each channel capable of carrying digitally encoded voice-band telephone signals or data. The transmission bit rate (*line speed*) for a T1 carrier is 1.544 Mbps, including an 8-kbps framing bit. The lengths of T1 carrier systems typically range from about 1 mile to over 50 miles.

T1 carriers use BPRZ-AMI encoding with regenerative repeaters placed every 3000, 6000, or 9000 feet. The transmission medium for T1 carriers is generally 19- to 22-gauge twisted-pair metallic cable. Because T1 carriers use BPRZ-AMI encoding, they are susceptible to losing clock synchronization on long strings of consecutive logic 0s. Ensuring

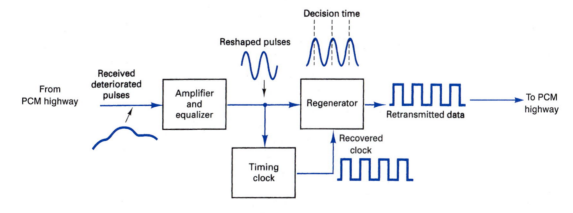

FIGURE 6-11 Regenerative repeater block diagram

sufficient transitions occur in the data stream is sometimes called *ones density*. Early T1 carrier systems provided measures to ensure that no single eight-bit byte was transmitted without at least one bit being a logic 1 or that 15 or more consecutive logic 0s were not transmitted. With modern T1 carriers, a technique called *binary eight zero substitution* (B8ZS) is used to ensure that sufficient transitions occur in the data to maintain clock synchronization. With B8SZ, whenever eight consecutive 0s are encountered, one of two special patterns is substituted for the eight 0s, either $+ - 0 - + 0\,0\,0$ or $- + 0 + - 0\,0\,0$. The + (plus) and − (minus) represent bipolar logic 1 conditions, and a 0 (zero) indicates a logic 0 condition. The eight-bit pattern substituted for the eight consecutive 0s is the one that purposely induces bipolar violations in the fourth and seventh bit positions. Ideally, the receiver will detect the bipolar violations and the substituted pattern and then substitute the eight 0s back into the data signal. During periods of low usage, eight logic 1s are substituted into idle channels. Two examples of B8ZS are illustrated here with their corresponding waveforms shown in Figures 6–12a and 6–12b, respectively.

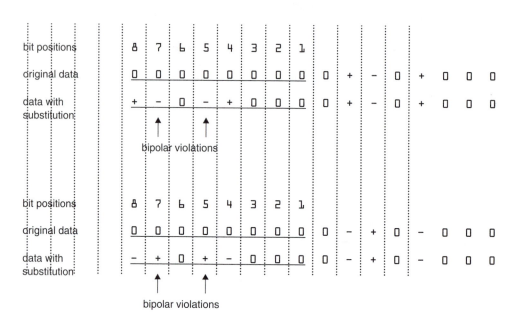

6-6-2   T2 Carrier System

T2 carriers time-division multiplex 96 64-kbps voice or data channels into a single 6.312-Mbps data signal for transmission over twisted-pair copper wire up to 500 miles over a special LOCAP (low capacitance) metallic cable. T2 carriers also use BPRZ-AMI encoding; however, because of the higher transmission rate, clock synchronization is even more critical than with a T1 carrier. A sequence of six consecutive logic 0s could be sufficient to cause loss of clock synchronization. Therefore, T2 carrier systems use an alternative method of ensuring that ample transitions occur in the data. This method is called *binary six zero substitution* (B6ZS).

With B6ZS, whenever six consecutive logic 0s occur, one of the following binary codes is substituted in its place: $0 - + 0 + -$ or $0 + - 0 - +$. Again, + and − represent logic 1s, and 0 represents a logic 0. The six-bit code substituted for the six consecutive 0s is selected to purposely cause a bipolar violation. If the violation is detected in the receiver, the original six 0s can be substituted back into the data signal. The substituted patterns produce bipolar violations (i.e., consecutive pulses with the same polarity) in the second and

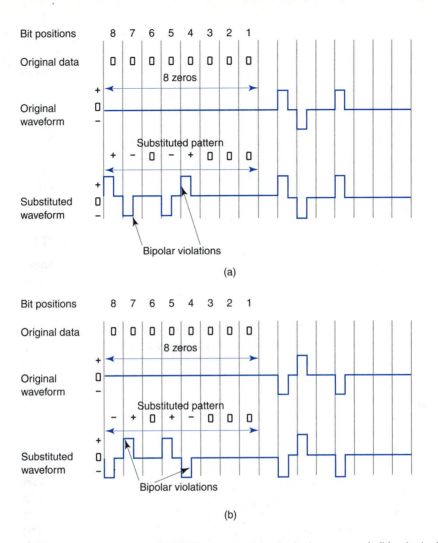

(a)

(b)

FIGURE 6-12 Waveforms for B8ZS example: (a) substitution pattern 1; (b) substitution pattern 2

fourth bits of the substituted patterns. If DS-2 signals are multiplexed to form DS-3 signals, the B6ZS code must be detected and stripped off from the DS-2 signal prior to DS-3 multiplexing. An example of B6ZS is illustrated here with its corresponding waveform shown in Figure 6-13.

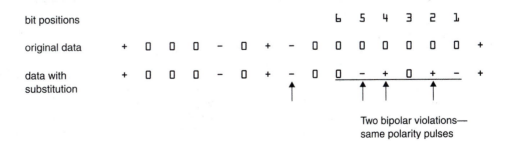

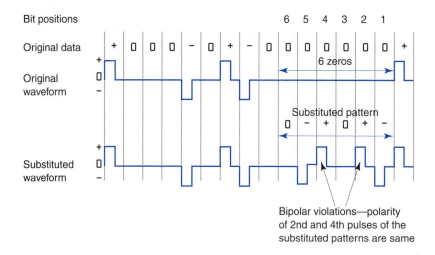

FIGURE 6-13 Waveform for B6ZS example

6-6-3 T3 Carrier System

T3 carriers time-division multiplex 672 64-kbps voice or data channels for transmission over a single 3A-RDS coaxial cable. The transmission bit rate for T3 signals is 44.736 Mbps. The coding technique used with T3 carriers is *binary three zero substitution* (B3ZS). Substitutions are made for any occurrence of three consecutive 0s.

6-6-4 T4M Carrier System

T4M carriers time-division multiplex 4032 64-kbps voice or data channels for transmission over a single T4M coaxial cable up to 500 miles. The transmission rate is sufficiently high that substitute patterns are impractical. Instead, T4M carriers transmit scrambled unipolar NRZ digital signals; the scrambling and descrambling functions are performed in the subscriber's terminal equipment.

6-6-5 T5 Carrier System

T5 carriers time-division multiplex 8064 64-kbps voice or data channels and transmits them at a 560.16-Mbps rate over a single coaxial cable.

6-7 EUROPEAN TIME-DIVISION MULTIPLEXING

In Europe, a different version of T carrier lines is used called *E lines*. Although the two systems are conceptually the same, they have different capabilities. Figure 6-14 shows the frame alignment for the E1 European standard PCM-TDM system. With the basic E1 system, a 125-μs frame is divided into 32 equal time slots. Time slot 0 is used for a frame alignment pattern and for an alarm channel. Time slot 17 is used for a *common signaling channel* (CSC). The signaling for all 30 voice-band channels is accomplished on the common signaling channel. Consequently, 30 voice-band channels are time-division multiplexed into each E1 frame.

With the European E1 standard, each time slot has eight bits. Consequently, the total number of bits per frame is

$$\frac{8 \text{ bits}}{\text{time slot}} \times \frac{32 \text{ time slots}}{\text{frame}} = \frac{256 \text{ bits}}{\text{frame}}$$

and the line speed for an E-1 TDM system is

Time slot 0	Time slot 1	Time slots 2–16	Time slot 17	Time slots 18–30	Time slot 31
Framing and alarm channel	Voice channel 1	Voice channels 2–15	Common signaling channel	Voice channels 16–29	Voice channel 30
8 bits	8 bits	112 bits	8 bits	112 bits	8 bits

(a)

Time slot 17

16 frames equal one multiframe; 500 multiframes are transmitted each second

Frame	Bits 1234	5678
0	0000	xyxx
1	ch 1	ch 16
2	ch 2	ch 17
3	ch 3	ch 18
4	ch 4	ch 19
5	ch 5	ch 20
6	ch 6	ch 21
7	ch 7	ch 22
8	ch 8	ch 23
9	ch 9	ch 24
10	ch 10	ch 25
11	ch 11	ch 26
12	ch 12	ch 27
13	ch 13	ch 28
14	ch 14	ch 29
15	ch 15	ch 30

x = spare
y = loss of multiframe alignment if a 1

4 bits per channel are transmitted once every 16 frames, resulting in a 500 words per second (2000 bps) signaling rate for each channel

(b)

FIGURE 6-14   CCITT TDM frame alignment and common signaling channel alignment: (a) CCITT TDM frame (125 μs, 256 bits, 2.048 Mbps); (b) common signaling channel

Table 6-4   European Transmission Rates and Capacities

Line	Transmission Bit Rate (Mbps)	Channel Capacity
E1	2.048	30
E2	8.448	120
E3	34.368	480
E4	139.264	1920

$$\text{line speed} = \frac{256 \text{ bits}}{\text{frame}} \times \frac{8000 \text{ frames}}{\text{second}} = 2.048 \text{ Mbps}$$

The European digital transmission system has a TDM multiplexing hierarchy similar to the North American Hierarchy except the European system is based on the 32-time-slot (30-voice-channel) E1 system. The *European Digital Multiplexing Hierarchy* is shown in Table 6-4. Interconnecting T carriers with E carriers is not generally a problem because most multiplexers and demultiplexers are designed to perform the necessary bit rate conversions.

Digital transmissions over a synchronous time-division multiplexing system often contain an abundance of time slots within each frame that contain no information (i.e., at any given instant, several of the channels may be idle). For example, TDM is commonly used to link remote data terminals or PCs to a common server or mainframe computer. A majority of the time, however, there are no data being transferred in either direction, even if all the terminals are active. The same is true for PCM-TDM systems carrying digital-encoded voice-grade telephone conversations. Normal telephone conversations generally involve information being transferred in only one direction at a time with significant pauses embedded in typical speech patterns. Consequently, there is a lot of time wasted within each TDM frame. There is an efficient alternative to synchronous TDM called *statistical time-division multiplexing*. Statistical time-division multiplexing is generally not used for carrying standard telephone circuits but is used more often for the transmission of data when they are called *asynchronous* TDM, *intelligent* TDM, or simply *stat muxs*.

A statistical TDM multiplexer exploits the natural breaks in transmissions by dynamically allocating time slots on a demand basis. Just as with the multiplexer in a synchronous TDM system, a statistical multiplexer has a finite number of low-speed data input lines with one high-speed multiplexed data output line, and each input line has its own digital encoder and buffer. With the statistical multiplexer, there are n input lines but only k time slots available within the TDM frame (where $k > n$). The multiplexer scans the input buffers, collecting data until a frame is filled, at which time the frame is transmitted. On the receive end, the same holds true—there are more output lines than time slots within the TDM frame. The demultiplexer removes the data from the time slots and distributes them to their appropriate output buffers.

Statistical TDM takes advantage of the fact that the devices attached to the inputs and outputs are not all transmitting or receiving all the time, and the data rate on the multiplexed line is lower than the combined data rates of the attached devices. In other words, statistical TDM multiplexers require a lower data rate than synchronous multiplexers need to support the same number of inputs. Alternately, a statistical TDM multiplexer operating at the same transmission rate as a synchronous TDM multiplexer can support more users.

Figure 6-15 shows a comparison between statistical and synchronous TDM. There are four data sources (A, B, C, and D) and four time slots, or epochs (t_1, t_2, t_3, and t_4). The synchronous multiplexer has an output data rate equal to four times the data rate of each of the input channels. During each sample time, data are collected from all four sources and transmitted regardless of whether there is any input. As the figure shows, during sample time t_1, channels C and D have no input data, resulting in a transmitted TDM frame void of information in time slots C_1 and D_1. With a statistical multiplexer, however, the empty time slots are not transmitted. A disadvantage of the statistical format is that the length of a frame varies and the positional significance of each time slot is lost. There is no way of knowing beforehand which channel's data will be in which time slot or how many time slots are included in each frame. Because data arrive and are distributed to receive buffers unpredictably, address information is required to ensure proper delivery. This necessitates more overhead per time slot for statistical TDM because each slot must carry an address as well as data.

The frame format used by a statistical TDM multiplexer has a direct impact on system performance. Obviously, it is desirable to minimize overhead to improve data throughput. Normally, a statistical TDM system will use a synchronous protocol such as HDLC (described in detail in a later chapter). With statistical multiplexing, control bits must be included within the frame. Figure 6-16a shows the overall frame format for a statistical TDM multiplexer. The frame includes a beginning flag and ending flag that indicate the beginning and end of the frame, an address field that identifies the transmitting device, a control field, a statistical TDM subframe, and a frame check sequence field (FCS), which provides error detection.

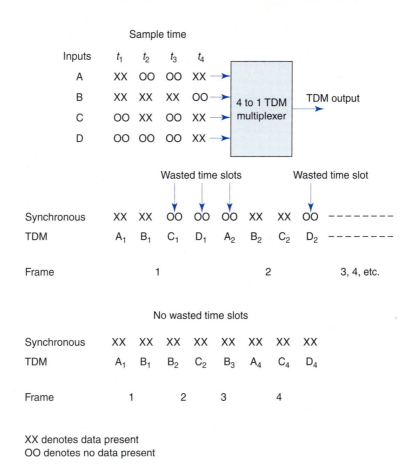

XX denotes data present
OO denotes no data present

FIGURE 6-15 Comparison between synchronous and statistical TDM

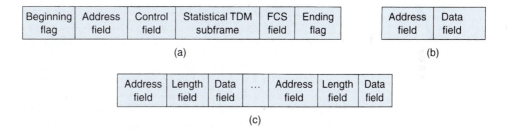

FIGURE 6-16 Statistical TDM frame format: (a) overall statistical TDM frame; (b) one source per frame; (c) multiple sources per frame

Figure 6-16b shows the frame when only one data source is transmitting. The transmitting device is identified in the address field. The data field length is variable and limited only by the maximum length of the frame. Such a scheme works well in times of light loads but rather inefficiently under heavy loads. Figure 6-16c shows one way to improve the efficiency by allowing more than one data source to be included within a single frame. With multiple sources, however, some means is necessary to specify the length of the data stream from each source. Hence, the statistical frame consists of sequences of data fields labeled with an address and a bit count. There are several techniques that can be used to further improve efficiency. The address field can be shortened by using relative addressing where each address specifies the position of the current source relative to the previously transmitted source and the total number of sources. With relative addressing, an eight-bit address field can be replaced with a four-bit address field.

Another method of refining the frame is to use a two-bit label with the length field. The binary values 01, 10, and 11 correspond to a data field of 1, 2, or 3 bytes, respectively, and no length field necessary is indicated by the code 00.

6-9   FRAME SYNCHRONIZATION

With TDM systems, it is imperative not only that a frame be identified but also that individual time slots (samples) within the frame be identified. To acquire frame synchronization, a certain amount of overhead must be added to the transmission. There are several methods used to establish and maintain frame synchronization, including added digit, robbed digit, added channel, statistical, and unique coding.

6-9-1   Added-Digit Framing

T1 carriers using D1, D2, or D3 channel banks use *added-digit framing*. A special *framing digit* (framing pulse) is added to each frame. Consequently, for an 8-kHz sample rate, 8000 digits are added each second. With T1 carriers, an alternating 1/0 frame-synchronizing pattern is used.

To acquire frame synchronization, the digital terminal in the receiver searches through the incoming data until it finds the framing bit pattern. This encompasses testing a bit, counting off 193 more bits, and then testing again for the opposite logic condition. This process continues until a repetitive alternating 1/0 pattern is found. Initial frame synchronization depends on the total frame time, the number of bits per frame, and the period of each bit. Searching through all possible bit positions requires N tests, where N is the number of bit positions in the frame. On average, the receiving terminal dwells at a false framing position for two frame periods during a search; therefore, the maximum average synchronization time is

$$\text{synchronization time} = 2NT = 2N^2 t_b \qquad \textbf{(6-13)}$$

where N = number of bits per frame
   T = frame period of $N\, t_b$
   t_b = bit time

For the T1 carrier, $N = 193$, $T = 125$ μs, and $t_b = 0.648$ μs; therefore, a maximum of 74,498 bits must be tested, and the maximum average synchronization time is 48.25 ms.

6-9-2   Robbed-Digit Framing

When a short frame is used, added-digit framing is inefficient. This occurs with single-channel PCM systems. An alternative solution is to replace the least significant bit of every nth frame with a framing bit. This process is called *robbed-digit framing*. The parameter n is chosen as a compromise between reframe time and signal impairment. For $n = 10$, the SQR is impaired by only 1 dB. Robbed-digit framing does not interrupt transmission but instead periodically replaces information bits with forced data errors to maintain frame synchronization.

6-9-3   Added-Channel Framing

Essentially, *added-channel framing* is the same as added-digit framing except that digits are added in groups or words instead of as individual bits. The European time-division multiplexing scheme previously discussed uses added-channel framing. One of the 32 time slots in each frame is dedicated to a unique synchronizing bit sequence. The average number of bits to acquire frame synchronization using added-channel framing is

$$\text{number of synchronization bits} = \frac{N^2}{2(2^K - 1)} \qquad \textbf{(6-14)}$$

where N = number of bits per frame
        K = number of bits in the synchronizing word

Example 6-5

For the European E1 32-channel system with $N = 256$, $K = 8$, and a 2.048-Mbps transmission rate, determine the average number of bits needed to synchronize and the average synchronization time.

Solution   Substituting into Equation 6-14 yields

$$\frac{256^2}{2(2^8 - 1)} = 128.5 \text{ bits}$$

Therefore, the average synchronization time is

$$\frac{128.5 \text{ bits}}{2.048 \text{ Mbps}} = 62.7 \text{ μs}$$

6-9-4   Statistical Framing

With *statistical framing*, it is not necessary to either rob or add digits. With the gray code, the second bit is a logic 1 in the central half of the code range and a logic 0 at the extremes. Therefore, a signal that has a centrally peaked amplitude distribution generates a high probability of a logic 1 in the second digit. Hence, the second digit of a given channel can be used for the framing bit.

6-9-5   Unique-Line Code Framing

With *unique-line code framing*, some property of the framing bit is different from the data bits. The framing bit is either made higher or lower in amplitude or with a different time duration. The earliest PCM-TDM systems used unique-line code framing. D1 channel banks used framing pulses that were twice the amplitude of normal data bits. With unique-line code framing, either added-digit or added-word framing can be used, or specified data bits can be used to simultaneously convey information and carry synchronizing signals. The advantage of unique-line code framing is that synchronization is immediate and automatic. The disadvantage is the additional processing requirements necessary to generate and recognize the unique bit.

6-10   FREQUENCY-DIVISION MULTIPLEXING

With *frequency-division multiplexing* (FDM), multiple sources that originally occupied the same frequency spectrum are each converted to a different frequency band and transmitted simultaneously over a single transmission medium, which can be a physical cable or the earth's atmosphere (i.e., wireless). Thus, many relatively narrow-bandwidth channels can be transmitted over a single wide-bandwidth transmission system without interfering with each other. FDM is used for combining many relatively narrowband sources into a single wideband channel, such as in public telephone systems. Essentially, FDM is taking a given bandwidth and subdividing it into narrower segments with each segment carrying different information.

FDM is an analog multiplexing scheme; the information entering an FDM system must be analog, and it remains analog throughout transmission. If the original source information is digital, it must be converted to analog before being frequency-division multiplexed.

A familiar example of FDM is the commercial AM broadcast band, which occupies a frequency spectrum from 535 kHz to 1605 kHz. Each broadcast station carries an information signal (voice and music) that occupies a bandwidth between 0 Hz and 5 kHz. If the information from each station were transmitted with the original frequency spectrum, it would be impossible to differentiate or separate one station's transmissions from another. Instead, each station amplitude modulates a different carrier frequency and produces a 10-kHz signal. Because the carrier frequencies of adjacent stations are separated by 10 kHz, the total commercial AM broadcast band is divided into 107 10-kHz frequency slots stacked next to each other in the frequency domain. To receive a particular station, a receiver is simply tuned to the frequency band associated with that station's transmissions.

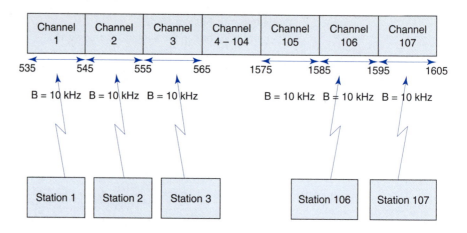

FIGURE 6-17 Frequency-division multiplexing of the commercial AM broadcast band

Figure 6-17 shows how commercial AM broadcast station signals are frequency-division multiplexed and transmitted over a common transmission medium (earth's atmosphere).

With FDM, each narrowband channel is converted to a different location in the total frequency spectrum. The channels are stacked on top of each other in the frequency domain. Figure 6-18a shows a simple FDM system where four 6-kHz channels are frequency-division multiplexed into a single 20-kHz combined channel. As the figure shows, channel 1 signals amplitude modulate a 100-kHz carrier in a balanced modulator, which inherently suppresses the 100-kHz carrier. The output of the balanced modulator is a double-sideband suppressed carrier waveform with a bandwidth of 10 kHz. The double-sideband waveform passes through a bandpass filter (BPF) where it is converted to a single-sideband signal. For this example, the lower sideband is blocked; thus, the output of the BPF occupies the frequency band between 100 kHz and 105 kHz (a bandwidth of 5 kHz).

Channel 2 signals amplitude modulate a 105-kHz carrier in a balanced modulator, again producing a double-sideband signal that is converted to single sideband by passing it through a bandpass filter tuned to pass only the upper sideband. Thus, the output from the BPF occupies a frequency band between 105 kHz and 110 kHz. The same process is used to convert signals from channels 3 and 4 to the frequency bands 110 kHz to 115 kHz and 115 kHz to 120 kHz, respectively. The combined frequency spectrum produced by combining the outputs from the four bandpass filters is shown in Figure 6-18b. As the figure shows, the total combined bandwidth is equal to 20 kHz, and each channel occupies a different 6-kHz portion of the total 20-kHz bandwidth.

There are many other applications for FDM, such as commercial FM and television broadcasting, high-volume telephone and data communications systems, and cable television and data distribution networks.

6-10-1 AT&T's FDM Hierarchy

Although AT&T is no longer the only long-distance common carrier in the United States, it still provides the vast majority of the long-distance services and, if for no other reason than its overwhelming size, has essentially become the standards organization for the telephone industry in North America. Figure 6-19 shows AT&T's FDM hierarchy. As the figure shows, voice channels are combined to form groups, groups are combined to form supergroups, and supergroups are combined to form mastergroups.

Message channel. The message channel is the basic building block of the FDM hierarchy. The basic message channel was originally intended for the analog transmission of voice signals, although it now includes any transmissions that utilize voice-band

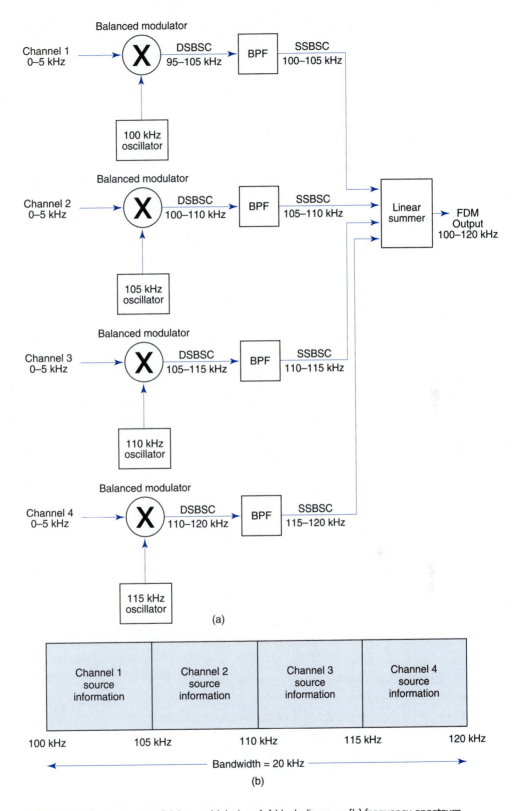

FIGURE 6-18 Frequency-division multiplexing: (a) block diagram; (b) frequency spectrum

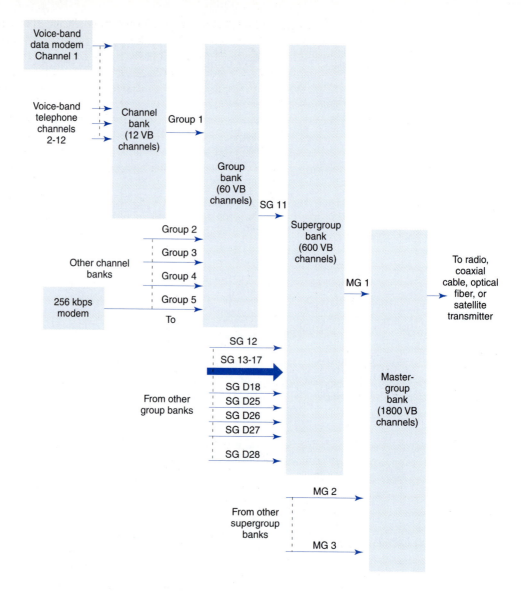

FIGURE 6-19 AT&T's FDM hierarchy

frequencies (0 kHz to 4 kHz), such as data transmission using voice-band data modems. The basic voice-band (VB) circuit is actually bandlimited to approximately a 300-Hz to 3000-Hz band, although for practical considerations it is considered a 4-kHz channel.

Basic group. A group is the next-higher level in the FDM hierarchy above the basic message channel and consequently is the first multiplexing step for combining message channels. A basic group consists of 12 voice-band message channels multiplexed together by stacking them next to each other in the frequency domain. Twelve 4-kHz voice-band channels occupy a combined bandwidth of 48 kHz (4 × 12).

Basic supergroup. The next-higher level in the FDM hierarchy is the supergroup, which is formed by frequency-division multiplexing five groups containing 12 channels each, for a combined bandwidth of 240 kHz (5 groups × 48 kHz/group or 5 groups × 12 channels/group × 4 kHz/channel).

Basic mastergroup. The next-higher level of multiplexing, shown in Figure 6-19, is the mastergroup, which is formed by frequency-division multiplexing 10 supergroups together for a combined capacity of 600 voice-band message channels occupying a bandwidth of 2.4 MHz (600 channels × 4 kHz/channel or 5 groups × 12/channels/group × 10 groups/supergroup). Typically, three mastergroups are frequency-division multiplexed together and placed on a single microwave or satellite radio channel. The capacity is 1800 VB channels utilizing a combined bandwidth of 7.2 MHz (3 mastergroups × 600 channels/mastergroup).

Mastergroups can be further multiplexed in mastergroup banks to form jumbogroups (3600 VB channels), multijumbogroups (7200 VB channels), and superjumbogroups (10,800 VB channels).

6-11 WAVELENGTH-DIVISION MULTIPLEXING

During the last two decades of the 20th century, the telecommunications industry witnessed an unprecedented growth in data traffic and the need for computer networking. The possibility of using *wavelength-division multiplexing* (WDM) as a networking mechanism for routing, switching, and selection based on wavelength began a new era in optical communications.

WDM promises to vastly increase the bandwidth capacity of optical transmission media. The basic principle behind WDM involves the transmission of multiple digital signals using several wavelengths without their interfering with one another. Digital transmission equipment currently being deployed utilizes optical fibers to carry only one digital signal per fiber per propagation direction. This technology enables many optical signals to be transmitted simultaneously by a single fiber cable.

Wavelength-division multiplexing is sometimes referred to as simply *wave-division multiplexing.* Since wavelength and frequency are closely related, wavelength-division multiplexing is similar to frequency-division multiplexing (FDM). WDM resembles FDM in that the idea is to send information signals that originally occupied the same band of frequencies through the same fiber at the same time without their interfering with each other. This is accomplished by modulating injection laser diodes, which are transmitting highly concentrated light waves at different wavelengths (i.e., at different optical frequencies). Therefore, WDM is coupling light at two or more discrete wavelengths into and out of an optical fiber. Each wavelength is capable of carrying vast amounts of information in either analog or digital form, and the information can already be time- or frequency-division multiplexed. Although the information used with lasers is almost always time-division multiplexed digital signals, the wavelength separation used with WDM is analogous to analog radio channels operating at different carrier frequencies. However, the carrier with WDM is in essence a wavelength rather than a frequency.

6-11-1 Wavelength-Division Multiplexing versus Frequency-Division Multiplexing

The basic principle of WDM is essentially the same as frequency-division multiplexing (FDM) where several signals are transmitted using different carriers, occupying nonoverlapping bands of a frequency or wavelength spectrum. In the case of WDM, the wavelength spectrum used is in the region of 1300 nm or 1500 nm, which are the two wavelength bands at which optical fibers have the least amount of signal loss. In the past, each window transmitted a single digit signal. With the advance of optical components, each transmitting window can be used to propagate several optical signals, each occupying a small fraction of the total wavelength window. The number of optical signals multiplexed with a window is limited only by the precision of the components used. Current technology allows over 100 optical channels to be multiplexed into a single optical fiber.

Although FDM and WDM share similar principles, they are not the same. The most obvious difference is that optical frequencies (in THz) are much higher than radio frequencies (in MHz and GHz). Probably the most significant difference, however, is in the way the two signals propagate through their respective transmission media. With FDM, signals propagate at the same time and through the same medium and follow the same transmission path. The basic principle of WDM, however, is somewhat different. Different wavelengths in a light pulse travel through an optical fiber at different speeds (e.g., blue light propagates slower than red light). In standard optical fiber communications systems, as the light propagates down the cable, wavelength dispersion causes the light waves to spread out and distribute their energy over a longer period of time. Thus, in standard optical fiber systems, wavelength dispersion creates problems, which impose limitations on the system's performance. With WDM, however, wavelength dispersion is the essence of how the system operates.

With WDM, information signals from multiple sources modulate lasers operating at different wavelengths. Hence, the signals enter the fiber at the same time and travel through the same medium. However, they do not take the same path down the fiber. Since each wavelength takes a different transmission path, each arrives at the receive end at slightly different times. The result is a series of rainbows made of different colors (wavelengths) each about 20 billionths of a second long, simultaneously propagating down the cable.

Figure 6-20 illustrates the basic principles of FDM and WDM signals propagating through their respective transmission media. As shown in Figure 6-20a, FDM channels all propagate at the same time and over the same transmission medium and take the same transmission path; however, they occupy different bandwidths. In Figure 6-20b, it can be seen that with WDM, each channel propagates down the same transmission medium at the same time; however, each channel occupies a different bandwidth (wavelength), and each wavelength takes a different transmission path.

6-11-2 Dense-Wave-Division Multiplexing, Wavelengths, and Wavelength Channels

WDM is generally accomplished at approximate wavelengths of 1550 nm (1.55 µm) with successive frequencies spaced in multiples of 100 GHz (e.g., 100 GHz, 200 GHz, 300 GHz, and so on). At 1550-nm and 100-GHz frequency separation, the wavelength separation is approximately 0.8 nm. For example, three adjacent wavelengths each separated by 100 GHz correspond to wavelengths of 1550.0 nm, 1549.2 nm, and 1548.4 nm. Using a multiplexing technique called *dense-wave-division multiplexing* (D-WDM), the spacing between adjacent frequencies is considerably less. Unfortunately, there does not seem to be a standard definition of exactly what D-WDM means. Generally, optical systems carrying multiple optical signals spaced more than 200 GHz or 1.6 nm apart in the vicinity of 1550 nm are considered standard WDM. WDM systems carrying multiple optical signals in the vicinity of 1550 nm with less than 200 GHz of separation are considered D-WDM. Obviously, the more wavelengths used in a WDM system, the closer they are to each other and the denser the wavelength spectrum.

Light waves are comprised of many frequencies (wavelengths), and each frequency corresponds to a different color. Transmitters and receivers for optical fiber systems have been developed that transmit and receive only a specific color (i.e., a specific wavelength at a specific frequency with a fixed bandwidth). WDM is a process in which different sources of information (channels) are propagated down an optical fiber on different wavelengths where the different wavelengths do not interfere with each other. In essence, each wavelength adds an optical lane to the transmission superhighway, and the more lanes there are, the more traffic (voice, data, video, and so on) can be carried on a single optical fiber cable. In contrast, conventional optical fiber systems have only one channel per cable, which is used to carry information over a relatively narrow bandwidth. A Bell Laboratories

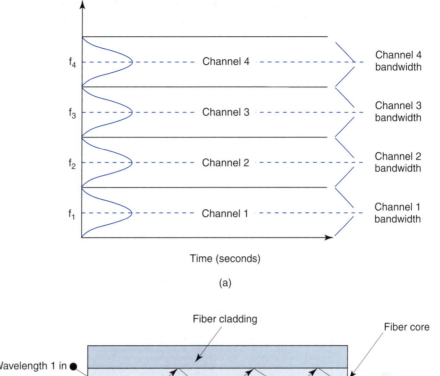

(a)

(b)

FIGURE 6-20 (a) Frequency-division multiplexing; (b) wave-length-division multiplexing

research team recently constructed a D-WDM transmitter using a single femtosecond, erbium-doped fiber-ring laser that can simultaneously carry 206 digitally modulated wavelengths of color over a single optical fiber cable. Each wavelength (channel) has a bit rate of 36.7 Mbps with a channel spacing of approximately 36 MHz.

Figure 6-21a shows the wavelength spectrum for a WDM system using six wavelengths, each modulated with equal-bandwidth information signals. Figure 6-21b shows how the output wavelengths from six lasers are combined (multiplexed) and then propagated over a single optical cable before being separated (demultiplexed) at the receiver with wavelength selective couplers. Although it has been proven that a single, ultrafast light source can generate hundreds of individual communications channels, standard WDM communications systems are generally limited to between 2 and 16 channels.

WDM enhances optical fiber performance by adding channels to existing cables. Each wavelength added corresponds to adding a different channel with its own information source and transmission bit rate. Thus, WDM can extend the information-carrying capacity of a fiber to hundreds of gigabits per second or higher.

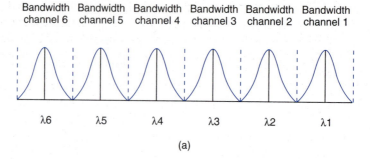

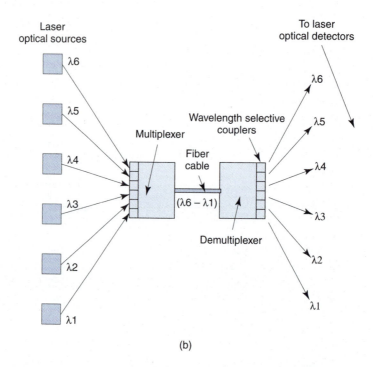

FIGURE 6-21 (a) Wavelength spectrum for a WDM system using six wavelengths; (b) multiplexing and demultiplying six lasers

6-11-3 Advantages and Disadvantages of WDM

An obvious advantage of WDM is enhanced capacity, and with WEM, full-duplex transmission is also possible with a single fiber. In addition, optical communications networks use optical components, which are simpler, more reliable, and often less costly than their electronic counterparts. WDM has the advantage of being inherently easier to reconfigure (i.e., adding or removing channels). For example, WDM local area networks have been constructed that allow users to access the network by simply tuning to a certain wavelength.

There are also limitations to WDM. Signals cannot be placed so close in the wavelength spectrum that they interfere with each other. Their proximity depends on system design parameters, such as whether optical amplification is used and what optical technique is used to combine and separate signals at different wavelengths. The International Telecommunications Union adopted a standard frequency grid for D-WDM with a spacing of 100 GHz or integer multiples of 100 GHz, which at 1550 nm corresponds to a wavelength spacing of approximately 0.8 nm.

With WDM, the overall signal strength should be approximately the same for each wavelength. Signal strength is affected by fiber attenuation characteristics and the degree of amplification, both of which are wavelength dependent. Under normal conditions, the wavelengths chosen for a system are spaced so close to one another that attenuation differs very little among them.

One difference between FDM and WDM is that WDM multiplexing is performed at extremely high optical frequencies, whereas FDM is performed at relatively low radio and baseband frequencies. Therefore, radio signals carrying FDM are not limited to propagating through a contained physical transmission medium, such as an optical cable. Radio signals can be propagated through virtually any transmission medium, including free space. Therefore, radio signals can be transmitted simultaneously to many destinations, whereas light waves carrying WDM are limited to a two-point circuit or a combination of many two-point circuits that can go only where the cable goes.

The information capacity of a single optical cable can be increased n-fold, where n represents how many different wavelengths the fiber is propagating at the same time. Each wavelength in a WDM system is modulated by information signals from different sources. Therefore, an optical communications system using a single optical cable propagating n separate wavelengths must utilize n modulators and n demodulators.

6-12 SYNCHRONOUS OPTICAL NETWORK

The *synchronous optical network* (SONET) is a multiplexing system similar to conventional time-division multiplexing except SONET was developed to be used with optical fibers. The initial SONET standard is OC-1. This level is referred to as *synchronous transport level 1* (STS-1). STS-1 has a 51.84-Mbps synchronous frame structure comprised of 28 DS-1 signals. Each DS-1 signal is equivalent to a single 24-channel T1 digital carrier system. Thus, one STS-1 system can carry 672 individual 64-kbps voice or data channels (24×28). With STS-1, it is possible to extract or add individual DS-1 signals without completely disassembling the entire frame.

OC-48 is the second level of SONET multiplexing. It combines 48 OC-1 systems for a total channel capacity of 32,256. OC-48 has a transmission bit rate of 2.48332 Gbps (2.48332 billion bits per second). A single-channel optical fiber can carry one OC-48 system; however, as many as 16 OC-48 systems can be combined using wave-division multiplexing and propagated over a single optical cable. The light spectrum is divided into 16 different wavelengths with an OC-48 system attached to each transmitter for a combined capacity of 516,096 64-kbps voice or data channels $(16 \times 32,256)$. The synchronous optical network and its relevance to computer networking are described in detail in a later chapter.

QUESTIONS

6-1. Define *multiplexing*.

6-2. Describe *time-division multiplexing*.

6-3. Describe a *T1 carrier system*.

6-4. What is the purpose of *signaling bits*?

6-5. What is *frame synchronization*? How is it achieved in a PCM-TDM system?

6-6. Describe the *superframe format*. Why is it used?

6-7. Describe the *extended superframe format*. Why is it used?

6-8. Describe a *fractional T carrier*.

6-9. Describe *digital line encoding*.

6-10. Briefly describe *unipolar* and *bipolar* transmissions.

6-11. Briefly describe *return-to-zero* and *nonreturn-to-zero* transmission.

6-12. Contrast the bandwidth considerations of *return-to-zero* and *nonreturn-to-zero* transmission.

6-13. Contrast the clock recovery capabilities of *return-to-zero* and *nonreturn-to-zero* transmission.

6-14. Contrast error detection and decoding capabilities of *return-to-zero* and *nonreturn-to-zero* transmission.

6-15. What is *digital biphase*?

6-16. What is a *regenerative repeater*?

6-17. Explain B8ZS, B6ZS, and B3ZS.

6-18. Briefly explain the following framing techniques: *added-digit framing*, *robbed-digit framing*, *added-channel framing*, *statistical framing*, and *unique-line code framing*.

6-19. Briefly explain *the European E1 time-division multiplexing format*.

6-20. Briefly describe *statistical time-division multiplexing*.

6-21. Describe *frequency-division multiplexing*.

6-22. Briefly describe a *basic message channel*.

6-23. Briefly describe the AT&T *FDM hierarchy*.

6-24. Describe the basic concepts of *wavelength-division multiplexing*.

6-25. What is the difference between WDM and D-WDM?

6-26. List the advantages and disadvantages of WDM.

6-27. Briefly describe the SONET standard, including OC-1 and OC-48.

PROBLEMS

6-1. A PCM-TDM system multiplexes 24 voice-band channels. Each sample is encoded into seven bits, and a framing bit is added to each frame. The sampling rate is 9000 samples per second. Determine the line speed in bps.

6-2. A PCM-TDM system multiplexes 32 voice-band channels each with a bandwidth of 0 kHz to 4 kHz. Each sample is encoded with an eight-bit PCM code. Determine

  a. Minimum sample rate
  b. Line speed in bps

6-3. For the following bit sequence, draw the timing diagram for UPRZ, UPNRZ, BPRZ, BPNRZ, and BPRZ-AMI: 1 1 1 0 0 1 0 1 0 1 1 0 0

6-4. A PCM-TDM system multiplexes 20 voice-band channels. Each sample is encoded into seven bits, and a framing bit is added to each frame. The sampling rate is 10,000 samples per second. Determine

  a. Maximum analog input frequency
  b. Line speed in bps

6-5. A PCM-TDM system multiplexes 30 voice-band channels each with a bandwidth of 0 kHz to 3 kHz. Each sample is encoded with a nine-bit PCM code. Determine

  a. Minimum sample rate
  b. Line speed in bps

CHAPTER 7

Wireless Communications Systems

CHAPTER OUTLINE

OBJECTIVES

- Define *free-space propagation*
- Define *electromagnetic polarization*
- Describe rays and wavefronts and the relationship between them
- Define *electromagnetic radiation*
- Explain a spherical wavefront and the inverse square law
- Describe wave attenuation and absorption and the relationship between them
- Describe the following optical properties of radio waves: refraction, reflection, diffraction, and interference
- Describe ground-wave, space-wave, and sky-wave propagation
- Explain skip distance
- Define *free-space path loss*
- Contrast the advantages and disadvantages of microwave radio communications
- Describe the components that make up a microwave radio link
- Describe a microwave repeater
- Define *satellite*
- List and describe the three satellite elevation categories
- List and describe the three satellite orbital patterns
- Define *geosynchronous satellite*
- Describe the Clark orbit

191

- Contrast the advantages and disadvantages of geosynchronous satellites
- Describe satellite look angles
- Define *satellite footprints*
- List and describe the three satellite multiple access arrangements

7-1 INTRODUCTION

With *wireless communications systems,* electromagnetic signals are emitted from an antenna, propagate through the *earth's atmosphere* (air) or *free space* (a vacuum), and are then received (captured) by another antenna. The signals are *unguided;* therefore, the direction they propagate depends on the direction in which they were emitted and any obstacles they may encounter while propagating. Signals propagating through an unguided transmission medium are available to anyone with a device capable of receiving them.

In Chapter 3 we explained *transverse electromagnetic* (TEM) *waves* and also described how metallic wires and optical fiber cables can be used as a transmission medium to transfer TEM waves from one point to another. However, very often in data communications systems and computer networks, it is impractical or impossible to interconnect two pieces of equipment with a physical cable. This is especially true when the equipment is separated by large spans of water, rugged mountains, or harsh desert terrain or when communicating with satellite transponders orbiting 22,300 miles above earth. Also, when the transmitters and receivers are mobile, such as with cellular telephones, it is impossible to provide a physical connection. Therefore, *free space* or earth's atmosphere is often used as the transmission medium.

Free-space propagation of electromagnetic waves is often called *radio-frequency* (RF) *propagation* or simply *radio propagation.* Although free space implies a vacuum, propagation through earth's atmosphere is often referred to as free-space propagation, the primary difference being that earth's atmosphere introduces losses to the signal that are not encountered in a vacuum. In addition, earth's atmosphere may cause the signal to undergo certain optical phenomena not possible in a vacuum. TEM waves will propagate through any dielectric material, including air. However, TEM waves do not propagate well through lossy conductors, such as seawater, because the electric fields cause currents to flow in the material that rapidly dissipate the wave's energy.

Wireless communications systems include both terrestrial and satellite microwave radio systems, broadcast radio systems (such as commercial FM, AM, and television), two-way mobile radio, and cellular telephone. Terrestrial microwave and satellite communications are discussed in this chapter, and the cellular telephone is described in Chapter 10.

Radio waves are electromagnetic waves and, like light, propagate through free space in a straight line with a velocity of 300,000,000 meters per second. Other forms of electromagnetic waves include infrared, ultraviolet, X rays, and gamma rays. To propagate radio waves through earth's atmosphere, it is necessary that the energy be radiated from the source and directed toward the receiver, and then the energy must be captured at the receiving end.

7-2 ELECTROMAGNETIC POLARIZATION

Electromagnetic waves are comprised of an electric and a magnetic field at 90 degrees to each other. The *polarization* of a plane electromagnetic wave is simply the orientation of the electric field vector in respect to earth's surface (i.e., looking at the horizon). If the polarization remains constant, it is described as *linear polarization.* Horizontal and vertical polarizations are two forms of linear polarization. A wave is *horizontally polarized* if the electric field propagates parallel to the earth's surface, and the wave is *vertically polarized* if the electric field propagates perpendicular to the earth's surface. The wave is described as having *circular polarization* if the polarization vector rotates 360 degrees as the wave moves one wavelength through space and the field strength is equal at all angles of polarization. When the field strength varies with changes in polarization, this is described as *elliptical polarization*. A rotating wave can turn in either direction. If the vector rotates in a clockwise direction, it is *right handed*, and if the vector rotates in a counterclockwise direction, it is considered *left handed*.

7-3 RAYS AND WAVEFRONTS

Electromagnetic waves are invisible; therefore, they must be analyzed by indirect methods using schematic diagrams. The concepts of *rays* and *wavefronts* are aids to illustrating the effects of electromagnetic wave propagation. A ray is a line drawn along the direction of propagation of an electromagnetic wave. Rays are used to show the relative direction of propagation; however, a ray does not necessarily represent a single electromagnetic wave. Several rays are shown in Figure 7-1 (R_a, R_b, R_c, and so on). A wavefront shows a surface of constant phase of electromagnetic waves. A wavefront is formed when points of equal phase on rays propagating from the same source are joined together. Figure 7-1 shows a wavefront with a surface that is perpendicular to the direction of propagation (rectangle *ABCD*). When a surface is plane, its wavefront is perpendicular to the direction of propagation.

Most wavefronts are more complicated than a simple plane wave. Figure 7-2 shows a point source, several rays propagating from it, and the corresponding wavefront. A *point*

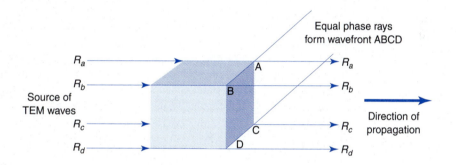

FIGURE 7-1   Plane wave comprised of rays R_a, R_b, R_c, and R_d forming wavefront ABCD

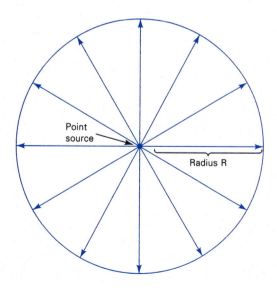

FIGURE 7-2   Wavefront from a point source

source is a single location from which rays propagate equally in all directions (i.e., an *isotropic source*). The wavefront generated from a point source is simply a sphere with radius R and its center located at the point of origin of the waves.

7-4 ELECTROMAGNETIC RADIATION

Electromagnetic radiation represents the flow of electromagnetic waves (energy) in the direction of propagation. The rate at which energy passes through a given surface area in free space is called *power density*. Therefore, power density is energy per unit of time per unit of area and is usually given in watts per square meter. Mathematically, power density is

$$\mathscr{P} = \mathscr{E}\mathscr{H} \qquad \text{(7-1)}$$

where $\mathscr{P}$ = power density (watts per meter squared)
  $\mathscr{E}$ = rms electric field intensity (volts per meter)
  $\mathscr{H}$ = rms magnetic field intensity (ampere turns per meter)

7-5 SPHERICAL WAVEFRONT AND THE INVERSE SQUARE LAW

7-5-1 Spherical Wavefront

Figure 7-3 shows a point source radiating power at a constant rate uniformly in all directions. Such a source is called an *isotropic radiator*. An isotropic radiator produces a spherical wavefront with radius R. All points at distance R from the source lie on the surface of the sphere and have equal power densities. For example, in Figure 7-3, points A and B are an equal distance from the source. Therefore, the power densities at points A and B are equal. At any instant of time, the total power radiated, P_{rad} watts, is uniformly distributed over the total surface of the sphere (assuming a lossless transmission medium). Therefore, the power density at any point on the sphere is the total radiated power divided by the total area of the sphere. Mathematically, the power density at any point on the surface of a spherical wavefront is

$$\mathscr{P} = \frac{P_{rad}}{4\pi R^2} \qquad \text{(7-2)}$$

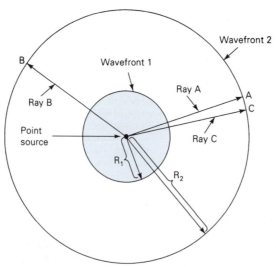

FIGURE 7-3 Spherical wavefront from an isotropic source

where $\mathcal{P}$ = power density at distance R from point source (watts per meter squared)
P_{rad} = total power radiated by point source (watts)
R = radius of the sphere (which is equal to the distance from any point on the surface of the sphere to the source)
$4\pi R^2$ = area of the sphere with radius R (square meters)

Example 7-1

For an isotropic radiator emitting 100 W of power, determine the power density 1000 meters from the source.

Solution   Substituting into Equation 7-2 yields

$$\mathcal{P} = \frac{100 \text{ watts}}{4\pi 1000^2}$$

$$= 7.96 \ \mu w$$

7-5-2 Inverse Square Law

From Equation 7-2, it can be seen that the farther the wavefront moves from the source, the smaller the power density. The total power distributed over the surface of the sphere remains the same. However, because the area of the sphere increases in direct proportion to the distance from the source squared (i.e., the radius of the sphere squared), the power density is inversely proportional to the square of the distance from the source. This relationship is called the *inverse square law*.

7-6 WAVE ATTENUATION AND ABSORPTION

Free space is a vacuum, so no loss of energy occurs as a wave propagates through it. However, as waves propagate through free space, they spread out, resulting in a reduction in power density. This is called *attenuation* and occurs in free space as well as the Earth's atmosphere. Since earth's atmosphere is not a vacuum, it contains particles that can absorb electromagnetic energy. This type of reduction of power is called *absorption loss* and does not occur in waves traveling outside earth's atmosphere.

7-6-1 Attenuation

The inverse square law for radiation (Equation 7-2) mathematically describes the reduction in power density with distance from the source. As a wavefront moves away from the source, the electromagnetic field radiated from the source spreads out. That is, the waves move farther away from each other and, consequently, the number of waves per unit area decreases. None of the radiated power is lost or dissipated; the wave simply spreads out or disperses over a larger area decreasing the power density. This reduction in power density with distance is equivalent to a power loss and is commonly called *wave attenuation*. Because the attenuation is due to the spherical spreading of the wave in space, it is sometimes called *space attenuation*. Wave attenuation is generally expressed in terms of the common logarithm of the power density ratio (dB loss). Mathematically, wave attenuation is

$$\gamma_A = 10 \log\left(\frac{P_1}{P_2}\right) \tag{7-3}$$

where γ_A = wave attenuation (dB loss)
P_1 = power density at point 1
P_2 = power density at point 2

The reduction of power density due to the inverse square law presumes free-space propagation (i.e., through a vacuum) and is called wave attenuation.

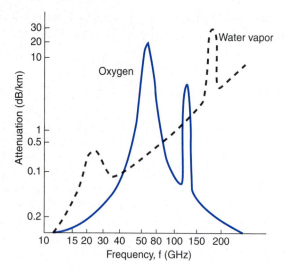

FIGURE 7-4 Atmospheric absorption of electromagnetic waves

Example 7-2

Determine the wave attenuation between two points with power densities of $P_1 = 10\ \mu\mu W/m^2$ and $P_2 = 0.1\ W/m^2$.

Solution Substituting into Equation 7-3 yields

$$\gamma_A = 10 \log\left(\frac{10\ \mu W/m^2}{0.1\ \mu W/m^2}\right)$$

$$= 20\ dB\ of\ loss$$

7-6-2 Absorption

Earth's atmosphere is not a vacuum. Rather, it is comprised of atoms and molecules of various substances, such as gases, liquids, and solids. Some of these materials are capable of absorbing electromagnetic waves. As an electromagnetic wave propagates through earth's atmosphere, energy is transferred from the wave to the atoms and molecules of substances in the atmosphere. This transfer of energy is called *wave absorption* and is analogous to an I^2R power loss. Once absorbed, the energy is lost forever, thus reducing the power density of the signal.

Absorption of radio frequencies in a normal atmosphere depends on frequency and is relatively insignificant at frequencies below approximately 10 GHz. Figure 7-4 shows atmosphere absorption in decibels per kilometer due to oxygen and water vapor for radio frequencies above 10 GHz. It can be seen that certain frequencies are affected more or less by absorption, creating peaks and valleys in the curves. Wave attenuation due to absorption depends not on the distance from the radiating source but rather on the total distance that the wave propagates through the atmosphere. In other words, for a homogeneous medium (one with uniform properties throughout), the absorption experienced during the first mile of propagation is the same as for the last mile. Also, abnormal atmospheric conditions, such as heavy rain or dense fog, absorb more energy than a normal atmosphere.

7-7 OPTICAL PROPERTIES OF RADIO WAVES

In earth's atmosphere, ray-wavefront propagation may be altered from free-space behavior by optical effects such as *refraction, reflection, diffraction*, and *interference*. Using rather unscientific terminology, refraction can be thought of as bending, reflection as bouncing, diffraction as scattering, and interference as colliding. Refraction, reflection, diffraction, and interference are called *optical properties* because they were first observed in the science of optics, which is the behavior of light waves. Because light waves are high-frequency electromagnetic waves, it stands to reason that optical properties will also apply to radio-wave propagation.

7-7-1 Refraction

Electromagnetic *refraction* is the change in direction of an electromagnetic wave as it passes obliquely from one medium to another medium with a different density (refractive index). The velocity at which an electromagnetic wave propagates is inversely proportional to the density of the medium in which it is propagating. Therefore, refraction occurs whenever a radio wave passes from one medium into another.

Figure 7-5 shows refraction of a wavefront at a plane boundary between two media with different densities. For this example, medium 1 is less dense than medium 2 ($n_1 > n_2$, thus $v_1 > v_2$). It can be seen that ray A enters the more dense medium before ray B. Therefore, ray B propagates more rapidly than ray A and travels distance from B to B′ during the same time that ray A travels the distance from A to A′. Therefore, wavefront A′B′ is tilted or bent in a downward direction. Because a ray is defined as being perpendicular to the wavefront at all points, the rays in Figure 7-3 have changed direction at the interface of the two media. Whenever a ray passes from a less dense to a more dense medium, it is effectively bent toward the normal (imaginary line drawn perpendicular to the interface at the point of incidence). Conversely, whenever a ray passes from a more dense to a less dense medium, it is effectively bent away from the normal. The *angle of incidence* is the angle formed between the incident wave and the normal, and the *angle of refraction* is the angle formed between the refracted wave and the normal.

As with optical frequencies, how an electromagnetic wave reacts when it meets the interface of two transmissive materials with different indexes of refraction can be explained with *Snell's law*. Mathematically, Snell's law states,

$$\sin \theta_1 \left(\frac{n_1}{n_2} \right) = \sin \theta_2 \tag{7-4}$$

where θ_1 = angle of incidence (degrees)
θ_2 = angle of refraction (degrees)
n_1 = refractive index of material 1 (unitless)
n_2 = refractive index of material 2 (unitless)

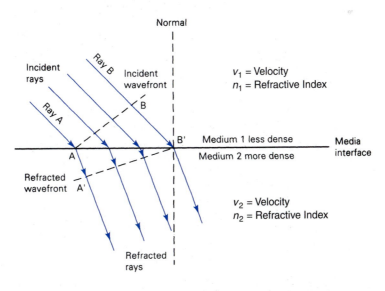

FIGURE 7-5 Refraction at a plane boundary between two media

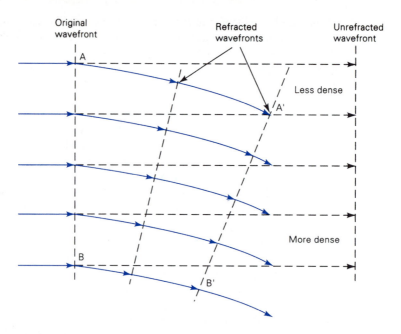

FIGURE 7-6 Wavefront refraction in a gradient medium

Refraction also occurs when a wavefront propagates in a medium that has a density gradient that is perpendicular to the direction of propagation (i.e., parallel to the wavefront). Figure 7-6 shows wavefront refraction in a transmission medium that has a gradient refractive index, such as earth's atmosphere. The medium is more dense near the bottom (close to earth's surface) and less dense near the top (in earth's upper atmosphere). Therefore, rays traveling in the upper layers of the atmosphere travel faster than rays traveling near earth's surface and, consequently, the wavefront tilts downward. The tilting occurs in a gradual fashion as the wave progresses.

7-7-2 Reflection

Electromagnetic wave *reflection* occurs when an incident wave strikes a boundary of two media and some or all of the incident power does not enter the second material (i.e., they are reflected). Figure 7-7 shows electromagnetic wave reflection at a plane boundary be-

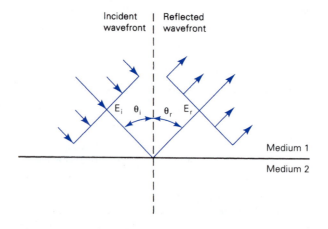

FIGURE 7-7 Electromagnetic reflection at a plane boundary of two media

tween two media. Because all the reflected waves remain in medium 1, the velocities of the reflected and incident waves are equal. Consequently, the angle of reflection equals the angle of incidence ($\theta_i = \theta_r$). The ratio of the reflected to the incident power is Γ, expressed mathematically as

$$\Gamma = \frac{P_r}{P_i} \tag{7-5}$$

where Γ = reflection coefficient (unitless)
           P_r = power reflected (watts)
           P_i = power incident (watts)

For perfect conductors, $\Gamma = 1$, and all the incident power is reflected. For imperfect conductors, Γ is a function of the angle of incidence, the electric field orientation, and the dielectric constants of the two materials. In addition, some of the incident waves penetrate medium 2 and are absorbed and converted to heat.

Reflection also occurs when the reflective surface is irregular; however, such a surface may destroy the shape of the wavefront. When an incident wavefront strikes an irregular surface, it is randomly scattered in many directions. Such a condition is called *diffuse reflection*, whereas reflection from a perfectly smooth surface is called *specular* (mirrorlike) *reflection*.

7-7-3 Diffraction

Diffraction is defined as the modulation or redistribution of energy within a wavefront when it passes near the edge of an opaque object. Diffraction is the phenomenon that allows light or radio waves to propagate (peek) around corners. The previous discussions on refraction and reflection assumed that the dimensions of the refracting and reflecting surfaces were large with respect to a wavelength. However, when a wavefront passes near an obstacle or discontinuity with dimensions comparable in size to a wavelength, simple geometric analysis cannot be used to explain the results. In these cases, *Huygen's principle* is necessary.

Huygen's principle states that every point on a given spherical wavefront can be considered as a secondary point source of electromagnetic waves from which other secondary waves (wavelets) are radiated outward. Huygen's principle is illustrated in Figure 7-8. Normal wave propagation considering an infinite plane is shown in Figure 7-8a. Each secondary point source (p_1, p_2, and so on) radiates energy outward in all directions. However, the wavefront continues in its original direction rather than spreading out because cancellation of the secondary wavelets occurs in all directions except straight forward. Therefore, the wavefront remains plane.

When a finite plane wavefront is considered, as in Figure 7-8b, cancellation in random directions is incomplete. Consequently, the wavefront spreads out or scatters. This scattering effect is called diffraction. Figure 7-8c shows diffraction around the edge of an obstacle. It can be seen that wavelet cancellation occurs only partially. Diffraction occurs around the edge of the obstacle, which allows secondary waves to "sneak" around the corner of the obstacle into what is called the *shadow zone*. For example, this phenomenon can be observed when a door is opened into a dark room. Light rays diffract around the edge of the door and illuminate the area behind the door.

7-7-4 Interference

Radio wave *interference* occurs when two or more electromagnetic waves combine in such a way that system performance is degraded. Refraction, reflection, and diffraction are categorized as geometric optics, which means that their behavior is analyzed primarily in terms of rays and wavefronts. Interference, on the other hand, is subject to the principle of *linear superposition* of electromagnetic waves and occurs whenever two or more waves simultaneously occupy the same point in space.

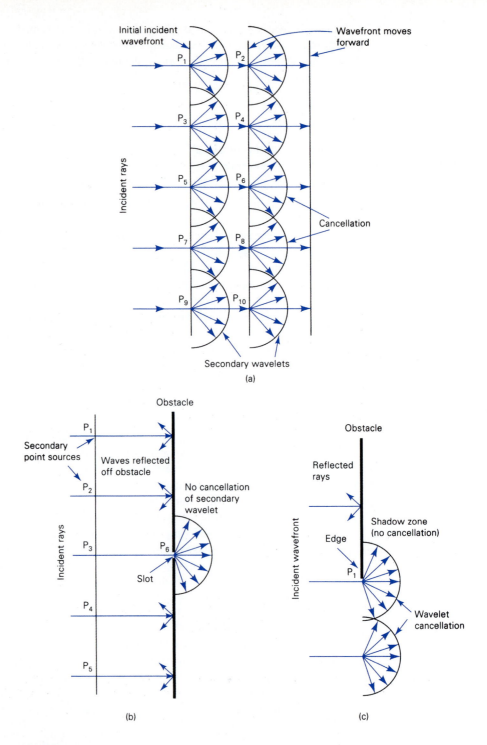

FIGURE 7-8 Electromagnetic wave diffraction: (a) Huygen's principle for a plane wavefront; (b) finite wavefront through a slot; (c) around an edge

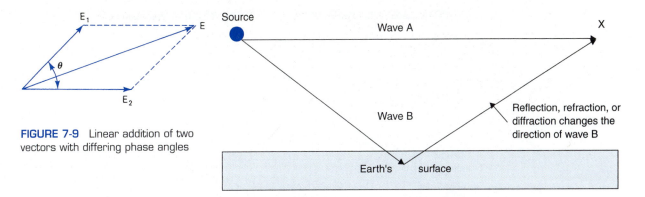

FIGURE 7-9 Linear addition of two vectors with differing phase angles

FIGURE 7-10 Electromagnetic wave interference

Figure 7-9 shows the linear addition of two instantaneous voltage vectors whose phase angles differ by angle θ. It can be seen that the total voltage is not simply the sum of the two vector magnitudes but rather the phasor addition of the two. Depending on the phase angles of the two vectors, either addition or subtraction can occur.

Figure 7-10 shows interference between two electromagnetic waves in free space. It can be seen that at point X the two waves occupy the same area in space. However, wave B has traveled a different path than wave A and, therefore, their relative phase angles may be different. If the difference in distance traveled is an odd-integral multiple of one-half wavelength, reinforcement takes place. If the difference is an even-integral multiple of one-half wavelength, total cancellation occurs. More than likely, the difference in distance falls somewhere between the two, and partial cancellation occurs.

7-8 TERRESTRIAL PROPAGATION OF ELECTROMAGNETIC WAVES

Electromagnetic radio waves traveling within earth's atmosphere are called *terrestrial waves*, and communications between two or more points on earth is called *terrestrial radio communications*. Earth's atmosphere and earth itself influence terrestrial waves. In terrestrial radio communications, waves can be propagated between points in several ways, depending on the type of system and the environment. Electromagnetic waves travel in straight lines except when earth and its atmosphere alter their path. Essentially, there are three modes of propagating electromagnetic waves within earth's atmosphere: ground wave propagation, space wave propagation, and sky wave propagation. Figure 7-11 shows the three modes of terrestrial propagation.

7-8-1 Ground Wave Propagation

Ground waves are electromagnetic waves that travel along the surface of earth. Therefore, ground waves are sometimes called *surface waves*. Ground waves must be vertically polarized because the electric field in a horizontally polarized wave is parallel to earth's surface and short-circuited by the conductivity of the ground. With ground waves, the changing electric field induces voltages in earth's surface, which cause currents to flow that are very similar to those in a transmission line. Earth's surface also has resistance and dielectric losses. Therefore, ground waves are attenuated as they propagate. Ground waves propagate best over a surface that is a good conductor, such as salt water, and poorly over dry desert areas. Ground wave losses increase rapidly with frequency; therefore, ground wave propagation is generally limited to frequencies below 2 MHz.

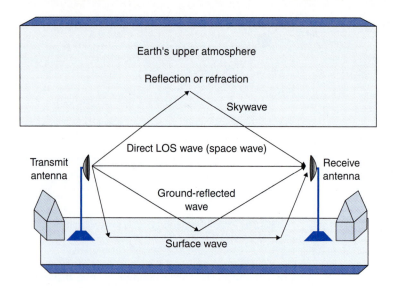

FIGURE 7-11 Normal modes of wave propagation

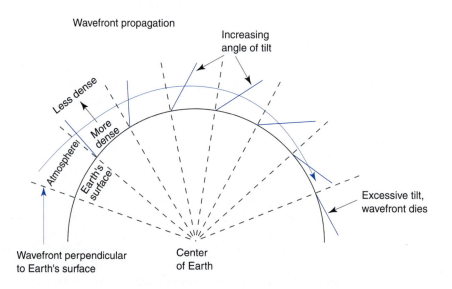

FIGURE 7-12 Surface (ground) wave propagation

Figure 7-12 shows ground wave propagation. Earth's atmosphere has a gradient density (i.e., the density decreases gradually with distance from earth's surface), which causes the wavefront to tilt progressively forward. Therefore, the ground wave propagates around earth, remaining close to its surface, and if enough power is transmitted, the wavefront could propagate beyond the horizon or even around the entire circumference of earth. However, care must be taken when selecting the frequency and the terrain over which the ground wave will propagate to ensure that the wavefront does not tilt excessively and simply turn over, lie flat on the ground, and cease to propagate.

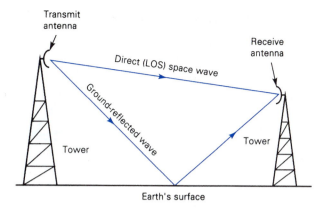

FIGURE 7-13 Space wave propagation

Ground wave propagation is commonly used for ship-to-ship and ship-to-shore communications, for radio navigation, and for maritime mobile communications.

7-8-2 Space Wave Propagation

Space wave propagation of electromagnetic energy includes radiated energy that travels in the lower few miles of earth's atmosphere. Space waves include both direct and ground-reflected waves (see Figure 7-13). Direct waves travel essentially in a straight line between transmit and receive antennas. Space wave propagation with direct waves is commonly called *line-of-sight* (LOS) *transmission*. Therefore, direct space wave propagation is limited by the curvature of the earth. *Ground-reflected waves* are waves reflected by earth's surface as they propagate between transmit and receive antennas.

Figure 7-13 shows space wave propagation between two antennas. It can be seen that the field intensity at the receive antenna depends on the distance between the two antennas (attenuation and absorption) and whether the direct and ground-reflected waves are in phase (interference).

The curvature of earth presents a horizon to space wave propagation commonly called the *radio horizon*. Because of atmospheric refraction, the radio horizon extends beyond the optical horizon for the common standard atmosphere. The radio horizon is approximately four-thirds that of the optical horizon. Refraction is caused by the troposphere because of changes in its density, temperature, water vapor content, and relative conductivity. The radio horizon can be lengthened by simply elevating the transmit or receive antennas (or both) above earth's surface with towers or by placing the antennas on top of mountains or tall buildings.

Because the conditions in earth's lower atmosphere are subject to change, the degree of refraction can vary with time. A special condition called *duct propagation* occurs when the density of the lower atmosphere is such that electromagnetic waves can propagate within the duct for great distances, causing them to propagate around earth following its natural curvature.

7-8-3 Sky Wave Propagation

Electromagnetic waves that are directed above the horizon level are called *sky waves*. Typically, sky waves are radiated in a direction that produces a relatively large angle with reference to earth. Sky waves are radiated toward the sky, where they are either reflected or refracted back to earth by the *ionosphere*. Because of this, sky wave propagation is

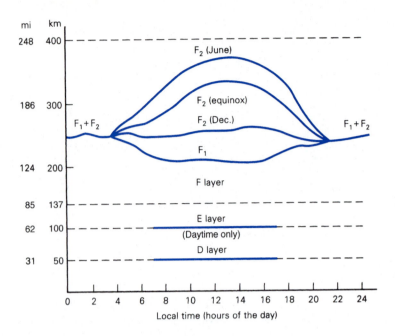

FIGURE 7-14 Ionospheric layers

sometimes called *ionospheric propagation*. The ionosphere is the region of space located approximately 50 km to 400 km (31 mi to 248 mi) above earth's surface.

The ionosphere is the upper portion of earth's atmosphere. Therefore, it absorbs large quantities of the sun's radiant energy, which ionizes the air molecules. The upper atmosphere has a higher percentage of ionized molecules than the lower atmosphere. The higher the ion density, the more the refraction. Also, because of the ionosphere's nonuniform composition and its temperature and density variations, it is *stratified*. Essentially, three layers make up the ionosphere (the D, E, and F layers) and are shown in Figure 7-14. It can be seen that all three layers of the ionosphere vary in location and in *ionization density* with the time of day. They also fluctuate in a cyclic pattern throughout the year and in accordance with the 11-year *sunspot cycle*. The ionosphere is most dense during times of maximum sunlight (i.e., during the daylight hours in the summer). Because the density and location of the ionosphere vary over time, the effects it has on electromagnetic radio wave propagation also vary.

7-9 SKIP DISTANCE

Skip distance is the minimum distance from a transmit antenna that a sky wave of given frequency will be returned to earth. Figure 7-15a shows several rays with different elevation angles being radiated from the same point on earth. It can be seen that the location where the wave is returned to earth moves closer to the transmitter as the elevation angle (ϕ) increases. Eventually, the angle of elevation is sufficiently high that the wave penetrates the ionosphere and continues into space totally escaping earth's atmosphere.

Figure 7-15b shows the effect on the skip distance of the disappearance of the D and E layers of the ionosphere during nighttime. Effectively, the ceiling formed by the ionosphere is raised, allowing sky waves to travel higher before being refracted back to earth. This effect explains how, when I was a young boy growing up in Oriskany, New York, I could listen to WCFL in Chicago broadcasting the White Sox baseball games.

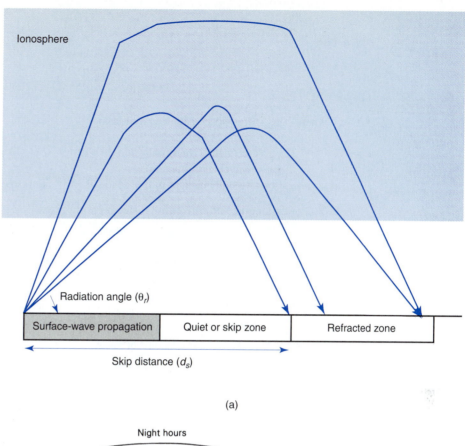

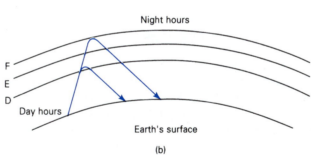

FIGURE 7-15 (a) Skip distance; (b) daytime-versus-nighttime propagation

7-10 FREE-SPACE PATH LOSS

Free-space path loss is often defined as the loss incurred by an electromagnetic wave as it propagates in a straight line through a vacuum with no absorption or reflection of energy from nearby objects. This is a misstated and often misleading definition. Free-space path loss is a fabricated engineering quantity that evolved from manipulating communications system link budget equations into a particular format. With free-space path loss, no electromagnetic energy is actually lost—it merely spreads out as it propagates away from the source resulting in a lower power density. A more appropriate term for the phenomenon is *spreading loss*. Spreading loss occurs simply because of the inverse square law. Spreading loss is a function of distance from the source and the wavelength (frequency) of the electromagnetic wave. Mathematically, free-space path loss is

$$L_p = \left(\frac{4\pi D}{\lambda} \right)^2 \qquad \text{(7-6a)}$$

$$= \left(\frac{4\pi f D}{c} \right)^2 \qquad \text{(7-6b)}$$

where L_p = free-space path loss (unitless)
D = distance (kilometers)
f = frequency (hertz)
λ = wavelength (meters)
c = velocity of light in free space (3×10^8 meters per second)

Rearranging and converting to decibel form gives

$$L_{p\text{(dB)}} = 32.4 + 20 \log f_{\text{(MHz)}} + 20 \log D_{\text{(km)}} \qquad \text{(7-6c)}$$

From Equations 7-6a, b, and c, it can be seen that free-space path loss increases with both frequency and distance.

Example 7-3

Determine the free-space path loss for a frequency of 6 GHz traveling a distance of 50 km.

Solution   Substituting into Equation 7-6c yields

$$L_{p\text{(dB)}} = 32.4 + 20 \log 6000 + 20 \log 50$$
$$L_{p\text{(dB)}} = 32.4 + 75.6 + 34$$
$$= 142 \text{ dB}$$

7-11   MICROWAVE COMMUNICATIONS SYSTEMS

Microwaves are generally described as electromagnetic waves with frequencies that range from approximately 500 MHz to 300 GHz. Therefore, microwave signals, because of their inherently high frequencies, have relatively short wavelengths. For example, a 100-GHz microwave signal has a wavelength of 0.3 cm as compared to a 10-MHz signal, which has a wavelength of 30 meters. Table 7-1 lists some of the microwave radio frequency bands available in the United States, and Table 7-2 lists several of the microwave radio frequency assignments.

The vast majority of the communications systems established since the mid-1980s are digital in nature and, thus, carry information in digital form. However, terrestrial (earth-

Table 7-1   Microwave Radio-Frequency Bands

Band Designation	Frequency Range (GHz)
L	1–2
S	2–4
C	4–8
X	8–12
K_u	12–18
K	18–27
K_a	27–40
Millimeter	40–300
Submillimeter	>300

Table 7-2 Microwave Radio-Frequency Assignments

Service	Frequency (MHz)	Band
Military	1710–1850	L
Operational fixed	1850–1990	L
Studio transmitter link	1990–2110	L
Common carrier	2110–2130	S
Operational fixed	2130–2150	S
Operational carrier	2160–2180	S
Operational fixed	2180–2200	S
Operational fixed television	2500–2690	S
Common carrier and satellite downlink	3700–4200	S
Military	4400–4990	C
Military	5250–5350	C
Common carrier and satellite uplink	5925–6425	C
Operational fixed	6575–6875	C
Studio transmitter link	6875–7125	C
Common carrier and satellite downlink	7250–7750	C
Common carrier and satellite uplink	7900–8400	X
Common carrier	10,700–11,700	X
Operational fixed	12,200–12,700	X
Cable television (CATV) studio link	12,700–12,950	Ku
Studio transmitter link	12,950–13,200	Ku
Military	14,400–15,250	Ka
Common carrier	17,700–19,300	Ka
Satellite uplink	26,000–32,000	K
Satellite downlink	39,000–42,000	Q
Satellite crosslink	50,000–51,000	V
Satellite crosslink	54,000–62,000	V

based) *microwave radio relay systems* using frequency modulation (FM) or digitally modulated carriers (phase-shift keying [PSK] or quadrative amplitude modulation [QAM]) still provide approximately 35% of the total information-carrying circuit mileage in the United States. Microwave systems are used for carrying long-distance voice telephone service, metropolitan area networks, wide area networks, and the Internet.

There are many different types of microwave systems operating over distances that vary from 15 miles to 4000 miles in length. *Intrastate* or *feeder service* microwave systems are generally classified as *short haul* because they are used to carry information for relatively short distances, such as between cities within the same state. *Long-haul* microwave systems are those used to carry information for relatively long distances, such as *interstate* and *backbone* route applications. Microwave radio system capacities range from less than 12 voice-grade telephone circuits to more than 22,000 voice and data channels.

Figure 7-16 shows a typical layout for a terrestrial microwave system showing two terminal stations and two repeater stations. A terminal station is one in which information can originate or terminate, whereas a repeater is simply used to receive information from one station and retransmit it to another station.

7-11-1 Advantages of Microwave Radio Communications

Microwave radios propagate signals through earth's atmosphere between transmitters and receivers often located on top of towers spaced about 15 miles to 30 miles apart. Therefore, microwave radio systems have the obvious advantage of having the capacity to carry thousands of individual information channels between two points without the need for physical facilities, such as metallic cables or optical fibers. This avoids the need for acquiring right-of-ways through private property. In addition, radio waves are better suited for spanning

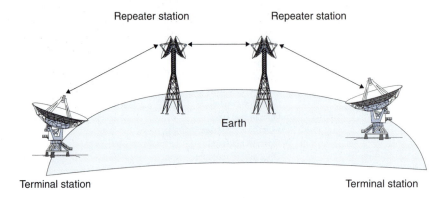

Repeater station   Repeater station

Earth

Terminal station   Terminal station

FIGURE 7-16   Typical terrestrial microwave system

large bodies of water, going over high mountains, or going through heavily wooded terrain that impose formidable barriers to physical cable systems. The advantages of microwave radio systems over cable facilities include the following:

1. Radio systems do not require a right-of-way acquisition between stations.
2. Each station requires the purchase or lease of only a small area of land.
3. Because of their high operating frequencies, microwave radio systems can carry large quantities of information.
4. High frequencies mean short wavelengths, which require relatively small antennas.
5. Radio signals are more easily propagated around physical obstacles, such as water and high mountains.
6. Microwave systems require fewer repeaters for amplification.
7. Distances between switching centers are less.
8. Underground facilities are minimized.
9. Minimum delay times are introduced.
10. Minimal crosstalk exists between voice channels.

Microwave radio systems also have several disadvantages, including the following:

1. The electronic circuits used with microwave frequencies are more difficult to analyze.
2. Conventional components, such as resistors, inductors, and capacitors, are more difficult to manufacture and implement at microwave frequencies.
3. Microwave components are more expensive.
4. Transistor transit time is a problem with microwave devices.
5. Signal amplification is more difficult with microwave frequencies.

7-11-2 Microwave Radio Link

Figure 7-17 shows the block diagram for a simplex (one-way only) microwave radio link comprised of a transmitter and a receiver separated by a distance of up to 40 miles. Full-duplex operation would require a duplicate set of equipment for the opposite direction.

A simplified block diagram for a microwave transmitter is shown in Figure 7-17a. The transmitter includes a modulator, mixer, and microwave generator as well as several stages of amplifications and filtering. The modulator may perform frequency modulation (FM) or some form of digital modulation such as PSK or QAM. The output of the modulator is an intermediate frequency (IF) carrier that has been modulated or encoded by the baseband input signal. The baseband signal is simply the information, which can be

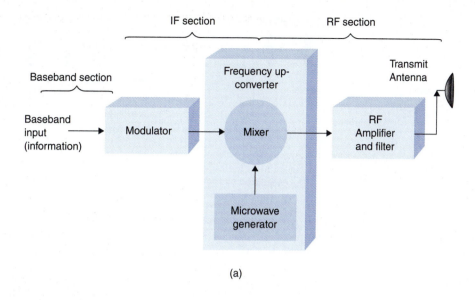

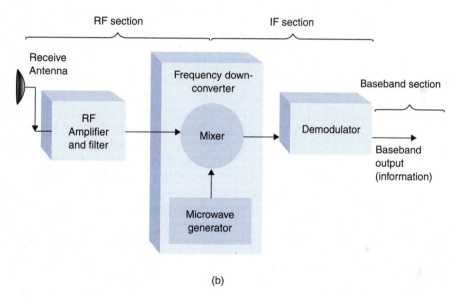

FIGURE 7-17 Simplified microwave radio link: (a) transmitter; (b) receiver

frequency-division multiplexed analog channels, time-division multiplexed digital voice channels, video, or high-speed digital data channels. The mixer and microwave generator (oscillator) combine to perform frequency up-conversion through nonlinear mixing (heterodyning). The purpose of the up-converter is to translate IF frequencies to RF microwave frequencies. Typical IF frequencies range between 60 MHz and 80 MHz, and typical RF frequencies range between 2 GHz and 18 GHz.

A simplified block diagram for a microwave receiver is shown in Figure 7-17b. A microwave receiver is comprised of a radio-frequency (RF) amplifier, a frequency down-converter (mixer and microwave generator), and a demodulator. The RF amplifier and filter increase the received signal level so that the down-converter can convert the RF signals to IF signals. The demodulator can be for FM, PSK, or QAM. The output of the demodulator is the original baseband (information) signals.

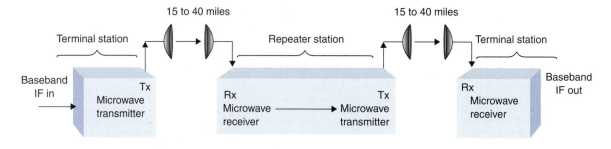

FIGURE 7-18 Microwave repeater

7-11-3 Microwave Radio Repeaters

The permissible distance between a microwave transmitter and receiver depends on several variables, such as transmitter output power level, receive noise threshold, terrain, atmospheric conditions, system capacity, reliability objectives, and performance expectations. Typically, the maximum distance is between 15 miles and 40 miles. Long-haul microwave systems span distances considerably longer than this. Consequently, a single-hop microwave system, such as the one shown in Figure 7-17, is inadequate for most practical system applications. With systems longer than 40 miles or when geographical obstructions, such as a mountain, block the transmission path, *repeaters* are needed. A microwave repeater is a receiver and a transmitter placed back to back in the system.

A simplified block diagram for a microwave repeater is shown in Figure 7-18. The repeater station receives a signal, amplifies and reshapes it, and then retransmits it to the next repeater or terminal station down line from it. A terminal station is simply a station at the end of a microwave system where information signals originate and terminate.

Microwave radio systems span distances from as short as a few miles up to several thousand miles, necessitating hundreds of repeater stations. The location of repeater sites is greatly influenced by the nature of the terrain between the surrounding sites. Preliminary route planning generally assumes relatively flat areas, and path (hop) lengths will average between 15 miles and 40 miles between stations. In relatively flat terrain, increasing path length will dictate increasing transmit powers. The exact distance between stations is determined primarily by line-of-sight path clearance and received signal strength. For frequencies above 10 GHz, local rainfall patterns could also have a significant bearing on path length.

7-12 SATELLITE COMMUNICATIONS SYSTEMS

In astronomical terms, a *satellite* is a celestial body that orbits around a planet (e.g., the moon is a satellite of earth). However, in aerospace terms, a satellite is a space vehicle launched by humans that orbits earth or another celestial body. Communications satellites are man-made satellites that orbit earth, providing a multitude of communications services to a wide variety of consumers, including military, governmental, private, and commercial subscribers.

Figure 7-19 shows a typical layout for a satellite microwave system showing two earth stations and one satellite. In essence, a communications satellite is a microwave repeater in the sky. The purpose of most communications satellites is to relay signals between two or more earth stations. Satellites use essentially the same frequencies as microwave systems. A satellite repeater is called a *transponder*, and a satellite may have many transponders. A satellite system consists of one or more satellite space vehicles, a ground-based control sta-

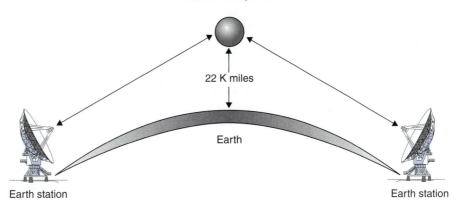

Satellite transponder

22 K miles

Earth

Earth station

Earth station

FIGURE 7-19   Typical satellite microwave system

Table 7-3   Satellite Radio-Frequency Bands

Band Designation	Frequency Range
P	225–490 MHz
J	350–530 MHz
L	1530–2700 MHz
S	2500–2700 MHz
C	3400–6425 MHz
X	7250–8400 MHz
K_u	10.95–14.5 GHz
K_c	17.7–21.1 GHz
K	27.5–31 GHz
Q	36–46 GHz
V	56–56 GHz
W	56–100 GHz

tion, and a network of earth station users that provide the interface facilities for the transmission, reception, and processing of terrestrial communications traffic through the satellite system. Transmissions to and from satellites are categorized as either *bus* or *payload*. The bus includes control mechanisms that support the payload operation. The payload is the actual user information.

Although there are many types of satellite systems, the most prevalent systems are used for communications, surveillance, weather, and navigation. Communications satellites are used extensively by the government, the military, and commercial communications companies for transferring voice, data, and video information between users located all over the world. Weather and surveillance satellites are used primarily by government and military agencies, while navigation satellites are used by virtually everyone, including the government, the military, civilians, and commercial companies. Satellites utilize many of the same frequency bands as terrestrial microwave radio systems. Table 7-3 lists the frequency bands commonly used for satellite systems.

7-12-1   Satellite Elevation Categories

Satellites are generally classified as having a low earth orbit (LEO), medium earth orbit (MEO), or geosynchronous earth orbit (GEO). Most LEO satellites operate in the 1.0-GHz to 2.5-GHz frequency range. Motorola's satellite-based mobile telephone system, Iridium, is an LEO system utilizing a 67-satellite constellation orbiting approximately 480 miles above earth's surface. The main advantage of LEO satellites is that the path loss between earth stations and space vehicles is much lower than for satellites revolving in medium- or high-altitude orbits. Less path loss equates to lower transmit powers, smaller antennas, and less weight.

MEO satellites operate in the 1.2-GHz to 1.67-GHz frequency band and orbit between 6000 miles and 12,000 miles above Earth. The Department of Defense's satellite-based global positioning system, NAVSTAR, is a MEO system with a constellation of 21 working satellites and six or more spares orbiting approximately 9500 miles above earth.

Geosynchronous satellites are high-altitude earth-orbit satellites operating primarily in the 2-GHz to 18-GHz frequency spectrum with orbits 22,300 miles above earth's surface. Geosynchronous, or *geostationary*, satellites are those that orbit in a circular pattern with an angular velocity equal to that of earth. Geostationary satellites have an orbital time of approximately 24 hours, the same as earth; thus, geosynchronous satellites appear to be stationary as they remain in a fixed position in respect to a given point on earth. Satellites in high-elevation, nonsynchronous circular orbits between 19,000 and 25,000 miles above earth are said to be in *near-synchronous* orbit.

7-12-2 Satellite Orbits and Orbital Patterns

Satellites can be generally classified as either *synchronous* or *nonsynchronous*. Synchronous satellites orbit earth above the equator with the same angular velocity as earth. Therefore, synchronous satellites appear to be stationary and remain in the same location with respect to a given point on earth. Most of the satellites used for commercial voice and data communications are synchronous. Synchronous satellites are discussed in more detail in a later section of this chapter.

Nonsynchronous satellites rotate around earth in circular or elliptical patterns as shown in Figures 7-20a and b, respectively. In a circular orbit, the speed or rotation is constant; however, in elliptical orbits, the speed depends on the altitude of the satellite. The velocity of a satellite in an elliptical orbit is greatest when the satellite is closest to earth. The point in an elliptical orbit located farthest from earth is called the *apogee*, and the point in an elliptical orbit located closest to earth is called the *perigee*.

If a satellite is orbiting in the same direction as earth's rotation (counterclockwise) and at an angular velocity greater than that of earth ($\omega_s > \omega_e$), the orbit is called a *prograde* or *posigrade* orbit. If the satellite is orbiting in the opposite direction as earth's rotation or in the same direction with an angular velocity less than that of earth ($\omega_s < \omega_e$), the orbit is called a *retrograde* orbit. Most nonsynchronous satellites revolve around earth in a prograde orbit. Therefore, the position of satellites in nonsynchronous orbits is continuously changing in respect to a fixed position on earth. Consequently, nonsynchronous satellites have to be used when available, which may be as little as 15 minutes per orbit. Another disadvantage of orbital satellites is the need for complicated and expensive tracking equipment at the earth stations so that they can locate the satellite as it comes into view on each orbit and then lock its antenna onto the satellite and track it as it passes overhead.

Although there are an infinite number of orbital paths possible, only three are useful for communications satellites. Figure 7-21 shows three paths a satellite can follow as it rotates around earth: inclined, equatorial, or polar. All satellites rotate around earth in an orbit that forms a plane that passes through the center of gravity of earth called the *geocenter*.

Inclined orbits are virtually all orbits except those that travel directly above the equator or directly above the North and South Poles. Figure 7-22a shows the *angle of inclination* of a satellite orbit. The angle of inclination is the angle between the earth's equatorial plane and the orbital plane of a satellite measured counterclockwise at the point in the or-

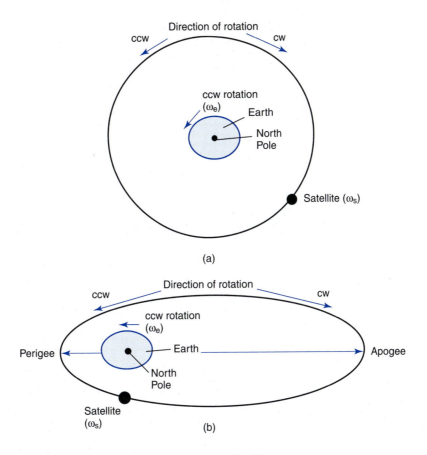

FIGURE 7-20   Satellite orbits: (a) circular; (b) elliptical

bit where it crosses the equatorial plane traveling from south to north. This point is called the *ascending node* and is shown in Figure 7-22b. The point where a polar or inclined or-bit crosses the equatorial plane traveling from north to south is called the *descending node*, and the line joining the ascending and descending nodes through the center of earth is called the *line of nodes*. Angles of inclination vary between 0 degrees and 90 degrees. To provide coverage to regions of high latitudes, inclined orbits are generally elliptical.

As shown in Figure 7-21, an *equatorial orbit* is when the satellite rotates in an orbit directly above the equator, usually in a circular path. With an equatorial orbit, the angle of inclination is 0 degrees, and there are no ascending or descending nodes and, hence, no line of nodes. All geosynchronous satellites are in equatorial orbits.

A *polar orbit* is when the satellite rotates in a path that takes it over the North and South Poles in an orbital pattern that is perpendicular to the equatorial plane. Polar orbiting satellites follow a low-altitude path close to earth's surface, passing over and very close to both the North and South Poles. The angle of inclination of a satellite in a polar orbit is nearly 90 degrees. It is interesting to note that 100% of earth's surface can be covered with a single satellite in a polar orbit. Satellites in polar orbits rotate around earth in a longitu-dinal orbit while earth is rotating on its axis in a latitudinal rotation. Consequently, the satel-lite's area of coverage is a diagonal area that forms a spiral pattern around the surface of earth resembling a barber pole. As a result, every location on earth lines within the area of coverage of a satellite in a polar orbit twice each day.

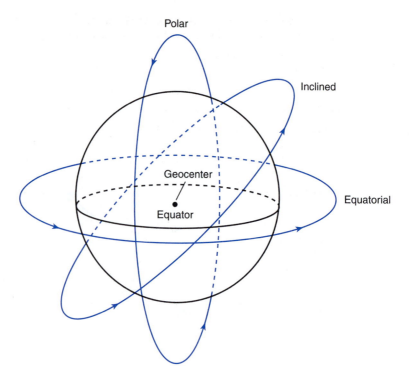

FIGURE 7-21   Satellite orbital patterns

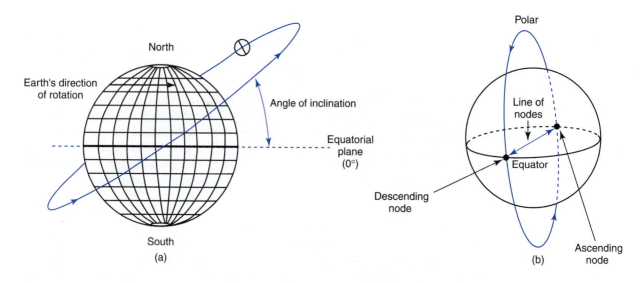

FIGURE 7-22   (a) Angle of inclination; (b) ascending node, descending node, and line of nodes

7-12-3   Geosynchronous Satellites

Since geosynchronous satellites orbit earth above the equator with the same angular veloc-ity as earth, they appear to remain in a fixed location above one spot on earth's surface. Therefore, no special antenna tracking equipment is necessary—earth station antennas are simply pointed at the satellite. A single high-altitude geosynchronous satellite can provide reliable communications to approximately 40% of the earth's surface.

Geosynchronous orbits are circular; therefore, the speed of rotation is constant throughout the orbit. There is only one geosynchronous earth orbit, and this orbit is occupied by a large number of satellites. In fact, geosynchronous orbits are the most widely used earth orbit.

7-12-3-1 Geosynchronous orbit requirements.

There are several requirements for satellites in geosynchronous orbits. The first and most obvious is that geosynchronous orbits must have a 0-degree angle of elevation (i.e., the space vehicle must be orbiting directly above earth's equatorial plane). Geosynchronous satellites must also orbit in the same direction as earth's rotation (eastward, toward the morning sun) with the same angular (rotational) velocity: one revolution per day. Using Kepler's third law, it can also be shown that geosynchronous satellites must revolve around earth in a circular pattern 42,164 kilometers from the center of earth. Because earth's equatorial radius is approximately 6378 kilometers, the height above mean sea level of a geosynchronous satellite is 35,768 km, or approximately 22,300 miles, above earth's surface. Therefore, the circumference (C) of a geosynchronous satellite orbit is

$$C = 2\pi(42,164 \text{ km})$$

$$= 264,790 \text{ km}$$

and the velocity (v) of a geosynchronous satellite is

$$v = \frac{264,790 \text{ km}}{24 \text{ hr}}$$

$$= 11,033 \text{ km/hr}$$

$$= 6840 \text{ mph}$$

7-12-3-2 Clarke orbit.

A geosynchronous earth orbit is sometimes referred to as the *Clarke orbit* or *Clarke belt*, after Arthur C. Clarke, who first suggested its existence in 1945 and proposed its use for communications satellites. Clarke was an engineer, a scientist, and a science fiction author who is probably best remembered for writing *2001: A Space Odyssey*. The Clarke orbit meets the concise setoff specifications for geosynchronous satellite orbits: (1) be located directly above the equator, (2) travel in the same direction as earth's rotation with a velocity of 6840 miles per hour, (3) have an altitude of 22,300 miles above earth, and (4) complete one revolution in 24 hours. As shown in Figure 7-23, three satellites in Clarke orbits separated by 120 degrees in longitude can provide communications over the entire globe except the polar regions.

7-12-3-3 Geosynchronous satellite summary.

The advantages and disadvantages of geosynchronous satellites are summarized here.

Advantages:
1. Geosynchronous satellites remain almost stationary in respect to a given earth station; therefore, expensive tracking equipment is not required at the earth stations.
2. Geosynchronous satellites are available to all earth stations within their *shadow* 100% of the time. The shadow of a satellite includes all the earth stations that have a line-of-sight path to the satellite.
3. Switching from one geosynchronous satellite to another as they orbit overhead is not necessary. Consequently, there are no transmission breaks due to switching times.

Disadvantages:
1. An obvious disadvantage of geosynchronous satellites is they require sophisticated and heavy propulsion devices on board to keep them in a fixed orbit.

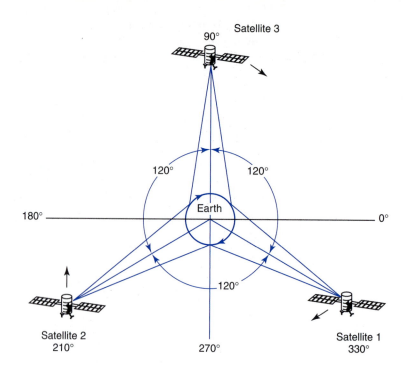

FIGURE 7-23 Three geosynchronous satellites in Clarke orbits

2. High-altitude geosynchronous satellites introduce much longer propagation delays. The round-trip propagation delay between two earth stations through a geosynchronous satellite is typically between 500 ms and 600 ms.
3. Geosynchronous satellites require higher transmit power levels and more sensitive receivers because of the longer distances and greater path losses.
4. High-precision spacemanship is required to place a geosynchronous satellite into orbit and to keep it there.

7-12-4 Satellite Look Angles

To optimize the performance of a satellite communications system, an earth station antenna must be pointed directly at the satellite. To ensure that the earth station antenna is aligned, two angles must be determined: the *azimuth* and the *elevation angle*. Azimuth and elevation angle are jointly referred to as the *look angles*. With geosynchronous satellites, the look angles of earth station antennas need to be adjusted only once, as the satellite will remain in a given position permanently, except for minor variations.

The location of a satellite is generally specified in terms of latitude and longitude similar to the way the location of a point on earth is described. However, because a satellite is orbiting many miles above the earth's surface, it has no latitude or longitude. Therefore, a point on the surface of earth directly below the satellite is used to identify its location. This point is called the *subsatellite point* (SSP), and for geosynchronous satellites the SSP must fall on the equator. Subsatellite points and earth station locations are specified using standard latitude and longitude coordinates. The standard convention specifies angles of longitude between 0 degrees and 180 degrees either east or west of the Greenwich prime meridian. Latitudes in the Northern Hemisphere are angles between 0° and 90°N and latitudes in the Southern Hemisphere are angles between 0° and 90°S. Since geosynchro-

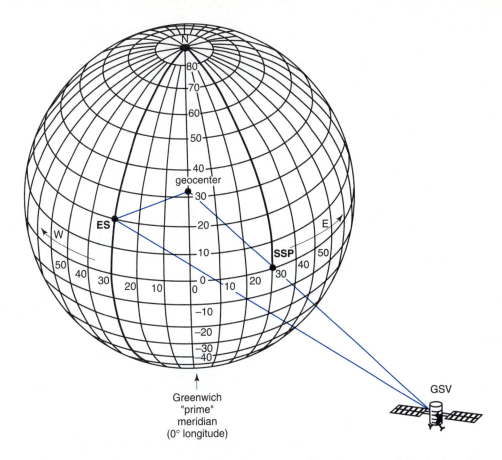

FIGURE 7-24 Geosynchronous satellite position, subsatellite point, and earth longitude and latitude coordinate system

nous satellites are all located directly above the equator, they all have a 0-degree latitude. Hence, geosynchronous satellite locations are normally given in degrees longitude east or west of the Greenwich meridian (e.g., 122°W or 78°E).

Figure 7-24 shows the position of a hypothetical geosynchronous satellite vehicle (GSV), its respective subsatellite point (SSP), and an arbitrarily selected earth station (ES) all relative to earth's geocenter. The SSP for the satellite shown is 30°E longitude and 0° latitude. The earth station has a location of 30°W longitude and 20°N latitude. Figure 7-25 shows the positions of several satellites in geosynchronous orbit.

7-12-5   Satellite Antenna Radiation Patterns: Footprints

The area of earth covered by a satellite depends on the location of the satellite in its orbit, its carrier frequency, and the directivity of its antenna. Satellite engineers select the antenna and carrier frequency for a particular spacecraft to concentrate the limited transmit power on a specific area of earth's surface. The geographical representation of the area on earth illuminated by the radiation from a satellite's antenna is called a *footprint* or sometimes a *footprint map*. In essence, a footprint of a satellite is the area on earth's surface that the satellite can receive from or transmit to. The shape of a satellite's footprint depends on the satellite's orbital path, height, and the type of antenna used. The higher the satellite, the more of the earth's surface it can cover. A typical satellite footprint is shown in Figure 7-26. The contour lines represent limits of equal receive power density.

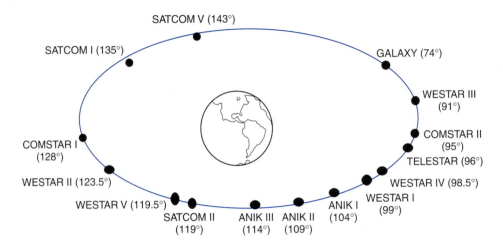

FIGURE 7-25 Satellites in geosynchronous earth orbits

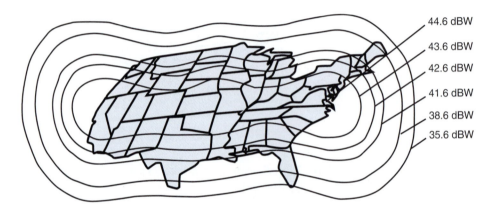

FIGURE 7-26 Satellite antenna radiation patterns (footprints)

The radiation pattern from a satellite's antenna is sometimes called a *beam* (see Figure 7-27). The smallest and most directive beam is called a *spot beam*, followed by *zonal beams*, *hemispherical beams*, and *earth (global) beams*. Spot beams concentrate their power to a very small geographical area. Spot and zonal beams blanket less than 10% of the earth's surface. Hemispherical beams typically target up to 20% of earth's surface, while global beams have beam widths of approximately 17° and are capable of covering approximately 42% of earth's surface, which is the maximum view of any one geosynchronous satellite.

7-12-6 Satellite Multiple-Accessing Arrangements

Satellite *multiple accessing* (sometimes called *multiple destination*) implies that more than one user has access to one or more transponders within a satellite's bandwidth allocation. Transponders are typically leased from a common carrier for the purpose of providing voice or data transmission to a multitude of users. The method by which a satellite transponder's bandwidth is used or accessed depends on the multiple-accessing method utilized.

Figure 7-28 illustrates the three most commonly used multiple-accessing arrangements: *frequency-division multiple accessing* (FDMA), *time-division multiple accessing* (TDMA), and *code-division multiple accessing* (CDMA).

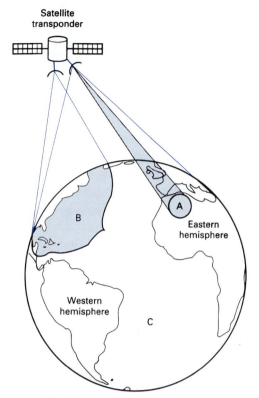

FIGURE 7-27 Beams: (a) spot; (b) zonal; (c) earth

7-12-6-1 Frequency-division multiple accessing. *Frequency-division multiple accessing* (FDMA) is a method of multiple accessing where a given RF bandwidth is divided into smaller frequency bands called *subdivisions*. Consequently, FDMA transmissions are separated in the frequency domain and, therefore, must share the total available transponder bandwidth as well as the total transponder power. A control mechanism is used to ensure that two or more earth stations do not transmit in the same subdivision at the same time. Essentially, the control mechanism designates a receive station for each of the subdivisions. Thus, with FDMA, transmission can occur from more than one station at the same time, but the transmitting stations must share the allocated power, and no two stations can utilize the same bandwidth.

7-12-6-2 Time-division multiple accessing. *Time-division multiple accessing* (TDMA) is the predominant multiple-accessing method used today. TDMA is a method of time-division multiplexing digitally modulated carriers between participating earth stations within a satellite network using a common satellite transponder. With TDMA, each earth station transmits a short burst of information during a specific time slot within a TDMA frame. The bursts must be synchronized so that each station's burst arrives at the satellite at a different time, thus avoiding a collision with another station's carrier. TDMA transmissions are separated in the time domain, and with TDMA, the entire transponder bandwidth and power are used for each transmission but for only a prescribed interval of time. Thus, with TDMA, transmission cannot occur from more than one station at the same time. However, the transmitting station can use all the allocated power and the entire bandwidth during its assigned time slot.

7-12-6-3 Code-division multiple accessing. With *code-division multiple accessing* (CDMA), there are no restrictions on time or bandwidth. Because there are no limitations on bandwidth, CDMA is sometimes referred to as *spread-spectrum multiple*

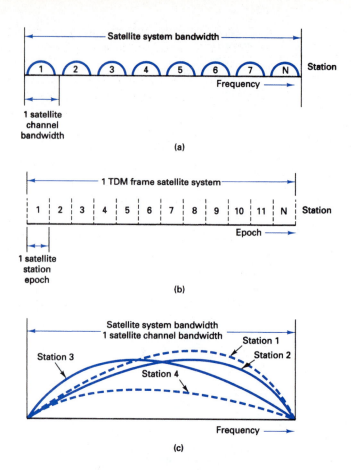

FIGURE 7-28 Multiple-accessing arrangements: (a) FDMA; (b) TDMA; (c) CDMA

accessing (SSMA). With CDMA, all earth stations transmit within the same frequency band and, for all practical purposes, have no limitations on when they may transmit or on which carrier frequency. Thus, with CDMA, the entire satellite transponder bandwidth is used by all stations on a continuous basis. Signal separation is accomplished with enve-lope encryption/decryption techniques.

QUESTIONS

7-1. Describe an *electromagnetic polarization*.

7-2. Describe an *electromagnetic ray*; a *wavefront*.

7-3. Describe *power density*.

7-4. Describe a *spherical wavefront*.

7-5. Explain the *inverse square law* and how it relates to electromagnetic waves.

7-6. Describe the difference between *wave attenuation* and *wave absorption*.

7-7. What are the *optical properties* of radio waves?

7-8. Describe the following terms and how they relate to radio-wave propagation: *refraction, re-flection, diffraction*, and *interference*.

7-9. Describe *ground wave propagation, space wave propagation*, and *sky wave propagation*.

7-10. Describe what is meant by the term *skip distance*.

7-11. Describe *free-space path loss*.

7-12. What is the approximate range for microwave frequencies?

7-13. List the advantages and disadvantages of *microwave radio communications* over *cable transmission facilities*.

7-14. What is the difference between a *terminal station* and a *repeater station*?

7-15. List and give examples of the three *satellite elevation categories*.

7-16. List and describe the three *orbital patterns* used by satellites.

7-17. Describe a *geosynchronous satellite*.

7-18. What is the *Clarke orbit*?

7-19. List the advantages and disadvantages of *geosynchronous satellites*.

7-20. What is meant by the term *look angles*?

7-21. Describe a *satellite footprint*.

7-22. What is a *satellite multiple-accessing arrangement*?

7-23. List and describe the three forms of *satellite multiple-accessing arrangements*.

PROBLEMS

7-1. Determine the power density for a radiated power of 1000 W at a distance 20 km from an isotropic antenna.

7-2. Determine the power density for Problem 7-1 for a point 30 km from the antenna.

7-3. Determine the change in power density when the distance from a source increases by a factor of 2.

7-4. If the distance from a source is reduced to one-fourth its value, what effect does this have on the power density?

7-5. The power density at a point from a source is 0.001 µW, and the power density at another point is 0.00001 µW. Determine the attenuation in dB.

7-6. Determine the wave attenuation between two points with power densities $P_1 = 25 \ \mu W/m^2$ and $P_2 = 0.05 \ \mu W/m^2$.

7-7. Determine the reflection coefficient for a reflected power of 0.001 W and an incident power of 0.008 W.

7-8. For a frequency of 2 GHz and a distance of 40 km, determine the free-space path loss.

7-9. Determine the path loss for the following frequencies and distances:

f (MHz)	D (km)
400	0.5
800	0.6
3000	10
5000	5
8000	20
18,000	15

C H A P T E R 8

Telephone Instruments and Signals

CHAPTER OUTLINE

OBJECTIVES

- Define *communications* and *telecommunications*
- Define and describe *subscriber loop*
- Describe the operation and basic functions of a standard telephone set
- Explain the relationship among telephone sets, local loops, and central office switching machines
- Describe the block diagram of a telephone set
- Explain the function and basic operation of the following telephone set components: ringer circuit, on/off-hook circuit, equalizer circuit, speaker, microphone, hybrid network, and dialing circuit
- Describe basic telephone call procedures
- Define *call progress tones* and *signals*
- Describe the following terms: *dial tone, dual-tone multifrequency, multifrequency, dial pulses, station busy, equipment busy, ringing, ring-back,* and *receiver on/off hook*
- Describe the basic operation of a cordless telephone
- Define and explain the basic format of caller ID
- Describe the operation of electronic telephones
- Describe the basic principles of paging systems

Communications is the process of conveying information from one place to another. Communications requires a source of information, a transmitter, a receiver, a destination, and some form of transmission medium (connecting path) between the transmitter and the receiver. The transmission path may be quite short, as when two people are talking face to face with each other or when a computer is outputting information to a printer located in the same room. *Telecommunications* is long-distance communications (from the Greek word *tele* meaning "distant" or "afar"). Although the word "long" is an arbitrary term, it generally indicates that communications is taking place between a transmitter and a receiver that are too far apart to communicate effectively using only sound waves.

Although often taken for granted, the telephone is one of the most remarkable devices ever invented. To talk to someone, you simply pick up the phone and dial a few digits, and you are almost instantly connected with them. The telephone is one of the simplest devices ever developed, and the telephone connection has not changed in nearly a century. Therefore, a telephone manufactured in the 1920s will still work with today's intricate telephone system.

Although telephone systems were originally developed for conveying human speech information (voice), they are now also used extensively to transport data. This is accomplished using modems that operate within the same frequency band as human voice. Anyone who uses a telephone or a data modem on a telephone circuit is part of a global communications network called the *public telephone network* (PTN). Because the PTN interconnects subscribers through one or more switches, it is sometimes called the *public switched telephone network* (PSTN). The PTN is comprised of several very large corporations and hundreds of smaller independent companies jointly referred to as *Telco*.

The telephone system as we know it today began as an unlikely collaboration of two men with widely disparate personalities: Alexander Graham Bell and Thomas A. Watson. Bell, born in 1847 in Edinburgh, Scotland, emigrated to Ontario, Canada, in 1870, where he lived for only six months before moving to Boston, Massachusetts. Watson was born in a livery stable owned by his father in Salem, Massachusetts. The two met in 1874 and invented the telephone in 1876. On March 10, 1876, one week after his patent was allowed, Bell first succeeded in transmitting speech in his lab at 5 Exeter Place in Boston. At the time, Bell was 29 years old and Watson only 22. Bell's patent, number 174,465, has been called the most valuable ever issued.

The telephone system developed rapidly. In 1877, there were only six telephones in the world. By 1881, 3,000 telephones were producing revenues, and in 1883, there were over 133,000 telephones in the United States alone. Bell and Watson left the telephone business in 1881, as Watson put it, "in better hands." This proved to be a financial mistake, as the telephone company they left evolved into the telecommunications giant known officially as the American Telephone and Telegraph Company (AT&T). Because at one time AT&T owned most of the local operating companies, it was often referred to as the *Bell Telephone System* and sometimes simply as "*Ma Bell.*" By 1982, the Bell System grew to an unbelievable $155 billion in assets ($256 billion in today's dollars), with over one million employees and 100,000 vehicles. By comparison, in 1998, Microsoft's assets were approximately $10 billion.

AT&T once described the Bell System as "the world's most complicated machine." A telephone call could be made from any telephone in the United States to virtually any other telephone in the world using this machine. Although AT&T officially divested the Bell System on January 1, 1983, the telecommunications industry continued to grow at an unbelievable rate. Some estimate that more than 1.5 billion telephone sets are operating in the world today.

8-2 THE SUBSCRIBER LOOP

The simplest and most straightforward form of telephone service is called *plain old telephone service* (POTS), which involves subscribers accessing the public telephone network through a pair of wires called the *local subscriber loop* (or simply *local loop*). The local loop is the most fundamental component of a telephone circuit. A local loop is simply an unshielded twisted-pair transmission line (cable pair), consisting of two insulated conductors twisted together. The insulating material is generally a polyethylene plastic coating, and the conductor is most likely a pair of 116- to 26-gauge copper wire. A subscriber loop is generally comprised of several lengths of copper wire interconnected at junction and cross-connect boxes located in manholes, back alleys, or telephone equipment rooms within large buildings and building complexes.

The subscriber loop provides the means to connect a telephone set at a subscriber's location to the closest telephone office, which is commonly called an *end office, local exchange office,* or *central office.* Once in the central office, the subscriber loop is connected to an *electronic switching system* (ESS), which enables the subscriber to access the public telephone network. The local subscriber loop is described in greater detail in Chapter 9.

8-3 STANDARD TELEPHONE SET

The word *telephone* comes from the Greek words *tele,* meaning "from afar," and *phone,* meaning "sound," "voice," or "voiced sound." The standard dictionary defines a telephone as follows:

> *An apparatus for reproducing sound, especially that of the human voice (speech), at a great distance, by means of electricity; consisting of transmitting and receiving instruments connected by a line or wire which conveys the electric current.*

In essence, *speech* is sound in motion. However, sound waves are acoustic waves and have no electrical component. The basic telephone set is a simple analog transceiver designed with the primary purpose of converting speech or acoustical signals to electrical signals. However, in recent years, new features such as multiple-line selection, hold, caller ID, and speakerphone have been incorporated into telephone sets, creating a more elaborate and complicated device. However, their primary purpose is still the same, and the basic functions they perform are accomplished in much the same way as they have always been.

The first telephone set that combined a transmitter and receiver into a single hand-held unit was introduced in 1878 and called the Butterstamp telephone. You talked into one end and then turned the instrument around and listened with the other end. In 1951, Western Electric Company introduced a telephone set that was the industry standard for nearly four decades (the rotary dial telephone used by your grandparents). This telephone set is called the Bell System 500-type telephone and is shown in Figure 8-1a. The 500-type telephone set replaced the earlier 302-type telephone set (the telephone with the hand-crank magneto, fixed microphone, handheld earphone, and no dialing mechanism). Although there are very few 500-type telephone sets in use in the United States today, the basic functions and operation of modern telephones are essentially the same. In modern-day telephone sets, the rotary dial mechanism is replaced with a Touch-Tone keypad. The modern Touch-Tone telephone is called a 2500-type telephone set and is shown in Figure 8-1b.

The quality of transmission over a telephone connection depends on the received volume, the relative frequency response of the telephone circuit, and the degree of interference. In a typical connection, the ratio of the acoustic pressure at the transmitter input to the corresponding pressure at the receiver depends on the following:

The translation of acoustic pressure into an electrical signal

The losses of the two customer local loops, the central telephone office equipment, and the cables between central telephone offices

FIGURE 8-1 (a) 500-type telephone set; (b) 2500-type telephone set

The translation of the electrical signal at the receiving telephone set to acoustic pressure at the speaker output

8-3-1 Functions of the Telephone Set

The basic functions of a telephone set are as follows:

1. Notify the subscriber when there is an incoming call with an audible signal, such as a bell, or with a visible signal, such as a flashing light. This signal is analogous to an interrupt signal on a microprocessor, as its intent is to interrupt what you are doing. These signals are purposely made annoying enough to make people want to answer the telephone as soon as possible.

2. Provide a signal to the telephone network verifying when the incoming call has been acknowledged and answered (i.e., the receiver is lifted off hook).

3. Convert speech (acoustical) energy to electrical energy in the transmitter and vice versa in the receiver. Actually, the microphone converts the acoustical energy to mechanical energy, which is then converted to electrical energy. The speaker performs the opposite conversions.

4. Incorporate some method of inputting and sending destination telephone numbers (either mechanically or electrically) from the telephone set to the central office switch over the local loop. This is accomplished using either rotary dialers (pulses) or Touch-Tone pads (frequency tones).

5. Regulate the amplitude of the speech signal the calling person outputs onto the telephone line. This prevents speakers from producing signals high enough in amplitude to interfere with other people's conversations taking place on nearby cable pairs (crosstalk).

6. Incorporate some means of notifying the telephone office when a subscriber wishes to place an outgoing call (i.e., handset lifted off hook). Subscribers cannot dial out until they receive a dial tone from the switching machine.

7. Ensure that a small amount of the transmit signal is fed back to the speaker, enabling talkers to hear themselves speaking. This feedback signal is sometimes called *sidetone* or *talkback*. Sidetone helps prevent the speaker from talking too loudly.

8. Provide an open circuit (idle condition) to the local loop when the telephone is not in use (i.e., on hook) and a closed circuit (busy condition) to the local loop when the telephone is in use (off hook).

9. Provide a means of transmitting and receiving call progress signals between the central office switch and the subscriber, such as on and off hook, busy, ringing, dial pulses, Touch-Tone signals, and dial tone.

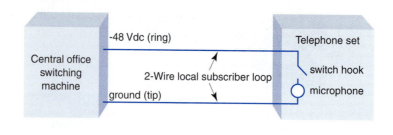

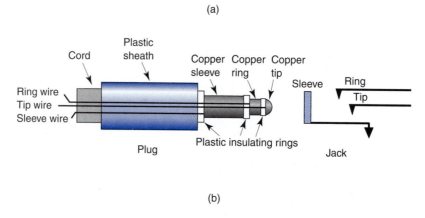

(a)

(b)

FIGURE 8-2 (a) Simplified two-wire loop showing telephone set hookup to a local switching machine; (b) plug and jack configurations showing tip, ring, and sleeve

FIGURE 8-3 RJ-11 connector

8-3-2 Telephone Set, Local Loop, and Central Office Switching Machines

Figure 8-2a shows how a telephone set is connected to a central office switching machine (local switch). As shown in the figure, a basic telephone set requires only two wires (one pair) from the telephone company to operate. Again, the pair of wires connecting a subscriber to the closest telephone office is called the *local loop*. One wire on the local loop is called the *tip,* and the other is called the *ring.* The names *tip* and *ring* come from the ¼-inch-diameter two-conductor phone plugs and patch cords used at telephone company switchboards to interconnect and test circuits. The tip and ring for a standard plug and jack are shown in Figure 8-2b. When a third wire is used, it is called the *sleeve.*

Since the 1960s, phone plugs and jacks have gradually been replaced in the home with a miniaturized plastic plug known as RJ-11 and a matching plastic receptacle (shown in Figure 8-3). *RJ* stands for *registered jacks* and is sometimes described as RJ-XX. RJ is a series of telephone connection interfaces (receptacle and plug) that are registered with the U.S. Federal Communications Commission (FCC). The term *jack* sometimes describes both the receptacle and the plug and sometimes specifies only the receptacle. RJ-11 is the most common telephone jack in use today and can have up to six conductors. Although an RJ-11 plug is capable of holding six wires in a ³⁄₁₆-inch-by-³⁄₁₆-inch body, only two wires (one pair) are necessary for a standard telephone circuit to operate. The other four wires can be used for a second telephone line and/or for some other special function.

As shown in Figure 8-2a, the switching machine outputs −48 Vdc on the ring and connects the tip to ground. A dc voltage was used rather than an ac voltage for several reasons: (1) to prevent power supply hum, (2) to allow service to continue in the event of a power outage, and (3) because people were afraid of ac. Minus 48 volts was selected to minimize electrolytic corrosion on the loop wires. The −48 Vdc is used for supervisory signaling and to provide talk battery for the microphone in the telephone set. On hook, off hook, and dial pulsing are examples of supervisory signals and are described in a later section of this chapter. It should be noted that −48 Vdc is the only voltage required for the operation of a standard telephone. However, most modern telephones are equipped with nonstandard (and often nonessential) features and enhancements and may require an additional source of ac power.

8-3-3 Block Diagram of a Telephone Set

A standard telephone set is comprised of a transmitter, a receiver, an electrical network for equalization, associated circuitry to control sidetone levels and to regulate signal power, and necessary signaling circuitry. In essence, a telephone set is an apparatus that creates an exact likeness of sound waves with an electric current. Figure 8-4 shows the functional block diagram of a *telephone set*. The essential components of a telephone set are the ringer circuit, on/off hook circuit, equalizer circuit, hybrid circuit, speaker, microphone, and a dialing circuit.

8-3-3-1 Ringer circuit.   The telephone *ringer* has been around since August 1, 1878, when Thomas Watson filed for the first ringer patent. The *ringer circuit,* which was originally an electromagnetic bell, is placed directly across the tip and ring of the local loop. The purpose of the ringer is to alert the destination party of incoming calls. The audible tone from the ringer must be loud enough to be heard from a reasonable distance and offensive enough to make a person want to answer the telephone as soon as possible. In modern telephones, the bell has been replaced with an electronic oscillator connected to the speaker. Today, ringing signals can be any imaginable sound, including buzzing, beeping, chiming, or your favorite melody.

8-3-3-2 On/off hook circuit.   The *on/off hook circuit* (sometimes called a *switch hook*) is nothing more than a simple single-throw, double-pole (STDP) switch placed across

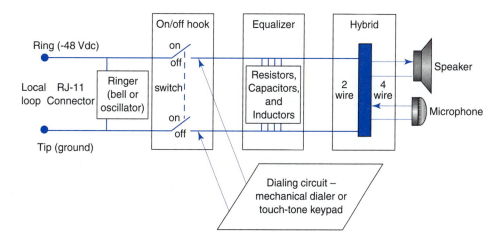

FIGURE 8-4   Functional block diagram of a standard telephone set

the tip and ring. The switch is mechanically connected to the telephone handset so that when the telephone is idle (on hook), the switch is open. When the telephone is in use (off hook), the switch is closed, completing an electrical path through the microphone between the tip and ring of the local loop.

8-3-3-3 Equalizer circuit. *Equalizers* are combinations of passive components (resistors, capacitors, and so on) that are used to regulate the amplitude and frequency response of the voice signals. The equalizer helps solve an important transmission problem in telephone set design, namely, the interdependence of the transmitting and receiving efficiencies and the wide range of transmitter currents caused by a variety of local loop cables with different dc resistances.

8-3-3-4 Speaker. In essence, the *speaker* is the receiver for the telephone. The speaker converts electrical signals received from the local loop to acoustical signals (sound waves) that can be heard and understood by a human being. The speaker is connected to the local loop through the hybrid network. The speaker is typically enclosed in the *handset* of the telephone along with the microphone.

8-3-3-5 Microphone. For all practical purposes, the *microphone* is the transmitter for the telephone. The microphone converts acoustical signals in the form of sound pressure waves from the caller to electrical signals that are transmitted into the telephone network through the local subscriber loop. The microphone is also connected to the local loop through the hybrid network. Both the microphone and the speaker are transducers, as they convert one form of energy into another form of energy. A microphone converts acoustical energy first to mechanical energy and then to electrical energy, while the speaker performs the exact opposite sequence of conversions.

8-3-3-6 Hybrid network. The *hybrid network* (sometimes called a *hybrid coil* or *duplex coil*) in a telephone set is a special balanced transformer used to convert a two-wire circuit (the local loop) into a four-wire circuit (the telephone set) and vice versa, thus enabling full-duplex operation over a two-wire circuit. In essence, the hybrid network separates the transmitted signals from the received signals. Outgoing voice signals are typically in the 1-V to 2-V range, while incoming voice signals are typically half that value. Another function of the hybrid network is to allow a small portion of the transmit signal to be returned to the receiver in the form of a *sidetone*. Insufficient sidetone causes the speaker to raise his voice, making the telephone conversation seem unnatural. Too much sidetone causes the speaker to talk too softly, thereby reducing the volume that the listener receives.

8-3-3-7 Dialing circuit. The *dialing circuit* enables the subscriber to output signals representing digits, and this enables the caller to enter the destination telephone number. The dialing circuit could be a rotary dialer, which is nothing more than a switch connected to a mechanical rotating mechanism that controls the number and duration of the on/off condition of the switch. However, more than likely, the dialing circuit is either an electronic dial-pulsing circuit or a Touch-Tone keypad, which sends various combinations of tones representing the called digits.

8-4 BASIC TELEPHONE CALL PROCEDURES

Figure 8-5 shows a simplified diagram illustrating how two telephone sets (subscribers) are interconnected through a central office dial switch. Each subscriber is connected to the switch through a local loop. The switch is most likely some sort of an electronic switching system (*ESS machine*). The local loops are terminated at the calling and called stations in telephone sets and at the central office ends to switching machines.

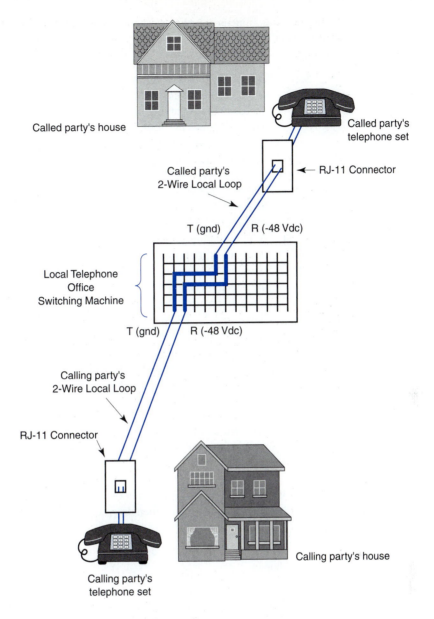

FIGURE 8-5 Telephone call procedures

When the calling party's telephone set goes off hook (i.e., lifting the handset off the cradle), the switch hook in the telephone set is released, completing a dc path between the tip and the ring of the loop through the microphone. The ESS machine senses a dc current in the loop and recognizes this as an off-hook condition. This procedure is referred to as *loop start operation* since the loop is completed through the telephone set. The amount of dc current produced depends on the wire resistance, which varies with loop length, wire gauge, type of wire, and the impedance of the subscriber's telephone. Typical loop resistance ranges from a few ohms up to approximately 1300 ohms, and typical telephone set impedances range from 500 ohms to 1000 ohms.

Completing a local telephone call between two subscribers connected to the same telephone switch is accomplished through a standard set of procedures that includes the 10 steps listed next. Accessing the telephone system in this manner is known as POTS (plain old telephone service):

Step 1   Calling station goes off hook.

Step 2   After detecting a dc current flow on the loop, the switching machine returns an audible dial tone to the calling station, acknowledging that the caller has access to the switching machine.

Step 3   The caller dials the destination telephone number using one of two methods: mechanical dial pulsing or, more likely, electronic dual-tone multifrequency (Touch-Tone) signals.

Step 4   When the switching machine detects the first dialed number, it removes the dial tone from the loop.

Step 5   The switch interprets the telephone number and then locates the local loop for the destination telephone number.

Step 6   Before ringing the destination telephone, the switching machine tests the destination loop for dc current to see if it is idle (on hook) or in use (off hook). At the same time, the switching machine locates a signal path through the switch between the two local loops.

Step 7a If the destination telephone is off hook, the switching machine sends a station-busy signal back to the calling station.

Step 7b If the destination telephone is on hook, the switching machine sends a ringing signal to the destination telephone on the local loop and at the same time sends a ring-back signal to the calling station to give the caller some assurance that something is happening.

Step 8   When the destination answers the telephone, it completes the loop, causing dc current to flow.

Step 9   The switch recognizes the dc current as the station answering the telephone. At this time, the switch removes the ringing and ring-back signals and completes the path through the switch, allowing the calling and called parties to begin their conversation.

Step 10 When either end goes on hook, the switching machine detects an open circuit on that loop and then drops the connections through the switch.

Placing telephone calls between parties connected to different switching machines or between parties separated by long distances is somewhat more complicated and is described in Chapter 9.

8-5   CALL PROGRESS TONES AND SIGNALS

Call progress tones and *call progress signals* are acknowledgment and status signals that ensure the processes necessary to set up and terminate a telephone call are completed in an orderly and timely manner. Call progress tones and signals can be sent from machines to machines, machines to people, and people to machines. The people are the subscribers (i.e., the calling and the called party), and the machines are the electronic switching systems in the telephone offices and the telephone sets themselves. When a switching machine outputs a call progress tone to a subscriber, it must be audible and clearly identifiable.

Signaling can be broadly divided into two major categories: *station signaling* and *interoffice signaling*. Station signaling is the exchange of signaling messages over local loops between stations (telephones) and telephone company switching machines. On the other hand, interoffice signaling is the exchange of signaling messages between switching machines. Signaling messages can be subdivided further into one of four categories: *alerting, supervising, controlling,* and *addressing.* Alerting signals indicate a request for service, such as going off hook or ringing the destination telephone. Supervising signals provide call status information, such as busy or ring-back signals. Controlling signals provide information in the form of announcements, such as number changed to another number, a number no longer in service, and so on. Addressing signals provide the routing information, such as calling and called numbers.

Examples of essential call progress signals are dial tone, dual-tone multifrequency tones, multifrequency tones, dial pulses, station busy, equipment busy, ringing, ring-back, receiver on hook, and receiver off hook. Tables 8-1 and 8-2 summarize the most important call progress tones and their direction of propagation, respectively.

8-5-1 Dial Tone

Siemens Company first introduced *dial tone* to the public switched telephone network in Germany in 1908. However, it took several decades before being accepted in the United

Table 8-1 Call Progress Tone Summary

Tone or Signal	Frequency	Duration/Range
Dial tone	350 Hz plus 440 Hz	Continuous
DTMF	697 Hz, 770 Hz, 852 Hz, 941 Hz, 1209 Hz, 1336 Hz, 1477 Hz, 1633 Hz	Two of eight tones On, 50-ms minimum Off, 45-ms minimum, 3-s maximum
MF	700 Hz, 900 Hz, 1100 Hz, 1300 Hz, 1500 Hz, 1700 Hz	Two of six tones On, 90-ms minimum, 120-ms maximum
Dial pulses	Open/closed switch	On, 39 ms Off, 61 ms
Station busy	480 Hz plus 620 Hz	On, 0.5 s Off, 0.5 s
Equipment busy	480 Hz plus 620 Hz	On, 0.2 s Off, 0.3 s
Ringing	20 Hz, 90 vrms (nominal)	On, 2 s Off, 4 s
Ring-back	440 Hz plus 480 Hz	On, 2 s Off, 4 s
Receiver on hook	Open loop	Indefinite
Receiver off hook	dc current	20-mA minimum, 80-mA maximum,
Receiver-left-off-hook alert	1440 Hz, 2060 Hz, 2450 Hz, 2600 Hz	On, 0.1 s Off, 0.1 s

Table 8-2 Call Progress Tone Direction of Propagation

Tone or Signal	Direction
Dial tone	Telephone office to calling station
DTMF	Calling station to telephone office
MF	Telephone office to telephone office
Dial pulses	Calling station to telephone office
Station busy	Telephone office to calling subscriber
Equipment busy	Telephone office to calling subscriber
Ringing	Telephone office to called subscriber
Ring-back	Telephone office to calling subscriber
Receiver on hook	Calling subscriber to telephone office
Receiver off hook	Calling subscriber to telephone office
Receiver-left-off-hook alert	Telephone office to calling subscriber

States. Dial tone is an audible signal comprised of two frequencies: 350 Hz and 440 Hz. The two tones are linearly combined and transmitted simultaneously from the central office switching machine to the subscriber in response to the subscriber going off hook. In essence, dial tone informs subscribers that they have acquired access to the electronic switching machine and can now dial or use Touch-Tone in a destination telephone number. After a subscriber hears the dial tone and begins dialing, the dial tone is removed from the line (this is called *breaking dial tone*). On rare occasions, a subscriber may go off hook and not receive dial tone. This condition is appropriately called *no dial tone* and occurs when there are more subscribers requesting access to the switching machine than the switching machine can handle at one time.

8-5-2 Dual-Tone Multifrequency

Dual-tone multifrequency (DTMF) was first introduced in 1963 with 10 buttons in Western Electric 1500-type telephones. DTMF was originally called *Touch-Tone.* DTMF is a more efficient means than dial pulsing for transferring telephone numbers from a subscriber's location to the central office switching machine. DTMF is a simple two-of-eight encoding scheme where each digit is represented by the linear addition of two frequencies. DTMF is strictly for signaling between a subscriber's location and the nearest telephone office or message switching center. DTMF is sometimes confused with another two-tone signaling system called *multifrequency signaling* (MF), which is a two-of-six code designed to be used only to convey information between two electronic switching machines.

Figure 8-6 shows the four-row-by-four-column keypad matrix used with a DTMF keypad. As the figure shows, the keypad is comprised of 16 keys and eight frequencies. Most household telephones, however, are not equipped with the special-purpose keys located in the fourth column (i.e., the A, B, C, and D keys). Therefore, most household telephones actually use two-of-seven tone encoding scheme. The four vertical frequencies (called the *low group frequencies*) are 697 Hz, 770 Hz, 852 Hz, and 941 Hz, and the four horizontal frequencies (called the *high group frequencies*) are 1209 Hz, 1336 Hz, 1477 Hz, and 1633 Hz. The frequency tolerance of the oscillators is ±.5%. As shown in Figure 8-6, the digits 2 through 9 can also be used to represent 24 of the 26 letters (Q and Z are omitted). The letters were originally used to identify one local telephone exchange from another, such as BR for Bronx, MA for Manhattan, and so on. Today, the letters are used to person-

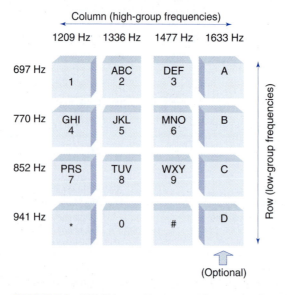

FIGURE 8-6 DTMF keypad layout and frequency allocation

Table 8-3 DTMF Specifications

Transmitter (Subscriber)	Parameter	Receiver (Local Office)
−10 dBm	Minimum power level (single frequency)	−25 dBm
+2 dBm	Maximum power level (two tones)	0 dBm
+4 dB	Maximum power difference between two tones	+4 dB
50 ms	Minimum digit duration	40 ms
45 ms	Minimum interdigit duration	40 ms
3 s	Maximum interdigit time period	3 s
Maximum echo level relative to transmit frequency level (−10 dB)		
Maximum echo delay (<20 ms)		

alize telephone numbers; for example, 1-800-UPS-MAIL equates to the telephone number 1-800-877-6245. When a digit (or letter) is selected, two of the eight frequencies (or seven for most home telephones) are transmitted (one from the low group and one from the high group). For example, when the digit 5 is depressed, 770 Hz and 1336 Hz are transmitted simultaneously. The eight frequencies were purposely chosen so that there is absolutely no harmonic relationship between any of them, thus eliminating the possibility of one frequency producing a harmonic that might be misinterpreted as another frequency.

The major advantages for the subscriber in using Touch-Tone signaling over dial pulsing is speed and control. With Touch-Tone signaling, all digits (and thus telephone numbers) take the same length of time to produce and transmit. Touch-Tone signaling also eliminates the impulse noise produced from the mechanical switches necessary to produce dial pulses. Probably the most important advantage of DTMF over dial pulsing is the way in which the telephone company processes them. Dial pulses cannot pass through a central office exchange (local switching machine), whereas DTMF tones will pass through an exchange to the switching system attached to the called number.

Table 8-3 lists the specifications for DTME. The transmit specifications are at the subscriber's location, and the receive specifications are at the local switch. Minimum power levels are given for a single frequency, and maximum power levels are given for two tones. The minimum duration is the minimum time two tones from a given digit must remain on. The interdigit time specifies the minimum and maximum time between the transmissions of any two successive digits. An echo occurs when a pair of tones is not totally absorbed by the local switch and a portion of the power is returned to the subscriber. The maximum power level of an echo is 10 dB below the level transmitted by the subscriber and must be delayed less than 20 ms.

8-5-3 Multifrequency

Multifrequency (MF) *tones* (codes) are similar to DTMF signals in that they involve the simultaneous transmission of two tones. MF tones are used to transfer digits and control signals between switching machines, whereas DTMF signals are used to transfer digits and control signals between telephone sets and local switching machines. MF tones are combinations of two frequencies that fall within the normal speech bandwidth so that they can be propagated over the same circuits as voice. This is called *in-band signaling.* In-band signaling is rapidly being replaced by *out-of-band signaling,* which is discussed in Chapter 9.

MF codes are used to send information between the control equipment that sets up connections through a switch when more than one switch is involved in completing a call. MF codes are also used to transmit the calling and called numbers from the originating telephone office to the destination telephone office. The calling number is sent first, followed by the called number.

Table 8-4 lists the two-tone MF combinations and the digits or control information they represent. As the table shows, MF tones involve the transmission of two of six possible

Table 8-4 Multifrequency Codes

Frequencies (Hz)	Digit or Control
700 + 900	1
700 + 1100	2
700 + 1300	4
700 + 1500	7
900 + 1100	3
900 + 1300	5
900 + 1500	8
1100 + 1300	6
1100 + 1500	9
1100 + 1700	Key pulse (KP)
1300 + 1500	0
1500 + 1700	Start (ST)
2600 Hz	IDLE

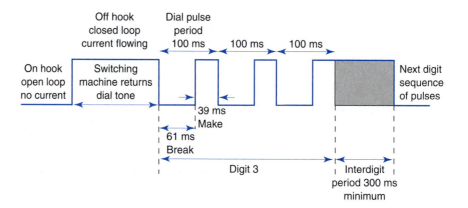

FIGURE 8-7   Dial pulsing sequence

frequencies representing the 10 digits plus two control signals. The six frequencies are 700 Hz, 900 Hz, 1100 Hz, 1300 Hz, 1500 Hz, and 1700 Hz. Digits are transmitted at a rate of seven per second, and each digit is transmitted as a 68-ms burst. The *key pulse* (KP) signal is a multifrequency control tone comprised of 1100 Hz plus 1700 Hz, ranging from 90 ms to 120 ms. The KP signal is used to indicate the beginning of a sequence of MF digits. The *start* (ST) signal is a multifrequency control tone used to indicate the end of a sequence of dialed digits. From the perspective of the telephone circuit, the ST control signal indicates the beginning of the processing of the signal. The IDLE signal is a 2600-Hz single-frequency tone placed on a circuit to indicate the circuit is not currently in use. For example, KP 3 1 5 7 3 6 1 0 5 3 ST is the sequence transmitted for the telephone number 315-736-1053.

8-5-4   Dial Pulses

Dial pulsing (sometimes called *rotary dial pulsing*) is the method originally used to transfer digits from a telephone set to the local switch. Pulsing digits from a rotary switch began soon after the invention of the automatic switching machine. The concept of dial pulsing is quite simple and is depicted in Figure 8-7. The process begins when the telephone set is lifted off hook, completing a path for current through the local loop. When the switching machine detects the off-hook condition, it responds with dial tone. After hearing the dial tone, the subscriber begins dial pulsing digits by rotating a mechanical dialing mechanism and then letting it return to its rest position. As the rotary switch returns to its rest position, it outputs a series of dial pulses corresponding to the digit dialed.

When a digit is dialed, the loop circuit alternately opens (breaks) and closes (makes) a prescribed number of times. The number of switch make/break sequences corresponds to the digit dialed (i.e., the digit 3 produces three switch openings and three switch closures). Dial pulses occur at 10 make/break cycles per second (i.e., a period of 100 ms per pulse cycle). For example, the digit 5 corresponds to five make/break cycles lasting a total of 500 ms. The switching machine senses and counts the number of make/break pairs in the sequence. The break time is nominally 61 ms, and the make time is nominally 39 ms. Digits are separated by an idle period of 300 ms called the *interdigit time.* It is essential that the switching machine recognize the interdigit time so that it can separate the pulses from successive digits. The central office switch incorporates a special *time-out circuit* to ensure that the break part of the dialing pulse is not misinterpreted as the phone being returned to its on-hook (idle) condition.

All digits do not take the same length of time to dial. For example, the digit 1 requires only one make/break cycle, whereas the digit 0 requires 10 cycles. Therefore, all telephone numbers do not require the same amount of time to dial or to transmit. The minimum time to dial pulse out the seven-digit telephone number 987-1234 is as follows:

digit	9	ID	8	ID	7	ID	1	ID	2	ID	3	ID	4
time (ms)	900	300	800	300	700	300	100	300	200	300	300	300	400

where ID is the interdigit time (300 ms) and the total minimum time is 5200 ms, or 5.2 seconds.

8-5-5 Station Busy

In telephone terminology, a *station* is a telephone set. A *station-busy signal* is sent from the switching machine back to the calling station whenever the called telephone number is off hook (i.e., the station is in use). The station-busy signal is a two-tone signal comprised of 480 Hz and 620 Hz. The two tones are on for 0.5 seconds, then off for 0.5 seconds. Thus, a busy signal repeats at a 60-pulse-per-minute (ppm) rate.

8-5-6 Equipment Busy

The *equipment-busy signal* is sometimes called a *congestion tone* or a *no-circuits-available tone.* The equipment-busy signal is sent from the switching machine back to the calling station whenever the system cannot complete the call because of equipment unavailability (i.e., all the circuits, switches, or switching paths are already in use). This condition is called *blocking* and occurs whenever the system is overloaded and more calls are being placed than can be completed. The equipment-busy signal uses the same two frequencies as the station-busy signal, except the equipment-busy signal is on for 0.2 seconds and off for 0.3 seconds (120 ppm). Because an equipment-busy signal repeats at twice the rate as a station-busy signal, an equipment busy is sometimes called a *fast busy,* and a station busy is sometimes called a *slow busy.* The telephone company refers to an equipment-busy condition as a *can't complete.*

8-5-7 Ringing

The *ringing signal* is sent from a central office to a subscriber whenever there is an incoming call. The purpose of the ringing signal is to ring the bell in the telephone set to alert the subscriber that there is an incoming call. If there is no bell in the telephone set, the ringing signal is used to trigger another audible mechanism, which is usually a tone oscillator circuit. The ringing signal is nominally a 20-Hz, 90-Vrms signal that is on for 2 seconds and then off for 4 seconds. The ringing signal should not be confused with the actual ringing sound the bell makes. The audible ring produced by the bell was originally made as annoying as possible so that the called end would answer the telephone as soon as possible, thus tying up common-usage telephone equipment in the central office for the minimum length of time.

8-5-8　Ring-Back

The *ring-back signal* is sent back to the calling party at the same time the ringing signal is sent to the called party. However, the ring and ring-back signals are two distinctively different signals. The purpose of the ring-back signal is to give some assurance to the calling party that the destination telephone number has been accepted and processed and is being rung. The ring-back signal is an audible combination of two tones at 440 Hz and 480 Hz that are on for 2 seconds and then off for 4 seconds.

8-5-9　Receiver On/Off Hook

When a telephone is *on hook,* it is not being used, and the circuit is in the *idle* (or *open*) *state.* The term *on hook* was derived in the early days of telephone when the telephone handset was literally placed on a hook (the hook eventually evolved into a cradle). When the telephone set is on hook, the local loop is open, and there is no current flowing on the loop. An on-hook signal is also used to terminate a call and initiate a disconnect.

When the telephone set is taken *off hook,* a switch closes in the telephone that completes a dc path between the two wires of the local loop. The switch closure causes a dc current to flow on the loop (nominally between 20 mA and 80 mA, depending on loop length and wire gauge). The switching machine in the central office detects the dc current and recognizes it as a receiver off-hook condition (sometimes called a *seizure* or *request for service*). The receiver off-hook condition is the first step to completing a telephone call. The switching machine will respond to the off-hook condition by placing an audible dial tone on the loop. The off-hook signal is also used at the destination end as an *answer signal* to indicate that the called party has answered the telephone. This is sometimes referred to as a *ring trip* because when the switching machine detects the off-hook condition, it removes (or trips) the ringing signal.

8-5-10　Other Nonessential Signaling and Call Progress Tones

There are numerous additional signals relating to initiating, establishing, completing, and terminating a telephone call that are nonessential, such as *call waiting tones, caller waiting tones, calling card service tones, comfort tones, hold tones, intrusion tones, stutter dial tone* (for voice mail), and *receiver off-hook tones* (also called *howler tones*).

8-6　CORDLESS TELEPHONES

Cordless telephones are simply telephones that operate without cords attached to the handset. Cordless telephones originated around 1980 and were quite primitive by today's standards. They originally occupied a narrow band of frequencies near 1.7 MHz, just above the AM broadcast band, and used the 117-vac, 60-Hz household power line for an antenna. These early units used frequency modulation (FM) and were poor quality and susceptible to interference from fluorescent lights and automobile ignition systems. In 1984, the FCC reallocated cordless telephone service to the 46-MHz to 49-MHz band. In 1990, the FCC extended cordless telephone service to the 902-MHz to 928-MHz band, which appreciated a superior signal-to-noise ratio. Cordless telephone sets transmit and receive over narrowband FM (NBFM) channels spaced 30 kHz to 100 kHz apart, depending on the modulation and frequency band used. In 1998, the FCC expanded service again to the 2.4-GHz band. Adaptive differential pulse code modulation and spread spectrum technology (SST) are used exclusively in the 2.4-GHz band, while FM and SST digital modulation are used in the 902-MHz to 928-MHz band. Digitally modulated SST telephones offer higher quality and more security than FM telephones.

In essence, a cordless telephone is a full-duplex, battery-operated, portable radio transceiver that communicates directly with a stationary transceiver located somewhere in

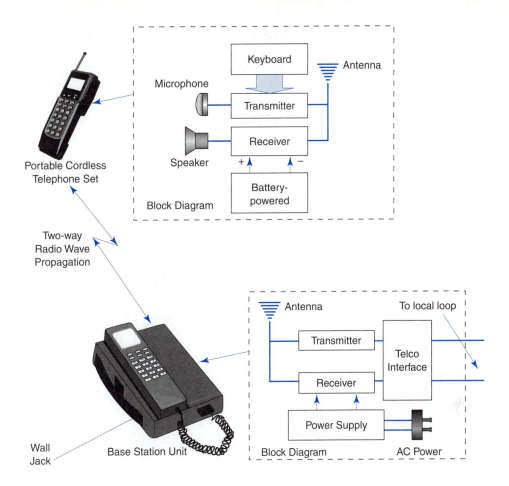

FIGURE 8-8 Cordless telephone system

the subscriber's home or office. The basic layout for a cordless telephone is shown in Figure 8-8. The base station is an ac-powered stationary radio transceiver (transmitter and receiver) connected to the local loop through a cord and telephone company interface unit. The interface unit functions in much the same way as a standard telephone set in that its primary function is to interface the cordless telephone with the local loop while being transparent to the user. Therefore, the base station is capable of transmitting and receiving both supervisory and voice signals over the subscriber loop in the same manner as a standard telephone. The base station must also be capable of relaying voice and control signals to and from the portable telephone set through the wireless transceiver. In essence, the portable telephone set is a battery-powered, two-way radio capable of operating in the full-duplex mode.

Because a portable telephone must be capable of communicating with the base station in the full-duplex mode, it must transmit and receive at different frequencies. In 1984, the FCC allocated 10 full-duplex channels for 46-MHz to 49-MHz units. In 1995 to help relieve congestion, the FCC added 15 additional full-duplex channels and extended the frequency band to include frequencies in the 43-MHz to 44-MHz band. Base stations transmit on high-band frequencies and receive on low-band frequencies, while the portable unit transmits on low-band frequencies and receives on high-band frequencies. The frequency assignments are listed in Table 8-5. Channels 16 through 25 are the original 10 full-duplex carrier frequencies. The maximum transmit power for both the portable unit and the base station is 500 mW. This stipulation limits the useful range of a cordless telephone to within 100 feet or less of the base station.

Table 8-5 43-MHz- to 49-MHz-Band Cordless Telephone Frequencies

	Portable Unit	
Channel	Transmit Frequency (MHz)	Receive Frequency (MHz)
1	43.720	48.760
2	43.740	48.840
3	43.820	48.860
4	43.840	48.920
5	43.920	49.920
6	43.960	49.080
7	44.120	49.100
8	44.160	49.160
9	44.180	49.200
10	44.200	49.240
11	44.320	49.280
12	44.360	49.360
13	44.400	49.400
14	44.460	49.460
15	44.480	49.500
16	46.610	49.670
17	46.630	49.845
18	46.670	49.860
19	46.710	49.770
20	46.730	49.875
21	46.770	49.830
22	46.830	49.890
23	46.870	49.930
24	46.930	49.970
25	46.970	49.990

Note. Base stations transmit on the 49-MHz band and receive on the 46-MHz band.

Cordless telephones using the 2.4-GHz band offer excellent sound quality utilizing digital modulation and twin-band transmission to extend their range. With twin-band transmission, base stations transmit in the 2.4-GHz band, while portable units transmit in the 902-MHz to 928-MHz band.

8-7 CALLER ID

Caller ID (identification) is a service originally envisioned by AT&T in the early 1970s, although local telephone companies have only recently offered it. The basic concept of caller ID is quite simple. Caller ID enables the destination station of a telephone call to display the name and telephone number of the calling party before the telephone is answered (i.e., while the telephone is ringing). This allows subscribers to screen incoming calls and decide whether they want to answer the telephone.

The caller ID message is a simplex transmission sent from the central office switch over the local loop to a caller ID display unit at the destination station (no response is provided). The caller ID information is transmitted and received using Bell System 202-compatible modems (ITU V.23 standard). This standard specifies a 1200-bps FSK (frequency-shift keying) signal with a 1200-Hz mark frequency (f_m) and a 2200-Hz space frequency (f_m). The FSK signal is transmitted in a burst between the first and second 20-Hz, 90-Vrms ringing signals, as shown in Figure 8-9a. Therefore, to ensure detection of the caller ID signal, the telephone must ring at least twice before being answered. The caller ID signal does not begin until 500 ms after the end of the first ring and must end 500 ms before the beginning of the second ring. Therefore, the caller ID signal has a 3-second window in which it must be transmitted.

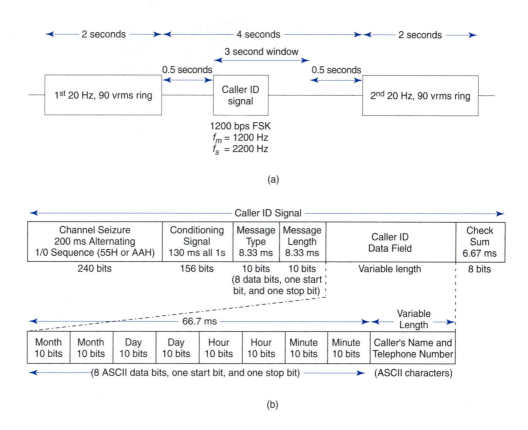

FIGURE 8-9   Caller ID: (a) ringing cycle; (b) frame format

The format for a caller ID signal is shown in Figure 8-9b. The 500-ms delay after the first ringing signal is immediately followed by the *channel seizure field,* which is a 200-ms-long sequence of alternating logic 1s and logic 0s (240 bits comprised of 120 pairs of alternating 1/0 bits, either 55 hex or AA hex). A *conditioning signal field* immediately follows the channel seizure field. The conditioning signal is a continuous 1200-Hz tone lasting for 130 ms, which equates to 156 consecutive logic 1 bits.

The protocol used for the next three fields—*message type field, message length field,* and *caller ID data field*—specifies asynchronous transmission of 16-bit characters (without parity) framed by one start bit (logic 0) and one stop bit (logic 1) for a total of 10 bits per character. The message type field is comprised of a 16-bit hex code, indicating the type of service and capability of the data message. There is only one message type field currently used with caller ID (04 hex). The message type field is followed by a 16-bit message length field, which specifies the total number of characters (in binary) included in the caller ID data field. For example, a message length code of 15 hex (0001 0101) equates to the number 21 in decimal. Therefore, a message length code of 15 hex specifies 21 characters in the caller ID data field.

The caller ID data field uses extended ASCII coded characters to represent a month code (01 through 12), a two-character day code (01 through 31), a two-character hour code in local military time (00 through 23), a two-character minute code (00 through 59), and a variable-length code, representing the caller's name and telephone number. ASCII coded digits are comprised of two independent hex characters (eight bits each). The first hex character is always 3 (0011 binary), and the second hex character represents a digit between 0 and 9 (0000 to 1001 binary). For example, 30 hex (0011 0000 binary) equates to the digit 0, 31 hex (0011 0001 binary) equates to the digit 1, 39 hex (0011 1001) equates to the digit

9, and so on. The caller ID data field is followed by a checksum for error detection, which is the 2's complement of the module 256 sum of the other words in the data message (message type, message length, and data words).

Example 8-1

Interpret the following hex code for a caller ID message (start and stop bits are not included in the hex codes):

04 12 31 31 32 37 31 35 35 37 33 31 35 37 33 36 31 30 35 33 xx

Solution 04—message type word

12—18 decimal (18 characters in the caller ID data field)

31, 31—ASCII code for 11 (the month of November)

32, 37—ASCII code for 27 (the 27th day of the month)

31, 35—ASCII code for 15 (the 15th hour—3:00 P.M.)

35, 37—ASCII code for 57 (57 minutes after the hour—3:57 P.M.)

33, 31, 35, 37, 33, 36, 31, 30, 35, 33—10-digit ASCII-coded telephone number
(315 736 1053)

xx—checksum (00 hex to FF hex)

8-8 ELECTRONIC TELEPHONES

Although 500- and 2500-type telephone sets still work with the public telephone network, they are becoming increasingly more difficult to find. Most modern-day telephones have replaced many of the mechanical functions performed in the old telephone sets with electronic circuits. Electronic telephones use integrated-circuit technology to perform many of the basic telephone functions as well as a myriad of new and, and in many cases, nonessential functions. The refinement of microprocessors has also led to the development of multiple-line, full-feature telephones that permit automatic control of the telephone set's features, including telephone number storage, automatic dialing, redialing, and caller ID. However, no matter how many new gadgets are included in the new telephone sets, they still have to interface with the telephone network in much the same manner as telephones did a century ago.

Figure 8-10 shows the block diagram for a typical electronic telephone comprised of one multifunctional integrated-circuit chip, a microprocessor chip, a Touch-Tone keypad, a speaker, a microphone, and a handful of discrete devices. The major components included in the multifunctional integrated circuit chip are DTMF tone generator, MPU (microprocessor unit) interface circuitry, random access memory (RAM), tone ringer circuit, speech network, and a line voltage regulator.

The Touch-Tone keyboard provides a means for the operator of the telephone to access the DTMF tone generator inside the multifunction integrated-circuit chip. The external crystal provides a stable and accurate frequency reference for producing the dual-tone multifrequency signaling tones.

The tone ringer circuit is activated by the reception of a 20-Hz ringing signal. Once the ringing signal is detected, the tone ringer drives a piezoelectric sound element that produces an electronic ring (without a bell).

The voltage regulator converts the dc voltage received from the local loop and converts it to a constant-level dc supply voltage to operate the electronic components in the telephone. The internal speech network contains several amplifiers and associated components that perform the same functions as the hybrid did in a standard telephone.

The microprocessor interface circuit interfaces the MPU to the multifunction chip. The MPU, with its internal RAM, controls many of the functions of the telephone, such as num-

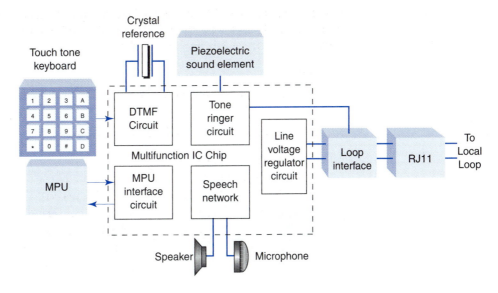

FIGURE 8-10 Electronic telephone set

ber storage, speed dialing, redialing, and autodialing. The bridge rectifier protects the telephone from the relatively high-voltage ac ringing signal, and the switch hook is a mechanical switch that performs the same functions as the switch hook on a standard telephone set.

8-9 PAGING SYSTEMS

Most *paging systems* are simplex wireless communications systems designed to alert subscribers of awaiting messages. Paging transmitters relay radio signals and messages from wire-line and cellular telephones to subscribers carrying portable receivers. The simplified block diagram of a paging system is shown in Figure 8-11. The infrastructure used with paging systems is somewhat different than the one used for cellular telephone systems. This is because standard paging systems are one way, with signals transmitted from the paging system to portable pager and never in the reverse direction. There are narrow-, mid-, and wide-area pagers (sometimes called local, regional, and national). Narrow-area paging systems operate only within a building or building complex, mid-area pagers cover an area of several square miles, and wide-area pagers operate worldwide. Most pagers are mid-area where one centrally located high-power transmitter can cover a relatively large geographic area, typically between 6 and 10 miles in diameter.

To contact a person carrying a pager, simply dial the telephone number assigned that person's portable pager. The paging company receives the call and responds with a query requesting the telephone number you wish the paged person to call. After the number is entered, a *terminating signal* is appended to the number, which is usually the # sign. The caller then hangs up. The paging system converts the telephone number to a digital code and transmits it in the form of a digitally encoded signal over a wireless communications system. The signal may be simultaneously sent from more than one radio transmitter (sometimes called *simulcasting* or *broadcasting*), as is necessary in a wide-area paging system. If the paged person is within range of a broadcast transmitter, the targeted pager will receive the message. The message includes a notification signal, which either produces an audible beep or causes the pager to vibrate, and the number the paged unit should call is shown on an alphanumeric display. Some newer paging units are also capable of displaying messages as well as the telephone number of the paging party.

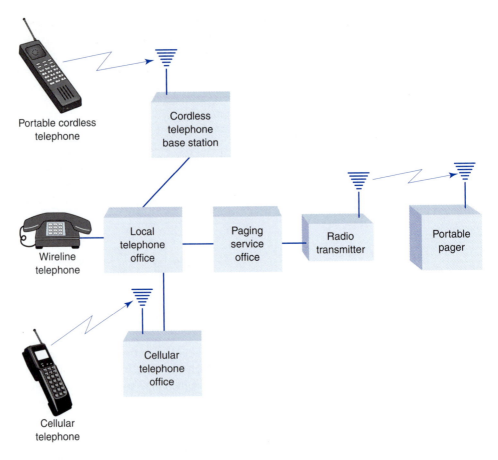

FIGURE 8-11 Simplified block diagram of a standard simplex paging system

Early paging systems used FM; however, most modern paging systems use FSK or PSK. Pagers typically transmit bit rates between 200 bps and 6400 bps with the following carrier frequency bands: 138 MHz to 175 MHz, 267 MHz to 284 MHz, 310 MHz to 330 MHz, 420 MHz to 470 MHz, and several frequency slots within the 900-MHz band.

Each portable pager is assigned a special code, called a *cap code,* which includes a sequence of digits or a combination of digits and letters. The cap code is broadcast along with the paging party's telephone number. If the portable paging unit is within range of the broadcasting transmitter, it will receive the signal, demodulate it, and recognize its cap code. Once the portable pager recognizes its cap code, the callback number and perhaps a message will be displayed on the unit. Alphanumeric messages are generally limited to between 20 and 40 characters in length.

Early paging systems, such as one developed by the British Post Office called Post Office Code Standardization Advisory Group (POCSAG), transmitted a two-level FSK signal. POCSAG used an asynchronous protocol, which required a long preamble for synchronization. The preamble begins with a long *dotting sequence* (sometimes called a *dotting comma*) to establish clock synchronization. Data rates for POCSAG are 512 bps, 1200 bps, and 2400 bps. With POCSAG, portable pagers must operate in the *always-on mode* all the time, which means the pager wastes much of its power resources on nondata preamble bits.

In the early 1980s, the European Telecommunications Standards Institute (ETSI) developed the ERMES protocol. ERMES transmitted data at a 6250 bps rate using four-level FSK (3125 baud). ERMES is a synchronous protocol, which requires less time to synchronize. ERMES supports 16 25-kHz paging channels in each of its frequency bands.

The most recent paging protocol, FLEX, was developed in the 1990s. FLEX is designed to minimize power consumption in the portable pager by using a synchronous time-slotted protocol to transmit messages in precise time slots. With FLEX, each frame is comprised of 128 data frames, which are transmitted only once during a 4-minute period. Each frame lasts for 1.875 seconds and includes two synchronizing sequences, a header containing frame information and pager identification addresses, and 11 discrete data blocks. Each portable pager is assigned a specific frame (called a *home frame*) within the frame cycle that it checks for transmitted messages. Thus, a pager operates in the high-power standby condition for only a few seconds every 4 minutes (this is called the *wakeup time*). The rest of the time, the pager is in an ultra–low power standby condition. When a pager is in the wakeup mode, it synchronizes to the frame header and then adjusts itself to the bit rate of the received signal. When the pager determines that there is no message waiting, it puts itself back to sleep, leaving only the timer circuit active.

QUESTIONS

8-1. Define the terms *communications* and *telecommunications.*

8-2. Define *plain old telephone service.*

8-3. Describe a *local subscriber loop.*

8-4. Where in a telephone system is the *local loop?*

8-5. Briefly describe the basic functions of a standard *telephone set.*

8-6. What is the purpose of the *RJ-11 connector?*

8-7. What is meant by the terms *tip* and *ring?*

8-8. List and briefly describe the essential components of a standard telephone set.

8-9. Briefly describe the steps involved in completing a local telephone call.

8-10. Explain the basic purpose of *call progress tones* and *signals.*

8-11. List and describe the two primary categories of *signaling.*

8-12. Describe the following signaling messages: *alerting, supervising, controlling,* and *addressing.*

8-13. What is the purpose of *dial tone,* and when is it applied to a telephone circuit?

8-14. Briefly describe *dual-tone multifrequency* and *multifrequency* signaling and tell where they are used.

8-15. Describe *dial pulsing.*

8-16. What is the difference between a *station-busy* signal and an *equipment-busy* signal?

8-17. What is the difference between a *ringing* and a *ring-back* signal?

8-18. Briefly describe what happens when a telephone set is taken *off hook.*

8-19. Describe the differences between the operation of a *cordless telephone* and a *standard telephone.*

8-20. Explain how *caller ID* operates and when it is used.

8-21. Briefly describe how a paging system operates.

C H A P T E R 9

The Telephone Circuit

CHAPTER OUTLINE

OBJECTIVES

- Define *telephone circuit, message,* and *message channel*
- Describe the transmission characteristics of a local subscriber loop
- Describe loading coils and bridge taps
- Describe loop resistance and how it is calculated
- Explain telephone message–channel noise and C-message noise weighting
- Describe the following units of power measurement: db, dBm, dBmO, rn, dBrn, dBrnc, dBrn 3-kHz flat, and dBrncO
- Define *psophometric noise weighting*
- Define and describe transmission parameters
- Define *private-line circuit*
- Explain bandwidth, interface, and facilities parameters
- Define *line conditioning* and describe C- and D-type conditioning
- Describe two-wire and four-wire circuit arrangements
- Explain hybrids, echo suppressors, and echo cancelers
- Define *crosstalk*
- Describe nonlinear, transmittance, and coupling crosstalk

A *telephone circuit* is comprised of two or more facilities, interconnected in tandem, to provide a transmission path between a source and a destination. The interconnected facilities may be temporary, as in a standard telephone call, or permanent, as in a dedicated private-line telephone circuit. The facilities may be metallic cable pairs, optical fibers, or wireless carrier systems. The information transferred is called the *message,* and the circuit used is called the *message channel.*

Telephone companies offer a wide assortment of message channels ranging from a basic 4-kHz voice-band circuit to wideband microwave, satellite, or optical fiber transmission systems capable of transferring high-resolution video or wideband data. The following discussion is limited to basic voice-band circuits. In telephone terminology, the word *message* originally denoted speech information. However, this definition has been extended to include any signal that occupies the same bandwidth as a standard voice channel. Thus, a message channel may include the transmission of ordinary speech, supervisory signals, or data in the form of digitally modulated carriers (FSK, PSK, QAM, and so on). The network bandwidth for a standard voice-band message channel is 4 kHz; however, a portion of that bandwidth is used for *guard bands* and signaling. Guard bands are unused frequency bands located between information signals. Consequently, the effective channel bandwidth for a voice-band message signal (whether it be voice or data) is approximately 300 Hz to 3000 Hz.

9-2 THE LOCAL SUBSCRIBER LOOP

The *local subscriber loop* is the only facility required by all voice-band circuits, as it is the means by which subscriber locations are connected to the local telephone company. In essence, the sole purpose of a local loop is to provide subscribers access to the public telephone network. The local loop is a metallic transmission line comprised of two insulated copper wires (a pair) twisted together. The local loop is the primary cause of *attenuation* and *phase distortion* on a telephone circuit. Attenuation is an actual loss of signal strength, and phase distortion occurs when two or more frequencies undergo different amounts of phase shift.

The *transmission characteristics* of a cable pair depend on the wire diameter, conductor spacing, dielectric constant of the insulator separating the wires, and the conductivity of the wire. These physical properties, in turn, determine the inductance, resistance, capacitance, and conductance of the cable. The resistance and inductance are distributed along the length of the wire, whereas the conductance and capacitance exist between the two wires. When the insulation is sufficient, the effects of conductance are generally negligible. Figure 9-1a shows the electrical model for a copper-wire transmission line.

The electrical characteristics of a cable (such as inductance, capacitance, and resistance) are uniformly distributed along its length and are appropriately referred to as *distributed parameters.* Because it is cumbersome working with distributed parameters, it is common practice to lump them into discrete values per unit length (i.e., millihenrys per mile, microfarads per kilometer, or ohms per 1000 feet). The amount of attenuation and phase delay experienced by a signal propagating down a metallic transmission line is a function of the frequency of the signal and the electrical characteristics of the cable pair.

There are seven main component parts that make up a traditional local loop:

Feeder cable (F1). The largest cable used in a local loop, usually 3600 pair of copper wire placed underground or in conduit.

Serving area interface (SAI). A cross-connect point used to distribute the larger feeder cable into smaller distribution cables.

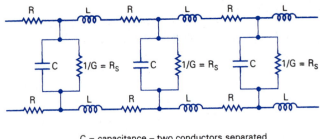

C = capacitance – two conductors separated
   by an insulator
R = resistance – opposition to current flow
L = self inductance
1/G = leakage resistance of dielectric
R_s = shunt leakage resistance

(a)

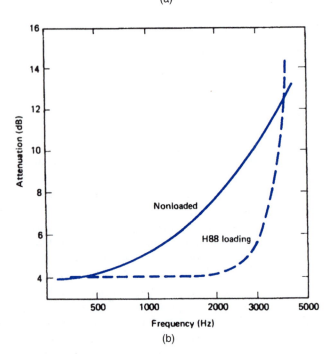

(b)

FIGURE 9-1   (a) Electrical model of a copper-wire transmission
line; (b) frequency-versus-attenuation characteristics for unloaded
and loaded cables

Distribution cable (F2). A smaller version of a feeder cable containing less wire pairs.

Subscriber or standard network interface (SNI). A device that serves as the demarcation point between local telephone company responsibility and subscriber responsibility for telephone service.

Drop wire. The final length of cable pair that terminates at the SNI.

Aerial. That portion of the local loop that is strung between poles.

Distribution cable and drop-wire cross-connect point. The location where individual cable pairs within a distribution cable are separated and extended to the subscriber's location on a drop wire.

Two components often found on local loops are loading coils and bridge taps.

9-2-1 Loading Coils

Figure 9-1b shows the effect of frequency on attenuation for a 12,000-foot length of 26-gauge copper cable. As the figure shows, a 3000-Hz signal experiences 6 dB more attenuation than a 500-Hz signal on the same cable. In essence, the cable acts like a low-pass filter. Extensive studies of attenuation on cable pairs have shown that a substantial reduction in attenuation is achieved by increasing the inductance value of the cable. Minimum attenuation requires a value of inductance nearly 100 times the value obtained in ordinary twisted-wire cable. Achieving such values on a uniformly distributed basis is impractical. Instead, the desired effect can be obtained by adding inductors periodically in series with the wire. This practice is called *loading,* and the inductors are called *loading coils.* Loading coils placed in a cable decrease the attenuation, increase the line impedance, and improve transmission levels for circuits longer than 18,000 feet. Loading coils allowed local loops to extend three to four times their previous length. A loading coil is simply a passive conductor wrapped around a core and placed in series with a cable creating a small electromagnet. Loading coils can be placed on telephone poles, in manholes, or on cross-connect boxes. Loading coils increase the effective distance that a signal must travel between two locations and cancels the capacitance that inherently builds up between wires with distance. Loading coils first came into use in 1900.

Loaded cables are specified by the addition of the letter codes A, B, C, D, E, F, H, X, or Y, which designate the distance between loading coils and by numbers, which indicate the inductance value of the wire gauge. The letters indicate that loading coils are separated by 700, 3000, 929, 4500, 5575, 2787, 6000, 680, or 2130 feet, respectively. B-, D-, and H-type loading coils are the most common because their separations are representative of the distances between manholes. The amount of series inductance added is generally 44 mH, 88 mH, or 135 mH. Thus, a cable pair designated 26H88 is made from 26-gauge wire with 88 mH of series inductance added every 6000 feet. The loss-versus-frequency characteristics for a loaded cable are relatively flat up to approximately 2000 Hz, as shown in Figure 9-1b. From the figure, it can be seen that a 3000-Hz signal will suffer only 1.5 dB more loss than a 500-Hz signal on 26-gauge wire when 88-mH loading coils are spaced every 6000 feet.

Loading coils cause a sharp drop in frequency response at approximately 3400 Hz, which is undesirable for high-speed data transmission. Therefore, for high-performance data transmission, loading coils should be removed from the cables. The low-pass characteristics of a cable also affect the phase distortion–versus–frequency characteristics of a signal. The amount of *phase distortion* is proportional to the length and gauge of the wire. Loading a cable also affects the phase characteristics of a cable. The telephone company must often add gain and delay equalizers to a circuit to achieve the minimum requirements. Equalizers introduce discontinuities or ripples in the bandpass characteristics of a circuit. Automatic equalizers in data modems are sensitive to this condition, and very often an overequalized circuit causes as many problems to a data signal as an underequalized circuit.

9-2-2 Bridge Taps

A *bridge tap* is an irregularity frequently found in cables serving subscriber locations. Bridge taps are unused sections of cable that are connected in shunt to a working cable pair, such as a local loop. Bridge taps can be placed at any point along a cable's length. Bridge taps were used for party lines to connect more than one subscriber to the same local loop. Bridge taps also increase the flexibility of a local loop by allowing the cable to go to more than one junction box, although it is unlikely that more than one of the cable pairs leaving a bridging point will be used at any given time. Bridge taps may or may not be used at some future time, depending on service demands. Bridge taps increase the flexibility of a cable by making it easier to reassign a cable to a different subscriber without requiring a person working in the field to cross connect sections of cable.

Bridge taps introduce a loss called *bridging loss.* They also allow signals to split and propagate down more than one wire. Signals that propagate down unterminated (open-circuited) cables reflect back from the open end of the cable, often causing interference with the original signal. Bridge taps that are short and closer to the originating or terminating ends often produce the most interference.

Bridge taps and loading coils are not generally harmful to voice transmissions, but if improperly used, they can literally destroy the integrity of a data signal. Therefore, bridge taps and loading coils should be removed from a cable pair that is used for data transmission. This can be a problem because it is sometimes difficult to locate a bridge tap. It is estimated that the average local loop can have as many as 16 bridge taps.

9-2-3 Loop Resistance

The dc resistance of a local loop depends primarily on the type of wire and wire size. Most local loops use 18- to 26-gauge, twisted-pair copper wire. The lower the wire gauge, the larger the diameter, the less resistance, and the lower the attenuation. For example, 26-gauge unloaded copper wire has an attenuation of 2.67 dB per mile, whereas the same length of 19-gauge copper wire has only 1.12 dB per mile. Therefore, the maximum length of a local loop using 19-gauge wire is twice as long as a local loop using 26-gauge wire.

The total attenuation of a local loop is generally limited to a maximum value of 7.5 dB with a maximum dc resistance of 1300 Ω, which includes the resistance of the telephone (approximately 120 Ω). The dc resistance of 26-gauge copper wire is approximately 41 Ω per 1000 feet, which limits the round-trip loop length to approximately 5.6 miles. The maximum distance for lower-gauge wire is longer of course.

The dc loop resistance for copper conductors is approximated by

$$R_{dc} = \frac{0.1095}{d^2} \tag{9-1}$$

where R_{dc} = dc loop resistance (ohms per mile)
       d = wire diameter (inches)

9-3 TELEPHONE MESSAGE–CHANNEL NOISE AND NOISE WEIGHTING

The *noise* that reaches a listener's ears affects the degree of annoyance to the listener and, to some extent, the intelligibility of the received speech. The total noise is comprised of room *background noise* and noise introduced in the circuit. Room background noise on the listening subscriber's premises reaches the ear directly through leakage around the receiver and indirectly by way of the sidetone path through the telephone set. Room noise from the talking subscriber's premises also reaches the listener over the communications channel. Circuit noise is comprised mainly of thermal noise, nonlinear distortion, and impulse noise, which are described in a later section of this chapter.

The measurement of interference (noise), like the measurement of volume, is an effort to characterize a complex signal. Noise measurements on a telephone message channel are characterized by how annoying the noise is to the subscriber rather than by the absolute magnitude of the average noise power. Noise interference is comprised of two components: annoyance and the effect of noise on intelligibility, both of which are functions of frequency. Noise signals with equal interfering effects are assigned equal magnitudes. To accomplish this effect, the American Telephone and Telegraph Company (AT&T) developed a weighting network called *C-message* weighting.

When designing the C-message weighting network, groups of observers were asked to adjust the loudness of 14 different frequencies between 180 Hz and 3500 Hz until the sound of each tone was judged to be equally annoying as a 1000-Hz reference tone in the absence of speech. A 1000-Hz tone was selected for the reference because empirical data indicated that 1000 Hz is the most annoying frequency (i.e., the best frequency response)

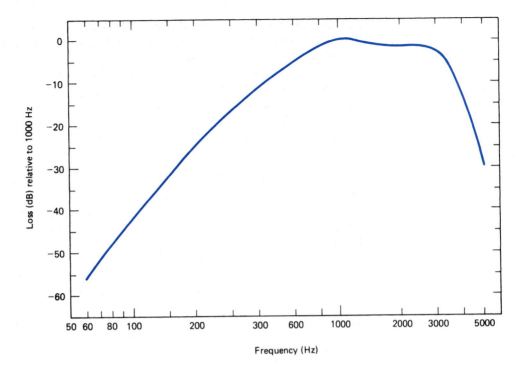

FIGURE 9-2 C-message weighting curve

to humans. The same people were then asked to adjust the amplitude of the tones in the presence of speech until the effect of noise on articulation (annoyance) was equal to that of the 1000-Hz reference tone. The results of the two experiments were combined, smoothed, and plotted, resulting in the C-message weighting curve shown in Figure 9-2. A 500-type telephone set was used for these tests; therefore, the C-message weighting curve includes the frequency response characteristics of a standard telephone set receiver as well as the hearing response of an average listener.

The significance of the C-message weighting curve is best illustrated with an example. From Figure 9-2, it can be seen that a 200-Hz test tone of a given power is 25 dB less disturbing than a 1000-Hz test tone of the same power. Therefore, the C-message weighting network will introduce 25 dB more loss for 200 Hz than it will for 1000 Hz.

When designing the C-message network, it was found that the additive effect of several noise sources combine on a root-sum-square (RSS) basis. From these design considerations, it was determined that a telephone message-channel noise measuring set should be a voltmeter with the following characteristics:

Readings should take into consideration that the interfering effect of noise is a function of frequency as well as magnitude.

When dissimilar noise signals are present simultaneously, the meter should combine them to properly measure the overall interfering effect.

It should have a transient response resembling that of the human ear. For sounds shorter than 200 ms, the human ear does not fully appreciate the true power of the sound. Therefore, noise-measuring sets are designed to give full-power indication only for bursts of noise lasting 200 ms or longer.

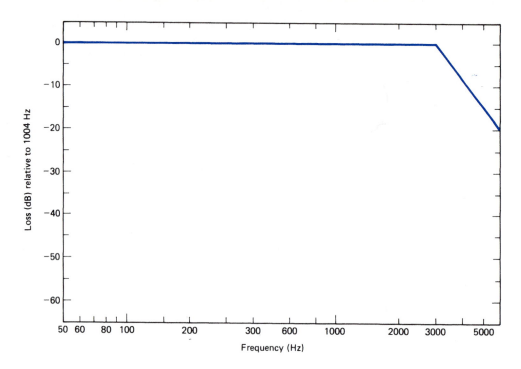

FIGURE 9-3 3-kHz flat response curve

When different types of noise cause equal interference as determined in subjective tests, use of the meter should give equal readings.

The reference established for performing message-channel noise measurements is −90 dBm (10^{-12} watts). The power level of −90 dBm was selected because, at the time, power levels could not measure levels below −90 dBm, and therefore it would not be necessary to deal with negative values when reading noise levels. Thus, a 1000-Hz tone with a power level of −90 dBm is equal to a noise reading of 0 dBrn. Conversely, a 1000-Hz tone with a power level of 0 dBm is equal to a noise reading of 90 dBrn, and a 1000-Hz tone with a power level of −40 dBm is equal to a noise reading of 50 dBrn.

When appropriate, other weighting networks can be substituted for C-message. For example, a *3-kHz flat* network is used to measure power density of white noise. This network has a nominal low-pass frequency response down 3 dB at 3 kHz and rolls off at 12 dB per octave. A 3-kHz flat network is often used for measuring high levels of low-frequency noise, such as power supply hum. The frequency response for a 3-kHz flat network is shown in Figure 9-3.

9-4 UNITS OF POWER MEASUREMENT

9-4-1 dB and dBm

To specify the amplitudes of signals and interference, it is often convenient to define them at some reference point in the system. The amplitudes at any other physical location can then be related to this reference point if the loss or gain between the two points is known. For example, sea level is generally used as the reference point when comparing elevations. By referencing two mountains to sea level, we can compare the two elevations regardless of where the mountains are located. A mountain peak in Colorado 12,000 feet above sea level is 4000 feet higher than a mountain peak in France 8000 feet above sea level.

The decibel (dB) is the basic yardstick used for making power measurements in communications. The unit dB is simply a logarithmic expression representing the ratio of one power level to another and expressed mathematically as

$$dB = 10 \log\left(\frac{P_1}{P_2}\right) \qquad (9\text{-}2)$$

where P_1 and P_2 are power levels at two different points in a transmission system.

From Equation 9-2, it can be seen that when $P_1 = P_2$, the power ratio is 0 dB; when $P_1 > P_2$, the power ratio in dB is positive; and when $P_1 < P_2$, the power ratio in dB is negative. In telephone and telecommunications circuits, power levels are given in dBm and differences between power levels in dB.

Equation 9-2 is essentially dimensionless since neither power is referenced to a base. The unit dBm is often used to reference the power level at a given point to 1 milliwatt. One milliwatt is the level from which all dBm measurements are referenced. The unit dBm is an indirect measure of absolute power and expressed mathematically as

$$dBm = 10 \log\left(\frac{P}{1 \text{ mW}}\right) \qquad (9\text{-}3)$$

where P is the power at any point in a transmission system.

From Equation 9-3, it can be seen that a power level of 1 mW equates to 0 dBm, power levels above 1 mW have positive dBm values, and power levels less than 1 mW have negative dBm values.

Example 9-1

Determine

a. The power levels in dBm for signal levels of 10 mW and 0.5 mW.
b. The difference between the two power levels in dB.

Solution a. The power levels in dBm are determined by substituting into Equation 9-3:

$$dBm = 10 \log\left(\frac{10 \text{ mW}}{1 \text{ mW}}\right) = 10 \text{ dBm}$$

$$dBm = 10 \log\left(\frac{0.5 \text{ mW}}{1 \text{ mW}}\right) = -3 \text{ dBm}$$

b. The difference between the two power levels in dB is determined by substituting into Equation 9-2:

$$dB = 10 \log\left(\frac{10 \text{ mW}}{0.5 \text{ mW}}\right) = 13 \text{ dB}$$

or $\qquad 10 \text{ dBm} - (-3 \text{ dBm}) = 13 \text{ dB}$

The 10-mW power level is 13 dB higher than a 0.5-mW power level.

Experiments indicate that a listener cannot give a reliable estimate of the loudness of a sound but can distinguish the difference in loudness between two sounds. The ear's sensitivity to a change in sound power follows a logarithmic rather than a linear scale, and the dB has become the unit of this change.

9-4-2 Transmission Level Point, Transmission Level, and Data Level Point

Transmission level point (TLP) is defined as the optimum level of a test tone on a channel at some point in a communications system. The numerical value of the TLP does not describe the total signal power present at that point—it merely defines what the ideal level should be.

The *transmission level* (TL) at any point in a transmission system is the ratio in dB of the power of a signal at that point to the power the same signal would be at a 0-dBm transmission level point. For example, a signal at a particular point in a transmission system measures −13 dBm. Is this good or bad? This could be answered only if it is known what the signal strength should be at that point. TLP does just that. The reference for TLP is 0 dBm. A −15-dBm TLP indicates that, at this specific point in the transmission system, the signal should measure −15 dBm. Therefore, the transmission level for a signal that measures −13 dBm at a −15-dBm point is −2 dB. A 0 TLP is a TLP where the signal power should be 0 dBm. TLP says nothing about the actual signal level itself.

Data level point (DLP) is a parameter equivalent to TLP except TLP is used for voice circuits, whereas DLP is used as a reference for data transmission. The DLP is always 13 dB below the voice level for the same point. If the TLP is −15 dBm, the DLP at the same point is −28 dBm. Because a data signal is more sensitive to nonlinear distortion (harmonic and intermodulation distortion), data signals are transmitted at a lower level than voice signals.

9-4-3 Units of Measurement

Common units for signal and noise power measurements in the telephone industry include dBmO, rn, dBrn, dBrnc, dBrn 3-kHz flat, and dBrncO.

9-4-3-1 dBmO. *dBmO* is dBm referenced to a zero transmission level point (0 TLP). dBmO is a power measurement adjusted to 0 dBm that indicates what the power would be if the signal were measured at a 0 TLP. dBmO compares the actual signal level at a point with what that signal level should be at that point. For example, a signal measuring −17 dBm at a −16-dBm transmission level point is −1 dBmO (i.e., the signal is 1 dB below what it should be, or if it were measured at a 0 TLP, it would measure −1 dBm).

9-4-3-2 rn (reference noise). *rn* is the dB value used as the reference for noise readings. Reference noise equals −90 dBm or 1 pW (1×10^{-12} W). This value was selected for two reasons: (1) early noise measuring sets could not accurately measure noise levels lower than −90 dBm, and (2) noise readings are typically higher than −90 dBm, resulting in positive dB readings in respect to reference noise.

9-4-3-3 dBrn. *dBrn* is the dB level of noise with respect to reference noise (−90 dBm). dBrn is seldom used by itself since it does not specify a weighting. A noise reading of −50 dBm equates to 40 dBrn, which is 40 dB above reference noise (−50 − [− 90]) = 40 dBrn.

9-4-3-4 dBrnc. *dBrnc* is similar to dBrn except dBrnc is the dB value of noise with respect to reference noise using C-message weighting. Noise measurements obtained with a C-message filter are meaningful, as they relate the noise measured to the combined frequency response of a standard telephone and the human ear.

9-4-3-5 dBrn 3-kHz flat. *dBrn 3-kHz flat* noise measurements are noise readings taken with a filter that has a flat frequency response from 30 Hz to 3 kHz. Noise readings taken with a 3-kHz flat filter are especially useful for detecting low-frequency noise, such as power supply hum. dBrn 3-kHz flat readings are typically 1.5 dB higher than dBrnc readings for equal noise power levels.

9-4-3-6 dBrncO. *dBrncO* is the amount of noise in dBrnc corrected to a 0 TLP. A noise reading of 34 dBrnc at a +7-dBm TLP equates to 27 dBrncO. dBrncO relates noise power readings (dBrnc) to a 0 TLP. This unit establishes a common reference point throughout the transmission system.

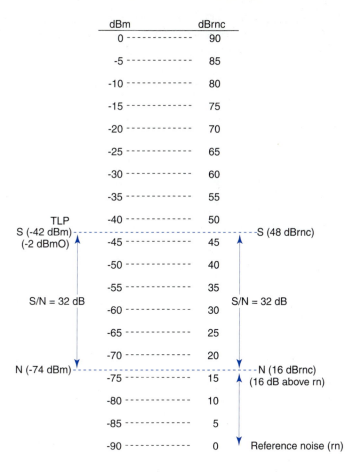

dBm		dBrnc
0	------------	90
-5	------------	85
-10	------------	80
-15	------------	75
-20	------------	70
-25	------------	65
-30	------------	60
-35	------------	55

TLP
S (-42 dBm) -------------------------------- S (48 dBrnc)
(-2 dBmO)

-40	------------	50
-45	------------	45
-50	------------	40
-55	------------	35

S/N = 32 dB   S/N = 32 dB

-60	------------	30
-65	------------	25
-70	------------	20

N (-74 dBm) -------------------------------- N (16 dBrnc)
(16 dB above rn)

-75	------------	15
-80	------------	10
-85	------------	5
-90	------------	0

Reference noise (rn)

FIGURE 9-4   Figure for Example 9-2

Example 9-2

For a signal measurement of -42 dBm, a noise measurement of 16 dBrnc, and a -40-dBm TLP, determine

a. Signal level in dBrnc.
b. Noise level in dBm.
c. Signal level in dBmO.
d. Signal-to-noise ratio in dB. (For the solutions, refer to Figure 9-4.)

Solution   a. The signal level in dBrnc can be read directly from the chart shown in Figure 9-4 as 48 dBrnc. The signal level in dBrnc can also be computed mathematically as follows:

$$-42 \text{ dBm} - (-90 \text{ dBrn}) = 48 \text{ dBrnc}$$

b. The noise level in dBm can be read directly from the chart shown in Figure 9-4 as -74 dBm. The noise level in dBm can also be calculated as follows:

$$-90 + 16 = -74 \text{ dBm}$$

c. The signal level in dBmO is simply the difference between the actual signal level in dBm and the TLP or 2 dBmO as shown in Figure 9-4. The signal level in dBmO can also be computed mathematically as follows:

$$-42 \text{ dBm} - (-40 \text{ dBm}) = -2 \text{ dBmO}$$

d. The signal-to-noise ratio is simply the difference in the signal power in dBm and the noise power in dBm or the signal level in dBrnc and the noise power in dBrnc as shown in Figure 9-4 as 32 dB. The signal-to-noise ratio is computed mathematically as

$$-42 \text{ dBm} - (-74 \text{ dBm}) = 32 \text{ dB}$$

or
$$48 \text{ dBrnc} - 16 \text{ dBrnc} = 32 \text{ dB}$$

9-4-4 Psophometric Noise Weighting

Psophometric noise weighting is used primarily in Europe. Psophometric weighting assumes a perfect receiver; therefore, its weighting curve corresponds to the frequency response of the human ear only. The difference between C-message weighting and psophometric weighting is so small that the same conversion factor may be used for both.

9-5 TRANSMISSION PARAMETERS AND PRIVATE-LINE CIRCUITS

Transmission parameters apply to dedicated *private-line data circuits* that utilize the private sector of the public telephone network—circuits with bandwidths comparable to those of standard voice-grade telephone channels that do not utilize the public switched telephone network. Private-line circuits are direct connections between two or more locations. On private-line circuits, transmission facilities and other telephone company–provided equipment are hardwired and available only to a specific subscriber. Most private-line data circuits use four-wire, full-duplex facilities. Signal paths established through switched lines are inconsistent and may differ greatly from one call to another. In addition, telephone lines provided through the public switched telephone network are two wire, which limits high-speed data transmission to half-duplex operation. Private-line data circuits have several advantages over using the switched public telephone network:

Transmission characteristics are more consistent because the same facilities are used with every transmission.

The facilities are less prone to noise produced in telephone company switches.

Line conditioning is available only on private-line facilities.

Higher transmission bit rates and better performance is appreciated with private-line data circuits.

Private-line data circuits are more economical for high-volume circuits.

Transmission parameters are divided into three broad categories: bandwidth parameters, which include attenuation distortion and envelope delay distortion; interface parameters, which include terminal impedance, in-band and out-of-band signal power, test signal power, and ground isolation; and facility parameters, which include noise measurements, frequency distortion, phase distortion, amplitude distortion, and nonlinear distortion.

9-5-1 Bandwidth Parameters

The only transmission parameters with limits specified by the FCC are attenuation distortion and envelope delay distortion. *Attenuation distortion* is the difference in circuit gain experienced at a particular frequency with respect to the circuit gain of a reference frequency. This characteristic is sometimes referred to as *frequency response, differential gain, and 1004-Hz deviation. Envelope delay distortion* is an indirect method of evaluating the phase delay characteristics of a circuit. FCC tariffs specify the limits for attenuation distortion and envelope delay distortion. To reduce attenuation and envelope delay distortion and improve the performance of data modems operating over standard message channels, it is often necessary to improve the quality of the channel. The process used to improve a basic telephone channel is called *line conditioning*. Line conditioning improves the high-frequency response of a message channel and reduces power loss.

The attenuation and delay characteristics of a circuit are artificially altered to meet limits prescribed by the *line conditioning* requirements. Line conditioning is available only to private-line subscribers at an additional charge. The *basic voice-band channel* (sometimes called a *basic 3002 channel*) satisfies the minimum line conditioning requirements. Telephone companies offer two types of special line conditioning for subscriber loops: C-type and D-type.

9-5-1-1 C-type line conditioning. *C-type conditioning* specifies the maximum limits for attenuation distortion and envelope delay distortion. C-type conditioning pertains to line impairments for which compensation can be made with filters and equalizers. This is accomplished with telephone company–provided equipment. When a circuit is initially turned up for service with a specific C-type conditioning, it must meet the requirements for that type of conditioning. The subscriber may include devices within the station equipment that compensate for minor long-term variations in the bandwidth requirements.

There are five classifications or levels of C-type conditioning available. The grade of conditioning a subscriber selects depends on the bit rate, modulation technique, and desired performance of the data modems used on the line. The five classifications of C-type conditioning are the following:

C1 and C2 conditioning pertain to two-point and multipoint circuits.

C3 conditioning is for access lines and trunk circuits associated with private switched networks.

C4 conditioning pertains to two-point and multipoint circuits with a maximum of four stations.

C5 conditioning pertains only to two-point circuits.

Private switched networks are telephone systems provided by local telephone companies dedicated to a single customer, usually with a large number of stations. An example is a large corporation with offices and complexes at two or more geographical locations, sometimes separated by great distances. Each location generally has an on-premise *private branch exchange* (PBX). A PBX is a relatively low-capacity switching machine where the subscribers are generally limited to stations within the same building or building complex. *Common-usage access lines* and *trunk circuits* are required to interconnect two or more PBXs. They are common only to the subscribers of the private network and not to the general public telephone network. Table 9-1 lists the limits prescribed by C-type conditioning for attenuation distortion. As the table shows, the higher the classification of conditioning imposed on a circuit, the flatter the frequency response and, therefore, a better-quality circuit.

Attenuation distortion is simply the frequency response of a transmission medium referenced to a 1004-Hz test tone. The attenuation for voice-band frequencies on a typical cable pair is directly proportional to the square root of the frequency. From Table 9-1, the attenuation distortion limits for a basic (unconditioned) circuit specify the circuit gain at any frequency between 500 Hz and 2500 Hz to be not more than 2 dB more than the circuit gain at 1004 Hz and not more than 3 dB below the circuit gain at 1004 Hz. For attenuation distortion, the circuit gain for 1004 Hz is always the reference. Also, within the frequency bands from 300 Hz and 499 Hz and from 2501 Hz to 3000 Hz, the circuit gain cannot be

Table 9–1   Basic and C-Type Conditioning Requirements

	Attenuation Distortion (Frequency Response Relative to 1004 Hz)		Envelope Delay Distortion	
Channel Conditioning	Frequency Range (Hz)	Variation (dB)	Frequency Range (Hz)	Variation (μs)
Basic	300–499	+3 to −12	800–2600	1750
	500–2500	+2 to −8		
	2501–3000	+3 to −12		
C1	300–999	+2 to −6	800–999	1750
	1000–2400	+1 to −3	1000–2400	1000
	2401–2700	+3 to −6	2401–2600	1750
	2701–3000	+3 to −12		
C2	300–499	+2 to −6	500–600	3000
	500–2800	+1 to −3	601–999	1500
	2801–3000	+2 to −6	1000–2600	500
			2601–2800	3000
C3 (access line)	300–499	+0.8 to −3	500–599	650
	500–2800	+0.5 to −1.5	600–999	300
	2801–3000	+0.8 to −3	1000–2600	110
			2601–2800	650
C3 (trunk)	300–499	+0.8 to −2	500–599	500
	500–2800	+0.5 to −1	600–999	260
	2801–3000	+0.8 to −2	1000–2600	80
			2601–3000	500
C4	300–499	+2 to −6	500–599	3000
	500–3000	+2 to −3	600–799	1500
	3001–3200	+2 to −6	800–999	500
			1000–2600	300
			2601–2800	500
			2801–3000	1500
C5	300–499	+1 to −3	500–599	600
	500–2800	+0.5 to −1.5	600–999	300
	2801–3000	+1 to −3	1000–2600	100
			2601–2800	600

greater than 3 dB above or more than 12 dB below the gain at 1004 Hz. Figure 9-5 shows a graphical presentation of basic line conditioning requirements.

Figure 9-6 shows a graphical presentation of the attenuation distortion requirements specified in Table 9-1 for C2 conditioning, and Figure 9-7 shows the graph for C2 conditioning superimposed over the graph for basic conditioning. From Figure 9-7, it can be seen that the requirements for C2 conditioning are much more stringent than those for a basic circuit.

Example 9-3

A 1004-Hz test tone is transmitted over a telephone circuit at 0 dBm and received at −16 dBm. Determine

a. The 1004-Hz circuit gain.
b. The attenuation distortion requirements for a basic circuit.
c. The attenuation distortion requirements for a C2 conditioned circuit.

Solution   **a.** The circuit gain is determined mathematically as

$$0 \text{ dBm} - (-16 \text{ dB}) = -16 \text{ dB (which equates to a loss of 16 dB)}$$

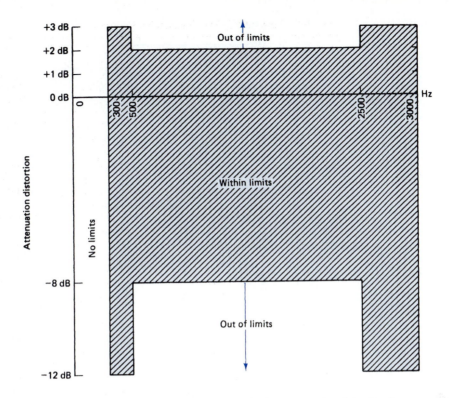

FIGURE 9-5 Graphical presentation of the limits for attenuation distortion for a basic 3002 telephone circuit

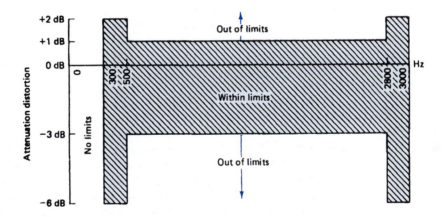

FIGURE 9-6 Graphical presentation of the limits for attenuation distortion for a C2 conditioned telephone circuit

b. Circuit gain requirements for a basic circuit can be determined from Table 9-1:

Frequency Band	Requirements	Minimum Level	Maximum Level
500 Hz and 2500 Hz	+2 dB and −8 dB	−24 dBm	−14 dBm
300 Hz and 499 Hz	+3 dB and −12 dB	−28 dBm	−13 dBm
2501 Hz and 3000 Hz	+3 dB and −12 dB	−28 dBm	−13 dBm

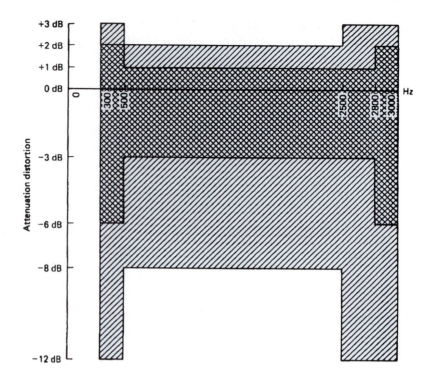

FIGURE 9-7 Overlay of Figure 9-5 over Figure 9-6 to demonstrate the more stringent requirements imposed by C2 conditioning compared to a basic (unconditioned) circuit

c. Circuit gain requirements for a C2 conditioned circuit can be determined from Table 9-1:

Frequency Band	Requirements	Minimum Level	Maximum Level
500 Hz and 2500 Hz	+1 dB and −3 dB	−19 dBm	−15 dBm
300 Hz and 499 Hz	+2 dB and −6 dB	−22 dBm	−14 dBm
2801 Hz and 3000 Hz	+2 dB and −6 dB	−22 dBm	−14 dBm

A linear phase-versus-frequency relationship is a requirement for error-free data transmission—signals are delayed more at some frequencies than others. Delay distortion is the difference in phase shifts with respect to frequency that signals experience as they propagate through a transmission medium. This relationship is difficult to measure because of the difficulty in establishing a phase (time) reference. Envelope delay is an alternate method of evaluating the phase–versus–frequency relationship of a circuit.

The time delay encountered by a signal as it propagates from a source to a destination is called *propagation time,* and the delay measured in angular units, such as degrees or radians, is called *phase delay.* All frequencies in the usable voice band (300 Hz to 3000 Hz) do not experience the same time delay in a circuit. Therefore, a complex waveform, such as the output of a data modem, does not possess the same phase–versus–frequency relationship when received as it possessed when it was transmitted. This condition represents a possible impairment to a data signal. The *absolute phase delay* is the actual time required for a particular frequency to propagate from a source to a destination through a communications channel. The difference between the absolute delays of all the frequencies is phase distortion. A graph of phase delay–versus–frequency for a typical circuit is nonlinear.

By definition, envelope delay is the first derivative (slope) of phase with respect to frequency:

$$\text{envelope delay} = \frac{d\theta(\omega)}{d\omega} \tag{9-4}$$

In actuality, envelope delay only closely approximates $d\theta(\omega)/d\omega$. Envelope delay measurements evaluate not the true phase–versus–frequency characteristics but rather the phase of a wave that is the result of a narrow band of frequencies. It is a common misconception to confuse true phase distortion (also called delay distortion) with envelope delay distortion (EDD). *Envelope delay* is the time required to propagate a change in an AM envelope (the actual information-bearing part of the signal) through a transmission medium. To measure envelope delay, a narrowband amplitude-modulated carrier, whose frequency is varied over the usable voice band, is transmitted (the amplitude-modulated rate is typically between 25 Hz and 100 Hz). At the receiver, phase variations of the low-frequency envelope are measured. The phase difference at the different carrier frequencies is *envelope delay distortion*. The carrier frequency that produces the minimum envelope delay is established as the reference and is normalized to zero. Therefore, EDD measurements are typically given in microseconds and yield only positive values. EDD indicates the relative envelope delays of the various carrier frequencies with respect to the reference frequency. The reference frequency of a typical voice-band circuit is typically around 1800 Hz.

EDD measurements do not yield true phase delays, nor do they determine the relative relationships between true phase delays. EDD measurements are used to determine a close approximation of the relative phase delay characteristics of a circuit. Propagation time cannot be increased. Therefore, to correct delay distortion, equalizers are placed in a circuit to slow down the frequencies that travel the fastest more than frequencies that travel the slowest. This reduces the difference between the fastest and slowest frequencies, reducing the phase distortion.

The EDD limits for basic and conditioned telephone channels are listed in Table 9-1. Figure 9-8 shows a graphical representation of the EDD limits for a basic telephone channel,

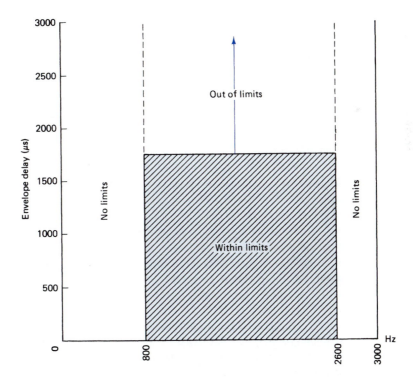

FIGURE 9-8 Graphical presentation of the limits for envelope delay in a basic telephone channel

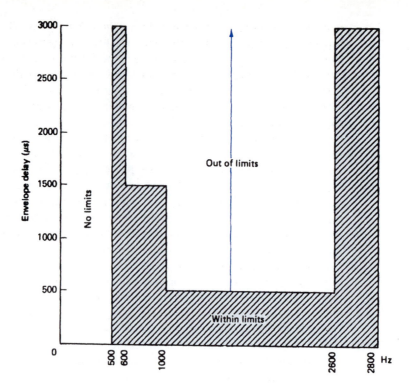

FIGURE 9-9 Graphical presentation of the limits for envelope delay in a telephone channel with C2 conditioning

and Figure 9-9 shows a graphical representation of the EDD limits for a channel meeting the requirements for C2 conditioning. From Table 9-1, the EDD limit of a basic telephone channel is 1750 μs between 800 Hz and 2600 Hz. This indicates that the maximum difference in envelope delay between any two carrier frequencies (the fastest and slowest frequencies) within this range cannot exceed 1750 μs.

Example 9-4

An EDD test on a basic telephone channel indicated that an 1800-Hz carrier experienced the minimum absolute delay of 400 μs. Therefore, it is the reference frequency. Determine the maximum absolute envelope delay that any frequency within the 800-Hz to 2600-Hz range can experience.

Solution The maximum envelope delay for a basic telephone channel is 1750 μs within the frequency range of 800 Hz to 2600 Hz. Therefore, the maximum envelope delay is 2150 μs (400 μs + 1750 μs).

The absolute time delay encountered by a signal between any two points in the continental United States should never exceed 100 ms, which is not sufficient to cause any problems. Consequently, relative rather than absolute values of envelope delay are measured. For the previous example, as long as EDD tests yield relative values less than +1750 μs, the circuit is within limits.

9-5-1-2 D-type line conditioning. *D-type conditioning* neither reduces the noise on a circuit nor improves the signal-to-noise ratio. It simply sets the minimum requirements for *signal-to-noise* (S/N) *ratio* and *nonlinear distortion.* If a subscriber requests D-type conditioning and the facilities assigned to the circuit do not meet the requirements, a different facility is assigned. D-type conditioning is simply a requirement and does not add anything to the circuit, and it cannot be used to improve a circuit. It simply places higher requirements on circuits used for high-speed data transmission. Only circuits that meet D-type conditioning requirements can be used for high-speed data transmission. D-type conditioning is sometimes referred to as *high-performance conditioning* and can be applied to private-line data circuits in addition to either basic or C-conditioned requirements. There are two categories for D-type conditioning: D1 and D2. Limits imposed by D1 and D2 are virtually identical. The only difference between the two categories is the circuit arrangement to which they apply. D1 conditioning specifies requirements for two-point circuits, and D2 conditioning specifies requirements for multipoint circuits.

D-type conditioning is mandatory when the data transmission rate is 9600 bps because without D-type conditioning, it is highly unlikely that the circuit can meet the minimum performance requirements guaranteed by the telephone company. When a telephone company assigns a circuit to a subscriber for use as a 9600-bps data circuit and the circuit does not meet the minimum requirements of D-type conditioning, a new circuit is assigned. This is because a circuit cannot generally be upgraded to meet D-type conditioning specifications by simply adding corrective devices, such as equalizers and amplifiers. Telephone companies do not guarantee the performance of data modems operating at bit rates above 9600 bps over standard voice-grade circuits.

D-type conditioned circuits must meet the following specifications:

Signal-to-C-notched noise ratio: ≥ 28 dB

Nonlinear distortion

Signal-to-second-order distortion: ≥ 35 dB

Signal-to-third-order distortion: ≥ 40 dB

The signal-to-notched noise ratio requirement for standard circuits is only 24 dB, and they have no requirements for nonlinear distortion.

Nonlinear distortion is an example of correlated noise and is produced from nonlinear amplification. When an amplifier is driven into a nonlinear operating region, the signal is distorted, producing multiples and sums and differences (cross products) of the original signal frequencies. The noise caused by nonlinear distortion is in the form of additional frequencies produced from nonlinear amplification of a signal. In other words, no signal, no noise. Nonlinear distortion produces distorted waveforms that are detrimental to digitally modulated carriers used with voice-band data modems, such as FSK, PSK, and QAM. Two classifications of nonlinear distortion are *harmonic distortion* (unwanted multiples of the transmitted frequencies) and *intermodulation distortion* (cross products [sums and differences] of the transmitted frequencies, sometimes called *fluctuation noise* or *cross-modulation noise*). Harmonic and intermodulation distortion, if of sufficient magnitude, can destroy the integrity of a data signal. The degree of circuit nonlinearity can be measured using either harmonic or intermodulation distortion tests.

Harmonic distortion is measured by applying a single-frequency test tone to a telephone channel. At the receive end, the power of the fundamental, second, and third harmonic frequencies is measured. Harmonic distortion is classified as second, third, or *n*th order or as total harmonic distortion. The actual amount of nonlinearity in a circuit is determined by comparing the power of the fundamental with the combined powers of the second and third harmonics. Harmonic distortion tests use a single-frequency (704-Hz) source (see Figure 9-10); therefore, no cross-product frequencies are produced.

Although simple harmonic distortion tests provide an accurate measurement of the nonlinear characteristics of analog telephone channel, they are inadequate for digital (T carrier) facilities. For this reason, a more refined method was developed that uses a multifrequency test-tone signal. Four test frequencies are used (see Figure 9-11): two designated

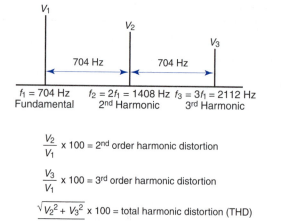

$f_1 = 704$ Hz $f_2 = 2f_1 = 1408$ Hz $f_3 = 3f_1 = 2112$ Hz
Fundamental   2nd Harmonic   3rd Harmonic

$$\frac{V_2}{V_1} \times 100 = \text{2nd order harmonic distortion}$$

$$\frac{V_3}{V_1} \times 100 = \text{3rd order harmonic distortion}$$

$$\frac{\sqrt{V_2{}^2 + V_3{}^2}}{V_1} \times 100 = \text{total harmonic distortion (THD)}$$

FIGURE 9-10 Harmonic distortion

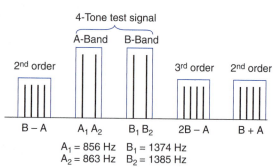

FIGURE 9-11 Intermodulation distortion

the A band ($A_1 = 856$ Hz, $A_2 = 863$ Hz) and two designated the B band ($B_1 = 1374$ Hz and $B_2 = 1385$ Hz). The four frequencies are transmitted with equal power levels, and the total combined power is equal to that of a normal data signal. The nonlinear amplification of the circuit produces multiples of each frequency (harmonics) and their cross-product frequencies (sum and difference frequencies). For reasons beyond the scope of this text, the following second- and third-order products were selected for measurement: B + A, B − A, and 2B − A. The combined signal power of the four A and B band frequencies is compared with the second-order cross products and then compared with the third-order cross products. The results are converted to dB values and then compared to the requirements of D-type conditioning.

Harmonic and intermodulation distortion tests do not directly determine the amount of interference caused by nonlinear circuit gain. They serve as a figure of merit only when evaluating circuit parameters.

9-5-2 Interface Parameters

The two primary considerations of the interface parameters are electrical protection of the telephone network and its personnel and standardization of design arrangements. The interface parameters include the following:

Station equipment impedances should be 600 Ω resistive over the usable voice band.

Station equipment should be isolated from ground by a minimum of 20 MΩ dc and 50 kΩ ac.

The basic voice-grade telephone circuit is a 3002 channel; it has an ideal bandwidth of 0 Hz to 4 kHz and a usable bandwidth of 300 Hz to 3000 Hz.

The circuit gain at 3000 Hz is 3 dB below the specified in-band signal power.

The gain at 4 kHz must be at least 15 dB below the gain at 3 kHz.

The maximum transmitted signal power for a private-line circuit is 0 dBm.

The transmitted signal power for dial-up circuits using the public switched telephone network is established for each loop so that the signal is received at the telephone central office at −12 dBm.

Table 9-2 summarizes interface parameter limits.

Table 9–2 Interface Parameter Limits

Parameter	Limit
1. Recommended impedance of terminal equipment	600 Ω resistive ± 10%
2. Recommended isolation to ground of terminal equipment	At least 20 MΩ dc
	At least 50 kΩ ac
	At least 1500 V rms breakdown voltage at 60 Hz
3. Data transmit signal power	0 dBm (3-s average)
4. In-band transmitted signal power	2450-Hz to 2750-Hz band should not exceed signal power in 800-Hz to 2450-Hz band
5. Out-of-band transmitted signal power	
Above voice band:	
(a) 3995 Hz–4005 Hz	At least 18 dB below maximum allowed in-band signal power
(b) 4-kHz–10-kHz band	Less than –16 dBm
(c) 10-kHz–25-kHz band	Less than –24 dBm
(d) 25-kHz–40-kHz band	Less than –36 dBm
(e) Above 40 kHz	Less than –50 dBm

Below voice band:

(f) rms current per conductor as specified by Telco but never greater than 0.35 A.

(g) Magnitude of peak conductor-to-ground voltage not to exceed 70 V.

(h) Conductor-to-conductor voltage shall be such that conductor-to-ground voltage is not exceeded. For an underground signal source, the conductor-to-conductor limit is the same as the conductor-to-ground limit.

(i) Total weighted rms voltage in band from 50 Hz to 300 Hz, not to exceed 100 V. Weighting factors for each frequency component (f) are $f^2/10^4$ for f between 50 Hz and 100 Hz and $f^{3.3}/10^{6.6}$ for f between 101 Hz and 300 Hz.

6. Maximum test signal power: same as transmitted data power.

9-5-3 Facility Parameters

Facility parameters represent potential impairments to a data signal. These impairments are caused by telephone company equipment and the limits specified pertain to all private-line data circuits using voice-band facilities, regardless of line conditioning. Facility parameters include 1004-Hz variation, C-message noise, impulse noise, gain hits and dropouts, phase hits, phase jitter, single-frequency interference, frequency shift, phase intercept distortion, and peak-to-average ratio.

9-5-3-1 1004-Hz variation.

The telephone industry has established 1004 Hz as the standard test-tone frequency; 1000 Hz was originally selected because of its relative location in the passband of a standard voice-band circuit. The frequency was changed to 1004 Hz with the advent of digital carriers because 1000 Hz is an exact submultiple of the 8-kHz sample rate used with T carriers. Sampling a continuous 1000-Hz signal at an 8000-Hz rate produced repetitive patterns in the PCM codes, which could cause the system to lose frame synchronization.

The purpose of the 1004-Hz test tone is to simulate the combined signal power of a standard voice-band data transmission. The 1004-Hz channel loss for a private-line data circuit is typically 16 dB. A 1004-Hz test tone applied at the transmit end of a circuit should be received at the output of the circuit at −16 dBm. Long-term variations in the gain of the transmission facility are called *1004-Hz variation* and should not exceed ±4 dB. Thus, the received signal power must be within the limits of −12 dBm to −20 dBm.

9-5-3-2 C-message noise.

C-message noise measurements determine the average weighted rms noise power. Unwanted electrical signals are produced from the random movement of electrons in conductors. This type of noise is commonly called *thermal noise* because its magnitude is directly proportional to temperature. Because the electron movement is completely random and travels in all directions, thermal noise is also called *random noise,* and because it contains all frequencies, it is sometimes referred to as *white noise.* Thermal noise is inherently present in a circuit because of its electrical makeup. Because thermal noise is additive, its magnitude is dependent, in part, on the electrical length of the circuit.

C-message noise measurements are the terminated rms power readings at the receive end of a circuit with the transmit end terminated in the characteristic impedance of the telephone line. Figure 9-12 shows the test setup for conducting terminated C-message noise readings. As shown in the figure, a C-message filter is placed between the circuit and the power meter in the noise measuring set so that the noise measurement evaluates the noise with a response similar to that of a human listening to the noise through a standard telephone set speaker.

There is a disadvantage to measuring noise this way. The overall circuit characteristics, in the absence of a signal, are not necessarily the same as when a signal is present. Using compressors, expanders, and automatic gain devices in a circuit causes this difference. For this reason, *C-notched noise* measurements were developed. C-notched noise measurements differ from standard C-message noise measurements only in the fact that a *holding tone* (usually 1004 Hz or 2804 Hz) is applied to the transmit end of the circuit while the noise measurement is taken. The holding tone ensures that the circuit operation simulates a loaded voice or data transmission. *Loaded* is a communications term that indicates the presence of a signal power comparable to the power of an actual message transmission. A narrowband notch filter removes the holding tone before the noise power is measured. The test setup for making C-notched noise measurements is shown in Figure 9-13. As the

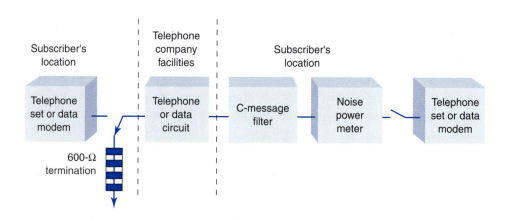

FIGURE 9-12   Terminated C-message noise test setup

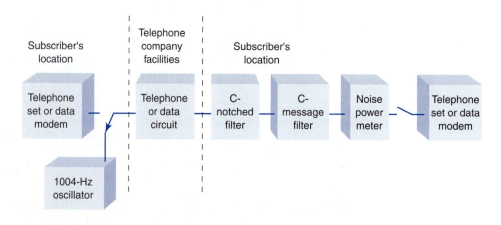

FIGURE 9-13   C-notched noise test setup

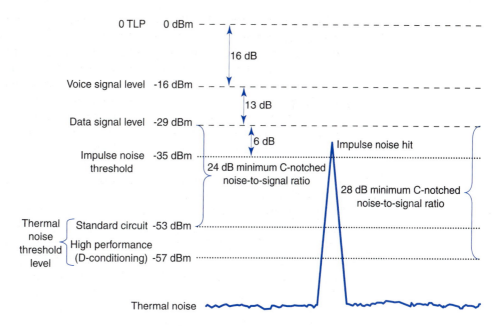

FIGURE 9-14 C-notched noise and impulse noise

figure shows, the notch filter is placed in front of the C-message filter, thus blocking the holding tone from reaching the power meter.

The physical makeup of a private-line data circuit may require using several carrier facilities and cable arrangements in tandem. Each facility may be analog, digital, or some combination of analog and digital. Telephone companies have established realistic C-notched noise requirements for each type of facility for various circuit lengths. Telephone companies guarantee standard private-line data circuits a minimum signal-to-C-notched noise ratio of 24 dB. A standard circuit is one operating at less than 9600 bps. Data circuits operating at 9600 bps require D-type conditioning, which guarantees a minimum signal-to-C-notched noise ratio of 28 dB. C-notched noise is shown in Figure 9-14. Telephone companies do not guarantee the performance of voice-band circuits operating at bit rates in excess of 9600 bps.

9-5-3-3 Impulse noise. *Impulse noise* is characterized by high-amplitude peaks (impulses) of short duration having an approximately flat frequency spectrum. Impulse noise can saturate a message channel. Impulse noise is the primary source of transmission errors in data circuits. There are numerous sources of impulse noise—some are controllable, but most are not. The primary cause of impulse noise is man-made sources, such as interference from ac power lines, transients from switching machines, motors, solenoids, relays, electric trains, and so on. Impulse noise can also result from lightning and other adverse atmospheric conditions.

The significance of impulse noise hits on data transmission has been a controversial topic. Telephone companies have accepted the fact that the absolute magnitude of the impulse hit is not as significant as the magnitude of the hit relative to the signal amplitude. Empirically, it has been determined that an impulse hit will not produce transmission errors in a data signal unless it comes within 6 dB of the signal level as shown in Figure 9-14. Impulse hit counters are designed to register a maximum of seven counts per second. This leaves a 143-ms lapse called a *dead time* between counts when additional impulse hits are not registered. Contemporary high-speed data formats transfer data in a block or frame format, and whether one hit or many hits occur during a single transmission is unimportant, as any error within a message generally necessitates retransmission of the entire message. It has been determined that counting additional impulses during the time of a single transmission does not correlate well with data transmission performance.

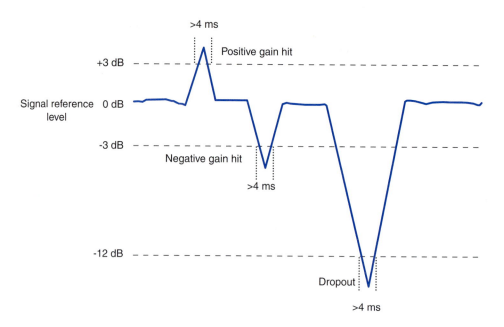

FIGURE 9-15 Gain hits and dropouts

Impulse noise objectives are based primarily on the error susceptibility of data signals, which depends on the type of modem used and the characteristics of the transmission medium. It is impractical to measure the exact peak amplitudes of each noise pulse or to count the number that occur. Studies have shown that expected error rates in the absence of other impairments are approximately proportional to the number of impulse hits that exceed the rms signal power level by approximately 2 dB. When impulse noise tests are performed, a 2802-Hz holding tone is placed on a circuit to ensure loaded circuit conditions. The counter records the number of hits in a prescribed time interval (usually 15 minutes). An impulse hit is typically less than 4 ms in duration and never more than 10 ms. Telephone company limits for recordable impulse hits is 15 hits within a 15-minute time interval. This does not limit the number of hits to one per minute but, rather, the average occurrence to one per minute.

9-5-3-4 Gain hits and dropouts.   A *gain hit* is a sudden, random change in the gain of a circuit resulting in a temporary change in the signal level. Gain hits are classified as temporary variations in circuit gain exceeding ±3 dB, lasting more than 4 ms, and returning to the original value within 200 ms. The primary cause of gain hits is noise transients (impulses) on transmission facilities during the normal course of a day.

A *dropout* is a decrease in circuit gain (i.e., signal level) of more than 12 dB lasting longer than 4 ms. Dropouts are characteristics of temporary open-circuit conditions and are generally caused by deep fades on radio facilities or by switching delays. Gain hits and dropouts are depicted in Figure 9-15.

9-5-3-5 Phase hits.   *Phase hits* (slips) are sudden, random changes in the phase of a signal. Phase hits are classified as temporary variations in the phase of a signal lasting longer than 4 ms. Generally, phase hits are not recorded unless they exceed ±20C° peak. Phase hits, like gain hits, are caused by transients produced when transmission facilities are switched. Phase hits are shown in Figure 9-16.

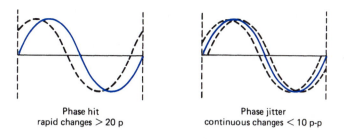

Phase hit
rapid changes > 20 p

Phase jitter
continuous changes < 10 p-p

FIGURE 9-16 Phase hits and phase jitter

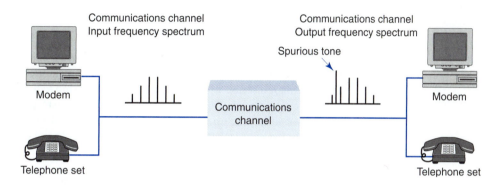

FIGURE 9-17 Single-frequency interference (spurious tone)

9-5-3-6 Phase jitter. *Phase jitter* is a form of incidental phase modulation—a continuous, uncontrolled variation in the zero crossings of a signal. Generally, phase jitter occurs at a 300-Hz rate or lower, and its primary cause is low-frequency ac ripple in power supplies. The number of power supplies required in a circuit is directly proportional to the number of transmission facilities and telephone offices that make up the message channel. Each facility has a separate phase jitter requirement; however, the maximum acceptable end-to-end phase jitter is 10° peak to peak regardless of how many transmission facilities or telephone offices are used in the circuit. Phase jitter is shown in Figure 9-16.

9-5-3-7 Single-frequency interference. *Single-frequency interference* is the presence of one or more continuous, unwanted tones within a message channel. The tones are called *spurious tones* and are often caused by crosstalk or cross modulation between adjacent channels in a transmission system due to system nonlinearities. Spurious tones are measured by terminating the transmit end of a circuit and then observing the channel frequency band. Spurious tones can cause the same undesired circuit behavior as thermal noise. Single-frequency interference is shown in Figure 9-17.

9-5-3-8 Frequency shift. *Frequency shift* is when the frequency of a signal changes during transmission. For example, a tone transmitted at 1004 Hz is received at 1005 Hz. Analog transmission systems used by telephone companies operate single-sideband suppressed carrier (SSBSC) and, therefore, require coherent demodulation. With coherent demodulation, carriers must be synchronous—the frequency must be reproduced exactly in the receiver. If this is not accomplished, the demodulated signal will be offset in frequency by the difference between transmit and receive carrier frequencies. The longer a circuit, the more analog transmission systems and the more likely frequency shift will occur. Frequency shift is shown in Figure 9-18.

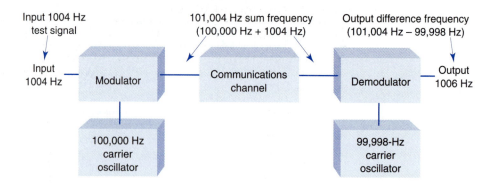

FIGURE 9-18 Frequency shift

9-5-3-9 Phase intercept distortion. *Phase intercept distortion* occurs in coherent SSBSC systems, such as those using frequency-division multiplexing when the received carrier is not reinserted with the exact phase relationship to the received signal as the transmit carrier possessed. This impairment causes a constant phase shift to all frequencies, which is of little concern for data modems using FSK, PSK, or QAM. Because these are practically the only techniques used today with voice-band data modems, no limits have been set for phase intercept distortion.

9-5-3-10 Peak-to-average ratio. The difficulties encountered in measuring true phase distortion or envelope delay distortion led to the development of *peak-to-average ratio* (PAR) tests. A signal containing a series of distinctly shaped pulses with a high peak voltage-to-average voltage ratio is transmitted. Differential delay distortion in a circuit has a tendency to spread the pulses, thus reducing the peak voltage-to-average voltage ratio. Low peak-to-average ratios indicate the presence of differential delay distortion. PAR measurements are less sensitive to attenuation distortion than EDD tests and are easier to perform.

9-5-3-11 Facility parameter summary. Table 9-3 summarizes facility parameter limits.

9-6 VOICE-FREQUENCY CIRCUIT ARRANGEMENTS

Electronic communications circuits can be configured in several ways. Telephone instruments and the voice-frequency facilities to which they are connected may be either *two wire* or *four wire*. Two-wire circuits have an obvious economic advantage, as they use only half as much copper wire. This is why most local subscriber loops connected to the public switched telephone network are two wire. However, most private-line data circuits are configured four wire.

9-6-1 Two-Wire Voice-Frequency Circuits

As the name implies, *two-wire transmission* involves two wires (one for the signal and one for a reference or ground) or a circuit configuration that is equivalent to using only two wires. Two-wire circuits are ideally suited to simplex transmission, although they are often used for half- and full-duplex transmission.

Figure 9-19 shows the block diagrams for four possible two-wire circuit configurations. Figure 9-19a shows the simplest two-wire configuration, which is a passive circuit consisting of two copper wires connecting a telephone or voice-band modem at one

Table 9-3 Facility Parameter Limits

Parameter	Limit		
1. 1004-Hz loss variation	Not more than ±4 dB long term		
2. C-message noise	Maximum rms noise at modem receiver (nominal − 16 dBm point)		
Facility miles		*dBm*	*dBrncO*
0–50		−61	32
51–100		−59	34
101–400		−58	35
401–1000		−55	38
1001–1500		−54	39
1501–2500		−52	41
2501–4000		−50	43
4001–8000		−47	46
8001–16,000		−44	49
3. C-notched noise	(minimum values)		
(a) Standard voice-band channel	24-dB signal to C-notched noise		
(b) High-performance line	28-dB signal to C-notched noise		
4. Single-frequency interference	At least 3 dB below C-message noise limits		
5. Impulse noise			
Threshold with respect to	*Maximum counts above threshold*		
1004-Hz holding tone	*allowed in 15 minutes*		
0 dB	15		
+4 dB	9		
+8 dB	5		
6. Frequency shift	±5 Hz end to end		
7. Phase intercept distortion	No limits		
8. Phase jitter	No more than 10° peak to peak (end-to-end requirement)		
9. Nonlinear distortion (D-conditioned circuits only)			
Signal to second order	At least 35 dB		
Signal to third order	At least 40 dB		
10. Peak-to-average ratio	Reading of 50 minimum end to end with standard PAR meter		
11. Phase hits	8 or less in any 15-minute period greater than ±20 peak		
12. Gain hits	8 or less in any 15-minute period greater than ±3 dB		
13. Dropouts	2 or less in any 15-minute period greater than 12 dB		

station through a telephone company interface to a telephone or voice-band modem at the destination station. The modem, telephone, and circuit configuration are capable of two-way transmission in either the half- or the full-duplex mode.

Figure 9-19b shows an active two-wire transmission system (i.e., one that provides gain). The only difference between this circuit and the one shown in Figure 9-19a is the addition of an amplifier to compensate for transmission line losses. The amplifier is unidirectional and, thus, limits transmission to one direction only (simplex).

Figure 9-19c shows a two-wire circuit using a digital T carrier for the transmission medium. This circuit requires a T carrier transmitter at one end and a T carrier receiver at the other end. The digital T carrier transmission line is capable of two-way transmission; however, the transmitter and receiver in the T carrier are not. The transmitter encodes the analog voice or modem signals into a PCM code, and the decoder in the receiver performs the opposite operation, converting PCM codes back to analog. The digital transmission medium is a pair of copper wires.

Figures 9-19a, b, and c are examples of *physical two-wire circuits,* as the two stations are physically interconnected with a two-wire metallic transmission line. Figure 9-19d shows an *equivalent two-wire circuit.* The transmission medium is earth's atmosphere, and there are no copper wires between the two stations. Although earth's atmosphere is capable

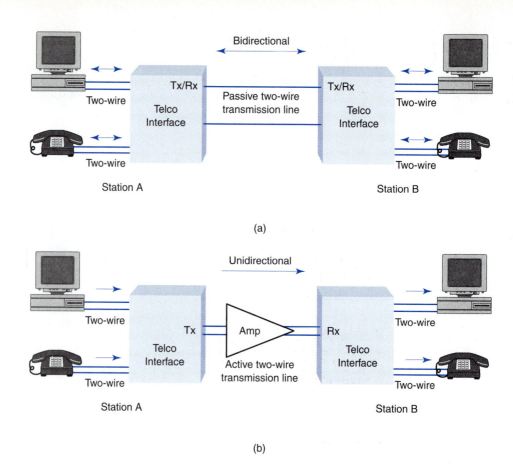

(a)

(b)

FIGURE 9-19 Two-wire configurations: (a) passive cable circuit; (b) active cable circuit
(*Continued*)

of two-way simultaneous transmission, the radio transmitter and receiver are not. Therefore, this is considered an equivalent two-wire circuit.

9-6-2 Four-Wire Voice-Frequency Circuits

As the name implies, *four-wire transmission* involves four wires (two for each direction—a signal and a reference) or a circuit configuration that is equivalent to using four wires. Four-wire circuits are ideally suited to full-duplex transmission, although they can (and very often do) operate in the half-duplex mode. As with two-wire transmission, there are two forms of four-wire transmission systems: *physical four wire* and *equivalent four wire.*

Figure 9-20 shows the block diagrams for four possible four-wire circuit configurations. As the figures show, a four-wire circuit is equivalent to two two-wire circuits, one for each direction of transmission. The circuits shown in Figures 9-20a, b, and c are physical four-wire circuits, as the transmitter at one station is hardwired to the receiver at the other station. Therefore, each two-wire pair is unidirectional (simplex), but the combined four-wire circuit is bidirectional (full duplex).

The circuit shown in Figure 9-20d is an equivalent four-wire circuit that uses earth's atmosphere for the transmission medium. Station A transmits on one frequency (f_1) and receives on a different frequency (f_2), while station B transmits on frequency f_2 and receives on frequency f_1. Therefore, the two radio signals do not interfere with one another, and simultaneous bidirectional transmission is possible.

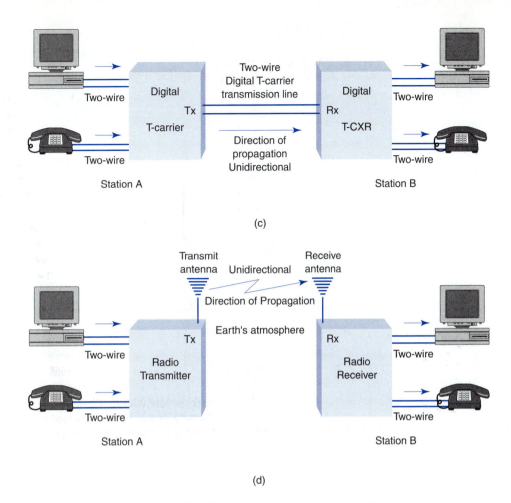

(c)

(d)

FIGURE 9-19 (Continued) (c) digital T carrier system; (d) wireless radio carrier system

9-6-3 Two Wire versus Four Wire

There are several inherent advantages of four-wire circuits over two-wire circuits. For instance, four-wire circuits are considerably less noisy, have less crosstalk, and provide more isolation between the two directions of transmission when operating in either the half- or the full-duplex mode. However, two-wire circuits require less wire, less circuitry, and thus less money than their four-wire counterparts.

Providing amplification is another disadvantage of four-wire operation. Telephone or modem signals propagated more than a few miles require amplification. A bidirectional amplifier on a two-wire circuit is not practical. It is much easier to separate the two directions of propagation with a four-wire circuit and install separate amplifiers in each direction.

9-6-4 Hybrids, Echo Suppressors, and Echo Cancelers

When a two-wire circuit is connected to a four-wire circuit, as in a long-distance telephone call, an interface circuit called a *hybrid,* or *terminating, set* is used to affect the interface. The hybrid set is used to match impedances and to provide isolation between the two directions of signal flow. The hybrid circuit used to convert two-wire circuits to four-wire circuits is similar to the hybrid coil found in standard telephone sets.

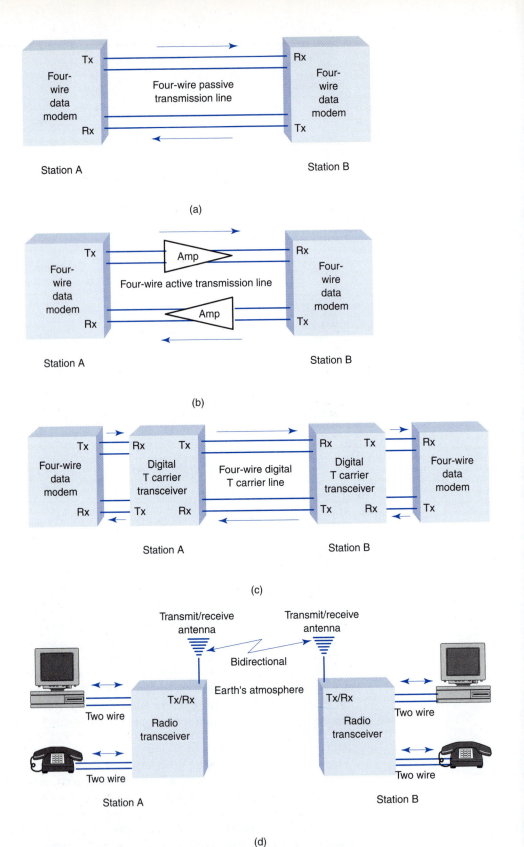

FIGURE 9-20 Four-wire configurations: (a) passive cable circuit; (b) active cable circuit; (c) digital T carrier system; (d) wireless radio carrier system

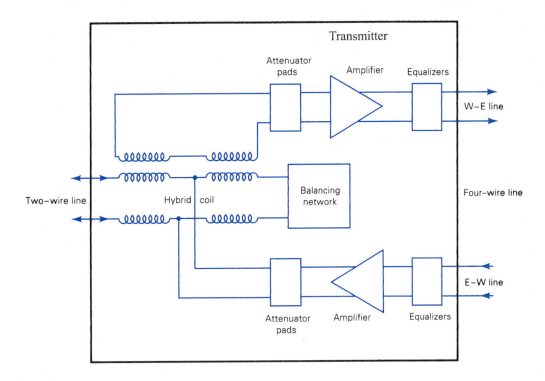

FIGURE 9-21 Hybrid (terminating) sets

Figure 9-21 shows the block diagram for a two-wire to four-wire hybrid network. The hybrid coil compensates for impedance variations in the two-wire portion of the circuit. The amplifiers and attenuators adjust the signal power to required levels, and the equalizers compensate for impairments in the transmission line that affect the frequency response of the transmitted signal, such as line inductance, capacitance, and resistance. Signals traveling west to east (W-E) enter the terminating set from the two-wire line, where they are inductively coupled into the west-to-east transmitter section of the four-wire circuit. Signals received from the four-wire side of the hybrid propagate through the receiver in the east-to-west (E-W) section of the four-wire circuit, where they are applied to the center taps of the hybrid coils. If the impedances of the two-wire line and the balancing network are properly matched, all currents produced in the upper half of the hybrid by the E-W signals will be equal in magnitude but opposite in polarity. Therefore, the voltages induced in the secondaries will be 180° out of phase with each other and, thus, cancel. This prevents any of the signals from being retransmitted to the sender as an echo.

If the impedances of the two-wire line and the balancing network are not matched, voltages induced in the secondaries of the hybrid coil will not completely cancel. This imbalance causes a portion of the received signal to be returned to the sender on the W-E portion of the four-wire circuit. Balancing networks can never completely match a hybrid to the subscriber loop because of long-term temperature variations and degradation of transmission lines. The talker hears the returned portion of the signal as an echo, and if the round-trip delay exceeds approximately 45 ms, the echo can become quite annoying. To eliminate this echo, devices called *echo suppressors* are inserted at one end of the four-wire circuit.

Figure 9-22 shows a simplified block diagram of an echo suppressor. The speech detector senses the presence and direction of the signal. It then enables the amplifier in the appropriate direction and disables the amplifier in the opposite direction, thus preventing the echo

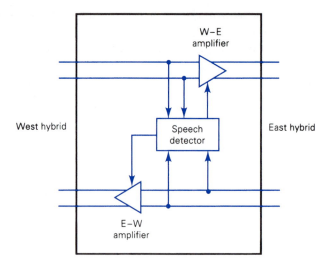

FIGURE 9-22 Echo suppressor

from returning to the speaker. A typical echo suppressor suppresses the returned echo by as much as 60 dB. If the conversation is changing direction rapidly, the people listening may be able to hear the echo suppressors turning on and off (every time an echo suppressor detects speech and is activated, the first instant of sound is removed from the message, giving the speech a choppy sound). If both parties talk at the same time, neither person is heard by the other.

With an echo suppressor in the circuit, transmissions cannot occur in both directions at the same time, thus limiting the circuit to half-duplex operation. Long-distance carriers, such as AT&T, generally place echo suppressors in four-wire circuits that exceed 1500 electrical miles in length (the longer the circuit, the longer the round-trip delay time). Echo suppressors are automatically disabled when they receive a tone between 2020 Hz and 2240 Hz, thus allowing full-duplex data transmission over a circuit with an echo suppressor. Full-duplex operation can also be achieved by replacing the echo suppressors with *echo cancelers*. Echo cancelers eliminate the echo by electrically subtracting it from the original signal rather than disabling the amplifier in the return circuit.

9-7 CROSSTALK

Crosstalk can be defined as any disturbance created in a communications channel by signals in other communications channels (i.e., unwanted coupling from one signal path into another). Crosstalk is a potential problem whenever two metallic conductors carrying different signals are located in close proximity to each other. Crosstalk can originate in telephone offices, at a subscriber's location, or on the facilities used to interconnect subscriber locations to telephone offices. Crosstalk is a subdivision of the general subject of interference. The term *crosstalk* was originally coined to indicate the presence of unwanted speech sounds in a telephone receiver caused by conversations on another telephone circuit.

The nature of crosstalk is often described as either *intelligible* or *unintelligible*. Intelligible (or near intelligible) crosstalk is particularly annoying and objectionable because the listener senses a real or fancied loss of privacy. Unintelligible crosstalk does not violate privacy, although it can still be annoying. Crosstalk between unlike channels, such as different types of carrier facilities, is usually unintelligible because of frequency inversion, frequency displacement, or digital encoding. However, such crosstalk often retains the syllabic pattern of speech and is more annoying than steady-state noise (such as thermal noise) with the same average power. Intermodulation noise, such as that found in multichannel frequency-division multiplexed telephone systems, is a form of interchannel crosstalk that is usually unintelligible. Unintelligible crosstalk is generally grouped with other types of noise interferences.

The use of the words *intelligible* and *unintelligible* can also be applied to non–voice circuits. The methods developed for quantitatively computing and measuring crosstalk between voice circuits are also useful when studying interference between voice circuits and data circuits and between two data circuits.

There are three primary types of crosstalk in telephone systems: nonlinear crosstalk, transmittance crosstalk, and coupling crosstalk.

9-7-1 Nonlinear Crosstalk

Nonlinear crosstalk is a direct result of nonlinear amplification (hence the name) in analog communications systems. Nonlinear amplification produces harmonics and cross products (sum and difference frequencies). If the nonlinear frequency components fall into the passband of another channel, they are considered crosstalk. Nonlinear crosstalk can be distinguished from other types of crosstalk because the ratio of the signal power in the disturbing channel to the interference power in the disturbed channel is a function of the signal level in the disturbing channel.

9-7-2 Transmittance Crosstalk

Crosstalk can also be caused by inadequate control of the frequency response of a transmission system, poor filter design, or poor filter performance. This type of crosstalk is most prevalent when filters do not adequately reject undesired products from other channels. Because this type of interference is caused by inadequate control of the transfer characteristics or transmittance of networks, it is called *transmittance crosstalk.*

9-7-3 Coupling Crosstalk

Electromagnetic coupling between two or more physically isolated transmission media is called *coupling crosstalk.* The most common coupling is due to the effects of near-field mutual induction between cables from physically isolated circuits (i.e., when energy radiates from a wire in one circuit to a wire in a different circuit). To reduce coupling crosstalk due to mutual induction, wires are twisted together (hence the name *twisted pair*). Twisting the wires causes a canceling effect that helps eliminate crosstalk. Standard telephone cable pairs have 20 twists per foot, whereas data circuits generally require more twists per foot. Direct capacitive coupling between adjacent cables is another means in which signals from one cable can be coupled into another cable. The probability of coupling crosstalk occurring increases with cable length, signal power, and frequency.

There are two types of coupling crosstalk: near end and far end. *Near-end crosstalk* (NEXT) is crosstalk that occurs at the transmit end of a circuit and travels in the opposite direction as the signal in the disturbing channel. *Far-end crosstalk* (FEXT) occurs at the far-end receiver and is energy that travels in the same direction as the signal in the disturbing channel.

9-7-4 Unit of Measurement

Crosstalk interference is often expressed in its own special decibel unit of measurement, dBx. Unlike dBm, where the reference is a fixed power level, dBx is referenced to the level on the cable that is being interfered with (whatever the level may be). Mathematically, dBx is

$$dBx = 90 - (\text{crosstalk loss in decibels}) \tag{9-5}$$

where 90 dB is considered the ideal isolation between adjacent lines.

For example, the magnitude of the crosstalk on a circuit is 70 dB lower than the power of the signal on the same circuit. The crosstalk is then 90 dB − 70 dBx = 20 dBx.

9-1. Briefly describe a *local subscriber loop*.

9-2. Explain what *loading coils* and *bridge taps* are and when they can be detrimental to the performance of a telephone circuit.

9-3. What are the designations used with *loading coils*?

9-4. What is meant by the term *loop resistance*?

9-5. Briefly describe *C-message noise weighting* and state its significance.

9-6. What is the difference between *dB* and *dBm*?

9-7. What is the difference between a *TLP* and a *DLP*?

9-8. What is meant by the following terms: *dBmO, rn, dBrn, dBrnc,* and *dBrncO*?

9-9. What is the difference between *psophometric noise weighting* and *C-message weighting*?

9-10. What are the three categories of *transmission parameters*?

9-11. Describe *attenuation distortion* and *envelope delay distortion*.

9-12. What is the reference frequency for attenuation distortion? Envelope delay distortion?

9-13. What is meant by *line conditioning*? What types of line conditioning are available?

9-14. What kind of circuits can have C-type line conditioning; D-type line conditioning?

9-15. When is D-type conditioning mandatory?

9-16. What limitations are imposed with D-type conditioning?

9-17. What is meant by *nonlinear distortion*? What are two kinds of nonlinear distortion?

9-18. What considerations are addressed by the *interface parameters*?

9-19. What considerations are addressed by *facility parameters*?

9-20. Briefly describe the following parameters: 1004-Hz variation, C-message noise, impulse noise, gain hits and dropouts, phase hits, phase jitter, single-frequency interference, frequency shift, phase intercept distortion, and peak-to-average ratio.

9-21. Describe what is meant by a *two-wire circuit* and a *four-wire circuit*.

9-22. Briefly describe the function of a two-wire-to-four-wire *hybrid set*.

9-23. What is the purpose of an *echo suppressor* and an *echo canceler*?

9-24. Briefly describe *crosstalk*.

9-25. What is the difference between *intelligible* and *unintelligible* crosstalk?

9-26. List and describe three types of crosstalk.

9-27. What is meant by *near-end crosstalk* and *far-end crosstalk*?

PROBLEMS

9-1. Describe what the following loading coil designations mean:

    a. 22B44
    b. 19H88
    c. 24B44
    d. 16B135

9-2. Frequencies of 250 Hz and 1 kHz are applied to the input of a C-message filter. Would their difference in amplitude be (greater, the same, or less) at the output of the filter?

9-3. A C-message noise measurement taken at a -22-dBm TLP indicates -72 dBm of noise. A test tone is measured at the same TLP at -25 dBm. Determine the following levels:

    a. Signal power relative to TLP (dBmO)
    b. C-message noise relative to reference noise (dBrn)
    c. C-message noise relative to reference noise adjusted to a 0 TLP (dBrncO)
    d. Signal-to-noise ratio

9-4. A C-message noise measurement taken at a -20-dBm TLP indicates a corrected noise reading of 43 dBrncO. A test tone at data level (0 DLP) is used to determine a signal-to-noise ratio of 30 dB. Determine the following levels:

    a. Signal power relative to TLP (dBmO)
    b. C-message noise relative to reference noise (dBrnc)
    c. Actual test-tone signal power (dBm)
    d. Actual C-message noise (dBm)

9-5. A test-tone signal power of -62 dBm is measured at a -61-dBm TLP. The C-message noise is measured at the same TLP at -10 dBrnc. Determine the following levels:

    a. C-message noise relative to reference noise at a O TLP (dBrncO)
    b. Actual C-message noise power level (dBm)
    c. Signal power level relative to TLP (dBmO)
    d. Signal-to-noise ratio (dB)

9-6. Sketch the graph for attenuation distortion and envelope delay distortion for a channel with C4 conditioning.

9-7. An EDD test on a basic telephone channel indicated that a 1600-Hz carrier experienced the minimum absolute delay of 550 μs. Determine the maximum absolute envelope delay that any frequency within the range of 800 Hz to 2600 Hz can experience.

9-8. The magnitude of the crosstalk on a circuit is 66 dB lower than the power of the signal on the same circuit. Determine the crosstalk in dBx.

C H A P T E R   10

The Public Telephone Network

CHAPTER OUTLINE

OBJECTIVES

- Define *public telephone company*
- Explain the differences between the public and private sectors of the public telephone network
- Define *telephone instruments, local loops, trunk circuits,* and *exchanges*
- Describe the necessity for central office telephone exchanges
- Briefly describe the history of the telephone industry
- Describe operator-assisted local exchanges
- Describe automated central office switches and exchanges and their advantages over operator-assisted local exchanges
- Define *circuits, circuit switches,* and *circuit switching*
- Describe the relationship between local telephone exchanges and exchange areas
- Define *interoffice trunks, tandem trunks,* and *tandem switches*
- Define *toll-connecting trunks, intertoll trunks,* and *toll offices*
- Describe the North American Telephone Numbering Plan Areas
- Describe the predivestiture North American Telephone Switching Hierarchy
- Define the five classes of telephone switching centers
- Explain switching routes

- Describe the postdivestiture North American Telephone Switching Hierarchy
- Define *Common Channel Signaling System No. 7 (SS7)*
- Describe the basic functions of SS7
- Define and describe SS7 signaling points

10-1 INTRODUCTION

The telecommunications industry is the largest industry in the world. There are over 1400 independent telephone companies in the United States, jointly referred to as the *public telephone network* (PTN). The PTN uses the largest computer network in the world to interconnect millions of subscribers in such a way that the myriad of companies function as a single entity. The mere size of the PTN makes it unique and truly a modern-day wonder of the world. Virtually any subscriber to the network can be connected to virtually any other subscriber to the network within a few seconds by simply dialing a telephone number. One characteristic of the PTN that makes it unique from other industries is that every piece of equipment, technique, or procedure, new or old, is capable of working with the rest of the system. In addition, using the PTN does not require any special skills or knowledge.

10-2 TELEPHONE TRANSMISSION SYSTEM ENVIRONMENT

In its simplest form, a telephone transmission system is a pair of wires connecting two telephones or data modems together. A more practical transmission system is comprised of a complex aggregate of electronic equipment and associated transmission medium, which together provide a multiplicity of channels over which many subscribers' messages and control signals are propagated.

In general, a telephone call between two points is handled by interconnecting a number of different transmission systems in tandem to form an overall *transmission path (connection)* between the two points. The manner in which transmission systems are chosen and interconnected has a strong bearing on the characteristics required of each system because each element in the connection degrades the message to some extent. Consequently, the relationship between the performance and the cost of a transmission system cannot be considered only in terms of that system. Instead, a transmission system must be viewed with respect to its relationship to the complete system.

To provide a service that permits people or data modems to talk to each other at a distance, the communications system (telephone network) must supply the means and facilities for connecting the subscribers at the beginning of a call and disconnecting them at the completion of the call. Therefore, switching, signaling, and transmission functions must be involved in the service. The *switching function* identifies and connects the subscribers to a suitable transmission path. *Signaling functions* supply and interpret control and supervisory signals needed to perform the operation. Finally, *transmission functions* involve the actual transmission of a subscriber's messages and any necessary control signals. New transmission systems are inhibited by the fact that they must be compatible with an existing multitrillion-dollar infrastructure.

10-3 THE PUBLIC TELEPHONE NETWORK

The public telephone network (PTN) accommodates two types of subscribers: *public* and *private.* Subscribers to the private sector are customers who lease equipment, transmission media (*facilities*), and services from telephone companies on a permanent basis. The leased circuits are designed and configured for their use only and are often referred to as *private-line* circuits or *dedicated* circuits. For example, large banks do not wish to share their communications network with other users, but it is not cost effective for them to construct their own networks. Therefore, banks lease equipment and facilities from public telephone companies and essentially operate a private telephone or data network within the PTN. The public telephone companies are sometimes called *service providers,* as they lease equipment and provide services to other private companies, organizations, and government agencies. Most metropolitan area networks (MANs) and wide area networks (WANs) utilize private-line data circuits and one or more service providers.

Subscribers to the public sector of the PTN share equipment and facilities that are available to all the public subscribers to the network. This equipment is appropriately called *common usage equipment,* which includes transmission facilities and telephone switches. Anyone with a telephone number is a subscriber to the public sector of the PTN. Since subscribers to the public network are interconnected only temporarily through switches, the network is often appropriately called the *public switched telephone network* (PSTN) and sometimes simply as the *dial-up network.* It is possible to interconnect telephones and modems with one another over great distances in fractions of a second by means of an elaborate network comprised of central offices, switches, cables (optical and metallic), and wireless radio systems that are connected by routing *nodes* (a node is a *switching point*). When someone talks about the public switched telephone network, they are referring to the combination of lines and switches that form a system of electrical routes through the network.

In its simplest form, data communications is the transmittal of digital information between two pieces of digital equipment, which includes computers. Several thousand miles may separate the equipment, which necessitates using some form of transmission medium to interconnect them. There is an insufficient number of transmission media capable of carrying digital information in digital form. Therefore, the most convenient (and least expensive) alternative to constructing an all-new all-digital network is to use the existing PTN for the transmission medium. Unfortunately, much of the PTN was designed (and much of it constructed) before the advent of large-scale data communications. The PTN was intended for transferring voice, not digital data. Therefore, to use the PTN for data communications, it is necessary to use a modem to convert the data to a form more suitable for transmission over the wireless carrier systems and conventional transmission media so prevalent in the PTN.

There are as many network configurations as there are subscribers in the private sector of the PTN, making it impossible to describe them all. Therefore, the intent of this chapter is to describe the public sector of the PTN (i.e., the public switched telephone network). Private-line data networks are described in later chapters of this book.

10-4 INSTRUMENTS, LOCAL LOOPS, TRUNK CIRCUITS, AND EXCHANGES

Telephone network equipment can be broadly divided into four primary classifications: instruments, local loops, trunk circuits, and exchanges.

10-4-1 Instruments

An *instrument* is any device used to originate and terminate calls and to transmit and receive signals into and out of the telephone network, such as a 2500-type telephone set, a cordless telephone, or a data modem. The instrument is often referred to as *station equipment* and the location of the instrument as the *station.* A *subscriber* is the operator or user of the instrument. If you have a home telephone, you are a subscriber.

10-4-2 Local Loops

As described in Chapters 8 and 9, the *local loop* is simply the dedicated cable facility used to connect an instrument at a subscriber's station to the closest telephone office. In the United States alone, there are several hundred million miles of cable used for local subscriber loops. Everyone who subscribes to the PTN is connected to the closest telephone office through a local loop. Local loops connected to the public switched telephone network are two-wire metallic cable pairs. However, local loops used with private-line data circuits are generally four-wire configurations.

10-4-3 Trunk Circuits

A *trunk circuit* is similar to a local loop except trunk circuits are used to interconnect two telephone offices. The primary difference between a local loop and a trunk is that a local loop is permanently associated with a particular station, whereas a trunk is a common-usage connection. A trunk circuit can be as simple as a pair of copper wires twisted together or as sophisticated as an optical fiber cable. A trunk circuit could also be a wireless communications channel. Although all trunk circuits perform the same basic function, there are different names given to them, depending on what types of telephone offices they interconnect and for what reason. Trunk circuits can be two wire or four wire, depending on what type of facility is used. Trunks are described in more detail in a later section of this chapter.

10-4-4 Exchanges

An *exchange* is a central location where subscribers are interconnected, either temporarily or on a permanent basis. Telephone company switching machines are located in exchanges. Switching machines are programmable matrices that provide temporary signal paths between two subscribers. Telephone sets and data modems are connected through local loops to switching machines located in exchanges. Exchanges connected directly to local loops are often called *local exchanges* or sometimes *dial switches* or *local dial switches*. The first telephone exchange was installed in 1878, only two years after the invention of the telephone. A central exchange is also called a *central telephone exchange, central office* (CO), *central wire center, central exchange, central office exchange,* or simply *central*.

The purpose of a telephone exchange is to provide a path for a call to be completed between two parties. To process a call, a switch must provide three primary functions:

Identify the subscribers
Set up or establish a communications path
Supervise the calling processes

10-5 LOCAL CENTRAL OFFICE TELEPHONE EXCHANGES

The first telephone sets were self-contained, as they were equipped with their own battery, microphone, speaker, bell, and ringing circuit. Telephone sets were originally connected directly to each other with heavy-gauge iron wire strung between poles, requiring a dedicated cable pair and telephone set for each subscriber you wished to be connected to. Figure 10-1a shows two telephones interconnected with a single cable pair. This is simple enough; however, if more than a few subscribers wished to be directly connected together, it became cumbersome, expensive, and very impractical. For example, to interconnect one subscriber to five other subscribers, five telephone sets and five cable pairs are needed, as shown in Figure 10-1b. To completely interconnect four subscribers, it would require six cable pairs, and each subscriber would need three telephone sets. This is shown in Figure 10-1c.

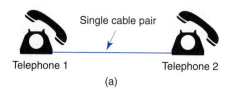

Telephone 1   Telephone 2

(a)

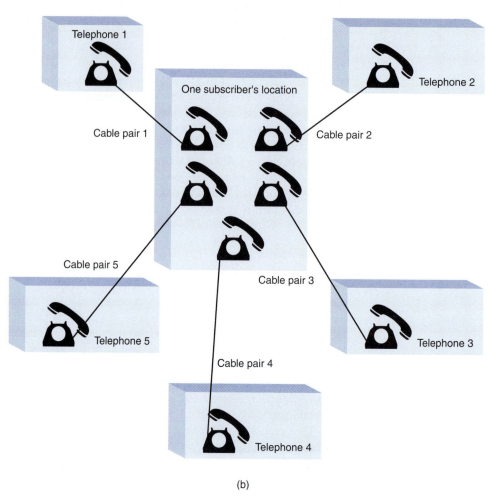

(b)

FIGURE 10-1 Dedicated telephone interconnections: (a) interconnecting two subscribers; (b) interconnecting one subscriber to five other telephone sets; (*Continued*)

The number of lines required to interconnect any number of stations is determined by the following equation:

$$N = \frac{n(n-1)}{2}$$ (10-1)

where n = number of stations (parties)
           N = number of interconnecting lines

The number of dedicated lines necessary to interconnect 100 parties is

$$N = \frac{100(100-1)}{2} = 4950$$

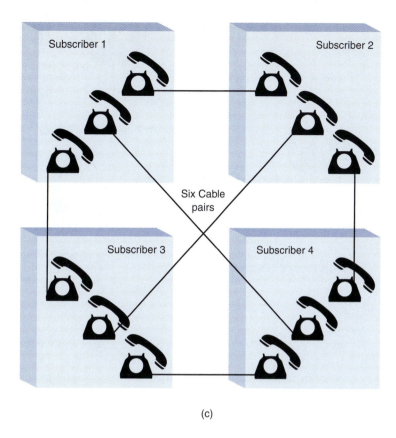

(c)

FIGURE 10-1 (*Continued*) (c) interconnecting four subscribers

In addition, each station would require either 100 separate telephones or the capability of switching one telephone to any of 99 lines.

These limitations rapidly led to the development of the *central telephone exchange.* A telephone exchange allows any telephone connected to it to be interconnected to any of the other telephones connected to the exchange without requiring separate cable pairs and telephones for each connection. Generally, a community is served by only one telephone company. The community is divided into zones, and each zone is served by a different central telephone exchange. The number of stations served and the density determine the number of zones established in a given community. If a subscriber in one zone wishes to call a station in another zone, a minimum of two local exchanges is required.

10-6 OPERATOR-ASSISTED LOCAL EXCHANGES

The first commercial telephone switchboard began operation in New Haven, Connecticut, on January 28, 1878, marking the birth of the public switched telephone network. The switchboard served 21 telephones attached to only eight lines (obviously, some were party lines). On February 17 of the same year, Western Union opened the first large-city exchange in San Francisco, California, and on February 21, the New Haven District Telephone Company published the world's first telephone directory comprised of a single page listing only 50 names. The directory was immediately followed by a more comprehensive listing by the Boston Telephone Dispatch Company.

The first local telephone exchanges were *switchboards* (sometimes called *patch panels* or *patch boards*) where manual interconnects were accomplished using *patchcords* and

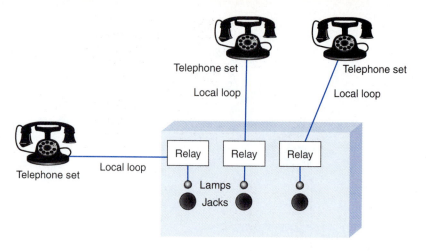

FIGURE 10-2 Patch panel configuration

jacks. All subscriber stations were connected through local loops to jacks on the switchboard. Whenever someone wished to initiate a call, they sent a ringing signal to the switchboard by manually turning a crank on their telephone. The ringing signal operated a relay at the switchboard, which in turn illuminated a supervisory lamp located above the jack for that line, as shown in Figure 10-2. Manual switchboards remained in operation until 1978, when the Bell System replaced their last cord switchboard on Santa Catalina Island off the coast of California near Los Angeles.

In the early days of telephone exchanges, each telephone line could have 10 or more subscribers (residents) connected to the central office exchange using the same local loop. This is called a *party line,* although only one subscriber could use their telephone at a time. Party lines are less expensive than private lines, but they are also less convenient. A private telephone line is more expensive because only telephones from one residence or business are connected to a local loop.

Connecting 100 private telephone lines to a single exchange required 100 local loops and a switchboard equipped with 100 relays, jacks, and lamps. When someone wished to initiate a telephone call, they rang the switchboard. An operator answered the call by saying, "Central." The calling party told the operator whom they wished to be connected to. The operator would then ring the destination, and when someone answered the telephone, the operator would remove her plug from the jack and connect the calling and called parties together with a special patchcord equipped with plugs on both ends. This type of system was called a *ringdown* system. If only a few subscribers were connected to a switchboard, the operator had little trouble keeping track of which jacks were for which subscriber (usually by name). However, as the popularity of the telephone grew, it soon became necessary to assign each subscriber line a unique telephone number. A switchboard using four digits could accommodate 10,000 telephone numbers (0000 to 9999).

Figure 10-3a shows a central office patch panel connected to four idle subscriber lines. Note that none of the telephone lines is connected to any of the other telephone lines. Figure 10-3b shows how subscriber 1 can be connected to subscriber 2 using a temporary connection provided by placing a patchcord between the jack for line 1 and the jack for line 2. Any subscriber can be connected to any other subscriber using patchcords.

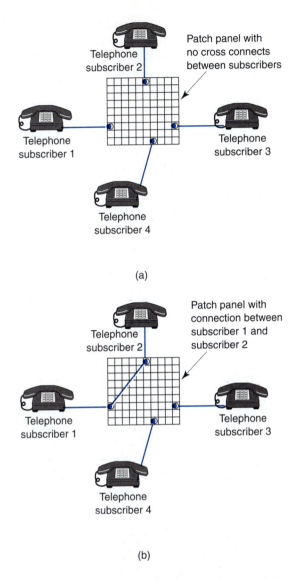

Telephone subscriber 2

Patch panel with no cross connects between subscribers

Telephone subscriber 1

Telephone subscriber 3

Telephone subscriber 4

(a)

Telephone subscriber 2

Patch panel with connection between subscriber 1 and subscriber 2

Telephone subscriber 1

Telephone subscriber 3

Telephone subscriber 4

(b)

FIGURE 10-3 Central office exchange: (a) without interconnects; (b) with an interconnect

10-7 AUTOMATED CENTRAL OFFICE SWITCHES AND EXCHANGES

As the number of telephones in the United States grew, it quickly became obvious that operator-assisted calls and manual patch panels could not meet the high demand for service. Thus, automated switching machines and exchange systems were developed.

An *automated switching system* is a system of sensors, switches, and other electrical and electronic devices that allows subscribers to give instructions directly to the switch without having to go through an operator. In addition, automated switches performed interconnections between subscribers without the assistance of a human and without using patchcords.

In 1890 an undertaker in Kansas City, Kansas, named Alman Brown Strowger was concerned that telephone company operators were diverting his business to his competitors. Consequently, he invented the first automated switching system using electromechanical relays. It is said that Strowger worked out his original design using a cardboard box, straight pins, and a pencil.

With the advent of the Strowger switch, mechanical dialing mechanisms were added to the basic telephone set. The mechanical dialer allowed subscribers to manually dial the telephone number of the party they wished to call. After a digit was entered (dialed), a relay in the switching machine connected the caller to another relay. The relays were called *stepping relays* because the system stepped through a series of relays as the digits were entered. The stepping process continued until all the digits of the telephone number were entered. This type of switching machine was called a *step-by-step* (SXS) switch, *stepper,* or, perhaps more commonly, a *Strowger* switch. A step-by-step switch is an example of a progressive switching machine, meaning that the connection between the calling and called parties was accomplished through a series of steps.

Between the early 1900s and the mid-1960s, the Strowger switch gradually replaced manual switchboards. The Bell System began using steppers in 1919 and continued using them until the early 1960s. In 1938, the Bell System began replacing the steppers with another electromechanical switching machine called the *crossbar* (XBAR) *switch.* The first No. 1 crossbar was cut into service at the Troy Avenue central office in Brooklyn, New York, on February 14, 1938. The crossbar switch used sets of contact points (called *crosspoints*) mounted on horizontal and vertical bars. Electromagnets were used to cause a vertical bar to cross a horizontal bar and make contact at a coordinate determined by the called number. The most versatile and popular crossbar switch was the #5XB. Although crossbar switches were an improvement over step-by-step switches, they were short lived, and most of them have been replaced with *electronic switching systems* (ESS).

In 1965, AT&T introduced the No. 1 ESS, which was the first computer-controlled central office switching system used on the PSTN. ESS switches differed from their predecessors in that they incorporate *stored program control* (SPC), which uses software to control practically all the switching functions. SPC increases the flexibility of the switch, dramatically increases its reliability, and allows for automatic monitoring of maintenance capabilities from a remote location. Virtually all the switching machines in use today are electronic stored program control switching machines. SPC systems require little maintenance and require considerably less space than their electromechanical predecessors. SPC systems make it possible for telephone companies to offer the myriad of services available today, such as three-way calling, call waiting, caller identification, call forwarding, call within, speed dialing, return call, automatic redial, and call tracing. Electronic switching systems evolved from the No. 1 ESS to the No. 5 ESS, which is the most advanced digital switching machine developed by the Bell System.

Automated central office switches paved the way for totally *automated central office exchanges,* which allow a caller located virtually anywhere in the world to direct dial virtually anyone else in the world. Automated central office exchanges interpret telephone numbers as an address on the PSTN. The network automatically locates the called number, tests its availability, and then completes the call.

10-7-1   Circuits, Circuit Switches, and Circuit Switching

A *circuit* is simply the path over which voice, data, or video signals propagate. In telecommunications terminology, a circuit is the path between a source and a destination (i.e., between a calling and a called party). Circuits are sometimes called *lines* (as in telephone lines). A *circuit switch* is a programmable matrix that allows circuits to be connected to one another. Telephone company circuit switches interconnect input loop or trunk circuits to output loop or trunk circuits. The switches are capable of interconnecting any circuit connected to it to any other circuit connected to it. For this reason, the switching process is called *circuit switching,* and therefore the public telephone network is considered a *circuit-switched network.* Circuit switches are *transparent.* That is, they interconnect circuits without altering the information on them. Once a circuit switching operation has been performed, a transparent switch simply provides continuity between two circuits.

10-7-2   Local Telephone Exchanges and Exchange Areas

Telephone exchanges are strategically placed around a city to minimize the distance between a subscriber's location and the exchange and also to optimize the number of stations connected to any one exchange. The size of the service area covered by an exchange depends on subscriber density and subscriber calling patterns. Today, there are over 20,000 local exchanges in the United States.

Exchanges connected directly to local loops are appropriately called *local exchanges.* Because local exchanges are centrally located within the area they serve, they are often called *central offices* (CO). Local exchanges can directly interconnect any two subscribers whose local loops are connected to the same local exchange. Figure 10-4a shows a local exchange with six telephones connected to it. Note that all six telephone numbers begin with 87. One subscriber of the local exchange can call another subscriber by simply dialing their seven-digit telephone number. The switching machine performs all tests and switching operations necessary to complete the call. A telephone call completed within a single local exchange is called an *intraoffice call* (sometimes called an *intraswitch* call). Figure 10-4b shows how two stations serviced by the same exchange (874-3333 to 874-4444) are interconnected through a common local switch.

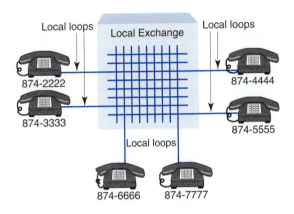

(a)

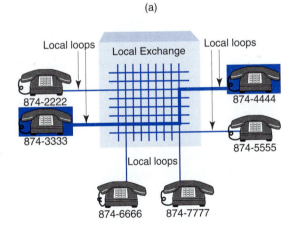

(b)

FIGURE 10-4 Local exchange: (a) no interconnections;
(b) 874-3333 connected to 874-4444

In the days of manual patch panels, to differentiate telephone numbers in one local exchange from telephone numbers in another local exchange and to make it easier for people to remember telephone numbers, each exchange was given a name, such as Bronx, Redwood, Swift, Downtown, Main, and so on. The first two digits of a telephone number were derived from the first two letters of the exchange name. To accommodate the names with dial telephones, the digits 2 through 9 were each assigned three letters. Originally, only 24 of the 26 letters were assigned (Q and Z were omitted); however, modern telephones assign all 26 letters to oblige personalizing telephone numbers (the digits 7 and 9 are now assigned four letters each). As an example, telephone numbers in the Bronx exchange begin with 27 (B on a telephone dial equates to the digit 2, and R on a telephone dial equates to the digit 7). Using this system, a seven-digit telephone number can accommodate 100,000 telephone numbers. For example, the Bronx exchange was assigned telephone numbers between 270-0000 and 279-9999 inclusive. The same 100,000 numbers could also be assigned to the Redwood exchange (730-0000 to 739-9999).

10-7-3   Interoffice Trunks, Tandem Trunks, and Tandem Switches

Interoffice calls are calls placed between two stations that are connected to different local exchanges. Interoffice calls are sometimes called *interswitch* calls. Interoffice calls were originally accomplished by placing special plugs on the switchboards that were connected to cable pairs going to local exchange offices in other locations around the city or in nearby towns. Today telephone-switching machines in local exchanges are interconnected to other local exchange offices on special facilities called *trunks* or, more specifically, *interoffice trunks*. A subscriber in one local exchange can call a subscriber connected to another local exchange over an interoffice trunk circuit in much the same manner that they would call a subscriber connected to the same exchange. When a subscriber on one local exchange dials the telephone number of a subscriber on another local exchange, the two local ex-

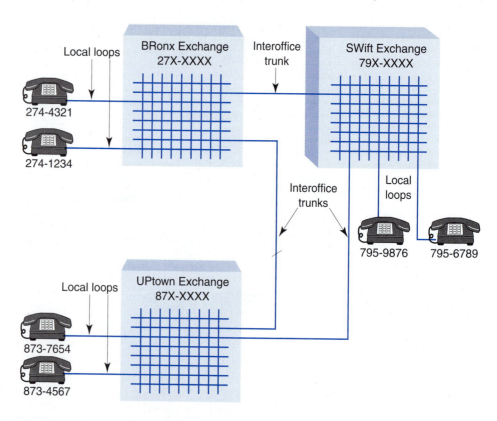

FIGURE 10-5   Interoffice exchange system

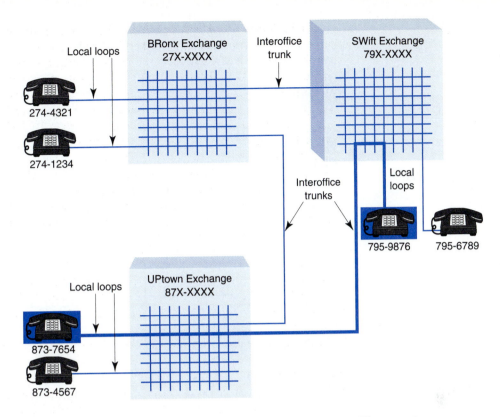

FIGURE 10-6 Interoffice call between subscribers serviced by two different exchanges

changes are interconnected with an interoffice trunk for the duration of the call. After either party terminates the call, the interoffice trunk is disconnected from the two local loops and made available for another interoffice call. Figure 10-5 shows three exchange offices with two subscribers connected to each. The telephone numbers for subscribers connected to the Bronx, Swift, and Uptown exchanges begin with the digits 27, 79, and 87, respectively. Figure 10-6 shows how two subscribers connected to different local exchanges can be interconnected using an interoffice trunk.

In larger metropolitan areas, it is virtually impossible to provide interoffice trunk circuits between all the local exchange offices. To interconnect local offices that do not have interoffice trunks directly between them, tandem offices are used. A *tandem office* is an exchange without any local loops connected to it (tandem meaning "in conjunction with" or "associated with"). The only facilities connected to the switching machine in a tandem office are trunks. Therefore, tandem switches interconnect local offices only. A *tandem switch* is called a *switcher's switch,* and trunk circuits that terminate in tandem switches are appropriately called *tandem trunks* or sometimes *intermediate trunks.*

Figure 10-7 shows two exchange areas that can be interconnected either with a tandem switch or through an interoffice trunk circuit. Note that tandem trunks are used to connect the Bronx and Uptown exchanges to the tandem switch. There is no name given to the tandem switch because there are no subscribers connected directly to it (i.e., no one receives dial tone from the tandem switch). Figure 10-8 shows how a subscriber in the Uptown exchange area is connected to a subscriber in the Bronx exchange area through a tandem switch. As the figure shows, tandem offices do not eliminate interoffice trunks. Very often, local offices have the capabilities to be interconnected with direct interoffice trunks as well as through a tandem office. When a telephone call is made from one local office to another, an interoffice trunk is selected if one is available. If not, a route through a tandem office is the second choice.

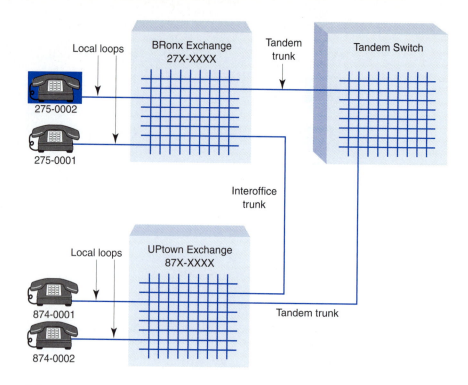

FIGURE 10-7 Interoffice switching between two local exchanges using tandem trunks and a tandem switch

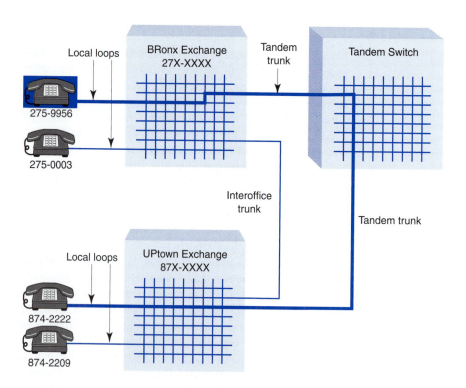

FIGURE 10-8 Interoffice call between two local exchanges through a tandem switch

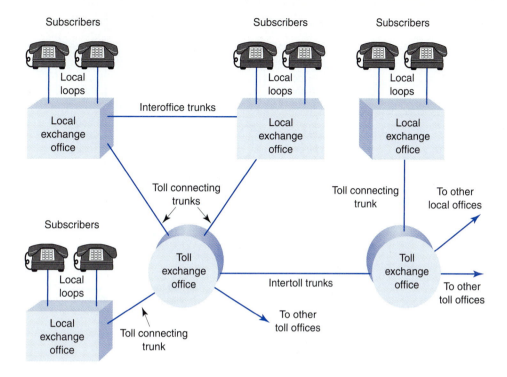

FIGURE 10-9 Relationship between local exchange offices and toll offices

10-7-4 Toll-Connecting Trunks, Intertoll Trunks, and Toll Offices

Interstate long-distance telephone calls require a special telephone office called a *toll office*. There are approximately 1200 toll offices in the United States. When a subscriber initiates a long-distance call, the local exchange connects the caller to a toll office through a facility called a *toll-connecting trunk* (sometimes called an *interoffice toll trunk*). Toll offices are connected to other toll offices with *intertoll trunks*. Figure 10-9 shows how local exchanges are connected to toll offices and how toll offices are connected to other toll offices. Figure 10-10 shows the network relationship between local exchange offices, tandem offices, toll offices, and their respective trunk circuits.

10-8 NORTH AMERICAN TELEPHONE NUMBERING PLAN AREAS

The *North American Telephone Numbering Plan* (NANP) was established to provide a telephone numbering system for the United States, Mexico, and Canada that would allow any subscriber in North America to direct dial virtually any other subscriber without the assistance of an operator. The network is often referred to as the DDD (*direct distance dialing*) network. Prior to the establishment of the NANP, placing a long-distance telephone call began by calling the long-distance operator and having her manually connect you to a trunk circuit to the city you wished to call. Any telephone number outside the caller's immediate area was considered a long-distance call.

North America is now divided into *numbering plan areas* (NPAs) with each NPA assigned a unique three-digit number called an *area code*. Each NPA is further subdivided into smaller service areas each with its own three-digit number called an *exchange code* (or

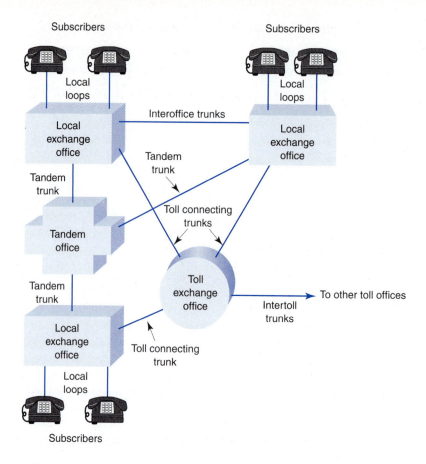

FIGURE 10-10 Relationship between local exchanges, tandem offices, and toll offices

prefix). Initially, each service area had only one central telephone switching office and one prefix. However, today a switching office can be assigned several exchange codes, depending on user density and the size of the area the office services. Each subscriber to a central office prefix is assigned a four-digit *extension number*. The three-digit area code represents the first three digits of a 10-digit telephone number, the three-digit prefix represents the next three digits, and the four-digit extension represents the last four digits of the telephone number.

Initially, within the North American Telephone Numbering Plan Area, if a digit could be any value from 0 through 9, the variable X designated it. If a digit could be any value from 2 through 9, the variable N designated it. If a digit could be only a 1 or a 0, it was designated by the variable 1/0 (one or zero). Area codes were expressed as N(1/0)N and exchange codes as NNX. Therefore, area codes could not begin or end with the digit 0 or 1, and the middle digit had to be either a 0 or a 1. Because of limitations imposed by electromechanical switching machines, the first two digits of exchange codes could not be 0 or 1, although the third digit could be any digit from 0 to 9. The four digits in the extension could be any digit value from 0 through 9. In addition, each NPA or area code could not have more than one local exchange with the same exchange code, and no two extension numbers within any exchanges codes could have the same four-digit number. The 18-digit telephone number was expressed as

$$N(1/0)N - NNX - XXXX$$

$$\underbrace{\qquad}_{\text{area code}} \quad \underbrace{\qquad}_{\text{prefix}} \quad \underbrace{\qquad}_{\text{extension}}$$

With the limitations listed for area codes, there were

$$N(1/0)N$$
$$(8)(2)(8) = 128 \text{ possibilities}$$

Each area code was assigned a cluster of exchange codes. In each cluster, there were

$$(N)(N)(X)$$
$$(8)(8)(10) = 640 \text{ possibilities}$$

Each exchange code served a cluster of extensions, in which there were

$$(X)(X)(X)(X)$$
$$(10)(10)(10)(10) = 10,000 \text{ possibilities}$$

With this numbering scheme, there were a total of $(128)(640)(10,000) = 819,200,000$ telephone numbers possible in North America.

When the NANP was initially placed into service, local exchange offices dropped their names and converted to their exchange number. Each exchange had 10 possible exchange codes. For example, the Bronx exchange was changed to 27 (B = 2 and r = 7). Therefore, it could accommodate the prefixes 270 through 279. Although most people do not realize it, telephone company employees still refer to local exchanges by their name. In January 1958, Wichita Falls, Texas, became the first American city to incorporate a true all-number calling system using a seven-digit number without attaching letters or names.

The popularity of cellular telephone has dramatically increased the demand for telephone numbers. By 1995, North America ran out of NPA area codes, so the requirement that the second digit be a 1 or a 0 was dropped. This was made possible because by 1995 there were very few electromagnetic switching machines in use in North America, and with the advent of SS7 signaling networks, telephone numbers no longer had to be transported over voice switching paths. This changed the numbering scheme to NXN-NNX-XXXX, which increased the number of area codes to 640 and the total number of telephones to 4,096,000,000. Figure 10-11 shows the North American Telephone Numbering Plan Areas as of January 2002.

The International Telecommunications Union has adopted an international numbering plan that adds a prefix in front of the area code, which outside North America is called a *city code*. The city code is one, two, or three digits long. For example, to call London, England, from the United States, one must dial 011-44-491-222-111. The 011 indicates an international call, 44 is the country code for England, 491 is the city code for London, 222 is the prefix for Piccadilly, and 111 is the three-digit extension number of the party you wish to call.

10-9 TELEPHONE SERVICE

A telephone connection may be as simple as two telephones and a single local switching office, or it may involve a multiplicity of communications links including several switching offices, transmission facilities, and telephone companies.

Telephone sets convert acoustic energy to electrical signals and vice versa. In addition, they also generate supervisory signals and address information. The subscriber loop provides a two-way path for conveying speech and data information and for exchanging ringing, switching, and supervisory signals. Since the telephone set and the subscriber loop are permanently associated with a particular subscriber, their combined transmission properties can be adjusted to meet their share of the total message channel objectives. For example, the higher efficiencies of new telephone sets and modems compensate for increased loop loss, permitting longer loop lengths or using smaller-gauge wire.

The small percentage of time (approximately 10% during busy hours) that a subscriber loop is utilized led to the development of line concentrators between subscribers and

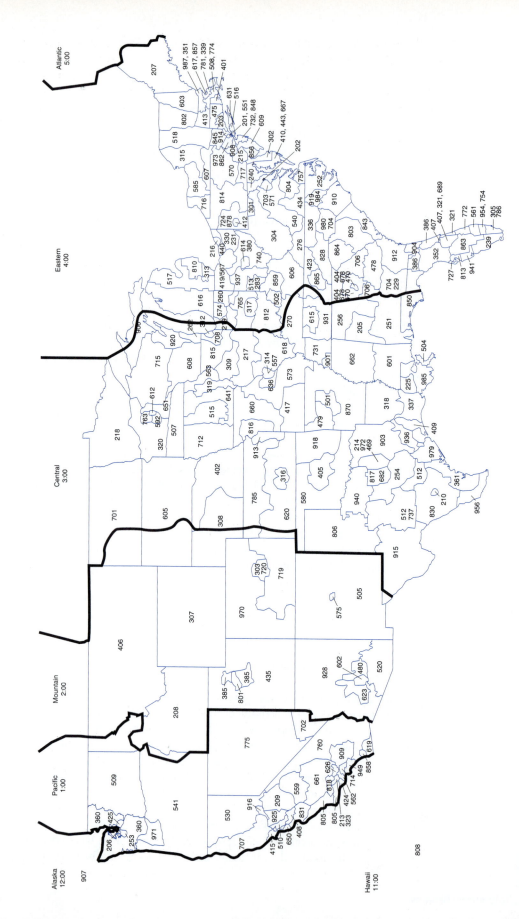

FIGURE 10-11 North American Telephone Numbering Plan Areas

294

central offices. A concentrator allows many subscribers to share a limited number of lines to a central office switch. For example, there may be 100 subscriber loops connected to one side of a concentrator and only 10 lines connected between the concentrator and the central office switch. Therefore, only 10 of 100 (10%) of the subscribers could actually access the local office at any one time. The line from a concentrator to the central office is essentially a trunk circuit because it is shared (common usage) among many subscribers on an "as needed" basis. As previously described, trunk circuits of various types are used to interconnect local offices to other local offices, local offices to toll offices, and toll offices to other toll offices.

When subscribers are connected to a toll office through toll-connecting trunks, the message signal is generally handled on a two-wire basis (both directions of transmission on the same pair of wires). After appropriate switching and routing functions are performed at toll offices, messages are generally connected to intertoll trunks by means of a two-wire-to-four-wire *terminating set* (*term set* or *hybrid*), which splits the two directions of signal propagation so that the actual long-distance segment of the route can be accomplished on a four-wire basis (separate cable pairs for each direction). Signals are connected through intertoll trunks to remote toll-switching centers, which may in turn be connected by intertoll trunks to other toll-switching centers and ultimately reach the recipient of the call through a toll-connecting trunk, a local office, another four-wire-to-two-wire term set, a local switching office, and a final subscriber loop as shown in Figure 10-12. A normal two-point

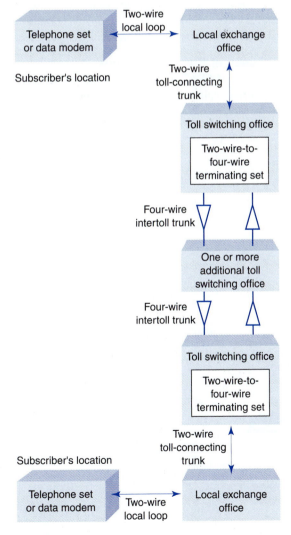

FIGURE 10-12 Long-distance telephone connection

telephone connection never requires more than two local exchange offices; however, there may be several toll-switching offices required, depending on the location of the originating and destination stations.

10-10 NORTH AMERICAN TELEPHONE SWITCHING HIERARCHY

With the advent of automated switching centers, a hierarchy of switching exchanges evolved in North America to accommodate the rapid increase in demand for long-distance calling. Thus, telephone company switching plans include a *switching hierarchy* that allows a certain degree of route selection when establishing a telephone call. A *route* is simply a path between two subscribers and is comprised of one or more switches, two local loops, and possibly one or more trunk circuits. The choice of routes is not offered to subscribers.

Telephone company switches, using software translation, select the best route available at the time a call is placed. The best route is not necessarily the shortest route. The best route is most likely the route requiring the fewest number of switches and trunk circuits. If a call cannot be completed because the necessary trunk circuits or switching paths are not available, the calling party receives an equipment (fast) busy signal. This is called *blocking*. Based on telephone company statistics, the likelihood that a call be blocked is approximately 1 in 100,000. Because software translations in automatic switching machines permit the use of alternate routes and each route may include several trunk circuits, the probability of using the same facilities on identical calls is unlikely. This is an obvious disadvantage of using the PSTN for data transmission because inconsistencies in transmission parameters occur from call to call.

10-10-1 Classes of Switching Offices

Before the divestiture of AT&T in 1984, the Bell System North American Switching Hierarchy consisted of five ranks or classes of switching centers as shown in Figure 10-13. The

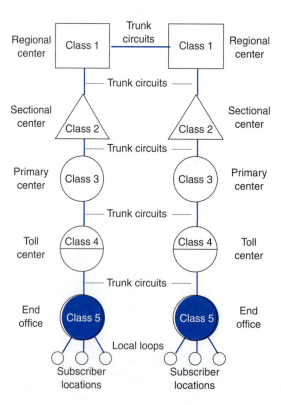

FIGURE 10-13 AT&T switching hierarchy prior to the 1984 divestiture

highest-ranking office was the regional center, and the lowest-ranking office was the end office. The five classifications of switching offices were as follows.

10-10-1-1 Class 5 end office. A class 5 office is a local exchange where subscriber loops terminated and received dial tone. End offices interconnected subscriber loops to other subscriber loops and subscriber loops to tandem trunks, interoffice trunks, and toll-connecting trunks. Subscribers received unlimited local call service in return for payment of a fixed charge each month, usually referred to as a *flat rate*. Some class 5 offices were classified as class 4/5. This type of office was called an *access tandem office,* as it was located in rural, low-volume areas and served as a dedicated class 5 office for local subscribers and also performed some of the functions of a class 4 toll office for long-distance calls.

10-10-1-2 Class 4 toll center. There were two types of class 4 offices. The class 4C toll centers provided human operators for both outward and inward calling service. Class 4P offices usually had only outward operator service or perhaps no operator service at all. Examples of operator-assisted services are person-to-person calls, collect calls, and credit card calls. Class 4 offices concentrated traffic in one switching center to direct outward traffic to the proper end office. Class 4 offices also provided centralized billing, provided toll customers with operator assistance, processed toll and intertoll traffic through its switching system, and converted signals from one trunk to another.

Class 3, 2, and 1 offices were responsible for switching intertoll-type calls efficiently and economically; to concentrate, collect, and distribute intertoll traffic; and to interconnect intertoll calls to all points of the direct distance dialing (DDD) network.

10-10-1-3 Class 3 primary center. This office provided service to small groups of class 4 offices within a small area of a state. Class 3 offices provided no operator assistance; however, they could serve the same switching functions as class 4 offices. A class 3 office generally had direct trunks to either a sectional or a regional center.

10-10-1-4 Class 2 sectional center. Sectional centers could provide service to geographical regions varying in size from part of a state to all of several states, depending on population density. No operator services were provided; however, a class 2 office could serve the same switching functions as class 3 and class 4 offices.

10-10-1-5 Class 1 regional center. Regional centers were the highest-ranking office in the DDD network in terms of the size of the geographical area served and the trunking options available. Ten regional centers were located in the United States and two in Canada. Class 1 offices provided no operator services; however, they could serve the same switching functions as class 2, 3, or 4 offices. Class 1 offices had direct trunks to all the other regional centers.

10-10-2 Switching Routes

Regional centers served a large area called a *region.* Each region was subdivided into smaller areas called *sections,* which were served by primary centers. All remaining switching centers that did not fall into these categories were toll centers or end offices. The switching hierarchy provided a systematic and efficient method of handling long-distance telephone calls using hierarchical routing principles and various methods of automatic *alternate routing.* Alternate routing is a simple concept: if one route (path) is not available, select an alternate route that is available. Therefore, alternate routing often caused many toll offices to be interconnected in tandem to complete a call. When alternate routing is used, the actual path a telephone call takes may not resemble what the subscriber actually dialed. For example, a call placed between Phoenix, Arizona, and San Diego, California, may be routed from Phoenix to Albuquerque to Las Vegas to Los Angeles to San Diego. Common switching control equipment may have to add to, subtract from, or change the dialed information when

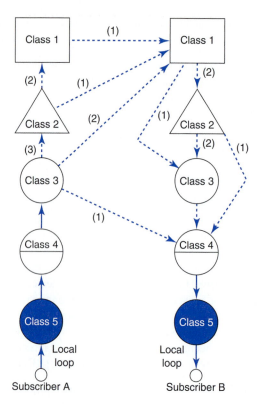

FIGURE 10-14 Choices of switching routes

routing a call to its destination. For example, an exchange office may have to add a prefix to a call with one, two, or three routing digits just to advance the call through an alternate route.

The five-class switching hierarchy is a *progressive switching scheme* that establishes an end-to-end route mainly through trial and error. Progressive switching is slow and unreliable by today's standards, as signaling messages are transported over the same facilities as subscriber's conversations using analog signals, such as multifrequency (MF) tones. Figure 10-14 shows examples of several choices for routes between subscriber A and subscriber B. For this example, there are 10 routes to choose from, of which only one requires the maximum of seven *intermediate links.* Intermediate links are toll trunks in tandem, excluding the two terminating links at the ends of the connection. In Figure 10-14, the first-choice route requires two intermediate links. Intermediate links are not always required, as in many cases a single *direct link,* which would be the first choice, exists between the originating and destination toll centers.

For the telephone office layout shown in Figure 10-15, the simplest connection would be a call between subscribers 1 and 2 in city A who are connected to the same end office. In this case, no trunk circuits are required. An interoffice call between stations 1 and 3 in city A would require using two tandem trunk circuits with an interconnection made in a tandem office. Consider a call originating from subscriber 1 in city A intended for subscriber 4 in city B. The route begins with subscriber 1 connected to end office 1 through a local loop. From the end office, the route uses a toll-connecting trunk to the toll center in city A. Between city A and city B, there are several route choices available. Because there is a high community of interest between the two cities, there is a direct intertoll trunk between city A and city B, which would be the first choice. However, there is an alternate route between city A and city B through the primary center in city C, which would probably be the second choice. From the primary center, there is a direct, high-usage intertoll trunk to both city A

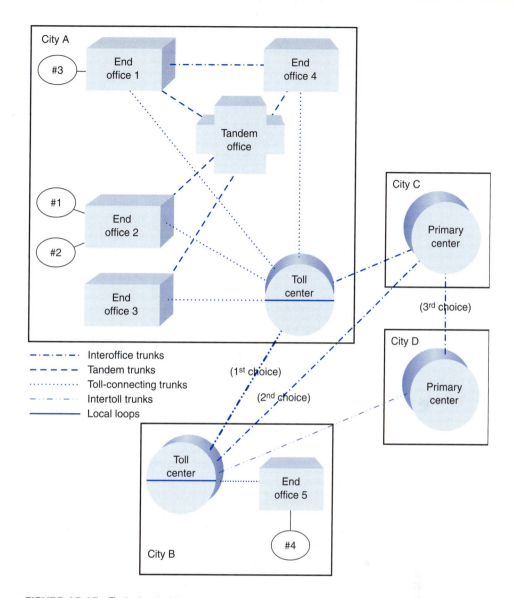

FIGURE 10-15 Typical switching routes

and city B (possibly the third choice), or, as a last resort, the toll centers in city A and city B could be interconnected using the primary centers in city C and city D (fourth choice).

The probability that a telephone call would require more than n links in tandem to reach the final destination decreases rapidly as n increases from 2 to 7. This is primarily because a large majority of long-distance toll calls are made between end offices associated with the same regional switching center, which of course would require fewer than seven toll trunks. Although the maximum number of trunks is seven, the average number for a typical toll call is only three. In addition, even when a telephone call was between telephones associated with different regional centers, the call was routed over the maximum of seven intermediate trunks only when all the normally available high-usage trunks are busy. The probability of this happening is only ρ^5, where ρ is the probability that all trunks in any one high-usage group are busy. Finally, many calls do not originate all the way down the hierarchy since each higher class of office will usually have class 5 offices homing on it that will act as class 4 offices for them.

10-11 COMMON CHANNEL SIGNALING SYSTEM NO. 7 (SS7) AND THE POSTDIVESTITURE NORTH AMERICAN SWITCHING HIERARCHY

Common Channel Signaling System No. 7 (i.e., SS7 or C7) is a global standard for telecommunications defined by the International Telecommunications Union (ITU) Telecommunications Sector (ITU-T). SS7 was developed as an alternate and much-improved means of transporting signaling information through the public telephone network. The SS7 standard defines the procedures and protocol necessary to exchange information over the PSTN using a separate digital signaling network to provide wireless (cellular) and wireline telephone call setup, routing, and control. SS7 determines the switching path before any switches are actually enabled, which is a much faster and more reliable switching method than the old five-class progressive switching scheme. The SS7 signaling network performs its functions by exchanging telephone control messages between the SS7 components that support completing the subscribers' connection.

The functions of the SS7 network and protocol are as follows:

1. Basic call setup, management, and tear-down procedures
2. Wireless services, such as personal communications services (PCS), wireless roaming, and mobile subscriber authentication
3. Local number portability (LNP)
4. Toll-free (800/888) and toll (900) wireline services
5. Enhanced call features, such as call forwarding, calling party name/number display, and three-way calling
6. Efficient and secure worldwide telecommunications service

10-11-1 Evolution of SS7

When telephone networks and network switching hierarchies were first engineered, their creators gave little thought about future technological advancements. Early telephone systems were based on transferring analog voice signals using analog equipment over analog transmission media. As a result, early telephone systems were not well suited for modern-day digital services, such as data, digitized voice, or digitized video transmission. Therefore, when digital services were first offered in the early 1960s, the telephone networks were ill prepared to handle them, and the need for an intelligent all-digital network rapidly became evident.

The ITU commissioned the Comiteé Consultatif International Téléphonique et Télégraphique (CCITT) to study the possibility of developing an intelligent all-digital telecommunications network. In the mid-1960s, the ITU-TS (International Telecommunications Union Telecommunications—Standardization Sector) developed a digital signaling standard known as *Signaling System No. 6* (SS6) that modernized the telephone industry. *Signaling* refers to the exchange of information between call components required to provide and maintain service. SS6, based on a proprietary, high-speed data communications network, evolved into *Signaling System No. 7* (SS7), which is now the telephone industry standard for most of the civilized world ("civilized world" because it was estimated that in 2002, more than half the people in the world had never used a telephone). High-speed packet data and out-of-band signaling characterize SS7. Out-of-band signaling is signaling that does not take place over the same path as the conversation. Out-of-band signaling establishes a separate digital channel for exchanging signaling information. This channel is called a *signaling link*.

The protocol used with SS7 uses a message structure, similar to X.25 and other message-based protocols, to request services from other networks. The messages propagate from one network to another in small bundles of data called *packets* that are independent of the subscriber voice or data signals they pertain to. In the early 1960s, the ITU-TS developed a *common channel signaling* (CCS) known as *Common Channel Interoffice Signaling System No. 6* (SS6). The basic concept of the common channel signaling is to use a facility (separate from the voice facilities) for transferring control and signaling information between telephone offices.

When first deployed in the United States, SS6 used a packet switching network with 2.4-kbps data links, which were later upgraded to 4.8 kbps. Signaling messages were sent as part of a data packet and used to request connections on voice trunks between switching offices. SS6 was the first system to use packet switching in the PSTN. Packets consisted of a block of data comprised of 12 signal units of 28 bits each, which is similar to the method used today with SS7.

SS7 is an architecture for performing out-of-band signaling in support of common telephone system functions, such as call establishment, billing, call routing, and information exchange functions of the PSTN. SS7 identifies functions and enables protocols performed by a telephone signaling network. The major advantages of SS7 include better monitoring, maintenance, and network administration. The major disadvantage is its complex coding.

Because SS7 evolved from SS6, there are many similarities between the two systems. SS7 uses variable-length signal units with a maximum length, therefore making it more versatile and flexible than SS6. In addition, SS7 uses 56-kbps data links (64 kbps for international links), which provide a much faster and efficient signaling network. In the future, data rates of 1.544 Mbps nationally and 2.048 Mbps internationally are expected.

In 1983 (just prior to the AT&T divestiture), SS6 was still widely used in the United States. When SS7 came into use in the mid-1980s, SS6 began to be phased out of the system, although SS6 was still used in local switching offices for several more years. SS7 was originally used for accessing remote databases rather than for call setup and termination. In the 1980s, AT&T began offering Wide Area Telephone Service (WATS), which uses a common 800 area code regardless of the location of the destination. Because of the common area code, telephone switching systems had a problem dealing with WATS numbers. This is because telephone switches used the area code to route a call through the public switched network. The solution involved adding a second number to every 800 number that is used by the switching equipment to actually route a call through the voice network. The second number is placed in a common, centralized database accessible to all central offices. When an 800 number is called, switching equipment uses a data link to access the database and retrieve the actual routing number. This process is, of course, transparent to the user. Once the routing number is known, the switching equipment can route the call using standard signaling methods.

Shortly after implementing the WATS network, the SS7 network was expanded to provide other services, such as call setup and termination. However, the database concept has proven to be the biggest advantage of SS7, as it can also be used to provide routing and billing information for all telephone services, including 800 and 900 numbers, 911 services, custom calling features, caller identifications, and a host of other services not yet invented.

In 1996, the FCC mandated *local number portability* (LNP), which requires all telephone companies to support the *porting* of a telephone number. Porting allows customers to change to a different service and still keep the same telephone number. For example, a subscriber may wish to change from *plain old telephone service* (POTS) to ISDN, which would have required changing telephone numbers. With LNP, the telephone number would remain the same because the SS7 database can be used to determine which network switch is assigned to a particular telephone number.

Today, SS7 is being used throughout the Bell Operating Companies telephone network and most of the independent telephone companies. This in itself makes SS7 the world's largest data communications network, as it links wireline telephone companies, cellular telephone companies, and long-distance telephone companies together with a common signaling system. Because SS7 has the ability to transfer all types of digital information, it supports most of the new telephone features and applications and is used with ATM, ISDN, and cellular telephone.

10-11-2 Postdivestiture North American Switching Hierarchy

Today, the North American telephone system is divided into two distinct functional areas: signaling and switching. The signaling network for the telephone system is SS7, which is used to determine how subscriber's voice and data signals are routed through the network. The switching network is the portion of the telephone network that actually transports the voice and data from one subscriber to another. The signaling part of the network establishes and disconnects the circuits that actually carry the subscriber's information.

After the divestiture of AT&T, technological advances allowed many of the functions distributed among the five classes of telephone offices to be combined. In addition, switching equipment was improved, giving them the capability to act as local switches, tandem switches, or toll switches. The new North American Switching Hierarchy consolidated many of the functions of the old hierarchy into two layers, with many of the functions once performed by the higher layers being located in the end office. Therefore, the postdivestiture telephone network can no longer be described as a hierarchy of five classes of offices. It is now seen as a system involving two decision points. The postdivestiture North American Switching Hierarchy is shown in Figure 10-16. Long-distance access is now accomplished through an access point called the *point-of-presence* (POP). The term *point-of-presence* is a telecommunications term that describes the legal boundaries for the responsibility of maintaining equipment and transmission lines. In essence, it is a demarcation point separating two companies.

After the divestiture of AT&T, calling areas were redefined and changed to *Local Access and Transport Areas* (LATAs) with each LATA having its own three-level hierarchy. Although the United States was originally divided into only 160 local access and transport areas, there are presently over 300 LATA dispersed throughout the United States. Within these areas, local telephone companies provide the facilities and equipment to interconnect

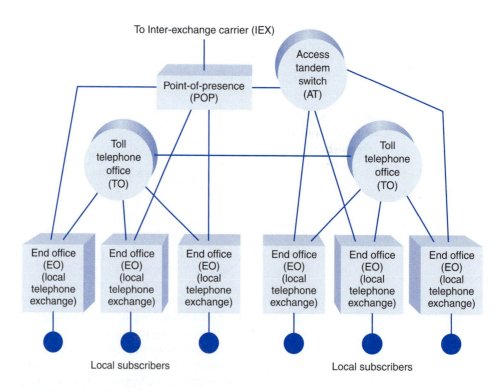

FIGURE 10-16 Postdivestiture North American Switching Hierarchy

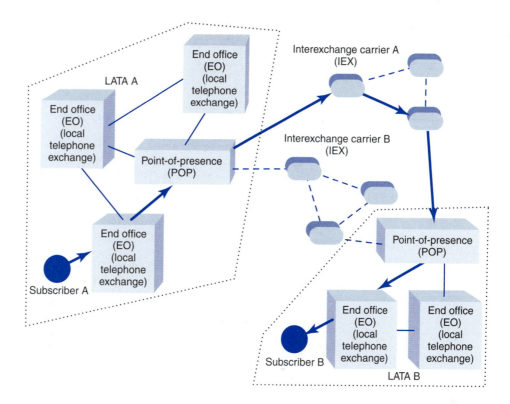

FIGURE 10-17 Example of an interexchange call between subscriber A in LATA A to subscriber B in LATA B

subscribers within the LATA. The telephone companies are called *local exchange carriers* (LECs), *exchange carriers* (ECs), *operating telephone companies* (OTCs), and *telephone operating companies* (TOCs). The areas serviced by local exchanges were redistributed by the Justice Department to provide telephone companies a more evenly divided area with equal revenue potential. Telephone calls made within a LATA are considered a function of the *intra-LATA network*.

Telephone companies further divided each LATA into a *local market* and a *toll market*. The toll market for a company is within its LATA but is still considered a long-distance call because it involves a substantial distance between the two local offices handling the call. These are essentially the only long-distance telephone calls local operating companies are allowed to provide, and they are very expensive. If the destination telephone number is in a different LATA than the originating telephone number, the operating company must switch to an *interexchange carrier* (IC, IEC, or IXC) selected by the calling party. In many cases, a direct connection is not possible, and an interexchange call must be switched first to an access tandem (AT) switch and then to the interexchange carrier point-of-presence. Figure 10-17 shows an example of an interexchange call between subscriber A in LATA A through interexchange carrier A to subscriber B in LATA B.

10-11-3 SS7 Signaling Points

Signaling points provide access to the SS7 network, access to databases used by switches inside and outside the network, and the transfer of SS7 messages to other signaling points within the network.

Every network has an addressing scheme to enable a node within the network to exchange signaling information with nodes it is not connected to by a physical link. Each node

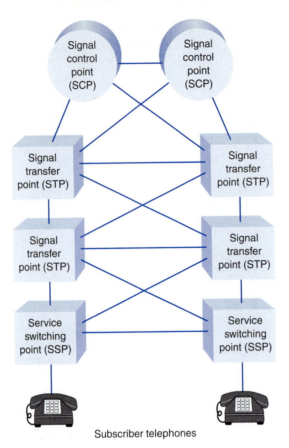

Subscriber telephones

FIGURE 10-18 SS7 signaling point topology

is uniquely identified by a numeric *point code.* Point codes are carried in signaling messages exchanged between signaling points to identify the source and destination of each message (i.e., an *originating point code* and a *destination point code*). Each signaling point is identified as a *member* part of a cluster of signaling points. Similarly, a cluster is defined as being part of a complete network. Therefore, every node in the American SS7 network can be addressed with a three-level code that is defined by its network, cluster, and member numbers. Each number is an eight-bit binary number between 0 and 255. This three-level address is called the *point code.* A point code uniquely identifies a signaling point within the SS7 network and is used whenever it is addressed. A neutral party assigns network codes on a nationwide basis. Because there are a limited number of network numbers, networks must meet a certain size requirement to receive one. Smaller networks may be assigned one or more cluster numbers within network numbers 1, 2, 3, and 4. The smallest networks are assigned point codes within network number 5. The cluster they are assigned to is determined by the state where they are located. Network number 0 is not available for assignment, and network number 255 is reserved for future use.

The three types of signaling points are listed here, and a typical SS7 topology is shown in Figure 10-18.

10-11-3-1 Service switching points (SSPs). Service switching points (sometimes called *signal switching points*) are local telephone switches (in either end or tandem offices) equipped with SS7-compatible software and terminating signal links. The SSP provides the functionality of communicating with the voice switch by creating the packets or signal units necessary for transmission over the SS7 network. An SSP must convert signaling information from voice switches into SS7 signaling format. SSPs are basically local access points that send signaling messages to other SSPs to originate, terminate, or switch calls. SSPs may also send query messages to centralized databases to determine how to route a call.

10-11-3-2 Signal transfer points (STPs). Signal transfer points are the packet switches of the SS7 network. STPs serve as routers in the SS7 network, as they receive and route incoming signaling messages to the proper destination. STPs seldom originate a message. STPs route each incoming message to an outgoing signaling link based on routing information contained in the SS7 message. Because an STP acts like a network router, it provides improved utilization of the SS7 network by eliminating the need for direct links between all signaling points.

10-11-3-3 Service control points (SCPs). Service control points (sometimes called *signal control points*) serve as an interface to telephone company databases. The databases store information about subscriber's services, routing of special service numbers (such as 800 and 900 numbers), and calling card validation for fraud protection and provide information necessary for advanced call-processing capabilities. SCPs also perform protocol conversion from SS7 to X.25, or they can provide the capability of communicating with the database directly using an interface called a *primitive,* which provides access from one level of the protocol to another level. SCPs also send responses to SSPs containing a routing number(s) associated with the called number.

SSPs, STPs, and SCPs are interconnected with digital carriers, such as T1 or DS-0 links, which carry the signaling messages between the SS7 network devices.

10-11-4   SS7 Call Setup Example

A typical call setup procedure using the SS7 signaling network is as follows:

1. Subscriber A goes off hook and Touch-Tones out the destination telephone number of subscriber B.
2. The local telephone translates the tones to binary digits.
3. The local telephone exchange compares the digits to numbers stored in a routing table to determine whether subscriber B resides in the same local switch as subscriber A. If not, the call must be transferred onto an outgoing trunk circuit to another local exchange.
4. After the switch determines that subscriber B is served by a different local exchange, an SS7 message is sent onto the SS7 network. The purposes of the message are as follows:
    i. To find out if the destination number is idle.
   ii. If the destination number is idle, the SS7 network makes sure a connection between the two telephone numbers is also available.
  iii. The SS7 network instructs the destination switch to ring subscriber B.
5. When subscriber B answers the telephone, the switching path is completed.
6. When either subscriber A or subscriber B terminates the call by hanging up, the SS7 network releases the switching path, making the trunk circuits and switching paths available to other subscribers of the network.

QUESTIONS

10-1. What are the purposes of telephone network *signaling functions*?

10-2. What are the two types of subscribers to the public telephone network? Briefly describe them.

10-3. What is the difference between *dedicated* and *switched* facilities?

10-4. Describe the term *service provider.*

10-5. Briefly describe the following terms: *instruments, local loops, trunk circuits,* and *exchanges.*

10-6. What is a *local office telephone exchange?*

10-7. What is an *automated central office switch?*

10-8. Briefly describe the following terms: *circuits, circuit switches,* and *circuit switching.*

10-9. What is the difference between a *local telephone exchange* and an *exchange area?*

10-10. Briefly describe *interoffice trunks, tandem trunks,* and *tandem switches.*

10-11. Briefly describe *toll-connecting trunks, intertoll trunks,* and *toll offices.*

10-12. Briefly describe the *North American Telephone Numbering Plan.*

10-13. What is the difference between an *area code,* a *prefix,* and an *extension?*

10-14. What is meant by the term *common usage?*

10-15. What does *blocking* mean? When does it occur?

10-16. Briefly describe the *predivestiture North American Telephone Switching Hierarchy.*

10-17. Briefly describe the five *classes* of the predivestiture North American Switching Hierarchy.

10-18. What is meant by the term *switching route?*

10-19. What is meant by the term *progressive switching scheme?*

10-20. What is *SS7?*

10-21. What is *common channel signaling?*

10-22. What is meant by the term *local number portability?*

10-23. What is meant by the term *plain old telephone service?*

10-24. Briefly describe the *postdivestiture North American Switching Hierarchy.*

10-25. What is a LATA?

10-26. What is meant by the term *point-of-presence?*

10-27. Describe what is meant by the term *local exchange carrier.*

10-28. Briefly describe what is meant by SS7 *signaling points.*

10-29. List and describe the three SS7 signaling points.

10-30. What is meant by the term *point code?*

C H A P T E R   11

Cellular Telephone Concepts

CHAPTER OUTLINE

OBJECTIVES

- Give a brief history of mobile telephone service
- Define *cellular telephone*
- Define *cell* and explain why it has a honeycomb shape
- Describe the following types of cells: macrocell, microcell, and minicell
- Describe edge-excited, center-excited, and corner-excited cells
- Define *service areas, clusters,* and *cells*
- Define *frequency reuse*
- Explain frequency reuse factor
- Define *interference*
- Describe cochannel and adjacent channel interference
- Describe the processes of cell splitting, sectoring, segmentation, and dualization
- Explain the differences between cell-site controllers and mobile telephone switching offices
- Define *base stations*
- Define and explain *roaming* and *handoffs*
- Briefly describe the purpose of the IS-41 protocol standard
- Define and describe the following cellular telephone network components: *electronic switching center, cell-site controller, system interconnects, mobile and portable telephone units,* and *communications protocols*
- Describe the cellular call procedures involved in making the following types of calls: mobile to wireline, mobile to mobile, and wireline to mobile

11-1 INTRODUCTION

The basic concepts of *two-way mobile telephone* are quite simple; however, mobile telephone systems involve intricate and rather complex communications networks comprised of analog and digital communications methodologies, sophisticated computer-controlled switching centers, and involved protocols and procedures. Cellular telephone evolved from two-way mobile FM radio. The purpose of this chapter is to present the fundamental concepts of cellular telephone service. Cellular services include standard *cellular telephone service* (CTS), *personal communications systems* (PCS), and *personal communications satellite systems* (PCSS).

11-2 MOBILE TELEPHONE SERVICE

Mobile telephone services began in the 1940s and were called MTSs (*mobile telephone systems* or sometimes *manual telephone systems,* as all calls were handled by an operator). MTS systems utilized frequency modulation and were generally assigned a single carrier frequency in the 35-MHz to 45-MHz range that was used by both the mobile unit and the base station. The mobile unit used a push-to-talk (PTT) switch to activate the transceiver. Depressing the PTT button turned the transmitter on and the receiver off, whereas releasing the PTT turned the receiver on and the transmitter off. Placing a call from an MTS mobile telephone was similar to making a call through a manual switchboard in the public telephone network. When the PTT switch was depressed, the transmitter turned on and sent a carrier frequency to the base station, illuminating a lamp on a switchboard. An operator answered the call by plugging a headset into a jack on the switchboard. After the calling party verbally told the operator the telephone number they wished to call, the operator connected the mobile unit with a patchcord to a trunk circuit connected to the appropriate public telephone network destination office. Because there was only one carrier frequency, the conversation was limited to half-duplex operation, and only one conversation could take place at a time. The MTS system was comparable to a party line, as all subscribers with their mobile telephones turned on could hear any conversation. Mobile units called other mobile units by signaling the operator who rang the destination mobile unit. Once the destination mobile unit answered, the operator disconnected from the conversation, and the two mobile units communicated directly with one another through the airways using a single carrier frequency.

MTS mobile identification numbers had no relationship to the telephone numbering system used by the public telephone network. Local telephone companies in each state, which were generally Bell System Operating Companies, kept a record of the numbers assigned to MTS subscribers in that state. MTS numbers were generally five digits long and could not be accessed directly through the public switched telephone network (PSTN).

In 1964, the Improved Mobile Telephone System (IMTS) was introduced, which used several carrier frequencies and could, therefore, handle several simultaneous mobile conversations at the same time. IMTS subscribers were assigned a regular PSTN telephone number; therefore, callers could reach an IMTS mobile phone by dialing the PSTN directly, eliminating the need for an operator. IMTS and MTS base station transmitters outputted powers in the 100-W to 200-W range, and mobile units transmitted between 5 W and 25 W. Therefore, IMTS and MTS mobile telephone systems typically covered a wide area using only one base station transmitter.

Because of their high cost, limited availability, and narrow frequency allocation, early mobile telephone systems were not widely used. However, in recent years, factors such as technological advancements, wider frequency spectrum, increased availability, and improved reliability have stimulated a phenomenal increase in people's desire to talk on the telephone from virtually anywhere, at any time, regardless of whether it is necessary, safe, or productive.

Today, mobile telephone stations are small handsets, easily carried by a person in their pocket or purse. In early radio terminology, the term *mobile* suggested any radio transmitter, receiver, or transceiver that could be moved while in operation. The term *portable* described a relatively small radio unit that was handheld, battery powered, and easily carried by a person moving at walking speed. The contemporary definition of mobile has come to mean moving at high speed, such as in a boat, airplane, or automobile, or at low speed, such as in the pocket of a pedestrian. Hence, the modern, all-inclusive definition of mobile telephone is any wireless telephone capable of operating while moving at any speed, battery powered, and small enough to be easily carried by a person.

Cellular telephone is similar to two-way mobile radio in that most communications occur between base stations and mobile units. Base stations are fixed-position transceivers with relatively high-power transmitters and sensitive receivers. Cellular telephones communicate directly with base stations. Cellular telephone is best described by pointing out the primary difference between it and two-way mobile radio. Two-way mobile radio systems operate half-duplex and use PTT transceivers. With PTT transceivers, depressing the PTT button turns on the transmitter and turns off the receiver, whereas releasing the PTT button turns on the receiver and turns off the transmitter. With two-way mobile telephone, all transmissions (unless scrambled) can be heard by any listener with a receiver tuned to that channel. Hence, two-way mobile radio is a *one-to-many* radio communications system. Examples of two-way mobile radio are *citizens band* (CB), which is an AM system, and *public land mobile radio,* which is a two-way FM system such as those used by police and fire departments. Most two-way mobile radio systems can access the public telephone network only through a special arrangement called an *autopatch,* and then they are limited to half-duplex operation where neither party can interrupt the other. Another limitation of two-way mobile radio is that transmissions are limited to relatively small geographic areas unless they utilize complicated and expensive repeater networks.

On the other hand, cellular telephone offers full-duplex transmissions and operates much the same way as the standard wireline telephone service provided to homes and businesses by local telephone companies. Mobile telephone is a *one-to-one* system that permits two-way simultaneous transmissions, and, for privacy, each cellular telephone is assigned a unique telephone number. Coded transmissions from base stations activate only the intended receiver. With mobile telephone, a person can virtually call anyone with a telephone number, whether it be through a cellular or a wireline service.

Cellular telephone systems offer a relatively high user capacity within a limited frequency spectrum providing a significant innovation in solving inherent mobile telephone communications problems, such as spectral congestion and user capacity. Cellular telephone systems replaced mobile systems serving large areas (cells) operating with a single base station and a single high-power transmitter with many smaller areas (cells), each with its own base station and low-power transmitter. Each base station is allocated a fraction of the total channels available to the system, and adjacent cells are assigned different groups of channels to minimize interference between cells. When demand for service increases in a given area, the number of base stations can be increased, providing an increase in mobile-unit capacity without increasing the radio-frequency spectrum.

11-3 EVOLUTION OF CELLULAR TELEPHONE

In the July 28, 1945, *Saturday Evening Post,* E. K. Jett, then the commissioner of the FCC, hinted of a cellular telephone scheme that he referred to as simply a *small-zone* radio-telephone system. On June 17, 1946, in St. Louis, Missouri, AT&T and Southwestern Bell introduced the first American commercial mobile radio-telephone service to private customers. In the same year, similar services were offered to 25 major cities throughout the United States. Each city utilized one base station consisting of a high-powered transmitter and a sensitive receiver that were centrally located on a hilltop or tower that covered an area within a 30- to 50-mile radius of the base station. In 1947, AT&T introduced a radio-telephone service they called *highway service* between New York and Boston. The system operated in the 35-MHz to 45-MHz band.

The first half-duplex, PTT FM mobile telephone systems introduced in the 1940s operated in the 35-MHz to 45-MHz band and required 120-kHz bandwidth per channel. In the early 1950s, the FCC doubled the number of mobile telephone channels by reducing the bandwidth to 60 kHz per channel. In 1960, AT&T introduced direct-dialing, full-duplex mobile telephone service with other performance enhancements, and in 1968, AT&T proposed the concept of a cellular mobile system to the FCC with the intent of alleviating the problem of spectrum congestion in the existing mobile telephone systems. Cellular mobile telephone systems, such as the *Improved Mobile Telephone System* (IMTS), were developed, and recently developed miniature integrated circuits enabled management of the necessarily complex algorithms needed to control network switching and control operations. Channel bandwidth was again halved to 30 kHz, increasing the number of mobile telephone channels by twofold.

In 1966, Don Adams, in a television show called *Get Smart*, unveiled the most famous mobile telephone to date: the fully mobile shoe phone. Some argue that the 1966 *Batphone supra* was even more remarkable, but it remained firmly anchored to the Batmobile, limiting Batman and Robin to vehicle-based telephone communications.

In 1974, the FCC allocated an additional 40-MHz bandwidth for cellular telephone service (825 MHz to 845 MHz and 870 MHz to 890 MHz). These frequency bands were previously allocated to UHF television channels 70 to 83. In 1975, the FCC granted AT&T the first license to operate a developmental cellular telephone service in Chicago. By 1976, the Bell Mobile Phone service for metropolitan New York City (approximately 10 million people) offered only 12 channels that could serve a maximum of 543 subscribers. In 1976, the FCC granted authorization to the American Radio Telephone Service (ARTS) to install a second developmental system in the Baltimore–Washington, D.C., area. In 1983, the FCC allocated 666 30-kHz half-duplex mobile telephone channels to AT&T to form the first U.S. cellular telephone system called Advanced Mobile Phone System (AMPS).

In 1991, the first digital cellular services were introduced in several major U.S. cities, enabling a more efficient utilization of the available bandwidth using voice compression. The calling capacity specified in the U.S. Digital Cellular (USDC) standard (EIA IS-54) accommodates three times the user capacity of AMPS, which used conventional frequency modulation (FM) and frequency-division multiple accessing (FDMA). The USDC standard specifies digital modulation, speech coding, and time-division multiple accessing (TDMA). Qualcomm developed the first cellular telephone system based on code-division multiple accessing (CDMA). The Telecommunications Industry Association (TIA) standardized Qualcomm's system as Interim Standard 95 (IS-95). On November 17, 1998, a subsidiary of Motorola Corporation implemented Iridium, a satellite-based wireless personal communications satellite system (PCSS).

11-4 CELLULAR TELEPHONE

The key principles of *cellular telephone* (sometimes called *cellular radio*) were uncovered in 1947 by researchers at Bell Telephone Laboratories and other telecommunications companies throughout the world when they developed the basic concepts and theory of cellular telephone. It was determined that by subdividing a relatively large geographic market area, called a *coverage zone*, into small sections, called *cells*, the concept of *frequency reuse* could be employed to dramatically increase the capacity of a mobile telephone channel. Frequency reuse is described in a later section of this chapter. In essence, cellular telephone systems allow a large number of users to share the limited number of *common-usage* radio

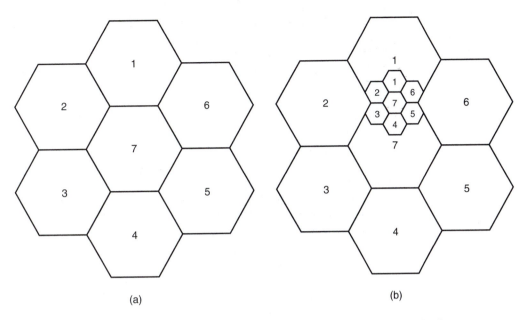

FIGURE 11-1 (a) Honeycomb cell pattern; (b) honeycomb pattern with two sizes of cells

channels available in a region. In addition, integrated-circuit technology, microprocessors and microcontroller chips, and the implementation of Signaling System No. 7 (SS7) have recently enabled complex radio and logic circuits to be used in electronic switching machines to store programs that provide faster and more efficient call processing.

11-4-1 Fundamental Concepts of Cellular Telephone

The fundamental concepts of cellular telephone are quite simple. The FCC originally defined geographic cellular radio coverage areas on the basis of modified 1980 census figures. With the cellular concept, each area is further divided into hexagonal-shaped cells that fit together to form a *honeycomb* pattern as shown in Figure 11-1a. The hexagon shape was chosen because it provides the most effective transmission by approximating a circular pattern while eliminating gaps inherently present between adjacent circles. A cell is defined by its physical size and, more importantly, by the size of its population and traffic patterns. The number of cells per system and the size of the cells are not specifically defined by the FCC and has been left to the providers to establish in accordance with anticipated traffic patterns. Each geographical area is allocated a fixed number of cellular voice channels. The physical size of a cell varies, depending on user density and calling patterns. For example, large cells (called *macrocells*) typically have a radius between 1 mile and 15 miles with base station transmit powers between 1 W and 6 W. The smallest cells (called *microcells*) typically have a radius of 1500 feet or less with base station transmit powers between 0.1 W and 1 W. Figure 11-1b shows a cellular configuration with two sizes of cell.

Microcells are used most often in high-density areas such as found in large cities and inside buildings. By virtue of their low effective working radius, microcells exhibit milder propagation impairments, such as reflections and signal delays. Macrocells may overlay clusters of microcells with slow-moving mobile units using the microcells and faster-moving units using the macrocells. The mobile unit is able to identify itself as either fast or slow moving, thus allowing it to do fewer cell transfers and location updates. Cell transfer algorithms can be modified to allow for the small distances between a mobile unit and the

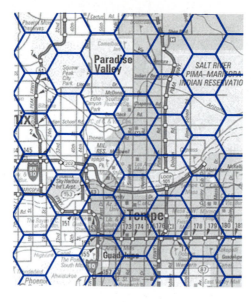

FIGURE 11-2 Hexagonal cell grid superimposed over a metropolitan area

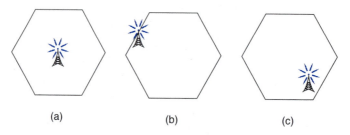

    (a)   (b)   (c)

FIGURE 11-3 (a) Center excited cell; (b) edge excited cell; (c) corner excited cell

microcellular base station it is communicating with. Figure 11-2 shows what a hexagonal cell grid might look like when superimposed over a metropolitan area.

Occasionally, cellular radio signals are too weak to provide reliable communications indoors. This is especially true in well-shielded areas or areas with high levels of interference. In these circumstances, very small cells, called *picocells,* are used. Indoor picocells can use the same frequencies as regular cells in the same areas if the surrounding infrastructure is conducive, such as in underground malls.

When designing a system using hexagonal-shaped cells, base station transmitters can be located in the center of a cell (*center-excited cell* shown in Figure 11-3a) or on three of the cells' six vertices (*edge-* or *corner-excited cells* shown in Figures 11-3b and c). *Omnidirectional* antennas are normally used in center-excited cells, and sectored directional antennas are used in edge- and corner-excited cells (omnidirectional antennas radiate and receive signals equally well in all directions).

Cellular telephone is an intriguing mobile radio concept that calls for replacing a single, high-powered fixed base station located high above the center of a city with multiple, low-powered duplicates of the fixed infrastructure distributed over the coverage area on sites placed closer to the ground. The cellular concept adds a spatial dimension to the simple cable-trunking model found in typical wireline telephone systems.

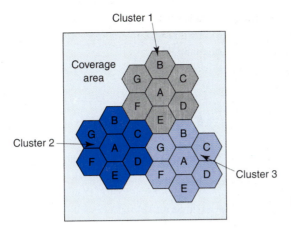

Cluster 1

Coverage area

Cluster 2

Cluster 3

FIGURE 11-4 Cellular frequency reuse concept

11-5 FREQUENCY REUSE

Frequency reuse is the process in which the same set of frequencies (channels) can be allocated to more than one cell, provided the cells are separated by sufficient distance. Reducing each cell's coverage area invites frequency reuse. Cells using the same set of radio channels can avoid mutual interference, provided they are properly separated. Each cell base station is allocated a group of channel frequencies that are different from those of neighboring cells, and base station antennas are chosen to achieve a desired coverage pattern within its cell. However, as long as a coverage area is limited to within a cell's boundaries, the same group of channel frequencies may be used in different cells without interfering with each other, provided the two cells are sufficient distance from one another.

Figure 11-4 illustrates the concept of frequency reuse in a cellular telephone system. The figure shows a geographic cellular radio coverage area containing three groups of cells called *clusters*. Each cluster has seven cells in it, and all cells are assigned the same number of full-duplex cellular telephone channels. Cells with the same letter use the same set of channel frequencies. As the figure shows, the same sets of frequencies are used in all three clusters, which essentially increases the number of usable cellular channels available threefold. The letters A, B, C, D, E, F, and G denote the seven sets of frequencies.

The frequency reuse concept can be illustrated mathematically by considering a system with a fixed number of full-duplex channels available in a given area. Each service area is divided into clusters and allocated a group of channels, which is divided among N cells in a unique and disjoint channel grouping where all cells have the same number of channels but do not necessarily cover the same size area. Thus, the total number of cellular channels available in a cluster can be expressed mathematically as

$$F = GN \tag{11-1}$$

where F = number of full-duplex cellular channels available in a cluster
G = number of channels in a cell
N = number of cells in a cluster

The cells that collectively use the complete set of available channel frequencies make up the cluster. When a cluster is duplicated m times within a given service area, the total number of full-duplex channels can be expressed mathematically as

$$C = mGN$$

or
$$= mF \tag{11-2}$$

where C = total channel capacity in a given area
        m = number of clusters in a given area
        G = number of channels in a cell
        N = number of cells in a cluster

Example 11-1

Determine the number of channels per cluster and the total channel capacity for a cellular telephone area comprised of 10 clusters with seven cells in each cluster and 10 channels in each cell.

Solution Substituting into Equation 11-1, the total number of full-duplex channels is

$$F = (10)(7)$$
$$= 70 \text{ channels per cluster}$$

Substituting into Equation 11-3, the total channel capacity is

$$C = (10)(7)(10)$$
$$= 700 \text{ channels total}$$

From Example 11-1, it can be seen that through frequency reuse, 70 channels (frequencies), reused in 10 clusters, produce 700 usable channels within a single cellular telephone area.

From Equations 11-1 and 11-2, it can be seen that the channel capacity of a cellular telephone system is directly proportional to the number of times a cluster is duplicated in a given service area. The factor N is called the *cluster size* and is typically equal to 3, 7, or 12. When the cluster size is reduced and the cell size held constant, more clusters are required to cover a given area, and the total channel capacity increases. The frequency reuse factor of a cellular telephone system is inversely proportional to the number of cells in a cluster (i.e., $1/N$). Therefore, each cell within a cluster is assigned $1/N$th of the total available channels in the cluster.

The number of subscribers who can use the same set of frequencies (channels) in non-adjacent cells at the same time in a small area, such as a city, is dependent on the total number of cells in the area. The number of simultaneous users is generally four, but in densely populated areas, that number may be significantly higher. The number of users is called the frequency reuse factor (FRF). The frequency reuse factor is defined mathematically as

$$FRF = \frac{N}{C} \tag{11-3}$$

where FRF = frequency reuse factor (unitless)
        N = total number of full-duplex channels in an area
        C = total number of full-duplex channels in a cell

Meeting the needs of projected growth in cellular traffic is accomplished by reducing the size of a cell by splitting it into several cells, each with its own base station. Splitting cells effectively allows more calls to be handled by the system, provided the cells do not become too small. If a cell becomes smaller than 1500 feet in diameter, the base stations in adjacent cells would most likely interfere with one another. The relationship between frequency reuse and cluster size determines how cellular telephone systems can be rescaled when subscriber density increases. As the number of cells per cluster decreases, the possibility that one channel will interfere with another channel increases.

Cells use a hexagonal shape, which provides exactly six equidistant neighboring cells, and the lines joining the centers of any cell with its neighboring cell are separated by multiples of 60. Therefore, a limited number of cluster sizes and cell layouts is possible. To connect cells without gaps in between (*tessellate*), the geometry of a hexagon is such that the number of cells per cluster can have only values that satisfy the equation

$$N = i^2 + ij + j^2 \tag{11-4}$$

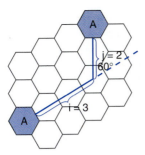

FIGURE 11-5 Locating first-tier cochannel cells

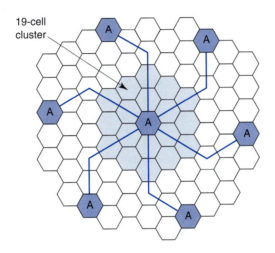

FIGURE 11-6 Determining first-tier cochannel cells for Example 11-2

where N = number of cells per cluster
            i and j = nonnegative integer values

The process of finding the tier with the nearest cochannel cells (called the *first tier*) is as follows and shown in Figure 11-5:

1. Move i cells through the center of successive cells.
2. Turn 60° in a counterclockwise direction.
3. Move j cells forward through the center of successive cells.

Example 11-2

Determine the number of cells in a cluster and locate the first-tier cochannel cells for the following values: $j = 2$ and $i = 3$.

Solution   The number of cells in the cluster is determined from Equation 11-4:

$$N = 3^2 + (3)(2) + 2^2$$
$$N = 19$$

Figure 11-6 shows the six nearest first-tier 1 cochannel cells for cell A.

11-6   INTERFERENCE

The two major kinds of interferences produced within a cellular telephone system are *cochannel interference* and *adjacent-channel interference*.

11-6-1 Cochannel Interference

When frequency reuse is implemented, several cells within a given coverage area use the same set of frequencies. Two cells using the same set of frequencies are called *cochannel cells,* and the interference between them is called *cochannel interference.* Unlike thermal noise, cochannel interference cannot be reduced by simply increasing transmit powers because increasing the transmit power in one cell increases the likelihood of that cell's transmissions interfering with another cell's transmission. To reduce cochannel interference, a certain minimum distance must separate cochannels.

Figure 11-7 shows cochannel interference. The base station in cell A of cluster 1 is transmitting on frequency f_1, and at the same time, the base station in cell A of cluster 2 is transmitting on the same frequency. Although the two cells are in different clusters, they both use the A group of frequencies. The mobile unit in cluster 2 is receiving the same frequency from two different base stations. Although the mobile unit is under the control of the base station in cluster 2, the signal from cluster 1 is received at a lower power level as cochannel interference.

Interference between cells is proportional not to the distance between the two cells but rather to the ratio of the distance to the cell's radius. Since a cell's radius is proportional to transmit power, more radio channels can be added to a system by either (1) decreasing the transmit power per cell, (2) making cells smaller, or (3) filling vacated coverage areas with new cells. In a cellular system where all cells are approximately the same size, cochannel interference is dependent on the radius (R) of the cells and the distance to the center of the nearest cochannel cell (D) as shown in Figure 11-8. Increasing the D/R ratio (sometimes called *cochannel reuse ratio*) increases the spatial separation between cochannel

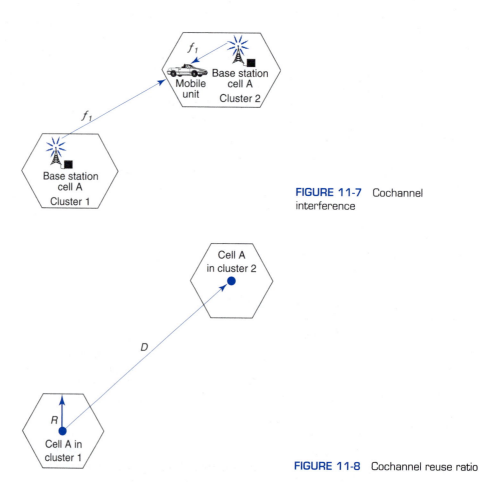

FIGURE 11-7 Cochannel interference

FIGURE 11-8 Cochannel reuse ratio

cells relative to the coverage distance. Therefore, increasing the cochannel reuse ratio (Q) can reduce cochannel interference. For a hexagonal geometry,

$$Q = \frac{D}{R} \tag{11-5}$$

where Q = cochannel reuse ratio (unitless)
D = distance to center of the nearest cochannel cell (kilometers)
R = cell radius (kilometers)

The smaller the value of Q, the larger the channel capacity since the cluster size is also smaller. However, a large value of Q improves the cochannel interference and, thus, the overall transmission quality. Obviously, in actual cellular system design, a trade-off must be made between the two conflicting objectives.

11-6-2 Adjacent-Channel Interference

Adjacent-channel interference occurs when transmissions from *adjacent channels* (channels next to one another in the frequency domain) interfere with each other. Adjacent-channel interference results from imperfect filters in receivers that allow nearby frequencies to enter the receiver. Adjacent-channel interference is most prevalent when an adjacent channel is transmitting very close to a mobile unit's receiver at the same time the mobile unit is trying to receive transmissions from the base station on an adjacent frequency. This is called the *near-far effect* and is most prevalent when a mobile unit is receiving a weak signal from the base station.

Adjacent-channel interference is depicted in Figure 11-9. Mobile unit 1 is receiving frequency f_1 from base station A. At the same time, base station A is transmitting frequency f_2 to mobile unit 2. Because mobile unit 2 is much farther from the base station than mobile unit 1, f_2 is transmitted at a much higher power level than f_1. Mobile unit 1 is located very close to the base station, and f_2 is located next to f_1 in the frequency spectrum (i.e., the adjacent channel); therefore, mobile unit 1 is receiving f_2 at a much higher power level than f_1. Because of the high power level, the filters in mobile unit 1 cannot block all the energy from f_2, and the signal intended for mobile unit 2 interferes with mobile unit 1's reception

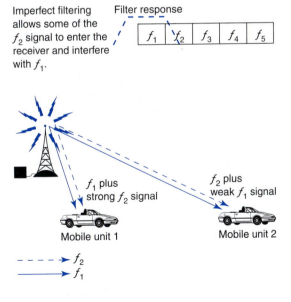

FIGURE 11-9 Adjacent-channel interference

of f_1. f_1 does not interfere with mobile unit 2's reception because f_1 is received at a much lower power level than f_2.

Using precise filtering and making careful channel assignments can minimize adjacent-channel interference in receivers. Maintaining a reasonable frequency separation between channels in a given cell can also reduce adjacent-channel interference. However, if the reuse factor is small, the separation between adjacent channels may not be sufficient to maintain an adequate adjacent-channel interference level.

11-7 CELL SPLITTING, SECTORING, SEGMENTATION, AND DUALIZATION

The Bell System proposed cellular telephone systems in the early 1960s as a means of alleviating congested frequency spectrums indigenous to wide-area mobile telephone systems using line-of-sight, high-powered transmitters. These early systems offered reasonable coverage over large areas, but the available channels were rapidly used up. For example, in the early 1970s, the Bell System could handle only 12 simultaneous mobile telephone calls at a time in New York City. Modern-day cellular telephone systems use relatively low-power transmitters and generally serve a much smaller geographical area.

Increases in demand for cellular service in a given area rapidly consume the cellular channels assigned the area. Two methods of increasing the capacity of a cellular telephone system are cell splitting and sectoring. Cell splitting provides for an orderly growth of a cellular system, whereas sectoring utilizes directional antennas to reduce cochannel and adjacent-channel interference and allow channel frequencies to be reassigned (reused).

11-7-1 Cell Splitting

Cell splitting is when the area of a cell, or independent component coverage areas of a cellular system, is further divided, thus creating more cell areas. The purpose of cell splitting is to increase the channel capacity and improve the availability and reliability of a cellular telephone network. The point when a cell reaches maximum capacity occurs when the number of subscribers wishing to place a call at any given time equals the number of channels in the cell. This is called the *maximum traffic load* of the cell. Splitting cell areas creates new cells, providing an increase in the degree of frequency reuse, thus increasing the channel capacity of a cellular network. Cell splitting provides for orderly growth in a cellular system. The major drawback of cell splitting is that it results in more *base station transfers* (handoffs) per call and a higher processing load per subscriber. It has been proven that a reduction of a cell radius by a factor of 4 produces a 10-fold increase in the handoff rate per subscriber.

Cell splitting is the resizing or redistribution of cell areas. In essence, cell splitting is the process of subdividing highly congested cells into smaller cells each with their own base station and set of channel frequencies. With cell splitting, a large number of low-power transmitters take over an area previously served by a single, higher-powered transmitter. Cell splitting occurs when traffic levels in a cell reach the point where channel availability is jeopardized. If a new call is initiated in an area where all the channels are in use, a condition called *blocking* occurs. A high occurrence of blocking indicates that a system is overloaded.

Providing wide-area coverage with small cells is indeed a costly operation. Therefore, cells are initially set up to cover relatively large areas, and then the cells are divided into smaller areas when the need arises. The area of a circle is proportional to its radius squared. Therefore, if the radius of a cell is divided in half, four times as many smaller cells could be created to provide service to the same coverage area. If each new cell has the same number of channels as the original cell, the capacity is also increased by a factor of 4. Cell splitting allows a system's capacity to increase by replacing large cells with several smaller cells while not disturbing the channel allocation scheme required to prevent interference between cells.

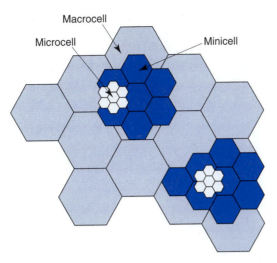

Microcell   Macrocell   Minicell

FIGURE 11-10   Cell splitting

Figure 11-10 illustrates the concept of cell splitting. Macrocells are divided into mini-cells, which are then further divided into microcells as traffic density increases. Each time a cell is split, its transmit power is reduced. As Figure 11-10 shows, cell splitting increases the channel capacity of a cellular telephone system by rescaling the system and increasing the number of channels per unit area (channel density). Hence, cell splitting decreases the cell radius while maintaining the same cochannel reuse ratio (D/R).

Example 11-3

Determine

a. The channel capacity for a cellular telephone area comprised of seven macrocells with 10 channels per cell.
b. Channel capacity if each macrocell is split into four minicells.
c. Channel capacity if each minicell is further split into four microcells.

Solution   **a.**
$$\frac{10 \text{ channels}}{\text{cell}} \times \frac{7 \text{ cells}}{\text{area}} = 70 \text{ channels/area}$$

b. Splitting each macrocell into four minicells increases the total number of cells in the area to $4 \times 7 = 28$. Therefore,
$$\frac{10 \text{ channels}}{\text{cell}} \times \frac{28 \text{ cells}}{\text{area}} = 280 \text{ channels/area}$$

c. Further splitting each minicell into four microcells increases the total number of cells in the area to $4 \times 28 = 112$. Therefore,
$$\frac{10 \text{ channels}}{\text{cell}} \times \frac{112 \text{ cells}}{\text{area}} = 1120 \text{ channels/area}$$

From Example 11-3, it can be seen that each time the cells were split, the coverage area appreciated a fourfold increase in channel capacity. For the situation described in Example 11-3, splitting the cells twice increased the total capacity by a factor of 16 from 70 channels to 1120 channels.

11-7-2   Sectoring

Another means of increasing the channel capacity of a cellular telephone system is to decrease the *D/R* ratio while maintaining the same cell radius. Capacity improvement can be achieved by reducing the number of cells in a cluster, thus increasing the frequency reuse. To accomplish this, the relative interference must be reduced without decreasing transmit power.

In a cellular telephone system, cochannel interference can be decreased by replacing a single omnidirectional antenna with several directional antennas, each radiating within a

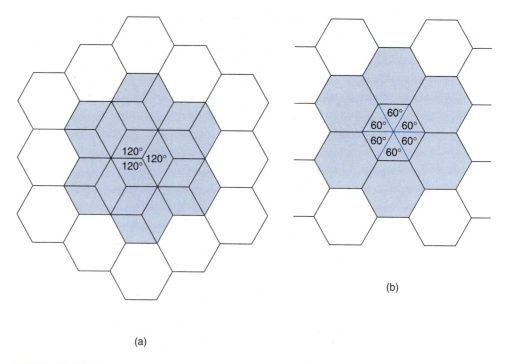

(a)

(b)

FIGURE 11-11 Sectoring: (a) 120-degree sectors; (b) 60-degree sectors

smaller area. These smaller areas are called *sectors,* and decreasing cochannel interference while increasing capacity by using directional antennas is called *sectoring.* The degree in which cochannel interference is reduced is dependent on the amount of sectoring used. A cell is normally partitioned either into three 60° or six 120° sectors as shown in Figure 11-11. In the three-sector configuration shown in Figure 11-11b, three antennas would be placed in each 120° sector—one transmit antenna and two receive antennas. Placing two receive antennas (one above the other) is called *space diversity.* Space diversity improves reception by effectively providing a larger target for signals radiated from mobile units. The separation between the two receive antennas depends on the height of the antennas above the ground. This height is generally taken to be the height of the tower holding the antenna. As a rule, antennas located 30 meters above the ground require a separation of eight wavelengths, and antennas located 50 meters above the ground require a separation of 11 wavelengths.

When sectoring is used, the channels utilized in a particular sector are broken down into sectored groups that are used only within a particular sector. With seven-cell reuse and 120° sectors, the number of interfering cells in the closest tier is reduced from six to two. Sectoring improves the signal-to-interference ratio, thus increasing the system's capacity.

11-7-3 Segmentation and Dualization

Segmentation and *dualization* are techniques incorporated when additional cells are required within the reuse distance. Segmentation divides a group of channels into smaller groupings or segments of mutually exclusive frequencies; cell sites, which are within the reuse distance, are assigned their own segment of the channel group. Segmentation is a means of avoiding cochannel interference, although it lowers the capacity of a cell by enabling reuse inside the reuse distance, which is normally prohibited.

Dualization is a means of avoiding full-cell splitting where the entire area would otherwise need to be segmented into smaller cells. When a new cell is set up requiring the same channel group as an existing cell (cell 1) and a second cell (cell 2) is not sufficiently far from cell 1 for normal reuse, the busy part of cell 1 (the center) is converted to a primary cell, and the same channel frequencies can be assigned to the new competing cell (cell 2). If all available channels need to be used in cell 2, a problem would arise because the larger secondary cell in cell 1 uses some of these, and there would be interference. In practice, however, cells are assigned different channels, so this is generally not a problem. A drawback of dualization is that it requires an extra base station in the middle of cell 1. There are now two base stations in cell 1: one a high-power station that covers the entire secondary cell and one a low-power station that covers the smaller primary cell.

11-8 CELLULAR SYSTEM TOPOLOGY

Figure 11-12 shows a simplified cellular telephone system that includes all the basic components necessary for cellular telephone communications. The figure shows a wireless radio network covering a set of geographical areas (cells) inside of which mobile two-way radio units, such as cellular or PCS telephones, can communicate. The radio network is defined by a set of radio-frequency transceivers located within each of the cells. The locations of these radio-frequency transceivers are called *base stations*. A base station serves as central control for all users within that cell. Mobile units (such as automobiles and pedestrians) communicate directly with the base stations, and the base stations communicate

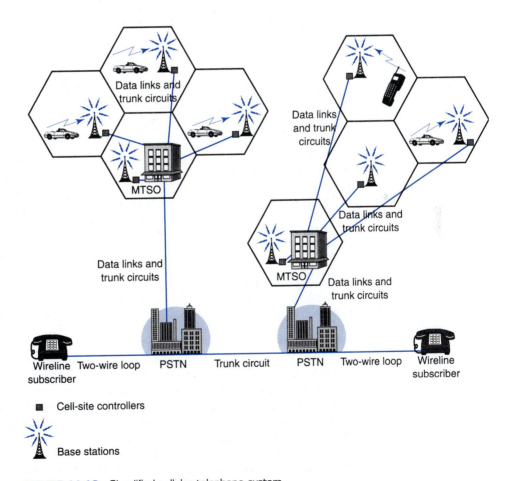

FIGURE 11-12 Simplified cellular telephone system

directly with a *Mobile Telephone Switching Office* (MTSO). An MTSO controls channel assignment, call processing, call setup, and call termination, which includes signaling, switching, supervision, and allocating radio-frequency channels. The MTSO provides a centralized administration and maintenance point for the entire network and interfaces with the public telephone network over wireline voice trunks and data links. MTSOs are equivalent to class 4 toll offices, except smaller. Local loops (or the cellular equivalent) do not terminate in MTSOs. The only facilities that connect to an MTSO are trunk circuits. Most MTSOs are connected to the SS7 signaling network, which allows cellular telephones to operate outside their service area.

Base stations can improve the transmission quality, but they cannot increase the channel capacity within the fixed bandwidth of the network. Base stations are distributed over the area of system coverage and are managed and controlled by an on-site computerized *cell-site controller* that handles all cell-site control and switching functions. Base stations communicate not only directly with mobile units through the airways using control channels but also directly with the MTSO over dedicated data control links (usually four wire, full duplex). Figure 11-12 shows how trunk circuits interconnect cell-site controllers to MTSOs and MTSOs with exchange offices within the PSTN.

The base station consists of a low-power radio transceiver, power amplifiers, a control unit (computer), and other hardware, depending on the system configuration. Cellular and PCS telephones use several moderately powered transceivers over a relatively wide service area. The function of the base station is to provide an interface between mobile telephone sets and the MTSO. Base stations communicate with the MTSO over dedicated data links, both metallic and nonmetallic facilities, and with mobile units over the airwaves using control channels. The MTSO provides a centralized administration and maintenance point for the entire network, and it interfaces with the PSTN over wireline voice trunks to honor services from conventional wireline telephone subscribers.

To complicate the issue, an MTSO is known by several different names, depending on the manufacturer and the system configuration. *Mobile Telephone Switching Office* (MTSO) is the name given by Bell Telephone Laboratories, *Electronic Mobile Xchange* (EMX) by Motorola, *AEX* by Ericcson, *NEAX* by NEC, and *Switching Mobile Center* (SMC) and *Master Mobile Center* (MMC) by Novatel. In PCS networks, the mobile switching center is called the MCS.

Each geographic area or cell can generally accommodate many different user channels simultaneously. The number of user channels depends on the accessing technique used. Within a cell, each radio-frequency channel can support up to 20 mobile telephone users at one time. Channels may be statically or dynamically assigned. Statically assigned channels are assigned a mobile unit for the duration of a call, whereas dynamically assigned channels are assigned a mobile unit only when it is being used. With both static and dynamic assignments, mobile units can be assigned any available radio channel.

11-9 ROAMING AND HANDOFFS

Roaming is when a mobile unit moves from one cell to another—possibly from one company's service area into another company's service area (requiring *roaming agreements*). As a mobile unit (car or pedestrian) moves away from the base station transceiver it is communicating with, the signal strength begins to decrease. The output power of the mobile unit is controlled by the base station through the transmission of up/down commands, which depends on the signal strength the base station is currently receiving from the mobile unit. When the signal strength drops below a predetermined threshold level, the electronic switching center locates the cell in the honeycomb pattern that is receiving the strongest signal from the particular mobile unit and then transfers the mobile unit to the base station in the new cell.

One of the most important features of a cellular system is its ability to transfer calls that are already in progress from one cell-site controller to another as the mobile unit moves

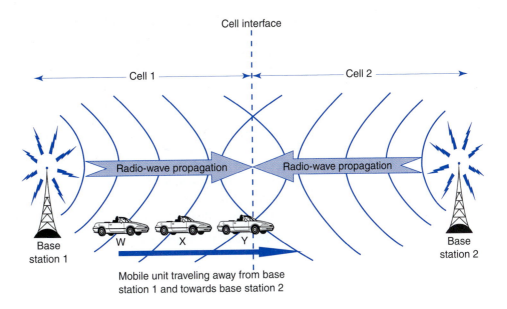

Cell 1   Cell 2

Radio-wave propagation   Radio-wave propagation

Base
station 1

Base
station 2

W   X   Y

Mobile unit traveling away from base
station 1 and towards base station 2

FIGURE 11-13   Handoff

from cell to cell within the cellular network. The base station transfer includes converting the call to an available channel within the new cell's allocated frequency subset. The transfer of a mobile unit from one base station's control to another base station's control is called a *handoff* (or *handover*). Handoffs should be performed as infrequently as possible and be completely *transparent* (*seamless*) to the subscriber (i.e., the subscribers cannot perceive that their facility has been switched). A handoff consists of four stages: (1) initiation, (2) resource reservation, (3) execution, and (4) completion. A connection that is momentarily broken during the cell-to-cell transfer is called a *hard handoff*. A hard handoff is a *break-before-make process*. With a hard handoff, the mobile unit breaks its connection with one base station before establishing voice communications with a new base station. Hard handoffs generally occur when a mobile unit is passed between disjointed systems with different frequency assignments, air interface characteristics, or technologies. A flawless handoff (i.e., no perceivable interruption of service) is called a *soft handoff* and normally takes approximately 200 ms, which is imperceptible to voice telephone users, although the delay may be disruptive when transmitting data. With a soft handoff, a mobile unit establishes contact with a new base station before giving up its current radio channel by transmitting coded speech signals to two base stations simultaneously. Both base stations send their received signals to the MTSO, which estimates the quality of the two signals and determines when the transfer should occur. A complementary process occurs in the opposite direction. A soft handoff requires that the two base stations operate synchronously with one another.

Figure 11-13 shows how a base station transfer is accomplished when a mobile unit moves from one cell into another (the figure shows a soft handoff). The mobile unit is moving away from base station 1 (i.e., toward base station 2). When the mobile unit is at positions W and X, it is well within the range of base station 1 and very distant from base station 2. However, when the mobile unit reaches position Y, it receives signals from base station 1 and base station 2 at approximately the same power level, and the two base stations should be setting up for a handoff (i.e., initiation and resource reservation). When the mobile unit crosses from cell 1 into cell 2, the handoff should be executed and completed.

Computers at cell-site controllers should transfer calls from cell to cell with minimal disruption and no degradation in the quality of transmission. The computers use *handoff decision algorithms* based on variations in signal strength and signal quality. When a call is in progress, the switching center monitors the received signal strength of each user channel. Handoffs can be initiated when the signal strength (or signal-to-interference ratio), measured by either the base station or the mobile unit's receiver, falls below a predetermined threshold level (typically between -90 dBm and -100 dBm) or when a network resource management needs to force a handoff to free resources to place an emergency call. During a handoff, information about the user stored in the first base station is transferred to the new base station. A condition called *blocking* occurs when the signal level drops below a usable level and there are no usable channels available in the target cell to switch to. To help avoid blocking or loss of a call during a handoff, the system employs a load-balancing scheme that frees channels for handoffs and sets handoff priorities. Programmers at the central switching site continually update the switching algorithm to amend the system to accommodate changing traffic loads.

The handoff process involves four basic steps:

1. *Initiation.* Either the mobile unit or the network determines the need for a handoff and initiates the necessary network procedures.
2. *Resource reservation.* Appropriate network procedures reserve the resources needed to support the handoff (i.e., a voice and a control channel).
3. *Execution.* The actual transfer of control from one base station to another base station takes place.
4. *Completion.* Unnecessary network resources are relinquished and made available to other mobile units.

11-9-1 IS-41 Standard

In the United States, roaming from one company's calling area into another company's calling area is called *interoperator roaming* and requires prior agreements between the two service providers. To provide seamless roaming between calling areas served by different companies, the Electronics Industries Association/Telecommunications Industry Association (EIA/TIA) developed the IS-41 protocol, which was endorsed by the Cellular Telecommunication Industry Association (CITA). IS-41 aligns with a subprotocol of the SS7 protocol stack that facilitates communications among databases and other network entities. The IS-41 standard is separated into a series of recommendations.

The principal purposes of IS-41 are to allow mobile units to roam and to perform handoffs of calls already in progress when a mobile unit moves from one cellular system into another without subscriber intervention. Before deployment of SS7, X.25 provided the carrier services for data messages traveling from one cell (the *home location register* [HLR]) to another cell (the *visitor location register* [VLR]). IS-41 provides the information and exchanges necessary to establish and cancel registration in various databases. IS-41 aligns with the ANSI version of SS7 to communicate with databases and other network functional entities.

IS-41 relies on a feature called *autonomous registration,* the process where a mobile unit notifies a serving MTSO of its presence and location through a base station controller. The mobile unit accomplishes autonomous registration by periodically transmitting its identity information, thus allowing the serving MTSO to continuously update its customer list. IS-41 allows MTSOs in neighboring systems to automatically register and validate locations of roaming mobile units so that users no longer need to manually register as they travel.

11-10 CELLULAR TELEPHONE NETWORK COMPONENTS

There are six essential components of a cellular telephone system: (1) an electronic switching center, (2) a cell-site controller, (3) radio transceivers, (4) system interconnections, (5) mobile telephone units, and (6) a common communications protocol.

11-10-1 Electronic Switching Centers

The *electronic switching center* is a digital telephone exchange located in the MTSO that is the heart of a cellular telephone system. The electronic switch performs two essential functions: (1) it controls switching between the public wireline telephone network and the cell-site base stations for wireline-to-mobile, mobile-to-wireline, and mobile-to-mobile calls, and (2) it processes data received from the cell-site controllers concerning mobile unit status, diagnostic data, and bill-compiling information. Electronic switches communicate with cell-site controllers using a data link protocol, such as X.25, at a transmission rate of 9.6 kbps or higher.

11-10-2 Cell-Site Controllers

Each cell contains one *cell-site controller* (sometimes called *base station controller*) that operates under the direction of the switching center (MTSO). Cell-site controllers manage each of the radio channels at each site, supervise calls, turn the radio transmitter and receiver on and off, inject data onto the control and voice channels, and perform diagnostic tests on the cell-site equipment. Base station controllers make up one part of the *base station subsystem.* The second part is the *base transceiver station* (BTS).

11-10-3 Radio Transceivers

Radio transceivers are also part of the base station subsystem. The radio transceivers (combination transmitter/receiver) used with cellular telephone system voice channels can be either narrowband FM for analog systems or either PSK or QAM for digital systems with an effective audio-frequency band comparable to a standard telephone circuit (approximately 300 Hz to 3000 Hz). The control channels use either FSK or PSK. The maximum output power of a cellular transmitter depends on the type of cellular system. Each cell base station typically contains one radio transmitter and two radio receivers tuned to the same channel (frequency). The radio receiver that detects the strongest signal is selected. This arrangement is called *receiver diversity.* The radio transceivers in base stations include the antennas (both transmit and receive). Modern cellular base station antennas are more aesthetically appealing than most antennas and can resemble anything from a window shutter to a palm tree to an architectural feature on a building.

11-10-4 System Interconnects

Four-wire leased lines are generally used to connect switching centers to cell sites and to the public telephone network. There is one dedicated four-wire trunk circuit for each of the cell's voice channels. There must also be at least one four-wire trunk circuit to connect switching centers to each cell-site controller for transferring control signals.

11-10-5 Mobile and Portable Telephone Units

Mobile and *portable telephone units* are essentially identical. The only differences are that portable units have a lower output power, have a less efficient antenna, and operate exclusively on batteries. Each mobile telephone unit consists of a control unit, a multiple-frequency radio transceiver (i.e., multiple channel), a logic unit, and a mobile antenna. The control unit houses all the user interfaces, including a built-in handset. The transceiver uses a frequency synthesizer to tune into any designated cellular system channel. The logic unit interrupts subscriber actions and system commands while managing the operation of the transceiver (including transmit power) and control units.

11-10-6 Communications Protocol

The last constituent of a cellular telephone system is the *communications protocol,* which governs the way telephone calls are established and disconnected. There are several layers of protocols used with cellular telephone systems, and these protocols differ between cellular networks. The protocol implemented depends on whether the voice and control channels are analog or digital and what method subscribers use to access the network. Examples of cellular communications protocols are IS-54, IS-136.2, and IS-95.

Telephone calls over cellular networks require using two full-duplex radio-frequency channels simultaneously, one called the *user channel* and one called the *control channel*. The user channel is the actual voice channel where mobile users communicate directly with other mobile and wireline subscribers through a base station. The control channel is used for transferring control and diagnostic information between mobile users and a central cellular telephone switch through a base station. Base stations transmit on the *forward control channel* and *forward voice channel* and receive on the *reverse control channel* and *reverse voice channel*. Mobile units transmit on the reverse control channel and reverse voice channel and receive on the forward control channel and forward voice channel.

Completing a call within a cellular telephone system is similar to completing a call using the wireline PSTN. When a mobile unit is first turned on, it performs a series of start-up procedures and then samples the receive signal strength on all user channels. The mobile unit automatically tunes to the control channel with the strongest receive signal strength and synchronizes to the control data transmitted by the cell-site controller. The mobile unit interprets the data and continues monitoring the control channel(s). The mobile unit automatically rescans periodically to ensure that it is using the best control channel.

11-11-1 Call Procedures

Within a cellular telephone system, three types of calls can take place involving mobile cellular telephones: (1) mobile (cellular) to wireline (PSTN), (2) mobile (cellular) to mobile (cellular), and (3) wireline (PSTN) to mobile (cellular). A general description is given for the procedures involved in completing each of the three types of calls involving mobile cellular telephones.

11-11-1-1 Mobile (cellular)-to-wireline (PSTN) call procedures.

1. Calls from mobile telephones to wireline telephones can be initiated in one of two ways:
   a. The mobile unit is equivalently taken off hook (usually by depressing a talk button). After the mobile unit receives a dial tone, the subscriber enters the wireline telephone number using either a standard Touch-Tone keypad or with speed dialing. After the last digit is depressed, the number is transmitted through a reverse control channel to the base station controller along with the mobile unit's unique identification number (which is not the mobile unit's telephone number).
   b. The mobile subscriber enters the wireline telephone number into the unit's memory using a standard Touch-Tone keypad. The subscriber then depresses a send key, which transmits the called number as well as the mobile unit's identification number over a reverse control channel to the base station switch.
2. If the mobile unit's ID number is valid, the cell-site controller routes the called number over a wireline trunk circuit to the MTSO.
3. The MTSO uses either standard call progress signals or the SS7 signaling network to locate a switching path through the PSTN to the destination party.
4. Using the cell-site controller, the MTSO assigns the mobile unit a nonbusy user channel and instructs the mobile unit to tune to that channel.
5. After the cell-site controller receives verification that the mobile unit has tuned to the selected channel and it has been determined that the called number is on hook, the mobile unit receives an audible call progress tone (ring-back) while the wireline caller receives a standard ringing signal.
6. If a suitable switching path is available to the wireline telephone number, the call is completed when the wireline party goes off hook (answers the telephone).

11-11-1-2 Mobile (cellular)-to-mobile (cellular) call procedures.

1. The originating mobile unit initiates the call in the same manner as it would for a mobile-to-wireline call.
2. The cell-site controller receives the caller's identification number and the destination telephone number through a reverse control channel, which are then forwarded to the MTSO.
3. The MTSO sends a page command to all cell-site controllers to locate the destination party (which may be anywhere in or out of the service area).
4. Once the destination mobile unit is located, the destination cell-site controller sends a page request through a control channel to the destination party to determine if the unit is on or off hook.
5. After receiving a positive response to the page, idle user channels are assigned to both mobile units.
6. Call progress tones are applied in both directions (ring and ring-back).
7. When the system receives notice that the called party has answered the telephone, the switches terminate the call progress tones, and the conversation begins.
8. If a mobile subscriber wishes to initiate a call and all user channels are busy, the switch sends a directed retry command, instructing the subscriber's unit to reattempt the call through a neighboring cell.
9. If the system cannot allocate user channels through a neighboring cell, the switch transmits an intercept message to the calling mobile unit over the control channel.
10. If the called party is off hook, the calling party receives a busy signal.
11. If the called number is invalid, the calling party receives a recorded message announcing that the call cannot be processed.

11-11-1-3 Wireline (PSTN)-to-mobile (cellular) call procedures.

1. The wireline telephone goes off hook to complete the loop, receives a dial tone, and then inputs the mobile unit's telephone number.
2. The telephone number is transferred from the PSTN switch to the cellular network switch (MTSO) that services the destination mobile number.
3. The cellular network MTSO receives the incoming call from the PSTN, translates the received digits, and locates the base station nearest the mobile unit, which determines if the mobile unit is on or off hook (i.e., available).
4. If the mobile unit is available, a positive page response is sent over a reverse control channel to the cell-site controller, which is forwarded to the network switch (MTSO).
5. The cell-site controller assigns an idle user channel to the mobile unit and then instructs the mobile unit to tune to the selected channel.
6. The mobile unit sends verification of channel tuning through the cell-site controller.
7. The cell-site controller sends an audible call progress tone to the subscriber's mobile telephone, causing it to ring. At the same time, a ring-back signal is sent back to the wireline calling party.
8. The mobile answers (goes off hook), the switch terminates the call progress tones, and the conversation begins.

QUESTIONS

11-1. What is the contemporary mobile telephone meaning for the term *mobile*?

11-2. Contrast the similarities and differences between *two-way mobile radio* and *cellular telephone*.

11-3. Describe the differences between a cellular telephone *service area*, a *cluster*, and a *cell*.

11-4. Why was a *honeycomb pattern* selected for a cell area?

11-5. What are the differences between *macrocells*, *minicells*, and *microcells*?

11-6. What is meant by a *center-excited* cell? *Edge-excited* cell? *Corner-excited* cell?

11-7. Describe *frequency reuse*. Why is it useful in cellular telephone systems?

11-8. What is meant by *frequency reuse factor*?

11-9. Name and describe the two most prevalent types of interference in cellular telephone systems.

11-10. What significance does the *cochannel reuse factor* have on cellular telephone systems?

11-11. What is meant by the *near-far effect*?

11-12. Describe the concept of *cell splitting*. Why is it used?

11-13. Describe the term *blocking*.

11-14. Describe what is meant by *channel density*.

11-15. Describe *sectoring* and state why it is used.

11-16. What is the difference between a *MTSO* and a *cell-site controller*?

11-17. Explain what the term *roaming* means.

11-18. Explain the term *handoff*.

11-19. What is the difference between a *soft* and a *hard* handoff?

11-20. Explain the term *break before make* and state when it applies.

11-21. Briefly describe the purpose of the *IS-41 standard*.

11-22. Describe the following terms: *home location register, visitor location register,* and *autonomous registration*.

11-23. List and describe the six essential components of a *cellular telephone network*.

11-24. Describe what is meant by the terms *forward channel* and *reverse channel*.

PROBLEMS

11-1. Determine the number of channels per cluster and the total channel capacity for a cellular telephone area comprised of 12 clusters with seven cells in each cluster and 16 channels in each cell.

11-2. Determine the number of cells in clusters for the following values: $j = 4$ and $i = 2$ and $j = 3$ and $i = 3$.

11-3. Determine the cochannel reuse ratio for a cell radius of 0.5 miles separated from the nearest cochannel cell by a distance of 4 miles.

11-4. Determine the distance from the nearest cochannel cell for a cell radius of 0.4 miles and a cochannel reuse factor of 12.

11-5. Determine

    a. The channel capacity for a cellular telephone area comprised of seven macrocells with 16 channels per cell.

    b. Channel capacity if each macrocell is split into four minicells.

    c. Channel capacity if each minicell is further split into four microcells.

11-6. A cellular telephone company has acquired 150 full-duplex channels for a given service area. The company decided to divide the service area into 15 clusters and use a seven-cell reuse pattern and use the same number of channels in each cell. Determine the total number of channels the company has available for its subscribers at any one time.

C H A P T E R   12

Cellular Telephone Systems

CHAPTER OUTLINE

OBJECTIVES

- Define *first-generation analog cellular telephone systems*
- Describe and outline the frequency allocation for the Advanced Mobile Telephone System (AMPS)
- Explain frequency-division multiple accessing (FDMA)
- Describe the operation of AMPS control channels
- Explain the AMPS classification of cellular telephones
- Describe the concepts of personal communications systems (PCS)
- Outline the advantages and disadvantages of PCS compared to standard cellular telephone
- Describe second-generation cellular telephone systems
- Explain the operation of N-AMPS cellular telephone systems
- Define *digital cellular telephone*
- Describe the advantages and disadvantages of digital cellular telephone compared to analog cellular telephone
- Describe time-division multiple accessing (TDMA)
- Describe the purpose of IS-54 and what is meant by dual-mode operation
- Describe IS-136 and explain its relationship to IS-54
- Describe the format for a USDC digital voice channel
- Explain the classifications of USDC radiated power
- Describe the basic concepts and outline the specifications of IS-95
- Describe code-division multiple accessing (CDMA)

- Outline the CDMA frequency and channel allocation for cellular telephone
- Explain the classifications of CDMA radiated power
- Summarize North American cellular and PCS systems
- Describe global system for mobile communications (GSM)
- Describe the services provided by GSM
- Explain GSM system architecture
- Describe the GSM radio subsystem
- Describe the basic concepts of a Personal Communications Satellite System (PCSS)
- Outline the advantages and disadvantages of PCSS over terrestrial cellular telephone systems

12-1 INTRODUCTION

Like nearly everything in the modern world of electronic communications, cellular telephone began as a relatively simple concept. However, the increased demand for cellular services has caused cellular telephone systems to evolve into complicated networks and internetworks comprised of several types of cellular communications systems. New systems have evoked new terms, such as *standard cellular telephone service* (CTS), *personal communications systems* (PCS), and *Personal Communications Satellite System* (PCSS), all of which are full-duplex mobile telephone systems that utilize the cellular concept.

Cellular telephone began as a relatively simple two-way analog communications system using frequency modulation (FM) for voice and frequency-shift keying (FSK) for transporting control and signaling information. The most recent cellular telephone systems use higher-level digital modulation schemes for conveying both voice and control information. In addition, the Federal Communications Commission (FCC) has recently assigned new frequency bands for cellular telephone. The following sections are intended to give the reader a basic understanding of the fundamental meaning of the common cellular telephone systems and the terminology used to describe them.

12-2 FIRST-GENERATION ANALOG CELLULAR TELEPHONE

In 1971, Bell Telephone Laboratories in Murry Hill, New Jersey, proposed the cellular telephone concept as the *Advanced Mobile Telephone System* (AMPS). The cellular telephone concept was an intriguing idea that added a depth or spatial dimension to the conventional wireline trunking model used by the public telephone company at the time. The cellular plan called for using many low-profile, low-power cell-site transceivers linked through a central computer-controlled switching and control center. AMPS is a standard cellular telephone service (CTS) initially placed into operation on October 13, 1983, by Illinois Bell that incorporated several large cell areas to cover approximately 2100 square miles in the Chicago area. The original system used omnidirectional antennas to minimize initial equipment costs and employed low-power (7-watt) transmitters in both base stations and mobile units. Voice-channel radio transceivers with AMPS cellular telephones use narrowband frequency modulation (NBFM) with a usable audio-frequency band from 300 Hz to 3 kHz and a maximum frequency deviation of ± 12 kHz for 100% modulation. Using Carson's rule, this corresponds to an approximate bandwidth of 30 kHz. Empirical information determined that an AMPS 30-kHz telephone channel requires a minimum signal-to-interference ratio (SIR) of 18 dB for satisfactory performance. The smallest reuse factor that satisfied this requirement utilizing 120° directional antennas was 7. Consequently, the AMPS system uses a seven-cell reuse pattern with provisions for cell splitting and sectoring to increase channel capacity when needed.

12-2-1 AMPS Frequency Allocation

In 1980, the FCC decided to license two common carriers per cellular service area. The idea was to eliminate the possibility of a monopoly and provide the advantages that generally

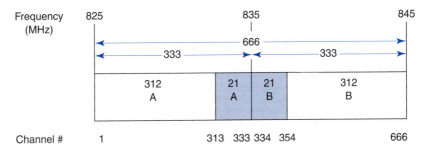

Reverse channels — mobile unit transmit and base station receive frequencies

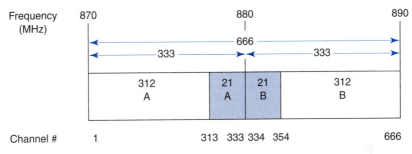

Forward channels — mobile unit transmit and base station receive frequencies

Shaded areas denote control channels (A system: 313 to 333 and B system: 334 to 354)

FIGURE 12-1 Original Advanced Mobile Phone Service (AMPS) frequency spectrum

accompany a competitive environment. Subsequently, two frequency allocation plans emerged—system A and system B—each with its own group of channels that shared the allocated frequency spectrum. System A is defined for the non-wireline companies (i.e., cellular telephone companies) and system B for existing wireline companies (i.e., local telephone companies). The FCC initially assigned the AMPS system a 40-MHz frequency band consisting of 666 two-way channels per service area with 30-kHz spacing between adjacent channels.

Figure 12-1 shows the original frequency management system for the AMPS cellular telephone system. The A channels are designated 1 to 333, and the B channels are designated 334 to 666. For mobile units, channel 1 has a transmit frequency of 825.03 MHz, and channel 666 has a transmit frequency of 844.98 MHz. For base stations, channel 1 has a transmit frequency of 870.03 MHz, and channel 666 has a transmit frequency of 889.98 MHz. The receive frequencies are, of course, just the opposite.

Simultaneous transmission in both directions is a transmission mode called *full duplex* (FDX) or simply *duplexing*. Duplexing can be accomplished using frequency- or time-domain methods. *Frequency-division duplexing* (FDD) is used with AMPS and occurs when two distinct frequency bands are provided to each user. In FDD, each duplex channel actually consists of two simplex (one-way) channels (base station to mobile and mobile to base station). A special device called a duplexer is used in each mobile unit and base station to allow simultaneous transmission and reception on duplex channels.

Transmissions from base stations to mobile units are called *forward links*, whereas transmissions from mobile units to base stations are called *reverse links*. (Forward links are

Cellular Telephone Systems

331

Reverse channels — mobile unit transmit and base station receive frequencies

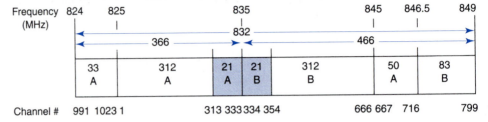

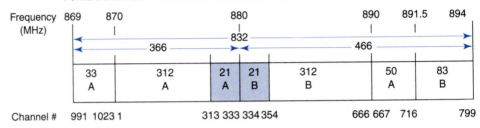

Forward channels — mobile unit transmit and base station receive frequencies

Shaded areas denote control channels (A system: 313 to 333 and B system: 334 to 354)

FIGURE 12-2 Complete Advanced Mobile Phone Service (AMPS) frequency spectrum

sometimes called *downlinks,* and reverse links are sometimes called *uplinks.*) The receiver for each channel operates 45 MHz above the transmit frequency. Consequently, every two-way AMPS radio channel consists of a pair of simplex channels separated by 45 MHz. The 45-MHz separation between transmit and receive frequencies was chosen to make use of inexpensive but highly selective duplexers in the mobile units.

In 1989, the FCC added an additional 10-MHz frequency spectrum to the original 40-MHz band, which increased the number of simplex channels by 166 for a total of 832 (416 full duplex). The additional frequencies are called the *expanded spectrum* and include channels 667 to 799 and 991 to 1023. The complete AMPS frequency assignment is shown in Figure 12-2. Note that 33 of the new channels were added below the original frequency spectrum and that the remaining 133 were added above the original frequency spectrum. With AMPS, a maximum of 128 channels could be used in each cell.

The mobile unit's transmit carrier frequency in MHz for any channel is calculated as follows:

$$f_t = 0.03\,N + 825 \qquad\qquad \text{for } 1 \le N \le 866 \qquad\qquad \textbf{(12-1)}$$

$$f_t = 0.03(N - 1023) + 825 \qquad \text{for } 990 \le N \le 1023 \qquad \textbf{(12-2)}$$

where f_t = transmit carrier frequency (MHz)
       N = channel number

The mobile unit's receive carrier frequency is obtained by simply adding 45 MHz to the transmit frequency:

$$f_r = f_t + 45 \text{ MHz} \qquad\qquad\qquad \textbf{(12-3)}$$

The base station's transmit frequency for any channel is simply the mobile unit's receive frequency, and the base station's receive frequency is simply the mobile unit's transmit frequency.

Example 12-1

Determine the transmit and receive carrier frequencies for

a. AMPS channel 3.
b. AMPS channel 991.

Solution

a. The transmit and receive carrier frequencies for channel 3 can be determined from Equations 12-1 and 12-3:

transmit
$$f_t = 0.03N + 825$$
$$= 0.03(3) + 825$$
$$= 825.09 \text{ MHz}$$

receive
$$f_r = 825.09 \text{ MHz} + 45 \text{ MHz}$$
$$= 870.09 \text{ MHz}$$

b. The transmit and receive carrier frequencies for channel 991 can be determined from Equations 12-2 and 12-3:

transmit
$$f_t = 0.03(991 - 1023) + 825$$
$$= 824.04 \text{ MHz}$$

receive
$$f_r = 824.04 \text{ MHz} + 45 \text{ MHz}$$
$$= 869.04 \text{ MHz}$$

Table 12-1 summarizes the frequency assignments for AMPS. The set of control channels may be split by the system operator into subsets of dedicated control channels, paging channels, or access channels.

The FCC controls the allocation of cellular telephone frequencies (channels) and also issues licenses to cellular telephone companies to operate specified frequencies in geographic areas called *cellular geographic serving areas* (CGSA). CGSAs are generally designed to lie within the borders of a standard metropolitan statistical area (SMSA), which defines geographic areas used by marketing agencies that generally correspond to the area covered by a specific wireline LATA (local access and transport area).

12-2-2   Frequency-Division Multiple Accessing

Standard cellular telephone subscribers access the AMPS system using a technique called *frequency-division multiple accessing* (FDMA). With FDMA, transmissions are separated in the frequency domain—each channel is allocated a carrier frequency and channel bandwidth within the total system frequency spectrum. Subscribers are assigned a pair of voice channels (forward and reverse) for the duration of their call. Once assigned a voice channel, a subscriber is the only mobile unit using that channel within a given cell. Simultaneous transmissions from multiple subscribers can occur at the same time without interfering with one another because their transmissions are on different channels and occupy different frequency bands.

12-2-3   AMPS Identification Codes

The AMPS system specifies several identification codes for each mobile unit (see Table 12-2). The *mobile identification number* (MIN) is a 34-bit binary code, which in the United States represents the standard 10-digit telephone number. The MIN is comprised of a three-digit area code, a three-digit prefix (exchange number), and a four-digit subscriber (extension) number. The exchange number is assigned to the cellular operating company. If a subscriber changes service from one cellular company to another, the subscriber must be assigned a new cellular telephone number.

Table 12-1 AMPS Frequency Allocation

			AMPS
Channel spacing			30 kHz
Spectrum allocation			40 MHz
Additional spectrum			10 MHz
Total number of channels			832

System A Frequency Allocation		
AMPS		
Channel Number	Mobile TX, MHz	Mobile RX, MHz
1	825.030	870.030
313[a]	834.390	879.390
333[b]	843.990	879.990
667	845.010	890.010
716	846.480	891.480
991	824.040	869.040
1023	825.000	870.000
System B Frequency Allocation		
334[c]	835.020	880.020
354[d]	835.620	880.620
666	844.980	890.000
717	846.510	891.000
799	848.970	894.000

[a]First dedicated control channel for system A.
[b]Last dedicated control channel for system A.
[c]First dedicated control channel for system B.
[d]Last dedicated control channel for system B.

Table 12-2 AMPS Identification Codes

Notation	Name	Length (Bits)	Description
MIN	Mobile identifier	34	Directory number assigned by operating company to a subscriber (telephone number)
ESN	Electronic serial number	32	Assigned by manufacturer to a mobile station (telephone)
SID	System identifier	15	Assigned by regulators to a geographical service area
SCM	Station class mark	4	Indicates capabilities of a mobile station
SAT	Supervisory audio tone	*	Assigned by operating company to each base station
DCC	Digital color code	2	Assigned by operating company to each base station

Another identification code used with AMPS is the *electronic serial number* (ESN), which is a 32-bit binary code permanently assigned to each mobile unit. The ESN is similar to the VIN (vehicle identification number) assigned to an automobile or the MAC address on a network interface card (NIC) in that the number is unique and positively identifies a specific unit.

Table 12-3 AMPS Mobile Phone Power Levels

Power Level	Class I		Class II		Class III	
	dBm	mW	dBm	mW	dBm	mW
0	36	4000	32	1600	28	640
1	32	1600	32	1600	28	640
2	28	640	28	640	28	640
3	24	256	24	256	24	256
4	20	102	20	102	20	102
5	16	41	16	41	16	41
6	12	16	12	16	12	16
7	8	6.6	8	6.6	8	6.6

The third identification code used with AMPS is the four-bit *station class mark* (SCM), which indicates whether the terminal has access to all 832 AMPS channels or only 666. The SCM also specifies the maximum radiated power for the unit (Table 12-3).

The *system identifier* (SID) is a 15-bit binary code issued by the FCC to an operating company when it issues it a license to provide AMPS cellular service to an area. The SID is stored in all base stations and all mobile units to identify the operating company and MTSO and any additional shared MTSO. Every mobile unit knows the SID of the system it is subscribed to, which is the mobile unit's *home system*. Whenever a mobile unit initializes, it compares its SID to the SID broadcast by the local base station. If the SIDs are the same, the mobile unit is communicating with its home system. If the SIDs are different, the mobile unit is roaming.

Local operating companies assign a two-bit *digital color code* (DCC) and a *supervisory audio tone* (SAT) to each of their base stations. The DCC and SAT help the mobile units distinguish one base station from a neighboring base station. The SAT is one of three analog frequencies (5970 Hz, 6000 Hz, or 6030 Hz), and the DCC is one of four binary codes (00, 01, 10, or 11). Neighboring base stations transmit different SAT frequencies and DCCs.

12-2-4 AMPS Control Channels

The AMPS channel spectrums are divided into two basic sets or groups. One set of channels is dedicated for exchanging control information between mobile units and base stations and is appropriately termed *control channels* (shaded areas in Figures 12-1 and 12-2). Control channels cannot carry voice information; they are used exclusively to carry service information. There are 21 control channels in the A system and 21 control channels in the B system. The remaining 790 channels make up the second group, termed *voice* or *user channels*. User channels are used for propagating actual voice conversations or subscriber data.

Control channels are used in cellular telephone systems to enable mobile units to communicate with the cellular network through base stations without interfering with normal voice traffic occurring on the normal voice or user channels. Control channels are used for call origination, for call termination, and to obtain system information. With the AMPS system, voice channels are analog FM, while control channels are digital and employ FSK. Therefore, voice channels cannot carry control signals, and control channels cannot carry voice information. Control channels are used exclusively to carry service information. With AMPS, base stations broadcast on the *forward control channel* (FCC) and listen on the *reverse control channel* (RCC). The control channels are sometimes called *setup* or *paging channels.* All AMPS base stations continuously transmit FSK data on the FCC so that idle cellular telephones can maintain a lock on the strongest FCC regardless of their location. A subscriber's unit must be *locked* (sometimes called *camped*) on an FCC before it can originate or receive calls.

Each base station uses a control channel to simultaneously page mobile units to alert them of the presence of incoming calls and to move established calls to a vacant voice channel. The forward control channel transmits a 10-kbps data signal using FSK. Forward

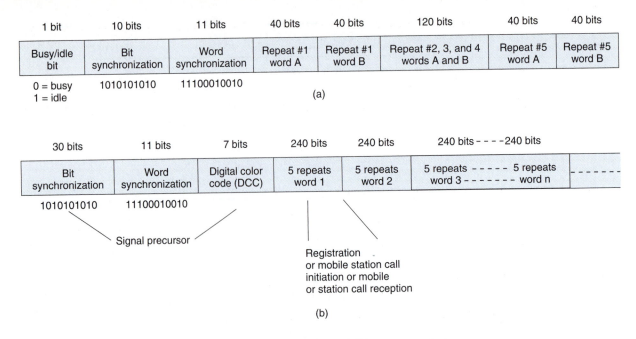

1 bit	10 bits	11 bits	40 bits	40 bits	120 bits	40 bits	40 bits
Busy/idle bit	Bit synchronization	Word synchronization	Repeat #1 word A	Repeat #1 word B	Repeat #2, 3, and 4 words A and B	Repeat #5 word A	Repeat #5 word B

0 = busy
1 = idle
1010101010
11100010010

(a)

30 bits	11 bits	7 bits	240 bits	240 bits	240 bits - - - -240 bits	
Bit synchronization	Word synchronization	Digital color code (DCC)	5 repeats word 1	5 repeats word 2	5 repeats - - - - - 5 repeats word 3 - - - - - - - word n	

1010101010
11100010010

Signal precursor

Registration or mobile station call initiation or mobile or station call reception

(b)

FIGURE 12-3 Control channel format: (a) forward control channel; (b) reverse control channel

control channels from base stations may contain overhead data, mobile station control information, or control file information.

Figure 12-3a shows the format for an AMPS forward control channel. As the figure shows, the control channel message is preceded by a 10-bit *dotting scheme,* which is a sequence of alternating 1s and 0s. The dotting scheme is followed by an 11-bit *synchronization word* with a unique sequence of 1s and 0s that enables a receiver to instantly acquire synchronization. The sync word is immediately followed by the message repeated five times. The redundancy helps compensate for the ill effects of fading. If three of the five words are identical, the receiver assumes that as the message.

Forward control channel data formats consist of three discrete information streams: stream A, stream B, and the busy-idle stream. The three data streams are multiplexed together. Messages to the mobile unit with the least-significant bit of their 32-bit *mobile identification number* (MIN) equal to 0 are transmitted on stream A, and MINs with the least-significant bit equal to 1 are transmitted on stream B. The busy-idle data stream contains *busy-idle bits,* which are used to indicate the current status of the reverse control channel (0 = busy and 1 = idle). There is a busy-idle bit at the beginning of each dotting sequence, at the beginning of each synchronization word, at the beginning of the first repeat of word A, and after every 10 message bits thereafter. Each message word contains 40 bits, and forward control channels can contain one or more words.

The types of messages transmitted over the FCC are the *mobile station control message* and the *overhead message train.* Mobile station control messages control or command mobile units to do a particular task when the mobile unit has not been assigned a voice channel. Overhead message trains contain *system parameter overhead messages, global action overhead messages,* and *control filler messages.* Typical mobile-unit control messages are *initial voice channel designation messages, directed retry messages, alert messages,* and *change power messages.*

Figure 12-3b shows the format for the reverse control channel that is transmitted from the mobile unit to the base station. The control data are transmitted at a 10-kbps rate and include *page responses, access requests,* and *registration requests.* All RCC messages begin with the RCC seizure precursor, which consists of a 30-bit *dotting sequence,* an 11-bit *synchronization word,* and the coded *digital color code* (DCC), which is added so that the control channel is not confused with a control channel from a nonadjacent cell that is reusing the same frequency. The mobile telephone reads the base station's DCC and then returns a coded version of it, verifying that the unit is locked onto the correct signal. When the call is finished, a 1.8-second *signaling time-out signal* is transmitted. Each message word contains 40 bits and is repeated five times for a total of 200 bits.

12-2-5 Voice-Channel Signaling

Analog cellular channels carry both voice using FM and digital signaling information using binary FSK. When transmitting digital signaling information, voice transmissions are inhibited. This is called *blank and burst:* the voice is blanked, and the data are transmitted in a short burst. The bit rate of the digital information is 10 kbps. Figure 12-4a shows the voice-channel signaling format for a forward voice channel, and Figure 12-4b shows the format for the reverse channel. The digital signaling sequence begins with a 101-bit dotting sequence that readies the receiver to receive digital information. After the dotting sequence, a synchronization word is sent to indicate the start of the message. On the forward voice channel, digital signaling messages are repeated 11 times to ensure the integrity of the message, and on the receive channel they are repeated five times. The forward channel uses 40-bit words, and the reverse channel uses 48-bit words.

12-3 PERSONAL COMMUNICATIONS SYSTEM

The Personal Communications System (PCS) is a relatively new class of cellular telephony based on the same basic philosophies as standard cellular telephone systems (CTSs), such as AMPS. However, PCS systems are a combination of cellular telephone networks and the *Intelligent Network,* which is the entity of the SS7 interoffice protocol that distinguishes the physical components of the switching network, such as the signal service point (SSP), signal control point (SCP), and signal transfer point (STP), from the services provided by the SS7 network. The services provided are distinctly different from the switching systems and protocols that promote and support them. PCS was initially considered a new service, although different companies have different visions of exactly what PCS is and what services it should provide. The FCC defines PCS mobile telephone as "a family of mobile or portable radio communications services, which provides services to individuals and business and is integrated with a variety of competing networks." In essence, PCS is the North American implementation of the European GSM standard.

Existing cellular telephone companies want PCS to provide broad coverage areas and fill in service gaps between their current service areas. In other words, they want PCS to be an extension of the current first- and second-generation cellular system to the 1850-MHz to 2200-MHz band using identical standards for both frequency bands. Other companies would like PCS to compete with standard cellular telephone systems but offer enhanced services and better quality using extensions of existing standards or entirely new standards. Therefore, some cellular system engineers describe PCS as a third-generation cellular telephone system, although the U.S. implementation of PCS uses modifications of existing cellular protocols, such as IS-54 and IS-95. Most cellular telephone companies reserve the designation third-generation PCS to those systems designed for transporting data as well as voice.

Although PCS systems share many similarities with first-generation cellular telephone systems, PCS has several significant differences that, most agree, warrant the use of a different name. Many of the differences are transparent (or at least not obvious) to the users of the networks. Probably the primary reason for establishing a new PCS cellular telephone system was because first-generation cellular systems were already overcrowded, and it was obvious that they would not be able to handle the projected demand

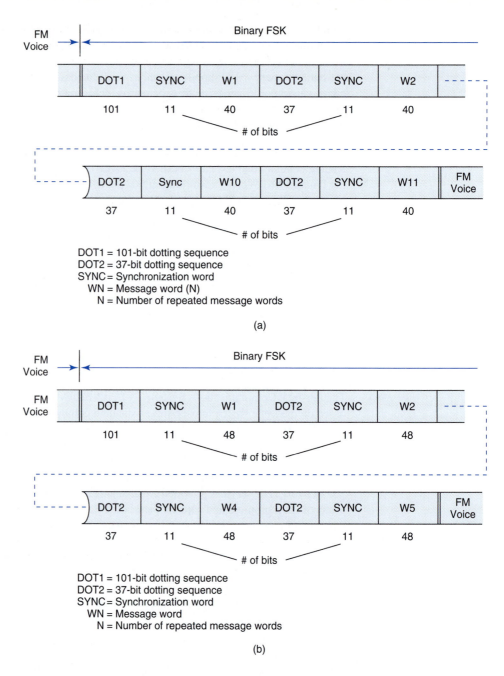

FIGURE 12-4 Voice channel format: (a) forward channel; (b) reverse channel

for future cellular telephone services. In essence, PCS services were conceived to provide subscribers with a low-cost, feature-rich wireless telephone service.

Differences between PCS systems and standard cellular telephone systems generally include but are certainly not limited to the following: (1) smaller cell size, (2) all digital, and (3) additional features. Cellular systems generally classified as PCS include IS-136 TDMA, GSM, and IS-95 CDMA.

The concept of *personal communications services* (also PCS) originated in the United Kingdom when three companies were allocated a band of frequencies in the 1.8-GHz band to develop a *personal communications network* (PCN) throughout Great Britain. The terms *PCS* and *PCN* are often used interchangeably. However, PCN refers to a wireless networking concept where any user can initiate or receive calls regardless of where they are using a portable, personalized transceiver. PCS refers to a new wireless system that incorporates enhanced network features and is more personalized than existing standard cellular telephone systems but does not offer all the features of an ideal PCN.

In 1990, the FCC adopted the term PCS to mean *personal communications services,* which is the North American implementation of the *global system for mobile communications.* However, to some people, PCS means *personal communications system,* which specifies a category or type of cellular telephone system. The exact nature of the services provided by PCS is not completely defined by the cellular telephone industry. However, the intention of PCS systems is to provide enhanced features to first- and second-generation cellular telephone systems, such as messaging, paging, and data services.

PCS is more of a concept than a technology, the concept being to assign everyone a *personal telephone number* (PTN) that is stored in a database on the SS7 network. This database keeps track of where each mobile unit can be reached. When a call is placed from a mobile unit, an *artificial intelligence network* (AIN) in SS7 determines where and how the call should be directed. The PCS network is similar to the D-AMPS system in that the MTSO stores three essential databases: *home location register, visitor location register,* and *equipment identification registry.*

> *Home location register* (HLR). The HLR is a database that stores information about the user, including home subscription information and what supplementary services the user is subscribed to, such as call waiting, call hold, call forwarding, and call conferencing (three-way calling). There is generally only one HLR per mobile network. Data stored on the HLR are semipermanent, as they do not usually change from call to call.
>
> *Visitor location register* (VLR). The VLR is a database that stores information about subscribers in a particular MTSO serving area, such as whether the unit is on or off and whether any of the supplementary services are activated or deactivated. There is generally only one VLR per mobile switch. The VLR stores permanent data, such as that found in the HLR, plus temporary data, such as the subscriber's current location.
>
> *Equipment identification registry* (EIR). The EIR is a database that stores information pertaining to the identification and type of equipment that exists in the mobile unit. The EIR also helps the network identify stolen or fraudulent mobile units.

Many of the services offered by PCS systems are not currently available with standard cellular telephone systems, such as available mode, screen, private, and unavailable.

> *Available mode.* The available mode allows all calls to pass through the network to the subscriber except for a minimal number of telephone numbers that can be blocked. The available mode relies on the delivery of the calling party number, which is checked against a database to ensure that it is not a blocked number. Subscribers can update or make changes in the database through the dial pad on their PCS handset.
>
> *Screen mode.* The screen mode is the PCS equivalent to caller ID. With the screen mode, the name of the calling party appears on the mobile unit's display, which allows PCS users to screen calls. Unanswered calls are automatically forwarded to a *forwarding destination* specified by the subscriber, such as voice mail or another telephone number.
>
> *Private mode.* With the private mode, all calls except those specified by the subscriber are automatically forwarded to a forwarding destination without ringing the subscriber's handset. Subscribers can make changes in the list of allowed calling numbers through the dial pad on their handset.

Unavailable mode. With the unavailable mode, no calls are allowed to pass through to the subscriber. Hence, all incoming calls are automatically forwarded to a forwarding destination.

PCS telephones are intended to be small enough to fit into a shirt pocket and use digital technology, which is quieter than analog. Their transmit power is relatively low; therefore, PCS systems utilize smaller cells and require more base stations than standard cellular systems for a given service area. PCS systems are sometimes called *microcellular* systems. The fundamental concept of PCS is to assign each mobile unit a PTN that is stored in a database on the SS7 common signaling network. The database keeps track of where mobile units are. When a call is placed for a mobile unit, the SS7 artificial intelligence network determines where the call should be directed.

The primary disadvantage of PCS is network cost. Employing small cells requires using more base stations, which equates to more transceivers, antennas, and trunk circuits. Antenna placement is critical with PCS. Large towers typically used with standard cellular systems are unacceptable in neighborhoods, which is where a large majority of PCS antennas must be placed.

PCS base stations communicate with other networks (cellular, PCS, and wireline) through a PCS switching center (PSC). The PSC is connected directly to the SS7 signaling network with a link to a signaling transfer point. PCS networks rely extensively on the SS7 signaling network for interconnecting to other telephone networks and databases.

PCS systems generally operate in a higher frequency band than standard cellular telephone systems. The FCC recently allocated an additional 160-MHz band in the 1850-MHz to 2200-MHz range. PCS systems operating in the 1900-MHz range are often referred to as *personal communications system 1900* (PCS 1900).

12-4 SECOND-GENERATION CELLULAR TELEPHONE SYSTEMS

First-generation cellular telephone systems were designed primarily for a limited customer base, such as business customers and a limited number of affluent residential customers. When the demand for cellular service increased, manufacturers searched for new technologies to improve the inherent problems with the existing cellular telephones, such as poor battery performance and channel unavailability. Improved batteries were also needed to reduce the size and cost of mobile units, especially those that were designed to be handheld. Weak signal strengths resulted in poor performance and a high rate of falsely initiated handoffs (*false handoffs*).

It was determined that improved battery performance and higher signal quality were possible only by employing digital technologies. In the United States, the shortcomings of the first-generation cellular systems led to the development of several second-generation cellular telephone systems, such as narrowband AMPS (N-AMPS) and systems employing the IS-54, IS-136, and IS-95 standards. A second-generation standard, known as *Global System for Mobile Communications* (GSM), emerged in Europe.

12-5 N-AMPS

Because of uncertainties about the practicality and cost effectiveness of implementing digital cellular telephone systems, Motorola developed a narrowband AMPS system called N-AMPS to increase the capacity of the AMPS system in large cellular markets. N-AMPS was originally intended to provide a short-term solution to the traffic congestion problem in the AMPS system. N-AMPS allows as many as three mobile units to use a single 30-kHz cellular channel at the same time. With N-AMPS, the maximum frequency deviation is reduced, reducing the required bandwidth to 10 kHz and thus providing a threefold increase in user capacity. One N-AMPS channel uses the carrier frequency for the existing AMPS channel, and, with the other two channels, the carrier frequencies are offset by ±10 kHz. Each 10-kHz subchannel is capable of handling its own calls. Reducing the bandwidth degrades speech quality by lowering the signal-to-interference ratio. With narrower bandwidths, voice channels are more vulnerable to interference than standard AMPS channels and would generally require a higher frequency reuse factor. This is compensated for with the addition of an *interference avoidance scheme* called *Mobile Reported Interference* (MRI), which uses voice companding to provide synthetic voice channel quieting.

N-AMPS systems are *dual mode* in that mobile units are capable of operating with 30-kHz channels or with 10-kHz channels. N-AMPS systems use standard AMPS control channels for call setup and termination. N-AMPS mobile units are capable of utilizing four types of handoffs: *wide channel to wide channel* (30 kHz to 30 kHz), *wide channel to narrow channel* (30 kHz to 10 kHz), *narrow channel to narrow channel* (10 kHz to 10 kHz), and *narrow channel to wide channel* (10 kHz to 30 kHz).

12-6 DIGITAL CELLULAR TELEPHONE

Cellular telephone companies were faced with the problem of a rapidly expanding customer base while at the same time the allocated frequency spectrum remained unchanged. As is evident with N-AMPS, user capacity can be expanded by subdividing existing channels (band splitting), partitioning cells into smaller subcells (cell splitting), and modifying antenna radiation patterns (sectoring). However, the degree of subdivision and redirection is limited by the complexity and amount of overhead required to process handoffs between cells. Another serious restriction is the availability and cost of purchasing or leasing property for cell sites in the higher-density traffic areas.

Digital cellular telephone systems have several inherent advantages over analog cellular telephone systems, including better utilization of bandwidth, more privacy, and incorporation of error detection and correction.

AMPS is a first-generation analog cellular telephone system that was not designed to support the high-capacity demands of the modern world, especially in high-density metropolitan areas. In the late 1980s, several major manufacturers of cellular equipment determined that digital cellular telephone systems could provide substantial improvements in both capacity and performance. Consequently, the *United States Digital Cellular* (USDC) system was designed and developed with the intent of supporting a higher user density within a fixed-bandwidth frequency spectrum. Cellular telephone systems that use digital modulation, such as USDC, are called *digital cellular.*

The USDC cellular telephone system was originally designed to utilize the AMPS frequency allocation scheme. USDC systems comply with IS-54, which specifies dual-mode operation and backward compatibility with standard AMPS. USDC was originally designed to use the same carrier frequencies, frequency reuse plan, and base stations. Therefore, base stations and mobile units can be equipped with both AMPS and USDC channels within the same telephone equipment. In supporting both systems, cellular carriers are able to provide new customers with digital USDC telephones while still providing service to existing customers with analog AMPS telephones. Because the USDC system maintains compatibility with AMPS systems in several ways, it is also known as *Digital AMPS* (D-AMPS or sometimes DAMPS).

The USDC cellular telephone system has an additional frequency band in the 1.9-GHz range that is not compatible with the AMPS frequency allocation. Figure 12-5 shows the frequency spectrum and channel assignments for the 1.9-GHz band (sometimes called the PCS band). The total usable spectrum is subdivided into subbands (A through F); however, the individual channel bandwidth is limited to 30 kHz (the same as AMPS).

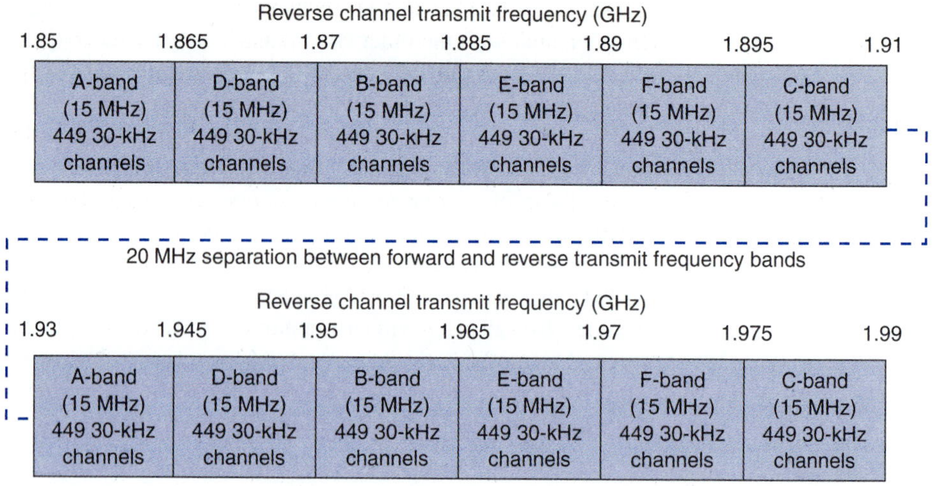

Reverse channel transmit frequency (GHz)

| 1.85 | 1.865 | 1.87 | 1.885 | 1.89 | 1.895 | 1.91 |

A-band (15 MHz) 449 30-kHz channels	D-band (15 MHz) 449 30-kHz channels	B-band (15 MHz) 449 30-kHz channels	E-band (15 MHz) 449 30-kHz channels	F-band (15 MHz) 449 30-kHz channels	C-band (15 MHz) 449 30-kHz channels

20 MHz separation between forward and reverse transmit frequency bands

Reverse channel transmit frequency (GHz)

| 1.93 | 1.945 | 1.95 | 1.965 | 1.97 | 1.975 | 1.99 |

A-band (15 MHz) 449 30-kHz channels	D-band (15 MHz) 449 30-kHz channels	B-band (15 MHz) 449 30-kHz channels	E-band (15 MHz) 449 30-kHz channels	F-band (15 MHz) 449 30-kHz channels	C-band (15 MHz) 449 30-kHz channels

FIGURE 12-5   1.9-GHz cellular frequency band

12-6-1   Time-Division Multiple Accessing

USDC uses *time-division multiple accessing* (TDMA) as well as frequency-division multiple accessing (FDMA). USDC, like AMPS, divides the total available radio-frequency spectrum into individual 30-kHz cellular channels (i.e., FDMA). However, TDMA allows more than one mobile unit to use a channel at the same time by further dividing transmissions within each cellular channel into time slots, one for each mobile unit using that channel. In addition, with AMPS FDMA systems, subscribers are assigned a channel for the duration of their call. However, with USDC TDMA systems, mobile-unit subscribers can only *hold* a channel while they are actually talking on it. During pauses or other normal breaks in a conversation, users must relinquish their channel so that other mobile units can use it. This technique of *time-sharing* channels significantly increases the capacity of a system, allowing more mobile-unit subscribers to use a system at virtually the same time within a given geographical area.

A USDC TDMA transmission frame consists of six equal-duration time slots enabling each 30-kHz AMPS channel to support three full-rate or six half-rate users. Hence, USDC offers as much as six times the channel capacity as AMPS. The original USDC standard also utilizes the same 50-MHz frequency spectrum and frequency-division duplexing scheme as AMPS.

The advantages of digital TDMA multiple-accessing systems over analog AMPS FDMA multiple-accessing systems are as follows:

1. Interleaving transmissions in the time domain allows for a threefold to sixfold increase in the number of mobile subscribers using a single cellular channel. Time-sharing is realized because of digital compression techniques that produce bit rates approximately one-tenth that of the initial digital sample rate and about one-fifth the initial rate when error detection/correction (EDC) bits are included.
2. Digital signals are much easier to process than analog signals. Many of the more advanced modulation schemes and information processing techniques were developed for use in a digital environment.
3. Digital signals (bits) can be easily encrypted and decrypted, safeguarding against eavesdropping.
4. The entire telephone system is compatible with other digital formats, such as those used in computers and computer networks.
5. Digital systems inherently provide a quieter (less noisy) environment than their analog counterparts.

12-6-2 EIA/TIA Interim Standard 54

In 1990, the Electronics Industries Association and Telecommunications Industry Association (EIA/TIA) standardized the dual-mode USDC/AMPS system as Interim Standard 54 (IS-54), *Cellular Dual Mode Subscriber Equipment. Dual mode* specifies that a mobile station complying with the IS-54 standard must be capable of operating in either the analog AMPS or the digital (USDC) mode for voice transmissions. Using IS-54, a cellular telephone carrier could convert any or all of its existing analog channels to digital. The key criterion for achieving dual-mode operation is that IS-54 digital channels cannot interfere with transmissions from existing analog AMPS base and mobile stations. This goal is achieved with IS-54 by providing digital control channels and both analog and digital voice channels. Dual-mode mobile units can operate in either the digital or the analog mode for voice and access the system with the standard AMPS digital control channel. Before a voice channel is assigned, IS-54 mobile units use AMPS forward and reverse control channels to carry out user authentications and call management operations. When a dual-mode mobile unit transmits an access request, it indicates that it is capable of operating in the digital mode; then the base station will allocate a digital voice channel, provided one is available. The allocation procedure indicates the channel number (frequency) and the specific time slot (or slots) within that particular channel's TDMA frame. IS-54 specifies a 48.6-kbps rate per 30-kHz voice channel divided among three simultaneous users. Each user is allocated 13 kbps, and the remaining 9.6 kbps is used for timing and control overhead.

In many rural areas of the United States, analog cellular telephone systems use only the original 666 AMPS channels (1 through 666). In these areas, USDC channels can be added in the extended frequency spectrum (channels 667 through 799 and 991 through 1023) to support USDC telephones that roam into the system from other areas. In high-density urban areas, selected frequency bands are gradually being converted one at a time to the USDC digital standard to help alleviate traffic congestion. Unfortunately, this gradual changeover from AMPS to USDC often results in an increase in the interference and number of dropped calls experienced by subscribers of the AMPS system.

The successful and graceful transition from analog cellular systems to digital cellular systems using the same frequency band was a primary consideration in the development of the USDC standard. The introduction of N-AMPS and a new digital spread-spectrum standard has delayed the widespread deployment of the USDC standard throughout the United States.

12-6-3 USDC Control Channels and IS-136.2

The IS-54 USDC standard specifies the same 42 *primary control channels* as AMPS and 42 additional control channels called *secondary control channels*. Thus, USDC offers twice as many control channels as AMPS and is, therefore, capable of providing twice the capacity of control traffic within a given market area. Carriers are allowed to dedicate the secondary control channels for USDC-only use since AMPS mobile users do not monitor and cannot decode the new secondary control channels. In addition, to maintain compatibility with existing AMPS cellular telephone systems, the primary forward and reverse control channels in USDC cellular systems use the same signaling techniques and modulation scheme (FSK) as AMPS. However, a new standard, IS-136.2 (formerly IS-54, Rev.C), replaces FSK with $\pi/4$ DQPSK modulation for the 42 dedicated USDC secondary control channels, allowing digital mobile units to operate entirely in the digital domain. The IS-136.2 standard is often called North American–Time Division Multiple Accessing (NA-TDMA). IS-54 Rev.C was introduced to provide PSK (phase-shift keying) rather than FSK on dedicated USDC control channels to increase the control data rates and provide additional specialized services, such as paging and short messaging between private mobile user groups. *Short message service* allows for brief paging-type messages and short e-mail messages (up to 239 characters) that can be read on the mobile phone's display and entered using the keypad.

IS-136 was developed to provide a host of new features and services, positioning itself in a competitive market with the newer PCS systems. Because IS-136 specifies short messaging capabilities and private user-group features, it is well suited as a wireless paging system. IS-136 also provides an additional *sleep mode,* which conserves power in the mobile units. IS-136 mobile units are not compatible with IS-54 units, as FSK control channels are not supported.

The digital control channel is necessarily complex, and a complete description is beyond the scope of this book. Therefore, the following discussion is meant to present a general overview of the operation of a USDC digital control channel.

The IS-54 standard specifies three types of channels: analog control channels, analog voice channels, and a 10-kbps binary FSK digital control channel (DCCH). The IS-54 Rev.C standard (IS-136) provides for the same three types of channels plus a fourth—a digital control channel with a signaling rate of 48.6 kbps on USDC-only control channels. The new digital control channel is meant to eventually replace the analog control channel. With the addition of a digital control channel, a mobile unit is able to operate entirely in the digital domain, using the digital control channel for system and cell selection and channel accessing and the digital voice channel for digitized voice transmissions.

IS-136 details the exact functionality of the USDC digital control channel. The initial version of IS-136 was version 0, which has since been updated by revision A. Version 0 added numerous new services and features to the USDC digital cellular telephone system, including enhanced user services, such as short messaging and displaying the telephone number of the incoming call; sleep mode, which gives the telephone set a longer battery life when in the standby mode; private or residential system service; and enhanced security and validation against fraud. The newest version of IS-136, revision A, was developed to provide numerous new features and services by introducing an enhanced vocoder, over-the-air activation where the network operators are allowed to program information into telephones directly over the air, calling name and number ID, and enhanced hands-off and priority access to control channels.

IS-136 specifies several private user-group features, making it well adapted for wireless PBX and paging applications. However, IS-136 user terminals operate at 48.6 kbps and are, therefore, not compatible with IS-54 FSK terminals. Thus, IS-136 modems are more cost effective, as it is necessary to include only the 48.6-kbps modem in the terminal equipment.

12-6-3-1 Logical channels. The new digital control channel includes several *logical channels* with different functions, including the *random access channel* (RACH); the *SMS point-to-point, paging, and access response channel* (SPACH); the *broadcast control channel* (BCCH); and the *shared channel feedback* (SCF) channel. Figure 12-6 shows the logical control channels for the IS-136 standard.

12-6-3-2 Random access channel (RACH). RACH is used by mobile units to request access to the cellular telephone system. RACH is a unidirectional channel specified for transmissions from mobile-to-base units only. Access messages, such as origination, registration, page responses, audit confirmation, serial number, and message confirmation, are transmitted on the RACH. It also transmits messages that provide information on authentication, security parameter updates, and *short message service* (SMS) point-to-point messages. RACH is capable of operating in two modes using contention resolution similar to voice channels. RACH can also operate in a *reservation mode* for replying to a base-station command.

12-6-3-3 SMS point-to-point, paging, and access response channel (SPACH). SPACH is used to transmit information from base stations to specific mobile stations. RACH is a unidirectional channel specified for transmission from base stations to mobile units only and is shared by all mobile units. Information transmitted on the SPACH channel includes three separate logical subchannels: *SMS point-to-point messages, paging messages,* and *access response messages.* SPACH can carry messages related to a single mo-

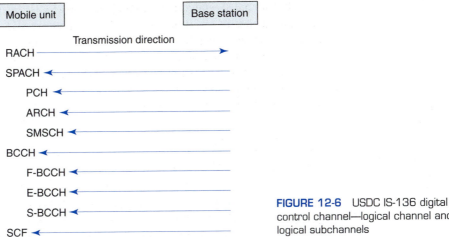

FIGURE 12-6 USDC IS-136 digital control channel—logical channel and logical subchannels

bile unit or to a small group of mobile units and allows larger messages to be broken down into smaller blocks for transmission.

The paging channel (PCH) is a subchannel of the logical channel of SPACH. PCH is dedicated to delivering pages and orders. The PCH transmits *paging messages, message-waiting messages,* and *user-alerting messages.* Each PCH message can carry up to five mobile identifiers. Page messages are always transmitted and then repeated a second time. Messages such as *call history count updates* and *shared secret data updates* used for the authentication and encryption process also are sent on the PCH.

The access response channel (ARCH) is also a logical subchannel of SPACH. A mobile unit automatically moves to an ARCH immediately after successful completion of contention- or reservation-based access on a RACH. ARCH can be used to carry assignments to another resource or other responses to the mobile station's access attempt. Messages assigning a mobile unit to an analog voice channel or a digital voice channel or redirecting the mobile to a different cell are also sent on the ARCH along with registration access (accept, reject, or release) messages.

The SMS channel (SMSCH) is used to deliver short point-to-point messages to a specific mobile station. Each message is limited to a maximum of 200 characters of text. Mobile-originated SMS is also supported; however, SMS, where a base station can broadcast a short message designated for several mobile units, is not supported in IS-136.

12-6-3-4 Broadcast control channel (BCCH).
BCCH is an acronym referring to the F-BCCH, E-BCCH, and S-BCCH logical subchannels. These channels are used to carry generic, system-related information. BCCH is a unidirectional base station–to–mobile unit transmission shared by all mobile units.

The *fast broadcast control channel* (F-BCCH) broadcasts digital control channel (DCCH) structure parameters, including information about the number of F-BCCH, E-BCCH, and S-BCCH time slots in the DCCH frame. Mobile units use F-BCCH information when initially accessing the system to determine the beginning and ending of each logical channel in the DCCH frame. F-BCCH also includes information pertaining to access parameters, including information necessary for authentication and encryptions and information for mobile access attempts, such as the number of access retries, access burst size, initial access power level, and indication of whether the cell is barred. Information addressing the different types of registration, registration periods, and system identification information, including network type, mobile country code, and protocol revision, is also provided by the F-BCCH channel.

The *extended broadcast control channel* (E-BCCH) carries less critical broadcast information than F-BCCH intended for the mobile units. E-BCCH carries information about neighboring analog and TDMA cells and optional messages, such as emergency information, time and date messaging, and the types of services supported by neighboring cells.

The *SMS broadcast control channel* (S-BCCH) is a logical channel used for sending short messages to individual mobile units.

12-6-3-5 Shared channel feedback (SCF) channel. SCF is used to support random access channel operation by providing information about which time slots the mobile unit can use for access attempts and also if a mobile unit's previous RACH transmission was successfully received.

12-6-4 USDC Digital Voice Channel

Like AMPS, each USDC voice channel is assigned a 30-kHz bandwidth on both the forward and the reverse link. With USDC, however, each voice channel can support as many as three full-rate mobile users simultaneously by using digital modulation and a TDMA format called *North American Digital Cellular* (NADC). Each radio-frequency voice channel in the total AMPS FDMA frequency band consists of one 40-ms TDMA frame comprised of six time slots containing 324 bits each, as shown in Figure 12-7. For full-speech rate, three users

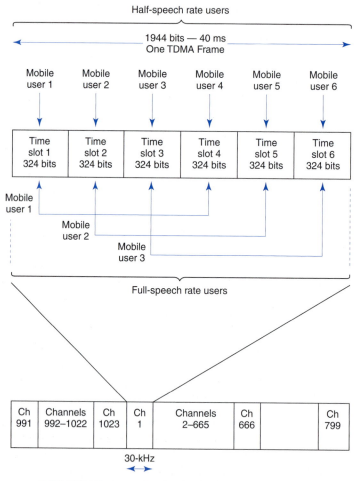

FIGURE 12-7 North American Digital Cellular TDMA frame format

share the six time slots in an equally spaced manner. For example, mobile user 1 occupies time slots 1 and 4, mobile user 2 occupies time slots 2 and 5, and mobile user 3 occupies time slots 3 and 6. For half-rate speech, each user occupies one time slot per frame. During their respective time slots, mobile units transmit short bursts (6.67 ms) of a digital-modulated carrier to the base station (i.e., *uplink transmissions*). Hence, full-rate users transmit two bursts during each TDMA frame. In the *downlink* path (i.e., from base stations to mobile units), base stations generally transmit continuously. However, mobile units listen only during their assigned time slot. The average cost per subscriber per base station equipment is lower with TDMA since each base station transceiver can be shared by up to six users at a time.

General Motors Corporation implemented a TDMA scheme called E-TDMA, which incorporates six half-rate users transmitting at half the bit rate of standard USDC TDMA systems. E-TDMA systems also incorporate *digital speech interpolation* (DSI) to dynamically assign more than one user to a time slot, deleting silence on calls. Consequently, E-TDMA can handle approximately 12 times the user traffic as standard AMPS systems and four times that of systems complying with IS-54.

Each time slot in every USDC voice-channel frame contains four data channels—three for control and one for digitized voice and user data. The full-duplex *digital traffic channel* (DTC) carries digitized voice information and consists of a *reverse digital traffic channel* (RDTC) and a *forward digital traffic channel* (FDTC) that carry digitized speech information or user data. The RDTC carries speech data from the mobile unit to the base station, and the FDTC carries user speech data from the base station to the mobile unit. The three supervisory channels are the *coded digital verification color code* (CDVCC), the *slow associated control channel* (SACCH), and the *fast associated control channel* (FACCH).

12-6-4-1 Coded digital verification color code. The purpose of the CDVCC color code is to provide cochannel identification similar to the SAT signal transmitted in the AMPS system. The CDVCC is a 12-bit message transmitted in every time slot. The CDVCC consists of an eight-bit digital voice color code number between 1 and 255 appended with four additional coding bits derived from a shortened Hamming code. The base station transmits a CDVCC number on the forward voice channel, and each mobile unit using the TDMA channel must receive, decode, and retransmit the same CDVCC code (handshake) back to the base station on the reverse voice channel. If the two CDVCC values are not the same, the time slot is relinquished for other users, and the mobile unit's transmitter will be automatically turned off.

12-6-4-2 Slow associated control channel. The SACCH is a signaling channel for transmission of control and supervision messages between the digital mobile unit and the base station while the mobile unit is involved with a call. The SACCH uses 12 coded bits per TDMA burst and is transmitted in every time slot, thus providing a signaling channel in parallel with the digitized speech information. Therefore, SACCH messages can be transmitted without interfering with the processing of digitized speech signals. Because the SACCH consists of only 12 bits per frame, it can take up to 22 frames for a single SACCH message to be transmitted. The SACCH carries various control and supervisory information between the mobile unit and the base station, such as communicating power-level changes and handoff requests. The SACCH is also used by the mobile unit to report signal-strength measurements of neighboring base stations so, when necessary, the base station can initiate a *mobile-assisted handoff* (MAHO).

12-6-4-3 Fast associated control channel. The FACCH is a second signaling channel for transmission of control and specialized supervision and traffic messages between the base station and the mobile units. Unlike the CDVCC and SACCH, the FACCH does not have a dedicated time slot. The FACCH is a *blank-and-burst* type of transmission that, when transmitted, replaces digitized speech information with control and supervision messages within a subscriber's time slot. There is no limit on the number of speech frames that can be replaced with FACCH data. However, the digitized voice information is somewhat

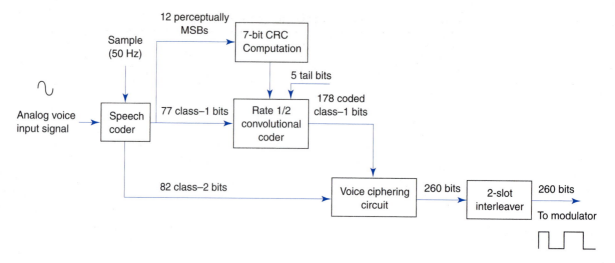

FIGURE 12-8 USDC digital voice-channel speech coder

protected by preventing an entire digitized voice transmission from being replaced by FACCH data. The 13-kbps net digitized voice transmission rate cannot be reduced below 3250 bps in any given time slot. There are no fields within a standard time slot to identify it as digitized speech or an FACCH message. To determine if an FACCH message is being received, the mobile unit must attempt to decode the data as speech. If it decodes in error, it then decodes the data as an FACCH message. If the cyclic redundancy character (CRC) calculates correctly, the message is assumed to be an FACCH message. The FACCH supports transmission of dual-tone multiple-frequency (DTMF) Touch-Tones, call release instruction, flash hook instructions, and mobile-assisted handoff or mobile-unit status requests. The FACCH data are packaged and interleaved to fit in a time slot similar to the way digitized speech is handled.

12-6-5 Speech Coding

Figure 12-8 shows the block diagram for a USDC digital voice-channel speech encoder. Channel error control for the digitized speech data uses three mechanisms for minimizing channel errors: (1) a rate one-half convolutional code is used to protect the more vulnerable bits of the speech coder data stream, (2) transmitted data are interleaved for each speech coder frame over two time slots to reduce the effects of Rayleigh fading, and (3) a cyclic redundancy check is performed on the most perceptually significant bits of the digitized speech data.

With USDC, incoming analog voice signals are sampled first and then converted to a binary PCM in a special *speech coder* (*vocoder*) called a *vector sum exciter linear predictive* (VSELP) *coder* or a *stochastically excited linear predictive* (SELP) *coder.* Linear predictive coders are time-domain types of vocoders that attempt to extract the most significant characteristics from the time-varying speech waveform. With linear predictive coders, it is possible to transmit good-quality voice at 4.8 kbps and acceptable, although poorer-quality, voice at lower bit rates.

Because there are many predictable orders in spoken word patterns, it is possible, using advanced algorithms, to compress the binary samples and transmit the resulting bit stream at a 13-kbps rate. A consortium of companies, including Motorola, developed the VSELP algorithm, which was subsequently adopted for the IS-54 standard. Error-detection and -correction (EDC) bits are added to the digitally compressed voice signals to reduce the effects of interference, bringing the final voice data rate to 48.6 kbps. Compression/expansion and error-detection/correction functions are implemented in the telephone handset by a special microprocessor called a digital signal processor (DSP).

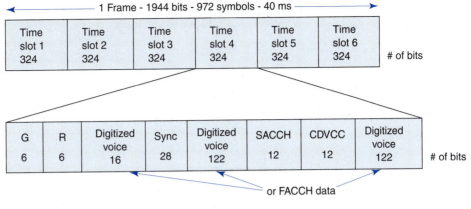

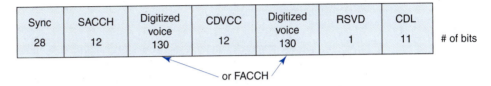

Mobile station-to-base (reverse) channel

Sync	SACCH	Digitized voice	CDVCC	Digitized voice	RSVD	CDL	
28	12	130	12	130	1	11	# of bits

or FACCH

Base-to-mobile station (forward) channel

FIGURE 12-9 USDC digital voice-channel slot and frame format

The VSELP coders output 7950 bps and produce a speech frame every 20 ms, or

$$\frac{7950 \text{ bits}}{\text{second}} \times \frac{20 \text{ ms}}{\text{frame}} = 159 \text{ bits per frame}$$

Fifty speech frames are outputted each second containing 159 bits each, or

$$\frac{50 \text{ frames}}{\text{second}} \times \frac{159 \text{ bits}}{\text{frame}} = 7950 \text{ bps}$$

The 159 bits included in each speech coder frame are divided into two classes according to the significance in which they are perceived. There are 77 class 1 bits and 82 class 2 bits. The class 1 bits are the most significant and are, therefore, error protected. The 12 most-significant class 1 bits are block coded using a seven-bit CRC error-detection code to ensure that the most-significant speech coder bits are decoded with a low probability of error. The less significant class 2 bits have no means of error protection.

After coding the 159 bits, each speech code frame is converted in a 1/2 convolution coder to 260 channel-coded bits per frame, and 50 frames are transmitted each second. Hence, the transmission bit rate is increased from 7950 bps for each digital voice channel to 13 kbps:

$$\frac{260 \text{ bits}}{\text{frame}} \times \frac{50 \text{ frames}}{\text{second}} = 13 \text{ kbps}$$

Figure 12-9 shows the time slot and frame format for the forward (base station to mobile unit) and reverse (mobile unit to base station) links of a USDC digital voice channel. USDC voice channels use frequency-division duplexing; thus, forward and reverse channel time slots operate on different frequencies at the same time. Each time slot carries interleaved digital voice data from the two adjacent frames outputted from the speech coder.

G1	RSDSDVSDWSDXSDYS	G2

Where G1 = 6-bit guard time

     R = 6-bit length ramp time

     S = 28-bit synchronization word

     D = 12-bit CDVCC code

     G2 = 44-bit guard time

     V = 0000

     W = 00000000

     X = 000000000000

     Y = 0000000000000000

FIGURE 12-10 USDC shortened burst digital voice channel format

In the reverse channel, each time slot contains two bursts of 122 digitized voice bits and one burst of 16 bits for a total of 260 digitized voice bits per frame. In addition, each time slot contains 28 synchronization bits, 12 bits of SACCH data, 12 bits of CDVCC bits, and six guard bits to compensate for differences in the distances between mobile units and base stations. The guard time is present in only the reverse channel time slots to prevent overlapping of received bursts due to radio signal transit time. The ramp-up time consists of six bits that allow gradual rising and falling of the RF signal energy within the time slot. Thus, a reverse channel time slot consists of 324 bits. If an FACCH is sent instead of speech data, one time slot of speech coding data is replaced with a 260-bit block of FACCH data.

In the forward channel, each time slot contains two 130-bit bursts of digitized voice data (or FACCH data if digitized speech is not being sent) for a total of 260 bits per frame. In addition, each forward channel frame contains 28 synchronization bits, 12 bits of SACCH data, 12 CDVCC bits, and 12 reserved bits for a total of 324 bits per time slot. Therefore, both forward and reverse voice channels have a data transmission rate of

$$\frac{324 \text{ bits}}{\text{time slot}} \times \frac{6 \text{ time slots}}{40 \text{ ms}} = 48.6 \text{ kbps}$$

A third frame format, called a *shortened burst,* is shown in Figure 12-10. Shortened bursts are transmitted when a mobile unit begins operating in a larger-diameter cell because the propagation time between the mobile and base is unknown. A mobile unit transmits shortened burst slots until the base station determines the required time offset. The default delay between the receive and transmit slots in the mobile is 44 symbols, which results in a maximum distance at which a mobile station can operate in a cell to 72 miles for an IS-54 cell.

12-6-6 USDC Digital Modulation Scheme

To achieve a transmission bit rate of 48.6 kbps in a 30-kHz AMPS voice channel, a *bandwidth (spectral) efficiency* of 1.62 bps/Hz is required, which is well beyond the capabilities of binary FSK. The spectral efficiency requirements can be met by using conventional pulse-shaped, four-phase modulation schemes, such as QPSK and OQPSK. However, USDC voice and control channels use a *symmetrical differential, phase-shift keying* technique known as π/4 DQPSK, or π/4 *differential quadriphase shift keying* (DQPSK), which offers several advantages in a mobile radio environment, such as improved cochannel rejection and bandwidth efficiency.

A 48.6-kbps data rate requires a symbol (baud) rate of 24.3 kbps (24.3 kilobaud per second) with a symbol duration of 41.1523 μs. The use of pulse shaping and π/4 DQPSK supports the transmission of three different 48.6-kbps digitized speech signals in a 30-kHz

Table 12-4 NA-TDMA Mobile Phone Power Levels

Power Level	Class I dBm	Class I mW	Class II dBm	Class II mW	Class III dBm	Class III mW	Class IV dBm
0	36	4000	32	1600	28	640	28
1	32	1600	32	1600	28	640	28
2	28	640	28	640	28	640	28
3	24	256	24	256	24	256	24
4	20	102	20	102	20	102	20
5	16	41	16	41	16	41	16
6	12	16	12	16	12	16	12
7	8	6.6	8	6.6	8	6.6	8
8	—	—	Dual mode only		—		— 4 dBm ± 3 dB
9	—	—	Dual mode only		—		— 0 dBm ± 6 dB
10	—	—	Dual mode only		—		— − 4 dBm ± 9 dB

bandwidth with as much as 50 dB of adjacent-channel isolation. Thus, the bandwidth efficiency using π/4 DQPSK is

$$\eta = \frac{3 \times 48.6 \text{ kbps}}{30 \text{ kHz}}$$

$$= 4.86 \text{ bps/Hz}$$

where η is the bandwidth efficiency.

In a π/4 DQPSK modulator, data bits are split into two parallel channels that produce a specific phase shift in the analog carrier, and, since there are four possible bit pairs, there are four possible phase shifts using a quadrature I/Q modulator. The four possible differential phase changes, π/4, −π/4, 3π/4, and −3π/4, define eight possible carrier phases. Pulse shaping is used to minimize the bandwidth while limiting the intersymbol interference. In the transmitter, the PSK signal is filtered using a square-root raised cosine filter with a roll-off factor of 0.35. PSK signals, after pulse shaping, become a linear modulation technique, requiring linear amplification to preserve the pulse shape. Using pulse shaping with π/4 DQPSK allows for the simultaneous transmission of three separate 48.6-kbps speech signals in a 30-kHz bandwidth.

12-6-7 USDC Radiated Power

NA-TDMA specifies 11 radiated power levels for four classifications of mobile units, including the eight power levels used by standard AMPS transmitters. The fourth classification is for dual-mode TDMA/analog cellular telephones. The NA-TDMA power classifications are listed in Table 12-4. The highest power level is 4 W (36 dBm), and successive levels differ by 4 dB, with the lowest level for classes I through III being 8 dBm (6.6 mW). The lowest transmit power level for dual-mode mobile units is −4 dBm (0.4 mW) ± 9 dB. In a dual-mode system, the three lowest power levels can be assigned only to digital voice channels and digital control channels. Analog voice channels and FSK control channels transmitting in the standard AMPS format are confined to the eight power levels in the AMPS specification. Transmitters in the TDMA mode are active only one-third of the time; therefore, the average transmitted power is 4.8 dB below specifications.

12-7 INTERIM STANDARD 95

FDMA is an access method used with standard analog AMPS, and both FDMA and TDMA are used with USDC. Both FDMA and TDMA use a frequency channelization approach to frequency spectrum management; however, TDMA also utilizes a time-division accessing approach. With FDMA and TDMA cellular telephones, the entire available cellular radio-frequency spectrum is subdivided into narrowband radio channels to be used for one-way communications links between cellular mobile units and base stations.

In 1984, Qualcomm Inc. proposed a cellular telephone system and standard based on spread-spectrum technology with the primary goal of increasing capacity. Qualcomm's new system enabled a totally digital mobile telephone system to be made available in the United States based on *code-division multiple accessing* (CDMA). The U.S. Telecommunications Industry Association recently standardized the CDMA system as Interim Standard 95 (IS-95), which is a mobile–to–base station compatibility standard for dual-mode wideband spread-spectrum communications. CDMA allows users to differentiate from one another by a unique code rather than a frequency or time assignment and, therefore, offers several advantages over cellular telephone systems using TDMA and FDMA, such as increased capacity and improved performance and reliability. IS-95, like IS-54, was designed to be compatible with existing analog cellular telephone system (AMPS) frequency band; therefore, mobile units and base stations can easily be designed for dual-mode operation. Pilot CDMA systems developed by Qualcomm were first made available in 1994.

NA-TDMA channels occupy exactly the same bandwidth as standard analog AMPS signals. Therefore, individual AMPS channel units can be directly replaced with TDMA channels, which are capable of carrying three times the user capacity as AMPS channels. Because of the wide bandwidths associated with CDMA transmissions, IS-95 specifies an entirely different channel frequency allocation plan than AMPS.

The IS-95 standard specifies the following:

1. Modulation—digital OQPSK (uplink) and digital QPSK (downlink)
2. 800-MHz band (IS-95A)
   45-MHz forward and reverse separation
   50-MHz spectral allocation
3. 1900-MHz band (IS-95B)
   90-MHz forward and reverse separation
   120-MHz spectral allocation
4. 2.46-MHz total bandwidth
   1.23-MHz reverse CDMA channel bandwidth
   1.23-MHz forward CDMA channel bandwidth
5. Direct-sequence CDMA accessing
6. 8-kHz voice bandwidth
7. 64 total channels per CDMA channel bandwidth
8. 55 voice channels per CDMA channel bandwidth

12-7-1   CDMA

With IS-95, each mobile user within a given cell, and mobile subscribers in adjacent cells use the same radio-frequency channels. In essence, frequency reuse is available in all cells. This is made possible because IS-95 specifies a direct-sequence, spread-spectrum CDMA system and does not follow the channelization principles of traditional cellular radio communications systems. Rather than dividing the allocated frequency spectrum into narrow-bandwidth channels, one for each user, information is transmitted (spread) over a very wide frequency spectrum with as many as 20 mobile subscriber units simultaneously using the same carrier frequency within the same frequency band. Interference is incorporated into the system so that there is no limit to the number of subscribers that CDMA can support. As more mobile subscribers are added to the system, there is a *graceful degradation* of communications quality.

With CDMA, unlike other cellular telephone standards, subscriber data change in real time, depending on the voice activity and requirements of the network and other users of the network. IS-95 also specifies a different modulation and spreading technique for the forward and reverse channels. On the forward channel, the base station simultaneously transmits user data from all current mobile units in that cell by using different spreading sequences (codes) for each user's transmissions. A pilot code is transmitted with the user data at a higher power level, thus allowing all mobile units to use coherent detection. On the reverse link, all mobile units respond in an asynchronous manner (i.e., no time or duration limitations) with a constant signal level controlled by the base station.

The speech coder used with IS-95 is the Qualcomm 9600-bps *Code-Excited Linear Predictive* (QCELP) coder. The vocoder converts an 8-kbps compressed data stream to a 9.6-kbps data stream. The vocoder's original design detects voice activity and automatically reduces the data rate to 1200 bps during silent periods. Intermediate mobile user data rates of 2400 bps and 4800 bps are also used for special purposes. In 1995, Qualcomm introduced a 14,400-bps vocoder that transmits 13.4 kbps of compressed digital voice information.

12-7-1-1 CDMA frequency and channel allocations.

CDMA reduces the importance of frequency planning within a given cellular market. The AMPS U.S. cellular telephone system is allocated a 50-MHz frequency spectrum (25 MHz for each direction of propagation), and each service provider (system A and system B) is assigned half the available spectrum (12.5 MHz). AMPS common carriers must provide a 270-kHz guard band (approximately nine AMPS channels) on either side of the CDMA frequency spectrum. To facilitate a graceful transition from AMPS to CDMA, each IS-95 channel is allocated a 1.25-MHz frequency spectrum for each one-way CDMA communications channel. This equates to 10% of the total available frequency spectrum of each U.S. cellular telephone provider. CDMA channels can coexist within the AMPS frequency spectrum by having a wireless operator clear a 1.25-MHz band of frequencies to accommodate transmissions on the CDMA channel. A single CDMA radio channel takes up the same bandwidth as approximately 42 30-kHz AMPS voice channels. However, because of the frequency reuse advantage of CDMA, CDMA offers approximately a 10-to-1 channel advantage over standard analog AMPS and a 3-to-1 advantage over USDC digital AMPS.

For reverse (downlink) operation, IS-95 specifies the 824-MHz to 849-MHz band and forward (uplink) channels the 869-MHz to 894-MHz band. CDMA cellular systems also use a modified frequency allocation plan in the 1900-MHz band. As with AMPS, the transmit and receive carrier frequencies used by CDMA are separated by 45 MHz. Figure 12-11a shows the frequency spacing for two adjacent CDMA channels in the AMPS frequency band. As the figure shows, each CDMA channel is 1.23 MHz wide with a 1.25-MHz frequency separation between adjacent carriers, producing a 200-kHz guard band between CDMA channels. Guard bands are necessary to ensure that the CDMA carriers do not interfere with one another. Figure 12-11b shows the CDMA channel location within the AMPS frequency spectrum. The lowest CDMA carrier frequency in the A band is at AMPS channel 283, and the lowest CDMA carrier frequency in the B band is at AMPS channel 384. Because the band available between 667 and 716 is only 1.5 MHz in the A band, A band operators have to acquire permission from B band carriers to use a CDMA carrier in that portion of the frequency spectrum. When a CDMA carrier is being used next to a non-CDMA carrier, the carrier spacing must be 1.77 MHz. There are as many as nine CDMA carriers available for the A and B band operator in the AMPS frequency spectrum. However, the A and B band operators have 30-MHz bandwidth in the 1900-MHz frequency band, where they can facilitate up to 11 CDMA channels.

With CDMA, many users can share common transmit and receive channels with a transmission data rate of 9.6 kbps. Using several techniques, however, subscriber information is spread by a factor of 128 to a channel chip rate of 1.2288 Mchips/s, and transmit and receive channels use different spreading processes.

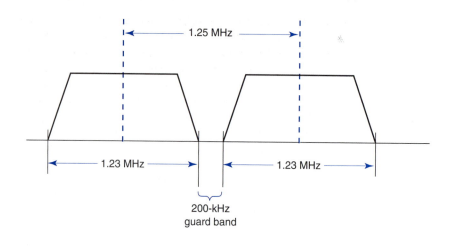

(a)

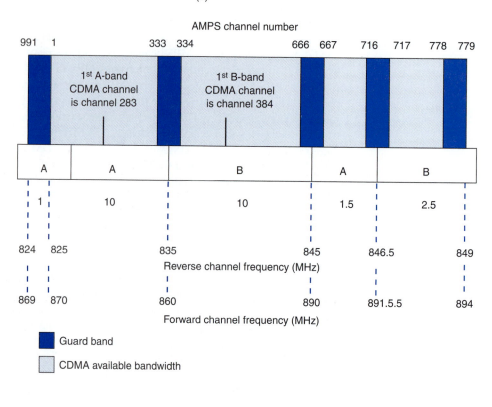

(b)

FIGURE 12-11 (a) CDMA channel bandwidth, guard band, and frequency separation; (b) CDMA channel location within the AMPS frequency spectrum

In the uplink channel, subscriber data are encoded using a rate 1/2 convolutional code, interleaved, and spread by one of 64 orthogonal spreading sequences using Walsh functions. Orthogonality among all uplink cellular channel subscribers within a given cell is maintained because all the cell signals are scrambled synchronously.

Downlink channels use a different spreading strategy since each mobile unit's received signal takes a different transmission path and, therefore, arrives at the base station at a different time. Downlink channel data streams are first convolutional encoded with a rate

1/3 convolution code. After interleaving, each block of six encoded symbols is mapped to one of the available orthogonal Walsh functions, ensuring 64-ary orthogonal signaling. An additional fourfold spreading is performed by subscriber-specified and base station–specific codes having periods of 2^{14} chips and 2^{15} chips, respectively, increasing the transmission rate to 1.2288 Mchips/s. Stringent requirements are enforced in the downlink channel's transmit power to avoid the near-far problem caused by varied receive power levels.

Each mobile unit in a given cell is assigned a unique spreading sequence that ensures near perfect separation among the signals from different subscriber units and allows transmission differentiation between users. All signals in a particular cell are scrambled using a pseudorandom sequence of length 2^{15} chips. This reduces radio-frequency interference between mobiles in neighboring cells that may be using the same spreading sequence and provides the desired wideband spectral characteristics even though all Walsh codes do not yield a wideband power spectrum.

Two commonly used techniques for spreading the spectrum are *frequency hopping* and *direct sequencing*. Both of these techniques are characteristic of transmissions over a bandwidth much wider than that normally used in narrowband FDMA/TDMA cellular telephone systems, such as AMPS and USDC. For a more detailed description of frequency hopping and direct sequencing, refer to Chapter 25.

12-7-1-2 Frequency-hopping spread spectrum.

Frequency-hopping spread spectrum was first used by the military to ensure reliable antijam and to secure communications in a battlefield environment. The fundamental concept of frequency hopping is to break a message into fixed-size blocks of data with each block transmitted in sequence except on a different carrier frequency. With frequency hopping, a pseudorandom code is used to generate a unique frequency-hopping sequence. The sequence in which the frequencies are selected must be known by both the transmitter and the receiver prior to the beginning of the transmission. The transmitter sends one block on a radio-frequency carrier and then switches (hops) to the next frequency in the sequence and so on. After reception of a block of data on one frequency, the receiver switches to the next frequency in the sequence. Each transmitter in the system has a different hopping sequence to prevent one subscriber from interfering with transmissions from other subscribers using the same radio channel frequency.

12-7-1-3 Direct-sequence spread spectrum.

In direct-sequence systems, a high-bit-rate pseudorandom code is added to a low-bit-rate information signal to generate a high-bit-rate pseudorandom signal closely resembling noise that contains both the original data signal and the pseudorandom code. Again, before successful transmission, the pseudorandom code must be known to both the transmitter and the intended receiver. When a receiver detects a direct-sequence transmission, it simply subtracts the pseudorandom signal from the composite receive signal to extract the information data. In CDMA cellular telephone systems, the total radio-frequency bandwidth is divided into a few broadband radio channels that have a much higher bandwidth than the digitized voice signal. The digitized voice signal is added to the generated high-bit-rate signal and transmitted in such a way that it occupies the entire broadband radio channel. Adding a high-bit-rate pseudorandom signal to the voice information makes the signal more dominant and less susceptible to interference, allowing lower-power transmission and, hence, a lower number of transmitters and less expensive receivers.

12-7-2 CDMA Traffic Channels

CDMA traffic channels consist of a downlink (base station to mobile unit) channel and an uplink (mobile station to base station) channel. A CDMA downlink traffic channel is shown in Figure 12-12a. As the figure shows, the downlink traffic channel consists of up to 64 channels, including a broadcast channel used for control and traffic channels used to carry subscriber information. The broadcast channel consists of a pilot channel, a synchronization channel, up to seven paging channels, and up to 63 traffic channels. All these channels share the same 1.25-MHz CDMA frequency assignment. The traffic channel is identified

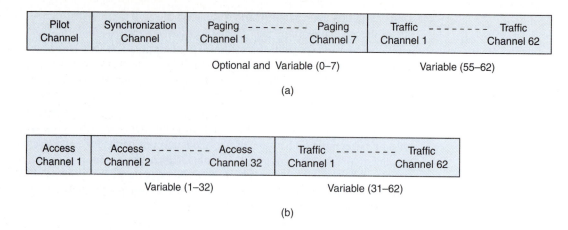

Pilot Channel	Synchronization Channel	Paging ──────── Paging Channel 1 Channel 7	Traffic ──────── Traffic Channel 1 Channel 62

Optional and Variable (0–7)   Variable (55–62)

(a)

Access Channel 1	Access ──────── Access Channel 2 Channel 32	Traffic ──────── Traffic Channel 1 Channel 62

Variable (1–32)   Variable (31–62)

(b)

FIGURE 12-12 IS-95 traffic channels: (a) downlink; (b) uplink

by a distinct user-specific long-code sequence, and each access channel is identified by a distinct access channel long-code sequence.

The pilot channel is included in every cell with the purpose of providing a signal for the receiver to use to acquire timing and provide a phase reference for coherent demodulation. The pilot channel is also used by mobile units to compare signal strengths between base stations to determine when a handoff should be initiated. The synchronization channel uses a Walsh W32 code and the same pseudorandom sequence and phase offset as the pilot channel, allowing it to be demodulated by any receiver that can acquire the pilot signal. The synchronization channel broadcasts synchronization messages to mobile units and operates at 1200 bps. Paging channels convey information from the base station to the mobile station, such as system parameter messages, access parameter messages, CDMA channel list messages, and channel assignment messages. Paging channels are optional and can range in number between zero and seven. The paging channel is used to transmit control information and paging messages from the base station to the mobile units and operates at either 9600 bps, 4800 bps, or 2400 bps. A single 9600-bps pilot channel can typically support about 180 pages per second for a total capacity of 1260 pages per second.

Data on the downlink traffic channel are grouped into 20-ms frames. The data are first convolutionally coded and then formatted and interleaved to compensate for differences in the actual user data rates, which vary. The resulting signal is spread with a Walsh code and a long pseudorandom sequence at a rate of 1.2288 Mchips/s.

The uplink radio channel transmitter is shown in Figure 12-12b and consists of access channels and up to 62 uplink traffic channels. The access and uplink traffic channels use the same frequency assignment using direct-sequence CDMA techniques. The access channels are uplink-only, shared, point-to-point channels that provide communications from mobile units to base stations when the mobile unit is not using a traffic channel. Access channels are used by the mobile unit to initiate communications with a base station and to respond to paging channel messages. Typical access channel messages include acknowledgments and sequence number, mobile identification parameter messages, and authentication parameters. The access channel is a random access channel with each channel subscriber uniquely identified by their pseudorandom codes. The uplink CDMA channel can contain up to a maximum of 32 access channels per supported paging channel. The uplink traffic channel operates at a variable data rate mode, and the access channels operate at a fixed 4800-bps rate. Access channel messages consist of registration, order, data burst, origination, page response, authentication challenge response, status response, and assignment completion messages.

Table 12-5 CDMA Power Levels

Class	Minimum EIRP	Maximum EIRP
I	-2 dBW (630 mW)	3 dBW (2.0 W)
II	-7 dBW (200 mW)	0 dBW (1.0 W)
III	-12 dBW (63 mW)	-3 dBW (500 mW)
IV	-17 dBW (20 mW)	-6 dBW (250 mW)
V	-22 dBW (6.3 mW)	-9 dBW (130 mW)

Subscriber data on the uplink radio channel transmitter are also grouped into 20-ms frames, convolutionally encoded, block interleaved, modulated by a 64-ary orthogonal modulation, and spread prior to transmission.

12-7-2-1 CDMA radiated power. IS-95 specifies complex procedures for regulating the power transmitted by each mobile unit. The goal is to make all reverse-direction signals within a single CDMA channel arrive at the base station with approximately the same signal strength (± 1 dB), which is essential for CDMA operation. Because signal paths change continuously with moving units, mobile units perform power adjustments as many as 800 times per second (once every 1.25 ms) under control of the base station. Base stations instruct mobile units to increase or decrease their transmitted power in 1-dB increments (± 0.5 dB).

When a mobile unit is first turned on, it measures the power of the signal received from the base station. The mobile unit assumes that the signal loss is the same in each direction (forward and reverse) and adjusts its transmit power on the basis of the power level of the signal it receives from the base station. This process is called *open-loop power setting*. A typical formula used by mobile units for determining their transmit power is

$$P_t \, \text{dBm} = -76 \, \text{dB} - P_r \qquad \text{(12-4)}$$

where P_t = transmit power (dBm)
  P_r = received power (dBm)

Example 12-2

Determine the transmit power for a CDMA mobile unit that is receiving a signal from the base station at -100 dBm.

Solution Substituting into Equation 12-4 gives

$$P_t = -76 - (-100)$$
$$P_t = 24 \text{ dBm, or } 250 \text{ mW}$$

With CDMA, rather than limit the maximum transmit power, the minimum and maximum effective isotropic radiated power (EIRP) is specified (EIRP is the power radiated by an antenna times the gain of the antenna). Table 12-5 lists the maximum EIRPs for five classes of CDMA mobile units. The maximum radiated power of base stations is limited to 100 W per 1.23-MHz CDMA channel.

12-8 NORTH AMERICAN CELLULAR AND PCS SUMMARY

Table 12-6 summarizes several of the parameters common to North American cellular and PCS telephone systems (AMPS, USDC, and PCS).

Table 12-6 Cellular and PCS Telephone Summary

Parameter	Cellular System		
	AMPS	USDC (IS-54)	IS-95
Access method	FDMA	FDMA/TDMA	CDMA/FDMA
Modulation	FM	$\pi/4$ DQPSK	BPSK/QPSK
Frequency band			
Base station	869–894 MHz	869–894 MHz	869–894 MHz
Mobile unit	824–849 MHz	824–849 MHz	824–849 MHz
Base station	—	1.85–1.91 GHz	1.85–1.91 GHz
Mobile unit	—	1.93–1.99 GHz	1.93–1.99 GHz
RF channel bandwidth	30 kHz	30 kHz	1.25 MHz
Maximum radiated power	4 W	4 W	2 W
Control channel	FSK	PSK	PSK
Voice channels per carrier	1	3 or 6	Up to 20
Frequency assignment	Fixed	Fixed	Dynamic

12-9 GLOBAL SYSTEM FOR MOBILE COMMUNICATIONS

In the early 1980s, analog cellular telephone systems were experiencing a period of rapid growth in western Europe, particularly in Scandinavia and the United Kingdom and to a lesser extent in France and Germany. Each country subsequently developed its own cellular telephone system, which was incompatible with everyone else's system from both an equipment and an operational standpoint. Most of the existing systems operated at different frequencies, and all were analog. In 1982, the *Conference of European Posts and Telegraphs* (CEPT) formed a study group called *Groupe Spécial Mobile* (GSM) to study the development of a pan-European (*pan* meaning "all") public land mobile telephone system using ISDN. In 1989, the responsibility of GSM was transferred to the *European Telecommunications Standards Institute* (ETSI), and phase I of the GSM specifications was published in 1990. GSM had the advantage of being designed from scratch with little or no concern for being backward compatible with any existing analog cellular telephone system. GSM provides its subscribers with good quality, privacy, and security. GSM is sometimes referred to as the *Pan-European cellular system.*

Commercial GSM service began in Germany in 1991, and by 1993 there were 36 GSM networks in 22 countries. GSM networks are now either operational or planned in over 80 countries around the world. North America made a late entry into the GSM market with a derivative of GSM called *PCS-1900.* GSM systems now exist on every continent, and the acronym GSM now stands for *Global System for Mobile Communications.* The first GSM system developed was GSM-900 (phase I), which operates in the 900-MHz band for voice only. Phase 2 was introduced in 1995, which included facsimile, video, and data communications services. After implementing PCS frequencies (1800 MHz in Europe and 1900 MHz in North America) in 1997, GSM-1800 and GSM-1900 were created.

GSM is a second-generation cellular telephone system initially developed to solve the fragmentation problems inherent in first-generation cellular telephone systems in Europe. Before implementing GSM, all European countries used different cellular telephone standards; thus, it was impossible for a subscriber to use a single telephone set throughout Europe. GSM was the world's first totally digital cellular telephone system designed to use the services of SS7 signaling and an all-digital data network called *integrated services digital network* (ISDN) to provide a wide range of network services. With between 20 and 50 million subscribers, GSM is now the world's most popular standard for new cellular telephone and personal communications equipment.

12-9-1 GSM Services

The original intention was to make GSM compatible with ISDN in terms of services offered and control signaling. Unfortunately, radio-channel bandwidth limitations and cost prohibited GSM from operating at the 64-kbps ISDN basic data rate.

GSM telephone services can be broadly classified into three categories: *bearer services, teleservices,* and *supplementary services.* Probably the most basic bearer service provided by GSM is telephony. With GSM, analog speech signals are digitally encoded and then transmitted through the network as a digital data stream. There is also an emergency service where the closest emergency service provider is notified by dialing three digits similar to 911 services in the United States. A wide variety of data services is offered through GSM, where users can send and receive data at rates up to 9600 bps to subscribers in POTS (plain old telephone service), ISDN networks, Packet Switched Public Data Networks (PSPDN), and Circuit Switched Public Data Networks (CSPDN) using a wide variety of access methods and protocols, such as X.25. In addition, since GSM is a digital network, a modem is not required between the user and the GSM network.

Other GSM data services include Group 3 facsimile per ITU-T recommendation T.30. One unique feature of GSM that is not found in older analog systems is the *Short Message Service* (SMS), which is a bidirectional service for sending alphanumeric messages up to 160 bytes in length. SMS can be transported through the system in a store-and-forward fashion. SMS can also be used in a *cell-broadcast mode* for sending messages simultaneously to multiple receivers. Several supplemental services, such as *call forwarding* and *call barring,* are also offered with GSM.

12-9-2 GSM System Architecture

The system architecture for GSM as shown in Figure 12-13 consists of three major interconnected subsystems that interact among one another and with subscribers through

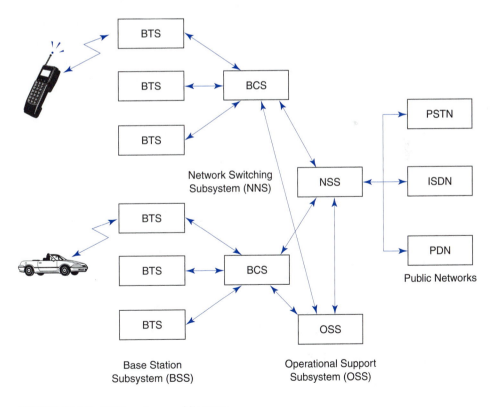

FIGURE 12-13 GSM system architecture

specified network interfaces. The three primary subsystems of GSM are *Base Station Subsystem* (BSS), *Network Switching Subsystem* (NSS), and *Operational Support Subsystem* (OSS). Although the mobile station is technically another subsystem, it is generally considered part of the base station subsystem.

The BSS is sometimes known as the radio *subsystem* because it provides and manages radio-frequency transmission paths between mobile units and the mobile switching center (MSC). The BSS also manages the radio interface between mobile units and all other GSM subsystems. Each BSS consists of many base station controllers (BSCs), which are used to connect the MCS to the NSS through one or more MSCs. The NSS manages switching functions for the system and allows the MSCs to communicate with other telephone networks, such as the public switched telephone network and ISDN. The OSS supports operation and maintenance of the system and allows engineers to monitor, diagnose, and troubleshoot every aspect of the GSM network.

12-9-3 GSM Radio Subsystem

GSM was originally designed for 200 full-duplex channels per cell with transmission frequencies in the 900-MHz band; however, frequencies were later allocated at 1800 MHz. A second system, called DSC-1800, was established that closely resembles GSM. GSM uses two 25-MHz frequency bands that have been set aside for system use in all member companies. The 890-MHz to 915-MHz band is used for mobile unit–to–base station transmissions (reverse-link transmissions), and the 935-MHz to 960-MHz frequency band is used for base station–to–mobile unit transmission (forward-link transmissions). GSM uses frequency-division duplexing and a combination of TDMA and FDMA techniques to provide base stations simultaneous access to multiple mobile units. The available forward and reverse frequency bands are subdivided into 200-kHz wide voice channels called *absolute radio-frequency channel numbers* (ARFCN). The ARFCN number designates a forward/reverse channel pair with 45-MHz separation between them. Each voice channel is shared among as many as eight mobile units using TDMA.

Each of the ARFCN channel subscribers occupies a unique time slot within the TDMA frame. Radio transmission in both directions is at a 270.833-kbps rate using binary Gaussian minimum shift keying (GMSK) modulation with an effective channel transmission rate of 33.833 kbps per user.

The basic parameters of GSM are the following:

1. GMSK modulation (Gaussian MSK)
2. 50-MHz bandwidth:
   890-MHz to 915-MHz mobile transmit band (reverse channel)
   935-MHz to 960-MHz base station transmit band (forward channel)
3. FDMA/TDMA accessing
4. Eight 25-kHz channels within each 200-kHz traffic channel
5. 200-kHz traffic channel
6. 992 full-duplex channels
7. Supplementary ISDN services, such as call diversion, closed user groups, caller identification, and *short messaging service* (SMS), which restricts GSM users and base stations to transmitting alphanumeric pages limited to a maximum of 160 seven-bit ASCII characters while simultaneously carrying normal voice messages

12-10 PERSONAL COMMUNICATIONS SATELLITE SYSTEM

Mobile Satellite Systems (MSS) provide the vehicle for a new generation of wireless telephone services called *personal communications satellite systems* (PCSS). Universal wireless telephone coverage is a developing MSS service that promises to deliver mobile subscribers both traditional and enhanced telephone features while providing wide-area global coverage.

MSS satellites are, in essence, radio repeaters in the sky, and their usefulness for mobile communications depends on several factors, such as the space-vehicle altitude, orbital pattern, transmit power, receiver sensitivity, modulation technique, antenna radiation pattern (footprint), and number of satellites in its constellation. Satellite communications systems have traditionally provided both narrowband and wideband voice, data, video, facsimile, and networking services using large and very expensive, high-powered earth station transmitters communicating via high-altitude, geosynchronous earth-orbit (GEO) satellites. Personal communications satellite services, however, use low earth-orbit (LEO) and medium earth-orbit (MEO) satellites that communicate directly with small, low-power mobile telephone units. The intention of PCSS mobile telephone is to provide the same features and services offered by traditional, terrestrial cellular telephone providers. However, PCSS telephones will be able to make or receive calls anytime, anywhere in the world. A simplified diagram of a PCSS system is shown in Figure 12-14.

The key providers in the PCSS market include American Mobile Satellite Corporation (AMSC), Celsat, Comsat, Constellation Communications (Aries), Ellipsat (Ellipso), INMARSAT, LEOSAT, Loral/Qualcomm (Globalstar), TMI communications, TWR (Odysse), and Iridium LLC.

12-10-1 PCSS Advantages and Disadvantages

The primary and probably most obvious advantage of PCSS mobile telephone is that it provides mobile telephone coverage and a host of other integrated services virtually anywhere in the world to a truly global customer base. PCSS can fill the vacancies between land-based cellular and PCS telephone systems and provide wide-area coverage on a regional or global basis.

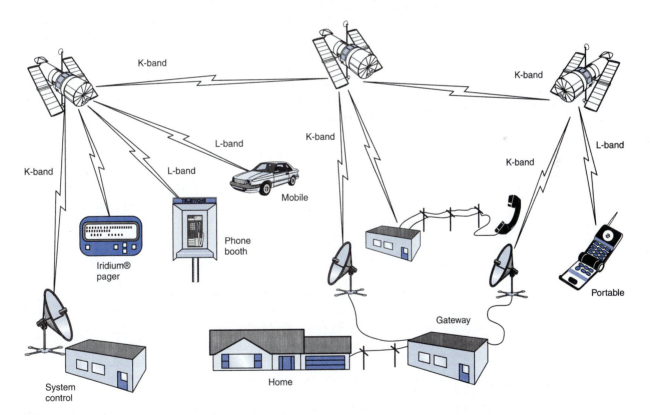

FIGURE 12-14 Overview of Iridium PCSS mobile telephone system

PCSS is ideally suited to fixed cellular telephone applications, as it can provide a full complement of telephone services to places where cables can never go because of economical, technical, or physical constraints. PCSS can also provide complementary and backup telephone services to large companies and organizations with multiple operations in diverse locations, such as retail, manufacturing, finance, transportation, government, military, and insurance.

Most of the disadvantages of PCSS are closely related to economics, with the primary disadvantage being the high risk associated with the high costs of designing, building, and launching satellites. There is also a high cost for the terrestrial-based networking and interface infrastructure necessary to maintain, coordinate, and manage the network once it is in operation. In addition, the intricate low-power, dual-mode transceivers are more cumbersome and expensive than most mobile telephone units used with terrestrial cellular and PCS systems.

12-10-2 PCSS Industry Requirements

PCSS mobile telephone systems require transparent interfaces and feature sets among the multitude of terrestrial networks currently providing mobile and wireline telephone services. In addition, the interfaces must be capable of operating with both ANSI and CCITT network constraints and be able to provide interpretability with AMPS, USDC, GMS, and PCS cellular telephone systems. PCSS must also be capable of operating dual-mode with *air-access protocols,* such as FDMA, TDMA, or CDMA. PCSS should also provide unique MSS feature sets and characteristics, such as inter-/intrasatellite handoffs, land-based–to–satellite handoffs, and land-based/PCSS dual registration.

12-10-3 Iridium Satellite System

Iridium LLC is an international consortium owned by a host of prominent companies, agencies, and governments, including the following: Motorola, General Electric, Lockheed, Raytheon, McDonnell Douglas, Scientific Atlanta, Sony, Kyocera, Mitsubishi, DDI, Kruchinew Enterprises, Mawarid Group of Saudi Arabia, STET of Italy, Nippon Iridium Corporation of Japan, the government of Brazil, Muidiri Investments BVI, LTD of Venezuela, Great Wall Industry of China, United Communications of Thailand, the U.S. Department of Defense, Sprint, and BCE Siemens.

The *Iridium project,* which even sounds like something out of *Star Wars,* is undoubtedly the largest commercial venture undertaken in the history of the world. It is the system with the most satellites, the highest price tag, the largest public relations team, and the most peculiar design. The $5 billion, gold-plated *Iridium* mobile telephone system is undoubtedly (or at least intended to be) the Cadillac of mobile telephone systems. Unfortunately (and somewhat ironically), in August 1999, on Friday the 13th, Iridium LLC, the beleaguered satellite-telephone system spawned by Motorola's Satellite Communications Group in Chandler, Arizona, filed for bankruptcy under protection of Chapter 11. However, Motorola Inc., the largest stockholder in Iridium, says it will continue to support the company and its customers and does not expect any interruption in service while reorganization is under way.

Iridium is a satellite-based wireless personal communications network designed to permit a wide range of mobile telephone services, including voice, data, networking, facsimile, and paging. The system is called Iridium after the element on the periodic table with the atomic number 77 because Iridium's original design called for 77 satellites. The final design, however, requires only 66 satellites. Apparently, someone decided that element 66, dysprosium, did not have the same charismatic appeal as Iridium, and the root meaning of the word is "bad approach." The 66-vehicle LEO interlinked satellite constellation can track the location of a subscriber's telephone handset, determine the best routing through a network of ground-based gateways and intersatellite links, establish the best path for the telephone call, initiate all the necessary connections, and terminate the call on completion. The system also provides applicable revenue tracking.

With Iridium, two-way global communications is possible even when the destination subscriber's location is unknown to the caller. In essence, the intent of the Iridium system is to provide the best service in the telephone world, allowing telecommunication anywhere, anytime, and any place. The FCC granted the Iridium program a full license in January 1995 for construction and operation in the United States.

Iridium uses a GSM-based telephony architecture to provide a digitally switched telephone network and global dial tone to call and receive calls from any place in the world. This global roaming feature is designed into the system. Each subscriber is assigned a personal phone number and will receive only one bill, no matter in what country or area they use the telephone.

The Iridium project has a satellite network control facility in Landsdowne, Virginia, with a backup facility in Italy. A third engineering control complex is located at Motorola's SATCOM location in Chandler, Arizona.

12-10-3-1 System layout. Figure 12-14 shows an overview of the Iridium system. Subscriber telephone sets used in the Iridium system transmit and receive L band frequencies and utilize both frequency- and time-division multiplexing to make the most efficient use of a limited frequency spectrum. Other communications links used in Iridium include EHF and SHF bands between satellites for telemetry, command, and control as well as routing digital voice packets to and from gateways. An Iridium telephone enables the subscriber to connect either to the local cellular telephone infrastructure or to the space constellation using its *dual-mode* feature.

Iridium gateways are prime examples of the advances in satellite infrastructures that are responsible for the delivery of a host of new satellite services. The purpose of the gateways is to support and manage roaming subscribers as well as to interconnect Iridium subscribers to the public switched telephone network. Gateway functions include the following:

1. Set up and maintain basic and supplementary telephony services
2. Provide an interface for two-way telephone communications between two Iridium subscribers and Iridium subscribers to subscribers of the public switched telephone network
3. Provide Iridium subscribers with messaging, facsimile, and data services
4. Facilitate the business activities of the Iridium system through a set of cooperative mutual agreements

12-10-3-2 Satellite constellation. Providing full-earth coverage is the underlying basis of the Iridium satellite system. Iridium uses 66 operational satellites (there are also some spares) configured at a mean elevation of 420 miles above earth in six nearly polar orbital planes (86.4° tilt), in which 11 satellites revolve around earth in each orbit with an orbital time of 100 minutes, 28 seconds. This allows Iridium to cover the entire surface area of earth, and whenever one satellite goes out of view of a subscriber, a different one replaces it. The satellites are phased appropriately in north–south necklaces forming *corotating planes* up one side of earth, across the poles, and down the other side. The first and last planes rotate in opposite directions, creating a virtual *seam*. The corotating planes are separated by 31.6°, and the seam planes are 22° apart.

Each satellite is equipped with three L band antennas forming a honeycomb pattern that consists of 48 individual spot beams with a total of 1628 cells aimed directly below the satellite, as shown in Figure 12-15. As the satellite moves in its orbit, the footprints move across earth's surface, and subscriber signals are switched from one beam to the next or from one satellite to the next in a handoff process. When satellites approach the North or South Pole, their footprints converge, and the beams overlap. Outer beams are then turned off to eliminate this overlap and conserve power on the spacecraft. Each cell has 174 full-duplex voice channels for a total of 283,272 channels worldwide.

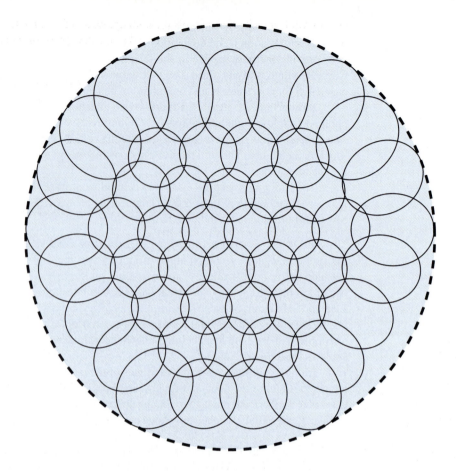

FIGURE 12-15 Iridium system spot beam footprint pattern

Using satellite *cross-links* is the unique key to the Iridium system and the primary differentiation between Iridium and the traditional satellite *bent-pipe system,* where all transmissions follow a path from earth to satellite to earth. Iridium is the first mobile satellite to incorporate sophisticated, onboard digital processing on each satellite and cross-link capability between satellites.

Each satellite is equipped with four satellite-to-satellite cross-links to relay digital information around the globe. The cross-link antennas point toward the closest spacecraft orbiting in the same plane and the two adjacent corotating planes. *Feeder link* antennas relay information to the terrestrial gateways and the system control segment located at the earth stations.

12-10-3-3 Frequency plan and modulation.   On October 14, 1994, the FCC issued a report and order Dockett #92-166 defining L band frequency sharing for subscriber units in the 1616-MHz to 1626.5-MHz band. Mobile satellite system cellular communications are assigned 5.15 MHz at the upper end of this spectrum for TDMA/FDMA service. CDMA access is assigned the remaining 11.35 MHz for their service uplinks and a proportionate amount of the S band frequency spectrum at 2483.5 MHz to 2500 MHz for their downlinks. When a CDMA system is placed into operation, the CDMA L band frequency spectrum will be reduced to 8.25 MHz. The remaining 3.1 MHz of the frequency spectrum will then be assigned to either the Iridium system or another TDMA/FDMA system.

All Ka band uplinks, downlinks, and cross-links are packetized TDM/FDMA using quadrature phase-shift keying (QPSK) and FEC 1/2 rate convolutional coding with Viterbi decoding. Coded data rates are 6.25 Mbps for gateways and satellite control facility links and 25 Mbps for satellite cross-links. Both uplink and downlink transmissions occupy 100 MHz of bandwidth, and intersatellite links use 200 MHz of bandwidth. The frequency bands are as follows:

L band subscriber-to-satellite voice links = 1.616 GHz to 1.6265 GHz

Ka band gateway downlinks = 19.4 GHz to 19.6 GHz

Ka band gateway uplinks = 29.1 GHz to 29.3 GHz

Ka intersatellite cross-links = 23.18 GHz to 23.38 GHz

QUESTIONS

12-1. What is meant by a *first-generation* cellular telephone system?

12-2. Briefly describe the AMPS system.

12-3. Outline the AMPS *frequency allocation*.

12-4. What is meant by the term *frequency-division duplexing*?

12-5. What is the difference between a *wireline* and *non-wireline* company?

12-6. Describe a *cellular geographic serving area*.

12-7. List and describe the three *classifications* of AMPS cellular telephones.

12-8. What is meant by the *discontinuous transmission mode*?

12-9. List the features of a *personal communications system* that differentiate it from a *standard cellular telephone network*.

12-10. What is the difference between a *personal communications network* and *personal communications services*?

12-11. Briefly describe the functions of a *home location register*.

12-12. Briefly describe the functions of a *visitor location register*.

12-13. Briefly describe the functions of an *equipment identification registry*.

12-14. Describe the following services: *available mode, screen mode, private mode*, and *unavailable mode*.

12-15. What is meant by a *microcellular system*?

12-16. List the advantages of a *PCS cellular system* compared to a *standard cellular system*.

12-17. List the disadvantages of a PCS cellular system.

12-18. What is meant by the term *false handoff*?

12-19. Briefly describe the N-AMPS cellular telephone system.

12-20. What is an *interference avoidance scheme*?

12-21. What are the four types of *handoffs* possible with N-AMPS?

12-22. List the advantages of a *digital cellular system*.

12-23. Describe the *United States Digital Cellular* system.

12-24. Describe the *TDMA scheme* used with USDC.

12-25. List the advantages of *digital* TDMA over *analog* AMPS FDMA.

12-26. Briefly describe the EIA/TIA *Interim Standard IS-54*.

12-27. What is meant by the term *dual mode*?

12-28. Briefly describe the EIA/TIA *Interim Standard IS-136*.

12-29. What is meant by the term *sleep mode*?

12-30. Briefly describe the *North American Digital Cellular* format.

12-31. Briefly describe the E-TDMA scheme.

12-32. Describe the differences between the radiated power classifications for USDC and AMPS.

12-33. List the IS-95 specifications.

12-34. Describe the *CDMA format* used with IS-95.

12-35. Describe the differences between the *CDMA radiated power procedures* and AMPS.

12-36. Briefly describe the GSM cellular telephone system.

12-37. Outline and describe the *services* offered by GSM.

12-38. Briefly describe the GSM *system architecture*.

What are the three *primary subsystems* of GSM?

Briefly describe the GSM *radio subsystem*.

List the *basic parameters* of GSM.

Briefly describe the *architecture* of a PCSS.

List the *advantages* and *disadvantages* of PCSS.

Outline the *industry requirements* of PCSS.

C H A P T E R 13

Data Communications Codes, Error Control, and Data Formats

CHAPTER OUTLINE

OBJECTIVES

- Define the terms *data, information,* and *data transmission*
- Define the terms *character framing* and *message framing*
- Describe several of the popular data communications codes
- Describe bar codes and explain several common formats
- Define *error control* and relate it to probability of error and bit error rate
- Define *error detection* and describe several common techniques of achieving it
- Define *error correction* and describe the Hamming code
- Describe character synchronization and explain the differences between asynchronous and synchronous data formats

13-1 INTRODUCTION

Data can be generally defined as *information* that is stored in digital format, usually in the form of binary digits. Although information can originate from either analog or digital sources, data communications networks are involved primarily with transporting information from digital sources, such as computers and computer-related equipment.

Data may be *alphabetic, numeric,* or *symbolic* in nature. Therefore, information can be encoded in many formats, such as binary-weighted numbers, binary-coded digits (BCD), alphanumeric characters, microprocessor op-codes, digitized analog signals such as voice or video, control codes, source and destination addresses, program data, graphics, and many more. When a key is depressed on a standard keyboard, the symbol (digit, letter, and so on) that the key represents is converted to a unique binary code. Each key has a different code. In essence, the keyboard is an *encoder*. There are different types of keyboards. For example, ASCII keyboard produces ASCII codes, hex keyboards produce hex codes, and so on. Information can also be encoded with analog signal codes, such as with the Touch-Tone keypad on a standard telephone set. When a key is depressed, a two-tone signal is transmitted representing the digit depressed. Each digit has a different two-tone code.

Information sources generate signals that are not in a form suitable for transmission over a data communications network. Therefore, the information must be formatted (encoded) to comply with the requirements of the network prior to transmission and then unformatted (decoded) at the destination. Information converted to a binary code is sometimes called a *data transmission code*. Data transmission codes allow people to communicate with people and machines to communicate with machines over data communications networks.

In essence, binary codes add meaning to bits. Hex codes are simply shorthand notations for binary sequences and by themselves add no meaning to the data. The hex code 31 represents the binary sequence 0011 0001, which could mean different things in different situations. For example, 31 hex represents the digit 1 in ASCII and the number 39 in BCD and in decimal carries a binary weighting equivalent to 49.

In addition to sending data transmission codes, additional information may be added to the code to perform error control (i.e., error detection and/or correction [EDC]). Error control enables receivers to determine whether errors have occurred during transmission and, when they do occur, to correct them. Additional data may also be added to the original information to perform specific network functions, such as indicating the beginning and end of a character (*character framing*) or at the beginning and end of a message (*message framing*). Data added to the original information is called *overhead*, and sometimes a message contains more overhead than user information.

13-2 DATA COMMUNICATIONS CHARACTER CODES

Data communications codes are often used to represent characters and symbols, such as letters, digits, and punctuation marks. Therefore, these types of data communications codes are called *character codes, character sets, symbol codes*, or *character languages*. In essence, there are only three types of characters used with character codes: data-link control characters, alphanumeric characters, and graphic control characters. *Data-link control characters* are used to facilitate the orderly flow of data from source to destination. *Graphic control characters* involve the syntax or presentation of the data at the receiver, and *alphanumeric characters* represent the various symbols used for letters, numbers, and punctuation. Data-link control characters include special function characters for framing messages, such as STX (start of text) and ETX (end of text). Graphic control characters include special characters for displaying text on printers and monitors, such as VT (vertical tab), HT (horizontal tab), and NL (new line). Data-link and graphic control characters are described in more detail in a later chapter of this book.

The first character set was the Morse code. However, today the three most common character sets are the Baudot code, the American Standard Code for Information Interchange (ASCII), and the Extended Binary Decimal Interchange Code (EBCDIC).

13-2-1 Morse Code

The first character code that saw widespread usage was the *Morse code*, developed in 1840 by Samuel F. B. Morse. The Morse code was first used in 1844 when Morse established the world's first digital communications link between Washington, D.C., and Baltimore,

Table 13-1 International Morse Code

Letter	Code	Letter	Code	Letter	Code
A	• –	N	– •	1	• – – – –
B	– • • •	O	– – –	2	• • – – –
C	– • – •	P	• – – •	3	• • • – –
D	– • •	Q	– – • –	4	• • • • –
E	•	R	• – •	5	• • • • •
F	• • – •	S	• • •	6	– • • • •
G	– – •	T	–	7	– – • • •
H	• • • •	U	• • –	8	– – – • •
I	• •	V	• • • –	9	– – – – •
J	• – – –	W	• – –	0	– – – – –
K	– • –	X	– • • –		
L	• – • •	Y	– • – –		
M	– –	Z	– – • •		
Period	• – • – • –	Fraction bar (slash)	– • • – •		
Comma	– – • • – –	Wait	• – • • •		
?	• • – – • •	End of message	• – • – •		
Error	• • • • • • • •	Invitation to transmit	– • –		
:	– – – • • •	End of transmission	• • • – • –		
;	– • – • – •	Double dash (break)	– • • • –		
( )	– • – – • –				
SOS (International distress signal: "save our ship")			• • • – – – • • •		

Maryland, using the telegraph system. Morse's code soon became known as the *International Morse Code*. The Morse code is an example of a *variable-length source code*, where the length of the code varies from character to character. Today, for the most part, amateur radio operators are the only people who use the Morse code. The Morse code is a digital encoding scheme that uses three unequal-length symbols (*dot, dash,* and *space*) to encode alphanumeric characters, punctuation marks, and an interrogation word. Spaces separate words and are approximately the length of two dashes. Table 13-1 lists the International Morse Code character set. Note that the code can be used to represent only letters, digits, and punctuation marks.

The Morse code is inadequate for use in modern digital equipment because all characters do not have the same number of symbols or take the same length of time to send, and each Morse code operator transmits code at a different rate. It literally requires the reasoning ability of a human brain to decode Morse code. Morse code also has an insufficient selection of graphic and data-link control characters to facilitate the transmission and presentation of the data used in contemporary computer applications.

13-2-2 Baudot Code

The *Baudot code* (sometimes called the *Telex code*) was the first *fixed-length character code* developed for machines rather than for people. A French postal engineer named Thomas Murray developed the Baudot code in 1875 and named the code after Emile Baudot, an early pioneer in telegraph printing. The Baudot code (pronounced *baw-dough*) is a *fixed-length source code* (sometimes called a *fixed-length block code*). With fixed-length source codes, all characters are represented in binary and have the same number of symbols (bits). The Baudot code is a five-bit character code that was used primarily for low-speed teletype equipment, such as the TWX/Telex system and radio teletype (RTTY). With a five-bit code, there are only 2^5, or 32, combinations possible, which is insufficient to represent the 26 letters of the alphabet, 10 digits, and the various punctuation marks and control characters. Therefore, the Baudot code uses two special control characters called *figure shift* (FS) and *letter shift* (LS) to expand its capabilities to 58 characters. Characters that follow a letter shift are interpreted from the letter column, and characters that follow the

Table 13-2 Baudot Code

Letter	Figure	Bit 4	3	2	1	0
A	—	1	1	0	0	0
B	?	1	0	0	1	1
C	:	0	1	1	1	0
D	$	1	0	0	1	0
E	3	1	0	0	0	0
F	!	1	0	1	1	0
G	&	0	1	0	1	1
H	#	0	0	1	0	1
I	8	0	1	1	0	0
J	'	1	1	0	1	0
K	(	1	1	1	1	0
L	)	0	1	0	0	1
M	.	0	0	1	1	1
N	,	0	0	1	1	0
O	9	0	0	0	1	1
P	0	0	1	1	0	1
Q	1	1	1	1	0	1
R	4	0	1	0	1	0
S	bel	1	0	1	0	0
T	5	0	0	0	0	1
U	7	1	1	1	0	0
V	;	0	1	1	1	1
W	2	1	1	0	0	1
X	/	1	0	1	1	1
Y	6	1	0	1	0	1
Z	"	1	0	0	0	1
Figure shift		1	1	1	1	1
Letter shift		1	1	0	1	1
Space		0	0	1	0	0
Line feed (LF)		0	1	0	0	0
Blank (null)		0	0	0	0	0

figure shift are interpreted from the figure column. For example, the letter shift followed by 1 0 0 0 0 represents the letter E, whereas the figure shift followed by 1 0 0 0 0 represents the digit 3. Each time you switch from the letter column to the figure column and vice versa, you must insert either a letter shift or a figure shift character.

In 1910, Howard Krum introduced a method to indicate the start and end of a character. Incorporating Krum's start/stop technology with the Baudot code led to the development of machines that could read holes punched in a paper tap, convert the readings to electrical signals, and transmit the electrical signals over a wire to a distant machine that could duplicate the original paper tape. The U.S. Postal Service adopted the technology and began the first teleprinter service.

The latest version of the Baudot code is recommended by the CCITT as the International Alphabet No. 2 and is still used worldwide by the international Telex network. Today, the only widespread application for the Baudot code is with *Telecommunications Display Devices* (TDD) used by the deaf (sometimes called *Teletype for the Deaf* [TTD]). The most recent version of the Baudot code is shown in Table 13-2.

13-2-3 ASCII Code

In 1963, in an effort to standardize data communications codes, the United States adopted the Bell System model 33 teletype code as the *United States of America Standard Code for Information Exchange* (USASCII), better known simply as ASCII-63. ASCII is an extension of an earlier code known as the *six-bit trans code,* which was used extensively from 1910 through the mid-1950s. Since its adoption, ASCII (pronounced *as-key*) has progressed through the 1965, 1967, and 1977 versions, with the 1977 version being recommended by the ITU as the International Alphabet No. 5, in the United States as ANSI Standard X3.4–1986 (R1997), and by the International Organization for Standardization as ISO-14962 (1997).

ASCII is the standard character set for source-coding the alphanumeric character set that human beings understand but computers do not (computers understand only 1s and 0s). ASCII is a seven-bit fixed-length character set that can represent 2^7, or 128, possible combinations. With the ASCII code, the least-significant bit (LSB) is designated b_0, and the most-significant bit (MSB) is designated b_7 as shown here:

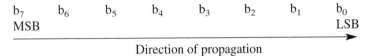

The terms *least-significant bit* and *most-significant bit* are somewhat of a misnomer because character codes do not represent weighted binary numbers, and all bits are equally significant. Bit b_7 is not part of the ASCII code but is generally reserved for the parity bit, which is explained later in this chapter. With character codes, it is more meaningful to refer to bits by their order than by their position; b_0 is the zero-order bit, b_1 the first-order bit, b_7 the seventh-order bit, and so on. With serial transmission, the bit transmitted first is generally called the LSB. With ASCII, the low-order bit (b_0) is the LSB and is transmitted first. ASCII is probably the code most often used today.

The 1977 version of the ASCII code with odd parity is shown in Table 13-3 (note that the parity bit is not included in the hex code). To convert the code shown in Table 13-3 to even parity, simply complement the parity bit (b^7). Note in Table 13-3 that characters are grouped by function—all uppercase letters together, all digits together, all control characters together, and so on. The logical placement of characters within the code provides programming as well as hardware development and testing advantages.

13-2-4 Extended ASCII Code

Soon after development of the IBM PC in 1981, computer and software engineers realized that the standard ASCII code with only 128 possibilities was insufficient. Therefore, IBM developed a new code called the *extended ASCII code.* The extended ASCII character set forbids using a parity bit. Therefore, all eight bits can be used to represent characters. Adding an eighth bit doubles the number of possible codes ($2^8 = 256$ codes). Codes 00 hex through 7F hex are backward compatible with the standard ASCII code. Codes 8F through FF are used to represent additional scientific, mathematical, graphic, and foreign characters. Several other companies, such as Microsoft Windows, created their own specialized versions of extended ASCII.

13-2-5 EBCDIC Code

The *extended binary-coded decimal interchange code* (EBCDIC) is an eight-bit fixed-length character set developed in 1962 by the International Business Machines Corporation (IBM). EBCDIC is used almost exclusively with large-scale IBM computers and peripheral equipment. With eight bits, 2^8, or 256, codes are possible, although only 139 of the 256 codes are assigned characters. Unspecified codes can be assigned to specialized characters and functions. The term *binary coded decimal* was selected because the second hex character for all letter and digit codes contains only the hex values from 0 to 9, which have the same binary sequence as BCD codes.

With the EBCDIC code, the LSB is designated b_7, and the MSB is designated b_0. Therefore, the high-order bit (b_7) is transmitted first, and the low-order bit (b_0) is transmitted last. The EBCDIC code does not facilitate the use of a parity bit. The EBCDIC code is shown in Table 13-4.

Table 13-3 ASCII-77—Odd Parity

Bit	7	6	5	4	3	2	1	0	Hex	Bit	7	6	5	4	3	2	1	0	Hex
NUL	1	0	0	0	0	0	0	0	00	@	0	1	0	0	0	0	0	0	40
SOH	0	0	0	0	0	0	0	1	01	A	1	1	0	0	0	0	0	1	41
STX	0	0	0	0	0	0	1	0	02	B	1	1	0	0	0	0	1	0	42
ETX	1	0	0	0	0	0	1	1	03	C	0	1	0	0	0	0	1	1	43
EOT	0	0	0	0	0	1	0	0	04	D	1	1	0	0	0	1	0	0	44
ENQ	1	0	0	0	0	1	0	1	05	E	0	1	0	0	0	1	0	1	45
ACK	1	0	0	0	0	1	1	0	06	F	0	1	0	0	0	1	1	0	46
BEL	0	0	0	0	0	1	1	1	07	G	1	1	0	0	0	1	1	1	47
BS	0	0	0	0	1	0	0	0	08	H	1	1	0	0	1	0	0	0	48
HT	1	0	0	0	1	0	0	1	09	I	0	1	0	0	1	0	0	1	49
NL	1	0	0	0	1	0	1	0	0A	J	0	1	0	0	1	0	1	0	4A
VT	0	0	0	0	1	0	1	1	0B	K	1	1	0	0	1	0	1	1	4B
FF	1	0	0	0	1	1	0	0	0C	L	0	1	0	0	1	1	0	0	4C
CR	0	0	0	0	1	1	0	1	0D	M	1	1	0	0	1	1	0	1	4D
SO	0	0	0	0	1	1	1	0	0E	N	1	1	0	0	1	1	1	0	4E
SI	1	0	0	0	1	1	1	1	0F	O	0	1	0	0	1	1	1	1	4F
DLE	0	0	0	1	0	0	0	0	10	P	1	1	0	1	0	0	0	0	50
DC1	0	0	0	1	0	0	0	1	11	Q	0	1	0	1	0	0	0	1	51
DC2	1	0	0	1	0	0	1	0	12	R	0	1	0	1	0	0	1	0	52
DC3	0	0	0	1	0	0	1	1	13	S	1	1	0	1	0	0	1	1	53
DC4	1	0	0	1	0	1	0	0	14	T	0	1	0	1	0	1	0	0	54
NAK	0	0	0	1	0	1	0	1	15	U	1	1	0	1	0	1	0	1	55
SYN	0	0	0	1	0	1	1	0	16	V	1	1	0	1	0	1	1	0	56
ETB	1	0	0	1	0	1	1	1	17	W	0	1	0	1	0	1	1	1	57
CAN	1	0	0	1	1	0	0	0	18	X	0	1	0	1	1	0	0	0	58
EM	0	0	0	1	1	0	0	1	19	Y	1	1	0	1	1	0	0	1	59
SUB	0	0	0	1	1	0	1	0	1A	Z	1	1	0	1	1	0	1	0	5A
ESC	1	0	0	1	1	0	1	1	1B	[	0	1	0	1	1	0	1	1	5B
FS	0	0	0	1	1	1	0	0	1C	\	1	1	0	1	1	1	0	0	5C
GS	1	0	0	1	1	1	0	1	1D	]	0	1	0	1	1	1	0	1	5D
RS	1	0	0	1	1	1	1	0	1E	∧	0	1	0	1	1	1	1	0	5E
US	0	0	0	1	1	1	1	1	1F	-	1	1	0	1	1	1	1	1	5F
SP	0	0	1	0	0	0	0	0	20	`	1	1	1	0	0	0	0	0	60
!	1	0	1	0	0	0	0	1	21	a	0	1	1	0	0	0	0	1	61
"	1	0	1	0	0	0	1	0	22	b	0	1	1	0	0	0	1	0	62
#	0	0	1	0	0	0	1	1	23	c	1	1	1	0	0	0	1	1	63
$	1	0	1	0	0	1	0	0	24	d	0	1	1	0	0	1	0	0	64
%	0	0	1	0	0	1	0	1	25	e	1	1	1	0	0	1	0	1	65
&	0	0	1	0	0	1	1	0	26	f	1	1	1	0	0	1	1	0	66
'	1	0	1	0	0	1	1	1	27	g	0	1	1	0	0	1	1	1	67
(	1	0	1	0	1	0	0	0	28	h	0	1	1	0	1	0	0	0	68
)	0	0	1	0	1	0	0	1	29	i	1	1	1	0	1	0	0	1	69
*	0	0	1	0	1	0	1	0	2A	j	1	1	1	0	1	0	1	0	6A
+	1	0	1	0	1	0	1	1	2B	k	0	1	1	0	1	0	1	1	6B
,	0	0	1	0	1	1	0	0	2C	l	1	1	1	0	1	1	0	0	6C
-	1	0	1	0	1	1	0	1	2D	m	0	1	1	0	1	1	0	1	6D
.	1	0	1	0	1	1	1	0	2E	n	0	1	1	0	1	1	1	0	6E
/	0	0	1	0	1	1	1	1	2F	o	1	1	1	0	1	1	1	1	6F
0	1	0	1	1	0	0	0	0	30	p	0	1	1	1	0	0	0	0	70
1	0	0	1	1	0	0	0	1	31	q	1	1	1	1	0	0	0	1	71
2	0	0	1	1	0	0	1	0	32	r	1	1	1	1	0	0	1	0	72
3	1	0	1	1	0	0	1	1	33	s	0	1	1	1	0	0	1	1	73
4	0	0	1	1	0	1	0	0	34	t	1	1	1	1	0	1	0	0	74
5	1	0	1	1	0	1	0	1	35	u	0	1	1	1	0	1	0	1	75
6	1	0	1	1	0	1	1	0	36	v	0	1	1	1	0	1	1	0	76
7	0	0	1	1	0	1	1	1	37	w	1	1	1	1	0	1	1	1	77
8	0	0	1	1	1	0	0	0	38	x	1	1	1	1	1	0	0	0	78

(Continued)

Table 13-3 (*Continued*)

Bit	7	6	5	4	3	2	1	0	Hex	Bit	7	6	5	4	3	2	1	0	Hex
				Binary Code										Binary Code					
9	1	0	1	1	1	0	0	1	39	y	0	1	1	1	1	0	0	1	79
:	1	0	1	1	1	0	1	0	3A	z	0	1	1	1	1	0	1	0	7A
;	0	0	1	1	1	0	1	1	3B	{	1	1	1	1	1	0	1	1	7B
<	1	0	1	1	1	1	0	0	3C	|	0	1	1	1	1	1	0	0	7C
=	0	0	1	1	1	1	0	1	3D	}	1	1	1	1	1	1	0	1	7D
>	0	0	1	1	1	1	1	0	3E	~	1	1	1	1	1	1	1	0	7E
?	1	0	1	1	1	1	1	1	3F	DEL	0	1	1	1	1	1	1	1	7F

NUL = null
SOH = start of heading
STX = start of text
ETX = end of text
EOT = end of transmission
ENQ = enquiry
ACK = acknowledge
BEL = bell
BS = back space
HT = horizontal tab
NL = new line

VT = vertical tab
FF = form feed
CR = carriage return
SO = shift-out
SI = shift-in
DLE = data link escape
DC1 = device control 1
DC2 = device control 2
DC3 = device control 3
DC4 = device control 4
NAK = negative acknowledge

SYN = synchronous
ETB = end of transmission block
CAN = cancel
SUB = substitute
ESC = escape
FS = field separator
GS = group separator
RS = record separator
US = unit separator
SP = space
DEL = delete

Table 13-4 EBCDIC Code

Bit	0	1	2	3	4	5	6	7	Hex	Bit	0	1	2	3	4	5	6	7	Hex
				Binary Code										Binary Code					
NUL	0	0	0	0	0	0	0	0	00		1	0	0	0	0	0	0	0	80
SOH	0	0	0	0	0	0	0	1	01	a	1	0	0	0	0	0	0	1	81
STX	0	0	0	0	0	0	1	0	02	b	1	0	0	0	0	0	1	0	82
ETX	0	0	0	0	0	0	1	1	03	c	1	0	0	0	0	0	1	1	83
	0	0	0	0	0	1	0	0	04	d	1	0	0	0	0	1	0	0	84
PT	0	0	0	0	0	1	0	1	05	e	1	0	0	0	0	1	0	1	85
	0	0	0	0	0	1	1	0	06	f	1	0	0	0	0	1	1	0	86
	0	0	0	0	0	1	1	1	07	g	1	0	0	0	0	1	1	1	87
	0	0	0	0	1	0	0	0	08	h	1	0	0	0	1	0	0	0	88
	0	0	0	0	1	0	0	1	09	i	1	0	0	0	1	0	0	1	89
	0	0	0	0	1	0	1	0	0A		1	0	0	0	1	0	1	0	8A
	0	0	0	0	1	0	1	1	0B		1	0	0	0	1	0	1	1	8B
FF	0	0	0	0	1	1	0	0	0C		1	0	0	0	1	1	0	0	8C
	0	0	0	0	1	1	0	1	0D		1	0	0	0	1	1	0	1	8D
	0	0	0	0	1	1	1	0	0E		1	0	0	0	1	1	1	0	8E
	0	0	0	0	1	1	1	1	0F		1	0	0	0	1	1	1	1	8F
DLE	0	0	0	1	0	0	0	0	10		1	0	0	1	0	0	0	0	90
SBA	0	0	0	1	0	0	0	1	11	j	1	0	0	1	0	0	0	1	91
EUA	0	0	0	1	0	0	1	0	12	k	1	0	0	1	0	0	1	0	92
IC	0	0	0	1	0	0	1	1	13	l	1	0	0	1	0	0	1	1	93
	0	0	0	1	0	1	0	0	14	m	1	0	0	1	0	1	0	0	94
NL	0	0	0	1	0	1	0	1	15	n	1	0	0	1	0	1	0	1	95
	0	0	0	1	0	1	1	0	16	o	1	0	0	1	0	1	1	0	96
	0	0	0	1	0	1	1	1	17	p	1	0	0	1	0	1	1	1	97
	0	0	0	1	1	0	0	0	18	q	1	0	0	1	1	0	0	0	98
EM	0	0	0	1	1	0	0	1	19	r	1	0	0	1	1	0	0	1	99
	0	0	0	1	1	0	1	0	1A		1	0	0	1	1	0	1	0	9A
	0	0	0	1	1	0	1	1	1B		1	0	0	1	1	0	1	1	9B
DUP	0	0	0	1	1	1	0	0	1C		1	0	0	1	1	1	0	0	9C
SF	0	0	0	1	1	1	0	1	1D		1	0	0	1	1	1	0	1	9D
FM	0	0	0	1	1	1	1	0	1E		1	0	0	1	1	1	1	0	9E

(*Continued*)

Table 13-4 (Continued)

Bit	0	1	2	3	4	5	6	7	Hex	Bit	0	1	2	3	4	5	6	7	Hex
					Binary Code										Binary Code				
ITB	0	0	0	1	1	1	1	1	1F		1	0	0	1	1	1	1	1	9F
	0	0	1	0	0	0	0	0	20		1	0	1	0	0	0	0	0	A0
	0	0	1	0	0	0	0	1	21	~	1	0	1	0	0	0	0	1	A1
	0	0	1	0	0	0	1	0	22	s	1	0	1	0	0	0	1	0	A2
	0	0	1	0	0	0	1	1	23	t	1	0	1	0	0	0	1	1	A3
	0	0	1	0	0	1	0	0	24	u	1	0	1	0	0	1	0	0	A4
	0	0	1	0	0	1	0	1	25	v	1	0	1	0	0	1	0	1	A5
ETB	0	0	1	0	0	1	1	0	26	w	1	0	1	0	0	1	1	0	A6
ESC	0	0	1	0	0	1	1	1	27	x	1	0	1	0	0	1	1	1	A7
	0	0	1	0	1	0	0	0	28	y	1	0	1	0	1	0	0	0	A8
	0	0	1	0	1	0	0	1	29	z	1	0	1	0	1	0	0	1	A9
	0	0	1	0	1	0	1	0	2A		1	0	1	0	1	0	1	0	AA
	0	0	1	0	1	0	1	1	2B		1	0	1	0	1	0	1	1	AB
	0	0	1	0	1	1	0	0	2C		1	0	1	0	1	1	0	0	AC
ENQ	0	0	1	0	1	1	0	1	2D		1	0	1	0	1	1	0	1	AD
	0	0	1	0	1	1	1	0	2E		1	0	1	0	1	1	1	0	AE
	0	0	1	0	1	1	1	1	2F		1	0	1	0	1	1	1	1	AF
	0	0	1	1	0	0	0	0	30		1	0	1	1	0	0	0	0	B0
	0	0	1	1	0	0	0	1	31		1	0	1	1	0	0	0	1	B1
SYN	0	0	1	1	0	0	1	0	32		1	0	1	1	0	0	1	0	B2
	0	0	1	1	0	0	1	1	33		1	0	1	1	0	0	1	1	B3
	0	0	1	1	0	1	0	0	34		1	0	1	1	0	1	0	0	B4
	0	0	1	1	0	1	0	1	35		1	0	1	1	0	1	0	1	B5
	0	0	1	1	0	1	1	0	36		1	0	1	1	0	1	1	0	B6
BOT	0	0	1	1	0	1	1	1	37		1	0	1	1	0	1	1	1	B7
	0	0	1	1	1	0	0	0	38		1	0	1	1	1	0	0	0	B8
	0	0	1	1	1	0	0	1	39		1	0	1	1	1	0	0	1	B9
	0	0	1	1	1	0	1	0	3A		1	0	1	1	1	0	1	0	BA
	0	0	1	1	1	0	1	1	3B		1	0	1	1	1	0	1	1	BB
RA	0	0	1	1	1	1	0	0	3C		1	0	1	1	1	1	0	0	BC
NAK	0	0	1	1	1	1	0	1	3D		1	0	1	1	1	1	0	1	BD
	0	0	1	1	1	1	1	0	3E		1	0	1	1	1	1	1	0	BE
SUB	0	0	1	1	1	1	1	1	3F		1	0	1	1	1	1	1	1	BF
SP	0	1	0	0	0	0	0	0	40	{	1	1	0	0	0	0	0	0	C0
	0	1	0	0	0	0	0	1	41	A	1	1	0	0	0	0	0	1	C1
	0	1	0	0	0	0	1	0	42	B	1	1	0	0	0	0	1	0	C2
	0	1	0	0	0	0	1	1	43	C	1	1	0	0	0	0	1	1	C3
	0	1	0	0	0	1	0	0	44	D	1	1	0	0	0	1	0	0	C4
	0	1	0	0	0	1	0	1	45	E	1	1	0	0	0	1	0	1	C5
	0	1	0	0	0	1	1	0	46	F	1	1	0	0	0	1	1	0	C6
	0	1	0	0	0	1	1	1	47	G	1	1	0	0	0	1	1	1	C7
	0	1	0	0	1	0	0	0	48	H	1	1	0	0	1	0	0	0	C8
	0	1	0	0	1	0	0	1	49	I	1	1	0	0	1	0	0	1	C9
¢	0	1	0	0	1	0	1	0	4A		1	1	0	0	1	0	1	0	CA
.	0	1	0	0	1	0	1	1	4B		1	1	0	0	1	0	1	1	CB
<	0	1	0	0	1	1	0	0	4C		1	1	0	0	1	1	0	0	CC
(	0	1	0	0	1	1	0	1	4D		1	1	0	0	1	1	0	1	CD
+	0	1	0	0	1	1	1	0	4E		1	1	0	0	1	1	1	0	CE
		0	1	0	0	1	1	1	1	4F		1	1	0	0	1	1	1	1	CF
&	0	1	0	1	0	0	0	0	50	}	1	1	0	1	0	0	0	0	D0
	0	1	0	1	0	0	0	1	51	J	1	1	0	1	0	0	0	1	D1
	0	1	0	1	0	0	1	0	52	K	1	1	0	1	0	0	1	0	D2
	0	1	0	1	0	0	1	1	53	L	1	1	0	1	0	0	1	1	D3
	0	1	0	1	0	1	0	0	54	M	1	1	0	1	0	1	0	0	D4
	0	1	0	1	0	1	0	1	55	N	1	1	0	1	0	1	0	1	D5
	0	1	0	1	0	1	1	0	56	O	1	1	0	1	0	1	1	0	D6

(Continued)

Table 13-4 (*Continued*)

Bit	0	1	2	3	4	5	6	7	Hex	Bit	0	1	2	3	4	5	6	7	Hex
	0	1	0	1	0	1	1	1	57	P	1	1	0	1	0	1	1	1	D7
	0	1	0	1	1	0	0	0	58	Q	1	1	0	1	1	0	0	0	D8
	0	1	0	1	1	0	0	1	59	R	1	1	0	1	1	0	0	1	D9
!	0	1	0	1	1	0	1	0	5A		1	1	0	1	1	0	1	0	DA
$	0	1	0	1	1	0	1	1	5B		1	1	0	1	1	0	1	1	DB
*	0	1	0	1	1	1	0	0	5C		1	1	0	1	1	1	0	0	DC
)	0	1	0	1	1	1	0	1	5D		1	1	0	1	1	1	0	1	DD
:	0	1	0	1	1	1	1	0	5E		1	1	0	1	1	1	1	0	DE
¬	0	1	0	1	1	1	1	1	5F		1	1	0	1	1	1	1	1	DF
−	0	1	1	0	0	0	0	0	60	\	1	1	1	0	0	0	0	0	E0
/	0	1	1	0	0	0	0	1	61		1	1	1	0	0	0	0	1	E1
−	0	1	1	0	0	0	1	0	62	S	1	1	1	0	0	0	1	0	E2
	0	1	1	0	0	0	1	1	63	T	1	1	1	0	0	0	1	1	E3
	0	1	1	0	0	1	0	0	64	U	1	1	1	0	0	1	0	0	E4
	0	1	1	0	0	1	0	1	65	V	1	1	1	0	0	1	0	1	E5
	0	1	1	0	0	1	1	0	66	W	1	1	1	0	0	1	1	0	E6
	0	1	1	0	0	1	1	1	67	X	1	1	1	0	0	1	1	1	E7
	0	1	1	0	1	0	0	0	68	Y	1	1	1	0	1	0	0	0	E8
	0	1	1	0	1	0	0	1	69	Z	1	1	1	0	1	0	0	1	E9
	0	1	1	0	1	0	1	0	6A		1	1	1	0	1	0	1	0	EA
	0	1	1	0	1	0	1	1	6B		1	1	1	0	1	0	1	1	EB
%	0	1	1	0	1	1	0	0	6C		1	1	1	0	1	1	0	0	EC
	0	1	1	0	1	1	0	1	6D		1	1	1	0	1	1	0	1	ED
>	0	1	1	0	1	1	1	0	6E		1	1	1	0	1	1	1	0	EE
?	0	1	1	0	1	1	1	1	6F		1	1	1	0	1	1	1	1	EF
	0	1	1	1	0	0	0	0	70	0	1	1	1	1	0	0	0	0	F0
	0	1	1	1	0	0	0	1	71	1	1	1	1	1	0	0	0	1	F1
	0	1	1	1	0	0	1	0	72	2	1	1	1	1	0	0	1	0	F2
	0	1	1	1	0	0	1	1	73	3	1	1	1	1	0	0	1	1	F3
	0	1	1	1	0	1	0	0	74	4	1	1	1	1	0	1	0	0	F4
	0	1	1	1	0	1	0	1	75	5	1	1	1	1	0	1	0	1	F5
	0	1	1	1	0	1	1	0	76	6	1	1	1	1	0	1	1	0	F6
	0	1	1	1	0	1	1	1	77	7	1	1	1	1	0	1	1	1	F7
	0	1	1	1	1	0	0	0	78	8	1	1	1	1	1	0	0	0	F8
▲	0	1	1	1	1	0	0	1	79	9	1	1	1	1	1	0	0	1	F9
:	0	1	1	1	1	0	1	0	7A		1	1	1	1	1	0	1	0	FA
#	0	1	1	1	1	0	1	1	7B		1	1	1	1	1	0	1	1	FB
@	0	1	1	1	1	1	0	0	7C		1	1	1	1	1	1	0	0	FC
▲	0	1	1	1	1	1	0	1	7D		1	1	1	1	1	1	0	1	FD
−	0	1	1	1	1	1	1	0	7E		1	1	1	1	1	1	1	0	FE
"	0	1	1	1	1	1	1	1	7F		1	1	1	1	1	1	1	1	FF

DLE = data-link escape
DUP = duplicate
EM = end of medium
ENQ = enquiry
EOT = end of transmission
ESC = escape
ETB = end of transmission block
ETX = end of text
EUA = erase unprotected to address
FF = form feed
FM = field mark
IC = insert cursor

ITB = end of intermediate transmission block
NUL = null
PT = program tab
RA = repeat to address
SBA = set buffer address
SF = start field
SOH = start of heading
SP = space
STX = start of text
SUB = substitute
SYN = synchronous
NAK = negative acknowledge

Bar codes are those omnipresent black and white striped stickers that seem to appear on every consumer item found in every store in the United States and most of the rest of the world. Although bar codes were developed in the early 1970s, they were not used extensively until the mid-1980s. A bar code is a series of vertical black bars separated by vertical white bars (called spaces). The widths of the bars and spaces, along with their reflective abilities, represent binary 1s and 0s that identify a specific item. In addition, bar codes may contain information regarding cost, inventory management and control, security access, shipping and receiving, production counting, document and order processing, automatic billing, and many other applications. A typical bar code is shown in Figure 13-1a.

Figure 13-1b shows the layout of the fields found on a typical bar code. The start field consists of a unique sequence of bars and spaces used to identify the beginning of the data field. The data characters correspond to the bar code symbology or format used. Serial data encoded in the data character field is extracted from the card with an optical scanner. The scanner reproduces logic conditions that correspond to the difference in reflectivity of the printed bars and underlying white spaces. To read the information, simply scan over the printed bar with a smooth, uniform motion. A photodetector in the scanner senses the reflected light and converts it to electrical signals for decoding.

There are several standard bar code formats used today in industry. The format is selected on the basis of the type of data being stored, how the data is being stored, system performance, and which format is most popular with business and industry. Bar codes are generally classified as being discrete, continuous, or 2D.

> *Discrete code.* A discrete bar code has spaces or gaps between characters. Therefore, each character within the bar code is independent of every other character. Code 39 is an example of a discrete bar code.
>
> *Continuous code.* A continuous bar code does not include spaces between characters. An example of a continuous bar code is the Universal Product Code (UPC).
>
> *2D code.* A 2D (two-dimensional) bar code stores data in two dimensions instead of in conventional linear bar codes, which store data along only one axis. A 2D bar code has a larger storage capacity than one-dimensional bar codes (typically 1 kilobyte or more per data symbol).

(a)

Start field	Data characters	Check characters	Stop field

(b)

FIGURE 13-1   (a) Bar code; (b) bar code layout

Table 13-5   Code 39 Character Set

Character	Binary Code									Bars $b_8b_6b_4b_2b_0$	Spaces $b_7b_5b_3b_1$	Check Sum Value
	b_8	b_7	b_6	b_5	b_4	b_3	b_2	b_1	b_0			
0	0	0	0	1	1	0	1	0	0	00110	0100	0
1	1	0	0	1	0	0	0	0	1	10001	0100	1
2	0	0	1	1	0	0	0	0	1	01001	0100	2
3	1	0	1	1	0	0	0	0	0	11000	0100	3
4	0	0	0	1	1	0	0	0	1	00101	0100	4
5	1	0	0	1	1	0	0	0	0	10100	0100	5
6	0	0	1	1	1	0	0	0	0	01100	0100	6
7	0	0	0	1	0	0	1	0	1	00011	0100	7
8	1	0	0	1	0	0	1	0	0	10010	0100	8
9	0	0	1	1	0	0	1	0	0	01010	0100	9
A	1	0	0	0	0	1	0	0	1	10001	0010	10
B	0	0	1	0	0	1	0	0	1	01001	0010	11
C	1	0	1	0	0	1	0	0	0	11000	0010	12
D	0	0	0	0	1	1	0	0	1	00101	0010	13
E	1	0	0	0	1	1	0	0	0	10100	0010	14
F	0	0	1	0	1	1	0	0	0	01100	0010	15
G	0	0	0	0	0	1	1	0	1	00011	0010	16
H	1	0	0	0	0	1	1	0	0	10010	0010	17
I	0	0	1	0	0	1	1	0	0	01010	0010	18
J	0	0	0	0	1	1	1	0	0	00110	0010	19
K	1	0	0	0	0	0	0	1	1	10001	0001	20
L	0	0	1	0	0	0	0	1	1	01001	0001	21
M	1	0	1	0	0	0	0	1	0	11000	0001	22
N	0	0	0	0	1	0	0	1	1	00101	0001	23
O	1	0	0	0	1	0	0	1	0	10100	0001	24
P	0	0	1	0	1	0	0	1	0	01100	0001	25
Q	0	0	0	0	0	0	1	1	1	00011	0001	26
R	1	0	0	0	0	0	1	1	0	10010	0001	27
S	0	0	1	0	0	0	1	1	0	01010	0001	28
T	0	0	0	0	1	0	1	1	0	00110	0001	29
U	1	1	0	0	0	0	0	0	1	10001	1000	30
V	0	1	1	0	0	0	0	0	1	01001	1000	31
W	1	1	1	0	0	0	0	0	0	11000	1000	32
X	0	1	0	0	1	0	0	0	1	00101	1000	33
Y	1	1	0	0	1	0	0	0	0	10100	1000	34
Z	0	1	1	0	1	0	0	0	0	01100	1000	35
–	0	1	0	0	0	0	1	0	1	00011	1000	36
.	1	1	0	0	0	0	1	0	0	10010	1000	37
space	0	1	1	0	0	0	1	0	0	01010	1000	38
*	0	1	0	0	1	0	1	0	0	00110	1000	—
$	0	1	0	1	0	1	0	0	0	00000	1110	39
/	0	1	0	1	0	0	0	1	0	00000	1101	40
+	0	1	0	0	0	1	0	1	0	00000	1011	41
%	0	0	0	1	0	1	0	1	0	00000	0111	42

13-3-1   Code 39

There are numerous bar code formats available using codes that vary from numeric symbols only to full ASCII code. One of the most common codes was developed in 1974 and is called *Code 39*. Because Code 39 was one of the first of the bar codes used for postal routing, it is sometimes called the USS Code 39. Code 39 (also called *Code 3 of 9* or *3 of 9 Code*) uses an alphanumeric code similar to ASCII and is shown in Table 13-5. Code 39 consists of 36 unique codes representing the 10 digits and 26 uppercase letters. There are seven additional codes used for special characters and an exclusive start/stop character coded as an asterisk (*). Code 39 bar codes are ideally suited for making labels, such as name badges.

(a)   (b)

FIGURE 13-2 Code 39 bar code. (a) human readable; (b) normal

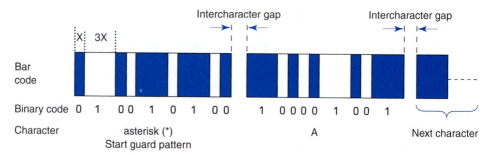

FIGURE 13-3 Code 39 bar code

Each Code 39 character contains nine vertical elements (five bars and four spaces). The logic condition (1 or 0) of each element is encoded by varying the width of the bar or space (i.e., width modulation). A wide element, whether it be a bar or a space, represents a logic 1, and a narrow element represents a logic 0. A 3:1 wide-to-narrow width ratio is generally used to distinguish logic 1s from logic 0s (i.e., a bar or space representing a logic 1 is three times wider than a bar or space representing a logic 0). In some high-resolution cases, the wide-to-narrow ratio is reduced to 2.2:1. Three of the nine elements in a Code 39 character must be logic 1s, and the rest must be logic 0s. In addition, of the three logic 1s, two must be bars and one a space. Each character begins and ends with a black bar with alternating white bars in between. Since Code 39 is a discrete code, all characters are separated with an intercharacter gap, which is usually one character wide. Figure 13-2 shows an example of an actual Code 39 bar code. The asterisks at the beginning and end of the code are start and stop characters, respectively. Figure 13-3 shows the Code 39 representation of the start/stop code (*) followed by an intercharacter gap and then the Code 39 representation of the letter A.

Code 39 bar codes can be produced with almost any size font as long as a wide-to-narrow width ratio of 3:1 is maintained. Table 13-6 lists Code 39 specifications for several common point sizes, including the width of a narrow bar and the approximate width of a character.

Table 13-6 Code 39 Specifications

Point size	Approximate Width of Narrow Bar or Space		Approximate Character Width	
24	0.42 mm	0.016 in.	6.72 mm	0.265 in.
16	0.28 mm	0.011 in.	4.48 mm	0.177 in.
12	0.21 mm	0.008 in.	3.36 mm	0.130 in.
8	0.14 mm	0.006 in.	2.24 mm	0.090 in.
6	0.105 mm	0.004 in.	1.68 mm	0.066 in.

The United States Department of Defense recommends an application of Code 39 called Logistics Applications of Automated Marking and Reading Symbols (LOGMARS) that uses an error-checking sequence that includes an optional error-checking character. Each character in the Code 39 character set, except the start/stop character (*), has a check character value assigned to it as shown in Table 13-5. To compute the check character, the value of the characters within a message are arithmetically summed and then divided by a constant. The remainder is converted to a Code 39 character that is appended to the end of the character field. With Code 39, the constant 43 was chosen because it produces the widest range of checksum possibilities without producing a remainder with a value larger than the highest possible number that can be represented with a single Code 39 character. The highest checksum value is 42, which is the % character.

Example 13-1

Compute the check characters for the Code 39 message * C O D E space 3 9 *.

Solution   The check values for all the characters except the start/stop character (*) are added together

$$C \qquad O \qquad D \qquad E \qquad space \qquad 3 \qquad 9$$
$$12 \; + \; 24 \; + \; 13 \; + \; 14 \; + \; 38 \; + \; 3 \; + \; 9 \; = \; 113$$

Dividing the sum by the constant 43 yields

$$\frac{113}{43} = 2 \text{ with a remainder of } 27$$

Therefore, the final message is

$$*C \quad O \quad D \quad E \quad space \quad 3 \quad 9 \quad R*$$

where R is the character with a check value of 27

13-3-2   Extended Code 39

The *extended Code 39* format encodes the full 128 character ASCII code set by defining specific two-character Code 39 sequences for each ASCII character that is not included in the normal Code 39 character set. However, it is not possible to encode the extended ASCII character set for hex codes between 80 and FF since they are not included in the extended Code 39 character set. The extended Code 39 character set is shown in Table 13-7.

13-3-3   Universal Product Code

The grocery industry developed the *Universal Product Code* (UPC) sometime in the early 1970s to identify their products. The *National Association of Food Chains* officially adopted the UPC code in 1974. Today, UPC codes are found on virtually every grocery item, from a candy bar to a can of beans.

There are three versions of UPC code: A, D, and E. Version A is considered the regular version as it encodes a 12-digit number. Version E uses a technique called *zero suppression* to compress a 12-digit code into a six-digit code, which allows the label to fit onto small packages. Version D uses a variable-length code and is limited to special applications such as ATM cards, check guarantee cards, and credit cards.

Figures 13–4a, b, and c show the character set, label format, and sample bit patterns for the standard Version A UPC code. Unlike Code 39, the UPC code is a continuous code since there are no intercharacter gaps. Each UPC label contains a 12-digit number. The two long bars shown in Figure 13-4b on the outermost left- and right-hand sides of the label are called the *start guard pattern* and the *stop guard pattern,* respectively. The start and stop guard patterns consist of a 101 (bar-space-bar) sequence, which is used to frame the 12-digit UPC number. The left and right halves of the label are separated by a *center guard pattern,* which consists of two *long bars* in the center of the label (they are called long bars because they are physically longer than the other bars and spaces on the label). The two long bars are separated with a space between them and spaces on both sides of the bars. Therefore, the UPC center guard pattern is 01010 as shown in Figure 13-4b. The first six digits

Table 13-7　Extended Code 39 Character Set

ASCII	Hex	Extended Code 39	ASCII	Hex	Extended Code 39	ASCII	Hex	Extended Code 39	ASCII	Hex	Extended Code 39
NUL	00	%U	SP	20	space	@	40	%V	`	60	%W
SOH	01	$A	!	21	/A	A	41	A	a	61	+A
STX	02	$B	"	22	/B	B	42	B	b	62	+B
ETX	03	$C	#	23	/C	C	43	C	c	63	+C
EOT	04	$D	$	24	/D‡	D	44	D	d	64	+D
ENQ	05	$E	%	25	/E‡	E	45	E	e	65	+E
ACK	06	$F	&	26	/F	F	46	F	f	66	+F
BEL	07	$G	'	27	/G	G	47	G	g	67	+G
BS	08	$H	(	28	/H	H	48	H	h	68	+H
HT	09	$I	)	29	/I	I	49	I	i	69	+I
LF	0A	$J	*	2A	/J‡	J	4A	J	j	6A	+J
VT	0B	$K	+	2B	/K‡	K	4B	K	k	6B	+K
FF	0C	$L	,	2C	/L	L	4C	L	l	6C	+L
CR	0D	$M	-	2D	-‡	M	4D	M	m	6D	+M
SO	0E	$N	.	2E	.‡	N	4E	N	n	6E	+N
SI	0F	$O	/	2F	/O‡	O	4F	O	o	6F	+O
DLE	10	$P	0	30	0‡	P	50	P	p	70	+P
DC1	11	$Q	1	31	1‡	Q	51	Q	q	71	+Q
DC2	12	$R	2	32		R	52	R	r	72	+R
DC3	13	$S	3	33	3‡	S	53	S	s	73	+S
DC4	14	$T	4	34	4‡	T	54	T	t	74	+T
NAK	15	$U	5	35	5‡	U	55	U	u	75	+U
SYN	16	$V	6	36	6‡	V	56	V	v	76	+V
ETB	17	$W	7	37	7‡	W	57	W	w	77	+W
CAN	18	$X	8	38	8‡	X	58	X	x	78	+X
EM	19	$Y	9	39	9‡	Y	59	Y	y	79	+Y
SUB	1A	$Z	:	3A	/Z	Z	5A	Z	z	7A	+Z
ESC	1B	%A	;	3B	%F	[	5B	%K	{	7B	%P
FS	1C	%B	<	3C	%G	\	5C	%L	|	7C	%Q
GS	1D	%C	=	3D	%H	]	5D	%M	}	7D	%R
RS	1E	%D	>	3E	%I	^	5E	%N	~	7E	%S
US	1F	%E	?	3F	%J	_	5F	%O	DEL	7F	%T‡

of the UPC code are encoded on the left half of the label (called the *left-hand characters*), and the last six digits of the UPC code are encoded on the right half (called the *right-hand characters*). Note in Figure 13-4a that there are two binary codes for each character. When a character appears in one of the first six digits of the code, it uses the left-hand code, and when a character appears in one of the last six digits, it uses the right-hand code. Note that the right-hand code is simply the complement of the left-hand code. For example, if the second and ninth digits of a 12-digit code UPC are both 4s, the digit is encoded as a 0100011 in position 2 and as a 1011100 in position 9. The UPC code for the 12-digit code 012345 543210 is

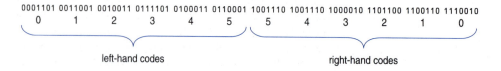

The first left-hand digit in the UPC code is called the *UPC number system character*, as it identifies how the UPC symbol is used. Table 13-8 lists the 10 UPC number system characters. For example, the UPC number system character 5 indicates that the item is intended to be used with a coupon. The other five left-hand characters are data characters. The

UPC Character Set

Left-hand character	Decimal digit	Right-hand character
0001101	0	1110010
0011001	1	1100110
0010011	2	1101100
0111101	3	1000010
0100011	4	1011100
0110001	5	1001110
0101111	6	1010000
0111011	7	1000100
0110111	8	1001000
0001011	9	1110100

(a)

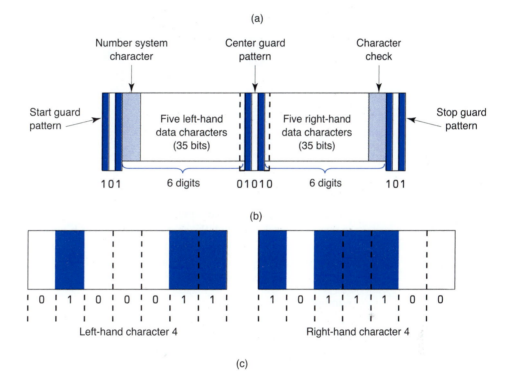

(b)

(c)

FIGURE 13-4 (a) UPC Version A character set; (b) UPC label format; (c) left- and right-hand bit sequence for the digit 4

first five right-hand characters are data characters, and the sixth right-hand character is a check character that is used for error detection. The decimal value of the number system character is always printed to the left of the UPC label, and on most UPC labels the decimal value of the check character is printed to the right of the UPC label.

With UPC codes, the width of the bars and spaces does not correspond to logic 1s and 0s. Instead, the digits 0 through 9 are encoded into a combination of two variable-width bars and two variable-width spaces that occupy the equivalent of seven bit positions. Figure 13-4c shows the variable-width code for the UPC character 4 when it is one of the first six digits of the code (i.e., left-hand bit sequence) and when it is one of the last six digits of the code (i.e., right-hand bit sequence). A single bar (one bit position) represents a logic 1, and

Table 13-8 UPC Number System Characters

Character	Intended Use
0	Regular UPC codes
1	Reserved for future use
2	Random-weight items that are symbol marked at the store
3	National Drug Code and National Health Related Items Code
4	Intended to be used without code format restrictions and with check digit protection for in-store marking of nonfood items
5	For use with coupons
6	Regular UPC codes
7	Regular UPC codes
8	Reserved for future use
9	Reserved for future use

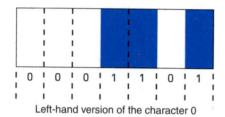

Left-hand version of the character 0

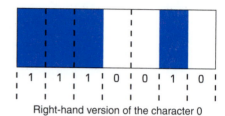

Right-hand version of the character 0

FIGURE 13-5 UPC Character 0

a single space represents a logic 0. However, close examination of the UPC character set in Table 13-8 will reveal that all UPC digits are made up of bit patterns that yield two variable-length bars and two variable-width spaces, with the bar and space widths ranging from one to three bits long. For the UPC character 4 shown in Figure 13-4c, the left-hand character is comprised of a one-bit space followed in order by a one-bit bar, a three-bit space, and a two-bit bar. The right-hand character is comprised of a one-bit bar followed in order by a one-bit space, a three-bit bar, and a two-bit space.

Example 13-2

Determine the UPC label structure for the digit 0.

Solution From Figure 13-4a, the binary sequence for the digit 0 in the left-hand character field is 0001101, and the binary sequence for the digit 0 in the right-hand character field is 1110010.

The left-hand sequence is comprised of three successive 0s, followed by two 1s, one 0, and one 1. The three successive 0s are equivalent to a space three bits long. The two 1s are equivalent to a bar two bits long. The single 0 and single 1 are equivalent to a space and a bar each one bit long.

The right-hand sequence is comprised of three 1s followed by two 0s, a 1, and a 0. The three 1s are equivalent to a bar three bits long. The two 0s are equivalent to a space two bits long. The single 1 and single 0 are equivalent to a bar and a space each one bit long each. The UPC pattern for the digit 0 is shown in Figure 13-5.

13-3-4 POSTNET

The U.S. Postal Service developed the bar code called *POSTNET* (*POST*al *N*umeric *E*ncoding *T*echnique) to increase the accuracy, decrease sorting time, and move mail through the system based on the destination's ZIP code.

POSTNET is the bar code used to encode the standard five-digit ZIP code as well as the nine-digit ZIP+4 and the 11-digit Delivery Point Bar Code (DPBC). The 11-digit DPBC code adds two additional digits to the nine-digit ZIP+4 code for a street address or

FIGURE 13-6 POSTNET bar code

Table 13-9 POSTNET Bar Code

Decimal Number	Binary code and position weighting b_4 7	b_3 4	b_2 2	b_1 1	b_0 0	POSTNET Bar Code
0	1	1	0	0	0	
1	0	0	0	1	1	
2	0	0	1	0	1	
3	0	0	1	1	0	
4	0	1	0	0	1	
5	0	1	0	1	0	
6	0	1	1	0	0	
7	1	0	0	0	1	
8	1	0	0	1	0	
9	1	0	1	0	0	

post office box number. The 11-digit DPBC code enhances automation by eliminating sorting mail by hand. Figure 13-6 shows a POSTNET nine-digit ZIP+4 bar code.

POSTNET uses a unique method of encoding binary 1s and 0s into a bar code format. With POSTNET, the logic conditions are encoded by height. Both logic 1s and 0s are represented with vertical black bars of the same width. However, the vertical bars that represent logic 1s are taller than the vertical bars representing logic 0s.

13-3-4-1 POSTNET Bar Code Format. With the POSTNET five-digit ZIP, nine-digit ZIP+4, and 11-digit DPBC bar codes, each of the 10 digits is assigned a unique five-bit (five bar) code as shown in Table 13-9. The code is a two-of-five code, as two of the five bits are always logic 1s (tall bars) and three of the five bits are always logic 0s (short bars). The values assigned the binary positions in the five-bit code do not follow standard binary weighting. The most-significant bit (b_4) has a weighting of 7, followed by weightings of 4,

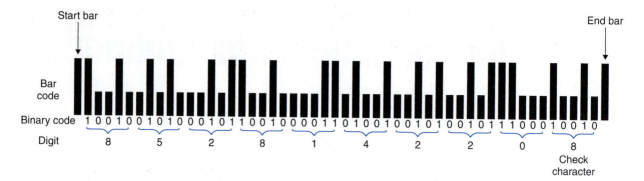

FIGURE 13-7   POSTNET bar code for the nine-digit ZIP+4 code for the ZIP code 85281–4220

2, and 1, and the least-significant bit (b_0) has a weighting of 0. All digits follow the weighting scheme except the digit 0, which is assigned the code 11000. For example, the binary code for the digits 8 is determined as follows:

$$
\begin{array}{llllll}
\text{weighting} & 7 & 4 & 2 & 1 & 0 \\
\text{bit position} & b_4 & b_3 & b_2 & b_1 & b_0 \\
\text{digit 8} & 1 & 0 & 0 & 1 & 0 \\
& 7 & + & & 1 & = \quad 8 \\
\end{array}
$$

All POSTNET bar codes begin with a tall start bar and end with a tall stop bar. The bits in the bar codes themselves are arranged in contiguous order from left to right to complete the numeric portion of the code. A special error-detection character called a check character is appended to the end of each bar code. The check character is one digit long (five bits) and, when added to the sum of the bar code digits, causes the sum of all the digits (including the check character) to equal the next-highest multiple of 10.

Example 13-3

Determine the check character for the nine-digit ZIP+4 number 85281–4220.

Solution   The sum of all the digits in the number 85281–4220 is 32. The next-highest multiple of 10 is 40, so the checksum is 8 (32 + 8 = 40). The POSTNET bar code for the number 85281–4220 is shown in Figure 13-7.

13-4 ERROR CONTROL

A data communications circuit can be as short as a few feet or as long as several thousand miles, and the transmission medium can be as simple as a pair of wires or as complex as a microwave, satellite, or optical fiber communications system. Therefore, because of the nonideal transmission characteristics associated with any communications system, it is inevitable that errors will occur, and it is necessary to develop and implement procedures for error control. Transmission errors are caused by electrical interference from natural sources, such as lightning, as well as from man-made sources, such as motors, generators, power lines, and fluorescent lights.

Data communications errors can be generally classified as *single bit, multiple bit*, or *burst*. Single-bit errors are when only one bit within a given data string is in error. Single-bit errors affect only one character within a message. A multiple-bit error is when two or more nonconsecutive bits within a given data string are in error. Multiple-bit errors can affect one or more characters within a message. A burst error is when two or more consecutive bits within a given data string are in error. Burst errors can affect one or more characters within a message.

Error performance is the rate in which errors occur, which can be described as either an expected value or an empirical value. The theoretical (mathematical) expectation of the rate at which errors will occur is called *probability of error* ($P[e]$), whereas the actual historical record of a system's error performance is called *bit error rate* (BER). For example, if a system has a $P(e)$ of 10^{-5}, this means that mathematically the system can expect to experience one bit error for every 100,000 bits transported through the system ($10^{-5} = 1/10^5$ = 1/100,000). If a system has a BER of 10^{-5}, this means that in the past there was one bit error for every 100,000 bits transported. Typically, a BER is measured and then compared with the probability of error to evaluate a system's performance.

Error control can be divided into two general categories: *error detection* and *error correction*.

13-5 ERROR DETECTION

Error detection is the process of monitoring data and determining when transmission errors have occurred. Error-detection techniques neither correct errors nor identify which bits are in error; they indicate only when an error has occurred. The purpose of error detection is not to prevent errors from occurring but to prevent undetected errors from occurring. How data communications systems react to transmission errors is system dependent and varies considerably.

The most common error-detection techniques used for data communications networks are redundancy, echoplex, exact-count coding, and redundancy checking, which includes vertical redundancy checking, checksum, longitudinal redundancy checking, and cyclic redundancy checking. Generally speaking, the higher the transmission bit rate, the more complicated the form of error detection used.

13-5-1 Redundancy

Redundancy is a form of error detection where each data unit is sent multiple times, usually twice. At the receive end, the two units are compared, and if they are the same, it is assumed that no transmission errors have occurred. When the data unit is a single character, it is called *character redundancy*, whereas if the data unit is the entire message, it is called *message redundancy*. Character redundancy is the most common form of redundancy. If the exact same character is not received twice in succession, a transmission error must have occurred. With message redundancy, if the exact same sequence of characters is not received twice in succession, in exactly the same order, a transmission error must have occurred.

Another type of redundancy used with short messages is to transmit the same message several times. At the receive end, if a given number of the messages are the same, it is assumed to be a successful transmission. For example, a message might be sent 10 times, and if seven or more of the received messages are the same, it is considered a successful transmission.

13-5-2 Echoplex

Echoplex (sometimes called *echo checking*) is a relatively simple form of error-detection scheme used almost exclusively with data communications systems involving human operators working in real time at computer terminals or PCs. With echoplex, receiving devices retransmit received data back to the transmitting device; therefore, echoplex requires full-duplex operation. Each character is transmitted immediately after it has been typed on the keyboard. At the receive end, once a character has been received, it is immediately transmitted back to the originating terminal, where it appears on that terminal's screen. When the character appears on the screen, the operator has verification that the character has been received at the destination terminal. If a transmission error occurs, the wrong character will be displayed on the transmit terminal's screen. When this happens, the operator can send a backspace and remove the erroneous character and then type and resend the correct character.

Table 13-10 ARQ Exact-Count Code

	Binary Code							Character	
Bit:	1	2	3	4	5	6	7	Letter	Figure
	0	0	0	1	1	1	0	Letter shift	
	0	1	0	0	1	1	0	Figure shift	
	0	0	1	1	0	1	0	A	–
	0	0	1	1	0	0	1	B	?
	1	0	0	1	1	0	0	C	:
	0	0	1	1	1	0	0	D	(WRU)
	0	1	1	1	0	0	0	E	3
	0	0	1	0	0	1	1	F	%
	1	1	0	0	0	0	1	G	@
	1	0	1	0	0	1	0	H	£
	1	1	1	0	0	0	0	I	8
	0	1	0	0	0	1	1	J	(bell)
	0	0	0	1	0	1	1	K	(
	1	1	0	0	0	1	0	L	)
	1	0	1	0	0	0	1	M	.
	1	0	1	0	1	0	0	N	,
	1	0	0	0	1	1	0	O	9
	1	0	0	1	0	1	0	P	0
	0	0	0	1	1	0	1	Q	1
	1	1	0	0	1	0	0	R	4
	0	1	0	1	0	1	0	S	'
	1	0	0	0	1	0	1	T	5
	0	1	1	0	0	1	0	U	7
	1	0	0	1	0	0	1	V	=
	0	1	0	0	1	0	1	W	2
	0	0	1	0	1	1	0	X	/
	0	0	1	0	1	0	1	Y	6
	0	1	1	0	0	0	1	Z	+
	0	0	0	0	1	1	1		(blank)
	1	1	0	1	0	0	0		(space)
	1	0	1	1	0	0	0		(line feed)
	1	0	0	0	0	1	1		(carriage return)

Echoplex is a simple concept requiring relatively simple circuitry. However, a disadvantage of echoplex is evident if a transmitted character has been received correctly but a transmission error occurs while it is being sent back to the originator. This would impose an unnecessary retransmission. Another disadvantage of echoplex is that it relies on human operators to detect and correct transmission errors. Echoplex also requires full-duplex operation when useful information is actually being transmitted in only one direction, and since echoplex requires all data to be repeated back to the source, it more than doubles the time necessary to transmit the information.

13-5-3 Exact-Count Encoding

With *exact-count encoding,* the number of binary 1s (and binary 0s) in each character is the same. The Code 39 bar code is an example of an exact-count code where each nine-bit code contains three logic 1s. The POSTNET bar code is another example of an exact-count code where each five-bit character includes two logic 1 bits.

Table 13-10 shows a data transmission code called the *ARQ code,* which is a seven-bit character set based on exact-count encoding. Each ARQ character contains three logic 1s and four logic 0s. Therefore, a simple count of the number of 1s (or 0s) received in each character can determine if a transmission error has occurred. The ARQ code is sometimes called the *3 of 7 code* and like the Baudot code uses letter and figure shift characters to expand the code's capabilities to 56 unique characters.

13-5-4 Redundancy Checking

Duplicating each data unit for the purpose of detecting errors, as with redundancy, is an effective but rather costly means of detecting errors, especially with long messages. It is much more efficient to add bits to data units to check for transmission errors. Adding bits for the sole purpose of detecting errors is called *redundancy checking*. There are four basic types of redundancy checks used with data communications: vertical redundancy checking (VRC), checksums (CS), longitudinal redundancy checking (LRC), and cyclic redundancy checking (CRC).

13-5-4-1 Vertical Redundancy Checking. *Vertical redundancy checking* (VRC) is probably the simplest error-detection scheme used for asynchronous data communications systems and is generally referred to as *character parity* or simply *parity*. Vertical redundancy checking received its name from ticker tapes and IBM punch cards that used a vertical orientation to encode the bits in each character. The bits were lined up from top to bottom in a vertical column. With character parity, each character has its own parity bit. Since the parity bit is not actually part of the character, it is considered a redundant bit. An *n*-character message would have *n* redundant parity bits.

With character parity, a single bit (called the *parity bit*) is added to each character to force the total number of logic 1s in the character, including the parity bit, to be either an odd number (*odd parity*) or an even number (*even parity*). For example, the ASCII code for the letter C is 43 hex or P1000011 binary, where the P bit is the parity bit. There are three logic 1s in the code, not counting the parity bit. If odd parity is used, the P bit is made a logic 0, keeping the total number of logic 1s at three, an odd number. If even parity is used, the P bit is made a logic 1, making the total number of logic 1s four, an even number.

A closer examination of parity indicates that the parity bit is independent of the number of logic 0s in the code and unaffected by pairs of logic 1s. For the letter C, if all the 0 bits are dropped, the code is P 1 − − − − 1 1. For odd parity, the P bit is still a 0, and for even parity, the P bit is still a logic 1. If pairs of 1s are also excluded, the code is either P 1 − − − − − −, P − − − − − − 1, or P − − − − − 1 −. Again, for odd parity, the P bit is a logic 0, and for even parity the P bit is a logic 1.

The definition of parity is *equivalence* or *equality*. A logic gate that determines when all its inputs are equal is the exclusive OR (XOR) gate, which is a type of equivalency gate. With XOR gates, if all inputs are equal (either all 0s or all 1s), the output is 0. If all inputs are not equal, the output is a 1. Figure 13-8 shows the logic diagrams for two circuits that are commonly used to generate a parity bit. Essentially, both circuits go through a comparison process eliminating 0s and pairs of 1s. The circuit shown in Figure 13-8a uses *sequential* (*serial*) comparisons, whereas the circuit shown in Figure 13-8b uses *combinational* (*parallel*) comparisons. With the sequential parity generator, b_0 is XORed with b_1, the result is XORed with b_2, and so on. The result of the last XOR operation is compared with a *bias bit*. If even parity is desired, the bias bit is made a logic 0. If odd parity is desired, the bias bit is made a logic 1. The output of the circuit is the parity bit, which is appended to the character code. With the parallel parity generator, comparisons are made in layers or levels. Pairs of bits (b_0 and b_1, b_2 and b_3, and so on) are XORed. The results of the first-level XOR gates are then XORed together. The process continues until only one bit is left, which is XORed with the bias bit. Again, if even parity is desired, the bias bit is made a logic 0, and if odd parity is desired, the bias bit is made a logic 1.

The circuits shown in Figure 13-8 can also be used for the parity checker in the receiver. A parity checker uses the same procedure as a parity generator except that the logic condition of the final comparison is used to determine if a parity violation has occurred (for odd parity, a logic 1 indicates an error and a logic 0 indicates no error; for even parity, a logic 1 indicates an error and a logic 0 indicates no error).

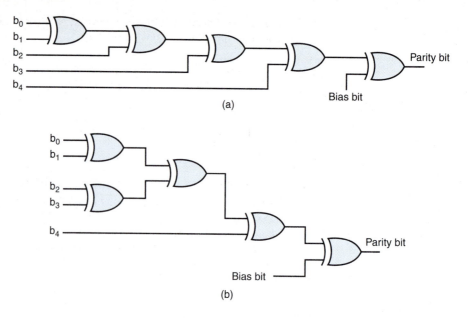

FIGURE 13-8   Parity generators: (a) serial; (b) parallel

The primary advantage of parity is its simplicity. The disadvantage is that when an even number of bits are received in error, the parity checker will not detect it because when the logic condition of an even number of bits is changed, the parity remains the same. Consequently, over a long period of time, parity will theoretically detect only 50% of the transmission errors (this assumes an equal probability that an even or an odd number of bits could be in error).

Example 13-4

Determine the odd and even parity bits for the ASCII character R.

Solution   The hex code for the ASCII character R is 52, which is P1010010 binary, where P designates the parity bit.

For odd parity, the parity bit is made a 0 because 52 hex contains three logic 1s, which is already an odd number. Therefore, the odd-parity bit sequence for the ASCII character R is 01010010.

For even parity, the parity bit is 1, making the total number of logic 1s in the eight-bit sequence four, an even number. Therefore, the even-parity bit sequence for the ASCII character R is 11010010.

Other forms of parity include *space parity* (the parity bit is always a 0), *marking parity* (the parity bit is always a 1), *no parity* (the parity bit is not sent or checked), and *ignored parity* (the parity bit is always a 0 and is ignored). Space and mark parity are capable of detecting errors only in the parity bit, which is generally about one-eighth (12.5%) of the bits transmitted. Space and marking parity are useful only when errors occur in a large number of bits. Ignored parity allows receivers that are incapable of checking parity to communicate with devices that use parity.

13-5-4-2 Checksum.   *Checksum* is another relatively simple form of redundancy error checking where the data within a message is summed together to produce an error-checking character (checksum). The checksum is appended to the end of the message. The receiver replicates the summing operation and determines its own sum and checksum character for the message. The receiver's checksum is compared to the checksum appended to the message, and if they are the same, it is assumed that no transmission errors have occurred. If the two checksums are different, a transmission error has definitely occurred.

There are five primary ways of calculating a checksum: *check character, single precision, double precision, Honeywell,* and *residue.*

Check character checksum. With a check character checksum, a decimal value is assigned to each character. The decimal values for each character of the message are added together (summed) to produce the checksum character, which is appended to the end of the message as redundant bits and transmitted. The Code 39 and POSTNET bar codes use modified forms of check characters to determine a checksum.

Single-precision checksum. Single-precision checksum is probably the most common method of calculating checksums. With single precision, the checksum is calculated by simply performing binary addition of the data within the message. However, with n-bit characters (where n equals the number of bits in each character), if the sum of the data exceeds $2^n - 1$, a carryout occurs. The carry bit is ignored, and only the n-bit checksum is appended to the message. Therefore, the checksum with single-precision addition is the LSB of the arithmetic sum of the binary data being transmitted.

Example 13-5

Determine the single-precision checksum for the following five-character ASCII message: HELLO.

Solution   From Table 13-3, the hex codes (excluding the parity bit) for the message HELLO are

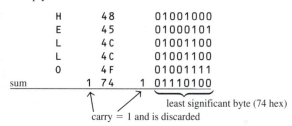

```
direction  ←─────────────────────
of transmission    H    E    L    L    0
                   48   45   4C   4C   4F
```

The checksum is simply the sum of the hex characters

```
        H      48       01001000
        E      45       01000101
        L      4C       01001100
        L      4C       01001100
        0      4F       01001111
 sum        1  74     1 01110100
```

carry = 1 and is discarded → least significant byte (74 hex)

The least significant byte of the addition process is 74 hex, which is appended to the message producing the following data stream:

```
direction  ←──────────────────────────
of transmission   48   45   4C   4C   4F   74
                  H    E    L    L    0    checksum
```

Double-precision checksum. A double-precision checksum is computed in the same manner as with single-precision except the checksum is $2n$ bits long. For example, if the data is comprised of eight-bit characters, the checksum would contain 16 bits, thereby reducing the probability of producing an erroneous checksum. If a double-precision checksum were used in Example 13-5, the checksum would have been 0174 hex, which, when appended to the message, produces the following data stream:

```
direction  ←──────────────────────────
of transmission   48   45   4C   4C   4F   01   74
                  H    E    L    L    0    checksum
```

Honeywell checksum. The Honeywell checksum is another form of double-precision checksum. The Honeywell checksum is $2n$ bits long; however, the checksum is based on interleaving consecutive data words to form double-length words. The double-length words are summed together to produce a double-precision checksum.

Example 13-6

Determine the Honeywell checksum for the following four-character ASCII message: HELP.

Solution From Table 13-3, the hex codes (excluding the parity bit) for the message HELLO are

```
        direction    ←————————————————————
    of transmission     H     E     L     P
                       48    45    4C    50
```

The checksum is the sum of groups of two hex characters

```
                4C  50
                48  45
                94  95
```

16-bit sum of the addition process is 9495 hex, which is appended to the message, producing the following data stream:

```
        direction    ←————————————————————————————
    of transmission     48    45    4C    50    95    94
                         H     E     L     P     checksum
```

Residue checksum. The residue checksum is virtually identical to the single-precision checksum except for the way the carry bit is handled. With the residue checksum, the carry bit is wrapped around and added to the LSB of the sum, adding complexity. If the single-precision checksum calculated in Example 13-5 were performed with residue checksum, the checksum would have been 75 and the transmitted message changed to

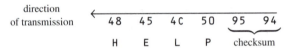
```
        direction    ←————————————————————————————————
    of transmission     48    45    4C    4C    4F    75
                         H     E     L     L     O     ＼checksum
```

13-5-4-3 Longitudinal redundancy checking. *Longitudinal redundancy checking* (LRC) is a redundancy error-detection scheme that uses parity to determine if a transmission error has occurred within a message and is therefore sometimes called *message parity*. With LRC, each bit position has a parity bit. In other words, b_0 from each character in the message is XORed with b_0 from all of the other characters in the message. Similarly, b_1, b_2, and so on are XORed with their respective bits from all the other characters in the message. Essentially, LRC is the result of XORing the "characters" that comprise the message, whereas VRC is the XORing of the bits within a single character. With LRC, even parity is generally used, whereas with VRC, odd parity is generally used.

The LRC bits are computed in the transmitter while the data is being sent and then appended to the end of the message as a redundant character. In the receiver, the LRC is recomputed from the data, and the recomputed LRC is compared to the LRC appended to the message. If the two LRC characters are the same, most likely no transmission errors have occurred. If they are different, one or more transmission errors have occurred.

Example 13-7 shows how VRC and LRC are calculated and how they can be used together.

Example 13-7

Determine the VRCs and LRC for the following ASCII-encoded message: THE CAT. Use odd parity for the VRCs and even parity for the LRC.

Solution

Character		T	H	E	sp	C	A	T	LRC
Hex		54	48	45	20	43	41	54	2F
ASCII code	b_0	0	0	1	0	1	1	0	1
	b_1	0	0	0	0	1	0	0	1
	b_2	1	0	1	0	0	0	1	1
	b_3	0	1	0	0	0	0	0	1
	b_4	1	0	0	0	0	0	1	0
	b_5	0	0	0	1	0	0	0	1
	b_6	1	1	1	0	1	1	1	0
Parity bit (VRC)	b_7	0	1	0	0	0	1	0	0

The LRC is 00101111 binary (2F hex), which is the character / in ASCII. Therefore, after the LRC character is appended to the message, it would read T H E C A T /.

In Example 13-7, the VRC bit for each character is computed in the vertical direction, and the LRC bits are computed in the horizontal direction. Therefore, LRC is sometimes called *horizontal redundancy checking* (HRC). The group of characters that comprise a message (i.e., THE CAT) is often called a *block* or *frame* of data. Therefore, the bit sequence for the LRC is often called a *block check character* (BCC) or *frame check character* or sometimes a *block check sequence* (BCS) or *frame check sequence* (FCS). BCS and FCS are probably more appropriate terms because the LRC has no function as a character (i.e., it does not represent an alphanumeric, graphic control, or data-link control character); the LRC is simply a sequence of bits used for error detection.

Historically, LRC detects between 95% and 98% of all transmission errors. LRC will not detect transmission errors when an even number of characters has an error in the same bit position. For example, if b_4 in an even number of characters is in error, the LRC is still valid even though multiple transmission errors have occurred.

If VRC and LRC are used simultaneously, the only time an error would go undetected is when an even number of bits in an even number of characters were in error and the same bit positions in each character were in error, which is highly unlikely to happen. VRC does not identify which bit is in error in a character, and LRC does not identify which character has an error in it. However, for single-bit errors, VRC used together with LRC will identify which bit is in error. Otherwise, VRC and LRC reliably identify only that a transmission error has occurred.

13-5-4-4 Cyclic redundancy checking. Probably the most reliable redundancy checking technique for error detection is a convolutional coding scheme called *cyclic redundancy checking* (CRC). With CRC, approximately 99.999% of all transmission errors are detected. There are several popular versions of CRC, including CRC-12, CRC-ITU, CRC-16, and CRC-32. CRC-12 is a 12-bit redundancy code used for transmission of data streams comprised of six-bit characters. CRC-32 is specified as an option for some point-to-point synchronous data transmission standards. CRC-ITU is a European-standard 16-bit CRC-generating polynominal. CRC-16 (sometimes called *cyclical parity*) is generally used with eight-bit codes, such as EBCDIC and extended ASCII, or seven-bit codes using character parity (VRC).

In the United States, the most common CRC code is CRC-16. With CRC-16, 16 bits are used for the block check sequence. With CRC, the entire data stream is treated as a long continuous binary number. Because the BCC is separate from the message but transported within the same transmission, CRC is considered a *systematic code*. Cyclic block codes are often written as (n, k) cyclic codes where n = bit length of transmission and k = bit length of message. Therefore, the length of the BCC in bits is BCC = $n - k$.

A CRC-16 block check character is the remainder of a binary division process. A data message polynominal $G(x)$ is divided by a unique generator polynominal function $P(x)$, the quotient is discarded, and the remainder is truncated to 16 bits and appended to the message as a BCC. The generator polynominal must be a prime number (i.e., a number divisible by only itself and 1). CRC-16 detects all single-bit errors, all double-bit errors provided that the divisor contains at least three logic 1s, all odd number of bit errors provided that the division contains a factor 11, all error bursts of 16 bits or less, and 99.9% of error bursts greater than 16 bits long. For randomly distributed errors, it is estimated that the likelihood of CRC-16 not detecting an error is 10^{-14}, which equates to one undetected error every two years of continuous data transmission at a rate of 1.544 Mbps.

With CRC generation, the division is not accomplished with standard arithmetic division. Instead, modulo-2 division is used, where the remainder is derived from an exclusive OR (XOR) operation. In the receiver, the data stream, including the CRC code, is divided by the same generating function $P(x)$. If no transmission errors have occurred, the remainder will be zero. In the receiver, the message and CRC character pass through a block check register. After the entire message has passed through the register, its contents should be zero if the receive message contains no errors.

Mathematically, CRC can be expressed as

$$\frac{G(x)}{P(x)} = Q(x) + R(x) \qquad \text{(13-1)}$$

where $G(x)$ = message polynominal
$P(x)$ = generator polynominal
$Q(x)$ = quotient
$R(x)$ = remainder

The generator polynomial for several common CRC standards are listed here:

CRC-12 $P(x) = x^{12} + x^{11} + x^3 + x^2 + x^1 + x^0$

CRC-ITU $P(X) = x^{16} + x^{12} + x^5 + x^0$

CRC-16 $P(x) = x^{16} + x^{15} + x^2 + x^0$

CRC-32 $P(x) = x^{32} + x^{26} + x^{23} + x^{22} + x^{16} + x^{12} + x^{11} + x^{10} + x^8$
$+ x^7 + x^5 + x^4 + x^2 + x^1 + x^0$

where $x^0 = 1$

The number of bits in the CRC code is equal to the highest exponent of the generating polynomial. The exponents identify the bit positions in the generating polynomial that contain a logic 1. Therefore, for CRC-16, b_{16}, b_{15}, b_2, and b_0 are logic 1s, and all other bits are logic 0s.

Figure 13-9 shows the block diagram for a circuit that will generate a CRC-16 BCC. A CRC-generating circuit requires one shift register for each bit in the BCC. Note that there are 16 shift registers in Figure 13-9. Also note that an XOR gate is placed at the output of the shift registers for each bit position of the generating polynomial that contains a logic 1, except for x^0. The BCC is the content of the 16 registers after the entire message has passed through the CRC-generating circuit.

Example 13-8

Determine the BCS for the following data- and CRC-generating polynomials:

$$\text{Data } G(x) = x^7 + x^5 + x^4 + x^2 + x^1 + x^0$$
$$= 10110111$$
$$\text{CRC } P(x) = x^5 + x^4 + x^1 + x^0$$
$$= 110011$$

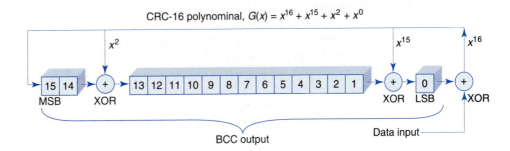

CRC-16 polynominal, $G(x) = x^{16} + x^{15} + x^2 + x^0$

FIGURE 13-9 CRC-16 generating circuit

Solution First, $G(x)$ is multiplied by the number of bits in the CRC code, which is 5.

$$x^5(x^7 + x^5 + x^4 + x^2 + x^1 + x^0) = x^{12} + x^{10} + x^9 + x^7 + x^6 + x^5 = 1011011100000$$

Then divide the result by $P(x)$:

```
                                      1 1 0 1 0 1 1 1
          1 1 0 0 1 1 | 1 0 1 1 0 1 1 1 0 0 0 0 0
                        1 1 0 0 1 1
                        1 1 1 1 1 0 1
                        1 1 0 0 1 1
                          1 1 1 0 1 0
                          1 1 0 0 1 1
                            1 0 0 1 0 0
                            1 1 0 0 1 1
                              1 0 1 1 1 0
                              1 1 0 0 1 1
                                1 1 1 0 1 0
                                1 1 0 0 1 1
                                  0 1 0 0 1  = CRC
```

The CRC is appended to the data to give the following data stream:

$$\overbrace{1\ 0\ 1\ 1\ 0\ 1\ 1\ 1}^{G(x)}\ \overbrace{0\ 1\ 0\ 0\ 1}^{CRC}$$

At the receiver, the data is again divided by $P(x)$:

```
                                      1 1 0 1 0 1 1 1
          1 1 0 0 1 1 | 1 0 1 1 0 1 1 1 0 1 0 0 1
                        1 1 0 0 1 1
                        1 1 1 1 1 0 1
                        1 1 0 0 1 1
                          1 1 1 0 1 0
                          1 1 0 0 1 1
                            1 0 0 1 1 0
                            1 1 0 0 1 1
                              1 0 1 0 1 0
                              1 1 0 0 1 1
                                1 1 0 0 1 1
                                1 1 0 0 1 1
                                0 0 0 0 0 0  Remainder = 0,
```

which means there were no transmission errors

Although detecting errors is an important aspect of data communications, determining what to do with data that contains errors is another consideration. There are two basic types of errors: *lost message* and *damaged message*. A lost message is one that never arrives at the destination or one that arrives but is damaged to the extent that it is unrecognizable. A damaged message is one that is recognized at the destination but contains one or more transmission error.

Data communications network designers have developed two basic strategies for handling transmission errors: *error-detecting codes* and *error-correcting codes*. Error-detecting codes include enough redundant information with each transmitted message to enable the receiver to determine when an error has occurred. Parity bits, block check characters, and cyclic redundancy characters are examples of error-detecting codes. Error-correcting codes include sufficient extraneous information along with each message to enable the receiver to determine when an error has occurred and which bit is in error.

Transmission errors can occur as single-bit errors or as bursts of errors, depending on the physical processes that caused them. Having errors occur in bursts is an advantage when data is transmitted in blocks or frames containing many bits. For example, if a typical block size is 10,000 bits and the system has a probability of error of 10^{-4} (one bit error in every 10,000 bits transmitted), independent bit errors would most likely produce an error in every block. However, if errors occur in bursts of 1000, only one or two blocks out of every 1000 transmitted would contain errors. The disadvantage of bursts of errors is they are more difficult to detect and even more difficult to correct than isolated single-bit errors.

In the early days of data communications, virtually all communications took place between human operators sitting in front of terminals communicating with mainframe computers. Because of the low bit rates available at the time, about the most effective means of providing error correction was a technique called *symbol substitution*. Symbol substitution was designed to be used in a human environment—when there is a human being at a terminal to analyze the received data and make decisions on its integrity. With symbol substitution, if a character is received in error, rather than revert to a higher level of error correction or display the incorrect character, a unique character that is undefined by the character code, such as a reverse question mark, is substituted for the bad character. If the operator cannot discern the flawed character, a retransmission is called for (i.e., symbol substitution is a form of *selective retransmission*). For example, if the message involved alpha characters, the operator could probably figure out what the incorrect character is. However, if the message involved numbers, the operator would probably request a retransmission.

In the modern world of data communications, there are two primary methods used for error correction: *retransmission* and *forward error correction*.

13-6-1 Retransmission

Retransmission, as the name implies, is when a receive station requests the transmit station to resend a message (or a portion of a message) when the message is received in error. Because the receive terminal automatically calls for a retransmission of the entire message, retransmission is often called ARQ, which is an old two-way radio term that means *automatic repeat request* or *automatic retransmission request*. ARQ is probably the most reliable method of error correction, although it is not always the most efficient. Impairments on transmission media often occur in bursts. If short messages are used, the likelihood that impairments will occur during transmission is small. However, short messages require more *acknowledgments* and *line turnarounds* than do long messages. Acknowledgments are when the recipient of data sends a short message back to the sender acknowledging receipt of the last transmission. The acknowledgment can indicate a successful transmission (positive acknowledgement) or an unsuccessful transmission (negative acknowledgment). Line turnarounds are when a receive station becomes the transmit station, such as when acknowledgments are sent or when retransmissions are sent in response to a negative acknowledgment. Acknowledgments and line turnarounds for error control are forms of over-head (data other than user information that must be transmitted). With long messages, less turnaround time is needed, although the likelihood that a transmission error will occur is higher than for short messages. It can be shown statistically that messages between 256 and 512 characters long are the optimum size for ARQ error correction.

There are two basic types of ARQ: discrete and continuous. *Discrete ARQ* uses *acknowledgments* to indicate the successful or unsuccessful reception of data. There are two basic types of acknowledgments: positive and negative. The destination station responds with a *positive acknowledgment* when it receives an error-free message. The destination station responds with a *negative acknowledgment* when it receives a message containing errors to call for a retransmission. If the sending station does not receive an acknowledgment after a predetermined length of time (called a *time-out*), it retransmits the message. This is called *retransmission after time-out*.

Another type of ARQ, called *continuous ARQ*, is used when messages are divided into smaller blocks or frames that are sequentially numbered and transmitted in succession without waiting for acknowledgments between blocks. Continuous ARQ allows the destination station to asynchronously request the retransmission of a specific block (or blocks) of data and still be able to reconstruct the entire message once all blocks have been successfully transported through the system. This technique is sometimes called *selective repeat*, as it can be used to call for a retransmission of an entire message or only a portion of a message.

13-6-2 Forward Error Correction

Forward error correction (FEC) is the only error-correction scheme that actually detects and corrects transmission errors when they are received without requiring a retransmission. With FEC, redundant bits are added to the message before transmission. When an error is detected, the redundant bits are used to determine which bit is in error. Correcting the bit is a simple matter of complementing it. The number of redundant bits necessary to correct errors is much greater than the number of bits needed to simply detect errors. Therefore, FEC is generally limited to one-, two-, or three-bit errors.

FEC is ideally suited for data communications systems when acknowledgments are impractical or impossible, such as when simplex transmissions are used to transmit messages to many receivers or when the transmission, acknowledgment, and retransmission time is excessive, as when communicating to faraway places, such as deep-space vehicles. The purpose of FEC codes is to eliminate the time wasted for retransmissions. However, the addition of the FEC bits to each message wastes time itself. Obviously, a trade-off is made between ARQ and FEC, and system requirements determine which method is best suited to a particular application. Probably the most popular error correction code is the Hamming code.

13-6-2-1 Hamming code.   A mathematician named Richard W. Hamming, who was an early pioneer in the development of error-detection/correction procedures, developed the *Hamming code* while working at Bell Telephone Laboratories. The Hamming code is an *error-correcting code* used for correcting transmission errors in synchronous data streams. However, the Hamming code will correct only single-bit errors. It cannot correct multiple-bit errors or burst errors, and it cannot identify errors that occur in the Hamming bits themselves. The Hamming code, as with all FEC codes, requires the addition of overhead to the message, thus increasing the length of a transmission.

Hamming bits (sometimes called *error bits*) are inserted into a character at random locations. The combination of the data bits and the Hamming bits is called the Hamming code. The only stipulation on the placement of the Hamming bits is that both the sender and the receiver must agree on where they are placed. To calculate the number of redundant Hamming bits necessary for a given character length, a relationship between the character bits and the Hamming bits must be established. As shown in Figure 13-10, a data unit contains m character bits and n Hamming bits. Therefore, the total number of bits in one data unit is

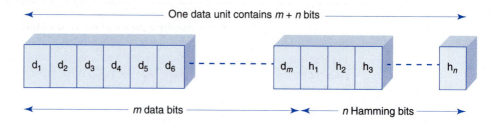

One data unit contains $m + n$ bits

m data bits n Hamming bits

FIGURE 13-10 Data unit comprised of m character bits and n Hamming bits

$m + n$. Since the Hamming bits must be able to identify which bit is in error, n Hamming bits must be able to indicate at least $m + n + 1$ different codes. Of the $m + n$ codes, one code indicates that no errors have occurred, and the remaining $m + n$ codes indicate the bit position where an error has occurred. Therefore, $m + n$ bit positions must be identified with n bits. Since n bits can produce 2^n different codes, 2^n must be equal to or greater than $m + n + 1$. Therefore, the number of Hamming bits is determined by the following expression:

$$2^n \geq m + n + 1 \qquad (13\text{-}2)$$

where n = number of Hamming bits
          m = number of bits in each data character

A seven-bit ASCII character requires four Hamming bits ($2^4 > 8 + 4 + 1$) that could be placed at the end of the character bits, at the beginning of the character bits, or interspersed throughout the character bits. Therefore, to include the Hamming bits requires transmitting 11 bits per ASCII character, which equates to a 57% increase in the message length.

The Hamming code uses parity to determine the logic condition of the Hamming bits. Each Hamming bit equates to the even-parity bit for a different combination of data bits. For example, the data unit representing ASCII character A shown in Figure 13-11a uses four Hamming (parity) bits. The seven data bits are placed in bit positions 3, 5, 6, 7, 9, 10, and 11, and the four Hamming bits are placed in bit positions 1, 2, 4, and 8 (all powers of 2) and designated n_1, n_2, n_4, and n_8, respectively. The data bits included in the calculation of each Hamming bit are shown next. Also shown in the figure are the logic conditions for the data bits for ASCII character A and the logic condition of each of the Hamming bits.

	Data bits					Hamming bit (even parity)
bit positions	3	5	7	9	11	n_1
logic conditions	1	0	0	0	1	0
bit positions	3	6	7	10	11	n_2
logic conditions	1	0	0	0	1	0
bit positions	5	6	7			n_4
logic conditions	0	0	0			0
bit positions	9	10	11			n_8
logic conditions	0	0	1			1

Figure 13-11b shows ASCII character A after the Hamming bits have been added. If an error occurs in one data bit, one or more of the Hamming bits will indicate a parity error. To determine the data bit in error, simply add the numbers of the parity bits that failed. For example, if bit 6 were received in error, the received bit sequence would be 0 0 1 0 0 *1* 0 1 0 0 1. Parity checks for n_1 and n_8 would pass, but parity checks for n_2 and n_4 would fail. To determine the bit position in error (called the *syndrome*), simply add the positions of the Hamming bits that are in error. In this case, $n_2 + n_4$ equates to $2 + 4 = 6$. Thus, bit 6 is in error.

An alternate method of determining the Hamming bits is shown in Example 13-9.

ASCII code for upper case letter A = **1000001**

Bit #	1	2	3	4	5	6	7	8	9	10	11
logic condition	n_1	n_2	**1**	n_4	**0**	**0**	**0**	n_8	**0**	**0**	**1**

Hamming bits n_1, n_2, n_4, and n_8

(a) Before determining logic condition of Hamming bits

- -

Bit #	1	2	3	4	5	6	7	8	9	10	11
logic condition	**0**	**0**	1	**0**	0	0	0	**1**	0	0	1

Hamming bits **0, 0, 0, and 1**

(b) After determining logic condition of Hamming bits

FIGURE 13-11 (a) Before determining logic condition of Hamming bits; (b) after determining logic condition of Hamming bits

Example 13-9

For a 12-bit data string of 101100010010, determine the number of Hamming bits required, arbitrarily place the Hamming bits into the data string, determine the logic condition of each Hamming bit, assume an arbitrary single-bit transmission error, and prove that the Hamming code will successfully detect the error.

Solution Substituting $m = 12$ into Equation 13-2, the number of Hamming bits is

$$\text{for } n = 4 \qquad 2^4 = 16 \geq 12 + 4 + 1 = 17$$

$16 < 17$; therefore, four Hamming bits are insufficient:

$$\text{for } n = 5 \qquad 2^5 = 32 \geq 12 + 5 + 1 = 18$$

$32 > 18$; therefore, five Hamming bits are sufficient, and a total of 17 bits make up the data stream (12 data plus 5 Hamming).

Arbitrarily place five Hamming bits into bit positions 4, 8, 9, 13, and 17:

bit position	17	16	15	14	13	12	11	10	9	8	7	6	5	4	3	2	1
	H	1	0	1	H	1	0	0	H	H	0	1	0	H	0	1	0

To determine the logic condition of the Hamming bits, express all bit positions that contain a logic 1 as a five-bit binary number and XOR them together:

Bit position	Binary number	
2	00010	
6	00110	
XOR	00100	
12	01100	
XOR	01000	
14	01110	
XOR	00110	
16	10000	
XOR	10110	= Hamming bits

$$b_{17} = 1, \qquad b_{13} = 0, \qquad b_9 = 1, \qquad b_8 = 1, \qquad b_4 = 0$$

The 17-bit Hamming code is

	H				H			H	H			H				
1	1	0	1	0	1	0	0	1	1	0	1	0	0	0	1	0

Assume that during transmission, an error occurs in bit position 14. The received data stream is

$$1 \ 1 \ 0 \ \underline{0} \ 0 \ 1 \ 0 \ 0 \ 1 \ 1 \ 0 \ 1 \ 0 \ 0 \ 0 \ 1 \ 0$$
$$\underbrace{}$$
error

At the receiver, to determine the bit position in error, extract the Hamming bits, and XOR them with the binary code for each data bit position that contains a logic 1:

Bit position	Binary number
Hamming bits	10110
2	00010
XOR	10100
6	00110
XOR	10010
12	01100
XOR	11110
16	10000
XOR	01110 = 14

Therefore, bit position 14 contains an error.

13-7 CHARACTER SYNCHRONIZATION

In essence, *synchronize* means to harmonize, coincide, or agree in time. There are several levels of synchronization necessary for successful data communications to occur, including modem (carrier) synchronization, clock synchronization, character synchronization, and message synchronization. *Modem synchronization* involves recovering the carrier reference (both frequency and phase), which is necessary for coherent demodulation. *Clock synchronization* is simply duplicating the transmit clock in the receiver. *Message synchronization* is identifying the beginning and end of the actual message, which may be embedded within a more complex data transmission. *Character synchronization* involves identifying the beginning and end of a character within a message. The purpose of this section is to describe how character synchronization is achieved. The three other forms of synchronization are described in more detail in subsequent chapters of this book.

Clock synchronization ensures that the transmitter and receiver agree on a precise time slot for the occurrence of a bit and a precise rate at which bits will be transported through the system. When a continuous string of data is received, it is also necessary to identify which bits belong to which characters and which bits are the MSBs and LSBs of the character. In essence, this is *character synchronization:* identifying the beginning and end of a character code. In data communications circuits, there are two formats commonly used to achieve character synchronization: asynchronous and synchronous.

13-7-1 Asynchronous Serial Data

The term *asynchronous* literally means "without synchronism," which in data communications terminology means "without a specific time reference." *Asynchronous serial data* is typically used with so called *dumb terminals*, which are basically inexpensive, limited-capability computer terminals attached to relatively low-speed data networks. Dumb terminals can receive data but usually cannot modify the data once it is displayed on the screen. Dumb terminals generally send data in real time. Therefore, unless the operator is Superman (or Superwoman), there is always a slight pause (sometimes called *idle time*) between characters, and the pause is not always the same length of time. Consequently, it is necessary for a receiver to resynchronize to each character. Asynchronous data transmis-

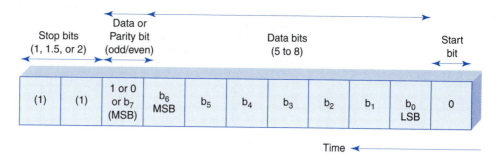

FIGURE 13-12 Asynchronous data format

sion is sometimes called *start/stop transmission* because each data character is framed between *start* and *stop bits*. The start and stop bits identify the beginning and end of the character so that the time gaps between characters do not present a problem. For asynchronously transmitted serial data, framing characters individually with start and stop bits is sometimes said to occur on a *character-by-character* basis.

Figure 13-12 shows the format used to frame a character for asynchronous serial data transmission. The first bit transmitted is the start bit, which is always a logic 0. The character bits are transmitted next beginning with the LSB and ending with the MSB. The data character can contain between five and eight bits. The parity bit (if used) is transmitted directly after the MSB of the character. The last bit transmitted is the stop bit, which is always a logic 1, and there can be either one, one and a half, or two stop bits. Therefore, a data character may be comprised of between seven and 11 bits.

A logic 0 is used for the start bit because an idle line condition (no data transmission) on a data communications circuit is identified by the transmission of continuous logic 1s (called *idle line ones*). Therefore, the start bit of a character is identified by a high-to-low transition in the received data, and the bit that immediately follows the start bit is the LSB of the character code. All stop bits are logic 1s, which guarantees a high-to-low transition at the beginning of each character. After the start bit is detected, the data and parity bits are clocked into the receiver. If data is transmitted in real time (i.e., as the operator types data into the computer terminal), the number of idle line 1s between each character will vary. During this *dead time*, the receive will simply wait for the occurrence of another start bit (i.e., high-to-low transition) before clocking in the next character. Obviously, both slipping over and slipping under produce errors. However, the errors are somewhat self-inflicted, as they occur in the receiver and are not a result of an impairment that occurred during transmission.

With asynchronous data, it is not necessary that the transmit and receive clocks be continuously synchronized; however, their frequencies should be close, and they should be synchronized at the beginning of each character. When the transmit and receive clocks are substantially different, a condition called *clock slippage* may occur. If the transmit clock is substantially lower than the receive clock, *underslipping* occurs. With underslipping, the receive clock samples the receive data faster than the bit rate. Consequently, each successive sample occurs earlier in the bit time until finally a bit is sampled twice. Slipping under is shown in Figure 13-13a. If the transmit clock is substantially higher than the receive clock, a condition called *overslipping* occurs. With overslipping, the receive clock samples the receive data slower than the bit rate. Consequently, each successive sample occurs later in the bit time until finally a bit is completely skipped. Slipping over is shown in Figure 13-13b.

For the situations illustrated in Figures 13-13a and b, the difference in the transmit and receive clocks rates were purposely exaggerated to illustrate the concept of clock slippage.

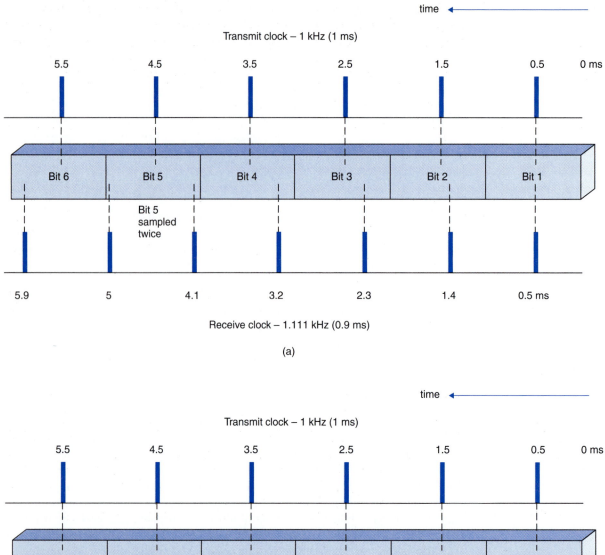

FIGURE 13-13 Clock slippage: (a) slipping under; (b) slipping over

In an actual asynchronous transmission system, it may take several hundred clock cycles before clock slippage occurs. When start and stop bits are used to frame each data character, the receiver essentially resynchronizes at the beginning of each character when the start bit is detected. Therefore, as long as the transmit and receive clock frequencies are reasonably close, the possibility of clock slippage occurring with an 11-bit character is remote.

Example 13-10

For the following sequence of bits, identify the ASCII-encoded character, the start and stop bits, and the parity bits (assume even parity and two stop bits):

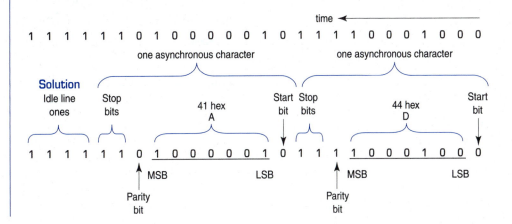

13-7-2 Synchronous Serial Data

Synchronous serial data is usually associated with *smart* or *intelligent* computer terminals, which are usually PCs that possess the capacity to modify or alter information they receive. Synchronous data generally involves transporting serial data at relatively high speeds in groups of characters called *blocks* or *frames*. Therefore, synchronous data is not sent in real time. Instead, a message is composed or formulated, then the entire message is transmitted as a single entity with no time lapses between characters. With *synchronous data*, rather than frame each character independently with start and stop bits, a unique sequence of bits, sometimes called a *synchronizing* (SYN) *character*, is transmitted at the beginning of each message. For synchronously transmitted serial data, framing characters in blocks is sometimes said to occur on a *block-by-block* basis. For example, with ASCII code, the SYN character is 16 hex, and with EBCDIC the SYN character is 32 hex. The receiver disregards incoming data until it receives one or more SYN characters. Once the synchronizing sequence is detected, the receiver clocks in the next eight bits and interprets them as the first character of the message. The receiver continues clocking in bits, interpreting them in groups of eight until it receives another unique character that signifies the end of the message. The end-of-message character varies with the type of protocol being used and what type of message it is associated with. With synchronous data, the transmit and receive clocks must be synchronized because character synchronization occurs only once at the beginning of a message.

With synchronous data, each character has two or three bits added to each character (one start and either one, one and a half, or two stop bits). These bits are additional overhead and, thus, reduce the efficiency of the transmission (i.e., the ratio of information bits to total transmitted bits). Synchronous data generally has two SYN characters (16 bits of overhead) added to each message. Therefore, asynchronous data is more efficient for short messages, and synchronous data is more efficient for long messages.

Example 13-11

For the following string of ASCII-encoded characters, identify each character (assume odd parity):

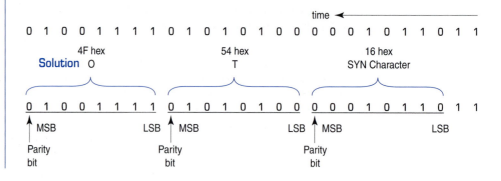

QUESTIONS

13-1. Define *data*.

13-2. What is meant by the term *overhead*?

13-3. What are some of the other names for *data communications codes*?

13-4. Explain why the *Morse code* is inadequate for modern-day data communications networks.

13-5. What is the purpose of the *figure shift* and *letter shift* in the Baudot code?

13-6. What is the difference between the *ASCII* and *extended ASCII* codes?

13-7. What is a *bar code*, and when is it commonly used?

13-8. Describe what is meant by *discrete bar code*, *continuous bar code*, and *2D bar code*.

13-9. What is the *check character* used for in a bar code?

13-10. How are logic 0s and 1s indicated with the *Code 39 bar code*?

13-11. How are logic 0s and 1s indicated with the *UPC bar code*?

13-12. How are logic 0s and 1s indicated with the *POSTNET bar code*?

13-13. Describe what is meant by *error control*.

13-14. What is the difference between *error detection* and *error correction*?

13-15. Briefly describe *redundancy error detection*.

13-16. Briefly describe *echoplex*.

13-17. How does *exact-count encoding* detect errors?

13-18. Briefly describe how *vertical redundancy checking* accomplishes error detection.

13-19. What is meant by the terms *odd parity*, *even parity*, *marking parity*, and *spacing parity*?

13-20. What is the difference between *no parity* and *ignored parity*?

13-21. Briefly describe the following error-detection schemes: *checksum*, *check character checksum*, *single-precision checksum*, *double-precision checksum*, *Honeywell checksum*, and *residue checksum*.

13-22. Give a brief description of *longitudinal redundancy checking*.

13-23. What is the difference between *character* and *message parity*?

13-24. Briefly describe *cyclic redundancy checking*.

13-25. List and describe two methods of *error correction*.

13-26. Give a brief explanation of the *Hamming code*.

13-27. What is meant by *character synchronization*?

13-28. Briefly compare and contrast *asynchronous* and *synchronous serial data formats*.

13-29. What is meant by a *start bit* and a *stop bit*?

13-30. What is a *SYN character*?

PROBLEMS

13-1. Determine the bit sequences for the letter H for the Baudot code, ASCII code, EBCDIC code, and ARQ code.

13-2. Compute the check character for the following Code 39 message: *BIG FOOT*.

13-3. Determine left- and right-hand label structure for the UPC code for the digit 7.

13-4. Determine the check character for the POSTNET nine-digit ZIP+4 number 13424–7654.

13-5. Determine the number of bits in error for a BER of 10^{-6} and a message that is 23 million bits long.

13-6. A probability of error of 10^{-5} indicates a projected error rate of 1 bit in every _____ bits transmitted.

13-7. Determine the single-precision checksum for the following five-character ASCII message: HOUSE.

13-8. Determine the double-precision checksum for the following five-character ASCII message: HORSE.

13-9. Determine the Honeywell checksum for the following four-character ASCII message: YOUR.

13-10. Determine the residue checksum for the following five-character ASCII message: TABLE.

13-11. Determine the LRC and VRC for the following message (use even parity for LRC and odd parity for VRC):

D A T A sp C O M M U N I C A T I O N S

13-12. Determine the LRC and VRC for the following message (use even parity for LRC and odd parity for VRC):

A S C I I sp C O D E

13-13. Determine the BCC for the following data- and CRC-generating polynomials:

$G(x) = x^7 + x^4 + x^2 + x^0 = 10010101$
$P(x) = x^5 + x^4 + x^1 + x^0 = 110011$

13-14. Determine the BCC for the following data- and CRC-generating polynomials:

$G(x) = x^8 + x^5 + x^2 + x^0$
$P(x) = x^5 + x^4 + x^1 + x^0$

13-15. How many Hamming bits are required for a single EBCDIC character?

13-16. How many Hamming bits are required for a 13-bit character?

13-17. How many Hamming bits are required for an 11-bit asynchronous ASCII character?

13-18. Determine the Hamming bits for the ASCII character B. Insert the Hamming bits into every other bit location starting from the left.

13-19. Determine the Hamming bits for the ASCII character C (use odd parity and two stop bits). Insert the Hamming bits into every other location starting at the right.

C H A P T E R   14

Data Communications Hardware

CHAPTER OUTLINE

OBJECTIVES

- Define the term *data communications hardware*
- List and describe the three basic elements of a data communications system
- Describe the term *data terminal equipment* and explain several examples
- Describe the term *data communications equipment* and list several examples
- List and describe the seven components that make up a two-point data communications circuit
- Explain the differences between terms *primary stations* and *secondary stations*
- Describe the terms *line control unit* and *front-end processor* and explain the differences between the two
- Describe the operation of UARTs, USRTs, and USARTs and outline the differences between them
- Describe the functions of a serial interface
- List the parameters for the RS-232, RS-449, and RS-530 interfaces
- Explain the operation of the RS-232 serial interface
- Compare and contrast the RS-232, RS-449, and RS-530 serial interfaces
- Define the terms *analog* and *digital loopback* and explain the differences between them
- Describe the X.21 Recommendation

14-1 INTRODUCTION

Data communications involves the distribution and exchange of information between people and machines. Therefore, the topic of *data communications hardware* encompasses an extremely wide range of data communications equipment, including both digital and analog devices. Although data communications hardware is utilized in all seven layers of the OSI protocol hierarchy and in virtually all types of data communications systems and networks, the intent of this chapter is to describe only hardware associated with the lowest layer of the OSI protocol hierarchy: the physical layer.

14-2 DATA COMMUNICATIONS HARDWARE

Digital information sources, such as personal computers, communicate with each other using the POTS (plain old telephone service) telephone network in a manner very similar to the way analog information sources, such as human conversations, communicate with each other using the POTS telephone network. With both digital and analog information sources, special devices are necessary to interface the sources to the telephone network.

Figure 14-1 shows a comparison between human speech (analog) communications and computer data (digital) communications using the POTS telephone network. Figure 14-1a shows how two humans communicate over the telephone network using standard analog telephone sets. The telephone sets interface human speech signals to the telephone network and vice versa. At the transmit end, the telephone set converts acoustical energy (information) to electrical energy, and at the receive end, the telephone set converts electrical energy back to acoustical energy. Figure 14-1b shows how digital data is transported over the telephone network. At the transmitting end, a telco interface converts digital data from the transceiver to analog electrical energy that is transported through the telephone network. At the receiving end, a telco interface converts the analog electrical energy received from the telephone network back to digital data.

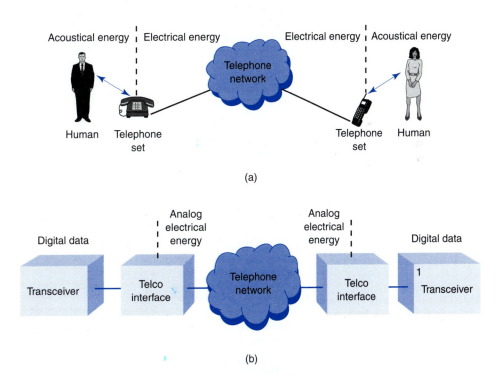

FIGURE 14-1 Telephone communications network: (a) human communications; (b) digital data communications

In simplified terms, a data communications system is comprised of three basic elements: a transmitter (source), a transmission path (data channel), and a receiver (destination). For two-way communications, the transmission path would be bidirectional and the source and destination interchangeable. Therefore, it is usually more appropriate to describe a data communications system as connecting two *endpoints* (sometimes called *nodes*) through a common communications channel. The two endpoints may not possess the same computing capabilities; however, they must be configured with the same basic components. Both endpoints must be equipped with special devices that perform unique functions, make the physical connection to the data channel, and process the data before it is transmitted and after it has been received. Although the special devices are sometimes implemented as a single unit, it is generally easier to describe them as separate entities. In essence, all endpoints must have three fundamental components: *data terminal equipment* (DTE), *data communications equipment* (DCE), and a *serial interface.*

14-2-1 Data Terminal Equipment

Data terminal equipment (DTE) can be virtually any binary digital device that generates, transmits, receives, or interprets data messages. In essence, a DTE is where information originates or terminates. DTEs are the data communications equivalent to the person in a telephone conversation. DTEs contain the hardware and software necessary to establish and control communications between endpoints in a data communications system; however, DTEs seldom communicate directly with other DTEs. Examples of DTEs include video display terminals, printers, and personal computers.

Over the past 50 years, data terminal equipment has evolved from simple on-line printers to sophisticated high-level computers. Data terminal equipment includes the concept of *terminals, clients, hosts,* and *servers.* Terminals are devices used to input, output, and display information, such as keyboards, printers, and monitors. A client is basically a modern-day terminal with enhanced computing capabilities. Hosts are high-powered, high-capacity mainframe computers that support terminals. Servers function as modern-day hosts except with lower storage capacity and less computing capability. Servers and hosts maintain local databases and programs and distribute information to clients and terminals.

Data terminal equipment is a general term that describes the digital equipment used to adapt digital signals to a format more suitable for transmission. Examples of DTEs include the following:

Teletypewriter (TTY). A relatively simple device that accepts keyboard-generated characters and commands, prints transmit and receive information, and temporarily stores information that may be forwarded to other destinations later. Teletypewriters serve as an interface between human operators and the data communications system and generally do not possess any computing or programming capabilities.

Video display terminal (VDT). A device that accepts keyboard-generated characters and commands and displays keyboard generated characters as well as received information on a video screen. The combination of a video display and a teletypewriter is sometimes called a *keyboard display* (KD).

Transactional terminal. A device that transmits, receives, and processes real-time data transactions. Examples of transactional terminals include automatic teller machines (ATMs), bar code readers, and data acquisition devices, such as thermometers, heat sensors, and intrusion alarms.

Specialized terminal. A relatively simple device that sources data from bar code readers and other types of optical scanners.

Intelligent terminal. A programmable keyboard and display device capable of performing higher-level tasks generally assigned to more sophisticated computers, such as screen formatting, word processing, text editing, database manipulating, and so on. An intelligent terminal can be a personal computer.

Smart terminal. A device that possesses sufficient memory and logic circuitry to enable it to perform limited and very specific tasks, such as facsimile machines and order entry machines.

Dumb terminal. An extremely simple device with little or no computing capacity or memory. Dumb terminals rely almost entirely on higher-level computers to perform even the most routine functions.

Workstation. A device that contains a significant amount of computing and processing power, such as a personal computer. Workstations can function on their own but generally access other computers and processors when performing their functions.

Line control unit (LCU). A device that performs virtually all the data communications–related functions at a remote location. LCUs direct the flow of data traffic between the data communications channel and local terminals. LCUs functions include performing serial-to-parallel and parallel-to-serial data conversion, formatting data, inserting and deleting data-link control characters, and performing error detection and correction.

Front-end processor (FEP). A device that performs virtually all the data communications–related functions at a host location. An FEP relieves a host computer of the relatively slow processes inherent with data communications circuits. FEPs are similar to LCUs except FEPs serve as an interface between a host computer and all the data circuits that it serves. One FEP at a host location can communicate with LCUs at hundreds of remote locations.

14-2-2 Data Communications Equipment

Data communications equipment (DCE) is a general term used to describe equipment that interfaces data terminal equipment to a transmission channel, such as a digital T1 carrier or an analog telephone circuit. The output of a DTE can be digital or analog, depending on the application. In essence, a DCE is a *signal conversion device,* as it converts signals from a DTE to a form more suitable to be transported over a transmission channel. A DCE also converts those signals back to their original form at the receive end of a circuit. DCEs are transparent devices responsible for transporting bits (1s and 0s) between DTEs through a data communications channel. The DCEs neither know nor care about the content of the data.

There are several types of DCEs, depending on the type of transmission channel used. Common DCEs are *channel service units* (CSUs), *digital service units* (DSUs), and *data modems.* CSUs and DSUs are used to interface DTEs to digital transmission channels. Data modems are used to interface DTEs to analog telephone networks. Because data communications channels are terminated at each end in a DCE, DCEs are sometimes called *data circuit-terminating equipment* (DCTE). CSUs and DSUs are described in a later chapter of this book, whereas data modems are described in subsequent sections of this chapter.

14-3 DATA COMMUNICATIONS CIRCUITS

A data modem is a DCE used to interface a DTE to an analog telephone circuit commonly called POTS. Figure 14-2a shows a simplified diagram for a two-point data communications circuit using a POTS link to interconnect the two endpoints (endpoint A and endpoint B). As shown in the figure, a two-point data communications circuit is comprised of the seven basic components:

1. DTE at endpoint A
2. DCE at endpoint A
3. DTE/DCE interface at endpoint A
4. Transmission path between endpoint A and endpoint B

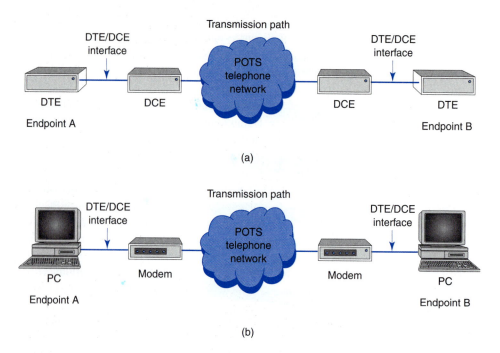

FIGURE 14-2   Two-point data communications circuit: (a) DTE/DCE representation; (b) device representation

5. DCE at endpoint B
6. DTE at endpoint B
7. DTE/DCE interface at endpoint B

The DTEs can be terminal devices, personal computers, mainframe computers, front-end processors, printers, or virtually any other pierce of digital equipment. If a digital communications channel were used, the DCE would be a CSU or DSU. However, because the communications channel is a POTS link, the DCE is a data modem.

Figure 14-2b shows the same equivalent circuit as is shown in Figure 14-2a except the DTE and DCE have been replaced with the actual devices they represent—the DTE is a PC and the DCE is a modem. In most modern-day personal computers for home use, the modem is simply a card installed inside the computer.

Figure 14-3 shows the block diagram for a centralized multipoint data communications circuit using several POTS data communications links to interconnect three endpoints. The circuit is arranged in a bus topology with central control provided by a mainframe computer (host) at endpoint A. The host station is sometimes called the *primary station*. Endpoints B and C are called *secondary stations*. The primary station is responsible for establishing and maintaining the data link and for ensuring an orderly flow of data between itself and each of the secondary stations. Data flow is controlled by an applications program stored in the mainframe computer at the primary station.

At the primary station, there is a mainframe computer, a front-end processor (DTE), and a data modem (DCE). At each secondary station, there is a modem (DCE), a line control unit (DTE), and a *cluster* of terminal devices (PCs, printers, and so on). The line control unit at the secondary stations is referred to as a *cluster controller,* as it controls data flow between several terminal devices and the data communications channel. Line control units at secondary stations are sometimes called *station controllers* (STACOs), as they control data flow to and from all the data communications equipment located at that station.

For simplicity, Figure 14-3 only shows one data circuit served by the mainframe computer at the primary station. However, there can be dozens of different circuits served

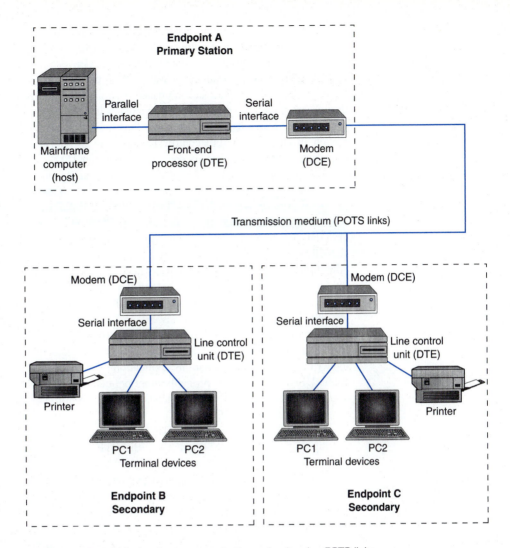

FIGURE 14-3 Multipoint data communications circuit using POTS links

by one mainframe computer. Therefore, the primary station line control unit (i.e., the front-end processor) must have enhanced capabilities for storing, processing, and retransmitting the data it receives from all secondary stations on all the circuits it serves. The primary station stores software for database management of all the circuits it serves. Obviously, the duties performed by the front-end processor at the primary station are much more involved than the duties performed by the line control units at the secondary stations. The FEP directs data traffic to and from many different circuits, which could all have different parameters (i.e., different bit rates, character codes, data formats, protocols, and so on). The LCU at the secondary stations directs data traffic between one data communications link and a relative few terminal devices, which all transmit and receive data at the same speed and use the same data-link protocol, character code, data format, and so on.

14-4   LINE CONTROL UNIT

As previously stated, a line control unit (LCU) is a DTE and DTEs have several important functions. At the primary station, the LCU is often called a FEP because it processes information and serves as an interface between the host computer and all the data communications circuits it serves. Each circuit served is connected to a different port on the FEP. The FEP directs the flow of input and output data between data communications circuits and their respective applications programs. The data interface between the mainframe computer and the FEP transfers data in parallel at relatively high bit rates. However, data transfers between the modem and the FEP are accomplished in serial and at a much lower bit rate. The FEP at the primary station and the LCU at the secondary stations perform parallel-to-serial and serial-to-parallel conversions. They also house the circuitry that performs error detection and correction. In addition, data-link control characters are inserted and deleted in the FEP and LCUs (data-link control characters are described in Chapter 15).

Within the FEP and LCUs, a single special-purpose integrated circuit performs many of the fundamental data communications functions. This integrated circuit is called a *universal asynchronous receiver/transmitter* (UART) if it is designed for asynchronous data transmission, a *universal synchronous receiver/transmitter* (USRT) if it is designed for synchronous data transmission, and a *universal synchronous/asynchronous receiver/transmitter* (USART) if it is designed for either asynchronous or synchronous data transmission. Although technically a UART (pronounced *u-art*), a USRT (pronounced *u-sart*), and a USART (also pronounced *u-sart*) are not the same and do not offer the same features, the acronyms are often used interchangeably to describe peripheral devices that perform similar data communication's functions. All three types of circuits specify general-purpose integrated-circuit chips located in an LCU or FEP that allow DTEs to interface with DCEs. For illustration purposes, UARTs and USRTs are described separately. However, in modern-day data communications applications, the two circuits are often combined into a single USART chip that is probably more popular today simply because it can be adapted to either asynchronous or synchronous data transmission. USARTs are available in 24- to 64-pin dual in-line packages (DIPs).

UARTS, USRTS, and USARTS are devices that operate external to the central processing unit (CPU) in a DTE that allow the DTE to communicate serially with other data communications equipment, such as DCEs. They are also essential data communications components in terminals, workstations, personal computers, and many other types of serial data communications devices. In most modern computers, USARTs are normally included on the motherboard and connected directly to the serial port. UARTs, USRTs, and USARTs designed to interface to specific microprocessors often have unique manufacturer-specific names. For example, Motorola manufactures a special purpose UART chip called an *asynchronous communications interface adapter* (ACIA).

The following descriptions are intended to describe the general operation of UARTs and USRTs and do not necessarily explain the operation of any particular device manufactured by a specific company.

14-4-1 Universal Asynchronous Receiver/Transmitter

A *universal asynchronous receiver/transmitter* (UART) is used for asynchronous transmission of serial data between a DTE and a DCE. Asynchronous data transmission means that an asynchronous data format is used and there is no clocking information transferred between the DTE and the DCE. The primary functions performed by a UART are the following:

1. Serial-to-parallel data conversion in the transmitter and parallel-to-serial data conversion in the receiver
2. Error detection by inserting parity bits in the transmitter and checking parity bits in the receiver
3. Inserting start and stop bits in the transmitter and detect and remove start and stop bits in the receiver
4. Formatting of data in the transmitter and receiver (i.e., combining items 1 through 3 in a meaningful sequence)
5. Providing transmit and receive status information to the CPU

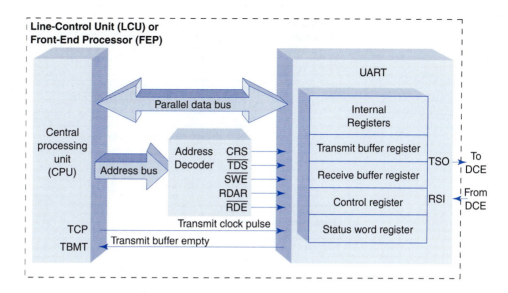

FIGURE 14-4 Line control unit UART interface

6. Allowing voltage-level conversion between the DTE and the serial interface and vice versa
7. Providing a means of achieving bit and character synchronization

Transmit and receive functions can be performed by a UART simultaneously because the transmitter and receiver have separate control signals and clock signals and share a bidirectional data bus, which allows them to operate virtually independently of one another. In addition, input and output data are double buffered, which allows for continuous data transmission and reception.

Figure 14-4 shows a simplified block diagram of a line control unit showing the relationship between the UART and the CPU that controls the operation of the UART. The CPU coordinates data transfer between the line control unit (or FEP) and the modem. The CPU is responsible for programming the UART's control register, reading the UART's status register, transferring parallel data to and from the UART transmit and receive buffer registers, providing clocking information to the UART, and facilitating the transfer of serial data between the UART and the modem.

A UART can be divided into two functional sections: the transmitter and the receiver. Figure 14-5 shows a simplified block diagram of a UART. Before transferring data in either direction, an eight-bit control word must be programmed into the UART control register to specify the nature of the data. The control word specifies the number of data bits per character; whether a parity bit is included with each character and, if so, whether it is odd or even parity; the number of stop bits inserted at the end of each character; and the receive clock frequency relative to the transmit clock frequency. Essentially, the start bit is the only bit in the UART that is not optional or programmable, as there is always one start bit, and it is always a logic 0. Table 14-1 shows the control-register coding format for a typical UART.

As specified in Table 14-1, the parity bit is optional and, if used, can be either odd or even. To select parity, NPB is cleared (logic 0), and to exclude the parity bit, NBP is set (logic 1). Odd parity is selected by clearing POE (logic 0), and even parity is selected by setting POE (logic 1). The number of stop bits is established with the NSB1 and NSB2 bits and can be one, one and a half, or two. The character length is determined by NDB1 and NDB2 and can be five, six, seven, or eight bits long. The maximum character length

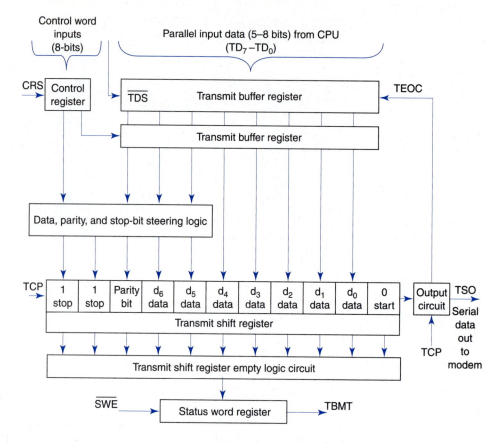

FIGURE 14-5 UART transmitter block diagram

Table 14-1 UART Control Register Inputs

D_7 and D_6

Number of stop bits

NSB1	NSB2	No. of Bits
0	0	Invalid
0	1	1
1	0	1.5
1	1	2

D_5 and D_4

NPB (parity or no parity)

1	No parity bit (RPE disabled in receiver)
0	Insert parity bits in transmitter and check parity bits in receiver

POE (parity odd or even)

1	Even parity
0	Odd parity

D_3 and D_2

Character length

NDB1	NDB2	Bits per Word
0	0	5
0	1	6
1	0	7
1	1	8

D_1 and D_0

Receive clock (baud rate factor)

RC1	RC2	Clock Rate
0	0	Synchronous mode
0	1	1X
1	0	16X
1	1	32X

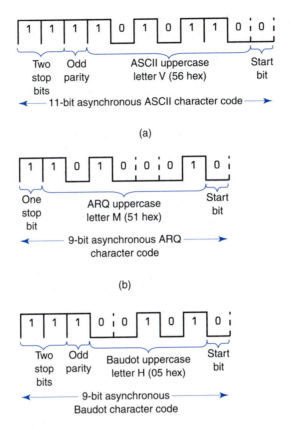

(a)

(b)

(c)

FIGURE 14-6 Asynchronous characters: (a) ASCII character; (b) ARQ character; (c) Baudot character

is 11 bits (i.e., one start bit, eight data bits, and two stop bits or one start bit, seven data bits, one parity bit, and two stop bits). Using an 11-bit character format with ASCII encoding is sometimes called *full ASCII*.

Figure 14-6 shows three of the character formats possible with a UART. Figure 14-6a shows an 11-bit data character comprised of one start bit, seven ASCII data bits, one odd-parity bit, and two stop bits (i.e., full ASCII). Figure 14-6b shows a nine-bit data character comprised of one start bit, seven ARQ data bits, and one stop bit, and Figure 14-6c shows another nine-bit data character comprised of one start bit, five Baudot data bits, one odd-parity bit, and two stop bits.

14-4-1-1 UART control register.   Table 14-1 shows the programming options for a typical UART control register (sometimes called the *command register*). Because there is only one control register, it determines the data characteristics for both the UART transmitter and the UART receiver, which must be the same. The control word is sometimes called the *mode instruction word*. A control word format for a typical UART is shown here:

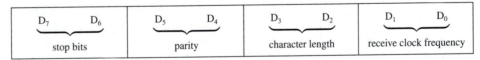

D_7 D_6	D_5 D_4	D_3 D_2	D_1 D_0
stop bits	parity	character length	receive clock frequency

Bits D_7 and D_6 specify the number of stop bits inserted in the transmitter and how many stop bits the receiver expects to see at the end of each received character. D_5 and D_4 specify whether a parity bit is inserted at the end of each character in the transmitter and the type of parity used (i.e., odd or even). D_5 and D_4 also specify the type of parity the receiver will check for. D_3 and D_2 determine the number of data bits included in each transmit and receive character. D_1 and D_0 specify the clock rate (sometimes called the *baud rate factor*) for the UART receiver in respect to the transmit clock rate. For example, $1\times$ (times one) specifies that the receive clock rate is the same as the transmit clock rate, $16\times$ (times 16) specifies that the receive clock rate is 16 times faster than the transmit clock rate, and $32\times$ (times 32) specifies 32 times faster. However, this parameter is often misunderstood, as the receive clock rate is not the actual receive data rate, as will be seen later.

Example 14-1

Determine the mode instruction word for the following parameters:

odd parity	seven-bit characters
two stop bits	$16\times$ receive clock rate

Solution   From the control register specifications listed in Table 14-1,

D_7	D_6	D_5	D_4	D_3	D_2	D_1	D_0
1	1	0	0	1	0	1	0
stop bits (2)		parity (odd)		character length (7-bits)		receive clock frequency (16X)	

The mode instruction word for Example 14–1 is CA hex. The CPU places a control word on the control register input lines and then clocks them into the UART with the *control register strobe* (CRS), which may be active high or active low, depending on the UART.

14-4-1-2 Status word register.   The *status word register* is an *n*-bit data register inside the UART that keeps track of the status of the UART's transmit and receive buffer registers. The number of bits in the status register is determined by the complexity of the UART. The CPU reads the status word register by activating the $\overline{\text{SWE}}$ lead. Typical status conditions compiled by the status word register are shown here and include the following status conditions:

TBMT   Transmit buffer empty
RPE   Receive parity error
RFE   Receive framing error
RDA   Receive data available
ROR   Receiver overrun
DSR   Data set ready

The significance of the status word is described in subsequent sections of this chapter. The bit assignments for a typical UART status word are shown here:

D_7	D_6	D_5	D_4	D_3	D_2	D_1	D_0
DSR	RFE	ROR	RPE	TBMT	RDA	not used	not used

Example 14-2

Determine the status word for a UART with the following parameters:

active high-status signals	overrun error
DSR active	no parity error
no framing error	receive data available
unused bits = 0	transmit buffer empty active

Solution

DSR active	$D_7 = 1$
no framing error	$D_6 = 0$

overrun error $D_5 = 1$
no parity error $D_4 = 0$
receive data available $D_3 = 1$
TMBT $D_2 = 1$
unused bits = 0 D_1 and $D_0 = 0$
Status word 10101100 or AC hex

14-4-1-3 UART transmitter. The operation of the typical UART transmitter shown in Figure 14-5 is quite logical. However, before the UART can send or receive data, the UART control register must be loaded with the desired mode instruction word. This is accomplished by the CPU in the DTE, which applies the mode instruction word to the control word bus and then activates the control register strobe (CRS).

Figure 14-7 shows the signaling sequence that occurs between the CPU and the UART transmitter. On receipt of an active *status word enable* ($\overline{SWE}$) signal, the UART sends a *transmit buffer empty* (TBMT) signal from the status word register to the CPU to indicate that the transmit buffer register is empty and the UART is ready to receive more data. When the CPU senses an active condition of TBMT, it applies a parallel data character to the transmit data lines (TD_7 through TD_0) and strobes them into the transmit buffer register with an active signal on the *transmit data strobe* signal ($\overline{TDS}$). The contents of the transmit buffer register are transferred to the transmit shift register when the *transmit end-of-character* (TEOC) signal goes active (the TEOC signal is internal to the UART and simply tells the transmit buffer register when the transmit shift register is empty and available to receive data). The data passes through the steering logic circuit, where it picks up the appropriate start, stop, and parity bits. After data has been loaded into the transmit shift register, it is serially outputted on the *transmit serial output* (TSO) pin at a bit rate equal to the transmit clock (TCP) frequency. While the data in the transmit shift register is serially clocked out of the UART, the CPU applies the next character to the input of the transmit buffer register. The process repeats until the CPU has transferred all its data.

14-4-1-4 UART receiver. A simplified block diagram for a UART receiver is shown in Figure 14-8. The number of stop bits and data bits and the parity bit parameters specified for the UART receiver must be the same as those of the UART transmitter. The UART receiver ignores the reception of idle line 1s. When a valid start bit is detected by the start bit verification circuit, the data character is clocked into the receive shift register.

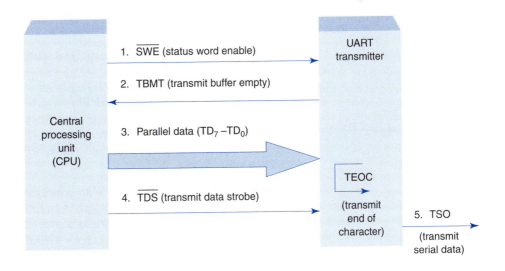

FIGURE 14-7 UART transmitter signal sequence

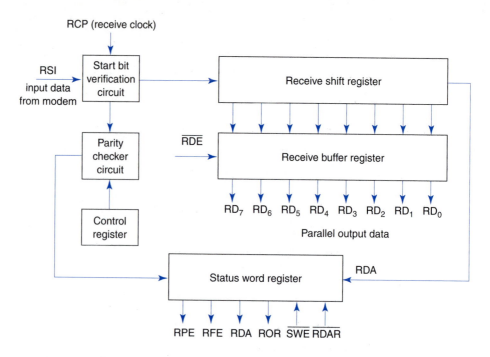

FIGURE 14-8 UART receiver block diagram

If parity is used, the parity bit is checked in the parity checker circuit. After one complete data character is loaded into the shift register, the character is transferred in parallel into the receive buffer register, and the *receive data available* (RDA) flag is set in the status word register. The CPU reads the status register by activating the $\overline{\text{SWE}}$ signal, and if RDA is active, the CPU reads the character from the receive buffer register by placing an active signal on the receive data enable (RDE) pin. After reading the data, the CPU places an active signal on the *receive data available reset* ($\overline{\text{RDAR}}$) pin, which resets the RDA pin. Meanwhile, the next character is received and clocked into the receive shift register, and the process repeats until all the data has been received. Figure 14-9 shows the receive signaling sequence that occurs between the CPU and the UART.

The CPU determines the status of the receive data by reading the status word register. The status flags relevant to the UART receiver are the following:

1. *Receive parity error (RPE)*. The RPE flag is set when a received character has a parity error in it.
2. *Receive framing error (RFE)*. The RFE flag is set when a character is received without any or with an improper number of stop bits.
3. *Receiver overrun (ROR)*. The ROR flag is set when a character in the receive buffer register is written over by another receive character because the CPU failed to service an active condition on REA before the next character was received from the receive shift register.

14-4-1-5 Start-bit verification circuit. With asynchronous data transmission, precise timing is less important than following an agreed-on format or pattern for the data. Each transmitted data character must be preceded by a start bit and end with one or more stop bits. Because data received by a UART has been transmitted from a distant UART whose clock is asynchronous to the receive UART, bit synchronization is achieved by establishing a timing reference at the center of each start bit. Therefore, it is imperative that

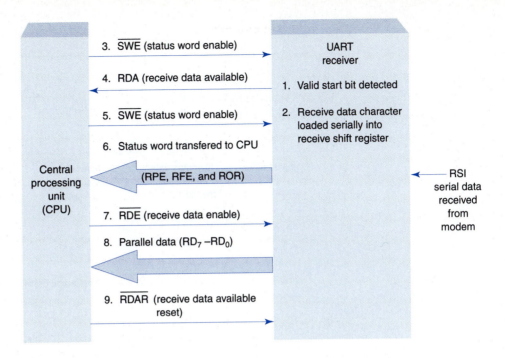

FIGURE 14-9   UART receiver signal sequence

a UART detect the occurrence of a valid start bit early in the bit cell and establish a timing reference before it begins to accept data.

The primary function of the start bit verification circuit is to detect valid start bits, which indicate the beginning of a data character. Figure 14-10a shows an example of how a noise hit can be misinterpreted as a start bit. The input data is a continuous string of idle line 1s, which are typically transmitted when there is no information. Idle line 1s are interpreted by a receiver as continuous stop bits (i.e., no data). If a noise impulse occurs that causes the receive data to go low at the same time the receiver clock is active, the receiver will interpret the noise impulse as a start bit. If this happens, the receiver will misinterpret the logic condition present during the next clock as the first data bit (b_0) and the following clock cycles as the remaining data bits (b_1, b_2, and so on). The likelihood of misinterpreting noise hits as start bits can be reduced substantially by clocking the UART receiver at a rate higher than the incoming data. Figure 14-10b shows the same situation as shown in Figure 14-10a, except the receive clock pulse (RCP) is 16 times (16X) higher than the receive serial data input (RSI). Once a low is detected, the UART waits seven clock cycles before resampling the input data. Waiting seven clock cycles places the next sample very near the center of the start bit. If the next sample detects a low, it assumes that a valid start bit has been detected. If the data has reverted to the high condition, it is assumed that the high-to-low transition was simply a noise pulse and, therefore, is ignored. Once a valid start bit has been detected and verified (Figure 14-10c), the start bit verification circuit samples the incoming data once every 16 clock cycles, which essentially makes the sample rate equal to the receive data rate (i.e., 16 RCP/16 = RCP). The UART continues sampling the data once every 16 clock cycles until the stop bits are detected, at which time the start bit verification circuit begins searching for another valid start bit. UARTs are generally programmed for receive clock rates of 16, 32, or 64 times the receive data rate (i.e., 16X, 32X, and 64X).

Another advantage of clocking a UART receiver at a rate higher than the actual receive data is to ensure that a high-to-low transition (valid start bit) is detected as soon as

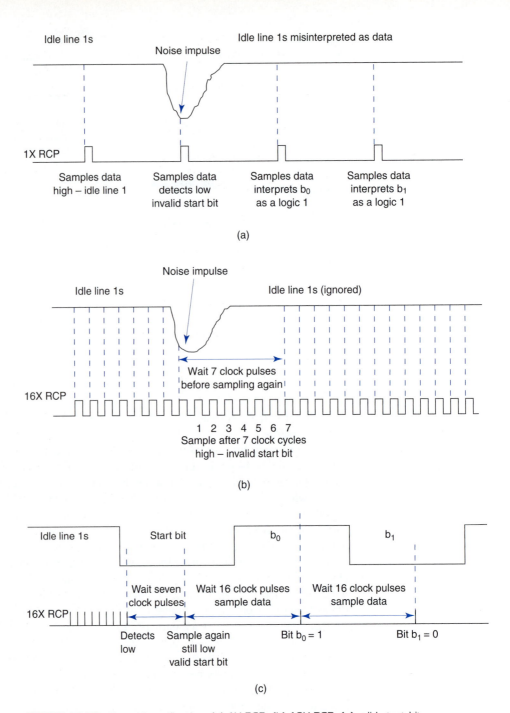

FIGURE 14-10 Start bit verification: (a) 1X RCP; (b) 16X RCP; (c) valid start bit

possible. This ensures that once the start bit is detected, subsequent samples will occur very near the center of each data bit. The difference in time between when a sample is taken (i.e., when a data bit is clocked into the receive shift register and the actual center of a data bit) is called the *sampling error*. Figure 14-11 shows a receive data stream sampled at a rate 16 times higher (16 RCP) than the actual data rate (RCP). As the figure shows, the start bit is not immediately detected. The difference in time between the beginning of a start bit and when it is detected is called the *detection error*. The maximum

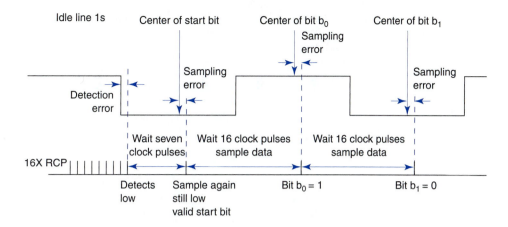

FIGURE 14-11   16× receive clock rate

detection error is equal to the time of one receive clock cycle ($t_{cl} = 1/R_{cl}$). If the receive clock rate equaled the receive data rate, the maximum detection error would approach the time of one bit, which would mean that a start bit would not be detected until the very end of the bit time. Obviously, the higher the receive clock rate, the earlier a start bit would be detected.

Because of the detection error, successive samples occur slightly off from the center of the data bit. This would not present a problem with synchronous clocks, as the sampling error would remain constant from one sample to the next. However, with asynchronous clocks, the magnitude of the sampling error for each successive sample would increase (the clock would slip over or slip under the data), eventually causing a data bit to be either sampled twice or not sampled at all, depending on whether the receive clock is higher or lower than the transmit clock.

Figure 14-12 illustrates how sampling at a higher rate reduces the sampling error. Figures 14–12a and b show data sampled at a rate eight times the data rate (8×) and 16 times the data rate (16×), respectively. It can be seen that increasing the sample rate moves the sample time closer to the center of the data bit, thus decreasing the sampling error.

Placing stop bits at the end of each data character also helps reduce the *clock slippage* (sometimes called *clock skew*) problem inherent when using asynchronous transmit and receive clocks. Start and stop bits force a high-to-low transition at the beginning of each character, which essentially allows the receiver to resynchronize to the start bit at the beginning of each data character. It should probably be mentioned that with UARTs, the data rates do not have to be the same in each direction of propagation (e.g., you could transmit data at 1200 bps and receive at 600 bps). However, the rate at which data leaves a transmitter must be the same as the rate data entering the receiver at the other end of the circuit. If you transmit at 1200 bps, it must be received at the other end at 1200 bps.

Example 14-3

Determine the bit time, receive clock rate, and maximum detection error for a UART receiving data at 1000 bps (f_b) with a receiver clock 16 times (16×) faster than the incoming data.

Solution   The time of one bit (t_b) is the reciprocal of the bit rate, or

$$t_b = \frac{1}{f_b}$$

Therefore,

$$t_b = \frac{1}{1000} = 1 \text{ ms, or } 1000 \text{ μs}$$

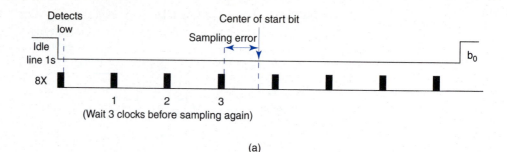

(a)

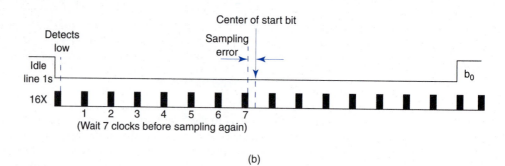

(b)

FIGURE 14-12   Sampling error: (a) 8× RCP; (b) 16× RCP

The receive clock rate is simply

$$R_{cl} = 16 f_b$$
$$= 16,000 \text{ Hz}$$

The time of one receive clock cycle is the reciprocal of the receive clock rate (R_{cl}), or

$$t_{cl} = \frac{1}{R_{cl}}$$

Therefore,

$$t_{cl} = \frac{1}{16,000} = 62.5 \text{ μs}$$

Therefore, the maximum detection error is equal to the time of one receive clock cycle, or 62.5 μs.

14-4-1-6 Dual universal asynchronous receiver/transmitter.   A *dual universal asynchronous receiver/transmitter* (DUART) is a single-chip data communications device that provides two full-duplex asynchronous receiver/transmitter channels in a single package. The operating mode and data format of each channel can be programmed independently. In addition, each DUART contains an internal programmable counter/timer that allows the two transmitter/receiver pairs to operate at a variety of fixed receive clock rates. DUARTs are particularly adaptable to dual-speed data-channel applications, such as required with clustered terminal systems (i.e., LCUs with multiple terminals, workstations, or PCs). Four-channel (quad) UARTs (QARTs), which provide four full-duplex asynchronous channels, and eight-channel UARTs (Octal UARTs), are also available.

14-4-2   Universal Synchronous Receiver/Transmitter

A *universal synchronous receiver/transmitter* (USRT) is used for synchronous transmission of data between a DTE and a DCE. Synchronous data transmission means that a synchronous data format is used and that clocking information is generally transferred between the DTE and the DCE. A USRT performs the same basic functions as a UART, except for synchronous data (i.e., the start and stop bits are omitted and replaced by unique synchronizing characters). The primary functions performed by a USRT are the following:

1. Serial-to-parallel and parallel-to-serial data conversions
2. Error detection by inserting parity bits in the transmitter and checking parity bits in the receiver
3. Inserting and detecting unique data synchronization (SYN) characters
4. Formatting data in the transmitter and receiver (i.e., combining items 1 through 3 in a meaningful sequence)
5. Providing transmit and receive status information to the CPU
6. Voltage level conversion between the DTE and the serial interface and vice versa
7. Providing a means of achieving bit and character synchronization

The block diagram for a typical USRT is shown in Figure 14-13. The USRT operates very similar to a UART; therefore, only the differences are explained. With a USRT, start and stop bits are not allowed. Instead, unique SYN characters are loaded into the transmit and receive SYN registers before transmitting or receiving data.

Like the UART, a USRT has only one control register, and it determines the data characteristics for both the USRT transmitter and the USRT receiver, which must be the same. The control signals for a USRT are essentially the same as for the UART except for the single *sync character signal* (SCS). SCS specifies the number of SYN characters inserted in the transmitter before transmitting data.

The *status word register* for a USRT is also essentially the same as for a UART except for the addition of a *SYN detect/break detect* flag (SYNDET/BD). SYNDET/BD can be used as either an input or an output status flag. When it is programmed as an input, it signals the USRT to start assembling data characters on the next receive clock pulse. When used as an output, it indicates when a valid SYN character has been received.

14-4-2-1 USRT transmitter.
For the following explanation, refer to the transmit section of the block diagram for a USRT shown in Figure 14-13. Prior to loading or transmitting data received from the CPU, the USRT control register must be programmed, the transmit clock signal (TCP) for the USRT must be set at the desired transmission rate, and the desired SYN character must be placed on the parallel input data bus (DB_7–DB_0). The SYN character is loaded into the transmit SYN register when the CPU pulses the *transmit SYN strobe* (TSS). Once the SYN character is initially loaded, it need not be loaded again unless, of course, it is changed. Anytime after the transmit SYN character has been loaded, the CPU can place a data character onto the parallel data bus. Data is loaded into the transmit data register by pulsing the *transmit data strobe* (TDS).

At the beginning of each data transmission, a SYN character is copied from the transmit sync register to the transmit shift register and then transmitted serially to the modem on the *transmit serial output* (TSO) pin. For reasons that will be explained in a later chapter, two SYN characters are generally transmitted at the beginning of each message. After the SYN characters have been transmitted, the first data character is transferred from the transmit data register to the transmit shift register, and the next data character is loaded into the transmit data register. Characters are transferred from the transmit data register to the transmit shift register provided that the TDS pulse occurs during the presently transmitted character. If TDS is not pulsed, the next transmitted character loaded into the transmit shift register is taken from the transmit sync register. The multiplexer provides the steering logic and circuitry that determines whether data is loaded into the transmit shift register from the transmit data register or the transmit sync register. When TDS is pulsed, data is transferred from the transmit data register to the transmit shift register. After each transmitted SYN character, the *SYN character transmit* (SCT) control signal is set. The *transmit buffer empty* (TBMT) signal is used by the USRT to request the next character from the CPU. Figure 14-14a shows the signaling sequence that occurs between the CPU and the USRT transmitter.

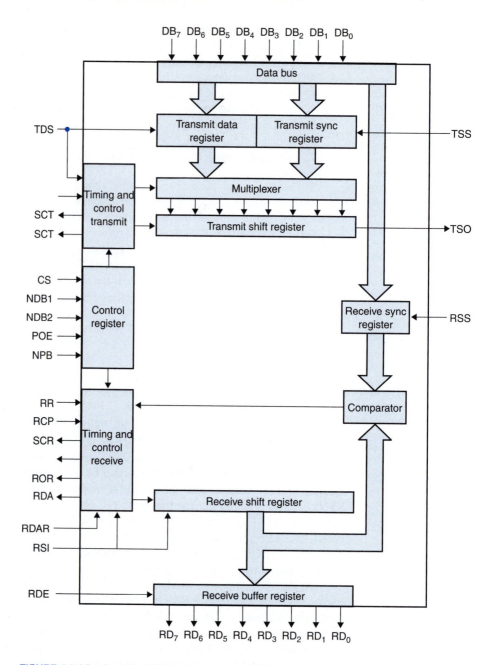

FIGURE 14-13 Simplified block diagram of a USRT

14-4-2-2 USRT receiver. For the following explanation, refer to the receive section of the block diagram for a USRT shown in Figure 14-13. The receive clock signal for a USRT is the same as the *transmit clock signal* (TCP); therefore, the transmit and receive data rates must be the same. The receive SYN character is loaded into the USRT receiver by the CPU by placing the SYN character onto the parallel input data bus (DB_7–DB_0) and then activating the *receive sync strobe* (RSS) signal. Again, the receive SYN character needs to be loaded only once, and it is always the same character as the transmit SYN character. Normally, the transmit and receive SYN characters are loaded one after the other and, of course, before attempting to transmit or receive data.

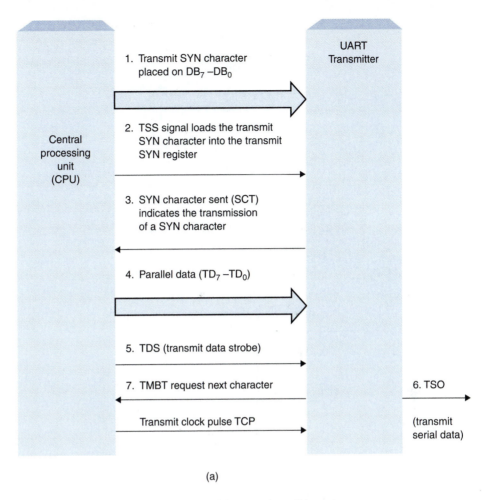

CPU side steps:

1. Transmit SYN character placed on DB₇–DB₀

2. TSS signal loads the transmit SYN character into the transmit SYN register

3. SYN character sent (SCT) indicates the transmission of a SYN character

4. Parallel data (TD₇–TD₀)

5. TDS (transmit data strobe)

7. TMBT request next character

Transmit clock pulse TCP

UART Transmitter

Central processing unit (CPU)

6. TSO

(transmit serial data)

(a)

FIGURE 14-14 USRT signal sequence: (a) transmitter; (b) receiver

In the USRT receiver, reception of data begins with a high-to-low transition on the *receive rest* (RR) input signal, which places the receiver into the *search mode* (sometimes called the *bit phase*). In the search mode, serially received data arriving on *receive serial input* (RSI) are examined in groups of eight bits; however, the data is examined after each bit is received until a valid SYN character is found. After each bit is clocked into the receive shift register, the contents of the receive shift register are compared to the contents of the receive SYN register. If the contents of the two registers are the same, it is assumed that a valid SYN character has been received, and the *SYN character receive* (SCR) output signal is set. Once a valid SYN character has been received, it is transferred into the receive buffer register, and the receive is placed into the *character mode*. In the character mode, the receive no longer searches for a SYN character, and subsequent received data is examined on a character-by-character basis (i.e., in successive groups of eight). Each receive character is examined, and the appropriate information is transferred to the status register, such as *receive data available* (RDA), *receive parity error* (RPE), *receiver overrun* (ROR), and reception of other SYN characters. Parallel receive data is outputted to the CPU on the parallel data output bus (RD₇–RD₀). Figure 14-14b shows the signaling sequence that occurs between the CPU and the USRT receiver.

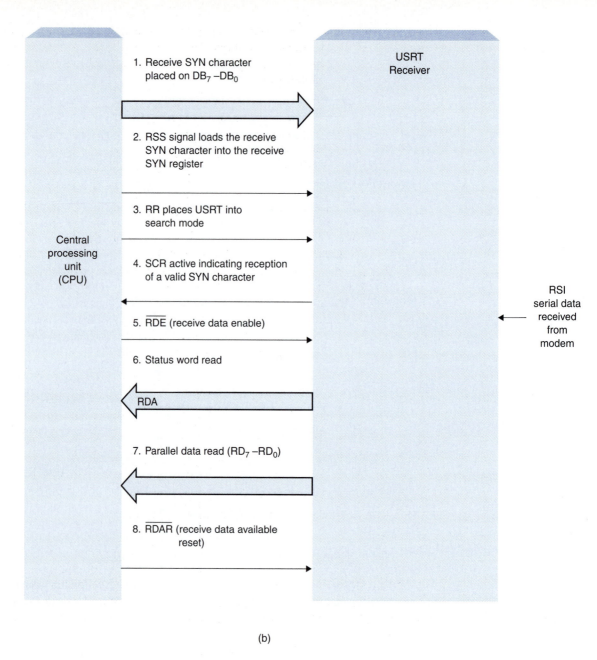

1. Receive SYN character placed on $DB_7 - DB_0$

2. RSS signal loads the receive SYN character into the receive SYN register

3. RR places USRT into search mode

4. SCR active indicating reception of a valid SYN character

5. $\overline{RDE}$ (receive data enable)

6. Status word read

RDA

7. Parallel data read ($RD_7 - RD_0$)

8. $\overline{RDAR}$ (receive data available reset)

Central processing unit (CPU)

USRT Receiver

RSI serial data received from modem

(b)

FIGURE 14-14 *(Continued)*

14-5 SERIAL INTERFACES

To ensure an orderly flow of data between a DTE and a DCE, a standard serial interface is used to interconnect them. The serial interface coordinates the flow of data, control signals, and timing information between the DTE and the DCE.

Before serial interfaces were standardized, every company that manufactured data communications equipment used a different interface configuration. More specifically, the cable arrangement between the DTE and the DCE, the type and size of the connectors, and the voltage levels varied considerably from vendor to vendor. To interconnect equipment manufactured by different companies, special level converters, cables, and connectors had to be designed, constructed, and implemented for each application. A serial interface standard should provide the following:

1. A specific range of voltages for transmit and receive signal levels
2. Limitations for the electrical parameters of the transmission line, including source and loads impedance, cable capacitance, and other electrical characteristics outlined later in this chapter
3. Standard cable and cable connectors
4. Functional description of each signal on the interface

In 1962, the Electronics Industries Association (EIA), in an effort to standardize interface equipment between data terminal equipment and data communications equipment, agreed on a set of standards called the *RS-232 specifications* (*RS* meaning "recommended standard"). The official name of the RS-232 interface is *Interface Between Data Terminal Equipment and Data Communications Equipment Employing Serial Binary Data Interchange*. In 1969, the third revision, RS-232C, was published and remained the industrial standard until 1987, when the RS-232D was introduced, which was followed by the RS-232E in the early 1990s. The RS-232D standard is sometimes referred to as the EIA-232 standard. Versions D and E of the RS-232 standard changed some of the pin designations. For example, data set ready was changed to DCE ready, and data terminal ready was changed to DTE ready.

The RS-232 specifications identify the mechanical, electrical, functional, and procedural descriptions for the interface between DTEs and DCEs. The RS-232 interface is similar to the combined ITU-T standards V.28 (electrical specifications) and V.24 (functional description) and is designed for serial transmission up to 20 kbps over a maximum distance of 50 feet (approximately 15 meters).

14-5-1 RS-232 Serial Interface Standard

The mechanical specification for the RS-232 interface specifies a cable with two connectors. The standard RS-232 cable is a sheath containing 25 wires with a DB25P-compatible male connector (plug) on one end and a DB25S-compatible female connector (receptacle) on the other end. The DB25P-compatible and DB25S-compatible connectors are shown in Figures 14–15a and b, respectively. The cable must have a plug on one end that connects to the DTE and a receptacle on the other end that connects to the DCE. There is also a special PC nine-pin version of the RS-232 interface cable with a DB9P-compatible male connector on one end and a DB9S-compatible connector at the other end. The DB9P-compatible and DB9S-compatible connectors are shown in Figures 14–15c and d, respectively (note that there is no correlation between the pin assignments for the two connectors). The nine-pin

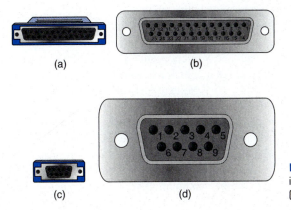

(a)   (b)

(c)   (d)

FIGURE 14-15 RS-232 serial interface connector: (a) DB25P; (b) DB25S; (c) DB9P; (d) DP9S

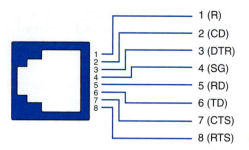

1 (R)
2 (CD)
3 (DTR)
4 (SG)
5 (RD)
6 (TD)
7 (CTS)
8 (RTS)

FIGURE 14-16 EIA-561 modular connector

Table 14-2   RS-232 Voltage Specifications

	Data Signals		Control Signals	
	Logic 1	Logic 0	Enable (On)	Disable (Off)
Driver (output)	−5 V to −15 V	+5 V to +15 V	+5 V to +15 V	−5 V to −15 V
Terminator (input)	−3 V to −25 V	+3 V to +25 V	+3 V to +25 V	−3 V to −25 V

version of the RS-232 interface is designed for transporting asynchronous data between a DTE and a DCE or between two DTEs, whereas the 25-pin version is designed for transporting either synchronous or asynchronous data between a DTE and a DCE. Figure 14-16 shows the eight-pin EIA-561 modular connector, which is used for transporting asynchronous data between a DTE and a DCE when the DCE is connected directly to a standard two-wire telephone line attached to the public switched telephone network. The EIA-561 modular connector is designed exclusively for dial-up telephone connections.

Although the RS-232 interface is simply a cable and two connectors, the standard also specifies limitations on the voltage levels that the DTE and DCE can output onto or receive from the cable. The DTE and DCE must provide circuits that convert their internal logic levels to RS-232-compatible values. For example, a DTE using TTL logic interfaced to a DCE using CMOS logic are not compatible. *Voltage-leveling circuits* convert the internal voltage levels from the DTE and DCE to RS-232 values. If both the DCE and the DTE output and accept RS-232 levels, they are electrically compatible regardless of which logic family they use internally. A voltage leveler is called a *driver* if it outputs singles onto the cable and a *terminator* if it accepts signals from the cable. In essence, a driver is a transmitter, and a terminator is a receiver. Table 14-2 lists the voltage limits for RS-232-compatible drivers and terminators. Note that the data and control lines use *non–return to zero, level* (NRZ-L) bipolar encoding. However, the data lines use negative logic, while the control lines use positive logic.

From examining Table 14-2, it can be seen that the voltage limits for a driver are more inclusive than the voltage limits for a terminator. The output voltage range for a driver is between +5 V and +15 V or between −5 V and −15 V, depending on the logic level. However, the voltage range in which a terminator will accept is between +3 V and +25 V or between −3 V and −25 V. Voltages between ±3 volts are undefined and may be interpreted by a terminator as a high or a low. The difference in the voltage levels between the driver output and the terminator input is called *noise margin* (NM). The noise margin reduces the susceptibility to interface caused by noise transients induced into the cable. Figure 14-17a shows the relationship between the driver and terminator voltage ranges. As shown in Figure 14-17a, the noise margin for the minimum driver output voltage is 2 V (5 − 3), and the noise margin for the maximum driver output voltage is 10 V (25 − 15). (The minimum noise margin of 2 V is called the *implied noise margin*.) Noise margins will vary, of course,

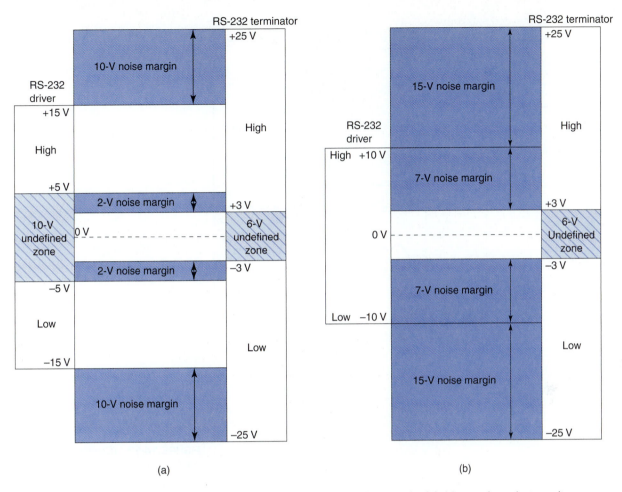

(a)

(b)

FIGURE 14-17 RS-232 logic levels and noise margin: (a) driver and terminator voltage ranges; (b) noise margin with a +10 V high and −10 V low

depending on what specific voltages are used for highs and lows. When the noise margin of a circuit is a high value, it is said to have *high noise immunity*, and when the noise margin is a low value, it has *low noise immunity*. Typical RS-232 voltage levels are +10 V for a high and −10V for a low, which produces a noise margin of 7 V in one direction and 15 V in the other direction. The noise margin is generally stated as the minimum value. This relationship is shown in Figure 14-17b. Figure 14-17c illustrates the immunity of the RS-232 interface to noise signals for logic levels of +10 V and −10 V.

The RS-232 interface specifies single-end (unbalanced) operation with a common ground between the DTE and DCE. A common ground is reasonable when a short cable is used. However, with longer cables and when the DTE and DCE are powered from different electrical busses, this may not be true.

Example 14-4

Determine the noise margins for an RS-232 interface with driver signal voltages of ±6 V.

Solution The noise margin is the difference between the driver signal voltage and the terminator receive voltage, or

$$NM = 6 - 3 = 3V \quad \text{or} \quad NM = 25 - 6 = 19V$$

The minimum noise margin is 3 V.

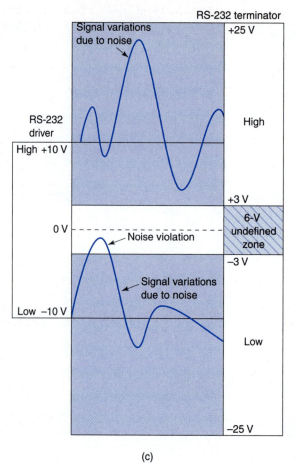

FIGURE 14-17 *(Continued)* (c) noise violation

(c)

14-5-1-1 RS-232 electrical equivalent circuit.
Figure 14-18 shows the equivalent electrical circuit for the RS-232 interface, including the driver and terminator. With these electrical specifications and for a bit rate of 20 kbps, the nominal maximum length of the RS-232 interface cable is approximately 50 feet.

14-5-1-2 RS-232 functional description.
The pins on the RS-232 interface cable are functionally categorized as either ground (signal and chassis), data (transmit and receive), control (handshaking and diagnostic), or timing (clocking signals). Although the RS-232 interface as a unit is bidirectional (signals propagate in both directions), each individual wire or pin is unidirectional. That is, signals on any given wire are propagated either from the DTE to the DCE or from the DCE to the DTE but never in both directions. Table 14-3 lists the 25 pins (wires) of the RS-232 interface and gives the direction of signal propagation (i.e., either from the DTE toward the DCE or from the DCE toward the DTE). The RS-232 specification designates the first letter of each pin with the letters A, B, C, D, or S. The letter categorizes the signal into one of five groups, each representing a different type of circuit. The five groups are the following:

A: ground

B: data

C: control

D: timing (clocking)

S: secondary channel

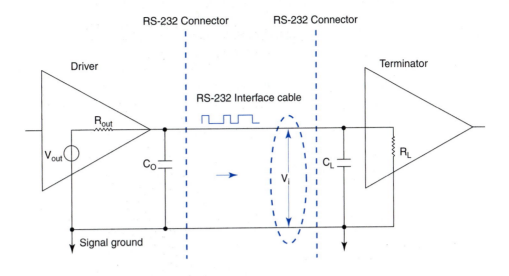

V_{out} — open-circuit voltage at the output of a driver (±5 V to ±15 V)
V_i — terminated voltage at the input to a terminator (±3 V to ±25 V)
C_L — load capacitance associated with the terminator, including the cable (2500 pF maximum)
C_O — capacitance seen by the driver including the cable (2500 pF maximum)
R_L — terminator input resistance (3000 Ω to 7000 Ω)
R_{out} — driver output resistance (300 Ω maximum)

FIGURE 14-18 RS-232 electrical specifications

Table 14-3 EIA RS-232 Pin Designations and Direction of Propagation

Pin Number	Pin Name	Direction of Propagation
1	Protective ground (frame ground or chassis ground)	None
2	Transmit data (send data)	DTE to DCE
3	Receive data	DCE to DTE
4	Request to send	DTE to DCE
5	Clear to send	DCE to DTE
6	Data set ready (modem ready)	DCE to DTE
7	Signal ground (reference ground)	None
8	Receive line signal detect (carrier detect or data carrier detect)	DCE to DTE
9	Unassigned	None
10	Unassigned	None
11	Unassigned	None
12	Secondary receive line signal detect (secondary carrier detect or secondary data carrier detect)	DCE to DTE
13	Secondary clear to send	DCE to DTE
14	Secondary transmit data (secondary send data)	DTE to DCE
15	Transmit signal element timing—DCE (serial clock transmit—DCE)	DCE to DTE
16	Secondary receive data	DCE to DTE
17	Receive signal element timing (serial clock receive)	DCE to DTE
18	Unassigned	None
19	Secondary request to send	DTE to DCE
20	Data terminal ready	DTE to DCE
21	Signal quality detect	DCE to DTE
22	Ring indicator	DCE to DTE
23	Data signal rate selector	DTE to DCE
24	Transmit signal element timing—DTE (serial clock transmit—DTE)	DTE to DCE
25	Unassigned	None

Table 14-4 EIA RS-232 Pin Designations

Pin Number	Pin Name	EIA Nomenclature	Common U.S. Acronyms
1	Protective ground (frame ground or chassis ground)	AA	GWG, FG, or CG
2	Transmit data (send data)	BA	TD, SD, TxD
3	Receive data	BB	RD, RxD
4	Request to send	CA	RS, RTS
5	Clear to send	CB	CS, CTS
6	Data set ready (modem ready)	CC	DSR, MR
7	Signal ground (reference ground)	AB	SG, GND
8	Receive line signal detect (carrier detect or data carrier detect)	CF	RLSD, CD, DCD
9	Unassigned	—	—
10	Unassigned	—	—
11	Unassigned	—	—
12	Secondary receive line signal detect (secondary carrier detect or secondary data carrier detect)	SCF	SRLSD, SCD, SDCD
13	Secondary clear to send	SCB	SCS, SCTS
14	Secondary transmit data (secondary send data)	SBA	STD, SSD, STxD
15	Transmit signal element timing—DCE (serial clock transmit—DCE)	DB	TSET, SCT-DCE
16	Secondary receive data	SBB	SRD, SRxD
17	Receive signal element timing (serial clock receive)	DD	RSET, SCR
18	Unassigned	—	—
19	Secondary request to send	SCA	SRS, SRTS
20	Data terminal ready	CD	DTR
21	Signal quality detect	CG	SQD
22	Ring indicator	CE	RI
23	Data signal rate selector	CH	DSRS
24	Transmit signal element timing—DTE (serial clock transmit—DTE)	DA	TSET, SCT-DTE
25	Unassigned	—	

Because the letters are nondescriptive designations, it is more practical and useful to use acronyms to designate the pins that reflect the functions of the pins. Table 14-4 lists the EIA signal designations plus the nomenclature more commonly used by industry in the United States to designate the pins.

Twenty of the 25 pins on the RS-232 interface are designated for specific purposes or functions. Pins 9, 10, 11, 18, and 25 are unassigned (unassigned does not necessarily imply unused). Pins 1 and 7 are grounds; pins 2, 3, 14, and 16 are data pins; pins 15, 17, and 24 are timing pins; and all the other pins are used for control or handshaking signals. Pins 1 through 8 are used with both asynchronous and synchronous modems. Pins 15, 17, and 24 are used only with synchronous modems. Pins 12, 13, 14, 16, and 19 are used only when the DCE is equipped with a secondary data channel. Pins 20 and 22 are used exclusively when interfacing a DTE to a modem that is connected to a standard dial-up telephone circuits on the public switched telephone network.

There are two full-duplex data channels available with the RS-232 interface; one channel is for *primary data* (actual information), and the second channel is for *secondary data* (diagnostic information and *handshaking* signals). The secondary channel is sometimes used as a reverse or backward channel, allowing the receive DCE to communicate with the transmit DCE while data is being transmitted on the primary data channel.

The functions of the 25 RS-232 pins are summarized here for a DTE interfacing with a DCE where the DCE is a data communications modem:

Pin 1: Protective ground, frame ground, or chassis ground (GWG, FG, or CG). Pin 1 is connected to the chassis and used for protection against accidental electrical shock. Pin 1 should be connected to the third-wire ground of an ac electrical system at one end of the cable (either at the DTE end or at the DCE end but not at both ends). If shielded cable is used, protective ground is also connected to the outer shield to prevent interference from external signals from reaching the inside conductors. Pin 1 is generally connected to signal ground (pin 7).

Pin 2: Transmit data or send data (TD, SD, or TxD). Pin 2 is one of two primary data pins. Serial data on the primary data channel is transported from the DTE to the DCE on pin 2. Primary data is the actual source information transported over the interface. The transmit data line is a transmit line for the DTE but a receive line for the DCE. The terms *send* and *receive* are relative with the reference being the DTE or the analog communications line when the DCE is a modem (i.e., the DTE is sending data on pin 2, which is converted to analog in the modem and outputted onto the analog communications channel). The DTE may hold the TD line at a logic 1 voltage level when no data is being transmitted and between characters when asynchronous data is being transmitted. Otherwise, the TD driver is enabled by an active condition on pin 5 (clear to send).

Pin 3: Receive data (RD or RxD). Pin 3 is the second primary data pin. Serial data is transported from the DCE to the DTE on pin 3. Again, the analog communications line or the DTE is the reference. Pin 3 is the receive data pin for the DTE and the transmit data pin for the DCE. The DCE may hold the TD line at a logic 1 voltage level when no data is being transmitted or when pin 8 (RLSD) is inactive. Otherwise, the RD driver is enabled by an active condition on pin 8.

Pin 4: Request to send (RS or RTS). For half-duplex data transmission, the DTE uses pin 4 to request permission from the DCE to transmit data on the primary data channel. When the DCE is a modem, an active condition on RTS turns on the modem's analog carrier. The RTS and CTS signals are used together to coordinate half-duplex data transmission between the DTE and DCE. For full-duplex data transmission, RTS can be held active permanently. The RTS driver is enabled by an active condition on pin 6 (data set ready).

Pin 5: Clear to send (CS or CTS). The CTS signal is a handshake from the DCE to the DTE (i.e., modem to LCU) in response to an active condition on RTS. An active condition on CTS enables the TD driver in the DTE. There is a predetermined time delay between when the DCE receives an active condition on the RTS signal and when the DCE responds with an active condition on the CTS signal. If the DCE is a modem, it uses the RTS/CTS time delay to modulate its carrier with a unique bit pattern called a *training sequence.* The purpose of the training sequence is to initialize the communications channel and synchronize the receive modem located at the distant end of the communications channel before any user information is transmitted. Synchronization in asynchronous modems is little more than initializing the modem's receiver circuitry and preparing it to receive data. However, synchronization for synchronous modems includes receiver initialization, carrier recovery, and clock synchronization. RTS/CTS delays vary with the type of modem. Asynchronous modems require a much shorter initialization time than synchronous modems. RTS/CTS delays vary from a few milliseconds for asynchronous modems to 500 ms or more for synchronous modems. CTS goes active when the modem is ready to transmit data and inactive to indicate that the DCE has completed data transmission. CTS is used to enable or disable the transmit data driver. When two digital devices are connected with an RS-232 interface, RTS and CTS can be connected together at each device, thus eliminating the RTS/CTS time delay.

Pin 6: Data set ready or modem ready (DSR or MR). DSR is a signal sent from the DCE to the DTE to indicate the availability of the communications channel. DSR is active only when the DCE and the communications channel are available. Under normal operation, the modem and the communications channel are always available. However, there are five situations when the modem or the communications channel are not available:

1. The modem is shut off (i.e., has no power).
2. The modem is disconnected from the communications line so the line can be used for normal telephone voice traffic (i.e., in the voice rather than the data mode).
3. The modem is in one of the self-test modes (i.e., analog or digital loopback).
4. The telephone company is testing the communications channel.
5. On dial-up circuits, DSR is held inactive while the telephone switching system is establishing a call and when the modem is transmitting a specific response (answer) signal to the calling station's modem.

An active condition on the DSR lead enables the request to send driver in the DTE, thus giving the DSR lead the highest priority of the RS-232 control leads.

Pin 7: Signal ground or reference ground (SG or GND). Pin 7 is the signal reference (return line) for all data, control, and timing signals (i.e. all pins except pin 1, chassis ground). Pin 7 is generally strapped to frame ground (pin 1).

Pin 8: Receive line signal detect, carrier detect, or data carrier detect (RLSD, CD, or DCD). The DCE uses this pin to signal the DTE when it determines that it is receiving a valid analog carrier (data carrier). An active RLSD signal enables the RD terminator in the DTE, allowing it to accept data from the DCE. An inactive RLSD signal disables the terminator for the DTE's receive data pin, preventing it from accepting invalid data. On half-duplex data circuits, RLSD is held inactive whenever RTS is active. RLSD typically remains active for a short time after detecting a loss of analog carrier to ensure that all the receive data has been demodulated and transferred to the DTE before disabling the RD line. The delay time is called the *RLSD turn-off delay.*

Pin 9: Unassigned. Pin 9 is non–EIA specified; however, it is often held at +12 Vdc for test purposes (+P).

Pin 10: Unassigned. Pin 10 is non–EIA specified; however, it is often held at –12 Vdc for test purposes (–P).

Pin 11: Unassigned. Pin 11 is non–EIA specified; however, it is often designated as equalizer mode (EM) and used by the modem to signal the DTE when the modem is self-adjusting its internal equalizers because error performance is suspected to be poor. When the carrier detect signal is active and the circuit is inactive, the modem is retraining (resynchronizing), and the probability of error is high. When receive line signal detect (pin 8) is active and EM is inactive, the modem is trained, and the probability of error is low.

Pin 12: Secondary receive line signal detect, secondary carrier detect, or secondary data carrier detect (SRLSD, SCD, or SDCD). Pin 12 is the same as RLSD (pin 8) except for the secondary data channel. SRLSD is active when the DCE is receiving an analog carrier on the secondary (diagnostic) data channel. An active condition on SRLSD enables the secondary receive data terminator (pin 16), allowing the DTE to accept secondary data from the DCE.

Pin 13: Secondary clear to send. The SCTS signal is sent from DCE to the DTE as a response (handshake) to the secondary request to send signal (pin 19). SCTS enables the driver in the DTE for secondary transmit data (pin 14).

Pin 14: Secondary transmit data or secondary send data (STD or STxD). Diagnostic data is transmitted from the DTE to the DCE on this pin. STD is enabled by an active condition on SCTS. Pin 14 can also be used for the new *sync signal*, which is non–EIA specified. New sync is optional and intended to be used with a modem at the primary location of a multipoint data circuit. When the primary is communicating with several secondary stations, rapid resynchronization of the modem receiver to many secondary transmitters is required. The receiver clock normally maintains the timing information of the previous message for a short time after the message has ended. However, this may interfere with resynchronizing to the carrier from the next station's modem. The DTE should provide an active condition on this pin for at least 1 ms but no longer than the intermessage interval to squelch the existing timing information in the modem.

Pin 15: Transmission signal element timing or serial clock transmit (TSET or SCT-DCE). With synchronous modems, the transmit clocking signal is sent from the DCE to the DTE on this pin. The clock is used in the DTE to synchronize the output data rate to the DCE's clock.

Pin 16: Secondary received data (SRD or SRxD). Diagnostic data is transmitted from the DCE to the DTE on this pin. The SRD driver is enabled by an active condition on secondary receive line signal detect.

Pin 17: Receiver signal element timing or serial clock receive (RSET or SCR). When synchronous modems are used, clocking information recovered by the DCE is sent to the DTE on this pin. The receive clock is used to clock data out of the DCE and into the DTE on the receive data line. The clock frequency is equal to the bit rate on the primary data channel.

Pin 18: Unassigned. Pin 11 is non–EIA specified; however, it is often used for the *local loopback* (LL) signal. Local loopback is a control signal sent from the DTE to the DCE placing the DCE (modem) into an analog loopback condition. Analog and digital loopbacks are described in a later section of this chapter.

Pin 19: Secondary request to send (SRS or SRTS). SRTS is used by the DTE to bid for the secondary data channel from the DCE. SRTS and SCTS coordinate the flow of data on the secondary data channel.

Pin 20: Data terminal ready (DTR). The DTE sends signals to the DCE on the DTR line concerning the availability of the data terminal equipment. DTR is used primarily with dial-up circuits to handshake with ring indicator (pin 22). The DTE disables DTR when it is unavailable, thus instructing the DCE not to answer an incoming call.

Pin 21: Signal quality detector (SQD). The DCE sends signals to the DTE on this line indicating the quality of the received analog carrier. An inactive (low) signal on SQD tells the DTE that the incoming signal is marginal and that there is a high likelihood that errors are occurring. Pin 21 is sometimes used for remote loopback, which is non–EIA specified. An active condition on pin 21 tells the DTE that the DCE is in a digital loopback, preventing the DCE from transporting data to the DTE on the RD line.

Pin 22: Ring indicator (RI). The RI line is used primarily on dial-up data circuits for the DCE to inform the DTE that there is an incoming call. If the DTE is ready to receive data, it responds to an active condition on RI with an active condition on DTR. DTR is a handshaking signal in response to an active condition on RI.

Pin 23: Data signal rate selector (DSRS). The DTE used this line to select one of two transmission bit rates when the DCE is equipped to offer two rates. (The data rate selector line can be used to change the transmit clock frequency.)

Pin 24: Transmit signal element timing or serial clock transmit (TSET or SCT-DTE). When synchronous modems are used, the transmit clocking signal is sent from the DTE to the DCE on this pin. Pin 24 is used only when the master clock is located in the DTE, in which case the modem synchronizes its transmit clock to the DTE's clock. When the DCE is providing the master clock, pin 24 is not used, and the DTE synchronizes its transmit clock to the master clock received from the DCE on pin 15. Sometimes the terms *internal* and *external* are used to indicate which device (the DTE or the DCE) is providing the master transmit clock. Whichever device is providing the master clock is optioned for internal timing, and the other device is optioned for external timing. The two devices can never have the same timing option. The receive clock is not optional—the DCE always provides a clock to the DTE that has been recovered from data received from the distant DCE.

Pin 25: Unassigned. Pin 5 is non–EIA specified; however, it is sometimes used as a control signal from the DCE to the DTE to indicate that the DCE is in either the remote or the local loopback mode.

For asynchronous transmission using either the DB9P/S-modular connector, only the following nine pins are provided:

1. Receive line signal detect
2. Receive data
3. Transmit data
4. Data terminal ready
5. Signal ground
6. Data set ready
7. Request to send
8. Clear to send
9. Ring indicator

The eight-pin EIA-561-modular connector pin assignments are the following:

1. Ring indicator
2. Receive line signal detect
3. Data terminal ready
4. Signal ground
5. Receive data
6. Transmit data
7. Clear to send
8. Request to send

14-5-1-3 RS-232 signals. Figure 14-19 shows the timing diagram for the transmission of one asynchronous data character over the RS-232 interface. The character is comprised of one start bit, one stop bit, seven ASCII character bits, and one even-parity bit. The transmission rate is 1000 bps, and the voltage level for a logic 1 is -10 V and for a logic 0 is $+10$V. The time of one bit is 1 ms; therefore, the total time to transmit one ASCII character is 10 ms.

14-5-1-4 RS-232 asynchronous data transmission. Figures 14–20a and b show the functional block diagram for the drivers and terminators necessary for transmission of asynchronous data over the RS-232 interface between a DTE and a DCE that is a modem. As shown in the figure, only the first eight pins of the interface are required, which includes the following signals: signal ground and chassis ground, transmit data and receive data, request to send, clear to send, data set ready, and receive line signal detect.

Figure 14-21 shows a timing diagram for control and data signals for a typical asynchronous data transmission over an RS-232 interface with the following parameters:

Modem RTS-CTS delay = 50 ms

DTE primary data message length = 100 ms

Modem training sequence = 50 ms

Propagation time = 10 ms

Modem RLSD turn-off delay time = 5 ms

When the DTE wishes to transmit data on the primary channel, it enables request to send ($t = 0$). After a predetermined RTS/CTS time delay time, which is determined by the mo-

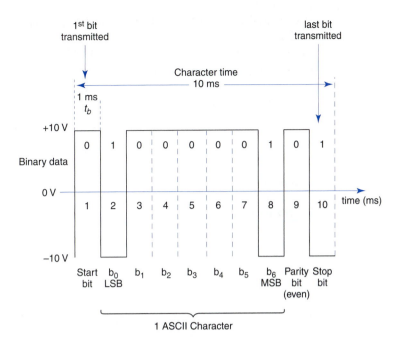

FIGURE 14-19 RS-232 data-timing diagram—ASCII uppercase letter A, one start bit, even parity, and one stop bit

dem (50 ms for this example), CTS goes active. During the 50-ms RTS/CTS delay, the modem outputs an analog carrier that is modulated by a unique bit pattern called a *training sequence*. The training sequence for asynchronous modems is generally nothing more than a series of logic 1s, which produce 50 ms of continuous mark frequency. The analog carrier is used to initialize the communications channel and the distant receive modem (with synchronous modems, the training sequence is more involved, as it would also synchronize the carrier and clock recovery circuits in the distant modem). After the RTS/CTS delay, the transmit data (TD) line is enabled, and the DTE begins transmitting user data. When the transmission is complete ($t = 150$ ms), RTS goes low, which turns off the modem's analog carrier. The modem acknowledges the inactive condition of RTS with an inactive condition on CTS.

At the distant end, the receive modem receives a valid analog carrier after a 10 ms propagation delay (P_d) and enables RLSD. The DCE sends an active RLSD signal across the RS-232 interface cable to the DT, which enables the receive data line (RD). However, the first 50 ms of the receive data is the training sequence, which is ignored by the DTE, as it is simply a continuous stream of logic 1s. The DTE identifies the beginning of the user data by recognizing the high-to-low transition caused by the first start bit ($t = 60$ ms). At the end of the message, the DCE holds RLSD active for a predetermined RLSD turn-off delay time (10 ms) to ensure that all the data received has been demodulated and outputted onto the RS-232 interface.

14-5-1-5 DTE/DCE loopback tests.
Loopbacks are temporary circuit configurations that connect a transmit line to a receive line for the purpose of performing tests and isolating troubles. There are two basic types of loopbacks: analog and digital.

Analog loopback. An analog loopback (ALB) is actually a modem feature that disconnects the analog side of a modem from the telephone line and connects the output of the transmitter to the input of the receiver. Analog loopbacks can be performed on two-wire

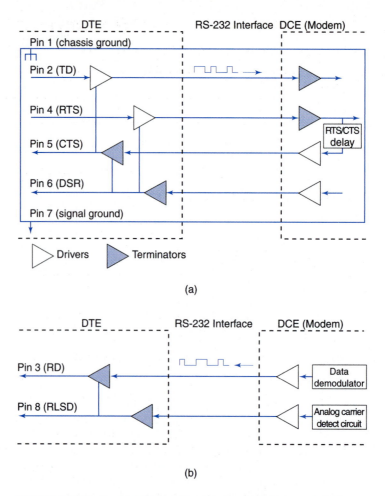

FIGURE 14-20 Functional block diagram for the drivers and terminators necessary for transmission of asynchronous data over the RS-232
interface between a DTE and a DCE (modem): (a) transmit circuits;
(b) receive circuits

or four-wire modems. Analog loopbacks disrupt the normal flow of user data and, therefore, should be performed only after a station has been properly taken off line.

A four-wire analog loopback is shown in Figure 14-22. Data from the DTE is
transferred to the modem over an RS-232 interface following standard interface procedures. The data modulates the carrier in the modem, which is then directed back into the
modem's receiver through a special loopback circuit. The modem may be two wire or
four wire. The loopback circuit is basically an analog switch with an attenuator that converts the transmit signal power to the receive signal power. The analog signal is demodulated in the receiver and then transported back to the DTE over the RS-232 interface. Because analog loopbacks enable a DTE to perform local tests by transmitting data
signals to itself through the local modem, they are sometimes called *local loopbacks*.

Analog loopbacks test only the local DTE and local DCE and do not test the
communications line or the destination DCE or DTE. Pin 18 on the RS-232 interface
is sometimes used for the loopback signal. An active signal on pin 18 places the modem into the analog loopback mode.

Digital loopback. A digital loopback (DLB) is another modem feature that disconnects
the digital side of a distant modem from the distant DTE and connects the digital output

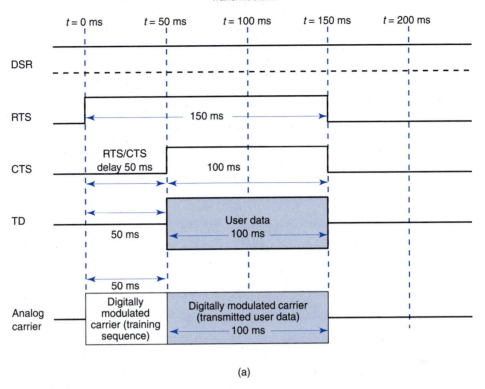

(a)

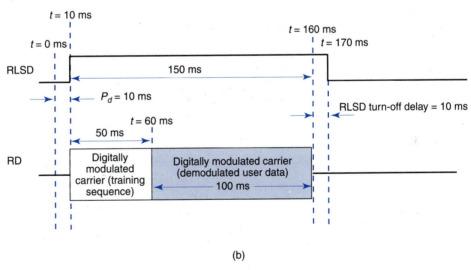

(b)

FIGURE 14-21 Typical timing diagram for control and data signals for asynchronous data transmission over the RS-232 interface between a DTE and a DCE (modem): (a) transmit timing diagram; (b) receive timing diagram

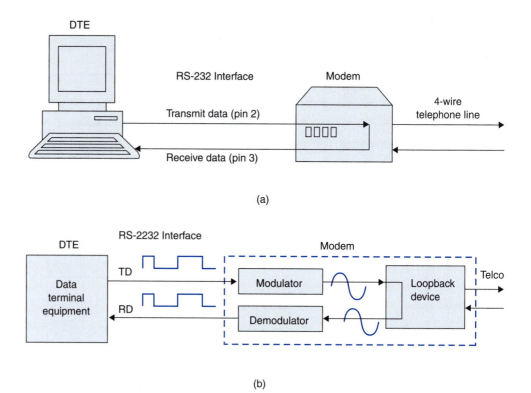

(a)

(b)

FIGURE 14-22 Analog loopback test: (a) functional diagram; (b) circuit block diagram

from the distant modem receiver to the digital input to the modem transmitter. Digital loopbacks test the transmit and receive sections of the local DTE, the local modem, and the distant modem. In addition, both directions of the communications channel are tested. Digital loopbacks are generally performed on four-wire circuits and can be initiated from the local or distant end. Because digital loopbacks break the normal transmission path, they disrupt the normal flow of user data.

A four-wire digital loopback is shown in Figure 14-23. Data from the DTE is transferred to the modem over an RS-232 interface following standard interface procedures. After the data modulates the carrier in the modem, the modulated carrier is transported through the telephone circuit to the distant modem. The distant modem receives and demodulates the data and then loops the data back to the transmitter, where it modulates the carrier. The modulated carrier is transported back to the source modem over the telephone circuit, where it is again demodulated and returned via the RS-232 interface to the source DTE. Digital loopbacks allow local DTEs to perform remote tests on local and distant (remote) modems; therefore, they are sometimes called *remote loopbacks*. Pin 21 on the RS-232 interface is sometimes used by the distant DCE to signal the distant DTE that the modem is in remote loopback and cannot send or receive user data.

14-5-1-6 DTE-to-DTE transmission using the RS-232 interface. The RS-232 interface can be used to interconnect two digital devices such as one PC to another PC or a PC to a peripheral device without using a modem. This is often desirable when the distance between the two devices is relatively short. In this case, one PC is assuming the role of DCE and the other PC or the peripheral the role of DTE—no modem is involved in the trans-

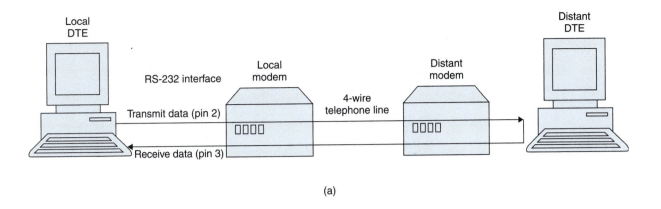

(a)

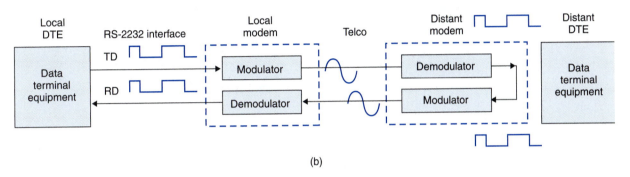

(b)

FIGURE 14-23 Digital loopback test: (a) functional diagram; (b) circuit block diagram

mission (in essence, the RS-232 interface performs the functions of the modem). Connecting two digital devices together with a standard RS-232 interface causes an interesting situation where two devices transmit and receive data on the same wires, as shown in Figure 14-24a. Obviously, a standard RS-232 interface will not work. However, a simple solution is to rewire the interface so that conflicts do not occur, as shown in Figure 14-24b. As shown in the figure, several pins on the interface are flipped over (*crisscrossed*) at one end of the interface (this is also called a *crossover*). The transmit data on one device is connected to the receive data on the other device and vice versa. The RTS and CTS wires are also reversed, as are the DTR and DSR wires. Such an arrangement is called a *null modem*. Null modems can be used to establish a digital-to-digital interconnection without the need for a modem. Figure 14-24c shows an alternate wiring configuration for a null modem. With synchronous transmission, the RxC and TxC (receive and transmit clocks) wires must also be flipped over.

14-5-2 RS-449 Serial Interface Standards

In the mid-1970s, it appeared that data rates had exceeded the capabilities of the RS-232 interface. Consequently, in 1977, the EIA introduced the RS-449 serial interface with the intention of replacing the RS-232 interface. The RS-449 interface specifies a 37-pin primary connector (DB37) and a nine-pin secondary connector (DB9) for a total of 46 pins, which provide more functions and faster data transmission rates and span greater distances than the RS-232 interface. The RS-449 is essentially an updated version of the RS-232 interface except the RS-449 standard outlines only the mechanical and functional specifications of the interface.

The RS-449 primary cable is for serial data transmission, while the secondary cable is for diagnostic information. Table 14-5a lists the 37 pins of the RS-449 primary cable and

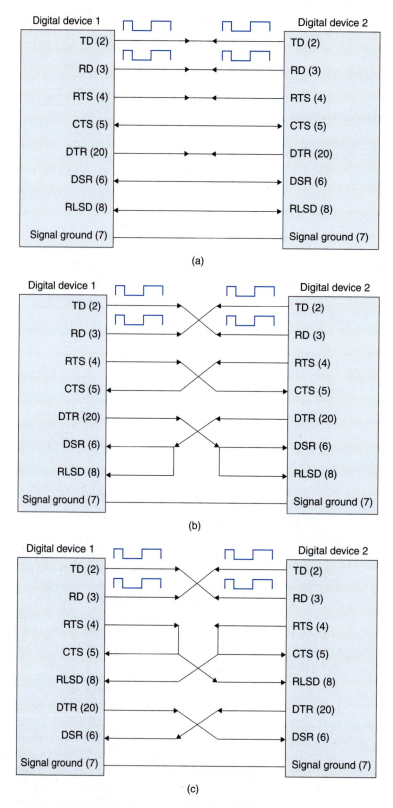

FIGURE 14-24 Interconnecting two digital devices with the RS-232 interface: (a) conflicting signals; (b) crossovers (null modem); (c) optional null modem configuration

Table 14-5a RS-449 Pin Designations (37-pin connector)

Pin Number	Pin Name	EIA Nomenclature
1	Shield	None
19	Signal	SG
37	Send common	SC
20	Receive common	RC
28	Terminal in service	IS
15	Incoming call	IC
12, 30	Terminal ready	TR
11, 29	Data mode	DM
4, 22	Send data	SD
6, 24	Receive data	RD
17, 35	Terminal timing	TT
5, 23	Send timing	ST
8, 26	Receive timing	RT
7, 25	Request to send	RS
9, 27	Clear to send	CS
13, 31	Receiver ready	RR
33	Signal quality	SQ
34	New signal	NS
16	Select frequency	SF
2	Signal rate indicator	SI
10	Local loopback	LL
14	Remote loopback	RL
18	Test mode	TM
32	Select standby	SS
36	Standby indicator	SB

Table 14-5b RS-449 Pin Designations (nine-pin connector)

Pin Number	Pin Name	EIA Nomenclature
1	Shield	None
5	Signal ground	SG
9	Send common	SC
2	Receive common	RC
3	Secondary send data	SSD
4	Secondary receive data	SRD
7	Secondary request to send	SRS
8	Secondary clear to send	SCS
6	Secondary receiver ready	SRF

their designations, and Table 14-5b lists the nine pins of the diagnostic cable and their designations. Note that the acronyms used with the RS-449 interface are more descriptive than those recommended by the EIA for the RS-232 interface. The functions specified by the RS-449 are very similar to the functions specified by the RS-232. The major difference between the two standards is the separation of the primary data and secondary diagnostic channels onto two separate cables.

The electrical specifications for the RS-449 were specified by the EIA in 1978 as either the RS-422 or the RS-423 standard. The RS-449 standard, when combined with RS-422A or RS-423A, was intended to replace the RS-232 interface. The primary goals of the new specifications are listed here:

1. Be compatible with the RS-232 interface standard
2. Replace the set of circuit names and mnemonics used with the RS-232 interface with more meaningful and descriptive names

3. Provide separate cables and connectors for the primary and secondary data channels
4. Provide single-ended or balanced transmission
5. Reduce cross-talk between signal wires
6. Offer higher data transmission rates
7. Offer longer distances over twisted-pair cable
8. Provide loopback capabilities
9. Improve performance and reliability
10. Specify a standard connector

Figure 14-25a shows the balanced digital interface circuit for the RS-422A. The RS-422A standard specifies a balanced interface cable capable of operating up to 10 Mbps and spanning distances up to 1200 m. However, this does not mean that 10 Mbps can be transmitted 1200 m. At 10 Mbps, the maximum distance is approximately 15 m, and 90 kbps is the maximum bit rate that can be transmitted 1200 m. Balanced implies that the data signal is outputted and propagates over two separate signal paths. The two signal paths see the same impedance with respect to ground. Note in Figure 14-25a that the data signals on the two balanced lines are 180° out of phase with each other (i.e., complements). The balanced terminator amplifies the difference between the two signals.

The prominent advantage of balanced operation is noise cancellation. With balanced transmission, both conductors carry signal current except the currents in the two wires travel in opposite directions. Currents that flow in opposite directions in a balanced pair are called *metallic circuit currents*. Currents that flow in the same direction are called *longitudinal currents*. External noise is generally induced into both signal paths with equal amplitudes and produces currents that travel in the same direction (longitudinal). Therefore, the noise is inherently canceled in the balanced terminator. The noise is common to both signal paths; therefore, it is referred to as *common-mode noise*. A measure of the terminator's ability to reject common-mode noise is called *common-mode rejection ratio* (CMRR).

The RS-423A standard specifies an unbalanced interface cable capable of operating at a maximum transmission rate of 100 kbps and spanning a maximum distance of 90 m. The RS-442A and RS-443A standards are similar to ITU-T V.11 and V.10, respectively. Figure 14-25b shows the unbalanced digital interface circuit for the RS-423A. With a bidirectional unbalanced line, one wire is at ground potential, and the currents in the two wires may be different. With an unbalanced line, interference is induced into only one signal path and, therefore, does not cancel in the terminator.

The primary objective of establishing the RS-449 interface standard was to maintain compatibility with the RS-232 interface standard. To achieve this goal, the EIA divided the RS-449 into two categories: *category I* and *category II* circuits. Category I circuits include only circuits that are compatible with the RS-232 standard. The remaining circuits are classified as category II. Category I and category II circuits are listed in Table 14-6.

Category I circuits can function with either the RS-422A (balanced) or the RS-423A (unbalanced) specifications. Category I circuits are allotted two adjacent wires for each RS-232 compatible signal, which facilitates either balanced or unbalanced operation. Category II circuits are assigned only one wire and, therefore, can facilitate only unbalanced (RS-423A) specifications.

The RS-449 interface provides 10 circuits not specified in the RS-232 standard:

1. *Local loopback (LL—pin 10).* Used by the DTE to request a local (analog) loopback from the DCE
2. *Remote loopback (RL—pin 14).* Used by the DTE to request a remote (digital) loopback from the distant DCE
3. *Select frequency (SF—pin 16).* Allows the DTE to select the DCE's transmit and receive frequencies

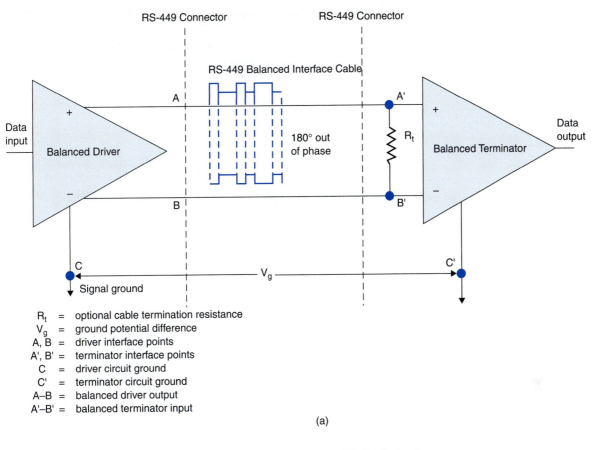

R_t = optional cable termination resistance
V_g = ground potential difference
A, B = driver interface points
A', B' = terminator interface points
C = driver circuit ground
C' = terminator circuit ground
A–B = balanced driver output
A'–B' = balanced terminator input

(a)

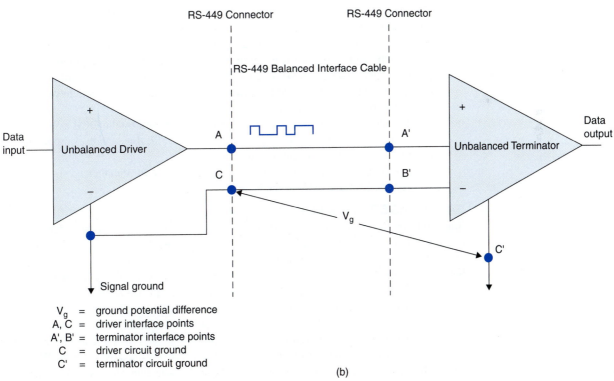

V_g = ground potential difference
A, C = driver interface points
A', B' = terminator interface points
C = driver circuit ground
C' = terminator circuit ground

(b)

FIGURE 14-25 (a) RS-422A balanced interface circuit; (b) RS-423A unbalanced interface circuit

Table 14-6 RS-449 Category I and Category II Circuits

Category I	
SD	Send data (4, 22)
RD	Receive data (6, 24)
TT	Terminal timing (17, 35)
ST	Send timing (5, 23)
RT	Receive timing (8, 26)
RS	Request to send (7, 25)
CS	Clear to send (9, 27)
RR	Receiver ready (13, 31)
TR	Terminal ready (12, 30)
DM	Data mode (11, 29)

Category II	
SC	Send common (37)
RC	Receive common (20)
IS	Terminal in service (28)
NS	New signal (34)
SF	Select frequency (16)
LL	Local loopback (10)
RL	Remote loopback (14)
TM	Test mode (18)
SS	Select standby (32)
SB	Standby indicator (36)

4. *Test mode (TM—pin 18).* Used by the DTE to signal the DCE that a test is in progress
5. *Receive common (RC—pin 20).* Common return wire for unbalanced signals propagating from the DCE to the DTE
6. *Terminal in service (IS—pin 28).* Used by the DTE to signal the DCE whether it is operational
7. *Select standby (SS—pin 32).* Used by the DTE to request that the DCE switch to standby equipment in the event of a failure on the primary equipment
8. *New signal (NS—pin 34).* Used with a modem at the primary location of a multipoint data circuit so that the primary can resynchronize to whichever secondary is transmitting at the time
9. *Standby indicator (SB—pin 36).* Intended to be used by the DCE as a response to the select standby (SS) signal to notify the DTE that standby equipment has replaced the primary equipment
10. *Send common (SC— pin 37).* Common return wire for unbalanced signals propagating from the DTE to the DCE

14-5-3 RS-530 Serial Interface Standards

Since industry did not readily adopt the RS-449 interface, it came and went virtually unnoticed by most of the data communications industry. Consequently, in 1987, the EIA introduced another new standard, the RS-530 serial interface, which was intended to operate at data rates between 20 kbps and 2 Mbps using the same 25-pin DB-25 connector used by the RS-232 interface. The pin functions of the RS-530 interface are essentially the same as the RS-449 category I pins with the addition of three category II pins: local loopback, remote loopback, and test mode. Table 14-7 lists the 25 pins for the RS-530 interface and their designations.

Like the RS-449 standard, the RS-530 interface standard does not specify electrical parameters. The electrical specifications for the RS-530 are outlined by either the RS-422A or the RS-423A standard. The RS-232, RS-449, and RS-530 interface standards provide specifications for answering calls but do not provide specifications for initiating calls (i.e.,

Table 14-7 RS-530 Pin Designations

Signal Name	Pin number
Shield	1
Transmit data[a]	2, 14
Receive data[a]	3, 16
Request to send[a]	4, 19
Clear to send[a]	5, 13
DCE ready[a]	6, 22
DTE ready[a]	20, 23
Signal ground	7
Receive line signal detect[a]	8, 10
Transmit signal element timing (DCE source)[a]	15, 12
Receive signal element timing (DCE source)[a]	17, 9
Local loopback[b]	18
Remote loopback[b]	21
Transmit signal element timing (DTE source)[a]	24, 11
Test mode[b]	25

[a]Category I circuits (RS-422A).
[b]Category II circuits (RS-423A).

dialing). The EIA has a different standard, RS-366, for automatic calling units. The principal use of the RS-366 is for dial backup of private-line data circuits and for automatic dialing of remote terminals.

14-5-4 CCITT X.21 Recommendation

In 1976, the CCITT (now the ITU-T) introduced the *X.21 recommendation*, which includes specifications for placing and receiving calls and for sending and receiving data using full-duplex synchronous transmission. The X.21 recommendation presumes a direct digital connection to a digital telephone network. Thus, all data transmissions must be synchronous, and the data communications equipment must provide both bit and character synchronization. The minimum data rate used with X.21 is 64 kbps because this is the bit rate currently used to encode voice in digital form onto the public telephone network. The ITU-T has a comparable standard for asynchronous data communications designated X.20.

A significant number of the pins on the RS-232, RS-449, and RS-530 interfaces are used for transporting control signals. Control signals are represented with positive or negative voltage levels. However, if control signals are encoded using control characters, such as those available with ASCII, they can be transported over the data lines. For this reason, the X.21 standard eliminates many of the control lines and directs control signals over data lines. This allows the X.21 standard to use significantly fewer wires than the EIA standards and to be used to propagate control signals between devices over digital telecommunications network. Therefore, X.21 is useful as an interface between DTEs and DCEs as well as between digital computers and digital interfaces, such as ISDN and X.25 (both described in later chapters).

X.21 uses a 15-pin (DB15) connector and specifies only 12 signals, which are listed in Table 14-8. Data and control signals are transmitted toward the DCE on the transmit line, and the DCE returns data and control signals on the receive line. The two control and two indication lines are control channels for the two transmission directions. The signal element timing line carries the bit-timing signal (clock), and the byte-timing line carries the character synchronization information. The electrical specifications for X.21 are listed in either recommendation X.26 (balanced) or recommendation X.27 (unbalanced).

The major advantage of the X.21 recommendation over the RS-232 and RS-530 standards is that X.21 signals are encoded in serial digital form, which sets the stage for providing special new services in computer communications.

Table 14-8 X.21 Pin Designations

Pin number	Function
1	Shield
2	Transmit data or control
3	Control
4	Receive data or control
5	Indication
6	Signal element timing
7	Byte timing
8	Signal ground
9	Transmit data or control
10	Control
11	Receive data or control
12	Indication
13	Signal element timing
14	Byte timing
15	Reserved

QUESTIONS

14-1. Define *data communications hardware*.

14-2. Define *data terminal equipment*.

14-3. List several types of data terminal equipment.

14-4. Briefly describe the following pieces of equipment: *teletypewriter*, *video display terminal*, *transactional terminal*, *specialized terminal*, *intelligent terminal*, *smart terminal*, *dumb terminal*, *workstation*, *line-control unit*, *front-end processor*.

14-5. Define *data communications equipment*.

14-6. List several types of data communications equipment.

14-7. What are the seven components that make up a two-point data communications circuit?

14-8. What is a *cluster*? A *cluster controller*?

14-9. Describe the differences between a line-control unit and a front-end processor.

14-10. Briefly describe the operation of a *UART transmitter*. A UART receiver.

14-11. Briefly describe the operation of a *USRT transmitter*. A UART receiver.

14-12. What is the function of the *control register* in a UART?

14-13. List and describe the standard inputs to a UART control register.

14-14. What is the purpose of the *status register* in a UART?

14-15. List and describe the *standard flags* of a UART status register.

14-16. List the primary functions performed by a *UART*. A *USRT*.

14-17. What is the purpose of a *start-bit verification circuit*?

14-18. Explain why the receive clock in a UART is much faster than the receive data rate.

14-19. List the status flags in a USRT that are not included in a UART.

14-20. List the features provided by a *serial interface*.

14-21. What is the purpose of a serial interface?

14-22. What is the nominal maximum length for the RS-232 interface ?

14-23. What are the general classifications of the pins on the RS-232 interface?

14-24. What is the difference between a *driver* and a *terminator*?

14-25. Explain how a *noise margin* is achieved with an RS-232 interface.

14-26. Briefly describe the functions of each pin on the RS-232 interface.

14-27. Describe the primary difference between the *RS-232 interface* and the RS-449 interface.

14-28. List and describe the two types of *loopbacks* associated with DTEs and DCEs.

14-29. List the primary goals for establishing the RS-449 interface.

14-30. Describe the differences between *category I* and *category II* circuits on the RS-449 interface.

14-31. Briefly describe the *X.21 recommendation*.

PROBLEMS

14-1. Determine the mode instruction word for a UART with the following parameters:

Even parity	Six-bit characters
1 stop bit	1× receive clock rate

14-2. Determine the mode instruction word for a UART with the following parameters:

No parity	Seven-bit characters
1.5 stop bits	16× receive clock rate

14-3. Determine the mode instruction word for a UART with the following parameters:

Odd parity	Five-bit characters
2 stop bits	32× receive clock rate

14-4. Determine the status word for a UART with the following parameters:

Active high-status signals	No overrun error
DSR inactive	Parity error
No framing error	Receive data available
Unused bits = 0	Transmit buffer empty active

14-5. Determine the status word for a UART with the following parameters:

Active high-status signals	Overrun error
DSR active	No parity error
Framing error	Receive data available
Unused bits = 0	Transmit buffer empty active

14-6. Determine the bit time, receive clock rate, and maximum detection error for a UART receiving data at 2000 bps (f_b) with a receiver clock 16 times (16×) faster than the incoming data.

14-7. Determine the bit time, receive clock rate, and maximum detection error for a UART receiving data at 600 bps (f_b) with a receiver clock 32 times (32×) faster than the incoming data.

14-8. Determine the noise margins for an RS-232 interface with driver output signal voltages of ±12 V.

14-9. Determine the noise margins for an RS-232 interface with driver output signal voltages of ±11 V.

14-10. Sketch the timing diagram for control and data signals for an asynchronous data transmission over an RS-232 interface with the following parameters:

Modem RTS-CTS delay = 200 ms

DTE primary data message length = 200 ms

Modem training sequence = 200 ms

Propagation time = 50 ms

Modem RLSD turn-off delay time = 10 ms

14-11. Sketch the timing diagram for control and data signals for an asynchronous data transmission over an RS-232 interface with the following parameters:

Modem RTS-CTS delay = 150 ms

DTE primary data message length = 2000 ms

Modem training sequence = 150 ms

Propagation time = 10 ms

Modem RLSD turn-off delay time = 20 ms

14-12. Sketch the timing diagram for control and data signals for an asynchronous data transmission over an RS-232 interface with the following parameters:

Modem RTS-CTS delay = 100 ms

DTE primary data message length = 50 ms

Modem training sequence = 100 ms

Propagation time = 100 ms

Modem RLSD turn-off delay time = 10 ms

CHAPTER 15

Data Communications Equipment

OBJECTIVES

- Define *data terminal equipment* and *data communications equipment*
- Describe the basic functions of a channel service unit
- Describe the basic functions of a digital service unit
- Define *voiceband data communications modem*
- Explain the difference between *bits per second* and *baud*
- List and describe the basic blocks of a voice-band modem
- Describe the two classifications for voice-band modems
- Describe the characteristics of *asynchronous voice-band modems*
- List and describe the basic blocks of an asynchronous voice-band modem
- Describe the characteristics of synchronous voice-band modems
- Explain modem synchronization
- Describe the function of a modem equalizers
- Explain modem training
- Describe the purpose of scrambler and descrambler circuits
- Describe the meaning of the terms *bis* and *terbo*
- Describe the basic characteristics of the following ITU-T voice-band modem specifications: ITU-T V.29, V.32, V.32bis, V.32terbo, V.33, V.34 fast, V.34+, V.42, V.42bis, V.90, and V.92

- List and describe the most common modulation methods used with 56K voiceband modems
- Describe and explain the difference between probability of error and bit error rate

15-1 INTRODUCTION

Data communications equipment (DCE) describes the hardware that allows *digital terminal equipment* (DTE) to access a transmission facility, such as a metallic or optical fiber cable. Because DCEs are used to terminate the digital portion of a data communications circuit, they are sometimes called *data circuit-terminating equipment* (DCTE).

Digital terminal equipment are always some type of digital device; however, the transmission facility they connect to may have been designed for either digital or analog signals. Therefore, there are two basic types of data communications equipment. The first type of DCE is called a *modem*. Modems allow digital terminal equipment to access and interface to analog transmission facilities. In essence, the purpose of a modem is to convert digital signals to analog signals and vice versa. Therefore, modems include both an *analog-to-digital converter* (ADC) and a *digital-to-analog converter* (DAC). The ADC is used in the modem's receiver, and the DAC is used in the modem's transmitter. The second type of DCE actually involves two devices that allow digital equipment to access and interface with digital transmission facilities. The two devices are called *channel service units* (CSUs) and *digital service units* (DSUs). Although CSUs and DSUs perform different functions, they are often combined into one unit.

15-2 DIGITAL SERVICE UNIT AND CHANNEL SERVICE UNIT

Digital service units (DSUs) and *channel service units* (CSUs) are *customer premise equipment* (CPE) used to terminate a digital circuit at a subscriber's location and allow the subscriber to connect to a local central telephone office. CSUs and DSUs allow devices connected to local area networks to access and connect to digital transmission facilities leased from service providers, such as a local telephone company. The digital transmission facilities vary in capacity from switched or dedicated 56 kbps *dataphone digital service* (DDS) lines to dedicated T1 carrier systems with bit rates between 64 kbps and 1.544 Mbps.

A typical CSU/DSU configuration could involve attaching a local area network's bridge or router to a telecommunications line or connecting a mainframe computer to a telecommunications line. The DSU is connected to a CSU, which in turn is connected to the telecommunications line. Since the divesture of AT&T, most vendor-provided routers and bridges are manufactured with built-in combination CSU/DSUs, which allows customer premise equipment to access the telecommunications line directly. Figure 15-1 shows two examples of how DSUs and CSUs are used to interconnect customer provided equipment to a telephone company central office.

15-2-1 Digital Service Unit

At the transmit end, a digital service unit (DSU) converts unipolar digital signals (such as TTL) from a local area network's digital terminal equipment into self-clocking bipolar digital signals that are capable of being transmitted more efficiently over a telecommunications line. At the receive end, a DSU removes any special codes inserted by the transmitting DSU and converts the bipolar digital signals back to unipolar. A DSU may also provide timing (clock) recovery, control signaling, and synchronous sampling.

15-2-2 Channel Service Unit

A channel service unit (CSU) serves as the demarcation point between the digital station equipment and the telecommunications line. A CSU physically terminates the telecommunications line, performs signal regeneration and reshaping, performs zero substitution, and converts digital signals to a format more suitable for transmission over the digital

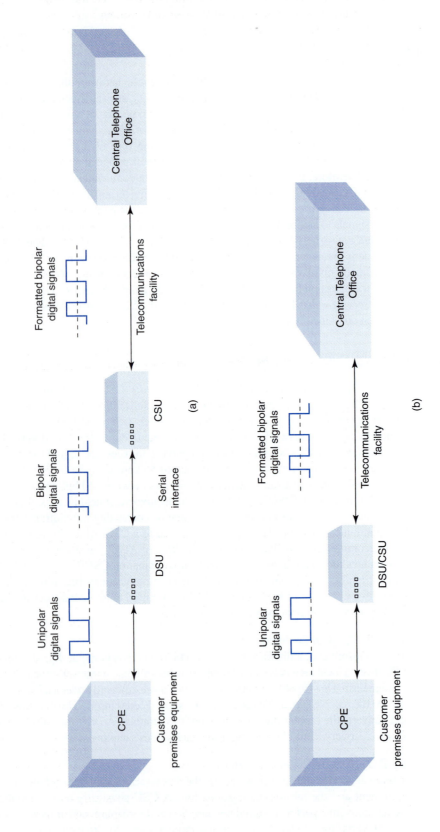

FIGURE 15-1 Channel service units and digital service units: (a) separate DSU and CSU; (b) combined DSU/CSU

transmission facility, such as AMI, 2B1Q encoding. CSUs may also perform other functions, such as channel multiplexing and line conditioning (equalization); execute certain carrier-controlled tests, such as local and remote loopbacks; and provide performance measurement capabilities and statistical compilations. Common CSU interfaces use a standard RJ-48 modular connector or a V.35 connection.

15-3 VOICE-BAND DATA COMMUNICATIONS MODEMS

The most common type of data communications equipment (DCE) is the *data communications modem*. Alternate names include *datasets, dataphones,* or simply *modems*. The word *modem* is a contraction derived from the words *modulator* and *demodulator*.

In the 1960s, the business world recognized a rapidly increasing need to exchange digital information between computers, computer terminals, and other computer-controlled digital equipment that were separated by substantial distances. The only transmission facilities available at the time were analog voice-band telephone circuits. Telephone circuits were designed for transporting analog voice signals within a bandwidth between approximately 300 Hz and 3000 Hz. In addition, telephone circuits often included amplifiers and other analog devices that would not propagate digital signals. Therefore, modems were designed to communicate with each other using analog signals that occupied the same bandwidth used for standard voice telephone communications. Data communications modems designed to operate over the limited bandwidth of the public telephone network are called *voice-band modems.*

Because digital information cannot be transported directly over analog transmission media (at least not in digital form), the primary purpose of a *data communications modem* is to interface computers, computer networks, and other digital terminal equipment to analog communications facilities. Modems are also used when computers are too far apart to be directly interconnected using standard computer cables. In the transmitter (modulator) section of a modem, digital signals are encoded onto an analog carrier signal. The digital signals modulate the carrier, producing digitally modulated analog signals that are capable of being transported through the analog communications media. Therefore, the output of a modem is an analog signal that is carrying digital information. In the receiver section of a modem, digitally modulated analog signals are demodulated. Demodulation is the reverse process of modulation. Therefore, modem receivers (demodulators) simply extract digital information from digitally modulated analog carriers.

The most common (and simplest) modems available are those intended to be used to interface DTEs through a serial interface to standard voice-band telephone lines and provide reliable data transmission rates between 300 bps and 56 kbps. These types of modems are sometimes called *telephone-loop modems* or *POTS modems,* as they are connected to the telephone company through the same local loops used for standard voice telephone circuits. More sophisticated modems (sometimes called *broadband modems* or *cable modems*) are also available that are capable of transporting data at much higher bit rates over wideband communications channels, such as those available with optical fiber, coaxial cable, microwave radio, and satellite communications systems. Broadband modems operate using a different set of standards and protocols than telephone loop modems.

A modem is, in essence, a transparent repeater that converts electrical signals in digital form to electrical signals in analog form and vice versa. A modem is transparent, as it does not interpret or change the information contained in the data. It is a repeater, as it is not a destination for data—it simply repeats or retransmits it. Modems are physically located between digital terminal equipment (DTE) and the analog communications channel. Modems work in pairs with one located at each end of a data communications circuit. The two modems do not need to be manufactured by the same company; however, they must use compatible protocols, modulation schemes, data encoding formats, and transmission rates.

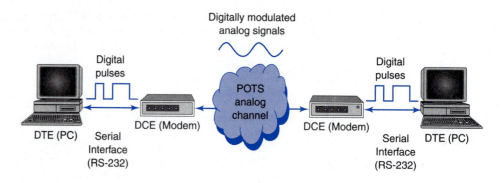

FIGURE 15-2 Data communications modems—POTS analog channel

Figure 15-2 shows how a typical voice-band modem is used to facilitate the transmission of digital data between DTEs over a standard voice-grade telephone circuit. At the transmit end, the modem receives discrete digital pulses (which are usually in binary form) from a DTE through a serial digital interface (such as the RS-232). The DCE converts the digital pulses to analog signals. In essence, a modem transmitter is a *digital-to-analog converter* (DAC), as it converts digital pulses to analog representations. The analog signals are outputted onto an analog communications channel where they are transported through the system to a distant receiver. At the destination end of a data communications system, a modem receives analog signals from the communications channel and converts them to digital pulses. In essence, a modem receiver is an *analog-to-digital converter* (ADC). The demodulated digital pulses are then outputted onto a serial digital interface and transported to the DTE.

15-3-1   Bits per Second versus Baud

The parameters *bits per second* (bps) and *baud* are often misunderstood and, consequently, misused. Baud, like bit rate, is a rate of change; however, baud refers to the rate of change of the signal on the transmission medium after encoding and modulation have occurred. Bit rate refers to the rate of change of a digital information signal, which is usually binary. Baud is the reciprocal of the time of one output *signaling element,* and a signaling element may represent several information bits. A signaling element is sometimes called a *symbol,* which could encode the data as a change in the amplitude, frequency, or phase. For example, binary signals are generally encoded and transmitted one bit at a time in the form of two discrete voltage levels that represent logic 1s (highs) and logic 0s (lows). A baud is also transmitted one at a time; however, a baud may represent more than one information bit. Thus, the baud of a data communications system may be considerably less than the bit rate.

15-4   BELL SYSTEM–COMPATIBLE VOICE-BAND MODEMS

At one time, Bell System modems were virtually the only modems in existence. This is because AT&T operating companies once owned 90% of the telephone systems in the United States and the AT&T operating tariff allowed only equipment manufactured by Western Electric Company and furnished by Bell System operating companies to be connected to AT&T telephone lines. However, in 1968, AT&T lost a landmark Supreme Court decision, the *Carterfone decision,* which allowed equipment manufactured by non-Bell companies to interconnect to the vast AT&T communications network, providing the equipment met Bell System specifications. The *Carterfone* decision began the *interconnect industry,* which has led to competitive data communications offerings by a large number of independent companies.

The operating parameters for Bell System modems are the models from which the international standards specified by the ITU-T evolved. Bell System modem specifications apply only to voice-band modems that existed in 1968; therefore, their specifications pertain only to modems operating at data transmission rate of 9600 bps or less.

15-5 VOICE-BAND MODEM BLOCK DIAGRAM

Figure 15-3 shows a simplified block diagram for a data communications modem. For simplicity, only the primary functional blocks of the transmitter and receiver are shown. The basic principle behind a modem transmitter is to convert information received from a DTE in the form of binary digits (bits) to digitally modulated analog signals. The reverse process is accomplished in the modem receiver.

The primary blocks of a modem are described here:

1. *Serial interface circuit.* This interfaces the modem transmitter and receiver to the serial interface. The transmit section accepts digital information from the serial interface, converts it to the appropriate voltage levels, and then directs the information to the modulator. The receive section receives digital information from the demodulator circuit, converts it to the appropriate voltage levels, and then directs the information to the serial interface. In addition, the serial interface circuit manages the flow of control, timing, and data information transferred between the DTE and the modem, which includes handshaking signals and clocking information.

2. *Modulator circuit.* This receives digital information from the serial interface circuit. The digital information modulates an analog carrier producing a digitally modulated analog signal. In essence, the modulator converts digital changes in the information to analog changes in the carrier. The output from the modulator is directed to the transmit bandpass filter and equalizer circuit.

3. *Bandpass filter and equalizer circuit.* There are bandpass filter and equalizer circuits in both the transmitter and the receiver sections of the modem. The transmit bandpass filter limits the bandwidth of the digitally modulated analog signals to a bandwidth appropriate for transmission over a standard telephone circuit. The receive bandpass filter limits the bandwidth of the signals allowed to reach the demodulator circuit, thus reducing noise

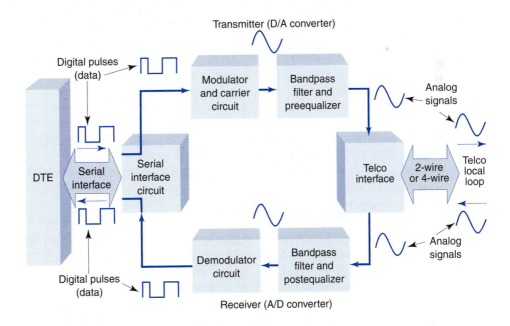

FIGURE 15-3 Simplified block diagram for a modem

and improving system performance. Equalizer circuits compensate for bandwidth and gain imperfections typically experienced on voice-band telephone lines.

4. *Telco interface circuit.* The primary functions of the telco interface circuit are to match the impedance of the modem to the impedance of the telephone line and regulate the amplitude of the transmit signal. The interface also provides electrical isolation and protection and serves as the demarcation (separation) point between subscriber equipment and telephone company–provided equipment. The telco line can be two wire or four wire, and the modem can operate half or full duplex. When the telephone line is two wire, the telco interface circuit would have to perform four-wire-to-two-wire and two-wire-to-four-wire conversions.

5. *Demodulator circuit.* This receives modulated signals from the bandpass filter and equalizer circuit and converts the digitally modulated analog signals to digital signals. The output from the demodulator is directed to the serial interface circuit, where it is passed on to the serial interface.

6. *Carrier and clock generation circuit.* The carrier generation circuit produces the analog carriers necessary for the modulation and demodulation processes. The clock generation circuit generates the appropriate clock and timing signals required for performing transmit and receive functions in an orderly and timely fashion.

15-6 VOICE-BAND MODEM CLASSIFICATIONS

Data communications modems can be generally classified as either *asynchronous* or *synchronous* and use one of the following digital modulation schemes: amplitude-shift keying (ASK), frequency-shift keying (FSK), phase-shift keying (PSK), or quadrature amplitude modulation (QAM). However, there are several additional ways modems can be classified, depending on which features or capabilities you are trying to distinguish. For example, modems can be categorized as internal or external; low speed, medium speed, high speed, or very high speed; wideband or voice band; and personal or commercial. Regardless of how modems are classified, they all share a common goal, namely, to convert digital pulses to analog signals in the transmitter and analog signals to digital pulses in the receiver.

Some of the common features provided data communications modems are listed here:

- Automatic dialing, answering, and redialing
- Error control (detection and correction)
- Caller ID recognition
- Self-test capabilities, including analog and digital loopback tests
- Fax capabilities (transmit and receive)
- Data compression and expansion
- Telephone directory (telephone number storage)
- Adaptive transmit and receive data transmission rates (300 bps to 56 kbps)
- Automatic equalization
- Synchronous or asynchronous operation

15-7 ASYNCHRONOUS VOICE-BAND MODEMS

Asynchronous voiceband modems can be generally classified as low-speed voice-band modems, as they are typically used to transport asynchronous data (i.e., data framed with start and stop bits). Synchronous data is sometimes used with an asynchronous modem; however, it is not particularly practical or economical. Synchronous data transported by asynchronous modems is called *isochronous transmission.* Asynchronous modems use relatively simple modulation schemes, such as ASK or FSK, and are restricted to relatively low-speed applications (generally less than 2400 bps), such as telemetry and caller ID.

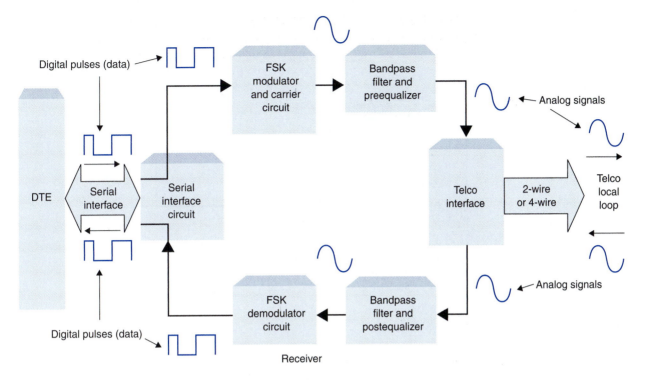

FIGURE 15-4 Simplified block diagram for an asynchronous FSK modem

Figure 15-4 shows a simplified block diagram for an asynchronous voice-band FSK modem. In the transmitter, serial data received from the DTE is applied directly to the FSK modulator. Clocking information is not required because FSK modulators are typically voltage-controlled oscillators, which react to voltage levels independent of any timing reference. The voltage for a binary 1 produces a mark frequency, and the voltage for a binary 0 produces a space frequency. Thus, FSK modulators use single-bit encoding, which means that each input bit modulates the carrier independent of any other data bits. Single-bit encoding produces an output frequency that is changing (shifting) at the same rate as the input data. The rate of change of the digital input data is called the bit rate and the rate of change of output analog signal is called baud (or baud rate). With FSK, the output frequency changes at the same rate as the input data rate; therefore, with FSK the bit rate is equal to the baud.

The minimum bandwidth necessary to propagate an FSK signal is dependent on the mark and space frequencies and the bit rate. The modulated FSK signal is band-limited and preequalized and then outputted onto the telephone line. Asynchronous receivers use noncoherent demodulation, which means that the carrier and clocking information is not recovered in the receiver.

There are several standard asynchronous voice-band modems designed for low-speed data applications using the switched public telephone network. To operate full duplex with a two-wire dial-up circuit, it is necessary to divide the usable bandwidth of a voice-band circuit in half, creating two equal-capacity data channels. A popular modem that does this is the Bell System 103 or compatible modem.

15-7-1 Bell System 103–Compatible Modem
The 103 modem is capable of full-duplex operation over a two-wire telephone line at bit rates up to 300 bps. With the 103 modem, there are two data channels, each with their own mark and space frequencies. One data channel is called the *low-band channel* and occupies a bandwidth from 300 Hz to 1650 Hz (i.e., the lower half of the usable voice band). A second data

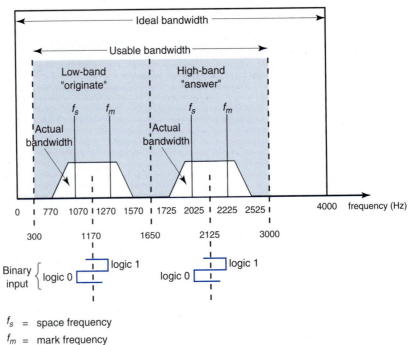

f_s = space frequency
f_m = mark frequency

FIGURE 15-5 Frequency and bandwidth allocation for a Bell System 103–compatible modem

channel, called the *high-band channel,* occupies a bandwidth from 1650 Hz to 3000 Hz (i.e., the upper half of the usable voice band). The mark and space frequencies for the low-band channel are 1270 Hz and 1070 Hz, respectively. The mark and space frequencies for the high-band channel are 2225 Hz and 2025 Hz, respectively. Separating the usable bandwidth into two narrower bands is called *frequency-division multiplexing* (FDM). FDM allows full duplex (FDX) transmission over a two-wire circuit, as signals can propagate in both directions at the same time without interfering with each other because the frequencies for the two directions of propagation are different. FDM allows full-duplex operation over a two-wire telephone circuit. Because FDM reduces the effective bandwidth in each direction, it also reduces the maximum data transmission rates. A 103 modem operates at 300 baud and is capable of simultaneous transmission and reception of 300 bps.

The frequency spectrum for a 103 modem is shown in Figure 15-5. The low-band channel is called the *originate band,* and the high-band channel is called the *answer band.* The designations originate and answer were chosen because a modem that originates a call transmits on the low-band (originate) frequencies and receives on the high-band frequencies, and the modem that answers the call transmits on the high-band (answer) frequencies and receives on the low-band frequencies.

As shown in Figure 15-5, as the binary input to the FSK modulator switches from a logic 1 to a logic 0, the output frequency switches from a mark frequency to a space frequency and vice versa. The receiver converts mark frequencies to logic 1 voltage levels and space frequencies to logic 0 voltage levels.

The ideal bandwidth of a standard voice-grade telephone circuit is 4000 Hz (0 Hz to 4000 Hz); however, the usable bandwidth is limited to approximately 2700 Hz (300 Hz to 3000 Hz). The minimum bandwidth for an FSK can be approximated by the formula

$$B = 2(\Delta f + f_b) \tag{15-1}$$

where B = bandwidth (hertz)

Δf = frequency deviation $\left(\dfrac{|f_m - f_s|}{2} \text{ hertz} \right)$

f_b = bit rate (bps)

For a 103 modem, the minimum bandwidth for the originate and answer bands are determined as follows:

originate band   answer band

$$\Delta f = \frac{|1270 - 1070|}{2} = 100 \qquad \Delta f = \frac{|2225 - 2025|}{2} = 100$$

Therefore,

$$B = 2(100 + 300)$$
$$= 800 \text{ Hz } (\pm 400 \text{ Hz})$$

As shown in Figure 15-5, the passbands for the originate and answer bands are symmetrical around the frequency located halfway between the mark and space frequencies. Therefore, the bandwidths for the originate and answer channels are

$$\text{originate band} = 1170 \text{ Hz} \pm 400 \text{ Hz} = 770 \text{ Hz to } 1570 \text{ Hz}$$
$$\text{answer band} = 2125 \text{ Hz} \pm 400 \text{ Hz} = 1725 \text{ Hz to } 2525 \text{ Hz}$$

With a data transmission rate of 300 bps and a minimum bandwidth of 800 Hz, the bandwidth efficiency for a 103 modem is

$$\text{bandwidth efficiency} = \frac{300 \text{ bps}}{800 \text{ Hz}} \times 100 = 37.5\%$$

15-7-2 Bell System 202–Compatible Modem

The 202T and 202S modem are identical except the 202T modem specifies four-wire, full-duplex operation and the 202S modem specifies two-wire, half-duplex operation. Therefore, the 202T is utilized on four-wire private-line data circuits, and the 202S modem is designed for the two-wire switched public telephone network. Probably the most common application of the 202 modem today is caller ID, which is a simplex system with the transmitter in the telephone office and the receiver at the subscriber's location.

The 202 modem is an asynchronous 1200-baud transceiver utilizing frequency-shift keying with a transmission bit rate of 1200 bps over a standard voice-grade telephone line. The frequency spectrum for a 202 modem is shown in Figure 15-6. The mark frequency is 1200 Hz, and the space frequency is 2200 Hz. You might note that the mark frequency for the 202 modem is the lower frequency, while the space frequency for the 103 modem is the lower frequency. There is no existing standard that specifies that the mark or space frequency must be the higher or lower frequency.

From Equation 15-1, the minimum bandwidth for a 202 modem is

$$\Delta f = \frac{|1200 - 2200|}{2} = 500$$

and

$$B = 2(500 + 1200)$$
$$B = 3400 \text{ Hz}$$

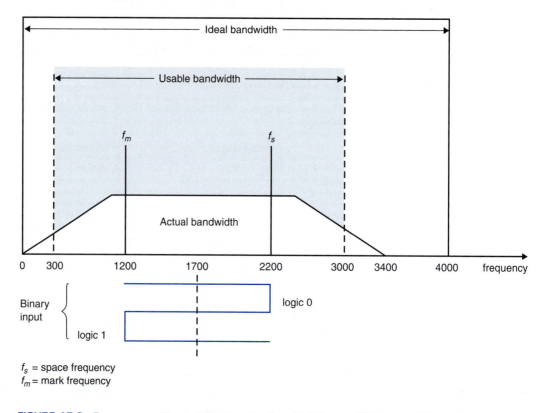

FIGURE 15-6 Frequency and bandwidth allocation for a Bell System 202–compatible modem

From the previous calculation and Figure 15-6, it can be seen that the minimum bandwidth for a 202 modem operating at 1200 bps exceeds the usable bandwidth for a standard voice-grade telephone circuit by 700 Hz. However, Equation 15-1 is an approximation and assumes the worst-case condition, which for FSK is an alternating 1/0 data pattern. Since the majority of the time the energy is concentrated within the 300-Hz to 3000-Hz bandwidth, data transmission rates of 1200 bps with acceptable bit error rates are possible with the 202 modem.

With a data transmission rate of 1200 bps and an available bandwidth of 2700 Hz, the bandwidth efficiency for a 202 modem is

$$\text{bandwidth efficiency} = \frac{1200 \text{ bps}}{2700 \text{ Hz}} \times 100 = 44.4\%$$

15-8 SYNCHRONOUS VOICE-BAND MODEMS

Synchronous modems use phase-shift keying (PSK) or quadrature amplitude modulation (QAM) to transport synchronous data (i.e., data preceded by unique SYN characters) at transmission rates between 2400 bps and 56,000 bps over standard voice-grade telephone lines. The modulated carrier is transmitted to the distant modem, where a coherent carrier is recovered and used to demodulate the data. The transmit clock is recovered from the data and used to clock the received data into the DTE. Because of the addition of clock and carrier recovery circuits, synchronous modems are more complicated and, thus, more expensive than asynchronous modems.

PSK is commonly used in medium speed synchronous voice-band modems, typically operating between 2400 bps and 4800 bps. More specifically, QPSK is generally used with 2400-bps modems and 8-PSK with 4800-bps modems. QPSK has a bandwidth efficiency of 2 bps/Hz; therefore, the baud rate and minimum bandwidth for a 2400-bps synchronous modem are 1200 baud and 1200 Hz, respectively. The standard 2400-bps synchronous modem is the Bell System 201C or equivalent. The 201C modem uses a 1600-Hz carrier frequency and has an output spectrum that extends from approximately 1000 Hz to 2200 Hz. Because 8-PSK has a bandwidth efficiency of 3 bps/Hz, the baud rate and minimum bandwidth for 4800-bps synchronous modems are 1600 baud and 1600 Hz, respectively. The standard 4800-bps synchronous modem is the Bell System 208A. The 208A modem also uses a 1600-Hz carrier frequency but has an output spectrum that extends from approximately 800 Hz to 2400 Hz. Both the 201C and the 208A are full-duplex modems designed for use with four-wire private-line circuits. The 201C and 208A modems can operate over two-wire dial-up circuits but only in the simplex mode. There are also half-duplex two-wire versions of both modems: the 201B and 208B.

High-speed synchronous voice-band modems operate at 9600 bps and use 16-QAM modulation. 16-QAM has a bandwidth efficiency of 4 bps/Hz; therefore, the baud and minimum bandwidth for 9600-bps synchronous modems are 2400 baud and 2400 Hz, respectively. The standard 9600-bps modem is the Bell System 209A or equivalent. The 209A uses a 1650-Hz carrier frequency and has an output spectrum that extends from approximately 450 Hz to 2850 Hz. The Bell System 209A is a four-wire synchronous voice-band modem designed to be used on full-duplex private-line circuits. The 209B is the two-wire version designed for half-duplex operation on dial-up circuits.

Table 15-1 summarizes the Bell System voice-band modem specifications. The modems listed in Table 15-1 are all relatively low speed by modern standards. Today, the Bell System–compatible modems are used primarily on relatively simple telemetry circuits, such as remote alarm systems, and on metropolitan and wide-area private-line data networks, such

Table 15-1 Bell System Modem Specifications

Bell System Designation	Transmission Facility	Operating Mode	Circuit Arrangement	Synchronization Mode	Modulation	Transmission Rate
103	Dial-up	FDM/FDX	Two wire	Asynchronous	FSK	300 bps
113A/B	Dial-up	FDM/FDX	Two wire	Asynchronous	FSK	300 bps
201B	Dial-up	HDX	Two wire	Synchronous	QPSK	2400 bps
201C	Private line	FDX	Four wire	Synchronous	QPSK	2400 bps
202S	Dial-up	HDX	Two wire	Asynchronous	FSK	1200 bps
202T	Private line	FDX	Four wire	Asynchronous	FSK	1800 bps
208A	Private line	FDX	Four wire	Synchronous	8-PSK	4800 bps
208B	Dial-up	HDX	Two wire	Synchronous	8-PSK	4800 bps
209A	Private line	FDX	Four wire	Synchronous	16-QAM	9600 bps
209B	Dial-up	HDX	Two wire	Synchronous	16-QAM	9600 bps
212A	Dial-up	HDX	Two wire	Asynchronous	FSK	600 bps
212B	Private line	FDX	Four wire	Synchronous	QPSK	1200 bps

Dial-up = switched telephone network
Private line = dedicated circuit
FDM = frequency-division multiplexing
HDX = half duplex
FDX = full duplex
FSK = frequency-shift keying
QPSK = four-phase PSK
8-PSK = eight-phase PSK
16-QAM = 16-state QAM

as those used by department stores to keep track of sales and inventory. The more advanced, higher-speed data modems are described in a later section of this chapter.

15-9 MODEM SYNCHRONIZATION

During the request-to-send/clear-to-send (RTS/CTS) delay, a transmit modem outputs a special, internally generated bit pattern called a *training sequence*. This bit pattern is used to synchronize (train) the receive modem at the distant end of the communications channel. Depending on the type of modulation, transmission bit rate, and modem complexity, the training sequence accomplishes one or more of the following functions:

1. Initializes the communications channel, which includes disabling echo and establishing the gain of automatic gain control (AGC) devices
2. Verifies continuity (activates RLSD in the receive modem)
3. Initializes descrambler circuits in the receive modem
4. Initializes automatic equalizers in the receive modem
5. Synchronizes the receive modem's carrier to the transmit modem's carrier
6. Synchronizes the receive modem's clock to the transmit modem's clock

15-9-1 Modem Equalizers

Equalization is the compensation for phase delay distortion and amplitude distortion inherently present on telephone communications channels. One form of equalization provided by the telephone company is C-type conditioning, which is available only on private-line circuits. Additional equalization may be performed by the modems themselves. *Compromise equalizers* are located in the transmit section of a modem, and they provide *preequalization*—they shape the transmitted signal by altering its delay and gain characteristics before the signal reaches the telephone line. It is an attempt by the modem to compensate for impairments anticipated in the bandwidth parameters of the communications line. When a modem is installed, the compromise equalizers are manually adjusted to provide the best error performance. Typically, compromise equalizers affect the following:

1. Amplitude only
2. Delay only
3. Amplitude and delay
4. Neither amplitude nor delay

Compromise equalizer settings may be applied to either the high- or the low-voice-band frequencies or symmetrically to both at the same time. Once a compromise equalizer setting has been selected, it can only be changed manually. The setting that achieves the best error performance is dependent on the electrical length of the circuit and the type of facilities that make it up (i.e., one or more of the following: twisted-pair cable, coaxial cable, optical fiber cable, microwave, digital T-carriers, and satellite).

Adaptive equalizers are located in the receiver section of a modem, where they provide postequalization to the received signals. Adaptive equalizers automatically adjust their gain and delay characteristics to compensate for phase and amplitude impairments encountered on the communications channel. Adaptive equalizers may determine the quality of the received signal within its own circuitry, or it may acquire this information from the demodulator or descrambler circuits. Whatever the case, the adaptive equalizer may continuously vary its settings to achieve the best overall bandwidth characteristics for the circuit.

Figure 15-7 shows where compromise and adaptive equalizers are placed in a modem to achieve pre- and post-equalization.

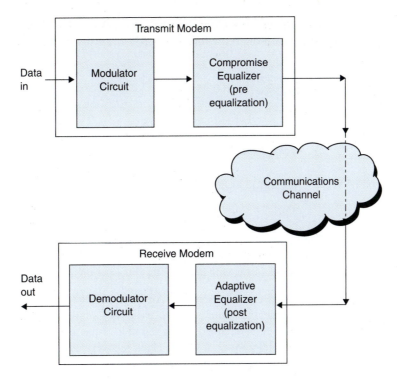

FIGURE 15-7 Compromise and adaptive equalizers

15-9-2 Low-Speed Modem Synchronization

Because low-speed modems are generally asynchronous and use noncoherent FSK, the transmit carrier frequency and the clock frequency are not recovered in the receive modem. Therefore, scrambler and descrambler circuits are unnecessary. If pre- or postequalization is used, they are generally adjusted manually and do not require initialization. Therefore, the training sequence for low-speed asynchronous voice-band modems is generally nothing more than a string of consecutive logic 1s (called *idle line ones*) and is used primarily to initialize the communications line by verifying continuity, setting the gain of AGC amplifiers, and disabling echo suppressors.

15-9-3 Medium- and High-Speed Modem Synchronization

Medium- and high-speed voice-band modems are used for transmission rates of 2400 bps and higher. To transmit at these bit rates, PSK or QAM modulation is used, which requires the receive carrier oscillators to be at least frequency coherent (and possibly phase coherent) with the transmit modem's carrier oscillator. Because these modems are synchronous, clock recovery must also be accomplished in the receive modem. Therefore, synchronous voice-band modems contain scrambler and descrambler circuits and adaptive (automatic) equalizers.

   15-9-3-1 Training.   The type of modulation and encoding technique used in a modem determines the number of bits required in the training sequence and, therefore, the duration of the training sequence. A 208-compatible modem, for example, is a synchronous, 4800-bps modem that uses 8-PSK. The training sequence for a 208-compatible modem is shown in Figure 15-8. Each symbol contains three bits (one tribit) of information and is 0.625 ms in duration. The four-phase idle code sequences through four of the eight possible signal phases. This allows the receive to recover the carrier and the clock timing information rapidly. The four-phase test word allows the adaptive equalizer in the receive

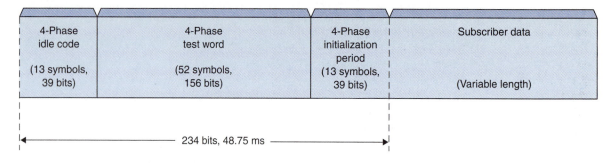

| 4-Phase idle code (13 symbols, 39 bits) | 4-Phase test word (52 symbols, 156 bits) | 4-Phase initialization period (13 symbols, 39 bits) | Subscriber data (Variable length) |

234 bits, 48.75 ms

FIGURE 15-8 Training sequence for a 208 modem

modem to adjust its final settings. The eight-phase initialization period prepares the descrambler circuits for eight-phase operation. The entire training sequence (234 bits) requires only 48.75 ms with a 4800-bps transmission rate.

15-9-3-2 Clock recovery. Although timing (clock) synchronization is initially established during the training sequence (i.e., before user data is propagated through the modem), it must be maintained for the duration of the transmission.

Figure 15-9 shows a simple clock recovery circuit and how it extracts clocking information from a received data stream. In Figure 15-9a, the demodulated data is applied directly to one input of an exclusive OR gate. The same data is delayed by one-half of a bit time and applied to a second input of the same exclusive OR gate. Figure 15-9b shows the timing diagram for the exclusive OR gate. From the figure, it can be seen that an alternating 1/0 data sequence produces a clock at the output of the exclusive OR gate with a frequency equal to the data rate. However, a prolonged sequence of consecutive logic 1s or logic 0s produces an output void of transitions and loss of clock synchronization. This could be prevented by placing restrictions on the customer's protocol and message format to forbid long strings of consecutive logic 1s or logic 0s from occurring, but this is a poor solution to the problem. A better method would be to scramble the data in the transmit modem before modulation and descramble the data in the receive modem after demodulation.

15-9-3-3 Scrambler and descrambler circuits. The purpose of a scrambler circuit is to detect undesirable sequences of 1s and 0s and convert them to a sequence more conducive to clock recovery. The job of a descramble is to detect scrambled data and convert it back to its original sequence. If a scrambler circuit simply randomized data, it would be impossible to convert the data back to its original sequence in the receiver. Therefore, scrambler circuits pseudorandomize data and alter it in a logical manner, allowing it to be descrambled in the receiver. For a scrambler circuit to be effective, the descrambler must contain the appropriate descrambling algorithm to recover the original bit sequence before data is sent to the DTE. The purpose of the scrambler is not simply to randomize the bit sequence but also to detect the occurrence of an undesirable bit sequence and convert it to a more acceptable pattern (i.e., one containing transitions).

Figure 15-10 illustrates the basic idea behind scrambler and descrambler circuits. The scrambler shown detects eight consecutive logic 0s and converts them to an alternating 1/0 pattern, which is more conducive to clock recovery. The descrambler detects the alternating 1/0 pattern and converts the sequence back to the original eight consecutive 0s. Actual scrambler and descrambler circuits obviously use more complex scrambling and descrambling algorithms.

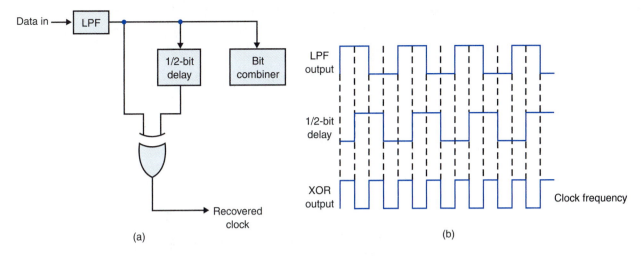

FIGURE 15-9   Clock recovery: (a) circuit; (b) timing diagram

Modem Transmitter

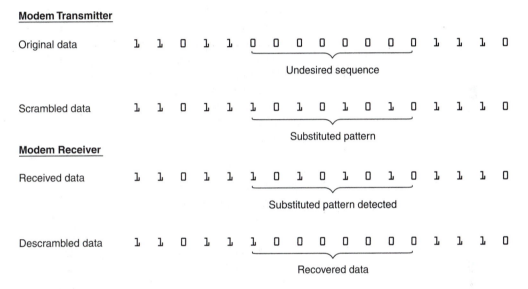

FIGURE 15-10   Scrambler and descrambler

15-10   ITU-T VOICE-BAND MODEM SPECIFICATIONS

Since the late 1980s, the International Telecommunications Union–Telephone Sector (ITU-T, formerly CCITT) has developed transmission standards for data communications modems outside the United States. The ITU-T specifications are known as the V-series, which include a number indicating the standard (V.21, V.32, and so on). Sometimes the V-series is followed by the French word *bis,* meaning "second," which indicates that the standard is a revision of an earlier standard. If the standard includes the French word *terbo,* meaning "third," the bis standard has been modified.

ITU-T specifications include Bell System–compatible standards as well as a number of ultra-high-speed voice-band modems that have been developed in the past decade or so with transmission rates well in excess of 9600 bps. For example, ITU-T standards V.21,

Table 15-2 ITU-T V-Series Modem Standards

ITU-T Designation	Specification
V.1	Defines binary 0/1 data bits as space/mark line conditions
V.2	Limits output power levels of modems used on telephone lines
V.4	Sequence of bits within a transmitted character
V.5	Standard synchronous signaling rates for dial-up telephone lines
V.6	Standard synchronous signaling rates for private leased communications lines
V.7	List of modem terminology in English, Spanish, and French
V.10	Unbalanced high-speed electrical interface specifications (similar to RS-423)
V.11	Balanced high-speed electrical interface specifications (similar to RS-422)
V.13	Simulated carrier control for full-duplex modem operating in the half-duplex mode
V.14	Asynchronous-to-synchronous conversion
V.15	Acoustical couplers
V.16	Electrocardiogram transmission over telephone lines
V.17	Application-specific modulation scheme for Group III fax (provides two-wire, half-duplex trellis-coded transmission at 7.2 kbps, 9.6 kbps, 12 kbps, and 14.4 kbps)
V.19	Low-speed parallel data transmission using DTMF modems
V.20	Parallel data transmission modems
V.21	0- to 300-bps full-duplex two-wire modems similar to Bell System 103
V.22	1200/600-bps full-duplex modems for switched or dedicated lines
V.22bis	1200/2400-bps two-wire modems for switched or dedicated lines
V.23	1200/75-bps modems (host transmits 1200 bps and terminal transmits 75 bps). V.23 also supports 600 bps in the high channel speed. V.23 is similar to Bell System 202. V.23 is used in Europe to support some videotext applications.
V.24	Known in the United States as RS-232. V.24 defines only the functions of the interface circuits, whereas RS-232 also defines the electrical characteristics of the connectors.
V.25	Automatic answering equipment and parallel automatic dialing similar to Bell System 801 (defines the 2100-Hz answer tone that modems send)
V.25bis	Serial automatic calling and answering—CCITT equivalent to the Hayes AT command set used in the United States
V.26	2400-bps four-wire modems identical to Bell System 201 for four-wire leased lines
V.26bis	2400/1200-bps half-duplex modems similar to Bell System 201 for two-wire switched lines
V.26terbo	2400/1200-bps full-duplex modems for switched lines using echo canceling
V.27	4800-bps four-wire modems for four-wire leased lines similar to Bell System 208 with manual equalization
V.27bis	4800/2400-bps four-wire modems same as V.27 except with automatic equalization
V.28	Electrical characteristics for V.24
V.29	9600-bps four-wire full-duplex modems similar to Bell System 209 for leased lines
V.31	Older electrical characteristics rarely used today
V.31bis	V.31 using optocouplers
V.32	9600/4800-bps full-duplex modems for switched or leased facilities
V.32bis	4.8-kbps, 7.2-kbps, 9.6-kbps, 12-kbps, and 14.4-kbps modems and rapid rate regeneration for full-duplex leased lines
V.32terbo	Same as V.32bis except with the addition of adaptive speed leveling, which boosts transmission rates to as high as 21.6 kbps
V.33	12.2 kbps and 14.4 kbps for four-wire leased communications lines
V.34	(V. fast) 28.8-kbps data rates without compression
V.34+	Enhanced specifications of V.34
V.35	48-kbps four-wire modems (no longer used)
V.36	48-kbps four-wire full-duplex modems
V.37	72-kbps four-wire full-duplex modems
V.40	Method teletypes use to indicate parity errors
V.41	An older obsolete error-control scheme
V.42	Error-correcting procedures for modems using asynchronous-to-synchronous conversion (V.22, B.22bis, V.26terbo, V.32, and V.32bis, and LAP M protocol)
V.42bis	Lempel-Ziv-based data compression scheme used with V.42 LAP M
V.50	Standard limits for transmission quality for modems
V.51	Maintenance of international data circuits

Table 15-2 *(Continued)*

ITU-T Designation	Specification
V.52	Apparatus for measuring distortion and error rates for data transmission
V.53	Impairment limits for data circuits
V.54	Loop test devices of modems
V.55	Impulse noise-measuring equipment
V.56	Comparative testing of modems
V.57	Comprehensive tests set for high-speed data transmission
V.90	Asymmetrical data transmission—receives data rates up to 56 kbps but restricts transmission bit rates to 33.6 kbps
V.92	Asymmetrical data transmission—receives data rates up to 56 kbps but restricts transmission bit rates to 48 kbps
V.100	Interconnection between public data networks and public switched telephone networks
V.110	ISDN terminal adaptation
V.120	ISDN terminal adaptation with statistical multiplexing
V.230	General data communications interface, ISO layer 1

V.23, and V.26 describe modem specifications similar to the Bell System 103, 202, and 201 modems, respectively. V.22 describes modem specifications similar to the Bell System 212A, and V.29 outlines specifications similar to the Bell System 209. Table 15-2 lists some of the current ITU-T modem specifications and recommendations.

15-10-1   ITU-T Modem Specification V.29

The ITU-T V.29 specification is the first internationally accepted standard for a 9600-bps transmission rate. The V.29 standard is intended to provide synchronous data transmission over four-wire leased telephone lines. V.29 uses 16-QAM modulation of a 1700-Hz carrier frequency. Data is clocked into the modem in groups of four bits called *quadbits,* resulting in a 2400-baud transmission rate. Occasionally, V.29 modems are used in the half-duplex mode over two-wire switched telephone lines. Pseudo-full-duplex operation can be achieved over the two-wire lines using a method called *ping-pong*. With ping-pong, data sent to the modem at each end of the circuit by their respective DTEs is buffered and automatically exchanged over the data link by rapidly turning the carriers on and off in succession.

Pseudo-full-duplex operation over two-wire transmission lines can also be accomplished using *statistical duplexing*. Statistical duplexing utilizes a 300-bps reverse data channel. The reverse channel allows a data operator to enter keyboard data while simultaneously receiving a file from the distant modem. By monitoring the data buffers inside the modem, the direction of data transmission can be determined, and the high- and low-speed channels can be reversed.

15-10-2   ITU-T Modem Specification V.32

The ITU-T V.32 specification provides for a 9600-bps transmission rate with true full-duplex operation over four-wire leased private line or two-wire switched telephone lines. V.32 also provides for data rates of 2400 bps and 4800 bps. V.32 specifies 16-QAM similar to V.29 except with a carrier frequency of 1800 Hz. V.32 is similar to V.29, except with V.32 an advanced coding technique called *trellis encoding* is specified. Trellis encoding produces a superior signal-to-noise ratio by dividing the incoming data stream into groups of five bits called *quintbits* (M-ary, where $M = 2^5 = 32$). The constellation diagram for V.32 encoding was developed by Dr. Ungerboeck at IBM Zuerich Research Laboratory and combines coding and modulation to improve bit error performance. The basic idea behind trellis encoding is to introduce controlled redundancy, which reduces channel error rates by doubling the number of signal points on the QAM constellation. The trellis encoding constellation diagram for V.32 is shown in Figure 15-11.

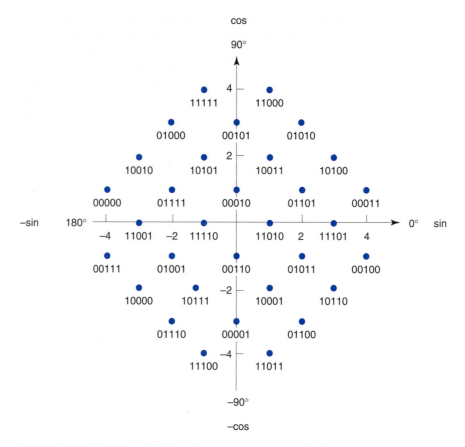

FIGURE 15-11 V.32 constellation diagram using trellis encoding

Full-duplex operation over two-wire switched telephone lines is achieved with V.32 using a technique called *echo cancellation*. Echo cancellation involves adding an inverted replica of the transmitted signal to the received signal. This allows the data transmitted from each modem to simultaneously use the same carrier frequency, modulation scheme, and bandwidth.

15-10-3 ITU-T Modem Specification V.32bis and V.32terbo

ITU-T recommendation V.32bis was introduced in 1991 and created a new benchmark for the data modem industry by allowing transmission bit rates of 14.4 bps over standard voice-band telephone channels. V.32bis uses a 64-point signal constellation with each signaling condition representing six bits of data ($M = 2^6 = 54$). The transmission rate for V.32bis is

$$\frac{6 \text{ bits}}{\text{code}} \times \frac{2400 \text{ codes}}{\text{second}} = \frac{14{,}400 \text{ bits}}{\text{second}}$$

The signaling rate is

$$\frac{14{,}400 \text{ bits/second}}{6 \text{ bits/symbol}} = \frac{2400 \text{ symbols}}{\text{second}} = 2400 \text{ baud}$$

V.32bis also includes automatic *fall-forward* and *fall-back* features, which allow the modem to change its transmission rate to accommodate changes in the quality of the communications channel. The fall-back feature slowly reduces the transmission bit rate to 12.2 kbps, 9.6 kbps, or 4.8 kbps. The fall-forward feature gives the modem the ability to return to a higher transmission rate when the quality of the communications channel improves. V.32bis supports Group III fax, which is the transmission standard that outlines the connection procedures used between two fax machines or fax modems. V.32bis also specifies the data compression procedure used during transmissions.

In August 1993, U.S. Robotics introduced V.32terbo. V.32terbo includes all the features of V.32bis plus a proprietary technology called *adaptive speed leveling*. V.32terbo includes two categories of new features: increased data rates and enhanced fax capabilities. V.32terbo also outlines the new 19.2-kbps data transmission rate developed by AT&T.

15-10-4 ITU-T Modem Specification V.33

ITU-T specification V.33 is intended for modems that operated over four-wire dedicated two-point private-line circuits. V.33 uses trellis coding and is similar to V.32, except a V.33 signal element includes six information bits and one redundant bit, resulting in a data transmission rate of 14.4 kbps and a signal rate of 2400 baud using an 1800-Hz carrier. The 128-point constellation diagram used with V.33 is shown in Figure 15-12.

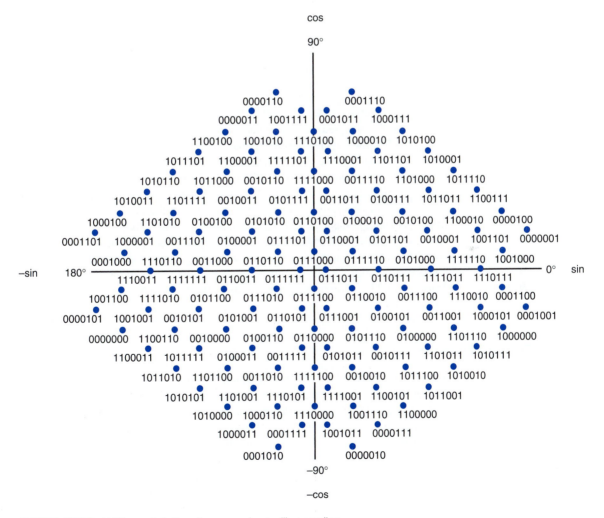

FIGURE 15-12 V.33 constellation diagram using trellis encoding

15-10-5   ITU-T Modem Specification V.34 Fast (V.fast)

Officially adopted in 1994, V.fast is considered the next generation in data transmission. Data rates of 28.8 kbps without compression are possible using V.34. Using current data compression techniques, V.34 automatically adapts to changes in transmission-line characteristics and dynamically adjusts data rates either up or down, depending on the quality of the communications channel.

V.34 innovations include the following:

1. Nonlinear coding, which offsets the adverse effects of system nonlinearities that produce harmonic and intermodulation distortion and amplitude proportional noise
2. Multidimensional coding and constellation shaping, which enhance data immunity to channel noise
3. Reduced complexity in decoders found in receivers
4. Precoding data for better utilization of the available bandwidth of the communications channel, which improves transmission of data in the outer limits of the channel, where amplitude, frequency, and phase distortion are at their worst
5. Line probing, which is a technique that modem receivers use to rapidly determine the best correction to compensate for transmission-line impairments

15-10-6   ITU-T Modem Specification V.34+

V.34+ is an enhanced standard adopted by the ITU in 1996. V.34+ adds 31.2 kbps and 33.6 kbps to the V.34 specification. Theoretically, V.34+ adds 17% to the transmission rate; however, many modem manufacturers do not consider it significant enough to warrant serious consideration at this time.

15-10-7   ITU-T Modem Specification V.42 and V.42bis

In 1988, the ITU adopted the V.42 standard *error-correcting procedures for DCEs* (modems). V.42 specifications address asynchronous-to-synchronous transmission conversions and error control that includes both detection and correction. V.42's primary purpose specifies a relatively new modem protocol called *Link Access Procedures for Modems* (LAP-M). LAP-M is almost identical to the packet-switching protocol used with the X.25 standard.

V.42bis is a specification designed to enhance the error-correcting capabilities of modems that implement the V.42 standard. Modems employing data compression schemes have proven to significantly surpass the data throughput performance of their predecessors. The V.42bis standard is capable of achieving somewhere between 3-to-1 and 4-to-1 compression ratios for ASCII-coded text. The compression algorithm specified is British Telecom's BTLZ. Throughput rates of up to 56 kbps can be achieved using V.42bis data compression.

15-11   56K MODEMS

The ITU-T developed the V.90 specification in February 1998 during a meeting in Geneva, Switzerland. The V.90 recommendation is similar to 3COM's x2 and Lucent's K56flex in that it defines an asymmetrical data transmission technology where the upstream and downstream data rates are not the same. V.90 allows modem downstream (receive) data rates up to 56 kbps and upstream (transmit) data rates up to 33.6 kbps. These data rates are inappropriate in the United States and Canada, as the Federal Communications Commission and Canadian Radio-television and Telecommunications Commission limit transmission rates offered by telephone companies to no more than 53 kbps. The reason for this restriction is to reduce crosstalk on telephone company feeder cables.

In 2000, the ITU-T approved a new modem standard called V.92. V.92 offers three improvements over V.90 that can be achieved only if both the transmit and the receive modems and the Internet Service Provider (ISP) have V.92 compliant modems. V.92 offers the following:

1. An upstream transmission rate of 48 kbps accomplished through improved signaling and compression techniques.
2. Faster call setup capabilities because V.92 modems remember information from previous calls to the same destination.
3. Incorporation of a hold option that can put a data call on hold for up to 16 minutes when voice calls are received. Subscribers must subscribe to call waiting from their local service provider to use the hold option.

15-12 MODEM CONTROL: THE AT COMMAND SET

First-generation data communications modems, such as the Bell System line of datasets, are often called *dumb modems* because they consisted of little more than a modulator, a demodulator, and a serial interface. It was awkward to use dumb modems on dial-up circuits because they were originally designed for private-line data circuits. Human operators at both ends of the circuit were required to perform most of the functions necessary to initiate, complete, and terminate a call.

Intelligent (smart) modems were introduced in the early 1970s. Smart modems include the features of standard modems; however, they also include built-in microprocessors that perform routine functions, such as automatic answering, call initiating and dialing, busy signal recognition, and error correction. Smart modems are controlled through a system of commands sent in ASCII code over an RS-232 interface. The most common system of modem commands is the *AT command set,* which is also known as the *Hayes modem command set.* Hayes Microcomputer Products originally developed the AT command set for its own line of modems, but other manufacturers soon adopted the system, and it rapidly became the de facto standard in the United States.

15-12-1 Modem Transmission Modes

There are two *transmission modes* used by data communications modems: *originate* and *answer.* The two modes are physically different, as they use different carrier frequencies and modulation schemes. The modem that initiates communications typically uses the originate mode, and the modem that receives the call uses the answer mode. However, the modem that initiates a call does not have to remain in the originate mode unless the modem is capable of operating only in that mode. If the destination modem can operate only in the originate mode, the initiating modem must switch to the answer mode.

15-12-2 Modem Operational Modes

There are four operational modes used by computers to communicate with data communications modems using the AT command set: *local command mode, handshake mode, online mode,* and *off-line mode.* Depending on which operational mode a modem is in, it will interpret data it receives from its local computer either as a modem command or as user data intended to be forwarded to the distant computer.

Local command mode. All modem commands in the AT command set begin with the ASCII characters A and T (ATtention) followed by the appropriate command or set of commands. Whenever a modem is disconnected from the telephone line and not communicating directly with another modem, it is in the command mode. When in the command mode, a modem monitors information sent to it from the local computer looking for the ASCII characters AT. After detecting the AT sequence, the modem interprets the characters immediately following the AT sequence as commands. For example, the ASCII character T is the command to use dual-tone multifrequency

(DTMF) signals rather than dial pulses, and the ASCII character D is the command to begin dialing out. The command to call the telephone number 1-602-461-7777 using DTMF is the character sequence ATDT16022467777. When in the command mode, the modem also accepts other instructions and data communications parameters from the computer, such as bit rate and parity.

Handshake mode. In the handshake mode, the two modems are in controlled (rather than the computer), and no data is actually transferred. Each modem confirms reception of a valid analog carrier from the distant modem. Once a connection is established, the calling modem transmits a training sequence, which the distant modem uses to adjust it adaptive equalizers. The modems also check for compatible bit rates and negotiate error correction and data compression procedures. Once the handshake procedures are completed, the calling modem changes to the on-line mode and sends a *connected* message to its local computer, which includes the negotiated transmission rate and compression protocol. If the handshake is unsuccessful, the modem sends a *no carrier* message to the computer, provides an on-hook indication to the telephone line, and returns to the local command mode.

On-line mode. Once communications has been established with a remote modem, the local modem switches to the on-line mode, which is also called the *data transfer mode*. In the on-line mode, a modem is transparent and interprets characters received from its local computer and the distant modem, including the letters A and T, as user data. The local modem simply accepts the characters and allows them to modulate its carrier before sending them to the remote location. The local terminal can switch the modem from the on-line mode to the command mode by momentarily pausing the transmission of data, sending three consecutive plus signs $(+++)$, and then pausing again. This sequence is called the *escape code*. In response to the escape code, the modem switches to the command mode and begins monitoring data for the ASCII AT command code. If a modem detects a loss of carrier from the distant modem, it will automatically change from the on-line mode to the off-line mode.

Off-line mode. When in this mode, the modem connection remains open in the background, allowing the local computer to send commands to the modem to perform certain functions that are necessary before any user data can be exchanged. For example, the computer can instruct the modem to open or close a memory buffer, enable local echo, read the local director, or terminate a call.

15-12-3   Command Types

The Hayes AT command set includes a set of *basic commands, command extensions, proprietary commands,* and *register commands*:

Basic commands. Basic commands include a fundamental core of modem commands used to place a modem off hook, dial a number, answer an incoming call, and echo characters to the local computer. Basic commands are universally supported by all Hayes modems and virtually all modems manufactured by most other companies. However, some modems will not accept the entire basic command set or every possible setting for each command. Basic commands all begin with the letters A through Z followed by an optional variable that further defines a feature within the command that supports two or more conditions. Table 15-3 lists and describes the basic Hayes command set.

Command extensions. Command extensions enable additional functions (extensions) that can be added to basic commands. The ampersand (&) is the prefix that precedes the letters A through Z to identify command extensions. Table 15-4 lists and describes the extended Hayes command set.

Table 15-3 Hayes AT Basic Command Set

Command	Description
A	Answer call
A/	Repeat last command
B	Select the method of modem modulation
C	Turn modem's carrier on or off
D	Dial a telephone number
E	Enable or inhibit echo of characters to the screen
F	Switch between half- and full-duplex operation
H	Hang up telephone (go on hook) or pick up telephone (go off hook)
I	Request identification code or request checksum
L	Select the speaker volume
M	Turn speaker off or on
N	Negotiate handshake options
O	Place modem on-line
P	Dial pulse out
Q	Request modem to send or inhibit sending result code
R	Change modem mode to "originate only"
S	Set modem register values
T	Touch-Tone out
V	Send result codes as digits or words
W	Select negotiation progress message
X	Use basic or extended result code set
Z	Reset the modem

Table 15-4 Hayes AT Command Extensions

Command	Description
&B	Control automatic retraining
&C	Carrier detect signal
&D	Data terminal ready signal
&F	Factory default control
&G	Guard tone control
&K	Flow control
&L	Select dial-up mode
&M	Select error control mode
&P	Select pulse dialing make/break ratio
&Q	Select data compression and error-control mode
&S	Data set ready signal
&T	Perform modem test
&U	Control trellis coding for V.32
&W	Store modem configuration
&Y	Set configuration at power on
&Z	Store phone number

Proprietary commands. Proprietary commands vary widely among modem manufacturers and are generally identified by either a percent sign (%) or a backslash (\) character.

Registered commands. Registered commands pertain to a specific location in a modem's onboard memory. Modem functions are defined by values associated with different memory (register) locations. The value of each register can be programmed (set); therefore, the registers are often called *S registers*. Table 15-5 lists and describes the first 13 (0 through 12) S register control parameters, which are the standardized parameters.

Table 15-5 Hayes AT Command S Register Parameters

Command	Function	Default Value	Range
S0	Ring to answer on	*	0–255
S1	Count number of rings	0	0–255
S2	Escape code character	ASCII 43	ASCII 0–127
S3	Carriage return character	ASCII 13	ASCII 0–127
S4	Line feed character	ASCII 10	ASCII 0–127
S5	Backspace character	ASCII 8	ASCII 0–127
S6	Dial tone wait time (seconds)	2	2–255
S7	Carrier wait time (seconds)	30	1–255
S8	Pause time caused by comma (seconds)	2	0–255
S9	Carrier detect response time (1/10ths of seconds)	6	1–255
S10	Time delay between loss of carrier and hang up (1/10ths of seconds)	7	1–255
S11	Touch-Tone duration and spacing time (milliseconds)	70	50–255
S12	Escape sequence guard time (1/20th of seconds)	50	0–255

15-13 CABLE MODEMS

Cable modems are similar to standard voice-band modems, except cable modems operate at higher frequencies, operate at higher bit rates, use more sophisticated modulation and demodulation schemes, and require more bandwidth than conventional voice-band modems.

Cable modems connect subscribers to cable TV (CATV) facilities, such as coaxial cables and optical fibers, and provide high-speed Internet access and video services. A single CATV channel can support multiple individual subscribers or a local area network using a shared network protocol capable of supporting many users, such as Ethernet.

Cable modems are broadband modems, which simply means that they require wider bandwidths than standard voice-band modems. Another example of a broadband modem is an Asymmetric Digital Subscriber Line (ADSL) modem. DSL is described in a later chapter.

Subscribers are connected through a cable modem over a TV cable to a CATV network *headend,* which is the originating point for the CATV audio, video, and data signals. With cable modems, the upstream (subscriber to headend) and downstream (headend to subscriber) frequency spectrums and transmission rates are not the same (asymmetrical transmission). For obvious reasons, the bandwidths are considerably wider and transmission rates considerably higher on the downstream connection. Cable modem frequency assignments and transmission rates are the following:

Upstream

    Carrier frequency: 5 MHz to 40 MHz

    Transmission rate: 19.2 kbps to 3 Mbps

Downstream

Carrier frequency: 250 MHz to 850 MHz

Transmission rate: 10 Mbps to 30 Mbps

Individual data channels are assigned frequency slots in 250-kHz increments with a maximum bandwidth of 6 MHz.

15-14 PROBABILITY OF ERROR AND BIT ERROR RATE

Probability of error ($P[e]$) and *bit error rate* (BER) are often used interchangeably, although in practice they do have slightly different meanings. $P(e)$ is a theoretical (mathematical) expectation of the bit error rate for a given system. BER is an empirical (historical) record of a system's actual bit error performance. For example, if a system has a $P(e)$ of 10^{-5}, this means that mathematically you can expect one bit error in every 100,000 bits transmitted $(1/10^{-5} = 1/100,000)$. If a system has a BER of 10^{-5}, this means that in past performance there was one bit error for every 100,000 bits transmitted. A bit error rate is measured and then compared with the expected probability of error to evaluate a system's performance.

QUESTIONS

15-1. Describe *data communications modems* and tell where they are used in data communications circuits.

15-2. What is meant by a *Bell System–compatible* modem?

15-3. What is the difference between *asynchronous* and *synchronous* modems?

15-4. What are meant by the terms *data terminal equipment* and *data communications equipment*?

15-5. Describe the basic functions of a *channel service unit*.

15-6. What is the purpose of a *digital service unit*?

15-7. What is meant by the term *voice-band data communications modem*?

15-8. Explain the difference between *bits per second* and *baud*.

15-9. List and describe the basic blocks of a voice-band modem.

15-10. What are the two classifications for voice-band modems?

15-11. List and describe the characteristics of asynchronous voice-band modems.

15-12. List and describe the basic blocks of an asynchronous voice-band modem.

15-13. List and describe the characteristics of synchronous voice-band modems.

15-14. What is *modem synchronization*?

15-15. Define *modem synchronization* and list its functions.

15-16. What is meant by *modem training*?

15-17. What are the purposes of *scrambler* and *descrambler circuits*?

15-18. What is meant by the terms *bis* and *terbo*?

15-19. List and describe the most common modulation methods used with 56K voice-band modems.

15-20. Explain the difference between the terms *probability of error* and *bit error rate*.

15-21. Describe the basic characteristics of the following ITU-T voice-band modem specifications: ITU-T V.29, V.32, V.32bis, V.32terbo, V.33, V.34 fast, V.34+, V.42, V.42bis, V.90, and V.92.

15-22. Describe the *AT command set*.

15-23. List and describe the two transmission modes used with data communications modems.

15-24. List and describe the four modem operational modes.

15-25. List and describe the four types of commands used with the *Hayes AT command set*.

15-26. Describe the differences between *cable modems* and *standard voice-band modems*.

C H A P T E R 16

Data-Link Protocols

CHAPTER OUTLINE

OBJECTIVES

- Define *protocol* and *data-link protocol*
- Describe data-link protocol functions
- Define and describe the following data-link protocol functions: *line discipline, flow control*, and *error control*
- Define *character-* and *bit-oriented protocols*
- Describe asynchronous data-link protocols
- Describe character and block transmission modes
- Describe synchronous data-link protocols
- Explain the operation of XMODEM, YMODEM, ZMODEM, and KERMIT
- Describe IBM's 83B asynchronous data-link protocol
- Define *synchronous data-link protocols*
- Explain binary synchronous communications
- Define and describe *synchronous data-link control*
- Define and describe *high-level data-link control*

16-1 INTRODUCTION

A *network architecture* outlines the way in which a data communications network is arranged or structured and generally includes the concepts of levels or layers within the ar-

474

chitecture. Each layer within the network consists of specific *protocols*, or rules, for communicating that perform a given set of functions. *Layered network architectures* consist of two or more independent levels where each level has a specific set of responsibilities and functions, including data transfer, flow control, data segmentation and reassembly, sequence control, error detection, and error correction.

Computer networks communicate using *protocols*, which define the procedures that the systems involved in the communications process will use. Protocols are used today in every network architectural layer to provide networking capabilities such as how much data can be sent, how it will be sent, how it will be addressed, and what procedure will be used to ensure that there are no undetected errors.

Protocols are arrangements between people or processes. In essence, a protocol is a set of customs, rules, or regulations dealing with formality or precedence, such as diplomatic or military protocol. Each functional layer of a network is responsible for providing a specific service to the data being transported through the network by providing a set of rules (protocols) that perform a specific function (or functions) within the network. *Data communications protocols* are sets of rules governing the orderly exchange of data within the network or a portion of the network, whereas network architecture is a set of layers and protocols that govern the operation of the network. The list of protocols used by a system is called a *protocol stack*, which generally includes only one protocol per layer.

A *data-link protocol* is a set of rules implementing and governing an orderly exchange of data between layer two devices, such as line control units and front-end processors. Protocols outline precise character sequences that ensure an orderly exchange of data between two layer two devices, such as line control units and front-end processors.

16-2 DATA-LINK PROTOCOL FUNCTIONS

For communications to occur over a data network, there must be at least two devices working together (one transmitting and one receiving). In addition, there must be some means of controlling the exchange of data. For example, most communication between computers on networks is conducted half duplex even though the circuits that interconnect them may be capable of operating full duplex. Most data communications networks, especially local area networks, transfer data half duplex where only one device can transmit at a time. Half-duplex operation requires coordination between stations. Data-link protocols perform certain network functions that ensure a coordinated transfer of data. Some data networks designate one station as the *control station* (sometimes called the *primary station*). This is sometimes referred to as *primary–secondary* communications. In centrally controlled networks, the primary station enacts procedures that determine which station is transmitting and which is receiving. The transmitting station is sometimes called the *master station*, whereas the receiving station is called the *slave station*. In primary–secondary networks, there can never be more than one master at a time; however, there may be any number of slave stations. In other networks, all stations are equal, and any station can transmit at any time. This type of network is sometimes called a *peer-to-peer network*. In a peer-to-peer network, all stations have equal access to the network, but when they have a message to transmit, they must contend with the other stations on the network for access.

Data-link protocol *functions* include line discipline, flow control, and error control. *Line discipline* coordinates hop-to-hop data delivery where a hop is a computer, a network controller, or some type of network-connecting device, such as a router. *Line discipline* determines which device is transmitting and which is receiving at any point in time. *Flow control* coordinates the rate at which data is transported over a link and generally provides an acknowledgment mechanism that ensures that data is received at the destination. *Error control* specifies a means of detecting and correcting transmission errors.

16-2-1 Line Discipline

In essence, line discipline is coordinating half-duplex transmission on a data communications network. There are two fundamental ways that line discipline is accomplished in a data communications network: *enquiry/acknowledgment* (ENQ/ACK) and *poll/select*.

16-2-1-1 ENQ/ACK.

Enquiry/acknowledgment (ENQ/ACK) is a relatively simple data-link layer line discipline that works best in simple network environments where there is no doubt as to which station, computer, or device is the intended receiver. An example is a network comprised of only two stations (i.e., a two-point network) where the stations may be interconnected permanently or on a temporary basis through a switched network, such as the public telephone network.

Before data can be transferred between stations, procedures must be invoked to establish continuity between the source and destination stations and ensure that the destination station is ready and capable of receiving data. These are the underlying purposes of line discipline procedures. ENQ/ACK line discipline procedures determine which device on a network can initiate a transmission and whether the intended receiver is available and ready to receive the message. Assuming that all stations on the network have equal access to the transmission medium, a data session can be initiated by any station using the ENQ/ACK. An exception would be a printer, of course, which cannot initiate a session with a computer.

The initiating station begins a session by transmitting a frame of data called an enquiry (ENQ), which identifies the receiving station. In essence, the ENQ sequence solicits the receiving station to determine whether it is ready to receive a message. With half-duplex operation, after the initiating station sends the ENQ, it waits for a response from the destination station indicating whether it is ready to receive a message. If the destination station is ready to receive a message, it responds with a *positive acknowledgment* (ACK), and if it is not ready to receive a message, it responds with a *negative acknowledgment* (NAK). If the destination station does not respond with an ACK or a NAK within a specified period of time, the initiating station retransmits the ENQ. How many inquiries are made varies from network to network, but generally after three unsuccessful attempts to establish communications, the initiating station gives up (this is sometimes called a *time-out*). The initiating station may attempt to establish a session later; however, after several unsuccessful attempts, the problem is generally referred to a higher level of authority (such as a human).

A negative acknowledgment (NAK) transmitted by the destination station in response to an ENQ generally indicates a temporary unavailability, and the initiating station will simply attempt to establish a session later. A positive acknowledgment (ACK) from the destination station indicates that it is ready to receive data and tells the initiating station it is free to transmit its message. All transmitted message frames end with a unique terminating sequence, such as end of transmission (EOT), which indicates the end of the message frame. The destination station acknowledges all message frames received with either an ACK or a NAK. An ACK transmitted in response to a received message indicates that the message received without errors, and a NAK indicates that the message was received containing errors. A NAK transmitted in response to a message is an automatic request for retransmission of the rejected message.

Figure 16-1 shows how a session is established and how data is transferred using ENQ/ACK procedures. Station A initiates the session by sending an ENQ to station B. Station B responds with a positive acknowledgment (ACK), indicating that it is ready to receive a message. Station A transmits message frame 1, which is acknowledged by station B with an ACK. Station A then transmits message frame 2, which is rejected by station B with a NAK, indicating that the message was received with errors. Station A then retransmits message frame 2, which is received without errors and acknowledged by station B with an ACK.

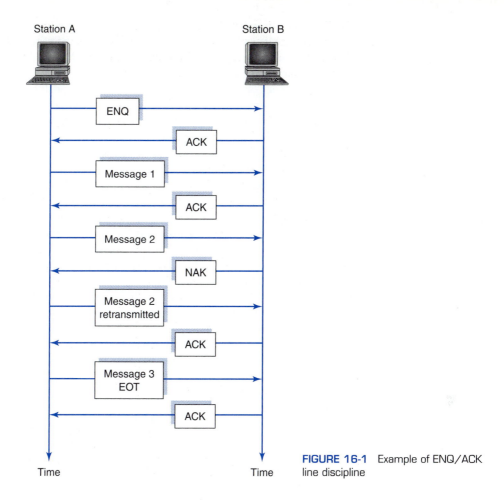

FIGURE 16-1 Example of ENQ/ACK line discipline

16-2-1-2 Poll/Select. The poll/select line discipline is best suited to *centrally controlled* data communications networks using a multipoint topology, such as a bus, where one station or device is designated as the primary station and all other stations are designated as secondary stations. Multipoint data communications networks using a single transmission medium must coordinate access to the network to prevent more than one station from attempting to transmit data at the same time. In addition, all exchanges of data must occur through the primary station. Therefore, if one secondary station wishes to transmit data to another secondary station, it must do so through the primary station. This is analogous to transferring data between memory devices in a computer using a central processing unit (CPU) where all data is read into the CPU from the source memory and then written to the destination memory.

In a poll/select environment, the primary station controls the data link, while secondary stations simply respond to instructions from the primary. The primary determines which device or station has access to the transmission channel (medium) at any given time. Hence, the primary initiates all data transmissions on the network with polls and selections.

A *poll* is a solicitation sent from the primary station to a secondary station to determine if the secondary station has data to transmit. In essence, the primary designates a secondary as a transmitter (i.e., the master) with a poll. A *selection* is when the primary designates a secondary as a destination or recipient of data. A selection is also a query from the primary to determine if the secondary is ready to receive data. With two-point networks using ENQ/ACK procedures, there was no need for addresses because transmissions from one station were obviously intended for the other station. On multipoint networks, however, addresses are necessary because all transmissions from the primary go to all secondaries to identify which secondary is being polled or selected. All secondary stations receive all polls and selections transmitted from the primary. With poll/select procedures, each secondary station is assigned one or more addresses for identification. It is up to the secondaries to examine the address to determine if the poll or selection is intended for them. The primary has no address because transmissions from all secondary stations go only to the primary. A primary can poll only one station at a time; however, it can select more than one secondary at a time using group (more than one station) and broadcast (all stations) addresses.

When a primary polls a secondary, it is soliciting the secondary for a message. If the secondary has a message to send, it responds to the poll with the message. This is called a positive acknowledgment to a poll. If the secondary has no message to send, it responds with a negative acknowledgment to the poll, which confirms that it received the poll but has no messages to send at that time. This is called a negative acknowledgment to a poll.

When a primary selects a secondary, it is identifying the secondary as a receiver. If the secondary is available and ready to receive data, it responds with an ACK. If it is not available or ready to receive data, it responds with a NAK.

Figure 16-2 shows how polling and selections are accomplished using poll/select procedures. The primary polls station A, which responds with a negative acknowledgment to a poll (NAK), indicating that it received the poll but has no message to send. Then the primary polls station B, which responds with a positive acknowledgment to a poll (i.e., a mes-

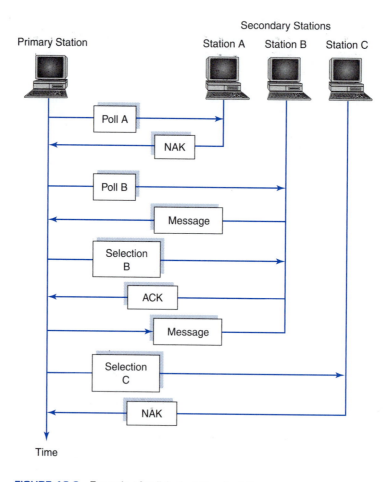

FIGURE 16-2   Example of poll/select line discipline

sage). The primary then selects station B to see if it ready to receive a message. Station B responds with a positive acknowledgment to the selection (ACK), indicating that it is ready to receive a message. The primary transmits the message to station B. The primary then selects station C, which responds with a negative acknowledgment to the selection (NAK), indicating that it is not ready to receive a message.

16-2-2 Flow Control

Flow control defines a set of procedures that tells the transmitting station how much data it can send before it must stop transmitting and wait for an acknowledgment from the destination station. The amount of data transmitted must not exceed the storage capacity of the destination station's buffer capacity. Therefore, the destination station must have some means of informing the transmitting station when its buffers are nearly at capacity and to temporarily stop sending data or to send data at a slower rate. There are two common methods of flow control: stop and wait and sliding window.

16-2-2-1 Stop-and-wait flow control. With *stop-and-wait* flow control, the transmitting station sends one message frame and then waits for an acknowledgment before sending the next message frame. After it receives an acknowledgment, it transmits the next frame. The transmit/acknowledgment sequence continues until the source station sends an end-of-transmission sequence. The primary advantage of stop-and-wait flow control is simplicity. The primary disadvantage is speed, as the time lapse between each frame is wasted time. Each frame takes essentially twice as long to transmit as necessary because both the message and the acknowledgment must traverse the entire length of the data link before the next frame can be sent.

Figure 16-3 shows an example of stop-and-wait flow control. The source station sends message frame 1, which is acknowledged by the destination station. After stopping

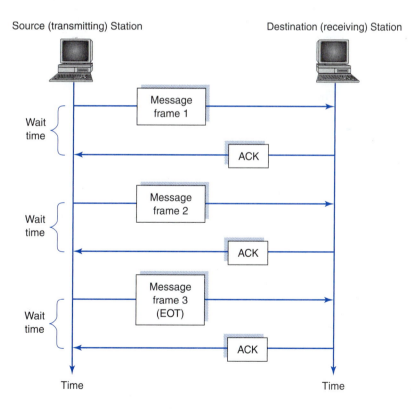

FIGURE 16-3 Example of stop-and-wait flow control

transmission and waiting for the acknowledgment, the source station transmits the next frame (message frame 2). After sending the second frame, there is another lapse in time while the destination station acknowledges reception of frame 2. The time it takes the source station to transport three frames equates to at least three times as long as it would have taken to send the message in one long frame.

16-2-2-2 Sliding window flow control. With *sliding window* flow control, a source station can transmit several frames in succession before receiving an acknowledgment. There is only one acknowledgment for several transmitted frames, thus reducing the total transmission time considerably over the stop-and-wait technique.

The term *sliding window* refers to imaginary boxes at the source and destination stations with the capacity of holding several frames of data. Message frames can be acknowledged any time before the window is filled with data. To keep track of which frames have been acknowledged and which have not, sliding window procedures require a modulo-n numbering scheme where each transmitted frame is identified with a unique sequence number between 0 and $n - 1$. With a three-bit binary numbering scheme, there are eight numbers possible (0, 1, 2, 3, 4, 5, 6, and 7), and, therefore, the windows must have the capacity of holding $n - 1$ (seven) frames of data. The reason for this is explained later in this chapter.

The primary advantage of sliding window flow control is network utilization. With fewer acknowledgments (i.e., fewer line turnarounds), considerably less network time is wasted acknowledging messages and more time can be spent actually sending messages. The primary disadvantages of sliding window flow control are complexity and hardware capacity. Each secondary station on a network must have sufficient buffer space to hold $2(n - 1)$ frames of data ($n - 1$ transmit and $n - 1$ receive frames), and the primary station must have buffer space to hold $m[2(n - 1)]$, where m equals the number of secondary stations on the network. In addition, each secondary must keep track of the number of each unacknowledged frame it transmits and each unacknowledged frame it receives, and the primary station must keep track of all unacknowledged frames it transmits and receives for each secondary station.

16-2-3 Error Control

Error control includes both error detection and error correction. However, with the data-link layer, error control is concerned primarily with error detection and message retransmission, which is a method of error correction.

With poll/select line disciplines, all polls, selections, and message transmissions end with some type of end-of-transmission sequence. In addition, all messages transported from the primary to a secondary or from a secondary to the primary are acknowledged with ACK or NAK sequences to verify the validity of the message. An ACK sequence means the message was received with no transmission errors, and a NAK sequence means the message was received with errors. A NAK is an automatic call for retransmission of the last message.

Error detection at the data-link layer can be accomplished with VRC, LCR, or CRC, and error correction is generally accomplished with *automatic repeat request* (ARQ), sometimes called *automatic request for retransmission*. With ARQ, any time a transmission error is detected, the destination station sends a negative acknowledgment (NAK) back to the source station requesting retransmission of the last message frame or frames. ARQ also calls for retransmission of missing or lost frames—those frames that are damaged so severely that the destination station does not recognize them and frames where the acknowledgments (both ACKs and NAKs) are lost.

There are two types of ARQ: stop and wait and sliding window. Stop-and-wait flow control generally incorporates *stop-and-wait ARQ*, and sliding window flow control usually implements ARQ in one of two variants: *go-back-n frames* or *selective reject* (SREJ). With go-back-n frames, the destination station tells the source station to go back n frames and retransmit all of them, even if all of them did not contain errors. With selective reject, the destination station tells the source station to retransmit only the frame (or frames) received in error. Go-back-n frames is easier to implement; however, it also wastes more time, as most of the frames retransmitted were not received in error. Selective reject is more complicated to implement but saves transmission time, as only those frames that are actually damaged are retransmitted.

16-3 CHARACTER- AND BIT-ORIENTED DATA-LINK PROTOCOLS

All data-link protocols transmit control information either in separate control frames or in the form of overhead that is added to the data and included in the same frame. Data-link protocols can be generally classified as either character or bit oriented.

16-3-1 Character-Oriented Protocols

Character-oriented protocols interpret a frame of data as a group of successive bits combined into predefined patterns of fixed length, usually eight bits each. Each group of bits represents a unique character. Control information is included in the frame in the form of standard characters from an existing character set, such as ASCII. Control characters convey important information pertaining to line discipline, flow control, and error control.

With character-oriented protocols, unique data-link control characters, such as start of text (STX) and end of text (ETX), no matter where they occur in a transmission, warrant the same action or perform the same function. For example, the ASCII code 02 hex represents the STX character. Start of text, no matter where 02 hex occurs within a data transmission, indicates that the next character is the first character of the text or information portion of the message. Care must be taken to ensure that the bit sequences for data-link control characters do not occur within a message unless they are intended to perform their designated data-link functions.

Character-oriented protocols are sometimes called *byte-oriented* protocols. Examples of character-oriented protocols are XMODEM, YMODEM, ZMODEM, KERMIT, BLAST, IBM's 83B asynchronous data-link protocol, and IBM's binary synchronous communications (BSC [bisync]). Bit-oriented protocols are more efficient than character-oriented protocols.

16-3-2 Bit-Oriented Protocols

A *bit-oriented protocol* (BOP) is a discipline for serial-by-bit information transfer over a data communications channel. With bit-oriented protocols, data-link control information is transferred as a series of successive bits that may be interpreted individually on a bit-by-bit basis or in groups of several bits rather than in a fixed-length group of n bits, where n is usually the number of bits in a data character. In a bit-oriented protocol, there are no dedicated data-link control characters. With bit-oriented protocols, the control field within a frame may convey more than one control function.

Bit-oriented typically convey more information into shorter frames than character-oriented protocols. The most popular bit-oriented protocols are synchronous data-link communications (SDLC) and high-level data-link communications (HDLC).

16-4 DATA TRANSMISSION MODES

Data transmission modes describe how human-operated terminals and computers transmit and receive alphanumeric data characters between other terminals and computers. There are only two basic transmission modes available: character and block.

16-4-1 Character Mode

When operating in the *character mode*, character codes are transmitted immediately after an operator depresses a key. The character is sent asynchronously because transmissions are not synchronized with the speed of the operator's keystrokes and operators type at different speeds. When an operator is not typing, the terminal is in the *idle state*. Data characters transmitted from the source station to the destination station are displayed on the screen at the current location of the cursor. Non–data characters, such as Bell (BEL), CR (carriage return), LF (line feed), and so on, are acted on accordingly.

16-4-2 Block Mode

In the *block mode*, data characters are not transmitted immediately as they are typed. Instead, operators enter characters into their terminals and PCs, where they are stored in buffers and displayed on the screen. When an operator is ready to transmit the information displayed on the screen, he or she depresses the Enter, Send, or Return key, which transmits all the data characters previously entered into memory. The assortment of characters transmitted as a group is called a *block* or *frame* of data. The format of the data within the block or frame depends on the system protocol used. Most modern terminals and PCs are capable of operating in either the character or the block mode. The character mode of transmission is more common when terminals and PCs located in remote stations are communicating directly to a host or mainframe computer through a direct communications channel, such as a dial-up telephone line. The block mode of transmission is more appropriate for multidrop data communications circuits operating in a polling environment.

16-5 ASYNCHRONOUS DATA-LINK PROTOCOLS

Asynchronous data-link protocols are relatively simple protocols generally used with asynchronous data and asynchronous modems. Asynchronous protocols, such as XMODEM, YMODEM, and ZMODEM, are commonly used to facilitate communications between two personal computers over the public switched network.

16-5-1 XMODEM

In 1979, a man named Ward Christiansen developed the first *file transfer protocol* designed to facilitate transferring data between two PCs over the public switched telephone network. Christiansen's protocol is now called *XMODEM*. XMODEM is a relatively simple data-link protocol intended for low-speed applications using asynchronous data and asynchronous modems. Although XMODEM was designed to be used between two PCs, it can also be used between a PC and a mainframe or host computer.

XMODEM specifies a half-duplex stop-and-wait protocol using a data frame comprised of four fields. The frame format for XMODEM contains four fields, as shown in Figure 16-4. The four fields for XMODEM are the SOH field, header field, data field, and error-detection field. The first field of an XMODEM frame is simply a one-byte start-of-heading

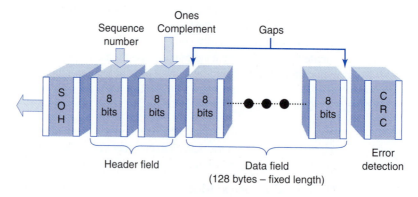

*Each 8-bit character contains start and stop bits (white bars) and characters are separated from each other with gaps.

FIGURE 16-4 XMODEM frame format

(SOH) field. SOH is a data-link control character indicating the beginning of the header. SOH simply indicates that the next byte is the first byte of the header. The second field is a two-byte sequence that is the actual header for the frame. The first header byte is called the *sequence number*, as it contains the number of the current frame being transmitted. The second header byte is simply the 2-s complement of the first byte, which is used to verify the validity of the first header byte (this is sometimes called *complementary redundancy*). The next field is the information field, which contains the actual user data. The information field has a maximum capacity of 128 bytes (e.g., 128 ASCII characters). The last field of the frame is an eight-bit frame check sequence (CRC-8) that is used for error detection.

Data transmission and control are quite simple with the XMODEM protocol—too simple for most modern-day data communications networks. Since XMODEM was designed to be used with the dial-up telephone network, the modems at the originate and destination ends of the circuit establish continuity, as described in Chapter 15. The process of transferring data begins when the answer (destination) station sends a NAK character to the originate (source) station; however, when transmitted by the destination station at the beginning of a data transfer, it simply indicates that the destination station is ready to receive data. After the source station receives the initial NAK character, it sends the first data frame and then waits for an acknowledgment from the destination station. If the data is received without errors, the destination station responds to the frame with an ACK character. If the data is received with errors, the destination station responds with a NAK character. However, when the NAK character is sent in response to an information frame, it indicates a negative acknowledgment to the Federal Communications Commission (FCC), which calls for a retransmission of the data. Whenever the originate station receives a NAK character, it retransmits the same frame. Each time the destination station receives a data frame, it responds with a NAK or an ACK, depending on whether the FCC indicates a transmission error has occurred. If the source station does not receive an ACK or a NAK after a predetermined length of time, it retransmits the last frame. When the destination station fails to respond to a frame, it is called a *time-out*. Time-outs are treated the same as NAKs. If the destination station wishes to terminate transmissions, it sends a cancel (CAN) signal to the source station.

16-5-2 YMODEM

YMODEM is a protocol similar to XMODEM except with the following exceptions:

1. The information field has a maximum capacity of 1024 bytes.
2. Two CAN characters are required to abort a transmission.
3. ITU-T-CRC 16 is used to calculate the frame check sequence.
4. Multiple frames can be sent in succession and then acknowledged with a single ACK or NAK character.

16-5-3 ZMODEM

ZMODEM is a newer protocol that simply combines the features of XMODEM and YMODEM.

16-5-4 KERMIT

The *KERMIT* protocol was developed at Columbia University and may be the most widely used asynchronous protocol today. KERMIT is a terminal emulation program as well as a file transfer protocol similar to XMODEM. With KERMIT, the sender waits for a NAK before it begins transmitting. However, KERMIT allows transmitting control characters as text. The control character is transformed into a printable character by adding a fixed number to its ASCII code and adding a pound sign (#) to the front of the transformed character. Consequently, a control character used as text is actually comprised of two characters. When the receiver detects a # character, it discards it and interprets the next character as a control character. When it is necessary to transmit a # character, two #s are sent.

16-5-5 IBM's 83B Asynchronous Data-Link Protocol

IBM's *83B asynchronous data-link protocol* is virtually identical to the old Western Electric Company's 8A1/8B1 *Selective Calling System*. The 83B protocol, unlike the protocols described in the previous sections of this chapter, is an asynchronous data-link protocol designed for multidrop private-line data circuits using an asynchronous data format and asynchronous modems. The 83B asynchronous data-link protocol was one of the first protocols designed for a central-controlled multipoint data circuit with a polling environment. There are probably few practical applications for the 83B protocol today; however, it provides a good example of the concepts of polling, selecting, and acknowledging described in the beginning sections of this chapter.

With the 83B asynchronous data-link protocol, the primary station is the host, and the remote stations are the secondaries. The XMODEM, YMODEM, ZMODEM, and KERMIT protocols described previously were designed primarily for two-point data communications circuits using the public telephone network (i.e., dial-up telephone network). The 83B asynchronous data-link protocol uses vertical redundancy checking (character parity) as the error-detection technique and either symbol substitution or ARQ (retransmission) for error correction.

Remote stations with the 83B protocol may be in one of four operating modes: *line monitoring*, *transmit*, *receive*, or *local*. When a secondary station is in the line-monitoring mode, it simply monitors messages on the circuit looking for a transmission with its polling or selection address. When a station is in the line-monitoring mode, it is neither transmitting nor receiving data. When a station is in the transmit mode, it has been designated the master. In the transmit mode, the station can send formatted messages or acknowledgments. When a station is in the receive mode, it has been selected by the primary station and designated as a receiver (slave). In the receive mode, the secondary can receive formatted messages, polls, or acknowledgments from the primary station. For a terminal operator to enter information into his or her computer or terminal, it must be in the local mode. A terminal can be placed into the local mode through software commands sent from the primary, or the operator can do it manually from the keyboard. When in the line-monitoring mode, a secondary station is simply monitoring the communications channel, looking for polls or selections.

16-5-5-1 83B polling sequence.
A primary station sends a polling sequence to identify a secondary station as a transmitter. The polling sequence for most asynchronous protocols designed for multipoint, private-line data circuits is quite simple and usually encompasses sending one or two data-link control characters and then a *station polling address* (SPA). A general poll using IBM's 83B asynchronous data-link protocol is the three-character sequence shown here:

$$
\begin{array}{ccc}
E & D & S \\
O & C & P \\
T & 3 & A
\end{array}
$$

where

$$
\begin{array}{l}
E \\
O = \text{end of transmission} \\
T \\
D \\
C = \text{device control three} \\
3 \\
S \\
P = \text{station polling address} \\
A
\end{array}
$$

The EOT character is a data-link control character called a *clearing character*. An EOT character precedes all polling and selection sequences. EOT places all secondary stations into the line-monitoring mode. When in the line-monitoring mode, a secondary station listens to the line for its polling or selection address. DC3 is a data-link control character that can mean several different things, depending on where it occurs. When DC3 follows an EOT, it indicates that the next character is a station polling address. A typical polling sequence is shown here:

```
E   D
O   C   A
T   3
```

For this example, the station polling address is the ASCII character *A*. Station A has been designated the master and must respond with either a formatted message or an acknowledgment to the poll.

16-5-5-2 83B responses to polls.   With the 83B asynchronous protocol, there are two acknowledgment sequences that may be transmitted in response to a poll: a positive acknowledgment and a negative acknowledgment. A positive acknowledgment to a poll simply means that the secondary station received the poll and has no messages to send but is ready to receive formatted messages. A negative acknowledgment to a poll means that the secondary station received the poll and has no messages to receive but is not ready to receive. The sequences for positive and negative acknowledgments to a poll and their functions are listed here:

Acknowledgments to a poll	Function
Positive	
A	No messages to transmit
\ C	Ready to receive
K	
Negative	
\ \	No messages to transmit
	Not ready to receive

If a secondary station has a formatted message to send when it is polled, it simply responds with the message format shown here:

```
S           E
T  message  O
X           T
```

```
        S
where   T  = start of text
        X
```

The EOT and STX characters are not part of the message. They are data-link characters inserted by the controller to frame the message. STX indicates that the actual message (user data) begins with the character that immediately follows it. The EOT character at the end of the sequence signals the end of the message and designates the primary station as the master so that it can acknowledge whether the transmission was successful.

Sometimes it is necessary or desirable to transmit coded data, in addition to the message, that is used only for data-link management, such as date, time, message number, message priority, or routing information. This bookkeeping information is not part of the message; it is overhead and is transmitted as heading information. To identify the heading, the message begins with a start-of-heading (SOH) character. SOH is transmitted first, followed by the heading information, STX, and then the message. The entire sequence is terminated with an EOT character. When a heading is included, STX terminates the heading and indicates the beginning of the message. The format for transmitting heading information together with data is shown here:

```
S                                S              E
O  heading (date, time, etc.)   T  message data  O
H                                X              T
```

16-5-5-3 83B selection sequence. A primary station sends a selection sequence to identify a secondary station as a receiver. The selected receiver must respond to a selection to indicate whether it is ready to receive a message. The selection sequence for the 83B protocol is very similar to the polling sequence and is shown here:

```
E    S
O    S    D
T    A    A
```

```
        E
where   O  = end of transmission
        T

        S
        S  = station selection address
        A

        D
        A  = device address
```

As with the polling sequence, a selection sequence begins with the transmission of an EOT character, which ensures that all secondary stations are in the line-monitoring mode. Following the EOT character is a two-character selection address, which includes a station selection address (SSA) and a device address (DA). The SSA identifies the station being selected, and the DA identifies the specific device at the designated station.

A typical selection sequence is shown here:

```
E
O    A    X
T
```

For this example, the station selection address is the ASCII character A, and the device address is the ASCII character X (i.e., station A, device X). Station A has been selected by the primary to receive a message and must respond with some form of an acknowledgment to indicate whether it is ready to receive a message.

16-5-5-4 83B responses to a selection. Once selected, a secondary station must respond with one of three acknowledgment sequences indicating its status. The three acknowledgments to a selection are listed here along with their statuses:

Negative Acknowledgments to a Selection		
\	\	Not ready to receive (terminal in local or printer out of paper)
*	*	Not ready to receive (have a formatted message to transmit)

```
          Positive Acknowledgment to a Selection

              A
          \   C   Ready to receive
              K
```

More than one secondary station can be selected simultaneously with group or broad-cast addresses. Group addresses are used when the primary desires to select more than one secondary station but not all of them. A single broadcast address is used to simul-taneously select all the secondary stations. With asynchronous protocols, acknowledg-ments for group and broadcast selections are somewhat involved and for this reason are seldom used.

16-5-5-5 83B primary message format.   Messages transmitted from primary sta-tions use exactly the same data format as messages transmitted from secondary stations as shown here:

```
          S              E
          T   message    O
          X              T
```

STX, when transmitted by a primary station, is called a *blinding character*, as it causes all previously unselected secondary stations to ignore the transmission. Consequently, only the previously selected secondary station receives the message transmitted by the primary. The unselected secondary stations remain blinded until they receive an EOT character, at which time they will return to the line-monitoring mode and begin looking for polls or selections addressed for their station. The ETO character at the end of the sequence signals the end of the message and also designates the secondary station that received the message as the mas-ter so that it can acknowledge whether the transmission was successful.

16-6   SYNCHRONOUS DATA-LINK PROTOCOLS

With *synchronous data-link protocols*, remote stations can have more than one terminal, PC, or printer. A group of terminals, computers, printers, and other digital devices is some-times called a *cluster*. A single line control unit (LCU) can serve a cluster with as many as 50 devices (terminals, PCs, and printers). Synchronous data-link protocols generally are used with synchronous data and synchronous modems and can be either character or bit ori-ented. One of the most commonly used synchronous data-link protocols is IBM's binary synchronous communications (BSC).

16-6-1   Binary Synchronous Communications

Binary synchronous communications (BSC) is a synchronous character-oriented data-link protocol developed by IBM. BSC is sometimes called *bisync* or *bisynchronous communi-cations*. With BSC, each data transmission is preceded by a unique synchronization (SYN) character as shown here:

```
          S   S
          Y   Y   message block
          N   N
```

The message block can be a poll, a selection, an acknowledgment, or a message containing user information.

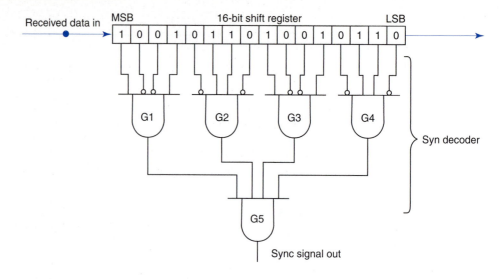

FIGURE 16-5 Bisync detector circuit

The SYN character for ASCII is 16 hex and for EBCDIC 32 hex. The SYN character places the USART receiver in the character (byte) mode and prepares it to receive data in eight-bit groupings. With BSC, SYN characters are always transmitted in pairs (hence the name *bisync* or *bisynchronous communications*). Figure 16-5 shows the logic diagram for an even-parity ASCII SYN character detection circuit. Received data are shifted serially one bit at a time through the detection circuit, where they are monitored in groups of 16 bits looking for the occurrence of two successive SYN characters (1616 hex). When two successive SYN characters are detected, the SYNC output status signal goes active.

If eight successive bits are received in the middle of a message that is equivalent to a SYN character, they are ignored. For example, the characters A and b have the following hex codes:

$$A = 41H \qquad b = 62H$$

If the ASCII characters A and b occur successively during a heading field, data field, or error-detection field, the following bit sequence would occur:

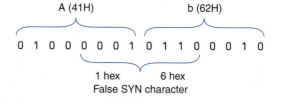

As you can see, it appears that a SYN character has been received when in fact it has not. To avoid this situation with BSC, SYN characters are always transmitted in pairs, and, consequently, if only one is received and detected, it is ignored. The likelihood of two false SYN characters occurring one immediately after the other is remote.

When synchronous data-link protocols are used in multipoint private-line circuits, the concepts of polling, selecting, and acknowledging are identical to asynchronous protocols; however, with bisync, group and broadcast selections are not allowed.

16-6-1-1 BSC polling sequences. There are two polling formats used with bisync: general and specific. The format for a general poll is the following:

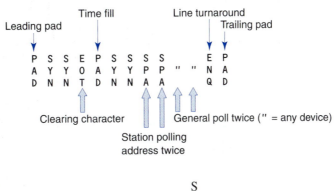

where
P		S	
A = pad		Y = synchronization character	
D		N	
E		S	
O = end of transmission		P = station polling address	
T		A	
S			
Y = synchronization character		" = general poll	
N			
E			
N = inquiry			
Q			

The PAD character at the beginning of the sequence is called a *leading pad* and is either a 55 hex or an AA hex (01010101 or 10101010). As you can see, a leading pad is simply a string of alternating logic 1s and logic 0s. The purpose of the leading pad is to ensure that transitions occur in the data before transmission of the actual message. The transitions are needed for clock recovery in the receive modem to maintain bit synchronization. Immediately following the leading pad are two SYN characters that establish character synchronization. The EOT character is again used as a clearing character that places all secondary stations into the line-monitoring mode. The PAD character immediately following the second SYN character is simply a string of successive logic 1s that serves as a time fill, giving each of the secondary stations time to clear. The number of logic 1s transmitted during this time fill may not be a multiple of eight bits. Consequently, the two SYN characters are repeated to reestablish character synchronization. Two station polling address (SPA) characters are transmitted for error detection (character redundancy). A secondary will not recognize or respond to a poll unless its SPA appears twice in succession. The two quotation marks signify that the poll is a general poll for any device at that station that has a formatted message to send. If two or more devices have messages to transmit when a general poll is received, the station controller determines which device's message is transmitted. This allows the controller to prioritize the devices at the station. The enquiry (ENQ) character is sometimes called a *format* or *line turnaround* character because it simply completes the polling sequence and initiates a line turnaround (i.e., the secondary station identified by the SPA is designated the master and must respond to the poll).

The PAD character at the end of the polling sequence is called a *trailing pad* and is simply a 7F (DEL, or delete character). The purpose of the trailing pad is to ensure that the

Table 16-1 BSC Station and Device Addresses

Station or Device Number	SPA	SSA	DA	Station or Device Number	SPA	SSA	DA
0	sp	—	sp	16	&	0	&
1	A	/	A	17	J	1	J
2	B	S	B	18	K	2	K
3	C	T	C	19	L	3	L
4	D	U	D	20	M	4	M
5	E	V	E	21	N	5	N
6	F	W	F	22	O	6	O
7	G	X	G	23	P	7	P
8	H	Y	H	24	Q	8	Q
9	I	Z	I	25	R	9	R
10	[	-	[	26	]	:	]
11	.	,	.	27	$	#	$
12	<	%	<	28	*	@	*
13	(	—	(	29	)	'	)
14	+	>	+	30	;	=	;
15	!	?	!	31	^	"	^

RLSD signal in the receive modem is held active long enough for the entire received message to be demodulated. If the carrier were shut off immediately at the end of the polling sequence, RLSD would go inactive and disable the receive data pin. If the last character of the polling sequence were not completely demodulated, the end of it would be cut off. Trailing pads are inserted at the end of all transmissions for the same purpose.

With BSC, there is a second form of polling sequence called a *specific poll*. The format for a specific poll is the following:

Device address
twice

The character sequence for a specific poll is similar to a general poll except that two device address (DA) characters are substituted for the two quotation marks. With a specific poll, both the station and the device address are included. Therefore, a specific poll is an invitation for only one specific device at a given secondary station to transmit its message. Again, two DA characters are transmitted for redundancy error detection.

Table 16-1 gives a list of station polling addresses, station selection addresses, and device addresses for a BSC system with a maximum of 32 stations and 32 devices.

Example 16-1

Determine the character sequence for the following:

a. General poll of station 8
b. A specific poll of device 10 at station 6.

Solution

a. From Table 16-1, the SPA for station 8 is H; therefore, the sequence for a general poll is

```
P  S  S  E  P  S  S              E  P
A  Y  Y  O  A  Y  Y  H  H  "  "  N  A
D  N  N  T  D  N  N              Q  D
```

b. From Table 16-1, the SPA for station 6 is F and the DA for device 10 is [; therefore, the sequence for a specific poll is

```
P   S   S   E   P   S   S                           E   P
A   Y   Y   O   A   Y   Y   F   F   [   [   N   A
D   N   N   T   D   N   N                           Q   D
```

c. With bisync, there are only two ways in which a secondary station can respond to a poll: with a formatted message or with a *handshake*. A handshake is simply a response from the secondary that indicates it has no formatted messages to transmit (a handshake is a negative response to a poll). The character sequence for a handshake is

```
P   S   S   E   P
A   Y   Y   O   A
D   N   N   T   D
```

16-6-1-2 BSC selection sequence.   The format for a selection with BSC is the following:

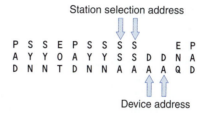

Station selection address

```
P   S   S   E   P   S   S   S   S           E   P
A   Y   Y   O   A   Y   Y   S   S   D   D   N   A
D   N   N   T   D   N   N   A   A   A   A   Q   D
```

Device address

The sequence for a selection is very similar to that of a specific poll except two SSA characters are substituted for the two SPA characters. SSA stands for station selection address. All selections are specific; they are for a specific device at a specific station.

Example 16-2

Determine the character sequence for a BSC selection of device 22 at station 18.

Solution From Table 16-1, the DA for device 22 is O, and the SSA for station 18 is 2; therefore, the selection sequence is

```
P   S   S   E   P   S   S                           E   P
A   Y   Y   O   A   Y   Y   2   2   O   O   N   A
D   N   N   T   D   N   N                           Q   D
```

A secondary station can respond to a selection with either a positive or a negative acknowledgment. A positive acknowledgment to a selection indicates that the device selected is ready to receive. The character sequence for a positive acknowledgment is

```
P   S   S   D       P
A   Y   Y   L   0   A
D   N   N   E       D
```

A negative acknowledgment to a selection indicates that the selected device is not ready to receive. A negative acknowledgment is called a *reverse interrupt* (RVI). The character sequence for a negative acknowledgment to a selection is

```
P   S   S   D       P
A   Y   Y   L   <   A
D   N   N   E       D
```

16-6-1-3 BSC message sequence. With bisync, formatted messages are sent from secondary stations to the primary station in response to a poll and sent from primary stations to secondary stations after the secondary has been selected. Formatted messages use the following format:

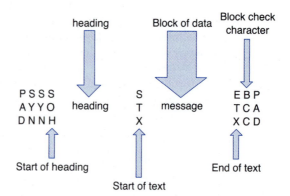

The block check character (BCC) is an error detection character. Longitudinal redundancy checking (LRC) is used for error detection with ASCII-coded messages, and cyclic redundancy checking (CRC-16) is used for EBCDIC-coded messages (when CRC-16 is used, there are two block check characters). The block check character is sometimes called a block check sequence (BCS) because it does not represent a character; it is simply a sequence of bits used for error detection.

The BCC is computed beginning with the first character after SOH and continues through and includes the end-of-text (ETX) character. (If there is no heading, the BCC is computed beginning with the first character after start of text.) With synchronous protocols, data are transmitted in blocks or frames. Blocks of data are generally between 256 and 1500 bytes long. ETX is used to terminate the last block of a message. End of block (ETB) is used for multiple block messages to terminate all message blocks except the last one. The last block of a message is always terminated with ETX. The receiving station must acknowledge all BCCs.

A positive acknowledgment to a BCC indicates that the BCC was good, and a negative acknowledgment to a BCC indicates that the BCC was bad. A negative acknowledgment is an automatic request for retransmission (ARQ). The character sequences for positive and negative acknowledgments are as follow:

Positive responses to a block check character (message):

	P	S	S	D		P	
	A	Y	Y	L	0	A	(even numbered blocks)
	D	N	N	E		D	

or

	P	S	S	D		P	
	A	Y	Y	L	1	A	(odd numbered blocks)
	D	N	N	E		D	

Negative response to a block check character (message):

P	S	S	N	P
A	Y	Y	A	A
D	N	N	K	D

where $\begin{matrix} N \\ A \\ K \end{matrix}$ = negative acknowledgment

16-6-1-4 BSC transparency. It is possible that a device attached to one or more of the ports of a station controller is not a computer terminal or printer, such as a microprocessor-controlled system that is used to monitor environmental conditions (temperature, humidity, and so on) or a security alarm system. If so, the data transferred between it and the primary station does not consist of ASCII- or EBCDIC-coded characters; they could be microprocessor op-codes or binary encoded data. Consequently, it would be possible that an eight-bit sequence could occur within the message that is equivalent to a data-link control character. For example, if the binary code 00000011 (03 hex) occurred in a message, the controller would misinterpret it as the ASCII code for the data-link control character ETX. If this happened, the controller would terminate the message and interpret the next 8- or 16-bit sequence as the block check character. To prevent this from occurring, the controller is made *transparent* to the data. With bisync, a *data-link escape* (DLE) character is used to achieve transparency. To place a controller into the transparent mode, STX is preceded by a DLE. This causes the controller to transfer the data to the selected device without searching through the message looking for data-link control characters. To come out of the transparent mode, DLE ETX is transmitted. To transmit a bit sequence equivalent to DLE as part of the text, it must be preceded by a DLE character (i.e., DLE DLE). There are only six instances when it is necessary to precede a character with a DLE:

1. *DLE STX.* Places the receive controller into the transparent mode.
2. *DLE ETX.* Used to terminate the last block of transparent text and take the controller out of the transparent mode.
3. *DLE ETB.* Used to terminate blocks of transparent text in all blocks of data except the final block.
4. *DLE ITB.* Used to terminate blocks of transparent text other than the final block when ITB (end-of-intermittent block) is used for a block-terminating character. ITB is used occasionally when consecutive blocks are transmitted in succession without acknowledgments for each block. Only one acknowledgment occurs at the end of the last block. This eliminates handshakes for all blocks except the final block of a message. The disadvantage of using ITB characters is that if an error occurs in any of the blocks, all the blocks must be retransmitted.
5. *DLE SYN.* Used only with transparent messages more than 1 second long. With bisync, two SYN characters are inserted in the text to ensure that the receive controller does not lose character synchronization. In a multipoint circuit with a polling environment, it is highly unlikely that any block of data would exceed 1 second in duration. SYN character insertion is used almost exclusively with two-point data circuits.
6. *DLE DLE.* Used to transmit a bit sequence equivalent to DLE as part of the text (i.e., DLE DLE).

Several examples of including data-link escape characters in transparent BSC messages are shown here:

P	S	S	D	S		D	E	B	P
A	Y	Y	L	T	message	L	T	C	D
D	N	N	E	X		E	X	C	D

D   E   D   I
L   T   L   T
E   B   E   B
D   S   D   S
L   Y   L   Y
E   N   E   N
D   D
L   L
E   E

16-7 SYNCHRONOUS DATA-LINK CONTROL

Synchronous data-link control (SDLC) is a synchronous bit-oriented protocol developed in the 1970s by IBM for use in *system network architecture* (SNA) environments. SDLC was the first link-layer protocol based on synchronous, bit-oriented operation. The International Organization for Standardization modified SDLC and created high-level data-link control (HDLC), and the International Telecommunications Union–Telecommunications Standardization Sector (ITU-T) subsequently modified HDLC to create link access procedures (LAPs). The Institute of Electrical and Electronic Engineers (IEEE) later modified HDLC and created IEEE 802.2. Although each of these protocol variations is important in its own domain, SDLC remains the primary SNA link-layer protocol for wide-area data networks.

SDLC can transfer data simplex, half duplex, or full duplex and can operate over a bus or ring (loop) topology. With a BOP, there is a single control field within a message *frame* that performs essentially all the data-link control functions. SDLC frames are similar to blocks of data and with SDLC generally limited to 256 characters in length. EBCDIC was the original character language used with SDLC.

There are two types of stations defined by SDLC: *primary stations* and *secondary stations*. There is only one primary station in an SDLC circuit that controls data exchange on the communications channel and issues *commands*. All the other stations on an SDLC circuit are secondary stations, which receive commands and return (transmit) *responses* to the primary station.

There are three transmission states with SDLC: transient, idle, and active. The *transient state* exists before and after an initial transmission and after each line turnaround. A secondary station assumes that the circuit is in an idle state after 15 or more consecutive logic 1s have been received. The *active state* exists whenever either the primary or one of the secondary stations is transmitting information or control signals.

16-7-1 SDLC Frame Format

Figure 16-6 shows the frame format used with SDLC. Frames transmitted from the primary and secondary stations use exactly the same format. There are five *fields* included in an SDLC frame:

1. Flag field
2. Address field
3. Control field
4. Information (or text) field
5. Frame check character (FCC) field

16-7-1-1 SDLC flag field.   There are two *flag fields* per frame, each with a minimum length of one byte. The two flag fields are the *beginning flag* and the *ending flag*. Flags are used for the *delimiting sequence* for the frame and to achieve *frame and charac-*

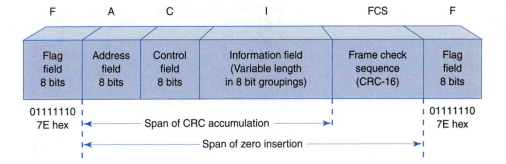

FIGURE 16-6   SDLC frame format

ter synchronization. The delimiting sequence sets the limits of the frame (i.e., when the frame begins and when it ends). The flag is used with SDLC in a manner similar to the way SYN characters are used with bisync—to achieve character synchronization. The bit sequence for a flag is 01111110 (7E hex), which is the character "=" in the EBCDIC code. There are several variations to how flags are used with SDLC:

1. One beginning and one ending flag for each frame:

Beginning flag					Ending flag
01111110	address	control	text	FCC	01111110

2. The ending flag from one frame is used for the beginning flag for the next frame:

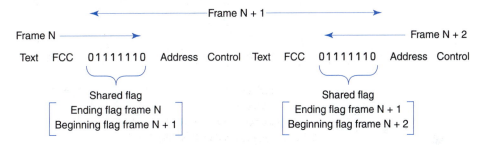

3. The last zero of an ending flag can be the first zero of the beginning flag of the next frame:

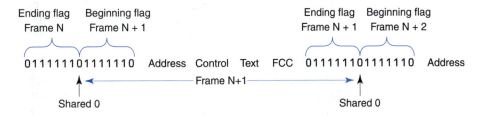

4. Flags are transmitted continuously during the time between frames in lieu of idle line 1s:

16-7-1-2 SDLC address field. The address field for SDLC contains eight bits; thus, 256 addresses are possible. The address 00 hex (00000000) is called the *null* address and is never assigned to a secondary station. The null address is used for network testing. The address FF hex (11111111) is called the *broadcast* address and is common to all secondary stations. The primary station is the only station that can transmit the broadcast address. When a frame is transmitted with the broadcast address, it is simultaneously sent to all secondary stations. The remaining 254 addresses can be used as unique station addresses intended for one secondary station only or as group addresses that are intended for more than one secondary station but not all of them.

In frames sent by the primary station, the address field contains the address of the secondary station (i.e., the address of the destination). In frames sent from a secondary station, the address field contains the address of the secondary (i.e., the address of the station sending the message). The primary station has no address because all transmissions from secondary stations go to the primary station.

16-7-1-3 SDLC information field. All information transmitted in an SDLC frame must be in the information field (I field), and the number of bits in the information field must be a multiple of eight. An information field is not allowed with all SDLC frames; however, the data within an information field can be user information or control information.

16-7-1-4 SDLC control field. The control field is an eight-bit field that identifies the type of frame being transmitted. The control field is used for polling, confirming previously received frames, and several other data-link management functions. There are three frame formats with SDLC: information, supervisory, and unnumbered.

With an *information frame*, there must be an information field, and the information field must contain user data. Information frames are used for transmitting sequenced information that must be acknowledged by the destination station. The bit pattern for the control field of an information frame is

A logic 0 in the least-significant bit position identifies an information frame (I frame), which is b_7 with the EBCDIC code. With information frames, the primary can select a secondary station, send formatted information, confirm previously received information frames, and poll a secondary station with a single transmission.

Bit b_3 of an information frame is called a *poll* (P) or *not-a-poll* ($\overline{P}$) bit when sent by the primary station and a *final* (F) or *not-a final* ($\overline{F}$) bit when sent by a secondary station. In frames sent from a primary, if the primary desires to poll the secondary (i.e., solicit it for information), the P bit in the control field is set (logic 1). If the primary does not wish to poll the secondary, the P bit is reset (logic 0). A secondary cannot transmit frames unless it receives a frame addressed to it with the P bit set. When the primary is transmitting multiple frames to the same secondary, b_3 is a logic 0 in all but the last frame. In the last frame, b_3 is set, which demands a response from the secondary. When a secondary is transmitting multiple frames to the primary, b_3 in the control field is a logic 0 in all but the last frame. In the last frame, b_3 is set, which simply indicates that frame is the last one in the sequence.

In an information frame, bits b_4, b_5, and b_6 of the control field are the ns bits, which are used for numbering transmitted frames (*ns* stands for "number sent"). All information frames must be numbered. With three bits, the binary numbers 000 through 111 (0 through 7) can be represented. The first frame transmitted by each station is designated frame 000, the second frame 001, and so on up to frame 111 (the eighth frame), at which time the count cycles back to 000 and repeats.

In an information frame, bits b_0, b_1, and b_2 in the control field are the nr bits, which are used to indicate the status of previously received information frames (*nr* stands for "number received"). The nr bits are used to confirm frames received without errors and to automatically request retransmission of information frames received with errors. The nr is the number of the next information frame that the transmitting station expects to receive or the number of the next information frame that the receiving station will transmit. The nr confirms received frames through nr $-$ 1. Frame nr $-$ 1 is the last information frame received without a transmission error. For example, when a station transmits an nr $= 5$, it is confirming successful reception of previously unconfirmed frames up through 4. Together, the ns and nr bits are used for error correction (ARQ). The primary station needs to keep track of an ns and an nr for each secondary station. Each secondary station must keep track of only its ns and nr. After all frames have been confirmed, the primary station's ns must agree with the secondary station's nr and vice versa.

For the following example, both the primary and the secondary station begin with their nr and ns counters reset to 000. The primary begins the information exchange by sending three information frames numbered 0, 1, and 2 (i.e., the ns bits in the control character for the three frames are 000, 001, and 010). In the control character for the three frames, the primary transmits an nr $= 0$ (i.e., 000). An nr $= 0$ is transmitted because the next frame the primary expects to receive from the secondary is frame 0, which is the secondary's present ns. The secondary responds with two information frames (ns $= 0$ and 1). The secondary received all three frames from the primary without any errors; therefore, the nr transmitted in the secondary's control field is 3, which is the number of the next frame the primary will send. The primary now sends information frames 3 and 4 with an nr $= 2$, which confirms the correct reception of frames 0 and 1 from the secondary. The secondary responds with frames ns $= 2$, 3, and 4 with an nr $= 4$. The nr $= 4$ confirms reception of only frame 3 from the primary (nr $-$ 1). Consequently, the primary retransmits frame 4. Frame 4 is transmitted together with four additional frames (ns $= 5, 6, 7$, and 0). The primary's nr $= 5$, which confirms frames 2, 3, and 4 from the secondary. Finally, the secondary sends information frame 5 with an nr $= 1$, which confirms frames 4, 5, 6, 7, and 0 from the primary. At this point, all frames transmitted have been confirmed except frame 5 from the secondary. The preceding exchange of information frames is shown in Figure 16-7.

With SDLC, neither the primary nor the secondary station can send more than seven numbered information frames in succession without receiving a confirmation. For example, if the primary sent eight frames (ns $= 0, 1, 2, 3, 4, 5, 6$, and 7) and the secondary responded with an nr $= 0$, it is ambiguous which frames are being confirmed. Does nr $= 0$ mean that all eight frames were received correctly, or does it mean that frame 0 had an error in it and all eight frames must be retransmitted? (All frames beginning with nr $-$ 1 must be retransmitted.)

Example 16-3

Determine the bit pattern for the control field of an information frame sent from the primary to a secondary station for the following conditions:

a. Primary is sending information frame 3 (ns $= 3$)
b. Primary is polling the secondary (P $= 1$)
c. Primary is confirming correct reception of frames 2, 3, and 4 from the secondary (nr $-$ 5)

Solution

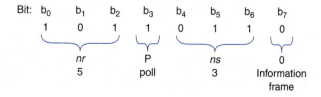

Primary Station

Status		ns	0	1	2		3	4			4	5	6	7	0
		nr:	0	0	0		2	2			5	5	5	5	5
		P/P̄	0	0	1		0	1			0	0	0	0	1

Control Field	b_0	0	0	0		0	0			1	1	1	1	1
	b_1	0	0	0		1	1			0	0	0	0	0
	b_2	0	0	0		0	0			1	1	1	1	1
	b_3	0	0	1		0	1			0	0	0	0	1
	b_4	0	0	0		0	1			1	1	1	1	0
	b_5	0	0	1		1	0			0	0	1	1	0
	b_6	0	1	0		1	0			0	1	0	1	0
	b_7	0	0	0		0	0			0	0	0	0	0
hex code		00	02	14		46	58			A8	AA	AC	AE	B0

Secondary Station

Status	ns:	0	1		2	3	4			5
	nr:	3	3		4	4	4			1
	F/F̄	0	1		0	0	1			1

Control Field	b_0	0	0		1	1	1			0
	b_1	1	1		0	0	0			0
	b_2	1	1		0	0	0			1
	b_3	0	1		0	0	1			1
	b_4	0	0		0	0	1			1
	b_5	0	0		1	1	0			0
	b_6	0	1		0	1	0			1
	b_7	0	0		0	0	0			0
hex code		60	72		84	86	98			3A

FIGURE 16-7   SDLC exchange of information frames

Example 16-4

Determine the bit pattern for the control field of an information frame sent from a secondary station to the primary for the following conditions:

a. Secondary is sending information frame 7 (ns = 7)
b. Secondary is not sending its final frame (F = 0)
c. Secondary is confirming correct reception of frames 2 and 3 from the primary (nr – 4)

Solution

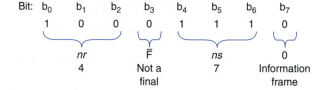

With *supervisory frames*, an information field is not allowed. Consequently, supervisory frames cannot be used to transfer numbered information; however, they can be used to assist in the transfer of information. Supervisory frames can be used to confirm previously received information frames, convey ready or busy conditions, and for a primary to poll a secondary station when the primary does not have any numbered information to send to the secondary. The bit pattern for the control field of a supervisory frame is

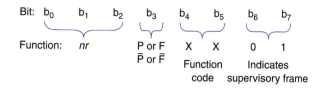

A supervisory frame is identified with a 01 in bit positions b_6 and b_7, respectively, of the control field. With the supervisory format, bit b_3 is again the poll/not-a-poll or final/not-a-final bit, and b_0, b_1, and b_2 are the nr bits. Therefore, supervisory frames can be used by a primary to poll a secondary, and both the primary and the secondary stations can use supervisory frames to confirm previously received information frames. Bits b_4 and b_5 in a supervisory are used either to indicate the receive status of the station transmitting the frame or to request transmission or retransmission of sequenced information frames. With two bits, there are four combinations possible. The four combinations and their functions are the following:

b_4	b_5	Receive Status
0	0	Ready to receive (RR)
0	1	Ready not to receive (RNR)
1	0	Reject (REJ)
1	1	Not used with SDLC

When a primary station sends a supervisory frame with the P bit set and a status of ready to receive, it is equivalent to a general poll. Primary stations can use supervisory frames for polling and also to confirm previously received information frames without sending any information. A secondary uses the supervisory format for confirming previously received information frames and for reporting its receive status to the primary. If a secondary sends a supervisory frame with RNR status, the primary cannot send it numbered information frames until that status is cleared. RNR is cleared when a secondary sends an information frame with the F bit = 1 or a supervisory frame indicating RR or REJ with the F bit = 0. The REJ command/response is used to confirm information frames through nr − 1 and to request transmission of numbered information frames beginning with the frame number identified in the REJ frame. An information field is prohibited with a supervisory frame, and the REJ command/response is used only with full-duplex operation.

Example 16-5

Determine the bit pattern for the control field of a supervisory frame sent from a secondary station to the primary for the following conditions:

a. Secondary is ready to receive (RR)
b. It is a final frame
c. Secondary station is confirming correct reception of frames 3, 4, and 5 (nr = 6)

Solution

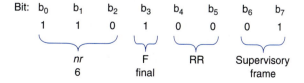

An *unnumbered frame* is identified by making bits b_6 and b_7 in the control field both logic 1s. The bit pattern for the control field of an unnumbered frame is

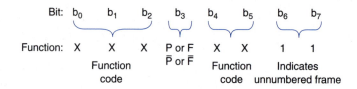

With an unnumbered frame, bit b_3 is again either the poll/not-a-poll or final/not-a-final bit. There are five X bits (b_0, b_1, b_2, b_4, and b_5) included in the control field of an unnumbered frame and are used for various unnumbered commands and responses. With five bits available, there are 32 unnumbered commands/responses possible. The control field in an unnumbered frame sent from a primary station is called a *command*, and the control field in an unnumbered frame sent from a secondary station is called a *response*. With unnumbered frames, there are neither ns nor nr bits included in the control field. Therefore, numbered information frames cannot be sent or confirmed with the unnumbered format. Unnumbered frames are used to send network control and status information. Two examples of control functions are (1) placing a secondary station on- or off-line and (2) initializing a secondary station's line control unit (LCU). Table 16-2 lists several of the more common unnumbered commands and responses. Numbered information frames are prohibited with all unnumbered frames. Therefore, user information cannot be transported with unnumbered frames, and, thus, the control field for unnumbered frames does not include nr and ns bits. However, information fields containing control information are allowed with the following unnumbered commands and responses: UI, FRMR, CFGR, TEST, and XID.

A secondary station must be in one of three modes: *initialization mode, normal response mode*, or *normal disconnect mode*. The procedures for placing a secondary station into the initialization mode are system specified and vary considerably. A secondary in the normal response mode cannot initiate unsolicited transmissions; it can transmit only in response to a frame received with the P bit set. When in the normal disconnect mode, a sec-

Table 16-2 SDLC Unnumbered Commands and Responses

Binary Configuration							
b_0		b_7	Acronym	Command	Response	1 Field Prohibited	Resets ns and nr
000	P/F	0011	UI	Yes	Yes	No	No
000	F	0111	RIM	No	Yes	Yes	No
000	P	0111	SIM	Yes	No	Yes	Yes
100	P	0011	SNRM	Yes	No	Yes	Yes
000	F	1111	DM	No	Yes	Yes	No
010	P	0011	DISC	Yes	No	Yes	No
011	F	0011	UA	No	Yes	Yes	No
100	F	0111	FRMR	No	Yes	No	No
111	F	1111	BCN	No	Yes	Yes	No
110	P/F	0111	CFGR	Yes	Yes	No	No
010	F	0011	RD	No	Yes	Yes	No
101	P/F	1111	XID	Yes	Yes	No	No
001	P	0011	UP	Yes	No	Yes	No
111	P/F	0011	TEST	Yes	Yes	No	No

ondary is off-line. In this mode, a secondary station will accept only the TEST, XID, CFGR, SNRM, or SIM commands from the primary station and can respond only if the P bit is set. The unnumbered commands and responses are summarized here:

1. *Unnumbered information (UI).* UI can be a command or a response that is used to send unnumbered information. Unnumbered information transmitted in the I field is acknowledged with an unnumbered acknowledgment (UA) frame.

2. *Set initialization mode (SIM).* SIM is a command that places a secondary station into the initialization mode. The initialization procedure is system specified and varies from a simple self-test of the station controller to executing a complete IPL (initial program logic) program. SIM resets the ns and nr counters at the primary and secondary stations. A secondary is expected to respond to a SIM command with an unnumbered acknowledgment (UA) response.

3. *Request initialization mode (RIM).* RIM is a response sent by a secondary station to request the primary to send a SIM command.

4. *Set normal response mode (SNRM).* SNRM is a command that places a secondary into the normal response mode (NRM). A secondary station cannot send or receive numbered information frames unless it is in the normal response mode. Essentially, SNRM places a secondary station on-line. SNRM resets the ns and nr counters at both the primary and the secondary stations. UA is the normal response to a SNRM command. Unsolicited responses are not allowed when a secondary is in the NRM. A secondary station remains in the NRM until it receives a disconnect (DISC) or SIM command.

5. *Disconnect mode (DM).* DM is a response transmitted from a secondary station if the primary attempts to send numbered information frames to it when the secondary is in the normal disconnect mode.

6. *Request disconnect (RD).* RD is a response sent by a secondary when it wants the primary to place it in the disconnect mode.

7. *Disconnect (DISC).* DISC is a command that places a secondary station in the normal disconnect mode (NDM). A secondary cannot send or receive numbered information frames when it is in the normal disconnect mode. When in the NDM, a secondary can receive only SIM or SNRM commands and can transmit only a DM response. The expected response to a DISC is UA.

8. *Unnumbered acknowledgment (UA).* UA is an affirmative response that indicates compliance to SIM, SNRM, or DISC commands. UA is also used to acknowledge unnumbered information frames.

9. *Frame reject (FRMR).* FRMR is for reporting procedural errors. The FRMR response is an answer transmitted by a secondary after it has received an invalid frame from the primary. A received frame may be invalid for any one of the following reasons:

   a. The control field contains an invalid or unassigned command.
   b. The amount of data in the information field exceeds the buffer space in the secondary station's controller.
   c. An information field is received in a frame that does not allow information fields.
   d. The nr received is incongruous with the secondary's ns. For example, if the secondary transmitted ns frames 2, 3, and 4 and then the primary responded with an nr = 7. A secondary station cannot release itself from the FRMR condition, nor does it act on the frame that caused the condition. The secondary repeats the FRMR response until it receives one of the following mode-setting commands: SNRM, DISC, or SIM. The information field for a FRMR response must contains three bytes (24 bits) and has the following format:

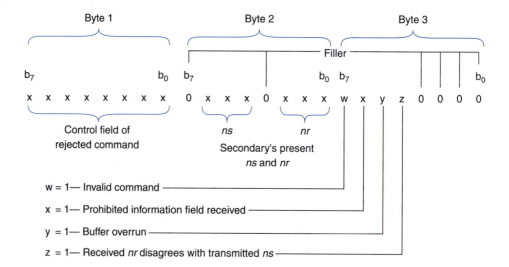

Byte 1 · Byte 2 · Byte 3

Filler

b_7 ... b_0 b_7 ... b_0 b_7 ... b_0

x x x x x x x x 0 x x x 0 x x x w x y z 0 0 0 0

Control field of rejected command

ns   *nr*

Secondary's present *ns* and *nr*

w = 1— Invalid command
x = 1— Prohibited information field received
y = 1— Buffer overrun
z = 1— Received *nr* disagrees with transmitted *ns*

10. *TEST.* The TEST command/response is an exchange of frames between the primary station and a secondary station. An information field may be included with the TEST command; however, it cannot be sequenced (numbered). The primary sends a TEST command to a secondary in any mode to solicit a TEST response. If an information field is included with the command, the secondary returns it with its response. The TEST command/response is exchanged for link testing purposes.

11. *Exchange station identification (XID).* XID can be a command or a response. As a command, XID solicits the identification of a secondary station. An information field can be included in the frame to convey the identification data of either the primary or the secondary. For dial-up circuits, it is often necessary that the secondary station identify itself before the primary will exchange information frames with it, although XID is not restricted to only dial-up circuits.

16-7-1-5 Frame check character field.   The FCC field contains the error detection mechanism for SDLC. The FCC is equivalent to the BCC used with binary synchronous communications (BSC). SDLC uses CRC-16 and the following generating polynomial: $x^{16} + x^{12} + x^5 + x^1$. Frame check characters are computed on the data in the address, control, and information fields.

16-7-2   SDLC Loop Operation

An SDLC loop operates half duplex. The primary difference between the loop and bus configurations is that in a loop, all transmissions travel in the same direction on the communications channel. In a loop configuration, only one station transmits at a time. The primary station transmits first, then each secondary station responds sequentially. In an SDLC loop, the transmit port of the primary station controller is connected to the receive port of the controller in the first down-line secondary station. Each successive secondary station is connected in series with the transmission path with the transmit port of the last secondary station's controller on the loop connected to the receive port of the primary station's controller. Figure 16-8 shows the physical layout for an SDLC loop.

In an SDLC loop, the primary transmits sequential frames where each frame may be addressed to any or all of the secondary stations. Each frame transmitted by the primary station contains an address of the secondary station to which that frame is directed. Each secondary station, in turn, decodes the address field of every frame and then serves as a repeater for all stations that are down-loop from it. When a secondary station detects a frame with its address, it copies the frame, then passes it on to the next down-loop station. All frames transmitted by

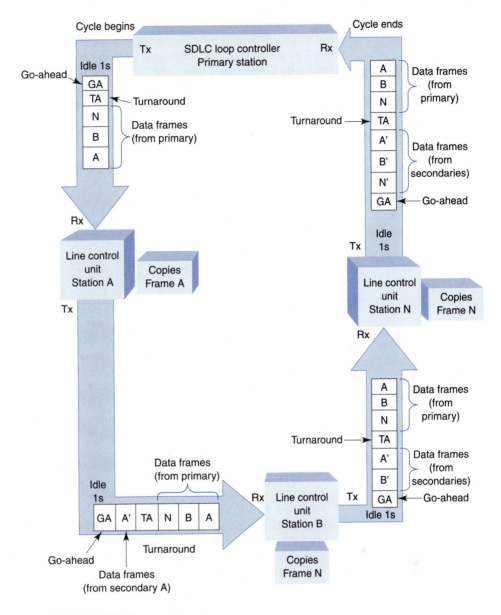

FIGURE 16-8 SDLC loop configuration

the primary are returned to the primary. When the primary has completed transmitting, it follows the last flag with eight consecutive logic 0s. A flag followed by eight consecutive logic 0s is called a *turnaround sequence*, which signals the end of the primary's transmissions. Immediately following the turnaround sequence, the primary transmits continuous logic 1s, which is called the *go-ahead sequence*. A secondary station cannot transmit until it receives a frame address to it with the P bit set, a turnaround sequence, and then a go-ahead sequence. Once the primary has begun transmitting continuous logic 1s, it goes into the receive mode.

The first down-loop secondary station that receives a frame addressed to it with the P bit set changes the go-ahead sequence to a flag, which becomes the beginning flag of the secondary station's response frame or frames. After the secondary station has transmitted its last frame, it again becomes a repeater for the idle line 1s from the primary. These idle line 1s again become the go-ahead sequence for the next down-loop secondary station. The next down-loop secondary station that receives a frame addressed to it with the P bit set detects the turnaround sequence, any frames transmitted from up-loop secondary stations, and then the go-ahead sequence. Each secondary station inserts its response frames immediately after the last repeated frame. The cycle is competed when the primary station receives its own turnaround sequence, a series of response frames, and then the go-ahead sequence.

The previously described sequence is summarized here:

1. Primary transmits sequential frames to one or more secondary stations.
2. Each transmitted frame contains a secondary station's address.
3. After a primary has completed transmitting, it follows the last flag of the last frame with eight consecutive logic zeros (turnaround sequence) followed by continuous logic ones (go-ahead sequence -0111111111111 – – – –).
4. The turnaround sequence alerts secondary stations of the end of the primary's transmissions.
5. Each secondary, in turn, decodes the address field of each frame and removes frames addressed to them.
6. Secondary stations serve as repeaters for any down-line secondary stations.
7. Secondary stations cannot transmit frames of their own unless they receive a frame with the P bit set.
8. The first secondary station that receives a frame addressed to it with the P bit set changes the seventh logic one in the go-ahead sequence to a logic zero, thus creating a flag. The flag becomes the beginning flag for the secondary station's response frames.
9. The next down-loop secondary station that receives a frame addressed to it with the P bit set, detects the turnaround sequence, any frames transmitted by other up-loop secondary stations, and then the go-ahead sequence.
10. Each secondary station's response frames are inserted immediately after the last repeated frame.
11. The cycle is completed when the primary receives its own turnaround sequence, a series of response frames, and the go-ahead sequence.

16-7-2-1 SDLC loop configure command/response. The configure command/response (CFGR) is an unnumbered command/response that is used only in SDLC loop configurations. CFGR contains a one-byte *function descriptor* (essentially a subcommand) in the information field. A CFGR command is acknowledged with a CFGR response. If the low-order bit of the function descriptor is set, a specified function is initiated. If it is reset, the specified function is cleared. There are six subcommands that can appear in the configure command/response function field:

1. *Clear—(00000000).* A *clear* subcommand causes all previously set functions to be cleared by the secondary. The secondary's response to a clear subcommand is another clear subcommand, 00000000.
2. *Beacon test (BCN)—(0000000X).* The *beacon test* subcommand causes the secondary receiving it to turn on (00000001) or turn off (00000000) its carrier. The beacon response is called a carrier, although it is not a carrier in the true sense of the word. The beacon test command causes a secondary station to begin transmitting a beacon response, which is not a carrier. However, if modems were used in the circuit, the beacon response would cause the modem's carrier to turn on. The beacon test is used to isolate open-loop continuity problems. In addition, whenever a secondary station detects a loss of signal (either data or idle line 1s), it automatically begins to transmit its beacon response. The secondary will continue transmitting the beacon until the loop resumes normal status.

3. *Monitor mode—(0000010X)*. The *monitor* command (00000101) causes the addressed secondary station to place itself into the monitor (receive-only) mode. Once in the monitor mode, a secondary cannot transmit until it receives either a monitor mode clear (00000100) or a clear (00000000) subcommand.

4. *Wrap—(0000100X)*. The *wrap* command (00001001) causes a secondary station to loop its transmissions directly to its receiver input. The wrap command places the secondary effectively off-line for the duration of the test. A secondary station takes itself out of the wrap mode when it receives a wrap clear (00001000) or clear (00000000) subcommand.

5. *Self-test—(0000101X)*. The self-test subcommand (00001011) causes the addressed secondary to initiate a series of internal diagnostic tests. When the tests are completed, the secondary will respond. If the P bit in the configure command is set, the secondary will respond following completion of the self-test or at its earliest opportunity. If the P bit is reset, the secondary will respond following completion of the test to the next poll-type frame it receives from the primary. All other transmissions are ignored by the secondary while it is performing a self-test; however, the secondary will repeat all frames received to the next down-loop station. The secondary reports the results of the self-test by setting or clearing the low-order bit (X) of its self-test response. A logic 1 means that the tests were unsuccessful, and a logic 0 means that they were successful.

6. *Modified link test—(0000110X)*. If the *modified link test* function is set (X bit set), the secondary station will respond to a TEST command with a TEST response that has an information field containing the first byte of the TEST command information field repeated *n* times. The number *n* is system specified. If the X bit is reset, the secondary station will respond with a zero-length information field. The modified link test is an optional subcommand and is only used to provide an alternative form of link test to that previously described for the TEST command.

16-7-2-2 Transparency. With SDLC, the flag bit sequence (01111110) can occur within a frame where it is not intended to be a flag. For instance, within the address, control, or information fields, a combination of one or more bits from one character combined with one or more bits from an adjacent character could produce a 01111110 pattern. If this were to happen, the receive controller would misinterpret the sequence for a flag, thus destroying the frame. Therefore, the pattern 01111110 must be prohibited from occurring except when it is intended to be a flat.

One solution to the problem would be to prohibit certain sequences of characters from occurring, which would be difficult to do. A more practical solution would be to make a receiver transparent to all data located between the beginning and ending flags. This is called *transparency*. The *transparency mechanism* used with SDLC is called *zero-bit insertion* or *zero stuffing*. With zero-bit insertion, a logic 0 is automatically inserted after any occurrence of five consecutive logic 1s except in a designated flag sequence (i.e., flags are not zero inserted). When five consecutive logic 1s are received and the next bit is a 0, the 0 is automatically deleted or removed. If the next bit is a logic 1, it must be a valid flag. An example of zero insertion/deletion is shown here:

Original frame bits at the transmit station:

01111110	01101111	11010011	1110001100110101	01111110
Beginning flag	Address	Control	Frame check character	Ending flag

After zero insertion but prior to transmission:

01111110 01101111 101010011 1110000110011010101 01111110

| Beginning flag | Address | Control | Frame check character | Ending flag |

Inserted zeros

After zero deletion at the receive end:

01111110 01101111 11010011 1110001100110101 01111110

| Beginning flag | Address | Control | Frame check character | Ending flag |

16-7-2-3 Message abort. Message abort is used to prematurely terminate an SDLC frame. Generally, this is done only to accommodate high-priority messages, such as emergency link recovery procedures. A message abort is any occurrence of 7 to 14 consecutive logic 1s. Zeros are not inserted in an abort sequence. A message abort terminates an existing frame and immediately begins the higher-priority frame. If more than 14 consecutive logic 1s occur in succession, it is considered an idle line condition. Therefore, 15 or more contiguous logic 1s place the circuit into the idle state.

16-7-3 Invert-on-Zero Encoding

With binary synchronous transmission such as SDLC, transmission and reception of data must be time synchronized to enable identification of sequential binary digits. Synchronous data communications assumes that bit or time synchronization is provided by either the DCE or the DTE. The master transmit clock can come from the DTE or, more likely, the DCE. However, the receive clock must be recovered from the data by the DCE and then transferred to the DTE. With synchronous data transmission, the DTE receiver must sample the incoming data at the same rate that it was outputted from the transmit DTE. Although minor variations in timing can exist, the receiver in a synchronous modem provides data clock recovery and dynamically adjusted sample timing to keep sample times midway between bits. For a DCE to recover the data clock, it is necessary that transitions occur in the data. Traditional unipolar (UP) logic levels, such as TTL (0 V and +5 V), do not provide transitions for long strings of logic 0s or logic 1s. Therefore, they are inadequate for clock recovery without placing restrictions on the data. *Invert-on-zero coding* is the encoding scheme used with SDLC because it guarantees at least one transition in the data for every seven bits transmitted. Invert-on-zero coding is also called *NRZI* (*nonreturn-to-zero inverted*).

With NRZI encoding, the data are encoded in the controller at the transmit end of and then decoded in the controller at the receive end. Figure 16-9 shows examples of NRZI en-

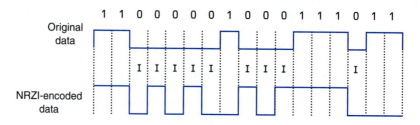

*I indicates to invert

FIGURE 16-9 NRZI encoding

coding and decoding. The encoded waveform is unchanged by 1s in the NRZI encoder. However, logic 0s cause the encoded transmission level to invert from its previous state (i.e., either from a high to a low or from a low to a high). Consequently, consecutive logic 0s are converted to an alternating high/low sequence. With SDLC, there can never be more than six logic 1s in succession (a flag). Therefore, a high-to-low transition is guaranteed to occur at least once every seven bits transmitted except during a message abort or an idle line condition. In a NRZI decoder, whenever a high/low or low/high transition occurs in the received data, a logic 0 is generated. The absence of a transition simply generates a logic 1. In Figure 16-9, a high level is assumed prior to encoding the incoming data.

NRZI encoding was intended to be used with asynchronous modems that do not have clock recovery capabilities. Consequently, the receive DTE must provide time synchronization, which is aided by using NRZI-encoded data. Synchronous modems have built-in scrambler and descrambler circuits that ensure transitions in the data, and thus NRZI encoding is unnecessary. The NRZI encoder/decoder is placed between the DTE and the DCE.

Figure 16-8 shows the logical location of the zero insertion/deletion and NRZI encoding/decoding circuits in an SDLC circuit.

16-8 HIGH-LEVEL DATA-LINK CONTROL

In 1975, the International Organization for Standardization (ISO) defined several sets of substandards that, when combined, are called *high-level data-link control* (HDLC). HDLC is a superset of SDLC; therefore, only the added capabilities are explained.

HDLC comprises three standards (subdivisions) that outline the frame structure, control standards, and class of operation for a bit-oriented data-link control (DLC). The three standards are the following:

1. ISO 3309-1976(E)
2. ISO 4335-1979(E)
3. ISO 7809-1985(E)

16-8-1 ISO 3309-1976(E)

The ISO 3309 standard defines the frame structure, delimiting sequence, transparency mechanism, and error-detection method used with HDLC. With HDLC, the frame structure and delimiting sequence are essentially the same as with SDLC. An HDLC frame includes a beginning flag field, an address field, a control field, an information field, a frame check character field, and an ending flag field. The delimiting sequence with HDLC is a binary 01111110, which is the same flag sequence used with SDLC. However, HDLC computes the frame check characters in a slightly different manner. HDLC uses CRC-16 for error detection with a generating polynominal specified by CCITT V.41. At the transmit station, the CRC characters are computed such that when included in the FCC computations at the receive end, the remainder for an errorless transmission is always FOB8.

HDLC has extended addressing capabilities. HDLC can use an eight-bit address field or an *extended addressing* format that is virtually limitless. With extended addressing, the address field may be extended recursively. If b_0 in the address field is a logic 1, the seven remaining bits are the secondary's address (the ISO defines the low-order bit as b_0, whereas SDLC designates the high-order bit as b_0). If b_0 is a logic 0, the next byte is also part of the address. If b_0 of the second byte is a logic 0, a third address byte follows and so on until an address byte with a logic 1 for the low-order bit is encountered. Essentially, there are seven bits available in each address byte for address encoding. An example of a three-byte extended addressing scheme is shown in the following. Bit b_0 in the first two bytes of the address field are logic 0s, indicating that one or more additional address bytes follow. However, b_0 in the third address byte is a logic 1, which terminates the address field. There are a total of 21 address bits (seven in each byte):

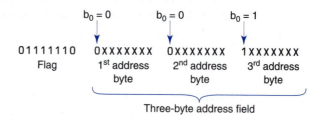

16-8-2 ISO 4335-1979(E)

The ISO 3309 standard defines the elements of procedure for HDLC. The control and information fields have increased capabilities over SDLC and there are two additional operational modes allowed with HDLC.

16-8-2-1 Control field. With HDLC, the control field can be extended to 16 bits. Seven bits are for the ns, and seven bits are for the nr. Therefore, with the extended control format, there can be a maximum of 127 outstanding (unconfirmed) frames at any given time. In essence, a primary station can send 126 successive information frames to a secondary station with the P bit = 0 before it would have to send a frame with the P bit = 1.

With HDLC, the supervisory format includes a fourth status condition: *selective reject* (SREJ). SREJ is identified by two logic 1s in bit positions b_4 and b_5 of a supervisory control field. With SREJ, a single frame can be rejected. A SREJ calls for the retransmission of only one frame identified by the three-bit nr code. A REJ calls for the retransmission of all frames beginning with frames identified by the three-bit nr code. For example, the primary sends I frames ns = 2, 3, 4, and 5. If frame 3 were received in error, a REJ with an nr of 3 would call for a retransmission of frames 3, 4, and 5. However, a SREJ with an nr of 3 would call for the retransmission of only frame 3. SREJ can be used to call for the retransmission of any number of frames except only one at a time.

16-8-2-2 Information field. HDLC permits any number of bits in the information field of an information command or response. With HDLC, any number of bits may be used for a character in the I field as long as all characters have the same number of bits.

HDLC has two operational modes not specified in SDLC: asynchronous response mode and asynchronous disconnect mode:

1. *Asynchronous response mode (ARM).* With the ARM, secondary stations are allowed to send unsolicited responses. To transmit, a secondary does not need to have received a frame from the primary with the P bit set. However, if a secondary receives a frame with the P bit set, it must respond with a frame with the F bit set.
2. *Asynchronous disconnect mode (ADM).* An ADM is identical to the normal disconnect mode except that the secondary can initiate a DM or RIM response at any time.

16-8-3 ISO 7809-1985(E)

The ISO 7809 standard combines previous standards 6159(E) (unbalanced) and 6256(E) (balanced) and outlines the class of operation necessary to establish the link-level protocol.

16-8-3-1 Unbalanced operation. This class of operation is logically equivalent to a multipoint private-line circuit with a polling environment. There is a single primary station responsible for central control of the network. Data transmission may be either half or full duplex.

16-8-3-2 Balanced operation. This class of operation is logically equivalent to a two-point private-line circuit where each station has equal data-link responsibilities. Channel access is accomplished through contention on a two-wire circuit using the asynchronous response mode. Data transmission is half duplex on a two-wire circuit and full duplex on a four-wire circuit.

QUESTIONS

16-1. Define *data-link protocol*.

16-2. What is meant by a *primary station*? A *secondary station*?

16-3. What is a *master station*? *Slave station*?

16-4. List and describe the three data-link protocol functions.

16-5. Briefly describe the *ENQ/ACK line discipline*.

16-6. Briefly describe the *poll/select line discipline*.

16-7. Briefly describe the *stop-and-wait method* of flow control.

16-8. Briefly describe the *sliding window method* of flow control.

16-9. What is the difference between *character-* and *bit-oriented* protocols?

16-10. Describe the difference between *asynchronous* and *synchronous* protocols.

16-11. Briefly describe how the *XMODEM protocol* works.

16-12. Why is IBM's 3270 protocol called *bisync*?

16-13. Briefly describe the *polling sequence* for *BSC*, including the difference between a general and specific poll.

16-14. Briefly describe the *selection sequence* for BSC.

16-15. Describe how BSC achieves transparency.

16-16. What is the difference between a *command* and a *response* with SDLC?

16-17. What are the three *transmission states* used with SDLC?

16-18. What are the five fields used with SDLC?

16-19. What is the *delimiting sequence* used with SDLC?

16-20. What are the three frame formats used with SDLC?

16-21. What are the purposes of the *ns* and *nr* bit sequences?

16-22. What is the difference in the *P* and *F* bits?

16-23. With SDLC, which frame types can contain an information field?

16-24. With SDLC, which frame types can be used for error correction?

16-25. What SDLC command/response is used for reporting procedural errors?

16-26. When is the *configure* command/response used with SDLC?

16-27. What is the *go-ahead sequence*? The *turnaround sequence*?

16-28. What is the *transparency mechanism* used with SDLC?

16-29. What supervisory condition exists with HDLC that is not included in SDLC?

16-30. What are the transparency mechanism and delimiting sequence for HDLC?

16-31. Briefly describe *invert-on-zero encoding*.

16-32. List and describe the HDLC *operational modes*.

PROBLEMS

16-1. Determine the BSC sequence for a general poll of station 7.

16-2. Determine the BSC sequence for a specific poll of station 7, device 3.

16-3. Determine the BSC sequence for a selection of device 12 at station 16.

16-4. Determine the hex code for the control field in an SDLC frame for the following conditions (assume P/F bit = 1):

a. information frame; transmitting frame 4; confirming reception of frames 2, 3, and 4
b. information frame; transmitting frame 3; confirming reception of frames 3, 4, and 5
c. information frame; transmitting frame 7; confirming reception of frames 7, 0, 1, and 2
d. information frame; transmitting frame 2; confirming reception of frames 6 and 7
e. supervisory frame; confirming reception of frames 2 and 3; ready to receive
f. supervisory frame; confirming reception of frames 7, 0, and 1; not ready to receive
g. supervisory frame; confirming reception of frames 2, 3, 4, 5, and 6; ready to receive

16-5. Determine the hex code for the control field for a SNRM command

16-6. Insert 0s into the following SDLC data stream:

111 001 000 011 111 111 100 111 110 100 111 101 011 111 111 111 001 011

16-7. Insert 0s into the following SDLC data stream:

001 011 000 000 111 111 111 111 110 100 111 101 011 111 111 111 001 011

16-8. Delete 0s from the following SDLC data stream:

111 110 000 001 111 110 100 111 110 100 111 110 011 111 010 111 001 011

001 110 000 111 110 110 101 111 100 100 101 111 101 111 010 111 110 011

16-9. Sketch the NRZI waveform from the following data stream:

1 0 0 1 1 1 0 0 1 0 1 0 1 0 0 0 1 0 1 1 1 0 0 0 1 1 1

16-10. Sketch the NRZI waveform from the following data stream:

0 0 1 1 1 0 0 1 0 1 1 1 1 0 0 0 1 0 1 0 1 1 0 1 1 1 1

C H A P T E R 17

Network Topologies and Connectivity Devices

CHAPTER OUTLINE

OBJECTIVES

- Define *local area network*
- Define *transmission formats*
- Explain baseband transmission
- Explain broadband transmission
- Compare the advantages and disadvantages of baseband and broadband transmission
- Describe the following LAN topologies: star, bus, tree bus, and ring
- Describe the characteristics of the following LAN topologies: star, bus, tree bus, and ring
- Describe the advantages and disadvantages of the following LAN topologies: star, bus, tree bus, and ring
- Describe the listen and transmit mode used with a ring topology
- Explain data collision
- Define *collision domain*
- Define *broadcast domain*
- Describe two- and four-wire network media
- Define *connectivity device*
- Explain what is meant by a layer 1 connectivity device
- Describe the characteristics of a repeater
- Describe the characteristics of a hub
- Describe how layer 1 connectivity devices affect collision and broadcast domains
- Explain the differences between a repeater and a hub
- Explain what is meant by a layer 2 connectivity device

- Describe the characteristics of a bridge
- Describe the characteristics of a switch
- Explain the differences between a bridge and a switch
- Describe how layer 2 connectivity devices affect collision and broadcast domains
- Explain what is meant by a layer 3 connectivity device
- Describe the characteristics of a router
- Describe the characteristics of a gateway
- Explain the differences between a router and a gateway
- Describe the characteristics of a layer 3 switch
- Describe how layer 3 connectivity devices affect collision and broadcast domains
- Describe the characteristics of a Brouter

17-1   INTRODUCTION

A network is formed when two or more devices are interconnected in such a manner that information and/or resources can be shared among two or more devices. A network can be further defined by the number of devices interconnected, the topography used to interconnect them, the physical size of the network (i.e., how close together or far apart the devices are), and several other factors that will be explained later. A *local area network* (LAN) is a private computer network with a relatively small number of devices located within a relatively small geographical area, such as a single room or a floor of a building.

The Institute of Electrical and Electronic Engineers (IEEE) defines a LAN as "a data communications system allowing a number of independent devices to communicate directly with each other, within a moderately-sized geographic area over a physical communications channel of moderate data rate." Therefore, in simple terms, a LAN is a relatively small data communications network comprised of computing resources connected together in such a way that the devices can share information with each other. Computing resources include computers, peripherals, and connecting devices. Connecting devices are devices that connect devices to the transmission medium and devices that connect segments of a network together, such as repeaters and bridges.

When two or more networks are interconnected for the purpose of exchanging information and services, an internetwork is formed. Internetworks are sometimes called internets (lowercase *i*) and should not be confused with the global Internet (uppercase *I*).

All networks require some form of addressing mechanism to identify one device from another. Addressing can also be used to identify the network itself or a portion of a network (i.e., a subnetwork). Network, subnetwork, and device addresses can be identified at two levels of the OSI protocol hierarchy. Hardware addresses identify devices at the data-link layer (layer 2) of the OSI protocol hierarchy. Hardware addresses are the six-byte media access control (MAC) addresses permanently burned into a network interface card (NIC). Every network device and interface with a NIC has a unique hardware address. The hardware address may also be called the physical, node, Ethernet, or LAN address. Internet Protocol (IP) addresses identify devices at the network layer (layer 3) of the OSI protocol hierarchy. IP addresses are sometimes called *logical addresses*.

The types of networks and network protocols described in Chapter 16 are more appropriately adapted to private networks that provide limited and very explicit services to a specific user (or users). The user is often a single company or organization or perhaps a department within an organization. For example, private networks are used to interconnect branches of a bank to the main office. Private networks are also appropriate for supporting systems of automated data entry devices, such as the card readers used with *automatic teller machines* (ATMs). Private networks generally do not provide Internet access to their subscribers. Therefore, private networks usually communicate at the data-link level using relatively simple protocols, such as SDLC or HDLC. With SDLC and HDLC, it is often not necessary to use addressing above the data-link level.

17-2 TRANSMISSION FORMATS

There are two transmission formats used with computer networks: baseband and broadband. The transmission format determines how communications between a large number of data terminals can utilize a single transmission medium. Transmission formats determine how signals from two or more devices can share a single transmission medium. Sharing a transmission medium is called *multiplexing*.

17-2-1 Baseband Transmission Format

Baseband transmission formats are defined as transmission formats that use digital signaling (i.e., discrete-level pulses). In addition, baseband formats use the transmission medium as a single-channel medium, as they allow only one station to transmit at a time and all stations must transmit and receive the same types of signals (encoding schemes, bit rates, and so on). Baseband networks time-division multiplex (TDM) digital signals onto the transmission medium. With TDM, all stations connected to the transmission medium have access to and can use the medium, but only one station can transmit at a time. If more than one station transmits at the same time, their transmissions will collide, and collisions destroy the integrity of the data.

The entire frequency spectrum (bandwidth) of a transmission line is used (or at least made available to) whichever station is transmitting at the time. With a baseband transmission format, when a signal is inserted at any point on the transmission medium, it propagates in all directions until it reaches the ends of the media where it is absorbed. This type of transmission is sometimes referred to as *bidirectional* in the sense that when a signal leaves a device and passes through the medium interface (node) onto the transmission medium, it propagates away from the node in all directions, such as on a linear bus. *Unidirectional* transmission is when signals propagate in only one direction on the transmission medium, which is the case in a ring topology, where signals leave the source and propagate around the loop in one direction (either clockwise or counterclockwise).

Because digital signals are attenuated as they propagate down a transmission line, LANs using a baseband transmission format are limited to a relatively short length, which limits their capacity.

Baseband transmission is summarized here:

1. It uses digital signaling (i.e., discrete pulses).
2. The transmission medium is used as a single-channel device, as only one station can transmit at a time.
3. All stations must transmit and receive the same types of signals (i.e., encoding schemes, bit rates, and so on).
4. Only one station can transmit at a time.
5. The entire frequency spectrum is used by (or made available to) whichever station is presently transmitting.
6. It has a limited capacity and length.
7. It can support bidirectional and unidirectional transmission, depending on the topography.

17-2-2 Broadband Transmission Format

Broadband transmission formats generally utilize frequency-division multiplexing (FDM) and use the connecting media as a multichannel medium. With FDM, each channel occupies a different frequency band within the total available bandwidth. Consequently, each channel can contain different modulation and encoding schemes and operate at different transmission bit rates. A broadband network permits voice, data, and video to be transmitted

Table 17-1 Baseband versus Broadband Transmission Formats

Baseband	Broadband
Digital signaling	Analog signaling
Single-channel medium	Multichannel medium
All stations must transmit and receive the same type of signals	All stations do not need to transmit and receive the same types of signals
Only one station can transmit at a time	More than one station can transmit at a time
Entire bandwidth used by whichever station is transmitting	Bandwidth is divided among multiple users
Limited capacity and length	Greater capacity and length
Time-division multiplexing transmissions	Frequency-division multiplexing
Bidirectional or undirectional	Unidirectional
Baseband advantages:	Broadband advantages:
Less expensive	High capacity
Simpler technology	Multiple traffic types
Easier and quicker to install	More flexible circuit configurations
	Cover larger area
Baseband disadvantages:	Broadband disadvantages:
Single channel	Require RF modems and amplifiers
Limited capacity	Complex installation and maintenance
Limited length	

simultaneously over the same transmission medium. However, broadband systems are unidirectional, as they require RF modems, amplifiers, and more complicated transceivers than baseband systems.

With a broadband transmission format, signals can propagate in opposite directions on the same transmission line at the same time, provided that they occupy different frequency bands (i.e., different channels). Because broadband transmission uses analog signals to carry the information, broadband networks can span greater distances than baseband networks. Broadband systems can extend for hundreds of miles, and the circuit components used with broadband LANs easily facilitate splitting and joining operations; consequently, the most common topologies used with them are bus and tree.

The layout for a broadband system is usually much more complex than for a baseband system and, therefore, more difficult and expensive to implement. The primary disadvantages of baseband systems are its limited capacity and length. Broadband systems can carry a wide variety of different kinds of signals on a number of channels. Table 17-1 summarizes baseband and broadband transmission formats.

The broadband transmission format is summarized here:

1. The transmission medium is used as a multichannel device.
2. Each station can transmit at the same time but on different frequency bands within the total allocated bandwidth (i.e., FDM).
3. Each channel may contain different modulation and encoding schemes and operate at different transmission rates.
4. It permits simultaneous transmission of voice, data, and video signals.
5. It requires RF modems, amplifiers, and more complicated transceivers than baseband.
6. It spans greater distances than baseband systems.
7. It has a complex circuit layout.
8. It is difficult and expensive to implement.

There are two topologies associated with local area networks: *physical* and *logical*. The logical topology of a LAN refers to the relationship between nodes or devices (servers, terminals, personal computers, workstations, printers, modems, and so on) as viewed by the software that controls data delivery between the devices. The physical topology of a LAN identifies how the nodes or devices are geometrically (physically) interconnected. Two or more nodes are interconnected by a link, and when two or more links are combined, they form a *topology*. The transmission media used with LANs include twisted-pair cable, coaxial cable, optical fiber cable, and wireless. Wireless involves the transmission of digitally modulated radio-frequency signals and uses the earth's atmosphere for the transmission medium. Presently, most LANs use twisted-pair cable; however, optical cable systems are being installed in many new networks and used to interconnect two or more networks. Fiber systems operate at higher transmission bit rates and have a larger capacity to transfer information than metallic cable systems. The latest transmission medium trend for LANs is wireless. However, there are security issues with wireless that need to be resolved before they receive widespread acceptance.

Messages are transported across LANs in data units called *frames*. Each frame contains a header, a trailer, and the actual message. The header generally includes the source and destination addresses and any pertinent network control information, such as flow control and sequencing information. The trailer is generally some form of error-detection mechanisms, such as a CRC-16 or CRC-32 frame check sequence.

LAN addressing may include unicast (unique for each device), multicast (more than one device), and broadcast (all devices) addressing, depending on the type of LAN and what topography is used. The most common LAN topologies are the star, bus, and ring.

17-3-1 Star Topology

The primary feature of the *star topology* is that each station is radially linked to a central node through a direct point-to-point connection, as shown in Figure 17-1. All devices (computers, bridges, and so on) are connected to the transmission medium (ring) through

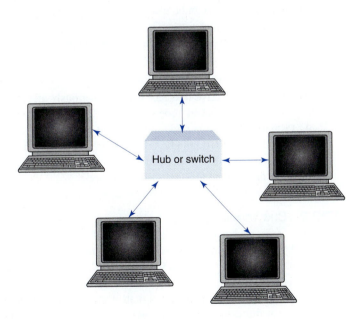

FIGURE 17-1 Star topology

medium interfaces. A medium interface is an active device either on the NIC or on a separate piece of hardware used to interface the computer to the transmission line. With a star topology, the propagation time between devices is relatively constant because the devices are generally located in the same room approximately the same distance from the central node. Some star networks use a hub for the central node. A hub is a layer 1 device; therefore, transmissions from all stations enter the central node, where they are retransmitted on all the outgoing links. Therefore, although the circuit arrangement physically resembles a star, it is logically equivalent to a bus because a transmission from any station is received by all the other stations. However, in most modern-day high-speed LAN applications, the central node is a switch, which is a layer-2 device. Therefore, transmissions are forwarded based on hardware addresses and are forwarded only onto the outgoing link, where the destination device resides. In this case, the other devices on the network do not receive the transmission unless it has a multicast or broadcast destination address. With a star topology, the destination station removes all messages.

Central nodes offer a convenient location for system or station troubleshooting because all traffic between outlying devices must flow through the central node. The central node is sometimes referred to as a *star coupler* or a *central switch*. The star configuration is best adapted to applications where most of the communications occur between the central node and the outlying devices or between outlying devices and a server connected to the central node. The star arrangement is also well suited to systems where there is a large demand to communicate with a relatively small number of devices. Time-sharing systems are generally configured with a star topology, and a star configuration is also well suited for word processing and database management applications.

One disadvantage of a star topology is that the network is only as reliable as the central node. When the central node fails, the system fails. If one or more outlying devices fail, the rest of the devices can continue to use the remainder of the network. When failure of any single entity within a network is critical to the point that it will disrupt service on the entire network, that entity is referred to as a *critical resource*. Thus, the central node in a star configuration is a critical resource.

The transmission mode of choice for most star LANs is baseband. Virtually any physical transmission medium can be used with the ring topology. Twisted-pair wires offer low cost and high transmission rates. The highest transmission rates, however, are achieved with optical fiber cables, except at a higher installation cost.

Characteristics of star topologies are the following:

1. Failure of the transmission medium does not seriously affect the network.
2. Failure of an active interface does not seriously affect the network.
3. Failure of the central node seriously affects the network.
4. Propagation delay is relatively constant between interfaces (nodes) on the network.
5. Unicast, multicast, and broadcast addresses are possible.
6. Twisted-pair wire, coaxial cable, or optical fibers are possible.
7. They operate in half- or full-duplex mode.
8. All messages are removed by the destination station.

17-3-2 Bus Topology

In essence, a *linear bus topology* is a multipoint or multidrop circuit configuration where individual devices are interconnected by a common, shared communications channel (transmission medium), as shown in Figure 17-2. The transmission medium is a two-wire facility, such as twisted-pair or coaxial cable designed to provide half-duplex communications between all the nodes connected to it. With a bus topology, all devices connect, using appropriate medium-interfacing hardware, directly to a common linear transmission medium, generally referred to as a bus. The media-interfacing hardware is generally a passive device that does not repeat, amplify, or regenerate signals. Because the signal attenu-

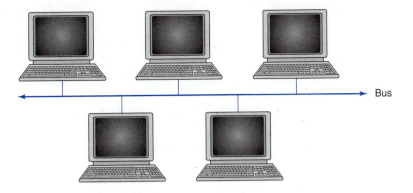

FIGURE 17-2 Bus topology

ates as it travels down the transmission line, the length of the bus is limited. However, the length can be increased by placing active connectivity devices, such as repeaters and bridges, between cable segments of the LAN.

In a bus LAN topology, network control is not centralized to a particular device. In fact, the most distinguishing feature of a bus LAN is that control is distributed among all the devices connected to the LAN. Data transmissions on a bus are usually in the form of small packets containing layer 2 source and destination addresses and data. When one device desires to transmit data to another device, it monitors the bus first to determine if it is currently being used. If no other devices are communicating over the network (i.e., the network is idle), the monitoring station can commence to transmit its data. When one device begins transmitting, all devices become receivers, including the transmitting device. Each receiver must monitor all transmission on the LAN and determine which are intended for them. When a device identifies its own address as the destination address in a received frame, the device must act on the frame; otherwise, the device ignores the frame.

One advantage of a bus topology is that no special routing or circuit switching is required, and, therefore, it is not necessary to store and retransmit messages intended for other devices. This advantage eliminates a considerable amount of message identification overhead and processing time. However, with heavy-usage LAN systems, there is a high likelihood that more than one device may desire to transmit at the same time. When transmissions from two or more devices occur simultaneously, a data collision occurs, disrupting data communications on the entire network. Obviously, a prioritized contention plan is required to handle data collisions. Such a plan is called *carrier sense, multiple access with collision detect* (CSMA/CD), which is discussed in Chapter 18.

Because network control is not centralized in a bus LAN, a device failure will not disrupt data flow on the entire LAN. The critical resource in this case is not a device but instead the bus itself. A failure on the transmission medium anywhere along the bus opens the network and, depending on the location of the break and the versatility of the communications channel, may disrupt communications on the entire network.

The addition of a new node to a bus LAN can sometimes be a problem because gaining access to the bus cable may be a cumbersome task, especially if it is enclosed within a wall, floor, or ceiling. One means of reducing installation problems is to add secondary buses to the primary communications channel. By branching off into other buses, a multiple-bus structure called a *tree bus* is formed. Figure 17-3 shows a tree bus LAN configuration.

With a bus topology, devices do not forward the frames it receives from the bus, and a destination device must remove all frames addressed to it. Because multiple transmissions on a bus result in collisions, a bus must operate either simplex or half duplex. In addition, the propagation time between devices is not constant, as it depends on the length of the bus,

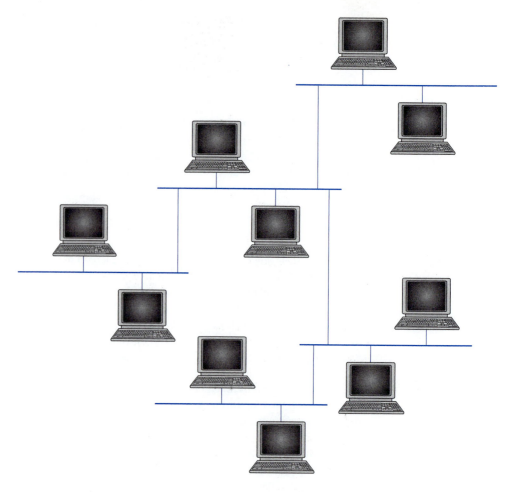

FIGURE 17-3 Tree bus topology

the number of devices attached to the bus, and how far apart the source and destination devices are from each other.

Characteristics of bus topologies are the following:

1. Failure of the transmission medium may seriously affect the network, depending on where the failure occurs.
2. Failure of a passive medium interface does not seriously affect the network.
3. Propagation delay is not the same between all devices.
4. Unicast, multicast, and broadcast addresses are possible.
5. Twisted-pair, coaxial cable, or optical fibers can be used.
6. They must operate simplex or half duplex.
7. All frames are removed by the destination station.

17-3-3   Ring Topology

With a *ring topology,* adjacent devices are interconnected in a closed-loop configuration, as shown in Figure 17-4. Adjacent devices on a ring topology are interconnected through direct, point-to-point connections. All devices are connected to the transmission medium (ring) through medium interfaces. A medium interface is an active device either on the NIC or on a separate piece of hardware that is used to interface the computer to the transmission

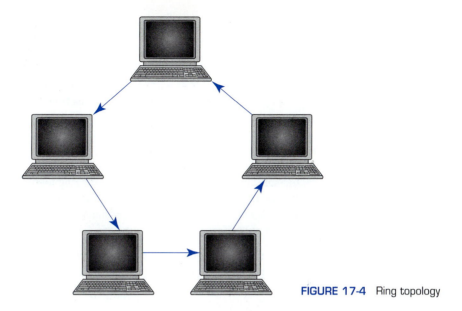

FIGURE 17-4 Ring topology

line. There are two interface states possible in a ring: operational and bypass. A device in the operational mode is part of the ring and may be in either the listen or the transmit mode, and the medium interface can switch from the device between the listen and transmit modes on demand.

Listen mode is used when a device either has no messages to send or it is off-line (not allowed to send messages). A device in the listen mode can receive messages, but they are simply copied, delayed one bit, and then regenerated and sent down-ring.

When a device receives a packet and the destination address does not match the physical address of the device, the device discards the packet, and the medium interface regenerates the packet and sends it to the next down-ring device. When a device receives a packet and the destination address matches the physical address of the device, the interface checks the packet for transmission errors, transfers the data portion of the packet to the device, regenerates the frame, and retransmits it to the next down-ring device.

Transmit mode is used when a device is in the process of transmitting a message. The interface receives incoming data, delivers it to the device, and accepts any outgoing messages from the device. The received message is discarded and replaced by the new message created by the device, or the received message is forwarded down-ring along with the new messages.

A device in the bypass state is removed from the ring without disrupting traffic to and from other devices. In the bypass mode, a device's receiver is connected directly to its transmitter, essentially routing messages around the device.

Each device on a ring participates as a repeater between two adjacent links (transmission line segments) within the ring. The repeaters are relatively simple devices capable of receiving data packets from one link and retransmitting the data packets on a second link. Messages are propagated in the simplex mode (one way only) from node to node around the ring (either clockwise or counterclockwise). All nodes have a predecessor (previous node) and a successor (next node).

Messages can be removed from a ring by the originating (source) or terminating (destination) device. If removal is performed by the source, messages propagate around the ring until they have circled the entire loop and return to the originating node. The originating device verifies that the data frame has circled the entire loop and returned exactly as it was sent. Hence, the ring configuration serves as an inherent error-detection mechanism. If frames are removed by the destination, they propagate only through devices located between the source and the destination devices. With destination removal, multicast and broadcast addressing are not possible.

The destination device can acknowledge reception of the data by setting or clearing appropriate bits within the control segment of the data frames. Data frames contain source and destination address fields as well as additional network control information and user data. Each device examines incoming frames, copies frames designated for them, and acts as a repeater for all frames by retransmitting them (bit by bit) to the next down-line device. A device should neither alter the content of the received frame nor change the transmission rate.

The transmission mode of choice for most ring LANs is baseband. Virtually any physical transmission medium can be used with the ring topology. Twisted-pair wires offer low cost and high transmission rates. The highest transmission rates, however, are achieved with optical fiber cables, except at a higher installation cost.

Characteristics of ring topologies are the following:

1. Failure of the transmission medium seriously affects the network.
2. Failure of an active interface seriously affects the network.
3. When active interfaces are used, there is virtually no limit to the length of the network.
4. Propagation delay is dependent on the number of interfaces (nodes) on the network.
5. Unicast, multicast, and broadcast addresses are possible.
6. Twisted-pair wires, coaxial cable, or optical fibers are possible.
7. They operate in simplex mode.
8. Messages can be removed by the source or destination station.

17-4 COLLISION AND BROADCAST DOMAINS

To improve performance and reliability, networks are often separated into smaller segments called *domains*. Although domains can be described physically, it is generally more meaningful to describe their boundaries logically on the basis of groups of device addresses. The physical size of a domain, how many domains are established on a network, and how many devices reside on a domain are determined by a variety of factors, such as the following:

1. Number of devices on the network
2. Local operating system
3. Network operating system
4. Network size (physical length and area covered)
5. Type of traffic and traffic patterns
6. Topography or physical layout
7. Addressing mechanism (protocol layer)

There are two basic types of network addressing domains: collision domains and broadcast domains.

17-4-1 Collision Domains

Collisions take place when transmissions (data frames) sent from two or more sources occur on the same network medium at the same time (i.e., when electrical signals from two or more sources appear at the same point at the same time). When electrical signals collide, their integrity is destroyed. Since messages are sent in packets or frames containing streams of contiguous bits, collisions are more likely to occur on high-density networks than on relatively inactive ones. A network with only a few devices or one that has little traffic would have little difficulty with collisions, as it would be highly unlikely that more than one device would have the need to transmit at the same time. However, on large, high-volume, high-density networks, the likelihood that two or more devices may transmit at the same time is relatively high.

Collisions destroy all data frames involved in the collision. To avoid collisions, the network must incorporate a system for managing competition for the medium. When devices compete for a transmission medium, it is called *contention,* as the devices contend (compete) for the medium. A section of a transmission medium common to more than one device is called a *shared medium* because more than one device can transmit and receive on it. Each section of a network where data can originate and collide is called a *collision domain,* and all shared-medium environments are collision domains. A shared medium can be a transmission medium, such as a cable or a patchcord, or it can be an interconnecting device, such as a hub or a repeater. When devices communicate with one another on a segment of a network populated by devices using a shared medium, the segment is a collision domain. A network with a bus topology has only one collision domain, as all devices are logically connected to the same transmission medium. If two or more stations transmit at the same time onto a bus topography, a collision will inevitably occur.

Collisions are independent of addressing layers. For example, if one device transmits a layer 2 frame of data onto a shared medium at the same time another device transmits a layer 3 datagram onto the same transmission medium, a collision will result because the two transmissions occur on the same collision domain at the same time. Figure 17-5 shows a network using a bus topology with a shared medium having only one collision domain. If any two devices transmit onto the bus at the same time, their transmissions collide, as shown in the figure. A network with only one collision domain is called a *flat network;* where *flat* simply means that there are no connectivity devices in the network that separate the network into protocol hierarchies (layers) that could isolate transmissions to a specific segment. In a network with a flat collision domain, all transmissions are propagated onto all segments at the same protocol layer.

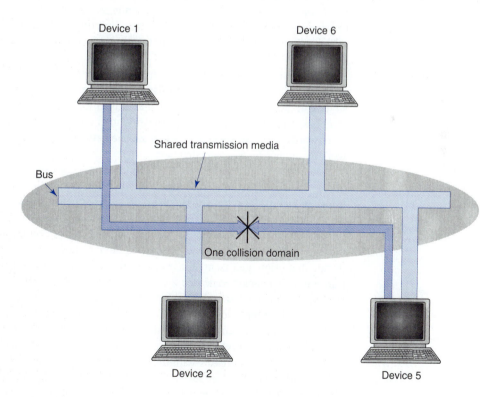

FIGURE 17-5 Data collision on a shared media

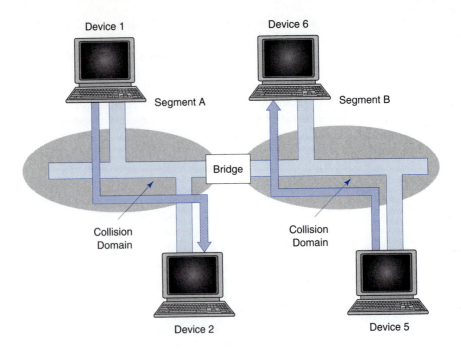

Device 1

Device 6

Segment A

Segment B

Bridge

Collision
Domain

Collision
Domain

Device 2

Device 5

FIGURE 17-6 Separating collision domains with a bridge

Some LANs use contention to determine which device has permission to transmit. In such networks, a device monitors "listens" to the transmission medium, and if it senses what appears to be an idle medium, it sends its message. However, because data transmissions are not instantaneous, a silent medium does not necessarily mean it is idle. Another device may be transmitting, but the transmission may not have reached the listening device yet.

Special devices, called *connectivity devices,* can be used to prevent (or at least minimize) collisions by isolating transmissions to a specific segment of a network. Transmissions can be isolated with layer 2 and higher connectivity devices, such as bridges, switches, and routers, that isolate transmissions on the basis of destination addresses. In other words, a transmission is passed from one network segment to another or from one network to another on an as-needed basis.

Isolating transmissions to segments of a network establishes multiple collision domains. Figure 17-6 shows a network divided into two collision domains separated by a bridge, which isolates transmissions on the two segments on the basis of destination hardware addresses. When a network is divided into more than one collision domain, the likelihood of a collision occurring is reduced because transmissions are passed from one segment to another only when the destination device resides on a different segment than the device that originated (sent) the transmission. For example, in Figure 17-6, device 1 can transmit a data packet to device 2 on segment A of the network at the same time device 5 transmits a data packet to device 6 on segment B. A collision did not occur because there was no need for the bridge to forward either data pack to the other segment.

Boundaries for collision domains can be established on the basis of either hardware or IP addresses. On LANs without Internet access, collision domains are established on the basis of hardware addresses with their boundaries established with bridges and switches. On networks with Internet access, collision domains are based on IP addresses with their boundaries established with routers or gateways.

The advantages to using flat networks are the following:

1. They are exceptionally easy for devices to communicate with each other.
2. They are less expensive to install and maintain.
3. They are easy to install and maintain.
4. They are less complicated and time consuming to manage.

The disadvantages to using flat networks are the following:

1. More devices can be connected to a flat network, which means that network resources are consumed faster.
2. Protocols that broadcast extensively compromise the performance of the network and the devices connected to the network.
3. Collisions are more likely to occur.
4. Devices have to contend with all other devices on the network for the transmission medium.

17-4-1-1 Two-wire versus four-wire network media. The type of transmission media used in a network can affect collisions. A two-wire transmission medium is when a single pair of metallic conductors or one optical fiber cable is used to propagate signals between devices. With a two-wire transmission medium, signals can propagate in both directions on the medium but not necessarily at the same time. Signals transmitted from a device propagate on the same transmission medium that the device receives signals from. Consequently, a device cannot send and receive signals at the same time. The exception is when frequency-division multiplexing is used and signals propagating in the two directions occupy different bandwidths, which is possible only with analog transmission. When devices on a network transmit and receive digital signals over a two-wire transmission medium, they must operate either simplex (one way only, either transmit or receive) or half duplex (both directions, but not at the same time). Figure 17-7a shows a two-wire network. Note that the transmitter and receiver for each device are connected to the same transmission medium. Therefore, a device is capable of receiving only when it is not transmitting, and only one device on a network can transmit at a time. Whenever two or more devices transmit at the same time, a collision will occur, as shown in Figure 17-7b.

A four-wire transmission medium is when two pairs of metallic conductors or two optical fiber cables are used to propagate signals between devices. With a four-wire transmission medium, signals propagate in one direction on one pair of wires and in the other direction on the other pair of wires. That is, a device transmits on one pair of wires and receives on the other pair. Four-wire networks must interconnect devices through an interface that accepts signals on one pair of wires and retransmits the signals on another, as shown in Figure 17-8. The interface is a connectivity device, such as a hub, a switch, or a router. Note in Figure 17-8 that the transmit pair on each device is connected to the receive pair on the interface and vice versa. With such an arrangement, a device can operate full duplex and transmit on one pair of wires and receive on the other pair at the same time without causing a collision. The figure shows device 1 transmitting on one pair of wires to the connectivity device, which receives the transmission and retransmits it to device 2. At the same time, device 3 is transmitting on one pair of wires to the connectivity device, which receives the transmission and retransmits it to device 1. Device 1 is transmitting and receiving at the same time (full duplex), while device 2 is receiving and device 3 is transmitting. Although two devices are transmitting at the same time, there is no collision. With four-wire full duplex networks, the likelihood of a collision occurring is much less than with two-wire networks. This will become more evident in later sections of this chapter.

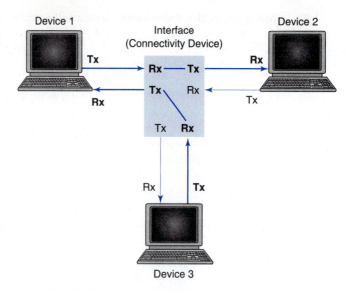

Each line represents one pair of wires

FIGURE 17-8 Four-wire network

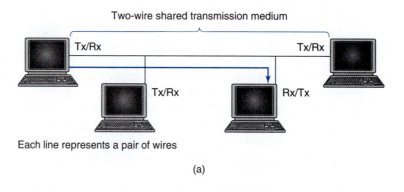

Each line represents a pair of wires

(a)

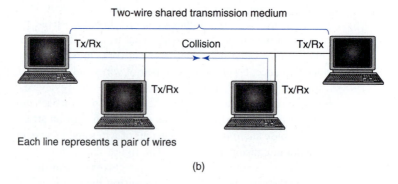

Each line represents a pair of wires

(b)

FIGURE 17-7 Two-wire network: (a) configuration; (b) collision

17-4-2 Broadcast Domains

Data frames can be transmitted from a single source to a single destination or from a single source simultaneously to more than one destination. Thus, data transmissions can be categorized as *unicast,* multicast, or *broadcast*. A unicast transmission is when a data packet is sent from a single source to a single destination. A multicast transmission is when a data packet is sent from a single source to more than one destination but not to all the other devices on the network. A broadcast transmission is when a data packet is sent from a single source with a destination address that identifies all the other devices on the network (i.e., all the other devices on the network are destinations for the same data packet).

A *broadcast domain* is a segment of a network serving a group of devices that will receive broadcasts sent from any device in the group using the same layer protocol or addressing mechanism. The boundary for a broadcast can be based on hardware (layer 2) or IP (layer 3) addresses. However, broadcast domains are typically bounded by routers based on IP addresses because routers do not forward broadcast messages and switches do. Broadcast messages must be restricted and isolated because if they were not, whenever a device transmitted a broadcast message, it would propagate uninhibited from one network to the next network, essentially flooding the networks with data frames, which would severely hinder network performance. To prevent broadcasts from flooding a network, special devices are included in networks to block broadcast messages, thus preventing them from propagating onto subnetworks or adjoining networks. Special devices, such as a router, logically separate networks and internetworks into broadcast domains, limiting where and how far a broadcast message can propagate. Broadcasts are generally allowed on small LANs and subnetworks of larger networks; however, they occur at the data-link layer. Therefore, layer 2 switches pass all frames where the destination hardware address is the MAC broadcast address, which is an address containing all logic 1s.

Figure 17-9 shows a network that has been separated with a router into three broadcast domains based on layer 3 IP addresses. Broadcasts can occur on all three segments at the same time without interfering with each other because the router will not pass broadcasts. Thus, each broadcast is isolated to only one segment of the network.

A network where all devices share the same broadcast domain is called a *flat broadcast network*. Here, *flat* simply means that there are no hierarchical devices, such as routers, in the network to block broadcasts and isolate them to a specific segment of the network. Thus, a network with no connectivity devices has a flat collision hierarchy and a flat broadcast hierarchy (i.e., the collision and broadcast domains share the same boundaries). A network configured with layer 2 and layer 3 connectivity devices is not flat for collisions or for broadcasts and is not likely to have the same boundaries for collision and broadcast domains.

17-4-2-1 LAN broadcasts. Broadcasts on LANs can occur at two different protocol layers: the network layer (layer 3) using a network protocol, such as IP, or at the data-link layer (layer 2) using hardware (MAC) addresses. When a device (host or router) wishes to broadcast network-specific information, such as ARP requests (address resolution protocol), onto a LAN, it must send a frame onto the network using the MAC broadcast address for the destination address. The MAC broadcast address ensures that the layer 2 connectivity devices (i.e., switches and bridges) on the network will pass the frame and that all network devices will accept the frame and decode it at the data-link layer. However, if the broadcast is carrying network- or transport-layer information, such as found in an IP datagram, the frame must also contain an encapsulated IP header with an IP broadcast address for the destination address. The broadcast IP address ensures that the each device will pass the IP datagram to the IP protocol so that the network-specific information can be decoded.

17-5 CONNECTIVITY DEVICES

Linking devices and networks together and separating collision and broadcast domains requires special components called *connectivity devices*. Connectivity devices include both networking and internetworking devices, as shown in Figures 17–10a and b. *Networking devices* operate at the physical and data-link layer (layers 1 and 2) of the OSI protocol

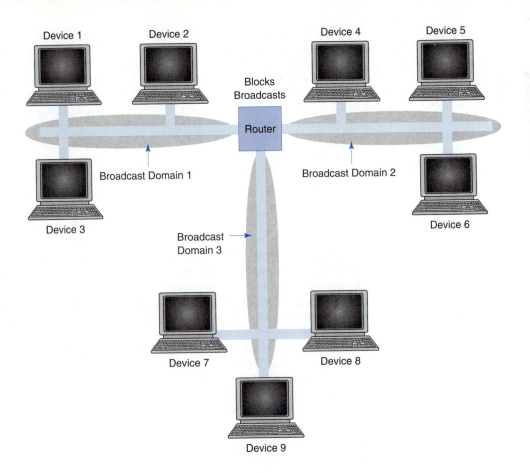

FIGURE 17-9 Separating a network into broadcast domains

hierarchy. Networking devices include repeaters and hubs at the physical layer and bridges and switches at the network layer. Networking devices are used to transport data frames between devices on a network or to interconnect and transport data frames between two or more segments of a network. Repeaters and hubs operate at the physical layer, while bridges and switches operate at the data-link layer. *Internetworking devices,* such as routers and gateways, operate at the network layer (layer 3) of the OSI protocol hierarchy, where they are used to interconnect two or more networks or subnetworks. Internetworking devices are used to route messages across a network, between networks, across a subnetwork, between subnetworks, and throughout the Internet. Connectivity devices can also be used to separate collision and broadcast domains.

17-5-1 Layer 1 Connectivity Devices

Connectivity devices that operate at layer 1 of the OSI protocol hierarchy include repeaters and hubs. Layer 1 devices are address independent, as they are not concerned with hardware or IP addresses. In fact, they are not concerned with any addresses. Layer 1 connectivity devices are transparent to the data traffic they transport, and they cannot separate collision or broadcast domains. In fact, because layer 1 devices pass collisions, they actually extend collision and broadcast domains. Networks using Layer 1 connectivity devices are flat networks for both collisions and broadcasts.

17-5-1-1 Repeaters. The transmission medium used with most LANs is CAT5 UTP cable, which typically limits the maximum length of a network segment to 100 meters (approximately 333 feet) or possibly less. This is because signals deteriorate with cable length,

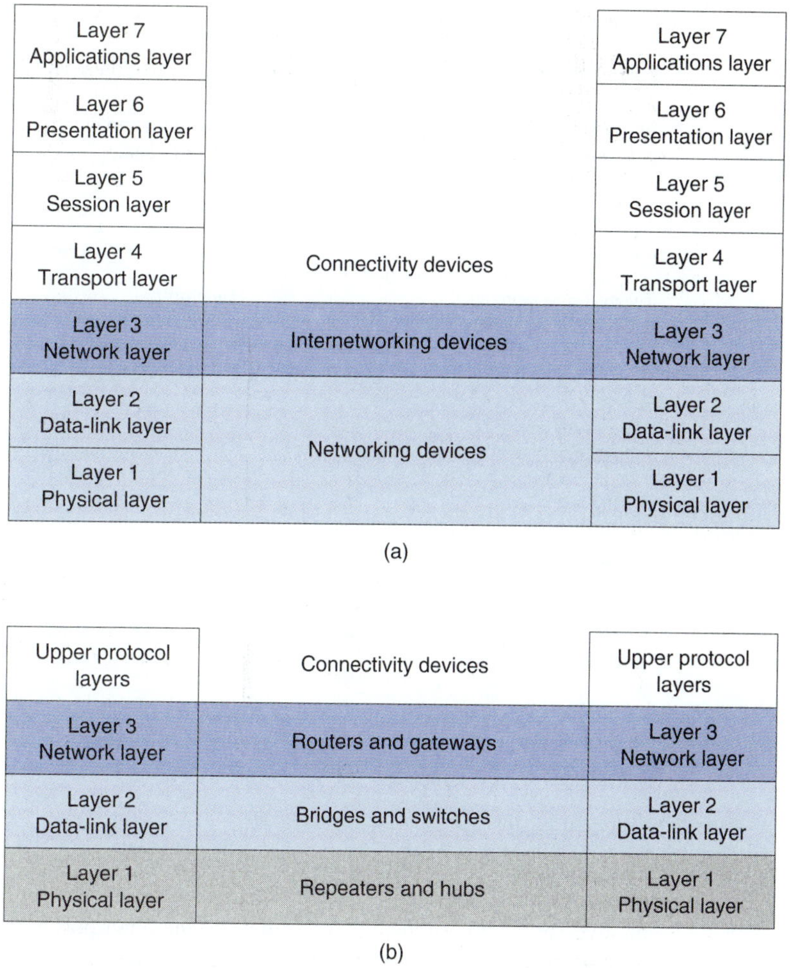

Layer 7 Applications layer		Layer 7 Applications layer
Layer 6 Presentation layer		Layer 6 Presentation layer
Layer 5 Session layer		Layer 5 Session layer
Layer 4 Transport layer	Connectivity devices	Layer 4 Transport layer
Layer 3 Network layer	Internetworking devices	Layer 3 Network layer
Layer 2 Data-link layer	Networking devices	Layer 2 Data-link layer
Layer 1 Physical layer		Layer 1 Physical layer

(a)

Upper protocol layers	Connectivity devices	Upper protocol layers
Layer 3 Network layer	Routers and gateways	Layer 3 Network layer
Layer 2 Data-link layer	Bridges and switches	Layer 2 Data-link layer
Layer 1 Physical layer	Repeaters and hubs	Layer 1 Physical layer

(b)

FIGURE 17-10   Connectivity devices and the OSI 7-layer protocol hierarchy: (a) networking and internetworking devices; (b) connectivity devices

and 100 meters is the maximum length the electrical signals typically used with computer networks can propagate over a copper transmission medium without deteriorating to the point that they become unrecognizable. A *repeater* is a connectivity device that is used to extend the physical length of a network. The term *repeater* is an old radio term for a transceiver (transmitter and receiver), which is an analog device that simply receives weak radio signals and then retransmits them at a higher power. There is a limit to how many segments can be interconnected with repeaters. The maximum number of repeaters is always one less than the maximum number of segments. Typical networks are limited to five segments interconnected with four repeaters. This is sometimes called the *four-repeater rule*.

Network repeaters operate at the bit level, as they simply *repeat* data (i.e., 1s and 0s), which allows them to travel longer distances on a transmission medium. Repeaters regenerate signals and transport them from one network segment to another at the physical layer (layer 1). Repeaters should not be confused with amplifiers, as amplifiers are linear devices that amplify both the input signal and any noise present with the signal. Repeaters regenerate digital signals. Regeneration prevents signals from being attenuated to the point that they become unrecognizable by accepting attenuated signals that may be contaminated with noise and reproducing the original signals without the noise. Figure 17-11a shows how a repeater regenerates signals.

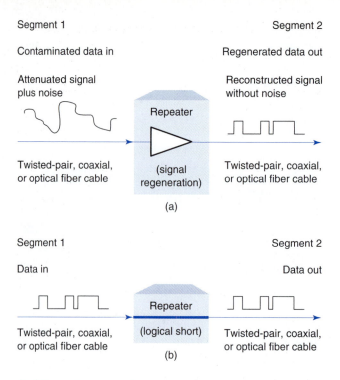

Segment 1   Segment 2

Contaminated data in   Regenerated data out

Attenuated signal   Reconstructed signal
plus noise   without noise

Repeater

Twisted-pair, coaxial,   (signal   Twisted-pair, coaxial,
or optical fiber cable   regeneration)   or optical fiber cable

(a)

Segment 1   Segment 2

Data in   Data out

Repeater

Twisted-pair, coaxial,   (logical short)   Twisted-pair, coaxial,
or optical fiber cable   or optical fiber cable

(b)

FIGURE 17-11   Repeater: (a) as a regenerator; (b) logical
diagram

A repeater (sometimes called a *regenerator*) is a two-port device that accepts signals from a source segment on one port, regenerates the signals and the timing information at the bit level, and then retransmits the signals on the other port to the adjoining segment (an intermediate, or a destination, segment). Although a repeater regenerates signals, it is logically equivalent to a short circuit, as shown in Figure 17-11b. A repeater simply passes data from input to output without changing the content or format of the message in any way.

To pass data through the repeater in a meaningful way, the data frames and protocols on the source side of the repeater must be identical (or at least compatible) to those on the destination side of the repeater. Repeaters operate independent of addresses. That is, they repeat all signals (including collisions) regardless of the source and destination addresses and regardless of the whether the data packet is carrying a layer 2 (physical) or layer 3 (IP) protocol. A repeater does not enable communications between incompatible or dissimilar networks, such as between an 802.3 LAN (Ethernet) and an 802.5 LAN (Token Ring).

Repeaters are basically dumb devices, as they do not translate data from one form to another or filter data in any way, and they do not change the functionality of the network. However, for a repeater to work, the two segments it joins together must use the same control methodology to access the transmission medium (access methodologies are discussed later in this chapter). Repeaters can move from one physical medium to another, such as an Ethernet packet from a coaxial cable segment to an optical fiber segment, provided that the repeater is capable of accepting both physical connections.

Repeaters are typically placed at the end of a network segment to increase the physical length of the network. A repeater is a good choice to extend the physical network beyond its distance or device limitations when neither segment is generating much traffic or when cost is a primary design constraint. When repeaters are used on segments in a star network, it is referred to as an *extended star* network. Repeaters can be used to extend the phys-

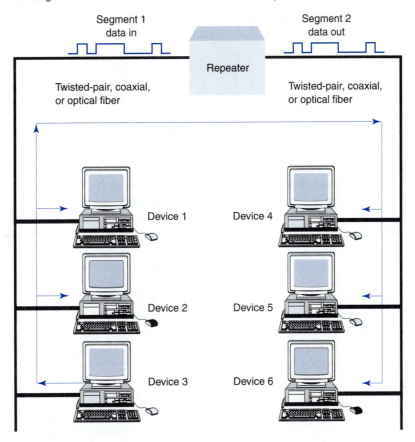

Segments 1 and 2 must use same frame format and protocol

Segment 1
data in

Repeater

Segment 2
data out

Twisted-pair, coaxial,
or optical fiber

Twisted-pair, coaxial,
or optical fiber

Device 1

Device 4

Device 2

Device 5

Device 3

Device 6

FIGURE 17-12   Network extension using a repeater

ical length of a network; however, at the same time they also extend collision and broadcast domains. Although the network is physically segmented with a repeater, collisions and broadcasts pass through the repeater just as though it were not there.

Figure 17-12 shows a network with two segments (1 and 2) using a repeater to interconnect the two segments. Segment 1 can be any transmission medium (i.e., twisted-pair copper cable, coaxial cable, or optical fiber), and segment 2 can be any transmission medium. However, the two transmission media do not have to be the same. All transmissions on segment 1 propagate through the repeater to segment 2 and vice versa. Therefore, if a device on one segment sends a broadcast message, the devices on both segments receive it. The figure shows a signal transmitted from device 2 on segment 1 passing through the repeater onto segment 2 destined for device 6. The repeater is a transparent device with only one function: to regenerate signals received from one segment and retransmit them onto an adjoining segment.

Because repeaters are address independent and forward all data frames, they extend both collision and broadcast domains. Figure 17-13a shows three separate networks where each network is a collision domain. Figure 17-13b shows the same three networks interconnected with two repeaters to form one larger network. When the networks are interconnected with repeaters, they form one larger physical network with one collision domain that extends across all three segments. The combined network also has only one broadcast domain that extends across all three segments.

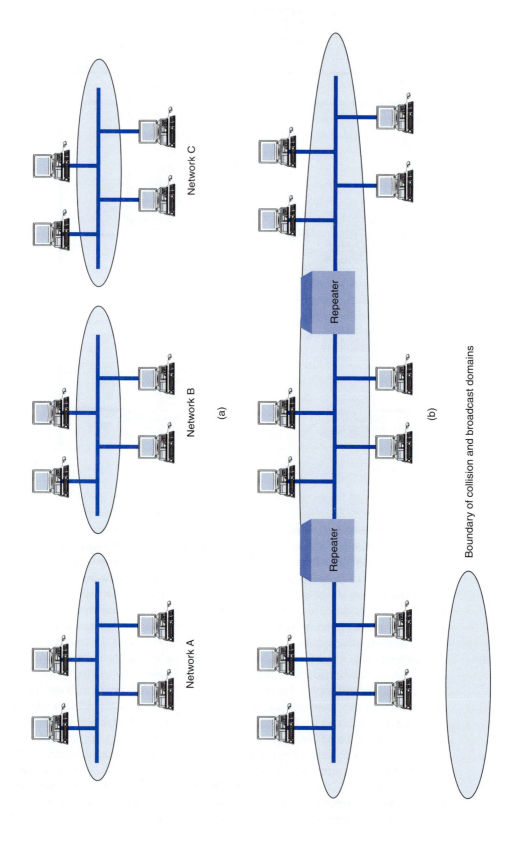

FIGURE 17-13 (a) Three independent networks with three collision and three broadcast domains; (b) one large network with one collision domain and one broadcast domain

Network A

Network B

Network C

(a)

Repeater

Repeater

(b)

Boundary of collision and broadcast domains

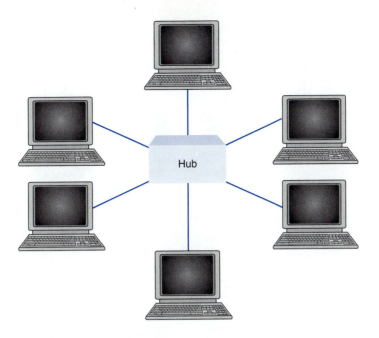

FIGURE 17-14   Star with a central hub

17-5-1-2 Hubs.   A *hub* is a repeater with more than two ports (i.e., multiport) that is used to interconnect LAN devices. A hub is generally placed in the center of a network, as shown in Figure 17-14, essentially converting the network topology from a linear bus to a star. Hubs are sometimes called *concentrators* or *wiring concentrators* when they are the central component in a star topology because they provide a common connection among devices. Since a hub is a layer 1 device, it simply repeats everything it receives. When a device transmits a frame into one port of the hub, shown in Figure 17-15a, the hub repeats the transmission out all the other ports. For example, if device 1 transmits a frame destined to device 4, the frame is also sent to devices 2 and 3, as shown in Figure 17-15b. Since all devices receive all frames, it is up to the devices to decode all frames and determine which frames are intended for them.

In a bus topology, a hub can be used to interconnect two or more segments of a network or to interconnect two or more devices on the network, creating a star within a bus, as shown in Figure 17-16. One obvious advantage of using a hub-based star configuration rather than a linear bus topography is reliability. With a linear bus, all devices are attached directly to the transmission medium. Therefore, a break in the transmission medium can severely affect network operation. With a hub-based star topology, a break in a transmission medium affects only one device. However, if the hub itself fails, the entire network is out of service.

A hub, like a repeater, is a layer 1 device that operates at the bit level and simply regenerates and repeats data. Hubs can be passive or active. Passive hubs have no electrical power and, therefore, cannot regenerate signals; they simply repeat them. Active hubs are capable of regenerating and repeating data. Because hubs are address independent, they can neither filter data traffic nor select the best path (route) to send data. An intelligent hub is a special kind of active hub that regenerates and repeats signals and also has an onboard processor enabling it to perform diagnostic tests and detect when there is a problem with a port.

Because hubs generally have 8 to 20 ports, they are sometimes called *multiport repeaters*. A hub is a multiport device that generally serves as the center of a network, whereas repeaters are two-port devices that are placed at the end of a network to extend its length. A hub does not enable communications between incompatible networks. For a hub to work,

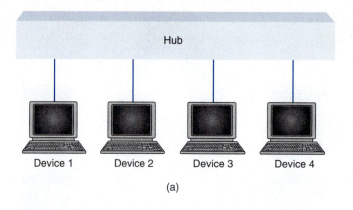

(a)

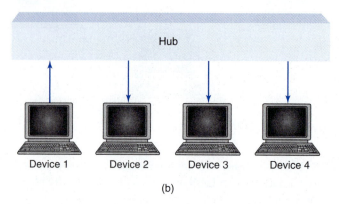

(b)

FIGURE 17-15 (a) Hub-based network; (b) transmission through a hub

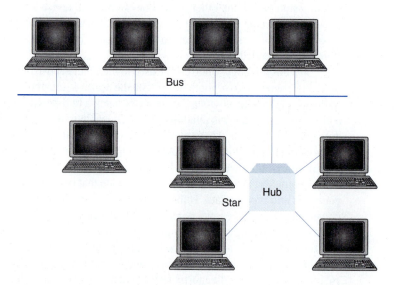

FIGURE 17-16 Star within a bus

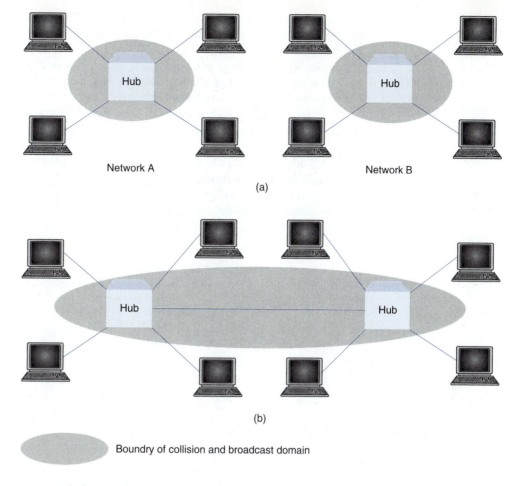

Network A   Network B

(a)

(b)

Boundry of collision and broadcast domain

FIGURE 17-17 (a) Two separate hub-based star networks with two separate collision and broadcast domains; (b) one larger network with one collision and one broadcast domain

all segments that it joins must have the same access method (i.e., CSMA/CD or token passing). Hybrid hubs are hubs (either active or passive) that can accommodate several different types of cabling (i.e., twisted-pair wire, coaxial cable, or optical fiber).

A hub can be used to extend the physical length of a network. However, a hub also extends collision and broadcast domains. Figure 17-17a shows two hub-based star networks, each with one collision and one broadcast domain. Figure 17-17b shows the same two networks interconnected with a hub, forming one larger network that has only one collision and one broadcast domain.

Hubs are versatile and offer several advantages, such as the following:

1. Changing or expanding wiring systems is as simple as plugging in another computer or another hub.
2. Using different types of physical ports allows hubs to accommodate a variety of cable types.
3. A centralized connection point for wiring the media can be created.
4. A centralized monitoring and troubleshooting point for network activity and traffic is provided. Many active hubs contain diagnostic capabilities to indicate whether a connection is working.
5. The reliability of a network is increased.

Two-wire hubs are, for the most part, a thing of the past, as most modern LANs use four-wire transmission media with either a hub- or a switch-based star topology. The block diagram for a four-wire hub is shown in Figure 17-18. Each device's transmitter is connected directly to a receiver port on the hub, and each device's receiver is connected directly to a transmitter port on the hub. This can be accomplished by using a crossover cable between each of the devices and the hub or by simply reversing the transmit and receive connections in the hub with an internal crossover. When a hub provides an internal crossover, they are generally labeled accordingly. For example, ports without crossovers (straight-through connections) are labeled 1, 2, 3, 4, and so on, and ports with internal crossovers are labeled 1X, 2X, 3X, and so on. On some hubs, one or more straight-through ports can be crossed over by simply pressing a button (usually labeled *uplink*).

Figure 17-19a shows how a device can be connected to a hub using a crossover cable connected to straight-through ports on the hub. Figure 17-19b shows the same connection using a straight-through cable and one crossover port. Two hubs can be connected to-

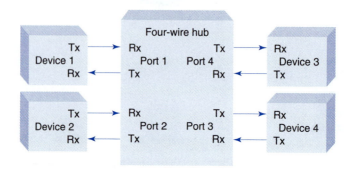

FIGURE 17-18 Four-wire hub

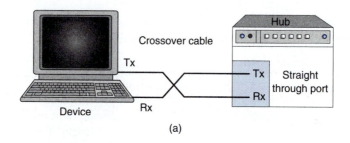

(a)

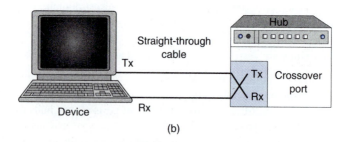

(b)

FIGURE 17-19 (a) Device-to-hub with crossover cable to straight-through port; (b) device-to-hub with straight-through cable to crossover port

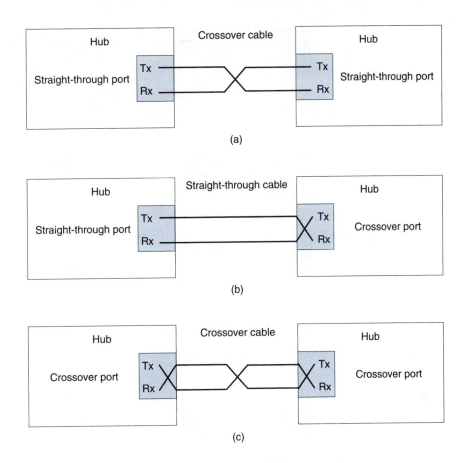

FIGURE 17-20 (a) Hub-to-hub with crossover cable and straight-through ports; (b) hub-to-hub with straight-through cable and one crossover port; (c) hub-to-hub with crossover cable and two crossover ports

gether with a crossover cable connected between two straight-through ports, as shown in Figure 17-20a, or by using a straight-through cable to connect a straight-through port on one hub to a crossover port on the other hub, as shown in Figure 17-20b. If all the ports on the two hubs are crossover ports, they can be interconnected with a crossover cable (i.e., three crossovers is equivalent to one), as shown in Figure 17-20c.

17-5-1-3 Layer 1 collision and broadcast domain summary. A network that does not utilize connectivity devices to interconnect devices or network segments, such as the one shown in Figure 17-21a, has only one collision and one broadcast domain that encompasses the entire network. If two or more devices transmit at the same time, a collision will occur, as there are no devices to filter, block, or separate transmissions. Likewise, when a device sends a broadcast packet, it will propagate across the entire network and be received by all devices.

When two network segments are interconnected with a repeater, as shown in Figure 17-21b, there is only one collision and one broadcast domain. Because the repeater is logically equivalent to a piece of wire, it ignores all addresses regardless of what layer protocol is used. Therefore, the repeater will simply forward all the data frames it receives on one port out the other port. If a device transmits on one segment, the packet will propagate across that segment, through the repeater, and onto the other segment. Therefore, regardless of from which segment a transmission originates or on which segment the destination resides, the data

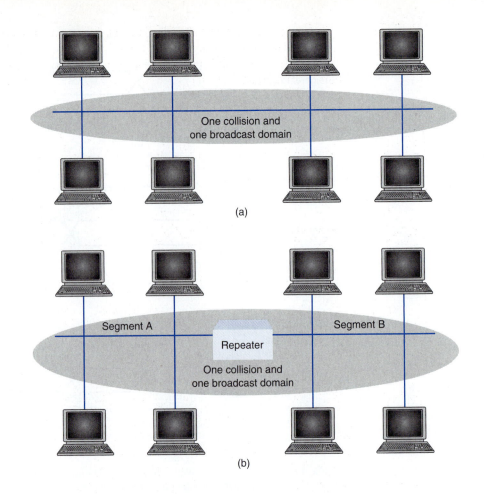

(a)

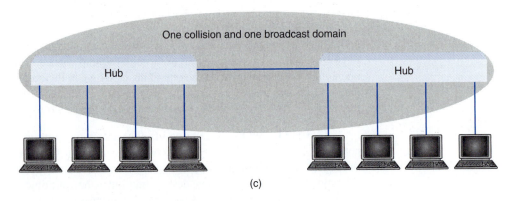

(b)

One collision and one broadcast domain

Hub

Hub

(c)

FIGURE 17-21 Layer 1 collision domains: (a) network with no connectivity devices; (b) network segments interconnected with a repeater; (c) segments interconnected with hubs

packet will propagate onto both segments of the network. The repeater combines the collision and broadcast domains of the two segments into one collision and one broadcast domain.

A network that uses hubs to interconnect devices, as shown in Figure 17-21c, also has only one collision domain. Because a hub ignores all addresses regardless of their protocol layer, it simply forwards all data frames it receives on one port out all the other ports, including the port connected to the other hub. If a device transmits on one segment, the packet will propagate across that segment, through both hubs, and onto the other segment. Therefore, regardless of which device originates a transmission, the data packet will propagate through the hubs onto the other segments. The hub extends the collision and broadcast domains of each segment to include the other segment, thus creating one large collision and one large broadcast domain that encompasses the entire network.

17-5-2 Layer 2 Connectivity Devices

Connectivity devices that operate at layer 2 of the OSI protocol hierarchy include bridges and switches. Layer 2 devices use hardware (MAC) addresses to forward data frames over LANs and deliver them to their intended destination. Layer 2 devices separate collision domains; however, they do not separate broadcast domains. Layer 2 devices are transparent to higher-layer protocols, such as IP. Bridges and switches can be used to join two dissimilar networks, such as an Ethernet LAN and a Token Ring LAN, provided that they are equipped with the proper interfacing hardware.

17-5-2-1 Bridges. A *bridge* is an intelligent two-port connectivity device used on LANs to direct data frames between two LAN segments based on hardware (MAC) addresses that are burned into the NICs of every device connected to the network. In essence, using a bridge creates network segments on the basis of groups of hardware addresses. The functionality of a bridge relies on its ability to make intelligent decisions about when it should and when it should not forward messages to the next segment. A bridge reads the hardware addresses of data frames it receives and compares them to an *address table* (sometimes called a *forwarding table*) to determine if the destination device is on the same local segment or an adjoining segment. The address table contains a list of all the devices on both network segments and their hardware addresses. If the destination device is on the same segment as the source device, the bridge ignores the packet. If the packet is destined for a device on an adjoining segment (i.e., a segment directly connected to the bridge), the bridge forwards the packet to that segment. Therefore, a bridge uses a simple pass/no pass logic to determine if it drops or forwards a data frame.

In what is sometimes called a *simple* or *primitive bridge,* addresses are entered manually into the bridge table, which contains lists of hardware addresses for all devices on each network segment. More sophisticated bridges called *transparent* or *learning bridges* use bridging protocols to dynamically update address tables. Bridges filter and either pass or block frames on the basis of hardware addresses. Therefore, bridges are not concerned with network-layer protocols, such as TCP/IP, or with IP addresses.

A bridge improves network performance by eliminating unnecessary traffic, thus reducing the possibility of collisions. A bridge is a device that can be used to join two LANs or two LAN segments that use identical protocols at the hardware level to form one larger network where each segment is a separate collision domain. All devices connected to a bridge must use the same protocol; therefore, the amount of processing required is minimal. Bridges allow devices on either LAN to access resources on the other LAN. Therefore, bridges can be used to increase the length or number of nodes on a network. Bridges filter traffic, keeping local traffic local and forwarding only traffic intended for a device on another segment of the network. Bridges separate collision domains; however, they pass broadcasts and extend broadcast domains.

To know when to forward a frame and when not to, bridges use tables to keep track of which hardware addresses reside on each segment. When a bridge receives a data frame, it checks the destination hardware address. If the destination address is for a device on the other segment, the bridge forwards the frame; otherwise, the bridge discards it. Because bridges look only at hardware addresses, they are not concerned with higher-layer network protocols that may be encapsulated in the data frame. Bridges collect, direct, and forward

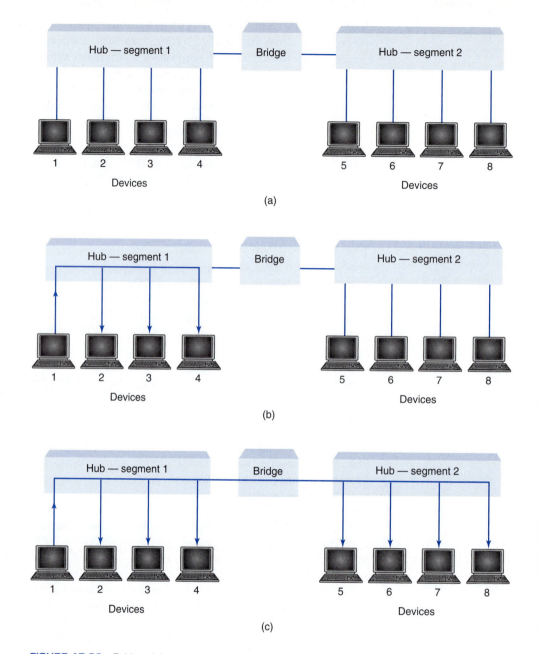

FIGURE 17-22 Bridge: (a) network configuration; (b) single-segment transmission; and (c) segment-to-segment transmission

frames between two or more LAN segments. Therefore, bridges must be more intelligent than hubs, as they examine the source and destination addresses in received frames and forward or discard frames on the basis of intelligent addressing decisions. Bridges divide networks into smaller collision domains, allowing more than one device to transmit at the same time, provided that they transmit on different collision domains.

Figure 17-22a shows two LAN segments interconnected with a bridge. The devices on each LAN segment are interconnected with hubs; however, the two segments are interconnected with a bridge. The bridge filters traffic and does not forward data frames unless the destination hardware address is for a device on a different network segment than the source device. For example, if device 1 sends a frame to device 2 (single-segment transmission), as shown in Figure 17-22b, the bridge would not pass the frame. However, if device 1 sends a frame to device 5, as shown in Figure 17-22c, the bridge would pass the frame from segment 1 to segment 2 (segment-to-segment transmission).

17-5-2-2 Switches. A *switch* is a multiport bridge that provides bridging functionality, except with greater efficiency. A switch operates at the data-link layer and filters and forwards data frames on the basis of hardware addresses. However, switches can be connected directly to devices and operate much faster than bridges. Switches can also support new functionalities, such as virtual LANs. A *transparent* or *learning bridge* is a layer 2 switch that constructs its own forwarding table as it performs its switching functions. When initially installed, a switch does not have any entries in its forwarding table. However, as it receives data frames, it learns where devices are located (i.e., which port they are connected to). When a switch receives a frame for a destination that is not listed in its forwarding table, it sends a broadcast frame to every device connected to it. The device in question responds to the broadcast, enabling the switch to add it to its forwarding table. Some networks use bridging protocols to construct forwarding tables. With bridging protocols, switches and bridges communicate with each other during idle network time and exchange information about the devices that are connected to their ports. Bridging protocols allow layer 2 devices to dynamically adjust their forwarding tables as changes occur on the network.

Bridges and switches are sometimes installed redundantly, which means that there may be more than one bridge or switch connecting two LAN segments together. When this type of configuration is used with transparent devices, it may create a network loop where a data frame may be passed around the loop over and over and never reach its destination. To avoid this situation, spanning tree algorithms are used.

A switch is a connectivity device similar to a hub, except switches make intelligent routing decisions on the basis of layer 2 hardware addresses, whereas hubs are dumb and do not make any routing decisions. The differences between a hub and a switch are virtually the same as the differences between a repeater and a bridge. A switch regenerates signals and forwards data frames similar to a hub. However, switching occurs at the data-link layer, and switches can perform several additional functions, such as controlling data flow, error detection and correction, physical addressing, and managing access to the physical medium.

Switches alleviate congestion in LANs and render them more efficient, as they concentrate connectivity by segmenting networks into multiple collision domains. Switches can divide a network into virtually one segment per device, as shown in Figure 17-23a. Although the configuration resembles that of a hub, there is one very important difference: the only data transported on a given segment is data that either originated from or is destined to the device on that segment. For example, the only transmissions that should ever occur on segment 1 are transmissions to or from device 1. Figure 17-23b shows device 1 sending a frame to device 3. As can be seen, the only two switch segments the frame appears on are the segments between device 1 and the switch and between device 3 and the switch.

Switches isolate transmissions to only two segments of a network: the segment from which the frame originated and the segment on which the destination device resides. Therefore, in essence, a switch can be used to convert a hub-based network from a logical bus topology to a logical star topology. A switch effectively creates a point-to-point connection from each data packet.

Some switches are multilayered devices in that they can provide interconnections to more than two pairs of devices at the same time. For example, in Figure 17-23c, device 1 can send a data frame to device 4 at the same time that device 2 sends a data frame to device 3, without a collision occurring. Therefore, an N-port switch can theoretically accommodate $N/2$ simultaneous connections.

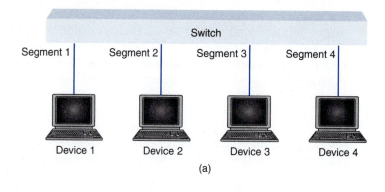

(a)

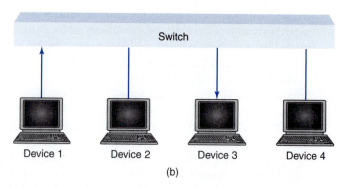

(b)

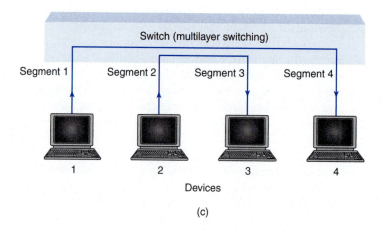

(c)

FIGURE 17-23   Switch: (a) configuration; (b) switching operation

There are three basic physical designs used with layer 2 switches: shared memory, matrix, and bus architecture:

Shared memory. A switch that stores all incoming data frames in a common memory buffer area and then transmits them to only the port connected to the segment where the destination device resides.

Matrix. A switch with an internal grid with the input ports and outputs crossing each other. When a data packet is received on an input port, the hardware address is compared to the lookup table to locate the appropriate output port. The switch then makes a connection at the grid where the source and destination ports intersect, thus reducing the possibility of collisions.

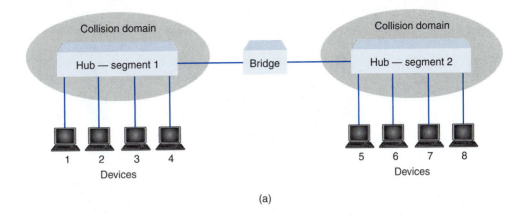

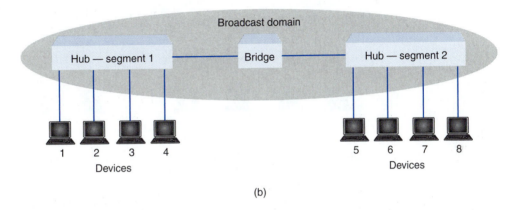

FIGURE 17-24 Bridge: (a) collision domains; (b) broadcast domains

Bus architecture. A switch with a bus architecture shares a common bus or internal transmission path with all the ports using time-division multiple accessing (TDMA). TDMA divides the transmission path into time slots that can be shared by several ports. If a bus architecture is not managed properly, collisions may occur.

Switches are designed with two types of *strategies* (sometimes called *fabrics*): store and forward and *cut-through*. A *store-and-forward switch* stores data on its onboard buffers until it receives the entire frame and computes the CRC, looks up the destination address, determines which interface to output the data on, and then forwards the frame. A *cut-through switch* copies only the destination address into its onboard buffers, computes the CRC, looks up the destination address, determines which interface to output the frame on, and then forwards. However, a cut-through switch begins forwarding the frame as soon as it determines the destination address.

17-5-2-3 Layer 2 collision and broadcast domain summary. Collisions do not propagate through layer 2 connectivity devices. Therefore, a bridge can be used to extend the physical length of a network without extending the collision domain. Figure 17-24a shows how a bridge can be installed between two LANs to divide a flat network with one collision domain into a segmented network with two collision domains. The only traffic that propagates through the bridge is traffic intended for a device on the other network. Local traffic on each network stays local (i.e., on that network). LANs that use bridges to divide the LAN into smaller segments can operate more efficiently because there are fewer devices

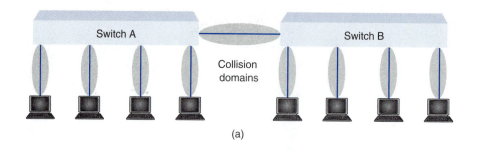

(a)

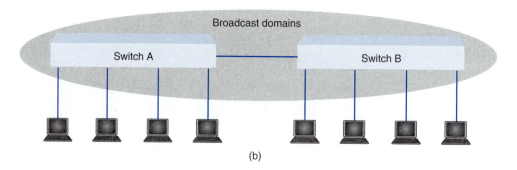

(b)

FIGURE 17-25 Switch configuration: (a) collision domains; (b) broadcast domains

in each collision domain. Bridges pass broadcast frames; therefore, bridges extend broadcast domains, as shown in Figure 17-24b.

Switches also block collisions. A 10-Mbps four-wire LAN that uses layer 2 switches with only one device per segment essentially creates a network that performs as though it has only two devices on it: a transmitting device and a receiving device. In addition, each segment is its own collision domain, as the only transmissions on it are frames sent from that device or frames intended for that device. Figure 17-25a shows two switches that are populated with devices that have been interconnected. As can be seen, the switch separates collision domains. However, switches pass broadcast frames; therefore, switches extend broadcast domains, as shown in Figure 17-25b.

Figure 17-26 shows a network containing both layer 1 (hub) and layer 2 (switch) connectivity devices. Figure 17-26a shows the collision domains for the network. As can be seen, the switch (L2) separates the network into five collision domains, and the hubs (L1) have no bearing on the boundaries of collision domains. Figure 17-26b shows how the same network reacts to broadcast frames. As can be seen, neither the switch nor the hubs have any effect on the boundaries for broadcasts. Therefore, the network is flat for broadcasts but not for collisions.

17-5-3 Layer 3 Connectivity Devices

Connectivity devices that operate at layer 3 of the OSI protocol hierarchy are sometimes called *internetworking devices*, as they interconnect networks and subnetworks at the network layer. Therefore, layer 3 connectivity devices forward data frames on the basis of IP addresses. Layer 3 connectivity devices include routers, gateways, and layer 3 switches. Layer 3 devices forward data frames between subnetworks, networks, and the Internet. Therefore, they are the most sophisticated means of interconnecting network components.

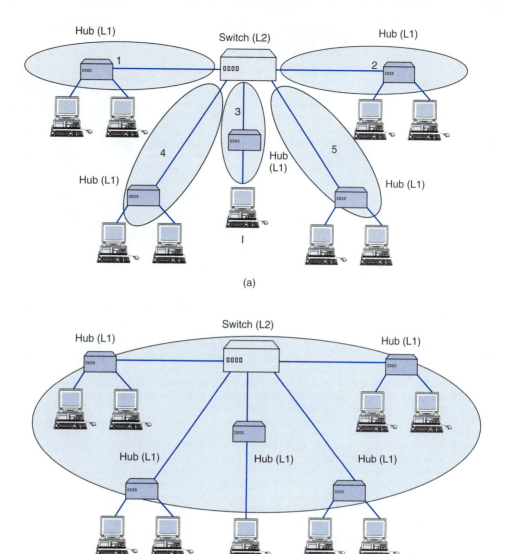

FIGURE 17-26 Layers 1 and 2 collision and broadcast domains: (a) collision domains; (b) broadcast domains

17-5-3-1 Routers. *Routers* operate at the network layer (layer 3) and, therefore, make forwarding decisions on the basis of network addresses, which are called IP (Internet Protocol) or *logical addresses*. Routers are the backbone of internetworks and the Internet, as they can be used to interconnect different types of networks, such as those using different architectures and protocols. Routers can switch and route packets across multiple networks and networks that use different layer 2 technologies, such as Ethernet, Token Ring, and FDDI. Therefore, routers can be used to interconnect any two devices in the world that are logically connected to the Internet either directly or through a LAN. Routers are also used to divide networks into subnetworks and to separate collision and broadcast domains.

The primary purpose of a router is to examine the source and destination IP (layer 3) addresses of data packets it receives and to direct those packets out the appropriate port and

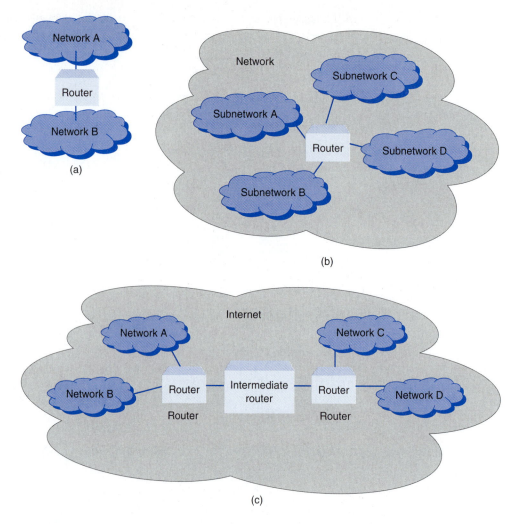

FIGURE 17-27 Routers: (a) interconnecting networks; (b) interconnecting subnetworks within a network; (c) intermediate router

over the best path available at the time. Routers accomplish this by using routing protocols to exchange protocol-specific information between networks. There are several routing algorithms available, including distance vector routing and link state routing. The routing information is used to formulate routing tables, which are used to determine the best route between two networks.

Each port on a router where a network or subnetwork is attached is called a *router interface,* and every interface has an IP address. Router interfaces connected to LANs or subnetworks are commonly designated E0, E1, and so on (E for Ethernet). Interfaces connected to WANs (wide area networks) are often designated W0, W1, and so on (W for WAN).

Figure 17-27a shows how a router can be used to interconnect two networks, and Figure 17-27b shows how a router can be used to segment a network into subnets. Figure 17-27c shows how a router is used as an intermediate routing device to interconnect networks that are separated by great distances.

17-5-3-2 Gateways. A *gateway* is basically a special-purpose router that allows local networks access to the Internet. A gateway is generally a software package installed in a router. In essence, a gateway is a doorway into and out of a LAN. Therefore, the primary difference between a standard router and a gateway is that gateways can perform protocol conversions, whereas standard routers operate on network protocols only, such as IP. Gateways have software that can link different programs or protocols by examining the entire data packet and perform any necessary translations of incompatible protocols. For example, a gateway can receive an Ethernet frame from a device on a local network, decapsulate the frame, remove the Ethernet heading, and then forward only the IP portion of the frame, which contains Internet-specific information, such as source and destination IP addresses. Conversely, gateways can receive Internet-specific information (such as found in an IP) from a distant network, encapsulate it into an Ethernet frame, and pass it onto the local network for delivery to the destination device. An intermediate router on the Internet is capable of interpreting and forwarding only Internet-specific information.

Figure 17-28a shows how an intermediate router forwards IP datagrams over the Internet. As can be seen, an intermediate router receives IP datagrams and forwards IP datagrams without converting from one protocol to another. Figure 17-28b shows how a gateway can be used to interface a LAN to the Internet. As can be seen in the figure, the IP protocol on the LAN side of the gateway is encapsulated in a layer 2 Ethernet protocol. The Ethernet frame is decapsulated in the gateway where the Ethernet header is removed. Therefore, only the IP datagram is forwarded onto the Internet. Figure 17-24c shows how a gateway forwards datagrams between two LANs using different protocols. LAN 1 is an Ethernet LAN, and LAN 2 uses AppleTalk. The gateway interface decapsulates the LAN header on incoming frames and removes the header. The gateway then transfers the IP datagram to the outgoing interface, where it is encapsulated in the LAN header of the destination LAN. In Figure 17-28c, an IP datagram is transported from network A across an Ethernet LAN to the gateway. The gateway converts the frame to AppleTalk before forwarding it onto network 2.

17-5-3-3 Layer 3 switches.   *Layer 3 switching* is a relatively new form of switching. Layer 3 switching is designed to replace routers on local networks. Layer 3 switches are sometimes called *Internet Protocol switches* or *IP switches*. Layer 3 switches are basically a cross between a switch and a router. The fundamental difference between a router and a layer 3 switch is that layer 3 switches are hardware controlled and have optimized their hardware to pass data as fast as a layer 2 switch, yet they make decisions on how to forward traffic on the basis of layer 3 network addresses. In a layer 3 switch, each port is a separate LAN port. However, the switch's forwarding engine computes switching routes on the basis of IP addresses rather than hardware addresses. Layer 3 switches are faster than routers; however, routers make intelligent decisions on how to route packets and have more extensive routing tables. Therefore, routers are more appropriate for interconnecting networks with a large number of subnetworks or for large internetworks, such as the Internet.

17-5-3-4 Brouters.   A *brouter* is a connectivity device that combines the best qualities of a switch: a multiport bridge and a router. Actually, it would probably be more appropriate to call the device a *swouter* (switch/router) because *switch* has probably become a more common term for a layer 2 device than *bridge*. Either way, a brouter can act like a router for one protocol and a switch (or bridge) for all others. Brouters can route selected routable protocols (e.g., IP), bridge nonroutable protocols (e.g., Ethernet), and provide more cost-effective and manageable internetworking than switches and routers. A brouter is a good choice in an environment that mixes several homogeneous LAN segments.

17-5-3-5 Layers 1, 2, and 3 collision and broadcast domain summary.   Figure 17-29 shows an internetwork comprised of two networks (network A and network B). The

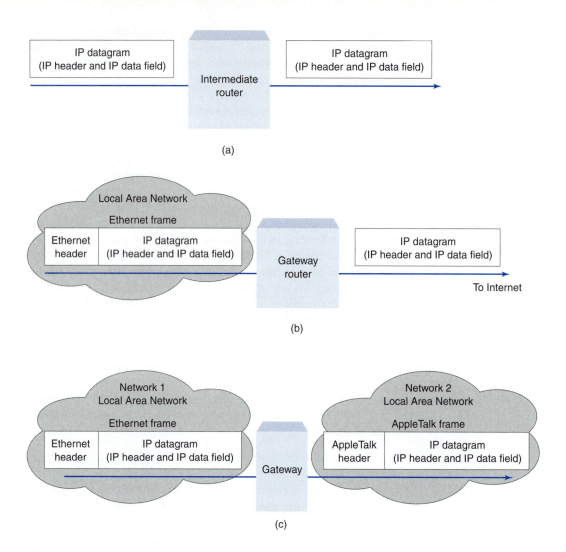

FIGURE 17-28 Gateway (a) forwarding IP datagrams over the Internet; (b) interfacing a LAN to the Internet; (c) forwarding datagrams between two LANs

devices on network A are interconnected with layer 1 hubs, which are connected to a layer 2 switch. The switch is connected directly to a layer 1 repeater, which simply extends the physical length of the link and forwards data frames to a layer 3 router. The devices attached to each hub use a shared media; therefore, each hub comprises a single collision domain. The layer 2 switch blocks collisions; however, the repeater does not. Therefore, the link between the switch and the router represents one collision domain.

The devices on network B are interconnected with a layer 2 switch, which is connected directly to a layer 3 router. The switch separates each device segment into separate collision domains and also isolates the link between the switch and the router to a separate collision domain.

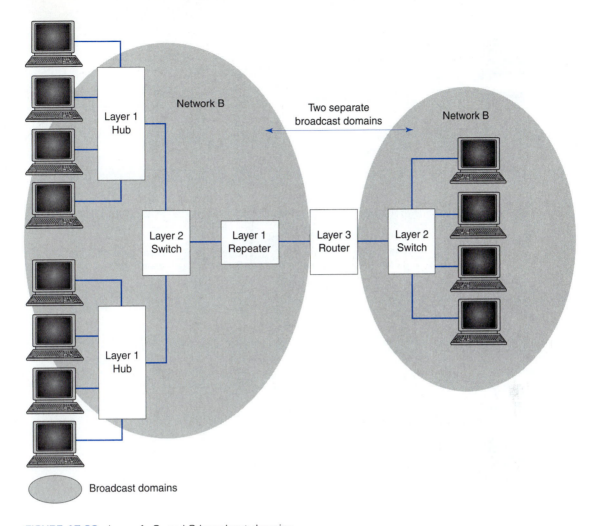

Broadcast domains

FIGURE 17-29 Layer 1, 2, and 3 broadcast domains

Figure 17-30 shows the same internetwork, except divided into broadcast domains. As can be seen, the only device that blocks broadcasts is the layer 3 router. Therefore, there are only two broadcast domains, one on each side of the router.

Table 17-2 is a summary of the connectivity devices described in the previous sections comparing the collision and broadcast characteristics of the devices.

17-6 STANDARD CONNECTIVITY DEVICE LOGIC SYMBOLS

For simplicity, the figures in this chapter use block diagram types of symbols to depict connectivity devices. The reason for this is because standards organizations seem to be having a difficult time agreeing on (i.e., standardizing) logic symbols for connectivity devices. Figure 17-31 shows what are probably the most commonly used logic symbols by industry today. However, keep in mind that there are several variations for each of the symbols shown.

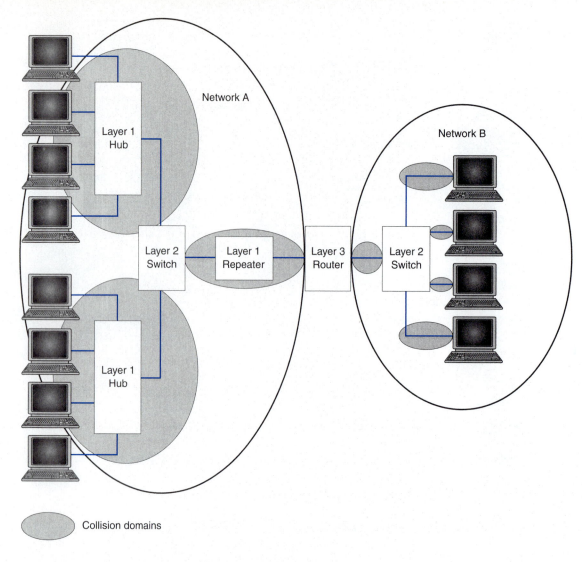

Collision domains

FIGURE 17-30 Layer 1, 2, and 3 collision domain summary

Table 17-2 Connectivity Device Summary

Device	Number of Ports	Protocol Layer	Passes Collisions	Passes Broadcasts
Repeater	Two	1	Yes	Yes
Hub	Three or more	1	Yes	Yes
Bridge	Two	2	No	Yes
Switch	Three or more	2	No	Yes
Router	Two or more	3	No	No
Gateway	Two or more	3	No	No

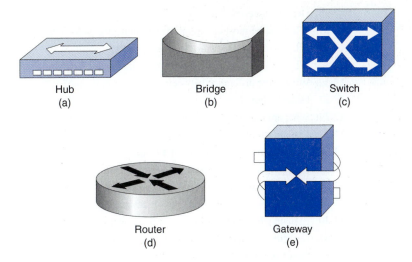

FIGURE 17-31 Standard connectivity device logic symbols: (a) hub; (b) bridge; (c) switch; (d) router; (e) gateway

QUESTIONS

17-1. How is a *network* formed?

17-2. Define *local area network* (LAN).

17-3. How does the IEEE define a local area network?

17-4. How is an *internetwork* formed?

17-5. What type of mechanism do all networks use to identify one device from another?

17-6. At what two levels can addresses be identified?

17-7. Define *transmission formats*.

17-8. Explain what is meant by *baseband transmission*.

17-9. Explain what is meant by *broadband transmission*.

17-10. Compare the advantages and disadvantages of baseband and broadband transmission.

17-11. Describe the differences between *physical* and *logical topologies*.

17-12. List and describe the three types of addresses used on LANs.

17-13. Briefly describe the following LAN topologies: *star, bus, tree bus,* and *ring.*

17-14. Describe the characteristics of the following LAN topologies: star, bus, tree bus, and ring.

17-15. What is the primary disadvantage of a star topology?

17-16. What is the primary disadvantage of a bus topology?

17-17. What is the primary disadvantage of a ring topology?

17-18. Describe the listen and transmit mode used with a ring topology.

17-19. Explain what is meant by a *data collision*.

17-20. Define *collision domain*.

17-21. What type of devices can separate collision domains?

17-22. What types of addressing can be used to establish collision boundaries?

17-23. What is meant by a *flat network* relative to collision domains?

17-24. Briefly describe the differences between *two-* and *four-wire networks*.

17-25. Define *broadcast domain*.

17-26. What is meant by a flat network relative to broadcast domains?

17-27. What is the difference between networking and internetworking connectivity devices?

17-28. Define *connectivity device*.

17-29. What are the most common layer 1 connectivity devices?

17-30. What is the primary purpose of a *repeater*?

17-31. Why are repeaters typically placed at the end of a network?

17-32. What is the difference between a *repeater* and a *hub*?

17-33. Describe how layer 1 connectivity devices affect collision and broadcast domains.

17-34. What are the most common layer 2 connectivity devices?

17-35. Explain what is meant by a layer 2 connectivity device.

17-36. Describe the characteristics of a *bridge*.

17-37. Describe the characteristics of a *switch*.

17-38. What is the difference between a bridge and a switch?

17-39. Describe how layer 2 connectivity devices affect collision and broadcast domains.

17-40. Explain what is meant by a layer 3 connectivity device.

17-41. Describe the characteristics of a *router*.

17-42. Describe the characteristics of a *gateway*.

17-43. What are the differences between a router and a gateway?

17-44. Describe the characteristics of a layer 3 switch.

17-45. Describe how layer 3 connectivity devices affect collision and broadcast domains.

17-46. Describe the characteristics of a *brouter*.

C H A P T E R   18

Local Area Networks

CHAPTER OUTLINE

OBJECTIVES

- Define *local area network* (LAN)
- Describe the advantages of using LANs
- Describe the purposes of IEEE Project 802
- Define *access control methodologies*
- Define *medium access method*
- Define *medium access control*
- Name and describe the two broad categories of medium access
- Describe what the word *contention* means in respect to a LAN
- Define *multiple access*
- Describe carrier sense multiple access
- Describe carrier sense multiple access with collision detect
- Describe carrier sense multiple access with collision avoidance
- Describe the term *controlled access*
- Describe the basic operation of a token ring LAN
- Describe the functions of the logical link control sublayer
- List and describe the LLC Service Classes
- Describe the differences between connectionless and connection-oriented service
- Describe the differences between acknowledged and unacknowledged service
- Describe protocol data unit

- Describe a LLC PDU header
- Define *source and destination service access points*
- Describe an IEEE 802.3 MAC frame format
- Explain MAC addressing
- Explain the format used with Ethernet standard notations
- Describe the differences between an 802.3 frame and an Ethernet II frame
- Describe the following 10-Mbps Ethernets: 10Base-5, 10Base-2, 10Base-T, 10Base-F, and 10Broad-36
- Describe the differences between a bridged Ethernet and a bus Ethernet
- Describe the differences between a switched Ethernet and a bus Ethernet
- Describe the operation of full-duplex and full-duplex switched Ethernet
- Describe MAC control packets
- Explain what is meant by the term *fast Ethernet*
- Describe the fast Ethernet MAC sublayer
- Describe the following 100-Mbps Ethernets: 100Base-TX, 10Base-T4, and 10Base-FX
- Define *gigabit Ethernet*

18-1 INTRODUCTION

Studies have indicated that approximately 80% of the communications among data terminals, such as personal computers (PCs), and other data terminal equipment occurs within a relatively small local environment. A local area network (LAN) provides the most economical and effective means of handling local data communications needs. A LAN is typically a privately owned data communications system in which the users share resources, including hardware- and software-based processes. LANs provide two-way communications between a large variety of data communications terminals within a limited geographical area, such as within the same room, building, or building complex. Most LANs link equipment that is within a few hundred yards of each other.

Figure 18-1 shows how several PCs can be connected to a LAN to share common resources, such as a modem, printer, or server. The server may be a more powerful computer than the other PCs sharing the network, or it may simply have more disk storage space. A server "serves" information to the other PCs on the network in the form of software and data information files. A PC server is analogous to a mainframe computer, except on a much smaller scale.

LANs allow for a roomful or more of computers to share common resources, such as printers, databases, files, and modems. The average PC uses these devices only a small percentage of the time, so there is no need to dedicate individual printers and servers to each

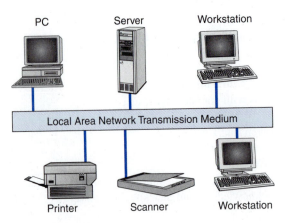

FIGURE 18-1 Local area network

PC. To print a document or file, a PC simply sends the information over the network to the print server. The print server organizes and prioritizes the documents and then sends them, one document at a time, to a common-usage printer. Meanwhile, the PCs are free to continue performing other useful tasks. Likewise, when a PC needs information from a database, it requests the information from a server, and the server sends the information over the LAN to the PC that requested the information. When a PC needs a modem, the network establishes a *virtual connection* between the modem and the PC. The network is transparent to the virtual connection, which allows the PC to communicate with the modem as if they were connected directly to each other. If a PC needs a connection to the global Internet, the network establishes a connection through a router, essentially allowing the PC to communicate directly with any device in the world with an Internet Protocol (IP) address.

LANs allow people and machines to send and receive messages and documents through the network much quicker than they could be sent through the paper mail system. *Electronic mail* (e-mail) is a communications system that allows users to send messages to each other through their computers. E-mail enables any PC on a network to send or receive information from any other PC on the network as long as the PCs and the mail server use the same or compatible software. E-mail can also be used to interconnect users on different networks in different cities, states, countries, or even continents. To send an e-mail message, a user at one PC sends its address and a message along with the destination address to a server. The server effectively "relays" the message to the destination PC if they are subscribers to the same network. If the destination PC is busy or not available for whatever reason, a server stores the message and delivers it at a later time. The server is the only computer that has to keep track of the locations and addresses of all the other PCs on the network. To send e-mail to subscribers on other networks, the mail server relays the message to the server on the destination user's network, which in turn relays the e-mail to the destination PC. E-mail can be used to send text information (letters) as well as program files, graphics, audio, and even video. This is referred to as *multimedia data communications*.

LANs are used extensively to interconnect a wide range of data services and computing resources, including the following:

Data terminals

Laser printers

Graphic plotters

Large-volume disk and tape storage devices

Facsimile machines

Personal computers

Workstations

Mainframe computers

Supercomputers

Data modems

Databases

Word processors

Public switched telephone networks

Digital carrier systems (T carriers)

E-mail servers

The capabilities of a LAN are established primarily by four factors: transmission format, topology, transmission medium, and access control protocols. Together, these four factors determine the type of data, transmission rate, efficiency, and applications that a network can effectively support.

18-2 IEEE PROJECT 802

When computer manufacturers developed the first LANs, they were expensive and worked only with certain types of computers and then with only a limited assortment of software. Early LANs also required a high degree of technical knowledge and expertise to install, maintain, and use. In 1980, the IEEE, in an effort to resolve problems with LANs, formed the 802 Local Area Network Standards Committee, which is informally called IEEE Project 802. The primary purpose of IEEE 802 is to set standards to enable intercommunications between equipment from a variety of manufacturers. Project 802 embraces the first two layers of the OSI protocol hierarchy (the physical and data-link layers) and to a lesser extent the network layer (layer 3). The strength of Project 802 is modularity, as it subdivides the tasks necessary to manage a LAN, allowing the designers to standardize the functions that are common and to isolate those that are not.

In 1983, the IEEE committee established several recommended standards for LANs. The most prominent of the 802 standards include 802.1 internetworking, 802.2 logical link control (LLC), 802.3 carrier sense multiple access with collision detect, 802.4 token bus, and 802.5 token ring. In 1985, the American National Standards Institute (ANSI) adopted the IEEE 802 standards, and in 1987 the standards were revised and reissued by the International Organization for Standardization (ISO) as ISO 8802.

18-3 ACCESS CONTROL METHODOLOGIES

Devices on LANs share the transmission medium, permitting simultaneous attempts to access the transmission medium by two or more devices. In a typical LAN, it is likely that more than one device may wish to use the communications channel at the same time. For two or more devices to share the same transmission medium, a means of controlling access is necessary. Media-sharing methods are known as *access control methodologies.* Network access methodologies include a *medium access method* (MAM) and some form of *medium access control* (MAC). Medium access method describes the means that devices use to gain access to the transmission medium, whereas medium access control describes how devices access the communications channel in a LAN without allowing two or more stations to transmit at the same time.

Access control methodologies are especially important on two-wire LANs or four-wire LANs using a bus topology, both of which have flat collision domains. Two-wire LANs are becoming a thing of the past; however, four-wire LANs configured as a hub-based star network (logical bus) are still fairly common. In four-wire LANs where devices gain access to the network through switches, collisions are virtually nonexistent or at least highly unlikely to occur.

18-4 MEDIUM ACCESS CONTROL

There are two broad categories of medium access: *random access* and *controlled access.* IEEE Standard 802.3 focuses on a random access method designed for bus topologies called *carrier sense, multiple access with collision detection* (CSMA/CD), and IEEE Standard 802.5 describes a controlled access method for ring topologies called *token passing.*

Medium access describes how devices access the communications channel in a LAN. Implementing access methods minimizes the likelihood of a collision occurring, although they cannot eliminate the possibility. However, access control ensure that collisions do not go undetected.

18-4-1 Random Access

With random access, all devices have equal access to the transmission medium, and there is no central control. That is, every device has the right to access the medium without being controlled by any other device. With random access, devices use *contention* to contend or compete with each other for the right to use the transmission media. However, no device is guaranteed access to the network. There are two primary features of random access:

1. There are no scheduled times for a device to transmit. If a device has something to say, it says it. If a collision occurs, it is dealt with later.
2. There are no specific rules concerning which device should transmit next. If the transmission medium is idle or appears to be idle, any device can use it.

With random access, if more than one device transmits at the same time, an *access conflict* (better known as a *collision*) occurs, and collisions destroy data. To avoid collisions or to resolve one when it happens, each device must incorporate a procedure that answers six fundamental questions:

1. When should a device attempt to access the medium?
2. What should a device do if it wishes to send a data frame and the medium is busy?
3. How should a device determine whether the transmission was successful?
4. How does a device know when a collision has occurred?
5. When a collision occurs, how does a device know if it was involved in it?
6. What should the device do if it detects a collision and knows that the frame it just transmitted was involved in it?

Random access methods evolved over the years in the following order:

Multiple access (MA)

Carrier sense multiple access (CSMA)

Carrier sense multiple access with collision detect (CSMA/CD)

Carrier sense multiple access with collision avoidance (CSMA/CA)

Multiple access by itself promotes a network atmosphere that borders on chaotic. The introduction of carrier sense improved multiple access significantly. With CSMA, a device listens to the transmission line before attempting a transmission, significantly reducing the probability of a collision. However, CSMA does not define what action should take place when a collision does occur. CSMA/CD involves sensing an idle transmission medium before transmitting and then implementing procedures to deal with a collision when one does occur. CSMA/CA involves listening to the medium and implementing procedures that avoid collisions.

18-4-1-1 Multiple access.   With multiple access (MA), any device can send a frame whenever it has one to send, and if there is a collision, it is hoped that someone will notice. ALOHA was the first random access method used. The University of Hawaii developed ALOHA in 1970 for a broadband wireless LAN operating at a 9600-bps transmission rate. The original ALOHA system (called *pure ALOHA*) was based on the following set of rather rudimentary rules:

1. Any device can send a message whenever it has a message to send.
2. Before sending a message, a station does not listen to the transmission medium to see whether it is idle.
3. After transmitting a message, a station does not listen to the medium to see if a collision has occurred.
4. After transmitting a message, the sending device waits for an acknowledgment. If it does not receive one after a predetermined time, which is approximately equal to the round-trip propagation delay, the station assumes that the message was lost, so it retransmits the message.

Pure ALOHA later evolved into *slotted ALOHA*. With slotted ALOHA, each device is allocated a precise time slot in which it must send its messages. Thus, it is theoretically impossible for a collision to occur, as no two devices are assigned the same time slot.

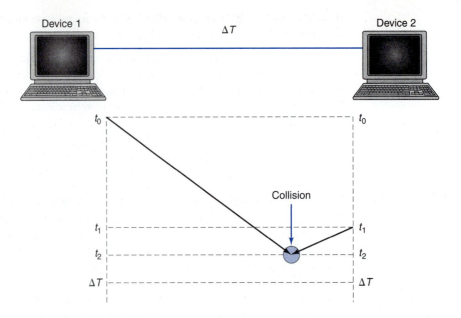

FIGURE 18-2 Collision detection with carrier sense multiple access

18-4-1-2 Carrier sense multiple access. *Carrier sense multiple access* (CSMA) is an access method used primarily with LANs configured in a bus topology. With CSMA, any device can send a message to any other device as long as the transmission medium is free of transmissions (idle). If two devices transmit at the same time, a collision may occur. To avoid collisions, devices monitor (listen to) the transmission medium to determine if the line is busy. In other words, CSMA employs the concept, *sense before transmitting* or, in human terms, *listen before talking.* This is the essence of carrier sense—to sense or detect whether a carrier is present. The term *carrier* probably came from the ALOHA multiple access system, which propagated analog signals. Most modern computer networks use a base-band format and transmit digital signals; therefore, the term *carrier* is technically incorrect, as *carrier* implies the transmission of analog signals. A more appropriate term today is probably *pulse sense* (i.e., detecting digital pulses).

If a device has a message to transmit when the line is busy, it waits for an idle medium before transmitting its message. If the medium appears quiet, the device sends its message. However, transmissions are not instantaneous; therefore, a quiet medium does not necessarily mean that the line is idle, as another device may have already begun to transmit a frame that has not reached the monitoring device yet. CSMA reduces the likelihood of a collision; however, it does not eliminate the possibility, and it does not provide procedures to deal with them when they occur.

Figure 18-2 illustrates how a collision can occur even when CSMA is employed. At time t_0, device 1 senses (listens) to the transmission medium. The medium is idle, so device 1 sends its message. It will take the message ΔT seconds to reach device 2, which is located at the distant end of the network. At time t_1, device 2 senses the medium and finds it idle because device 1's transmission has not reached it yet (i.e., $t_1 - t_0 < \Delta T$). Because the transmission medium appears idle to device 2, it transmits a frame. The two frames collide at time t_2.

The propagation delay ΔT is called the *vulnerable time,* as it is the time a transmission is vulnerable or susceptible to collisions. The vulnerable time is the length of time it takes a message to reach the most distant device on the network. The vulnerable time for a network is generally defined as the maximum propagation time between two devices on the network, which is the propagation time between the two devices separated by the greatest distance.

18-4-1-3 Carrier sense multiple access with collision detect.

Carrier sense multiple access with collision detect (CSMA/CD) is essentially the same as CSMA, except CSMA/CD includes procedures to detect collisions and deal with them when they occur. With CSMA/CD, devices listen to the transmission medium before transmitting and continue listening while they are transmitting. Therefore, CSMA/CD requires full-duplex capabilities, as a device must be able to transmit and receive at the same time. Consequently, if two devices transmit at the same time and a collision occurs, both devices detect garbled data resulting from the collision and impose recovery procedures.

The IEEE developed standard 802.3 to increase the performance of networks by minimizing the chances of collisions occurring. IEEE 802.3 outlines procedures for CSMA/CD for bus topologies. IEEE 802.3 specifies that any device (node) can send a message to any other device (or devices) as long as the transmission medium is free of transmissions from other stations. All devices continuously listen to the line. If a device has a message to send but the medium is busy, it waits until the line is idle. If two devices transmit at the same time and a collision occurs, the first device sensing the collision sends a special jamming signal onto the transmission medium. The device that sends the jamming signal is the device closest to where the collision occurred and may not be either of the devices involved in the collision. The jamming signal is made long enough that all devices connected to the medium detect it. Therefore, the devices that are transmitting at the time know they are involved in the collision. Devices involved in the collision cease transmitting (*back off*) and wait a random period of time before attempting a retransmission. The random delay time for each device is different, thereby allowing for prioritizing the device's retransmissions on the network. The device with the shortest *back-off period* has the highest priority and is allowed to retransmit first. If the difference between the back-off periods of the two devices is not long enough, their transmissions may collide again. When successive collisions occur, the back-off period for each device is doubled, which doubles the difference between them. The back-off period increases exponentially between 0 and 2^n times the maximum propagation time, where n increments with each successive attempt with a typical maximum value of 15. For this reason, the back-off period is sometimes called *exponential back-off*. The back-off period increments as follows:

$$\text{First back-off delay} = 2^1 (\Delta T)$$

$$\text{Second back-off delay} = 2^2 (\Delta T)$$

$$\text{Third back-off delay} = 2^4 (\Delta T)$$

where ΔT = the maximum network propagation time, which is the propagation time between the two devices on the network located the farthest distance from each other

A device with a frame to transmit sets its back-off exponent n to 1 and then senses the medium. If the line appears to be idle, it sends its frame. If it does not detect a collision while it is transmitting, it assumes that the transmission was successful. However, if the transmitting device detects a collision while it is sending its message, it terminates the frame and sends the jamming signal onto the medium for a specified length of time to inform the other devices on the network that a collision has occurred. The device increments its back-off value, and if it does not exceed the limit, it waits the new back-off period and then senses the line again. If the line is idle, it retransmits the message.

With CSMA/CD, devices contend for network access; therefore, no device is guaranteed access to the network at any specified time, and there is no prescribed sequence that specifies which device has access next. Also, to detect collisions, each station must be capable of operating full duplex (i.e., transmitting and receiving simultaneously).

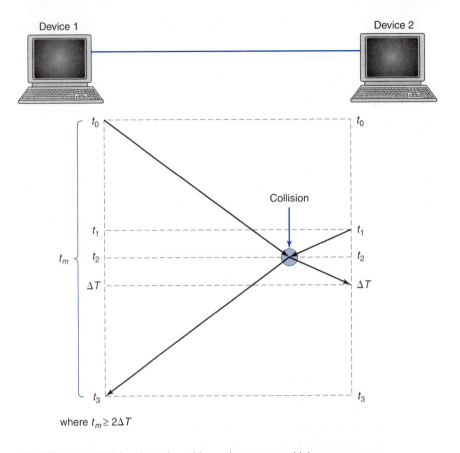

Device 1   Device 2

where $t_m \geq 2\Delta T$

FIGURE 18-3   Collision detection with carrier sense multiple access

Figure 18-3 illustrates how a collision is detected with CSMA/CD. At time t_0, device 1 senses (listens) to the transmission medium. The medium is idle, so device 1 sends its message. It takes the message ΔT seconds to reach device 2, which is located at the distant end of the network. However, at time t_1, device 2 senses the medium and finds it idle because device 1's transmission has not reached it yet (i.e., $t_1 - t_0 < \Delta T$). Because the transmission medium appears idle to device 2, it transmits a frame. The two frames collide at time t_2. The collision propagates in both directions on the transmission medium and reaches device 2 at the same time the message from device 1 would have had the collision not occurred (ΔT). The collision reaches device 1 at time t_3, and because the collision propagates at the same velocity as a normal transmission, $t_2 - t_0 = t_3 - t_2$. Although devices 1 and 2 both sense the collision, they would not know they were involved in it unless they were still transmitting when the collision reached them. Therefore, the minimum length of a device's transmission (t_m) must be at least as long as the maximum round-trip delay between the two devices, which is twice the one-way propagation time (i.e., $t_m \geq 2\Delta T$).

18-4-1-4 Carrier sense multiple access with collision avoidance. *Carrier sense multiple access with collision avoidance* (CSMA/CA) is similar to CSMA/CD except CSMA/CA implements steps to prevent collisions from occurring. Therefore, with CSMA/CA, there theoretically should be no collisions. CSMA/CA was designed to be used with wireless LANs. The procedure for transmitting a message with CSMA/CA is as follows:

1. The device resets its back-off exponent to 1.
2. If the device has a message to send, it listens to the transmission medium. If it detects an idle line, it waits a predetermined length of time, called the *interframe gap* (IFG).
3. The device waits another random amount of time and then transmits its frame.
4. The device waits a predetermined length of time for an acknowledgment.
5. If the device receives an acknowledgement within a predetermined time, it assumes that the transmission was successful.
6. If the device does not receive an acknowledgment within the prescribed time, something must be wrong (either the frame or the acknowledgment was lost or contaminated during transmission).
7. If it is determined that the transmission was unsuccessful, the device increments its back-off exponent, senses the line again, and then repeats steps 2 through 6.

18-4-2   Controlled Access

Controlled access does not use contention as a means of accessing the transmission media. Devices exchange information with each other to determine which device has the right to transmit. Therefore, each device is guaranteed access to the network. The IEEE outlines two procedures for controlled access: 802.4 *token bus* and 802.5 *token ring*. Token bus is seldom used today; however, token ring is the choice of most network designers for LANs using a ring (loop) topology.

18-4-2-1 Token ring.   In 1969, Olaf Soderblum was given credit for developing the first token passing ring network architecture. IBM, however, became the driving force behind the standardization and adoption of the token ring, and a prototype developed by IBM in Zurich, Switzerland, served as a model for IEEE standard 802.5.

With token ring (sometimes called *token passing*), devices do not contend for the right to transmit. Instead, a specific frame, called a *token,* is circulated around the ring from device to device, always in the same direction. The logical token is generated by a device designated the *active monitor.* The only function of the active monitor is to start the token circulating around the ring. Other than that, there is no central control on a token ring network. Each device has equal access to the media; however, before a device can transmit a frame, it must possess the token. Each device, in turn, acquires the token and examines the frame to determine if it is carrying a data packet addressed to it. If the frame contains a frame with the destination address of the receiving device, the device copies the contents of the frame into memory, appends any messages it has to transmit to the token, and then relinquishes the token by retransmitting all frames and the token to the next downlink device on the network.

Each data packet contains a source and destination address. A destination device confirms successful delivery of a frame by setting frame status flags and then forwarding the frame around the ring until it reaches the originating device, where the frame is removed. After a transmitted frame has propagated around the loop and returned to the originating device, the originating device removes the frame and relinquishes the token to the next down-loop device. A device cannot use a token twice in succession, and there is a time limitation on how long a token can be held. This prevents one device from disrupting data transmissions on the network by transmitting frames indefinitely or holding the token until it has a packet to send. When a device does not posses the token, it cannot originate a transmission—it can transmit a message only if it receives a frame addressed to it. However, the device can repeat frames destined to other down-loop devices.

Some token ring networks use a modified form of token passing where the token is relinquished as soon as a frame has been transmitted instead of waiting until the transmitted frame has been returned. This is known as an *early token release mechanism.*

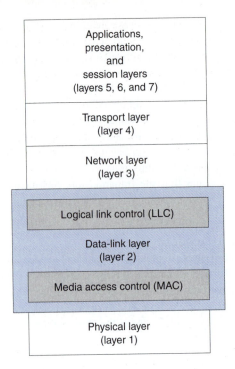

FIGURE 18-4   IEEE 802 data-link layer subdivisions

18-5   LAN DATA-LINK LAYER

The data-link layer of a LAN controls access to the network transmission medium and defines how upper-layer data is encapsulated into layer 2 frames. To separate these two functions, the IEEE 802 project subdivided the data-link layer into two sublayers: media access control (MAC) and logical link control (LLC), as shown in Figure 18-4.

18-5-1   Data-Link Addressing

The ISO LAN architecture specifies two levels of addressing for LANs: *service access point* (SAP) addressing and *medium access control* (MAC) addressing. The logical link control sublayer provides SAP addressing. SAP addresses denote a specific mechanism, process, or upper-layer protocol that is requesting services from the LLC sublayer. Each mechanism, process, or upper-layer protocol used by a device attached to a given LAN must use a different SAP address. MAC addressing uniquely identifies each device physically connected to a given LAN.

18-6   LOGICAL LINK CONTROL SUBLAYER

The basic idea behind the IEEE 802 project is to have one generic sublayer at the data-link layer that all LANs use; this allows interoperability between LANs carrying different upper-layer protocols. The IEEE developed the 802.2 LLC protocol in an attempt to standardize the LLC sublayer. The primary responsibility of the LLC sublayer is to control the exchange of messages between individual users of the LLC sublayer services, or virtually every device on the network. If two otherwise incompatible LANs use the same LLC sublayer, they can communicate with each other. Sharing the same LLC provides a common interface and offers reliability and flow control features. The LLC also allows LAN-specific aspects, such as media access, encoding, signaling, and the type of transmission media to be transparent to the upper-layer protocol. In essence, the LLC acts like a software bus, allowing multiple higher-layer protocols the ability to access one or more lower-layer networks. For example, a server may provide services to more than one type of network, which requires more than one type of NIC (e.g., one Ethernet and one token ring). The LLC transfers packets between upper-layer protocols through the appropriate network interface, allowing upper-layer protocols to operate without knowing the specifics of the lower-layer networks.

The LLC is non-architecture-specific in that it is the same for all IEEE-compliant LANs defined at the MAC sublayer. The MAC sublayer contains a number of distinct modules where each module carries proprietary information specific to the LAN product being used. Therefore, the MAC sublayer is virtual and unreliable.

LLC adds reliability to the link by supervising MAC frames until they reach their destination and acknowledging them when they are received. All networks do not use the LLC layer. For example, when a network uses the services of a reliable higher-layer protocol (such as TCP), it does not need the services provided by LLC.

ISO standards specify two features of the logical link control sublayer operation: a service definition and a protocol specification. The service definition specifies three service classes that define the interface between the LLC sublayer and the LAN data-link user and outline the services the LLC sublayer provides to the LAN data-link layer. It also outlines the information the user must provide the LLC sublayer that is requesting the services. The LLC protocol specification outlines the means by which it provides the LLC sublayer service.

18-6-1   LLC Service Classes

The LLC sublayer is responsible for supplying services to the users of the LAN, where the users are the upper-layer protocols encapsulated in the LLC frame, such as network-layer protocols (i.e., TCP [transmission control protocol]).

IEEE 802.2 defines three classes of services for the LLC sublayer: *connectionless acknowledged, connectionless unacknowledged* and *connection oriented*. Not all LANs provide connection-oriented service; however, they all provide connectionless service, and for many LANs it is the only service provided.

18-6-1-1 Connectionless service.   A connectionless service is similar to sending standard mail through the U.S. Postal Service. The information is put into an envelope and sent, and we hope it reaches the intended destination. There is no feedback from the destination (unacknowledged) unless we request one (acknowledged).

Connectionless service requires less protocol overhead than connection-oriented service because no communication takes place between the source and destination devices prior to sending a message. Each message is treated as a separate entity and processed independent from all other messages. Connectionless service provides no sequencing mechanism and may not provide any means of acknowledging messages. In addition, there is no flow control or error recovery. These services must be provided by the upper-layer protocols encapsulated in the LLC frame. Connectionless service is sometimes called *datagram service* and is used extensively to multicast or broadcast network-specific information, such as address resolution and domain name resolution.

Connectionless service supports point-to-point, multipoint, and broadcast communications and is suitable for carrying higher-level protocols that do the segmenting, addressing, routing, and error recovery, such as TCP/IP. There are two types of connectionless service: acknowledged and unacknowledged.

Connectionless acknowledged service. Connectionless acknowledged service falls somewhere between the connectionless unacknowledged and the connection-oriented services. With connectionless acknowledged service, the user protocol at the source is notified with an acknowledgment when the data is received by the destination's user protocol; therefore, it is considered a guaranteed delivery service. The acknowledgment can be positive (ACK) or negative (NAK). An ACK indicates that the data was received without errors and no retransmission is necessary, and a NAK indicates that the data was received with errors and a retransmission is requested. Connectionless acknowledged service operates half duplex, which means that transmissions can occur in both directions but not at the same time and that no connection is established before data is transferred. The connectionless acknowledged services implementation is seldom used. However, when it is used, the choices of services available are data transfer, polling, and selecting.

Connectionless unacknowledged service. With connectionless unacknowledged service, the data units are transmitted and received without being acknowledged. The only service available with connectionless unacknowledged service is data transfer, which simply delivers data packets between the user protocol and the LLC in the source device and from the LLC to the user protocol at the destination device. Connectionless unacknowledged services are simplex (one way only), as no connection is established prior to sending a message and no acknowledgment is returned when the message is received. Connection unacknowledged service is not considered a guaranteed delivery service.

18-6-1-2 Connection-oriented service.   A connection-oriented service is similar to carrying on a normal telephone conversation. A connection is established, information is exchanged (i.e., a conversation takes place), and then the connection is disconnected. In a network environment, a connection-oriented service provides a logical point-to-point link between two LLCs. A connection-oriented service responds to the higher-layer protocols by establishing the link prior to transferring data and disconnecting the link once data transfer is completed. Connection-oriented service is considered a guaranteed delivery service as long as the LLC connection is maintained between the source and destination.

There are five choices provided with a connection-oriented service: connection, data transfer, flow control, disconnection, and connection resetting. A connection is established between the LLCs in the source and destination devices before any data is exchanged. Flow control is used to specify the maximum byte length for a single data packet and also the maximum length of the message. After a source receives confirmation that the connection is established, data is transferred, and then the connection is released or disconnected. The user protocol at either the source or the destination can reset (reestablish) the connection if necessary.

Connection-oriented service provides a means of sequencing messages and detecting transmission errors. When errors are detected, the corrupted portion of the message is retransmitted. Connection-oriented service does not provide a means of multicasting. Therefore, for a source to send a message to more than one destination, the message must be sent individually to each destination.

18-6-2   LLC Protocol Specifications

The format specified by 802.2 for transporting data packets between two LLCs is a subset of high-level data-link control (HDLC). LLCs use essentially the same structure as an HDLC frame, where the responsibilities are divided into two sets of functions. One set of functions is performed by the LLC sublayer and includes logical addresses, control information, and data. This is considered the upper sublayer of the IEEE 802 data-link layer and is common to all LAN protocols. The second set of functions is specified by the lower sublayer of the IEEE 802 data-link layer, the medium access control (MAC) sublayer, and includes synchronization, flow control, and error control.

18-6-3   Protocol Data Unit

Data units transmitted from an entity in a protocol layer (N + 1) in one device to the same entity in the corresponding protocol layer (N + 1) in another device is called a *protocol data unit* (PDU). Each layer of the OSI protocol hierarchy has a PDU. The PDU used at the physical layer (layer 1) is called a 1-PDU, the PDU used at the data-link layer (layer 2) is called a 2-PDU, and so on. N-PDUs in the originating device communicate with the N-PDU in the destination device. Figure 18-5 shows PDU-to-PDU communications between a de-

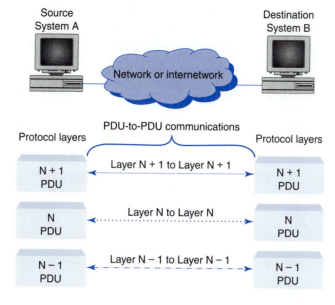

FIGURE 18-5 PDU-to-PDU communications between systems

vice on system A and a device on system B. As shown in the figure, protocol PDUs on one system communicate directly with the corresponding PDU on the other system.

Once an N-layer PDU leaves the N layer, it may have to pass through several protocol layers before reaching the N layer in the destination. PDUs flow downward through the protocol layers in the source, first upward and then downward through the protocol layers in intermediate systems and then upward through the protocol layers at the destination. N-PDUs should pass transparently through all other PDU layers. To ensure transparency, N-layer PDUs are encapsulated as they pass downward through lower-layer protocols in the source and decapsulated as they pass upward through the layers in the destination. Passing PDUs between protocol layers is made possible using interfaces between each pair of adjacent layers. Interfaces encapsulate the PDU in the lower-layer protocol. Encapsulation involves adding a header to each PDU as it moves downward through the layers in the source system. Headers identify in which layer the PDU resides. Headers are decapsulated in the destination as the PDU moves upward through the protocol layers. PDU encapsulation and decapsulation is shown in Figure 18-6. As shown in the figure, as a PDU moves downward through the layers in the source system, each protocol layer adds a PDU header to the PDU passed down to it, forming a new PDU. As PDUs move upward through the protocol layers in the destination system, each protocol removes its corresponding PDU.

18-6-4 LLC PDU Header

An IEEE 802.2 defined LLC PDU header has four fields: a one-byte DSAP field (destination service access point), a one-byte SSAP field (source service access point), a one- or two-byte control field, and a variable-length information field. The format for the IEEE 802.2 LLC PDU is shown in Figure 18-7. As can be seen in the figure, the PDU contains no flags (either beginning or ending), no CRC (error detection) field, and no address field. These three fields are added later in the MAC sublayer.

18-6-4-1 DSAP and SSAP. LAN data-link users request services through a LLC service access points (SAPs or LSAPs) into the LLC sublayer. There are two SAPs—one in the source called the source service access point (SSAP) and one in the destination called the destination service point (DSAP). SAPs are addresses used by the LLC to identify the upper-layer protocols on the source and destination devices (i.e., the devices that generate

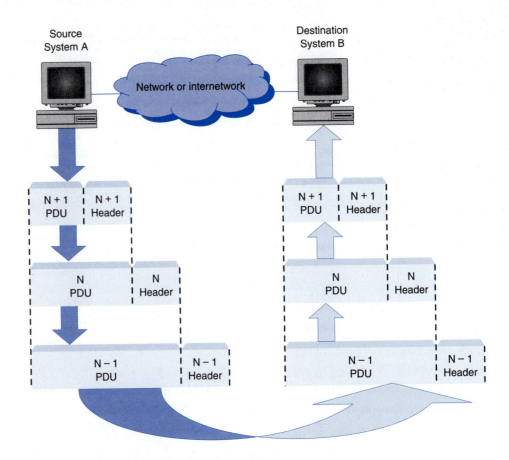

FIGURE 18-6 PDU Encapsulation and decapsulation

DSAP Field 1 Byte	SSAP Field 1 Byte	Control Field 1 or 2 Byte	Variable-Length Information Field

FIGURE 18-7 IEEE 802.2 LLC PDU format

and use the data). The DSAP and SSAP are similar to TCP/IP port numbers, as they identify a memory buffer and a logical link to the network layer protocol. The port numbers should not be confused with physical ports, which are connectors or jacks, such as the serial port on a PC or a port on a connectivity device. Port numbers identify the processes running on the source and destination devices that wish to communicate with each other. For example, the port number for NetBIOS is F0 hex, and the port number for the ARPANET IP is 06 hex. All SAP addresses where the second bit is set to a logic 1 are defined by the IEEE. Some of the more common SAP addresses are listed in Table 18-1. In essence, the DSAP and SSAP allow the LLC to identify which higher-layer protocol is encapsulated in its information field. DSAP and SSAP have the following characteristics:

DSAP. The DSAP identifies the LLC sublayer user that is to receive the LLC PDU. The DSAP can be for a single SAP (i.e., unique address), for more than one SAP (i.e., group address), or for all active SAPs (i.e., global broadcast address). The format for

Table 18-1 SAP Address Summary

Service Access Point (DSAP or SSAP)		
Binary	Hex	Protocol
00000000	00	Null SAP
00000010	02	Individual LLC Sublayer Management
00000011	03	Group LLC Sublayer Management
00000100	04	IBM SNA Path Control (Individual)
00000101	05	IBM SNA Path Control (Group)
00000110	06	ARPNET Dod TCP/IP Protocol Stack
00001000	08	IBM SNA
00001100	0C	IBM SNA
00001110	0E	PROWAY (DEC 955) Network Management and Installation
00011000	18	Texas Instruments
01000010	42	IEEE 802.1 Bridge Spanning Tree Protocol
01001110	4E	EIA RS-511 Manufacturing Message Service
01011110	5E	ISI IP
01111110	7E	ISO 802.6 (X.25 over IEEE 802.2 Type 2 LLC)
10000000	80	Xerox Network Systems (XNS)
10000110	86	Nestar
10001110	8E	PROWAY (DEC 955) Active Station List Management
10011000	98	ARPNET Address Resolution Protocol (ARP)
10101010	AA	Subnetwork Network Access Protocol (SNAP)
10111100	BC	Banyon VINES
11100000	E0	Novell Netware (IPX)
11110000	F0	IBM NetBIOS
11110100	F4	IBM LAN Management (Individual)
11110101	F5	IBM LAN Management (Group)
11111000	F8	IBM Remote Program Load (RPL)
11111010	FA	Ungermann-Bass
11111110	FE	ISO CLNS ISO-8473 Network-Layer Protocol
11111111	FF	Global DSAP

the DSAP is shown in Figure 18-8. The low-order bit identifies whether the address is intended for an individual device or a group of devices. The second bit is reserved for the IEEE, and the remaining bits are used for upper-level addressing.

SSAP. The SSAP identifies the LLC sublayer user that originated the LLC PDU. Therefore, the SSAP must be a unique address. The format for a SSAP is very similar to the DSAP and is shown in Figure 18-9. As can be seen, the only difference between the two formats is the meaning of the low-order bit. In the SSAP, the low-order bit identifies whether the frame is a command (0) or a response (1). A command is the initial frame sent by the originating device, and a response is the reply to the command.

18-6-4-2 Control field. The LLC protocol data unit is the unit of data in the LLC sublayer. The control field of the PDU describes the PDU type and includes other administrative information, such as sequencing and flow control information. LLC specifies three kinds of PDUs: information, supervisory, and unnumbered. An information PDU (I-PDU) is used to transport user data over a connection-oriented service. A supervisory PDU (S-PDU) provides flow and error control but cannot carry any data. An unnumbered PDU (U-PDU) is used to transport user data over a connectionless service or to transport management information over a connection-oriented service.

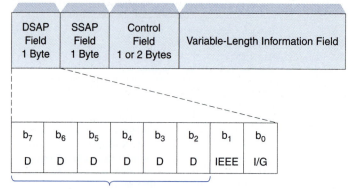

Used for upper-layer addressing

D = DSAP bits
b_0 = 0 individual (I) address
b_0 = 1 group (G) address

FIGURE 18-8 DSAP field format

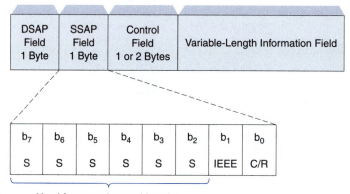

Used for upper-layer addressing

S = SSAP bits
b_0 = 0 command
b_0 = 1 response

FIGURE 18-9 SSAP field format

The control field for I- and S-PDUs can contain eight bits (standard) or 16 bits (extended); however, the control field for a U-PDU can have only eight bits. Figure 18-10 shows the control fields for the three PDU types specified for LLC PDUs. As can be seen in the figure, I-PDUs are identified with a 0 in bit position b_0, S-PDUs are identified with a 1 in bit position b_0 and a 0 in bit position b_1, and U-PDUs are identified with a 1 in both b_0 and b_1.

The format for standard eight- and 16-bit I-PDU control fields are shown in Figures 18–11a and b, respectively. As can be seen, the format for the control field is virtually identical to that of HDLC. The ns bits identify the number of the information PDU being sent, and the nr bits identify the number of the next information PDU the sending device expects to receive (i.e., one more than the last error-free PDU received). With three ns and three nr bits, PDU numbers can range between 0 and 7, and with seven ns and seven nr bits, PDU numbers can range between 0 and 127. As with HDLC, the ns and nr bits are used for error

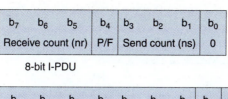

8-bit I-PDU

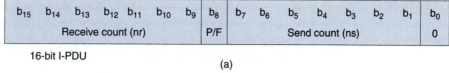

16-bit I-PDU

(a)

8-bit S-PDU

16-bit S-PDU

(b)

8-bit U-PDU

(c)

FIGURE 18-10 LLC PDU control field: (a) I-PDU; (b) S-PDU; (c) U-PDU

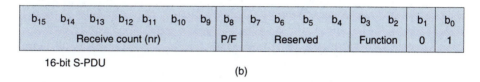

(a)

(b)

ns = number of the PDU being sent
nr= number of the next PDU the sending device expects to receive
P/F
  P = 1 poll
  P = 0 not a poll
  F = 1 final
  F = 0 not a final

FIGURE 18-11 I-PDU control field: (a) eight-bit standard; (b) 16-bit extended

b_7	b_6	b_5	b_4	b_3	b_2	b_1	b_0
nr	nr	nr	P/F	S	S	0	1

(a)

b_{15}	b_{14}	b_{13}	b_{12}	b_{11}	b_{10}	b_9	b_8	b_7	b_6	b_5	b_4	b_3	b_2	b_1	b_0
nr	nr	nr	nr	nr	nr	nr	P/F	R	R	R	R	S	S	0	1

(b)

nr = number of the next PDU the sending device expects to receive
S = supervisory function
P/F
  P = 1 poll
  P = 0 not a poll
  F = 1 final
  F = 0 not a final

FIGURE 18-12 S-PDU control field: (a) eight-bit standard; (b) 16-bit extended

control and for sequencing user data frames. Bit b_4 (eight-bit control field) or b_8 (16-bit control field) is the P/F bit. The P/F bit has two functions. In a PDU sent from a primary, it is the P bit and when it is a 1 indicates that the primary is polling the secondary and when it is 0 indicates that the primary is not polling the secondary. In a PDU sent from a secondary, the P/F bit is the F (final) bit and when it is a 1 indicates that this PDU is the last PDU in a sequence and when it is a 0 indicates that this is not the last PDU in a sequence.

The format for standard eight- and 16-bit S-PDU control fields are shown in Figures 18-12a and b, respectively. As can be seen, S-PDUs contain nr bits but no ns bits. Therefore, S-PDUs can be used to confirm reception of I-PDUs but forbid the transmission of sequenced information. Therefore, information fields are prohibited with S-PDUs. The R bits are reserved for future use, and the S bits constitute a two-bit subfield used to identify three types of S-PDUs:

b_3	b_2	
0	0	ready to receive
0	1	not ready to receive
1	0	reject

The format for a U-PDU is shown in Figure 18-13. U-PDUs contain neither ns nor nr bits. Therefore, they cannot be used to send or confirm reception of received I-PDUs. The five M bits identify up to 32 unnumbered commands and responses that are used to exchange unnumbered user information or management and control information between devices. Unnumbered commands and responses include: unnumbered acknowledgments (UA), disconnect (DISC), frame reject (FMRJ), loopback test (LT), and set asynchronous balanced mode (SABM).

For a more detailed explanation of the three PDU control formats, refer to the sections on SDLC and HDLC in Chapter 16.

18-7 MAC SUBLAYER

The MAC sublayer is where individual shared-media LAN technologies, such as Ethernet, are defined. The MAC sublayer resolves contention on a shared-media LAN, which basically determines what devices can access the LAN media and when. The MAC sublayer

b_7	b_6	b_5	b_4	b_3	b_2	b_1	b_0
M	M	M	P/F	M	M	1	1

m = command/response indicators
P/F
   P = 1 poll
   P = 0 not a poll
   F = 1 final
   F = 0 not a final

FIGURE 18-13 U-PDU control field

Preamble	Start frame delimiter (SFD)	Destination MAC address	Source MAC address	Length field	Data field IEEE 802.2 LLC and upper-layer network protocol	Frame check field CRC-32
10101010	10101011	DDDDDD	SSSSSS	002E to 05DC	Variable length	XXXX
7 bytes	1 byte	6 bytes	6 bytes	2 bytes		4 bytes

FIGURE 18-14 IEEE 802.3 MAC sublayer frame format

also includes synchronization, flow and error control procedures necessary for transporting information from one device to another, and physical (hardware) addresses of the next device to receive or route a frame. MAC protocols are LAN specific and include Ethernet, token ring, and token bus.

The MAC sublayer is where the LLC is encapsulated and passed on to the physical layer, where it is encoded and delivered to the transmission medium. At the destination, the MAC sublayer is removed from the transmission line, decoded, and passed on to the LLC for decapsulation. The MAC sublayer is also responsible for invoking CSMA/CD procedures. 803.2 MAC frames incorporate CRC-32 for error detection, which detects virtually all transmission errors. However, the MAC sublayer does not provide mechanisms for sequencing, acknowledging, or rejecting frames, making it an unreliable data-link layer transport medium. If a reliable network is desired, message sequencing and acknowledgments must be implemented in a higher-layer protocol encapsulated in the LLC PDU.

18-7-1   IEEE 802.3 MAC Frame Format

Figure 18-14 shows the format for the IEEE 802.3–specified MAC PDU frame, which is essentially the IEEE's version of Ethernet. The IEEE 802.2 LLC PDU is encapsulated in the data field of the MAC PDU. Thus, the MAC PDU transports the LLC PDU across the local network from the source device to the destination device (or devices). The MAC PDU frame is comprised of seven fields: preamble, start frame delimiter, destination address, source address, length field, data field, and frame check sequence field.

Preamble. The preamble field is comprised of seven bytes (56 bits) of alternating 1s and 0s that has two functions: (1) it alerts the destination device of an incoming frame, and (2) it provides the transitions necessary to establish clock synchronization between devices connected to the LAN medium. Because the preamble is actually added at the physical layer, it is technically not considered part of the 802.3 frame.

Start frame delimiter. The start frame delimiter (SFD) field is a one-octet field that marks the end of the preamble and beginning of the frame. The SFD is identical to a preamble byte, except it ends with two 1s (10101011). The SFD is sometimes

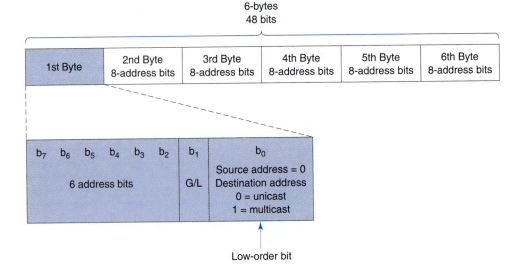

FIGURE 18-15   IEEE 802.3 MAC address field

considered the end of an eight-octet preamble. Reception of the SFD sequence indicates that the next byte is the destination address.

Destination address. The destination address field is a six-octet MAC address field that contains the hardware (physical) address of the frame's destination device (or devices).

Source address. The source address field is a six-octet MAC address field that contains the hardware (physical) address of the device that is sending the frame.

Length field. The length field is a two-octet field containing the number of bytes encapsulated in the LLC PDU data field. The range of values for the length field is typically 002E hex to 05DC hex (46 to 1500 decimal). The reason for this is explained later.

Data field. The data field typically carries between 46 and 1500 bytes, which is the payload for the frame. The payload is either a LLC PDU or one or more upper-layer protocols, such as Address Resolution Protocol (ARP) or Internet Protocol (IP) and either Transmission Control Protocol (TCP) or User Datagram Protocol (UDP).

Frame check sequence. The frame check sequence field is a four-octet field that carries a CRC-32 error-detection code derived from the header and data fields.

18-7-2   MAC Addressing

Hardware addresses are burned into NICs, and every device connected to a LAN has a MAC. The six-byte hardware address on the NIC is the MAC source address for the device. Source MAC addresses are always unicast addresses, as more than one device cannot originate a transmission. However, the destination MAC address can be unicast, multicast, or broadcast. A unicast destination address designates only one device as a recipient. Figure 18-15 shows the format for an 802.3 MAC source address field. In a unicast address, the low-order bit (b_0) is always 0. Therefore, the hex value for the first byte in a unicast address is always an even number. In the MAC destination address field, if the low-order bit is 0, the destination address is a unicast address, and if the low-order bit is a 1, it is a multicast address. Therefore, the first byte of a multicast address is always an odd number. If all 48 bits in the destination address field are 1s, it is a broadcast address (i.e., FF FF FF FF FF FF).

MAC address field
6-bytes
48 bits

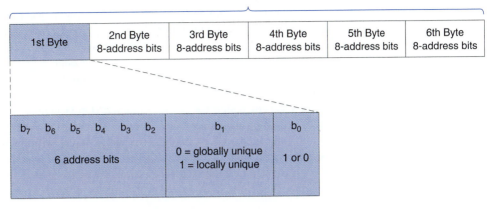

FIGURE 18-16 Globally and locally unique addresses

MAC source hardware addresses can be *globally* or *locally unique*. Globally unique means that the address is unique for all LANs in existence. Locally unique means that the address is unique on that LAN but that there is no guarantee that the address is not being used in another NIC in another LAN somewhere else in the world. The IEEE standard specifies the first three bytes of the MAC address as the *vendor address component* (sometimes called the *vendor block code*). The IEEE standard calls the unique block code an *organizationally unique identifier* (OUI). For example, if the first three bytes of the hardware address are 00 00 0C, Cisco manufactured the NIC card. The address 00 AA 00 identifies Intel as the manufacturer. A complete list of OUIs can be obtained from the Internet Assigned Numbers Authority (IANA). There are currently over 5000 venders registered with the IANA.

The last three bytes of the MAC address is the serial number of that vendor's interface card (sometimes called the *vendor-specific identifier*). The IEEE standard assigns each vendor that manufactures IEEE-compliant NICs a unique three-byte block code. With three bytes, there are 2^{24} unique vendor block codes possible. Each vendor uses its vendor block code for the first three bytes of every NIC it manufactures. The last three bytes are vendor specific and should be different on every NIC sold by a vendor. With three bytes, there are 2^{24} hardware addresses assigned to each NIC vendor.

The second bit (b_1) of the first byte of the MAC address is called the G/L bit (G = global and L = local). Bit b_1 identifies the address as either globally or locally unique. If $b_1 = 0$, the address is globally unique, and if $b_1 = 1$, the address is locally unique. Figure 18-16 shows the placement of the G/L bit in the MAC address. If network administrators want to ensure that each NIC on their LAN has a different hardware address, they should purchase NIC cards with a 0 in b_1 of the first byte.

18-8 ETHERNET

Ethernet is a baseband transmission system designed in 1972 by Robert Metcalfe and David Boggs of the Xerox Palo Alto Research Center. Metcalfe, who later founded 3COM Corporation, and his colleagues at Xerox developed the first experimental Ethernet system to interconnect a Xerox Alto personal workstation to a graphical user interface. The first experimental Ethernet system was later used to link Alto's workstations to each other and to link the workstations to servers and laser printers. The signal clock for the experimental Ethernet interface was derived from the Alto's system clock, which produced a data transmission rate of 2.94 Mbps.

Metcalfe's first Ethernet was called the Alto Aloha Network; however, in 1973, Metcalfe changed the name to Ethernet to emphasize the point that the system could support any computer (not just Alto's) and to stress the fact that the capabilities of his new network had evolved well beyond the original Aloha system. Metcalfe chose the name based on the word *ether*, meaning the air, the atmosphere, or the heavens, as an indirect means of describing a vital feature of the system: the physical medium (i.e., a cable). The physical medium carries data bits to all stations in much the same way that *luminiferous ether* was once believed to transport electromagnetic waves through space.

In July 1976, Metcalfe and Boggs published a landmark paper titled "Ethernet: Distributed Packet Switching for Local Computer." On December 13, 1977, Xerox Corporation received patent number 4,063,220 titled "Multipoint Data Communications System with Collision Detection." In 1979, Xerox joined forces with Intel and Digital Equipment Corporation in an attempt to make Ethernet an industry standard. In September 1980, the three companies jointly released the first version of the first Ethernet specification called the Ethernet Blue Book, DIX 1.0 (after the initials of the three companies), or Ethernet I.

18-8-1 Evolution of Ethernet

Ethernet I was replaced in November 1982 by the second version called Ethernet II (DIX 2.0), which remains the current standard. In 1983, the 802 Working Group of the IEEE released its first standard for Ethernet technology. The formal title of the standard was *IEEE 802.3 Carrier Sense Multiple Access with Collision Detection (CSMA/CD) Access Method and Physical Layer Specifications*. The IEEE subsequently reworked several sections of the original standard, especially in the area of the frame format definition, and in 1985 released the 802.3a standard, which was called *thin Ethernet, cheapernet,* or *10Base2 Ethernet*. In 1985, the IEEE also released the IEEE 802.3b 10Broad-36 standard, which defined a broadband transmission system with a transmission rate of 10 Mbps over a coaxial cable system.

In 1987, two additional standards were released: IEEE 802.3d and IEEE 802.3e. The 802.3d standard defined the *Fiber Optic Inter-Repeater Link* (FOIRL), which used two fiber optic cables to extend the maximum distance between 10-Mbps repeaters to 1000 meters. The IEEE 802.3e standard defined a 1-Mbps standard based on twisted-pair cable but was never widely accepted. In 1990, the IEEE introduced a major advance in Ethernet standards: IEEE 802.3i. The 802.3i standard defined 10Base-T, which permitted a 10-Mbps transmission rate over simple category 3 unshielded twisted-pair (UTP) cable. The widespread use of UTP cabling in existing buildings created a high demand for 10Base-T technology. 10Base-T also facilitated a star topology that made it much easier to install, manage, and troubleshoot. These advantages led to a vast expansion in the use of Ethernet.

In 1993, the IEEE released the 802.3j standard for 10Base-F (FP, FB, and FL), which permitted attachment over longer distances (2000 meters) through two optical fiber cables. This standard updated and expanded the earlier FOIRL standard. In 1995, the IEEE improved the performance of Ethernet technology by a factor of 10 when it released the 100-Mbps 802.3u 100Base-T standard. This version of Ethernet is commonly known as *fast Ethernet*. Fast Ethernet supported three media types: (1) 100Base-TX, which operates over two pairs of category 5 twisted-pair cable; (2) 100Base-T4, which operates over four pairs of category 3 twisted-pair cable; and (3) 100Base-FX, which operates over two multimode fibers.

In 1997, the IEEE released the 802.3x standard, which defined full-duplex Ethernet operation. Full-duplex Ethernet bypasses the normal CSMA/CD protocol and allows two stations to communicate over a point-to-point link, which effectively doubles the transfer rate by allowing each station to simultaneously transmit and receive separate data streams. In 1997, the IEEE also released the IEEE 802.3y 100Base-T2 standard for 100-Mbps operation over two pairs of category 3 balanced transmission line.

In 1998, IEEE once again improved the performance of Ethernet technology by a factor of 10 when it released the 1-Gbps 802.3z 1000Base-X standard, which is commonly called *gigabit Ethernet*. Gigabit Ethernet supports three media types: (1) 1000Base-SX, which operates with a 850-nm laser over multimode fiber; (2) 1000Base-LX, which operates with a 1300-nm laser over single and multimode fiber; and (3) 1000Base-CX, which operates over short-haul copper *twinax* shielded twisted-pair (STP) cable. In 1998, the IEEE also released the 802.3ac standard, which defines extensions to support virtual LAN (VLAN) tagging on Ethernet networks. In 1999, the release of the 802.3ab 1000Base-T standard defined 1-Gbps operation over four pairs of category 5 UTP cabling.

The topology of choice for Ethernet LANs is either a linear bus or a star, and all Ethernet systems employ CSMA/CD for the access method.

18-8-2 IEEE Ethernet Standard Notation

To distinguish the various implementations of Ethernet available, the IEEE 802.3 committee has developed a concise notation format that contains information about the Ethernet system, including such items as bit rate, transmission mode, transmission medium, and segment length. The IEEE 802.3 format is

<data rate in Mbps> <transmission mode> <maximum segment length in hundreds of meters>

or

<data rate in Mbps> <transmission mode> <transmission media>

The transmission rates specified for Ethernet are 10 Mbps, 100 Mbps, and 1 Gbps. There are only two transmission modes: baseband (base) or broadband (broad). The segment length can vary, depending on the type of transmission medium, which could be coaxial cable (no designation), twisted-pair cable (T), or optical fiber (F). For example, the notation 10Base-5 means a 10-Mbps transmission rate, a baseband mode of transmission, and a maximum segment length of 500 meters. The notation 100Base-T specifies a 100-Mbps transmission rate, a baseband mode of transmission, and a twisted-pair transmission medium. The notation 100Base-F means a 100-Mbps transmission rate, a baseband transmission mode, and an optical fiber transmission medium.

The IEEE currently supports nine 10-Mbps standards, six 100-Mbps standards, and five 1-Gbps standards. Table 18-2 lists some of the more common types of Ethernet, their cabling options, distances supported, and topology.

Table 18-2 Current IEEE Ethernet Standards

Transmission Rate	Ethernet System	Transmission Medium	Maximum Segment Length
10 Mbps	10Base-5	Coaxial cable (RG-8 or RG-11)	500 meters
	10Base-2	Coaxial cable (RG-58)	185 meters
	10Base-T	UTP/STP category 3 or better	100 meters
	10Broad-36	Coaxial cable (75-ohm)	Varies
	10Base-FL	Optical fiber	2000 meters
	10Base-FB	Optical fiber	2000 meters
	10Base-FP	Optical fiber	2000 meters
100 Mbps	100Base-T	UTP/STP category 5 or better	100 meters
	100Base-TX	UTP/STP category 5 or better	100 meters
	100Base-FX	Optical fiber	400–2000 meters
	100Base-T4	UTP/STP category 5 or better	100 meters
1000 Mbps	1000Base-LX	Long-wave optical fiber	Varies
	1000Base-SX	Short-wave optical fiber	Varies
	1000Base-CX	Short copper jumper	Varies
	1000Base-T	UTP/STP category 5 or better	Varies

Preamble	Start frame delimiter (SFD)	Destination MAC address	Source MAC address	Type field	Data field Upper-layer protocols	Frame check field CRC-32
10101010 7 bytes	10101011 1 byte	DDDDDD 6 bytes	SSSSSS 6 bytes	0600 to FFFF 2 bytes	(IP, TCP, UDP, etc.) 46 to 1500 bytes	XXXX 4 bytes

FIGURE 18-17 Ethernet II frame format

18-8-3 Ethernet Frame Formats

Over the years, four 10-Mbps Ethernet frame formats have emerged. Network environment generally dictates which format is implemented for a particular LAN. Network environment includes topography, device configuration, applications, and upper-layer protocols. The four Ethernet II formats are the following:

Ethernet II. The original format used with DIX. Includes a two-octet type field that indicates the higher-layer protocol carried inside the data field of the frame.

IEEE 802.3. The first generation of the IEEE standards committee, often referred to as a raw IEEE 802.3 frame. Novell was the only software vendor to use this format.

IEEE 802.3 with 802.2 LLC. Provides support for IEEE 802.2 LLC. IEEE 802.3 is the MAC format described earlier in this chapter.

IEEE 802.3 with SNAP. Similar to IEEE 802.3 but provides backward compatibility for 802.2 to Ethernet II formats and protocols.

Ethernet II and IEEE 802.3 with 802.2 LLC are the two most popular frame formats used on LANs today. Although they are sometimes thought to be the same thing, in actuality Ethernet II and IEEE 802.3 are not identical. However, the term *Ethernet* is generally used to refer to any IEEE 802.3–compliant network. Both Ethernet II and IEEE 802.3 specify that data be transmitted from one device to another in groups of data called frames. The frame formats for Ethernet II and IEEE 802.3 are very similar, and it is common to see both formats used on the same LAN, except for different applications.

18-8-4 Ethernet II Frame Formats

The 10-Mbps Ethernet II frame format is shown in Figure 18-17. The frame is comprised of seven fields: preamble, start frame delimiter, destination address, source address, type field, data field, and frame check sequence field. As can be seen, the Ethernet II MAC sublayer is identical to the IEEE 802.3 frame shown in Figure 18-14 except for the length field. Ethernet II specifies a type field in place of the length field. The type field is a two-byte field containing a number specifying the upper-layer protocol encapsulated in the data field in place of the IEEE 802.2 LLC PDU.

A hex value in the type field below 0600 identifies the frame as an IEEE 802.3–compliant frame carrying an IEEE 802.2 LLC PDU in the data field. In this case, the hex value specifies the length of the LLC PDU, which has been established for Ethernet as 1500 bytes. A type field of 0600 hex or higher specifies that an upper-layer protocol is encapsulated in the data field in place of the LLC PDU. Table 18-3 lists the *protocol types* specified by Ethernet II. The protocol type 0800 hex defines the Department of Defense IP, which is the network-layer protocol used in the global Internet.

18-8-5 10-Mbps Ethernet

From the mid-1980s to the late 1990s, 10 Mbps was the de facto standard transmission rate for Ethernet LANs using a bus topology regardless of the frame format used. Although re-

Table 18-3 Ethernet II Protocol Types

Hex Code	Protocol
0000–05DC	IEEE 802.3 LLC PDU
0600	Xerox XNS IDP
0800	Department of Defense IP (IPv4)
0801	X.75 Internet
0802	NBS Internet
0803	ECMA Internet
0804	CHAOSnet
0805	X.25 Level 3
0806	Address Resolution Protocol (ARP; for IP and CHAOSnet)
6001	DEC MOP Dump/Load Assistance
6002	DEC MOP Remote Console
6003	DEC DECnet Phase IV
6004	DEC LAT
6005	DEC DECnet Diagnostics
6010–6014	3COM Corporation
7000–7002	Ungermann-Bass download
7030	Proteon
7034	Cabletron
8035	Reverse ARP (RARP; for IP and CHAOSnet)
8046–8047	American Telephone and Telegraph Company (AT&T)
8088–808A	Xyplex
809B	Kinetics Ethertalk—Appletalk over Ethernet
80C0–80C3	Digital Communications Associates
80D5	IBM SNA Services over Ethernet
80F2	Retix
80F3–80F5	Kinetics
80F7	Apollo Computer
80FF–8103	Wellfleet Communications
8137–8138	Novell
8600	IPv6
8808	MAC control

cent innovations have led to the development of new LAN topologies with higher transmission rates, 10-Mbps LANs are still quite popular and include 10Base-5, 10Base-2, 10Base-T, 10Broad-36, and 10Base-F.

18-8-5-1 10Base-5 Ethernet. Figure 18-18 shows the physical layout for a 10Base-5 Ethernet system. The maximum number of cable segments supported with 10Base-5 Ethernet is five, interconnected with four repeaters or hubs. However, only three of the segments can be populated with nodes (computers). This is called the *5-4-3 rule:* five segments joined by four repeaters, but only three segments can be populated. The maximum segment length for 10Base-5 is 500 meters. Imposing maximum segment lengths are required for the CSMA/CD to operate properly. The limitations take into account Ethernet frame size, velocity of propagation on a given transmission medium, and repeater delay time to ensure that collisions that occur on the network are detected.

On 10Base-5 Ethernet, the maximum segment length is 500 meters with a maximum of five segments. Therefore, the maximum distance between any two nodes (computers) is $5 \times 500 = 2500$ meters. The worst-case scenario for collision detection is when the station at one end of the network completes a transmission at the same instant the station at the far end of the network begins a transmission. In this case, the station that transmitted first would not know that a collision had occurred. To prevent this from happening, a minimum frame length is imposed on Ethernet.

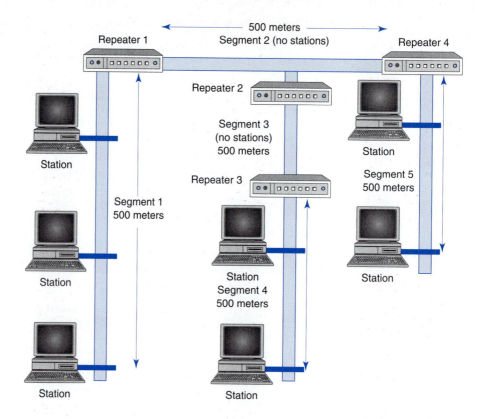

FIGURE 18-18 10 Mbps 5-4-3 Ethernet configuration

The minimum frame length for 10Base-5 is computed as follows. The velocity of propagation along the cable is assumed to be approximately two-thirds the speed of light, or

$$v_p = \frac{2}{3}v_c$$

$$v_p = \left(\frac{2}{3}\right)(3 \times 10^8 \text{ m/s})$$

$$v_p = 2 \times 10^8 \text{ m/s}$$

Thus, the length of a bit along a cable for a bit rate of 10 Mbps is

$$\text{bit length} = \frac{2 \times 10^8 \text{ m/s}}{10 \text{ Mbps}} = 20 \text{ m/bit}$$

and the maximum number of bits on a cable with a maximum length of 2500 meters is

$$\frac{2500 \text{ m}}{20 \text{ m/bit}} = 125 \text{ bits}$$

Therefore, the maximum time for a bit to propagate end to end is

$$\frac{2500 \text{ m}}{2 \times 10^8 \text{ m/s}} = 12.5 \text{ μs}$$

The round-trip delay time is sometimes called the *slot time*. Thus, the slot time is

$$\text{slot time} = 2 \times 12.5 \text{ μs} = 25 \text{ μs}$$

Therefore, the minimum length of an Ethernet message for a 10-Mbps transmission rate is

$$\frac{\text{round-trip delay}}{\text{bit time}} = \frac{25\ \mu s}{0.1\ \mu s}\ 250\ \text{bits}$$

where the time of a bit (t_b = 1/bit rate or 1/10 Mbps = 0.1 μs)

However, the minimum number of bits is doubled and rounded up to 512 bits (64 eight-bit bytes), which also increases the slot time to 512 μs.

The minimum length of an Ethernet frame is 64 bytes, which is based on the minimum transmission time at a 10-Mbps transmission rate. A minimum frame length of 64 bytes ensures that a device involved in the collision not only detects the collision but also knows that it is part of it. The maximum length of an Ethernet frame is 1518 bytes, which includes 18 bytes of header and 1500 bytes of data. An Ethernet frame shorter than 64 bytes is called a *runt,* and an Ethernet frame longer than 1518 bytes is called a *jabber.* Collisions that occur within the first 51.2 μs (512 bits or 64 bytes) of a transmission are called *early collisions,* and collisions that occur after that are called *late collisions.* When a collision is detected, the detecting device sends a special 32-bit *jam signal* to ensure that all the devices on the network are aware of the collision.

10Base-5 is the original Ethernet that specifies a *"thick"* 50-ohm double-shielded RG-11 coaxial cable for the transmission medium. Hence, this version is sometimes called *thicknet* or *thick Ethernet.* Because of its inflexible nature, 10Base-5 is sometimes called *"frozen yellow garden hose."* 10Base-5 Ethernet uses a bus topology with an external device called a *media access unit* (MAU) to connect terminals to the cable. The MAU is sometimes called a *vampire tap* because it connects to the cable simply by puncturing the cable with a sharp prong that extends into the cable until it makes contact with the center conductor. Each connection is called a *tap,* and the cable that connects the MAU to its terminal is called an *attachment unit interface* (AUI) or sometimes simply a *drop.* Within each MAU, a digital transceiver transfers electrical signals between the drop and the coaxial transmission medium. 10Base-5 supports a maximum of 100 nodes per segment. Repeaters are counted as nodes; therefore, the maximum capacity of a 10Base-5 Ethernet is 297 nodes. With 10Base-5, unused taps must be terminated in a 50-ohm resistive load. A drop left unterminated or any break in the cable will cause total LAN failure.

18-8-5-2 10Base-2 Ethernet. 10Base-5 Ethernet uses a 50-ohm RG-11 coaxial cable, which is thick enough to give it high noise immunity, thus making it well suited to laboratory and industrial applications. The RG-11 cable, however, is expensive to install. Consequently, the initial costs of implementing a 10Base-5 Ethernet system are too high for many small businesses. In an effort to reduce the cost, International Computer Ltd, Hewlett-Packard, and 3COM Corporation developed an Ethernet variation that uses thinner, less expensive 50-ohm RG-58 coaxial cable. RG-58 is less expensive to purchase and install than RG-11. In 1985, the IEEE 802.3 Standards Committee adopted a new version of Ethernet and gave it the name 10Base-2, which is sometimes called *cheapernet* or *thinwire* Ethernet.

10Base-2 Ethernet uses a bus topology and allows a maximum of five segments; however, only three can be populated. Each segment has a maximum length of 185 meters with no more than 30 nodes per segment. This limits the capacity of a 10Base-2 network to 96 nodes. 10Base-2 eliminates the MAU, as the digital transceiver is located inside the terminal and a simple BNC-T connector connects the NIC directly to the coaxial cable. This eliminates the expensive cable and the need to tap or drill into it. With 10Base-2 Ethernet, unused taps must be terminated in a 50-ohm resistive load and a drop left unterminated, or any break in the cable will cause total LAN failure.

It is possible to combine 10Base-5 and 10Base-2 segments in the same network by using a repeater that conforms to 10Base-5 on one side and 10Base-2 on the other side. The only restriction is that a 10Base-2 segment should not be used to bridge two 10Base-5 segments because a backbone segment should be as resistant to noise as the segments it connects.

18-8-5-3 10Base-T Ethernet. 10Base-T Ethernet is the most popular 10-Mbps Ethernet commonly used with PC-based LAN environments utilizing a star or bus topology. Because stations can be connected to a network hub through an internal transceiver; there is no need for an AUI. The T indicates unshielded twisted-pair cable. 10Base-T was developed to allow Ethernet to utilize existing voice-grade telephone wiring to carry Ethernet signals. Standard modular RJ-45 telephone jacks and four-pair UTP telephone wire are specified in the standard for interconnecting nodes directly to the LAN without an external AUI. The RJ-45 connector plugs directly into the NIC in the PC. 10Base-T operates at a transmission rate of 10 Mbps and uses CSMA/CD; however, it uses a multiport hub at the center of network to interconnect devices. This essentially converts each segment to a point-to-point connection into the LAN. The maximum segment length is 100 meters with no more than two nodes on each segment.

Nodes are added to the network through a port on the hub. When a node is turned on, its transceiver sends a direct current (dc) over the twisted-pair cable to the hub. The hub senses the current and enables the port, thus connecting the node to the network. The port remains connected as long as the node continues to supply dc to the hub. If the node is turned off or if an open or short circuit condition occurs in the cable between the node and the hub, dc stops flowing, and the hub disconnects the port from the network, allowing the remainder of the LAN to continue operating status quo. With 10Base-T Ethernet, a cable break affects only the nodes on that segment.

18-8-5-4 10Broad-36. 10Broad-36 is the only IEEE 802.3 specification for broadband transmission. 10Broad-36 operates at a 10-Mbps transmission rate over a 14-MHz bandwidth using a bus or tree topology. 10Broad-36 LANs use a 10-mm to 25-mm-diameter 75-ohm CATV cable for a transmission medium. The maximum length of an individual segment emanating from the headend is 1800 meters. 10Broad-36 uses differential PSK signaling where a change in phase between successive signaling elements indicates a logic 0 and no phase transition indicates a logic 1.

18-8-5-5 10Base-F Ethernet. With 10Base-F Ethernets, the F stands for fiber link, which is the transmission medium for all Base-F Ethernets. 10Base-F is an optical fiber medium specification that contains three subspecifications: 10Base-FP, 10Base-FL, and 10Base-FB.

10Base-FP. A passive star topology for interconnecting as many as 33 stations and repeaters to a central passive hub. Stations can be up to 1 kilometer from the hub. 10Base-FP uses asynchronous transmission.

10Base-FL. The most common 10-Mbps Ethernet that uses optical fiber for the transmission medium. 10Base-FL is arranged in a star topology where stations are connected directly to the network with a point-to-point link through an external AUI cable and an external transceiver called a *fiber-optic MAU.* Each transceiver is connected to the hub with two pairs of optical fiber cable, allowing for full-duplex operation. The cable specified is graded-index multimode cable with a 62.5-μm-diameter core with a maximum distance of 2 kilometers between a station and the central hub. 10Base-FL uses synchronous data transmission.

10Base-FB. A point-to-point link used as a backbone to connect up to 15 tandem repeaters with up to 2 kilometers between them. Each transceiver is connected to the hub with two pairs of optical fiber cable, allowing for full-duplex operation.

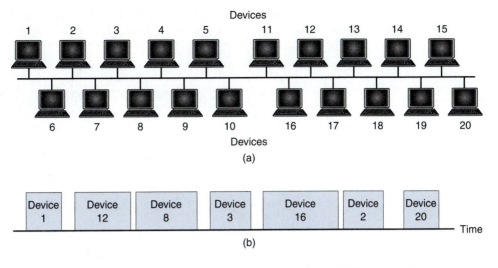

Devices

FIGURE 18-19   Ethernet LAN: (a) bus topology without bridges; (b) TDM on a LAN with a bus topology

18-8-6   Bridged Ethernet

Recent improvements in twisted-pair cable technology and the introduction of optical fibers to the LAN environment has facilitated increasing LAN transmission rates in excess of 2 Gbps (2 billion bits per second). In addition, recent improvements in four-wire bridging and switching technologies have led to the development of several innovative LAN topologies, such as *bridged Ethernet.*

As described in Chapter 17, bridges are two-port layer 2 connectivity devices. Therefore, bridges can be used to divide a network into segments where each segment is a separate collision domain. Since bridges forward frames on the basis of hardware addresses, they isolate transmissions to only those segments involved in the transmission (i.e., the segment or segments where the source and destination devices reside).

Figure 18-19a shows an Ethernet LAN with 20 devices connected to a common bus. Therefore, all 20 devices must share the same transmission medium. This is sometimes called *sharing the bandwidth;* however, in actuality, what is shared is time, not bandwidth. Whenever a device transmits onto the medium, it can use the entire bandwidth of the transmission medium. However, the total network time must be shared between 20 devices, where only one device can transmit at a time. Thus, assuming equal access, each device is theoretically allocated approximately 1/20th of the total network time. Sharing the time on a transmission medium is called *time-division multiplexing* (TDM). Figure 18-19b illustrates the concept of TDM. All 20 devices have an opportunity to transmit on the transmission medium, but only one can transmit at a time. On average, each device has access 1/20th of the time, which equivalently reduces the amount of data they can send to 1/20th of the total network data. This is equivalent to allocating 1/20th of the media bandwidth to each device, or reducing the average usable bandwidth by a factor of 20.

Figure 18-20 shows a network divided into two segments separated with a bridge where each segment is populated with 10 devices. The bridge prevents transmission between two devices on one segment from propagating onto the other segment and also prevents collisions on one segment from propagating onto the other segment. When traffic is not passing through the bridge, network time on each segment is now divided among only 10 devices, which equates to dividing the bandwidth by 10. Therefore, each device effectively has twice as much bandwidth as it would have without the bridge. This, of course,

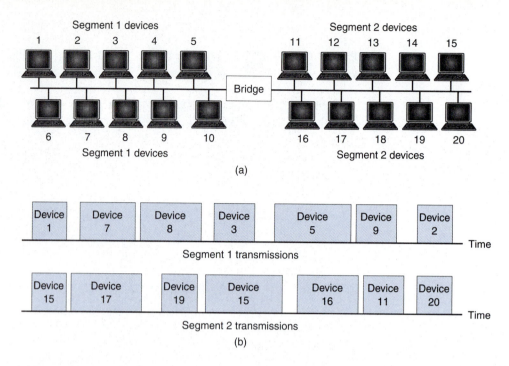

FIGURE 18-20   Ethernet LAN: (a) bus topology segmented with a bridge; (b) TDM on the two segments

assumes that most data exchanges occur between two devices on the same network segment. Figure 18-20b shows how transmissions on each segment are time-division multiplexed where transmission on one segment has no effect on transmission on the other segment, thereby allowing two devices on different segments to send messages at the same time, provided that the destination device is on the same segment. Thus, the bridge effectively increased data throughput on the network by a factor of 2.

18-8-7   Switched Ethernet

As described in the previous section, segmenting LANs with bridges provides networks with higher data throughputs, increased performance, and higher reliabilities. Therefore, it stands to reason that incorporating layer 2 switches into LAN architectures would increase the performance even more.

Isolating a network with N bridges divides the network into N + 1 segments. Therefore, a network with 20 devices using three bridges can facilitate four segments with five devices on each. However, since switches are multiport bridges, switches can be used to isolate individual devices, and the same network can be divided into 20 segments, one for each device. This also divides the network into 20 collision domains, making it virtually impossible for collisions to occur. Collisions are unlikely because the only device transmitting on a segment is the device connected to it, and since transmissions occur in sequence (one after the other), it is impossible for a frame transmitted from one device to collide with another frame transmitted from the same device. Collisions could theoretically occur on a device's receive pair, but only if the switch forwarded two or more frames onto the same segment at the same time, and that should never happen. If a collision should occur, a switch isolates the collision to one segment, making it impossible for devices attached to the other segments to detect it. Therefore, collisions go relatively unnoticed except for the device attached to the segment where the collision occurred. This

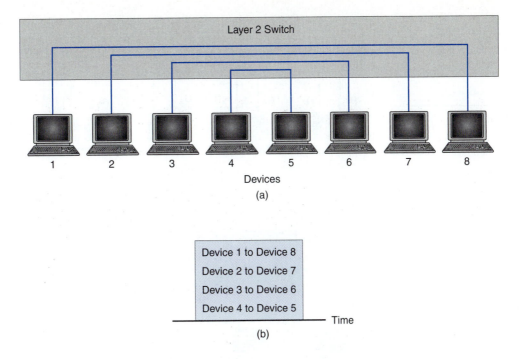

Device 1 to Device 8
Device 2 to Device 7
Device 3 to Device 6
Device 4 to Device 5

Time

(b)

FIGURE 18-21   Ethernet LAN: (a) switched Ethernet; (b) facilitating multiple transmissions

could be a disadvantage, as most likely the device involved in the collision that is not on the segment where the collision occurred would not know that one of its transmitted frames was involved in a collision. A layer 2 switch is a sophisticated N-port bridge, which allows it to isolate transmissions to only two segments and forward frames faster and more efficiently.

Figure 18-21a shows an Ethernet LAN populated with eight devices connected to the same switch. Theoretically, the switch can facilitate four transmissions between pairs of devices at the same time without interfering with each other. The figure shows how simultaneous transmissions could occur between four pairs of devices (1 and 8, 2 and 7, 3 and 6, and 4 and 5).

18-8-8   Full-Duplex Ethernet

Allowing the full-duplex exchange of information frames essentially doubles the capacity of a LAN, you might say equivalently doubling the bandwidth by providing four-wire transmission facilities and four-wire connectivity devices. Four-wire operation isolates a transmission to only one direction on the transmission medium, thus allowing a device to transmit on one medium and receive on a different medium at the same time.

Figure 18-22a shows a four-wire switched Ethernet LAN. Note that each device is attached to the switch with two transmission media, one connected to its transmitter and one connected to its receiver. Figure 18-22b shows device 1 transmitting to and receiving from device 4 at the same time.

With four-wire, full-duplex switched Ethernet, it is virtually impossible for collisions to occur. Therefore, employing a multiple access method, such as CSMA/CD, is unnecessary. Transmissions are separated by direction on segments that are isolated with switches, and the job of controlling access is left to the layer 2 switches, which receive, store, and forward frames when it is appropriate to do so.

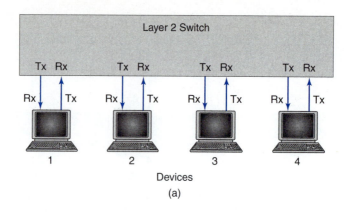

(a)

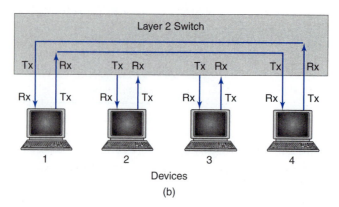

(b)

FIGURE 18-22 Four-wire switched Ethernet LAN: (a) configuration; (b) facilitating full-duplex operation

18-8-9 Full-Duplex Switched Ethernet

Because Ethernet was originally designed as a connectionless protocol, it does not include procedures at the MAC sublayer to control the flow of data or to recover frames received that contain errors. Ethernet uses CRC-32 for error detection, which is an extremely reliable error-detection mechanism. However, there is no means of acknowledging frames received successfully and no procedures to deal with transmission errors when they are detected. These shortcomings reduce the throughput, reliability, and effectiveness of traditional Ethernet LANs.

To eliminate these shortcomings from full-duplex switched Ethernet LANs, a new sublayer has been added between the MAC sublayer and the LLC sublayer to provide for flow and error control. The new sublayer is appropriately called *MAC control*. Thus, full-duplex switched Ethernet LANs utilize a data-link layer subdivided into three sublayers, as shown in Figure 18-23.

MAC control packets are encapsulated in a MAC frame in the same manner as the data packets handed down from upper-layer protocols; however, MAC control packets are transmitted between data packets to provide flow and error control, and they do not carry actual user information. To avoid unnecessary waste, MAC control packets are limited to 46 bytes.

Figure 18-24 shows how a MAC control packet is encapsulated in an Ethernet MAC frame. The format for a MAC frame with an encapsulated MAC control packet is the same as a standard MAC frame with the addition of the MAC control packet somewhere in the data field. However, the information in the fields is slightly different.

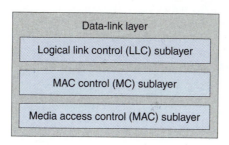

FIGURE 18-23 Full-duplex switched Ethernet data-link layer

Preamble	Start frame delimiter (SFD)	Destination MAC address	Source MAC address	Type/length field	Data field MAC control packet	Frame check field CRC-32
10101010 7 bytes	10101011 1 byte	01-08-C2-00-00-01 6 bytes	SSSSSS 6 bytes	8808 2 bytes	46 bytes	XXXX 4 bytes

FIGURE 18-24 Ethernet II frame carrying a MAC control packet

Destination address field. The recipient of a MAC control frame in a full-duplex switched Ethernet LAN is not the final destination of the frame; it is the switch the device is connected to (i.e., the device at the other end of the link). Therefore, the destination address is a special multicast address, 01-08-C2-00-00-01, accepted only by the switch. A multicast address is used so that all MAC control packets have the same destination address regardless of which device sends them, and there is no need for devices to learn or keep track of special addresses used only for frames carrying MAC control packets. In addition, bridges and switches will not pass multicast addresses, so this isolates each MAC control frame to only one segment of the network, namely, the segment for which it is carrying flow and error control information. Also, the special multicast address is accepted only by devices that have implemented the MAC control sublayer protocol and is ignored by all other devices.

Source address field. The source address field carries the hardware address of the source of the MAC control frame—either a device or a switch.

Type/length field. The length of a MAC control packet is fixed at 46 bytes. Therefore, in all frames carrying a MAC control packet, 8808 hex is placed in the type/length field to designate that the frame is carrying a MAC control packet.

Data field. Since MAC control packets cannot carry user information from an upper-layer protocol, the data field in a MAC control packet contains only information pertinent to MAC control functions.

Frame check sequence field. This field carries the familiar CRC-32 frame check sequence used in all Ethernet frames.

18-8-9-1 MAC control packets. As of this writing, there is only one type of packet defined for MAC control packets: the *pause* packet. Pause packets provide a relatively simple means of implementing flow control called *stop-start.* One device can send a pause packet to another device placing it in the *pause mode,* which instructs the receiving device to temporarily stop sending frames (pause) for a specified period of time called the *pause period.* When a device receives a pause packet, it sets a timer and ceases sending data frames until the pause time expires. If the paused device receives another pause packet with a new pause time, it sets its timer with the new pause period and continues waiting. If a device in the pause mode receives a pause packet with a pause period of 0, it removes

Code field	Parameter field
	Length of pause period in multiples of 512 bits
0001	
2 bytes	44 bytes

FIGURE 18-25   Pause packet format

itself from the pause mode and begins sending frames again. A device in the pause mode can send pause packets to temporarily halt frames traveling in the other direction on a full-duplex link.

The purpose of a pause packet is to slow down the rate at which frames are exchanged between two devices connected to opposite ends of a full-duplex link. For example, if a switch is receiving frames faster than it can forward them, its buffers become overloaded. The switch uses a pause packet to inform the sending device to slow down its delivery of data frames; otherwise, the switch will begin discarding them.

Figure 18-25 shows the format for a pause packet. The packet is 46 bytes long and includes a two-byte code field and a 44-byte parameter field. The code for a pause packet is 0001 hex. Thus far, the only parameter specified is the pause period, which defines how long a sender must wait before resuming transmissions. The parameter field actually contains a number that, when multiplied by the time it takes to transmit 512 bits, yields the actual pause time. The 512 bits equate to 64 bytes, which is the minimum length of an Ethernet frame including the 18-byte MAC header. Obviously, the higher the transmission rate, the shorter the pause period for a given multiplier.

18-8-10   Fast Ethernet

Over the past few years, it has become quite common for bandwidth-starved LANs to upgrade 10Base-T Ethernet LANs to 100Base-T (sometimes called *fast Ethernet*). The 100Base-T Ethernet includes a family of fast Ethernet standards offering 100-Mbps data transmission rates using CSMA/CD access methodology. The 100-Mbps Ethernet installations do not have the same design rules as 10-Mbps Ethernet, which allows several connections between hubs within the same segment (collision domain). Also, 100-Mbps Ethernet does not allow this flexibility. Essentially, the hub must be connected to an internetworking device, such as a switch or a router. This is called the 2-1 rule—two hubs minimum for each switch. The reason for this requirement is for collision detection within a domain. The transmission rate increased by a factor of 10; therefore, frame size, cable propagation, and hub delay are more critical.

18-8-10-1 Fast Ethernet MAC sublayer.   The intent of the IEEE 802 project was to develop a MAC sublayer for LANs that was the same for all Ethernet LANs with transmission rates between 10 Mbps and 100 Mbps. However, on full-duplex switched networks, implementing CSMA/CD is unnecessary other than for ensuring that the LAN is backward compatible with older Ethernet systems. The frame formats for 10-Mbps and 100-Mbps Ethernets are the same, as is the maximum and minimum frame lengths and the method of addressing. About the only difference between 10-Mbps and 100-Mbps LAN implementations is the slot time.

The slot time in bits is the same for 10-Mbps and 100-Mbps Ethernets: 512 bits, which equates to the minimum-length Ethernet frame (64 bytes). However, because 100-Mbps transmission is 10 times that of 10 Mbps, the slot time in seconds for 100 Mbps is 1/10th the time for 10 Mbps. The slot time for 10-Mbps transmission is 512 μs. Therefore, the slot time for 100 Mbps is 51.2 μs.

A shorter slot time reduces the likelihood of collisions occurring by a factor of 10. In addition, reducing the slot time reduces the maximum end-to-end length of a LAN. With 100 Mbps, the maximum length of an Ethernet segment is 250 meters.

18-8-10-2 100-Mbps Ethernet implementations. IEEE standard 802.3u details operation of the 100Base-T network. There are three media-specific physical layer standards for 100Base-TX: 100Base-T, 100Base-T4, and 100Base-FX.

100Base-TX Ethernet. 100Base-TX Ethernet is the most common of the 100-Mbps Ethernet standards and the system with the most technology available. 100Base-TX specifies a 100-Mbps data transmission rate over two pairs of category 5 UTP or STP cables with a maximum segment length of 100 meters and a 200-meter maximum network span. 100Base-TX uses a physical star topology (half duplex) or bus (full duplex) with the same media access method (CSMA/CD) and frame structures as 10Base-T; however, 100Base-TX requires a hub port and a NIC, both of which must be 100Base-TX compliant. 100Base-TX can operate full duplex in certain situations, such as from a switch to a server.

100Base-T4 Ethernet. 100Base-T4 is a physical-layer standard specifying 100-Mbps data rates using two pairs of category 3, 4, or 5 UTP or STP cable. 100Base-T4 was devised to allow installations that do not comply with category 5 UTP cabling specifications. 100Base-T4 will operate using category 3 UTP installation or better; however, there are some significant differences in the signaling. 100Base-T4 uses 8B6T signaling and requires a 100Base-T4 hub and NIC and has a 100-meter-maximum segment length and a 200-meter-maximum network span.

100Base-FX Ethernet. 100Base-FX is a physical-layer standard specifying 100-Mbps data rates over two optical fiber cables using a physical star topology. The logical topology for 100Base-FX can be either a star or a bus. 100Base-FX is generally used to interconnect 100Base-TX LANs to a switch or a router. 100Base-FX uses a full-duplex optical fiber connection with multimode cable that supports a variety of distances, depending on circumstances. 100Base-FX uses 4B5B, NRZI signaling and has a 100-meter-maximum segment length and a 400-meter-maximum network span.

18-8-11   1000-Mbps Ethernet (Gigabit Ethernet)

Although current switching technologies provide data throughput rates up to 2.4 Gbps, 1-gigabit Ethernet is the latest implementation of Ethernet that operates at a transmission rate of one billion bits per second and higher. The IEEE 802.3z Working Group is currently preparing standards for implementing gigabit Ethernet. Early deployments of gigabit Ethernet were used to interconnect 100-Mbps and gigabit Ethernet switches, and gigabit Ethernet is used to provide a *"fat pipe"* for high-density backbone connectivity. Gigabit Ethernet can use one of two approaches to medium access: half-duplex mode using CSMA/CD or full-duplex mode where there is no need for multiple accessing.

Gigabit Ethernet can be generally categorized as either two-wire 1000Base-X or four-wire 1000Base-T. 1000Base-T Ethernet is designed for systems using four twisted pairs of category 5 or higher UTP cable. Two-wire gigabit Ethernet can be either 1000Base-SX for short-wave optical laser over an optical fiber cable, 1000Base-LX for long-wave optical laser over an optical fiber, or 1000Base-CX for balanced 150-ohm short copper cables. The four-wire version of gigabit Ethernet is 1000Base-T. 1000Base-SX and 1000Base-LX use two optical fiber cables where the only difference between them is the wavelength (color) of the light waves propagated through the cable.

Although standards are currently in their development stage for gigabit Ethernet and the name sounds intriguing, it is very much a network of the future. Therefore, further discussion is not warranted at this time.

18-1. What feature distinguishes a *LAN* from other types of networks?

18-2. Describe the advantages of using LANs.

18-3. What is the primary purpose of the *IEEE project*?

18-4. What is an *access control methodology*?

18-5. What is a *medium access method*?

18-6. Explain *medium access control*.

18-7. List and describe the two broad categories of medium access.

18-8. Describe what the word *contention* means with respect to a LAN.

18-9. Define *multiple access*.

18-10. Explain *carrier sense multiple access*.

18-11. Explain *carrier sense multiple access with collision detect*.

18-12. Explain *carrier sense multiple access with collision avoidance*.

18-13. Explain what is meant by the term *controlled access*.

18-14. Describe the basic operation of a *token ring LAN*.

18-15. Describe functions of the *logical link control (LLC) sublayer*.

18-16. What kind of addressing is provided by the LLC?

18-17. List and describe the *LLC service classes*.

18-18. Describe the differences between *connectionless* and *connection-oriented service*.

18-19. Describe the differences between *acknowledged* and *unacknowledged service*.

18-20. What is a *protocol data unit*?

18-21. What is included in a *LLC PDU header*?

18-22. Define *source* and *destination service access points*.

18-23. Describe an *IEEE 802.3 MAC frame format*.

18-24. Describe *MAC addressing*.

18-25. Describe the format used with Ethernet standard notations.

18-26. Describe the differences between an *802.3 frame* and an *Ethernet II frame*.

18-27. List some of the characteristics of the following 10-Mbps Ethernets: 10Base-5, 10Base-2, 10Base-T, 10Base-F, and 10Broad-36.

18-28. Explain the differences between a *bridged Ethernet* and a *bus Ethernet*.

18-29. Explain the differences between a *switched Ethernet* and a *bus Ethernet*.

18-30. Explain the operation of *full-duplex* and *full-duplex switched Ethernet*.

18-31. Describe a *MAC control packet*.

18-32. Explain what is meant by the term *fast Ethernet*.

18-33. Describe the *fast Ethernet MAC sublayer*.

18-34. List some of the characteristics of the following 100-Mbps Ethernets: 100Base-TX, 10Base-T4, and 10Base-FX.

18-35. Define *gigabit Ethernet*.

18-36. With a 10-Mbps Ethernet, an early collision is one that occurs:

    a. in the first 512 μs of frame   **c.** in the first 9.6 μs of frame

    b. in the first 51.2 μs of frame   **d.** within the first 512 bytes of frame

18-37. A 10Base-5 Ethernet specifies:

    a. 5 Mbps   **c.** 10 meters per segment maximum

    b. broadband   **d.** 10 Mbps

18-38. A 10Base-T Ethernet specifies:

    a. fiber   **c.** twisted pair

    b. coax   **d.** barbed wire

18-39. A 100Base-F Ethernet specifies:

    a. 100-meter segments   **c.** twisted pair

    b. 100 Mbps   **d.** broadband

18-40. How long is the source address in an Ethernet frame?

    a. 37 inches   **c.** six bytes

    b. six bits   **d.** four bits

18-41. An Ethernet type field of 0600 or greater specifies:

    a. an embedded 802.2 LLC in the data field

    b. an Ethernet II frame

    c. 600 bytes in the data field

    d. cannot occur

18-42. What is the minimum length of the data field with Ethernet?

    a. 1500 bytes   **c.** 46 bytes

    b. 32 bytes   **d.** none of these

18-43. An Ethernet type field of 0800 specifies:

    a. 0800 bytes in the data field   **c.** a Department of Defense IP in the data field

    b. an IEEE 802.3 frame   **d.** a frame carrying LLC

18-44. The 5-4-3 rule specifies:

    a. no more than four segments   **c.** no more than three populated segments

    b. no more than five repeaters   **d.** no less than five segments

18-45. What is the maximum number of nodes on a 10Base-5 Ethernet?

    a. 10   **c.** 100

    b. 30   **d.** 297

18-46. Which type of Ethernet is called "thick Ethernet"?

    a. 10Base-2   **c.** 10Base-5

    b. 10Broad-36   **d.** 100Base-F

18-47. Which type of Ethernet is called "cheapernet"?

    a. 10Base-2   **c.** 10Base-5

    b. 10Broad-36   **d.** 100Base-F

18-48. Which type of Ethernet uses RJ-45 connectors?

    a. 10Base-T   **c.** 10Base-5

    b. 10Broad-36   **d.** 100Base-F

18-49. What type of encoding is used with Ethernet?

    a. B8ZS   **c.** Manchester

    b. BPRZ-AMI   **d.** TTL

PROBLEMS

18-1. Determine the LSAPs for the following protocols:

    a. ARPNET Dod TCP/IP protocol

    b. IEEE 802.1 Bridge Spanning Tree protocol

    c. Baynon VINES

    d. IBM LAN Management (individual)

18-2. Determine the contents of the length field for MAC frame for the following conditions:

    a. 1000-byte data field

    b. 600-byte data field

    c. data field carrying ARP for IP

    d. data field carrying Dod IP protocol

18-3. An Ethernet address of 07-01-02-03-04-05 is what type of address?

    a. unicast

    b. multicast

    c. broadcast

    d. plaster of Paris cast

18-4. An Ethernet address of 08-07-06-05-44-33 is what type of address?

    a. unicast

    b. multicast

    c. broadcast

    d. plaster of Paris cast

18-5. Which of the following could not be an Ethernet source address?

    a. 8A-7B-6C-DE-10-00

    b. EE-AA-C1-21-45-32

    c. 46-56-21-1A-DE-F4

    d. 8B-32-21-21-4D-34

18-6. Which of the following could not be an Ethernet unicast destination?

    a. 43-7B-6C-DE-10-10

    b. 44-AA-C1-23-45-32

    c. 46-56-21-1A-DE-F4

    d. 48-32-21-21-4D-34

18-7. Which of the following could not be an Ethernet multicast destination?

    a. B7-7B-6C-DE-10-00

    b. 7B-AA-C1-23-45-32

    c. 7C-56-21-1A-DE-F4

    d. 83-32-21-21-4D-34

18-8. Which of the following is a globally unique address?

    a. 13-B7-7B-6C-DE-10

    b. A1-7B-AA-C1-23-45

    c. AB-7C-56-21-1A-DE

    d. 3E-83-32-21-21-4D

C H A P T E R 19

TCP/IP Protocol Suite and Internet Protocol Addressing

CHAPTER OUTLINE

OBJECTIVES

- Briefly describe the history of the Internet
- Explain the relationships among the OSI seven-layer protocol hierarchy and the TCP/IP protocol hierarchy
- Explain the functions of each of the layers in the OSI seven-layer protocol hierarchy
- Describe the process for creating a Request for Comments document
- Explain the five status assignments for Requests for Comments
- Describe the primary purpose of the TCP/IP protocol suite
- Explain the three number systems used to represent IP addresses
- Explain how one can tell the class of an IP address
- Describe the basic concept of classful IP addressing
- Explain how IP addresses identify networks and hosts
- Describe the purpose of reserved IP addresses
- Compare the host, subnet, and network characteristics of class A, B, and C IP addresses
- Explain the purpose of class D IP addresses
- Explain the purpose of class E IP addresses
- Explain how loopback addresses are used
- Describe the three types of broadcast addresses
- Explain address masking and how masks are used to separate the network and host portions of an IP address

19-1 INTRODUCTION

Contrary to popular belief, the Internet has been around for quite some time. However, it was not widely used until relatively recently. The U.S. Department of Defense (DoD) Advanced Research Projects Agency (DARPA) developed TCP/IP in 1969 as a result of a resource-sharing experiment called Advanced Research Projects Agency Network (ARPANET). The original purpose of ARPANET (the original name given the Internet) was to provide a basis for early research into networking. Consequently, the Advanced Research Project Agency (ARPA) established a packet-switching network comprised of computers linked together with point-to-point leased communications lines.

The original conventions developed by ARPA to specify how individual computers communicated with each other across that network evolved into the Transmission Control Protocol/Internetwork Protocol (TCP/IP) protocol suites. The TCP/IP is a set of protocols (protocol suite) that defines how all transmissions are exchanged across the Internet. Although the TCP/IP protocol suite is named after its two most prominent protocols—Transmission Control Protocol (TCP) and Internet Protocol (IP)—it actually includes many other protocols.

The underlying purpose of TCP/IP was to provide a network of high-speed communications links that operate like a single network connecting many computers of any size and type. Since its introduction, ARPANET has evolved into a worldwide community of networks and internetworks known collectively as the Internet. The word *internet* (lowercase i) refers to a two or more TCP/IP networks interconnected with routers that form a computer internetwork. The word *Internet* (uppercase I) is more encompassing as it refers to the worldwide public Internet. Today, TCP/IP is an industry-standard protocol suite designed to interconnect networks and internetworks that span several wide area network (WAN) links.

19-2 TCP/IP PROTOCOL SUITE

The *TCP/IP protocol suite* was developed by the DoD before the inception of the seven-layer OSI model. Consequently, the TCP/IP layers do not completely match the OSI layers, which closely parallel IBM's System Network Architecture (SNA). TCP/IP is comprised of several interactive modules that provide specific functionality without necessarily operating independently of one another. The OSI seven-layer model specifies exactly which functions each layer performs, whereas TCP/IP is comprised of several relatively independent protocols that can be combined in many ways, depending on system needs. The term *hierarchical* simply means that the upper-level protocols are supported by one or more lower-level protocols.

Figure 19-1 shows the OSI 7-layer protocol hierarchy. Each layer is identified by name and by layer, and a brief description is given for each of the seven layers. Figure 19-2 shows a comparison of the OSI seven-layer protocol hierarchy and the TCP/IP protocol hierarchy. As can be seen in the figure, two of the TCP/IP protocol layers (the network and transport layers) are identical to the OSI layers. However, the TCP/IP model combines ISO layers 1 and 2 into a single layer and calls it the *network interface layer*. Some TCP/IP advocates refer to the TCP/IP protocol suite as a *three-layer model,* as they do not consider the physical-and data-link layers as part of the IP model. They refer to the lower two layers as the *subnetwork layer.* TCP/IP also combines OSI layers 5, 6, and 7 into a single layer called the *applications* or *process layer.* TCP/IP is concerned with transporting information from a process on one device to the same process on another device. However, TCP/IP is not concerned with how applications integrate layer 5, 6, and 7 functions into the process.

Layer	Name	Function
7	Applications	Provides a user interface: files, printing, messages, databases, and applications services
6	Presentation	Presents data: data encryption/decryption, compression, and translation services
5	Session	Separates applications: dialog control
4	Transport	End-to-end delivery: identifies processes with port numbers
3	Network	Logical (IP) addressing: routing
2	Data link	Hardware (physical or MAC) addressing: provides access to transmission media
1	Physical	Move bits: provides physical topology

FIGURE 19-1 OSI layers and functions

The OSI and TCP/IP protocol models are not the only protocol hierarchies in existence. For example, Figure 19-3 shows Cisco's three-layer protocol hierarchy, which is comprised of an access layer, a distribution layer, and a core layer.

Many Internet authorities consider the OSI model outdated and more of an academic tool than a practical or realistic model of how functions are performed on the Internet. Regardless of what model you prefer to use, the OSI model is generally accepted as the standard to which all other protocol hierarchies are referenced. Therefore, even when talking about the TCP/IP protocol model, it is common to reference network components and functions to the OSI seven-layer model, such as a layer N function or a layer N device. For example, a layer 2 connectivity device, such as a switch, is an OIS layer 2 device, as it operates at the data-link layer, and a layer 3 connectivity device, such as a router, is an OSI layer 3 device, as it operates at the network layer.

Figure 19-4 shows the TCP/IP protocol suite that is comprised of three functional layers (network layer, transport layer, and applications layer) and one sublayer, that provides the functions generally offered by the data-link and physical layers. The model shown is for IPv4, which is the network-layer protocol currently being used (Chapter 24 discusses the new network-layer protocol, which is called IPv6). As shown in Figure 19-4, each layer of the TCP/IP model includes two or more protocols. For example, network layer protocols include IP, ARP, RARP, IGRP, OSPF, and ICMP; transport-layer protocols include TCP and UDP; and applications-layer protocols include process protocols, such as SMTP, FTP, Telnet, HTTP, DNS, TFTP, BootP, DHCP, SNMP, and RIP.

Applications-layer protocols are referred to as being "over" either TCP or UDP, which are referred to as being "over" IP. As shown in Figure 19-4, "over" simply means

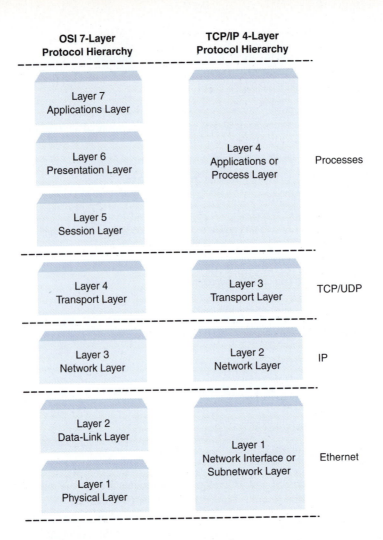

Layer 7
Applications Layer

Layer 6
Presentation Layer

Layer 5
Session Layer

Layer 4
Applications or
Process Layer

Processes

Layer 4
Transport Layer

Layer 3
Transport Layer

TCP/UDP

Layer 3
Network Layer

Layer 2
Network Layer

IP

Layer 2
Data-Link Layer

Layer 1
Physical Layer

Layer 1
Network Interface or
Subnetwork Layer

Ethernet

FIGURE 19-2 Protocol hierarchy comparison

above (a higher layer) in the protocol suite. As will be shown in later chapters, each TCP/IP protocol layer is encapsulated in a lower-layer protocol. For example, applications-layer protocols are encapsulated in a transport-layer protocol (either TCP or UDP), which is then encapsulated in a network-layer protocol (IP). IP is then encapsulated in a data-link protocol, such as the one used by Ethernet. Figure 19-5 illustrates the concept of protocol encapsulation. The figure shows the highest-layer protocol (DNS) encapsulated in the transport-layer protocol (TCP), which is encapsulated in the network-layer protocol (IP), which is encapsulated in the Ethernet data-link protocol. Each protocol has a different name for itself. For example, an Ethernet packet is called a frame, an IP packet is called a datagram, and a TCP packet is called a segment, which can sometimes be confusing. As shown, each layer adds a header to the message, which includes specific information pertinent to the data it is carrying, including identifying the upper-layer protocol it is transporting.

In the network world, it is common to refer to protocols as being routable, nonroutable, or routing protocols. A *routable protocol* is one that includes or is encapsulated in a protocol that contains source and destination IP addresses. For example, the header in an IP protocol includes the source and destination IP addresses; therefore, it is considered a

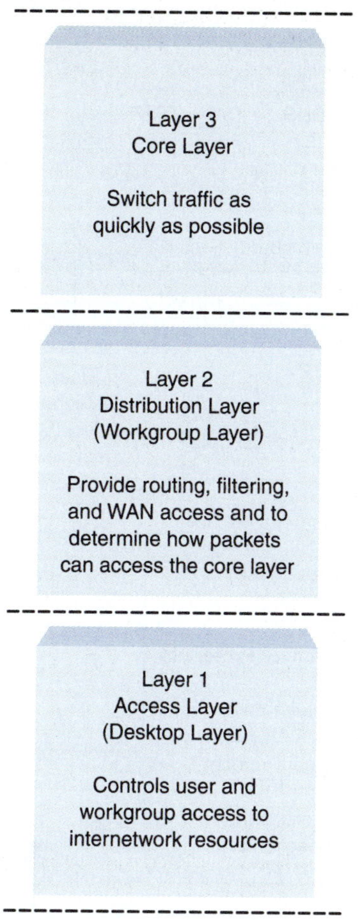

Layer 3
Core Layer

Switch traffic as
quickly as possible

Layer 2
Distribution Layer
(Workgroup Layer)

Provide routing, filtering,
and WAN access and to
determine how packets
can access the core layer

Layer 1
Access Layer
(Desktop Layer)

Controls user and
workgroup access to
internetwork resources

FIGURE 19-3   Cisco three-layer protocol hierarchy

routable protocol. ARP, on the other hand, does not contain IP addresses in its header, and it is not encapsulated in IP datagrams. Therefore, it is considered a *nonroutable protocol*. A *routing protocol* is one that is used by routers (layer 3 devices) to communicate with each other to perform routing functions. An example of a routing protocol is the Routing Information Protocol (RIP).

19-3   REQUEST FOR COMMENTS

The standards for TCP/IP are published as a series of documents called *Requests for Comments* (RFCs), which describe the internal workings of the Internet. Some RFCs describe network services and protocols and their implementations, whereas others simply summarize policies. All TCP/IP standards are published as RFCs; however, not all RFCs specify standards.

RFCs were introduced in 1969 for the purpose of documenting the functions of the Internet and the protocols that support it. The RFC series of documents is a set of technical and organizational annotations that deal directly with the Internet and Internet functions. RFC documents address numerous aspects of computer networking, including all protocol levels, specific procedures, software programs, and networking concepts.

The intent of the RFCs is to share information with the greater Internet community without the rigorous overview that each RFC undergoes. Most RFCs and Internet drafts

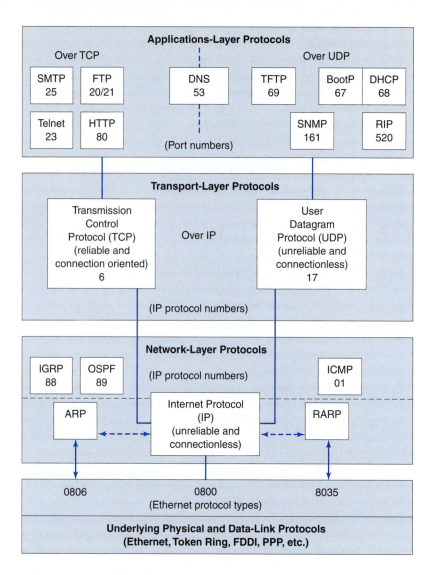

FIGURE 19-4 TCP/IP Protocol Stack

come from the IETF (Internet Engineering Task Force) working groups. As ideas are finalized and standards are developed, the Internet drafts become RFCs and are subject to a standard peer review process. The IETF is a large, open international community of network designers, operators, vendors, and researchers whose primary concern is the evolution of the Internet architecture and the efficient and effortless operation of the Internet. The IETF is open to any interested individual. Although the IETF holds meetings three times each year, most of the actual technical work is done in its working groups, which are organized by topic into several areas, such as routing, transport, and security. Much of the work of the IETF working groups is handled via mailing lists.

TCP/IP standards are developed by consensus, not by committees. Virtually anyone can submit a document for potential publication as an RFC. The submissions are reviewed by a task force comprised of several technical experts to determine their validity and pertinence relating to the underlying goals and purposes of the Internet. Once the document is determined to be worthy of consideration, it is assigned a status that specifies whether the document is being considered as a standard.

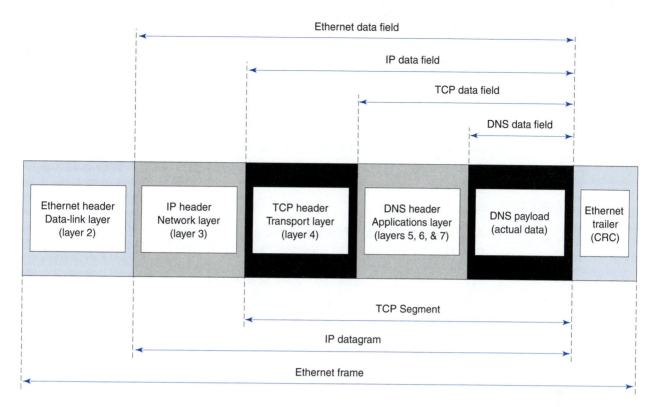

FIGURE 19-5 TCP/IP protocol encapsulation

There are five status assignments for RFC: required, recommended, elective, limited use, and not recommended.

Required. A required RFC must be implemented on all TCP/IP-based hosts and gateways.

Recommended. A recommended RFC is encouraged to be used on all TCP/IP hosts and gateways.

Elective. An elective RFC is optional, as its application has been agreed on but is not required.

Limited use. A limited-use RFC is not intended for general use.

Not recommended. A not-recommended RFC is simply not recommended for implementation.

When a document is being considered for a standard, it undergoes several stages of development, testing, and acceptance, called the *Internet standards process*. Each stage is formally labeled one of three maturity levels:

Proposed standard. A proposed standard is the lowest maturity level, as it is a proposal that seems to be relatively stable and believed to be well understood. It has also received significant review by the Internet community and appears to have sufficient community interest to be considered valuable.

Draft standard. A draft standard is the second level of maturity. Draft standards must be well understood and generally considered very stable concerning semantics and as a basis for developing a usable implementation.

Internet standard. The Internet standard specification is the highest maturity level, as it is characterized by a high degree of technical maturity and by a generally held belief that the specified protocol or service provides a significant benefit to the Internet community.

After passing a rigorous review process, a document is published and assigned an RFC number. The original RFC is never updated. If changes are necessary, a separate RFC is published with a different RFC number. Consequently, care must be taken to ensure that one has the most recent RFC for a particular topic. When an RFC is updated, it is labeled as "obsoleted by RFC XYZ," and the updated RFC is labeled "obsoletes RFC ABC."

19-4   IP ADDRESSING

A primary responsibility of the network layer of the TCP/IP protocol suite is to provide a method of establishing global communications between any two devices connected to the Internet. This is accomplished with an Internet addressing scheme that provides every device connected to the Internet with a unique identification code, which is analogous to assigning every telephone in the world a unique telephone number. The identification code used in the network layer of the TCP/IP protocol suite is called the *IP address* or the *Internet Protocol address*. IP addresses are unique; therefore, there is no duplication of IP addresses, as no two devices connected to the Internet can have the same IP address. However, a single device may have more than one IP address. IP addresses are also universal in the sense that IP addressing must be accepted and utilized by any device or network that wishes to be connected to the Internet.

TCP/IP currently uses IP version 4 addressing (IPv4), which is comprised of four groups of eight bits for a combined IP address length of 32 bits (four bytes or octets). With a 32-bit IP address, there are 2^{32}, or 4,294,967,296, addresses possible. In the early days of the Internet, there was an abundance of IP addresses. However, because of the popularity of the Internet, over the past few years IP addresses have been rapidly consumed, which has led to the development of the next generation IP addressing scheme called IP version 6 (IPv6). IPv6 is sometimes called IP *next generation* (IPng). IPng is currently in the design stage and not expected to be utilized worldwide for several years.

With IPv4, there are over four billion possible IP addresses, which might seem a sufficient number to accommodate all the users of the Internet. However, a variety of restrictions are imposed that limit the number of usable IP addresses to significantly less than four billion.

19-5   IP ADDRESS NOTATION

There are three methods commonly used to designate IP addresses: *binary* notation, *dotted-decimal* notation, and *hexadecimal* notation. With binary notation, an IP address is represented with a four-byte (32-bit) word, which is comprised of four eight-bit binary numbers separated with one or more spaces as shown here:

	First byte	Second byte	Third byte	Fourth byte
Binary	01001000	10001100	01000000	00000001

Because dealing with binary numbers is cumbersome, IP addresses are often represented in hexadecimal. A 32-bit binary word contains four groups of eight bits, and each eight-bit grouping can be represented with a two-character hexadecimal number separated with one or more space. Hexadecimal notation is also rather awkward to deal with and has limited usefulness. Therefore, it is generally used only in network programming environments. The 32-bit IP address shown previously can be represented in hexadecimal notation as shown here:

	First byte	Second byte	Third byte	Fourth byte
Binary	0100 1000	1000 1100	0100 0000	0000 0001
Hexadecimal	4 8	8 C	4 0	0 1

To make IP addresses more meaningful to humans and easier to read, they are generally written in what is called dotted-decimal notation. With dotted-decimal notation, each eight-bit binary grouping is replaced by its decimal equivalent value with a decimal point (dot) separating each number in the form N.N.N.N, where N is any decimal value between 0 and 255. Because each dotted-decimal number equates to eight bits, the range of values for each number is 0 to 255 (00000000 to 11111111 binary or 00 to FF hex). The dotted-decimal notation for the binary and hexadecimal notations shown previously is shown here:

	First byte	Second byte	Third byte	Fourth byte
Binary	0100 1000	1000 1100	0100 0000	0000 0001
Hexadecimal	4 8	8 C	4 0	0 1
Dotted-decimal	72 •	140 •	64 •	1

19-6 CLASSFUL IP ADDRESSING

When IP addressing began in the 1970s, the concept of classes was used to designate the different types of addresses. This addressing structure is called *classful addressing*. With classful addressing, the full range of 2^{32} IP addresses is divided into five classes: A, B, C, D, and E. Figure 19-6 shows the percentage of IP addresses allocated to each class.

From the figure, it can be seen that half (50%) of the total possible IP addresses are class A. One-quarter (25%) of the addresses are class B, one-eighth (12.5%) of the addresses are class C, and class D and class E make up one-sixteenth (6.25%) of the addresses.

Each IP address class can be identified by its binary, hexadecimal, or dotted-decimal notation. When the address is given in binary, the beginning bits are used to differentiate the classes as shown in Table 19-1. From the table, it can be seen that the IP address classes can be identified from the following criteria:

1. All IP addresses that begin with a logic 0 in the leftmost bit of the first byte are class A addresses.
2. All IP addresses that begin with a 10 in the first two bits of the leftmost byte are class B addresses.
3. All IP addresses that begin with a 110 in the first three bits of the leftmost byte are class C addresses.

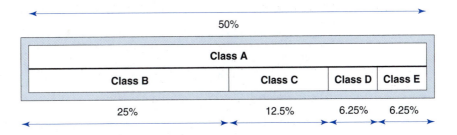

FIGURE 19-6   Percentage of IP addresses allocated each class

Table 19-1   Binary Representation of IP Addresses

Class	First Byte	Second Byte	Third Byte	Fourth Byte
A	*0*NNNNNNN	HHHHHHHH	HHHHHHHH	HHHHHHHH
B	*10*NNNNNN	NNNNNNNN	HHHHHHHH	HHHHHHHH
C	*110*NNNNN	NNNNNNNN	NNNNNNNN	HHHHHHHH
D	*1110*MMMM	MMMMMMMM	MMMMMMMM	MMMMMMMM
E	*1111*XXXX	XXXXXXXX	XXXXXXXX	XXXXXXXX
where	N designates bits associated with the network portion of the IP address (netids)			
	H designates bits associated with the host portion of the IP address (hostids)			
	M designates bits associated with a multicast address			
	X designates bits associated with addresses reserved for Internet research			

Note. Netids bits are used to designate the network address; hostids bits are used to designate the host address. Netids and hostids are described in a later section of this chapter.

4. All IP addresses that begin with a 1110 in the first four bits of the leftmost byte are class D addresses.
5. All IP addresses that begin with a 1111 in the first four bits of the leftmost byte are class E addresses.

The range of dotted-decimal numbers assigned to each IP address class can be found by converting each of the four binary numbers to their equivalent decimal value. The minimum value assigned to a particular class is determined by making all bits designated N, H, M, or X logic 0s, and the maximum number is determined by making all bits designated N, H, M, or X logic 1s. The corresponding dotted-decimal number is simply the decimal equivalent of each of the four eight-bit binary numbers.

For example, the minimum value for a class A address is

Binary	**0**0000000	00000000	00000000	00000000
Dotted-decimal	0 •	0 •	0 •	0

and the maximum value for a class A address is

Binary	**0**1111111	11111111	11111111	11111111
Dotted-decimal	127 •	255 •	255 •	255

The minimum and maximum values for class B addresses are

Minimum:

Binary	**10**000000	00000000	00000000	00000000
Dotted-decimal	128 •	0 •	0 •	0

Maximum:

Binary	**10**111111	11111111	11111111	11111111
Dotted-decimal	191 •	255 •	255 •	255

The minimum and maximum values for class C addresses are

Minimum:

Binary	**110**00000	00000000	00000000	00000000
Dotted-decimal	192 •	0 •	0 •	0

Maximum:

Binary	**110**11111	11111111	11111111	11111111
Dotted-decimal	223 •	255 •	255 •	255

Table 19-2 Range of Dotted-Decimal IP Addresses

Class	IP Address Range
A	0.0.0.0 to 127.255.255.255
B	128.0.0.0 to 191.255.255.255
C	192.0.0.0 to 223.255.255.255
D	224.0.0.0 to 239.255.255.255
E	240.0.0.0 to 255.255.255.255

The minimum and maximum values for class D addresses are

Minimum:

Binary	**1110**0000	00000000	00000000	00000000
Dotted-decimal	224 •	0 •	0 •	0

Maximum:

Binary	**1110**1111	11111111	11111111	11111111
Dotted-decimal	239 •	255 •	255 •	255

The minimum and maximum values for class E addresses are

Minimum:

Binary	**1111**0000	00000000	00000000	00000000
Dotted-decimal	240 •	0 •	0 •	0

Maximum:

Binary	**1111**1111	11111111	11111111	11111111
Dotted-decimal	255 •	255 •	255 •	255

Table 19-2 summarizes the ranges for the dotted-decimal IP addresses for the five address classifications.

19-7 NETWORK AND HOST IDENTIFICATION

With classful IP addressing, the IP address for class A, B, and C addresses is divided into two identification numbers: a *network identification* (netid) and a *host identification* (hostid). The netid identifies the network, and the hostid identifies the specific host (PC or other device) on the designated network. The length of the netid and hostid vary, depending on the address class. With class A addresses, the first eight bits represent the netid, and the remaining 24 bits make up the hostid. With class B addresses, the first 16 bits comprise the netid, and the remaining 16 bits make up the hostid. With class C addresses, the first 24 bits identify the netid, and the remaining eight bits identify the hostid. The network and host identification schemes for class A, B, and C addresses are summarized in Table 19-3.

With class A addresses, there are eight bits reserved for the network ID, and the first bit must be a logic 0. This leaves only seven bits for the netid; therefore, the theoretical maximum number of class A network addresses available is 2^7, or 128. With class A addresses, there are 24 bits reserved for the host ID; therefore, there are 2^{24}, or 16,777,216, host addresses mathematically possible for each class A network address. With class B addresses, there are 16 bits for the network ID, and the first two bits must be 10; therefore, the theoretical maximum number of class B network addresses is 2^{14}, or 16,384. With 16 hostid bits, there are 2^{16}, or 65,536, host addresses mathematically possible for each class B address. With class C addresses, there are 24 bits reserved for the network ID, and the first

Table 19-3   Network and Host Identification Schemes for Class A, B, and C Addresses

Class	First Byte	Second Byte	Third Byte	Fourth Byte
A	netid	hostid	hostid	hostid
B	netid	netid	hostid	hostid
C	netid	netid	netid	hostid

Table 19-4   Summary of Theoretically Possible Network and Host Addresses

Network Class	Number of Networks	Number of Hosts per Network	Total Number of Host Addresses	Percentage of Networks	Percentage of Total Hosts
A	128	16,777,216	2,147,483,648	0.006%	57.14%
B	16,384	65,536	1,073,741,824	0.78%	28.57%
C	2,097,152	256	536,870,912	99.2%	14.29%
Totals	2,113,664		3,758,096,384	100%	100%

three bits must be 110; therefore, the theoretical maximum number of class C network addresses is 2^{21}, or 2,097,152. With class C addresses, there are eight bits reserved for the hostid; therefore, the maximum number of host addresses mathematically possible for each class C address is 2^8, or 256.

From the preceding paragraph, it can be seen that the vast majority of the theoretically possible network addresses are class C (99.2%). However, the majority of host addresses theoretically possible are class A (57.14%). Table 19-4 lists the total theoretical number of network and host addresses for class A, B, and C networks. The table also lists the percentage of the total network and host addresses for class A, B, and C networks. Keep in mind that the table lists the total number of network and host addresses mathematically possible. Because there are restrictions that impose limitations on the number of network and host addresses, the actual number of usable network and host addresses is slightly less than the number shown in the table.

19-8   RESERVED IP ADDRESSES

With classful IP addressing, there are certain addresses and ranges of addresses that are restricted for specific uses and, therefore, are not allowed to be used on the Internet. For example, Class D addresses are reserved for multicast addresses, and class E addresses are reserved for research affiliations. The address 127.0.0.1 is reserved for a loopback address and is used only for self-testing the functionality of the TCP/IP protocol stack on a host machine. In addition to the loopback address, there are several network addresses (1 class A, 16 class B, and 256 class C) that are reserved for private network use only and cannot be used on the Internet. The private network addresses are listed in Table 19-5.

Table 19-5   Private IP Addresses

Class	Network Addresses
A (1)	10.0.0.0 to 10.0.0.255
B (16)	172.16.0.0 to 172.31.0.0
C (256)	192.168.0.0 to 192.168.255.0

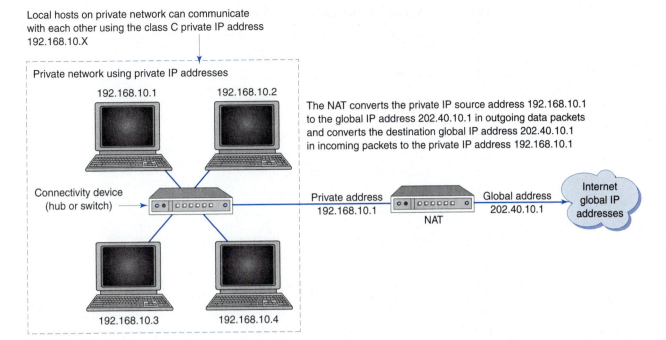

Local hosts on private network can communicate with each other using the class C private IP address 192.168.10.X

Private network using private IP addresses

192.168.10.1   192.168.10.2

The NAT converts the private IP source address 192.168.10.1 to the global IP address 202.40.10.1 in outgoing data packets and converts the destination global IP address 202.40.10.1 in incoming packets to the private IP address 192.168.10.1

Connectivity device (hub or switch)

Private address 192.168.10.1

Global address 202.40.10.1

Internet global IP addresses

NAT

192.168.10.3   192.168.10.4

FIGURE 19-7   Private-to-global IP address translation

Private IP addresses are not globally recognized and should never be used on the Internet. A global IP address is simply an IP address that is allowed to be used on the global Internet. Private IP addresses are intended for internal use in isolated networks, such as local area networks (LANs), without Internet access. However, private addresses can be used in networks with Internet access, provided that they undergo an address translation before reaching the Internet. With address translation, private IP source addresses are removed from outgoing data packets by a device called a *network address translator* (NAT) before the packets leave the network and are replaced with a global Internet IP address. The NAT is generally a router that isolates transmissions on the private network from the Internet. Conversely, destination global addresses in data packets received from the Internet are converted to private addresses before being placed on the local network. The NAT that performs the address translation is sometimes called a *proxy server.* Private-to-global address translation is illustrated in Figure 19-7.

The four hosts shown on the private network in Figure 19-7 use private IP addresses to communicate with each other over the local network using the TCP/IP protocol suite. However, if host 192.168.10.1 wishes to go out onto the Internet, the network address translator must convert the private IP address to a global address (202.40.10.1 in this example).

Private IP addresses allow a multitude of LANs to utilize the same IP addresses without interfering with each other's transmissions. Hosts on many different local networks can reuse the same private IP addresses, which significantly reduces the number of global IP addresses needed. Therefore, private IP addresses conserve the already depleted bank of global IP addresses.

19-9   IP ADDRESS SUMMARY

Figure 19-8 summarizes the IP address ranges for the network and host ID addresses for the five classes of IP addresses.

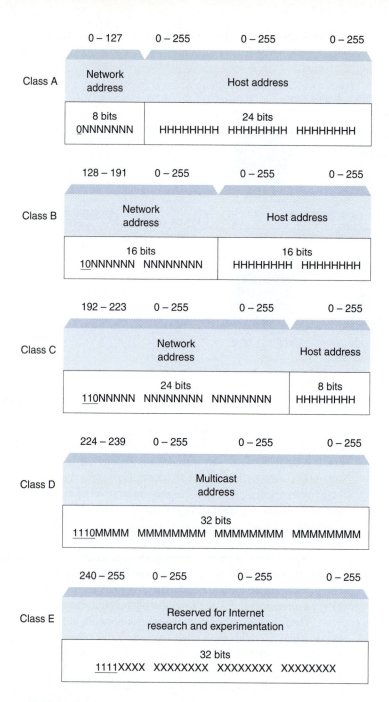

FIGURE 19-8 IP Address classifications

19-10 IP ADDRESS CLASSES

A major problem with classful addressing is that each address class is subdivided into a set number of address groups where each group has a fixed number of host identification numbers.

19-10-1 Class A Addresses

Class A addresses have eight netid bits; however, only seven can be used for addressing because the first bit must be a logic 0. Therefore, there are 128 class A address groups (2^7), and each group has a different network identification number. The first group has the netid 0 and includes addresses from 0.0.0.0. to 0.255.255.255. The second group has the netid 1 and includes addresses from 1.0.0.0 to 1.255.255.255. The last group has the netid 127 and includes addresses from 127.0.0.0 to 127.255.255.255. Each group of addresses has the same netid; however, the three hostid bytes can be any value between 0 and 255.

With classful IP addressing, the first class A address group (netid 0) and the last class A address group (netid 127) are reserved for specific purposes and, therefore, cannot be used for an Internet address. In addition, one group (netid 10) is reserved for private addresses. That leaves 125 class A network addresses that can actually be assigned and used on the Internet. However, each of the 125 class A netids can theoretically have as many as 16,777,216 hostids, which makes class A addresses attractive to large companies and organizations with a large number of hosts. However, it is highly unlikely that any organization would need 16,777,216 host addresses, which means that many of the class A addresses are wasted.

The first octet in a class A address group identifies the organization to the Internet. This address is called the *network address,* as it defines the network but not the individual hosts within the network. For example, the network address 62.0.0.0 identifies a class A network with the netid 62 but does not specify a host address, as a host cannot have the hostid 0.0.0. The address 62.0.0.5 identifies host 5 on class A network 62. Originally, the first and last addresses within each address group were reserved for special purposes, which meant that there are actually only 16,777,214 hostids available for each class A netid. However, in December 1995, the all-0s and all-1s host addresses were approved for use. This is appropriate, however, only if all the routers in the network use a contemporary routing protocol, such as version 2 of the Routing Internet Protocol (RIPv2) or Open Shortest Path First (OSPF). This may be a premature assumption, so if you want to err on the side of caution, these hostids should be avoided unless first checking your router documentation and consulting with your *Internet Service Provider* (ISP).

19-10-2   Class B Addresses

Class B addresses have 16 netid bits; however, only 14 can be used because the first two bits must be 10. Therefore, there are 16,384 class B address groups (2^{14}). However, 16 address groups (netids) are reserved for private addresses (172.16 to 172.31), leaving 16,368 netids available for Internet assignment. The first class B address group has the netid 128.0 and includes addresses from 128.0.0.0 to 128.0.255.255. The last class B address group has the netid 191.255 and includes addresses from 191.255.0.0 to 191.255.255.255. Each group of class B addresses has the same netid; however, the two hostid bytes can be any value between 0 and 255.

Because there are 16,368 class B address groups (netids), the total number of organizations that can acquire class B network addresses is 16,368. Each class B network address can theoretically have as many as 65,536 hostids, which makes class B addresses ideally suited to midsize companies and organizations.

The first two bytes in a class B address group identify the organization to the Internet. This address is called the network address, as it defines the network but not the hosts within the network. For example, the network address 170.8.0.0 identifies a class B network with the netid 170.8 but does not specify a host address, as a host cannot have the hostid 0.0. The address 170.8.0.10 identifies host 10 on class B network 170.8. Because the first and last addresses within each address group are reserved for special purposes, there are actually only 65,534 hostids available for each class B netid.

19-10-3   Class C Addresses

Class C addresses have 24 netid bits; however, only 21 can be used because the first three bits must be 110. Therefore, there are 2,097,152 class C addresses groups (2^{21}). However,

Table 19-6 Summary of Usable Network and Host Addresses

Network Class	Number of Networks	Number of Hosts per Network	Total Number of Host Addresses	Percentage of Networks	Percentage of Total Hosts
A	125	16,777,214	2,097,151,750	0.006%	56.64%
B	16,368	65,534	1,072,660,512	0.774%	28.97%
C	2,096,896	254	536,611,584	99.2%	14.39%
Totals	2,113,389		3,702,423,846	100%	100%

256 address groups (netids) are reserved for private addresses (192.168.0 to 192.168.255), leaving 2,096,896 netids available for assignment. The first class C address group has the netid 192.0.0 and includes addresses from 192.0.0.0 to 192.0.0.255, and the last class C address group has the netid 223.255.255 and includes addresses 223.255.255.0 to 223.255.255.255. Again, each group of addresses has the same netid, but the hostid byte can have any value between 0 and 255.

Because there are 2,096,896 class C address groups (netids), the total number of organizations that can acquire class C network addresses is 2,096,896. Each class C network address can theoretically have as many as 256 hostids, which makes class C addresses ideally suited for small companies and organizations with a small number of hosts.

The first three bytes of a class C address group identify the organization to the Internet. This address is called the network address, as it defines the network but not the hosts within the network. For example, the network address 210.10.10 identifies a class C network with the netid 210.10.10 but does not specify a host address, as a host cannot have the hostid 0. The address 210.10.10.8 identifies host 8 on class C network 210.10.10. Because the first and last addresses within each address group are reserved for special purposes, there are actually only 254 hostids available for each class C netid.

19-10-4   Class A, B, and C Address Summary
Table 19-6 summarizes the number of usable netids and hostids for class A, B, and C addresses.

19-10-5   Class D Addresses
Class A, B, and C addresses are *unicast,* which means that they originate from a single source and are transmitted to a single destination—a one-to-one transmission. Unicast addresses can be either a source address or a destination address, depending on which direction they are propagating. Class D addresses, however, are used for *multicasting* and can be only a destination address. Multicast communications is a one-to-many transmission. Multicasting is when a data packet is simultaneously sent from a single source to more than one destination, which allows multiple hosts to share a single destination IP address and receive the same message across a network with a single transmission. Multicast transmissions conserve network time, especially for high-volume data, such as video and teleconferencing applications. Multicasting is also used to communicate information that is important to more than one host, such as network routing information.

Class D addresses include only one address group, and each class D address defines a *group identification* address (groupid). A system connected to the Internet can have one or more class D multicast addresses in addition to one or more unicast addresses. Each multicast address identifies a group. If a system has 10 multicast addresses, it belongs to 10 groups.

Multicasting over the Internet can be performed at a local level (across a LAN) or at a global level (across the entire Internet). In a local environment, a multicast address identifies a group of hosts on a LAN that share a single destination multicast address. In a global environment, hosts on different LANs can form a group and be assigned a single multicast address.

Table 19-7 Category Multicast Addresses

Address	Assigned Group
224.0.0.0	Reserved
224.0.0.1	All systems on this subnet
224.0.0.2	All routers on this subnet
224.0.0.4	DVMRP routers
224.0.0.5	OSPFIGP routers
224.0.0.7	ST routers
224.0.0.8	ST hosts
224.0.0.9	RIP2 routers
224.0.0.10	IGRP routers
224.0.0.11	Mobile agents

Table 19-8 Conferencing and Teleconferencing Multicast Addresses

Address	Assigned Group
224.0.1.7	Audio news
224.0.1.10	IETF-1—low audio
224.0.1.11	IETF-1—audio
224.0.1.12	IETF-1—video
224.0.1.13	IETF-2—low audio
224.0.1.14	IETF-2—audio
224.0.1.15	IETF-2—video
224.0.1.16	Music service
224.0.1.17	SEANET telemetry
224.0.1.18	SEANET image

Some of the available multicast addresses have been assigned by the Internet authorities to specific categories of groups and, therefore, are not available to other users. The category multicast addresses all begin with the prefix 224.0.0, as shown in Table 19-7.

Some class D multicast addresses have been assigned by the Internet authorities for conferencing and teleconferencing. The conferencing and teleconferencing multicast addresses all begin with the prefix 224.0.1, as shown in Table 19-8.

When a host elects to employ multicast addressing, it must register itself on the network it is operating on, which gives it permission to "listen" for that address along with its unique IP address. The host must also notify the network gateway that it is registered for multicasting so that the gateway will forward multicast data packets onto that network. Otherwise, the gateway will block multicasts. A gateway is a router that provides the local network access to other networks, such as the Internet.

19-10-6   Class E Addresses

There is only one group of class E IP addresses that are designed for IP-related research and experimentation. Therefore, unless you work in a research and development environment, it is not likely that you will even encounter a Class E IP address.

Example 19-1

Determine the address class for the following binary, hexadecimal, and dotted-decimal IP addresses:

a. Binary: 10101000 00101011 00110010 10101010
b. Hexadecimal: E4 8A 26 71
c. Dotted-decimal: 140.24.10.1

Solution

a. Determining the address class for a binary number requires examining only the beginning bits of the leftmost octet:

If the first bit is a 0, the IP address is class A.
If the first two bits are 10, the IP address is class B.
If the first three bits are 110, the IP address is class C.
If the first four bits are 1110, the IP address is class D.
If the first four bits are 1111, the IP address is class E.

For the binary number *10*101000 00101011 00110010 10101010, the first two bits are 10; therefore, it must be a class B address:

b. Determining the address class for a hexadecimal number requires looking at only the first hex character of the leftmost octet.

If the first hex character is between 0 and 7, the IP address is class A.
If the first hex character is between 8 and B, the IP address is class B.
If the first hex character is C or D, the IP address is class C.
If the first hex character is E, the IP address is class D.
If the first hex character is F, the IP address is class E.

For the hexadecimal number E4 8A 26 71, the first hex character is E; therefore, it must be a class D address.

c. Determining the address class for a dotted-decimal number requires looking at only the first dotted-decimal number. The rules are simple:

If the first decimal number is between 0 and 127, the IP address is class A.
If the first decimal number is between 128 and 191, the IP address is class B.
If the first decimal number is between 192 and 223, the IP address is class C.
If the first decimal number is between 224 and 239, the IP address is class D.
If the first decimal number is between 240 and 255, the IP address is class E.

For the dotted-decimal number 140.24.10.1, the first decimal number falls within the range 128 and 191; therefore, it must be a class B address.

19-11 LOOPBACK ADDRESSES

The designated IP loopback address is 127.0.0.1; however, any IP address that begins with 127 can be used for the loopback address. The loopback address is used to test the local TCP/IP software on a host. A packet of data sent from a host with the destination network address of 127 never leaves the host machine—the packet is simply returned (looped back) to the host's protocol software. The loopback address can be the destination address in a data packet sent by a client process running an application program to a server process running on the same machine. The packet is transported from the process layer through the transport layer (TCP or UDP) to the network layer (IP), where it is looped back and returns through the network layer and transport layer to the process, as shown in Figure 19-9. It is important to note that a loopback address can be used only as a destination address.

19-12 BROADCAST ADDRESSES

As the name implies, a broadcast address is a one-to-all transmission. One-to-all transmissions are from a single host to every other device on the network with an IP address, which may include hosts, routers, switches, and a multitude of other devices. Broadcast addresses can be only destination addresses, and they identify every device that receives it as a recipient of that data packet. Broadcasts are intended to convey network-specific information onto a network. There are three types of broadcasts: global, direct, and limited.

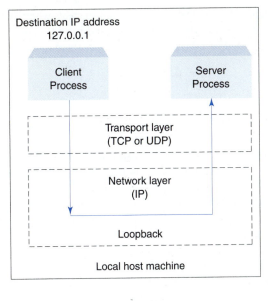

No access

Internet access

FIGURE 19-9   IP loopback operation

19-12-1   Global Broadcasts

Global broadcasts are transmissions from a single source to every device on the Internet that has an IP address. Global broadcasts are prohibited on the Internet. Therefore, a host cannot send data packets destined to all hosts and routers on the Internet for the simple reason that it would generate so much traffic that the Internet would be overwhelmed with broadcast transmissions and unable to route normal traffic. Routers block data packets with destination IP addresses containing a global broadcast address.

19-12-2   Direct Broadcasts

Direct broadcasts are used with class A, B, and C addresses, as they are transmissions sent by routers intended for all hosts on a specific local network. Direct broadcasting reduces the number of available IP addresses on each network by one, and direct broadcast IP addresses can be used only for a destination address. Direct broadcasts are data packets where the host portion of the destination IP address (hostid) contains all binary 1s. Examples of direct broadcast addresses are the following:

	netid	hostid
Class A	72.	255.255.255
Class B	170.72.	255.255
Class C	200.154.42.	255

19-12-3   Limited Broadcasts

Limited broadcasts use a class E address and are transmissions sent by hosts intended for all other hosts on a specific local network; therefore, they are received and processed by every host on the network with an IP address. A limited broadcast address can be used only for a destination address. Limited broadcasts are data packets where both the network and the host portion of the destination IP address (netid and hostid) contain all 1s (255.255.255.255). Routers block limited broadcasts, which confines them to local networks only.

19-13 ADDRESS MASKING

A mask is designed to cover something up. On the other hand, a mask can also be used to reveal certain things. An IP *address mask* works essentially the same way. For a router to deliver a data packet to the intended host, it must first direct the packet to the correct network. Therefore, a router must be capable of separating the network portion of a destination IP address (netid) from the total IP address. This is the purpose of an address mask. Once the network address is determined, the router can forward the data packet, including the entire source and destination IP addresses, to the destination network, where it is delivered to the intended host.

The simplest means of separating the netid from the total address is to first recognize what class address it is. Class identification is accomplished in routers by looking at the beginning bits in the IP address: first bit = 0, class A address; first two bits = 1 0, class B address; and so on. Once the router has determined the class of the address, it simply changes all hostid bits to 0s, which reveals the network address. For example, if a router receives a data packet with the address 130.20.64.1, it examines the beginning bits of the first octet and determines that the first two bits are 1 0, which indicates a class B address (130 = 10000010). The router then changes all host bits to logic 0s, which converts the IP address to the network address. Thus, the network portion (netid bits) of the IP address 130.20.64.1 is 130.20, which is converted to the network address 130.20.0.0.

The preceding example is shown here:

```
dotted decimal  ───────────────────►  binary

                         netid (130.20)              hostid (64.1)
                        ︷‾‾‾‾‾‾‾‾‾‾︷             ︷‾‾‾‾‾‾‾‾‾‾︷
   130.20.64.1  ─────►   10000010 00010100            01000000 00000001

                        10
                        ︷‾︷
                   class B
```

With class B addresses, the last 16 bits (two octets) represent the hostid, which are changed to logic 0s, giving the following network address:

```
                     netid (130.20)                 hostid (0.0)
                    ︷‾‾‾‾‾‾‾‾‾‾︷               ︷‾‾‾‾‾‾‾‾‾‾︷
                    10000010 00010100              00000000 00000000
Network address       130.      20.                  0.        0
```

Address masking is accomplished with IP addresses in the same manner in which it is accomplished in any binary manipulation—with a logical ANDing operation. When a bit (1 or 0) is logically ANDed with a 0, it produces a 0, and when a bit is logically ANDed with a 1, it retains its original logic condition (i.e., a 1 remains a 1, and a 0 remains a 0). For example,

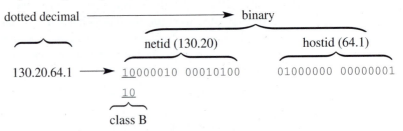

```
                        1011   1110        0011   0100
         ANDed with     1111   0000        0000   1111
                       ─────────────      ─────────────
                        1011   0000        0000   0100
                        ︷‾︷  ︷‾‾︷       ︷‾‾︷  ︷‾︷
                      remains becomes     becomes remains
                     the same  all 0s      all 0s  the same
```

Table 19-9 Class A, B, and C Default Masks

Class	Default Mask (Decimal)	Default Mask (Binary)
A	255.0.0.0	11111111 00000000 00000000 00000000
B	255.255.0.0	11111111 11111111 00000000 00000000
C	255.255.255.0	11111111 11111111 11111111 00000000

To extract the netid from an IP address, the address is ANDed with a specific binary sequence called a *mask*. The *default mask* is the mask that is applied to an IP address when subnetting is not being used (subnetting is described in a latter section of this chapter). Default masks have all logic 1s for the netid and all logic 0s for the hostid. The default mask for class A addresses is comprised of all logic 1s for the first octet (netid) and all logic 0s for the second, third, and fourth octets (hostid). With class B addresses, the default mask is comprised of all logic 1s for the first two octets (netid) and all logic 0s for the last two octets (hostid). With class C addresses, the default mask is comprised of all logic 1s for the first three octets (netid) and all logic 0s for the last octet (hostid). The default masks for class A, B, and C addresses are shown in Table 19-9.

Example 19-2

Determine the network address for the following IP addresses (assume that subnetting is not being used, and use the default mask):

a. 84.42.58.11
b. 144.62.12.9
c. 194.38.14.13

Solution

a. From observation, the IP address 84.42.58.11 is a class A address; therefore, the address is ANDed with the class A default mask to determine the network address:

IP address (decimal)	84	42	58	11
IP address (binary)	01010100	00101010	00111010	00001011
Default mask	11111111	00000000	00000000	00000000
Result of ANDing	01010100	00000000	00000000	00000000
Network address	84.	0.	0.	0

Netid 84.0.0.0

b. From observation, the IP address 144.62.12.9 is a class B address; therefore, the address is ANDed with the class B default mask to determine the network address:

IP address (decimal)	144	62	12	9
IP address (binary)	10010000	00111110	00001100	00001001
Default mask	11111111	11111111	00000000	00000000
Result of ANDing	01010100	00111110	00000000	00000000
Network address	144.	62.	0.	0

Netid 144.62.0.0

c. From observation, the IP address 194.38.14.13 is a class C address; therefore, the address is ANDed with the class C default mask to determine the network address:

IP address (decimal)	194	38	14	13
IP address (binary)	11000010	00100110	00001110	00001101
Default mask	11111111	11111111	11111111	00000000
Result of ANDing	11000010	00100110	00001110	00000000
Network address	194.	38.	14.	0.

Netid 194.38.14

QUESTIONS

19-1. What organization developed the Internet?

19-2. Briefly describe the functions of each layer of the *OSI seven-layer protocol hierarchy.*

19-3. Compare the layers of the OSI seven-layer protocol suite to the *TCP/IP protocol suite.*

19-4. Briefly describe the process of producing a *Request for Comments* (RFC) document.

19-5. Describe the five status assignments for an RFC.

19-6. Explain the meaning of the three RFC stages.

19-7. Describe the composition of an *IPv4 address.*

19-8. Describe the three numbering systems used to represent *IP addresses.*

19-9. Explain how each IP address class can be identified.

19-10. Compare the network and host limitations of *class A, B, and C IP addresses.*

19-11. Describe the purpose of *reserved IP addresses.*

19-12. Describe the purpose of *class D IP addresses.*

19-13. Describe the purpose of *class E IP addresses.*

19-14. How are *loopback addresses* identified?

19-15. What is the purpose of loopback addresses?

19-16. List and describe the three types of *broadcast addresses.*

19-17. What is the purpose of an *IP address mask?*

19-18. What are the default address masks for class A, B, and C IP addresses?

PROBLEMS

19-1. Determine the dotted-decimal notation for the following IP addresses:
   a. 00110101 11001010 00110111 00000111
   b. 10001110 00101110 11001100 00001110
   c. 11011110 10101110 00001111 01011101
   d. 11110000 10110001 00011101 10001110

19-2. Which of the following are class A addresses?
   a. 128.4.5.6
   b. 227.5.4.6
   c. 172.18.0.0
   d. 117.0.0.0
   e. 66.78.45.1
   f. 127.0.0.0
   g. 192.46.56.8
   h. 220.122.4.1
   i. 242.16.8.1

19-3. Which of the following are class B addresses?
   a. 128.4.5.6
   b. 227.5.4.6
   c. 172.18.0.0
   d. 117.0.0.0
   e. 66.78.45.1
   f. 127.0.0.0
   g. 192.46.56.8
   h. 220.122.4.1
   i. 242.16.8.1

19-4. Which of the following are class C addresses?
   a. 128.4.5.6
   b. 227.5.4.6
   c. 172.18.0.0

    d. 117.0.0.0

    e. 66.78.45.1

    f. 127.0.0.0

    g. 192.46.56.8

    h. 220.122.4.1

    i. 242.16.8.1

19-5. Which of the following are private IP addresses?

    a. 128.4.5.6

    b. 227.5.4.6

    c. 172.18.0.0

    d. 117.0.0.0

    e. 66.78.45.1

    f. 127.0.0.0

    g. 192.46.56.8

    h. 220.122.4.1

    i. 242.16.8.1

19-6. Which of the following are class D addresses?

    a. 128.4.5.6

    b. 227.5.4.6

    c. 172.18.0.0

    d. 117.0.0.0

    e. 66.78.45.1

    f. 127.0.0.0

    g. 192.46.56.8

    h. 220.122.4.1

    i. 242.16.8.1

19-7. Which of the following are class E addresses?

    a. 128.4.5.6

    b. 227.5.4.6

    c. 172.18.0.0

    d. 117.0.0.0

    e. 66.78.45.1

    f. 127.0.0.0

    g. 192.46.56.8

    h. 220.122.4.1

    i. 242.16.8.1

19-8. Which of the following are class A addresses?

    a. 78 F3 A2 B1

    b. 8C B3 32 18

    c. DC 12 08 01

    d. C8 12 9A 02

    e. E6 A4 12 10

    f. F8 10 10 11

    g. E1 BB 22 01

    h. 70 70 70 71

    i. 96 B2 18 AA

19-9. Which of the following are class B addresses?

    a. 62 F3 A2 B1

    b. C3 B3 32 18

    c. 7F 12 08 01

    d. BB 12 9A 02

    e. A1 A4 12 10

    f. 32 10 10 11

    g. 49 BB 22 01

    h. B8 70 70 71

    i. D5 B2 18 AA

19-10. Which of the following are class C addresses?

    a. 78 F3 A2 B1
    b. 8C B3 32 18
    c. DC 12 08 01
    d. C8 12 9A 02
    e. E6 A4 12 10
    f. F8 10 10 11
    g. E1 BB 22 01
    h. 70 70 70 71
    i. 96 B2 18 AA

19-11. Which of the following are class D addresses?

    a. 62 F3 A2 B1
    b. C3 B3 32 18
    c. 7F 12 08 01
    d. BB 12 9A 02
    e. A1 A4 12 10
    f. 32 10 10 11
    g. 49 BB 22 01
    h. B8 70 70 71
    i. D5 B2 18 AA

19-12. Which of the following are class E addresses?

    a. 78 F3 A2 B1
    b. 8C B3 32 18
    c. DC 12 08 01
    d. C8 12 9A 02
    e. E6 A4 12 10
    f. F8 10 10 11
    g. E1 BB 22 01
    h. 70 70 70 71
    i. 96 B2 18 AA

19-13. Which of the IP addresses listed in problem 19–17 are class A addresses?

19-14. Which of the IP addresses listed in problem 19–17 are class B addresses?

19-15. Which of the IP addresses listed in problem 19–17 are class C addresses?

19-16. Which of the IP addresses listed in problem 19–17 are class D addresses?

19-17. Which of the IP addresses listed are class E addresses?

    a. 10110101 11010000 11010101 01010001
    b. 00111010 11010000 11010101 01010001
    c. 11000101 11010000 11010101 01010001
    d. 11111010 11010000 11010101 01010001
    e. 00111010 11010000 11010101 01010001
    f. 11000001 11010000 11010101 01010001
    g. 10001010 11010000 11010101 01010001
    h. 01101010 11010000 11010101 01010001

19-18. Which of the following is the default mask for the IP address 198.0.4.201?

    a. 255.0.0.0
    b. 255.255.0.0
    c. 255.255.255.0
    d. 255.255.255.255

19-19. Which of the following is the default mask for the IP address 98.0.46.201?

    a. 255.0.0.0
    b. 255.255.0.0
    c. 255.255.255.0
    d. 255.255.255.255

19-20. Which of the following is the default mask for the IP address 172.14.6.8?

    a. 255.0.0.0

    b. 255.255.0.0

    c. 255.255.255.0

    d. 255.255.255.255

19-21. Determine the default mask and network address for the following IP host addresses:

    a. 70.45.121.33

    b. 155.67.80.2

    c. 205.35.66.12

    d. 192.56.78.111

    e. 50.12.12.12

C H A P T E R 20

Networks and Subnetworks

CHAPTER OUTLINE

OBJECTIVES

- Describe the essence of subnetting
- Explain the differences between default masks and subnet masks
- Show how subnet masks can be used to reveal subnet addresses
- Describe the differences between class A, B, and C subnet masks
- Describe the reason for using supernetting
- Describe the differences between subnetting and supernetting
- Explain how a supernet is formed using a supernet mask
- Explain the concept of variable-length blocks
- Explain the differences between classful and classless IP addressing
- List and describe the restrictions of classless addressing
- Describe slash notation for classless addressing
- Explain variable-length subnetting
- Describe classless interdomain routing

20-1   INTRODUCTION

Probably the most serious problem associated with classful IP addressing is that it wastes a significant portion of the available IP addresses, especially class A and class B host addresses. For example, each class A network address (netid) contains 16,777,216 host addresses (hostids), and it is highly unlikely that any network would ever need that many addresses; if it did, it would probably be unmanageable. One way to alleviate the problem of IP address squandering is to incorporate some means of dividing the available network addresses into smaller groups. This is the essence of *subnetting*—to partition a network into smaller groups, or *subnetworks,* with each *subnetwork* having its own *subnet address* and *subnet mask.* Subnetting allows a network to reorganize IP addresses into a more efficient and manageable system at both the global (Internet) and the local network level.

20-2   SUBNETTING

Figure 20-1 shows a simplified diagram of a network divided into four subnetworks logically separated with a single router. The router is a level 3 device that routes data packets based on IP addresses. Therefore, the only transmissions that pass through the router are those intended for a host on a different subnetwork or a different network than the host that originated the message. Thus, the router provides hosts on one subnetwork access to hosts on the other three subnetworks and allows hosts on all four subnetworks access to the global Internet.

Network designers created subnetting for a combination of reasons. Subnetting improves network performance and reliability by alleviating congestion on high-volume networks and separates broadcast and collision domains. Subnetting also provides a higher degree of security by allowing departments within an organization conveying sensitive information to be isolated onto separate subnetworks. Subnetting solves connectivity issues when a network is comprised of hosts that are on different floors of a building, in different buildings, in different cities, in different states, or in different countries. Subnetworks can also be used to connect two or more local area networks that use different media protocols, such as Ethernet, token ring, FDDI, or frame relay.

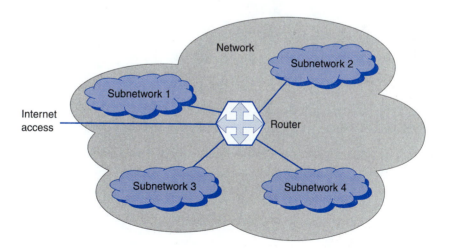

FIGURE 20-1   Network divided into four subnetworks

The original Internet was designed with five address classifications. The first three classifications (A, B, and C) utilized a hierarchy comprised of two layers: a *network layer* and a *host layer*. The network layer provides the network address, which identifies the network, and the host layer provides a host address, which identifies a specific host on the network. The two-layer address hierarchy is shown here:

> *Layer 1, network address (netid).* Identifies the network
>
> *Layer 2, host address (hostid).* Identifies the host

For example, an organization has the class B address 170.20.0.0, as shown in Figure 20-2a. Because it has only one network address, it can have only one physical network. However, the network can have as many as 65,536 hosts with addresses ranging from 170.20.0.0 to 170.20.255.255. This indeed would be a highly congested and truly unmanageable network.

Subnetting is a logical and relatively simple (although temporary) solution to the address allocation problem. In essence, subnetting adds a third layer to the addressing hierarchy—a *subnetwork layer*. Subnetting is permissible only with class A, B, and C addresses and, therefore, cannot be incorporated into multicast or private addresses (classes D and E). Broadcasts, however, are allowed on subnetworks.

The subnet layer further divides the network into smaller groupings called *subnetworks* (or *subnets*). The subnet layer is embedded between the network and host layers and identifies the subnetwork as shown here:

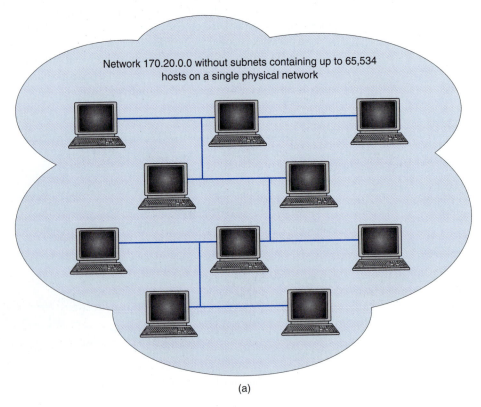

Network 170.20.0.0 without subnets containing up to 65,534 hosts on a single physical network

(a)

FIGURE 20-2 Class B network: (a) without subnets

Layer 1, network address (netid). Identifies the network

Layer 2, subnet address (subnetid). Identifies the subnetwork

Layer 3, host address (hostid). Identifies the host

In essence, hosts are combined to form subnetworks, and subnetworks are combined to form networks. Because adding subnetworks creates an intermediate layer in the addressing hierarchy, the IP addressing scheme must also include an intermediate identification number—the *subnet identification number (subnetid)*. Thus, when subnets are used, a host's IP address must contain three identification numbers: netid, subnetid, and hostid. The number of bits in the netid is fixed and depends on the address class (class A = eight bits, class B = 16 bits, and class C = 24 bits). However, when subnetting is incorporated, the bits previously designated as hostid bits are subdivided into two identification numbers: subnetid and hostid. Using subnets on an IP-based network is analogous to the 10-digit telephone number used on the public telephone network. The telephone number contains a three-digit area code, a three-digit prefix, and a four-digit extension number. An IP address is the complete address of a host, which contains a netid, a hostid, and a subnetid if subnetworks are used on the network. The three network identifications are shown in Table 20-1 for the class B address 130.24.32.1.

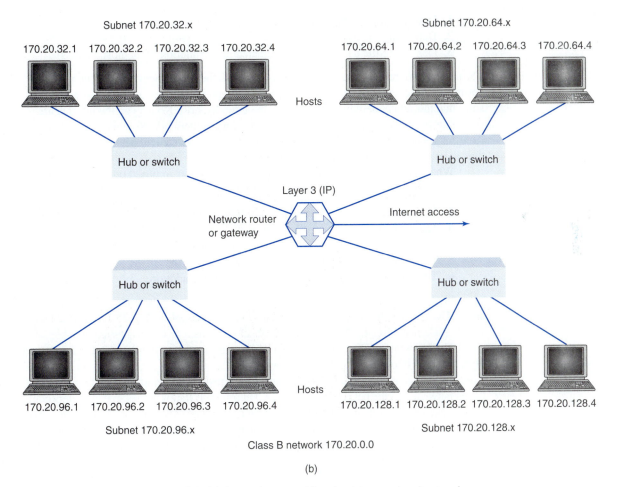

FIGURE 20-2   Class B network: (b) with four subnets and four hosts on each subnetwork

Table 20-1 Network Identifications for Class B Address 130.24.32.1

Network	Subnet	Host
netid	subnetid	hostid
130.24	32	1

Note. Network address 130.24.0.0, subnet address 130.24.32.0, host address 130.24.32.1.

Table 20-2 Subnet and Host Addresses for the Network Address 170.20.0.0 Divided into Four Subnets

Subnet Address	Host Addresses			
170.20.32.0	170.20.32.1	170.20.32.2	170.20.32.3	170.20.32.4
170.20.64.0	170.20.64.1	170.20.64.2	170.20.64.3	170.20.64.4
170.20.96.0	170.20.96.1	170.20.96.2	170.20.96.3	170.20.96.4
170.20.128.0	170.20.128.1	170.20.128.2	170.20.128.3	170.20.128.4

When subnetting is used, routing data packets to a destination host involves three steps:

1. Delivering the packet to the proper network (function of the network address)
2. Delivering the packet to the proper subnetwork (function of the subnet address)
3. Delivering the packet to the designated host (function of the host address)

The network address ensures that data packets reach the proper network, the subnet address ensures that the network delivers the packet to the proper subnet, and the host address ensures that the subnet delivers the packet to the intended host. Routers on the Internet are responsible for delivering data packets to the destination network. Routers and switches on local networks are responsible for forwarding data packets to the proper subnetwork, where they are delivered to their intended host.

When subnetting is used, there are several restrictions imposed on IP addresses:

1. The netid must be the leftmost bits of the address.
2. The netid must be a multiple of eight bits (eight, 16, or 24).
3. The subnetid must have at least two bits and cannot have more than 22.
4. The hostid must have at least two bits and cannot have more than 22.
5. The subnet bits must begin on the left end of an octet.
6. The hostid bits must be the rightmost bits of the address.

Figure 20-2b shows a class B network (170.20.0.0) with a three-layer addressing hierarchy. The network has been subdivided into four subnetworks—170.20.32.0, 170.24.64.0, 170.24.96.0, and 170.24.128.0—and each subnetwork has four hosts, as listed in Table 20-2.

The hosts on each subnetwork can be interconnected with either a layer 1 hub or a layer 2 switch, depending on how the network is configured. The subnetworks are separated from each other with a router. Routers separate transmissions and route data packets between networks on the basis of IP addresses. Therefore, it stands to reason that routers could also be used to separate and direct traffic between subnetworks that have different subnet addresses. Routers and switches prevent transmissions between hosts on one subnetwork from propagating onto other subnetworks. The only time a data packet passes through a router is if the packet contains a destination IP address of a host located on a different network or subnetwork. Routers separate collision and broadcast domains by segmenting the network into separate logical networks. Layer 2 switches separate collision domains; however, they limit the network to only one broadcast address.

The default mask used by routers to separate the netid from the hostid also sets the default configuration for the bits within an IP address. For example, with class A addresses, the default mask 255.0.0.0 designates the first octet (eight bits) as the netid and the remaining three octets (24 bits) as the hostid. With class B addresses, the default mask of 255.255.0.0 designates the first two octets (16 bits) as the netid and the remaining two octets (16 bits) as the hostid. With class C addresses, the default mask of 255.255.255.0 defines the first three octets (24 bits) as the netid and the remaining octet (eight bits) as the hostid. When subnetting is used, the netid is unaffected; however, the mask is changed so that some of the hostid bits are reallocated and designated subnetid bits. The same is true for class B and C masks. When a mask is changed from its default value to accommodate subnets, it is called a *subnet mask*.

As stated previously, the number of 1s in a default address mask is fixed (class A = eight, class B = 16, and class C = 24), and they must be placed in specific octets (class A—first octet, class B—first two octets, and class C—first three octets). In a subnet mask, however, the number of 1s must be more than the number of 1s in the corresponding default mask (i.e., more than eight with class A, more than 16 with class B, and more than 24 with class C). The first eight logic 1s in a class A subnet mask reveal the network address, and any additional logic 1s designate a subnet address. The first 16 logic 1s in a class B subnet mask reveal the network address, and any additional logic 1s designate a subnet address, and the first 24 logic 1s in a class C subnet mask reveal the network address, and any additional logic 1s designate a subnet address.

The default mask for a class A address is 255.0.0.0. When the default mask is applied to an IP address, the only address it uncovers is the network address. However, when the subnet mask 255.224.0.0 is applied to a class A address, it uncovers the network address and also indicates that the address contains three subnetid bits. The following diagram illustrates the concept of using default masks to extract the network address and subnet masks to extract the network and subnet addresses:

	255.	0.	0.	0
Default mask	11111111	00000000	00000000	00000000

reveals the network address   covers up the hostid bits

or	NNNNNNNN	HHHHHHHH	HHHHHHHH	HHHHHHHH

where   N = network address bits

	255.	224.	0.	0
Subnet mask	11111111	11100000	00000000	00000000

reveals the network address   reveals the subnet address   covers up the hostid bits

or	NNNNNNNN	SSSHHHHH	HHHHHHHH	HHHHHHHH

where   N = network address bits (netid)
         S = subnet address bits (subnetid)
         H = host address bits (hostid)

The default mask indicates that there are eight bits (one octet) in the netid portion of the address; however, it gives no indication of whether there are any subnet bits. The subnet mask, however, uncovers three subnet bits at the beginning of the second byte, which indicates that subnetting is being used, and the first three bits of the second octet represent the subnetid portion of the IP address. For example, ANDing the default mask with the address

70.64.10.5 reveals the network portion of the address (70) and indicates that there are no subnetid bits, as the remaining 24 bits are converted to logic 0s to produce the network address 70.0.0.0, as shown here:

		255.	0.	0.	0.
Default mask		11111111	00000000	00000000	00000000
Address		01000110	01000000	00001010	00000101
ANDing		01000110	00000000	00000000	00000000
		$\underbrace{\text{NNNNNNNN}}$			
Network address		70.	0.	0.	0

There are no hostid or subnetid bits revealed because the default subnet mask hides (masks) all bits except the netid bits.

ANDing the first octet of the subnet mask 255.224.0.0 with the address 70.64.10.5 reveals the network portion of the address (70). However, the second dotted-decimal number in the subnet mask 224 (11100000 binary) contains three leading 1s, which indicates that there are three subnetid bits in the address. The remaining hostid bits are changed to 0s, revealing three subnetid bits in the second octet (010) followed by five logic 0s, which indicates a subnetwork with the address 70.64.0.0, as shown here:

	255.	224.	0.	0
Subnet mask	11111111	11100000	00000000	00000000
Address	01000110	01000000	00001010	00000101
ANDing	01000110	01000000	00000000	00000000
	$\underbrace{\text{NNNNNNNN}}$	$\underbrace{\text{SSS}}$	all hostid bits are masked off	
Subnet address	70.	64.	0.	0
	netid	subnetid		

The network address is simply the netid followed by all 0s (70.0.0.0), and the subnet address is the network address plus the subnetid followed by all 0s (70.64.0.0). As far as the network and subnet addresses are concerned, the 21 0s that follow the subnetid are "don't care" bits, as they have nothing to do with either the network or the subnet address.

The number of bits included in the subnetid depends on the number of contiguous logic 1s in the subnet mask (contiguous simply means immediately following each other). In subnet masks, the first two bits must be logic 1s, and the logic 1s in the mask cannot be separated with 0s. All logic 1s in the subnet mask must be left justified. For the previous example, there were three contiguous logic 1s at the beginning of the second octet of the subnet mask. This indicated that there were three bits in the subnetid portion of the address. The default mask determines the netid, whereas the subnet mask identifies how many of the address bits are subnetid bits, which are used to determine the subnet address.

In any given octet, there are only eight binary combinations that contain contiguous 1s beginning on the left. The number of subnet bits in an address octet is equal to the number of leading 1s in the subnet mask. The eight combinations possible in a subnet mask, their equivalent dotted-decimal values, and the number of subnetid bits they reveal are shown in Table 20-3.

Example 20-1

Determine which of the following binary-coded octets are valid subnet masks and which are not and give the dotted-decimal equivalent of the valid masks:

a. 11101000
b. 11100000

Table 20-3 Subnet Mask Octet Possibilities

	One Octet of a Subnet Mask								Dotted-Decimal Value	Number of Subnet Bits
Binary Weighting	128	64	32	16	8	4	2	1		
	1	0	0	0	0	0	0	0	128	1
	1	1	0	0	0	0	0	0	192	2
	1	1	1	0	0	0	0	0	224	3
	1	1	1	1	0	0	0	0	240	4
	1	1	1	1	1	0	0	0	248	5
	1	1	1	1	1	1	0	0	252	6
	1	1	1	1	1	1	1	0	254	7
	1	1	1	1	1	1	1	1	255	8

c. 10000001

d. 11111000

Solution

a. 11101000—invalid, noncontiguous 1s

b. 11100000—valid, 224

c. 10000001—invalid, noncontiguous 1s

d. 11111000—valid, 248

Example 20-2

Determine the dotted-decimal notation and number of network and subnet address bits for the following addresses and subnet masks:

a. Class A address, subnet mask = 11111111 11110000 00000000 00000000

b. Class B address, subnet mask = 11111111 11111111 11100000 00000000

c. Class C address, subnet mask = 11111111 11111111 11111111 11000000

d. Class B address, subnet mask = 11111111 11111111 11111000 00000000

Solutions

a.

	11111111	11110000	00000000	00000000	
	NNNNNNNN	SSSS			
subnet mask	255.	240.	0.	0	eight network address bits and four subnet address bits

b.

	11111111	11111111	11100000	00000000	
	NNNNNNNN	NNNNNNNN	SSS		
subnet mask	255.	255.	224.	0	16 network address bits and three subnet address bits

c.

	11111111	11111111	11111111	11000000	
	NNNNNNNN	NNNNNNNN	NNNNNNNN	SS	
subnet mask	255.	255.	255.	192	24 network address bits and two subnet address bits

d.

	11111111	11111111	11111000	00000000	
	NNNNNNNN	NNNNNNNN	SSSSS		
subnet mask	255.	255.	248.	0	16 network address bits and five subnet address bits

Example 20-3

Determine the binary code and the number of subnet address bits for the following subnet masks:
a. Class A—255.248.0.0
b. Class B—255.255.192.0
c. Class C—255.255.255.224
d. Class B—255.255.240.0

Solution

a.   255.   248.   0.   0
    11111111. <u>11111</u>000. 00000000. 00000000   five subnet address bits
b.   255.   255.   192.   0
    11111111. 11111111. <u>11</u>000000. 00000000   two subnet address bits
c.   255.   255.   255.   224
    11111111. 11111111. 11111111. <u>111</u>00000   three subnet address bits
d.   255.   255.   240.   0
    11111111. 11111111. <u>1111</u>0000. 00000000   four subnet address bits

For a subnet mask to specify more than 256 subnet addresses, it must contain more than eight contiguous logic 1s and, therefore, must occupy more than one octet of the mask. The dotted-decimal address for each octet is determined independently using standard binary weightings: 128, 64, 32, 16, 8, 4, 2, and 1. For example, the class A subnet mask 255.255.192.0 is written in binary notation as

Binary	11111111	11111111	11000000	00000000
	NNNNNNNN	SSSSSSSS	SS	
Dotted-decimal	255.	255.	192.	0

Since this is a class A subnet mask, the first octet reveals the network address. The second octet extracts the first eight bits of the subnet address, and the first two bits of the third octet determine the remaining bits of the subnet address. For example, if the subnet mask 255.255.192.0 is applied to the class A address 50.68.172.121, the network and subnetwork addresses are determined as follows:

50.	68.	172.	121	00110010	01000100	10101100	01111001
255.	255.	192.	0	<u>11111111</u>	<u>11111111</u>	<u>11000000</u>	<u>00000000</u>
		ANDing		00110010	01000100	10000000	00000000
				NNNNNNNN	SSSSSSSS	SS	
	Network address			50.	0.	0.	0
	Subnetwork address			50.	68.	128.	0

20-5   THE ANATOMY OF AN IP ADDRESS

In essence, an IP address is the combination of three separate portions: a network portion, a subnet portion, and a host portion. Although dividing an IP address into its respective portions is not particularly useful in practice, it does lend itself to better understanding of the makeup of an IP address.

When a subnet mask is applied to an IP address, the following are revealed: netid, subnetid, and hostid. The netid is used to determine the network portion of the address, the subnetid is used to determine the subnetwork portion of the address, and the hostid is used to determine the host portion of the address. For example, if the subnet mask 255.255.224.0 is logically ANDed with the class B address 130.46.93.5, the following information is revealed:

		130	46	93	5
IP address	130.46.93.5	100000010	00101110	01011101	00000101
Subnet mask	255.255.224.0	<u>11111111</u>	<u>11111111</u>	<u>11100000</u>	<u>00000000</u>
	ANDing	10000010	00101110	010	00000000
		netid	netid	subnetid	

Thus, the network portion of the address is

$$10000010 \quad 00101110$$
$$130. \qquad 46.$$

And the subnet portion of the address is determined from the octet (or octets) containing subnetid bits:

$$\underline{01000000}$$
$$64.$$

The host portion is simply whatever is left (i.e., not included in the netid or subnetid). In this example, the hostid is comprised of the rightmost 13 bits of the address:

$$11101 \quad 00000101$$
$$29. \qquad 5$$

The IP address is found by simply adding the network, subnetwork, and host portions together:

Network portion	130	46	0.	0
Subnetwork portion	0	0	64	0
Host portion	0	0	29	5
Sum	130.	46.	93.	5

Apportioning IP addresses can be used to determine the network address by filling in the subnetid and hostid bits of the network portion with 0s and adding it to the network address, as shown here:

Network portion	130	46	—	—
	—	—	0	0
Network address	130.	46.	0.	0

The subnetwork address can also be determined by filling in the hostid bits of the subnet portion with 0s and adding it to the network portion, as shown here:

Network portion	130	46	—	—
Subnet portion	0	0	64	—
Hostid bits	—	—	0	0
Subnet address	130.	46.	64.	0

The host address can be determined by adding the network, subnetwork, and host portions, as shown here:

Network portion	130	46	—	—
Subnetwork portion	—	—	64	0
Host portion	—	—	29	5
Host address	130.	46.	93.	5

Example 20-4

For the following class A address and subnet mask, determine
a. the network, subnetwork, and host portions of the address
b. the network, subnetwork, and host addresses

IP address	42.56.29.13
Subnet mask	255.255.240.0

Solution

a.

		45	56	29	13
IP Address	42.56.29.13	00101010	00111000	00011101	00001101
Subnet mask	255.255.240.0	11111111	11111111	11110000	00000000
	ANDing	00101010	00111000	00010000	00000000
		netid		subnetid	

The network portion of the address is the leftmost eight bits:

$$\text{Network portion} \quad 00101010$$
$$42$$

The subnet portion of the address is the next 12 bits:

$$\text{Subnet portion} \quad 00111000 \quad 000100000$$
$$56. \qquad\qquad 16$$

The host portion of the address is the rightmost 12 bits:

$$\text{Host portion} \quad 00001101 \quad 00001101$$
$$13. \qquad\qquad 13$$

b. The network address is determined by filling in the subnetid and hostid bits of the network portion with 0s and adding it to the network address:

Network portion	42	—	—	—
	—	0	0	0
Network address	42.	0.	0.	0

The subnetwork address is determined by filling in the hostid bits of the subnet portion with 0s and adding it to the network portion, as shown here:

Network portion	42.	—	—	—
Subnet portion	—	56	16	—
Hostid bits	—	—	0	0
Subnet address	42.	56.	16.	0

The host address is determined by adding the network, subnetwork, and host portions:

Network portion	42.	—	—	—
Subnetwork portion	—	56.	16.	—
Host portion	—	—	13.	13
Subnet address	42.	56.	29.	13

20-6 SUBNET POSSIBILITIES

The number of subnet addresses possible depends on the subnet mask and how many subnetid bits are included in the address. For example, if there are two subnet bits, there are $2^2 = 4$ subnet addresses possible. With three subnet bits, there are $2^3 = 8$ subnet addresses possible. The number of subnet addresses possible is expressed mathematically as

$$\text{number of subnet} = 2^n \text{ addresses possible}$$

where n is the number of subnetid bits

With classful IP addressing, subnet addresses containing all 0s or all 1s cannot be used for subnet addresses. An address with all 0s is disallowed because it is the address of the subnet itself, and an address of all 1s is disallowed because this is the subnet broadcast address. Therefore, there are always two fewer subnet addresses allowed than there are subnet addresses possible. Thus, the number of usable subnet addresses with n subnetid bits is expressed mathematically as

$$\text{number of usable subnet addresses} = 2^n - 2$$

where n is the number of subnetid bits

Example 20-5

Determine the number of usable subnet addresses for the following number of subnetid bits:

a. 4
b. 5
c. 8
d. 11
e. 13

Solution

a. $2^4 - 2 = 14$
b. $2^5 - 2 = 30$
c. $2^8 - 2 = 254$
d. $2^{11} - 2 = 2046$
e. $2^{13} - 2 = 8190$

20-7 SUBNET ADDRESSES

With subnet addresses, the number of hosts allowed on a subnet and the range of host addresses are determined from the network address and the subnet mask. For example, the number of hosts and the range of host addresses on each subnet of a class C network with the address 200.40.14.0 and subnet mask 255.255.255.192 is determined as follows. The first three octets of the subnet mask (255.255.255) identify the network address (200.40.14.0). The last octet of the subnet mask (192) determines how many subnetid and hostid bits are included in the address and the range of subnet and host addresses. The number of subnetid and hostid bits is determined by converting the dotted-decimal value of the last octet (192) of the subnet mask to binary, as shown here:

				192				
Binary weighting	128	64	32	16	8	4	2	1
Binary code	1	1	0	0	0	0	0	0
	S	S	H	H	H	H	H	H

where   S = subnetid bits
        H = hostid bits

The binary code 11000000 indicates that the two leftmost bits of the last octet of the address are subnetid bits (S) and that the six rightmost bits are hostid bits (H). With two subnetid bits, there are $2^2 = 4$ subnet address possibilities: 00, 01, 10, 11. To determine the subnet address, all hostid bits are made 0, as shown in Table 20-4. From the table, it can be seen that the four subnet addresses are 200.40.14.0, 200.40.14.64, 200.40.14.128, and 200.40.14.192.

There are six hostids bits; thus, there are $2^6 = 64$ host addresses possible for each subnet address. The 64 host addresses range from all 0s (000000) to all 1s (111111) in each of the four subnet addresses. Keep in mind that with classful IP addressing, the first (all 0s)

Table 20-4 Subnet Address Calculations for Two Subnet Bits

	Binary Weightings								Decimal Value of
	128	64	32	16	8	4	2	1	Subnet Addresses
200.40.14	0	0	0	0	0	0	0	0	0
200.40.14	0	1	0	0	0	0	0	0	64
200.40.14	1	0	0	0	0	0	0	0	128
200.40.14	1	1	0	0	0	0	0	0	192
	subnetid bits			hostid bits					

Table 20-5 Subnets on Network 200.40.14.0

	128	64	32	Subnet 0 16	8	4	2	1	Decimal Value of Host Addresses 0
Binary weighting	0	0	0	0	0	0	0	0	0
	0	0	0	0	0	0	0	1	1
	0	0	0	0	0	0	1	0	2
	0	0	0	0	0	0	1	1	3
				4–61					
	0	0	1	1	1	1	1	0	62
	0	0	1	1	1	1	1	1	63

The host addresses for subnet 0 range from 200.40.14.0 to 200.40.14.63.

	128	64	32	Subnet 64 16	8	4	2	1	
Binary weighting	0	1	0	0	0	0	0	0	64
	0	1	0	0	0	0	0	1	65
	0	1	0	0	0	0	1	0	66
	0	1	0	0	0	0	1	1	67
				68–125					
	0	1	1	1	1	1	1	0	126
	0	1	1	1	1	1	1	1	127

The host addresses for subnet 64 range from 200.40.14.64 to 200.40.14.127.

	128	64	32	Subnet 128 16	8	4	2	1	
Binary weighting	1	0	0	0	0	0	0	0	128
	1	0	0	0	0	0	0	1	129
	1	0	0	0	0	0	1	0	130
	1	0	0	0	0	0	1	1	131
				132–189					
	1	0	1	1	1	1	1	0	190
	1	0	1	1	1	1	1	1	191

The host addresses for subnet 128 range from 200.40.14.128 to 200.40.14.191.

	128	64	32	Subnet 192 16	8	4	2	1	
Binary weighting	1	1	0	0	0	0	0	0	192
	1	1	0	0	0	0	0	1	193
	1	1	0	0	0	0	1	0	194
	1	1	0	0	0	0	1	1	195
				196–253					
	1	1	1	1	1	1	1	0	254
	1	1	1	1	1	1	1	1	255

The host addresses for subnet 192 range from 200.40.14.192 to 200.40.14.255.

and last (all 1s) host addresses are not allowed. In addition, the first (all 0s) and last (all 1s) subnet addresses are not allowed with classful IP addressing. However, as we will see later, these restrictions are not imposed on classless IP addressing. Therefore, the host addresses for the four subnets on network 200.40.14.0 are shown in Table 20-5.

For the example shown in Table 20-5 using classful addressing, there are only two subnet addresses allowed, and each subnet is allowed only 62 host addresses. The two subnets allowed are 200.40.14.64 and 200.40.14.128, and the range of host addresses allowed are 200.40.14.65 through 200.40.14.126 and 200.40.14.129 through 200.40.14.190.

Example 20-6

For a class C network with the classful address 192.192.14.0 and a subnet mask of 255.255.255.224, determine

a. the number of valid subnets
b. the number of valid hosts on each subnet
c. the total number of host addresses
d. the range of addresses for each valid subnet

Solution

a. With a class C address, only the last octet of the subnet mask is used to identify subnetid bits:

$$
\begin{array}{cccl}
192. & 192. & 14. & 0 \\
255. & 255. & 255. & \underline{224} \\
192. & 192. & 14. & \text{subnetids and hostids} \\
\end{array}
$$

Network address

To determine the number of subnetid bits, convert the dotted-decimal value for the last octet of the subnet mask to binary:

			224				
128	64	32	16	8	4	2	1

Binary   1   1   1   0   0   0   0   0

      S   S   S   H   H   H   H   H

A subnet mask of 224 indicates that there are three subnetid bits. Therefore, the number of classful subnets addresses allowed is

$$2^3 - 2 = 6$$

b. Since there are five hostid bits, the number of classful host addresses allowed is

$$2^5 - 2 = 30$$

c. The total number of classful host addresses allowed is simply the product of the number of subnet addresses and the number of host addresses, or

$$6 \times 30 = 180$$

d. The range of addresses are shown in Table 20-6.

20-8   CLASS A SUBNET MASKS

With class A networks, there are eight bits for the netid and 24 bits that can be divided between subnetid and hostid bits, depending on the subnet mask used. With classful IP addressing, class A subnet masks must have at least two logic 1s but not more than 22. One subnet bit is not allowed because it would produce only two subnets where one is all 0s and the other all 1s, both of which are not allowed. If there were 23 subnet bits, there is only one hostid bit and thus only two host addresses, one with all 0s and one with all 1s, both of which are also not allowed. If there were 24 subnet bits, there would not be any bits left in the address to designate hosts, which would yield a useless network.

Figure 20-3 summarizes classful class A subnetting. The figure shows all class A subnet masks possible and how many subnets bits (subnetids), subnetworks, hosts, and host address bits (hostids) are available for each mask. Note that the number of subnets equals $2^n - 2$, where n is the number of subnetid bits, and that the number of hosts is $2^a - 2$, where a is the number of hostid bits.

From Figure 20-3, it can be seen that the number of subnets is proportional to the number of subnetid bits and that the number of hosts on a subnet is inversely proportional to the number of subnets. Thus, as you increase the number of subnets on a network, you decrease the number of hosts allowed on each subnet.

Table 20-6 Range of Subnet Addresses for Example 20-6

		Subnetid	Hostid
Subnet 0	not allowed		
Subnet 32	192.192.14.32	001	00000—not allowed
	192.192.14.33	001	00001
	192.192.14.34–61	001	00010–11101
	192.192.14.62	001	11110
	192.192.14.63	001	11111—not allowed
Subnet 64	192.192.14.64	010	00000—not allowed
	192.192.14.65	010	00001
	192.192.14.66–93	010	00010–11101
	192.192.14.94	010	11110
	192.192.14.95	010	11111—not allowed
Subnet 96	192.192.14.96	011	00000—not allowed
	192.192.14.97	011	00001
	192.192.14.98–125	011	00010–11101
	192.192.14.126	011	11110
	192.192.14.127	011	11111—not allowed
Subnet 128	192.192.14.128	100	00000—not allowed
	192.192.14.129	100	00001
	192.192.14.130–157	100	00010–11101
	192.192.14.158	100	11110
	192.192.14.159	100	11111—not allowed
Subnet 160	192.192.14.160	101	00000—not allowed
	192.192.14.161	101	00001
	192.192.14.162–189	101	00010–11101
	192.192.14.190	101	11110
	192.192.14.191	101	11111—not allowed
Subnet 192	192.192.14.192	110	00000—not allowed
	192.192.14.193	110	00001
	192.192.14.194–221	110	00010–11101
	192.192.14.222	110	11110
	192.192.14.223	110	11111—not allowed
Subnet 224 not allowed			

Example 20-7

For the class A IP address 10.104.92.72 on a classful network with the subnet mask 255.255.240.0, determine the following:

a. IP address in binary
b. subnet addresses in binary
c. which bits are netids, subnetids, and hostids
d. network address
e. subnetwork address
f. host address

Solution

	10.	104.	92.	72
a. IP address (binary)	00001010.	01101000.	01011100.	01001000
	255.	255.	240.	0
b. Subnet mask (binary)	11111111.	11111111.	11110000.	00000000
	00001010.	01101000.	0101	
c. Bit meaning	NNNNNNNN.	SSSSSSSS.	SSSSHHHH.	HHHHHHHH
d. Network address	10.	0.	0.	0
e. Subnet address	10.	104.	80.	0
f. Host address	10.	104.	92.	72

0 to 127 | 0 to 255 | 0 to 255 | 0 to 255

Netid	Hostid or subnetid	Hostid or subnetid	Hostid or subnetid
Network unique 8 bits	0 to 255 8 bits	0 to 255 8 bits	0 to 255 8 bits

Subnet mask   255   0, 192, 224, 240, 248, 252, 254, 255

Number of subnets $(2^n - 2)$	Number of hosts $(2^a - 2)$	Number of host address bits $(a = 24 - n)$	Class A subnet mask	Number of subnet bits (n)
0	16,777,214	24	255.0.0.0	0
← not allowed →			255.128.0.0	1
2	4,194,302	22	255.192.0.0	2
6	2,097,150	21	255.224.0.0	3
14	1,048,574	20	255.240.0.0	4
30	524,286	19	255.248.0.0	5
62	262,142	18	255.252.0.0	6
126	131,070	17	255.254.0.0	7
254	65,534	16	255.255.0.0	8
510	32,766	15	255.255.128.0	9
1,022	16,382	14	255.255.192.0	10
2,046	8,190	13	255.255.224.0	11
4,094	4,094	12	255.255.240.0	12
8,190	2,046	11	255.255.248.0	13
16,382	1,022	10	255.255.252.0	14
32,766	510	9	255.255.254.0	15
65,534	254	8	255.255.255.0	16
131,070	126	7	255.255.255.128	17
262,142	62	6	255.255.255.192	18
524,286	30	5	255.255.255.224	19
1,048,574	14	4	255.255.255.240	20
2,097,150	6	3	255.255.255.248	21
4,194,302	2	2	255.255.255.252	22
← not allowed →			255.255.255.254	23
← not allowed →			255.255.255.255	24

FIGURE 20-3 Class A subnetting summary

Example 20-8

For a host on a class A network with the address 20.38.40.2 and a subnet mask of 255.248.0.0, determine

a. network address
b. subnet address where this host resides

Solution

a. The network address is determined by ANDing the default mask 255.0.0.0 with the host address:

Host address	20.	38.	40.	2
Default mask	255.	0.	0.	0
Network address	20.	0.	0.	0

Table 20-7 Subnet Addresses for Example 20-8

Binary Weighting	Subnetid Bits					Hostid Bits			Decimal Value
	128	64	32	16	8	4	2	1	
	0	0	0	0	0	0	0	0	0
	0	0	0	0	1	0	0	0	8
	0	0	0	1	0	0	0	0	16
	0	0	0	1	1	0	0	0	24
			32–216 in steps of eight						
	1	1	1	0	0	0	0	0	224
	1	1	1	0	1	0	0	0	232
	1	1	1	1	0	0	0	0	240
	1	1	1	1	1	0	0	0	248

b. The first step in determining the subnet address is to determine the number of subnetid bits by converting the second octet of the subnet mask to binary:

$$
\begin{array}{ccccccccc}
 & & & & 248 & & & & \\
\text{Binary weighting} & 128 & 64 & 32 & 16 & 8 & 4 & 2 & 1 \\
 & 1 & 1 & 1 & 1 & 1 & 0 & 0 & 0 \\
 & & \text{five subnetid bits} & & & & H & H & H
\end{array}
$$

The least significant of the subnetid bits has a binary weighting of eight. Therefore, the subnet addresses begin at 0 and increase by eight with each successive subnet address. The subnet addresses are shown in Table 20-7. The subnet addresses are 20.0.0.0, 20.8.0.0, 20.16.0.0, and so forth up to 20.248.0.0. To find the subnet address for the host address 20.38.40.2, simply AND the host address with the subnet mask. The result of ANDing the first octet reveals the network address, and the result of ANDing the three rightmost octets reveals the subnet address:

Host address	20.	38.	40.	2	00010010	00100110	00101000	00000010
Subnet mask	255.	248.	0.	0	11111111	11111000	00000000	00000000
	ANDing				00010010	00100000	00000000	00000000
					netid bits	subnetid bits		
Subnet address					20.	32.	0.	0

20-9 CLASS B SUBNET MASKS

With class B networks, there are 16 bits for the netid and 16 bits that can be divided between subnetid and hostid bits, depending on the subnet mask used. With classful IP addressing, class B subnet masks must have at least two logic 1s but not more than 14.

Figure 20-4 summarizes classful class B subnetting. The figure shows all class B subnet masks possible and how many subnet bits (subnetids), subnetworks, hosts, and host address bits (hostids) are available for each mask.

Example 20-9

For the IP address 172.25.12.45 and the subnet mask 255.255.255.0, determine the following:
a. IP address in binary
b. subnet addresses in binary
c. which bits are netids, subnetids, and hostids
d. network address
e. subnetwork address
f. host address

128 to 191	0 to 255	0 to 255	0 to 255
Netid	Netid	Hostid or Subnetid	Hostid or Subnetid
Network-unique 8 bits	Network-unique 8 bits	0 to 255 8 bits	0 to 255 8 bits

Subnet Mask   255   255   0, 192, 224, 240, 248, 252, 254, 255

Number of subnets ($2^n - 2$)	Number of hosts ($2^a - 2$)	Number of host address bits ($a = 16 - n$)	Class B subnet mask	Number of Subnet bits (n)
0	65,534	16	255.255.0.0	0
not allowed			255.255.128.0	1
2	16,382	14	255.255.192.0	2
6	8,190	13	255.255.224.0	3
14	4,094	12	255.255.240.0	4
30	2,046	11	255.255.248.0	5
62	1,022	10	255.255.252.0	6
126	510	9	255.255.254.0	7
254	254	8	255.255.255.0	8
510	126	7	255.255.255.128	9
1,022	62	6	255.255.255.192	10
2,046	30	5	255.255.255.224	11
4,094	14	4	255.255.255.240	12
8,190	6	3	255.255.255.248	13
16,382	2	2	255.255.255.252	14
not allowed			255.255.255.254	15
not allowed			255.255.255.255	16

FIGURE 20-4 Class B subnetting summary

Solution

	172.	25.	12.	45
a. IP address (binary)	10101100.	00011001.	00001100.	00101101
	255.	255.	255.	0
b. Subnet mask (binary)	11111111.	11111111.	11111111.	00000000
	10101100.	00011001.	00001100.	
c. Bit meaning	NNNNNNNN.	NNNNNNNN.	SSSSSSSS.	HHHHHHHH
d. Network address	172.	25.	0.	0
e. Subnet address	172.	25.	12.	0
f. Host address	172.	25.	12.	45

Example 20-10

For a host on a class B network with the address 140.38.40.2 and a subnet mask of 255.255.224.0, determine:

a. network address

b. subnet address where this host resides

Solution

a. The network address is determined by ANDing the default mask 255.255.0.0 with the host address:

Host address	140.	38.	40.	2
Default mask	255.	255.	0.	0
Network address	140.	38.	0.	0

Table 20-8   Subnet Addresses for Example 20-10

	Subnetid Bits			Hostid Bits					
	128	64	32	16	8	4	2	1	Decimal Value
Binary weighting	0	0	0	0	0	0	0	0	0
	0	0	1	0	0	0	0	0	32
	0	1	0	0	0	0	0	0	64
	0	1	1	0	0	0	0	0	96
	1	0	0	0	0	0	0	0	128
	1	0	1	0	0	0	0	0	160
	1	1	0	0	0	0	0	0	192
	1	1	1	0	0	0	0	0	224

b. The first step in determining the subnet address is to determine the number of subnetid bits by converting the third octet of the subnet mask to binary:

224

Binary weighting	128	64	32	16	8	4	2	1
	1	1	1	0	0	0	0	0
three subnetid bits				H	H	H	H	H

The least significant of the subnetid bits has a binary weighting of 32. Therefore, the subnet addresses begin at 0 and increase by 32 with each successive subnet address. The subnet addresses are shown in Table 20-8.

Therefore, the subnet addresses are 140.38.0.0, 140.38.32.0, 140.38.64.0, and so forth up to 140.38.224.0. To find the subnet address for the host address 140.38.40.2, simply AND the host address with the subnet mask. The result of ANDing the first and second octet reveals the network address, and the result of ANDing the third and fourth octets reveals the subnet address:

Host address	140.	38.	40.	2	10001100	00100110	00101000	00000010
Subnet mask	255.	255.	224.	0	11111111	11111111	11100000	00000000
			ANDing		10001100	00100110	00100000	00000000
					netid bits	netid bits	subnetid bits	
	Subnet address				140.	38.	32.	0

20-10   CLASS C SUBNET MASKS

With class C networks, there are 24 bits for the netid and eight bits that can be divided between subnetid and hostid bits, depending on the subnet mask used. With classful IP addressing, class C subnet masks must have at least two logic 1s but not more than six.

Figure 20-5 summarizes classful class C subnetting. The figure shows all class C subnet masks possible and how many subnets bits (subnetids), subnetworks, hosts, and host address bits (hostids) are available for each mask.

Example 20-11

For the IP address 192.168.40.112 and the subnet mask 255.255.255.192, determine the following:
a. IP address in binary
b. subnet addresses in binary
c. which bits are netids, subnetids, and hostids
d. network address
e. subnetwork address
f. host address

Solution

	192.	168.	40.	112
a. IP address (binary)	11000000.	10101000.	00101000.	01110000

	192 to 223	0 to 255	0 to 255	0 to 255
	Netid	Netid	Netid	Hostid or Subnetid
	Network-unique 8 bits	Network-unique 8 bits	Network-unique 8 bits	0 to 255 8 bits
Subnet Mask	255	255	255	0, 192, 224, 240, 248, 252, 254, 255

Number of subnets $(2^n - 2)$	Number of hosts $(2^a - 2)$	Number of host address bits $(a = 8 - n)$	Class C subnet mask	Number of Subnet bits (n)
0	254	8	255.255.255.0	0
← not allowed →			255.255.255.128	1
2	62	6	255.255.255.192	2
6	30	5	255.255.255.224	3
14	14	4	255.255.255.240	4
30	6	3	255.255.255.248	5
62	2	2	255.255.255.252	6
← not allowed →			255.255.255.254	7
← not allowed →			255.255.255.255	8

FIGURE 20-5 Class C subnetting summary

	255.	255.	255.	192
b. Subnet mask (binary)	11111111. 11000000.	11111111. 10101000.	11111111. 00101000.	11000000 01000000
c. Bit meaning	NNNNNNNN.	NNNNNNNN.	NNNNNNNN.	SSHHHHHH
d. Network address	192.	168.	40.	0
e. Subnet address	192.	168.	40.	64
f. Host address	192.	168.	40.	112

Example 20-12

For a host on a Class C network with the address 204.38.40.2 and a subnet mask of 255.255.255.240, determine:

a. network address
b. subnet address where this host resides

Solution

a. The network address is determined by ANDing the default mask 255.255.255.0 with the host address:

Host address	204.	38.	40.	2
Default mask	255.	255.	255.	0
Network address	204.	38.	40.	0

b. The first step in determining the subnet address is to determine the number of subnetid bits by converting the fourth octet of the subnet mask to binary:

	240							
Binary weighting	128	64	32	16	8	4	2	1
	1	1	1	1	0	0	0	0
four subnetid bits	H	H	H	H				

Table 20-9   Subnet Addresses for Example 20-12

	Subnetid Bits				Hostid Bits				
	128	64	32	16	8	4	2	1	Decimal Value
Binary weighting	0	0	0	0	0	0	0	0	0
	0	0	0	1	0	0	0	0	16
	0	0	1	0	0	0	0	0	32
			48–192 in steps of 16						
	1	1	0	1	0	0	0	0	208
	1	1	1	0	0	0	0	0	224
	1	1	1	1	0	0	0	0	240

The least significant of the subnetid bits has a binary weighting of 16. Therefore, the subnet addresses begin at 0 and increase by 16 with each successive subnet address. The subnet addresses are shown in Table 20-9.

Therefore, the subnet addresses are 204.38.40.0, 204.38.40.16, 204.38.40.32, and so forth up to 204.38.40.240. To find the subnet address for the host address 204.38.40.2, simply AND the host address with the subnet mask. The result of ANDing the first three octets reveals the network address, and the result of ANDing the last octet reveals the subnet address, as shown here:

Host address	204.	38.	40.	2	11001100	00100110	00101000	00000010
Subnet mask	255.	255.	255.	192	11111111	11111111	11111111	11000000
			ANDing		11001100	00100110	00101000	00000000
					netid bits	netid bits	netid bits	subnetid bits
Subnet address					204.	38.	40.	0

20-11   SUPERNETTING

Over the past several years, the Internet has experienced an exponential growth in utilization, which has raised a number of serious concerns. The foremost concern is facing the reality that the Internet is going to run out of addresses in the very near future. The other concerns are subtler, as they deal with the ability of the Internet's current routing system to adjust to the rapid and significant increase in traffic and whether it has the capability to sustain additional growth. The problem of running out of addresses manifested when the original Internet authorities assigned IP addresses on a network basis. When an organization received a class A network address, it automatically received all 16,777,216 host addresses whether they needed them all or not. Although class A and class B addresses are nearly depleted, approximately 15% of the class C addresses are still available.

20-11-1   Supernetworks

A temporary solution to the routing and address shortage problem is *supernetting,* which was developed to solve (or at least alleviate) the overcrowding problems on the Internet. With supernetting, an organization can combine several class C addresses to form a larger single block of addresses. In essence, supernetting combines small networks to create what is called a *supernetwork* where each network is treated like a subnetwork of the supernetwork. For example, rather than apply for a single class C address that supports 256 hosts, an organization can apply for a set of class C addresses. If an organization needs 900 addresses, it can apply for four class C addresses and then combine them into one larger network, which effectively forms a single supernetwork capable of supporting up to 1024 hosts.

Figure 20-6 illustrates the concept of supernetworking. Figure 20-6a shows the diagram for an organization with four separate class C networks: 210.18.8.0, 210.18.9.0, 210.18.10.0, and 210.18.11.0. Figure 20-6b shows the organization comprised of the same four class C networks that have been combined into one supernetwork with the supernet address 210.18.8.0.

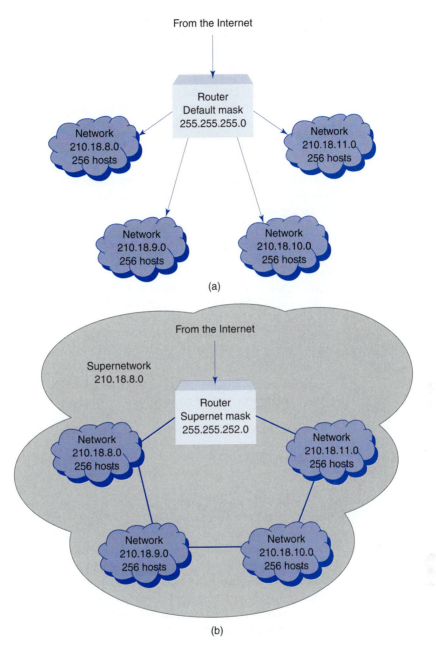

FIGURE 20-6 Supernetting: (a) four independent class C networks; (b) four class C networks combined into one supernetwork

When an organization receives multiple class C addresses, it can receive random addresses or addresses that occur in sequence. If the organization receives random addresses, routers on other networks (such as the Internet) will treat each address separately just as though they belonged to four different organizations. Random address assignment requires each router to have a separate entry in its routing table for each address, which increases the size of the routing table substantially.

A more logical solution is to assign an organization a continuous sequence of addresses, which allows routers to treat them as a single entity, necessitating only one entry in their routing table. To combine a group of class C addresses into a single supernetwork, the addresses must comply with the following set of rules:

1. The number of class C addresses must be a power of 2 (1, 2, 4, 8, 16, and so on).
2. The addresses must be contiguous with no gaps between them.
3. The third byte of the first network address in the supernetwork must be evenly divisible by the number of networks being combined (i.e., if the number of addresses combined is N, the third byte of the first address must be evenly divisible by N).

Supernetting reduces the amount of routing information stored in the backbone routers on the Internet, which reduces problems created when there are rapid changes in route availability due to irregular or unpredictable traffic variations (a condition called *route flapping*). Supernetting also eases the local administrative burden of updating external routing information.

Example 20-13

Which of the following sets of addresses meet the criteria necessary for supernetting four class C address groups?

a. 205.52.30.0	205.52.40.0	205.52.50.0	205.52.60.0
b. 202.12.16.0	202.12.17.0	202.12.18.0	202.12.19.0
c. 198.20.32.0	198.20.33.0	198.20.34.0	
d. 198.20.64.0	198.20.68.0	198.20.72.0	198.20.76.0
e. 210.40.64.0	210.40.65.0	210.40.66.0	210.40.67.0

Solution

a. Does not comply; third byte of first address is not evenly divisible by the number of address blocks (four) and the addresses are not contiguous.
b. Complies.
c. Does not comply; the number of addresses is not a power of 2.
d. Does not comply; the addresses are not contiguous.
e. Complies.

20-11-2 Supernet Masks

When an organization receives a class C network address, the first (lowest) host address in the group specifies the network address, and the default mask defines the range of addresses within the network. When an organization divides a group of addresses into subnets, the first (lowest) address in group (network address) and the subnet mask specify the subnet addresses and the range of host addresses within each subnet. Therefore, both the network address and the subnet mask must be known. Likewise, when an organization elects to combine several network addresses into a supernetwork, the first address of the supernet and the supernet mask must be known.

Subnet masks must contain more 1s than the default mask for that class. In essence, a subnet mask steals from the host identification bits (hostids) and converts them to subnet identification bits (subnetids). In contrast to subnetting, supernetting borrows bits from the network identification. Therefore, a supernet mask must contain fewer 1s than the default mask for that class. A comparison of default, subnet, and supernet masking is shown in Figure 20-7. The default mask is 255.255.255.0, which contains 24 contiguous 1s and does not support subnetting. The subnet mask shown has 27 contiguous 1s, which specifies three subnet bits (i.e., eight subnets). The supernet mask shown has 21 1s, which indicates that three of the network identification bits are used to define eight network addresses that have been combined into one supernetwork.

It is not possible to combine four class C networks using the default mask 255.255.255.0. However, if a supernet mask is used, the four networks can appear to a router as a single network. Figure 20-8 lists in binary form the network addresses for the four networks shown in

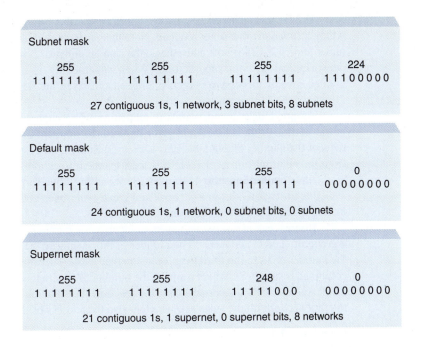

Subnet mask

255	255	255	224
1 1 1 1 1 1 1 1	1 1 1 1 1 1 1 1	1 1 1 1 1 1 1 1	1 1 1 0 0 0 0 0

27 contiguous 1s, 1 network, 3 subnet bits, 8 subnets

Default mask

255	255	255	0
1 1 1 1 1 1 1 1	1 1 1 1 1 1 1 1	1 1 1 1 1 1 1 1	0 0 0 0 0 0 0 0

24 contiguous 1s, 1 network, 0 subnet bits, 0 subnets

Supernet mask

255	255	248	0
1 1 1 1 1 1 1 1	1 1 1 1 1 1 1 1	1 1 1 1 1 0 0 0	0 0 0 0 0 0 0 0

21 contiguous 1s, 1 supernet, 0 supernet bits, 8 networks

FIGURE 20-7   Comparision of subnet, default, and supernet masks

Network Address	Network ID		Supernet ID	Host ID
	1 1 0 1 0 0 1 0	0 0 0 1 0 0 1 0	0 0 0 0 1 0 S S	H H H H H H H H
	1st octet	2nd octet	3rd octet	4th octet
210.18.8.0	1 1 0 1 0 0 1 0	0 0 0 1 0 0 1 0	0 0 0 0 1 0 0 0	H H H H H H H H
210.18.9.0	1 1 0 1 0 0 1 0	0 0 0 1 0 0 1 0	0 0 0 0 1 0 0 1	H H H H H H H H
210.18.10.0	1 1 0 1 0 0 1 0	0 0 0 1 0 0 1 0	0 0 0 0 1 0 1 0	H H H H H H H H
210.18.11.0	1 1 0 1 0 0 1 0	0 0 0 1 0 0 1 0	0 0 0 0 1 0 1 1	H H H H H H H H
Bit position	31 24	23 16	15 9 8	7 0

FIGURE 20-8   Binary addresses for the four networks shown in Figure 1a

Figure 20-6a. Each network has a different network address, and the combined range of IP address for the four class C networks is 210.18.8.0 to 210.18.11.255. Note that the four network addresses are exactly the same except for bit positions 9 and 8. Therefore, the same block of addresses can be assigned a single network address 210.18.8.0 with a supernet mask of 255.255.252.0. The supernet mask has two fewer 1s than the default mask, thus allowing two bits to be stolen from the network ID, creating a supernetwork comprised of four networks, as shown in Figure 20-6b. The addresses for the four networks are 210.18.8.0, 210.18.9.0, 210.18.10.0, and 210.18.11.0, which without knowing the supernet mask would appear to be four independent class C networks. However, using the supernet mask 255.255.252.0, the four networks appear as one larger network with the single address 210.18.8.0.

Unlike normal subnetting using subnet masks, the number of hosts on each network within a supernetwork is not reduced. Note that each network shown in Figure 20-6b has a maximum of 2^8 (256) host addresses—the same number of hosts each of the four original networks had without supernetting. When supernetting is utilized, an organization is granted several successive class C addresses and given the first address (the network address) and the supernet mask.

As previously stated, standard subnetting using subnet masks steals bits from the hostid portion of the address and gives them to the subnetid portion of the address, which reduces the number of available host addresses. Supernetting, on the other hand, steals bits from the netid portion of the address and uses them to establish a single larger supernetwork, thus conserving host addresses. Supernetting, therefore, increases the number of available host addresses on a network and greatly improves network performance by eliminating the need for internal routing.

Routers can use supernet masks to identify the number of bits used for supernetting a group of class C addresses. The number of bits is simply the difference between 24 and the number of 1s in the third octet of a supernet mask. Therefore, the number of networks combined is simply 2^n, where n equals the number of supernet bits. When an address is logically ANDed with a supernet mask, the result of the ANDing operation is the address of the first network on the supernet, which is also the address of the supernet.

Example 20-14

The first address of a supernet is 200.42.32.0, and the supernet mask is 255.255.248.0. Determine:

a. the number of networks in the supernet
b. the addresses of the networks
c. which of the following addresses belong to the supernets: 200.42.40.50, 200.42.35.18, and 200.43.33.17

Solution

a. The supernet mask has 21 leading 1s; therefore, there must be three supernet bits $(24 - 21 = 3)$. With three supernet bits, there are eight networks $(2^3 = 8)$.
b. The eight networks begin with the address 200.42.32.0 and end with the address 200.42.39.0.
c. Of the three addresses listed, only 200.42.35.18 is within the address range of this supernet.

Whenever a supernet mask is ANDed with a network address that is within its supernet, the result of the ANDing operation is the address of the first network within the supernet (i.e., the supernet address).

Example 20-15

The first address of a supernet is 200.42.32.0, and the supernet mask is 255.255.248.0. Determine which of the following addresses belongs to this supernetwork:

a. 200.42.40.50
b. 200.42.35.18
c. 200.42.18.17

Solution Each of the three addresses is ANDed with the supernet mask:

200.42.40.50 ANDed with 255.255.248.0 = 200.42.40.0 (not in this supernetwork)
200.42.35.18 ANDed with 255.255.248.0 = 200.42.32.0 (in this supernework)
200.42.42.23 ANDed with 255.255.248.0 = 200.42.16.0 (not in this supernetwork)

The only address that yielded the address of the first network in the supernet is 200.42.35.18; therefore, this is the only address that is part of this supernetwork.

20-12 CLASSLESS IP ADDRESSING

In 1996, in an effort to overcome the shortfalls of classful addressing, Internet authorities announced a new addressing architecture called *classless IP addressing*. The basic philosophy of classless addressing is to incorporate *variable-length blocks* of IP addresses that essentially belong to no class (hence the name *classless*). The underlying purpose for using variable-length blocks of addresses is to allow network designers the ability to design networks to accommodate virtually any company or organization regardless of its size. Variable-length blocks of addresses also allow networks to be subdivided in such a way that it is not necessary that all blocks have the same number of subnets or host addresses. The total number of IP addresses available with classless addressing is the same as with classful addressing (2^{32}, or approximately 4 billion). However, classless addressing allows the addresses to be allocated more discriminately and used more efficiently.

With classful addressing, the entire address allocation was divided into classes of addresses that were subdivided into smaller blocks of addresses that could be divided even further into subnets. However, the number of blocks within a class of addresses was a predetermined value (class A = 2^8, class B = 2^{16}, and class C = 2^{24}), and the number of addresses within a block was fixed (class A = 2^{24}, class B = 2^{16}, and class C = 2^8).

With classless addressing, the entire address allocation can be subdivided into blocks that can be subdivided further into subnets; however, the number of blocks is not fixed, and the size of the blocks can range from two addresses to literally thousands, depending on the need. Classless addressing is similar to classful addressing in the sense that the beginning address of a block and the network mask define the structure of the entire block. With classless addressing, blocks of addresses can be further subdivided into subnets, as will be shown later. However, small blocks of classless addresses should not be confused with subnets. A comparison of classful and classless addressing is shown in Figure 20-9.

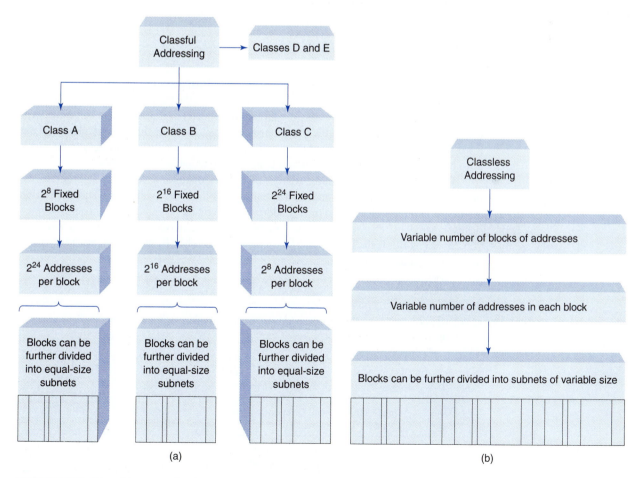

FIGURE 20-9 Classful versus classless addressing structure: (a) classful; (b) classless

Networks and Subnetworks

20-12-1   Restrictions on Classless Addresses

Humans like to think of IP addresses as dotted-decimal numbers; however, in actuality they are not. IP addresses are 32-bit binary weighted numbers, which are based on powers of 2. Therefore, one stipulation on the number of addresses in a block of classless addresses is that the number of addresses must be a power of 2 (1, 2, 4, 8, and so on). For example, a private household can be assigned two addresses, a small business eight addresses, and a large company or organization 2048 addresses. Another stipulation of classless addressing is that the beginning address of the block must be evenly divisible by the number of addresses in the block (the quotient must be a positive integer value). For example, if there are 16 addresses in a block, the beginning address must be a multiple of 16 (16, 32, 48, and so on).

Example 20-16

Determine which of the following IP addresses can be the beginning address for a classless block containing 16 addresses:

> 210.44.62.80
>
> 50.12.45.47
>
> 117.67.32.32
>
> 172.41.15.37

Solution

> 210.44.62.80 $80/16 = 5$ (can)
>
> 50.12.45.47 $47/16 = 2.9375$ (cannot)
>
> 117.67.32.32 $32/16 = 2$ (can)
>
> 172.41.15.37 $37/16 = 2.3125$ (cannot)

Of the four addresses listed, only 210.44.62.80 and 117.67.32.32, when divided by 16, yield a quotient that is a positive integer value.

20-12-2   Determining Classless Network Addresses

With a network that contains less than 256 host addresses, the beginning address of the block is relatively simple to determine, as only the rightmost octet of the address must be checked. Because the lowest address of the block must be a multiple of the number of addresses, the lowest bit position within the final octet that can be used to identify an address in the block must have a binary weighting equal to the number of addresses in the block.

To determine the beginning address of a block containing less than 256 addresses, only the rightmost octet is of concern. Therefore, only the binary weightings shown in Table 20-10 are of importance. If the lowest address in the block is eight bits b_7 through b_3 must be designated by the network mask as part of the network address (i.e., a mask of 255.255.255.248). If the lowest address in a block is 32, bits b_7 through b_5 must be part of the network address, and the network mask is 255.255.255.224.

Example 20-17

Determine

a. the network mask for a block of 32 addresses on a classless network with a beginning address of 200.40.10.0

Table 20-10   Binary Weightings for the Fourth Octet

First-Octet Netid	Second-Octet Netid	Third-Octet Netid	Fourth-Octet Netid or Hostid							
—	—	—	b_7	b_6	b_5	b_4	b_3	b_2	b_1	b_0
—	—	—	128	64	32	16	8	4	2	1

b. how many blocks of addresses are capable of containing 32 addresses

c. the block addresses

d. the range of addresses in each block

Solution

a. Since there are 32 addresses in the block, the network address bit with the lowest weighting must be 32. Therefore, b_7, b_6, and b_5 must be designated as network address bits, and b_4, b_3, b_2, b_1, and b_0 must be identified as hostid bits, as shown in Table 20-11. The mask must contain three bits in the fourth octet and is therefore 255.255.255.224.

b. Using three bits in the fourth octet of the mask allows for eight blocks with 32 addresses each.

c. The addresses for the eight blocks are determined as shown in Figure 20-10, which shows that the eight possible blocks of addresses begin at 200.40.10.0 and increment by 32, as shown in Table 20-12.

d. Since there are five hostid bits in each block, there are $2^5 = 32$ host addresses in each block, where the host addresses in each block range from 0 to 31. The range of host addresses for the eight blocks is shown in Table 20-13.

Table 20-11 Network Mask for Example 20-17

First-Octet Netid	Second-Octet Netid	Third-Octet Netid	Fourth-Octet							
			Netid			Hostid				
—	—	—	b_7	b_6	b_5	b_4	b_3	b_2	b_1	b_0
—	—	—	128	64	32	16	8	4	2	1
11111111.11111111.11111111.			1	1	1	0	0	0	0	0
255.	255.	255.	224							
			Network mask							

First-octet Netid Bits	Second-octet Netid Bits	Third-octet Netid Bits	Fourth-Octet							
			Netid Bits			Hostid Bits				
200	40	10	b_7	b_6	b_5	b_4	b_3	b_2	b_1	b_0
200	40	10	128	64	32	16	8	4	2	1
200.	40.	10.	0	0	0	0	0	0	0	0
			0							
200.	40.	10.	0	0	1	0	0	0	0	0
			32							
200.	40.	10.	0	1	0	0	0	0	0	0
			64							
200.	40.	10.	0	1	1	0	0	0	0	0
			96							
200.	40.	10.	1	0	0	0	0	0	0	0
			128							
200.	40.	10.	1	0	1	0	0	0	0	0
			160							
200.	40.	10.	1	1	0	0	0	0	0	0
			192							
200.	40.	10.	1	1	1	0	0	0	0	0
			224							

FIGURE 20-10 Address for eight blocks in Example 20-17

Table 20-12 Blocks of Addresses for Example 20-17

Block	Address
Block 0	200.40.10.0
Block 32	200.40.10.32
Block 64	200.40.10.64
Block 96	200.40.10.96
Block 128	200.40.10.128
Block 160	200.40.10.160
Block 192	200.40.10.192
Block 224	200.40.10.224

Table 20-13 Range of Host Addresses for Example 20-17

Block	Address Range
Block 0	200.40.10.0 to 200.40.10.31
Block 32	200.40.10.32 to 200.40.10.63
Block 64	200.40.10.64 to 200.40.10.95
Block 96	200.40.10.96 to 200.40.10.127
Block 128	200.40.10.128 to 200.40.10.159
Block 160	200.40.10.160 to 200.40.10.191
Block 192	200.40.10.192 to 200.40.10.223
Block 224	200.40.10.224 to 200.40.10.255

Table 20-14 Binary Weightings for Third and Fourth Octets in Networks Having between 256 and 16,536 Hosts

First-Octet Netid	Second-Octet Netid	Third-Octet Netid or Hostid								Fourth-Octet Netid or Hostid							
—	—	b_{15}	b_{14}	b_{13}	b_{12}	b_{11}	b_{10}	b_9	b_8	b_7	b_6	b_5	b_4	b_3	b_2	b_1	b_0
—	—	32768	16384	8192	4096	2048	1024	512	256	128	64	32	16	8	4	2	1

One feature that makes classless addressing so appealing is that the size of the block of addresses is relatively independent of where the addresses fall within the total IP address allocation. For instance, in the previous example the network required a block of less than 256 addresses, so it seemed appropriate to assign it a block of addresses in the class C range (0–127). However, it would have been just as appropriate to use a class A or B address because the mask (not the network number) defines the capability of the network. Unfortunately, most class A and B addresses are already assigned; therefore, classless addressing is utilized primarily with addresses that fall into the class C range.

To determine block addresses for a network having between 256 and 16,535 hosts, the two rightmost bytes must be checked, and for a network having between 16,536 and 16,777,215 hosts, the three rightmost bytes must be checked. These calculations are somewhat more involved, as they necessitate manipulating two or three octets of the address in binary. The weightings for the network and hostid bits for the third and fourth octets in networks having between 256 and 16,536 hosts are shown in Table 20-14.

Example 20-18

For a network that begins with the address 170.40.0.0, determine

a. the mask for establishing an address block containing 1024 addresses

b. how many blocks are capable of containing 1024 host addresses

1st and 2nd Octet Netid	3rd Octet Netid or Hostid								4th Octet Hostid							
	b_{15}	b_{14}	b_{13}	b_{12}	b_{11}	b_{10}	b_9	b_8	b_7	b_6	b_5	b_4	b_3	b_2	b_1	b_0
255.255	32768	16384	8192	4096	2048	1024	512	256	128	64	32	16	8	4	2	1
255.255	N	N	N	N	N	N	H	H	H	H	H	H	H	H	H	H
255.255	1	1	1	1	1	1	0	0	0	0	0	0	0	0	0	0
255.255	252.								0							

Note. N = netid bits; H = hosted bits.

FIGURE 20-11 Netid and hostid bits for Example 20-18

1st and 2nd Octets	3rd Octet Netid and Hostid Bits								4th Octet Hostid Bits							
	b_{15}	b_{14}	b_{13}	b_{12}	b_{11}	b_{10}	b_9	b_8	b_7	b_6	b_5	b_4	b_3	b_2	b_1	b_0
—																
170.40.	0	0	0	0	0	0	0	0	0	0	0	0	0	0	0	0
	0.								0							
170.40.	0	0	0	0	0	1	0	0	0	0	0	0	0	0	0	0
	4.								0							
---------- 170.40.8.0–170.40.244 ----------																
170.40.	1	1	1	1	1	0	0	0	0	0	0	0	0	0	0	0
	248.								0							
170.40.	1	1	1	1	1	1	0	0	0	0	0	0	0	0	0	0
	252.								0							

FIGURE 20-12 64 block addresses for Example 20-18

c. the block addresses

d. the range of host addresses in each block

Solution

a. Since there must be 1024 addresses in the block, the least-significant netid bit must have a binary weighting of 1024. Therefore, b_{15} through b_{10} must be designated netid bits and b_9 through b_0 designated as hostid bits, as shown in Figure 20-11. For the network mask to be able to identify six netid bits in the third octet, it must be 255.255.252.0.

b. Using six bits in the third octet for the netid enables 2^6 (64) blocks beginning with 170.40 capable of containing 1024 host addresses.

c. The lowest block address is the network address 170.40.0.0, and the beginning address of each successive block is incremented by four in dotted-decimal notation (which corresponds to 1024 binary). The 64 block addresses are derived as shown in Figure 20-12. The 64 possible block addresses begin at 170.40.0.0 and increment by four (1024 binary), as shown in Table 20-15.

d. Since each block contains 10 hostid bits, they each have 2^{10} host addresses with a dotted-decimal range from 0 to 3.255. The range of host addresses in each block is shown in Table 20-16.

Table 20-15 Block Addresses for Example 20-18

Block	Address
Block 0	170.40.0.0
Block 4	170.40.4.0
Block 8	170.40.8.0
Block 12	170.40.12.0
Blocks 16–244	
Block 248	170.40.248.0
Block 252	170.40.252.0

Table 20-16 Range of Host Addresses for Example 20-18

Block	Address Range
Block 0	170.40.0.0 to 170.40.3.255
Block 4	170.40.4.0 to 170.40.7.255
Block 8	170.40.8.0 to 170.40.11.255
Block 12	170.40.12.0 to 170.40.15.255
Blocks 16–244	170.40.16.0 to 170.40.247.255
Block 248	170.40.248.0 to 170.40.251.255
Block 252	170.40.252.0 to 170.40.255.255

20-12-3 Slash Notation for Classless Addressing

Writing the four-octet dotted-decimal notation for a network mask can be rather awkward. However, writing the entire address may not be necessary, especially since we know that the format for the mask is quite simple, as it is comprised of a series of left-justified contiguous 1s followed by a series of right-justified contiguous 0s. All we really need to know about the mask is how many leading 1s there are, which indirectly specifies how many trailing 0s it has. For example, the mask 255.255.248.0 has 21 leading 1s, so we could say that the mask contains 21 1s or simply append 21 to the network address separated by a slash, as shown here:

$$A.B.C.D/n$$

where $A.B.C.D$ = the dotted-decimal network address

n = number of 1s in the subnet mask

Writing a network address followed by a slash (/) and the number of 1s in the subnet mask (n) is called *slash notation*.

The number of 1s in a network mask defines the *prefix*, *prefix length*, *suffix*, and *suffix length* of a classless IP address. The prefix is the name given to the bits in a classless address that are common to all hosts within a given block of addresses. Prefix bits are similar to the netid bits in a classful address. The prefix length is simply the number of bits in the network portion of the address and is equal to n. The suffix is the part of a classless address that varies from host to host and is similar to hostid bits with classful addressing. The suffix length (s) is the number of bits in the host portion of the address and equals $32 - n$. The number of hosts in a block of addresses is simply 2^s, where s is the suffix length. Table 20-17 gives a summary of network masks for classless networks.

Example 20-19

Determine the network mask for the following classless addresses:

a. 160.40.211.13/21
b. 50.120.18.10/11
c. 210.144.40.70/26

Table 20-17 Classless Network Mask Summary

n	s	Network Mask	Number of Host Addresses
1	31	128.0.0.0	2^{31}
2	30	192.0.0.0	2^{30}
3	29	224.0.0.0	2^{29}
4	28	240.0.0.0	2^{28}
5	27	248.0.0.0	2^{27}
6	26	252.0.0.0	2^{26}
7	25	254.0.0.0	2^{25}
8	24	255.0.0.0	2^{24}
9	23	255.128.0.0	2^{23}
10	22	255.192.0.0	2^{22}
11	21	255.224.0.0	2^{21}
12	20	255.240.0.0	2^{20}
13	19	255.248.0.0	2^{19}
14	18	255.252.0.0	2^{18}
15	17	255.254.0.0	2^{17}
16	16	255.255.0.0	2^{16}
17	15	255.255.128.0	2^{15}
18	14	255.255.192.0	2^{14}
19	13	255.255.224.0	2^{13}
20	12	255.255.240.0	2^{12}
21	11	255.255.248.0	2^{11}
22	10	255.255.252.0	2^{10}
23	9	255.255.254.0	2^{9}
24	8	255.255.255.0	2^{8}
25	7	255.255.255.128	2^{7}
26	6	255.255.255.192	2^{6}
27	5	255.255.255.224	2^{5}
28	4	255.255.255.240	2^{4}
29	3	255.255.255.248	2^{3}
30	2	255.255.255.252	2^{2}
31	1	255.255.255.254	2^{1}
32	0	255.255.255.255	2^{0}

Solution

a. 160.40.211.13/21 $n = 21$, which specifies 21 1s in the subnet mask:

      11111111.11111111.11111000.00000000

        255.   255.   248.   0

b. 50.120.18.10/11 $n = 11$, which specifies 111s in the subnet mask:

      11111111.11100000.00000000.00000000

        255.   254.   0.   0

c. 210.144.40.70/26 $n = 26$, which specifies 26 1s in the subnet mask:

      11111111.11111111.11111111.11000000

        255.   255.   255.   192

Example 20-20

Given the address 210.18.35.24/28, determine

a. prefix length
b. suffix length
c. network mask
d. prefix
e. suffix
f. number of hosts in the block
g. minimum and maximum addresses of the block

h. network address

i. the range of address in the block

Solution

a. The prefix length $n = 28$.

b. The suffix length $s = 32 - n$, or $32 - 28 = 4$.

c. The subnet mask contains 28 logic 1s:

$$11111111.11111111.11111111.11110000$$
$$255. \qquad 255. \qquad 255. \qquad 240$$

d. The prefix is the first 28 bits of the address:

$$\underbrace{11010010.00010010.00100011.0001}_{\text{prefix}}1000$$

e. The suffix is the last four bits of the address:

$$11010010.00010010.00100011.0001\underline{1000}$$
$$\text{suffix}$$

f. The number of hosts is 2^s, or $2^4 = 16$.

g. The first 28 bits are common to all addresses within the block. Therefore, the minimum address for the block is determined by changing all suffix bits to 0, and the maximum address for the block is determined by changing all suffix bits to 1:

minimum address
$$11010010.00010010.00100011.0001\underline{0000}$$
$$210. \qquad 18. \qquad 35. \qquad 16$$

maximum address
$$11010010.00010010.00100011.0001\underline{1111}$$
$$210. \qquad 18. \qquad 35. \qquad 31$$

h. The network address is the minimum address in the block, or 210.18.35.16.

i. The range of addresses includes all addresses between the minimum and the maximum, or 210.18.35.16 to 210.18.35.31.

20-12-4 Subnetting on Classless Networks

Subnetting with classless networks is essentially the same as it is with classful networks except simpler. Each network has a subnet mask that identifies the subnetworks within its block of addresses. The subnet mask is simply the network mask with more 1s added to the prefix. The additional 1s identify how many subnets there are. For example, if the prefix for a network is 18, the prefix length can be increased to 21 bits, thus creating three subnet bits and eight (2^3) subnets.

Example 20-21

An organization is assigned a classless block of addresses with the address 135.24.10.64/26, and the organization wants to create four subnets. Determine

a. the network mask

b. the subnet mask

c. the four subnet addresses

d. the range of addresses for each subnet

Solution

a. The network mask contains 26 1s and is therefore

$$11111111.11111111.11111111.11000000$$
$$255. \qquad 255. \qquad 255. \qquad 192$$

b. To create four subnets, an additional two bits must be added to the network mask. Therefore, the subnet mask is

$$11111111.11111111.11111111.11110000$$
$$255. \qquad 255. \qquad 255. \qquad 240$$

1st Octet	2nd Octet	3rd Octet	\u2190 4th Octet \u2192						
					Subnetid Bits		Hostid Bits		
\u2190 Prefix \u2192					\u2192				
			b_7	b_6	b_5	b_4	b_3	b_2	b_1 b_0
135.	24.	10.	128	64	32	16	8	4	2 1
135.	24.	10.	0	1 (64)	0	0	H	H	H H
135.	24.	10.	0	1 (80)	0	1	H	H	H H
135.	24.	10.	0	1 (96)	1	0	H	H	H H
135.	24.	10.	0	1 (112)	1	1	H	H	H H

FIGURE 20-13 Subnet addresses for Example 20-21

c. The subnet addresses are determined by setting the two subnet bits to each of the four binary combinations possible with two bits, as shown in Figure 20-13. The four subnet addresses are

135.24.10.64/28   135.24.10.96/28
135.24.10.80/28   135.24.10.112/28

d. The ranges of addresses for the four subnets are found by first changing all the host (H) bits to 0s and finding the minimum address and then changing all host bits to 1s and finding the maximum address. The ranges of addresses for the four subnets are

135.24.10.64   to   135.24.10.79
135.24.10.80   to   135.24.10.95
135.24.10.96   to   135.24.10.111
135.24.10.112   to   135.24.10.127

20-12-5 Variable-Length Subnetting

Variable-length subnetting is when a network is divided into subnets, which further subdivide one or more of the original subnets into smaller subnets. The Internet allows variable-length subnetting on networks but not on the Internet itself. Variable-length subnetting involves using *variable-length subnet masks*. For example, an organization has a class C address 200.40.24.0 and wishes to subdivide the network into five subnets with 60, 55, 40, 30, and 25 hosts. Using two subnet bits on a classless network allows four subnets with 64 hosts each, which would accommodate the three subnets with 40, 55, and 60, hosts but there would be only one subnet address left for the remaining two subnets. Accommodation of five subnets requires using three subnet bits, which allows eight subnets, each accommodating only 32 host addresses.

The solution to this problem is to use variable-length subnet masks, which requires using two routers, as shown in Figure 20-14. The first router uses a subnet mask with 26 1s (255.255.255.192), as shown in Figure 20-15, to separate the network into four subnets each with a capacity of 64 hosts. Three of the subnets are used for the subnets requiring 40, 55, and 60 host addresses. The remaining subnet passes through a second router that applies a subnet mask with 27 1s (255.255.255.224). The second router further divides the subnet into two smaller subnets each capable of handling 32 hosts.

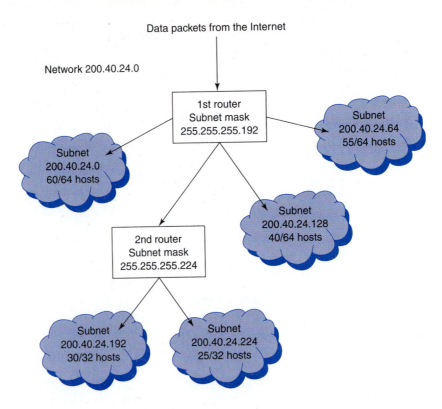

Data packets from the Internet

Network 200.40.24.0

1st router
Subnet mask
255.255.255.192

Subnet
200.40.24.64
55/64 hosts

Subnet
200.40.24.0
60/64 hosts

Subnet
200.40.24.128
40/64 hosts

2nd router
Subnet mask
255.255.255.224

Subnet
200.40.24.192
30/32 hosts

Subnet
200.40.24.224
25/32 hosts

FIGURE 20-14   Variable-length subnetting

Example 20-22

A small business has been assigned the classless network address 204.238.7.0. The proposed network structure for the business is shown in Figure 20-16. Determine

a. how many total subnets are needed for the network
b. how many total addresses are needed for the network
c. the subnet masks and classless IP subnet address for each of the complexes and the three wide area networks (WANs)
d. how many host addresses are used up to accommodate the entire network

Solution

a. Since each WAN requires its own subnet, the network must be divided into seven subnets: airport complex, design complex, home office complex, warehouse complex, and three WANs.
b. The total number of addresses required is the following:

airport complex	6
design complex	10
warehouse complex	33
home office complex	22
three WANs with two each	6
	77 total host addresses

c. Since there are seven total subnets required, it seems logical that on a classless network 3 subnet bits would be adequate, as that would accommodate 2^3, or eight, subnets each with the subnet mask 255.255.255.224. However, using three subnet bits leaves only five hostid bits, which provides for a maximum of 30 ($2^5 - 2$) hosts on each subnet. This allocation will not be adequate, as there are 33 hosts on the warehouse subnet. The only solution to the problem is to use variable-length subnet masks.

Net address	200.	40.	24.	0
1st subnet mask	255.	255.	255.	192

			4th Octet							
	Weightings	128	64	32	16	8	4	2	1
		H	H	H	H	H	H	H	H
		1	1	0	0	0	0	0	0
	After ANDing	S	S	0	0	0	0	0	0
Subnet 0			0	0	H	H	H	H	H	H
Subnet 64			0	1	H	H	H	H	H	H
Subnet 128			1	0	H	H	H	H	H	H
Subnet 192	200.	40.	24.	1	1	H	H	H	H	H	H

Subnet 192	200.	40.	24.	192
Second subnet mask	255.	255.	255.	224

			4th Octet							
	Weightings	128	64	32	16	8	4	2	1
		S	S	S	H	H	H	H	H
		1	1	1	0	0	0	0	0
	After ANDing	S	S	S	0	0	0	0	0
Subnet 192			1	1	0	H	H	H	H	H
Subnet 224			1	1	1	H	H	H	H	H

Subnet	Hosts	Address Range
0	64	200.40.24.0–200.40.24.63
64	64	200.40.24.64–200.40.24.127
128	64	200.40.24.128–200.40.24.191
192	32	200.40.24.192–200.40.24.223
224	32	200.40.24.224–200.40.24.255

FIGURE 20-15 Variable-length subnet masks

There are several acceptable solutions to the problem. However, it is general practice to assign the subnet with the largest number of hosts the lowest subnet address, the subnet with the second-largest number of hosts the next-higher subnet, and so forth. It is also customary to separate WAN subnet addresses from the populated subnets. Therefore, WAN subnets are generally assigned to the highest subnet addresses.

The warehouse complex is the most populated subnet with 33 hosts. To define 33 hosts requires six hostid bits, which leaves only two subnetid bits. Therefore, the subnet mask for the warehouse subnet is 255.255.255.192, and the warehouse subnet address is 204.238.7.0, as shown in Figure 20-17.

The home office subnet is populated with 22 hosts, which requires five hostid bits. Therefore, a subnet mask with three subnetid bits is required. Thus, the home office subnet mask is 255.255.255.224, and the subnet address is 204.238.7.64, as shown in Figure 20-17.

The design subnet has 10 hosts, which requires four hostid bits, which leaves five subnetid bits. Thus, the design subnet mask is 255.255.255.224, and the subnet address is 204.238.7.96, as shown in Figure 20-17.

The airport subnet must have six host addresses, which requires three hostid bits. Therefore, there are five subnetid bits, and the subnet mask is 255.255.255.248, and the subnet address is 204.238.7.112, as shown in Figure 20-17.

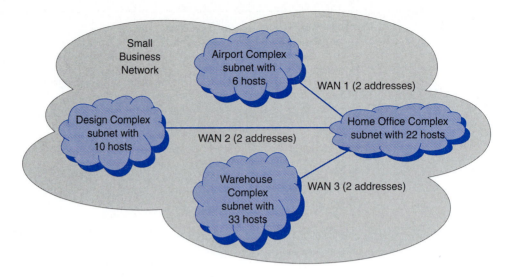

FIGURE 20-16 Network for Example 20-22

Each of the WANs requires two hostid bits, as shown in Figure 20-17. Therefore, all three WAN subnet masks are 255.255.255.252, and their subnet addresses are the following:

205.238.70.244
205.238.70.248
205.238.70.252

d. 77 out of 132 host addresses on the seven subnets are assigned, leaving 124 (120–243) host addresses available on unassigned subnets for future growth.

20-13   CLASSLESS INTERDOMAIN ROUTING

A disadvantage of classless addressing is routers have to change the organization of their routing tables and routing algorithms to accommodate variable-length subnetwork masks. The *Internet Engineering Task Force* (IETF) recently introduced a new technique of routing called *classless interdomain routing* (CIDR; pronounced "cider" as in apple cider). CIDR derived its name from the fact that it disregards the traditional class A, B, and C address designations and establishes the network–host identification boundary wherever it chooses and does it in a way that simplifies routing data packets through the Internet. CIDR outlines a new method of dividing up the available IP address allocation, which allows existing IP addresses to be combined into larger groupings called routing domains. CIDR allows each routing domain to contain more host addresses than the combined sums of the individual host addresses of all the domains it combines.

CIDR defines two fundamental conditions of the network: the address must be classless, and *interdomain routing* must be used. In addition, CIDR requires incorporating routers that are capable of dealing with it and a classless routing protocol, such as Routing Information Protocol version 2 (RIPv2). CIDR works best with blocks of contiguous addresses, which allows the addresses to be managed as one or more logical blocks of addresses that operate at specific bit boundaries that separate the host and network portions of the address. The primary idea behind CIDR is that it supports route aggregation, which permits existing addresses to be combined into larger routing domains. With CIDR, one entry in a routing table can represent the address space of literally thousands of traditional classful addresses. Internet Service Providers (ISPs) are overjoyed with CIDR, as it provides them with the additional address space they desperately need to continue to grow.

	Subnet address bits	WAN address bits	
255	1 1 1 1 1 1	1 1	WAN 3 (Subnet 204.238.7.252)
	2 out of 4	1 0	Address 204.238.7.254
		0 1	Address 204.238.7.253
252	1 1 1 1 1 1	0 0	Subnet Mask 255.255.255.252
251	1 1 1 1 1 0	1 1	WAN 2 (Subnet 204.238.7.248)
	2 out of 4	1 0	Address 204.238.7.250
		0 1	Address 204.238.7.249
248	1 1 1 1 1 0	0 0	Subnet Mask 255.255.255.252
247	1 1 1 1 0 1	1 1	WAN 1 (Subnet 204.238.7.244)
	2 out of 4	1 0	Address 204.238.7.246
		0 1	Address 204.238.7.245
244	1 1 1 1 0 1	1 0	Subnet Mask 255.255.255.252
243	1 1 1 1 0 0 1 1		120 Addresses are set aside for future growth
	– – – – – – not needed – – – –		
120	0 1 1 1 1 0 0 0		

	Subnet address bits	Host address bits	
119	0 1 1 1 0	1 1 1	Airport Complex (Subnet 204.238.7.112)
	6 out of 8	111 to 118	Hosts 204.238.7.113 – 118
112	0 1 1 1 0	0 0 0	Subnet Mask 255.255.255.248
111	0 1 1 0	1 1 1 1	Design Complex (Subnet 204.238.7.96)
	10 out of 16	97 to 110	Hosts 204.238.7.97 – 106
96	0 1 1 0	0 0 0 0	Subnet Mask 255.255.255.240
95	0 1 0	1 1 1 1 1	Home Office Complex (Subnet 204.238.7.64)
	22 out of 32	65 to 94	Hosts 204.238.7.65 – 86
64	0 1 0	0 0 0 0 0	Subnet Mask 255.255.255.224
63	0 0	1 1 1 1 1 1	Warehouse Complex (Subnet 204.238.7.0)
	33 out of 64	1 to 62	Hosts 204.238.7.1 – 33
0	0 0	0 0 0 0 0 0	Subnet Mask 255.255.255.192

FIGURE 20-17 Subnet masks for Example 20-22

The format for the routing table in a router configured to handle CIDR addresses includes two entities: an *address* and a *count*. The address is the beginning address, for a block of IP addresses, and the count specifies how many addresses are combined in the block. Creating a CIDR address is limited by the following:

1. All addresses included in the CIDR block of addresses must be contiguous. When multiple addresses are aggregated (combined), all addresses must be in numerical order. This allows the boundary between the network and host portion of the address to reflect the aggregation.
2. When addresses are aggregated, the CIDR address blocks should be combined in sets of addresses that are greater than 1 and equal to a lower-order bit pattern that corresponds to all 1s, such as 3 (11), 7 (111), 15 (1111), 31 (11111), and so on. This is necessary to make it possible to steal the corresponding number of contiguous bits (two, three, four, and so on) from the network portion of the CIDR address block and use them for the host portion of the address.
3. CIDR addresses are generally applied to class C addresses because they have fewer hosts assigned to them, and class C addresses are more plentiful than class B and C addresses. However, CIDR can be used with class B and C addresses.
4. To implement CIDR on a network, all routers in the routing domain must know CIDR notation, which is typically not a problem for routers manufactured since September 1993, when CIDR was approved for use on IP networks.

QUESTIONS

20-1. What is the primary purpose of using *subnetting*?

20-2. List and describe several reasons why network designers create subnetting.

20-3. Explain the differences between *default masks* and *subnet masks*.

20-4. List and describe the differences between *class A, B, and C* subnet masks.

20-5. Describe how a subnet mask separates the subnet address from an IP address.

20-6. How many subnet bits can be used with class A addresses? Class B? Class C?

20-7. Explain the reasons for using *supernetting*.

20-8. Describe the differences between *subnetting* and *supernetting*.

20-9. Describe how a *supernet* is formed using a *supernet mask*.

20-10. What is meant by *variable-length blocks*?

20-11. Describe the differences between *classful* and *classless IP addressing*.

20-12. What are the restrictions of classless addressing?

20-13. Describe *slash notation* for classless addressing.

20-14. Explain what is meant by *variable-length subnetting*.

20-15. Define *classless interdomain routing*.

20-16. List and describe the limitations of *CIDR*.

PROBLEMS

20-1. What is the default mask for network address 139.100.0.0?
   a. 255.0.0.0
   b. 255.255.255.0
   c. 255.255.0.0
   d. 255.255.255.255

20-2. Using the default mask, how many hosts can the classful network 109.0.0.0 support?

20-3. What subnet mask is necessary to establish six classful subnets on a class B network?

20-4. What subnet mask is necessary to establish 13 classful subnets on a class C network?

20-5. What subnet mask is necessary to establish 23 classful subnets on a class A network?

20-6. How many subnets can be supported on a classful C network with the following subnet mask: 255.255.255.192?

20-7. How many hosts can be supported on a classful class B network with the following subnet mask: 255.255.224.0?

20-8. How many hosts can be supported on a classful class B network with the following subnet mask: 255.255.255.240?

20-9. You have a host IP address of 40.150.73.10 and a subnet mask of 255.248.0.0:

   a. What subnet contains this host?
   b. What is the beginning host address for this subnet?
   c. What is the ending host address for this subnet?
   d. What is the broadcast address for this subnet?

20-10. You have been assigned the classless network address of 130.40.0.0.

   a. How many binary digits are required to define 13 subnets?
   b. What is the subnet mask?
   c. Identify the following subnets in dotted decimal notation: first subnet, second subnet, sixth subnet, and seventh subnet,

20-11. What is the broadcast address for the third subnet?

20-12. How many subnets are possible with the following classless network addresses?

   a. 75.0.0.0/8
   b. 144.8.0.0/20
   c. 222.8.1.0/26

20-13. Convert the following IP addresses to their classless notation:

   a. IP address = 80.0.0.0, subnet mask 255.240.0.0
   b. IP address = 168.22.0.0, subnet mask 255.255.192.0
   c. IP address = 220.10.1.0, subnet mask 255.255.255.224

20-14. What are the subnet addresses for the following IP addresses and masks?

   a. 80.231.4.2, 255.224.0.0
   b. 140.40.195.24, 255.255.192.0
   c. 210.40.30.147, 255.255.255.240

20-15. Determine the subnet addresses and subnet masks for a business network with the classful class C address 204.238.7.0 with six departments having separate subnetworks with the following number of hosts:

   a. Administrative subnet—22 hosts

                Subnet address _____
                Subnet mask _____

   b. Engineering subnet—17 hosts

                Subnet address _____
                Subnet mask _____

   c. Sales subnet—23 hosts

                Subnet address _____
                Subnet mask _____

   d. Maintenance subnet—18 hosts

                Subnet address _____
                Subnet mask _____

   e. Security subnet—19 hosts

                Subnet address _____
                Subnet mask _____

   f. Supervision subnet—28 hosts

                Subnet address _____
                Subnet mask _____

20-16. Determine the IP address for the following hosts on the network and subnetworks determined in problem 20-15:

  a. Twelfth host on the sales subnet
  b. Seventh host on the security subnet
  c. First host on the administrative subnet
  d. Last host on the supervision subnet

20-17. You have the classful network address 209.100.78.0:

  a. What is the default mask?
  b. Using the default mask, how many hosts can the network 209.100.78.0 support?
  c. What subnet mask is necessary to establish six usable subnets?
  d. List the six subnet addresses.
  e. How many hosts can be supported on each subnet?
  f. How many hosts can be supported on the entire network?
  g. List the address ranges for each of the six subnets and specify the broadcast address of each subnet.
  h. Which two subnets cannot be used?

Subnet Address	Start of Range	End of Range	Broadcast Address

20-18. You have a classful device IP address of 140.150.73.10 and a subnet mask of 255.255.224.0.

  a. What subnet contains this device?
  b. What is the beginning IP address for this subnet?
  c. What is the ending IP address for this subnet?
  d. What is the broadcast address for this subnet?

20-19. Which of the following sets of addresses meet the criteria necessary for supernetting four class C address groups?

  a. 200.52.32.0   200.52.33.0   200.52.34.0   200.52.35.0
  b. 212.12.30.0   212.12.45.0   212.12.60.0   212.12.75.0
  c. 195.20.32.0   195.20.33.0   195.20.34.0
  d. 198.20.4.0   198.20.8.0   198.20.12.0   198.20.16.0
  e. 110.40.64.0   110.40.65.0   110.40.66.0   110.40.67.0

20-20. You have been assigned the classless network address of 132.45.0.0 and you need to establish eight subnets:

  a. How many binary digits are required to define eight classless subnets?
  b. What is the classless address notation for this network?
  c. Identify the subnet addresses in dotted decimal notation.
  d. What is the broadcast address for subnet 3?

20-21. How many subnets are possible with the following classless network addresses?

  a. 75.0.0.0/16
  b. 144.8.0.0/19
  c. 222.8.1.0/27
  d. 178.10.0.0/18
  e. 88.0.0.0/12
  f. 170.24.0.0/22

20-22. Convert the following classful IP network addresses to their classless notation:

    a. IP address = 60.0.0.0, subnet mask 255.224.0.0
    b. IP address = 178.22.0.0, subnet mask 255.255.240.0
    c. IP address = 220.10.1.0, subnet mask 255.255.255.192
    d. IP address = 66.0.0.0, subnet mask 255.252.0.0
    e. IP address = 164.10.0.0, subnet mask 255.255.248.0

20-23. A small business has been assigned a classless network address 210.38.4.0. The proposed network structure for the business is shown here:

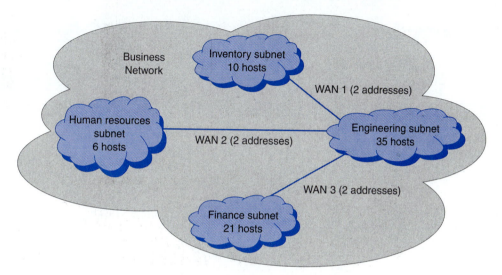

    a. How many total subnets are needed for the network?
    b. How many total addresses are needed for the network?

20-24. Determine the following for the network specified in problem 20-23:

    a. Classless IP subnet address for the airport subnet
    b. Subnet mask for the airport subnet
    c. Classless IP subnet address for the design subnet
    d. Subnet mask for the design subnet
    e. Classless IP subnet address for the warehouse subnet
    f. Subnet mask for the warehouse subnet
    g. Classless IP subnet address for the home office subnet
    h. Subnet mask for the home office subnet
    i. Classless IP subnet addresses for WAN 1
    j. Subnet mask for WAN 1
    k. Classless IP subnet addresses for WAN 2
    l. Subnet mask for WAN 2
    m. Classless IP subnet addresses for WAN 3
    n. Subnet mask for WAN 3
    o. How many host addresses are used for the entire network?

C H A P T E R 21

Network-Layer Protocols

CHAPTER OUTLINE

OBJECTIVES

- Explain address resolution
- Describe the differences in static and dynamic address mapping
- Explain the purpose of the address resolution protocol (ARP)
- Describe what an ARP cache is and what it is used for
- Explain proxy ARP
- Describe the ARP message format
- Explain the purpose of the reverse address resolution protocol (RARP)
- Describe what an RARP cache is and what it is used for
- Explain proxy RARP
- Describe the RARP message format
- Describe the primary functions of the Internet Protocol (IP)
- Describe the differences in hardware and IP addressing
- List and describe the functions of the fields in an IP header
- Describe fragmentation
- List and describe the options available with IP

21-1 INTRODUCTION

The *logical address* used with the TCP/IP protocol suite is the IP address, which is implemented in software. The logical address is an *internetworking address*, as it is globally unique and universally accepted. Logical addresses are necessary for transporting data between devices that reside on different physical networks or for transporting data across the Internet. However, on a physical network, such as an Ethernet LAN, each device (host, router, and so on) is identified with a locally unique *physical address* that is embedded in hardware, such as a network interface card (NIC). The physical address is sometimes called the *hardware address* or the *media access control* (MAC) *address*.

Transporting packets of data within IP-based networks requires two levels of addressing: logical and physical. When one device (host) wishes to communicate with another device (host), the source can acquire the IP address of the destination using a *domain name server* (DNS). However, to transport the packet across the physical network, the IP datagram must be encapsulated in a frame that includes the physical address of the destination device. Mapping an IP address to a physical address and vice versa is called *address resolution*.

21-2 ADDRESS RESOLUTION

Address resolution can be accomplished using either *dynamic* or *static mapping*. Static mapping is when an address table is created and updated manually on each machine. A table is stored individually on each machine on the network, creating some interesting limitations, such as the following:

1. If the NIC in a machine is changed, the physical address also changes.
2. In some local area networks (LANs; e.g., LocalTalk), the physical address changes each time a device is powered up.
3. On networks using a dynamic host configuration protocol (DHCP) to assign IP addresses, the IP address of a device can change each time the device is powered up.
4. A portable computer (such as a laptop) can move from one physical network to another, resulting in a different IP address on each network.

With static address mapping, changes in the address table must be made periodically. Manually updating an address table adds a substantial amount of administrative overhead to the network, seriously affecting network performance.

With dynamic address mapping, whenever a machine knows either the physical or the logical address of another host, it can use an address resolution protocol to automatically determine the other. The two most popular protocols used to perform dynamic address mapping are the *address resolution protocol* (ARP) and the *reverse address resolution protocol* (RARP). ARP is used to map an IP address to a physical (MAC) address, and RARP is used to map a physical address to an IP address. ARP and RARP operate at the network (IP) layer of the TCP/IP protocol suite.

21-3 ADDRESS RESOLUTION PROTOCOL

With IP networking, an IP address is mapped to a physical address using a process (protocol) called *address resolution protocol* (ARP). Figure 21-1 shows the location of ARP relative to the TCP/IP protocol suite. ARP is shown in the network layer, which is layer 3 of the OSI protocol hierarchy. However, as will be shown later, ARP is transported across a

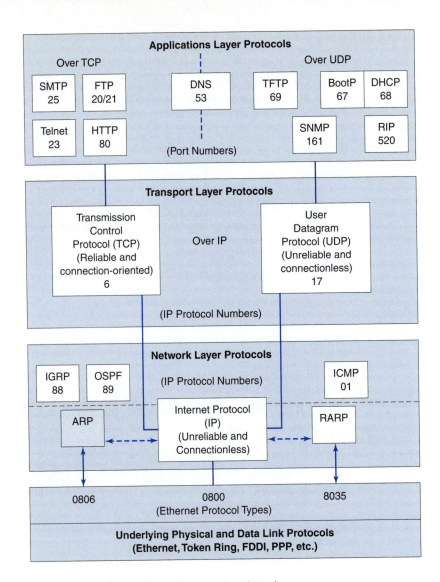

FIGURE 21-1 ARP and the TCP/IP protocol stack

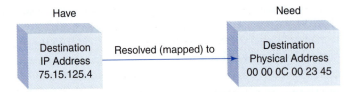

FIGURE 21-2 Address resolution protocol (ARP)

local network encapsulated in the information field of a layer 2 frame. Therefore, ARP relies on layer 2 hardware addressing to transport information across local networks. Since ARP messages are not encapsulated in the IP protocol, it is a nonroutable protocol.

ARP is used by a source device to resolve (map) the physical address of the destination device from its IP address. Figure 21-2 shows the process of mapping a physical address to a

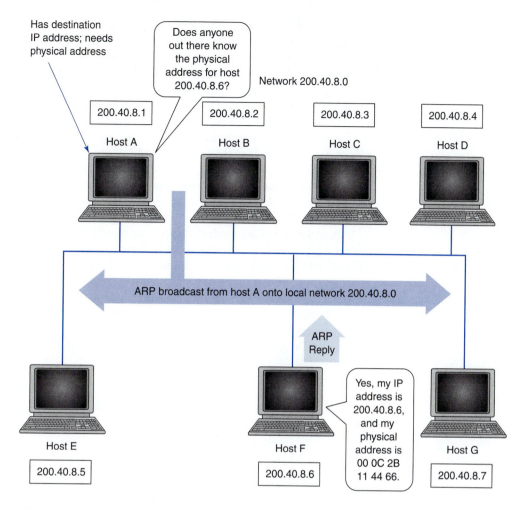

FIGURE 21-3 ARP broadcast onto local network

logical address using ARP. Whenever a host on a network wishes to locate the physical address of another host on the same network, it transmits an ARP query packet onto the local network. The query packet includes a broadcast destination address (FF FF FF FF FF FF for Ethernet) and the unique physical and logical addresses of the source. The packet also contains the IP address of the host it wishes to resolve. All devices (hosts and routers) on the network receive and process the ARP query. However, only the host with the IP address intended to be resolved replies to the query. The reply is a unicast transmission sent directly from the recipient to the inquiring host, which includes the recipient's IP and physical addresses.

Figure 21-3 shows host A sending an ARP broadcast onto the data link of its own physical network (200.40.8.0). The purpose of the broadcast is to resolve the hardware address of host F, which resides on the same network. Host A transmitted the ARP broadcast on the local network, querying whether any device on the network knows the physical address for host 200.40.8.6. Although all hosts on the network receive the broadcast, only host F has the IP address 200.40.8.6. Therefore, host F responds to the ARP request with an ARP reply, which contains its logical address as well as its local (hardware) address (00 0C 2B 11 44 66). Host A maps host F's IP and physical addresses and stores both of them in a location in volatile RAM called the ARP cache.

21-3-1 ARP Cache

Transporting messages between two hosts across a LAN in a TCP/IP networked system often requires sending more than one IP datagram. However, transmitting an ARP broadcast to determine the physical address for each datagram intended for the same destination is impractical and inefficient. The logical solution is for a host to maintain a table in volatile RAM, where it stores physical address to IP address matches. This volatile RAM is called the ARP *cache*. The ARP cache gives an IP host virtually instant access to the physical address it must provide the network interface (MAC) layer (i.e., Ethernet) with a datagram to a given destination. When a host resolves an IP address to a physical address, the information is stored temporarily in table form in the ARP cache. A host can use the ARP cache to resolve physical addresses intended for the same destination within a given period of time. However, because the capacity of an ARP cache is very limited, IP-to-MAC address mappings are not retained indefinitely.

An ARP cache table can include a variety of information, depending on the operating system. A typical ARP cache includes the IP (logical) address, the matching physical (MAC) address, and a timer. The *timer* (sometimes called the *time-out* or *time-to-live*) specifies the remaining lifetime of a cache entry in seconds. The initial timer setting varies from vendor to vendor with Microsoft using 120 seconds and Linux using 900 seconds. When the timer reaches 0, the ARP cache entry's validity has expired, and the resolution is flushed. Each time a host resolves a MAC address through its local cache, the timer is reset to the maximum value. On some systems, once the cache entry has expired, a new ARP query is automatically sent.

A typical ARP cache is shown here:

IP (Logical) Address	Physical (MAC) Address	Time Remaining
200.1.1.68	00 C0 93 A7 B1 EA	61
200.1.1.71	00 00 C0 IF 44 8D	109
200.1.1.82	00 00 0C FD 37 C2	110
200.1.1.79	00 A0 24 FF 2E 0B	95

An ARP cache may include a variety of additional information, such as attempt, state, and queue number:

Attempt. This entry stores the number ARP requests sent for this entry.

State. Each cache entry can have one of three states: *free, pending*, and *resolved*. A free state indicates that the time for this entry to remain in memory has expired and that the memory space can be used for a new entry. The pending state indicates that an ARP broadcast request has been sent to resolve the physical address but that a response has not been received yet. The resolved state indicates that the entry is complete and contains the IP address and that its corresponding physical address and packets intended for this destination can use information contained in this entry.

Queue number. ARP assigns a queue number to each datagram waiting for address resolution. Datagrams waiting for the same destination are generally given the same queue number.

21-3-2 Proxy ARP

Address resolution can be accomplished across network boundaries using a technique called *proxy* (promiscuous) ARP. Proxy ARP creates a subnetting effect in that it allows a layer 3 connectivity device (such as a router) to act on behalf of a set of hosts rather than for the benefit of an individual host. The IP addresses included in the set identify remote hosts (i.e., hosts that reside on a different subnet than the host requesting the address resolution). For example, when a router is running a proxy ARP software and receives an ARP

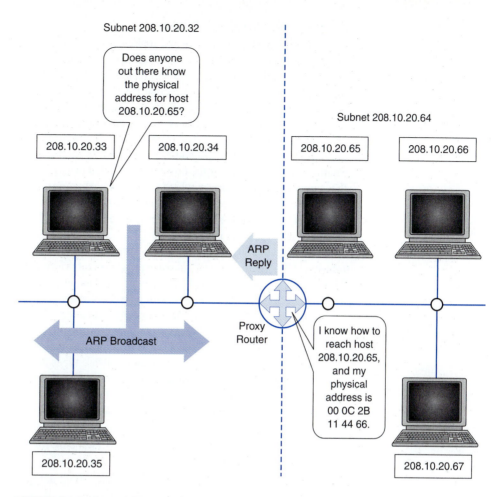

FIGURE 21-4 Proxy ARP

broadcast request for the physical address of a host that resides on a different subnet that is connected to the same router, the router responds with an ARP reply specifying its own physical address. Thus, the originating host maps the IP address of the remote host to the physical address of the router. Subsequent transmissions destined for the remote host are directed to the router, which simply forwards data packets to the intended destination IP address on the remote subnet.

Figure 21-4 shows host 208.10.20.33 sending an ARP broadcast onto its own local subnet (208.10.20.32) that is intended to resolve the physical address of host 208.10.20.65. The router recognizes that host 208.10.20.65 is included in a set of addresses for hosts that reside on a different subnet (208.10.20.64). Consequently, the router returns an ARP reply containing the router's physical address, which tells host 208.10.20.33 to resolve IP address 208.10.20.65 to the router's physical address (00 0C 2B 11 44 66). Subsequent packets sent from host 208.10.20.33 intended for host 208.10.20.65 are sent to the router, which forwards the packets to host 208.10.20.65.

The proxy router shown in Figure 21-4 replies in behalf of all hosts located on subnet 64 to ARP requests received from subnet 32 and replies in behalf of all hosts residing on subnet 64 to hosts residing on subnet 32. Hosts within a specific set of addresses are sometimes called *protégés* of that set. Thus, hosts residing on subnet 32 are protégés of subnet 64, and hosts residing on subnet 64 are protégés of subnet 32.

21-3-3 ARP Message Packet

ARP message packets (requests and replies) are transported across a physical network at the internetwork level (layer 2). The ARP message packets are embedded in the information field of the Internet-layer protocol (Ethernet). Figures 21–5a and b show an ARP message packet encapsulated in an Ethernet II frame (excluding the preamble), and Figure 21-5c shows how the frame would appear on a protocol analyzer.

As shown in Figures 21-5a, b, and c, the Ethernet header makes up the first 14 bytes of the frame. The first six bytes are the destination hardware address, which for ARP request packets is the broadcast address (FF FF FF FF FF FF hex) and, with ARP reply packets, the physical address of the host that originated the ARP request. The next six bytes are the source hardware address, which is the physical address of the originating host in a request packet and the physical address of the host or router responding to the request in a reply packet. The next two bytes are the type field, which is 0806 hex for Ethernet II frames carrying ARP message packets.

Immediately after the header is a variable-length information field followed by the four-byte CRC field. The ARP message packet is comprised of the nine fields listed in Table 21-1. Figure 21-5c shows the hex sequence for what a typical ARP packet encapsulated in an Ethernet frame might look like with a protocol analyzer.

The fields included in an ARP header are described as follows:

Hardware type field (HTYPE). A 16-bit field that defines the kind of network the ARP is running over. Table 21-2 lists the hardware types currently available for ARP. The hardware type field for the packet shown in Figure 21-5 is a 0001 hex, which indicates that it is running over an Ethernet II LAN.

Protocol type field (PTYPE). A 16-bit field that defines what network-layer protocol is running on the network. Although the protocol type field can be any higher-level protocol, thus far ARP has been implemented only for IP version 4 (IPv4). The protocol type value for IPv4 is 0800 hex.

Hardware length field (HLEN). An eight-bit field defining the number of bytes contained in the source and destination address fields. With Ethernet II, MAC addresses are used for the hardware address; therefore, the HLEN is 06 hex.

Protocol length field (PLEN). An eight-bit field defining the number of bytes contained in the source and destination protocol fields. With IPv4, the IP address is 32 bits (four bytes) long; therefore, the PLEN for IPv4 is 04 hex.

Operation field (OPER). A 16-bit field that defines the type (operation) of ARP packet being transported. ARP currently defines only two operations: request and reply. The hex codes for ARP request and ARP reply packets are the following:

    0001 hex: ARP request

    0002 hex: ARP reply

Source hardware address field (SHA). This is a variable-length field that contains the hardware (physical) address of the source of the ARP packet. With Ethernet II LANs, six-byte MAC addresses are used for hardware addresses.

Source protocol address field (SPA). A variable length field that contains the source protocol (logical) address, which is a four-byte IP address with IPv4.

Destination hardware address field (DHA). This is a variable-length field that contains the destination's hardware (physical) address, which with Ethernet II LANs is a six-byte MAC address. With an ARP request packet, the destination address field generally contains the MAC broadcast address (FF FF FF FF FF FF). However, since the reason for an ARP request is to find the physical address of the device that is intended to receive a message, vendors can fill the destination hardware address field in one of four ways:

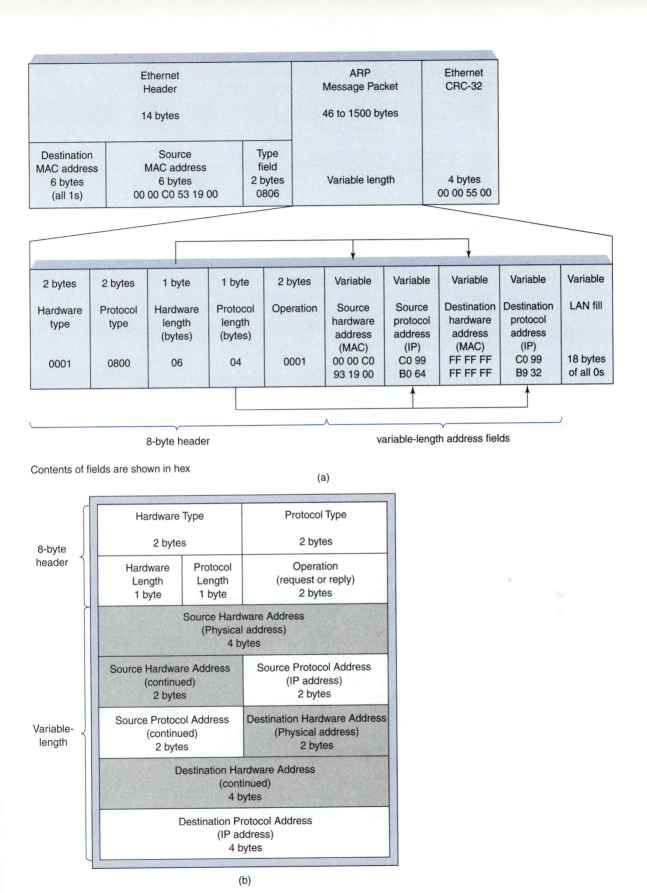

Contents of fields are shown in hex

(a)

(b)

FIGURE 21-5 ARP message packet: (a) encapsulated in an Ethernet II frame; (b) ARP message packet aligned on a four-byte boundary

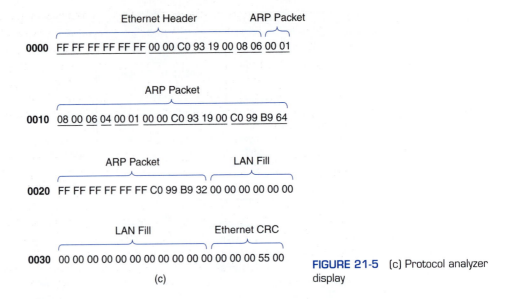

Ethernet Header   **ARP Packet**

0000 FF FF FF FF FF FF 00 00 C0 93 19 00 08 06 00 01

ARP Packet

0010 08 00 06 04 00 01 00 00 C0 93 19 00 C0 99 B9 64

ARP Packet   **LAN Fill**

0020 FF FF FF FF FF FF C0 99 B9 32 00 00 00 00 00 00

LAN Fill   **Ethernet CRC**

0030 00 00 00 00 00 00 00 00 00 00 00 00 00 00 55 00

(c)

FIGURE 21-5 (c) Protocol analyzer display

Table 21-1 ARP Fields

Field	Length (Bytes)
Hardware type	2
Protocol type	2
Hardware length	1
Protocol length	1
Operation	2
Source hardware address	Variable
Destination hardware address	Variable
Source protocol address	Variable
Destination protocol address	Variable

Table 21-2 Hardware Type Field

Type	Meaning	Type	Meaning
1	Ethernet	12	Localnet
2	Experimental Ethernet	13	Ultralink
3	Amateur radio AX.25	14	SMDS
4	Proteon ProNET token ring	15	Frame relay
5	ChaosNET	16	ATM
6	IEEE 802 Networks	17	HDLC
7	ARCnet	18	Unassigned
8	Hyperchannel	19	Asynchronous transmission mode
9	Lanstar	20	Serial line
10	Autonet short address	21	Asynchronous transmission mode
11	Local talks		

1. Six bytes of 1s: FF FF FF FF FF FF
2. Six bytes of 0s: 00 00 00 00 00 00
3. Six bytes from the sending host's buffer that look like a real address but are simply a buffer dump
4. The current or recently expired hardware address from the ARP cache

Destination protocol address field (DPA). A variable-length field that contains the destination's protocol (logical) address, which is a four-byte IP address with IPv4.

LAN fill. Ethernet-based ARP requests and replies do not contain enough characters to fulfill the minimum 46 bytes of user data required in an Ethernet frame. Therefore, to avoid losing the packet as runts, Ethernet drivers must add LAN fill (not to be confused with landfill) to make up the difference. Since the ARP message is only 28 bytes long, Ethernet requires 18 bytes of padding (or fill) to meet the 46-byte minimum.

An example of what an ARP request/reply exchange over an Ethernet LAN might look like with a network analyzer is shown in the following graphic. Note that the Ethernet destination address in the request frame is FF-FF-FF-FF-FF-FF, which is the broadcast address. The source address (00-04-76-39-B2-54) is the device that initiated the broadcast. The ARP header identifies the subnetwork protocols (i.e., physical and data-link levels) as Ethernet with a six-byte hardware address and a four-byte protocol (IP) address. The destination protocol address in the ARP header is the IP address the source wishes to resolve, and the source protocol address is the IP address of the originating device. In the reply packet, the Ethernet destination address is the source hardware address from the request packet, and the source address is the hardware address of the device that is responding to the request. The ARP packet includes the source and destination IP, which are reversed from the request packet. The responder hardware address is the address the request packet needed resolved, and the destination hardware address is the address of the originating device.

Request	
	Ethernet Header
ETHER	Destination Broadcast: FF-FF-FF-FF-FF-FF
ETHER	Source: 00-04-76-39-B2-54
ETHER	Protocol: ARP
ETHER	FCS: E5EC60E7
	ARP Header
ARP	Hardware address format = 1 (Ethernet)
ARP	Protocol address format = 2048 (IP, 0800 hex)
ARP	Hardware address length = 6
ARP	Protocol address length = 4
ARP	Operation = ARP request
ARP	Source protocol address = 10.10.4.189
ARP	Destination protocol address = 10.10.4.10
Reply	
	Ethernet Header
ETHER	Destination: 00-04-76-39-B2-54
ETHER	Source: 00-0C-95-E2-54-90
ETHER	Protocol: ARP
ETHER	FCS: BF615105
	ARP Header
ARP	Hardware address format = 1 (Ethernet)
ARP	Protocol address format = 2048 (IP, 0800 hex)
ARP	Hardware address length = 6
ARP	Protocol address length = 4
ARP	Operation = ARP reply
ARP	Responder hardware address = 00-0C-95-E2-54-90
ARP	Responder protocol address = 10.10.4.10
ARP	Destination hardware address = 00-04-76-39-B2-54
ARP	Destination protocol address = 10.10.4.189

Have

Destination
Physical Address
00 00 0C 00 23 45

Resolved (mapped) to →

Need

Destination
IP Address
75.15.125.4

FIGURE 21-6 Reverse address resolution protocol (RARP)

21-4 REVERSE ADDRESS RESOLUTION PROTOCOL

With IP networking, a physical address is mapped to an IP address using a process (protocol) called *reverse address resolution protocol* (RARP). RARP is used by a source to resolve (map) the physical address of the destination device from its IP address.

RARP is used primarily with diskless devices, which do not have a configuration file. A diskless device must boot from ROM, which usually supplies minimum booting information. In addition, the manufacturer, who has no idea what network the device will be on, generally installs ROM. Therefore, the manufacturer does not know the device's IP address. To acquire its IP address, a device reads its hardware address from its NIC card and then broadcasts an RARP request onto the network to resolve its IP address from whatever device on the network assigns or keeps track of device IP addresses.

Figure 21-6 shows the process of mapping a physical address to a logical address using ARP. The format for RARP is identical as ARP, except for the operation field. An RARP request is 3 and an RARP reply is 4.

21-5 INTERNET PROTOCOL

The *Internet Protocol* (IP) is the essence of the TCP/IP protocol suite. Without it, the Internet would not exist, at least not as we know it. IP is the mechanism that upper-layer protocols, such as TCP and UDP, use to transport data packets throughout the world.

Figure 21-7 shows the relative placement of the IP protocol relative to the TCP/IP protocol stack. As can be seen, IP is a layer 3 network protocol, which contains source and destination IP addresses. Therefore, IP is a routable protocol.

IP version 4 (IPv4) is currently the only network-layer protocol in the TCP/IP protocol suite. Therefore, in subsequent discussions, "IP" will imply IPv4. The primary function of a network protocol is to move packets of data called *datagrams* through a system of IP networks that are interconnected with routers. As shown in Figure 21-8, IP is a *host-to-host protocol* responsible for delivering datagrams from a source host to a destination host (i.e., end-to-end delivery) through an internetwork comprised of two or more networks and routers. To accomplish this, IP provides a layer 3 addressing mechanism called *IP addressing*, which provides IP addresses for all devices connected to the Internet. Therefore, IP is a routable OSI layer 3 protocol responsible for IP addressing and the fragmentation and reassembly of datagrams. In contrast, routing data frames across a LAN such as Ethernet requires using hardware (MAC) addresses employing layer 2 switching concepts.

IP is an unreliable and *connectionless* protocol, as it provides *best-effort delivery*. Best effort simply means it does its best to transport messages to the destination; however, it includes no means of error checking or tracking the messages it carries. Therefore, IP assumes the unreliability of the underlying protocol layers. If reliability is desired, IP must be combined with a reliable transport protocol such as the *transmission control protocol* (TCP). If reliability is not important, IP can be paired with an unreliable transport protocol, such as the *user datagram protocol* (UDP).

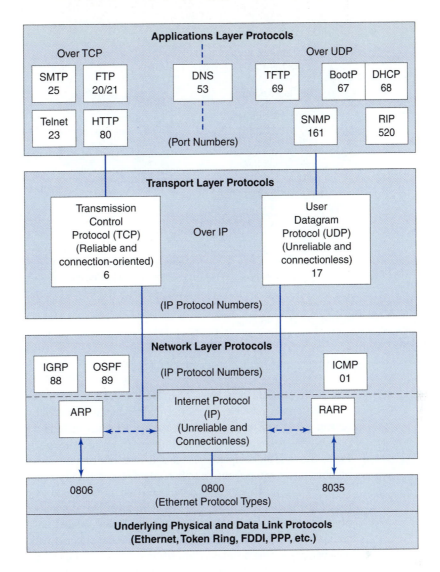

FIGURE 21-7 IP and the TCP/IP protocol stack

IP is a connectionless datagram protocol, as it does not create virtual circuits to deliver the datagrams and there is no call setup procedure to alert the destination of an incoming message. With a connectionless protocol, each datagram is transported independently of all other datagrams. Therefore, each datagram may take a different path through the network, and no record is kept of the route datagrams take. This implies that datagrams sent from the same source may arrive at the destination in a different order than they were sent. With IP, if one or more of the datagrams are lost or corrupted during transmission and never arrives at the destination, IP has no mechanism for calling for retransmissions of the lost or corrupted datagrams.

IP moves datagrams with the same functionality as the U.S. Postal Service delivers standard first-class letters. An IP datagram is placed on the network by the source computer, analogous to depositing a letter in a mailbox. IP attempts to deliver the datagram through physical and logical connections just as the Postal Service tries to deliver a letter using trucks, planes, buses, and mail carriers. Just like the Postal Service, IP makes no guarantee of delivery and gives no notification of failure. In the IP network, if you want reliability, you must incorporate a reliable higher-layer protocol. If you want reliability from the Postal Service, you have to upgrade to a higher class of service.

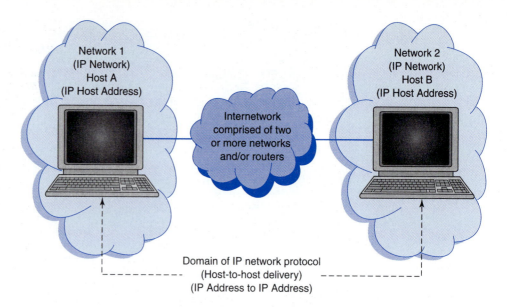

FIGURE 21-8 IP domain

21-6 HARDWARE ADDRESSING VERSUS IP ADDRESSING

The differences between layer-2 hardware addressing and layer-3 IP addressing are similar to the difference between interoffice mail and mail delivered by the U.S. Postal Service. To deliver a piece of interoffice mail, the sender must include the name of the recipient and possibly his room and building numbers. To ensure successful delivery through the U.S. Postal Service, the recipient's name and complete mailing are required (i.e., post office box number, street name, house number, city name, and a zip code).

To transport a data packet between two hosts located on different networks is similar to sending mail through the U.S. Postal Service. This is somewhat more involved than simply transporting data packets between two hosts located on the same network, which is analogous to sending interoffice mail. On a local area network, such as an Ethernet LAN, data packets are encapsulated into a frame, which contains the source and destination hardware (MAC) addresses. IP addresses are not necessary for two hosts to communicate with each other over a physical network because packet routing is based on layer-2 hardware (MAC) addresses. However, to transport data between two networks separated both physically and logically by a layer-3 router requires using some form of higher-layer addressing, such as IP. It should probably be noted that some Ethernet LANs route packets based on IP addresses using layer-3 switches and routers.

Figure 21-9a shows two Ethernet networks (LANs) interconnected with a router. Each LAN has its own network IP address. Host A on network 10.4.0.0 can transmit a message to a host on the same network, such as host B, by sending an Ethernet frame onto the network with the source hardware address 00 00 11 11 11 11 and a destination hardware address of 00 00 22 22 22 22 (no IP address is needed). However, if host A on network 10.4.0.0 wishes to send a message to host C on network 10.5.0.0 as shown in Figure 21-9b, the message must pass through the router. Any data packets that pass through a router must contain a layer-3 address, such as IP addresses. Therefore, host A must send an Ethernet frame to interface 1 on the router with a destination hardware address of 00 00 33 33 33 33 and a source hardware address of 00 00 11 11 11 11. The Ethernet frame must also contain an IP packet, which includes the source IP address of host A (10.4.0.1) and the destination IP address of host C (10.5.0.1). Once the router receives the Ethernet frame, it removes the IP packet and routes it to network 10.5.0.0 through interface 2 where the packet

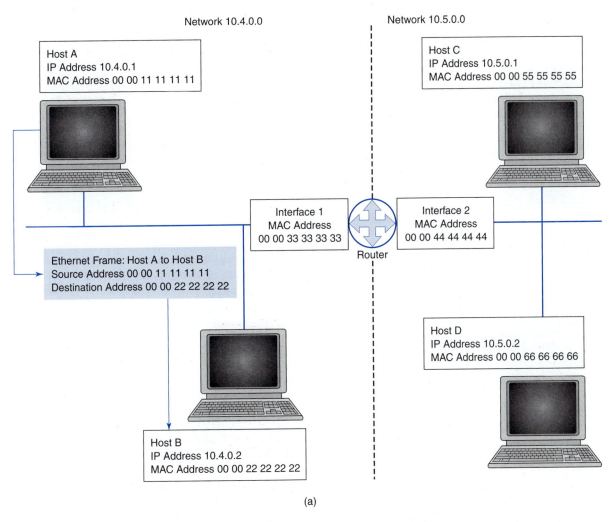

Network 10.4.0.0

Network 10.5.0.0

Host A
IP Address 10.4.0.1
MAC Address 00 00 11 11 11 11

Host C
IP Address 10.5.0.1
MAC Address 00 00 55 55 55 55

Interface 1
MAC Address
00 00 33 33 33 33

Router

Interface 2
MAC Address
00 00 44 44 44 44

Ethernet Frame: Host A to Host B
Source Address 00 00 11 11 11 11
Destination Address 00 00 22 22 22 22

Host D
IP Address 10.5.0.2
MAC Address 00 00 66 66 66 66

Host B
IP Address 10.4.0.2
MAC Address 00 00 22 22 22 22

(a)

FIGURE 21-9 MAC addressing versus IP addressing: (a) host-to-host same network

is encapsulated into another Ethernet frame. The Ethernet frame is transmitted onto network 10.5.0.0 with the source hardware address of interface 2 (00 00 44 44 44 44) and the destination hardware address of host C (00 00 55 55 55 55).

The message transmissions described in the previous paragraph are intended to illustrate the basic idea of local area network routing versus internetwork routing. The examples are somewhat oversimplified because they do not include the concept of address resolution (i.e., determining IP addresses from MAC addresses and vice versa) or the details of routing protocols.

21-7 IP DATAGRAM

A datagram is a variable-length packet of data comprised of two parts: a header and a data field. A standard IP header is 20 bytes long; however, it can be increased up to 60 bytes with the addition of special options. The header contains information vital to network routing and delivery. Some of the functions performed by the IP header are:

1. Logical addressing
2. Message fragmentation
3. Determining message length

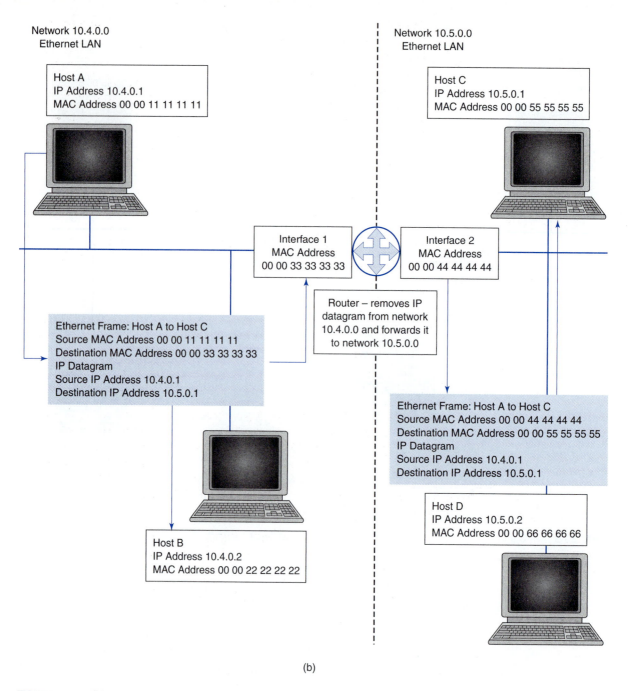

Network 10.4.0.0
Ethernet LAN

Host A
IP Address 10.4.0.1
MAC Address 00 00 11 11 11 11

Network 10.5.0.0
Ethernet LAN

Host C
IP Address 10.5.0.1
MAC Address 00 00 55 55 55 55

Interface 1
MAC Address
00 00 33 33 33 33

Interface 2
MAC Address
00 00 44 44 44 44

Router – removes IP
datagram from network
10.4.0.0 and forwards it
to network 10.5.0.0

Ethernet Frame: Host A to Host C
Source MAC Address 00 00 11 11 11 11
Destination MAC Address 00 00 33 33 33 33
IP Datagram
Source IP Address 10.4.0.1
Destination IP Address 10.5.0.1

Ethernet Frame: Host A to Host C
Source MAC Address 00 00 44 44 44 44
Destination MAC Address 00 00 55 55 55 55
IP Datagram
Source IP Address 10.4.0.1
Destination IP Address 10.5.0.1

Host D
IP Address 10.5.0.2
MAC Address 00 00 66 66 66 66

Host B
IP Address 10.4.0.2
MAC Address 00 00 22 22 22 22

(b)

FIGURE 21-9 (b) host-to-host different network

4. Determining the type of service
5. Higher-layer protocol identification
6. Message routing
7. Error checking

The IP header is divided into 13 fields of various lengths, where each field performs a specific function. The 13 fields are:

1. IP Version
2. IP Header length
3. Precedence
4. Differentiated services
5. Total IP length
6. Datagram identification number
7. Fragmentation area
8. Time to live
9. Protocol
10. Check sum
11. Source IP address
12. Destination IP address
13. Options (which are optional)

Figures 21-10a and b show an IP header (without options) encapsulated in an Ethernet frame and Figure 21-10c shows what the frame would look like with a network analyzer.

21-7-1 IP Version Field

The first hex character (first four bits of the first byte) identifies the version of IP that created the header. If the IP versions at the source and destination do not match, the two stations will not be able to communicate. However, the only version of IP currently being used on the global Internet is IPv4.

21-7-2 P Header Length Field

The header length field (HLEN) specifies the number of 32-bit data words (four bytes each) in the IP header. The HLEN can be any number between 5 and 15 (5 to F hex), which specifies between 20 (5 × 4) and 60 (15 × 4) bytes. A standard IP header begins with 45 hex, which indicates IPv4 with the default length of 20 bytes, and an IP header beginning with 4F indicates IPv4 with a 60-byte header. An Ethernet frame containing IP information has 0800 in its type field, and the first byte of the IP header is a hex number between 45 and 4F. The header length can be summarized as follows:

HLEN (hex)	Meaning
<5	illegal header length
5	minimum header length (20 bytes)
6 to F	indicates options are present and the header is between 24 and 60 bytes long

Example 21-1

Determine the IP header length for a HLEN field of A hex.

Solution The length of the IP header is

$$A \text{ times } 4 =$$
$$10 \text{ times } 4 = 40$$

where A hex = 10 decimal

21-7-3 Differentiated Services Field

The Internet Engineering Task Force (IETF) has recently redefined and renamed this field, which was formerly called the *service type field*. Both interpretations are currently being implemented; therefore, descriptions of both are given.

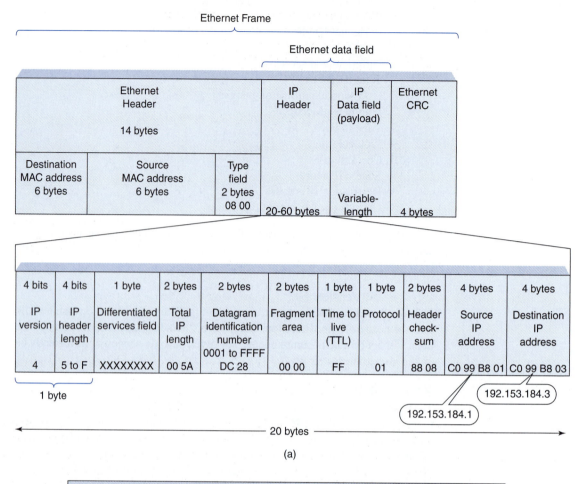

(a)

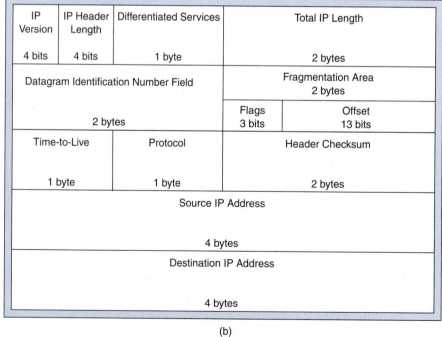

(b)

FIGURE 21-10 IP header (without options): (a) header format; (b) IP header aligned on four-byte boundary

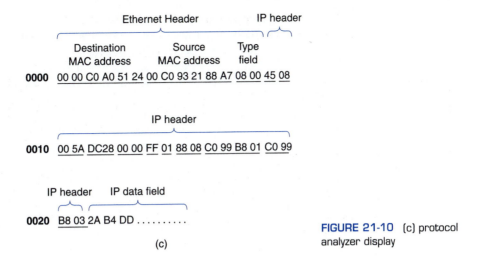

FIGURE 21-10 (c) protocol analyzer display

21-7-3-1 Service type field. The format for a service type field is shown in Table 21-3. As the table shows, the service type field is subdivided into two subfields: a three-bit precedence of data subfield and a four-bit type of service subfield and one additional bit reserved for future use. The precedence of data subfield is currently not widely used; however, in the future it will set the priority of datagrams by informing routers along the datagram's path about the importance of the data contained in the datagram.

The precedence-of-data subfield contains a three-bit word; therefore, there are eight codes possible, most of which were set for DARPA (Defense Advanced Research Projects Agency) or for military use. Table 21-4 lists the eight precedence-of-data codes and their priorities. Several of the codes are intended for clearing network congestion problems or as network management tools to explore problems in networks experiencing unusually high rates of errors. For example, the national network management (code 111) is considerably more urgent than routine traffic (code 000). The idea behind the precedence of data subfield

Table 21-3 Service Type Field (precedence and type of service subfields)

Precedence Subfield			Type of Service Subfield				
			Delay	Throughput	Reliability	Cost	Reserve
X	X	X	X	X	X	X	0

Note: X can be a 1 or a 0.

Table 21-4 Precedence of Data Subfield

Precedence Code			Priority
1	1	1	National network control
1	1	0	International network control
1	0	1	Critic/ECP
1	0	0	Flash override
0	1	1	Flash
0	1	0	Immediate
0	0	1	Priority
0	0	0	Routine

is to allow congested routers to selectively discard datagrams on the basis of their priority. Most commercial software packages are just beginning to use the precedence of data subfield. When the precedence of data subfield is used, normal traffic is rated as routine. When a router's buffer is close to full and the software decides to purge the buffer, the router can discard routine data to free buffer space, allowing it to forward the remaining higher-priority traffic.

Another use for the precedence-of-data subfield is when new users are seeking network access through busy circuits. To acquire access, the new users must have a higher precedence than at least one of the current users. It is assumed that the higher-precedence users have more important data than the lower-precedence users. Therefore, lower-level users are bumped off the circuit in favor of the higher-precedence users. This logic applies up to the flash level. When all circuits are busy handling flash traffic, a user must have a flash override code just to access the network.

Because the Internet itself has no direct knowledge of how to optimize the route for a particular application or user, the IP protocol attempts to provide a means for upper-layer protocols to convey information to the network-layer protocol concerning route selection criteria. This facility is called type of service (TOS), which is found in the four bits that follow the precedence in the service type field. The TOS field specifies the type of service the sender desires, such as the following:

1. Delay
2. Throughput
3. Reliability
4. Cost

Routes through the Internet vary considerably in the quality of service provided because some paths are more reliable than others, some are more expensive, some have longer propagation delays, and some have a higher throughput. Obviously, there are trade-offs because the route with the lowest cost may also be the least reliable route or the route with the longest propagation delay.

The delay, throughput, reliability, and cost bits are mutually exclusive (i.e., only one can be active at a time), and each has a special meaning:

Delay bit. Setting the delay bit requests the route with the least (minimum) propagation delay. IP routers along the path that support this feature select the shortest route and the fastest transmission path available to reach the destination.

Throughput bit. If the throughput bit is set, routers that support this function will route the datagram over the path experiencing the highest (maximum) throughput. For example, if a file transfer protocol, such as FTP, is attempting to transfer files with the greatest efficiency, it will set the throughput bit.

Cost bit. If the cost bit is set, the datagram will be routed over the path that provides the sending organization the least (minimum) expensive route available.

Reliability bit. When the reliability bit is set, the sending application is requesting that the datagram travel over the most (maximim) reliable route (i.e., the one with the least chance of losing or contaminating the data). However, this feature works only when IP routers along the route support it.

The terms *minimum* and *maximum* are relative terms. For example, the route with the minimum delay may or may not have a delay that the user considers low, and the route that incurs the maximum reliability may or may not be considered reliable to the user.

Table 21-5 lists the types of service available along with their codes and descriptions, and Table 21-6 lists the default type of service codes for several of the more common protocols.

Table 21-5   Type of Service Bits

Delay	Throughput	Reliability	Cost	Description
		TOS Bits		
0	0	0	0	Normal (default setting)
0	0	0	1	Minimize cost
0	0	1	0	Maximum reliability
0	1	0	0	Highest throughput
1	0	0	0	Minimum delay

Table 21-6   Protocol Default Type of Service

Protocol	TOS Bits	Type of Service
ICMP	0000	Routine
BOOTP	0000	Routine
IGP	0010	Maximum reliability
SNMP	0010	Maximum reliability
TELNET	1000	Minimum delay
FTP (data)	0100	Maximum throughput
FTP (control)	1000	Minimum delay
TFTP	1000	Minimum delay
SMTP (data)	0100	Maximum throughput
SMTP (command)	1000	Minimum delay
DNS (UDP query)	1000	Minimum delay
DNS (TCP query)	0000	Routine
DNS (zone)	0100	Maximum throughput

Example 21-2

Determine the precedence and type of service for the following hex values in the service type field:

a. 68
b. 24
c. 10

Solution   To determine the precedence and type of service, the hex values must be converted to binary.

a.

| | | | | | Type-of-Service-Subfield | | | |
Hex	Precedence Subfield			Delay	Throughput	Reliability	Cost	Reserve
68	0	1	1	0	1	0	0	0

Precedence: Flash, Type of Service: highest throughput

b.

| | | | | | Type-of-Service Subfield | | | |
Hex	Precedence Subfield			Delay	Throughput	Reliability	Cost	Reserve
24	0	0	1	0	0	1	0	0

Precedence: Priority, Type of Service: maximum reliability

c.

| | | | | | Type-of-Service Subfield | | | |
Hex	Precedence Subfield			Delay	Throughput	Reliability	Cost	Reserve
10	0	0	0	1	0	0	0	0

Precedence: Routine, Type of Service: minimum delay

21-7-3-2 Differentiated services field.   The differentiated services field uses the first six bits of the field for a codepoint subfield, and the last two bits are reserved for future growth. The codepoint subfield can be used in two ways:

1. If the three rightmost bits of the codepoint are logic 0s, the three leftmost bits are interpreted in the same way that the precedence bits were interpreted with the service type of format. Therefore, it is compatible with the service field in this aspect.

Table 21-7 Codepoint Assignments and Values

Category	Codepoint Bits	Range of Values	Assigning Authority
1	XXXXX0	All even values (0, 2, 4, up to 62)	Internet (IETF)
2	XXXX11	Every fourth value beginning with 3 (3, 7, 11, up to 63)	Local
3	XXXX01	Every fourth value beginning with 1 (1, 5, 9, up to 61)	Temporary

2. When the three rightmost bits are not all logic 0s, all six bits are used to define 2^6, or 64, possible service types based on priorities assigned by either Internet or local authorities as listed in Table 21-7. As shown in the table, the 64 types are divided into three categories with the first category containing 32 types and the second and third categories each containing 16 types. All service types specified by category 1 codepoints are assigned by Internet authorities (IETF) and have a logic 0 in the sixth bit. Therefore, category 1 types comprise all the even-numbered codepoints (i.e., 0, 2, 4, 6, and up through 62) as shown here:

Six-Bit Codepoint

	b_5	b_4	b_3	b_2	b_1	b_0	
Binary weightings	32	16	8	4	2	1	
Codepoint bit position	X	X	X	X	X	X	
	0	0	0	0	0	0	Codepoint 0
			Codes 2–60				
	1	1	1	1	1	0	Codepoint 62

All category 2 codepoints are assigned by local authorities and have logic 1s in both the b_1 and the b_0 bit positions. Therefore, category 2 codepoints comprise every fourth codepoint beginning with codepoint 3 (i.e., 3, 7, 11, and up through 63) as shown here:

Six-Bit Codepoint

	b_5	b_4	b_3	b_2	b_1	b_0	
Binary weightings	32	16	8	4	2	1	
Codepoint bit position	X	X	X	X	X	X	
	0	0	0	0	1	1	Codepoint 3
			Codes 7–59 in steps of 4				
	1	1	1	1	1	1	Codepoint 63

All category 3 service types are assigned on a temporary basis for experimentation and have a logic 0 in bit position b_1 and a logic 1 in bit position b_0. Therefore, category 3 codepoints comprise every fourth codepoint beginning with codepoint 1 (i.e., 1, 5, 9, and up to 61) as shown here:

Six-Bit Codepoint

	b_5	b_4	b_3	b_2	b_1	b_0	
Binary weightings	32	16	8	4	2	1	
Codepoint bit position	X	X	X	X	X	X	
	0	0	0	0	0	1	Codepoint 1
			Codes 5–57 in steps of 4				
	1	1	1	1	0	1	Codepoint 61

The codepoints are not contiguous because if they were, the first category's codes would range from 0 to 31, the second category's codes from 32 to 47, and the third category's code from 48 to 63. This would not be compatible with the type of service interpretation because all codes ending with three 0s (0, 8, 16, 24, 32, 40, 48, and 56) would fall into all three categories. With differentiated services assignments, these services belong to category 1. Keep in mind that these assignments have not been finalized.

Example 21-3

Determine the decimal value of the codepoint, the category, and the assigning authority for the following differentiated services fields:

a. 000010
b. 100101
c. 001011

Solution

a. codepoint decimal value = 2, category = 1, assigning authority = Internet
b. codepoint decimal value = 37, category = 3, assigning authority = temporary
c. codepoint decimal value = 11, category = 2, assigning authority = local

21-7-4   Total IP Length Field

This is a 16-bit field that specifies the total IP length of the datagram, which includes the IP header and all data in the IP data field (i.e., the payload). To determine the length of the data field (i.e., the data coming from the upper-layer protocol), simply subtract the IP header length from the total IP length. The two-byte IP length field can be used to determine the length of the data field being carried by the IP datagram. The number of bytes in the header and data field are found as follows:

$$\text{IP header length} = \text{HLEN times } 4$$

$$\text{Data field length} = \text{total IP length} - \text{IP header length}$$

With 16 bits (two bytes), the maximum datagram size is 65,535 bytes, which is a number much higher than is currently practical given the buffer space and error rates of data communications circuits currently available. For example, to determine how many bytes of data are being transported in an IP datagram carrying ICMP, simply subtracting the IP header length (20–60 bytes) and the ICMP header length (eight bytes) from the total IP length. This provides the number of bytes being sent and echoed as a result of a ping command, which is an ICMP command. The maximum size for this field when the IP datagram is traveling over an Ethernet data link is 1500 bytes (05DC hex), and when 802.3 LLC is carrying IP, its MTU (maximum transmission unit) is 1492 bytes (05D4).

Example 21-4

Determine the length of an ICMP data field for an Ethernet frame that is carrying an IP datagram with a 20-byte header, an ICMP message with a header length of eight bytes, and the Ethernet data field is 480 bytes long.

Solution   The total header length is the sum of the IP and ICMP headers:

$$\text{Total header length} = 20 + 8$$
$$= 28 \text{ bytes}$$

Thus, the length of the ICMP data field is

$$\text{ICMP data field} = \text{Ethernet length} - \text{IP and ICMP headers}$$
$$= 480 - 28$$
$$= 452 \text{ bytes}$$

21-7-5 Datagram ID Number Field

This is a two-byte (16-bit), host-specific field that carries the unique ID number of each datagram sent by a host. When the identification number is combined with the source IP address, it uniquely defines a datagram. With 16 bits, the range of identification numbers possible is 0 to 65,535. When TCP/IP software starts up on a host, the ID number is initialized and increments by one for each separate datagram sent. As will be seen later, the ID number is used to correlate logical errors and to help reassemble datagrams that have been fragmented. When fragmentation occurs, each datagram that is part of a larger message will have the same ID number. The ID number provides a common identification number for all the fragments so that the destination system can reassemble the fragments and recompile the original datagram before forwarding its payload to the next protocol up the stack.

21-7-6 Fragmentation Field

Because there is a high likelihood that an IP datagram will pass through several networks before reaching its final destination, it will undoubtedly pass through several routers along the way. Each time a datagram enters a router, it is decapsulated, processed, and then encapsulated into another frame before being forwarded to the next router or outputted onto the destination network. The protocol used by a physical network determines the format and length of a frame. Therefore, the format and length of a frame received from one physical network may not be the same as that of the next physical network. For example, one port on a router may be connected to a token ring LAN and another port connected to an Ethernet LAN. The router would be required to reformat and possibly resize any data packets transported between the two LANs. Whenever an IP datagram enters a network access layer, the interface reads the value of the fragmentation field.

When IP is required to transmit a datagram that exceeds the maximum length of the network access layer (such as Ethernet) allows, the datagram must be divided into smaller pieces (i.e., fragmented). Each network access layer specifies the maximum message length it can accept. This length is called the *maximum transfer unit* (MTU). Therefore, when an IP datagram is encapsulated in a frame, the total length of the datagram must either be less than the MTU of the physical network or will have to be divided into smaller segments (i.e., fragmented). The MTU of a protocol can vary from as few as 48 to as many as 65,535, as shown in the Table 21-8. Therefore, to make IP independent of physical network limitations, the maximum length of an IP datagram is equal to the largest value shown in the table 8 (65,536 bytes). Thus, whenever an IP datagram is placed on a network with an MTU that is less than the length of the datagram, the datagram must be fragmented.

When a datagram is fragmented, each of the fragments contains a header where some of the fields are duplicated (repeated) in each fragment and some are changed. In addition, a datagram may be fragmented more than once (refragmented). Because a datagram can be

Table 21-8 MTUs for Several Network Protocols

Protocol	MTU (Bytes)
Hyperchannel	65,535
16-Mbps token ring	17,914
4-Mbps token ring	4464
FDDI	4352
Ethernet	1500
IEEE 802.2 LLC	1492
X.25	576
PPP	296
ATM	48

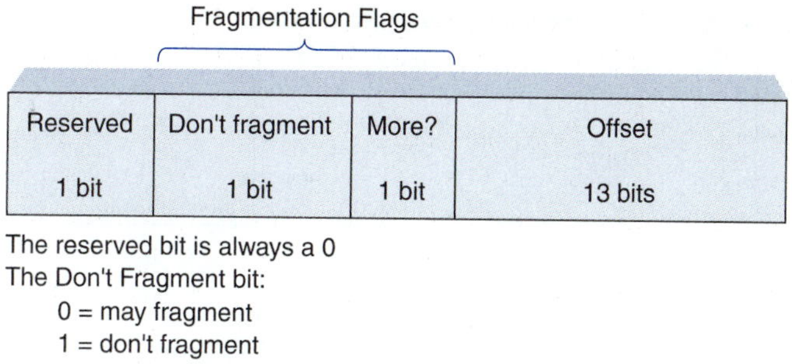

Reserved	Don't fragment	More?	Offset
1 bit	1 bit	1 bit	13 bits

The reserved bit is always a 0
The Don't Fragment bit:
    0 = may fragment
    1 = don't fragment

FIGURE 21-11 IP fragmentation field

fragmented by the source host or by any router in its path, it may be fragmented several times as it passes from one physical network to another en route to its final destination. Because each fragment becomes an independent datagram and each datagram may take a different route to its destination, the datagrams may not arrive at the destination at the same time or in the same order as they were transmitted. For these reasons, the datagram is not reassembled until it reaches the destination host. Losing one fragment means that the source would have to retransmit the entire datagram, and with so many routes available between the source and destination, a minor delay in getting all the fragments to the destination system could cause a reassembly time-out, which is described in a later section.

The *fragmentation field* in an IP header is a two-byte (16-bit) field that is subdivided into two subfields: a flag comprised of three bits and an offset comprised of 13 bits, as shown in Figure 21-11. The *don't fragment bit* determines whether the datagram may be fragmented. The rules are simple:

1. If the don't fragment bit equals 1, the application being transported is not allowing IP to fragment the message. If a router in the transmission path cannot forward the message without fragmenting it, the router discards the datagram and sends a destination-unreachable message back to the originating host.
2. If the don't fragment bit equals 0, the application is allowing IP to fragment the message.

The *more bit* is used in conjunction with the don't fragment bit to indicate whether a datagram is fragmented and, if it is fragmented, whether it is the first fragment, a middle fragment, or the last fragment of the message. The more bit is valid only if the don't fragment bit equals 0. The rules for the more bit are as follows:

1. If the more bit equals 0, this is the last fragment.
2. If the more bit equals 1, there are more fragments.

The *fragment offset field* is a 13-bit word that carries the relative position of the fragment with respect to the other fragments from the same datagram (i.e., the same group of fragments). The fragment offset is measured in groupings of eight bytes.

21-7-6-1 Fragmentation and reassembly. IP fragments are counted in units of eight bytes, and the fragment offset field identifies the fragment location relative to the beginning of the original unfragmented datagram. The basic fragmentation strategy is designed so that an unfragmented datagram has an all-0s fragment offset, and when an IP datagram is fragmented, its data portion must be broken on eight-byte boundaries. This format allows $2^{13} = 8192$ fragments of eight bytes each for a total of 65,536 bytes, which is consistent with the maximum length of an IP datagram. However, this is somewhat misleading because the header is counted in the total IP datagram length but is not counted in the fragments.

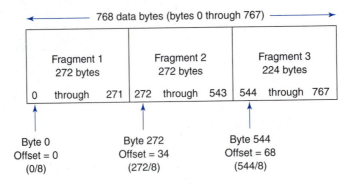

FIGURE 21-12 Fragmentation for Example 21-5

When fragmentation occurs, some of the IP options are copied into all fragments, but others are copied only in the first fragment. All Internet modules must be capable of forwarding datagrams carrying 68 bytes without further fragmentation because an IP header can be up to 60 bytes long and the minimum fragment size is eight bytes. In addition, all Internet destinations must be capable of receiving a datagram of 476 bytes either in one grouping or in fragments that require reassembly.

The IP header fields that may be affected by fragmentation include the options field, the more fragments flag, the fragment offset, the Internet header length field, and the total length field.

Example 21-5

Fragment a 768-byte datagram to send over a point-to-point (PPP) protocol.

Solution As shown in Figure 21-12, sending a 768-byte datagram with a PPP protocol requires dividing the datagram into three fragments. The first two fragments each contain 272 data bytes plus a 20-byte IP header for a total fragment size of 292 bytes. The length is limited to 272 data bytes rather than 276 because 272 is the largest multiple of 8 that when added to the 20-byte IP header does not exceed 296 bytes. The third fragment carries the remaining 224 data bytes plus the 20-byte IP header. The total payload transported in the three fragments is 272 + 272 + 224 = 768 bytes. However, counting the IP headers, the total length of each of the first two fragments is 292 bytes, and the total length of the third fragment is 244 bytes for a total length of 828 bytes. The offset in the first fragment is always 0, which indicates that the first eight-byte grouping begins with byte 0 (the first byte of the datagram). The offset in the second fragment is 34 because the second fragment begins with byte 272 (272/8 = 34). This also indicates that the first fragment contained 272 bytes of data. The offset in the third fragment is 68 because the fragment begins with byte 544 (272 + 272 = 544 and 544/8 = 68). The second fragment also carries 272 data bytes of data, leaving 224 data bytes of data for the third fragment. All three fragments have the same identification number, which enables the destination to know they are part of the same datagram.

The don't fragment bit, the more bit, and the offset can be used together to indicate whether a fragment is the first, a middle, or the last fragment of a datagram. The following relationships hold true and are summarized in Table 21-9:

Table 21-9 Fragmentation Field Flag Summary

More Bit	Offset	Datagram
0	0	Not fragmented
1	0	First fragment
1	>0	Middle fragment
0	>0	Last fragment

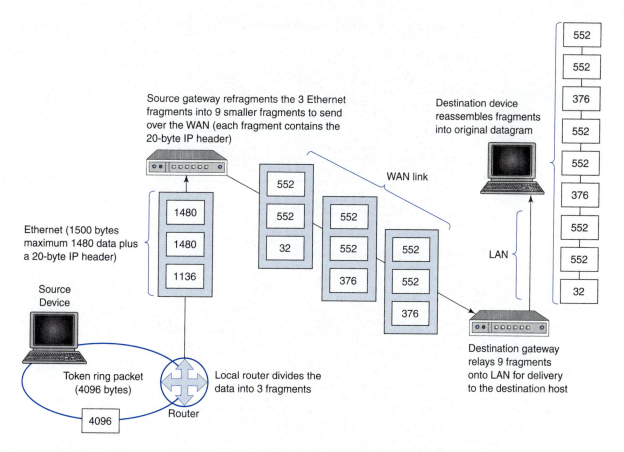

FIGURE 21-13 Fragmenting example

1. If the more bit is a 0 and the offset is 0, it must be the only fragment; therefore, the datagram has not been fragmented.
2. If the more bit is a 1 and the offset is 0, it must be the first of a series of fragments.
3. If the more bit is a 1 and the offset is greater than 0, it must be a middle fragment.
4. If the more bit is a 0 and the offset is greater than 0, it must be the last fragment.

Figure 21-13 shows a 4096-byte IP datagram sent from a device on a token ring LAN destined for the Internet. When the datagram reaches the local router, the router checks its routing table and determines that to reach the destination IP address, the datagram must be transported across the adjoining Ethernet LAN. Because Ethernet is limited to a 1500-byte data field, the router's IP entity fragments the data into three smaller packets fit into Ethernet frames. The token ring datagram is divided into two fragments carrying 1480 data bytes each and one fragment carrying 1136 data bytes. Each fragment also includes a copy of the 20-byte IP header, which brings the first two fragments up to the 1500-byte-maximum Ethernet limit.

When the fragments arrive at the gateway router, the destination IP address in the header directs the gateway to pass them on to a wide area network (WAN) link where there is no set MTU. However, IP specifications set an IP default datagram size of 576 bytes on the WAN. Thus, the gateway divides the arriving datagrams into smaller fragments containing a maximum of 572 bytes, which is comprised of a 20-byte IP header and 552 bytes of data (572 bytes total). The maximum length is 572 bytes rather than 576 because 572 is the highest multiple of 8 that when added to the 20-byte IP header does not exceed the maximum length of 576 bytes. The conditions of the fragmentation flags, offset, and identification number for the Ethernet segment shown in Figure 21-14 are described here:

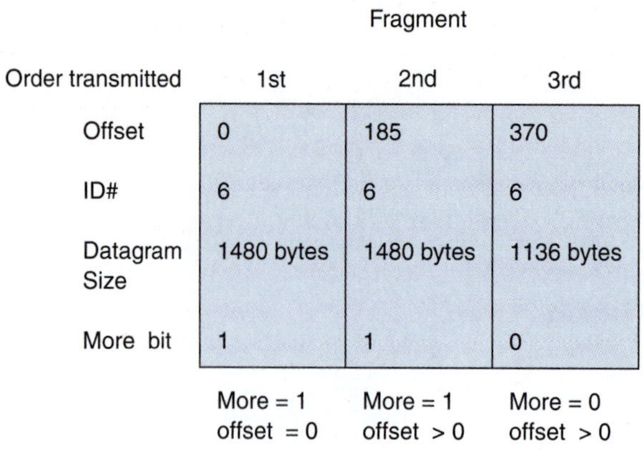

Fragment

Order transmitted	1st	2nd	3rd
Offset	0	185	370
ID#	6	6	6
Datagram Size	1480 bytes	1480 bytes	1136 bytes
More bit	1	1	0

	More = 1 offset = 0	More = 1 offset > 0	More = 0 offset > 0

FIGURE 21-14 Fragmentation flags, offset, and datagram identification for an Ethernet LAN carrying a 4096 byte IP datagram

First Ethernet Fragment

1. The more bit is 1, indicating that there is at least one more fragment to follow.
2. The offset is 0, as it is transmitting 1480 bytes beginning with byte 0.
3. The don't fragment bit is 0, indicating that the datagram can be fragmented.
4. The datagram identification number is 6.

Second Ethernet Fragment

1. The more bit is a 1, indicating that there is at least one more fragment to follow.
2. The offset is 185, indicating that the fragment begins with byte $185 \times 8 = 1480$.
3. The don't fragment bit is 0, indicating that the datagram can be fragmented.
4. The datagram identification number is 6.

Third Ethernet Fragment

1. The more bit is 0, indicating that it is the last fragment in the message.
2. The offset is 370, indicating that the fragment begins with byte $370 \times 8 = 2960$.
3. The don't fragment bit is 0, indicating that the datagram can be fragmented.
4. The datagram identification number is 6.

The nine refragmented WAN packets each contain the original IP header. However, the condition of the more and offset fields of the fragment field may change to reflect the refragmentation. Figure 21-13 shows the refragmented Ethernet frames leaving the source gateway on the same WAN. However, this is not always the case. Each of the refragmented WAN packets could take a different route to the destination gateway. Therefore, the refragmented WAN packets may arrive out of order. The destination gateway receives the WAN packets and simply forwards them onto its LAN, which transports them to the destination device, where they are reassembled in the proper sequence.

The conditions of the more and offset fields for the nine WAN fragments are summarized in the following table. The offset value is listed by octet and in its equivalent decimal value. The Range column lists the lowest and highest bytes contained in the fragment, and the Bytes column indicates how many bytes of data are contained in the fragment.

Ethernet Fragments					
Frame	More Bit	Offset (Octet)	Offset (Decimal)	Range (Decimal)	Bytes
1	1	0	0	0–1479	1480
2	1	185	1480	1480–2959	1480
3	0	370	2960	2960–4095	1136

		WAN Fragments			
Packet	More Bit	Offset (Octet)	Offset (Decimal)	Range (Decimal)	Bytes
1a	1	0	0	0–551	552
1b	1	69	552	552–1103	552
1c	1	138	1104	1104–1479	376
2a	1	185	1480	1480–2031	552
2b	1	254	2032	2032–2583	552
2c	1	323	2584	2584–2959	376
3a	1	370	2960	2960–3511	552
3b	1	439	3512	3512–4063	552
3c	0	508	4064	4064–4095	32

Example 21-6

For the following fragmentation conditions, determine

a. the order in which the fragments were transmitted
b. the number of bytes in each fragment

		Fragment	
Order received	First	Second	Third
Offset	240	0	120
ID #	14	14	14
More bit	0	1	1

Solution

	More = 0	More = 1	More = 1
	Offset > 0	Offset = 0	Offset > 0
Order transmitted	Third	First	Second
Number of bytes in the fragment	Cannot tell	960 bytes	960 bytes

Example 21-7

For the fragmentation conditions shown, determine

a. the order in which the fragments were transmitted
b. the number of bytes in each fragment

		Fragment	
Order received	First	Second	Third
Offset	80	0	160
ID #	156	156	156
More bit	1	1	0

Solution

	More = 1	More = 1	More = 0
	Offset > 0	Offset = 0	Offset > 0
Order transmitted	Second	First	Third
Number of bytes in the fragment	640 bytes	640 bytes	Cannot tell

Example 21-8

Show how a datagram carrying 65,536 bytes (the maximum length) can be transported over an Ethernet LAN where the maximum frame length is 1500 bytes, which includes the 20-byte IP header. Determine how many fragments are required; the offset value in binary, hex, and decimal for the first four and the last two fragments; and the number of bytes in the last fragment.

Solution The datagram must be separated into fragments where all fragments except the last fragment carry 1480 bytes of data plus a 20-byte IP header. If the maximum fragment length is

1480 bytes and the IP datagram length is 65,536, the offset in the first fragment is 0, the offset in the second fragment is 185 (1480 divided by 8), the offset in the third fragment is 370 (2960 divided by 8), and so on. This produces 44 fragments containing 1480 bytes each and one fragment (the last one) carrying only 416 bytes ([1480 × 44] + 416 = 65,536). Therefore, the offset in the last fragment is 8160 (1FCC hex), which, when multiplied by 8, equals 65,120. This allows 416 bytes in the last fragment with none of the fragments exceeding the 1500-byte-maximum Ethernet frame length and the sum of the fragmented data not exceeding the 65,536-byte-maximum IP datagram length.

The solution is shown here in tabular form. As can be seen, all fragments except the last fragment contain 1480 bytes of data, and the last fragment contains 416 bytes for a total of 65,536 bytes. Also, after each of the first 44 fragments, the total number of bytes increments by 1480.

Fragment	13-Bit Offset Value			Actual Offset Number (Offset × 8)	Bytes in Fragment	Total Number of Bytes
	Binary	Hex	Decimal			
1st	0 0000 0000 0000	0000	0	0	1480	1480
2nd	0 0000 1011 1001	00B9	185	1480	1480	2960
3rd	0 0001 0111 0010	0172	370	2960	1480	4440
4th	0 0010 0010 1011	022B	555	4440	1480	5920
5th–43rd	------				------	63,640
44th	1 1111 0001 0011	1F13	7955	63,640	1480	65,120
45th	1 1111 1100 1100	1FCC	8140	65,120	416	65,536

An example of what an IP datagram after it has been divided into three fragments and sent over an Ethernet LAN might look like with a network analyzer is shown in the following graphic. In all three fragments, the don't fragment bit equals 0 (false), indicating that the datagram can be fragments. The more bit in the first two fragments equals 1, indicating that there are additional fragments, and the more bit in the third fragment is 0, indicating that it is the last fragment. The offset in the first fragment is 0, indicating that it is the first fragment. The offset in the second fragment is B9 hex, which equates to 185 decimal, and 185 multiplied by 8 equals 1480 (the number of bytes contained in the first datagram). The offset in the third fragment is 172 hex, which equates to 370 decimal, and 370 multiplied by 8 equals 2960 (2960 − 1480 = 1480, the number of bytes in the second datagram).

First IP Datagram Fragment

                                            Ethernet Header
ETHER   Destination: 00-04-76-36-E5-7E
ETHER   Source: 00-04-76-36-E5-A4
ETHER   Protocol Type: IP
ETHER   FCS = 2B15

                                            IP Header
IP   Version = 4
IP   Header Length = 20
IP   Differentiated Services (DS) Field = 00
                0 0 0 0 0 0 . .   DS codepoint = Default PHB (0)
                            0 0   Unused
IP   Packet length = 1500
IP   Datagram Id = 1Fa
IP   Fragmentation Info = 2000 hex
                . 0 . . . .   Don't Fragment Bit = False
                . . 1 . . . .   More Fragments Bit = True
                .   . . . 0 0 0 0 0 0 0 0 0 0 0 0 0   Fragment Offset = 0
IP   Time to live = 123
IP   Protocol = ICMP (1)
IP   Header checksum = 90E7
IP   Source address = 192.168.0.131
IP   Destination address = 192.168.0.108

Second IP Datagram Fragment

		Ethernet Header
ETHER	Destination: 00-04-76-36-E5-7E	
ETHER	Source: 00-04-76-36-E5-A4	
ETHER	Protocol Type: IP	
ETHER	FCS = A23B	

IP Header

IP   Version = 4
IP   Header Length = 20
IP   Differentiated Services (DS) Field = 00

       0 0 0 0 0 0 0 . .   DS codepoint = Default PHB (0)
                  0 0 Unused

IP   Packet length = 1500
IP   Datagram Id = 1Fa
IP   Fragmentation Info = 20B9 hex

   . 0 . . . .   Don't Fragment Bit = False
   . . 1 . . . .   More Fragments Bit = True
   . . . 0 0 0 0 0 1 0 1 1 1 0 0 1   Fragment Offset = 1480

IP   Time to live = 125
IP   Protocol = ICMP (1)
IP   Header checksum = 902E
IP   Source address = 192.168.0.131
IP   Destination address = 192.168.0.108

Third IP Datagram Fragment

		Ethernet Header
ETHER	Destination: 00-04-76-36-E5-7E	
ETHER	Source: 00-04-76-36-E5-A4	
ETHER	Protocol Type: IP	
ETHER	FCS = C246	

IP Header

IP   Version = 4
IP   Header Length = 20

       0 0 0 0 0 0 0 . .   DS codepoint = Default PHB (0)
                  0 0 Unused

IP   Packet length = 68
IP   Datagram Id = 1Fa
IP   Fragmentation Info = 0172 hex

   . 0 . . . .   Don't Fragment Bit = False
   . . 0 . . . .   More Fragments Bit = False
   . . . 0 0 0 0 1 0 1 1 1 0 0 1 0   Fragment Offset = 2960

IP   Time to live = 121
IP   Protocol = ICMP (1)
IP   Header checksum = B50D
IP   Source address = 192.168.0.131
IP   Destination address = 192.168.0.108

21-7-7   Time-to-Live Field

The *time-to-live field* (TTL) is the lifetime control mechanism for an IP datagram. Before the days of routers, the TTL indicated how many seconds a datagram could exist "live" on the Internet. In essence, the TTL was the timestamp for the datagram. If all data fragments do not arrive at the destination before expiration of the TTL of the first-arriving fragment, the fragments are discarded, and an ICMP error message is sent back to the originating IP. TTLs are specified as hex values ranging from 01 to FF. Each unit of TTL represented 1 second. Therefore, a TTL of 255 means a datagram can exist for 255 seconds. After 255 seconds, if the entire message has not reached the destination, it is discarded.

Table 21-10 Protocol ID Numbers

Hex	Decimal	Protocol
01	1	ICMP
06	6	TCP
08	8	EGP
11	17	UDP
58	88	IGRP
59	89	OSPF

Today, because routers do not keep track of the amount of time a datagram spends in the network, the TTL represents how many routers a datagram can pass through before it is discarded. The TTL is initialized at the originating host, and each router the datagram passes through decrements the TTL by 1. If the TTL is not equal to 0, the router forwards the datagram to the next router, network, or subnet. When the TTL = 0, the datagram is discarded, and an ICMP message is sent back to the originating host indicating that the datagram was undeliverable. In essence, the TTL is now the hop (router) limit for a message.

21-7-8 Protocol Field

This is an eight-bit field that carries the ID number of the higher-level protocol residing in the IP data field. The ID number is necessary because IP datagrams can encapsulate data from a number of higher-level protocols such as TCP and UDP. The protocol field is similar to a shipping label because it identifies what transport protocol is intended to receive the datagram's payload. The most common value for the protocol field is 06 hex, which specifies TCP (transmission control protocol). The ID numbers for several other common protocols are listed in Table 21-10.

21-7-9 IP Header Checksum Field

Checksum is the primary means of error detection used by protocols in the TCP/IP protocol suite. The checksum provides error checking on the IP header only. The IP header checksum is calculated on the number of bytes specified in the header length field. However, the receiver performs the checksum calculation on the entire header, including the checksum.

The checksum is calculated at the sending end as follows:

1. The checksum is made all 0s.
2. The datagram is divided into k groups of n bits each, where n is usually 16.
3. The 16-bit groups are added together using one's complement arithmetic, which produces an n-bit sum.
4. The sum is complemented and placed in the checksum field.

The checksum is calculated at the receiving end as follows:

1. The received datagram is divided into k groups of n bits each.
2. The groups are added together.
3. The sum is 1s complemented.
4. If the result is all 0s, the packet is assumed to be error free; otherwise, the packet is rejected.

For example, if the sum of the groups at the source with the checksum equal to all 0s is 10110, the checksum sum would be its 1s complement, or 01001. Therefore, at the destination, assuming that no errors have occurred, the sum of the groups is the original sum plus the 1s complement of the sum, or

$$
\begin{array}{ll}
10110 & \text{sum} \\
+01001 & \text{1s complement} \\
\hline
11111 &
\end{array}
$$

The complement of 11111 is 00000, which is all 0s, indicating that no errors have occurred. An example of a checksum calculation is given in Chapter 22.

21-7-10   IP Address Fields

The last two fields of the IP header are the source and destination IP addresses (in hex). Unlike the network access layer protocols (such as Ethernet), the source address precedes the destination address:

> *Source IP address.* The sender's 32-bit IP address is identified in four bytes (four pairs of hex characters).
>
> *Destination IP address.* The destination's 32-bit IP address is identified in four bytes (four pairs of hex characters). For example,
>
> > CO 99 B8 01 hex equates to 192.153.184.1 in dotted-decimal notation
> >
> > CO 99 B8 03 hex equates to 192.153.184.3 in dotted-decimal notation

21-8   IP OPTIONS

The IP protocol specifies a two-part header: a required 20-byte header and a variable-length option header with a maximum length of 40 bytes. According to statistics released by several Internet Service Providers, less than 0.003% of all IP datagrams include options, which equates to less than 30 datagrams for every million datagrams sent. However, when included, the IP option header immediately follows the destination IP address in the required 20-byte IP header. Whenever IP options are used, they must be transmitted contiguously (with no separators between them) and occur in four-byte increments because the header length field in an IP datagram defines the header length in four-byte increments. Option fields are indicated when the value in the IP header length field in the required 20-byte IP header is greater than 5.

IP options may or may not appear in IP datagrams. However, when included, they must be implemented by all hosts, routers, and gateways. Therefore, transmission of IP options is optional in most IP datagrams; however, when included, their implementation is not optional. In certain environments, the security option is required in all datagrams.

IP options are not required; however, depending on the application and its implementation, IP users can access several useful IP options that can be used to measure propagation time, record network traffic routing, specify routing instructions to routing devices (routers and gateways) on the network, and specify the security level of the datagram. IP options are used primarily to provide additional IP routing control, such as recording the route an IP datagram takes or specifying the route an IP datagram must take.

21-8-1   IP Option Header

Figure 21-15 shows the format for an IP option header, which contains three fields: a one-byte option code field, which may be followed by a one-byte-length field and a variable-length data field with a maximum length of 39 bytes. To use an option, the correct option class and number must be present in the option code field.

The option field is variable in length and may contain no options or multiple options. There are two cases for the format of an IP option:

1. A single byte indicating the option type
2. An option type byte, an option length byte, and the actual option data bytes

21-8-2   Code Field

The code field is an eight-bit field containing three subfields: copy, class, and number.

> *Copy subfield.* The copy subfield (sometimes called the copy flag) contains only one bit that controls how routers treat options when IP datagrams must be fragmented.

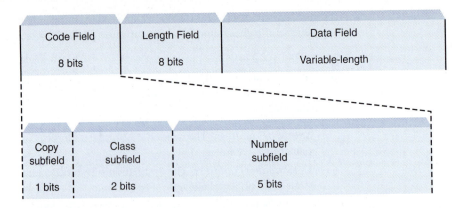

Code Field	Length Field	Data Field
8 bits	8 bits	Variable-length

Copy subfield	Class subfield	Number subfield
1 bits	2 bits	5 bits

Copy bit	Meaning		Number bits	Meaning
0	Copy only into 1st fragment		00000	End of option
1	Copy into all fragments		00001	No option
			00011	Loose source route
			00100	Timestamp
Class bits	Meaning		00111	Record route
0 0	Datagram control		01001	Strict source route
0 1	Reserved for future use		10100	Router Alert
1 0	Debugging and management		00010	Security
1 1	Reserved for future use		01000	Stream ID (obsolete)

FIGURE 21-15 IP option header

The condition of the copy flag indicates whether the option header will be copied into only the first fragment of a fragmented datagram (copy bit = 0) or copied into all fragments (copy bit = 1). However, the copy bit does not affect the required 20-byte IP header, which is always copied into all fragments. The copy bit is not itself optional in the sense that with some IP options it is always a 0 and with other IP options always a 1.

Class subfield. The class subfield contains two bits that define the general function of the option by specifying the class of the option and a specific option in that class. The two-bit class subfield is interpreted as follows:

Option Class	Meaning
00	The option contains network or datagram or network control information
01	Reserved
10	The option contains system management and debugging information
11	Reserved

The codes 01 or 11 currently have no meaning, as they have not been defined and are reserved for future use.

Number subfield. The option number subfield contains five bits that define the type of option invoked in the data field. With five bits, there are 32 possibilities; however, currently there are only nine being used.

The numerical equivalent of the binary code in the code field is sometimes called the *option type*. For example, the code field for an end of option is 00000000 (type 0), and the code field for a strict source routing option is 10001001 (type 137).

21-8-3   Length Field

The length field is not included in all option types. However, when present, the length field specifies the total length of the option header, including the code, length, and data fields. The length field also designates the maximum size of the option data field. For the current version of IP (version 4), the maximum length of the data field is 39 (27 hex) because the maximum length of an IP option is 40 bytes and the first byte contains the copy, class, and number subfields, which are not counted in the length.

21-8-4   Data Field

The data field contains data specific to the option type specified in the number subfield. The data field is not found in all option headers.

21-8-5   Option Types

Figure 21-16 shows the nine options currently being used, which include two single-byte options and seven multiple-byte options. The single-byte options do not require the length or the data fields; however, the multiple-byte options do.

Table 21-11 lists the IP options currently used and the contents of code and length fields and the code subfields. As shown in the table, with some options the copy bit is always a 1, and in other options it is always a 0. In addition, the class and number codes are specified for each option. The table also indicates when the length field is used and its value for the various options.

The IP software in a device must pass all the IP options it receives, except no operation and end of option, on to the transport-layer protocol or to the ICMP protocol when the datagram contains an ICMP message. The transport-layer protocol must interpret all IP options they support and silently ignore (discard) the rest.

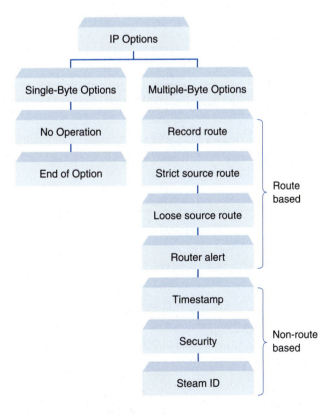

FIGURE 21-16 IP option categories

Table 21-11 IP Option Headers

First Byte Codefield						
Copy Bit Subfield	Class Subfield	Number Subfield	Type		Length Field (Bytes)	IP Option Name
			Hex	Decimal		
Non–route-based options						
0	00	00000	00	0	Not included	End of option
0	00	00001	01	1	Not included	No operation
0	10	00100	44	68	Variable length	Timestamp
1	00	00010	82	130	11	Security
0	00	01000	08	8	4	Stream ID (obsolete)
Route-based options						
0	00	00111	07	7	Variable length	Record route
1	00	01001	89	137	Variable length	Strict source routing
1	00	00011	83	131	Variable length	Loose source routing
1	00	10100	94	148	4	Router alert

21-8-5-1 Single-Byte Options   There are only two single-byte IP options, and their purposes are relatively straightforward.

No Operation. The no operation option is a padding character, as it has no function other than as a filler that is placed at the beginning of the options or between options to align the next option on a 32-bit boundary, as shown in Figures 21–17a and b. The no operation byte has no length field. The code field for a no option is 0 00 00001 (01 hex), which specifies a datagram control option and copy only into first fragment.

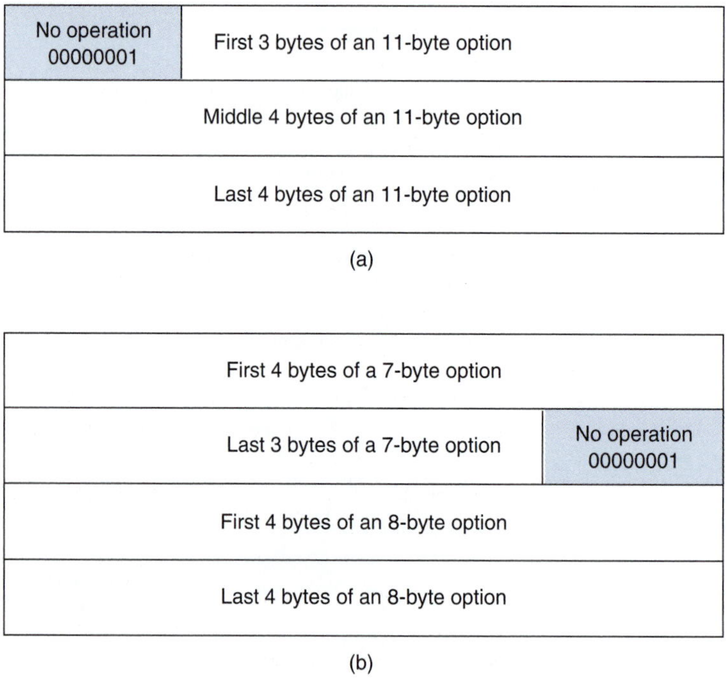

No operation 00000001	First 3 bytes of an 11-byte option
Middle 4 bytes of an 11-byte option	
Last 4 bytes of an 11-byte option	

(a)

First 4 bytes of a 7-byte option	
Last 3 bytes of a 7-byte option	No operation 00000001
First 4 bytes of an 8-byte option	
Last 4 bytes of an 8-byte option	

(b)

FIGURE 21-17 No operation option: (a) 11-byte option; (b) seven-byte option followed by an eight-byte option

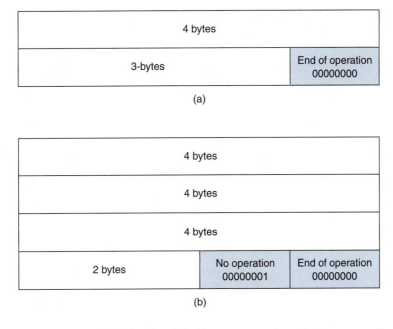

4 bytes	
3-bytes	End of operation 00000000

(a)

4 bytes		
4 bytes		
4 bytes		
2 bytes	No operation 00000001	End of operation 00000000

(b)

FIGURE 21-18   End of option: (a) without no operation; (b) with no operation

End of Option. The end of operation (sometimes called end of option list) must be used at the end of a variable-length option. Consequently, only one end of option can be used. The end of option is placed at the end of all options, not the end of each option, and need be used only if the end of the options does not otherwise coincide with the end of the IP option header as specified in the header length field of the IP datagram.

The end of option indicates the end of the option header and that the next byte is the beginning of the IP datagram data field, as shown in Figure 21-18a. If more than one byte is required to align the option field on a 16-bit boundary, no operation options must be used followed by the end of option, as shown in Figure 21-18b. The end of option byte is 00000000 (00 hex).

21-8-5-2 Multiple-Byte Options   There were seven multiple-byte options, which are comprised of the one-byte copy field, a one-byte length field, and a variable-length data field. As shown in Table 21-11, four of the multiple-byte options are considered route-based options, two of the multiple-byte options are considered non–route-based options, and one of the options (stream ID) is obsolete.

Record route option. The record route IP option provides a means to record the route an IP datagram takes as it propagates from the source device to the destination device. The record route option is used with record echoes (ping -r). The record route is similar to the traceroute feature. Traceroute can compile information from more routers than record route; however, record route records the route in both directions, whereas traceroute only provides the path in one direction.

Figure 21-19a shows the format for the record route option. The first three bytes are for the code, length, and pointer fields. The code for the record route option is 7, which specifies that if the datagram carrying the option is fragmented, the option is copied only into the first fragment. The length field specifies the total byte length of the option, including the header and data fields. In essence, the contents of the length field determine how many router addresses can be carried in the data field.

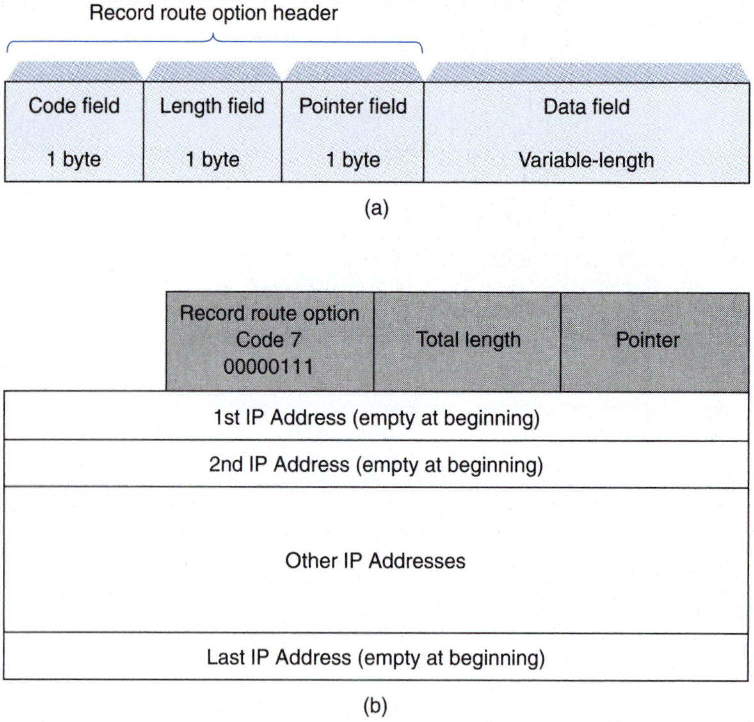

Code field	Length field	Pointer field	Data field
1 byte	1 byte	1 byte	Variable-length

(a)

Record route option Code 7 00000111	Total length	Pointer
1st IP Address (empty at beginning)		
2nd IP Address (empty at beginning)		
Other IP Addresses		
Last IP Address (empty at beginning)		

(b)

FIGURE 21-19 Record route option: (a) format; (b) format aligned on a four-byte boundary

A variable-length data field follows the three-byte header. When a record route option leaves the source device, an empty data field is created (i.e., a reserved space filled with 0s). The data field is filled (one IP address at a time) with the addresses of all the routers the datagram encounters until either the datagram reaches the destination or the data field is filled. The pointer field contains an offset that indicates the location of the next available byte (empty space) in the data field that can store an IP address. The initial value in the pointer field is 4, and its value increments by 4 every time another router IP address is placed into the data field.

IP addresses are four bytes long, the header is three bytes long, and the maximum option length is 40 bytes. Therefore, the maximum number of IP addresses the data field can carry is nine, and the maximum length of a record route option is 39 bytes (the three-byte header plus a 36-byte data field). The second, third, and fourth bytes are the code, length, and pointer fields, respectively. Figure 21-19b shows the format for the record route option. Although the diagram is shown in 32-bit groupings, in an actual datagram the option is not aligned on any octet boundary. The option begins with the header, which is immediately followed by the IP address of the outgoing interface of the first router to forward the datagram. The second four-byte block is the IP address of the second router interface to forward the datagram and so forth until the reserved space is filled. The number of bytes reserved for IP addresses in the data field does not change when addresses are inserted. If there is room in the data field but not enough for an entire IP address (four bytes), the original datagram is considered in error and is discarded.

Figure 21-20a shows an internetworking configuration, and Figures 21–20b through f show the record route option for a datagram as it travels from the source de-

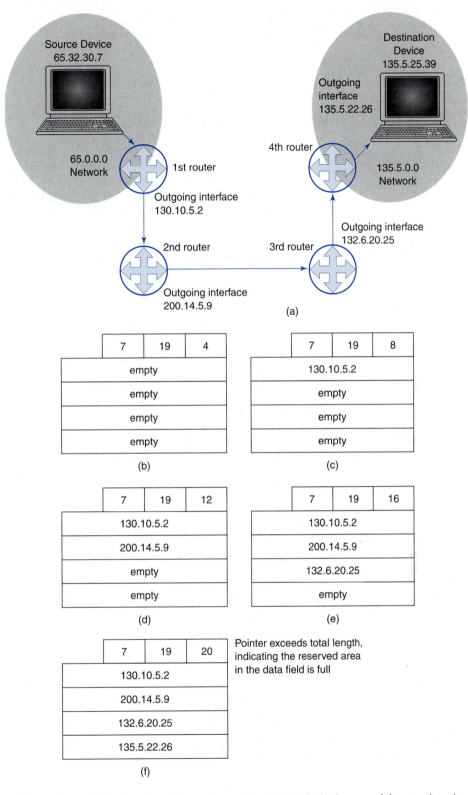

FIGURE 21-20 Record route option as it propagates through the Internet: (a) network and router configuration; (b) leaving source device; (c) leaving first router interface; (d) leaving second router interface; (e) leaving third router interface; (f) leaving fourth router interface

vice on network 65.0.0.0 to the destination device on network 135.5.0.0. The first three bytes in the option constitute the header, and the remaining 16 bytes are the reserved area in the data field. The first byte of the option is the code field for the record route option (7), the second byte is the length field (19), and the third byte is the pointer. Therefore, the total option length is 19 bytes. When the datagram leaves the source device, the 16-byte data field is empty, as shown in Figure 21-20b. The offset is 4, which means the first empty byte of the option data field is the fourth byte.

Figure 21-20c shows the option after the datagram passes through the first router. The code and length fields remain unchanged; however, the offset is incremented by 4. This is because the IP address of the router interface that forwarded the datagram (130.10.5.2) is placed in the first four bytes of the data field. Therefore, the first empty byte is now the eighth byte of the data field. Figures 21–20d, e, and f show the contents of the option after the datagram passes through the second, third, and fourth router interfaces. As shown, each time the datagram passes through a router, the pointer is incremented by 4, and another interface IP address is placed in the data field. The reserved area in the data field is full when the value in the pointer exceeds the total length of the option, as shown in Figure 21-20f. Any additional routers the datagram passes through are not recorded, as the reserved space in the option data field is filled. Once the datagram reaches the destination device, the data field in the record route option holds the record of the route the datagram took.

Strict source route option. The strict source route option provides a device the means to originate an IP datagram and specify the exact path the datagram must take to reach its destination even if the routers normally would have chosen a different route. This option is useful to the sender because it can choose to use a route with a specific type of service, such as minimum cost, maximum throughput, minimum delay, or most reliable. The option can also be used to select a route that does not go through a competitor's network.

With the strict source route option, the originating device provides a list of the hops (router interfaces) that the datagram must pass through to reach the destination, and those are the only routers the datagram can use. The path between two successive addresses in the list cannot have more than one physical network between them, and if it does, an error message is generated, as the datagram cannot follow the strict source route specified. If the datagram reaches a router not specified in the list, the datagram is discarded. If the datagram arrives at the destination and some of the specified routers were not visited, the datagram is discarded.

Figure 21-21 shows the format for the strict source route option. As can be seen, the format is very similar to the format for the record route. The code for the strict source

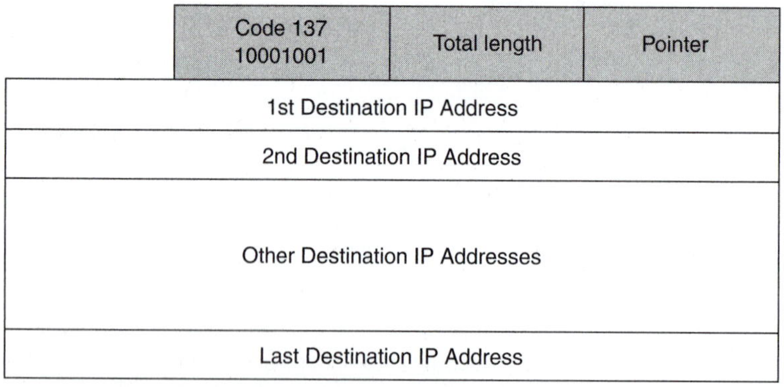

Code 137 10001001	Total length	Pointer
1st Destination IP Address		
2nd Destination IP Address		
Other Destination IP Addresses		
Last Destination IP Address		

FIGURE 21-21   Strict source route option format

route option is 137, which specifies that if the datagram is fragmented, the option must be copied into every fragment. Since the processing of IP options precedes reassembly, the original datagram will not be reassembled until the final destination is reached.

With the strict source route option, the originating device compiles a list of router interface IP addresses. All router interfaces must be visited in the same order as the addresses appear in the list. Before the datagram is sent, the first interface IP address is removed from the list and placed in the destination IP address field of the IP header. The remaining interface IP addresses are placed in sequence in the data field of the option. The original destination device IP address is the final entry in the list. The pointer field is initially set to 4, which points to the first interface address in the list (i.e., the second interface to be visited). When the datagram reaches the first router, the pointer is compared to the length field. If the value of the pointer is less than the header length, the router processes the datagram, replaces the first entry in the option data field with the IP destination address, replaces the contents of the IP destination address field with the next interface address, increments the pointer by 4, and sends the datagram to the next interface in the list. After the datagram has visited every router interface specified in the data field, the original destination IP address is placed into the IP destination address field, and the datagram is delivered to the final destination. However, if a router receives a datagram where the value in the pointer is greater than the value of the option length, the datagram has visited all the specified addresses and still has not reached the destination. When this happens, the datagram is discarded.

Figure 21-22a shows an internetworking configuration where device 65.32.30.7 is sending an IP datagram to device 145.5.25.39 using a strict source route option to specify the route. Figure 21-22b shows the original hop list compiled in the source, which is comprised of IP addresses for three router input interfaces. Figure 21-22c shows the source and destination IP addresses of the datagram and the contents of the option when they leave the source. The first byte is the code field for the record route option code (137), the second byte is the length field (15), and the third byte is the pointer (4). Therefore, the total option length is 15 bytes long. Three of the bytes constitute the header, and the remaining 12 bytes are filled with destination IP addresses. The offset is 4, which means the first interface address begins at the fourth byte of the option data field. Note that the destination IP address in the datagram is the first IP address specified in the hop list, the first destination IP address in the option data field is the IP address of the second router input interface, and the last destination IP address in the option data field is the IP address of the destination device.

Figure 21-22d shows the source and destination IP addresses for the datagram and the contents of the option when they leave the first router. The pointer is 8, which means the next destination interface address begins at the eighth byte of the option data field. The destination IP address is the second IP address specified in the hop list, and the first destination IP address in the option data field is the IP address of the input interface of the first router. The last two destination IP addresses in the option field are unchanged.

Figure 21-22e shows the source and destination IP addresses of the datagram and the contents of the option when they leave the second router. The pointer is 12, specifying the next interface address begins at the 12th byte of the option. The first two bytes of the option data field contain the IP addresses of the first two destination addresses specified in the hop list (the two interfaces the datagram has already visited).

Figure 21-22f shows the source and destination IP addresses of the datagram and the contents of the option when they leave the third (last) router. The destination address is now the address of the final destination. Also, note that the third address in the option data field is the IP address of the input interface of the third router. When the datagram reaches the destination device, the pointer will be larger than the header

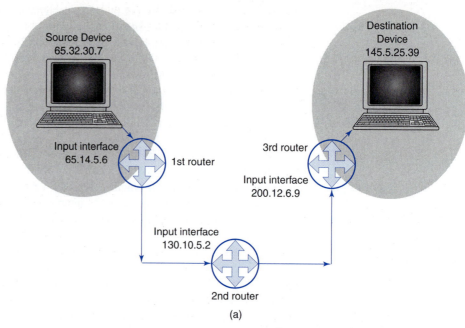

IP Datagram header leaving source:
Source IP address: 65.32.30.7
Destination IP address: 65.14.5.6

137	15	4
130.10.5.2		
200.12.6.9		
145.5.25.39		

Original hop list compiled in source:
1st destination interface = 65.14.5.6
2nd destination interface = 130.10.5.2
3rd destination interface = 200.12.6.9

(b)

(c)

IP Datagram header leaving 1st router:
Source IP address: 65.32.30.7
Destination IP address: 130.10.5.2

137	15	8
65.14.5.6		
200.14.5.9		
145.5.25.39		

(d)

IP Datagram header leaving 2nd router:
Source IP address: 65.32.30.7
Destination IP address: 200.14.5.9

137	15	12
65.14.5.6		
130.10.5.2		
145.5.25.39		

(e)

137	15	16
65.14.5.6		
130.10.5.2		
200.14.5.9		

(f)

IP Datagram header leaving 3rd router:
Source IP address: 65.32.30.7
Destination IP address: 145.5.25.39

FIGURE 21-22 Strict source route option: (a) network and router configuration; (b) original hop list; (c) option when it leaves the source; (d) option when it leaves the first router; (e) option when it leaves the second router; (f) option when it leaves the third router

length, and the option data field contains the addresses of the three router interfaces the datagram visited while moving through the Internet.

The strict source route option is sometimes called the *strict source and record route option* because once the datagram reaches the destination device, the option data field contains a sequential list of the router interfaces the datagram passed through while being transported through the Internet. The process of replacing the source route with the recorded route (even though it is recorded in the opposite direction) means that the option (and the IP header) stays a constant length as the datagram progresses through the Internet. When a device receives a strict source route option, it should respond to the datagram using a strict source route option, except passing through the routers in the reverse order.

Loose source route option. The loose source route option is very similar to the strict source route option, except more forgiving. With the loose source route option, every router interface listed in the option data field must be visited; however, the datagram can pass through other router interfaces. The loose source route option allows a host or router to use any route of any number of intermediate routers to reach the next required interface in the route. If routers replace the source route with the recorded route, the option is the called *loose source and record route option.*

Figure 21-23 shows the format for the loose source route option. The format is identical to that of the strict source route option except the code 131 is used.

Router alert option. The router alert option is optional. However, when used, it allows a device to specify to a router that it should pay more attention to the packet. Protocols that use the router alert option are RSVP path, RSVP path teardown, RSVP reservation confirmation, and IGMP.

Figure 21-24 shows the format for a router alert option, which includes an eight-bit code field, an eight-bit length field, and a 16-bit value field. The option code for the router alert option is 148, and the length is fixed at four bytes. Currently, there is only one code used in the 16-bit value field: 0000 hex, which simply indicates that the router must examine the datagram. The values 1 through 65,535 are currently reserved for future use.

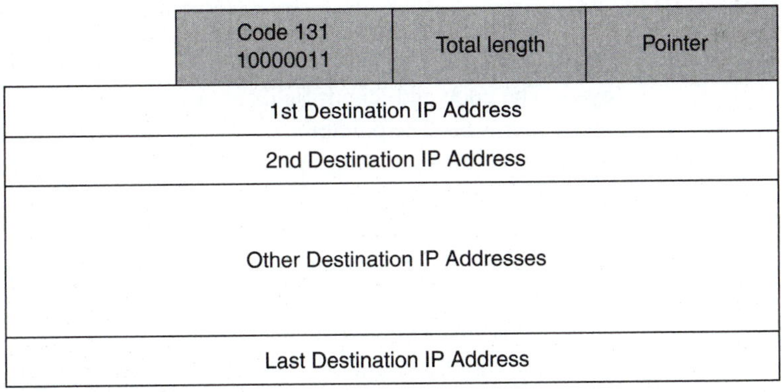

Code 131 10000011	Total length	Pointer
1st Destination IP Address		
2nd Destination IP Address		
Other Destination IP Addresses		
Last Destination IP Address		

FIGURE 21-23 Loose source route option format

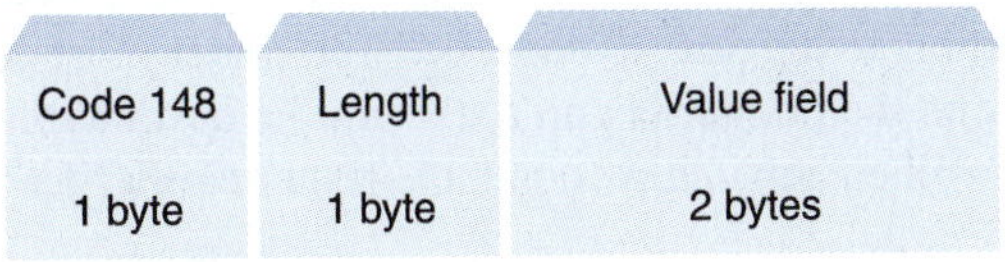

Code 148	Length	Value field
1 byte	1 byte	2 bytes

FIGURE 21-24 Router alert option format

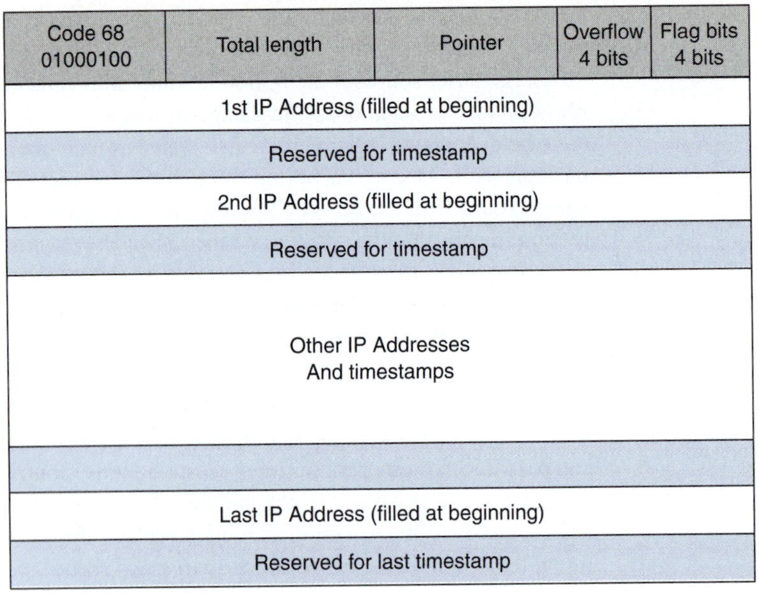

Code 68 01000100	Total length	Pointer	Overflow 4 bits	Flag bits 4 bits
1st IP Address (filled at beginning)				
Reserved for timestamp				
2nd IP Address (filled at beginning)				
Reserved for timestamp				
Other IP Addresses And timestamps				
Last IP Address (filled at beginning)				
Reserved for last timestamp				

FIGURE 21-25 Timestamp option format

Timestamp option. The timestamp option is used to record how long a router takes to process a datagram. The time is expressed in milliseconds from midnight, universal time. Network managers and users can use the timestamp option to track the behavior of Internet routers. The timestamp gives only an estimate of the time it takes a datagram to travel from one router to the next because the local clocks in the routers are probably not synchronized. The timestamp is a right-justified 32-bit value. If the time is not available in milliseconds or cannot be provided with respect to midnight universal time, any time can be inserted as a timestamp provided that the high-order bit is set to 1, which indicates that the time is a nonstandard value.

Figure 21-25 shows the format for the timestamp option, which is comprised of a four-byte header and a variable-length data field with a maximum length of 36 bytes. The code and length fields are identical to the other options. The code for a timestamp is 68. The pointer is the number of bytes from the beginning of the option to the end of the timestamps plus 1, which in essence points to the next available location to place a timestamp. The four-bit overflow field is used to keep track of how many routers could not add their timestamp to the data field because there were no fields available. The maximum number of routers the overflow field can count is 15 (1111 binary). If the overflow count overflows (increments beyond 15), the original datagram is considered to be in error and is discarded.

The flag field specifies the router responsibilities. When the flag contains a 0 (0000), each router must add a timestamp in the appropriate place in the data field. When the flag contains a 1 (0001), each router must add its outgoing interface IP address and a timestamp in the appropriate place in the data field. When the flag contains a 3 (0011), the IP addresses are already included in the data field, and the pointer points to the next IP address. Each router must compare the next IP address (i.e., the one the pointer is point at) with the destination IP address in the received datagram. When a match is found, the router overwrites the IP address with its out-

Code 68 01000100	Total length	Pointer	Overflow 4 bits	Flag bits 0000
Reserved for timestamp				
Reserved for timestamp				
Reserved for timestamp				

(a)

Code 68 01000100	Total length	Pointer	Overflow 4 bits	Flag bits 0001
Reserved for IP addresses				
Reserved for timestamp				
Reserved for IP addresses				
Reserved for timestamp				
Reserved for IP addresses				
Reserved for timestamp				

(b)

Code 68 01000100	Total length	Pointer	Overflow 4 bits	Flag bits 0011
Pre-specified IP addresses				
Reserved for timestamp				
Pre-specified IP addresses				
Reserved for timestamp				

(c)

FIGURE 21-26   Flags in the timestamp option: (a) flag = 0; (b) flag = 1; (c) flag = 3

going IP address and adds a timestamp. Figure 21-26 shows a comparison of the three flag conditions.

The following rules apply to the timestamp option:

1. The originating device must record a timestamp in a timestamp option whose Internet address fields are not prespecified or whose first prespecified address is the device's interface address.
2. The destination device must (if possible) add the current timestamp to a timestamp option before passing the option on to the transport layer or to the ICMP protocol for processing.
3. A timestamp value must follow the rules specified for the ICMP timestamp message.

Security option. The security option provides a way for a device to send security, compartmentation, handling restrictions, and TCC (closed user group) parameters.

Type	Length	Security field	Compartments field	Handling restriction field	Transmission control code field
1 byte	1 byte	2 bytes	2 bytes	2 bytes	3 bytes

FIGURE 21-27 Security option format

Table 21-12 IP Security Option Classifications

Binary Code		Hex Code	Classification
00000000	00000000	0000	Unclassified
11110001	00110101	F135	Confidential
01111000	10011010	785A	EFTO
10111100	01001101	BC4D	MMMM
10101111	00100110	AF26	PROG
10101111	00010011	AF13	Restricted
11010111	10001000	D788	Secret
01101011	11000101	6BC5	Top secret
00110101	11100010	35E2	Reserved for future use
10011010	11110001	9AF1	Reserved for future use
01001101	01111000	4D78	Reserved for future use
00100100	10111101	24BD	Reserved for future use
00010011	01011110	135E	Reserved for future use
10001001	10101111	89AF	Reserved for future use
11000100	11010110	C4D6	Reserved for future use
11100010	01101011	E26B	Reserved for future use

The format for the security option is shown in Figure 21-27. The security option type is 130, and the length field is the same as the other options. The 16-bit security field is used to specify 1 of 16 security classifications; however, currently only eight are being used. Table 21-12 lists the eight security classifications currently being used.

The compartments field contains all 0s when the information being transported is not compartmented. The other values for the compartments may be obtained from the Defense Intelligence Agency.

The handling restrictions (H) field contains values for the control and release markings, which are alphanumeric digraphs that are defined in the Defense Intelligence Agency Manual DIAM 65-19, Standard Security Markings.

The transmission control code (TCC) field provides a means to segregate traffic and define controlled communities of interest among subscribers. The TCC values are trigraphs and are available from the HQ DCA code 530.

QUESTIONS

21-1. Describe what is meant by a *globally unique IP address*.

21-2. Explain what is meant by *address resolution*.

21-3. What is the primary purpose of the *address resolution protocol* (ARP)?

21-4. List and describe the fields contained in ARP request and response packets.

21-5. Define *ARP cache* and explain what it is used for.

21-6. What is the difference between *dynamic* and *static address resolution?*

21-7. What information is generally stored in an ARP cache?

21-8. Describe the purpose of *proxy ARP.*

21-9. Describe the purpose of the *reverse address resolution protocol.*

21-10. Compare *connection-oriented* and *connectionless protocols.*

21-11. Describe the differences between *hardware* and *IP addressing.*

21-12. List the functions of the *Internet Protocol.*

21-13. List and describe the fields contained in the *IP header.*

21-14. Describe how the *checksum* is calculated in an IP header.

21-15. Describe the purposes of *IP options.*

21-16. Describe the fields contained in an *IP option header.*

21-17. List and describe the *single-byte IP options.*

21-18. List and describe the *multiple-byte IP options.*

PROBLEMS

21-1. Determine the contents of the first byte of an IP header if the IP protocol is IPv4 and the header has eight bytes of options.

21-2. Determine the contents of the first byte of an IP header if the IP protocol is IPv4 and the header has 20 bytes of options.

21-3. Determine the contents of the first byte of an IP header if the IP protocol is IPv4 and the header has 40 bytes of options.

21-4. Determine the precedence and type of service for the following hex values in an IP header type field:
   a. 70 hex
   b. 80 hex
   c. E4 hex
   d. 62 hex
   e. 28 hex
   f. 30 hex

21-5. Determine the decimal value for the codepoint, the category, and the assigning authority for the following differentiated service fields:
   a. 001110
   b. 100110
   c. 001111
   d. 011011
   e. 010101
   f. 101101

21-6. Determine the length of an ICMP data field for an Ethernet frame that is carrying an IP datagram with a 28-byte header and an eight-byte ICMP header and the Ethernet data field is 600 bytes long.

21-7. Use the following network analyzer display of an ARP packet to answer the following questions:

```
0000   00   07   08   00   06   04   00   02   00   00   BD   41   2A   1C   C1   98
0010   A0   14   00   00   40   A3   B1   CC   C1   98   0F   10
```

   a. What hardware type is being used?
   b. What protocol type is being used?
   c. How many bytes are in the hardware address?

d. How many bytes are in the protocol address?

e. What operation is being performed?

f. What is the source hardware address (in hex)?

g. What is the destination hardware address (in hex)?

h. What is the source protocol address (in dotted decimal)?

i. What is the destination protocol address (in dotted decimal)?

21-8. Use the following network analyzer display of an IP packet to answer the following questions:

```
0000   45   28   13   82   01   A2   21   34   F3   01   43   A1   C0   99   4A   01
0010   C0   99   4A   0C   --   --   --   --   --   --   --   --   --   --   --   --
```

a. What IP version is being used?

b. Are IP options being used?

c. What is the precedence?

d. What type of service is being requested?

e. What type of higher-level protocol is this datagram carrying?

f. How many data bytes were included in the original message (in decimal)?

g. What is the ID number for the datagram (in decimal)?

h. Is fragmenting allowed?

i. If yes, is this a first, last, or a middle fragment?

j. What is the decimal number of the first byte in this fragment?

k. Assuming the IP datagram is passing over an Ethernet, which fragment is this datagram in?

l. How many routers has this datagram propagated through?

m. What is the hex value for the checksum?

n. What is the source IP address (hex)?

o. What is the source IP address (dotted decimal)?

p. What is the destination IP address (hex)?

q. What is the destination IP address (dotted decimal)?

21-9. You have received the following three fragmented IP packets (X, Y, and Z). From the following information, answer the following questions.

Fragment	More Bit	Offset
X	1	007D
Y	0	00FA
Z	1	0000

a. Which fragment was sent first (X, Y, or Z)?

b. Which fragment was sent second (X, Y, or Z)?

c. Which fragment was sent last (X, Y, or Z)?

d. How many bytes does the second fragment contain (decimal)?

21-10. You have received the following three fragmented IP packets (X, Y, and Z). From the following information, answer the following questions.

Fragment	More Bit	Offset
X	0	0022
Y	1	0000
Z	1	0011

a. Which fragment was sent first (X, Y, or Z)?

b. Which fragment was sent last (X, Y, or Z)?

c. How many bytes does the first fragment contain (decimal)?

21-11. A maximum-length IP datagram carrying 65,536 bytes can be transported over a WAN with a maximum packet length of 292 bytes, which includes the 20-byte IP header. Determine

    a. How many fragments are required.

    b. The offset values in binary, hex, decimal for the first four and last two fragments and the number of bytes in the last fragment.

C H A P T E R   22

Internet Control Message Protocol

CHAPTER OUTLINE

OBJECTIVES

- Describe the major shortcomings of the Internet Protocol
- Name and describe the function of the three ICMP fields common to all ICMP messages
- List and describe the purposes of the two categories of ICMP messages
- Describe the operation of ICMP query requests and responses
- Describe the fields included in the headers for an echo request and an echo response
- Describe the fields included in the headers for a timestamp request and a timestamp response
- Describe the fields included in the headers for an address mask request and an address mask response
- Describe the fields included in the headers for a router discovery solicitation and a router advertisement
- Describe the operation of ICMP error-reporting messages
- Describe the fields included in the headers for a destination unreachable message
- Explain when a destination-unreachable message is sent
- Describe the fields included in the headers for a source quench message
- Explain when a source quench message is sent
- Describe the fields included in the headers for a redirect message
- Explain when a redirect message is sent
- Describe the fields included in the headers for a time-exceeded message
- Explain when a time-exceeded message is sent
- Describe the fields included in the headers for a parameter problem message
- Explain when a parameter problem is sent
- Describe how an ICMP checksum is calculated at the source and destination

704

22-1 INTRODUCTION

The Internet Protocol (IP) has several major shortcomings:

1. IP is an unreliable and connectionless datagram messaging service with no means of reporting procedural (logical) errors or performing error detection or correction on its own payload.
2. IP has no instrument that allows a host to perform queries to determine if a router or another host is active (alive).
3. IP also has no means of performing management queries concerning other hosts and routers.

Consequently, Internet authorities designed the *Internet Control Message Protocol* (ICMP) to compensate for the shortcomings of the IP. Figure 22-1 shows where ICMP is located in the TCP/IP protocol suite. ICMP is a layer 3 network protocol. However, ICMP messages are encapsulated in IP datagrams. Therefore, ICMP is a routable protocol.

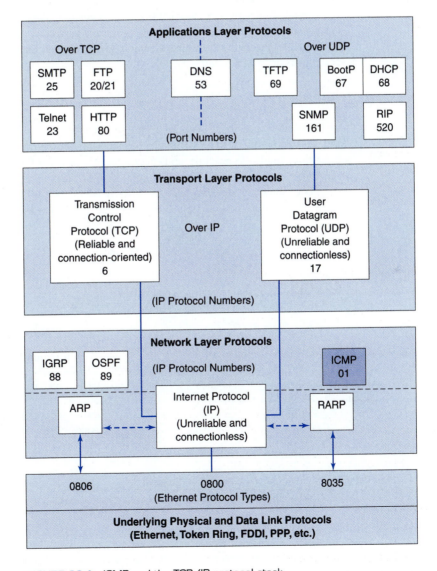

FIGURE 22-1   ICMP and the TCP/IP protocol stack

ICMP is the official gossipmonger (big mouth) of the Internet, as it provides feedback to hosts when problems are encountered while transporting datagrams through the Internet. ICMP is also the home of the *Packet Internet Groper* (*ping* command) and several other utilities that provide connection verification and time-based information. Hosts and gateways (routers) use ICMP as a control, messaging, and diagnostic protocol, which allows TCP/IP internetworks to handle logical errors that occur between the time a datagram leaves a source host and the time the datagram arrives at the destination host. ICMP provides hosts with the ability to initiate, transmit, receive, and process messages that help diagnose existing or potential connectivity problems on the Internet. Any device with a full IP address can send, receive, and process ICMP messages, which includes hosts, routers, gateways, and intelligent hubs.

22-2 ICMP ENCAPSULATION

Although ICMP is a layer 3 protocol, ICMP messages are encapsulated in the data field of layer 3 IP datagrams before they are passed down to the data-link layer. IP identifies that it is carrying ICMP packets by placing a type 1 code in the protocol field of the IP header.

Figure 22-2 shows an Ethernet frame carrying an IP datagram that has an ICMP message encapsulated in its data field. ICMP messages go wherever the destination address of the IP datagram carrying it designates, and ICMP messages can be sent from both client and server hosts or from a router.

22-3 ICMP MESSAGE FORMAT

All ICMP messages begin with an eight-byte header followed by a variable-length data field. The beginning of an ICMP header has the same format for all ICMP messages and includes three fields—type, code, and checksum—as shown in frame format in Figure 22-3a. Figure 22-3b shows an ICMP packet aligned to a four-byte boundary. What follows the type, code, and checksum fields in an ICMP header depends on what type of ICMP message is identified in the type field (i.e., what type of ICMP message is being sent).

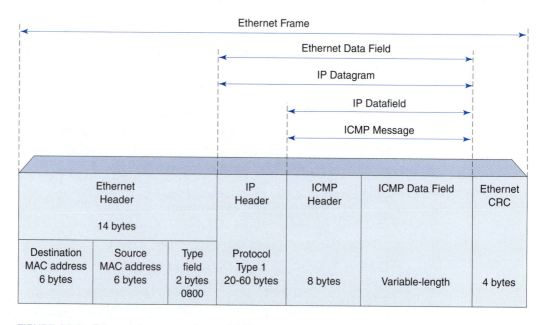

FIGURE 22-2 Ethernet frame carrying an ICMP message encapsulated in an IP datagram

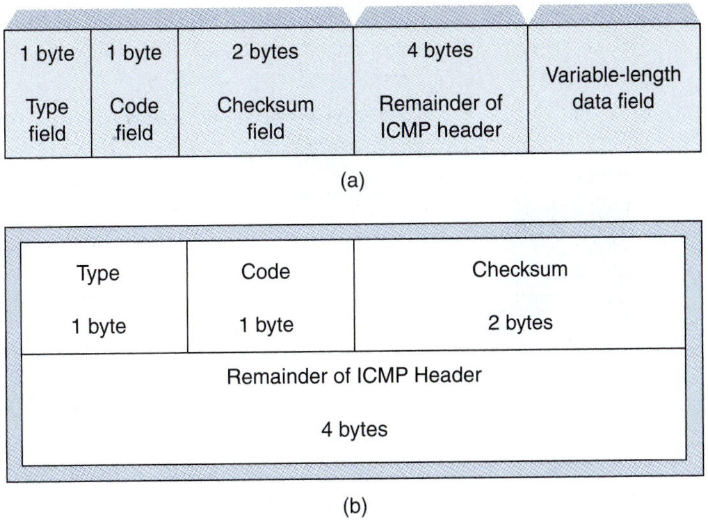

1 byte	1 byte	2 bytes	4 bytes	Variable-length data field
Type field	Code field	Checksum field	Remainder of ICMP header	

(a)

Type 1 byte	Code 1 byte	Checksum 2 bytes
Remainder of ICMP Header 4 bytes		

(b)

FIGURE 22-3 (a) ICMP message format; (b) ICMP packet aligned to a four-byte boundary

22-4   ICMP CATEGORIES, TYPES OF SERVICE, TYPES, AND CODES

TCP/IP specifies two broad categories of ICMP messages: *query* (*diagnostic*) and *variation* (*error reporting*). ICMP query messages occur in pairs—a request, which is usually sent by a host, followed by a response, which may be sent from a host or a router. ICMP variation messages are unsolicited transmissions that are sent by a host or a router to report procedural errors.

ICMP request messages can be transmitted in an IP datagram with any of the five type of service (TOS) priorities (i.e., minimum delay, maximum throughput, maximum reliability, minimize monetary cost, or normal service). ICMP reply messages are sent with the same TOS priority that was used in the corresponding ICMP request message. ICMP error-reporting messages are always sent with the default TOS (0000—normal service). ICMP messages are treated as interruptions by the TCP/IP stacks of many venders. Therefore, in an IP-based network, ICMP messages are generally the first to be lost or discarded, which limits their usefulness.

There are 16 basic types of ICMP messages of which only 14 are currently being used. The current ICMP messages include eight query (diagnostic) messages and six variation (error-reporting) messages. As shown in Figure 22-3, the first two bytes of all ICMP messages are the type and code fields. The type field identifies what kind of ICMP packet is being delivered, and the code field further defines the meaning of the ICMP packet. Each kind of message has a separate type to identify it.

22-5   ICMP QUERY (DIAGNOSTIC) MESSAGES

ICMP query messages can be sent from a host to another host or from a host to a router to acquire specific information about the other host or the router. Query messages help hosts acquire information about themselves and neighboring hosts and routers connected to their network. Hosts send ICMP queries to a specific destination for diagnostic purposes. ICMP queries can be used to diagnose certain network connectivity problems.

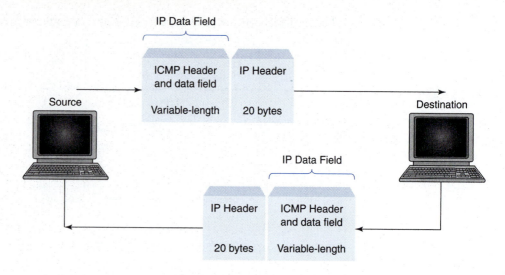

FIGURE 22-4 ICMP diagnostic message exchange (request/reply)

ICMP queries are transmitted from a source to a destination where the destination simply returns the ICMP query to the originating source. The format for the returned IP datagram is identical to the format of the original transmitted datagram; however, the data contained in the datagram may be different.

Figure 22-4 shows an example of an ICMP query request and its corresponding reply. The request includes an IP header followed by the ICMP message, which includes a header and a data field. The entire ICMP message is encapsulated in the data field of the IP datagram. As shown in the figure, the reply has exactly the same format as the request.

ICMP query messages can be further divided into two subcategories: ICMP request messages and ICMP response messages. Table 22-1 lists the four ICMP query messages currently being used, which include echo request, router solicitation, timestamp request, and address mask request. Table 22-2 lists the four ICMP response messages currently being used, which include echo response, router advertisement, timestamp reply, and address mask reply. Note that the code field for all eight types of query messages is 0.

Table 22-1 ICMP Query (Diagnostic) Requests

Type	Code	Message
8	0	Echo request (ping request used with type 0)
10	0	Router solicitation (used with type 9)
13	0	Timestamp request (used with type 14)
17	0	Address mask request (used with type 18)

Table 22-2 ICMP Query (Diagnostic) Responses

Type	Code	Message
0	0	Echo response (ping reply used with type 8)
9	0	Router advertisement (used alone or with type 10)
14	0	Timestamp reply (used with type 13)
18	0	Address mask reply (used with type 17)

22-5-1 Echo Request (Type 8) and Echo Response (Type 0)

ICMP echo request and echo response messages are a significant portion of a TCP/IP network manager's troubleshooting tools. An echo request combined with an echo reply can determine whether two systems (hosts or routers) are able to communicate with each other at the network (IP) layer. When someone issues a ping command, an echo request (type 8) message is generated and sent to the designated destination IP address. If the IP address is valid and the pinged destination's equipment supports the request function, an echo response (type 0) is returned. In its most basic form, the ping utility is used to verify network connectivity. The ping command can support multiple flags that expand its functionality, which may include using IP header options. For example, the ping command can include packet length (-l) and don't fragment (-f) flags. The ping command can be sent without any data, or it can include a variable-length data field that can be used to check for transmission errors.

Figures 22-5a and b show the header format for an ICMP type 8 (echo request) or type 0 (echo response) message, which includes the following fields:

1. Type field (one byte)
2. Code field (one byte)
3. Checksum (two bytes)
4. Identification number (two bytes)
5. Sequence number (two bytes)

The type field contains 08 in the request and a 00 in the response, both with a code of 00. Other than the type number and the checksum value, the formats for the request and response messages are identical. The identification number (ID #) in an ICMP ping request is randomly generated and changes with each new ping request issued. With 16 bits, there are 65,536 ID numbers possible (0–65,535). The sequence number is generally incremented

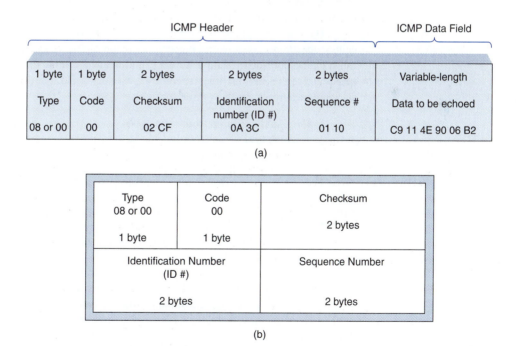

FIGURE 22-5 (a) ICMP type 8 (echo request) or type 0 (echo response) message format; (b) aligned to a four-byte boundary

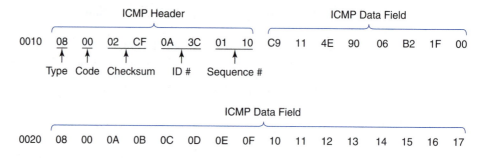

FIGURE 22-6   Example of an ICMP type 8 echo request message

by 1 in each successive echo request sent. Therefore, if the ping command calls for multiple echo requests to the same destination address, the sequence number changes for each request; however, the ID # stays the same. The data field is optional and may contain any data that the sender wishes with a length limited only by the IP maximum of 65,535. To allow for a variable-length data field, the end of the ICMP data field is left open. An ICMP echo response (type 0) looks identical to the echo request, except the contents of the type field is 00 hex instead of 08 and the checksum is different.

Figure 22-6 shows an example of a network analyzer display of an ICMP type 8 echo request with a checksum value of 02CF, an identification number of 0A3C, and a sequence number of 0110. The figure also shows the first 24 bytes of the ICMP data field.

22-5-2   Timestamp Request (Type 13) or Timestamp Reply (Type 14)

Two devices (either hosts or routers) can use the timestamp request and timestamp reply messages to resolve the round-trip delay time an IP datagram experiences propagating between the two devices. The clocks of two distant devices can also be synchronized using the timestamp request/reply.

The request/reply header for the timestamp ICMP message is identical to that of the echo request/reply as shown in Figures 22-7a and b. The header includes a type field, code field, checksum, identification number, and sequence number. However, timestamp requests and replies also include a specific 12-byte data field. The 12-byte data field in a timestamp request/reply contains three subfields:

1. Original timestamp (four bytes)
2. Receive timestamp (four bytes)
3. Transmit timestamp (four bytes)

The sequence for a timestamp request/response is shown in Figure 22-8 and described here:

1. The source machine creates a timestamp request message and fills the original timestamp subfield with the universal time determined at the time the request message is actually sent (time t_1).
2. The source machine fills the receive and transmit subfields with 0s.
3. The destination creates the timestamp reply message and copies the original timestamp from the request message (time t_1) and fills the receive timestamp subfield with the universal time shown by its clock at the time the request was actually received (time t_2).
4. The destination fills the transmit timestamp subfield with the universal time shown by its clock at the time the reply message is actually sent (time t_3).
5. The originating source receives the timestamp reply at time t_4.

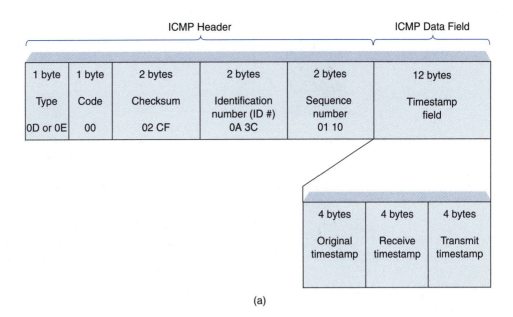

ICMP Header   ICMP Data Field

1 byte	1 byte	2 bytes	2 bytes	2 bytes	12 bytes
Type	Code	Checksum	Identification number (ID #)	Sequence number	Timestamp field
0D or 0E	00	02 CF	0A 3C	01 10	

4 bytes	4 bytes	4 bytes
Original timestamp	Receive timestamp	Transmit timestamp

(a)

Type	Code	Checksum	
0D or 0E 1 byte	00 1 byte	2 bytes	
Identification Number (ID #) 2 bytes		Sequence Number 2 bytes	
Original Timestamp 4 bytes			
Receive Timestamp 4 bytes			
Transmit Timestamp 4 bytes			

(b)

FIGURE 22-7 (a) ICMP type 13 (timestamp request) or type 14 (timestamp response) message format; (b) aligned to a four-byte boundary

The timestamp request and reply messages can be used at the source station to compute the round-trip delay time the message took to propagate between the source and destination stations as follows:

$$\text{Sending time } (t_s) = \text{receive timestamp } (t_2) - \text{original timestamp } (t_1)$$

$$\text{Return time } (t_r) = \text{time the datagram returned } (t_4) - \text{transmit timestamp } (t_3)$$

$$\text{Round-trip time } (t_{RT}) = \text{sending time } (t_s) + \text{return } (t_r)$$

The host-to-destination and destination-to-host one-way propagation times are accurate only if the source and destination clocks are synchronized. However, the round-trip propagation

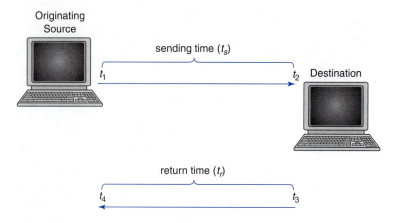

Originating
Source

sending time (t_s)

t_1 t_2 Destination

return time (t_r)

t_4 t_3

FIGURE 22-8 Timestamp request/response example

time is accurate even if the two clocks are not synchronized because each station's clock contributes twice to the round-trip calculation, thus canceling any error contributed to the differences between them.

Example 22-1

Original timestamp (t_1) = 42 time units

Receive timestamp (t_2) = 53 time units

Transmit timestamp (t_3) = 54 time units

Time datagram actually returned (t_r) = 61 time units

Solution

Sending time (t_s) = 53 − 42 = 11 time units

Return time (t_r) = 61 − 54 = 7 time units

Round-trip time (t_{RT}) = 11 + 7 = 18 units

The round-trip time calculated does not include the turnaround time at the destination. The total elapsed time (including the turnaround time at the receiver) is

Elapsed time = time datagram actually returned (t_4) − original timestamp (t_1)

= 61 − 42

= 19 time units (includes 1 unit of turnaround time)

Dividing the round-trip propagation time by 2 can approximate the one-way propagation time:

Round-trip time (t_{RT}) = 19/2

= 9.5 time units

22-5-3   Address Mask Request (Type 17) or Address Mask Reply (Type 18)

IP addresses include a network address, a host identifier, and in most cases a subnet address. A host generally knows its full IP address; however, it may not know if it is on a subnet, and if it is on a subnet, it may not know which subnet or the subnet address. It has no way of knowing what part of its IP address defines the network address and what part defines the subnet address. For example, a host may know that its IP address is 158.29.230.171, but it may not know that the 20 leftmost bits are the network and subnet identification bits and that the 12 rightmost bits are the host identification bits.

To resolve the network and subnet addresses, a host must know its subnet mask. For example, if you apply the default mask 255.255.255.0 and the subnet mask 255.255.255.240 to the host's IP address, the network, subnet, and host identification bits are revealed as shown here:

Host address	158 . 29 . 230 . 171	
	10011110.00011101.11100110.10101011	
Default mask	11111111.11111111.00000000.00000000	
Network address	158. 29. 0. 0	
Host address	158 . 29 . 230 . 171	
	10011110.00011101.11100110.10101011	
Default mask	11111111.11111111.11111111.11110000	
Network address	158. 29. 224. 0	
Host address	158 . 29 . 230 . 171	
	10011110.00011101.11100110.10101011	
Default mask	11111111.11111111.1111HHHH.HHHHHHHH	
Network address	6. 171	

To acquire its subnet mask, a host can send an addressmask request to the closest router on its local area network (LAN), and if the host knows the IP address of the router, it can send the request directly to it. If the host does not know the IP address of the router, it can broadcast an addressmask request onto the LAN. The router receiving the request responds with an addressmask reply message, which contains the subnet mask of the requesting host. The host can then apply the subnet mask to its IP address and derive its network and subnet addresses.

The format for an addressmask request and reply is shown in Figures 22–9a and b. The type field contains 17 (11 hex) in a request and 18 (12 hex) in the reply. The functions

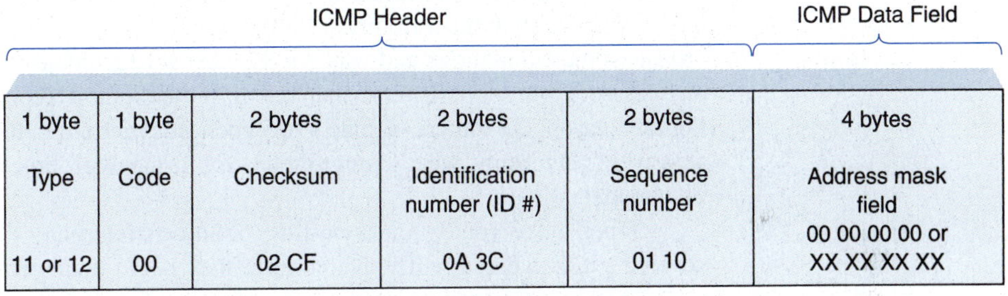

ICMP Header | ICMP Data Field

1 byte	1 byte	2 bytes	2 bytes	2 bytes	4 bytes
Type	Code	Checksum	Identification number (ID #)	Sequence number	Address mask field 00 00 00 00 or
11 or 12	00	02 CF	0A 3C	01 10	XX XX XX XX

(a)

Type 11 or 12 1 byte	Code 00 1 byte	Checksum 2 bytes
Identification Number (ID #) 2 bytes		Sequence Number 2 bytes
Address Mask 4 bytes		

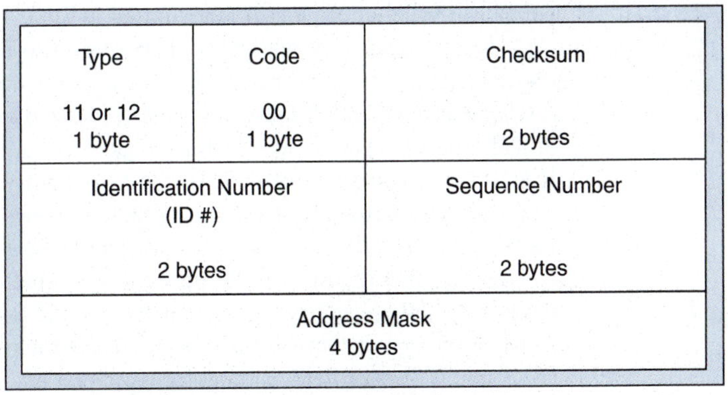

(b)

FIGURE 22-9 (a) ICMP type 17 (address-mask request) or type 18 (address-mask reply) message format; (b) aligned to a four-byte boundary

of the identification number, sequence number, and checksum are the same as the ICMP query messages previously discussed. In an addressmask request sent from a host, the address mask field is filled with all 0s (i.e., 00 00 00 00 hex), and in an addressmask reply sent from a router, the address mask field contains the host's subnet mask.

22-5-4 Router Discovery Messages (Type 9 and Type 10)

When a host on one network wishes to send a message to a host residing on another network, it must know the address of the routers connected to its own network. The host must also know if those routers are alive and functioning properly. ICMP router discovery messages are used to resolve the IP addresses of local routers.

ICMP router discovery messages help a host discover (determine) one or more of the operational routers on the host's subnet. There are currently two types of router discovery messages: router advertisement (type 9) and router solicitation (type 10). All routers on a subnet periodically send multicast router advertisements onto each of its multicast interfaces to announce the IP address of the interface. The default "*all hosts*" multicast IP address is 224.0.0.1. The default time between router advertisements is typically between 7 and 10 minutes.

A host on a subnet can passively discover the addresses of neighboring routers by simply listening for their advertisements. However, a host that boots up early in the delay period may not want to wait for one of the periodic advertisements. In this case, the host can send a multicast router solicitation onto the subnet requesting immediate advertisements. The default "*all-router*" multicast IP address is 224.0.0.2. In cases where a router will not process multicast addresses, the broadcast address is used. If the routers that receive the router solicitation message support IP multicast or broadcast addressing, they broadcast their routing information using a router advertisement messages. However, if the routers do not support the router portion of ICMP router discovery, the host's solicitation request is discarded. If a host does not receive a timely response, it may retransmit the solicitation a "small number of times," and if it still does not receive a response, it must cease broadcasting solicitations and wait for the next scheduled advertisement. If a host resides on a network that supports more than one IP router, it may receive more than one response. For this reason, the routers include a precedence level field in their advertisements that the host uses to determine what router to select for its gateway. Precedence levels are described in more detail later.

When a new router comes on-line (or an existing router that was not previously discovered), it is eventually discovered after they begin their periodic (and unsolicited) advertisements. If a network experiences a high number of lost packets, it should be accommodated by increasing the rate at which the routers advertise rather than by increasing the number of solicitations a host is permitted to send.

Router discovery messages enable hosts to discover the existence of neighboring routers. Therefore, router discovery messages are not considered routing protocol, as they do not determine which router a host should use to reach a particular destination. If a host initially makes a poor selection for its gateway router, the selected router should send an ICMP type 5 redirect message to the host to identify a more appropriate gateway router.

The format for a type 10 router solicitation message is shown in Figures 22-10a and b. As the figure shows, the router solicitation message is quite simple, as there is no need to convey information beyond the message type and code. Therefore, a router solicitation message is comprised of an eight-byte header and no data field. The header is actually only four bytes long and contains the three fields common to all ICMP messages: type, code, and checksum. The type field contains 10 (0A hex), and the code for a router solicitation is always 00 hex. To complete the eight-byte requirement for an ICMP header, four bytes of fill are appended to the solicitation in the area designated as reserved.

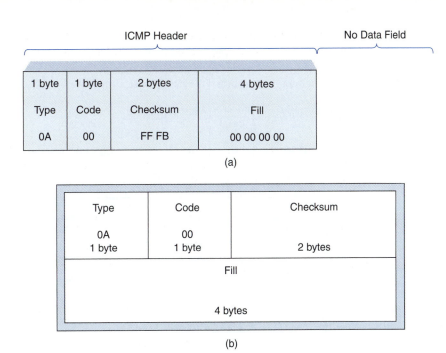

ICMP Header   No Data Field

1 byte	1 byte	2 bytes	4 bytes
Type	Code	Checksum	Fill
0A	00	FF FB	00 00 00 00

(a)

Type	Code	Checksum
0A	00	
1 byte	1 byte	2 bytes

Fill
4 bytes

(b)

FIGURE 22-10 (a) ICMP type 10 router solicitation message format; (b) aligned to a four-byte boundary

Figures 22-11a and b show the format for a router advertisement message, which includes an eight-byte header and a variable-length data field. The data field contains variable-length router addresses and a four-byte precedence (or preference) level for each router address (precedence fields are described later). The header for a type 9 router advertisement message includes the following six fields:

1. Type (one byte)
2. Code (one byte)
3. Checksum (two bytes)
4. Address count (one byte)
5. Address size (one byte)
6. Lifetime (two bytes)

The type, code, and checksum are the same as for all ICMP messages. The type for a router advertisement is 09 hex, and the code is 00 hex.

22-5-4-1 Address count field.　The address count field identifies the number of addresses advertised in the message, which determines the number of router address and address precedence fields included in the data field of the message. The number of router addresses and address precedence fields must be the same. For the header shown in Figure 22-11, the address count field is 01 hex; therefore, there is only one router address and one router precedence field.

22-5-4-2 Address size field.　The address size field sets the number of 32-bit (four-byte) addresses each router address field contains. For the header shown in Figure 22-11 and a network that is using IP addresses, the address size field is 01 hex, which means that each router address field contains one four-byte router IP address (C0 99 B9 64 or 192.153.185.100) followed by one four-byte address precedence field (00 00 00 00).

ICMP Header | ICMP Data Field

1 byte	1 byte	2 bytes	1 byte	1 byte	2 bytes	4 bytes	4 bytes	Additional addresses and preferences
Type	Code	Checksum	Address count	Address size	Lifetime	Router Address 1	Address Preference 1	
09	00	03 7B	01	01	07 08	C0 99 B9 64	00 00 00 00	

(a)

Type 09 1 byte	Code 00 1 byte	Checksum 2 bytes
Address Count 1 byte	Address Size 1 byte	Lifetime 2 bytes
Router Address 1 4 bytes		
Address Preference 1 4 bytes		
Additional addresses and preferences 4 bytes		

(b)

FIGURE 22-11   (a) ICMP type 9 router advertisement message format; (b) aligned to a four-byte boundary

22-5-4-3 Lifetime field.   The lifetime field identifies the number of seconds the router addresses are valid. For the header shown in Figure 22-11, 0708 hex equates to 1800 seconds, or 30 minutes.

A router can send a router advertisement message to advertise its own presence and also the presence of any additional routers or router interfaces on the network that it is aware of. Therefore, since each network may have more than one router and each router can have more than one IP address, the router address and address precedence fields of a type 9 router advertisement may be repeated as many times as needed to advertise all addresses. Each router and each router interface entry in the advertisement contains two fields: the IP address and the address precedence level.

The preference (or precedence) level defines the ranking of the router compared to other routers on the network. The preference level is used when a host is selecting a default router. The precedence level indicates which router is the preferred router to be selected as the default router. A higher preference level means a higher preference (more preferable). In other words, a host should select the router with the highest precedence level for its default router. Preference levels are defined with 2's complement (negative) numbers where the most positive number has the highest preference. Therefore, the router with an address preference level of 00000000 hex is considered the default router because 00000000 hex is the 2's complement of 0, which is the most positive value possible. If the address preference level is 80000000 hex, the router should never be selected as the default router because 80000000 hex is the 2's complement of $-2,147,483,648$, which is the most negative value possible.

The IP protocol is unreliable and is not concerned with error checking. Therefore, ICMP variation (error-reporting) messages were designed to compensate for the deficiencies of IP. ICMP query messages can be sent from a host or a router. However, variation messages can be sent only in response to an IP datagram that for some reason could not be delivered. Consequently, ICMP variation messages must be returned to the IP address of the source that originated the failed IP datagram, which is most often a host.

One significant limitation of ICMP variation messages is that if the problem relates to something that the host cannot control or change, the message cannot determine the actual cause of the problem. This limitation is a direct result of the flexibility of TCP/IP because IP datagrams take whatever route is available at the time, as only the source and destination IP addresses are included in the IP header. ICMP messages cannot be sent to intermediate routers (even if they are the cause of the problem) because there is no place in an IP datagram for IP addresses of intermediary routers.

There is one additional restriction on ICMP variation messages that is significant: if an ICMP variation messages experiences an error while returning to the original source of the failed IP datagram, the ICMP message is discarded, and the error report never reaches the originating source. This restriction is not imposed on ICMP diagnostic messages. If an error occurs with an ICMP diagnostic message, an ICMP variation message can be sent back to the source of the message to report the error.

ICMP error-reporting messages are responsible for reporting logical errors, not transmission errors. However, ICMP does not correct errors; it simply reports them. ICMP variation messages report errors to the host machine, not the operator of the machine. In other words, ICMP error-reporting messages are sent to the host machine and are invisible to humans except with a protocol analyzer.

Because only the source and destination IP addresses are included in an IP datagram, ICMP error-reporting messages are sent from the device that detected the error back to the source IP (host or router) that originated the message that was found in error. There are four instances when ICMP error-reporting messages are never generated:

1. In response to an IP datagram carrying another ICMP error reporting message
2. For a fragmented IP datagram that is not the first fragment
3. For an IP datagram having a multicast address
4. For an IP datagram having a special address, such as a loopback (127.0.0.0) or 0.0.0.0

Figure 22-12 shows the sequence and format for an ICM variation (error-reporting) message. As shown in the figure, the IP header in the variation message is necessary, as it is used to encapsulate the ICMP header and data field. All ICMP error-reporting messages contain a data field that includes the IP header of the failed IP datagram plus the first eight bytes of the data field from the failed IP datagram. The original IP datagram header is included to help the originating source identify what datagram was in error.

The first eight bytes of the original IP datagram's data field are included because they contain information about the UDP and TCP headers, such as the port and the sequence numbers. This information is necessary so that the source can inform the processes about the error. Table 22-3 lists the five kinds of variation messages and their types and codes.

22-6-1   Destination Unreachable (Type 3)

When a router cannot route a datagram to a network or to a host or when a host cannot deliver a datagram to a port or a protocol (process), the datagram is discarded, and the router or host sends a destination-unreachable (type 3) error message (variation) back to the originating source. A destination-unreachable message can be created by a router or by the destination host, depending on where the datagram failed.

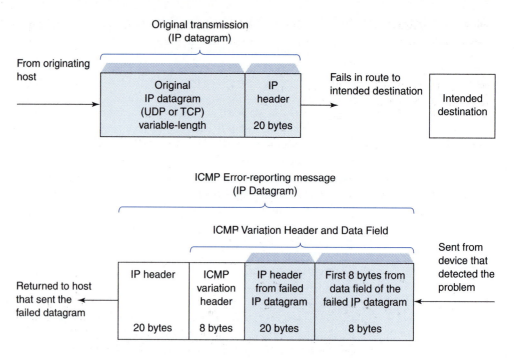

FIGURE 22-12   ICMP variation message exchange

Table 22-3   ICMP Variation (Error-Reporting) Messages

Type	Code	Message
3	0–15	Destination unreachable
4	0	Source quench (destination unreachable, fragment blocked)
5	0–3	Redirect
11	0–1	Time exceeded
12	0–1	Parameter problem
30	0–1	Traceroute

The format for a type 3 (destination unreachable) ICMP variation message is shown in Figures 22-13a and b. The type field contains 03 hex, which identifies it as an ICMP type 3 destination unreachable variation message. The code field in a destination unreachable message can contain any one of 16 different codes, which specify why the datagram could not be delivered. Table 22-4 lists the 16 destination-unreachable codes, what they indicate, and their intended purpose.

Codes 0 through 3 operate in a hierarchical arrangement with each other to indicate exactly how far the datagram traveled before it failed. Codes 6 and 7, 9 and 10, and 11 and 12 work together to identify the network or host that has the problem. Codes 13, 14, and 15 have been added by RFC 1812 to indicate the results of manager control of router functions.

Figure 22-14 shows an example of how ICMP error-reporting messages are used. The originating host (host 1, IP address X.X.X.X) sends an IP datagram intended for host 2 (IP address Y.Y.Y.Y). Somewhere along the way, the datagram fails, and an ICMP error-reporting message is sent back to host 1. If the datagram failed at the destination, host 2 is the device that would send the error-reporting message. If the datagram failed in a router, the router would send the error-reporting message.

Even when a destination-unreachable message is not generated, it does not necessarily mean that the datagram was successfully delivered. For example, if a datagram is trans-

ICMP Header ... ICMP Data Field

1 byte	1 byte	2 bytes	4 bytes	20 bytes	8 bytes
Type	Code	Checksum	Fill	Failed IP header	Failed IP data
03	00-0F	04 47	00 00 00 00	XX XX XX XX XX	YY YY YY

(a)

Type	Code	Checksum
03 1 byte	00–0F 1 byte	2 bytes

Fill 4 bytes
Failed IP Datagram Header 4 bytes
Failed IP Datagram Header 4 bytes
Failed IP Datagram Header 4 bytes
Failed IP Datagram Header 4 bytes
Failed IP Datagram Header 4 bytes
Failed IP Datagram Data 4 bytes
Failed IP Datagram Data 4 bytes

(b)

FIGURE 22-13 (a) ICMP type 3 (destination unreachable) message format; (b) aligned to a four-byte boundary

ported to the host over a LAN, such as Ethernet, it is impossible for a router to know whether the datagram was delivered.

An example of an ICMP type 3 error-reporting message is shown in Figure 22-15. The first 20 bytes shown (000B–001F) are the IP header in the datagram sent from the device detecting the problem back to the source of the failed datagram. The IP address (C0 99 B7 05) is the address of the detecting device, and the IP address (C0 99 B7 01) is the address of the source of the failed datagram, which is the destination of the ICMP message. The next eight bytes (0020–0027) are the ICMP header sent from the detecting device. Note that the ICMP type is 02 (destination unreachable) and that the code is 03 (port unreachable). The next 20 bytes are the IP header from the failed datagram. The source IP address is C0 99 B7 01, and the destination address is C0 99 B7 05 (note this is the same as the destination IP address of the failed IP datagram). The upper-layer protocol identified in the failed IP header is UDP (11 hex). The last eight bytes shown are the first eight bytes of the

Table 22-4 Destination-Unreachable Codes

Code	Indication	Purpose
0	Network unreachable	Cannot reach requested network, possibly because of hardware failure.
1	Host unreachable	Cannot reach requested host, possibly because of hardware failure.
2	Protocol unreachable	Destination host cannot find the upper-level protocol, such as UDP or TCP.
3	Port unreachable	Destination host's port is not available. The application program (process) that the datagram is destined for is not running at the moment.
4	Fragmentation blocked	DF (don't fragment) bit is set, but routing is impossible without fragmenting.
5	Source route failed	Source route is unavailable—one or more of the routers defined in the source routing option cannot be visited.
6	Destination network unknown	Network address is not in the routing table—router has no information about the destination network.
7	Destination host unknown	Host's address is not in the routing table—routing has no information about the destination host.
8	Source host isolated	Source is isolated and cannot talk to the Internet.
9	Destination network prohibited	Destination network is administratively blocking access.
10	Destination host prohibited	Destination host is administratively blocking access.
11	Network type of service (TOS) problem	Requested TOS prevents access because of unavailable TOS.
12	Host TOS problem	Requested TOS prevents access because of unavailable TOS.
13	Communications administratively prohibited	Prohibited because a router cannot forward a packet because of administrative filtering.
14	Host precedence violation	Sent by the first-hop router to a host to indicate that a requested precedence is not permitted for a particular combination of source/destination host or network, upper-layer protocol, and source/destination port.
15	Precedence cutoff in effect	Sent with a precedence level below that required by network operators.

Note: Codes 2 and 3 indicate problems at the destination host.

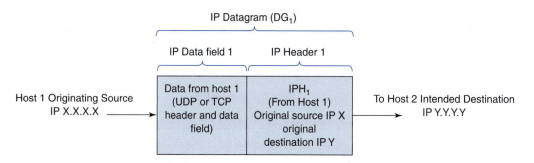

DG_1 = datagram from host 1
IPH_1 = IP header from host 1

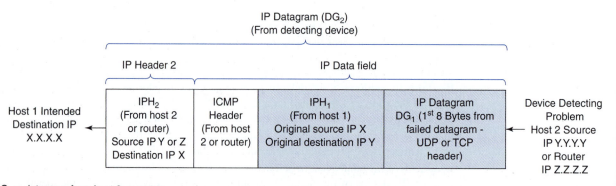

DG_2 = datagram from host 2 or router
IPH_2 = IP header from host 2 or router

FIGURE 22-14 ICMP error-reporting sequence

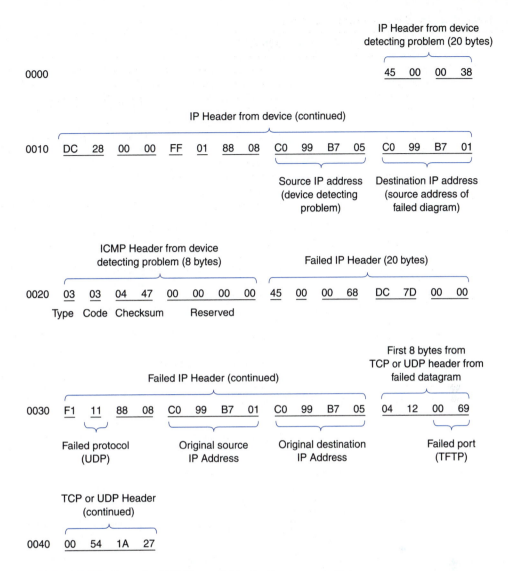

0000   IP Header from device detecting problem (20 bytes)   45   00   00   38

IP Header from device (continued)

0010   DC   28   00   00   FF   01   88   08   C0   99   B7   05   C0   99   B7   01

Source IP address (device detecting problem)   Destination IP address (source address of failed diagram)

ICMP Header from device detecting problem (8 bytes)   Failed IP Header (20 bytes)

0020   03   03   04   47   00   00   00   00   45   00   00   68   DC   7D   00   00

Type   Code   Checksum   Reserved

Failed IP Header (continued)   First 8 bytes from TCP or UDP header from failed datagram

0030   F1   11   88   08   C0   99   B7   01   C0   99   B7   05   04   12   00   69

Failed protocol (UDP)   Original source IP Address   Original destination IP Address   Failed port (TFTP)

TCP or UDP Header (continued)

0040   00   54   1A   27

FIGURE 22-15 Example of ICMP type 3 (destination unreachable) message

header in the transport-layer protocol (UDP) that was encapsulated in the data field of the failed IP datagram. The port number specified in the UDP header is 00 69, which is TFTP.

22-6-2 Source Quench (Type 4)

The IP protocol is a connectionless protocol, which means that there is no direct communications between the source host (which sends the datagram), the routers (which forwards the datagram), and the destination host (which processes the datagram). In addition, IP provides no mechanism for controlling the flow of the data encapsulated in the datagram (i.e., no organization to how packets of data are sent or acknowledged). Because of the lack of flow control, the source host is not informed if or when intermediate routers or the destination host have been overwhelmed with datagrams.

Lack of flow control can cause congestion in routers and hosts because they have a limited memory capacity (buffer space). If the memory is exceeded in a router or in a host, all or some of the received datagrams may be discarded. The ICMP source quench message was designed to add a degree of flow control to IP. When a router or host is overwhelmed with data packets and begins discarding IP datagrams, it sends a source quench message to the host that is sending the datagrams.

The source quench message has two purposes:

1. It informs the source that the datagram has been discarded.
2. It warns the source that there is a congestion problem somewhere in the path and that the source should slow down (quench) the rate at which it is sending data.

The device (router or host) experiencing congestion sends a source quench message back to the originating host for each datagram it discards. However, there is no mechanism to inform the originating host that the congestion problem is alleviated and the host can resume transmitting datagrams at its original rate. Instead, the source continues transmitting messages at the lower rate until it is no longer receiving source quench messages. Router or host congestion can be created from one-to-many or many-to-one communications:

1. With one-to-many communications, a single high-speed host can generate datagrams faster than a router or destination host can handle them. In this case, the source quench message is helpful because it tells the source to slow down its rate of transmissions.
2. With many-to-one communications, more than one host generates datagrams that must be routed by intermediate routers and processed by the destination host. In this case, because each source may be transmitting datagrams at different rates, the source quench message may not be particularly useful. The router or destination host has no clue what source is responsible for the congestion; therefore, it could discard datagrams from a slow source rather than dropping datagrams from the source that is actually creating the congestion problem.

The format for the source quench message is shown in Figures 22-16a and b. The ICMP header is comprised of the usual suspects: type, code, and checksum. The type is 04 hex, and the code is 00 hex. Because the last four bytes of the header are not used, they are filled with four bytes of 0s. The ICMP data field is comprised of the IP header from the discarded datagram and the first eight bytes of IP data field.

22-6-3 Redirect (Type 5)

When a router needs to forward a datagram to another network, the router must know the IP address of the next appropriate router. The same is true for a host because it must send a datagram to the correct router if it expects the datagram to be routed to the intended destination. Therefore, both routers and hosts have routing tables, which enables them to find the IP address of the next appropriate router. Routing of IP datagrams is executed dynamically (updated constantly) by routers and performed by routers (not hosts) because there are too many hosts. Hosts use static routing, which has a routing table with a limited number of entries.

Hosts usually know the routing table of only one router: its default router. For this reason, a host could forward a datagram destined for a remote network to the wrong router. When this happens, the router that erroneously receives the datagram should forward the datagram to the correct router (if known). When a router forwards a misdirected datagram, it also sends a redirection message to the originating host so the host can update its routing table.

Figure 22-17 illustrates the concept of redirection. Host A wants to send an IP datagram to host B, and host A generally communicates with network 1. Router R2 is obviously the most efficient routing choice. However, host A did not choose router R2 because it generally communicates directly with network 1. Consequently, the datagram goes to the only other choice: router 1. After consulting its routing table, router 1 determines that to deliver the datagram to host B, it must send it back into the same network datagram just arrived

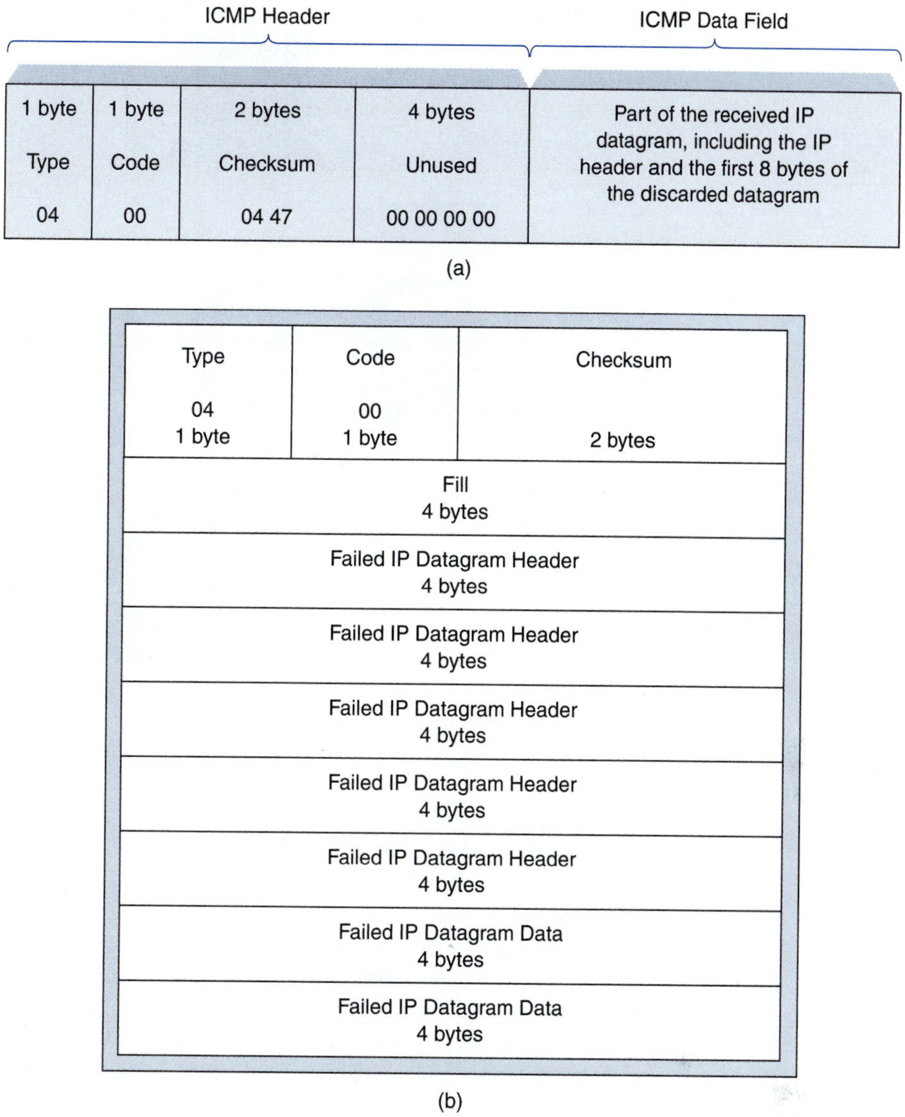

ICMP Header | ICMP Data Field

1 byte	1 byte	2 bytes	4 bytes	Part of the received IP datagram, including the IP header and the first 8 bytes of the discarded datagram
Type	Code	Checksum	Unused	
04	00	04 47	00 00 00 00	

(a)

Type 04 1 byte	Code 00 1 byte	Checksum 2 bytes
Fill 4 bytes		
Failed IP Datagram Header 4 bytes		
Failed IP Datagram Header 4 bytes		
Failed IP Datagram Header 4 bytes		
Failed IP Datagram Header 4 bytes		
Failed IP Datagram Header 4 bytes		
Failed IP Datagram Data 4 bytes		
Failed IP Datagram Data 4 bytes		

(b)

FIGURE 22-16 (a) ICMP type 4 (source quench) message format; (b) aligned to a four-byte boundary

from. Router 1 determines that the datagram should have gone to router 2, so it sends the datagram to router 2, which can forward it to host B. Router 1 also sends an ICMP redirection message to the originating IP address (host A) so that host A can update its routing table.

Figures 22-18a and b show the format for an ICMP redirection message. The type for a redirection message is always 5. There are four codes for a type 5 redirection message: 0, 1, 2, or 3:

Code 0. Redirection for the network-specific route

Code 1. Redirection for host-specific route

Code 2. Redirection for network-specific route based on the specific type of service

Code 3. Redirection for the host-specific route based on the specified type of service

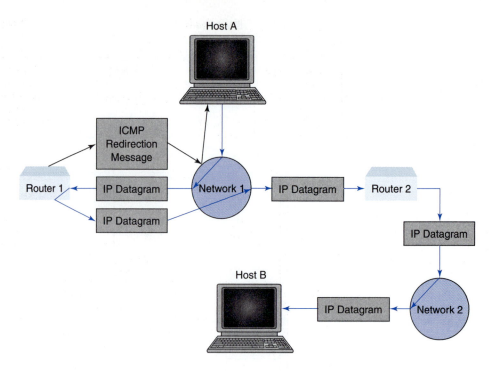

FIGURE 22-17 IP datagram redirection

Although a redirection message is considered an error-reporting message, the router does not discard the datagram but rather forwards it to the appropriate router. The four-byte gateway field identifies the IP address of the router the IP datagram should have been sent to so that the originating host can update its routing table. The failed IP header field contains the IP header from the failed datagram (i.e., the datagram that caused this ICMP redirection message to be sent). However, it does not mean that the IP datagram has actually failed; rather, the direction the datagram was sent failed, and the router has copied the datagram and sent it to the proper destination network. The failed IP data field contains the first eight bytes of the datagram's data field, which contains TCP or UDP header information.

22-6-4 Time Exceeded (Type 11)
There are two conditions when a time exceeded ICMP error-reporting message is sent:

1. Whenever a router or gateway receives a datagram whose time-to-live field falls to 0, the router discards the datagram and sends a time-exceeded message back to the originating source.
2. Whenever a final destination (host or server) does not receive all fragments from a message within a predetermined time interval, the destination discards any fragments it has received and sends a time-exceeded message back to the original source.

Figures 22-19a and b show the format for an ICMP time-exceeded message. The type field for all time-exceeded messages is 11 (0B hex), and the code field can be either a 0 or a 1:

Code 0. Used when a router discards a datagram because the time to live equals 0

Code 1. Used when a destination discards a datagram because it has not received all fragments of a datagram before the destination host's timer expires

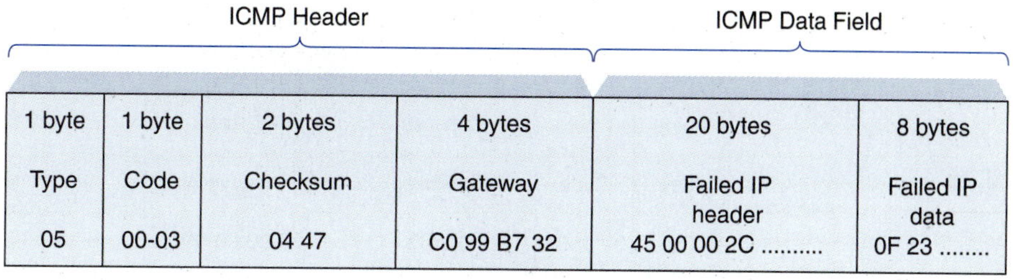

1 byte	1 byte	2 bytes	4 bytes	20 bytes	8 bytes
Type	Code	Checksum	Gateway	Failed IP header	Failed IP data
05	00-03	04 47	C0 99 B7 32	45 00 00 2C	0F 23 ....

(a)

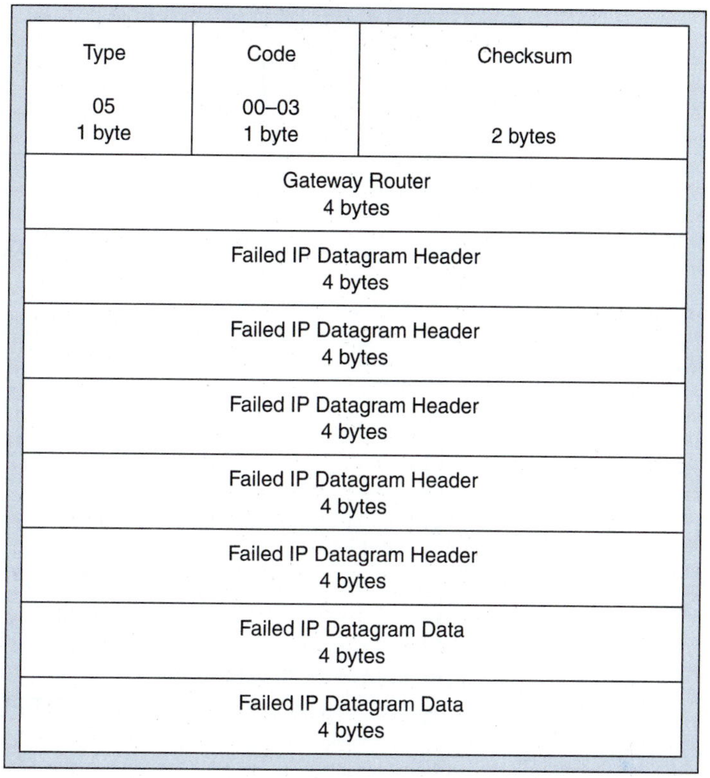

(b)

FIGURE 22-18 (a) ICMP type 5 (redirect) message format; (b) aligned to a four-byte boundary

The failed IP header and failed IP data fields contain the contents of the IP header and the first eight bytes of the data field from the datagram that was discarded.

22-6-5 Parameter Problem (Type 12)

Any ambiguity in the header portion of an IP datagram can create serious problems as the datagram propagates through the Internet. Whenever a router or a destination host discovers an ambiguous value or a missing field in a datagram, the router discards the datagram and sends an ICMP parameter problem-reporting message back to the originating source.

Figures 22-20a and b show the format for an ICMP parameter problem message. The type field of a parameter problem message is 12 (0C hex), and the code field can be either a 0 or a 1:

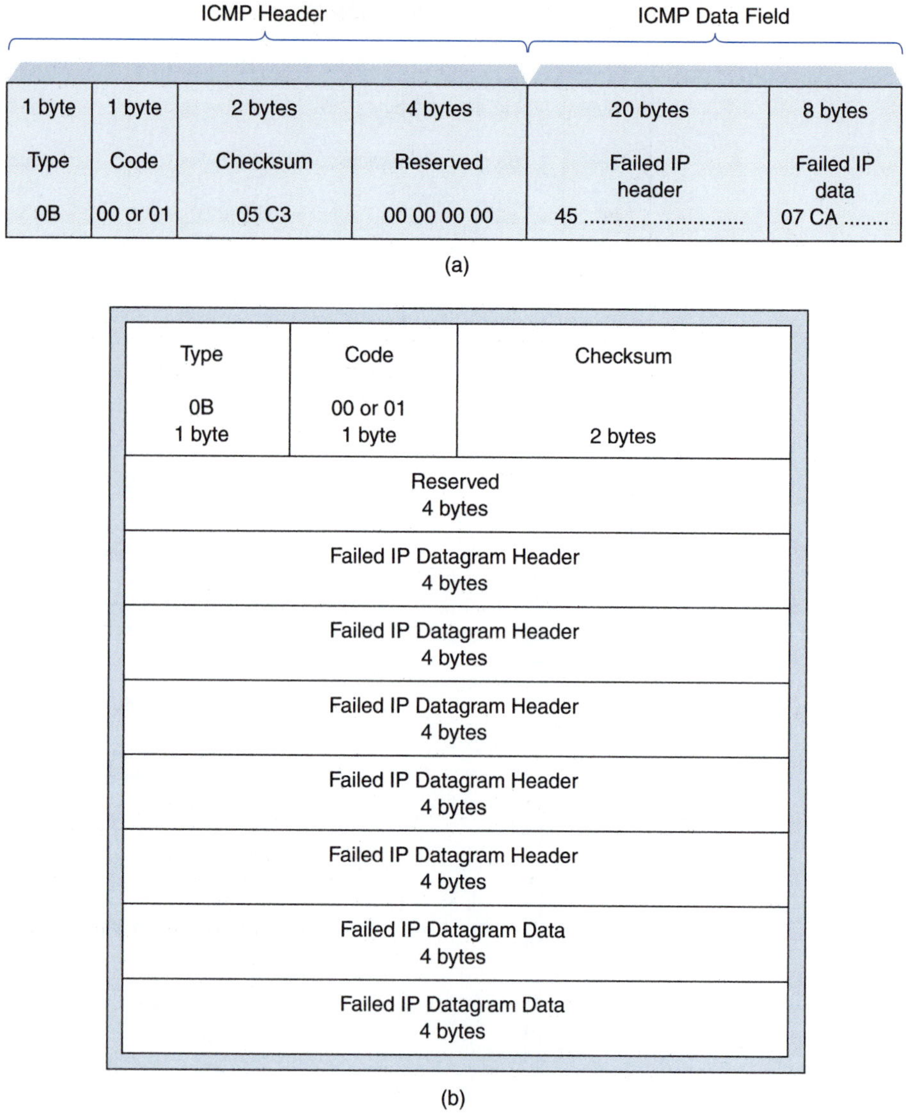

ICMP Header | ICMP Data Field

1 byte	1 byte	2 bytes	4 bytes	20 bytes	8 bytes
Type	Code	Checksum	Reserved	Failed IP header	Failed IP data
0B	00 or 01	05 C3	00 00 00 00	45	07 CA ....

(a)

Type	Code	Checksum
0B 1 byte	00 or 01 1 byte	2 bytes

Reserved
4 bytes

Failed IP Datagram Header
4 bytes

Failed IP Datagram Header
4 bytes

Failed IP Datagram Header
4 bytes

Failed IP Datagram Header
4 bytes

Failed IP Datagram Header
4 bytes

Failed IP Datagram Data
4 bytes

Failed IP Datagram Data
4 bytes

(b)

FIGURE 22-19 (a) ICMP type 11 (time exceeded) message format; (b) aligned to a four-byte boundary

Code 0. Used when there is an error or ambiguity in one of the header fields of an IP datagram and the value in the pointer field points to the byte with the suspected problem. For example, if the pointer field contains 01, the first byte of the header is invalid.

Code 1. Indicates that a required field of a datagram is missing. The pointer field is not used with a code 1 message and is reset to all 0s.

Code 2. Indicates that the original datagram had an invalid length. The pointer field is not used with a code 1 message and is reset to all 0s.

22-6-6 Future ICMP Message Formats

With the increased popularity of wireless networks, new ICMP messages are currently being developed and used on an experimental basis. Table 22-5 lists some of the future types of ICMP messages that may be adopted by the global Internet in the very near future. As

ICMP Header

1 byte	1 byte	2 bytes	1 byte	3 bytes
Type	Code	Checksum	Pointer	Unused
0C	00, 01, or 02	05 C3	00 to 14	00 00 00 00

(a)

Type 0B 1 byte	Code 00 1 byte	Checksum 2 bytes
Pointer 00 to 14 1 byte	Unused 3 bytes	

(b)

FIGURE 22-20 (a) ICMP type 12 (parameter problem) message format;
(b) aligned to a four-byte boundary

Table 22-5 Future ICM Message Types

Type	Message
19	Reserved for security
20–29	Reserved for robustness experiment
31	Datagram conversion error
32	Mobile host redirect
33	IPv6 where are you
34	IPv6 I am here
35	Mobile registration request
36	Mobile registration reply
37	Domain name request
38	Domain name reply
39	SKIP
40	Photuris
41–255	Reserved for future use

can be seen in the table, most of the proposed new ICMP messages are for wireless networks or for the future version of the Internet Protocol, IPv6.

22-7 ICMP CHECKSUM

The third and fourth bytes (the final two bytes common to all ICMP messages) are the ICMP checksum field. Since the IP checksum covers only its own header, ICMP must cover itself and any data it contains with a separate checksum. The checksum is the 16-bit 1s complement of the 1s complement of the sum of the ICMP message beginning with the ICMP type.

The ICMP checksum is calculated as follows:

1. The checksum field is initially reset to all 0s.
2. The ICMP message (header and data field) is divided into 16-bit words.
3. The 1s complement sum of the 16-bit words (including heading and data) is calculated; 1s complement addition simply means that any carryout bits are dropped. Therefore, the 1s complement sum is a 16-bit, multiple-precision sum.
4. The checksum is the complement of the sum, which is placed in the checksum field.

The ICMP checksum is tested as follows:

1. The ICMP message (header and data field) is divided into 16-bit words.
2. The sum of all 16-bit words received including the checksum is calculated.
3. The sum is complemented.
4. If the complemented value is all 0s, the message is considered good and is accepted. Otherwise, the message is rejected and discarded.

Figure 22-21 shows an example of the calculation and verification of an ICMP checksum. At the source of the ICMP message, the checksum field is filled with all 0s. The ICMP message is divided into 16-bit words, which are summed together. The sum is AFA3 hex, which is complemented (505C hex) and placed in the checksum field.

At the destination, the 1s complement sum is again calculated, except using the checksum value rather than 16 0s. If no transmission errors have occurred in the ICMP message, the 1s complement sum should be FFFF hex, which when 1s complemented produces all 0s (0000 hex).

22-8   PACKET INTERNET GROPER

The packet Internet groper (ping) command is an important troubleshooting tool. The ping command can be used to verify IP-level connectivity with remote TCP/IP hosts as well as with routers, servers, and so on. Ping uses ICMP echo requests and replies. Whenever you issue a ping command, a series of four ICMP echo requests are sent to the designated host. If connectivity is possible, the host returns an ICMP echo reply for each request. The ping command is used to troubleshoot connectivity, reachability, and name resolution.

The following command output displays the ping commands and responses from a successful ping:

Pinging 192.168.12.2
Reply from 192.168.12.2: bytes = 32 time <10 ms TTL = 126
Reply from 192.168.12.2: bytes = 32 time <20 ms TTL = 127
Reply from 192.168.12.2: bytes = 32 time <30 ms TTL = 125
Reply from 192.168.12.2: bytes = 32 time <40 ms TTL = 125

Ping Statistics for 192.168.12.2
Packets: sent = 4, received = 4, lost = 0 (0% loss),
Approximate round-trip times in milliseconds:
Minimum = 10 ms, maximum = 40 ms, average = 25 ms

If you issue the ping command and the ICMP echo replies are not returned, you receive the following output:

Pinging 192.168.12.2
Request timed out
Request timed out

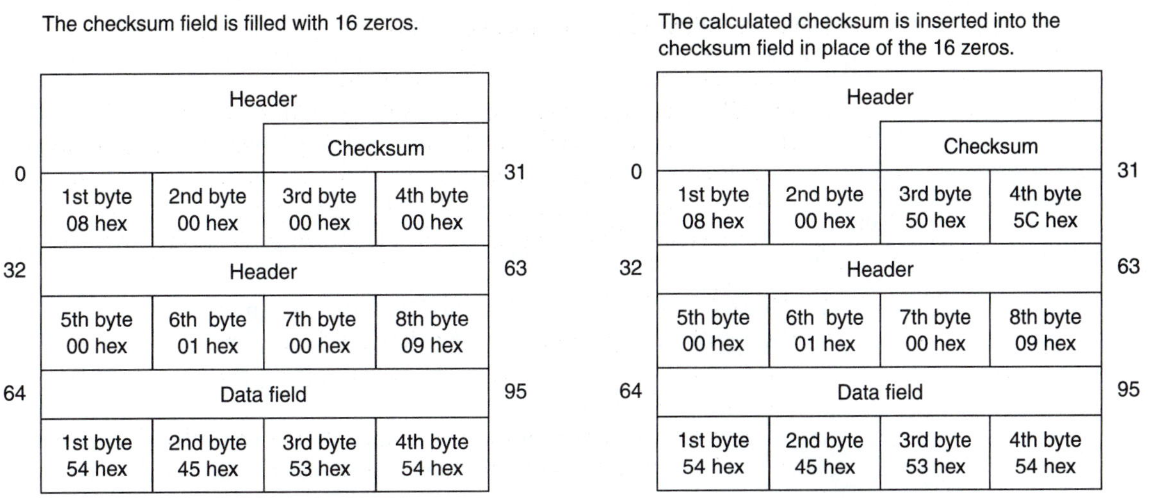

The checksum field is filled with 16 zeros.

Header			
		Checksum	
1st byte 08 hex	2nd byte 00 hex	3rd byte 00 hex	4th byte 00 hex

0 ... 31

Header			
5th byte 00 hex	6th byte 01 hex	7th byte 00 hex	8th byte 09 hex

32 ... 63

Data field			
1st byte 54 hex	2nd byte 45 hex	3rd byte 53 hex	4th byte 54 hex

64 ... 95

The calculated checksum is inserted into the checksum field in place of the 16 zeros.

Header			
		Checksum	
1st byte 08 hex	2nd byte 00 hex	3rd byte 50 hex	4th byte 5C hex

0 ... 31

Header			
5th byte 00 hex	6th byte 01 hex	7th byte 00 hex	8th byte 09 hex

32 ... 63

Data field			
1st byte 54 hex	2nd byte 45 hex	3rd byte 53 hex	4th byte 54 hex

64 ... 95

The checksum is calculated at the source:

	Hex	Binary	
1st 16 bits of header	0800	00001000	00000000
2nd 16 bits of header	0000	00000000	00000000
3rd 16 bits of header	0001	00000000	00000001
4th 16 bits of header	0009	00000000	00001001
1st 16 bits of data	5445	01010100	01000101
2nd 16 bits of data	5354	01010011	01010100
Sum	AFA3	10101111	10100011
Checksum	505C	01010000	01011100

(a)

Header and data field are received at the destination where the checksum is calculated:

	Hex	Binary	
1st 16 bits of header	0800	00001000	00000000
2nd 16 bits of header	505C	01010000	01011100
3rd 16 bits of header	0001	00000000	00000001
4th 16 bits of header	0009	00000000	00001001
1st 16 bits of data	5445	01010100	01000101
2nd 16 bits of data	5354	01010011	01010100
Sum	FFFF	11111111	11111111
Complemented Sum	0000	00000000	00000000

all zeros
(no errors)

(b)

FIGURE 22-21 ICMP checksum example: (a) checksum calculation at sender; (b) checksum calculation at destination

Request timed out
Request timed out

Ping Statistics for 192.168.12.2
Packets: sent = 4, received = 0, lost = 4 (100% loss),
Approximate round-trip times in milliseconds:
Minimum = 0 ms, maximum = 0 ms, average = 0 ms

When the ping fails, it is important to begin troubleshooting both the basic network configuration of the host (local and remote) and the IP configuration. Begin with simple troubleshooting by checking to see if the host has a network cable patched into a network jack. Also use the ipconfig command to verify IP settings.

The ping command, like the ipconfig command, has a large number of command "line switches." When you type "ping," you receive the following list of switch options:

Usage: ping [-t] [-a] [- n count] [-l size] [-f] [-i TTL] [-v TOS] [-r count] [-s count] [-j host-list] [-k host-list], [-w timeout] destination list, [-R], [-S SrcAddr], [-4] [-6] TargetName

Note that a given network does not necessarily support all the line switch options. Ping switch options are as follows:

-t   Ping the specified host until stopped (interrupted). To see statistics and continue, type "Control Break." To stop the ping sequence, type "Control C."

[ping −t <host name or IP address>]

-a   Resolve addresses to hostnames. Performs reverse name resolution on the destination IP address. If successful, the host name that matches the IP address is displayed.

[ping −a <IP address>]

-n count   Number of echo requests sent (the default value is 4).

[ping −n <number of requests><host name or IP address>]

-l size   Send buffer size. Specifies the length (in bytes) of the data field in the echo request messages sent. The default value is 32 bytes, and the maximum length is 65,527 bytes.

[ping −l <length of packet><host name or IP address>]

-f   Set the don't fragment flag in IP header (available with IPv4 only). When set (1), it specifies that the echo request message cannot be fragmented by routers in the path to the destination. When clear (0), it specifies that the echo request message can be fragmented. This parameter is used to troubleshoot the maximum transmission unit (MTU).

[ping −f <host name or IP address>]

-i TTL   Time to live. Specifies the value of the TTL field in the IP header for echo request messages sent. The default value is the default TLL value specified for the host. The maximum TTL value is 255.

[ping −i <number of routers><host name or IP address>]

-v TOS   Type of service. Specifies the type of service (TOS) field in the IP header for echo request messages sent (available with IPv4 only). TOS is specified as a decimal value between 0 and 255. The default value is 0 (routine any route).

[ping −v <TOS><host name or IP address>]

-r count   Record route for count hops. Specifies that the record route option in the IP header be used to record the path taken by the echo request messages sent (available with IPv4 only). Each hop (router) in the path uses an entry in the record route option. If possible, specify a count that is equal to or greater than the number of hops between the source and destination. The count must have a minimum value of 1 and a maximum value of 9.

[ping −r <number of routers><host name or IP address>]

-s count   Timestamp for count hops. Specifies that the Internet timestamp option in the IP header is used to record the time of arrival for the echo request message and cor-

responding echo reply message for each hop. The count must be a minimum value of 1 and a maximum value of 4. This is required for link-local destination addresses.

[ping −s <count><host name or IP address>]

-j host-list   Loose source route along host list. Specifies that the echo request messages use the loose source route option in the IP header with a set of intermediate destinations specified in host list (available with IPv4 only). With loose source routing, successive intermediate destinations can be separated by one or multiple routers. The maximum number of addresses or names in the host list is 9. The host list is a series of IP addresses (in dotted-decimal form) separated with spaces.

[ping −j <10.12.0.1 10.29.3.1 10.1.44.1 10.0.99.221>]

-k hostlist   Strict source route along host list. Specifies that the echo request messages use the strict source route option in the IP header with a set of intermediate destinations specified in hostlist (available with IPv4 only). With strict source routing, the next intermediate destination must be directly reachable (it must be on an interface of the router). The maximum number of addresses or names in the host list is 9. The host list is a series of IP addresses (in dotted-decimal form) separated with spaces.

[ping −k <10.12.0.1 10.29.3.1 10.1.44.1 10.0.99.221>]

-w timeout   Time-out in milliseconds to wait for each reply. Specifies the length of time (in milliseconds) to wait for the echo reply message that corresponds to a given echo request message. If the echo reply message is not received within the specified time-out period, the "request timed out" error message is displayed. The default time-out is 4000 milliseconds (4 seconds).

[ping −w <timeout in milliseconds><host name or IP address>]

-R   Specifies that the round-trip path be traced (available with IPv6 only).

[ping −R <host name or IP address>]

-S SrcAddr   Specifies the source address to use (available with IPv6 only).

[ping −S <source address>]

-4   Specifies that IPv4 is used to ping. This parameter is not required to identify the target host with an IPv4 address. It is only required to identify the target host by name.

[ping −4 <host name or IP address>]

-6   Specifies that IPv6 is used to ping. This parameter is not required to identify the target host with an IPv6 address. It is only required to identify the target host by name.

[ping −6 <host name or IP address>]

/?   Displays help from the command prompt.

QUESTIONS

22-1. List several shortcomings of the *Internet Protocol.*

22-2. Why is ICMP sometimes called the *official gossipmonger of the Internet?*

22-3. Describe how *ICMP messages* are encapsulated in Ethernet frames.

22-4. List and describe the three fields of an ICMP header that are common to all ICMP messages.

22-5. What are the two broad categories of ICMP messages?

22-6. List the eight query and six variation messages currently used with ICMP.

22-7. What is the purpose of *ICMP query messages?*

22-8. What are the two subcategories of ICMP query messages?

22-9. Describe the ICMP fields contained in *echo request* and *echo response messages*.

22-10. What are *timestamp request* and *timestamp replies* used for?

22-11. Describe the ICMP fields contained in timestamp request and timestamp reply messages.

22-12. What are *address mask request* and *address mask replies* used for?

22-13. Describe the ICMP fields contained in address mask request and address mask reply messages.

22-14. What are *router solicitation* and router *advertisements* used for?

22-15. Describe the ICMP fields contained in router solicitation and router advertisement messages.

22-16. What is the purpose of *ICMP error-reporting messages?*

22-17. What is a *destination-unreachable message* used for?

22-18. What is a *source quench message* used for?

22-19. What is a *redirect message* used for?

22-20. What is a *time-exceeded message* used for?

22-21. What is a *parameter problem message* used for?

PROBLEMS

22-1. Determine the ICMP message for the following type and code values:

   a. Type 9, code 0   **e.** Type 12, code 1
   b. Type 17, code 0   **f.** Type 3, code 1
   c. Type 3, code 8   **g.** Type 5, code 1
   d. Type 11, code 1   **h.** Type 4, code 0

22-2. Determine which of the following type numbers are for query messages and which are for variation messages, and for the query messages determine whether the message is a request or a reply:

   a. 4   **d.** 18
   b. 13   **e.** 12
   c. 9   **f.** 3

22-3. Determine the destination-unreachable code for the following conditions:

   a. A router cannot locate the destination network.
   b. Destination host blocks access.
   c. Destination network blocks access.
   d. The destination gateway cannot locate the destination host.
   e. The destination host does not support the process.
   f. Destination cannot locate the upper-level protocol.
   g. Fragmentation is blocked.

22-4. Use the following network analyzer display of an ICMP packet to answer the following questions:

0000   00 00 1C 2B B2 16 00 B0 AA BB CC DD EE FF 00 11

   a. What kind of ICMP message is this (query or error reporting)?
   b. What type of ICMP message is this?
   c. What is the code for this message?
   d. What is the identification number (hex)?
   e. What is the sequence number (hex)?
   f. Does the packet contain any information?

22-5. Use the following network analyzer display of an ICMP packet to answer the following questions:

```
0000   0D  00  C2  1A  84  A3  10  A3  00  43  A2  BC  A2  B1  3C  4A
0010   B2  1A  A3  C2
```

    a. Is this a request or a response?
    b. What type of ICMP message is this?
    c. What significance do bytes 9 to 12 have?
    d. What information is contained in the difference between the data contained in bytes 13 to 16 and the data contained in bytes 9 to 12?
    e. What data is contained in bytes 17 to 20?

22-6. Use the following network analyzer display of an ICMP packet to answer the following questions:

```
0000   12  00  B3  4A  A2  4B  03  B2  FF  FF  C0  00
```

    a. What kind of ICMP message is this (request or reply)?
    b. What is the type number for this ICMP message (decimal)?
    c. What information is contained in bytes 9 to 12?
    d. If the information in bytes 9 to12 is an IP address, what is its value (dotted decimal)?

22-7. Use the following network analyzer display of an ICMP router advertisement packet to answer the following questions:

```
0000   09  00  30  A2  02  01  03  84  46  10  4A  01  00  00  00  00
0010   46  10  4A  0A  80  00  00  00
```

    a. How many router addresses are contained in this message?
    b. What are they (dotted decimal)?
    c. What is the first address preference?
    d. What does the first address preference indicate?
    e. Is there a second address preference?
    f. If the answer to (e) is yes, what does it indicate?

22-8. Use the following network analyzer display of an ICMP destination-unreachable packet to answer the following questions:

```
0000   03  03  40  12  00  00  00  00  45  62  00  C8  DC  08  00  00
0010   F3  11  02  DC  32  14  10  01  C0  99  40  12  00  11  22  33
```

    a. Why was the destination unreachable?
    b. Was this message sent from a router or a host?
    c. What type of protocol was in the failed datagram?
    d. What was the time to live of the failed packet (decimal)?
    e. What is the address of the destination of the failed datagram (dotted decimal)?
    f. What is the address of the source of the failed datagram (dotted decimal)?

22-9. Use the following network analyzer display of an ICMP redirection packet to answer the following questions:

```
0000   05  01  51  10  20  04  10  08  4F  44  00  C8  DC  08  00  00
0010   F3  06  02  DC  32  14  10  01  C0  99  40  12  00  11  22  33
```

    a. What does the redirection code indicate?
    b. What is the IP address (dotted decimal) of the router this message should have been directed to?
    c. Does the failed IP datagram have options?
    d. What is the priority of the failed IP datagram?
    e. What is the upper-layer protocol contained in the failed IP datagram?

C H A P T E R   23

Transport-Layer Protocols

CHAPTER OUTLINE

OBJECTIVES

- Define the primary responsibilities of transport-layer protocols
- Describe port addresses and how they are used in the TCP/IP protocol suite
- Define *socket addresses*
- Define the fundamental concepts of user datagram protocol
- List and describe the UDP header fields
- Describe the operation of UDP
- Explain how UDP is encapsulated and decapsulated in an IP datagram
- Define the fundamental concepts of transmission control protocol
- List and describe the TCP header fields
- Describe the operation of TCP
- Explain how TCP is encapsulated and decapsulated in an IP datagram
- List and describe TCP options
- Explain how TCP connections are established and terminated

Figure 23-1 shows the relationship of the transport-layer protocols to the other protocols in the TCP/IP protocol stack. As shown in the figure, the transport layer falls between the application-layer protocols and the network-layer protocols. The primary function of transport-layer protocols is to serve as an interface between the processes running in the application layer and the internetworking mechanisms operating in the network layer. Transport-layer protocols do not operate at the process layer, and they are not process-layer protocols. However, transport-layer protocols are the mechanism (protocol) used by the TCP/IP protocol suite to ensure that processes are delivered to the proper protocol at the destination host. The TCP/IP protocol stack includes only two transport-layer protocols, UDP and TCP, with the primary purpose of providing process-to-process communications.

The Internet Protocol (IP) is a network-layer protocol responsible for providing host-to-host communications (i.e., from the source computer to the destination computer). The primary function of a layer 3 protocol, such as IP, is end-to-end delivery of entire data packets

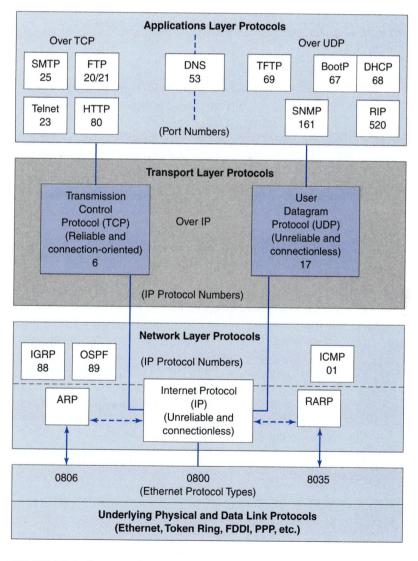

FIGURE 23-1 Transport-layer protocols and the TCP/IP protocol stack

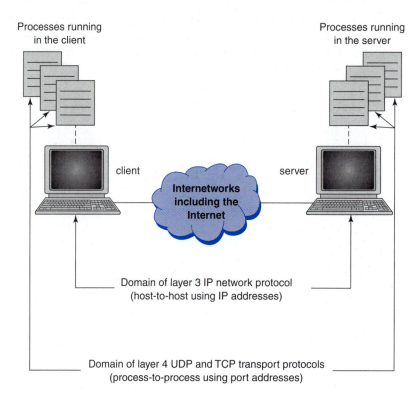

Processes running
in the client

Processes running
in the server

client

Internetworks
including the
Internet

server

Domain of layer 3 IP network protocol
(host-to-host using IP addresses)

Domain of layer 4 UDP and TCP transport protocols
(process-to-process using port addresses)

FIGURE 23-2 Domains of layer 3 (network) and layer 4 (transport) protocols

through an internetwork, such as the Internet, to a destination. Layer 3 protocols include IP addresses, which identify the source and destination devices. IP addresses also provide intermediate routers the information they need to enable them to direct data packets to the intended destination.

Once a data packet reaches the destination host, the data must be transferred to the appropriate process. Transport-layer protocols, such as user datagram protocol (UDP) and transport control protocol (TCP), provide the mechanisms necessary to transport messages of arbitrary length from a source to a destination. Therefore, transport protocols are responsible for the end-to-end delivery of individual packets of data between appropriate processes running on the host and destination computers. Consequently, transport-layer protocols include special numbers that identify these processes. The special numbers are called *port addresses*.

Figure 23-2 shows the domains of layer 3 network protocols (such as IP) and layer 4 transport protocols (such as UDP or TCP). As the figure shows, network-layer protocols ensure packet delivery between devices, whereas transport-layer protocols ensure message delivery between the processes running on the devices. The processes (i.e., port addresses) are embedded in the transport layer protocol, and the IP addresses are embedded in the network layer protocols.

23-2   PORT ADDRESSES

Probably the most common means of achieving process-to-process communications is through the client-server model where a process on a local host (client) requests a service from a process on a remote host (server). With the client-server model, the client and server processes have the same name. For example, for a client to acquire the day of the week and the time of the day from a server, the daytime client process must be running on the client, and the daytime server process must be running on the server.

Most modern operating systems support both multiuser and multiprogramming environments. In other words, computers (both clients and servers) have the capability of running several programs simultaneously. With process-to-process communications, four units must be defined:

1. Local host (client)
2. Local process (application running on the client)
3. Remote host (server)
4. Remote process (application running on the server)

IP addresses are used to uniquely define the local and remote hosts; however, local and remote processes must be defined with a second identifier called a *port number*. Port numbers are used with transport-layer protocols in a way very similar to how service access points (DSAPs and SSAPs) are used with logical link control. With the TCP/IP protocol suite, 16 bits are used to identify port numbers. Therefore, port numbers can be any integer value between 0 and 65,535. In the TCP/IP protocol suite, client and server tasks are uniquely identified with source and destination port numbers to enable information to be passed from a host to the proper service or user application. The transport protocol software running on the client host defines its process by randomly selecting a *dynamic* source port number. This is also called the *ephemeral* or *temporary* port number. The source port number defines the application or process in the client that originates and sends a data packet. The destination port number defines the application or process in the server that receives the data packet. The server's port number, however, cannot be randomly selected because if it were, the client host would have no way of knowing what the port number is. Therefore, TCP/IP uses universal port numbers for servers. These are called *well-known* port numbers.

The problem of port number selection could be resolved by using a two-step exchange where the client requests a port number from the server and the server responds. However, this would require additional overhead, so the designers of the TCP/IP protocol suite opted instead to assign permanent universal port numbers to servers. There are, however, exceptions to this rule where the source and destination are assigned the same well-known port numbers.

All client processes know the well-known port number for the corresponding server process. For example, the daytime client process can use the randomly generated temporary port number 53,000 to identify itself; however, the server process must use the well-known port number 13. Figure 23-3 illustrates the exchange of port numbers between a client and a server.

IP addresses and port numbers provide distinctly different functions in the process of selecting the final destination for a packet of data. The destination IP address defines one unique host among the myriad of hosts that exist in the global environment. Once a source host identifies a destination host, the destination port number uniquely defines the destination process.

23-3 INTERNET ASSIGNED NUMBER AUTHORITY PORT NUMBER RANGES

The Internet Assigned Numbers Authority (IANA) has divided the 65,536 port numbers into three categories, each with a specific range of numbers. The three IANA port categories are well known, registered, and dynamic.

Well-known ports. These are permanent port numbers in the range 0 through 1,023 that are assigned to specific processes. Well-known port numbers are controlled by the IANA as officially described by RFCS.

Registered ports. These are port numbers in the range 1,024 through 49,151 that are not assigned or controlled by the IANA. However, registered port numbers are recorded with the IANA to prevent duplication. Registered port numbers are listed by

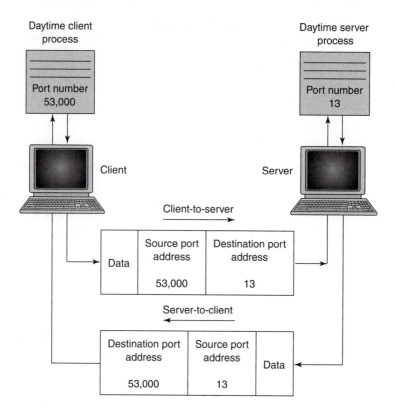

Daytime client
process

Port number
53,000

Client

Daytime server
process

Port number
13

Server

Client-to-server

Data	Source port address	Destination port address
	53,000	13

Server-to-client

Destination port address	Source port address	Data
53,000	13	

FIGURE 23-3 Exchange of port numbers

the IANA for the convenience of the Internet community. Registered port numbers are used by applications, services, and processes that are not documented by RFCs.

Dynamic (ephemeral) ports. These are port numbers in the range 49,152 through 65,535 that are randomly assigned by clients on a temporary basis for any process. The client that initiates the transport session identifies its process using dynamic port numbers. However, dynamic port numbers are released once the process is complete.

It should be noted that other operating systems could use different names for their categories with different number ranges than those specified by the IANA. For example, BSD Unix uses three ranges called reserved, ephemeral, and nonprivileged.

Table 23-1 lists several of the common well-known port numbers used with transport protocols, such as UDP and TCP.

23-4 SOCKET ADDRESSES

From the previous discussion, it was shown that transport-layer protocols require two identifiers to make a connection: the IP address and the port number. The combination of the IP address and the port number is called the *socket address*. The client socket address uniquely defines the client and the process it is running, and the server socket address uniquely defines the server and the process it is running. The concept of socket addresses is shown in Figure 23-4.

23-5 USER DATAGRAM PROTOCOL

The *user datagram protocol* (UDP) is an unreliable connectionless protocol that operates at the transport layer of the OSI seven-layer hierarchy (layer 4). UDP resides above IP, which is another unreliable connectionless protocol in the TCP/IP protocol suite. UDP

Table 23-1   Well-Known Port Numbers

Port Number	TCP/IP Protocol	Transport Layer Protocol	Protocol Description
7	Echo	UDP/TCP	Echoes a received UDP datagram back to the source host
9	Discard	UDP/TCP	Discards any UDP datagram that is received
11	Users	UDP/TCP	Active users
13	Daytime	UDP/TCP	Returns the date and time
15	Netstat	TCP	Retrieves information about routers and connections
17	Quote	UDP/TCP	Returns a quote of the date
19	Chargen	UDP/TCP	Returns a string of characters
20	FTP (data)	TCP	File transfer protocol (data)
21	FTP (control)	TCP	File transfer protocol (control)
23	TELNET	TCP	Terminal network
25	SMTP	TCP	Simple mail transfer protocol
53	Nameserver	UDP/TCP	Domain name service
67	Bootps	UDP/TCP	Server port to download boot-strap information
68	Bootpc	UDP/TCP	Client port to download boot-strap information
69	TFTP	UDP/TCP	Trivial file transfer protocol
79	Finger	TCP	Finger
80	HTTP	TCP	Hypertext transfer protocol
111	RPC	UDP/TPC	Remote procedure call
123	NTP	UDP	Network time protocol
161	SNMP (data)	UDP	Simple network management protocol
162	SNMP (control)	UDP	Simple network management protocol (trap)
520	RIP	TCP	Routing information protocol

Note. Port numbers are given in decimal.

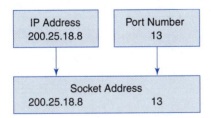

FIGURE 23-4   Formation of a socket address

transports data in units called *datagrams*, which are embedded in the data field of an IP packet with the protocol identifier 11 hex (17 decimal). UDP provides best effort delivery and, therefore, gives no guarantee of message delivery and provides no acknowledgment of datagrams received. About the only advantage UDP has over TCP (the other transport-layer protocol) is that UDP sometimes runs up to 40% faster than TCP.

UDP is unreliable in the sense that datagrams are neither sequenced (numbered) nor acknowledged. Therefore, datagrams are transported with no guarantee of delivery, and UDP provides no means of tracking lost or corrupted messages. If reliability is desired, UDP must be paired with a reliable higher-layer protocol. Since UDP is a connectionless protocol, each datagram is transported independent from all other datagrams. Consequently, each datagram may take a different path to the destination, making it impossible to provide a means of establishing, managing, or closing a connection between the source and destination.

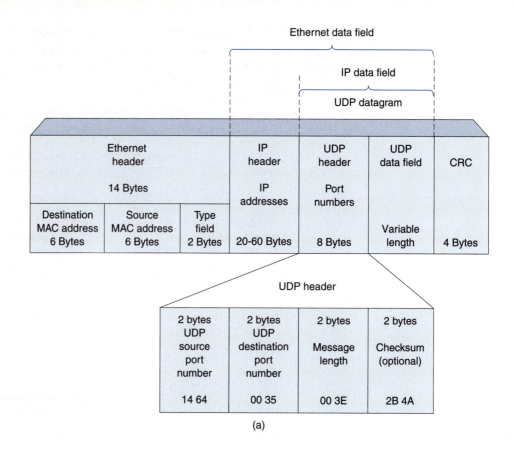

FIGURE 23-5 (a) Ethernet frame carrying a UDP datagram encapsulated in an IP data packet; (b) aligned to a four-byte boundary

23-5-1 UDP Header

UDP datagrams are encapsulated in the IP data field. However, to transport an IP packet across a physical network, such as Ethernet, the IP packets must be encapsulated in the frame of the physical-layer protocol, as shown in Figures 23–5a and b.

As shown in Figure 23-5, all UDP datagrams include a fixed-length header comprised of the four two-byte fields listed here:

1. UDP source port
2. UDP destination port
3. Variable-length message
4. Checksum

23-5-1-1 UDP source port field. The 16-bit UDP source port field contains the port number of the process running on the source host (i.e., the process that initiated the datagram). The source port number enables the destination host to return messages to the process at the source host. If the source host is a client sending a request, the port number is most likely a dynamic port number requested by the process and chosen by the UDP software running on the source host. If the source host is a server responding to a client's request, the port number is most likely a well-known port number.

23-5-1-2 UDP destination port field. The 16-bit UDP destination port field contains the port number of the process running on the destination host. If the destination host is a server (i.e., a client is sending a request), the destination port number is most likely a well-known port number. If the destination host is a client (i.e., a server is sending a response), the destination port number is most likely a dynamic port number that the server has copied from the request packet. In some cases, the source and destination port numbers are the same.

Although the client and server process port numbers are included in the UDP header, to utilize the services of UDP, a pair of socket addresses is also needed: the client socket address and the server socket address.

There are two kinds of applications (ports) that use UDP datagrams as their transport layer, and both need to transport the data packets to the destination host as quickly after the packet is created as possible. One type is the UDP applications that carry short messages that can be resent if they do not arrive as expected. These include Active Users, Quote of the Day, BootP, SNMP, and RIP. The second type of application that uses UDP datagrams involves longer messages that have been written with the reliability that is not typical of a UDP datagram. These include DNS, TFTP, and RPC (remote procedure call by Sun Microsystems).

23-5-1-3 Message length field. The message length field is 16 bits long and holds the length of the total UDP message in bytes, which includes the UDP header and any data in the variable-length UDP data field. To determine the number of data bytes, simply subtract 8 (the length of a UDP header) from the message length. For example, if the message length field contains 62, the UDP data field contains

$$
\begin{array}{rl}
62 & \text{UDP message length} \\
-8 & \text{UDP header} \\
\hline
54 & \text{UDP data field}
\end{array}
$$

The UDP total length field is relatively useless, as the length of the UDP data field can be determined by subtracting the sum of the IP and UDP headers from the number of bytes specified in the datagram total length field in the IP header. For example, if the total length field in an IP datagram carrying UDP is 82, the length of the UDP data field is

$$
\begin{array}{rl}
82 & \text{total length} \\
-(28) & \text{sum of IP (20) and UDP header (8)} \\
\hline
54 & \text{UDP data field}
\end{array}
$$

23-5-1-4 Checksum field. Since IP has no error-checking mechanism, UDP provides its own error checking. However, with UDP the checksum is optional; therefore, its reliability is reduced even further when it is not used. The checksum is derived from the UDP header, any data in the UDP message field, and part of the IP header. If network managers decide not to use a UDP checksum, the checksum field is filled with all 0s (0000 hex).

The UDP checksum includes three entities, as shown in Figure 23-6:

1. A 12-byte pseudoheader, which is comprised of a filler field containing eight bits of all 0s and information extracted from the IP header, including the 32-bit source

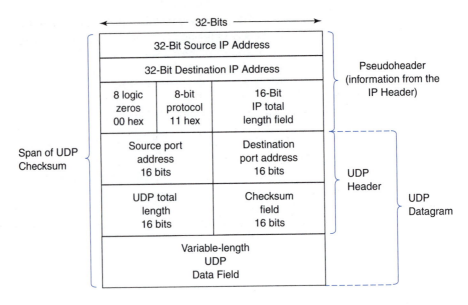

FIGURE 23-6 UDP checksum computation

IP address, the 32-bit destination IP address, the eight-bit protocol field containing 11 hex (17 decimal), and the 16-bit UDP message length field

2. The eight-byte UDP header
3. The variable-length UDP data field, which contains data handed down from the application layer

The reason for including the pseudoheader is to ensure that the datagram is delivered to the proper IP address if the IP header is corrupted. The protocol field is included to indicate that the message encapsulated in the IP datagram belongs to UDP and not TCP, which has a different protocol number.

The checksum is calculated at the source as follows:

1. The pseudoheader is created and then added to the UDP datagram.
2. The checksum field is filled with all 0s.
3. The total is divided into 16-bit (two-byte) words.
4. If the total number of bytes is an odd number, one byte of padding (eight logic 0s) is added. (The padding is used only for the purpose of calculating the checksum and is discarded after completion of the checksum computation.)
5. All the 16-bit words are summed together using 16-bit precision 1s complement arithmetic, which is standard binary addition where the carry bit is discarded. The checksum is actually only the least-significant 16 bits of the sum.
6. The checksum is the complement of the sum, which produces a 16-bit number that is inserted into the checksum field in place of the 0s.
7. The pseudoheading is removed along with any padding.
8. The UDP datagram is delivered to the IP software for encapsulation.

The checksum is calculated at the destination as follows:

1. The pseudoheader is re-created from the IP header and added to the UDP datagram.
2. Padding (eight-bit blocks of logic 0s) is added if necessary.
3. The total bit stream is divided into 16-bit words.
4. All 16-bit words are summed together using 1s complement arithmetic.

5. The sum is complemented, and if the result is all 0s, the pseudoheader and any added padding are dropped, and the datagram is accepted. If the result is not all 0s, the datagram is discarded.

23-5-2 UDP Operation

Since UDP is a connectionless service, each datagram transmitted is independent of every other datagram transmitted. There is no relationship between the different datagrams even if they originate from the same source process and are transported to the same destination process. User datagrams are not numbered, and there are no connection establishment or connection termination sequences. Therefore, each UDP datagram may be transported over a different route, and datagrams may not be received in the same sequence as they were transmitted.

One obvious shortcoming of UDP is that a higher-layer process using UDP cannot simply send a stream of data to the UDP software and expect it to subdivide the data into a sequence of related datagrams. Instead, each data stream must be small enough to fit into a single user datagram. Therefore, UDP is appropriate only for processes sending short messages.

23-5-3 UDP Flow and Error-Control Mechanism

UDP is an extremely simple and very unreliable transport protocol. UDP does not provide a packet numbering sequence or any mechanism for exchanging buffer capacity prior to transmitting a datagram. Therefore, there is no method of controlling the flow of data, and a destination could be overrun by an incoming message if the message exceeds the destination's buffer space.

The UDP protocol has no error-control mechanism other than the checksum. The checksum will detect most transmission errors; however, there are no procedures to call for retransmission of corrupted datagrams. Consequently, the corrupted datagram is discarded without informing the sender, and the sender is unaware that the datagram never reached its destination. If a process using UDP requires a flow and/or error-control mechanism, the process-layer protocol must provide it.

23-5-4 UDP Encapsulation and Decapsulation

A UDP datagram is virtually useless by itself, as it provides no means of identifying the source or destination hosts because neither logical (IP) nor physical (hardware) addresses are included in the UDP header. To transport a UDP datagram from a process on one host to a process on another host, the datagram must be encapsulated in a layer 3 protocol, such as IP, at the source and decapsulated at the destination. The processes of encapsulation and decapsulation are shown in Figure 23-7.

23-5-4-1 UDP encapsulation. There is a specific sequence of events that occur when a process on a host initiates a UDP message. The sequence begins at the application process and progresses through several stages of encapsulation before becoming a message suitable for transmission onto the local physical network, through an internetwork, and eventually to the destination host. The sequence is as follows:

1. The encapsulation procedure begins when the application process passes the message to the UDP software along with a pair of socket addresses (IP address and port number) and the length of the data field.
2. The UDP software prepares the UDP header and adds it to the message, thus creating a UDP datagram.
3. The datagram is passed on to the IP software along with the socket addresses.
4. The IP software prepares its own header and adds it to the message, adding 11 hex (17 decimal) to the IP protocol field. The protocol field indicates that the data originated from the UDP software.

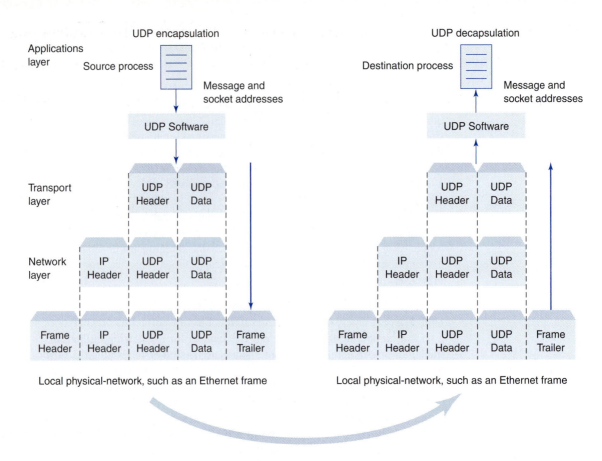

FIGURE 23-7 UDP encapsulation and decapsulation

5. The IP datagram is passed onto the data-link layer, which adds its own header and error-detection mechanism to the message, thus creating a frame.

6. The data-link layer passes the frame to the physical layer, where the data is encoded into appropriate electrical signals and transmitted onto the local network. For example, if the local physical network is an Ethernet, the IP datagram is placed into the information field of the Ethernet frame, and the entire frame is Manchester encoded and then placed onto the local transmission media.

23-5-4-2 UDP decapsulation. After a message arrives at the local physical network at the destination, a specific sequence of events occurs to ensure that the message is decapsulated and delivered to the proper process. The sequence is as follows:

1. The sequence begins at the physical layer, where the electrical signals are decoded and passed on to the data-link layer.

2. The data-link layer uses the header and error-detection bits to check for transmission errors.

3. If no errors are detected, the header and error-detection bits are removed, and the datagram is passed on to the IP software.

4. The IP software performs its own error checking, and if no transmission errors are detected, the IP header is removed, and the user datagram is passed on to the UDP software along with the socket addresses.

5. The UDP software verifies the checksum and tests the entire UDP datagram for errors.
6. If no errors are detected, the UDP header is removed, and the process data along with the socket addresses are passed on to the application process.
7. The source's socket address is passed to the process to enable the destination to respond to the message if necessary.

23-6 TRANSMISSION CONTROL PROTOCOL

Transmission control protocol (TCP), like UDP, is a transport-layer protocol responsible for providing a connection mechanism between the application program (process) at a source and the application program (process) at the destination; thus, TCP, unlike UDP, is a reliable, connection-oriented transport protocol. TCP adds reliability to the services provided by IP.

The application program at the source host sends streams of data to the transport layer, where TCP performs the following operations:

1. Opens a connection with the destination host using a sequence of handshakes
2. Segments the stream of data into transportable data units
3. Numbers each data unit
4. Transmits the data units one at a time

At the destination host, TCP performs the following operations:

1. Stores each data unit until all data units from a given data stream have arrived
2. Checks each data unit for errors
3. Passes only those data units with no errors on to the application layer
4. Delivers the data units to the destination application program as a continuous stream of data

After the entire data stream has been successfully transported from the source application program to the destination application program, it is the responsibility of the transport layer to close the connection. Because TCP is a reliable transport protocol, it uses an acknowledgment mechanism to verify successful transmission of data and also to call for retransmissions of corrupted or missing segments.

TCP offers full-duplex service, as data can flow in both directions at the same time. This, of course, presumes that the physical network and transmission media are capable of full-duplex operation.

TCP is a connection-oriented service, which means for a process at a source host to exchange data (transmit and receive) with a destination host, these steps must be followed:

a. The source host's TCP software must request a connection with the destination host's TCP software.
b. The source host receives approval from the destination host before any data is transmitted in either direction (i.e., a virtual connection is established between the source and destination hosts).
c. The source and destination's TCP software exchange data in both directions. This is called a *session*.
d. After the source and destination host's processes have completed sending and receiving data (i.e., the session is over), the TCP software at both ends discards the contents of their buffers, and the virtual connection is terminated.

23-6-1 TCP Services

TCP provides three primary services to the processes running at the application layer: stream delivery, buffering, and segmentation.

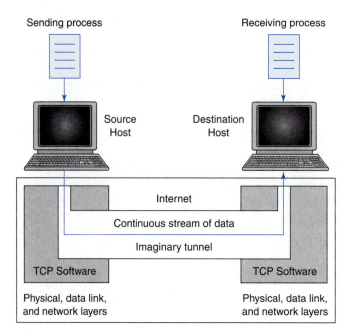

FIGURE 23-8 TCP tunneling

23-6-1-1 Stream delivery service. With UDP, a process running in the application program initiates a transmission by sending a packet of data to the UDP software. The UDP software adds a header to the packet, thus creating a datagram. The datagram is then delivered to the IP software, where it is encapsulated in an IP datagram, which is transmitted onto the local network. The process may deliver several packets of data to the UDP software; however, UDP treats each packet as an independent entity, without attempting to establish a relationship among the packets.

TCP is a stream-oriented protocol, which allows the source application to deliver data to the transport layer as a single continuous stream of data packets. Likewise, at the receiving end, TCP allows the receiving application to accept data as a single continuous stream of data. TCP accomplishes this by creating an environment where the source and destination processes appear to be interconnected by an imaginary "tunnel" that transports the data across the Internet. Transporting data through this imaginary tunnel is illustrated in Figure 23-8.

23-6-1-2 Buffering service. With typical applications, the source host may transmit data at a different rate than the destination host can receive it. For TCP to operate properly, it requires buffer space to store data at both the source and the destination ends for each direction of transmission. Therefore, with TCP, there are two sets of buffers, as shown in Figure 23-9.

As will be shown later, TCP uses the buffers to provide flow and error control. For TCP to operate properly, each end must know the buffer capacity of the other end, and TCP must keep each end informed on how much of that buffer space is available.

Essentially, the buffers provided by TCP ensure the following:

1. The source host does not transmit more data for which the buffer at the receive end has storage space.
2. The source buffer saves the data it has transmitted until the destination end acknowledges its reception.

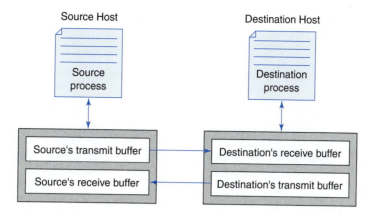

FIGURE 23-9   TCP buffers

23-6-1-3 Segmentation service. In essence, IP is the service provider to TCP. Because IP sends data in packets and not as a continuous stream of bytes, TCP must group the stream of data into smaller packets called *segments* where all segments do not need to be the same size. A TCP session is comprised of one or more segments. In other words, a message is made up of a stream of data bytes that are divided into segments. Each segment is transmitted independently; however, control information is transmitted in the TCP header that allows the destination host to reconstruct the original data stream. After TCP adds a header to a segment, it passes the segment to the IP layer for delivery. A TCP segment is encapsulated into an IP datagram and then transmitted.

At the receive end, the TCP segments can be received out of order, lost, or received in error and retransmitted without the destination process being aware (i.e., the entire operation is transparent to the receiving process).

23-6-2   TCP Header

Figure 23-10 shows a TCP segment encapsulated in an IP datagram that is transported across the local network in an Ethernet frame. The Ethernet frame begins with a 14-byte header and ends with a four-byte trailer. The Ethernet header contains the source and destination hardware addresses and a protocol field. The trailer contains only a four-byte CRC field used for error detection. The information field in the Ethernet frame contains an IP datagram that is comprised of an IP header and an IP data field. The TCP header and data field are encapsulated within the IP data field.

Figures 23–11a and b show a detailed layout for the TCP header. As shown in the figure, the TCP header is comprised of 10 fields of varying lengths:

1. Source port address
2. Destination port address
3. Source sequence number
4. Acknowledgment number
5. TCP header length
6. Reserved
7. Session flags
8. Sender window size
9. TCP checksum
10. Urgent data size

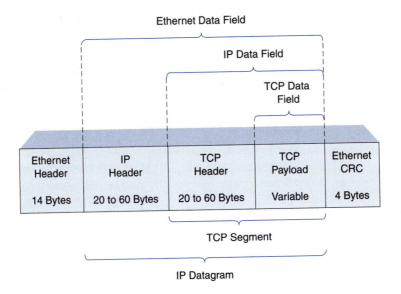

FIGURE 23-10 TCP encapsulated in an IP data field being transported in an Ethernet frame

23-6-2-1 Source port address. The first field of the TCP header is the 16-bit (two-byte) source port number, which serves the same purpose with TCP as it did with UDP: it identifies the process on the source that initiated the session.

23-6-2-2 Destination port address. The second field of the TCP header is the 16-bit (two-byte) destination port number, which serves the same purpose with TCP as it did with UDP: it identifies the intended process at the destination.

23-6-2-3 Source sequence number. The third field of the TCP header contain a four-byte source sequence number, which is a 32-bit number assigned to the first byte of data in the data field of this TCP segment. TCP is a stream-transport protocol; therefore, each byte transmitted is numbered to ensure connectivity. The source sequence number in successive TCP segments within a TCP session is incremented by the number of bytes contained in the previous segment.

23-6-2-4 Acknowledgment number. The fourth field of the TCP header contain a four-byte acknowledgment number, which is a 32-bit number defining the byte number that the receiver of a segment is expecting to receive next (i.e., the sequence number in the next segment it will receive). The source sequence number and acknowledgment number are used for flow control. Source and acknowledgment sequence numbers are used similar to the way the *ns* and *nr* fields are used with HDLC, except TCP acknowledges bytes, whereas HDLC acknowledges entire frames. If no transmission errors occur, the acknowledgment number sent should have a value one above the number of the last byte received.

Figure 23-12 shows an example of how source sequence numbers and acknowledgment numbers are used for controlling the flow of data between two hosts (a client and a server). Figure 23-12a shows the exchange in tabular form, and Figure 23-12b shows the same exchange in pictorial form. The example presumes that the session has already been established and does not include establishment and termination procedures. In addition, the initial source sequence and acknowledgment numbers have been arbitrarily chosen for this example.

The server has a 2500-byte data stream to send to the client. The server divides the data stream into three segments (segments 1, 2, and 3). The first two segments each carry 1000 bytes of data, and the third segment is carrying 500 bytes of data. The source se-

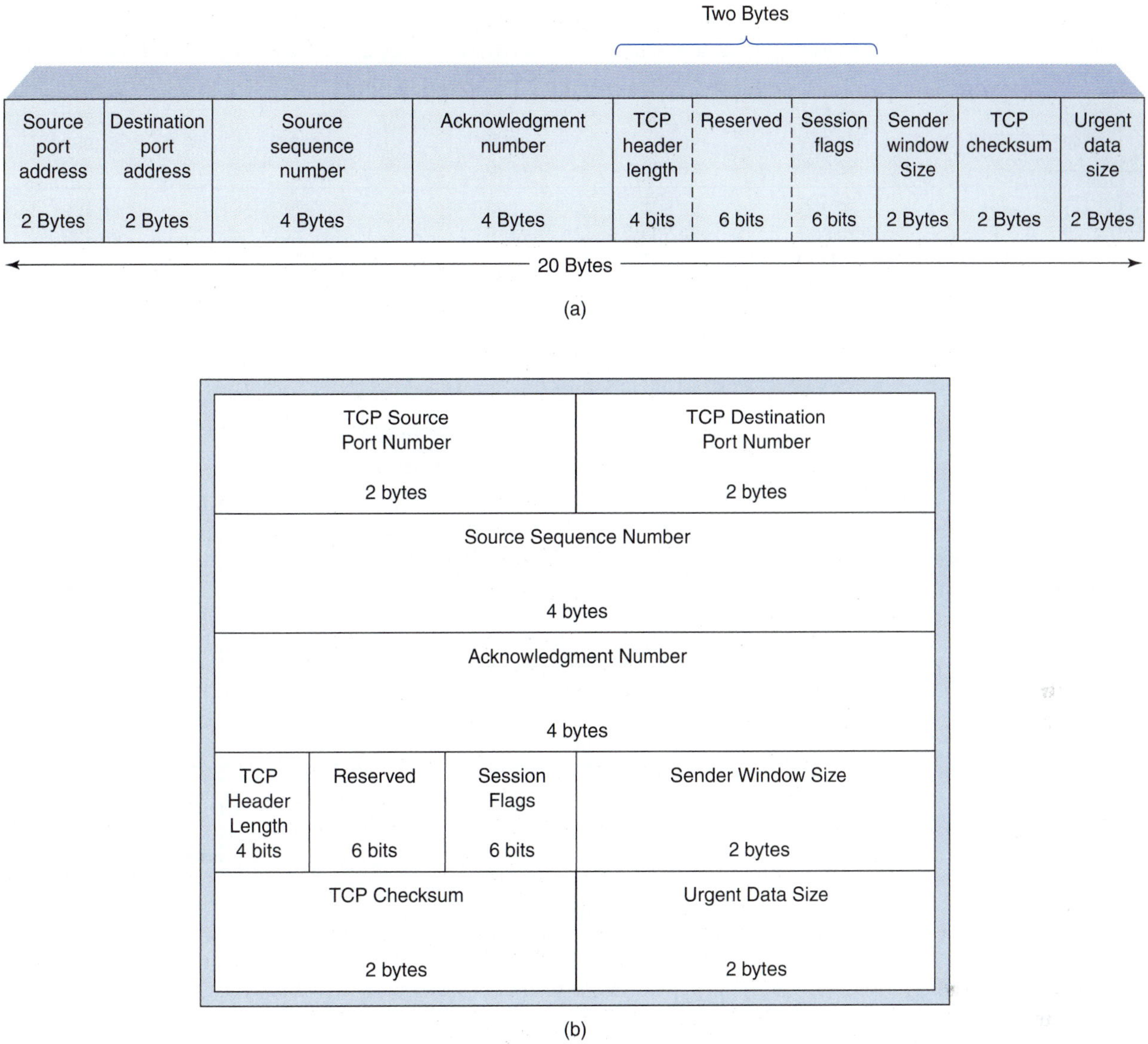

Two Bytes

Source port address	Destination port address	Source sequence number	Acknowledgment number	TCP header length	Reserved	Session flags	Sender window Size	TCP checksum	Urgent data size
2 Bytes	2 Bytes	4 Bytes	4 Bytes	4 bits	6 bits	6 bits	2 Bytes	2 Bytes	2 Bytes

20 Bytes

(a)

TCP Source Port Number 2 bytes			TCP Destination Port Number 2 bytes		
Source Sequence Number 4 bytes					
Acknowledgment Number 4 bytes					
TCP Header Length 4 bits	Reserved 6 bits	Session Flags 6 bits	Sender Window Size 2 bytes		
TCP Checksum 2 bytes			Urgent Data Size 2 bytes		

(b)

FIGURE 23-11   (a) TCP header format; (b) aligned to a four-byte boundary

quence number in the first segment is 50,000, which is the number of the first byte in the segment. Since the segment is carrying 1000 bytes of data, the number of the last byte in the segment is 50,999. The acknowledgment number in the first segment is 20,000, which indicates that the next byte the server expects to receive from the client is number 20,000. The second segment is carrying bytes 51,000 through 51,999, and the third byte is carrying bytes 52,000 through 52,499. Since the server has not received any new segments from the client, the acknowledgment number (20,000) is simply repeated in the second and third segments.

The client replies with two segments of data. Assuming that no transmission errors have occurred, the client sends a TCP segment with an acknowledgment sequence number of 52,500, which confirms reception of segments 1 through 3 (bytes 50,000–52,499). Byte number 52,500 should also be the number of the first byte in the next segment the server transmits to the client. The first segment is carrying 800 bytes (20,000–20,799), and the second segment

Segment	Host	Source Sequence Number	Acknowledge Number	Data Bytes in Segment	Total Number of Bytes in Data Stream (Accumulated)	Number of Last Byte in Segment
1	S	50,000	20,000	1000	1000	50,999
2	S	51,000	20,000	1000	2000	51,999
3	S	52,000	20,000	500	2500	52,499
1	C	20,000	52,500	800	800	20,799
2	C	20,800	52,500	400	1200	21,199
4	S	52,500	21,200	500	500	52,999

S = server, C = client

(a)

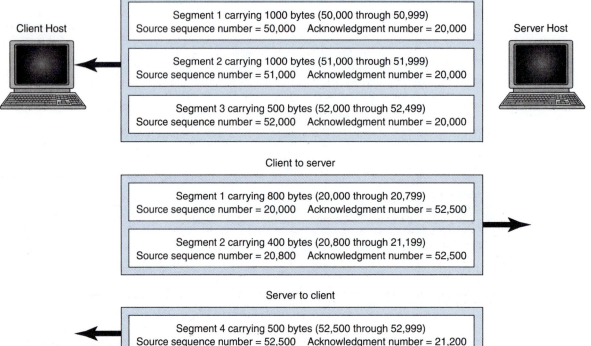

(b)

FIGURE 23-12 Using source and acknowledgment numbers: (a) segment information; (b) transmission sequence

is carrying 400 bytes (20,800–21,199). The acknowledgment number in both segments is 52,500, which confirms reception of bytes up through 52,499 and indicates that the next byte it receives should be 52,500 (i.e., the next byte the server sends should be number 52,500).

After receiving segments 1 and 2 from the client, the server sends frame 4 to the client carrying 500 bytes of data. The source sequence number is 52,500, and the acknowledgment number is 21,200, which confirms reception of the 1200 bytes of data in segments 1 and 2 (bytes 20,000–21,199).

23-6-2-5 TCP header length field. The first four bits of the 13th byte of the TCP header is the TCP header length field, which contains the number of four-byte words in the

1 bit	1 bit	1 bit	1 bit	1 bit	1 bit
URG	ACK	PSH	RST	SYN	FIN
1/0	1/0	1/0	1/0	1/0	1/0

FIGURE 23-13 TCP session field

TCP header. The header length can be between 20 and 60 bytes long. Therefore, the header length field can be any number between 5 and 15 (F hex). To determine how many bytes of data are contained in the TCP data field, simply subtract the number of bytes in the IP and TCP headers from the IP total length field as shown here:

$$N = \text{IP total length} - \text{IP header (20–60 bytes)} - \text{TCP header (20–60 bytes)}$$

where N = byte length of data field.

23-6-2-6 Reserved field. The last four bits of the 13th byte and the first two bits of the 14th byte of the TCP header are reserved for future use and are all logic 0s:

Session flag field. The last six bits of the 14th byte of the TCP header are the session bit flags, which are used for control. The format for the session field is shown in Figure 23-13.

Urgent Pointer (URG). The URG bit indicates whether the contents of the urgent data size field are valid. If the bit is a 1, the contents of the urgent data field are valid. If the bit is a 0, the contents of the urgent data field are invalid.

Valid acknowledgment (ACK). The ACK bit is set if the value in the acknowledgment field is a valid number.

Push request (PSH). The transmitting host uses the PSH request flag to request the receiving host to process the nonurgent data in the segment as soon as possible. This bit is set (1) when a rapid response is important, such as when using an interactive process protocol like Telnet or FTP. When active, the PSH bit tells the receive host to create and send a segment immediately without waiting until its window buffer is filled.

Reset session (RST). Setting the RST flag signifies the abnormal end to a session in which one host is not responding to the other. A RST is sent after retransmission has failed to restore the session and the retransmission timer has expired. It is sent every 30 seconds until the timer expires or the other host in the session acknowledges receiving it. A RST indicates that the connection must be reestablished.

Synchronize sequence number (SYN). The SYN flag is used to synchronize sequence numbers during establishment of a TCP connection, which is described later. In the first header sent from each host while establishing the session, the SYN flag is set to 1, which designates that each host is requesting that the opposite host in the session synchronize with the submitted source sequence number. To comply, the host that receives the flag adds 1 to the submitted source sequence number and returns it as the next TCP header's acknowledgment sequence number.

Final data (FIN). The FIN flag is used to terminate a connection. Talbe 23-2 lists some of the valid hex combinations that can be found in the session bit flag field.

23-6-2-7 Sender window size field. The 15th and 16th bytes of the TCP header contain the sender window size, which is the number of bytes the sender will accept from the

Table 23-2 Session-Bit Flag Field

Combination (Hex)	Meaning	Binary Sequence
02	SYN	00000010
04	RST	00000100
10	ACK	00010000
11	ACK/FIN	00010001
12	ACK/SYN	00010010
18	ACK/PSH	00011000
19	ACK/PSH/FIN	00011001
30	URG/ACK	00110000

other end of the session without requiring that the other end wait for an acknowledgment. This is a 16-bit field; therefore, the maximum window size is 65,535 bytes. In essence, the sender window size is the amount of receive buffer the sender has available for storage of TCP data. Since this field is part of each TCP header, this field can be used to constantly update the amount of buffer space available.

23-6-2-8 TCP checksum field. The 17th and 18th bytes of the TCP header contain the TCP checksum, which is a 16-bit field used for error detection. The TCP checksum is derived from the TCP header, the TCP application data, the IP header, and a pseudoheader following the same procedure as the UDP checksum. However, the UDP checksum is optional, whereas with TCP it is not.

23-6-2-9 Urgent data size field. The last two bytes of the TCP header contain the urgent data size. If the urgent data flag is 0, there is no urgent data in this segment. If the urgent data pointer (URG) in the session field is set, the urgent data size field contains a number that must be added to the sequence number to obtain the number of the last urgent data byte in the data section of the segment.

There are occasions when an application program must send urgent data. Urgent data is data that the application process at the source wants to be read out of order by the application process at the destination. Urgent data is placed at the beginning of the next segment sent. The remainder of the segment is normal data. The urgent pointer field defines how many urgent data bytes are at the beginning of the segment, which also indicates where the normal data begins.

For example, a TCP segment is carrying 1000 bytes of normal data beginning with byte 40,000. The urgent pointer is set, and the urgent data size field contains 0064 hex (100 decimal). Therefore, the first 100 bytes of data in the segment is urgent data (bytes 40,000–40,099), and the normal data begins with byte 40,100 and ends with byte 41,099.

23-7   TCP OPTIONS

A TCP header can contain up to 40 bytes of optional information that is used to convey additional information to the destination or to align other options. With TCP, the option field must be a multiple of 32 bits (four bytes). Although there are numerous options available with TCP, the most common options are shown in Figure 23-14. In general, TCP options can be divided into two basic categories: category 1 options, which are one-byte options, and category 2 options, which are multiple-byte options. There are two types of category 1 options: end of option and no operation. There are three types of category 2 options: maximum segment size, window scale factor, and timestamp. Several common TCP options and their codes are listed in Table 23-3.

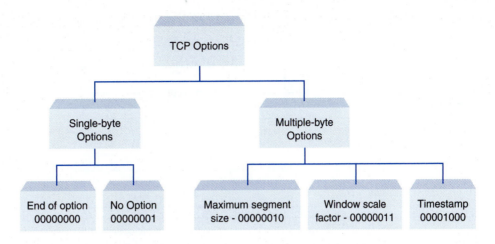

FIGURE 23-14 TCP options

Table 23-3 TCP Options

Code	Name of Option
0	End of option
1	No operation
2	Maximum segment size
3	Window scale factor
8	Timestamp

23-7-1 Category 1 Options

23-7-1-1 End of option.   This is a one byte option used for padding (i.e., fill), as shown in Figure 23-15. The code for an end of option is 00 hex (00000000). However, there can be only one end of option code, and it must be used in the last option if multiple options are used. When a receiver detects an end of option sequence, it begins looking for actual data (i.e., the payload). If more than one byte of filler is needed to align the option field, the no-operation option must be used followed by an end of option.

End of option conveys three items of information to the destination:

1. There are no more options in the header.
2. The remainder of the 32-bit word is garbage.
3. Data from the application program starts at the beginning of the next 32-bit word.

23-7-1-2 No operation.   This is another one-byte category 1 option that is used for filler between other options. The code for a no operation is 01 hex (00000001).

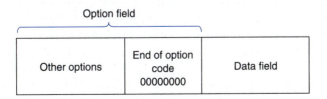

FIGURE 23-15 End of option format

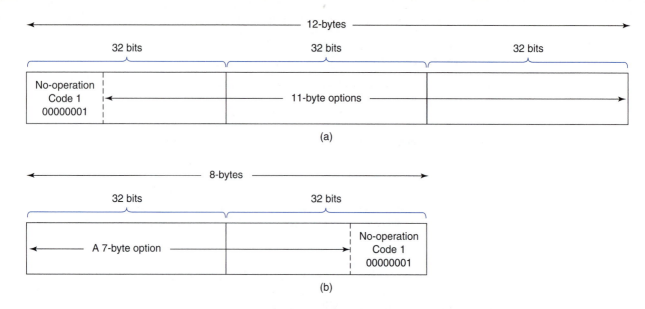

FIGURE 23-16 No operation option format: (a) used at beginning of an option; (b) used to align the next option

For example, a no operation can be used to align the next option on the 32-bit boundary, as shown in Figure 23-16. In Figure 23-16a, a no operation is inserted at the beginning of an 11-byte option, which completes the third 32-bit grouping. In Figure 23-16b, a no operation is used to complete the second 32-bit grouping.

23-7-2 Category 2 Options

23-7-2-1 *Maximum segment size* (MSS). MSS is a category 2 option with the code 02 hex (00000010). The format for the maximum segment size option is shown in Figure 23-17. The length field contains the byte length for this option, four bytes, which includes a one-byte code field, a one-byte length field, and a two-byte maximum segment size. MSS defines the largest packet of data that a destination can receive (not the longest segment). Since MSS is a 16-bit field, it can define packets up to 65,535 bytes long. However, the default MSS length is 536 bytes, which equates to a maximum segment length of 576 bytes (536 data + 20 bytes IP header + 20 bytes TCP header).

The MSS for a TCP session is determined during the TCP setup procedure, which is described later. The MSS is determined by the destination host's buffer capacity. Therefore, each host must inform the other host of its MSS before any data is transferred. If neither host defines the MSS, the default value is selected. The MSS option can be used only in setup segments and cannot be used in segments that are carrying actual data.

23-7-2-2 Window scale factor. The maximum scale factor is a category 2 option with the code 3 (00000011). The format for the window scale factor option is shown in Figure 23-18. The length field contains the byte length of this option, three bytes, which includes a one-byte code field, a one-byte length field, and a one-byte scale factor.

Code 2 00000010	Length 4 00000100	Maximum segment size 16 bits

FIGURE 23-17 Format for the maximum segment size (MSS) option

Code 3 00000011	Length 3 00000011	Scale factor 1 byte

FIGURE 23-18 Format for the window scale factor option

Code 8 00001000	Length 10 00001010	Timestamp value 4 bytes	Timestamp echo reply 4 bytes

FIGURE 23-19 Format for the timestamp option

The window size field in the TCP header defines the length of the host's sliding window, which is used for flow control. This field is 16 bits long with a range from 0 to 65,535, which may seem excessively long. However, this may not be the case for large messages that are transported with high-throughput, minimum delay transmission media, such as an optical fiber cable. The window size is increased by the factor 2^n, where n is the value specified in the scale factor field. The new window size is

$$NWS = OWS \times 2^n$$

where   NWS = new window size
           OWS = old window size
             n = scale factor

For example, if the window scale factor is 2, the actual window size is 2^2, or 4, times the window size specified in the TCP header. Although the window scale factor is an eight-bit number and can be any value between 0 and 255, TCP limits it to a maximum value of 16.

The window scale factor is used to increase the window size, which is determined during a TCP connection setup procedure. When data is being transferred, the size of the window specified in the TCP header can be changed, but only by the specified window scale factor.

23-7-2-3 Timestamp. The timestamp is a category 2 option with the code 8 (00001000). The format for the timestamp option is shown in Figure 23-19. The length field contains the byte length of this option, 10 bytes, which includes a one-byte code field, a one-byte length field, a four-byte timestamp value, and a four-byte timestamp echo reply.

The timestamp field is filled at the source when the segment containing it is transmitted. The destination receives the timestamp and stores its value. When the destination host acknowledges a received segment, it sends the previously stored timestamp value in an echo reply field. When the source receives the acknowledgment, it compares the current time with the time specified in the timestamp. The difference is the round-trip delay time, which includes the round-trip propagation time and the turnaround time.

23-8 TCP CONNECTION ESTABLISHMENT AND TERMINATION

TCP is a connection-oriented transport protocol. Therefore, a virtual path is established between the source and destination before any data segments are transmitted in either direction. Once a virtual path has been established, all data segments associated with a message are transported over the same virtual path. Using a singe virtual path allows messages to be acknowledged and retransmitted if necessary.

23-8-1 TCP Connection Establishment

TCP is capable of exchanging data full duplex (i.e., both directions at the same time). Therefore, each host must initialize communications and receive approval from the other host before transmitting data. There are four steps that must be taken before two hosts can exchange data with each other. For simplicity, the two hosts are referred to as host A and host B:

1. Host A begins the set up procedure by transmitting a segment to host B announcing its desire to establish a connection. The segment includes initialization information about traffic propagating from host A to host B.
2. Host B responds with a segment acknowledging the request from host A to establish a connection.
3. Host B transmits a segment that includes its initialization information about traffic propagating from host B to host A. Steps 2 and 3 are generally combined into one step and occur simultaneously. In other words, host B confirms the request from host A and transmits its own request at the same time.
4. Host A transmits a segment to acknowledge the request from host B.

If steps 2 and 3 are combined into one transmission, the connection procedure is called *three-way handshaking*. The process begins when a server tells its TCP software that it is ready to accept a connection. This is called a *request for a passive open*. At this time, the server is considered "open" and ready to accept connections from virtually any host that is connected to the global Internet. However, the server cannot make the connection itself. A client program must initiate a request for an active open by telling its own TCP software that it would like to be connected to a particular server. At that time, the client initiates the three-way handshaking procedure.

The three-way handshake sequence to establish a TCP connection is as follows:

1. The client initiates the procedure by sending the first segment, which is called a SYN segment, as the synchronize sequence number (SYN) flag in the session flag field is set. Setting the SYN flag tells the server to synchronize to the client's initialization sequence number (ISN), which is in the source sequence number field. The ISN is used for numbering subsequent bytes of information sent from the client to the server. The SYN segment also includes the source and destination port numbers. The source port number is probably a temporary (ephemeral) port number and the destination port number a well-known port number that clearly defines the process on the server the client wishes to communicate with. The SYN segment can also be used to establish the maximum segment size (MSS) that the client can receive from the server. In addition, if the client requires a large window size, it can define the window scale factor using the appropriate options. The acknowledgement sequence number in the SYN segment is 0, and the segment does not define the window size because a window size makes little sense in a segment that does not contain an acknowledgment number.
2. The second transmission in the three-step handshake is transmitted by the server and combines steps 2 and 3 of the connection establishment sequence. This segment is called SYN and ACK, as both the SYN and the ACK flags in the session flag field are set. The ACK flag, combined with the acknowledgment number field, confirms that the server has synchronized to the client's initialization sequence number. The server's acknowledgment sequence number is the client's source sequence number plus 1. The SYN flag requests the client to synchronize to the server's initialization sequence number, which is in the source sequence number field. The segment also contains the window scale factor option (if needed) and the MSS that the server can receive from the client.

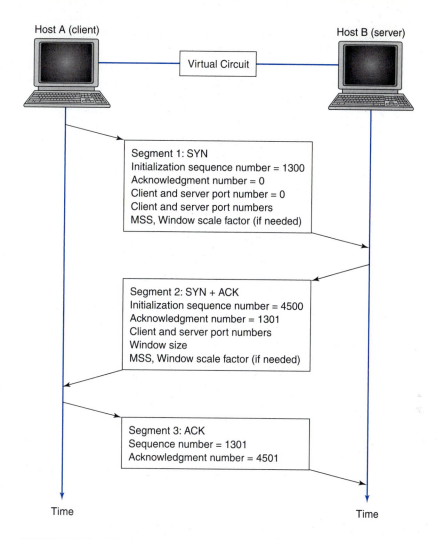

Host A (client)

Virtual Circuit

Host B (server)

Segment 1: SYN
Initialization sequence number = 1300
Acknowledgment number = 0
Client and server port number = 0
Client and server port numbers
MSS, Window scale factor (if needed)

Segment 2: SYN + ACK
Initialization sequence number = 4500
Acknowledgment number = 1301
Client and server port numbers
Window size
MSS, Window scale factor (if needed)

Segment 3: ACK
Sequence number = 1301
Acknowledgment number = 4501

Time

Time

FIGURE 23-20 TCP three-way connection handshake

3. The third and final transmission of the three-step handshake is when the client transmits an ACK segment, in which only the ACK flag is set in the session flag field. The client also adds one to the server's source sequence number and places that value in its acknowledgment sequence number field.

A three-way handshake to establish a TCP connection is shown in Figure 23-20. The client begins the sequence by sending a SYN segment with an initialization sequence number (ISN) of 1300 and an acknowledgment number of 0. The client responds with a SYN and ACK segment, which includes the initialization sequence number 4500 and the acknowledgment number 1301 (one more than the client's source sequence number). The final segment in the client's ACK segment includes the sequence number 1301 and the acknowledgment number 4501.

Figure 23-21 shows a client establishing a TCP session with a server using a TCP three-step handshake. The first segment sent from the client (segment 1) and the first segment sent from the server (segment 2) each have a 24-byte header (the 20-byte standard header plus four bytes of options). The maximum segment size option is sent in each of the first two segments, which sets the maximum limit to 1460 bytes. The client initially sets its window size to 16,348

	Segment 1 (client)	Segment 2 (server)	Segment 3 (client)
Ethernet header:			
Destination physical address:	00–C0–02–63–55–48	00–D0–59–05–C5–DB	00–C0–02–63–55–48
Source physical address:	00–D0–59–05–C5–DB	00–C0–02–63–55–48	00–D0–59–05–C5–DB
Protocol:	IP (0800)	IP (0800)	IP (0800)
FCS:	AC2D469F hex	2BB46348 hex	A8925B2F hex
IP header:			
IP version	4	4	4
Header length	20	20	20
Packet length	44	44	40
ID #	3E2E hex	D7CF hex	3E2F hex
Fragmentation field			
Don't fragment	1 = true	1 = true	1 = true
More fragments	0 = false	0 = false	0 = false
Fragment offset	0000000000000	0000000000000	0000000000000
TTL	124	115	122
Protocol	TCP (6)	TCP (6)	TCP (6)
Header checksum	A4BD hex	5820 hex	A4C4 hex
Source IP address	192.168.0.2	209.39.134.10	192.168.0.2
Destination IP address	209.39.134.10	192.168.0.2	209.39.134.10
TCP header:			
Source port number	1025 (blackjack)	FTP (21)	1025 (blackjack)
Destination port number	FTP (21)	1025 (blackjack)	FTP (21)
Sequence number	1379437769	2165615844	1379437770
Acknowledgment number	0	1379437770	2165615845
Data offset	24	24	20
Session flags	02 hex	12 hex	10 hex
URG flag	..0..... false	..0..... false	..0..... false
ACK flag	...0.... false	...1.... true	...1.... true
PSH flag	....0... false	....0... false	....0... false
RST flag	.....0.. false	.....0.. false	.....0.. false
SYN flag	......1. true	......1. true	......0. false
FIN flag	.......0 false	.......0 false	.......0 false
Window size	16348	8760	17520
Checksum	44F4 hex	49BF hex	3F44 hex
Urgent pointer	00000000	00000000	00000000
Option type 2			
MSS	1460	1460	No options

FIGURE 23-21 Example of TCP three-way handshake

bytes; however, in segment 3 it sends an updated window size of 17,520. The server sets its window size to 8760 bytes. Note the session flags in each segment:

Segment 1: SYN

Segment 2: ACK and SYN

Segment 3: ACK

Figure 23-22 shows a TCP session where several data segments are transmitted in each direction. Frames 1 through 3 are the three-step handshake, and frames 4 through 12 are data segments and acknowledgments exchanged between the client and server.

23-8-2   TCP Connection Termination

Either the client or the server can initiate terminating (closing) a TCP connection. However, terminating the connection in one direction does not automatically close the connection in

Frame Number	Host	Source Port Number	Destination Port Number	Source Sequence Number	Acknowledgment Sequence Number	Data Bytes in Segment	Data Bytes in Message
1	C	2513	80	246603657	0	0	
2	S	80	2513	692377068	246603658	0	
3	C	2513	80	246603658	692377069	0	
4	S	80	2513	692377069	246603658	400	400 total
5	S	80	2513	692377469	246603658	346	746 total
6	C	2513	80	246603658	692377815	0	
7	S	80	2513	692377815	246603659	1460	1460 total
8	S	80	2513	692379275	246603659	1460	2920 total
9	C	2513	80	246603659	692380735	0	
10	S	80	2513	692380735	246603660	1460	1460 total
11	C	2513	80	246603660	692382195	0	
12	S	80	2513	692382195	246603661	?	

C = client, S = server. Frames 1 to 3 are the three-step TCP connection handshake.

FIGURE 23-22   Example of a three-step handshake establishing a TCP connection and the exchange of several data segments between the client and server

the other direction. Therefore, the connection can be closed in one direction, and data transmission can still continue in the other direction.

There are four steps that must be taken to terminate a TCP connection between two hosts. Again, for simplicity, the client and server are designated host A and host B:

1. Host A begins the termination procedure by transmitting a segment to host B requesting a connection termination.
2. Host B sends a segment acknowledging host A's request to terminate, and the connection is closed between host A and host B. However, the connection between host B and host A is still open, and host B can continue sending data segments to host A.
3. Once host B has completed sending data segments, it transmits a segment to host A indicating that it wishes to close the connection.
4. Host A acknowledges the termination request from host B, at which time the connection is closed.

To terminate the connection in both directions requires four steps, as steps 2 and 3 cannot be combined as they were with the connection establishment. Therefore, the sequence described here is called *four-way handshaking*. However, steps 2 and 3 do not have to occur at the same time, as the connection may be closed in one direction and left open in the other direction. If both directions are terminated at the same time, which is highly likely, it is called a *three-step disconnect*.

A four-way handshake to terminate a connection is generally initiated by the client. The procedure begins when the client process tells its TCP software that it has completed transmitting data and wants to terminate the connection. This is called a *request for an active close*. After the client TCP software receives the request to terminate the connection, it closes the connection between the client and server. However, the connection between the server and the client remains open. When the process running on the server's TCP software has completed transmitting data to the client, it requests the TCP software to close the connection in the server-to-client direction. This is generally a *passive close*.

The four-way handshake sequence to terminate a TCP connection is as follows:

1. The client initiates the termination sequence by sending the first segment, which is called a FIN segment, as the FIN flag in the session flag field is set. Setting the FIN flag tells the server that the client wishes to terminate the TCP session in the client-to-server direction.

2. The TCP software in the server transmits the second segment in the sequence that confirms receipt of the FIN segment received from the client. The acknowledgment number in this segment is simply the sequence number received in the FIN segment plus 1. This essentially closes the connection in the client-to-server direction.

3. If the TCP software in the server still has data segments to send, it can continue transmitting in the server-to-client direction. After the server has completed sending all its data segments, it transmits the third segment of the sequence, which is also called a FIN segment. A FIN segment sent from the server to the client is a request to terminate the TCP session in the server-to-client direction.

4. The final segment in the four-step sequence is an ACK from the client TCP software to the server TCP software confirming the receipt of the FIN segment. The acknowledgment number in the ACK segment is the sequence number received from the server plus 1. At this time, the connections have been terminated in both directions.

A three-way handshake to terminate a TCP connection is shown in Figure 23-23. The client begins the sequence by sending an ACK + FIN segment with a sequence number of 3000 and an acknowledgment number that is equal to the last acknowledgment number it transmitted to the server. The server responds with an ACK segment, which includes the sequence number 5000 and the acknowledgment number 3001 (one more than the client's source sequence number). The third segment is the server's FIN segment with a source sequence number of 5001 and the acknowledgment number of 3001 repeated. The final segment in the sequence is the client's ACK segment, which includes the sequence number 3001 and the acknowledgment number 5002.

Figure 23-24 shows an example of a TCP three-step disconnect where the client initiates terminating a TCP session with a server using a three-step disconnect. In this example, sessions are terminated in both directions. Note the session flags in each segment:

Segment 1: ACK and FIN
Segment 2: ACK and FIN
Segment 3: ACK

23-9 TCP ERROR CONTROL

TCP is the only reliable transport-layer protocol in the TCP/IP protocol suite. Therefore, when a process delivers a stream of data to the TCP software, the software is expected to deliver the entire message to the process running on the destination host without errors and without any portion of the message lost or duplicated.

TCP error control includes mechanisms for detecting segments with corrupted data, missing segments, segments received out of order, lost segments, duplicate segments, and lost acknowledgments received in duplicate. TCP error control also includes a mechanism for correcting errors once they have been detected.

23-9-1 Corrupted Segments

TCP uses three relatively simple items to perform error detection: *checksum*, *acknowledgment*, and *time-out*. The checksum is used by the receive end to check for corrupted data. If a segment contains corrupted data, it is discarded. TCP uses source sequence and acknowledgment numbers to confirm reception of uncorrupted segments. TCP does not use

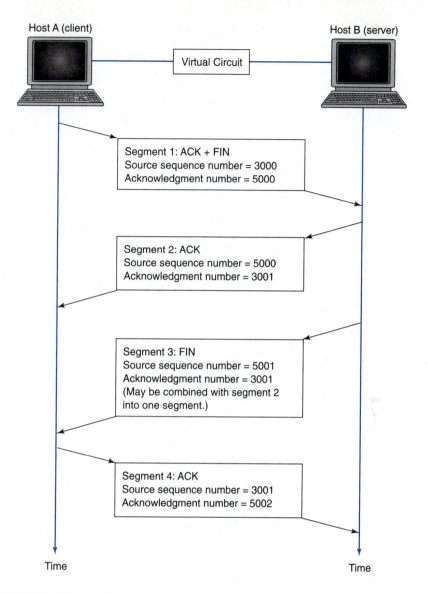

FIGURE 23-23 TCP four-way termination

a negative acknowledgment sequence. Instead, if a segment is not acknowledged before the time-out period expires, the segment is assumed either corrupted or lost.

TCP starts a timer for each segment transmitted. The counter remains open until the segment is acknowledged or the timer times out. Each open counter is checked periodically, and if the time-out period expires, it is assumed that the corresponding segment was corrupted or lost, and the segment is retransmitted.

How TCP deals with corrupted segments is best described with an example. The following example is illustrated in Figure 23-25:

1. Client transmits three segments (1, 2, and 3), and each segment contains 100 bytes of data. The three segment's transmissions are summarized in Table 23-4.
2. Server receives segments 1 and 2 without errors. However, segment 3 is received with corrupted data. Therefore, the server has successfully received data bytes 1000 through 1199.

	Segment 1 (client)	Segment 2 (server)	Segment 3 (client)
Ethernet header:			
Destination physical address	00–C0–02–63–55–48	00–D0–59–05–C5–DB	00–C0–02–63–55–48
Source physical address	00–D0–59–05–C5–DB	00–C0–02–63–55–48	00–D0–59–05–C5–DB
Protocol	IP (0800)	IP (0800)	IP (0800)
FCS	AC2D469F hex	2BB46348 hex	A8925B2F hex
IP header:			
IP version	4	4	4
Header length	20	20	20
Packet length	40	40	40
ID #	3E2E hex	D7CF hex	3E2F hex
Fragmentation field			
Don't fragment	1 = true	1 = true	1 = true
More fragments	0 = false	0 = false	0 = false
Fragment offset	0000000000000	0000000000000	0000000000000
TTL	124	115	122
Protocol	TCP (6)	TCP (6)	TCP (6)
Header checksum	A4BD hex	5820 hex	A4C4 hex
Source IP address	192.168.0.2	209.39.134.10	192.168.0.2
Destination IP address	209.39.134.10	192.168.0.2	209.39.134.10
TCP header:			
Source port number	1025 (blackjack)	FTP (21)	1025 (blackjack)
Destination port number	FTP (21)	1025 (blackjack)	FTP (21)
Sequence number	1379437769	2165615844	1379437770
Acknowledgment number	2165615844	1379437770	2165615845
Data offset	20	20	20
Session flags	11 hex	11 hex	10 hex
URG flag	. . 0 . . . . . false	. . 0 . . . . . false	. . 0 . . . . . false
ACK flag	. . . 1 . . . . true	. . . 1 . . . . true	. . . 1 . . . . true
PSH flag	. . . . 0 . . . false	. . . . 0 . . . false	. . . . 0 . . . false
RST flag	. . . . . 0 . . false	. . . . . 0 . . false	. . . . . 0 . . false
SYN flag	. . . . . . 0 . false	. . . . . . 0 . false	. . . . . . 0 . false
FIN flag	. . . . 1 true	. . . . 1 true	. . . . 0 false
Window size	16348	8760	17520
Checksum	44F4 hex	49BF hex	3F44 hex
Urgent pointer	00000000	00000000	00000000
Option: no options			

FIGURE 23-24 Example of TCP three-way disconnect

3. Server transmits a segment back to the client with an acknowledgment number of 1200, indicating that is has received and accepted 200 data bytes (1000–1199).

4. Since the server has not acknowledged segment 3, the client's timer expires. At that time, the client assumes that segment 3 (bytes 1200–1299) was either corrupted or lost, so it retransmits segment 3 with the same sequence number (1200).

5. The server receives the retransmitted without errors. Therefore, it responds with a segment containing the acknowledgment number 1300, which acknowledges segment 3 (bytes 1200–1299).

23-9-2 Lost or Missing Segments

Figure 23-26 shows how a lost segment is handled by TCP. TCP handles lost or missing segments in essentially the same way that it deals with corrupted segments. The destination host discards a corrupted segment, and an intermediate node, such as a router, discards lost segments. Therefore, lost segments never reach the final destination.

In Figure 23-26, the client transmits segments 1, 2, and 3. Segments 1 and 2 reach the destination; however, segment 3 never arrives at the server. The server transmits an ACK

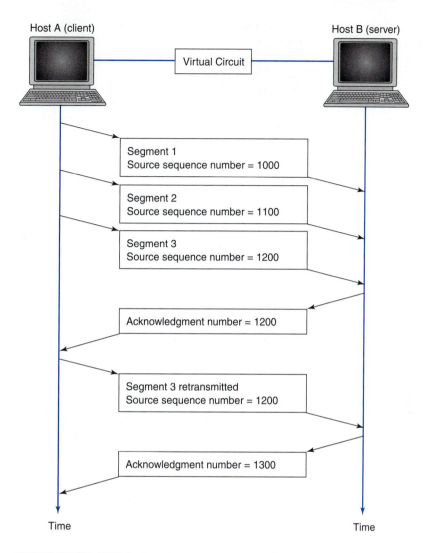

Host A (client)

Virtual Circuit

Host B (server)

Segment 1
Source sequence number = 1000

Segment 2
Source sequence number = 1100

Segment 3
Source sequence number = 1200

Acknowledgment number = 1200

Segment 3 retransmitted
Source sequence number = 1200

Acknowledgment number = 1300

Time

Time

FIGURE 23-25 TCP dealing with a corrupted segment

Table 23-4 TCP Corrupted Segment Example

Segment	Data Byte in Segment	Source Sequence Number	Range of Data in Segment
1	100	1000	1000–1099
2	100	1100	1100–1199
3	100	1200	1200–1299

segment back to the client acknowledging reception of segments 1 and 2 (bytes 1000–1199). After the client's time-out expires, it retransmits segment 3, which is received and acknowledged by the server.

23-9-3 Duplicate Received Segments

A duplicate segment may occur with TCP if an ACK segment does not arrive at the source before the time-out expires. The destination TCP software expects to receive a continuous stream of data. When a segment arrives with the same sequence number as a previously received segment, the TCP software discards the second (duplicate) segment.

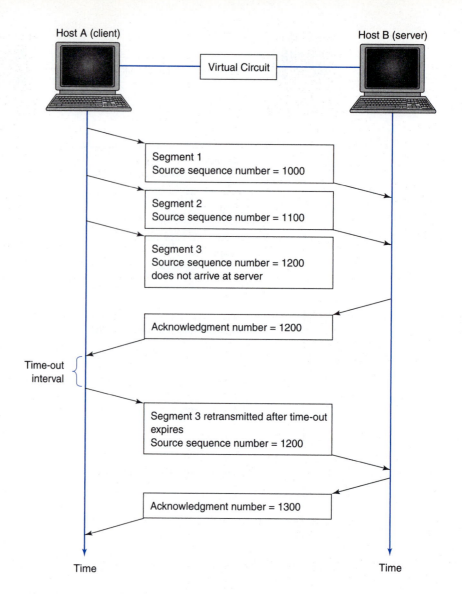

Host A (client)   Host B (server)

Virtual Circuit

Segment 1
Source sequence number = 1000

Segment 2
Source sequence number = 1100

Segment 3
Source sequence number = 1200
does not arrive at server

Acknowledgment number = 1200

Time-out
interval

Segment 3 retransmitted after time-out
expires
Source sequence number = 1200

Acknowledgment number = 1300

Time   Time

FIGURE 23-26   TCP dealing with a lost segment

23-9-4   Segments Received Out of Order

TCP uses IP to deliver it to the destination host. Since IP is an unreliable connectionless protocol, each datagram is treated as an independent entity. Network and internetwork routers are allowed to forward IP datagrams through whatever route they choose. One datagram may take the most economical path, another datagram may take the route with the highest throughput, and so on. Consequently, datagrams do not necessarily reach the destination in the same order in which they were transmitted.

TCP handles segments received out of order rather straightforwardly. TCP simply does not acknowledge a segment received out of order until it receives all the segments that precede it in the sequence. If the source times out before the segment arrives at the destination, a duplicate segment is transmitted, which is discarded at the receiver.

23-9-5   Lost Acknowledgment

Figure 23-27 shows how TCP deals with a lost acknowledgment. Because of the nature of TCP's error-control mechanism, a lost acknowledgment may not be noticed or detected. TCP

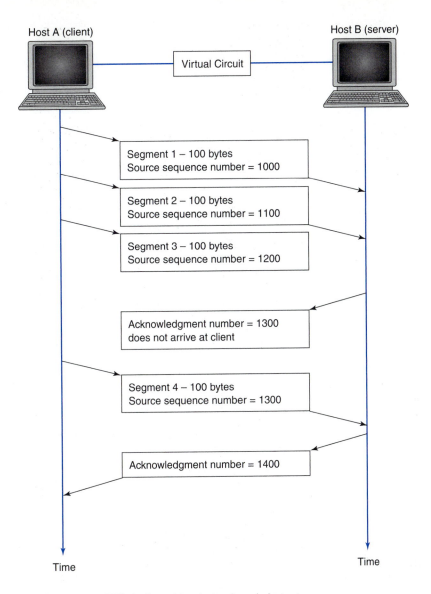

FIGURE 23-27 TCP dealing with a lost acknowledgment

uses an accumulative acknowledgment system. Therefore, the acknowledgment number in a segment confirms all segments (bytes) up to its value regardless of whether some of the segments have already been acknowledged. For example, in Figure 23-27, the client transmits segments 1, 2, and 3 (bytes 1000–1299). The acknowledgment sent by the server never arrives at the client. Later, the client transmits segment 4 (bytes 1300–1399), which is received and acknowledged by the server. When the server transmits the acknowledgment number 1400, it is acknowledging reception of bytes up through 1399 regardless of when they were transmitted.

23-10 TCP ENCAPSULATION AND DECAPSULATION

The TCP encapsulation and decapsulation procedure for processes is essentially the same as it is with UDP. Figure 23-28 shows a TCP process originating at a source that is encapsulated in a TCP segment. The TCP segment is encapsulated in an IP datagram, which is encapsulated in a physical-layer frame. The decapsulation procedure is identical except in reverse order.

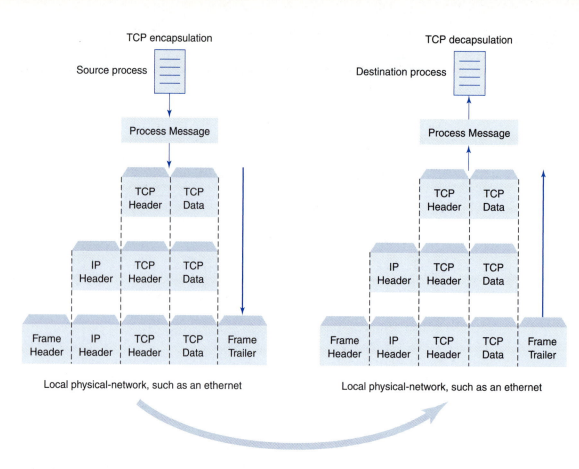

FIGURE 23-28 TCP encapsulation and decapsulation

QUESTIONS

23-1. What is the primary function of *transport-layer protocols*?

23-2. What is the primary function of *port addresses*?

23-3. List the three types of port addresses and give the range of numbers for each.

23-4. Describe *socket addresses*.

23-5. List the advantages of *UDP* over *TCP*.

23-6. List and describe the fields that comprise a *UDP header*.

23-7. Describe the operation of UDP, including the means of providing flow and error control.

23-8. Describe how UDP is encapsulated in an IP datagram.

23-9. List the operations performed by TCP at the source and destination host.

23-10. List and describe the services provided by TCP.

23-11. List and describe the fields that comprise a TCP header.

23-12. List and describe the fields in the reserved field of a TCP header.

23-13. List and describe the category 1 TCP options.

23-14. List and describe the category 2 TCP options.

23-15. Describe the steps for establishing and terminating a TCP connection.

23-16. Describe how TCP accomplishes error control.

23-17. Describe how TCP is encapsulated in an IP datagram.

23-1. Determine if the following port numbers are well known, registered, and dynamic:

    a. 851

    b. 31,657

    c. 43,301

    d. 62

    e. 55,601

    f. 81

23-2. Determine the TCP/IP protocol assigned the following port numbers:

    a. 19

    b. 7

    c. 162

    d. 25

    e. 111

    f. 520

23-3. Determine the length of a UDP data field if the message length field contains 0030 hex.

23-4. Determine the total message length field for a UDP datagram that is carrying 012C hex bytes of data.

23-5. Determine the meaning of the following session-bit flag fields:

    a. 11 hex

    b. 10 hex

    c. 18 hex

    d. 04 hex

    e. 02 hex

    f. 19 hex

23-6. Use the following network analyzer display of an IP datagram carrying UDP to answer the following questions:

```
0000 45  44  00  4A  A0  A2  04  00  F5  11  A2  35  C0  99  B4  01
0010 C0  99  B4  02  D6  D8  00  45  00  36  20  A3  B2  91  00  A2
```

    a. What IP version is being used?

    b. Are IP options being used?

    c. What is the precedence?

    d. What type of service is being requested?

    e. How many data bytes are included in the IP datagram (decimal)?

    f. Is fragmenting allowed?

    g. Is this the first, last, or a middle fragment?

    h. What is the decimal number of the first byte in this fragment?

    i. How many routers has this datagram propagated through (decimal)?

    j. What is the source IP address (dotted decimal)?

    k. What is the destination IP address (dotted decimal)?

    l. What is the source port number (decimal)?

    m. What is the category of the source port?

    n. Is the source most likely a client or a server?

    o. What is the destination port number (decimal)?

    p. What is the category of the destination port?

    q. Is the destination most likely a client or a server?

    r. What protocol is the destination process?

23-7. Use the following network analyzer display of an IP datagram carrying TCP to answer the following questions:

```
0000  45  44  00  4A  A0  A2  04  00  F5  06  A2  35  C0  99  40  B2
0010  C0  99  40  A1  00  35  D0  A2  04  A2  B1  22  00  1A  46  A2
0020  50  12  10  00  56  8F  00  00  00  00  00  00  00  00  00  00
```

a. What is the source port number (decimal)?
b. What is the category of the source port?
c. Is the source most likely a client or a server?
d. What is the destination port number (decimal)?
e. What is the category of the destination port?
f. Is the destination most likely a client or a server?
g. What protocol is the source process?
h. How many bytes are in the TCP header (decimal)?
i. Which session flags are set?
j. What is the sender's window size (decimal)?
k. Assuming that no errors have occurred, what should be the destination's acknowledgement number?
l. Assuming that no errors have occurred, what should be the destination's source sequence number?

C H A P T E R   24

Internet Protocol Version 6

CHAPTER OUTLINE

OBJECTIVES

- Describe the reasons for developing IPv6
- List and describe the advantages of IPv6 over IPv4
- Explain hexadecimal colon notation
- List and describe the three IPv6 address types
- List and describe the fields in provider-based unicast addresses
- Describe the following IPv6 special addresses: loopback, unspecified, and IPv4 over IPv6
- Define *local addresses*
- Describe link-local and site-local addresses
- List and explain the purpose of the fields in aggregatable global unicast addresses
- Describe how anycast addresses are used
- Describe how multicast addresses are used
- List and describe the functions of the fields contained in an IPv6 header
- Describe the purpose of IPv6 extensions and extension headers
- Describe ICMPv6 and compare it with ICMPv4
- Explain the two types of ICMPv6 messages
- Explain how an ICMPv6 checksum is calculated

24-1 INTRODUCTION

The original network-layer protocol in the TCP/IP protocol suite is Internetwork Protocol, version 4 (IPv4), which was designed to provide host-to-host communications across the Internet. However, the Internet has changed significantly since its inception in the early 1970s, leaving IPv4 with several inherent shortcomings, such as the following:

1. Limited number of IP addresses
2. Inefficient use of IP addresses
3. Inadequate delay and resource reservation strategies for real-time audio and video transmission
4. No provisions for encryption or decryption
5. No provision for authentication
6. Inefficient provisions for routing
7. Inefficient use of options
8. Inadequate types options

To overcome the shortcomings of IPv4, the Area Directors of the Internet Engineering Task Force at the Toronto IETF meeting on July 25, 1994, created a new generation of IP addressing called Internet Protocol version 6 (IPv6), which is sometimes called IPng (IP next generation). IPv6 was designed as an evolutionary and radical progression from IPv4. IPv6 can be installed as a normal software upgrade in Internet devices, as it is interoperable with the current IPv4. The deployment strategy for IPv6 is designed for immediate implementation, which is possible because IPv6 is flexible and capable of running on high-performance networks, such as ATM, and at the same time efficient enough to be used on low-bandwidth networks, such as wireless.

Some Internet users say IPv6 is overkill, as it provides capabilities that are well beyond the foreseeable future. This was done intentionally to provide a network-layer protocol capable of accommodating unforeseen changes and adaptations that are inevitable in a dynamic internetworking environment, such as the Internet.

IPv6 is a significant modification of IPv4. For example, with IPv6, the format and length of IP addresses and the format and capabilities of IP datagrams are dramatically different from those provided by IPv4. Modifying IP also necessitated modifying related protocols, such as ICMP and several other protocols, such as ARP, RARP, and IGMP, have been either omitted entirely or incorporated in a new version of ICMP called ICMPv6 or ICMPng. IPv6 also necessitates slight modifications to routing protocols, such as RIP and OSPF.

24-2 ADVANTAGES OF IPv6

IPv6 has several inherent advantages over IPv4, such as the following:

1. Longer addresses by increasing the IP address length from 32 bits to 128 bits
2. Extended address hierarchy
3. Flexible and improved header format
4. Provisions for protocol extensions
5. Support for autoconfiguration and renumbering
6. Improved support for resource allocation
7. Incorporation of extended options
8. Improved support for options
9. Encryption and decryption capabilities
10. Expanded routing and addressing capabilities
11. A new address type called anycast addresses

12. Adding a scope field to multicast addresses to improve the scalability of multi-cast routing
13. A simplified base header
14. Improved quality of service capabilities to enable the labeling of datagrams belonging to particular traffic flows for which the sender requests special handling
15. Improved authentication and privacy capabilities

24-3 IPv6 ADDRESSING FORMAT

IPv6 addresses are comprised of 128 bits (16 bytes) and identify individual interfaces and groups of interfaces. Therefore, there are 2^{128} addresses possible with IPv6, which is probably enough to provide multiple IP addresses to virtually every device in existence for the next several centuries.

With 128 address bits, representing an IP address in binary or dotted-decimal notation is rather cumbersome, as can be seen in Figures 24-1a and b, respectively. Therefore, a new notation scheme, called *hexadecimal colon notation*, is used with IPv6. An example of hexadecimal colon notation is shown in Figure 24-1c. Hexadecimal colon notation may not initially appear to be any less ominous than binary or dotted-decimal notation. However, as we will soon see, hexadecimal colon notation can be abbreviated, which greatly simplifies it.

24-3-1 Hexadecimal Colon Notation

Both IPv4 and IPv6 addresses are binary numbers. However, 32-bit IPv4 addresses are generally expressed in dotted-decimal notation comprised of four groups of eight bits where each group of eight bits is converted to its decimal equivalent value. With hexadecimal colon notation, 128-bit IP addresses are divided into eight segments where each segment is two bytes (four hex characters) long. Therefore, IP addresses are comprised of 32 hexadecimal digits where each group of four digits (16 bits) is separated with a colon (:), creating eight groups of four-character hexadecimal numbers, as shown here:

AF4B:2943:003E:17DC:0003:0000:2A01:0054

Representing an IP address with 32 hexadecimal digits is somewhat overwhelming, and 2^{128} IP addresses are significantly more than is needed in the foreseeable future. Therefore, many of the digits will initially be 0s or, better yet, groups of contiguous 0s. Therefore, the leading 0s in a hexadecimal segment can be omitted in groups of four bits, leaving only the trailing 0s (also in groups of four bits). Whenever a group of four hex characters has a value 1000 or less, the leading 0s can be dropped. For example, with the hexadecimal segment 003E, the first two hex 0s can be dropped, leaving the abbreviated value 3E. With the hexadecimal segment 0003, the first three hex 0s can be dropped, leaving the abbreviated value 3. The hexadecimal segment 0000 can be abbreviated as simply 0.

Abbreviating compresses IP addresses. Using abbreviations, the hexadecimal address

AF4B:2943:003E:17DC:0003:0000:2A01:0054

is abbreviated as

AF4B:2943:3E:17DC:3:0:2A01:54

Whenever a group of four hex characters has a value of 0000, the entire group can be dropped. Therefore, it is possible to abbreviate further when there are consecutive segments comprised of only 0s by removing the 0s and replacing them with a double colon, as shown here:

abbreviated: 4A3B:0:0:0:0:A1B5:0:A124
further abbreviated   4A3B::A1B5:0:A124

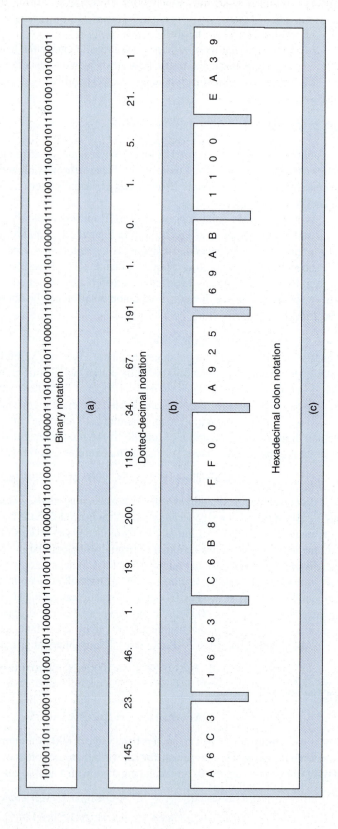

FIGURE 24-1 IPv6 Address notation: (a) binary; (b) dotted decimal; (c) hexadecimal colon

Replacing consecutive segments containing only 0s can be done only once per address. Therefore, when there are two or more occurrences of consecutive segments containing consecutive 0s, only one can be abbreviated with double colons. To expand the abbreviated address, simply align the unabbreviated segments with the ends of the address and then insert 0s until you complete the original expanded address length.

IPv6 addresses can use a slash notation to identify the network portion of the address, as shown here:

1085::8:800:321D:23F5:/60

2128:DFC0:0:8:80:800:11BB:B/24

The decimal number to the right of the slash indicates how many of the leftmost contiguous bits of the address are part of the network prefix.

24-3-2 IPv6 Address Types

IPv6 addresses are assigned to interfaces, not nodes. All interfaces belong to a single node; therefore, any of the node's interface addresses can be used to identify the node, and a single node may be assigned several IPv6 addresses.

With IPv6, there is no broadcast address. IPv6 specifies three address types: unicast, anycast, and multicast:

Unicast addresses. Unicast addresses are unique, as they define a single interface. A unicast address can be used for a source or a destination address.

Anycast addresses. Anycast addresses specify a group of devices whose addresses have the same prefix. All interfaces connected to the same physical network have the same prefix. Therefore, if a datagram is sent with an anycast address onto a physical network, it must be delivered to only one interface within the group, which is generally the one closest to the source or the one that is the most accessible. Anycast addresses can be used to address functions that are generally provided on the Internet at more than one network location, such as DHCP (dynamic host configuration protocol) servers, routers, or IRC (Internet relay chat) servers.

Multicast addresses. Multicast addresses can be used only for destination addresses. Multicast addresses specify a group of interfaces where each interface also has a unique unicast address. Multicast addresses do not necessarily share the same prefix, and the interfaces need not populate the same physical network. A datagram sent to a multicast address must be delivered to all interfaces in the group. Multicast addresses can be used for videoconferencing or router updates.

24-3-3 IPv6 Address Allocation

IPv6 addresses are divided into two parts. The first part is called the type *prefix*. The type prefix is a variable-length address that defines the purpose of the address. Type prefixes were designed so that no two address types begin with the same binary sequence. Therefore, there is nothing ambiguous about an address, as the type prefix is easily recognized. Figure 24-2 shows the IPv6 address format. As shown in the figure, all addresses begin with a variable-length type prefix and end with a variable-length code that represents the remainder of the IP address. Table 24-1 shows the type prefixes for each type of IP address

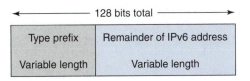

FIGURE 24-2 IPv6 address format

Table 24-1   IPv6 Type Prefixes

Prefix	Type	Fraction of IPv6 Addresses
0000 0000	Reserved	1/256
0000 0001	Reserved	1/256
0000 001	NSAP (Network Service Access Point)	1/128
0000 010	IPX (Novell)	1/128
0000 011	Reserved	1/128
0000 100	Reserved	1/128
0000 101	Reserved	1/128
0000 110	Reserved	1/128
0000 111	Reserved	1/128
0001	Reserved	1/16
001	Reserved	1/8
010	Provider-based unicast addresses	1/8
011	Reserved	1/8
100	Geographica unicast address	1/8
101	Reserved	1/8
110	Reserved	1/8
1110	Reserved	1/16
1111 0	Reserved	1/32
1111 10	Reserved	1/64
1111 110	Reserved	1/128
1111 1110 0	Reserved	1/512
1111 1110 10	Link-local address	1/1024
1111 1110 11	Site-local address	1/1024
1111 1111	Multicast addresses	1/256

and lists the prefix, the type, and what fraction of the total IP address allocation each type represents.

24-3-4   IPv6 Unicast Addresses

There are several kinds of unicast address assignments with IPv6: provider based, special (loopback, unspecified, and IPv4 capable), local (link-local use and site-local use), and aggregatable global.

24-3-4-1 Provider-based unicast addresses.   Provider-based unicast addresses are for global communications over the Internet and function in similar to classless IPv4 address using CIDR. Provider-based unicast addresses are the address type used by normal devices for their globally unique IP address. The format for provider-based unicast address is comprised of six fields, as shown in Figure 24-3:

1. Type identifier
2. Registry identifier

Type Field	Registry Field	Provider Identifier Field	Subscriber Identifier Field	Subnet Identifier Field	Note Identifier Field
3 bits	5 bits	16 bits 2 bytes	24 bits 3 bytes	32 bits 4 bytes	48 bits 6 bytes

FIGURE 24-3   Provider-based address format

3. Provider identifier
4. Subscriber identifier
5. Subnet identifier
6. Node identifier

Type identifier field. The type identifier is a three-bit type field that identifies the address as a provider-based unicast address with the three-bit prefix 010, as shown in Figure 24-4.

Registry identifier field. The registry identifier is a five-bit field that specifies the Internet address agency that the address is registered with. Internet address agencies assign provider identifiers to Internet Service Providers (ISPs), which in turn reassign portions of their allocated address space to their subscribers. At this writing, there are only three registry centers defined, as shown in Figure 24-4:

INTERNIC (11000): The center for North America

RIPNIC (01000): The center for European registration

APNIC (10100): The center for Asian and Pacific countries

Provider identifier field. The provider identifier is a variable-length field; however, 16 bits is the recommended and generally accepted length. Internet address agencies assign a provider identifier number to each ISP.

Subscriber identifier field. All organizations that subscribe to the Internet are assigned a 24-bit subscriber identification number. ISPs uniquely distinguish each of their subscribers with a subscriber identification number.

Subnet identifier field. The subnet identifier is a 32-bit field. The subnet identifier defines a specific physical network under the territory of a subscriber. This is necessary as a subscriber can have multiple subnetworks, and each subnetwork can have its own subnet identifier code. A specific subnet cannot span more than one physical link.

Node identifier. The node identifier (sometimes called interface identifier or interface ID) specifies a single device interface among a group of interfaces connected to a subnet and identified with the same subnet identifier. IPv6 specifies that interface identifiers comply with the IEEE EUI–64 format, which recommends a length of 64 bits. For Ethernet local area networks (LANS) the 48-bit hardware (physical) address embedded on the network interface card is used to derive the node identifier. The first 24 bits identify a specific manufacturer, and the last 24 bits guarantee uniqueness among a manufacturer's cards. The 48-bit MAC address is converted to a 64-bit node identifier by adding 16 bits (FFFE hex) between the two halves of the MAC number.

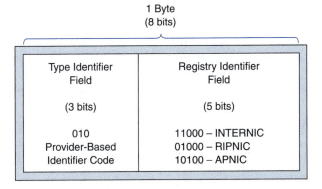

FIGURE 24-4 Type identifier and registry identifier fields

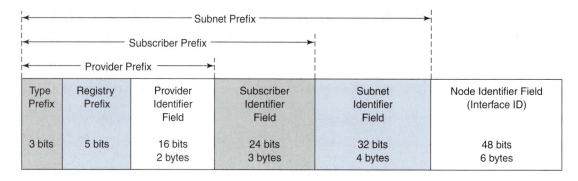

FIGURE 24-5 IPv6 address hierarchy

Type and Registry Fields	Provider, Subscriber, Subnet, and Node Identifiers
8 bits (all logic 0s)	120 bits (119 logic 0s and 1 logic 1)
00000000	00000000000000...............................000000001

FIGURE 24-6 IPv6 loopback address format

The provider-based address is a hierarchical identification comprised of several prefixes where each prefix defines a level of the hierarchy, as shown in Figure 24-5. The type prefix defines the address type, the registry prefix uniquely defines the registry agency for the address, the provider prefix uniquely defines the ISP, the subscriber prefix uniquely defines the subscriber, and the subnet prefix uniquely defines the subnet.

24-3-4-2 Special addresses. Special addresses are reserved addresses with prefixes that begin with eight contiguous logic 0s (00000000) in the type and registry fields. Special addresses include loopback addresses, unspecified addresses, and IPv4 addresses.

Loopback address. With IPv4, the entire class A address 127.x.x.x (2^{24} addresses) is dedicated to the loopback feature. With IPv6, there is only one loopback address that is comprised of 127 logic 0s followed by a logic 1, as shown in Figure 24-6. The loopback address is represented in hexadecimal colon notation with two colons and a 1 (::1). The loopback address is sent by a device to itself to test its own TCP/IP software without actually being connected to the local network. A loopback message never leaves the device that originates it. A test message is created in the applications layer and passed down through to the protocol stack and then back up the stack without ever reaching the local network. Loopbacks are used as diagnostic tools and cannot be used for source or destination addresses.

Unspecified address. An unspecified is, in essence, an address that is not really an address. An unspecified address is comprised of 128 logic 0s, as shown in Figure 24-7. Unspecified addresses can be represented in hexadecimal colon notation with two colons (::). The unspecified address is used when a host transmits an inquiry to learn its own IP address. When an inquiry is sent, the originator must specify its address, and since the originator does not know its address, it uses the unspecified address. The unspecified address cannot be used as a destination address.

Type and Registry Fields	Provider, Subscriber, Subnet, and Node Identifiers
8 bits (all logic 0s)	120 bits (all logic 0s)
00000000	00000000000000.............................000000000

FIGURE 24-7 IPv6 unspecified address format

Type and Registry Fields		IPv4 Address
8 bits (all logic 0s)	88 bits (all logic 0s)	32 bits
00000000	00000000000000.............................000000000	4.12.18.14

FIGURE 24-8 IPv6-compatible address format

IPv4 over IPv6 addresses. During the transition from IPv4 to IPv6, a device can embed its IPv4 address inside an IPv6 address. IPv6 includes two mechanisms that allow hosts and routers to dynamically tunnel IPv6 datagrams over an IPv4 routing infrastructure. One type of IPv6 address that will do this is called an *IPv4-compatible IPv6 address* (or simply *IP-compatible address*). The second type of IPv6 address is capable of carrying IPv4 addresses that represents the addresses of IPv4-only nodes (i.e., nodes that do not support IPv6). This type of address is called an *IPv4-mapped IPv6 address* (or simply an *IP-mapped address*). The two types are similar in that the first 80 bits of both types are reset to logic 0s, and the last 32 bits of each type contains the IPv4 IP address.

Compatible addresses are comprised of 96 logic 0s followed by the 32-bit IPv4 address. Compatible addresses are used when an IPv6 device sends a datagram to another IPv6 device and the datagram must pass through one or more IPv4 networks. For example, the IPv4 address 4.12.18.14 is translated to 0::040C:120E in hexadecimal colon notation. The IPv4 address is preceded by 96 logic 0s, as shown in Figure 24-8.

Mapped addresses are comprised of 80 logic 0s, followed by 16 logic 1s, followed by the 32-bit IPv4 address. Mapped addresses are used by IPv6 devices to send a datagram to another device using IPv4 when the datagram must pass through one or more IPv6 networks to an IPv4 network that delivers the datagram to the destination device. For example, the IPv4 address 4.12.18.14 is translated to 0::FFFF: 040C:120E in hexadecimal colon notation. The IPv4 address is preceded by 80 logic 0s and 16 logic 1s, as shown in Figure 24-9.

Because groups of 16 logic 0s or 16 logic 1s do not affect the value of a checksum, there is no need to recalculate the checksum when an IPv4 datagram is converted to an IPv6 datagram and vice versa.

24-3-4-3 Local addresses. Local addresses are reserved unicast addresses that begin with seven logic 1s and one logic 0 (11111110) in the type and registry fields. A local address is a unicast address with only a local routability scope (i.e., within a subnet or within a subscriber network). It is intended to use local addresses inside a specific network

Type and Registry Fields			IPv4 Address
8 bits (all logic 0s)	72 bits (all logic 0s)	16 bits (all logic 1s)	32 bits
00000000	00000000000000........000000000	1111111111111111	4.12.18.14

FIGURE 24-9 IPv6-mapped address format

10 bits (7 logic 1s, 1 logic 0, 1 logic 1, and 1 logic 0)	54 bits (all logic 0s)	64 bits (16 bits of fill plus a 48-bit MAC address)
1111111010	00000000000000........000000000	(node address)

FIGURE 24-10 IPv6 link-local address format

for local computing activities. Local addresses include link local addresses and site local addresses. With both types of addresses, the interface ID number must be unique in the domain in which it is being used. In most cases, the interface ID is the node's IEEE–802.2 48-bit MAC address. The subnet ID identifies a specific subnet at a site, and the combination of the subnet ID and interface ID form a unique local use address, which allows a large private internetwork to be constructed without requiring any additional IP address allocation.

Local use addresses allow organizations that are not connected to the global Internet to operate with local use addresses without having to acquire an address prefix from the global Internet address space. When an organization using local use addresses decides to connect to the global Internet, it can use its subnet ID and interface ID in combination with a global prefix (e.g., registry ID plus provider ID plus subscriber ID) to create a global address. This is a vast improvement over IPv4, which requires devices attached to networks using private (nonglobal) IP addresses to manually reassign a global IP address when they connect to the Internet. With IPv6, the number reassignment is done automatically.

Link-local addresses. Link-local addresses are similar to the private addresses used with IPv4, as they are addresses intended for use on a single link. Link-local addresses are used on local networks that utilize the TCP/IP protocol suite but for security reasons are not actually connected to the Internet. Link-local address prefixes begin with seven logic 1s, followed by one logic 0, followed by one logic 1, followed by one logic 0 (11111110 10), as shown in Figure 24-10. The next 54 bits are reset to logic 0s, leaving the remaining 64 bits for 16 bits of fill and the link-local (MAC) address. Link-local addresses are used in isolated networks; therefore, they have no global impact. Devices outside the isolated local network cannot send datagrams to devices attached to a network that uses link-local addresses.

Site-local addresses. Site-local addresses are addresses that are intended for use on a single site (location). Site-local addresses are used when several interconnected networks located in close proximity to each other utilize the TCP/IP protocol suite but for security reasons are not connected to the Internet. Site-local address prefixes be-

10 bits (7 logic 1s, 1 logic 0, and 2 logic 1s)	38 bits (all logic 0s)	16 bits	64 bits (MAC address)
1111111011	0000000........0000000	(subnet address)	(node address)

FIGURE 24-11 IPv6 site-local address format

Format Prefix Field	Top-Level Aggregation Identification Field	Reserved Field	Next-Level Aggregation Identification Field	Site-Level Aggregation Field	Interface Identification Field
3 bits 001	13 bits	8 bits	24 bits	16 bits	64 bits

FIGURE 24-12 Aggregatable global unicast address format

gin with seven logic 1s, followed by one logic 0, followed by two logic 1s (11111110 11), as shown in Figure 24-11. The next 38 bits are all logic 0s, which are followed by a 16-bit subnet address that identifies the site to which the address is local. The last 64 bits are for the MAC address, which includes a 16-bit filler. Site-local addresses are used in isolated internetworks; therefore, they have no global impact. Devices outside the isolated internetwork cannot send datagrams to devices attached to a network that uses site-local addresses.

24-3-4-4 Aggregatable global unicast addresses.   IPv6 created a specific kind of unicast address called the aggregatable global unicast address to aid in the administration and routing of IP addresses. Aggregatable global unicast addresses are comprised of six fields, as shown in Figure 24-12. The network portion (leftmost 64 bits) of the address is segmented into five explicit fields, which allows routes to these addresses to be aggregated (combined) into a single entry in a routing table. The rightmost 64 bits of the address are used for the interface ID field.

Format prefix field. The format prefix (FP) field is a three-bit identifier that shows which part of the IPv6 address space the address belongs to. Currently, all aggregatable addresses must have 001 in this field.

Top-level aggregation ID field. The top-level aggregation identification (TLA ID) field is a 13-bit field that allows 2^{13} (8192) top-level routes.

Reserved field. The reserved field contains eight logic 0s.

Next-level aggregation ID field. The next-level aggregation identification (NLA ID) field is 24 bits long and allows the organizations controlling any of the TLAs to divide their address blocks into any size they choose. The controlling organizations are most likely large ISPs, as they can share this space with other smaller users. For example, they might reserve half the space for themselves, which leaves the other half to share with smaller ISPs to allocate very large blocks of addresses. The smaller ISPs could then subdivide their blocks further.

Site-level aggregation ID field. The site-level aggregation identification field is one 16-bit field that permits 65,536 flat (same subnets) addresses. Or it could be set up as a hierarchy and allow 255 subnets each with 255 addresses. It is presumed that eventually IPv6 will allocate address blocks to individual sites of this size.

Interface ID field. The interface identification field uses the same IEEE EUI–64 format as described in a previous section.

24-3-5 Anycast Addresses

IPv6 anycast addresses are addresses that are assigned to more than one interface that typically belong to different nodes. A datagram transmitted to an anycast address is routed to the nearest interface with that anycast address (the nearest interface is determined by the routing protocol).

Anycast addresses can be used as part of a routing sequence that permits a device to choose the ISPs it wants to carry its datagrams. This capacity is sometimes referred to as *source-selected policies*. Source selection can be implemented by configuring anycast addresses that identify a set of routers that belong to a particular ISP where each anycast address identifies one service provider. The anycast addresses are used as intermediate addresses in an IPv6 routing header, which causes the datagram to be delivered over a route using a particular provider (or sequence of providers). Anycast addresses can also be used to identify the set of routers attached to a specific subnet or group of subnets, affording entry into a prescribed routing domain.

Anycast addresses are assigned from the unicast address space using any of the defined unicast addressing formats. Therefore, anycast addresses are syntactically indistinguishable from unicast addresses. Whenever a unicast address is assigned to more than one interface, it becomes an anycast address, and the nodes the address is assigned to must be explicitly configured; otherwise, it would be impossible to tell that it was an anycast address.

24-3-6 Multicast Addresses

Multicast addresses are used to send the same message to more than one destination. Multicast addresses define a group of interfaces, and an interface may belong to any number of multicast groups. Multicast addresses begin with the prefix 11111111 (eight logic 1s) in the first eight-bit field, as shown in Figure 24-13. The four-bit field immediately following the prefix is a flag where the first three bits are all 0s, as they are reserved for future use. The fourth bit of the flag field is the T bit, where T means "transient." When the T bit is a logic 0 (flag = 0000), the group address is permanently assigned (not transient). Permanent multicast group

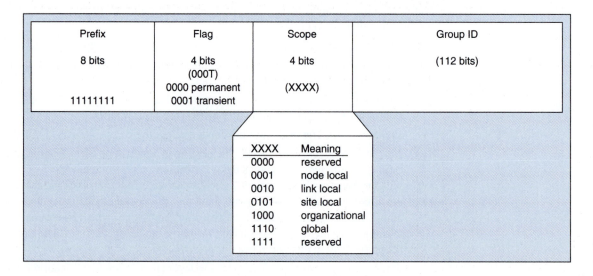

Prefix	Flag	Scope	Group ID
8 bits	4 bits (000T) 0000 permanent 0001 transient	4 bits (XXXX)	(112 bits)
11111111			

XXXX	Meaning
0000	reserved
0001	node local
0010	link local
0101	site local
1000	organizational
1110	global
1111	reserved

FIGURE 24-13 IPv6 multicast address format

Table 24-2 Multicast Scope Values

Scope	Meaning
0000	Reserved
0001	Node local
0010	Link local
0011	Unassigned
0100	Unassigned
0101	Site-local scope
0110	Unassigned
0111	Unassigned
1000	Organizational (local scope)
1001	Unassigned
1010	Unassigned
1011	Unassigned
1100	Unassigned
1101	Unassigned
1110	Global
1111	Reserved

addresses are sometimes called *well-known* multicast addresses. When the T bit is a logic 1 (flag = 0001), the multicast group address is not permanently assigned (temporary or transient). Permanent group addresses are defined by Internet authorities and can be accessed all the time. Transient group addresses are used only temporarily, such as during a teleconference.

The third field of a multicast address is the four-bit scope of the group address, which is used to limit the scope of the multicast group. The multicast scope values are listed in Table 24-2. The remaining 112 bits of the multicast address are used for the group ID number.

24-4 IPv6 HEADER

Although IPv6 has not been implemented as of this writing, a brief description of the IPv6 header format is given for comparison purposes. Figure 24-14 shows the format for an IPv6 datagram. The datagram is comprised of a fixed-length 40-byte mandatory header (called a basic header) and up to 65,535 bytes of payload, which may include one or more optional extension headers and a data packet carrying data from an upper-layer protocol. With IPv6, options do not affect the length of the base header because options are handled with extension headers, which are placed after the 40-byte base header.

24-4-1 IPv6 Base Header

The format for the 40-byte IPv6 base header is shown aligned to a 32-bit boundary in Figure 24-15. The changes in the IPv6 header reflect the changes in the protocol. The IPv6

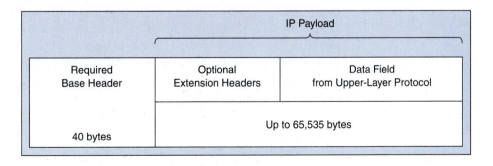

FIGURE 24-14 IPv6 datagram format

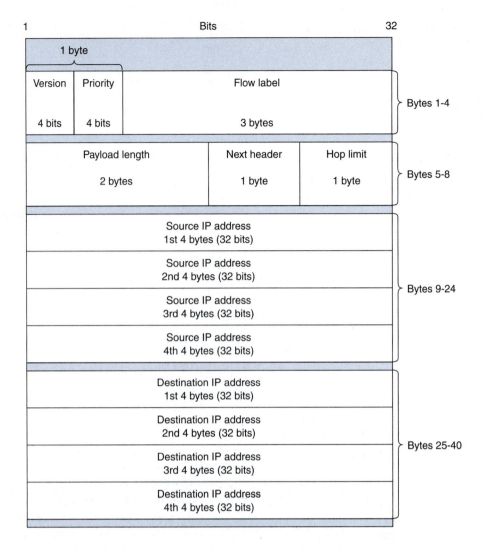

Top labels of figure:
1   Bits   32

| 1 byte |

| Version | Priority | Flow label | Bytes 1-4 |
| 4 bits | 4 bits | 3 bytes | |

| Payload length | Next header | Hop limit | Bytes 5-8 |
| 2 bytes | 1 byte | 1 byte | |

Source IP address
1st 4 bytes (32 bits)

Source IP address
2nd 4 bytes (32 bits)

Source IP address
3rd 4 bytes (32 bits)

Source IP address
4th 4 bytes (32 bits)

Bytes 9-24

Destination IP address
1st 4 bytes (32 bits)

Destination IP address
2nd 4 bytes (32 bits)

Destination IP address
3rd 4 bytes (32 bits)

Destination IP address
4th 4 bytes (32 bits)

Bytes 25-40

FIGURE 24-15 IPv6 base header format aligned to a 32-bit boundary

header is longer than the IPv4 header mainly because the length of the source and destination IP addresses has been extended from 32 bits (four bytes) with IPv4 to 128 bits (16 bytes) with IPv6. However, the IPv6 header has fewer fields (eight compared to 12 for IPv4). Therefore, the IPv6 header actually contains less information because some of the information previously contained in the IPv4 header has been moved to the options with IPv6. With the new version of IP, six of the header fields have been either eliminated or moved to the extension headers, three have been renamed or altered, and two new fields have been added. The header fields are listed in Table 24-3 and summarized here:

Version field. The version field is a four-bit field that defines the version of IP. For version IPv6, the value is 6.

Priority. The priority field is a four-bit field that defines the delivery priority of the datagram with respect to other traffic from the same source. The priority field is sometimes called the *traffic field,* the *class field,* or the *traffic class field.* The priority field is used when one of two consecutive datagrams must be discarded because of traffic congestion. When this happens, the datagram with the lowest priority is the one discarded.

Table 24-3 IP Fields Removed, Renamed, or Moved

Removed or moved fields
   Header length
   Type of service (and precedence)
   Identification
   Flags
   Fragment offset—moved to extension headers
   Header checksum
Renamed or altered fields
   Total length—replaced with payload length field
   Protocol—replaced with next header field
   Time to live—replaced with hop limit field
New fields
   Traffic class
   Flow label

Table 24-4 Priorities for Congestion-Controlled Traffic

Priority	Meaning
0	No specific traffic
1	Background data
2	Unattended data traffic
3	Reserved
4	Attended bulk data traffic
5	Reserved
6	Interactive traffic
7	Control traffic

IPv6 divides data traffic into two number ranges that specify a general category. Priority values from 0 through 7 specify the source is providing *congestion-controlled traffic,* such as TCP traffic that backs off in response to congestion. Priority values from 8 through 15 specify a priority of traffic called *noncongestion controlled,* which is traffic that does not back off in response to congestion. With noncongestion-controlled traffic, datagrams are sent at a constant rate in real time. *Congestion-controlled traffic* is data that adapts and adjusts itself to a slowdown in traffic. With congestion-controlled traffic, data packets may arrive at the destination delayed or out of order, or they could be lost and never arrive at the destination. A priority code between 0 and 7 is assigned to all congestion-control data, as listed in Table 24-4. The lowest priority is 0 (0000), and the highest priority is 7 (0111). *Noncongestion-controlled traffic* is traffic that expects minimum transmission delays, and discarding noncongestion-controlled traffic is not particularly desirable, as retransmitting the datagram is very often impossible. Examples of noncongestion-controlled traffic are real-time audio and real-time video. Noncongestion-controlled traffic is assigned a priority from 8 to 15, with 8 used for datagrams with the most redundancy, such as high-quality audio or video. A priority of 8 should be used for datagrams that the sender is most willing to have discarded under heavy congestion conditions. A priority of 15 should be used with datagrams having the least redundancy (i.e., datagrams that the sender is least willing to have discarded), such as low-quality audio and video.

Flow label. The flow label field is a three-byte field optional field that may be used by a source to label datagrams it wishes to request special handling by IPv6 routers. Flow instructions are designed to request a certain flow for the data transmitted from a particular source to a particular destination, such as nondefault or real-time quality of service. If the flow label field is not used, it is filled with 20 logic 0s.

A *flow* is a sequence of datagrams sent (unicast or multicast) from a specific source to a specific destination where the source is requesting special handling by the intermediate routers. The nature of the special handling instruction may be conveyed to the routers with a control protocol, such as resource reservations protocol, or by information contained within the flow packets themselves in a hop-by-hop option. More than one flow can be active at a time from a single source to a destination as well as traffic that is not associated with any flow. A flow of datagrams is uniquely identified by the combination of a source IP address and a nonzero flow label, and datagrams that are not associated with a flow have all 0s in the flow label.

A flow label is assigned to a flow by the source of the flow. New flow labels must be selected pseudorandomly and uniformly from a range of values between 000001hex and FFFFFF hex. The reason for the random allocation is to make any set of bits within the flow label field suitable to be used by routers when they look up the state associated with the flow. Flow labels can be used to reduce the processing delay in a router. When a router receives a datagram, rather than search through its routing table and then exercising a routing algorithm to define the address of the next hop, the router simply looks in a flow label table to determine the next hop. All datagrams from the same flow must be sent with the same source IP address, same destination IP address, and the same nonzero flow label.

Payload length. The payload length field is a two-byte field that defines the total length of the IPv6 datagram, excluding the base header. The maximum value for the payload length is 65,535. With IPv6, there is no need for a header length field, as the base header has a fixed length of 40 bytes and the extension headers are included in the payload length.

Next header. The next header field is an eight-bit field that defines the header that immediately follows the base header in the datagram, which is either one of the optional extension headers IP uses or a higher-layer protocol header, such as UDP or TCP. The next header field can also be used to specify an ICMP header and message follows. Table 24-5 lists the next header codes used with IPv6. Code 59 is called the *null header*, as it indicates that there are no additional headers.

Table 24-5 IP Version 6 Next Header Codes

Code	Next Header Indicated
0	Hop-by-hop option
1	ICMP (IPv4)
2	IGMP (IPv4)
3	Gateway-to-gateway protocol
4	IP in IP (IPv4 encapsulated in IPv6)
5	Stream
6	TCP
17	UDP
29	ISO TP4
43	Source routing
44	Fragmentation
45	Interdomain routing protocol (IDRP)
51	Authentication
52	Encrypted security payload
58	ICMPv6
59	Null (no additional headers)
60	Destination options
80	ISO CLNP
88	IGRP
89	OSPF
255	Reserved

Hop limit. The hop limit is an eight-bit field that serves the same purpose as the time to live (TTL) used with IPv4. The hop limit field holds an eight-bit unsigned integer that is decremented by 1 by each router that forwards the datagram.

Source address. The source address field is a 16-byte field that carries the 128-bit IP address of the device that originated the datagram.

Destination address. The destination address field is the only options header that can appear in more than one place. It is a 16-byte field that carries the 128-bit IP address of the device intended recipient of the datagram, which may not be the ultimate destination if an optional routing header is used.

24-5 IPv6 EXTENSIONS AND EXTENSION HEADERS

The 40-byte base header used with IPv6 includes all the necessary fields for delivering normal, unfragmented datagrams. Therefore, the designers of IPv6 decided that several of the fields contained in the IPv4 header were unnecessary for delivering many datagrams. Consequently, they elected to remove them from the base header but still provide their functions in optional *extension headers,* but only when it is necessary to do so. Extension headers can also be used to transport options. Only one of the IPv6 extension headers is examined or processed by routers along the datagram's delivery path. The rest of the extension headers are not processed until the datagram reaches its final destination. This facilitates a considerable improvement in router performance for datagrams containing options.

Although IPv4 allows intermediate routers to fragment datagrams, IPv4 does not. IPv6 datagrams must be fragmented by the originating source. However, IPv6, like IPv4, specifies that fragmented datagrams not be reassembled until they reach the final destination. The reason for using end-to-end fragmentation is to reduce router overhead, which permit routers to handle more datagrams in a given unit of time.

In addition to the IPv6 base header, there can be as many as six additional extension headers chained together one after the other following the base header. The extension headers must align on 64-bit boundaries and immediately follow the base header in the IPv6 datagram. Figure 24-16 shows the format for an IPv6 datagram when extension headers are used. As shown in the figure, there can be more than one extension header in an IPv6 datagram. Each extension header contains a header length field to identify when the header ends and a next header field that specifies if there is an additional header and, if so, which one.

There are currently six extension headers available with IPv6:

1. Hop-by-hop option header
2. Source routing
3. Fragmentation
4. Authentication
5. Encapsulating security payload
6. Destination option

An IPv6 base header can be followed by a transport-layer protocol, such as TCP or UDP, or by one or more extension headers. The extension headers are linked together after the IPv6 base header using the next header field in the base header and next header fields in the extensions themselves. IPv6 headers can be linked together over an Ethernet LAN, as shown in Figure 24-17.

IPv6 specifies the following recommended order for headers:

1. Data link header (such as Ethernet)
2. IPv6 base header
3. Hop-by-hop header

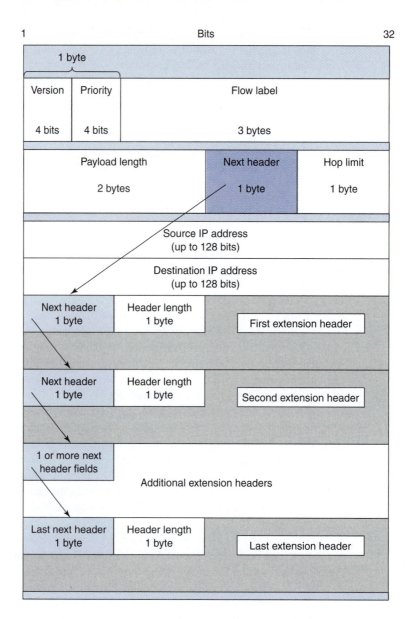

FIGURE 24-16 IPv6 datagram showing base header and extension headers

    4. Destination options header
    5. Fragment header
    6. Authentication header
    7. Encapsulated security payload (ESP) header
    8. Destination options header
    9. Transport-layer protocol header (TCP, UDP, or ICMP)
    10. Application-layer protocol header (FTP, HTTP, etc.)
    11. Application-layer payload
    12. Data-link trailer (such as Ethernet CRC field)

Note that the destination options header is shown in two locations, as it is the only options header that can appear more than time. The destination options header can appear in either or both of the positions shown. The reasons for this are explained later.

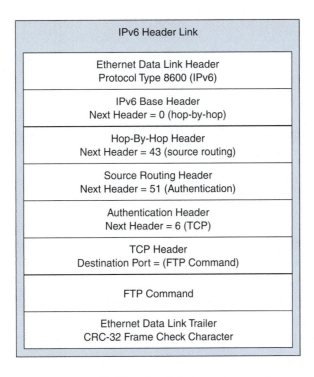

IPv6 Header Link
Ethernet Data Link Header Protocol Type 8600 (IPv6)
IPv6 Base Header Next Header = 0 (hop-by-hop)
Hop-By-Hop Header Next Header = 43 (source routing)
Source Routing Header Next Header = 51 (Authentication)
Authentication Header Next Header = 6 (TCP)
TCP Header Destination Port = (FTP Command)
FTP Command
Ethernet Data Link Trailer CRC-32 Frame Check Character

FIGURE 24-17 IPv6 headers linked together over an Ethernet LAN

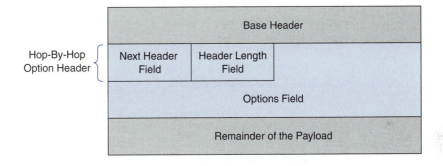

FIGURE 24-18 Hop-by-hop option header format

24-5-1 Hop-by-Hop Extension Header

The *hop-by-hop extension header* allows maximum flexibility in header definition and functionality, as it allows the source of the IP datagram to transfer information to all the routers visited by the datagram, such as management, debugging, and certain control functions. For example, if the original datagram exceeds 65,536 bytes, routers need to know this.

Figure 24-18 shows the format for the hop-by-hop option header, which contains only three fields. The first field is the next header field, which identifies whether there is an additional header and, if so, which one. The second field is a header length field that specifies the number of bytes in the header, including the next header and header length fields, as they are required for all extension and options headers. The remainder of the extension is the hop-by-hop information field, which contains option headers and their corresponding option information. Currently, there are only three options specified for IPv6: Pad1, PadN, and Jumbo Payload. Figure 24-19 shows the format for the options that may be included in

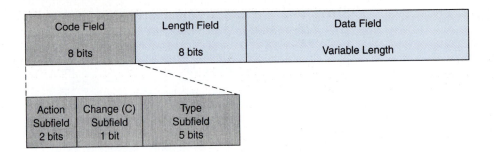

FIGURE 24-19 Hop-by-hop option header format

the hop-by-hop option header. The code field in the hop-by-hop option header contains three subfields: action, change (C), and type. The action subfield is two bits long and specifies the action to be taken if the option is not recognized. There are four actions:

00—skip over this option

01—discard the datagram, no additional action required

10—discard the datagram and send an error message

11—same as 10, except only if the destination address is not a multicast address

The change (C) subfield contains only one bit:

0—does not change in transmit

1—may be changed in transmit

The type subfield is five bits long and specifies the type of option contained in the variable-length data field. The type codes are the following:

00000—Pad1

00001—PadN

00010—Jumbo Payload

Pad1 option. The Pad1 is a one-byte option that can appear anywhere in the hop-by-hop option header. The purpose of the Pad1 option is to align other options that must begin at a specific place within a 32-bit boundary. If another option is off by exactly eight bits, the Pad1 option can be used for fill (i.e., a pad) to adjust the position of the other option. The Pad1 option has no length or optional data fields and is comprised of eight 0s. The code field for a Pad1 option is 0 (00000000), which indicates the following:

Action = 00 (skip)

Change = 0 (may be changed)

Type = 00000 (Pad1)

PadN option. The PadN shown in Figure 24-20 option is similar to the Pad1 option, except the PadN option is used when two or more bytes are needed to align another option on a 32-bit boundary. The PadN option is comprised of a one-byte option code field and a one-byte length field. The option length field specifies the number of padding bytes (00000000). The code field for a PadN option is 1 (00000001), which indicates the following:

Action = 00 (skip)

Change = 0 (may be changed)

Type = 00001 (PadN)

Code Field (00000001) 1 byte	Length Field 1 byte	Data Field (all 0s) Variable Length

FIGURE 24-20 PadN option format

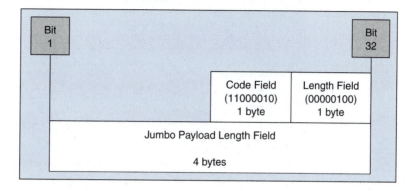

FIGURE 24-21 Jumbo payload option format

Jumbo payload option. The jumbo payload option is a six-byte option that defines the length of an IP datagram that contains more than 65,536 bytes (i.e., a jumbo datagram). The code field for a jumbo payload option is 194 (11000010), which indicates the following:

Action = 11 (discard the datagram and send an error message)

Change = 0 (may be changed)

Type = 00010 (jumbo payload)

The one-byte option length field specifies the length of the jumbo payload length field, which with IPv6 is four bytes (32 bits). The length of the jumbo payload is 2^n, where n is the number of bits in the jumbo payload length field. Therefore, the maximum length of a jumbo payload is $2^{32} - 1$, or 4,294,967,295.

The jumbo payload option must begin at a multiple of four bytes plus 2 from the beginning of the extension headers, as shown in Figure 24-21. Therefore, the jumbo payload option begins at the byte number $4x + 2$, where x is a small integer.

24-5-2 Source Routing Extension Header

The source routing option is essentially a combination of two IPv4 options: the strict source route option and the loose source route option. As shown in Figure 24-22, the source routing extension header is comprised of a minimum of seven fields. The next header and header length fields are the same as with the hop-by-hop option extension header. The type field indicates whether the option defines strict (1) or loose (0) routing. At the time of this writing, only loose routing has been defined. The addresses left field specifies the number of hops or segments remaining to reach the destination. The strict/loose mask field specifies the path over which the datagram should be routed. Strict routing must precisely follow the route indicated by the source. With loose routing, additional routers may be visited, provided that the specified routers are visited. The destination address field is updated in each router visited to indicate the address of the next router to visit (i.e., the immediate destination rather than the final destination).

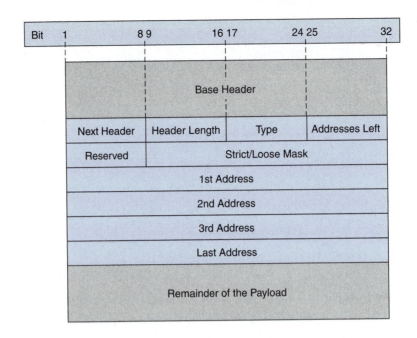

FIGURE 24-22 Source routing extension header format

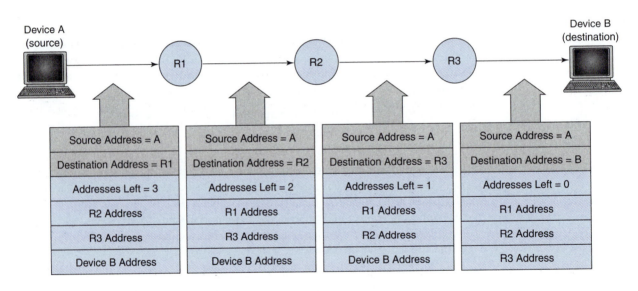

FIGURE 24-23 Example of source routing

Figure 24-23 shows an example of source routing. The initial source is device A, and the final destination is device B. The specified route is device A to router 1 to router 2 to router 3 to device B. Note that the destination address in the base header initially contains the address of the first router (R1). However, the destination address is constantly updated as the datagram propagates through a router with each router replacing its address with the address of the next router. Note also that the addresses in the extension header change from router to router. When the datagram arrives at the first router (R1), the router examines the packet to find its header. If all the information is correct, the router places the address of the next router from the list in the destination address field of the base header and replaces it with its own address. The process repeats in each successive router until the datagram reaches the final destination. The routing list can contain the addresses of up to 255 routers.

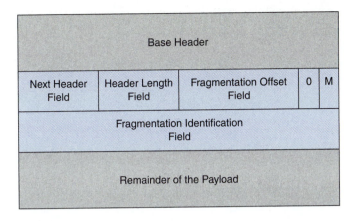

FIGURE 24-24 Fragmentation extension header format

24-5-3 Fragmentation Extension Header

With IPv6, the concept of fragmentation is identical to IPv4 in the sense that the length of a datagram cannot exceed the MTU (maximum transfer unit) of the network through which it is being transported. However, IPv6 mandates a new minimum MTU of 1280 bytes for Internet links. IPv6 does not support fragmentation by forwarding routers (all datagrams are treated as if the don't fragment bit is set). Therefore, with IPv6, the originating source is the only device that can fragment a datagram. The originating device can use a path MTU discovery procedure to determine the smallest MTU supported by the networks in the datagram's path. The datagram is then fragmented no larger than the length of the network with the smallest MTU. If an IPv6 datagram (or fragment) is longer than the MTU of the next hop, the datagram is discarded. If the source chooses not to use a path discovery procedure, it should fragment the datagram into 576-byte segments or smaller, as this is typically the minimum-length MTU for networks connected to the Internet.

Figure 24-24 shows the format for a fragmentation extension header. The next header and header length fields are identical to the other extension headers discussed. The fragmentation offset is used identical to IPv4 and specifies the fragment size in multiples of eight bytes. The M bit is the more bit, which is also used identical to IPv4, where M = 1 in all fragments except the last one. The don't fragment bit is unnecessary with IPv6, and the RS field is reserved for future use. The fragment identification field contains a 32-bit unique identification number that the destination device uses to regroup fragments from the same datagram (i.e., fragments with the same identification number).

24-5-4 Authentication Extension Header

The authentication extension header has two purposes: it validates the originator of a datagram, and it ensures the integrity of the actual data carried in the datagram by preventing address spoofing and connection theft. The recipient of a message must have some assurance that the implied sender is legitimate and not a phony. The recipient must also have some assurance that the data has not been altered in transit.

Figure 24-25a shows the format for an authentication extension header, which is comprised of six fields: an eight-bit next header, an eight-bit payload length, a 16-bit reserved field, a 32-bit security parameter index (SPI) field, a 32-bit sequence number field, and a variable-length authentication data field. The next header field points to the next header, and the payload length field specifies the number of four-byte words following the security parameter index field. The reserved field contains 16 logic 0s. The security parameter index identifies the algorithm used for authentication and also ensures that the destination recognizes old datagrams on the network. The authentication data field contains the actual data produced by the algorithm, which includes a cryptographic checksum of the payload data,

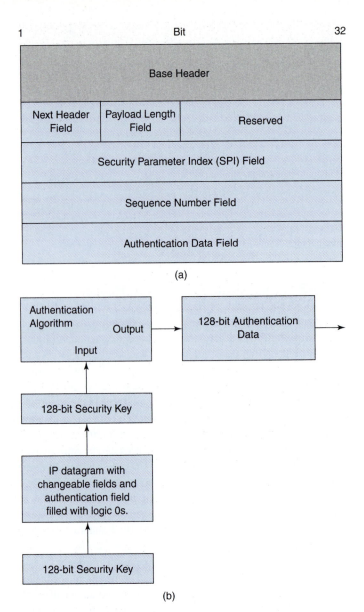

FIGURE 24-25   Authentication extension header: (a) format;
(b) authentication data calculation procedure

some of the extension header fields, and other privileged data shared by the authenticated devices.

Figure 24-25b shows how the contents of the authentication data field are determined. The originator of the datagram sends a 128-bit security key through the authentication algorithm followed by the IP datagram, and then the 128-bit security key is repeated. There are several authentication algorithms that can be used. Any fields in the datagram containing information that changes while being transported through the Internet (such as hop count) are filled with logic 0s. The datagram includes the authentication header extension with the authentication data field filled with all logic 0s. The authentication algorithm determines the authentication data, which is then inserted into the extension header in place of the logic 0s before the datagram is sent.

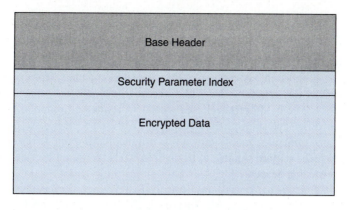

FIGURE 24-26   Encapsulated security payload extension header format

After the destination receives the secret key and the received datagram, it replaces any fields in the datagram that may have changed in transit with logic 0s and then passes the secret key and the datagram to the authentication algorithm. If the output of the authentication algorithm matches the contents of the authentication data field, the datagram is assumed to be authentic; otherwise, the datagram is discarded.

24-5-5   Encrypted Security Payload Extension Header

The encrypted security payload (ESP) extension header provides a means of providing confidentiality to a datagram by encrypting the data, thus preventing intruders from snooping. This header must always be the last header of the IP header chain, as it indicates the beginning of encrypted data.

Figure 24-26 shows the format for the encrypted security payload header, which is comprised of a 32-bit security parameter index. The security parameter index specifies the type of encryption and decryption used in the datagram. The variable-length encrypted data field contains the original data that is being encrypted plus any extra parameters required by the encrypting algorithm. There are two modes or methods of encrypting: transport mode and tunnel mode.

Transport mode. The transport mode is an encryption scheme where the transport protocol (either a TCP segment or a UDP datagram) is encrypted and then encapsulated in an IPv6 datagram. Transport mode encryption is used mostly to encrypt data transported between two hosts.

Tunnel mode. The tunnel mode is an encryption scheme where the entire IP datagram, including its base and extension headers, are encrypted and then encapsulated in a new IPv6 datagram using the ESP extension header. With this mode, there are actually two base headers transported: the original header and an encrypted header. Tunnel mode encryption is used mostly with gateways to encrypt data.

24-5-6   Destination Option Extension Header

The destination option is used by a source to pass information to the destination without giving intermediate routers access to the information. The format for the destination option is the same as the hop-by-hop option except the type numbers are different. The hop-by-hop and destination options separate the set of options that should be examined at each hop (router) from those that should be interpreted only by the final destination. When the destination option is placed early in the datagram, intermediate routers examine it, and when it appears after the ESP header, it can be examined only at the final destination.

ICMPv6 is used by IPv6 hosts and routers to report logical errors encountered while processing datagrams and to perform several other Internet-layer functions. ICMPv6 messages are grouped into two classes: error messages and query (informational) messages with distinctly different functions and different type ranges.

The ICMP protocol used with IPv4 is not well suited to IPv6. Therefore, it has been modified and appropriately renamed ICMPv6. The new version of ICMP complies with the same principles and strategy as the original ICMP protocol; however, it has been modified to make it more suitable to IPv6. ICMPv6 is identified by a next header value of 58 in the preceding header.

There are several protocols in IPv4 that operate independently and that are included in ICMPv6, such as ICMP, ARP, and IGMP. RARP has been dropped from IPv6, as BOOTP can do the same thing. ICMPv6 is used by IPv6 devices to report errors encountered in processing datagrams and to perform other Internet-layer functions, such as diagnostics and multicast membership reporting.

Figure 24-27 compares the network-layer protocols used with IPv4 to those used with IPv6, and Figure 24-28 shows the TCP/IP protocol suite for IPv6. Note that the Ethernet protocol type for IPv6 is 8600.

24-6-1   ICMPv6 Message Processing Rules

ICMPv6 implementations must observe several rules when processing ICMPv6 messages:

1. If a device receives an ICMPv6 error-reporting message of unknown type, the device must forward the message to the upper-layer protocol.
2. If a device receives an ICMPv6 informational message of unknown type, the device must silently discard it.
3. Every ICMPv6 error-reporting message (type 127 or lower) must include as much of the offending IPv6 datagram as possible without causing the error message to exceed the IPv6 MTU.

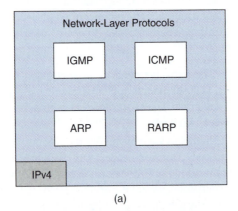

(a)

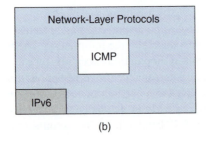

(b)

FIGURE 24-27   Network-layer protocols:
(a) IPv4; (b) IPv6

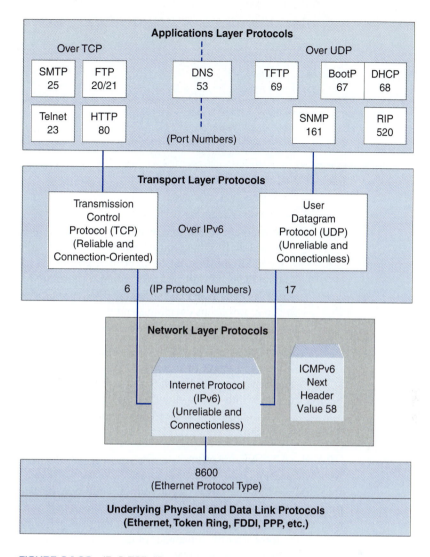

FIGURE 24-28 IPv6 TCP/IP protocol stack

4. In cases when the Internet-layer protocol is required to pass an ICMPv6 error message to the upper-layer process, the upper-layer protocol type is extracted from the original IP header (which is contained in the ICMP data field) and used to select the appropriate upper-layer process to handle the error. If the original datagram had an unusually large number of extension headers, it is possible that the upper-layer protocol type may not be present in the ICMPv6 message because the original datagram is truncated to meet the IPv6 MTU limit. In these cases, the error message is silently discarded after any IPv6-layer processing.

5. An ICMPv6 error-reporting message must never be sent as a result of receiving one of the following:

   a. An ICMPv6 error-reporting message

   b. A datagram destined to an IPv6 multicast address unless the error message is a datagram too big or a parameter problem

   c. A datagram sent as a lower-link multicast

   d. A datagram sent as a link-layer broadcast

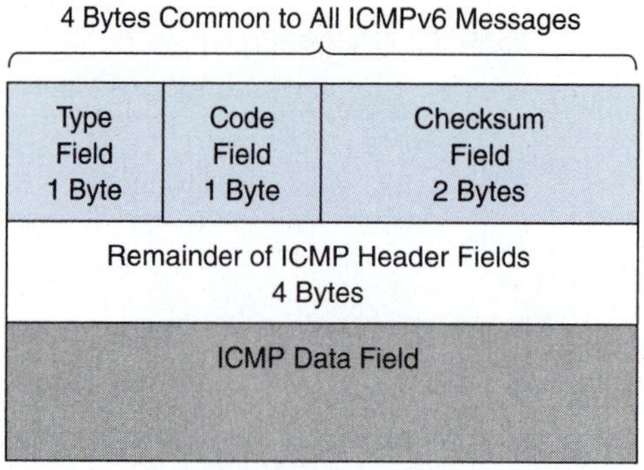

Type Field 1 Byte	Code Field 1 Byte	Checksum Field 2 Bytes
Remainder of ICMP Header Fields 4 Bytes		
ICMP Data Field		

FIGURE 24-29 ICMPv6 message format

 e. A datagram whose source address does not uniquely identify a single device, such as an IPv6 unspecified address, multicast address, or an address known by the ICMP message sender to be an IPv6 anycast address

6. To limit the bandwidth and forwarding costs, an IPv6 device must limit the rate at which it sends ICMPv6 messages. This is appropriate when a source sends a stream of erroneous datagrams and does not observe the resulting ICMPv6 error messages. There are several ways to implement the rate-limiting function:

 a. Timer-based—limit the rate at which error messages are sent to a given source or to any source to a maximum of once every T milliseconds, where T is an arbitrarily selected length with a common value of 1 second.

 b. Bandwidth-based—limit the rate at which error messages are sent from a specific interface to some fraction F of the attached link's bandwidth, where F is typically 2%.

24-6-2 ICMPv6 Header

An IPv6 header and possibly one or more IPv6 extension headers precede all ICMPv6 messages. All CMPv6 messages begin with an eight-byte header where the first four bytes (three fields) are the same in all message types, as shown in Figure 24-29. The first field is an eight-bit type field that defines what kind of ICMP message is being delivered and outlines the format for the remainder of the ICMP message. Having a logic 0 in the high-order bit uniquely identifies error-reporting messages. Therefore, error-reporting messages have type values ranging from 0 to 127. All query messages have a logic 1 in the high-order bit. Therefore, the type values for query messages range from 128 to 255. Table 24-6 lists the message types recommended for ICMPv6. Like ICMPv4, ICMPv6 divides messages into two broad categories: query (informational) messages and error-reporting messages.

 The second field of the ICMPv6 header is an eight-bit code field that defines the specific purpose of the message type. The third (and last) field common to all ICMPv6 message headers is the 16-bit checksum.

24-6-3 ICMPv6 Error-Reporting Messages

When an IPv6 datagram encounters a problem, the device that detects the problem formulates an ICMPv6 error-reporting messages. The ICMP message is encapsulated in an IPv6 datagram and sent back to the device that originated the datagram that encountered the problem. Currently, only five error-reporting messages are defined with IPv6 (destination unreachable, time exceeded, parameter problem, redirection, and datagram too big), of which four were provided with ICMPv4. The source quench message has been eliminated because the flow label field of the IPv6 header allows routers to control congestion and discard the lowest-priority messages. The packet too bit message has been added to ICMPv6 because fragmentation is the responsibility of the originator of an IPv6 datagram. If the originating

Table 24-6 ICMPv6 Types

Error-Reporting Messages		
Type	Code	Name
1	0-4	Destination unreachable
2	0	Packet too big
3	0-1	Time exceeded
4	0-2	Parameter problem
Query Messages		
Type	Code	Name
128	0	Echo request
129	0	Echo reply
130	0	Multicast listener query
131	0	Multicast listener report
132	0	Multicast listener done
133	0	Router advertisement
135	0	Neighbor solicitation
136	0	Neighbor advertisement
137	0	Redirect message
138	0, 1, and 255	Router renumbering
139	0	ICMP node information query
140	0	ICMP node information response
141	0	Inverse neighbor discovery solicitation message
142	0	Inverse neighbor discovery advertisement message

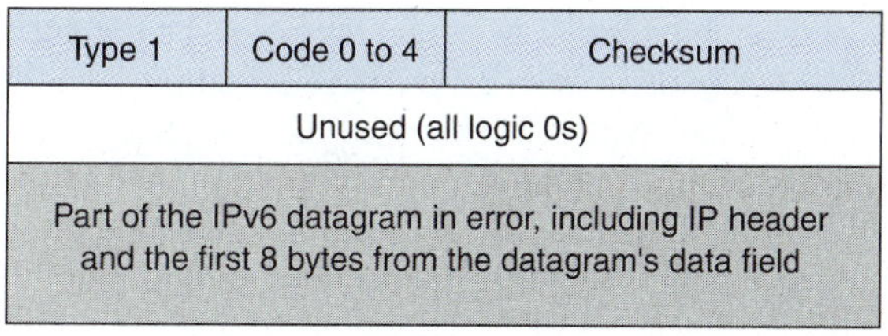

Type 1	Code 0 to 4	Checksum
Unused (all logic 0s)		
Part of the IPv6 datagram in error, including IP header and the first 8 bytes from the datagram's data field		

FIGURE 24-30 Destination-unreachable message format

device does not choose the correct datagram size, the routers have no choice other than to discard datagrams that exceed the maximum length and send an error message back to the sender.

24-6-3-1 Destination unreachable.   With ICMPv6, the destination-unreachable message has essentially the same meaning as it has with ICMPv4, except there are fewer codes (i.e., reasons why the datagram was unable to reach its destination). A destination-unreachable message should be generated in response to a datagram that cannot be delivered to its destination for any reasons except network congestion. The header is followed by part of the failed IP datagram, including the IP header and the first eight bytes of the datagram's data field.

Figure 24-30 shows the format for an ICMPv6 destination-unreachable message. As shown in the figure, the message begins with an eight-byte header comprised of four fields: type, code, checksum, and four bytes that are unused and filled with all logic 0s. The type for a destination-unreachable message is 1, and the code is one of five values: 0, 1, 2, 3, or 4. The code field specifies the reason why the datagram failed to reach its destination and was discarded. The five codes are the following:

Code 0. No route to destination (i.e., no matching entry in the forwarding nodes routing table).

Code 1. Communications with destination is administratively prohibited because of a firewall filter.

Code 2. The next destination address in the routing header is not a neighbor of the processing router, but the strict bit is set for that address (i.e., strict source routing is impossible).

Code 3. Destination address is unreachable for reasons not specified in codes 0 to 2, such as the inability to resolve the IPv6 destination address into a corresponding link address or some sort of link-specific problem.

Code 4. Destination port is unreachable, and the transport protocol has no alternative way to inform the sender.

24-6-3-2 Time exceeded. The time-exceeded message is virtually the same as the one used with ICMPv4. The only difference is that the type number in ICMPv6 is 3. Figure 24-31 shows the format for the time-exceeded message, which is comprised of the three essential fields, 32 bits of all 0s, and the standard ICMP data field.

Code 0 indicates that the datagram was discarded because the hop-limit field value became 0 before the datagram reached its final destination. If a router that receives a datagram with a hop limit of 0 or a router decrements a datagram's hop limit to 0, the datagram must be discarded and a time-exceeded message sent back to the source of the datagram. A code of 1 is used when a datagram is discarded because all the fragments did not arrive within a prescribed time limit.

24-6-3-3 Parameter problem. The ICMPv6 parameter-problem message is similar to its version counterpart. The format for a type 4 parameter-problem message is shown in Figure 24-32. As can be seen, the message is comprised of the three essential fields plus a 32-bit offset pointer. With ICMPv6, the type was changed to 4, and the size of the offset pointer field has been increased to four bytes.

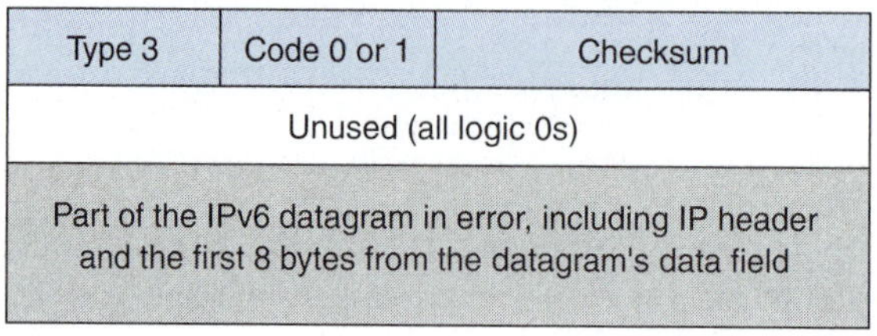

Type 3	Code 0 or 1	Checksum
Unused (all logic 0s)		
Part of the IPv6 datagram in error, including IP header and the first 8 bytes from the datagram's data field		

FIGURE 24-31 Time-exceeded message format

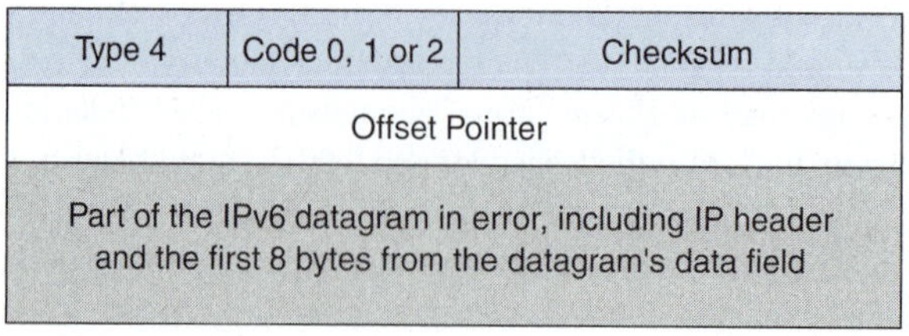

Type 4	Code 0, 1 or 2	Checksum
Offset Pointer		
Part of the IPv6 datagram in error, including IP header and the first 8 bytes from the datagram's data field		

FIGURE 24-32 Parameter-problem message format

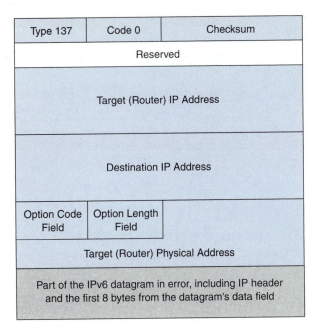

Type 137	Code 0	Checksum

Reserved

Target (Router) IP Address

Destination IP Address

Option Code Field	Option Length Field

Target (Router) Physical Address

Part of the IPv6 datagram in error, including IP header and the first 8 bytes from the datagram's data field

FIGURE 24-33   Redirection message format

When a datagram discovers a problem with a field in the IPv6 header or a field in an extension header that it cannot process, it discards the datagram and sends a parameter-problem message back to the datagram's source, indicating the type and location of the problem.

There are three code values with the ICMPv6 parameter problem:

Code 0. Indicates either that there is an error or that there is something ambiguous in one of the header fields. When a code 0 is used, the value in the offset pointer field points to the byte with the problem. For example, if the offset point is 2, the second byte is not a valid field.

Code 1. Indicates that the contents of a next header field is unrecognizable.

Code 2. Indicates the presence of an unrecognizable option.

24-6-3-4 Redirection.   The format for the type 137 redirection message is show in Figure 24-33. As can be seen, the format is essentially the same as ICMPv4 except it has been modified to accommodate longer IP addresses, and an additional option has also been added that contains the physical address of the target router.

24-6-3-5 Packet too big.   The packet-too-big message is a new message not included in ICMPv4. When a router receives a datagram that is longer than the MTU of the destination network, the router discards the datagram and then sends an ICMP error message back to the originating source. Figure 24-34 shows the format for the packet too big message, which is comprised of the three essential fields plus a 32-bit MTU field that contains the maximum-length packet the network will accept, which is information that is used as part of the path MTU discovery process.

A packet-too-big message is the only error message sent in response to a packet with an IPv6 multicast destination address, a link-layer multicast address, or a link-layer broadcast address.

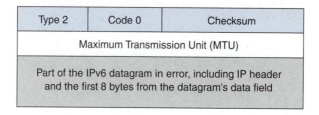

Type 2	Code 0	Checksum
Maximum Transmission Unit (MTU)		
Part of the IPv6 datagram in error, including IP header and the first 8 bytes from the datagram's data field		

FIGURE 24-34 Packet-too-big message format

Type Request 128 Reply 129	Code 0	Checksum
Identifier		Sequence Number
Optional Variable-Length Data Field (sent in request and returned in reply)		

FIGURE 24-35 Echo request and echo reply message format

24-6-4 ICMPv6 Query (Informational) Messages

As with ICMPv4, ICMPv6 messages can also be used to diagnose network problems using query messages. At this writing, four groups comprised of nine ICMPv6 query messages have been defined:

> Echo request and echo reply
> Router solicitation and router advertisement
> Neighbor solicitation and neighbor advertisement
> Group membership

Two pairs of ICMPv4 messages have been eliminated with ICMPv6: timestamp request and timestamp reply and address mask request and address mask reply. The timestamp messages were removed because they are implemented in other protocols, such as TCP. The address mask request and address reply messages were eliminated because the subnet field in an IPv6 address allows the subscriber to use up to $2^{32} - 1$ subnets, making the messages unnecessary.

24-6-4-1 Echo request and echo reply. Echo request and echo reply messages are used for diagnostic purposes, such as verifying continuity. An echo request is sent, and an echo reply is returned. Figure 24-35 shows the format for the ICMPv6 echo request and echo reply messages. As can be seen, the format is the same as it was for ICMPv4 except with new type numbers. The first three fields are standard for all ICMPv6 messages. The fourth and fifth fields are the identification and sequence numbers, which have the same purpose as ICMPv4.

24-6-4-1 Router solicitation and router advertisement. The router solicitation and advertisement messages perform essentially the same functions as they did with ICMPv4. The formats are shown in Figure 24-36a and b. As can be seen in Figure 24-36a, the format is the same for the solicitation except for the addition of an option (option 1)

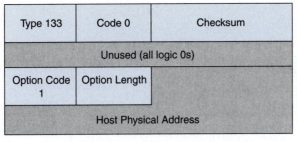

(a)

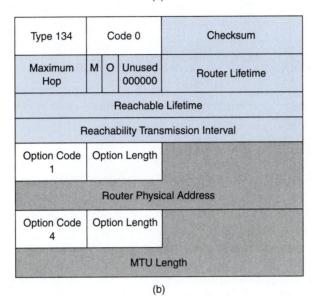

(b)

FIGURE 24-36   Router solicitation and router advertisement message formats: (a) solicitation, (b) advertisement

that allows the host to announce its physical address. However, the router advertisement is different with ICMPv6, as a router can advertise only itself and not any additional routers.

Figure 24-36b shows the placement of options in the router advertisement message, which includes three options (two defined and one proposed). Option 1 announces the router physical address, option 5 allows the router to announce the MTU size, and a third option that has been proposed allows the router to define the valid and preferred lifetime.

24-6-4-1 Neighbor solicitation and neighbor advertisement.   The address resolution protocol (ARP) has been eliminated with IPv6 and replaced with the ICMPv6 message pair: neighbor solicitation and neighbor advertisement messages. The basic idea is the same except the format for the message has changed. Figure 24-37 shows the formats for the neighbor solicitation and neighbor advertisement messages. The two formats are identical except for their codes. Both formats have a field that contains the IPv6 address to be resolved. There is only one option, and that is used by the destination to announce its physical address.

24-6-4-2 Group membership.   The Internet group management protocol (IGMP) has been eliminated with IPv6 and replaced with the ICMPv6 group membership message. With ICMPv6, there are three types of group membership messages—report, query,

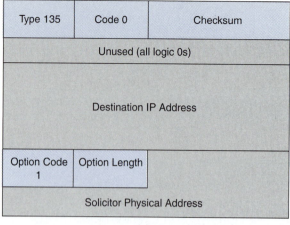

(a)

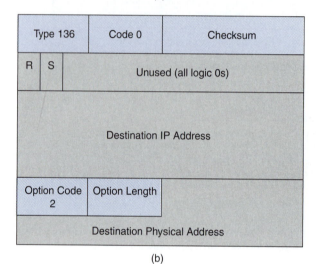

(b)

FIGURE 24-37 Neighbor solicitation and advertisement message formats: (a) solicitation; (b) advertisement

and termination—as shown in Figure 24-38. The report and termination messages are transmitted from a host to a router, and the query message is sent from a router to a host.

24-6-5   ICMP Checksum

The checksum for ICMPv6 is calculated in essentially the same manner as it is for ICMPv4. The checksum is the 16-bit 1s complement of the 1s complement sum of the entire ICMPv6 message beginning with the ICMPv6 message field prepended with a pseudoheader comprised of 24 bits of fill (all 0s) and the following IPv6 header fields: source IPv6 address, destination IPv6 address, upper-layer packet length, and next header. Consequently, to determine the checksum, a device (router or host) that sends an ICMPv6 message must determine the source and destination IP addresses from the IPv6 header.

If the device has more than one unicast address, it chooses the source address in one of four ways:

> **1.** If the ICMP message is a response to a message sent to one of the device's unicast addresses, the source address of the reply must be the same address.

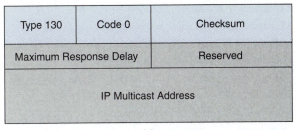

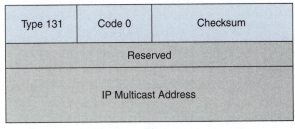

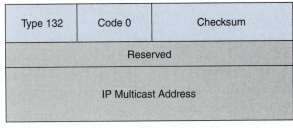

FIGURE 24-38 Group membership message formats:
(a) query; (b) report; (c) termination

2. If the ICMP message is a response to a message sent to a multicast or anycast group in which the device is a member, the source address of the reply must be the unicast address that belongs to the interface on which the multicast or anycast datagram was received.

3. If the ICMP message is a response to a message sent to an address that does not belong to the device, the source address should be the unicast address that belongs to the device that will be most helpful in diagnosing the cause of the error. For example, if the message is a response to a datagram forwarding action that cannot be successfully completed, the source address should be a unicast address belonging to the interface on which the datagram forwarding failed.

4. When the first three methods are inappropriate, the device's routing table must be examined to determine which interface should be used to transmit the message to its destination, and a unicast address belonging to that interface must be used as the source address of the message.

QUESTIONS

24-1. List the shortcomings of *IPv4*.

24-2. List the advantages of *IPv6* over IPv4.

24-3. Explain the *IPv6 address format*.

24-4. Explain *hexadecimal colon notation*.

24-5. Describe the following IPv6 address types: *unicast, anycast*, and *multicast*.

24-6. Describe the *IPv6 address allocation*.

24-7. List and describe the six fields contained in a *provider-based unicast address*.

24-8. List and describe the three *IPv6 special addresses*.

24-9. Explain *IPv6 local addresses*.

24-10. Describe the differences between *link-local* and *site-local addresses*.

24-11. Describe the purpose of *aggregatable global unicast addresses*.

24-12. List and describe the function of the fields contained in an aggregatable global unicast address.

24-13. Describe how anycast addresses are used with IPv6.

24-14. Describe how multicast addresses are used with IPv6.

24-15. List and describe the fields contained in an *IPv6 base header*.

24-16. Explain the differences between *congestion-controlled* and *noncongestion-controlled traffic*.

24-17. Explain the purpose of *IPv6 extensions* and *extension headers*.

24-18. List the recommended order for IPv6 headers.

24-19. Describe the purpose of the *hop-by-hop extension*.

24-20. Describe the purpose of the *source routing extension*.

24-21. Describe the purpose of the *fragmentation extension*.

24-22. Describe the purpose of the *authentication extension*.

24-23. Describe the purpose of the *encrypted security payload extension header*.

24-24. Explain the differences between the *transport* and *tunnel modes*.

24-25. Describe the purpose of the *destination option extension*.

24-26. List the IPv4 protocols that have been incorporated in IPv6.

24-27. Describe the *IPv6 message processing rules*.

24-28. List and describe the required fields in an ICMPv6 header.

24-29. Describe the purpose of *ICMPv6 error-reporting messages*.

24-30. Describe the *ICMPv6 destination-unreachable message*.

24-31. Describe the *ICMPv6 time-exceeded message*.

24-32. Describe the *ICMPv6 parameter-problem message*.

24-33. Describe the *ICMPv6 redirection message*.

24-34. Describe the *ICMPv6 packet-too-big message*.

24-35. Describe the purpose of *ICMPv6 informational messages*.

24-36. Compare the ICMPv6 informational messages with the ICMPv4 query messages.

24-37. Describe how the checksum is calculated with ICMPv6.

PROBLEMS

24-1. Identify the ICMPv6 type for the following prefixes:

    a. 0000 011
    b. 1111 0
    c. 1111 1110 11
    d. 0000 0001
    e. 011
    f. 1111 110
    g. 1111 1111
    h. 1111 111011

24-2. Determine the abbreviated form for the following IPv6 addresses:

    a. 1234:2CBA:228A:B000:0000:0000:0000:0000

    b. 0000:00BB:0000:0000:0000:0000:228A:B123

    c. 4320:0000:0000:0000:0000:228B:B001:0000

    d. 0000:0000:0000:5470:0000:0000:0000:0000

24-3. Expand the following abbreviated addresses:

    a. 0:0

    b. 0:B:B:0

    c. 0:5678::2

    d. 456::1:4

24-4. Determine the type for each of the following IPv6 addresses:

    a. FE90

    b. FEB1::34A2

    c. FF02::0

    d. 0::01

24-5. Determine the type for each of the following IPv6 addresses:

    a. 0::0

    b. 0::EEEE:0:0

    c. 581E:2345:3333

    d. 4820::15:33

24-6. Determine the hex contents of the hop limit field for the following limits:

    a. 14

    b. 42

    c. 31

    d. 250

24-7. Determine the contents of the priority field for the following classifications:

    a. 0

    b. 8 to 15

    c. 6

    d. 16

24-8. For the previous question, which datagram has the highest priority?

    a. 0

    b. 8 to 15

    c. 6

    d. 16

24-9. What extension header must be used to route a 5000-byte packet over an Ethernet LAN?

24-10. What is the maximum length of an IPv6 datagram?

24-11. What is the maximum length for a hop-by-hop extension header?

24-12. What are the only two possibilities for the first two bytes of a multicast address?

C H A P T E R   25

Configuration and Domain Name Protocols

CHAPTER OUTLINE

OBJECTIVES

- Define and describe the functions of configuration and domain name protocols
- Explain and describe the purposes of address acquisition and domain name resolution protocols
- Contrast the advantages and disadvantages of the BOOTP DHCP protocols
- List and explain the fields that constitute a BOOTP packet
- Describe the basic operation of BOOTP
- List and explain the fields that constitute a DHCP packet
- Describe the basic operation of DHCP
- Explain the structure and purpose of DHCP databases
- Describe an address lease
- List and explain the purpose and operation of DHCP options
- Describe DHCP states and procedures
- Explain the DHCP address renewal process
- Describe the DHCP address release process
- List and explain the fields that constitute a DNS packet
- Describe the basic operation of DNS
- Explain the concepts of name resolution and inverse name resolution
- Describe the structure and different types of domain names

- Define fully and partially qualified domain names
- Describe DNS question and answer sessions
- Describe DNS encapsulation

25-1 INTRODUCTION

Configuration protocols are process-layer protocols that assign IP addresses to hosts on local networks, and an *autoconfiguration protocol* is a configuration protocol that performs the assignments automatically. *Domain name protocols* are process-layer protocols that resolve IP addresses to domain names.

Although autoconfiguration and domain name protocols are application-layer protocols, they are not used to transport actual user files or information. They are used to exchange address and name information between servers and clients that clients need before they can use TCP/IP protocols to transport files and other information. Autoconfiguration protocols include the *bootstrap protocol* (BOOTP) and the *dynamic host configuration protocol* (DHCP), and the most common domain name resolution protocol is *domain name service* (DNS).

Figure 25-1 shows where these protocols are located in the TCP/IP protocol suite. As shown in the figure, BOOTP and DHCP are transported exclusively over UDP, which makes them unreliable and connectionless. Reliability is not a major concern with autoconfiguration protocols because the information is exchanged between a client host and a local server over a local network. DNS can be transported over UDP or TCP.

25-2 ADDRESS ACQUISITION AND DOMAIN NAME RESOLUTION

For a computer to communicate over a local area, it must know its own hardware address and the hardware address of the destination host. Hardware addresses are embedded on network interface cards (NICs) so that every device with a NIC has a hardware address. However, when a computer is connected to a TCP/IP network, it cannot send or receive IP datagrams unless it knows its own IP address and the IP address of the destination. IP addresses are not built into hardware; they are assigned to a device, usually on a temporary basis. IP addresses cannot be randomly selected or randomly assigned; they must be addresses that are part of a block of addresses belonging to a specific network or subnetwork. In addition, a computer connected to a TCP/IP network must know additional information (called *bootstrap information*), which includes the following:

1. Each device must know the subnet mask for the network or subnetwork where it resides.
2. Each device must know the IP address of at least one router that is connected to the network.
3. Each device must know the IP address of a device connected to the network that is capable of assigning IP addresses.
4. Each device must know the IP address of a device connected to the network that is capable of resolving names to IP addresses.

A device's IP address and items 1 through 3 in the previous list are generally found in a *configuration file* stored on a computer's hard drive, which the computer can access when it boots up. However, a diskless computer or a computer that is being booted up for the very first time has not been configured and, therefore, does not have a configuration file. In such cases, the computer must acquire the information from another computer using an autoconfiguration protocol. Autoconfiguration protocols are based on a client/server model using a request and response relationship where the client requests an IP address from a server and the server responds by sending the address. The client is the host that needs the information, and the server is responsible for providing it. The server assigns IP addresses

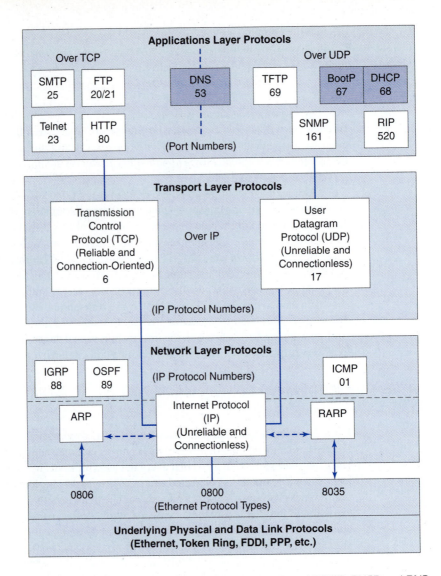

Applications Layer Protocols

Over TCP

| SMTP 25 | FTP 20/21 |
| Telnet 23 | HTTP 80 |

DNS 53

(Port Numbers)

Over UDP

| TFTP 69 | BootP 67 | DHCP 68 |
| SNMP 161 | RIP 520 |

Transport Layer Protocols

Transmission Control Protocol (TCP) (Reliable and Connection-Oriented) 6

Over IP

User Datagram Protocol (UDP) (Unreliable and Connectionless) 17

(IP Protocol Numbers)

Network Layer Protocols

| IGRP 88 | OSPF 89 |

(IP Protocol Numbers)

ICMP 01

ARP

Internet Protocol (IP) (Unreliable and Connectionless)

RARP

0806   0800   8035
(Ethernet Protocol Types)

Underlying Physical and Data Link Protocols
(Ethernet, Token Ring, FDDI, PPP, etc.)

FIGURE 25-1   TCP/IP protocol stack showing location of BOOTP, DHCP, and DNS

and provides each host with its subnet mask, the IP address of the default router, and the IP address of a server that is capable of assigning IP addresses. A host can acquire the information from the server statically (manually) or dynamically (automatically).

The two most common autoconfiguration protocols are BOOTP and DHCP, which communicate over UDP, which relies on IP for the delivery system. It may seem unusual that a host can use UDP to find its IP address when UDP itself relies on IP to transfer datagrams. How this is accomplished will soon become evident.

Computers are good at looking at IP addresses to determine where each datagram should be sent. However, humans are not as adept at remembering numbers or dealing with numeric addresses, whether they are in binary, hex, or dotted-decimal notation. When humans begin an application, such as e-mail or Web browsing, they typically supply a symbolic host name for the destination machine, such as *yahoo.com* or *msn.com*. Humans also prefer dealing with names for computers that indicate the person who owns or is using it, such as *johndoe*. For a name to be used in place of an IP address, there must be some means of cross-referencing names to IP addresses and vice versa. When the Internet was smaller and more manageable, a list of names and addresses were stored in a *host file* on the hard drive of a computer. The host file resolved names to IP addresses and could be updated periodically from a master host file on a server. Host files allowed computers to map computer names and Internet Web sites to IP addresses. However, in the modern Internet, it is virtually impossible for every computer to store every conceivable pair of names and addresses, as the file would have to be extremely large.

Assigning one centralized computer the job of storing and maintaining a comprehensive host file that virtually every computer in the Internet could use to map names to addresses would create a permanent traffic jam for that computer. A more realistic solution is to divide the host file and the task of resolving names to IP addresses among several strategically located computers called *name servers*. When a host needs to resolve a name to an IP address, it can simply contact the closest name serve, which invokes a DNS (*domain name system*) request to the local server (called a *domain name server* [DNS]). If the local DNS server does not have the requested information, it contacts other DNS servers that eventually locate the IP address for the host. DNS makes it considerably easier for a user to access applications without having to remember their IP addresses. DNS servers take advantage of the context-based memory clues that names offer to translate names into IP addresses. Network managers can also use a DNS server to control traffic to and from various servers connected to the local network.

25-3   BOOTP

BOOTP (bootstrap protocol) is an IP/UDP client/server applications protocol where the device sending a BOOTP request is called the *client* and the device sending the reply is called the *server*. BOOTP is designed to provide a diskless computer a unicast IP address, the IP address of the server, the gateway's IP address, and the path name of the boot file. The bootstrap process consists of two phases. The first phase determines the IP addresses and boot file selection. After the first phase is completed, control is passed to the second phase, where an actual file transfer occurs. BOOTP does not provide the client with a memory image; however, it provides the information necessary for the client to obtain the image. The actual transfer of the image file is typically accomplished using the *trivial file transfer protocol* (TFTP).

Figure 25-2 shows an Ethernet frame carrying a BOOTP message packet encapsulated in a UDP datagram that is in turn encapsulated in an IP datagram that is sufficiently short that fragmenting is not necessary. In the IP header in the request packet, the client supplies its own IP source address (if known); otherwise, the field is filled with all 0s. If the client does not know the server's IP address, it places the broadcast address (255.255.255.255) in the destination IP address field. The broadcast address simply means that the datagram is broadcasted onto the local cable facility, as the client does not know its own network address.

The UDP header in the client's request packet contains the source port number (BOOTP client 68) and the destination port number (BOOTP server 67). The two port numbers simply reverse their positions in the reply packet (i.e., destination 68 and source 67).

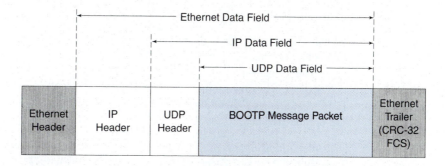

FIGURE 25-2   BOOTP encapsulated in an IP/UDP datagram being transported over an Ethernet

Operation Code (1Byte) 1 or 2	Hardware Type (1Byte) Ethernet = 1	Hardware Length (1Byte) Ethernet = 6	Hop Count (1 Byte) (00 to FF Hex)
Transaction Identification (4-Byte Integer)			
Number of Seconds (2 Bytes)		Unused (2 Bytes all 0s)	
Client IP Address (4 Bytes)			
Your IP Address (4 Bytes)			
Server IP Address 4 Bytes			
Gateway IP Address 4 Bytes			
Client Hardware Address 16 Bytes			
Server Name 64 Bytes			
Boot File Name 128 Bytes			
Options (64 Bytes)			

FIGURE 25-3 BOOTP packet format

Well-known port numbers are used for both port numbers to avoid confusing other hosts who happen to be listening on that port. The UDP length field contains the length of the UDP header plus the BOOTP packet. The UDP checksum is generally not used and filled with all 0s to reduce processing time.

The format for a BOOTP packet is shown in Figure 25-3. The formats for the request and response packets are the same. The BOOTP packet is comprised of 15 fixed-length fields containing a total of 300 bytes. Fixed-length fields are used for simplicity. The fields are summarized here:

Operation code field (op code). An eight-bit (one-byte) field that defines the type of BOOTP packet. There are two types: BOOTREQUEST (code 1) and BOOTREPLY (code 2). Clients send requests and the BOOTP name server sends replies.

Hardware type field. An eight-bit (one-byte) network interface protocol code field that defines the type of physical network the client is attached to. Each kind of local area network (LAN) has a unique hardware type. For example, for 10-Mbps Ethernet, the hardware is type 1, and for 100-Mbps Ethernet, the hardware is type 6.

Hardware length field. An eight-bit (one-byte) field that defines the byte length of the physical address. For example, standard NICs used on LANs, such as Ethernet, use a six-byte hardware address. The length of the client hardware address field is fixed at 16 bytes. Therefore, this field is necessary to identify how many of those bytes are actually used for the hardware address and how many are redundant.

Hop count field. An eight-bit (one-byte) field that specifies the number of routers between the client and the server. The hop count defines the maximum number of hops the BOOTP message packet can propagate through before it is discarded. A client fills this field with 0s; however, a gateway can use this field for cross-gateway booting (i.e., booting from a server on another network or subnetwork).

Transaction identification field. A four-byte field that contains a randomly generated integer number established by the client. Diskless machines use the transaction ID to match a BOOTP reply with the proper BOOTP request when multiple clients are requesting IP addresses at the same time. The reply packet sent from the BOOTP name server contains the same transaction ID that was in the client's request packet.

Number of seconds field. A two-byte field filled by the client that contains the number of seconds that have elapsed since the client started to boot or started the renewal process for a previously assigned IP address.

Client IP address field. A four-byte field that contains the client's current IP address (i.e., the IP address previously assigned to the client). Very often when a client initiates a boot, it does not have an IP address, so the client fills this field with all 0s. This field has no meaning when the server sends the reply packet.

Your IP address field. A four-byte field that the server fills with the client's current IP address in the reply packet. If the client does not know its IP address, the field is filled with all 0s in the request packet.

Server IP address field. A four-byte field that contains the server's IP address. This field is filled will all 0s in the request packet and replaced with the server's IP address in the reply packet.

Gateway IP address field. A four-byte field that contains the gateway router's IP address. This field is filled with all 0s in the request packet and replaced with the gateway router's IP address in the reply packet. This field is only used in cross-gateway booting.

Client hardware address field. A 16-byte field that contains the physical address of the client left justified. For a six-byte hardware address, the leftmost six bytes contain the address, and the 10 rightmost bytes are filled with 0s. This is redundant information, as the server can retrieve this information from the request packet; however, sending the hardware address in the packet simply speeds up the processing operation.

Server name field. An optional 64-byte field that contains the server's domain name. This field is filled with all 0s in the request packet and replaced with a null-terminating sequence and the server's domain name in the reply packet.

Boot filename field. An optional 128-byte field that points to where to find the boot file (i.e., additional bootstrap data). This field is filled with 128 bytes of logic 0s in the request packet and replaced with a null-terminating sequence and the fully qualified directory path name of the boot file in the reply packet. The client can then use the boot file path name to retrieve additional information.

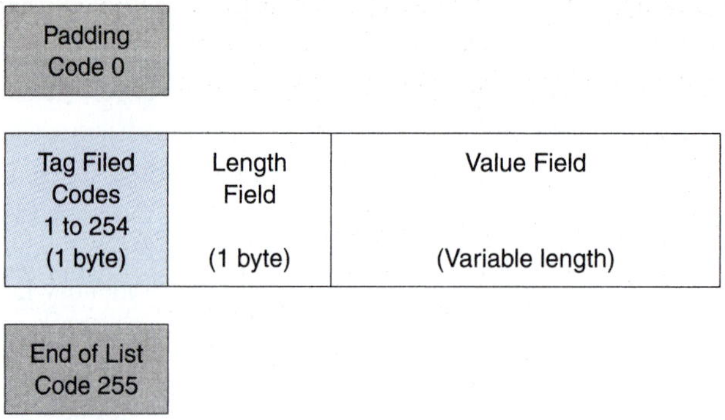

Padding
Code 0

Tag Filed Codes 1 to 254 (1 byte)	Length Field (1 byte)	Value Field (Variable length)

End of List
Code 255

FIGURE 25-4 BOOTP option format

Table 25-1 Partial List of BOOTP Options

Option	Code	Length (Bytes)	Contents
Padding	0	1	None—used only for padding
Subnet mask	1	4	Subnet mask of local network
Time of day	2	4	Time of day in universal time
Default routers	3	N	N/4 IP addresses of default router
Time servers	4	N	N/4 IP addresses of time server
IEN116 server[a]	5	N	N/4 IP address of IEN116 server
Domain servers	6	N	N/4 IP address of DNS server
Log server	7	N	N/4 IP address of log server
Quote server	8	N	N/4 IP address of quote server
Print servers	9	N	N/4 IP addresses of print server
Impress	10	N	N/4 IP address of impress server
RLP server	11	N	N/4 IP address of RLP server
Host name	12	N	DNS host name string
Boot size	13	2	Two-byte integer length of boot file
Merit dump file	14	Variable	Path file name
Extensions path	18	Variable	Extension path name
Vendor specific	128–254	Variable	Vendor-specific information
End of list	255	1	None—simply terminates option list

[a]IEN stands for Internet Engineering Note.

Options field. A 64-byte field that serves two purposes. It can be used to carry additional information, such as the subnet mask or the IP address of the default router, or more specific vendor information, such as hardware type or serial number. BOOTP options are generally called *vendor extensions*.

Figure 25-4 shows the format for BOOTP vender extensions. The extensions are comprised of three fields: a one-byte tag field, a one-byte length field, and a variable-length value field (sometimes called the TLV format). The tag field contains the extension code, which uniquely identifies the option. Fixed-length options without data consist of only a tag byte and no length byte. The length field does not include the tag or length fields, as it specifies only how many bytes are in the value field. Table 25-1 shows a partial list of the BOOTP vendor extensions. All fields that contain IP addresses are multiples of four-bytes long. The one-byte padding option is used only to align the options on a four-byte boundary. The one-byte end-of-list option terminates the extension field. Vendors must use tags 128 through 254 to provide additional information about vendor extensions in a reply packet. Extensions that contain NVT ASCII information in the data field should *not* include a trailing null character; however, destinations must be capable of deleting trailing null characters if they are received. The functions of several BOOTP vendor extensions listed in Table 25-1 are self-explanatory; however, many of them are defined in RFC 1497.

25-4 BOOTP OPERATIONS

BOOTP uses a two-step bootstrap procedure. Figure 25-5 depicts the steps involved in a BOOTP exchange. Before a client can initiate a bootstrap operation, the BOOTP server must activate UDP port number 67, which establishes a passive open enabling the server to accept BOOTP requests. After creating a passive open, the server waits for a request packet, which begins the two-step bootstrap procedure described here:

1. The client initiates the exchange by sending a BOOTP request packet to the server encapsulated in a UDP datagram that is encapsulated in an IP datagram. The UDP source port number is 68, and the UDP destination port number is 67. The client fills the source IP address field in the IP header with all 0s and fills the destination IP address field with the server's IP address (if known) or all 1s (broadcast).
2. The server answers the client's request packet with a BOOTP reply packet using the client's IP address for the source address or using the broadcast address for the destination address. The UDP source port number is 67, and the UDP destination port number is 68.

With BOOTP, the responsibility for providing reliable communications is placed on the client. However, BOOTP is encapsulated in UDP, which uses IP for delivery. Therefore, datagrams may be delayed, lost, delivered out of sequence, or duplicated. In addition, IP does not provide a checksum for the data it carries, which means that the UDP datagram could be delivered containing errors. To protect against delivering corrupted data, BOOTP requires that UDP use its checksum. BOOTP also requires that the don't fragment bit in the IP header be active to accommodate clients that do not have sufficient memory to re-assemble fragments. To handle lost or corrupted datagrams, BOOTP incorporates a system

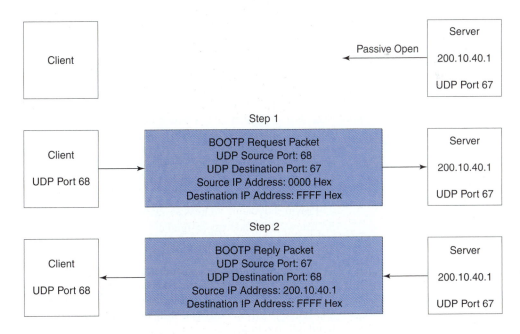

FIGURE 25-5 BOOTP operations

of time-outs and retransmissions. When a client transmits a BOOTP request, it starts a timer. If a response is not received within a specified time interval, the client retransmits the request.

BOOTP is a static configuration protocol designed for relatively static environments where each host is permanently connected to a network and the network manager creates a BOOTP configuration file that establishes the BOOTP parameters for each host. Whenever a client requests its IP address, the BOOTP server relies on a table to match the physical address of the client with its IP address. With the introduction of wireless networks and portable laptop computers, it is much faster and easier to move computers from one location to another. Unfortunately, BOOTP does not adapt well to mobility because it is capable only of statically mapping a host identifier to its parameters. In addition, the network manager must manually enter a set of parameters for each host on the network and then store it on the BOOTP server's configuration file.

Statically assigning BOOTP parameters is adequate if the computers remain stationary and there are sufficient IP addresses available to assign each computer a permanent IP address. However, when computers are mobile or there are more computers than IP addresses available, static assignment requires an excessive amount of overhead. In such networks, a dynamic (automated) method of address assignment, such as DHCP, is necessary.

25-5 DYNAMIC HOST CONFIGURATION PROTOCOL

The *dynamic host configuration protocol* (DHCP) is a service based on a clientserver model that allows network managers to set up servers that allocate and manage assemblies of IP addresses for client hosts that do not require permanent IP addresses. Like BOOTP, DHCP provides a framework for passing configuration information to hosts on a TCP/IP network. DHCP temporarily binds IP addresses and other configuration parameters to a DHCP client and provides a framework for passing configuration information to hosts residing on TCP/IP networks. Control information and configuration parameters are transported in data fields within the options field of the DHCP message.

Network engineers designed DHCP to provide dynamic configuration, which makes it more suitable than BOOTP for modern networks. DHCP is essentially an extension of BOOTP, as DHCP enhances the capabilities of BOOTP and is backward compatible with BOOTP. Therefore, a host running the BOOTP client software can request a static configuration from a DHCP server. DHCP was designed to provide computers with temporary IP addresses. Consequently, DHCP is well adapted to situations where hosts move from one network to another or when hosts are routinely connected and disconnected from the network, such as a subscriber to an Internet Service Provider.

DHCP includes the following characteristics:

1. Centralized IP address administration
2. Backward compatible with BOOTP
3. Supports multiple servers
4. Provides dynamic address assignment
5. Allows static address assignment
6. Does not interact with domain name service

There are three software elements defined by DHCP: client, server, and relay agent:

Client software. DHCP client software is installed in client machines to handle broadcast requests for automatic IP acquisition and for acquiring other essential configuration information.

Server software. DHCP server software is installed in servers designated to respond to client requests for IP addresses. DHCP servers also manage pools of IP addresses and related configuration information.

Relay agent software. DHCP clients broadcast requests onto local network segments to obtain an IP address. Routers block broadcasts, which means that responses from DHCP servers must come from the same network segment. DHCP relay agent software intercepts IP address requests on local network segments, repackages the requests, and then rebroadcasts them as unicast messages to DHCP servers with known addresses that reside on different network segments. The DHCP server sends its replies to the relay agent, which in turn forwards them to the client requesting the IP address.

25-5-1 DHCP Databases

DHCP servers use two databases. One database acquires IP addresses manually (statically) and binds them permanently to hardware addresses. This is the same kind of database that BOOTP uses to assign permanent IP addresses to DNS servers, e-mail servers, proxy servers, file servers, and routers, which require fixed IP addresses. Maintaining this type of database is a very time-consuming process for a network manager. The second database used with DHCP contains one or more blocks of IP addresses that are dynamically (automatically) assigned to clients on a first-come, first-serve basis. A block of addresses that is available for temporary assignment, it is called an *address pool*. If the address range is comprised of numeric IP addresses (usually contiguous) that fall under the control of DHCP, it is called an *address scope*. DHCP can exclude individual addresses or ranges of addresses within an address scope from dynamic assignment, which allows DHCP to manage existing IP address ranges that may have already been assigned to routers and servers. DHCP can allocate the remaining unassigned IP addresses to clients on demand. When a host no longer needs an IP address, it is released and returned to the scope so that it can be reissued to another host.

When a DHCP client requests a temporary IP address, the DHCP server assigns the client one of the unused (available) addresses from an address pool for a specified length of time. When a client sends a request to a DHCP server for a temporary IP address, the server checks its static database, and if it contains an entry that permanently binds an IP address to the hardware address of the requesting client, the server simply returns the permanent IP address to the client. If the static database does not contain a permanent assignment, the server assigns the client an IP address from the pool of available IP addresses and adds the entry to the dynamic database.

25-5-2 DHCP Leases

The IP addresses assigned from a pool of dynamic addresses are not permanent—DHCP servers issue a lease for a dynamic IP address that expires at the end of the lease time. The server's administrator can determine the length of the lease. Default lease times vary from one day to three weeks. For example, the default lease time for Windows 2000 is eight days, and for Windows NT 5.0 it is 36 hours. After half the lease time has expired, a client begins a lease renewal process that determines if the client can keep its current IP address beyond the lease time. If the client cannot renew its current address because the original DHCP server is not available, the client begins the process of renewing its address from another DHCP server. This process is called the *rebinding process*. If rebinding should fail, the client must totally release its IP address.

Once a lease has expired, the client must either stop using the IP address, request a lease renewal, or acquire a new IP address. The DHCP server has the option of renewing the lease or letting it expire, which essentially takes the client off the network. If there is more than one DHCP server on the network, each server may offer the client an IP address, and the client can select one lease offer (usually the first one) and then announce its selection to the other servers.

DHCP servers support three types of address leases: manual, automatic, and dynamic:

Manual lease. With manual leases, the network manager explicitly assigns all IP addresses manually (statically). Manual leases are used when network managers want DHCP to manage all IP addresses but do not want DHCP to dynamically assign them. Manual leases are too labor intensive for most large networks.

Automatic lease. With automatic leases, the DHCP server permanently assigns specific IP addresses and dynamically assigns the rest. This allows DHCP to maintain fixed addresses for routers and servers and dynamically assign addresses to clients and still enable DHCP to manage all the IP addresses on the network.

Dynamic lease. With dynamic leases, the DHCP server automatically assigns IP addresses to client hosts for a specified period of time when permanent addresses are not required. Dynamic leases are the least labor intensive and, therefore, probably the most prevalent type of lease.

25-5-3 DHCP Message Format

The format for DHCP message is based on the format used with BOOTP messages. In fact, the formats are so similar that DHCP is backward compatible with BOOTP, and in certain circumstances DHCP and BOOTP participants can exchange messages. Figure 25-6 shows the format for a DHCP message. There are only two differences: (1) tagged items with DHCP are called *options,* whereas in BOOTP they were called *vendor extensions,* and (2) DHCP includes a one-bit flag field that is not part of BOOTP. DHCP provides additional options to the option field; however, DHCP uses the same well-known port numbers as BOOTP: server 67 and client 68.

A one-bit flag field (sometimes called the *broadcast bit*) is located in the first bit of the 16-bit unused field that follows the number of seconds remaining field. The broadcast bit allows a client to specify a forced broadcast (rather than a unicast) reply from the server. With unicast replies, the destination IP address of the IP datagram is the address assigned to the client. Consequently, since the client may not know its IP address, it may very well ignore and discard the datagram. However, if the reply is an IP broadcast, every host (including the requesting computer) will receive and process the broadcast message. The rightmost 15 bits of the flag field are called *MBZ bits* (must be zero), as they are not used and are always 0.

25-5-4 DHCP Options

DHCP offers all the options as BOOTP along with several additional options that are not available with BOOTP. With DHCP, servers insert a special four-byte code called a *"magic cookie"* at the beginning of the option field. The magic cookie is in the form of an IP address with the dotted-decimal value 99.130.83.99 (63.82.53.53 hex). The magic cookie determines how the succeeding data should be interpreted. After a client reads the first 14 fields of the reply packet, it looks for the magic cookie. If it finds it, it knows that the next 60 bytes are options.

Table 25-2 lists the BOOTP and DHCP options. As you can see, the option list is somewhat overwhelming. However, all networks do not utilize all options. In fact, the only option required by DHCP is option 53, which provides the boot-up and discovery mechanisms a client needs to acquire a dynamic IP address.

The format for DHCP option 53 is shown in Figure 25-7. As can be seen, the option is comprised of three one-byte fields: code field (53), length field (1), and type field (1–8). The type code identifies the specific interaction currently taking place between the client and the server. Table 25-3 lists the eight message types available with the DHCP code 53 option.

DHCP option 53 messages are summarized here:

DHCP Discover. Client broadcast to locate available DHCP servers.

DHCP Offer. Server's response to DHCP Discover that contains an offer of an IP address and configuration parameters.

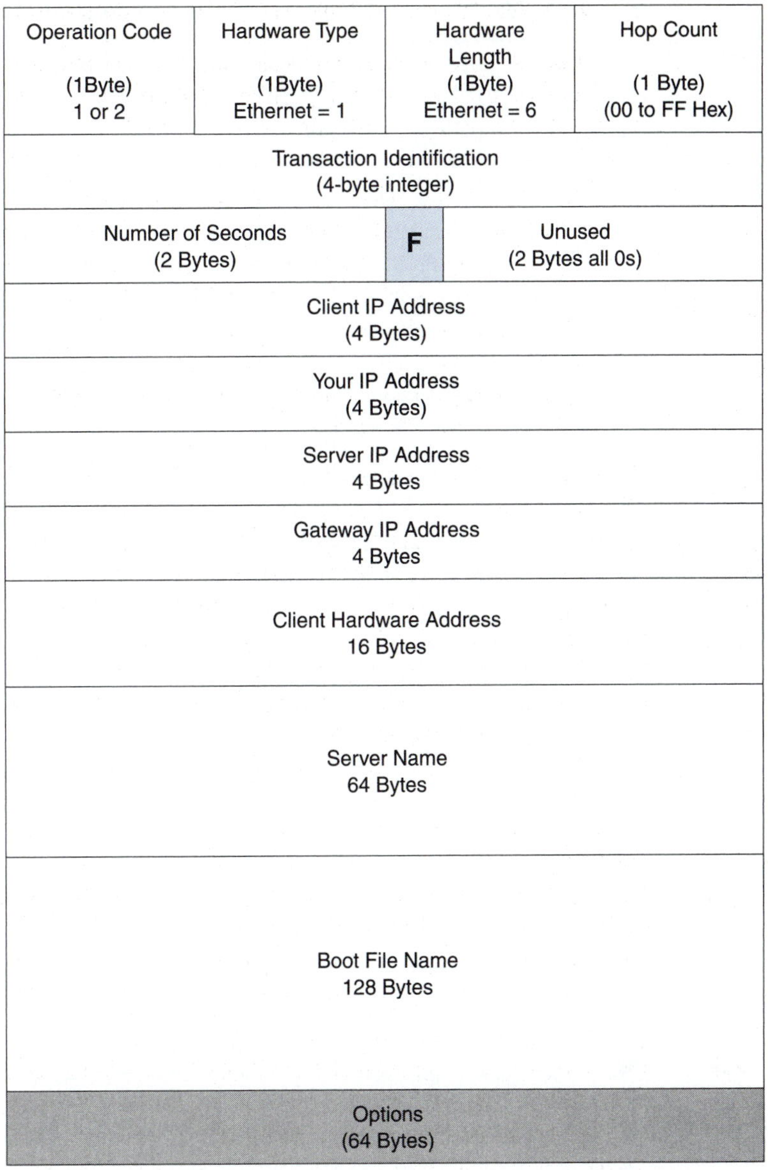

Operation Code (1Byte) 1 or 2	Hardware Type (1Byte) Ethernet = 1	Hardware Length (1Byte) Ethernet = 6	Hop Count (1 Byte) (00 to FF Hex)
Transaction Identification (4-byte integer)			
Number of Seconds (2 Bytes)	F	Unused (2 Bytes all 0s)	
Client IP Address (4 Bytes)			
Your IP Address (4 Bytes)			
Server IP Address 4 Bytes			
Gateway IP Address 4 Bytes			
Client Hardware Address 16 Bytes			
Server Name 64 Bytes			
Boot File Name 128 Bytes			
Options (64 Bytes)			

FIGURE 25-6   DHCP packet format

DHCP Request. Client sends to servers for one of three reasons: (1) requesting offered parameters from one DHCP server and implicitly declining offers from all other DHCP servers, (2) confirming the correctness of previously allocated IP address after a system reboot, or (3) extending the lease on a preassigned IP address.

DHCP Ack. Server sends with configuration parameters, which includes committed network address.

DHCP NAK. Server sends to indicate the client's notion of its network address is incorrect because the client has moved to a different network or subnet or the client's lease has expired.

DHCP Decline. Client sends to indicate that the network address is already in use.

DHCP Release. Client sends to relinquish its network address and cancel existing lease.

Table 25-2 BOOTP and DHCP Options and Vendor Extensions

Code	Data Length	Description
0	1	Pad
1	4	Subnet mask
2	4	Time offset
3	4+	Router
4	4+	Time server
5	4+	Name server
6	4+	Domain name server
7	4+	Log server
8	4+	Quote server
9	4+	LPR server
10	4+	Impress server
11	4+	Resource location server
12	1+	Host name
13	2	Boot file size
14	1+	Merit dump file
15	1+	Domain name
16	4	Swap server
17	1+	Root path
18	1+	Extensions path
19	1	IP forwarding enable/disable
20	1	Nonlocal source routing enable/disable
21	8+	Policy filter
22	2	Maximum datagram reassembly size
23	1	Default IP time to live
24	4	Path MTU aging time-out
25	2+	Path MTU plateau table
26	2	Interface MTU
27	1	All subnets are local
28	4	Broadcast address
29	1	Perform mask discovery
30	1	Mask supplier
31	1	Perform router discovery
32	4	Router solicitation address
33	8+	Static route (obsolete)
34	1	Trailer encapsulation
35	4	ARP cache time-out
36	1	Ethernet encapsulation
37	1	Default TCP TTL
38	4	TCP keep-alive interval
39	1	TCP keep-alive garbage
40	1+	Network information service domain
41	4+	Network information servers
42	4+	NTP servers
43	1+	Vendor-specific information
44	4+	NetBIOS over TCP/IP name server
45	4+	NetBIOS over TCP/IP datagram distribution server
46	1	NetBIOS over TCP/IP node type
47	1+	NetBIOS over TCP/IP scope
48	4+	X Window system font server
49	4+	X Window system display manager
50	4	Requested IP address
51	4	IP address lease time
52	1	Option overload
53	1	DHCP message type
54	4	Server identifier
55	1+	Parameter request list
56	1+	Message
57	2	Maximum DHCP message size
58	4	Renew time value
59	4	Rebinding time value
60	1+	Class identifier
61	2+	Client identifier

Table 25-2 BootP and DHCP Options and Vendor Extensions *(continued)*

Code	Data Length	Description
62	1–255	NetWare/IP domain name
63		NetWare/IP information
64	1+	Network information service + domain
65	4+	Network information service + servers
66	1+	TFTP server name
67	1+	Boot file name
68	0+	Mobile IP home agent
69	4+	Simple mail transport protocol server
70	4+	Post office protocol server
71	4+	Network news transport protocol server
72	4+	Default World Wide Web server
73	4+	Default finger server
74	4+	Default internet relay chat server
75	4+	StreetTalk server
76	4+	StreetTalk directory assistance server
77	Variable	User class information
78	Variable	SLP directory agent
79	Variable	SLP service scope
80	Variable	Naming authority
81	4+	FQDN (fully qualified domain name)
82	Variable	Agent circuit ID
83	Variable	Agent remote ID
84	Variable	Agent subnet mask
85	Variable	NDS servers
86	Variable	NDS tree name
87	Variable	NDS context
88	Variable	IEEE 1003.1 POSIX time zone
89	Variable	Fully qualified domain name
90	Variable	Authentication
91	Variable	Vines TCP/IP server option
92	Variable	Server selection option
93	Variable	Client system architecture
94	Variable	Client network device interface
95	Variable	LDAP (lightweight directory access protocol)
96	Variable	IPv6 transitions
97	Variable	UUID/GUID-based client identifier
98		Open group's user authentication
99		
100	Variable	Printer name
101	Variable	MDHCP multicast address
102–107		
108	Variable	Swap path option
109	Variable	Autonomous system number
110	Variable	IPX compatability
111		
112	Variable	NetInfo parent server address
113	Variable	NetInfo parent server tag
114	Variable	URL
115	Variable	DHCP fail-over protocol
116	1	Auto configure
117	2+	Name service search
118	4	Subnet selection
119	Variable	DNS domain search list
120	Variable	SIP servers DHCP option
121	5+	Classless static route option
122	Variable	CableLabs client configuration
123–125		
126	Variable	Extension
127	Variable	Extension
128–254		
255	0	End

Code Field 53 (1 byte)	Length Field 1 (1 byte)	Type Field 1 to 8 (1 byte)

FIGURE 25-7 DHCP option 53 format

Table 25-3 DHCP Option 53 Types

Type	Message
1	DHCP Discover
2	DHCP Offer
3	DHCP Request
4	DHCP Decline
5	DHCP Ack
6	DHCP NAK
7	DHCP Release
8	DHCP Inform

DHCP Inform. Client sends asking only for local configuration parameters, as client already has externally configured address.

With DHCP, the server host name and boot file name fields are rather long, and if the fields are not used, the entire space is wasted. However, DHCP option 52 defines an overload option that allows a DHCP server to use the space reserved for the server host name and boot file name in the DHCP message for other options. When a client receives the overload option, it instructs the client to ignore the usual meaning of the server host name and boot file name fields and to inspect those fields for other options.

25-6 DHCP STATES AND PROCEDURES

With DHCP, a client transitions among six states when acquiring, reacquiring, or releasing an IP address. The six states are initialization, selection, request, bound, renew, and rebind. DHCP uses four states to initially acquire an IP address (initialization, selection, request, and bound), and four states are used to rebind, renew, or release an IP address after a client has acquired an IP address (initialization, rebind, bound, and renew).

The six states are used to perform three essential DHCP processes: discovery, renewal, and release.

25-6-1 DHCP Address Discovery (Acquisition) Process

When a DHCP client initially boots up, it must execute a standard address discovery procedure to acquire an IP address. Once the client obtains an IP address, it tests the address by sending an ARP broadcast packet using a duplicate IP address.

A DHCP client does not have an IP address when it is booting up for the first time or after its current lease expires. Before a client can initiate a request for an IP address from a DHCP server, the server must activate UDP port number 67, which establishes a passive open enabling the server to accept DHCP requests. After creating a passive open, the server waits for a client to initiate the address discovery process.

DHCP servers use a four-step process to allocate IP addresses:

1. *Discovery process.* A DHCP client discovers all the DHCP servers on its local network segment by sending a broadcast message onto its segment.
2. *Offer process.* All DHCP servers that received the discovery message respond with a unicast transmission containing an offer for an IP address from the server's defined address scope.

3. *Request process.* The DHCP client broadcasts a second message accepting one of the addresses it was offered. The client can reject an offer by sending a DHCP decline packet, which is typically sent only if the client received more than one offer.

4. *Acknowledgment process.* The DHCP server that offered the IP address that the client accepted responds with an acknowledgment message. At this time, the client has an IP address, which begins the lease time. The acknowledgment packet essentially completes the four-packet DHCP discovery process.

To acquire an IP address, the client broadcasts a request for an IP address that identifies the client's hardware address. The discovery packet contains an IP header with the source IP address 0.0.0.0 because the client initially does not have an IP address. If the client was recently connected to the same network, it generally defines a preferred address, which is typically the last IP address it was assigned. This initial broadcast is called *DHCP discovery*. DHCP servers that receive the client's discovery broadcast respond with an IP address offer for a specific lease time (typically 1 hour). Because DHCP discovery relies on sending broadcast messages, the discovery process is restricted to one segment of a local network. Because it is impractical to have a DHCP server on every network segment, DHCP specifications include a process called the *relay agent process,* which allows DHCP broadcasts to be routed to another network segment. DHCP relay agent software located in a router intercepts address requests on a local cable segment, repackages the requests, and then retransmits them as unicast transmissions to one or more remote DHCP servers whose IP addresses are known. Relay agent software is generally installed on a router connected to segments containing DHCP clients. The relay agent that receives the client's broadcast request places its IP address in the appropriate field and forwards the request to the remote server. The remote server sends its reply to the relay agent, which forwards the message to the requesting client.

Figure 25-8a shows the four states used by a DHCP client to acquire an IP address. The figure also shows the state transitions that occur and the receive message that initiated the transition between states. In addition, the figure indicates whether the client or the server is sending the message. When a DHCP client initially boots up, it must perform a standard address discovery procedure to acquire an IP address that allows the computer to communicate over the network. The state transitions shown in Figure 25-8a are summarized here:

Initialization state to selection state. While a client is initially booting up, it enters the first state, which is the initialization state. While in the initialization state, the client broadcasts a DHCP Discover message onto the network that places the client in the selection state.

Selection state to request state. One or more of the DHCP servers on the network responds to the broadcast with a DHCP Offer message. The requesting client may receive more than one response; however, in rare cases, it may not receive any responses. The client chooses one of the responses (usually the first to arrive) and negotiates with the respective server for an address lease by sending a DHCP Request message, which puts the client in the request state. If a client does not receive an offer within a specified time, it tries four more times within a 2-second span. If there are still no replies, the client waits 5 minutes and then resends a request message.

Request state to bound state. The server responds to the request message with a positive acknowledgment (DHCP Ack), which begins the lease time and puts the client in the bound state. The client will remain in the bound state until the lease expires or the client releases the IP address.

Figure 25-8b shows the sequence of the four transmissions necessary for a host to acquire an IP address using DHCP: DHCP Discover, DHCP Offer, DHCP Request, and DHCP Ack.

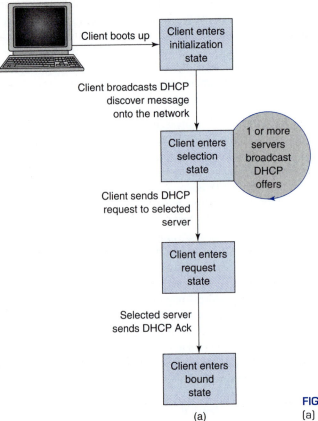

FIGURE 25-8 DHCP Discovery: (a) states and transitions

(a)

25-6-2 DHCP Address Renewal Process

When a client's lease nears its expiration time, the client must initiate an address renewal procedure if it wishes to maintain an IP address and continue communicating over the network. Figure 25-9 shows the four states used by a DHCP client to renew or release an IP address. The client is initially in the bound state, which is generally considered the normal operational state because this is the state the client is in while it is using a previously acquired IP address.

When a client enters the bound state, it sets three timers relating to lease renewal and records the time the address was received. The DHCP server determines explicit values for the timers when it assigns an IP address to a client. By default, the initial value loaded into the first timer is equal to one-half the lease time, and the value loaded into the second timer is equal to 87.5% of the lease time. After the first timer (the *renewal timer*) expires, the client must attempt to renew its lease if it wishes to keep the same IP address and communicate over the network. To request a lease renewal, the client must send a DHCP Request message to the server that assigned its current lease. The request message contains the client's current IP address and requests an extension of the current lease. After sending the DHCP Request message, the client enters the renew state and waits for a response from the server. Although a client can request a specific length for the lease extension, the server has the ultimate responsibility of determining its value. The server responds to the client in one of two ways: the server can instruct the client to cease using its current IP address, or it can renew the current lease. If the server approves the extension, it sends a DHCP Ack that places the client back into the bound state, as shown in Figure 25-10a. The DHCP Ack

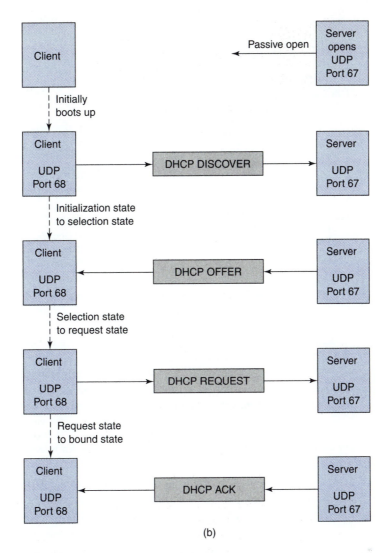

(b)

FIGURE 25-8 (b) message exchange sequence

may also include new values for the client's timers. When a server wishes to terminate the client's lease, it sends a DHCP NAK that instructs the client to immediately stop using the address and go into the initialize state, as shown in Figure 25-10b.

After sending a DHCP request message, a client remains in the renew state while awaiting a response. If the server is down or for some other reason unreachable, it will not respond, as shown in Figure 25-10c. To deal with this situation, DHCP uses the second timer set when the client initially entered the bound state. The second timer (called the *re-binding timer*) expires after 87.5% of the lease period, which causes the client to move from the renew state to the rebind state. While making the transition from the renew to the rebind state, the client assumes that the old DHCP server is still unavailable and begins broadcasting a DHCP Request message onto the network segment destined for any available DHCP server. Any server configured to provide DHCP services can respond to the broadcast. If the response is positive, as shown in Figure 25-10d, the client is granted an extension of its current lease, the two timers are reset, and the client returns to the bound state with the same IP address. If the response is negative, as shown in Figure 25-10e, the client

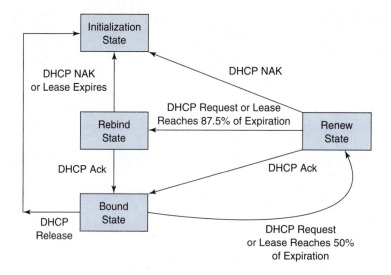

FIGURE 25-9 DHCP transitions among the bound, renew, rebind, and initialization states

is denied further use of IP address and must discontinue use of its IP address and move to the initialization state to acquire a new IP address.

If after a client moves to the rebind state and has been denied a lease renewal from all the servers on the network (including the original server) and after the third timer has expired, the client moves to the initialization state and begins acquiring a new IP address:

> *Bound state to renew state.* After the client's first timer reaches 50% of its initial value, it broadcasts a DHCP Request message for a lease renewal and moves to the renew state.
>
> *Renew state to bound state.* If a server approves the lease renewal, the client moves from the renew state back to the bound state and continues using its original IP address.
>
> *Renew state to initialization state.* If a server disapproves the lease renewal, the client moves from the renew state to the initialization state and begins the process of acquiring a new IP address.
>
> *Renew state to rebind state.* If a client does not receive a response from a server within the first 87.5% of the lease time, it moves to the rebind state and broadcasts a DHCP request message.
>
> *Rebind state to initialization state.* If the client receives a negative response from a server while in the rebind state, the client moves to the initialization state and begins the process of acquiring a new IP address.

25-6-3 DHCP Address Release (Termination) Process

When a client is in the normal state of operation (i.e., the bound state) and determines that it no longer needs an IP address, DHCP allows the client to release its current IP address without waiting for the lease time to expire. This is called an *early termination* and is useful when there are a limited number of IP addresses available. Early terminations allow servers to provide IP addresses to more clients than it has IP addresses.

A client transits a DHCP Release message to terminate an IP address before the timer expires. Terminating the IP address is final, as it prevents the client from using the address to communicate over the network. Figure 25-11 shows states and transitions involved in executing an early release. When a client sends a DHCP release message, it leaves the bound

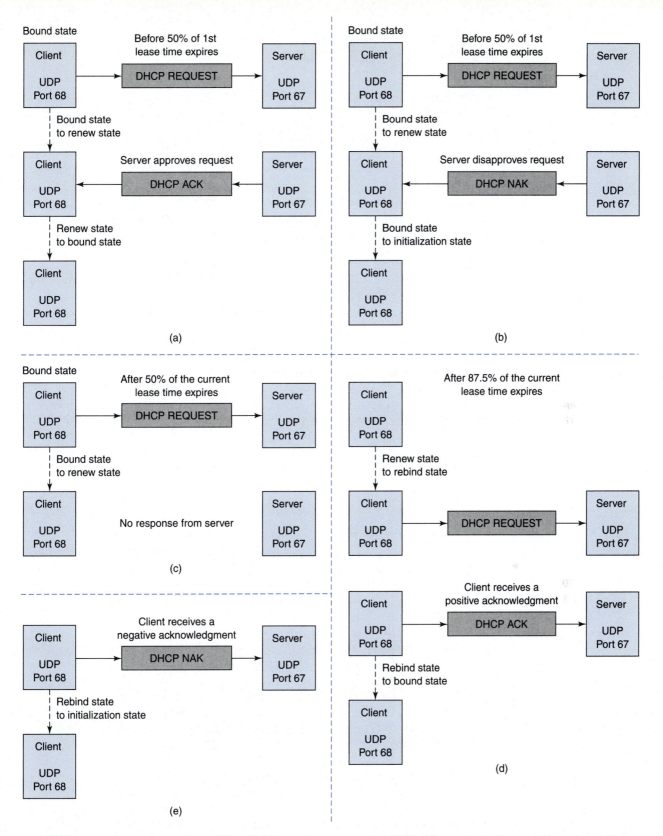

FIGURE 25-10 Renew states and transitions: (a) Renewal request before 50% of first timer expires—positive acknowledgment from server; (b) renewal attempt after 50% of first timer expires—negative response from server; (c) renewal attempt after 50% of first timer expires—no response from server; (d) renewal attempt after 87.5% of first timer expires—positive response; (e) renewal attempt after 87.5% of first timer expires—negative response

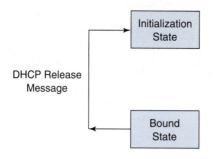

FIGURE 25-11   DHCP release states and transitions

state and returns to the initialization state, where it can request another IP address when it needs one. Servers do not respond to DHCP release messages.

25-6-4   Sample DHCP Boot Request Message

An example of what a DHCP boot request message, including the Ethernet, IP, and UDP headers, would look like with a network analyzer is shown here:

Ethernet Header

Destination Hardware Address	FF FF FF FF FF FF
Source Hardware Address	00 02 B9 E3 F9 40
Ethernet Protocol Type	0800 (IP)

IP Header

IP Version	4
Header Length	20 (no options)
Differentiated Services (DS) Field	00 hex
0000 00..	DS Codepoint = Default PHB (0)
00	Unused
Packet Length	276
ID	E58 hex
Fragmentation Information	00 hex
.0.	Don't Fragment = false
. .0.	More Fragments = false
. . . 0 0000 0000 0000	Fragment offset = 0
Time To Live	128
Protocol	UDP (17)
Header Checksum	1D5A
Source IP Address	10.10.4.30
Destination IP Address	255.255.255.255

UDP Header

Source Port	BOOTP (68)
Destination Port	BOOTP (67)
Length	256
Checksum	58A7

DHCP Header

Message Type	DHCP
Operation Code	Boot Request Message (1)
Hardware Address Type	1
Hardware Address Length	6

Hops	0
Transaction ID	0
Seconds Elapsed Since Client Started Trying to Reboot	2560
Flags	8000 hex
1...	Broadcast Flag Set
. 000 0000 0000 0000	Reserved
Client IP Address	10.10.4.30
Your IP Address	0.0.0.0
IP Address of Next Server to use BOOTP	0.0.0.0
Relay Agent IP Address	0.0.0.0
Client Hardware Address	000000000000000000000000000000000
Server Name	
BOOTP File Name	
DHCP Magic Cookie	63 82 53 63
DHCP Message Option	Code 53, Length 1
Message Type	DHCP INFORM (8)
Vendor Specific Information Option	Code 43, Length 2
Number of Bytes of Vendor Specific Information	2
End of Option	Code 255

If a DHCP server is configured to support BOOTP clients, it is probably configured to provide static as well as dynamic addresses. A DHCP server that supports BOOTP clients must interact with BOOTP clients according to the BOOTP protocol. For example, the DHCP server must formulate a BOOTP BOOTREPLY rather than a DHCP DHCPOF-FER. DHCP servers can send DHCP options only to BOOTP clients that are allowed by the joint DHCP and BOOTP vendor extensions.

25-7 DOMAIN NAME SYSTEM

The Internet is based on a hierarchical addressing system using numerical addresses, which allows routers to forward datagrams based on groups of addresses (numbers) rather than individual addresses. Numbers like 208.45.167.89 and 208.45.167.98 are awkward for humans to deal with and easy to get mixed up, as the only difference between the two numbers is the transposition of the last two digits. The domain naming system was developed to simplify the addressing system by equating numerical addresses to names. A domain naming is a string of characters, including letters and numbers, which are often abbreviations of words that are easily correlated with the owner of the address.

The *domain name system* (DNS) is a global network of name servers that translate host names into numerical IP addresses. Paul Mockapetris designed DNS in 1984 to help solve the rapidly increasing problems associated with the old name-to-address mapping system. The old name-to-address mapping system, called the *host table,* was maintained by the Stanford Research Institute's Network Information Center (SRI-NIC). New names were simply added to the table once or twice a week, and system administrators could download a file transfer protocol (FTP) containing the host table each week to update their local domain name servers. As the Internet expanded, the host table soon became completely unmanageable. The old system worked fine for name-to-address mapping; however, it simply was not practical or effective to manually update and distribute the vast amount of information contained in the file.

25-7-1   Name Resolution

DNS is the TCP/IP application-layer service that ties the entire Internet together by providing a means of resolving symbolic domain names, such as *microsoft.com,* to globally unique IP addresses, such as 207.46.249.27. Without this resolution service, humans would have to remember the numeric IP addresses for all their Internet destinations. DNS also supports other services, such as e-mail, by translating addresses like *jdoe@yahoo.com* into IP addresses so servers and routers can deal with them. Probably the most significant attribute of DNS is that no single organization is responsible for keeping it current because DNS is a distributed database that resides on multiple name servers located around the world. Distributed databases allow for an almost unlimited growth of the Internet. When the capabilities of all the DNS databases distributed throughout the Internet are combined, they have the ability to map every valid domain name in existence to its corresponding IP address. Each DNS database operates independent of all other DNS databases, which creates a widely distributed environment with no central structure. However, DNS does a very good job of providing a robust, dependable, and stable foundation for Internet addressing.

Host names are aliases assigned to IP addresses that identify the address as a specific TCP/IP host. Host names can be up to 255 characters long and may be comprised of alphabetic and numeric characters and the dash (-) and period (.) characters. More than one name can be assigned to a single host, and with most operating systems, host names do not have to be the same as the operating system name.

A DNS server is a computer that accepts requests from hosts to resolve domain names to IP addresses. No single DNS server needs to know the IP addresses for all names. The DNS system is distributed among numerous DNS servers where each server knows the names and IP addresses of the devices on their own network and the addresses of other DNS servers. A DNS server is simply a computer that is running DNS software. Since many of the DNS servers are Unix machines, the most popular DNS program is Berkeley Internet Name Domain (BIND). DNS software is generally comprised of two elements: the name server itself and a *resolver*. The name server supplies name–to–IP address translations in response to browser requests. If a server can resolve a name to an IP address without having to ask another DNS server, it is known as an *authoritative server*. If the name server cannot resolve the name itself, it asks other name servers for the information. The resolver is a client-side piece of software responsible for issuing DNS queries (requests) for TCP/IP applications and then relaying the responses back to the originating application. In essence, the resolver tracks down the IP address of a domain name by forwarding queries to other DNS servers. Usually a resolver will give a domain name to a server and requests the IP address. The server then consults the generic or country domains to try to locate the mapping. After a resolver receives a domain name request, it sends a query to the local DNS server for resolution. If the local server cannot resolve the request, it either refers the resolver to another server or forwards the query to another server.

25-7-2   Inverse Resolutions

Clients can also request names to IP addresses mapping, which is called a *pointer* or PTR query. DNS uses the inverse domain to perform this type of mapping by reversing the IP address and adding two labels, in-addr and arpa, creating a domain in which the inverse domain section can resolve. For example, if a resolver is asked to resolve the domain name for the IP address 123.43.54.131, it inverts the address and adds in-addr.arpa (which stands for "inverted address resolution"). Thus, the domain name 131.54.43.123.in-addr.arpa is sent to the local DNS server.

25-7-3   Recursive Resolution

A client resolver can request a recursive answer from a name server, which means the resolver expects the server to supply the final answer. If the server is an authoritative server for the domain name, it consults its own database and responds. However, if the server is not the authoritative server, it forwards the request to the parent server. If the parent server is the authoritative server, it responds; otherwise, it forwards the request to another server. This process continues until the name is resolved, at which time the response is returned to the requesting client.

25-7-4 DNS Resource Records

Data associated with domain names and address records are stored on DNS servers in special database records called *resource records* (sometimes called *address records* because it stores the name of a single IP address). There are two classes of resource records: *Internet* and *special*. Internet class resource records are the only records of interest to the general public, as the special resource records work only at the Massachusetts Institute of Technology (MIT).

Resource records within the Internet class belong to a named taxonomy of record types, of which many types are in use. There are only nine resource records typically used:

Address (A) record. Used to store domain name to IP address translation data

Canonical name (CNAME) record. Used to create aliases (contain the alias and the real name of the requested host)

Host information (HINFO) record. Used to store descriptive information about a specific Internet host (HINFO records may contain CPU [central processing unit] and operating system data, although it is usually left empty to discourage hackers)

Mail exchange (MX) record. Used to route SMTP (simple mail transfer protocol)-based e-mail on the Internet, to identify the IP address for a domain's master e-mail server, and mail filtering for organizations with firewalls

Name server (NS) record. Used to identify the DNS server for a requested zone

Pointer (PTR) record. Used to store in-addr.arpa information for IP address–to–domain name translation data and to support the operation known as a reserve DNS lookup

Start of authority (SOA) record. Used to identify the name server that is authoritative for a given DNS database segment and provides maintenance data about the zone to other servers

Text (TXT) record. May be used to add arbitrary text information to a DNS database for documentation

Well-known services (WKS) record. Used to list the IP-based services provided by the zone, such as FTP, HTTP, and Telnet, that an Internet host can supply

25-8 DOMAINS AND DOMAIN NAMES

A *domain name* is simply a symbolic (rather than numeric) name for a TCP/IP network resource. Domain names are easier for humans to remember, making it easier for them to navigate around the Internet. Domain names should be explicit and unique because the IP addresses they represent are explicit and unique. A *name space* is simply all the names assigned to machines on a network, internetwork, or the Internet. Name spaces map IP addresses to unique names. Name spaces can be organized two ways: flat and hierarchical.

With a *flat name space,* a name is assigned an IP address. Each name within a space is a sequence of unstructured characters that may or may not have a common component, and, if they do, it has no meaning. The primary disadvantage of a flat name space is that it must be centrally controlled to avoid ambiguity and replication, which makes a flat name space inadequate for large systems such as the Internet.

With a *hierarchical name space,* domain names are comprised of several parts where one part may define the nature of the organization, a second part the name of the organization, and a third part a specific department within an organization. With a hierarchical name space, the authority that assigns and controls the name space does not need to be centralized. A central authority must assign only the part of the name that defines the name and nature of the organization. The organization itself has the responsibility of assigning the rest

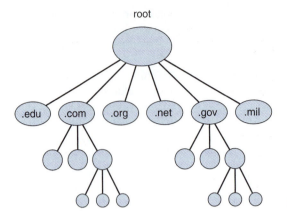

root

.edu .com .org .net .gov .mil

FIGURE 25-12 DNS server tree

of the name. Organizations may add suffixes or prefixes to the name to specifically define the hosts and resources that are part of the organization. The beauty of a hierarchical name space is that an organization can choose whatever prefixes or suffixes it wants, even if part of its name is the same as someone else's name, as long as the entire name is not the same. For example, two universities could designate the local name "library" to one of their computers. However, a central authority assigns one university the name "kansas.edu" and the other university the name "syracuse.edu." Even though the two computers share the same local name, the complete names of the two computers are different, as shown here:

library.kansas.edu

library.syracuse.edu

25-8-1 Domain Name Space

The concept of domain name space was developed to provide a structure for organizing hierarchical name spaces for designating domain names. With domain name space, names are defined in an inverted-tree structure, called a *tree*, a *DNS tree,* or a *DNS server tree.* The uppermost level of the tree where the DNS comes together is an artificial point called the *root* or sometimes the *core* of the tree, which never has a name. All name queries end at the root because the root is the top of the DNS hierarchy.

DNS trees can have four or more levels, as shown in Figure 25-12. Although the root binds the tree together, each level of the tree defines a distinctly different hierarchical level. Each node on a DNS tree represents a different domain, and all domains below a node are part of that domain. Therefore, one domain can be part of another domain, which may be part of still another domain. The DNS tree structure resembles the directory on a computer's hard drive, which has a drive letter for its root and various levels of folders and subfolders branching from the root.

A DNS tree for "ford" might look something like the tree shown in Figure 25-13. The domain "ranger" is in the "trucks" domain, which is in the "ford" domain, which is in the ".com" domain. Domain names are read from the node up the tree to the root. Therefore, the last label is the root label, which means that a full domain name will always end in a null label and that the last character is always a dot because the null string is nothing. Although the last label (the root label) is followed by a dot, the ending dot is generally omitted from the name. Therefore, the full domain name for "ranger" is "ranger.truck.ford.com.," which reads "ranger dot truck dot ford dot com dot." The top-level domain is ".com."

All domain names are comprised of character strings (called *labels*) separated by dots or periods. Each label must start with either a letter or a digit, and the label may contain a hyphen. Upper- and lowercase letters are not differentiated in labels. Each label is limited to 63 characters maximum, although most are much shorter. Every node must have a unique

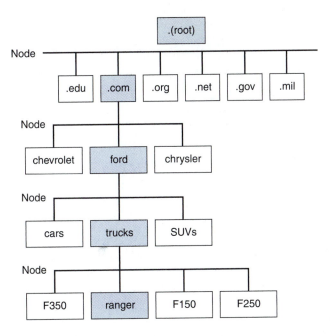

FIGURE 25-13 DNS tree

domain name; however, labels may be used more than once in a tree. Each node on the DNS tree has a label. However, the root label is a *null* (empty) string. DNS requires that nodes branching from the same node (sometimes called *children of a node*) have different labels, which guarantees that all domain names will be unique.

The rightmost label in a domain name is connected directly to the root and is called the *top-level domain* (TLD). The responsibility for operating each TLD, including maintaining a registry of the second-level domains, is delegated to particular organizations called *registry domain sponsors* or simply *delegees*. A sponsor is simply an organization that has been delegated to take on ICANN-TLD policy formulation responsibilities. The sponsor is also responsible for selecting the registry operator and establishing the roles played by registries and their relationship with the registry operator. Registry sponsors and operators must submit regular monthly reports to the Internet Corporation for Assigned Names and Numbers (ICANN). An unsponsored TLD operates under policies established by the global Internet directly through ICANN. Sponsored TLDs are specialized TLDs that have a sponsor representing the narrower community that is most affected by the TLD.

Internet authorities have created three types of top-level domains to reduce the time it takes a resolver to find the right network address: three- and four-letter generic codes (.edu, .gov, .com, .info), two-letter country codes (.uk, .us), and a special-purpose four-letter top-level domain for Internet infrastructure (.arpa). Grouping similar organizations means resolvers do not have to search through all possible Internet domains to resolve a domain name. The entire Internet domain space is located under the root, and 15 top-level root name servers constitute the source for all domain names that must be looked up. Each of those domains is discussed here:

Generic domains. Generic domains are intended to be used by the Internet general public. Generic domains are also referred to as gTLDs. There were originally seven generic domains (.com, .edu, .gov, .int, .mil, .net, and .org) created in the early 1980s of which three were registered without restrictions (.com, .net, and .org). The other four (.edu, .gov, .int, and .mil) have limited restrictions. Generic domains been subdivided into two types: *sponsored* TLDs (sTLDs) and *unsponsored* TLDs (uTLDs). In June 2001, seven additional domains (.biz, .info, .name, .pro, .aero, .coop, and

Table 25-4 Top-Level Domains

TDL	Introduced	Sponsored	Purpose	Sponsor/Operator
.aero	2001	Yes	Air transport industry	Societe Internationale de Telecommunications Aeronautiques SC (SITA)
.biz	2001	No	Businesses	NeuLevel
.com	1995	No	Unrestricted (intended to be used primarily for commercial organizations)	VeriSign, Inc.
.coop	2001	Yes	Cooperatives	DotCorperation, LLC
.edu	1995	Yes	U.S. educational institutions, such as schools, colleges, and universities	EDUCAUSE
.gov	1995	Yes	U.S. federal government (nonmilitary)	U.S. General Services Administration
.info	2001	No	Unrestricted use	Afilas, LLC
.int	1998	No	Multinational organizations (established by international treaties)	Internet Assigned Numbers Authority
.mil	1995	Yes	U.S. military	U.S. Department of Defense Network Information Center
.museum	2001	Yes	Museums	Museum Domain Management Association (MuseDoma)
.name	2001	No	Registration by individuals	Global Name Registry, Ltd
.net	1995	No	Unrestricted (intended to be used by network service providers, etc.)	VeriSign, Inc.
.org	1995	No	Unrestricted (intended for organizations that do not fit elsewhere and nonprofit organizations)	Public Interest Registry
.pro	2002	No	Accountants, lawyers, physicians, and other professionals	RegistryPro, Ltd

.museum) were introduced by ICANN. Four are unsponsored (.biz, .info, .name, and .pro), and three are sponsored (.aero, .coop, and .museum). Table 25-4 summarizes the generic domains.

Country code domains. Country codes, sometimes referred to as ccTLDs, have been created by the International Organization for Standardization (ISO) for over 240 countries and are intended for use as the individual countries deem necessary. Many countries form second-layer domains inside their country codes. For example, the United Kingdom uses .co for commercial organizations and .ac for academics. This gives colleges and universities domain names that end with .ac.uk and company domain names that end in .co.uk. Country codes are delegated to managers who operate the ccTLDs according to local policies that are adapted to best meet the cultural, linguistic, and legal circumstances of the specified country or territory.

.arpa domain. The .arpa domain is used for technical purposes. ICANN administers the .arpa TLD in cooperation with the Internet Architecture Board. The .arpa domain provides address-to-name mapping. For example, to resolve 192.136.118.123, the resolver will look for the name that matches 123.118.136.192 in-addr.arpa.

The Internet Assigned Numbers Authority (IANA) administers all top-level domains, including the ISO country-code domains that participate in the Internet. Each organization is divided into areas called *organizational zones* to simplify maintenance. An organizational zone is a separately administered section of the domain name server tree. Each zone has at least one domain name server, called the *primary server,* and most zones have multiple secondary servers that prevent a single point of failure. The DNS system is supported by distributed databases located in each organization domain and in each subdomain. Administrators at all domains and subdomains are responsible for maintaining their local databases and keeping them current so that when a server receives a request for domain name resolution, it can resolve the name.

25-8-2 Fully Qualified Domain Names

The full name of a system contains a local host name and its domain name, which includes a top-level domain. When a label is terminated in a null string, it is called a *fully qualified domain name* (FQDN). FQDNs contain the complete name of a host, which includes all labels from the least specific to the most specific (i.e., from the bottom of the tree to the top). An FQDN uniquely defines the name of a host. For example, ranger.trucks.ford.com is the FDQN for a computer named "ranger."

25-8-3 Partially Qualified Domain Names

When a domain name is not terminated in a null string, it is called a *partially qualified domain name* (PQDN). PQDNs begin from a node and go up the tree but do not reach the top (root). PQDNs are used only when the node resides on the same site as the client because the resolver can provide the missing labels to determine the FQDN. For example, for a host on the ford.com site to resolve the IP address of the "ranger" computer, the host can simply define the partial name "ranger," and the DNS client will add the suffix .truck.ford.com and then pass the address on to the DNS server. DNS clients generally maintain a list of suffixes.

25-8-4 Domain, Name Space Distribution, Zones, and Root Servers

A domain is a subtree of the domain name space, and the name of the domain is the domain name of the node at the top of the subtree. Domains themselves may be divided into smaller domains called *subdomains*. Figure 25-14 shows an example of a top-level domain name space with second-level domains and subdomains.

Domain name space information must be stored somewhere. However, if it were stored on one computer, it would be very inefficient because the server would have to distribute the

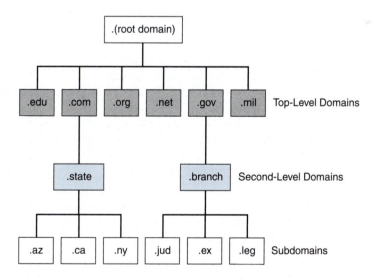

FIGURE 25-14 Top-level domain name space with second level domains and subdomains

vast amount of information to computers located virtually all over the world. It would also be unreliable because if the server failed, the data would be inaccessible.

To solve the efficiency and reliability problems, the information is distributed among several DNS servers. The root remains; however, one domain (subtree) is created for each first-level node. Domains created in this manner would be excessively large; therefore, DNS permits domains to be divided into smaller domains (subdomains). Each DNS server is responsible for one domain, creating a hierarchy of DNS servers similar to the hierarchy of domain names.

A DNS *zone* encompasses all the computers a server is responsible for. If a server is responsible for a domain that has not been divided into small domains, the domain and the zone are the same thing. The server constructs a database called a *zone file,* where it stores information about every node in that domain. If the server divides a domain into smaller subdomains and assigns some of its responsibility (and authority) to one or more other servers, the domain and the zone are no longer the same thing. Information stored in the lower-level servers is specific to the nodes in their subdomain. The original server also stores information that allows it to reference to the lower-level servers.

If a server has responsibility over a zone that consists of the entire DNS tree, it is called a *root server* or a *root-level server.* Root servers generally do not store information about specific domains because they delegate some of their authority to other subordinate servers. Because root servers are located at the top of the DNS hierarchy, they know how to reach any subdomain in the hierarchy.

25-9 DNS MESSAGE FORMATS

DNS only uses two types of messages: query and response. DNS messages are comprised of a header and one or more record sections. There are two types of records: *question records* and *resource records*. DNS query messages are comprised of a header and a question section that contains question records as shown in Figure 25-15a. DNS response messages are comprised of a header, a question section, and three kinds of resource sections that contain resource records. Resource records include answer records, authoritative records, and additional information records, as shown in Figure 25-15b.

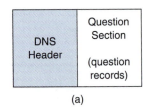

(a)

DNS Header	Question Section (question records)	Answer Section (answer records)	Authoritative Section (authoritative records)	Additional Information Section (additional information records)

Contain Resource Records

(b)

FIGURE 25-15 DNS message format: (a) query; (b) response

25-9-1 DNS Header

The format for the 12-byte DNS header is comprised of six two-byte fields, as shown in Figure 25-16. The six fields are summarized here:

Identification number field. A two-byte identification number used by a client to associate a response with the query. A client uses a different identification number for each query it sends. The server copies the client's identification number into the response header.

Flag field. A two-byte field containing eight subfields, as shown in Figure 25-17. The functions of the subfields are summarized here:

QR flag. A one-bit query/response subfield. If the QR bit is set (1), the message is a response, and if the QR bit is a 0, the message is a query.

Op code flag. A four-bit subfield used to define the type of query or response. Table 25-5 lists the op code types.

AA (authoritative answer) flag. A one-bit subfield that, when set (1), indicates that the name server is an authoritative server (the authority for the domain name). The AA flag is valid only in response messages.

TC (truncated) flag. A one-bit subfield that, when set (1), means that the response was more than 512 bytes and had to be truncated because it exceeded the MTU (maximum transmission unit) of the data field in the transport-layer protocol. The TC bit is generally used only when DNS is transported over UDP.

Identification Number Field	Flag Field	Question Count Field	Answer Count Field	Number of Authoritative Records Field	Number of Additional Records Field
2 bytes	2 bytes	2 bytes	2 bytes	2 bytes	2 bytes

FIGURE 25-16 DNS header format

QR	Op Code	AA	TC	RD	RA	Reserved (or included in the rCode subfield)	rCode
1 bit	4 bits	1 bit	1 bit	1 bit	1 bit	3 bits (all 0s)	4 bits

FIGURE 25-17 Flag field format

Table 25-5 Op Code Types

Op Code	Definition
0	Standard DNS query
1	Inverse DNS query
2	Server status report
3	Reserved
4	Notify
5	Update
6–15	Available for future assignment

RD (recursion desired) flag. A one-bit subfield that, when set (1), indicates that the client desires a recursive query if the target name server does not contain the information requested. The RD flag is set in the query message and repeated in the response message.

RA (recursion available) flag. A one-bit subfield that, when set (1), in a response, means that a recursive response is available in the name server. The RA bit is set only in response messages.

Reserved. A three-bit subfield filled with all 0s that is reserved for future use.

rCode (response code) flag. A four-bit subfield that indicates whether an error occurred in the response (only an authoritative server can determine if an error has occurred). In some advanced DNS servers, the rCode field is extended to seven bits by using the three-bit reserved field. Table 25-6 lists the rCode values and their meanings.

Question count field. A 16-bit field that specifies the number of question records in the question section of the message.

Answer count field. A 16-bit field that specifies the number of answer records in the answer section of the response message. This field is filled with all 0s in the query message.

Number of authoritative records field. A 16-bit field that specifies the number of authoritative records in the authoritative section of the response message. This field is filled with all 0s in the query message.

Number of additional records field. A 16-bit field that specifies the number of additional records in the additional section of the response message. This field is filled with all 0s in the query message.

Table 25-6   Response Code Values

rCode	Description
0	No error
1	Format error
2	Server failure
3	Nonexistent domain
4	Not implemented
5	Query refused
6	Name exists when it should not
7	Resource record set exists when it should not
8	Resource record set that should exist does not
9	Server not authoritative for zone
10	Name not contained in zone
11–15	Available for future assignment

The following rCodes are available only when a seven-bit rCode field is used:

16	Bad OPT version or TSIG signature failure
17	Key not recognized
18	Signature out of time window
19	Bad TKEY mode
20	Duplicate key name
21	Algorithm not supported
22–3840	Available for future assignment
3841–4095	Private use

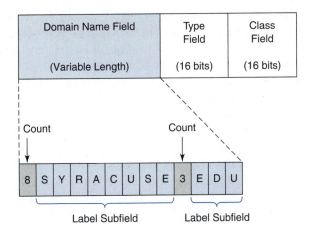

FIGURE 25-18 DNS question record format

25-9-2 DNS Question Section

The variable-length question section of a DNS message contains one or more question records and is present in both DNS query and response messages. Clients use question records to request information from a server. Figure 25-18 shows the format for a DNS question record, which contains three fields: a domain name field, a type field, and a class field.

25-9-3 DNS Answer Section

The DNS answer section is a variable-length section that contains resource records and is present only in response messages. The answer section contains answers sent from a server in response to questions asked by the resolver in a client.

25-9-4 DNS Authoritative Section

The DNS authoritative section is a variable-length section that also contains resource records and is present only in response messages. The authoritative section provides specific domain name information about one or more authoritative servers in response to a client's query.

25-9-5 DNS Additional Information Section

The variable-length additional information section contains one or more resource records and, like the answer and authoritative sections, is present only in response messages. The additional information section provides supplementary information to help the resolver locate domain names. For example, a server can use the additional information section to send the domain name of an authoritative server to the resolver and provide the authoritative server's IP address in the additional information section.

25-9-6 DNS Question and Resource Record Formats

There are two types of resource records used with DNS: question and resource. Question records are sent in the question section of both query and response messages. Resource records are sent in the answer, authoritative, and additional information sections in response messages.

25-9-6-1 Question record format   Clients use question (query) records to acquire information from a server. The question record is a variable-length field that consists of a series of variable-length address labels where each label is preceded by a one-byte length field, as shown in Figure 25-18. The question record is comprised of three fields: domain name field, type field, and class field:

Domain name field. The variable-length domain field holds domain names and is comprised of two or more length (count) bytes, and each length byte is followed by a label subfield. A length byte gives the count of how many characters are included in the label that immediately follows it. The domain name is formulated by simply combining the labels. For example, the domain name field shown in Figure 25-18 contains two length bytes and two labels. The first length field contains the number count 8, which means that the label immediately following it is comprised of eight characters. The second length field contains the number count 3, which indicates that the label immediately following it is comprised of three characters.

Type field. The 16-bit type field defines the type of query. Table 25-7 lists the most common query types, their mnemonic, and a brief description of each.

Class field. This is a 16-bit (two-byte) field that indicates the class for the query that is using DNS. Table 25-8 lists the classes currently being used.

25-9-6-2 Resource record format Each domain name can be associated with a record called, a *resource record,* as server databases are comprised of resource records. Servers return resource records to clients in response to client queries. Figure 25-19 shows the format for a resource record, which is comprised of six fields: domain name, domain type, domain class, time to live, resource data length, and resource data:

Domain name field. This is a variable-length field that holds the domain name to which the resource record belongs. The domain name is a duplicate of the domain name sent in the question record. Since DNS requires data compression capabilities wherever a domain name is repeated, the domain name field in a resource record is an offset that points to the corresponding domain name field in the question record.

Domain type field. This two-byte field is identical to the query type field in the question section except some of the types are not allowed, such as AXR and ANY.

Domain class field. This two-byte field is identical to the query class field in the question section.

Time-to-live field. This four-byte field indicates the number of seconds the data in the resource data field (i.e., the answer) is valid, which is how long it should be cached (saved) before it is discarded. If this field contains a 0, the resource data can be used for just one transaction and should not be cached.

Resource data length field. This is a two-byte field that defines the length of the resource data field (i.e., how many bytes it contains). This limits the maximum length that can be specified as 65,535, which should not be a problem because most resource records are 500 bytes long or less.

Resource data field. This variable-length field could contain several things: the actual answer to the query this message is responding to, the domain name of the authoritative server specified in the authoritative section, or additional information found in the additional information section. The format for the resource data field is specified in the type field and can be any of the following:

1. A number specified in bytes. For example, an IPv4 address is four bytes, and an IPv6 address is 16 bytes.
2. A domain name, which is expressed as a variable-length sequence of labels where a one-byte length field that defines the number of bytes in a label precedes each label. Because all domain names end with a null label, the last byte of all domain names is a length field containing the value 0. To prevent the length field from being confused with an offset pointer (explained next), the two high-order bits in a length field are always 0s. Therefore, the length field cannot hold a value greater than 63 (00111111 binary).

Table 25-7 Query Types

Type	Mnemonic	Description
1	A	A 32-bit host IP address used to convert a domain name to an IP address
2	NS	Identifies an authoritative name server
5	CNAME	Defines the canonical name for an alias for the official name
6	SOA	Marks the beginning of a zone of authority and is usually the first record in a zone file
7	MB	Mailbox domain name (experimental)
8	MG	Mail group member (experimental)
9	MR	Mail rename domain name (experimental)
10	Null	Null resource record (experimental)
11	WKS	Well-known service description that defines the network services a host provides
12	PTR	Domain name pointer used to convert an IP address to a domain name
13	HINFO	Host information that gives the description of the hardware and the operating system used by the host
14	MINFO	Mailbox or mail list information
15	MX	Mail exchange—redirects mail to a mail server
16	TXT	Text strings
17	RP	Responsible person
18	AFSDB	AFS data base location
19	X.25	X.25 public switched data network (PSDN) address
20	ISDN	Integrated Services Digital Network (ISDN) address
21	RT	Route through
22	NSAP	NSAP address, NSAP style A record
23	NSAP-PTR	NSAP resource record pointer
24	SIG	Security signature
25	KEY	Security key
26	PX	X.400 mail mapping information
27	GPOS	Geographical position
28	AAAA	IPv6 address
29	LOC	Location information
30	NXT	Next domain
31	EID	Endpoint identifier
32	NIMLOC	Nimrod locator
33	SRV	Server selection
34	ATMA	ATM address
35	NAPTR	Naming authority pointer
36	KX	Key exchange
37	CERT	CERT
38	A6	A6
39	DNAME	DNAME
40	SINK	SINK
41	OPT	OPT
100–103		IANA reserved
249	TKEY	Transaction key
250	TSIG	Transaction signature
251	IXFR	Incremental transfer
252	AXFR	Request for the transfer of an entire zone
253	MAILB	Mailbox-related resource record (MB, MG, or MR)
255	ANY	Request for all records

Table 25-8 Query Classes

Class	Mnemonic	Description
1	IN	Internet
3	CS	The COAS network
4	HS	The hesiod server developed by MIT

Domain Name Field	Domain Type Field	Domain Class Field	Time to Live Field	Resource Data Length Field	Resource Data
Variable Length	2 bytes	2 bytes	4 bytes	2 bytes	Variable Length

FIGURE 25-19 Resource record format

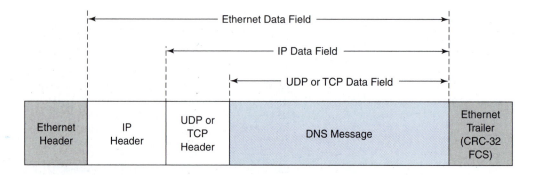

FIGURE 25-20 DNS encapsulated in an IP datagram being transported in UDP or TCP, which is encapsulated in an Ethernet frame

3. An offset pointer can replace the domain name. Offset pointers are two bytes long; however, the two high-order bits must both be 1s. Therefore, the highest value is $2^{14} - 1$, or 16,383.
4. A character string that contains a one-byte length field followed by the number of characters defined in the length field. The maximum length of a character string is 256 characters, which includes the length field.

25-10 DNS ENCAPSULATION

DNS can be encapsulated in either UDP or TCP. UDP is used when the DNS message is 512 bytes or less, and TCP is used for longer messages. Figure 25-20 shows the encapsulation of a DNS message. With either UDP or TCP, the well-known port number 53 is used for the server, and the client uses a temporary port number.

QUESTIONS

25-1. Define *configuration protocols*.

25-2. Define *domain name protocols*.

25-3. List and describe what is included in *bootstrap information*.

25-4. What are the two most common *autoconfiguration protocols*?

25-5. Explain when *BOOTP* is most appropriate.

25-6. Describe the format for encapsulating a BOOTP packet in an Ethernet frame.

25-7. List and describe the fields that comprise a BOOTP header.

25-8. Describe the operation of BOOTP.

25-9. What are the two most common *autoconfiguration protocols*?

25-10. Explain when *DHCP* is most appropriate.

25-11. Describe the format for encapsulating a DHCP packet in an Ethernet frame.

25-12. List and describe the fields that comprise a DHCP header.

25-13. Describe the operation of DHCP.

25-14. List the characteristics of DHCP.

25-15. Explain the three software elements of DHCP.

25-16. Describe the structure of the two DHCP databases.

25-17. Describe a DHCP *lease* and explain why it is used.

25-18. List and describe the three types of DHCP leases.

25-19. Describe the purpose of the *flag field* in a DHCP message.

25-20. List the DHCP options and describe their purposes.

25-21. Explain DHCP states and procedures.

25-22. Explain the DHCP address discovery process.

25-23. Explain the DHCP address renewal process.

25-24. Explain the DCHP address release process.

25-25. Describe the meaning and purpose of *DNS*.

25-26. Explain *name resolution* and *inverse resolution*.

25-27. List and describe the nine DNS resource records.

25-28. Describe the structure of *domain names*.

25-29. Explain *domain name space* and a *domain name tree*.

25-30. List and describe the three types of domains.

25-31. Explain the difference between *fully qualified* and *partially qualified domain names*.

25-32. Compare the terms *domain, name space distribution, zones,* and *root servers*.

25-33. Describe the two types of DNS message formats.

25-34. Describe the fields that comprise a DNS header.

25-35. Explain DNS *question, answer,* and *authoritative sections*.

25-36. Describe the DNS *additional information section*.

25-37. Describe DNS *question* and *resource record formats*.

25-38. Describe how DNS is encapsulated in an Ethernet frame.

PROBLEMS

25-1. What are the minimum and maximum lengths of a BOOTP packet?

25-2. What is the maximum time (in seconds) that can be stored in the number of seconds field of a BOOTP packet?

25-3. What is the least specific label in the domain name eng.lib.syr.edu?

25-4. What is the root label for the domain name eng.lib.syr.edu?

25-5. What is the maximum length of the query name subfield?

25-6. What is the maximum length of a name subfield in a DNS query message?

25-7. For the string 122.63.110.103.in-addr.arpa, what is the network address of the host?

25-8. Which of the following are FQDNs, and which are PQDNs?

    a. aaa

    b. aaa.bbb.

    c. aaa.bbb.net

    d. aaa.bbb.ccc.edu

25-9. Which of the following are FQDNs, and which are PQDNs?

    a. net.

    b. edu.

    c. aaa.net

    d. aaa.bb.ccc.net

25-10. Analyze the flag 8F80 hex.

25-11. Is the length of a question record fixed?

25-12. Is the length of a resource record fixed?

25-13. What is the length of a question record carrying the domain name: library.edu?

25-14. What is the length of a question record carrying an IP address?

25-15. What is the length of a resource record carrying the domain name: library.edu?

25-16. What is the length of a resource record carrying an IP address?

C H A P T E R   26

TCP/IP Application-Layer Protocols

CHAPTER OUTLINE

OBJECTIVES

- List the services provided by application-layer protocols
- Describe the purpose of the Telnet protocol
- List the services provided by the Telnet protocol
- Define a network virtual terminal
- Describe the NVT character set
- Describe Telnet encapsulation and embedding
- Describe the Telnet negotiations, options, and suboptions
- Describe the Telnet modes of operation
- Describe the purpose of the FTP protocol
- Explain FTP connections, file types, data structures, and transmission modes
- Explain how FTP commands and responses are processed
- Describe how an FTP file transfer occurs
- Describe the purpose of TFTP
- Explain the TFTP message types
- Describe TFTP connection establishment and termination
- Explain TFTP flow and error-control mechanisms
- Describe the purpose of SMTP
- Describe the makeup of an e-mail address
- List and describe SMTP commands and responses

- Describe the purpose of POP
- Describe the purpose of HTTP
- List the HTTP features
- Describe HTTP transactions
- Describe HTTP request and response message formats
- Describe a URL
- List and describe the HTTP headers

26-1 INTRODUCTION

Application-layer protocols allow users to access a network and provides user interfaces and support for many services, such as electronic mail (e-mail), shared database management, file access, file transfer, and so on. These services are an integral part of the TCP/IP protocol suite.

Application-layer protocols provide the following services:

Network virtual terminal (NVT). The NVT is the software version of a physical terminal that allows users to log on to remote hosts.

File Transfer, Access, and Management (FTAM). The FTAM application allows users to access, retrieve, manage, and control files stored in remote computers.

Electronic mail service. Electronic mail service applications provide a means for e-mail storage and forwarding.

Directory services. Directory service applications provide distributed database resources and a means to access global information concerning a multitude of objects and services.

The primary purpose of the Internet and the TCP/IP protocol suite is to provide users essential services that allow them to execute a variety of applications programs at a remote location. One possibility is to create clientserver application software for each service, which would be highly inefficient, expensive to implement, and a network manager's worst nightmare. Reliable stream delivery protocols, like TCP, make it possible for remote machines to interact with servers using keystrokes to send information and then reading the response.

A more reasonable solution is to provide a general-purpose clientserver protocol that allows users access to more than one application, especially since there are already several file transfer protocols, such as FTP and TFTP, and e-mail protocols, such as SMTP and PPP, available.

Figure 26-1 shows some of the application-layer protocols (processes) commonly used with TCP/IP and where they are located in the TCP/IP protocol suite relative to the lower-layer protocols. As can be seen, some application-layer protocols are transported over TCP and some over UDP. Since TCP is a reliable, connection-oriented transport-layer protocol and UDP is an unreliable, connectionless protocol, it is also true that the application-layer protocols transported over TCP are more reliable than those transported over UDP. Therefore, applications where reliability is essential, such as FTP and SMTP, are transported over TCP, and applications where reliability is not essential, such as TFTP, are transported over UDP.

Chapter 25 described the configuration and domain name protocols BOOTP, DHCP, and DNS, which are application-layer protocols. However, these protocols are used for exchanging addressing information and not for transferring actual user information. The purpose of this chapter is to introduce the reader to several of the more common application-layer protocols used with TCP/IP.

26-2 TELNET

Telnet, which stands for *TErminaL Network,* provides bidirectional byte-oriented communication that was originally designed to provide a communications method for terminal access. Telnet is generally thought of as a *remote terminal protocol* that provides a means for computers to accomplish local log-in or to complete remote logins using a lo-

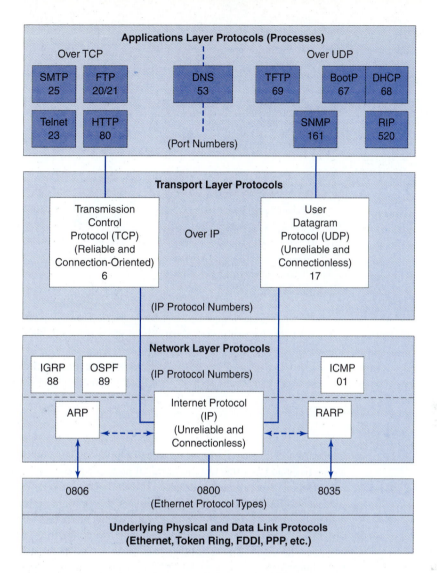

FIGURE 26-1 Application-layer protocols and the TCP/IP protocol stack

cal Telnet program to other computers over the Internet. Although this may have been its original intent, it can now be used for many other purposes. Telnet is rather unusual, as some consider it both a protocol and a program. The Telnet client program in one computer uses the Telnet protocol and TCP/IP to establish a virtual connection to a server program operating on another computer. The server side of the Telnet protocol allows the remote user to log in and operate like a dumb terminal attached directly to the server, which allows the terminal to perform remote log-ins and execute applications programs stored on the server using the Telnet server program to handle the communications needs. The Telnet server can pass data received from the client to other processes, including the remote login server.

Telnet allows hosts to exchange information about options they support while the connection between them is being established. After Telnet establishes a TCP connection between a user's host and a remote computer, it uses the connection to pass keystrokes from the user's keyboard directly to the remote computer just as if they had been typed on a keyboard attached directly to the remote machine. The remote computer is called a *network virtual terminal* (NVT). Telnet is considered a *transparent service*, as it appears that the user's

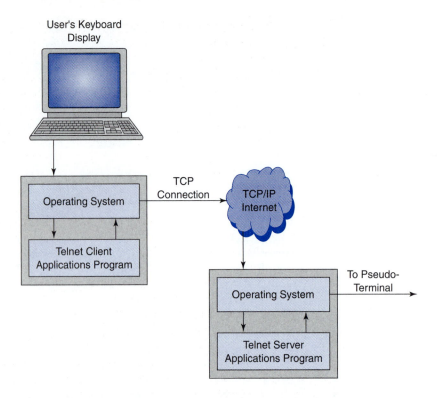

User's Keyboard
Display

TCP
Connection

TCP/IP
Internet

Operating System

Telnet Client
Applications Program

Operating System

To Pseudo-
Terminal

Telnet Server
Applications Program

FIGURE 26-2 Establishing a Telnet connection using a TCP connection

keyboard and display are attached directly to the remote computer. Although Telnet is not as sophisticated as some remote terminal protocols, however, it is widely available.

Telnet offers three basic services: it defines a network virtual terminal; it includes a mechanism for allowing clients and servers to negotiate options; and it provides a set of standard options. Figure 26-2 shows how a Telnet client and server implement applications programs. When client hosts summon Telnet, an application program on the user's machine becomes the client. The client then establishes a TCP connection with the server it wishes to communicate with. After the TCP connection has been established, the client program accepts keystrokes from the user's keyboard and transmits them to the server over the TCP connection. The server accepts the data and then relays it through the local operating system to a pseudoterminal.

26-2-1 The Network Virtual Terminal

To access remote computers operating on heterogeneous (diverse) systems, a host must know what type of computer it is connected to so the proper terminal emulator can be installed. For example, some systems requires that all lines of text be terminated with a carriage return character, while other systems require a two-character termination sequence (carriage return-line feed). Telnet solves this problem by defining how data, control, and command information is sent across the Internet using a universal interface known as the *network virtual terminal* (NVT) character set, which is a bidirectional character set. An NVT is defined as having a printer to respond to incoming data and a keyboard that generates outgoing data. An NVT keyboard has keys or key combinations and sequences for generating all 128 US ASCII codes, even if they have no effect on a NVT printer.

Using NVT, the Telnet software at the user or client end can translate (map) data and command characters received from the local terminal to their appropriate NVT characters and then deliver them to the network. The Telnet server receives NVT data and command

Table 26-1 NVT Control Codes

Name	Mnemonic	Value Hex	Value Decimal	Function
Null	NUL	0	0	No operation
Line feed	LF	0A	10	Moves the print head to the next print line with the same horizontal position
Carriage feed	CR	0D	13	Moves the print head to the left margin on the current print line
Bell	BEL	07	7	Produces an audible or visible signal
Backspace	BS	08	9	Moves the print head one space back
Horizontal tab	HT	09	9	Moves the print head to the next horizontal tab stop
Vertical tab	VT	0B	11	Moves the print head to the next vertical tab stop
Form feed	FF	0C	12	Moves the print head to the top of the next page, keeping the same horizontal position

characters from the network and translates (maps) them to a form accepted by the remote computer, which may or may not be the same character set used at the client. Comparable character translations must also be performed in the opposite direction.

26-2-2   NVT Character Set

The NVT character set includes two subsets: one subset is for data, and one subset is for control. NVT data characters use seven-bit codes, as display devices are required only to display, print, and process standard U.S. ASCII data and control characters, which are seven-bit codes. The seven-bit NVT characters are sent as eight-bit bytes with a logic 0 for the most-significant bit in each character. However, there is a Telnet option where it is possible to use the eight-bit extended ASCII code as long as the client and server agree on it.

There are three NVT control codes that all NVT terminals must understand: null (NUL), line feed (LF), and carriage return (CR). An end-of-line is transmitted as a carriage return followed by a line feed. An actual carriage return is transmitted as a carriage return followed by a null (NUL) character (all 0s). In addition, there are five NVT control codes that are optional: bell (BEL), backspace (BS), horizontal tab (HT), vertical tab (VT), and form feed (FF). The NVT control codes along with their mnemonic, hex and decimal values, and a brief description are listed in Table 26-1.

The Telnet protocol also specifies several remote control characters, called *commands*, that control the method and various details of the interaction between a client and a server. Commands are incorporated within the data stream and distinguished from NVT data characters by making the most-significant bit a logic 1. Commands are always introduced by a character with the hex code FF (decimal 255), known as an interpret as command (IAC) character. Telnet command codes along with their mnemonic, hex and decimal values, and a brief description of each are listed in Table 26-2.

26-2-3   Telnet Encapsulation and Embedding

Figure 26-3 shows an Ethernet frame carrying a Telnet message encapsulated in a TCP segment that is encapsulated in an IP datagram. The server uses the well-known port 23, and the client uses a random port. Telnet uses a single TCP connection for sending data and control information in both directions. Sending data and control information over the same TCP connection is made possible by embedding control characters in the data stream. Control characters are distinguished from data characters by preceding each control character with a special character called an *interpret as control* (IAC) character.

Table 26-2 NVT Command Codes

Name	Mnemonic	Hex	Decimal	Function
		Value		
End of file	EOF	EC	236	Designates the end of a file
End of record	EOR	EF	239	Designates the end of a record
Suboption end	SE	F0	240	End of a suboption or subnegotiation parameters
No operation	NOP	F1	241	Performs no function other than as a time fill
Data mark	DM	F2	242	Indicates the position of a synch event within a data stream The DM should always be accompanied by a TCP urgent notification
Break	BRK	F3	243	Indicates that the break or attention key was depressed
Interrupt process	IP	F4	244	Suspend, interrupt, or abort the process the NVT is connected to
Abort output	AO	F5	245	Allows the current process to complete but does not send its output to the user
Are you there	AYT	F6	246	Send back to the NVT some visible evidence that the AYT was received
Erase character	EC	F7	247	The receiver should delete the last undeleted character from the data stream
Erase line	EL	F8	248	The receiver should delete the last undeleted line from the data stream
Go ahead	GA	F9	249	Used under certain circumstances to tell the opposite end that it can now transmit
Subnegotiation begins	SB	FA	250	Subnegotiation of the indicated option follows
Will	WILL	FB	251	Indicates the desire to begin performing, or confirming that you are now performing, the indicated option
Won't	WONT	FC	252	Indicates the refusal to perform, or continue performing, the indicated option
Do	DO	FD	253	Indicates the request that the other party perform, or confirmation that you are expecting the other party to perform, the indicated option
Don't	DON'T	FE	254	Indicates the demand that the other party stop performing, or confirmation that you are no longer expecting the other party to perform, the indicated option
Interpret as command	IAC	FF	255	Interpret as a command

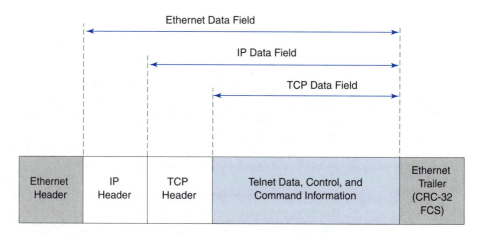

FIGURE 26-3 Telnet encapsulated in an IP/TCP datagram being transported over an Ethernet local area network

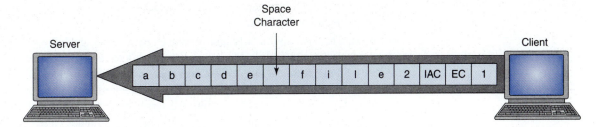

FIGURE 26-4 Embedding Telnet control characters

Table 26-3 Telnet Options

Name	Decimal Code	Meaning
Transmit binary	0	Change to an eight-bit code
Echo	1	Echo the data received from one side to the other side
Suppress go-ahead	3	Suppress (no longer send) go-ahead signals after data
Status	5	Request for status of a Telnet option from remote site
Timing mark	6	Request timing mark be inserted in return stream to synchronize the two ends of a connection
Terminal type	24	Exchange information about the make and model of a terminal
End of record	25	Terminate transmitted data with EOR character
Terminal speed	32	Set the terminal speed
Line mode	34	Change to the line mode (use local editing and send complete lines instead of individual characters)

Figure 26-4 shows an example of embedding Telnet control characters within a string of Telnet data characters. The client is sending a sequence of five data characters (abcde) and a file named (file1). However, the operator mistyped the file name as file2. With the default implementation of Telnet, a user cannot edit locally. Therefore, to correct the data, the user must send a backspace character and then the correct character, as shown here:

abcde file2<backspace>1

The backspace is converted to two NVT remote characters: interpret as control character (IAC) and erase character (EC), which are embedded between the mistyped message and the correct character and sent to the remote server, as shown in Figure 26-4.

26-2-4 Telnet Options

Like with other TCP/IP protocol, such as IP and TCP, Telnet options are parameters, conventions, and extra features and capabilities available to users who have terminals that are more sophisticated than the average terminal. However, users with less sophisticated terminals can still use the minimum features of Telnet. Telnet options allow clients and servers to reconfigure their connection.

Telnet clients and servers agree on options through a process of negotiation. The options can be established prior to utilizing the service or while the service is being used. Several of the NVT remote command characters discussed earlier are used to define options. Table 26-3 lists several of the more common Telnet options, their hex and decimal values, and the purpose of the option. A brief description of each option is given here:

Transmit binary option. The *transmit binary* option allows a receiver to accept all data characters as eight-bit binary characters, except the IAC character. When the IAC character is received, the following character or characters are interpreted as Telnet commands. If two IAC characters are received in succession, the first is discarded, and the second is interpreted as data.

Echo option. The *echo* option is generally enabled by a server and allows the server to echo (repeat) data received from a client. When using the echo option, every character sent by a client to a server will be sent back to the client's screen. Therefore, when a user depresses a key at the client end, the character is sent to the server but not displayed on the client's keyboard until the server sends the character back. The echo option operates in the same as an old error detection technique called *echoplex*. When a keystroke is displayed on the user's screen, it is a positive indication that the server received the character.

Suppress go-ahead option. The *suppress go-ahead* option simply suppresses the go-ahead (GA) character. The go-ahead character is sent from a server to a client to tell the client to accept a new line of characters from the user. The go-ahead character is discussed in more detail in the "Telnet Modes of Operation" section later in this chapter.

Status option. The *status* option is a means for a user or the process running on the client's machine to acquire status from the server concerning what Telnet options the server has enabled.

Timing mark option. The *timing mark* option allows the client or request that a timing mark be inserted in the return data stream for synchronization purposes. The timing mark confirms that all previously received data has been processed.

Terminal type option. The *terminal type* option gives the client a means of telling the server what terminal type the client is using, such as make and model. This allows programs to tailor their output, such as cursor positioning sequence, for a specific type of terminal.

Terminal speed option. The *terminal speed* option gives the client a means of telling the server its terminal speed.

End of record. The *end-of-record* (EOR) option terminates transmitted data with an EOR character.

Line mode option. The *line mode* option allows a client to switch to the line mode and use local editing and send complete lines instead of individual characters. The line mode is discussed in more detail in the "Telnet Modes of Operation" section later in this chapter.

26-2-5   Telnet Options Negotiations

A client and a server must negotiate before any of the Telnet options described in the previous section can be used. Initiating Telnet options is a *symmetrical protocol,* which means either the client or the server are given equal opportunity to request an option. One end requests an option, and the other end either accepts or rejects the option. A server can enable some Telnet options, the client can enable some options, and both the client and the server can enable some options. Options are agreed on before being used, which allows the client and server to share a common view of the extra capabilities that affect the interchange of data.

There are four control characters used with Telnet to negotiate options. The four characters, their decimal values, and their meanings are listed in Table 26-4. Either end of a Telnet dialogue can enable or disable an option either locally or remotely. Whichever end initiating an option sends a three-byte command of the form

IAC <operation> option

When IAC is interpreted as control, the operation is one of the control characters listed in Table 26-4, and the option is one of those listed in Table 26-3. Responses use the same format and the same control characters. Table 26-5 shows the control characters sent by the

Table 26-4 NVT Characters for Telnet Option Negotiation

Control Character	Decimal Value	Meaning of Character
WILL	251	Sender offering to enable an option or receiver accepting the request to enable an option
WONT	252	Sender rejecting an offer to enable an option, offering to disable an option, or accepting an offer to disable an option
DO	253	Sender approving an offer to enable an option or requesting to enable an option
DONT	254	Sender disapproving an offer to enable an option, approving an offer to disable an option, or offering to disable an option

Table 26-5 Telnet Option Enable/Disable Sequence

Sender Transmits	Receiver's Response	Implication
WILL	DO	Sender offers to enable an option, and receiver acknowledges that the option is now in effect
WILL	DONT	Sender offers to enable an option, and receiver rejects the option request
DO	WILL	Sender requests the receiver to enable an option, and receiver accepts
DO	WONT	Sender requests the receiver to enable an option, and receiver refuses
WONT	DONT	Sender offers to disable an option, and receiver confirms the offer
DONT	WONT	Sender requests the receiver not to use an option, and the receiver confirms it will not

initiating end (client or server) to offer to enable or disable an option and the corresponding control character returned by the other end (client or server) to confirm or refuse an offer. When a client sends a request, a server sends a response and vice versa.

26-2-6 Enabling and Disabling Telnet Options

As previously stated, some Telnet options can be enabled or disabled only by the client, some only by the server, and some by both the client and the server. All options are enabled or disabled through either an offer or a request. However, the Telnet protocol specifies that with some options only the client has the right to make an offer or request and that with some options only the server has the right.

Enable offer. A client or server can offer to enable an option, but only if the protocol gives it the right to do so. The receiving end can either approve or disapprove the offer. Figure 26-5a shows the process of offering to enable an option. The offering end sends a WILL control character, which asks will I enable this option, and the receiving end responds with either the DO command, which accepts the offer (i.e., please do), or the DONT command, which rejects the offer (i.e., please don't).

Request to enable. A client or server can request the other end to enable an option, and the request can be accepted or refused. Figure 26-5b shows the process of requesting an option to be enabled. The requesting end sends a DO command requesting the option be enabled (i.e., do enable the option). The receive end responds with the WILL command acknowledging the option has been enabled (i.e., I will) or the WONT command refusing to enable the option (i.e., I won't).

Disable offer. A client or server can offer to disable an option and the other end must approve the offer (it cannot disapprove the offer). Figure 26-5c shows the process of offering to disable an option. The offering end sends the WONT command offering

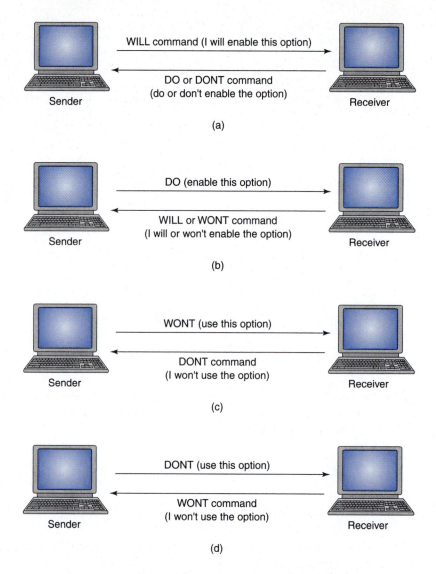

FIGURE 26-5 Telnet option negotiation: (a) offering to enable an option; (b) requesting to enable an option; (c) offering to disable an option; (d) requesting to disable an option

not to use the option (i.e., I won't use the option). The receive end must respond with the DONT command confirming the offer (i.e., don't use the option).

Request to disable. A client or server can request the other end disable an option, and the other end cannot reject the request. Figure 26-5d shows the process of requesting an option be disabled. The requesting end sends the DONT command requesting the other end not use an option (i.e., don't use this option). The receive end responds with the WONT command acknowledging it will not use the option (i.e., I won't use it).

26-2-7   Telnet Suboptions

With some Telnet options, additional information is necessary after the primary option has been enabled. For example, when the terminal type option is enabled, the type of terminal must be specified, and when the terminal speed option is enabled, the speed must be established. The additional information is sent in a suboption. A suboption begins with a SB (suboption begin, code 250) character and ends with a SE (suboption end, code 240) character.

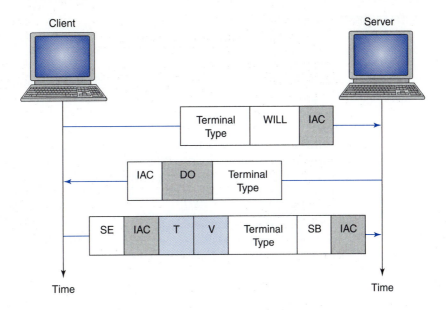

FIGURE 26-6 Suboption negotiation

Figure 26-6 shows an example of how Telnet suboptions are used. The client initially sends an offer to enable (WILL) command for the terminal type option. The server responds with an approval (DO) command. The client now sends a suboption begin command followed by the option name and then adds the characters "VT," meaning it will use the VT (virtual terminal) character set and keyboard layout.

26-2-8 Telnet Modes of Operation

There are three modes of operation specified for Telnet: default, character, and line.

26-2-8-1 Default mode.   The *default mode* is just that. If no other mode has been specified through option negotiations, the default mode is assumed. With the default mode of operation, the client does the echoing. The user types a character, and the client computer echoes the character back to the screen or the printer but does not actually transmit the character until an entire line of text has been entered. After a complete line of text has been transmitted to the server, the client waits for the go-ahead (GA) command before allowing the user to enter another line of text. Therefore, the default mode of operation is half duplex, which is not an efficient way to use a TCP connection, as TCP is capable of full-duplex communications.

26-2-8-2 Character mode.   With the *character mode,* after a user types each character, the client immediately sends it to the server, which normally echoes the character back to the client where it is displayed on the screen or printed on the printer. With the character mode, the character may not be immediately echoed, which creates additional overhead in the form of the user wasting time waiting for the echo before continuing. Additional overhead is created because each character entered by the user requires a three-step TCP sequence before it is sent.

26-2-8-3 Line mode.   With the *line mode,* editing (echoing, character erasing, line erasing, and so on) is performed by the client. After the editing is complete, the client sends the entire line to the server using a single TCP connection. The line mode is a relatively new mode that closely resembles the default mode. However, with the line mode, communications occurs full duplex with the client transmitting one line after another, without having to wait for the go-ahead character from the server.

File transfer protocol (FTP) provides a means for transferring files over a reliable connection-oriented transport protocol, such as TCP. FTP is the standard software mechanism used in the TCP/IP protocol suite for transferring a file from one host computer to another, which is probably the most common tasks performed in a networking environment. FTP client software is used to transfer files between hard drives and remote servers. Since servers are computers where Web pages are stored and distributed, FTP is also used to transfer Web browser files from remote Web servers to computers to graphically display Web information on the monitor. With FTP, you can download very large files and resume the transfer where it left off after an interruption, which saves time and aggravation. FTP can be used with text-based clients or with client software that uses a *graphical user interface* (GUI).

FTP is ideally suited to transferring files between two systems that use different file name conventions and when the two systems use different methods to represent text and data. FTP is also a good choice when the two systems have different directory structures. FTP has four essential objectives:

1. To promote file sharing
2. To encourage indirect or implicit (through software) use of remote computers
3. To protect users from differences in file storage systems among hosts
4. To transfer data and control information reliably and efficiently

FTP establishes two full-duplex (both directions simultaneously) connections between host computers. One connection is for transferring data, and the other connections are for transferring control information (i.e., commands and responses). Separating data and control functions is what makes FTP more efficient than other clientserver file transfer applications. FTP is encapsulated in TCP and uses two well-known port numbers. Port 21 is for transferring control information, and port 20 is for transferring data information. Figure 26-7 shows an Ethernet frame carrying FTP. As shown in the figure, the FTP packet is encapsulated in a TCP segment, which is encapsulated in an IP datagram, which is then encapsulated in the Ethernet frame.

Figure 26-8 shows the basic model for FTP. The client host has several essential components: a user interface, client control process software, client data transfer process software, a file system, and a set of FTP commands. The server host has two components: server control process software and server data transfer process software. The control con-

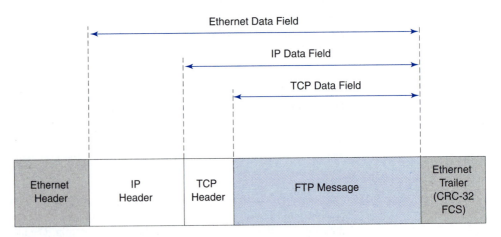

FIGURE 26-7   FTP encapsulated in an IP/TCP datagram being transported over an Ethernet local area network

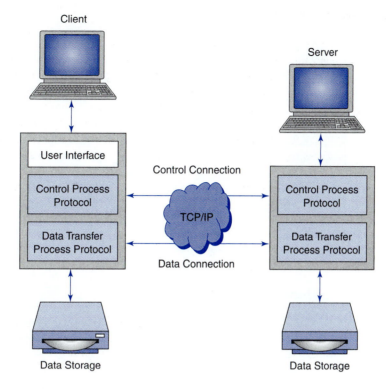

FIGURE 26-8 FTP model

nection is between the control process software in the two computers, and the data connection is between the data process software in the two computers. The control connection remains connected during the entire FTP file transfer session, whereas the data connection is opened before each file transfer and closed after each file transfer. The data connection is opened whenever commands are used that involve transferring files and closed when the file transfer is completed. Therefore, when a user begins an FTP session, the control connection opens and remains open while the data connection is opened and closed for each file transfer.

26-3-1 FTP Connections

There are two types of FTP connections—data transfer and control—that use different strategies and are connected to different ports. Files are transferred only over the data connection, as the control connection is used exclusively to transfer commands and to describe the functions to be performed.

26-3-1-1 FTP control connections. There are two steps to establishing a control connection:

1. The server creates a passive open on well-known port 21 and then waits for a client to request service.
2. The client issues an active open using a temporary port, which opens the control connection.

Once the control connection is opened, it remains open for the entire process. FTP control information uses the minimum delay IP service type because the control connection is an interactive real-time connection between the user (a human) and the server. When the user enters a command, it expects to receive immediate responses. Figure 26-9 illustrates how the initial control connection is established between a client using random port 60000 and

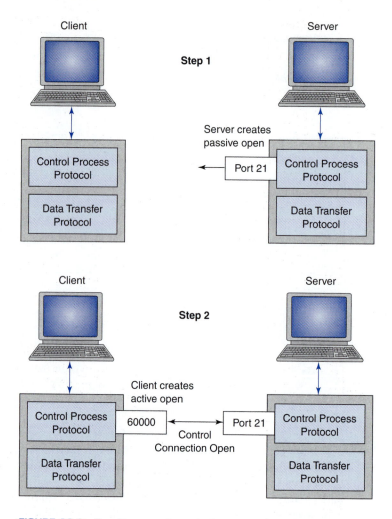

FIGURE 26-9 Opening control connecting

a server using the well-known port 21. After the initial control connection is established, the server process creates a child process, which is assigned the task of serving the client using a temporary port.

26-3-1-2 FTP Data-Transfer Connections At the server, the data connection uses the well-known port 20. There are three steps to establish a data connection:

1. The client initiates the connection by issuing a passive open using a temporary port. The client initiates this process because the client issues the commands necessary to transfer files.
2. The client sends the temporary port number to the server using a PORT command, which is described later.
3. After the server receives the temporary port number, it issues an open active using the well-known port 20 and the temporary port number provided by the client.

After the initial data transfer connection is established, the server process creates a child process, which is assigned the task of serving the client using a temporary port. Figure 26-10 illustrates how the initial data connection is established between a client using random port 61000 and a server using the well-known port 20.

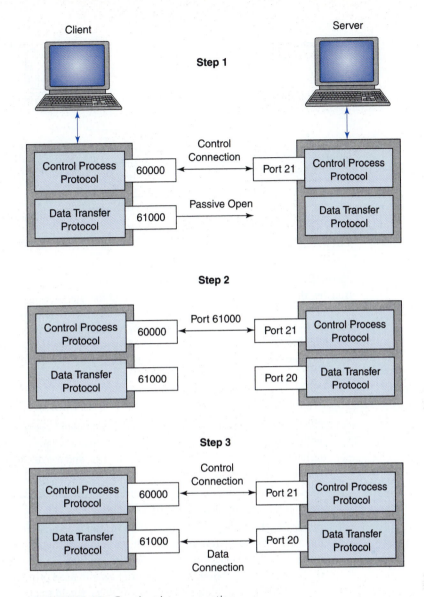

Client
Server

Step 1

Control Process Protocol | 60000 ←→ Port 21 | Control Process Protocol
Control Connection

Data Transfer Protocol | 61000 → Passive Open | Data Transfer Protocol

Step 2

Control Process Protocol | 60000 ←→ Port 21 | Control Process Protocol
Port 61000

Data Transfer Protocol | 61000 | Port 20 | Data Transfer Protocol

Step 3

Control Process Protocol | 60000 ←→ Port 21 | Control Process Protocol
Control Connection

Data Transfer Protocol | 61000 ←→ Port 20 | Data Transfer Protocol
Data Connection

FIGURE 26-10   Opening data connecting

26-3-1-3 Control channel communications.   FTP uses the *network virtual termi-nal* (NVT) ASCII character set, which is defined in the ARPA-Internet Protocol Handbook, to communicate across the control connection by exchanging commands and responses, as shown in Figure 26-11. FTP clients send commands, and FTP servers send responses. With FTP, ASCII characters are defined as the lower half of an extended ASCII eight-bit code set where the most-significant bit is 0. ASCII code is sufficient for the control connection be-cause commands and responses are sent one at a time. Since each command or response is comprised of only one short message, file format and structure is not a concern. Each mes-sage is terminated with a two-character end-of-token sequence, which is simply a carriage return character followed by a line feed character.

26-3-1-4 Data channel communications.   Data connections have different pur-poses and use a different implementation than control connections because data connections are used to transfer data files rather than control information between the client and the

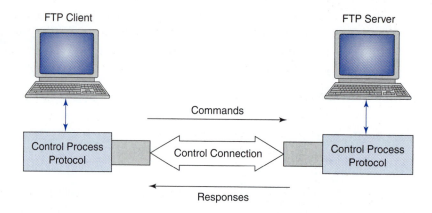

FIGURE 26-11 FTP communications using commands and responses

server. Before a file can be transferred, the client must specify what type of file it intends to transfer, the structure of the data in the file, and the transmission mode it wants to use.

There are two byte sizes allowed with FTP: the logical byte size of the file and the transfer byte size used for the transmission of data. The transfer byte size is always eight bits, which is not necessarily the byte size of the stored data in the system or the logical byte size for interpretation of the structure of the data.

26-3-2 FTP File Types

There are three types of files that can be transferred with FTP: *ASCII, EBCDIC,* and *image*.

ASCII file. The NVT ASCII code is the default character set for transferring text files with FTP. The source must convert its files from whatever representation it uses internally to NVT ASCII before sending the file, and the destination must convert NVT ASCII code it receives to whatever representation it uses internally.

EBCDIC file. If either host (or both) uses the EBCDIC code internally, the file can be transferred directly using the EBCDIC code.

Image file. The image file is the default format for transferring continuous streams of bits, which for transfer are packaged into eight-bit bytes. Therefore, the destination must store the data as contiguous bits. Image files are transferred without using any type of code conversion. The image file type is used primarily for the efficient storage and retrieval of binary files, such as those found in compiled programs.

If the file is encoded in either the ASCII or the EBCDIC code, the printability of the file must be defined as either *nonprint* or *TELNET*:

Nonprint. Nonprint is the default format for transferring text files when the format parameter is omitted. Because the file does not contain any vertical printing specifications (vertical tab, line feed, and so on) to direct the print head, the file cannot be printed without processing the information further. Generally, a printer will assume standard values for spacing and margins. The nonprint format is used when the file is going to be stored and processed at a later time.

TELNET. With the TELNET format, the file contains NVT ASCII–encoded vertical format control characters, such as CR (carriage return), LF (line feed), NL (new line—carriage return and line feed), and VT (vertical tab). With the TELNET format, the file can be printed immediately after being received without requiring any further processing.

26-3-3 FTP Data Structures   FTP is capable of transferring data using one of three structures: *file, record,* or *page*.

File structure. The file structure, which is the default structure, has essentially no internal structure, and the data is transferred as a continuous stream of data bytes.

Record structure. With the record structure, the file is divided into sequential records, which can be used only with text files. Record structures must be accepted for text files (i.e., files using the ASCII or EBCDIC code).

Page structure. The page structure is designed to transmit files that are discontinuous, sometimes known as *random access files* or as *holey* files. The page structure divides the file into separate indexed pages where each page has a number and a header. The pages can be stored or accessed either randomly or sequentially.

26-3-4   FTP Transmission Modes

The *transmission mode* defines how data is transported between the FTP and TCP software over the data connection. There are three transmission modes used with FTP: *stream* (default mode), *block,* and *compressed:*

Stream mode. With the stream transmission mode, data is transported from the FTP software to the TCP software as a continuous stream of bytes (i.e., file structure). The TCP software divides the data stream into groups of data called *segments.* Therefore, an end-of-file sequence is not required because simply closing the data connection terminates the file. If the record structure is used to transfer the data, the data is divided into records, and each record ends with a one-byte end-of-record (EOR) character and the file terminated with an end-of-file (EOF) character.

EOR and EOF are indicated with a two-byte control code. The first byte of the control code is all 1s, which is the escape (ESC) character. If the second byte is an EOR, it will have the low-order bit set (000000001), and if the second byte is an EOF, it will have the second low-order bit set (00000010). Thus, if the second byte is a binary 1, it is an EOR, and if it is a binary 2, it is an EOF. Both EOR and EOF can be set on the last byte transmitted, which is a binary 3 (00000011). To send an all-1s byte of data, it should be repeated in the second byte of the control code. The stream transmission mode is inherently somewhat unreliable because it is difficult to tell if the connection closed prematurely.

Block mode. With the block transmission mode, data is transported between the FTP software to the TCP software in groups of data called *blocks.* Each block is preceded by three header bytes. The first byte in the header is called the block *descriptor* code. There are currently four descriptor codes, whose decimal values listed in Table 26-6.

If the descriptor is an EOF, this is the last block in the file, and if the descriptor is an EOR, this is the last block in the record. A *suspected data* code indicates that the data being transferred is suspected of errors and not reliable. A *restart marker* code is provided to protect users from gross system failures, which includes failure of a host, an FTP process, or the underlying network.

The last two bytes of the header is a *count field,* which specifies the total length of the data block in bytes, thus indicating the beginning of the next block. The maximum block size is 65,535.

Table 26-6   FTP Descriptor codes

Decimal	Binary	Code Description
128	10000000	End-of-record block
64	01000000	End-of-file block
32	00100000	Suspected errors in data block
16	00010000	Data block is a restart marker

Compressed mode. The compressed transmission mode can be used if the file is excessively long. The compression method is generally *run-length encoding.* With run-length encoding, if a data unit (such as a space character) appears two or more times in succession, the unit is removed and replaced by one occurrence and a replication or filler byte that indicates the number of repetitions. To replace a string of six replications of the ASCII space character (20 hex), the following two bytes are sent:

2	6	20	hex value
1 0	00110	00100000	binary code

where the 1 0 in the two most-significant bits in the run-length code define the format for the replication byte

With the NVT ASCII code, the most-significant bit is 0 for the 120 valid characters. Therefore, run-length encoding is easily detected by the presence of a logic 1 in the most-significant bit, which is an invalid NVT ASCII character. The compressed transmission mode, like the block mode, does not close the connection to indicate the end of the file. Therefore, using the block or compressed transmission mode can leave the data connection for multiple file transfers.

26-3-5 Processing FTP Commands and Responses

An FTP control connection is used to establish a communications link between the client and server control processes. Clients send FTP commands, and servers send FTP responses.

26-3-5-1 FTP commands. *FTP commands* are sent as a sequence of uppercase ASCII characters that may be followed by an argument. FTP commands follow directly after the TCP header. There are essentially six groups of commands: access, file management, data formatting, port defining, file transferring, and miscellaneous. FTP commands specify the data connection (data port, transfer mode, representation type, and structure) parameters and the nature of the file system operation (store, retrieval, append, delete, and so on). Users should listen on the specified data port while servers initiate the data connection and data transfer in accordance with the specified parameters. However, it is not necessary that the data port be in the same host that initiates the FTP commands using the control connection.

26-3-5-2 Access commands. *Access commands* allow users to access the remote system. The most common access commands are listed in Table 26-7.

User (USER) command. The user command identifies the user name, which is required by the server to access its file system and is generally the first command transmitted after the control connection has been established.

Password (PASS) and account (ACCT) commands. The pass command is used for additional identification information, such as the user's password and account number. Some servers may also require the PASS and ACCT commands. The PASS command

Table 26-7 FTP Access Commands

Command	Argument	Description
USER	User ID	User information
PASS	User password	Password
ACCT	Account to be changed	Account information
REIN	None	Reinitialize
QUIT	None	Log out of system
ABOR	None	Abort the previous command

must be immediately preceded by the USER command, as it completes the user's identification for access control.

Reinitialize (REIN) command. The reinitialize command is used to terminate a user and flush all I/O and account information. However, the REIN command will not be implemented until any transfers in progress are completed.

Quit (QUIT) command. The user sends a quit (log-out) command to log out of the system. The quit command terminates a user, and if the file transfer is not in progress, the server closes the control connection. If a file transfer is in progress, the connection remains open until the transfer ends, at which time the server will close the connection.

Abort (ABOR) command. The abort command tells the server to abort the previous FTP service command and any associated data transfer. The ABOR command may require special action before being recognized by a server.

26-3-5-3 File management commands. *File management commands* allow users to access the file system on a remote computer, which allows the user to navigate through the directory structure, create new directories, delete files, and so on. Table 26-8 lists the most common file management commands.

Change working directory (CWD) command. The CWD command allows users to change to a different directory or data set for file storage or retrieval without altering log-in or accounting information. The argument is a path name, which specifies a directory or other system-dependent file group designator.

Change to parent directory (CDUP) command. The CDUP command is essentially a special case of the CWD command included to simplify the implementation of programs for transferring directory trees between operating systems using different syntaxes for naming their parent directory.

Delete (DELE) command. The delete command causes the file specified in the path name to be deleted from the server. The DELE command sometimes requires a second level of protection, such as a query, before the file is actually deleted.

List (LIST) command. The list command causes a list of subdirectories to be sent from the server to the passive data transfer process over the data connection. If the path name specifies a directory or other group of files, the server should transfer a list of files in the specified directory. If the path name specifies a file, the server should send current information on the file.

Name list (NLIST) command. The name list command causes a directory listing of subdirectories or files without other attributes to be sent from the server to the user site. The path name should specify a directory or other system-specific file group descriptor.

Table 26-8 FTP File Management Commands

Command	Argument	Description
CWD	Directory name	Change to a different directory
CDUP		Change to the parent directory
DELE	File name	Delete a file
LIST	Directory name	List subdirectories or files
NLIST	Directory name	List the names of the subdirectories or files without other attributes
MKD	Directory name	Creates a new directory
PWD	None	Display the name of the current directory
RMD	Directory name	Delete a directory
RNFR	Old file name	Identify a file to be renamed
RNTO	New file name	Rename a file
SMNT	File system name	Mount a file system

Table 26-9 FTP Data Formatting Commands

Command	Argument	Description
TYPE	A (ASCII), E (EBCDIC), I (Image), N (Nonprint), or T (Telnet)	Defines the file type and, if necessary, the print format
STRU	F (File), R (Record), or P (Page)	Defines the organization of the data
MODE	S (Stream), R (Record), or C (Compressed)	Defines the transmission mode

Make directory (MKD) command. The make directory command causes the directory specified in the path name to be created as a directory or a subdirectory of the current working directory.

Structure mount (SMNT) command. The structure mount command allows users to mount different file system data structures without altering their log-in or accounting information. The argument is a path name specifying a directory or other system-dependent file group designator.

26-3-5-4 Data formatting commands. *Data formatting commands* allow the user to define the data structure, file type, and transmission mode, which are then used by the file transfer commands. Table 26-9 lists the most common FTP data formatting commands.

Type (TYPE) command. The TYPE command is means for a user to specify a representation type, which defines the file type (first byte) and sometimes the print format (second byte). The TYPE commands are listed here, where the default representation type is ASCII nonprint:

A—ASCII code

E—EBCDIC code

N—Nonprint

C—Carriage control

T—Telnet format effectors

I—Image

L—Local byte size

File structure (STRU) command. The STRU command is a single Telnet character that specifies the file structure. The STRU codes are listed here, where the default structure is file:

F—File (no record structure)

R—Record structure

P—Page structure

Transfer mode (MODE) command. The MODE command is a single Telnet character code specifying the data transfer modes. The MODE codes are listed here, where the default mode is stream:

S—Stream

B—Block

C—Compressed

26-3-5-5 Port defining commands. *Port defining commands* specify the port number for the client-side data connection. There are two ways to accomplish this, as listed in Table 26-10. One method uses the port (PORT) command, and the other method uses the passive (PASV) command:

Table 26-10 FTP Port Defining Commands

Command	Argument	Description
PORT	Six-digit identifier	Client chooses a temporary port
PASV	None	Server chooses a temporary port

Port (PORT) command. The PORT command is a client-port specification that specifies the data port used for a data connection. There are defaults for both the client and the server data ports, and normally this command and its reply are not needed. However, if the PORT command is used, the command argument is the concatenation (interconnection) of the 32-bit IP address and the 16-bit TCP port number. The address information is divided into eight-bit fields where the value of each field is represented as a decimal number in character string representations where the fields are separated by commas, as shown here:

$$\text{PORT } h1, h2, h3, h4, h5, h6$$

where $h1$ is the high-order eight bits of the IP address and $h5$ is the high-order eight bits of the port number.

The client uses the PORT command to choose a temporary port number, which the server uses to create an active open.

Passive (PASV) command. A client sends the PASV command to request the server to choose a port number that is not the default port. After selecting a port number and creating a passive open, the server sends the port number with the response code 227 (from Table 26-10). After receiving the response, the client issues an active open using the assigned port number.

26-3-5-6 File transferring commands. *File transferring commands* (sometimes called *file service commands*) define the file transfer or the file system function requested by the user. FTP file transfer commands are listed in Table 26-11. The argument is a path name whose syntax must conform to server-site conventions using standard defaults and the language conventions of the control connection. The suggested default is the last specified device, directory, or file name or the standard default defined for local users. File transferring commands can be in any order, except a rename-from (RNFR) command must always be followed by a rename-to (RNTO) command, and the restart (REST) command must always be followed by the store (STOR) or retrieve (RETR) command

Table 26-11 FTP File Transfer (Service) Commands

Command	Argument	Description
RETR	Path name(s)	Retrieve files, and file(s) are transferred from the server to the client
STOR	Path name(s)	Stores files, and files(s) are transferred from the client to the server
APPE	Path name(s)	Similar to STOR except if the file already exists, the data must be appended to it
STOU	Path name(s)	Same as STOR except that the file name will be unique in the directory, and the existing file should not be overwritten
ALLO	Path name(s)	Allocate storage space for the files at the server
REST	Path name(s)	Position the file marker at a specified data point
STAT	Path name(s)	Return the status of files
RNFR	Path name(s)	Return a copy of the file
RNTO	Path name(s)	Specifies a new path name

Retrieve (RETR) command. The retrieve command causes the server to transfer a copy of the file specified in the path name to the client. This command does not affect the contents of the file in the server.

Store (STOR) command. The store command causes the server to accept the data transferred over the data connection and store the data as a new file in the server. If the file specified in the path name already exists in the server, its contents will be replaced by the new data.

Appended (APPE) command. The appended command causes the server to accept the data transferred over the data connection. If the file specified in the path name already exists in the server, the data will be appended to that file; otherwise, a new file will be created.

Store unique (STOU) command. The store unique command behaves like the STOR command, except the file created in the current directory will have a unique name, and the existing file is not overwritten.

Allocate (ALLO) command. The allocate command may be required by some servers to reserve sufficient storage space to accommodate the new file. The argument will be a decimal integer value representing the number of bytes of storage reserved for the file. For files sent with record or page structure, a maximum record or page size may also be necessary, which is indicated by a decimal integer in a second optional argument field. When present, the second argument must be separated from the first with three Telnet characters: <space><R><space>. The ALLO command must be followed by STOR and APPE commands.

Restart (REST) command. The argument field in the restart command represents the marker where the file transfer will be started. This command does not cause a file transfer but skips over the file to the specified data checkpoint.

Status (STAT) command. The status command is used to return file status responses. Status commands can be sent during a file transfer along with Telnet IP and synch signals. The server will respond with the status of the operation in progress, or the status can be sent between file transfers where the command may have an argument field. If the argument is a path name, the command is analogous to the list command, except data is transferred over the control connection.

Rename-from (RNFR) command. The rename-from command specifies the old path name of the file that is being renamed. The rename-from command must be immediately followed by the rename-to command to specify the new file path name.

Rename-to (RNTO) command. The rename-to command specifies the new path name of the file specified in the rename-from command.

26-3-5-7 Miscellaneous commands.   Miscellaneous commands are used to deliver information to FTP users at the client's site. Table 26-12 lists the common FTP miscellaneous commands.

Table 26-12   FTP File Miscellaneous Commands

Command	Argument	Description
HELP		Ask for information about the server
NOOP		Check to see if the server is alive
SITE	Commands	Specify the site-specific commands
SYST		Ask about the operating system used by the server

Help (HELP) command. The help command causes the server to send helpful information over the control connection to the client regarding its implementation status. The HELP command may accept an argument (any command name) and return more specific information in a response. The expected reply is type 211 or 214. The server can use the HELP reply to specify site-dependent parameters.

No operation (NOOP) command. The NOOP command does not affect any parameters or previously transmitted commands. It specifies no action from the server other than sending an okay reply.

Site-specific (SITE) command. Site-specific commands are used by servers to provide service specific to its system, essential to file transfer but not sufficiently universal to be included as commands themselves.

System (SYST) command. System commands are sent by clients to ask servers about their operating system. The first word of the reply to a system command must be one of the system names listed in the current version of the Assigned Numbers document.

26-3-5-8 FTP responses.   FTP responses are replies to FTP commands that ensure that requests and actions are synchronized during file transfers. All FTP commands warrant one or more FTP response, and each response must be easily distinguished. Responses indicate that intermediate states have been successful. A failure at any point in the sequence of exchanging commands and responses necessitates the retransmission of the entire sequence.

All responses have two parts: a three-digit number (transmitted as three alphanumeric characters) followed by some text. The number is interpreted logically to determine what state to enter next, whereas a human is intended to interpret the text. The three digits contain sufficient encoded information, so the user process will not need to examine the text and may either discard it or pass it on to the user (human).

Responses begin with a three-digit code, followed by a space and one line of text, which is terminated by the Telnet end-of-line code. In cases where the text is longer than one line, the complete text must be bracketed to let the user process know when it should stop reading the reply (i.e., stop processing data received on the control connection).

The first digit of the three-digit code defines the status of the command and must have a value between 1 and 5:

1yz (positive preliminary reply). The action requested is being initiated, and the server will transmit another reply before accepting the next command. This type of reply can be used to indicate the command was accepted and the user process may now pay attention to the data connection. The server process may send only one 1yz reply per command.

2yz (positive completion reply). The action requested has been completed, and another request may be initiated, as the server is ready to accept another command.

3yz (positive intermediate reply). The command has been received and accepted; however, more information is required. The command will be held in abeyance, pending receipt of additional information, and the user needs to transmit another command specifying the missing information.

4yz (transient negative completion reply). The action requested was not accepted, and the action requested did not take place because of a minor error. However, the error that prevents it from being completed is temporary. The user must return to the beginning of the command sequence and retransmit the same command later. Because it is difficult to assign a precise meaning to "transient," the user and server processes must agree on the interpretation. Therefore, each 4yz reply could have a slightly different time value, and the intent is that the user process is encouraged to try again later.

5yz (permanent negative completion reply). The command was received but was not accepted, and the action did not take place. The user should not reattempt the same command in the same sequence again even if a "permanent" error condition is corrected.

The second digit of the three-digit code further defines the status of the command and must have a value between 0 and 6:

x0z. This reply refers to syntax errors, syntactically correct commands that do not fit into any of the functional categories, and unimplemented or superfluous commands.

x1z. This reply is in response to requests for information, such as status or help.

x2z. This reply refers to the control and data connections.

x3z. This reply is used for authentication and accounting for the log-in and accounting procedures.

x4z. The purpose of this reply has not been specified.

x5z. This reply refers to the status of the server file system—the requested transfer or other file system actions.

The third digit of the three-digit code gives a more specific definition of the meaning of each of the functional categories specified by the second digit. Table 26-13 shows a brief list of the responses possible with the third digit.

26-3-6 FTP File Transfer

The underlying purpose of FTP is to transfer files using a variety of formats from one host to another host. FTP is capable of maintaining multiple sessions with multiple hosts, moving multiple files during each of the sessions. FTP uses two TCP sessions for each file transfer. The first session is called the FTP control session. TCP, using well-known port number 21, transmits commands to start, stop, and control the session. The second session uses well-known port number 20 to move the actual file between hosts.

Although file transfers with FTP occur over the data connection, they are under the control of commands sent over the control connection. File transfer with FTP could encompass one of the three things listed and shown in Figure 26-12:

1. A file can be copied from a server and sent to a client under the supervision of the RETR (retrieve) command. This type of file transfer is called *retrieving* a file.
2. A file can be copied from a client and sent to a server under the supervision of the store (STOR) command. This type of file transfer is called *storing* a file.
3. A list of directory or file names can be copied from a server and sent to a client under the supervision of the list (LIST) command. FTP treats this type of transfer in exactly the same way it treats a file, and the lists are transmitted over the data connection.

Figure 26-13 illustrates the commands sent when a client is retrieving a list of items stored in a directory in a server. The operation is summarized here:

Step 1—After a control connection is established to port 21, the FTP server transmits a *service ready* response (code 220) over the control connection.

Step 2—The client transmits the USER access command.

Step 3—The server transmits with the *user name ok, password required* (code 331) response.

Step 4—The client transmits the PASS access command.

Step 5—The server transmits the user *log-in is okay* response (code 230).

Table 26-13 FTP Responses

Code	Description
Positive Preliminary Reply	
120	Service will be ready shortly
125	Data connection is open, and data transfer will start shortly
150	File status is okay, and data connection will be opened shortly
Positive Completion Reply	
200	Command okay
202	Command not implemented; superfluous at this location
211	System status or help reply
212	Directory status
213	File status
214	Help message
215	Naming the operating system type
220	Service is ready
221	Service is closing control connection
225	Data connection is open; no transfer currently in progress
226	Closing data connection
227	Entering passive mode; server sends it IP address and port number
230	User logged in; proceed
250	Request file action okay
Positive Intermediate Reply	
331	User name okay; password needed
332	Need account for log-in
350	The file action is pending, and more information is needed
Transient Negative Completion Reply	
421	Service not available; closing control connection
425	Cannot open data connection
426	Connection is closed; transfer is aborted
450	File action not taken; file is not available
451	Action aborted because of local error
452	Action aborted because of insufficient storage
Permanent Negative Completion Reply	
500	Syntax error; command not recognized
501	Syntax error in parameters or arguments
502	Command not implemented
503	Bad sequence of commands
504	Command parameter not implemented
530	User not logged in
532	Need account for storing file
550	Action is not done; file unavailable
552	Requested action aborted; exceeded storage allocation
553	Requested action not taken; file name not allowed

Step 6—The client issues a passive open on a temporary port for the data connection and then transmits the PORT defining command, which contains the temporary port number, over the control connection.

Step 7—After the server prepares itself for issuing an active open over the data connection between port 20 on the server side of the connection and the temporary port it received from the client, the server transmits the *data connection will open shortly* response (code 150).

Step 8—The client transmits the LIST file management command.

Step 9—The server transmits the *data connection open, data transfer will start shortly* response (code 125).

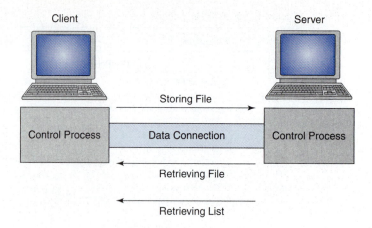

Client   Server

Control Process   Data Connection   Control Process

Storing File →

← Retrieving File

← Retrieving List

FIGURE 26-12 File transfer types

Step 10—The server transmits the list in file form over the data connection.

Step 11—After the entire list file has been transmitted, the server sends the *closing data connection* response (code 226) over the control connection.

Step 12—The client can now react in one of two ways. It can request closing the control connection using the QUIT access command, or it can transmit a different command that begins another activity (eventually opening another data connection).

Step 13—After the server receives the QUIT command, the server responds with the *service closing* response (code 221) and then close the control connection.

Figure 26-14 illustrates the commands sent when a binary image file is copied and stored in a server. The operation is summarized next (note that the first seven steps are identical to how a client retrieves a list of items stored in a directory in a server):

Step 1—After a control connection is established to port 21, the FTP server transmits a *service ready* response (code 220) over the control connection.

Step 2—The client transmits the USER access command.

Step 3—The server transmits with the *user name ok, password required* (code 331) response.

Step 4—The client transmits the PASS access command.

Step 5—The server transmits the user *log-in is okay* response (code 230).

Step 6—The client issues a passive open on a temporary port for the data connection and then transmits the PORT defining command, which contains the temporary port number, over the control connection.

Step 7—After the server prepares itself for issuing an active open over the data connection between port 20 on the server side of the connection and the temporary port it received from the client, the server transmits the *data connection will open shortly* response (code 150).

Step 8—Client transmits the TYPE command.

Step 9—Server transmits the *command okay* response (code 210).

Step 10—Client transmits STRU command.

Step 11—Server transmits the *command okay* response (code 210).

Step 12—Client transmits the STOR command.

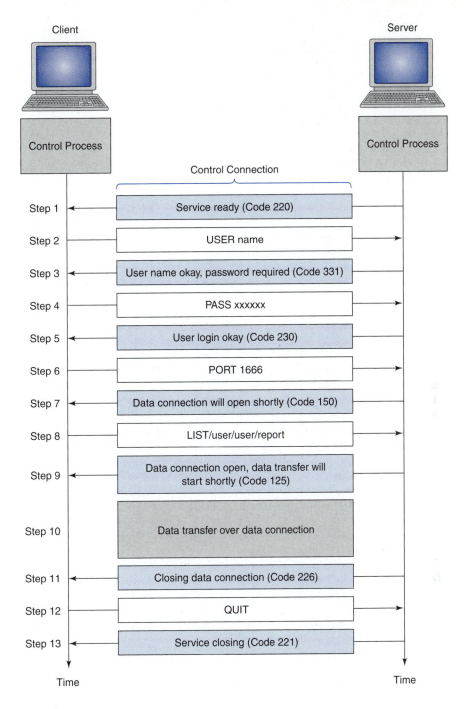

FIGURE 26-13 File retrieval—control connection

Step 13—Server opens data connection and transmits the *request file action okay* response (code 250).

Step 14—Client transmits the file over the data connection. After the entire file is sent, the data connection is closed, which means it is the end of the file.

Step 15—Server transmits the *closing data connection* response (code 226) over the control connection.

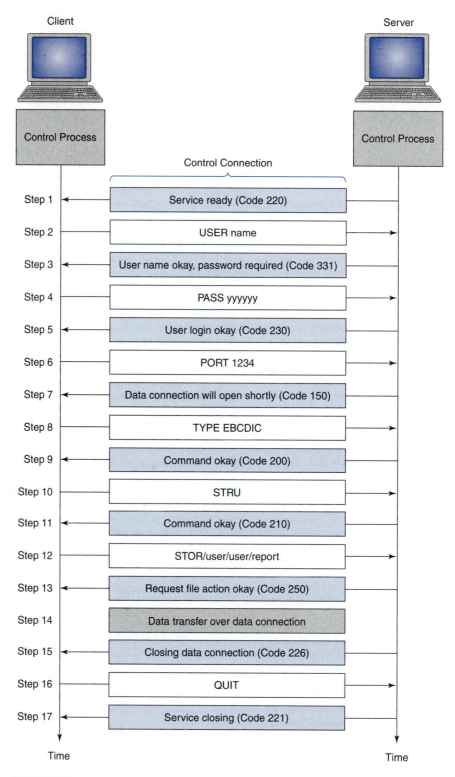

FIGURE 26-14 Storing a binary image file—control connection

Step 16—Client transmits the QUIT command or uses another command to open another data connection to transfer another file.

Step 17—Server transmits the *service closing* response (code 221), which closes the control connection.

26-4 TRIVIAL FILE TRANSFER PROTOCOL

The *trivial file transfer protocol* (TFTP) is exactly what the name implies: a trivial (frivolous) version of FTP used to transfer files. TFTP is appropriate when applications either do not need or cannot handle the complexity or full set of functions provided by FTP. TFTP provides a simple, unsophisticated, and inexpensive means of transferring files between hosts. TFTP is so simple that the entire software program can be stored on a single read-only memory chip on a diskless workstation. TFTP is capable of *reading* or *writing* files between a client and a remote server. The terms *reading* and *writing* are in reference to the client. Therefore, reading simply means copying from the server to the client, and writing means copying a file from the client to the server. TFTP cannot list directories and has no provision for user authentication.

TFTP is transported over UDP using well-known port 69. Therefore, TFTP is unreliable and connectionless. TFTP is used when speed is more important than efficiency (overhead) and reliability and there is no need for all the functionalities offered by FTP. For example, an FTP file would be cumbersome to load into a batch file for updating multiple hosts. Since TFTP uses UDP for its transport layer, it is small enough to do the job.

UDP is connectionless and does not guarantee delivery. To compensate for this, TFTP uses a time-out retransmission on both ends of the communications link. If a participating host does not receive either an acknowledgment or data within a specific period of time, another copy of the data or acknowledgment is sent. In other words, for each block of data transmitted, there must be an acknowledgment before the transfer can continue. Thus, all packets except the final packet are acknowledged separately.

26-4-1 TFTP Message Types

There are five types of TFTP message packets:

> Read request (RRQ)
> Write request (WRQ)
> Data (DATA)
> Acknowledge (ACK)
> Error (ERROR)

The RRQ and WRQ packets begin a request and determine what file is to be transferred. The DATA packets actually transfer the requested data, and ACK packets are used to acknowledge the receipt of each block of data received during a file transfer. The ERROR packet acknowledges any of the other packet types and signifies when an error has occurred.

26-4-1-1 Read request message type. A client uses a *read request* (RRQ) message type to establish a connection that is used for reading data from a server. The format for a RRQ message type is shown in Figure 26-15. An RRQ message is comprised of five fields, of which two are single-byte fields filled with all logic 0s. All-0s fields are used to indicate the end of the variable-length fields that immediately precedes them:

Opcode field. The opcode (operation code) is a two-byte field used to convey the type of TFTP message being delivered. The opcode for a RRQ message is 1.

File name field. The file name field is a variable-length field containing ASCII code that defines the name of the file. The end of the file name field is detected by the

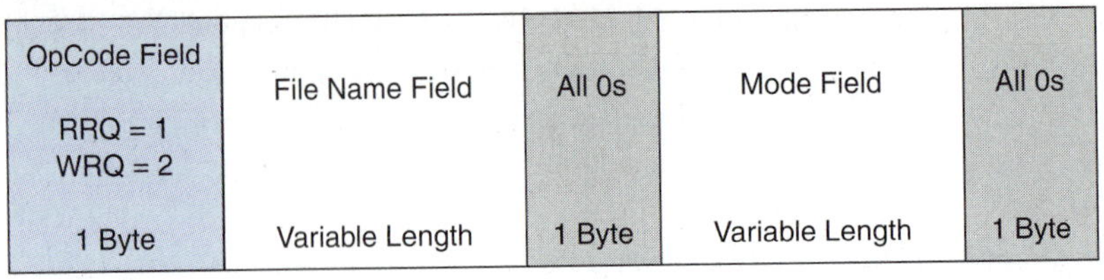

OpCode Field RRQ = 1 WRQ = 2 1 Byte	File Name Field Variable Length	All 0s 1 Byte	Mode Field Variable Length	All 0s 1 Byte

FIGURE 26-15 RRQ (WRQ) message format

occurrence of the one-byte all-0s field that immediately follows the file name. The file name field can use upper- or lowercase characters or a mixture of the two.

Mode field. The mode field is a variable-length field that defines the transfer mode. The end of the mode field is detected by the occurrence of the one-byte all-0s field that immediately follows the mode. The mode can be a *netascii* string for an ASCII file or an *octet* string for a binary file. The mode field can use upper- or lowercase characters or a mixture of the two.

26-4-1-2 Write request message type.   A client uses a *write request* (WRQ) message type to establish a connection for writing data from a server. The format for a RRQ message type is identical to the RRQ message type shown in Figure 26-15 except the opcode is 2.

26-4-1-3 Data message type.   A client or server uses a *data* (DATA) message type for sending blocks of data. Each data packet contains one block of data and must be acknowledged before the next block can be sent. The format for a DATA message is shown in Figure 26-16. A DATA message is comprised of the three fields described here:

Opcode field. The opcode (operation code) is a two-byte field used to convey the type of TFTP message being delivered. The opcode for a DATA message is 3.

Block number field. The block number field is a two-byte field containing the number of the data block being transmitted. Block numbers are used for sequencing and acknowledging. All data blocks contain sequence numbers that begin with 1.

Data field. The data field is a variable-length field with a maximum length of 512 bytes. All blocks except the last block of a message must contain exactly 512 bytes of data (bytes 0–511). When the client or server receives a block of data containing less than 512 bytes, it assumes that block is the last block of the message. This eliminates the necessity of sending an end-of-message character. The only time there is confusion is when the last block of data coincidently contains exactly 512 bytes of data. When this happens, an additional block filled with all 0s must be transmitted to signify the end of the message. Data can be sent in either NVT ASCII (netascii) or in binary octet form.

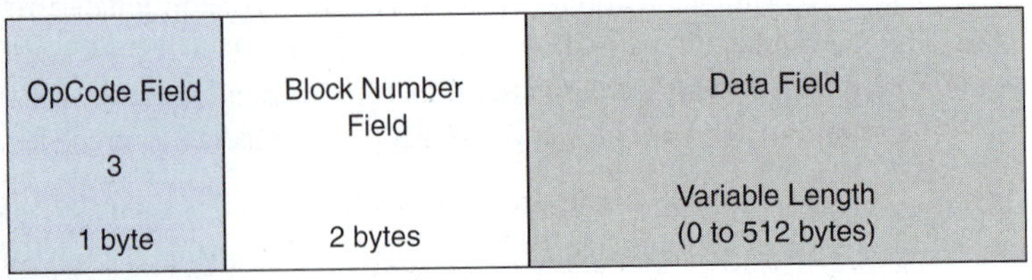

OpCode Field 3 1 byte	Block Number Field 2 bytes	Data Field Variable Length (0 to 512 bytes)

FIGURE 26-16 DATA message format

OpCode Field	Block Number Field
4	
1 byte	2 bytes

FIGURE 26-17 ACK message format

26-4-1-4 Acknowledgment message type. A client or server uses an *acknowledgment* (ACK) message type to acknowledge receipt of a block of data. The format for an ACK message type is shown in Figure 26-17. An ACK message is comprised of two fields:

Opcode field. The opcode (operation code) is a two-byte field used to convey the type of TFTP message being delivered. The opcode for an ACK message is 4.

Block number field. The block number field is a two-byte field that contains the number of the last block received. The ACK message can also be used as a response to a WRO message and transmitted by the server to indicate that it is ready to receive data, in which case the block number field has no meaning as is filled with all 0s.

26-4-1-5 Error message type. An *error* (ERROR) packet can be the acknowledgment for virtually any type of packet. A client or server can use an ERROR message type when a connection cannot be established or when a problem arises during data transmission (other than a transmission error). An ERROR message can also be transmitted as a negative acknowledgment in response to RRQ or WRQ messages. ERROR messages can be used during the actual data transfer phase when the next block of data cannot be transferred. It should be pointed out that the ERROR message is not used to indicate that a block of data is damaged or duplicated. Transmission errors like this are resolved with different error-control mechanisms.

The format for an ERROR message is shown in Figure 26-18. An ERROR message is comprised of four fields, of which one field is a single-byte field filled with all logic 0s used to terminate the variable-length error data field:

Opcode field. The opcode (operation code) is a two-byte field used to convey the type of TFTP message being delivered. The opcode for an ERROR message is 5.

Error number field. The error number field is a two-byte field that contains a number that defines the type of error the message is reporting. Table 26-14 lists the error numbers and their meanings.

Error data field. The error data field is a variable-length field that contains the data that was received in error. The error data field is terminated with a single byte of all 0s.

Most errors cause termination of the TFTP connection. Transmitting an error message that is not acknowledged and is not retransmitted signals an error. Therefore, time-outs are also used to detect a termination when the error packet is lost. The only error condition that does not cause termination is number 5 (unknown port number). In this case, an error packet is sent back to the originating host.

OpCode Field	Error Number Field	Error Data Field	All 0s
5			
1 byte	2 bytes	Variable Length	1 byte

FIGURE 26-18 ERROR message format

Table 26-14 ERROR Message Type Numbers

Error Number	Meaning
0	Not yet defined
1	File not found
2	Access violation
3	Disk full or quota on disk exceeded
4	Illegal operation
5	Unknown port number
6	File already exists
7	No such user

26-4-2   TFTP Connection Establishment and Termination

Because TFTP is transported over UDP, there is no connection establishment or disconnect sequence. UDP treats each datagram as an independent message. Multiple-block TFTP file transfers require some means of identifying which datagrams carrying blocks of data belong to the same message. TFTP uses the RRQ, WRQ, ACK, and ERROR message types to establish connections and a DATA message type containing less than 512 bytes to terminate connections.

Figures 26-19a and b show how connections are established for reading and writing files, respectively. To establish a connection for reading a file, the TFTP client transmits an RRQ message containing the file name and the mode of transmission. If the server can transmit the requested file, it responds with a DATA message containing the first block of data. If the server cannot open the file or if it encounters a permission restriction, the server sends a negative acknowledgment in the form of an ERROR message.

To establish a connection for writing, the TFTP client transmits a WRQ message, which includes the name of the file and the transmission mode. Provided that the server can accept a copy of the file, it responds to the WRQ with an ACK message (positive response) and a 0 for the block number. If there is a problem, the server sends an ERROR message, which is a negative response.

Once the entire file has been transferred, the connection is terminated. TFTP does not have a special character or sequence of characters to terminate a session. The session is automatically terminated when the last block of data is sent containing less than 512 bytes of data.

26-4-3   TFTP Flow and Error Control

With TFTP, the actual data transfer process occurs between establishing and disconnecting the connection. Because UDP does not provide a means of flow or error control, TFTP must provide its own mechanism to ensure an orderly transfer of a file containing more than one block of data.

26-4-3-1 TFTP flow control.   TFTP transmits blocks of data using the DATA message format. After a block has been sent, the sender waits for an ACK message. Providing the sender receives a positive acknowledgment before the time-out period expires, it sends the next block of data. Therefore, flow control is achieved by simply numbering blocks of data when they are transmitted and then waiting for a positive acknowledgment before transmitting the next block.

When a client wishes to read (retrieve) a file from a server, the client transmits an RRQ message. The server responds with DATA message block number 1, which contains the first block of data.

When a client wishes to write (store) a file to a server, the client transmits a WRQ message. The server responds with ACK message block number 0. After receiving the ACK message, the client transmits the first DATA message with block number 1.

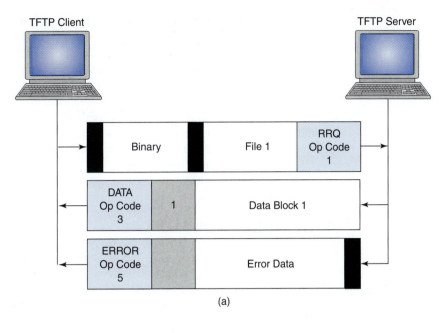

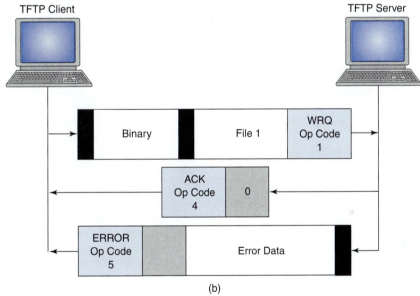

FIGURE 26-19 TFTP connection establishment: (a) for reading; (b) for writing

26-4-3-2 TFTP error control. TFTP uses an error-control mechanism that is different from the mechanism used by other protocols. The TFTP mechanism is symmetric, which simply means that both the sender and the receiver utilize time-outs for error control. Senders use time-outs for data messages and receivers use time-outs for acknowledgments. When a data message block is lost or corrupted beyond recognition, the sender retransmits the block after the time-out period has expired. If an acknowledgement is lost or corrupted beyond recognition, the receiver retransmits the acknowledgment after the time-out period has expired. Error control is necessary in four situations: when a message block has been damaged, when a message block is lost, when an acknowledgment message is lost, and for duplicated message.

With TFTP, there is no provision for sending negative acknowledgments. If a block is received in error, the receiver simply discards it. The sender will wait for an acknowledgment until the time-out period has expired, and then it will resend the message. There is no checksum in the DATA message. Therefore, the only way the receiver detects transmission errors is through the checksum field in the UDP header.

With TFTP, when a block is lost and never reaches the destination, no acknowledgment is sent. Consequently, after the time-out period has expired, the message block is retransmitted. If an acknowledgment is lost, one of two situations may occur. If the receiver's timer expires before the sender's timer, the receiver retransmits the acknowledgement; otherwise, the sender retransmits the message.

Duplicate message blocks are detected with TFTP through the block numbers. Whenever a duplicate message block is received, it is simply discarded.

26-4-4   TFTP Applications

TFTP includes no provision for security, as there is no user identification or password. Therefore, TFTP is useful for simple file transfers when security is not an important issue. TFTP can be used to initialize connectivity devices, such as bridges and routers. However, the primary application of TFTP is with BOOTP and DHCP. This is because TFTP requires only a small amount of memory and uses UDP over IP. Consequently, TFTP can be configured easily in ROM. When a device is powered up, TFTP is connected to a server to retrieve the configuration files. The powered-up device uses BOOPT or DHCP to acquire the name of the configuration file. The device then passes the name of the file to the TFTP client to retrieve the contents of the configuration file from the TFTP server.

26-5   SIMPLE MAIL TRANSFER PROTOCOL

Electronic mail (e-mail) is one of the most popular network services offered by the Internet. *Simple mail transfer protocol* (SMTP) is the standard protocol used by the TCP/IP protocol suite to support (e-mail). SMTP uses TCP to establish a connection and reliably transport e-mail messages through well-known port 25.

The purpose of SMTP is to transfer e-mail reliably and efficiently across the Internet between clients and servers. However, SMTP does not specify how the mail system will accept mail from a user or how the user interface will present received mail to the user. SMTP is independent of the transmission subsystem, as it requires only a reliable ordered data stream channel. SMTP does not specify how mail is stored or how frequently the mail system will attempt to transmit mail messages. SMTP is extremely straightforward, as communication between clients and servers occurs in the form of ASCII files that are easily read by humans.

One of the most important features of SMTP is its capability of relaying e-mail across virtually any transport service environment. The transport service provides an interprocess communications environment (IPCE) that can include one network, several networks, a subset of a network, or the Internet. Transport systems do not operate one to one with networks, as it is important that a process can communicate directly with another process through any mutual IPCE. E-mail is simply one of many applications of interprocess communications. It is important that e-mail be capable of communicating between processes in different IPCEs. Therefore, e-mail can be relayed between two hosts operating on different transport systems by another host that is common to both systems.

SMTP is a system that uses e-mail addresses to transfer electronic messages to other users both reliably and efficiently. SMTP provides a means of exchanging mail between two hosts using the same type of computer or two hosts using different types of computers. SMTP supports the transmission of the following types of messages:

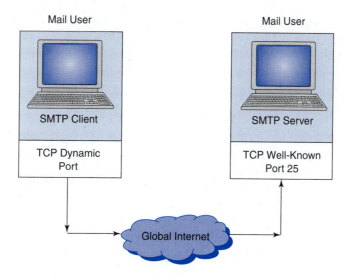

FIGURE 26-20 Fundamental idea of SMTP

Single messages to one or more recipients

Messages that are comprised of text, voice, video, or graphics

Messages to users who reside on networks outside the Internet

The SMTP model is initiated on a user mail request and is based on the following features:

The SMTP sender establishes a two-way communications channel with a SMTP receiver.

The SMTP receiver may be the ultimate destination or an intermediate destination.

SMTP commands are initiated by the SMTP sender and sent to an SMTP receiver.

SMTP responses are sent from the SMTP receiver to the SMTP sender in response to commands.

Figure 26-20 illustrates the fundamental ideal behind SMTP where the SMTP server uses the well-known TCP port 25 to transfer e-mail over the Internet to a SMTP client using a temporary port.

The SMTP client and server can be broken down into two components: a *user agent* (UA) and a *mail transfer agent* (MTA). User agents are remote client and server mail applications (software) used to send and receive e-mail, such as MH, Berkeley Mail, Zmail, and Mush. MTAs are the actual mail message. MTAs are defined by SMTP, but the details of their implementation are not. A client MTA sends e-mail, and a server MTA receives e-mail. A common MTA used by Unix is *sendmail*.

A client UA prepares the mail message, creates an envelope, and places the mail into the envelope. The envelope is given to the MTA, which transfers the envelope across the Internet to another MTA at the server end. At the server end, the MTA removes the mail from the envelope and gives it to the UA, as shown in Figure 26-21. Sometimes additional MTAs, acting like clients or servers, are required to relay mail between the client and server MTAs, as shown in Figure 26-22. The relaying MTAs allow sites that are not using TCP/IP to send e-mail to users at other sites that may or may not use TCP/IP. This is accomplished using a *mail gateway,* which is a relay MTA capable of receiving mail prepared by a protocol other than SMTP and converting it to the SMTP format before transmitting it. Mail gateways can also receive mail in SMTP format and convert to a different form before transmitting it, as shown in Figure 26-23.

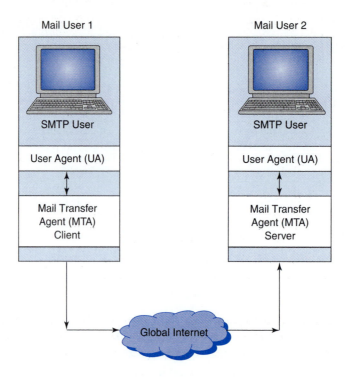

FIGURE 26-21   UA/MTA interaction

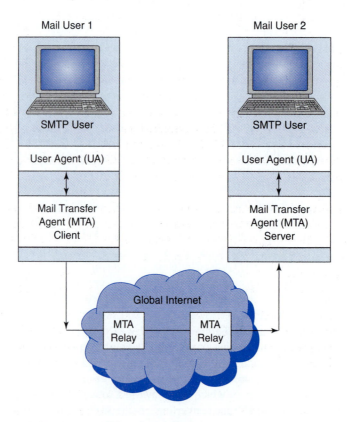

FIGURE 26-22   MTAs relaying mail

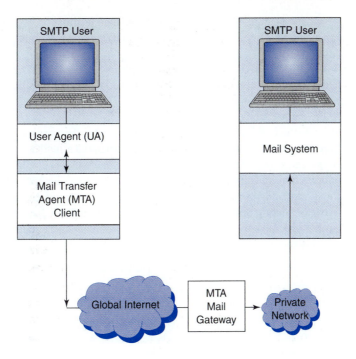

FIGURE 26-23 MTAs relaying mail

26-5-1 Sending and Receiving Mail Using SMTP

SMTP sends and receives e-mail messages using a sender SMPT process and a receiver SMTP process. The sender and receiver SMTP are processes that support the SMTP command list. SMTP senders communicate with SMTP receivers through a series of SMTP requests and responses that essentially control the delivery of the e-mail message. To send mail with SMTP, a user creates mail through the user agent that resembles standard postal mail, as it has an envelope that contains the actual mail message. The envelope generally includes the sender's address, the recipient's address, and possibly some additional information. The message includes a header (which defines the sender), the recipient, the subject of the message, and the body of the message (which contains the actual information intended for the recipient).

The user agent checks the mailboxes periodically, and if a user has received mail, the user agent informs the user by giving a notice. When the user is ready to read the mail, a list is displayed where each line contains a brief summary of a message. The summary generally contains the mail address of the sender, the subject of the message, and the time the mail was sent or received. To read a message, the user simply selects it, which displays it on the screen.

26-5-2 E-Mail Addresses

To identify users and deliver mail, a mail handling system must incorporate some type of addressing system that uniquely identifies each user. Within the global Internet, SMTP uses a relatively simple two-part addressing system comprised of a local part and a domain name separated by an ampersand (@), as shown here:

local-part @domain-name

The local part of the address defines the name of the a mailbox, which is a special file that stores all the user's receive mail until the user agent retrieves it.

The domain name is the name of the mail destination, which is a *mail exchanger* and not a host machine. An example of an e-mail address within the Internet is

tomasi@syracuse.edu

The domain name is assigned to the host mail exchanger designated to send and receive e-mail. The domain name can come from the DNS database, or it can be a logical name, such as the name of the organization.

E-mail addresses can be more complicated. For example, when mail gateways are used to connect outside networks to the Internet, the e-mail address must define the address of the gateway as well as the address of the recipient. The domain name part of the e-mail address must specify the name of the mail gateway in the DNS database. The local part of the e-mail address must specify the local physical network, the computer attached to the network, and the user mailbox. When a mailing system does not use the mailing address format specified by SMTP, it may cause problems and create confusion. An example of non-Internet e-mail address is

tomasi%syracuse.edu@forward.bs.net

Once the mail arrived at machine forward.bs.net, the mail gateway software would remove the local part, change the % sign to an @ sign, and forward the e-mail to tomasi@syracuse.edu. Some mail gateways use a colon or some other character in place of the % sign.

Domains is a concept recently introduced to the Internet mail system. Using domains changes the available address space from a flat global space comprised of simple character string host names to a hierarchically structured rooted tree of global addresses. The domain and host designator replaces the host name, which is a sequence of domain element strings separated by periods. The domain elements are ordered from the most specific to the least specific. For example, john.doe.uk might be host and domain identifiers. Whenever domain names are used with SMTP, only the official names are used, as nicknames and aliases are not allowed.

26-5-3 SMTP Commands

SMTP exchanges commands and responses when transferring messages between an MTA client and an MTA server. All commands and responses are character steams where each line is terminated with a two-character end-of-token sequence, which is simply a carriage return and line feed <CRLF>. The command codes themselves are alphanumeric characters that are followed by a space character if the command includes additional parameters.

Commands are comprised of a keyword (command code) possibly followed by one or more arguments followed by an end-of-token sequence (carriage return and line feed), as shown here:

keyword: argument(s)<CRLF>

The SMTP keyword consists of four upper- or lowercase alphabetic characters separated from the argument by one or more space characters.

SMTP commands define or classify the mail transfer or the mail system function requested by the user. SMTP defines 14 commands, which are listed in Table 26-15 and described next. The first five commands are mandatory, and all SMTP implementations must support them. The next three commands are commonly used and highly recommended. The last six commands are seldom used. As can be seen in Table 26-15, some the SMTP commands are inherent and require no argument, such as QUIT and RST, and some SMTP commands require an argument containing additional information, such as HELO and MAIL FROM. For SMTP to be workable, the following commands must be implemented by all receivers: HELO, MAIL, RCPT, DATA, RSET, NOOP, and QUIT.

The SMTP commands are summarized here:

Table 26-15 SMTP Commands

Keyword	Argument
HELO	Mail sender's host name
MAIL FROM	Sender of the mail message
RCPT TO	Intended recipient of the mail message
DATA	Body of the mail message
QUIT	None
RSET	None
VRFY	Name of recipient to be verified
NOOP	None
TURN	None
EXPN	Mailing list to be expanded
HELP	Command name
SEND FROM	Intended recipient of the mail message
SMOL FROM	Intended recipient of the mail message
SMAL FROM	Intended recipient of the mail message

HELO. The HELO command is used to identify the SMTP sender to the SMTP receiver. The argument is the domain name of the sending host followed by the end-of-token sequence (carriage return and line feed). The format is

> HELO: <sender's domain name><CRLF>
> HELO: syracuse.edu<CRLF>

CRLF indicates the carriage return and line feed characters.

MAIL FROM. The SMTP MAIL FROM command is the first step in the procedure as it identifies the sender of the message. The MAIL FROM command is sent by the sender to tell the SMTP receiver that a new mail transaction is beginning and to reset all state tables and buffers. The argument is the e-mail address of the sender (source address), which includes both the local part and domain name. The format is

> MAIL <space> FROM: <sender's email address><CRLF>
> MAIL FROM: wtomasi@devry.edu <CRLF>

RCPT TO. The RCPT command is the second step in the procedure. The client uses the RCPT (recipient) TO command to identify the intended recipient of the e-mail. The argument is the e-mail address of the recipient. When sending to multiple recipients, the command is repeated once for each recipient. The format is

> RCPT <space> TO: <recipient's email address><CRLF>
> RCPT TO: ballinger@devry.edu<CRLF>

DATA. The DATA command is used to transmit the actual message as well as the memo header, which may include Date, Subject, To, Cc, From, and so on. All lines that follow the DATA command are considered part of the mail message and may contain any of the 128 ASCII character codes. The message is terminated with a line containing only a period (i.e., CRLF.CRLF). The maximum total length of a command line, including the command word and the CRLC, is 512 characters. The format is

> DATA<CRLF>
> <Memo header and actual email message><CRLF>
> .<CRLF>

Without providing a provision for data transparency, the character sequence CRLF.CRLF ends the mail text and cannot be sent by a user. Users, in general, are not aware of such "forbidden" sequences. Therefore, to allow all user text to be transmitted transparently, the following procedures are implemented:

The sender must check the first character of a line before sending a line of mail text. If it is a period, one additional period is inserted at the beginning of the line.

A receiver checks the first line of mail text. If the line contains a single period, it must be the end of the mail message. If the first character is a period and the line contains additional characters, the first character is deleted.

All data characters are delivered to the recipient's mailbox, including control characters and format effectors. If the transmission channel provides an eight-bit data stream, the seven-bit ASCII codes are transmitted right justified with the high-order bit a logic 0.

QUIT. The QUIT command terminates the message. The QUIT command requires that the receiver return an OK reply and then close the transmission channel. The QUIT command has no argument and is sent as simply

QUIT<CRLF>

The HELO and QUIT commands are used to open and close an SMTP connection.

RSET. The RSET (reset) command causes the current mail transaction to abort and reset the connection. Any information stored about the sender, recipient, or mail data is discarded, and all buffers and state tables cleared. The recipient must reply with an OK. The RSET command has no argument and is sent as simply

RSET<CRLF>

VRFY. The client uses the VRFY (verify) command to verify the address of the recipient. The VRFY command asks the recipient to confirm that the argument identifies a user. If it is a user name, the full name of the user (if known) and the fully specified mailbox are returned. The e-mail address is the argument:

VRFY: <recipient's address><CRLF>
VRFY: wtomasi@devry.edu <CRLF>

NOOP. The client uses the NOOP (no operation) command to check the recipient's status. The NOOP command demands an OK response. The NOOP command does not affect any parameters or previously sent commands, and it specifies no further action other than an OK reply. The NOOP command has no argument and is sent as simply

NOOP<CRLF>

TURN. The TURN command is used to allow the sender and recipient to switch positions (i.e., the sender becomes the recipient and vice versa). However, most SMTP implementations do not support this command. The TURN command has no argument and is sent as simply

TURN<CRLF>

EXPN. The EXPN command requests the receiving host to expand the mailing list and to return the mailbox addresses of the recipients in the list. The full name of the users (if known) and the fully specified mailboxes are returned in a multiline reply. The format is

EXPN: <x y z><CRLF>

HELP. The HELP command requests the recipient to return "helpful" information about the command sent as the argument. The command may take an argument, which can be any command name, and return more specific information as a response. The format is

HELP: <mail><CRLF>

SEND FROM. The SEND FROM command is used to specify whether the mail is to be delivered to only the recipient's terminal and not the mailbox. However, if the recipient is not logged in, the mail is returned. The argument is the address of the sender:

SEND FROM: <sender's address><CRLF>
SEND FROM: wtomasi@devry.edu <CRLF>

SMOL FROM. The SMOL FROM command specifies whether the mail is to be delivered to the recipient's terminal if the user is active (open) or the recipient's mailbox if the user is not active. Therefore, if the recipient is logged in, the mail is delivered to the terminal, and if the recipient is not logged in, the mail is delivered to the mailbox. The argument is the address of the sender:

SMOL FROM: <sender's address><CRLF>
SMOL FROM: wtomasi@devry.edu <CRLF>

SMAL FROM. The SMAL FROM command specifies whether the mail is to be delivered to the recipient's terminal and one or more mailboxes. If the recipient is logged in, the mail is delivered to the terminal and the mailbox. If the recipient is not logged in, the mail is delivered only to the mailbox. The argument is the address of the sender:

SMAL FROM: <sender's address><CRLF>
SMAL FROM: wtomasi@devry.edu <CRLF>

There are restrictions on the order in which the SMTP commands may be used. The first command in a session must be the HELO command (which can also be used later in the session). The NOOP, HELP, EXPN and VRFY commands can be used at any time during a session. The MAIL, SEND, SMOL, or SMAL commands begin a mail transaction. The QUIT command must be the last command in a session and cannot be used at any other time.

26-5-4   SMTP Responses

SMTP replies acknowledge receipt of SMTP messages and error notifications. SMTP responses are sent from the server to the client as a three-digit number code sent as three alphanumeric characters. The number code is followed by additional text information. An SMTP reply has the form

<three-digit number code><space character><CRLF>

The three-digit codes are in the form xyz, where x is a digit between 1 and 5 that describes the category (state) of the reply and y and z are digits that further define the nature of the reply. The text included in the response is intended for the human user.

There is a hierarchy to the three-digit code. The first digit conveys general information. However, the second and third digits further define the reply, giving the receiver (it is hoped) sufficient information to determine if an error has occurred.

The first digit denotes whether the response is good, bad, or incomplete. The sender can determine its next action (proceed, redo, retrench, and so on) from the first digit. Although there are five possibilities for the first digit (x digit), only the last four are currently being used. The meanings of the first digit are the following:

1yz. Positive preliminary reply (currently not used)

2yz. Positive completion reply

3yz. Positive intermediate reply

4yz. Transient negative completion reply

5yz. Permanent negative completion reply

The second digit (y digit) indicates the type of error that occurred (i.e., mail system error, command syntax error, and so on), and the third digit (z digit) further defines the nature of the problem. The meanings of the second digit are the following:

x0z. Syntax

x1z. Information

x2z. Connection

x3z. Unspecified

x4z. Unspecified

x5z. Mail system

The SMTP responses are listed in Table 26-16 and described here:

2yz (positive completion reply) When the response code begins with the digit 2, the requested command has been successfully completed, and a new command may begin.

3yz (positive intermediate reply) When the response code begins with the digit 3, the command has been accepted; however, the recipient requires some additional information before completion can occur. The sender should send another command containing the specified information.

Table 26-16   SMTP Responses

Code	Description
Positive Completion Reply	
211	System status or help reply
214	Help message
220	Service ready
221	Service closing transmission channel
250	Request command completed
251	User not local; the message will be forwarded
Positive Intermediate Reply	
354	Start mail input
Transient Negative Completion Reply	
421	Service not available
450	Mailbox not available
451	Command aborted: local error
452	Command aborted: insufficient storage
Permanent Negative Completion Reply	
500	Syntax error; unrecognized command
501	Syntax error in parameters or arguments
502	Command not implemented
503	Bad sequence of commands
504	Command temporarily not implemented
550	Command is not executed; mailbox unavailable
551	User not local
552	Requested action aborted; exceeded storage location
553	Requested action not taken; mailbox name not allowed
554	Transaction failed

4yz (transient negative completion reply) When the response code begins with the digit 4, the requested command has been rejected, and the requested action did not occur. However, the error condition that caused the rejection is temporary, and the command may be sent again.

5yz (permanent negative completion reply) When the response code begins with the digit 5, the command has been rejected, and the requested action did not occur. The error condition that caused the rejection is permanent; therefore, the command cannot be resent until the condition has received higher-level attention, such as from a human.

Following is a sample exchange of SMTP commands and responses showing how user bill at host asu.edu (bill@asu.edu) sends an e-mail message to user joe at uoa.edu (joe@uoa.edu). The exchange includes three steps: connection establishment, message transfer, and connection termination. In the example, S denotes "from the server," and C denotes "from the client."

Connection establishment sequence:

Response	S:	220 UOA.edu Simple Mail Transfer Service Ready
Command	C:	HELO asu.edu
Response	S:	250 okay

Message transfer:

Command	C:	MAIL FROM: <bill@asu.edu)
Response	S:	250 okay
Command	C:	RCPT TO: <joe@uoa.edu
Response	S:	250 okay
Command	C:	DATA
	S:	354 start mail input; end with <CRLF>.<CRLF>
	C:	Actual email message, which
	C:	may continue for several lines.
	C:	This is the last line of the text.
	C:	<CRLF>.<CRLF>
Response	S:	250 okay

Connection termination:

Command	C:	QUIT
Response	S:	221 UOA.edu Service closing communications channel

26-5-5 Multipurpose Internet Mail Extension

Because of its simplicity, SMTP has the following restrictions:

SMTP messages can contain only ASCII characters.

The maximum line length for SMTP is 1000 characters or less.

SMTP messages must not exceed a predefined maximum length.

Multipurpose Internet mail extension (MIME) defines the method in which files are attached to an SMTP mail message. It does this by defining additional fields in the mail message header that describe new content types and a specific message body organization. MIME extensions allow data transmissions that were previously unsupported by Internet mail by encoding messages into readable ASCII to create a standard e-mail message.

MIME enhances the capabilities of normal Internet mail, including SMTP. MIME enables e-mail messages to carry the following message types:

Character sets other than ASCII

Multimedia messages, such as images, audio, and video

Multiple types within a single message

Multi-font messages

Messages with unlimited lengths

Binary files

26-6 POST OFFICE PROTOCOL

SMTP is used to transport e-mail across the Internet. After a SMTP mail server receives e-mail, it adds the e-mail message to the recipient's permanent mailbox. However, the recipient uses a different protocol to extract or delete e-mail messages from a mailbox. The most popular protocol for transferring e-mail messages from a permanent mailbox in a mail server to a local computer is called *Post Office Protocol version 3* (POP3). A user summons a POP3 client, which creates a TCP connection to a POP3 server residing on the mail server. After sending log-in and password information for authentication, the user client sends commands to retrieve a copy of one or more e-mail messages from the mailbox or to delete mail from the mailbox. The computer with the permanent mailbox must operate two servers: an SMTP server to accept e-mail and add it to the recipient's mailbox and a POP3 server, which allows the user to extract or delete messages from the mailbox.

26-7 HYPERTEXT TRANSFER PROTOCOL

Hypertext transfer protocol (HTTP) is an application-layer protocol for distributed two-way hypermedia information systems. HTTP has been used on the World Wide Web global information initiative since 1990. HTTP is used primarily to access data on the Web in a multitude of forms, including plain text, hypertext, audio, video, and many other forms. HTTP functions as a communications link between a browser and a Web server when Web pages are opened. Therefore, every time someone clicks a link or types a query, HTTP is being used.

HTTP derives its name because its efficiency allows it to be used in a hypertext environment that is rapidly changing from one document to another. HTTP can also be used as a generic protocol to provide communications between users and proxies (such as gateways) to other Internet systems.

HTTP operates similar to FTP but simpler because it does not require a control connection. Therefore, HTTP requires only one TCP connection to transfer data between the client and the server. The basic concept of HTTP is quite simple. A client transmits a request to the server that resembles mail, and the server responds with a reply that looks like mail to the client. Therefore, both the request and the response carry data in the form of a letter using a MIME format. Commands sent from the client to the server and the contents of the file sent from the server to the client are embedded in responses that also resemble letterlike messages.

26-7-1 HTTP Features

HTTP has several features or characteristics that separate it from other application-layer protocols:

HTTP operates at the application (process) layer and provides a communications link and message forwarding. However, HTTP is not reliable, and it does not perform retransmissions.

HTTP provides *bidirectional transfer* in the sense that the both servers and browsers can transfer copies of a Web page to each other.

HTTP uses *capability negotiation,* which is a feature that allows browsers and servers to figure out details together, such as the character set they will use to transfer requests and responses.

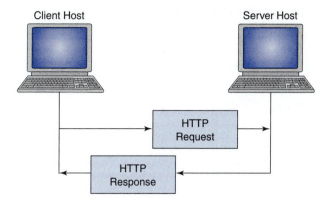

FIGURE 26-24 HTTP transaction

HTTP supports *caching,* which means that a browser can cache a copy of each Web page it retrieves for you. Caching speeds things up and saves time.

HTTP enables support for *intermediaries* so that any machine in the path between a browser and a server can be a proxy server. Proxy servers cache Web pages.

Servers do not maintain a history of HTTP sessions or HTTP requests.

26-7-2 HTTP Transactions

Because HTTP is a stateless protocol, transactions between clients and servers can occur only using the services of TCP, which provides a guaranteed delivery of messages using well-known port 80. Figure 26-24 illustrates an HTTP transaction, which like most protocols occurs through a series of requests and responses. An HTTP transaction is initiated when the client sends a request message (sometimes called a *request claim*). The server replies with a response (sometimes called a *request chain*).

The general formats for requests (claims) and responses (chains) are essentially the same and may include the following:

- A generic start line called a *request line* for request messages and a status line for responses
- A general header
- A message header
- One empty line
- A message body

26-7-2-1 Request messages format.   A request message is comprised of a request line, a header, and sometimes a body, as shown in Figure 26-25. The request line is comprised of three fields that define the type of request, uniform resource locator, and what version of HTTP is being used followed by a carriage return and line feed (CR/LF), as shown in Figure 26-26:

Request Type. HTTP version 1.1 defines several request types that categorize the request message into kinds of messages called *methods* or *method tokens*. The request method is the actual command or request a client issues to a server. Table 26-17 lists the method tokens and their functions:

GET. The GET method is used by a client when it wants to retrieve (or get) a document (file or resource) from a server. The address of the document is defined in the *uniform resource locator* (URL), which is described next. A server will generally reply to a GET request with the contents of the document in the body of the response message unless an error has occurred.

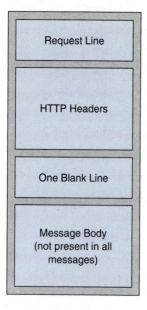

FIGURE 26-25 HTTP request message format

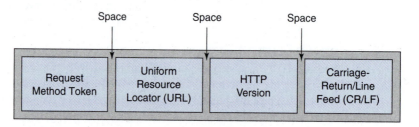

FIGURE 26-26 HTTP request line format

Table 26-17 Method Tokens

Token	Function
GET	Requests a document
HEAD	Requests information about a document
POST	Creates a new document on the server
PUT	Provide a new or replacement document on the server
PATCH	Provides only a list of differences that should be implemented in an existing file
COPY	Copy a document to another location
MOVE	Move a document to another location
DELETE	Remove a document from the server
LINK	Create a link or links from a document to another location
UNLINK	Delete links created by the LINK method
OPTIONS	Request information about available options
CONNECT	Used by proxy agents to set up for tunneling

HEAD. The HEAD method is used by a client when it wants information about a document but does not actually want the resource. The HEAD request is similar to GET request except the response from the server does not contain a body.

POST. The POST method is used by a client to provide information (input) to the server, for example, to create a new document on the server, such as posting a new message on a bulletin board.

PUT. The PUT method is sent by a client when it wants to provide a new document or replace an existing document. The document is included in the body of the request and stored in the location specified by the URL.

PATCH. The PATCH method is similar to the PUT method except the request contains a list of only the differences that should be implemented in an existing file.

COPY. The COPY method is sent by the client when it wishes to copy a document to another location.

MOVE. The MOVE method is sent by a client when it wishes to move a document from one location in the server to another location. The location of the source document is specified in the request line (URL), and the location of the destination is specified in the entity header.

DELETE. The DELETE method is sent by a client to remove a document from the server.

LINK. The LINK method is sent by a client to create a link (or links) from a document to another location. The location of the document is specified in the request line (URL), and the location of the destination is specified in the entity header.

UNLINK. The UNLINK method is sent by a client to delete a link (or links) created by the LINK method.

OPTION. The OPTION method is sent by a client to request available options from a server.

CONNECT. The Connect method is used by proxy agents to set up for tunneling.

Uniform resource locator. All Web pages are assigned a unique name to identify them. The name is called a *uniform resource locator* (URL), which is a specific type of *uniform resource identifier* (URI) that is used for specifying virtually all information on the Internet. The URL specifies five items—scheme, host computer, port number, path, and query—as shown in Figure 26-27. The scheme specifies the transfer protocol (such as HTTP) used to retrieve the document, such as Gopher, FTP, HTTP, News, and Telnet. The scheme also defines the remainder of the URL. For example a URL that complies with the HTTP scheme has the following format:

HTTP:// hostname [:port] / path [parameters] [? query]

where bracketed items are optional.

The host name string specifies the domain name or IP address of the computer where the server resides. Web pages are generally stored in computers that have been given alias names that often begin with the characters "www". [:port] is an optional protocol port number, which is needed only when the server does not use the well-known port number 80. Path is a string that identifies the path name of the file where the information is stored. The path itself can contain slashes that separate directories from subdirectories and files. [? Query] is an optional string that is used when a browser sends a question.

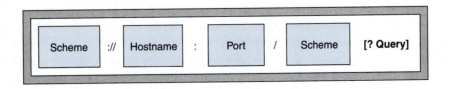

FIGURE 26-27 Uniform resource locator (URL) format

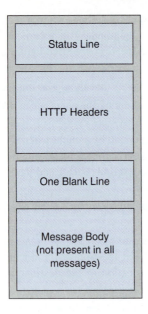

FIGURE 26-28 HTTP reply message format

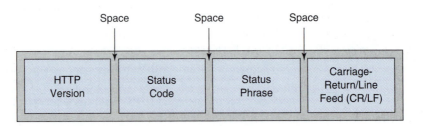

FIGURE 26-29 HTTP status line format

The URL shown next identifies a Web page where the server operates on computer www.nc.syracuse.edu and the document is named /sports/basketball/:

HTTP://www.nc.syracuse.edu/sports/basketball/

URLs have been known by many names, including WWW addresses, universal document identifiers, and universal resource identifiers.

Version. The version field specifies the version of HTTP used. Version 1.1 is the current version; however, versions 1.0 and 0.9 are still in use.

26-7-2-2 Response messages format. A response message is comprised of a status line, a header, and sometimes a body, as shown in Figure 26-28. The status line is comprised of three fields that define the version of HTTP being used, a status code, and a status phrase followed by a carriage return and line feed (CR/LF), as shown in Figure 26-29:

Version. The version field is identical to the request line.

Status code. The status code field is comprised of three digits. Table 26-18 lists the status codes and a description of each. The codes can be described in general terms as follows:

Codes in the 100 range—informational only

Codes in the 200 range—indicate a successful request

Codes in the 300 range—redirect the client to another URL

Table 26-18 HTTP Response Message Status Codes

Code	Phrase	Description
		Informational Codes
100	Continue	The beginning of the request has been received so the client can continue sending its request
101	Switching	The server is complying to a client's request to switch protocols defined in the upgrade header
		Success Codes
200	OK	The request is successful
201	Created	A new URL is created
202	Accepted	The request is accepted, but it is not immediately acted on
204	No content	There is no content in the body of the heading
		Redirection Codes
301	Multiple choices	The requested URL refers to more than one document
302	Moved permanently	The requested URL is no longer used by the server
304	Moved temporarily	The requested URL has moved temporarily
		Client Error Codes
400	Bad request	The request contains syntax errors
401	Unauthorized	The request lacks proper authorization
403	Forbidden	Service is denied
404	Not found	The document is not found
405	Method not allowed	The method requested is not supported by the URL
406	Not acceptable	The format requested is not acceptable
		Server Error Codes
500	Internal server error	There is an error in the server
501	Not implemented	The action requested cannot be performed
503	Service unavailable	The service requested is temporarily unavailable but may be requested later

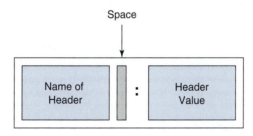

FIGURE 26-30 HTTP header format

Codes in the 400 range—indicate an error at the client's location

Codes in the 500 range—indicate an error at the server's location

Status phrase. The status phrase explains the status code in text form, as shown in Table 26-18.

26-7-3 HTTP Headers

HTTP headers are used to exchange additional information between the server and the client, such as the client requesting that a document be sent in some special format, or the server can send additional information about a document. HTTP headers can be one or more header lines where each header line is comprised of a header name, a colon, a space, and a header value, as shown in Figure 26-30.

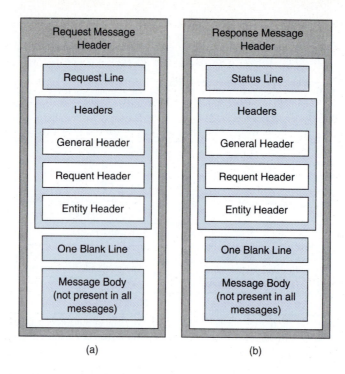

FIGURE 26-31 HTTP headers: (a) request message header; (b) response message header

There are four header categories: general, request, response, and entity. Request messages are allowed to contain only general, request, and entity headers. Response messages are allowed to contain only general, response, and entity headers. Figure 26-31 shows a comparison of request message headers and response message headers.

26-7-3-1 General header. All HTTP message headers begin with the same general header. As the name implies, a general header gives general information about the message. Some of the common general headers are listed in Table 26-19 and described here:

Cache-control. Specifies information about the caching operation of HTTP user agents, intermediaries, and servers. The intent of caching is to improve efficiency and reduces latency and network traffic by eliminating unnecessary data transfers. Cache control passes information through the request/response path and specifies when to cache or store transmitted or received data, how long the information should remain cached, and whether the cached information remains public.

Table 26-19 HTTP General Headers

Header
Cache-control
Connection
Date
MIME version
Upgrade
Pragma
Trailer
Upgrade
Via

Connection. Specifies whether the connection should be open or closed.

Date. Contains the current date and time of the request or response.

MIME version. The MIME version header specifies the MIME version being used.

Upgrade. Specifies the preferred communications protocol.

Pragma. Defines what directions an optional set of vendors should include in the request and response chains.

Trailer. Indicates what set of header fields will be in the trailer of a message.

Upgrade. Settle protocol and version type discrepancies and resolve compatibility issues between communicating devices.

Via. Enables proxies and gateways (intermediary devices) to keep track of forwarded messages and to enable them to identify various protocols and capabilities implemented by all devices involved in the request/reply chain.

26-7-3-2 Request header. Request headers are allowed only in request messages. A request header allows clients to send the server additional information about the request. Request headers specify the configuration and document format preferred by the client. Some of the common request headers are listed in Table 26-20 and described here:

Accept. Shows the media format a client is capable of accepting

Accept-charset. Specifies the character sets that may be used for the response (i.e., what the client can handle)

Accept-encoding. Sets a limit on the content encoding values (i.e., what the client can handle)

Accept-language. Sets a limit on the number of natural languages in the set (i.e., what the client can handle)

Authorization. Indicates what permissions the client has

From. The user's e-mail address

Host. The client's host and port number

If-modified-since. Ensures that the cached information is current by transmitting documents that are more recent than the specified date

If-unmodified-since. Transmits documents that have not changed since the specified date

If-match. Transmits the document only if it matches the given tag

Table 26-20 HTTP Request Headers

Header
Accept
Accept-charset
Accept-encoding
Accept-language
Authorization
From
Host
If-modified-since
If-unmodified-since
If-match
If-non-match
If-range
Range
Referrer
User Agent

Table 26-21 HTTP Response Headers

Header
Age
Accept-range
Public
Retry-after
Server

Table 26-22 HTTP Entity Headers

Header
Allow
Content-encoding
Content-language
Content-length
Content-range
Content-type
Etag
Expires
Last-modified
Location

If-non-match. Transmits the document only if it does not match the given tag

If-range. Transmits only the portion of the document that is missing

Range. Defines a resource

Referrer. Identifies the URL of the linked document

User agent. Specifies the client's program

26-7-3-3 Response header.   Response headers are allowed only in response messages. Response headers specify the configuration and other special information about the request. Some of the common response headers are listed in Table 26-21 and described here:

Age. Indicates whether a server accepts the range requested by the client

Accept-range. Specifies if a server has accepted the range requested by the client

Public. Shows the supported list of methods

Retry-after. Specifies the data in which a server is available

Server. Gives the server's name and version number

26-7-3-4 Entity header.   Entity headers give information concerning the body of the message document. Entity headers are sent primarily in response messages: however, request messages, such as POST and PUT methods, may also use entity headers. Some of the common response headers are listed in Table 26-22 and described here:

Allow. Indicates what methods a resource URL can support (i.e., which methods are valid)

Content-encoding. Specifies the encoding scheme

Content-language. Specifies the language

Content-length. Indicates the document length

Content-range. Specifies the document range

Content-type. Specifies the media type

Etag. Provides an entity tag

Expires. Specifies the time and date that the contents can be changed

Last-modified. Specifies the time and date the last change was made

Location. Specifies the location of the moved or created document

QUESTIONS

26-1. List and describe the services provided by *application-layer protocols.*

26-2. What does *Telnet* provide?

26-3. What does a *remote terminal protocol* provide?

26-4. What is a *network virtual terminal*?

26-5. List the three basic services provided by Telnet.

26-6. Describe the *NVT character set.*

26-7. Describe *Telnet encapsulation* and *decapsulation.*

26-8. List and describe each of the *Telnet options.*

26-9. Describe how *Telnet option negotiations* are accomplished.

26-10. Describe how Telnet options are enabled and disabled.

26-11. Describe the Telnet modes of operation.

26-12. What does the *file transfer protocol* (FTP) provide?

26-13. List the four essential objectives of FTP.

26-14. Construct and explain the basic FTP model.

26-15. List the two types of FTP *connections.*

26-16. Describe an FTP *data transfer connection.*

26-17. Describe how communications is established with an FTP *control channel.*

26-18. Describe how communications occurs over an FTP *data channel.*

26-19. List and describe the FTP *file types.*

26-20. List and describe the FTP *data structures.*

26-21. List and describe the FTP *transmission modes.*

26-22. How are FTP commands used?

26-23. List and describe the FTP *access commands.*

26-24. List and describe the FTP *file management commands.*

26-25. List and describe the FTP *data formatting commands.*

26-26. List and describe the FTP *port defining commands.*

26-27. List and describe the FTP *file transferring commands.*

26-28. Describe the format used with FTP *responses.*

26-29. List the steps involved when a client retrieves a list of items stored in a directory in a server.

26-30. Explain when the trivial file transfer protocol is appropriate.

26-31. What are the five types of TFTP *message packets*?

26-32. Describe the format for a *read request message* type.

26-33. Describe the format for a *write request message* type.

26-34. Describe the format for a *data message* type.

26-35. Describe the format for an *acknowledgment message* type.

26-36. Describe the format for an *error message* type.

26-37. Describe how a TFTP connection is established and terminated.

26-38. Explain how TFTP accomplishes *flow and error control.*

26-39. What are some of the applications for TFTP?

26-40. What is the purpose of the *simple mail transfer protocol* (SMTP)?

26-41. List the type of messages supported by SMTP.

26-42. Describe the fundamental ideal behind SMTP.

26-43. Describe the sending and receiving process for SMTP.

26-44. Describe the makeup of an *e-mail address*.

26-45. List and describe the *SMTP commands*.

26-46. List and describe the *SMTP responses*.

26-47. Describe a *multipurpose Internet mail extension* (MIME).

26-48. List the message types available with MIME.

26-49. What is the application of the *Post Office Protocol*?

26-50. What is the purpose of the *hypertext transfer protocol* (HTTP)?

26-51. List and describe the features of HTTP.

26-52. Describe the general format for HTTP *requests* and *responses*.

26-53. Describe the HTTP *request message format*.

26-54. Explain the format for a *uniform resource locator*.

26-55. Describe the HTTP *response message format*.

26-56. List the headers for HTTP.

26-57. Describe the format for an HTTP *request header*.

26-58. Describe the format for an HTTP *response header*.

26-59. Describe the format for an HTTP *entity header*.

26-60. Describe the format for an HTTP *general header*.

C H A P T E R 27

Integrated Services Data Networks

OBJECTIVES

- Define *integrated services* and *integrated services data networks*
- Define *broadband technology*
- Define a *public switched data network*
- Explain value-added networks
- Describe three packet switching techniques
- Define *X.25*
- Define the three types of X.25 network devices
- Describe the X.25 protocol layers
- Describe the purpose of the integrated services digital network
- Describe the conceptual view of ISDN
- Describe the ISDN architecture
- Describe the functional groupings and system connections used with ISDN
- Explain broadband ISDN
- Describe the principles of digital subscriber line
- Describe the XDSL family
- Explain the modulation schemes used with DSL
- Explain the ADSL frame format
- Describe the principles of asynchronous transfer mode
- Describe the ATM network components
- Describe the ATM protocol layers
- Describe the fundamental principle of frame relay

The communications industry is continually changing to meet the demands of contemporary telephone, video, and computer communications systems. Today, more and more people have the need or desire to communicate with each other than ever before. Meeting these needs requires updating and modernizing the infrastructure of the worldwide communications network. In addition, old standards are being updated and new standards developed and implemented on almost a daily basis.

Integrated services data networks are networks initiated by the telecommunications, data, and video networking industries to combine their services and transport virtually all information in digital format over a common communications system. From a networking perspective, the transmission media merely transport data in digital form. Once analog information, such as voice and video, has been digitized, it cannot be discerned from normal computer data. Therefore, it is treated the same. It makes no difference whether bits of information actually represent voice, data, or video streams. However, the telecommunications and video networking industries must figure out how to efficiently transport their information as packets of data over the existing communications system infrastructure. To provide integrated services, it is necessary to provide facilities (transmission media) that provide users with access to the integrated network. Protocols must also be designed to organize and control the flow of data across the network.

Integrated services data networks include *broadband technologies* that allow remote users to access data networks (including the Internet) through the public telephone system and transport data among numerous existing digital networks. Broadband access technologies route communication between networks or nodes within a network over wideband (high-bit-rate) transmission facilities. Because broadband access technologies have wider bandwidths, they allow users to communicate at data rates that exceed those of conventional dial-up voice-grade telephone connections (i.e., over 56 Kbps).

Integrated services data networks provide access to and transport information between private and public switched data networks and include the following:

X.25 User-to-network interface protocol

Integrated services digital network (ISDN)

Digital subscriber line (DSL)

Asynchronous transfer mode (ATM)

Frame relay (FR)

27-2 PUBLIC SWITCHED DATA NETWORKS

A *public switched data network* (PDN or PSDN) is a switched data communications network similar to the public telephone network except a PDN is designed for transferring data that may include digitized voice and video signals. A public switched data network is comprised of one or more wide-area data networks designed to provide access to a large number of subscribers with a wide variety of computer equipment.

Interconnecting computing devices and networks is best accomplished through *interlinked nodes* called *switches*. Switches are pieces of hardware that are capable of creating temporary connections between two or more devices or networks using hardware or software control. The basic principle behind a PDN is to transport data from a source to a destination through a network of intermediate *switching nodes* and transmission media. The switching nodes are not concerned with the content of the data, as their purpose is to provide *end stations* with access to transmission media and other switching nodes that will transport data from node to node until it reaches its final destination.

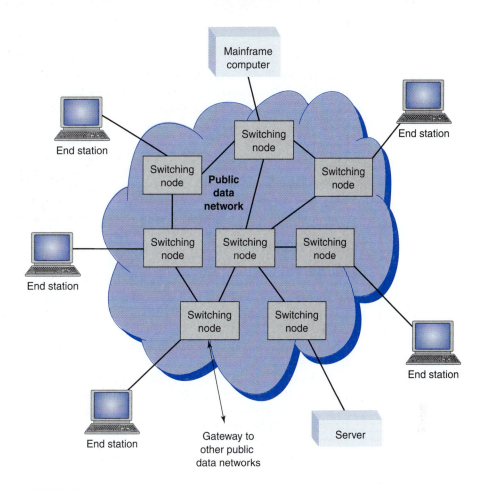

FIGURE 27-1 Public switched data network

Figure 27-1 shows a public switched data network comprised of several switching nodes interconnected with *transmission links* (channels). As shown in the figure, each switch is connected to multiple transmission links. The end-station devices can be personal computers, servers, mainframe computers, or any other piece of computer hardware capable of sending or receiving data. End stations are connected to the network through switching nodes. Data enter the network where they are routed through one or more intermediate switching nodes until they reach the intended destination.

As shown in Figure 27-1, some switching nodes connect only to other switching nodes (sometimes called *tandem switching nodes* or *switchers switches*), while some switching nodes are connected to end stations as well. Node-to-node communications links generally carry time-division multiplexed data. Public data networks are not *direct connected;* that is, they do not provide direct communications links between every possible pair of nodes.

Public switched data networks combine the concepts of value-added networks and packet switching networks.

27-2-1 Value-Added Networks

Value-added networks (VANs) "add value" to the services or facilities provided by common carriers to provide new types of communication services and enhanced capabilities to existing services. Examples of added values are error control, enhanced connection reliability, dynamic routing, failure protection, logical multiplexing, and data format conversions. A VAN comprises an organization that leases communications lines from common carriers, such as AT&T and MCI, and adds new types of communications services to those lines. Examples of VANs are GTE Telnet, DATAPAC, TRANSPAC, and Tymnet Inc.

27-2-2 Packet Switching Techniques

Packet switching involves dividing data messages into small bundles of information and transporting them through communications networks to their intended destinations using computer-controlled switches. There are three common switching techniques used with public data networks: circuit switching, message switching, and packet switching.

27-2-2-1 Circuit switching. *Circuit switching* was designed for making standard telephone calls on the public telephone network. Circuit switching creates a direct physical connection between two end stations, such as telephones or computers. Circuit switching is not as well-suited to data and other nonvoice transmissions, as they tend to transmit in bursts (spurts of data with idle gaps between them). Therefore, there is a lot of wasted time when nothing is being sent.

With circuit switching, the call is established, information is transferred (i.e., a conversation takes place), and then the call is terminated. The time required to establish the call is called the *setup time*. Once the call has been established, the circuits interconnected by the network switches are allocated to a single connection between two users for the duration of the call. Once a call has been established, information can be transferred in *real time*. When a call is terminated, the circuits and switches used to provide the interconnection are relinquished and made available for another user.

A circuit switch is a device with multiple inputs and multiple outputs that creates a temporary connection between an input and an output. Because there are a limited number of circuits and switching paths available, *blocking* can occur. Blocking is the inability to complete a call because there are no facilities or switching paths available between the source and destination locations. When circuit switching is used for data transfer, the terminal equipment at the source and destination must be compatible; they must use compatible modems and the same bit rate, character set, and protocol.

Figure 27-2 shows a *circuit-switched network* with three switches providing access to end stations and one switch that simply provides a transmission path between other switches. Any end station can be temporarily interconnected to any other end station through one or more switches.

A circuit switch is a *transparent switch*. A transparent switch interconnects the source and destination terminal equipment. Transparent switches do not add any value to the circuit, as they simply transport information.

Transmitting data in real time presents three additional drawbacks of using circuit switching for data:

1. The transmission rate is limited to that of the slowest device.
2. Circuit switching is inflexible in that once a circuit has been established, all data takes the same path regardless of whether that path is the most efficient means.
3. Circuit switching treats all transmission equally. A connection is established on demand using whatever transmission links and switching paths are available at the time. Therefore, it is difficult to prioritize messages and select transmission paths.

27-2-2-2 Message switching. *Message switching* is a form of *store-and-forward network,* as there is no direct connection between the sender and receiver. Data, including source and destination identification codes, is transmitted into the network and stored in a switch. Each switch within the network has message storage capabilities. Data is delivered to a switch along one transmission path and then rerouted to the destination or next switch along another transmission path. The network transfers data from one switch to the next

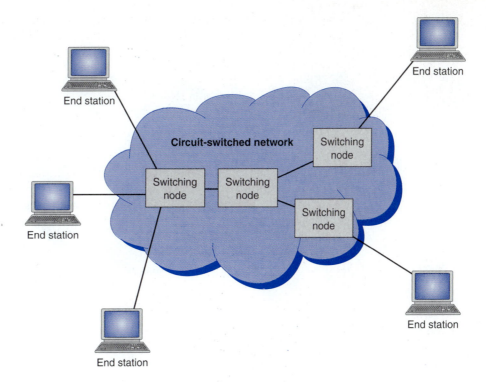

FIGURE 27-2 Circuit-switched network

switch when it is convenient to do so. Consequently, data is not transferred in real time, as there can be a delay at each switch.

With message switching, blocking cannot occur. However, the delay time from message transmission to reception can vary from call to call and can be quite long (possibly as long as 24 hours). With message switching, once information has entered the network, it is converted to a form more suitable for transmission through the network. At the receive end, the data is converted to a format compatible with the receiving data terminal equipment. Therefore, with message switching, the source and destination data terminal equipment do not need to be compatible. Message switching is more efficient than circuit switching because data that enters the network during busy times can be held in a switch and transmitted later when the demand has decreased.

Figure 27-3 shows a message-switched network. Each switch stores and relays data from a secondary storage device, which makes storing and switching times rather long. The primary application of message-switched networks is providing high-level network services for unintelligent devices where delayed delivery is acceptable.

A message switch is a *transactional switch* because it does more than simply provide a switching path to interconnect two segments of a communications circuit. A message switch can store data or change its format and bit rate and then convert the data back to its original form or an entirely different form at the destination. Message switching multiplexes data from different sources onto common transmission facilities.

27-2-2-3 Packet switching. With *packet switching,* data is divided into smaller units, called *packets,* and then transmitted through the network. The maximum length of the packet is determined by the network and can vary from network to network. In addition to the data, each packet contains a header with control information and source and destination

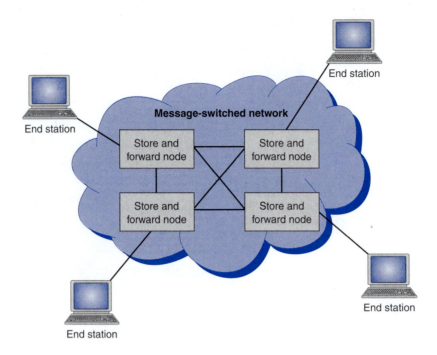

FIGURE 27-3 Message-switched network

addresses. Packets are transported over the network from node to node where each node stores the packet for a brief time while it determines the route the packet should take to reach the next node.

Because a packet can be held in memory at a switch for a short period of time, packet switching is sometimes called a *hold-and-forward network*. With packet switching, a message is divided into datagrams, and each datagram is treated as a separate entity and can take a different path through the network. Consequently, all packets do not necessarily arrive at the destination at the same time or in the same order in which they were transmitted.

Because packets are small, the hold time is generally quite short, message transfer is near real time, and blocking cannot occur. However, packet switching networks require complex and expensive switching arrangements and complicated protocols. A packet switch is also a transactional switch.

Packet switching uses two approaches: datagram and virtual circuit.

Datagram. With the datagram approach, each data packet is treated independently from all other packets, even when the packets belong to the same message. With datagram transmission, packets do not necessarily take the same route or arrive at the destination in the same order that they were transmitted. Therefore, each packet must be numbered, and all packets that comprise the same message must have a common identification code.

Virtual circuit. With the virtual circuit approach, the relationship between all packets that belong to the same message is preserved. All packets are transmitted one after the other and take the same route between the source and destination. Therefore, the packets arrive at the destination in the same order in which they were sent.

Table 27-1 summarizes circuit, message, and packet switching.

Table 27-1 Public Data Network Switching Summary

Circuit Switching	Message Switching	Packet Switching
Dedicated transmission path	No dedicated transmission path	No dedicated transmission path
Continuous transmission of data	Transmission of messages	Transmission of packets
Operates in real time	Not real time	Near real time
Messages not stored	Messages stored	Messages held for short time
Path established for entire message	Route established for each message	Route established for each packet
Call setup delay	Message transmission delay	Packet transmission delay
Busy signal if called party busy	No busy signal	No busy signal
Blocking may occur	Blocking cannot occur	Blocking cannot occur
User responsible for message-loss protection	Network responsible for lost messages	Network may be responsible for each packet but not for entire message
No speed or code conversion	Speed and code conversion	Speed and code conversion
Fixed bandwidth transmission (i.e., fixed information capacity)	Dynamic use of bandwidth	Dynamic use of bandwidth
No overhead bits after initial setup delay	Overhead bits in each message	Overhead bits in each packet

27-3 X.25 USER-TO-NETWORK INTERFACE PROTOCOL

X.25 is a packet-switched wide area network (WAN) developed in 1976 by the CCITT to provide a common interface protocol between public data networks. The X.25 interface is designated the *international standard for packet network access.* X.25 is a packet-switched data network interface protocol that defines an international recommendation for establishing connections and exchanging data and control information between user devices and network devices. X.25 is intended for terminal operation in the packet mode over public data networks.

There are three types of X.25 network devices: data terminal equipment (DTE), data circuit-terminating equipment (DTE), and packet switching exchanges (PSE), as shown in Figure 27-4. DTE equipment are end systems that communicate with one another across the X.25 data network and include terminals, PCs, and network hosts. DCE equipment are communications devices, such as modems and packet switches, that provide the interface between DTEs and a PSE. PSEs are switches that constitute the majority of the network. PSEs transfer data from one DTE to another through the X.25 network.

Another device commonly found in an X.25 network between DTEs and DCEs is a *packet assembler/disassembler* (PAD). PADs provide buffering (data storage), packet assembly (including the X.25 header), and packet disassembly (including the X.25 header). At the source, a PAD converts the user's data format to X.25 packets, and at the destination a PAD converts X.25 packets to the user's data format. The source and destination user's data can have different formats.

X.25 is a connection-oriented end-to-end protocol; however, because it uses the virtual circuit concept, the actual movement of data packets through the networks is transparent to the user, who sees the networks as a cloud through which packets pass on their way to the destination DTE. X.25 is a *subscriber network interface* (SNI), as it defines how a user's DTE communicates with the network and how data packets are transmitted over the network using DCEs.

27-3-1 X.25 Protocol Layers

Figure 27-5 shows the relationship between the X.25 protocol layers and the OSI seven-layer protocol hierarchy. As shown in the figure, X.25 addresses only the physical, data-link (frame), and network (packet) layers of the ISO seven-layer protocol hierarchy. X.25 defines how packet mode terminals can be connected to a packet network. It also describes the procedures required to establish, maintain, and terminate a connection as well as a set

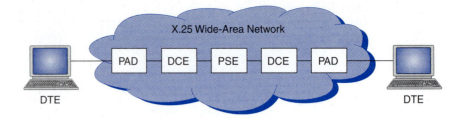

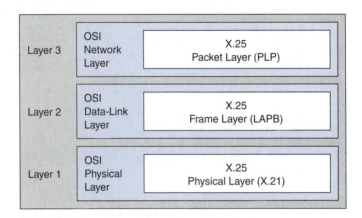

FIGURE 27-4 Conceptual view of X.25

FIGURE 27-5 X.25 layers compared to OSI layers

of services (sometimes called *facilities*) that provide additional functions, such as reverse charge, call direct, and delay control.

X.25 uses existing standards whenever possible. For example, X.25 specifies the X.21, X.24, and X.27 standards as the physical interfaces, which correspond to EIA RS-232, RS-423A, and RS 422A standards, respectively.

27-3-1-1 X.25 physical layer protocol. The *physical-layer protocol* used by X.25 is called the X.21 *digital interface.* The ITU-T designed the X.21 interface to enable all-digital communications between DTEs and DCEs and to address the problems inherent in many of the preexisting EIA interface standards. X.21 specifies how a DTE and DCE exchange signals to set up and clear calls.

X.21 specifies the 15-pin DB-15 connector as the physical connection, as shown in Table 27-2. Although the physical connector includes 15 pins, generally only eight are used. Many of the EIA standards utilize most of their pins for transporting control signals using positive and negative voltage levels. Therefore, each pin is limited to a single function. However, X.21 eliminates most of the control function pins by transporting meaningful serial codes, such as ASCII, over a single control lead. This consolidation of signals eliminates many of the control lines and also allows control signals to be propagated from network to network as well as between DTEs and DCEs. X.21 is designed for balanced transmission lines at a 64-kbps transmission rate. X.21 is appropriate for interfacing digital devices to modems or to other digital interfaces, such as ISDN and X.25.

The X.21 pin functions are summarized as follows:

Pins 1 and 8 are used for chassis and signal grounds, respectively.

Pins 2 and 3 of the X.21 interface are used to transport data and control information from the DTE to the DCE.

Table 27-2 X.21 Pin Designations

Pin Number	Function
1	Shield ground
2	Transmit data or control
3	Control
4	Receive data or control
5	Indication
6	Signal element timing
7	Byte timing
8	Signal ground
9	Transmit data or control
10	Control
11	Receive data or control
12	Indication
13	Signal element timing
14	Byte timing
15	Reserved for future use

Pins 4 and 5 are used to transport data information from the DCE to the DTE. The control and initiation pins are used for accomplishing an initial handshake between the DTE and DCE. Pin 3 (control) is equivalent to the request to send (RTS) pin used with several EIA standards. Pin 5 (indication) is comparable to clear to send.

Pin 6 is used to transport timing (clocking) information from the DCE to the DTE.

Pin 7 is used to send a pulse from the DCE to the DTE to group bits into byte frames. This is sometimes called *byte synchronization.* There is no equivalent signal on any of the EIA interface standards.

27-3-1-2 X.25 frame-layer protocol. The *frame* (or link)-*level protocol* is a layer 2 protocol intended to provide reliable data transfer between the DTE and the DCE by transmitting data as a sequence of frames. The link layer provides the following functions:

Transfer data efficiently and in a timely manner

Synchronize the link, ensuring that the receive is synchronized to the transmitter

Provide error detection and recovery

Identify and report procedural errors to a higher layer for recovery

There are several link protocols that may be used with X.25, including the following:

Link access protocol, balanced (LAPB). Derived from HDLC and currently the most popular link layer protocol for X.25, LAPB forms a logical link connection as well as all the other features of HDLC.

Link access protocol (LAP). An earlier version of LAPB that is outdated and seldom used today.

Link access procedure, D channel (LAPD). Derived from LAPB and used primarily for integrated services digital networks (ISDN). LAPD enables data transmission between a DTE and an ISDN node through special channels called *D channels.*

IEEE 802 logical link control (LLC). Used with X.25 to enable the packets to be transported across a Ethernet local area networks (LANs).

Most X.25 applications provide data-link control at the frame layer using a bit-oriented protocol called *link access procedure, balanced* (LAPB), which defines a subset of high-level data-link control (HDLC) as the international standard for the data-link layer. At the data-link level, LAPB provides for two-way, full-duplex communications between DTE and DCE at the packet network gateway. The American National Standards Institute

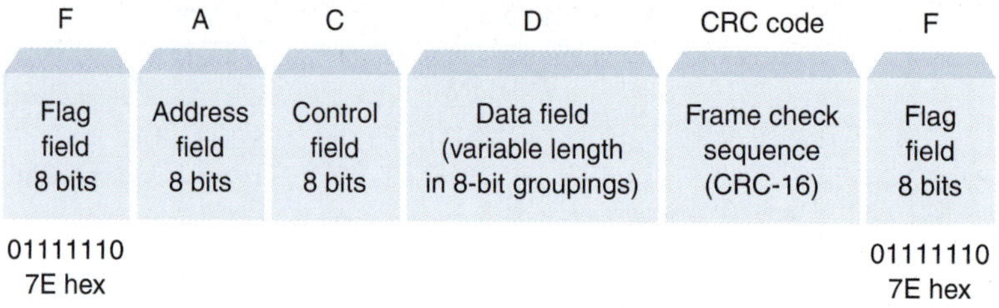

F	A	C	D	CRC code	F
Flag field 8 bits	Address field 8 bits	Control field 8 bits	Data field (variable length in 8-bit groupings)	Frame check sequence (CRC-16)	Flag field 8 bits

01111110
7E hex

01111110
7E hex

FIGURE 27-6 X.25 frame format

(ANSI) 3.66 Advanced Data Communications Control Procedures (ADCCP) is the U.S. standard. ANSI 3.66 and HDLC were designed for private-line data circuits using a centrally controlled polling environment. Consequently, the addressing and control procedures outlined by them are not appropriate for packet data networks. ANSI 3.66 and HDLC were selected for the data-link layer because of their frame format, delimiting sequence, transparency mechanism, and error-detection method.

Figure 27-6 shows the X.25 frame format. As shown in the figure, the frame format for X.25 is virtually identical to that of HDLC and includes flag, address, control, data, and checksum fields. The flag field contains the delimiting sequence 01111110, which is used to indicate the beginning and end of a frame. The address field contains the address of the DTE or DCE. The address field refers to a link address, not a network address. Communications between DTEs and DCEs with X.25 is point to point using the asynchronous balanced mode. Therefore, only two addresses are needed. A command issued by a DTE uses the address 00000001 (01 hex), and a response to the command uses the address 00000011 (03 hex). The network address of the destination terminal is contained in the packet header, which is encapsulated in the information field. The control field contains eight bits that specify the sequence numbers and the commands and responses for controlling data flow between the DTE and DCE. Like HDLC, LAPB has three types of frames: information (I frames), supervisory (S frames), and unnumbered (U frames), which is essentially the same as with HDLC.

Communications between a DTE and a DCE involve three phases: setup, data transfer, and disconnect. The data link between a DTE and a DCE must be established (set up) before data packets can be transferred. Either the DTE or the DCE can initiate establishing a link. Once the link has been established, both parties can transmit and receive network-layer packets, which includes both user data and control information. Once the network layer no longer needs the link, either the DTE or the DCE can initiate a disconnect frame to request disconnection, and the other end simply acknowledges the request.

Tables 27-3 and 27-4 show the data-link commands and responses, respectively, for an LAPB frame. During LAPB operation, most frames are commands. A response frame is compelled only when a command frame is received containing a poll (P bit) = 1. SABM/UA

Table 27-3 LAPB Commands

	Bit Number			
Command	8 7 6	5	4 3 2	1
1 (information)	nr	P	ns	0
RR (receiver ready)	nr	P	0 0 0	1
RNR (receiver not ready)	nr	P	0 1 0	1
REJ (reject)	nr	P	1 0 0	1
SABM (set asynchronous balanced mode)	0 0 1	P	1 1 1	1
DISC (disconnect)	0 1 0	P	0 0 1	1

Table 27-4 LAPB Responses

Response	Bit Number			
	8 7 6	5	4 3 2	1
RR (receiver ready)	*nr*	F	0 0 0	1
RNR (receiver not ready)	*nr*	F	0 1 0	1
REJ (reject)	*nr*	F	1 0 0	1
UA (unnumbered acknowledgment)	0 1 1	F	0 0 1	1
DM (disconnect mode)	0 0 0	F	1 1 1	1
FRMR (frame rejected)	1 0 0	F	0 1 1	1

is a command/response pair used to initialize all counters and timers at the beginning of a session. Similarly, DISC/DM is a command/response pair used at the end of a session. FRMR is a response to a procedural error or an illegal command for which there is no indication of transmission errors according to the frame check field.

Information (I) commands are used to transmit sequenced packets of user information. Packets are never sent as responses. Packets are acknowledged just as they are with HDLC. RR (receive ready) is sent by a station when it needs to respond (acknowledge) something but has no information packets to send. A response to an information command could be RR with F = 1. This procedure is called *checkpointing*. With X.25, the F (final bit) is called the M (more) bit, although it performs essentially the same function.

REJ (reject) is another way of requesting retransmission of frames. RNR (ready not to receive) is used for flow control to indicate a busy condition and prevent further transmissions until cleared with RR.

27-3-1-3 X.25 packet-layer protocol. The *packet* (network)-*layer protocol* is a layer 3 protocol. The layer 3 protocol used by X.25 is called simply the *packet-layer protocol* (PLP). The packet layer creates network data units called *packets* that contain user information as well as control information. The PLP is also responsible for establishing a connection, transferring data over the connection, and then terminating the connection. The packet-layer protocol is responsible for creating virtual circuits and negotiating network services between a DTE and DCE. The frame layer is responsible for establishing the link between the DTE and DCE, whereas the packet layer is responsible for establishing an end-to-end connection between two DTEs.

With X.25, flow and error control are responsibilities of both the frame and the packet layer. The frame layer provides link flow and error control between a DTE and a DCE, whereas the packet layer provides end-to-end flow and error control between two DTEs. Figure 27-7 illustrates the domains for error and flow control of the frame and packet layer.

The network layer of X.25 specifies two types of switching services: virtual circuits and datagrams.

Virtual circuits. X.25 is a *packet-switched virtual circuit network* that creates *virtual circuits* at the network layer. The physical connection established between a DTE and DCE is capable of establishing and simultaneously carrying more than one virtual circuit at a time, and a virtual circuit can carry either data or control information.

Figure 27-8 shows a single DTE (A) simultaneously connected to three distant DTEs (B, C, and D) using three separate virtual circuits that access the X.25 network through a single physical link. The virtual circuits are the following:

DTE A to DTE B

DTE A to DTE C

DTE A to DTE D

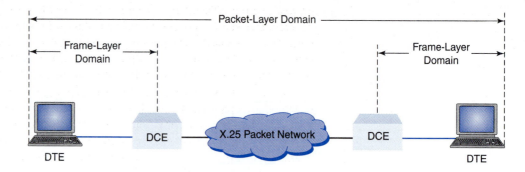

FIGURE 27-7 Error and flow control domains of X.25 frame and packet layers

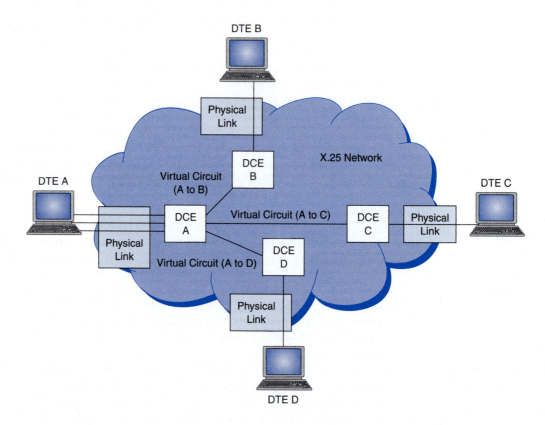

FIGURE 27-8 X.25 virtual circuits

Each virtual circuit originating from a DTE must be identified with two *virtual circuit identifiers,* which X.25 calls *logical channel numbers* (LCNs). LCNs enable the DTEs to keep track of what virtual circuit is connected to what destination DTE. There are two pairs of LCNs—one between the local DTE and the local DCE and the other between the distant DCE and the distant DTE.

There are two types of virtual circuits: permanent and switched.

Permanent virtual circuit. A permanent virtual circuit (PVC) is logically equivalent to a two-point private leased circuit except slower. A PVC is slower because a hardwired, end-to-end connection is not provided. The first time a connection is requested, the appropriate switches and circuits must be established through the network to provide the interconnection. A PVC identifies the routing between two predetermined subscribers of the network that is used for all subsequent messages. Therefore, subsequent messages do not require call setup or call clearing procedures. With a PVC, source and destination addresses are unnecessary because the two users are fixed and the network provider assigns permanent LCNs.

Switched virtual circuit. A switched virtual circuit (SVC) is logically equivalent to making a telephone call through the public switched telephone network except no direct end-to-end connect is made. An SVC is established at the beginning of each session. After a SVC is established, both DTE-DCE links are assigned an LCN. After completion of the data transfer, the virtual circuit is disconnected, and the LCNs are relinquished. An SVC is a one-to-many arrangement. Any VC subscriber can access any other SVC subscriber through a network of switches and communications channels. Switched virtual circuits are temporary virtual connections that use common-usage equipment and circuits. The source must provide its address and the address of the destination before an SVC can be completed. Switched virtual circuits are the most frequently used of the X.25 services.

A service called *fast select* enables the control packet that sets up the virtual circuit to carry data as well.

Datagram. A *datagram* (DG) is, at best, vaguely defined by X.25 and, until it is completely outlined, has very limited usefulness. With a DG, users send small packets of data into the network. Each packet is self-contained and travels through the network independent of other packets of the same message by whatever means available. The network does not acknowledge packets, nor does it guarantee successful transmission. However, if a message will fit into a single packet, a datagram is somewhat reliable. This is called a *single-packet-per-segment protocol.*

27-3-2 X.25 Packet Formats

A virtual call is the most efficient service offered for a packet network. There are two packet formats used with virtual calls: a call request packet and a data transfer packet.

27-3-2-1 X.25 call request packet. Whenever a DTE wishes to communicate with another DTE, it must first set up a connection with a *call request packet* (sometimes called a *control packet*). The virtual circuit is established when the DTE receives a call accepted packet. After the virtual circuit is established, the two DTEs use a full-duplex connection to exchange data packets.

Figure 27-9 shows the format for an X.25 call request packet. The delimiting sequence is 01111110 (the same as an HDLC flag), and the error-detection/correction mechanism is CRC-16 with ARQ (automatic request for retransmission). The link address field and the control field have little use and, therefore, are seldom used with packet networks. The remaining fields are defined in sequence:

Format identifier field. The format identifier (sometimes called a *general format identifier* [GFI]) is a four-bit field, as shown in Figure 27-10. The first bit is called the qualifier (Q) bit, which defines the source of control information. A logic 0 specifies PLP, and a logic 1 specifies other higher-level protocols. The second bit is the delivery (D) bit. The D bit defines which device should acknowledge the packet. A logic 0 specifies the local DCE, and a logic 1 specifies the remote DTE. The remaining two bits of the GRI specify the packet numbering sequence for flow control by specifying the length of the sequence number subfields contained in the packet type field. A 01 indicates there are only three bits in the sequence number fields (000–111 or 0–7), and a 10 indicates there are seven bits in the sequence number fields (0000000–1111111 or 0–127). There are two sequence numbers, P(r) and P(s), which

Flag	Link address field	Link control field	Format identifier	Logical channel identifier	Packet type	Calling address length	Called address length	Called address up to	Calling address up to
8 bits	8 bits	8 bits	4 bits	12 bits	8 or 16 bits	4 bits	4 bits	60 bits	60 bits

Null field	Facilities length field	Facilities field up to	Protocol ID	User data up to	Frame check sequence	Flag
2 zeros	6 bits	512 bits	32 bits	96 bits	16 bits	8 bits

FIGURE 27-9 X.25 call request packet format

Q-bit (Qualifier)	D-bit (Delivery)	Packet numbering sequence (2 bits) 01 = modular 8 (0 to 7) 10 = modular 128 (0 to 127)

FIGURE 27-10 Format indentifier format

are used the same way nr and ns are used with HDLC, that is; for flow control. P(s) is the number of the packet being sent by the transmitting device. P(r) is the number of the next packet the transmitting device expects to receive or the number of the next packet the receiving device should transmit.

Logical channel number field. The logical channel number (LCN) is a 12-bit binary number that identifies the source and destination users for a given virtual call. To provide a sense of hierarchy, X.25 originally divided the LCN field into a four-bit group field and an eight-bit channel field that defined a logical group channel number (LGCN) and an eight-bit channel number. However, today the 12-bit sequence is simply referred to as a single logical channel number (LCN). After a source user has gained access to the network and has identified the destination user, the source and destination users are assigned an LCN. In subsequent packets, the source and destination addresses are unnecessary, as only the LCN is needed. When two users disconnect, the LCN is relinquished and can be reassigned to new users. There are 4096 (2^{12}) LCNs available. Therefore, there may be as many as 4096 virtual calls established at any given time.

Packet type identifier field. The packet type identifier (PTI) is used to identify the function and content of the packet (i.e., its type), such as new request, call clear, call reset, and so on. There are two formats used for the packet type identifier field. One format is used to identify whether a three-bit number sequence or a seven-bit number sequence is being used, as shown in Figure 27-11. With the three-bit number sequence scheme, the packet type field contains either three P(r) bits or a three-bit packet type field. With three bits, eight different packet types (functions) can be defined. The remaining two bits are 01. With the seven-bit number sequence scheme, the packet type field contains a six-bit packet type field followed by 01 and a second byte containing seven P(r) bits followed by a 0. With six bits, 64 different packet types can be defined.

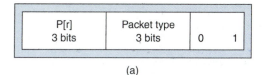

(a)

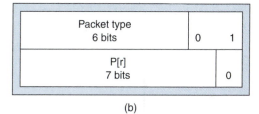

(b)

FIGURE 27-11   Packet type field for RR, RNR, and REJ: (a) three-bit number sequencing; (b) seven-bit number sequencing

FIGURE 27-12   Packet type field for call request packets carrying control data

There are three types of packets: *receive ready* (RR), *receive not ready* (RNR), and *reject* (REJ):

> **RR** *(000 or 000000)*. RR indicates that the device (DTE or DCE) is ready to receive more data packets. RR can also be used with P(r) to acknowledge the receipt of previously received packets.
>
> **RNR** *(001 or 000001)*. RNR indicates that the device (DTE or DCE) cannot accept any additional data packets at this time.
>
> **REJ** *(010 or 000010)*. REJ indicates that there were transmission errors and to retransmit all previously unconfirmed frames beginning with the frame identified by P(r).

The second type of packet identifier, shown in Figure 27-12, is used when the call request packet is also carrying data. The data, however, is control information, not user data. Because the packet is not carrying sequenced user information and cannot be used to confirm previously received sequenced information packets, there are no P(r) or P(s) bits. The packet type field contains eight bits, which includes a six-bit packet type field followed by 11. With six bits, 64 different packet types can be defined:

> *Calling address length field.* The calling address length is a four-bit binary number that specifies how many BCD digits appear in the calling address field (i.e., how long the calling address is). With four bits, a maximum of 15 digits can be specified.
>
> *Called address length field.* The called address length is identical to the calling address length except it identifies the number of BCD digits that appear in the called address field (i.e., how long the called address is). The most commonly used address length is 14 digits; therefore, the contents of the calling address length field is typically 1110 binary (E hex).
>
> *Called address field.* The called address contains the BCD code for the destination address. The number of BCD digits is specified in the called address length field. There can be a maximum of 15 BCD digits (60 bits) assigned to a destination user.
>
> *Calling address field.* The calling address contains the BCD codes for the source address. The number of BCD digits is specified in the calling address length field. There can be a maximum of 15 BCD digits (60 bits) assigned to a destination user.

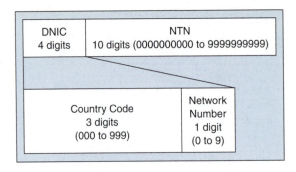

FIGURE 27-13 X.121 address format

Facilities length field. The facilities length identifies in binary the number of eight-bit octets present in the facilities field.

Facilities field. The facilities field contains up to 512 bits of optional network facility information, such as reverse billing information, closed user groups, and whether it is a simplex or half-duplex transmit or simplex or half-duplex receive connection. The facilities field can also be used to specify the maximum window size rather than using the default value of 128 bytes and two packets.

Protocol identifier field. The protocol identifier field contains a 32-bit code that is reserved for the subscriber to insert user-level protocol functions, such as log-on procedures and user identification practices.

User data field. The user data field of a call request packet can contain up to 96 bits of user data. The data is unnumbered (unsequenced) and is not confirmed. Therefore, there is no means of flow control or error correction. This field is generally used for user passwords.

X.25 uses an addressing system defined by X.121, which is similar to the system used by the public switched telephone network (area code, prefix, and extension). Figure 27-13 shows the X.121 protocol format. The address contains 14 digits. The first four digits are called the data network identification code (DNIC), which identifies the network. Three digits are used to identify the country, and one digit is used to define the network within that country. There are 1000 country codes and 10 network codes for a total of 10,000 combinations. Countries with more than one public data network may have more than one country code. The remaining 10 digits are called the national terminal number (NTN) and are used to identify a specific DTE within the specified network. The allocation of the NTN digits is not defined by X.121, which allows each network to allocate their 10 billion addresses any way they wish.

27-3-2-2 X.25 Data transfer packet.   Figure 27-14 shows the format for an X.25 data transfer packet. A data transfer packet is similar to a call request packet except that a data transfer packet can carry user data, and the data transfer packet contains send-and-receive packet sequence fields that were not included with the call request format.

The flag, link address, link control, format identifier, LCI, and FCS fields are identical to those used with the call request packet. The send and receive packet sequence fields are described as follows:

Send packet sequence field. Each successive data transfer packet sent is assigned the next P(s) number in sequence. P(s) can be a three- or seven-bit binary number and, thus, can number packets from either 0 through 7 (000–111) or 0 through 127 (0000000–1111111). The numbering sequence is identified in the format identifier. The send packet field always contains eight bits. Therefore, the unused bits are reset to 0s.

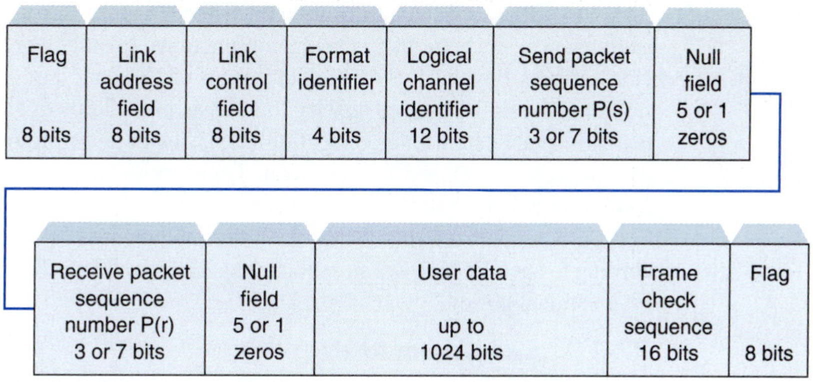

FIGURE 27-14 X.25 data transfer packet format

Receive packet sequence field. The receive packet sequence field is used to confirm received data transfer packets and call for retransmission of packets received in error using automatic request for retransmission (ARQ). The user data field in a data transfer packet can carry considerably more source information than an I field in a call request packet (i.e., 1024 bits as opposed to 96 bits).

P(s) and P(r) are used to sequence data transfer packets and ensure an orderly flow of packets containing user data. P(s) and P(r) are also used for error correction.

User data field. The user data field of a data transfer packet can contain up to 1024 bits of user data. The data is numbered (sequenced) and therefore provides a means of flow control and error correction.

27-4 INTEGRATED SERVICES DIGITAL NETWORK

The *integrated services digital network* (ISDN) is a network proposed and designed by the major telephone companies in conjunction with the International Telecommunications Union–Telephony (ITU-T) with the intent of providing worldwide telecommunications support of voice, data, video, and facsimile information within the same network (in essence, ISDN is the integrating of a wide range of services into a single multipurpose network). ISDN is a network that proposes to replace existing public telecommunications networks and deliver a wide variety of services by interconnecting an unlimited number of independent users through a common communications network.

To date, only a small number of ISDN facilities have been developed; however, the telephone industry is presently implementing an ISDN system so that in the near future, subscribers will access the ISDN system using existing public telephone and data networks. The ITU-T in its recommendation ITU-T I.120 (1984) defined the basic principles and evolution of ISDN.

27-4-1 Principles of ISDN

The main feature of the ISDN concept is to support a wide range of voice (telephone) and nonvoice (digital data) applications in the same network using a limited number of standardized facilities. ISDNs support a wide variety of applications, including both switched and nonswitched (dedicated) connections. Switched connections include both circuit- and packet-switched connections and their concatenations. Whenever practical, new services introduced into an ISDN should be compatible with 64-kbps switched digital connections. The 64-kbps digital connection is the basic building block of ISDN.

An ISDN will contain intelligence for the purpose of providing service features, maintenance, and network management functions. In other words, ISDN is expected to provide services beyond the simple setting up of switched circuit calls. A layered protocol structure should be used to specify the access procedures to an ISDN and can be mapped into the open system interconnection (OSI) seven-layer protocol hierarchy model. Standards already developed for OSI-related applications can be used for ISDN, such as X.25 level 3 for access to packet switching services.

It is recognized that ISDNs may be implemented in a variety of configurations according to specific national situations. This accommodates both single-source and competitive national policy.

27-4-2 Evolution of ISDN

ISDNs will be based on the concepts developed for telephone ISDNs and may evolve by progressively incorporating additional functions and network features, including those of any other dedicated networks, such as circuit and packet switching for data, so as to provide for existing and new services.

The transition from an existing network to a comprehensive ISDN may require a period of time extending over one or more decades. During this period, arrangements must be developed for internetworking of services on ISDNs and services on other networks.

In the evolution toward ISDN, digital end-to-end connectivity will be obtained via plant and equipment used in existing networks, such as digital transmission, time-division multiplexed switching, and/or space-division multiplexed switching. Existing relevant recommendations for these constituent elements of an ISDN are contained in the appropriate series of recommendations of ITU-T and CCIR.

In the early stages of the evolution of ISDNs, some interim user-network arrangements may need to be adopted in certain countries to facilitate early penetration of digital service capabilities. An evolving ISDN may also include at later stages switched connections at bit rates higher and lower than 64 kbps.

27-4-3 Conceptual View of ISDN

ISDN is a symmetrical broadband access technology, as the uplink and downlink transmission rates are the same. Uplink is from the subscriber to the network, and downlink is from the network to the subscriber. Broadband simply means the transmission rates exceed those typically transported over standard voice-grade telephone circuits (i.e., >56 kbps).

Figure 27-15 shows a view of how ISDN can be conceptually viewed by a subscriber (customer) of the system. Customers gain access to the ISDN system through a local interface connected to a digital transmission medium called a *digital pipe*. There are several sizes of pipe available with varying capacities (i.e., bit rates), depending on customer need. For example, a residential customer may require only enough capacity to accommodate a telephone and a personal computer. However, an office complex may require a pipe with sufficient capacity to handle a large number of digital telephones interconnected through a nonpremise private branch exchange (PBX) or a large number of computers on one or more LANs.

Figure 27-16 shows the ISDN user network, which illustrates the variety of network users and the need for more than one capacity pipe. A single residential telephone is at the low end of the ISDN demand curve, followed by a multiple-drop arrangement serving a telephone, a personal computer, and a home alarm system. Industrial complexes would be at the high end of the demand curve, as they require sufficient capacity to handle hundreds of telephones and several LANs. Although a digital pipe has a fixed capacity, the traffic on the pipe can be comprised of data from a dynamic variety of sources with varying signal types and bit rates that have been multiplexed into a single high-capacity pipe. Therefore, a customer can gain access to both circuit- and packet-switched services through the same pipe. Because of the obvious complexity of ISDN, it requires a rather complex control system to facilitate multiplexing and demultiplexing data to provide the required services.

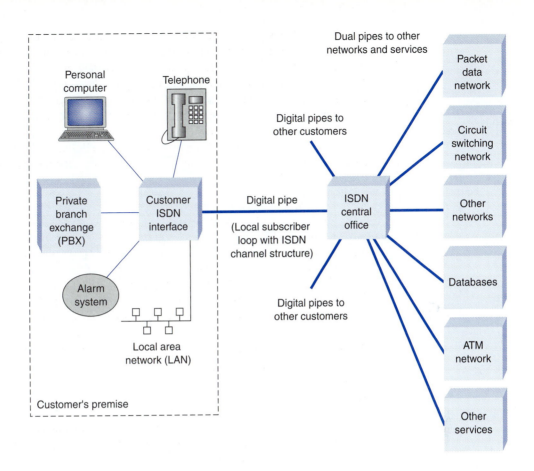

FIGURE 27-15   Subscriber's conceptual view of ISDN

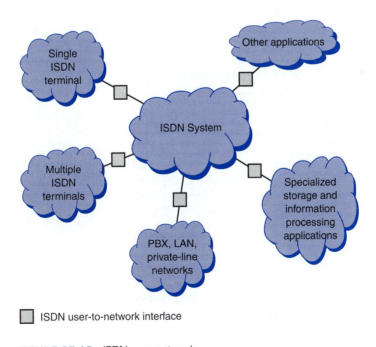

☐ ISDN user-to-network interface

FIGURE 27-16   ISDN user network

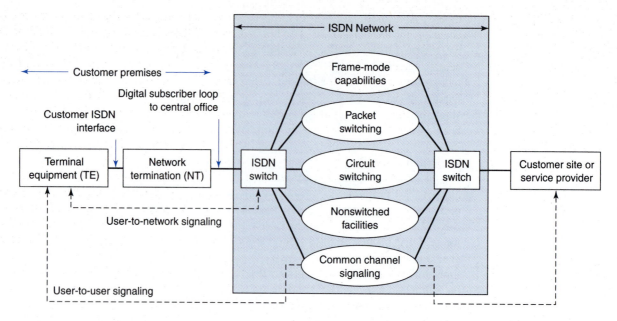

FIGURE 27-17 ISDN architecture

27-4-4 Objectives of ISDN

The key objectives of developing a worldwide ISDN are the following:

1. *System standardization*— ensures universal access to the network.
2. *Achieving transparency*— allows customers to use a variety of protocols and applications.
3. *Separating functions*— ISDN should provide services that preclude competitiveness.
4. *Variety of configurations*— provide both private-line (leased) and switched services.
5. *Addressing cost-related tariffs*— ISDN service should be directly related to cost and independent of the nature of the data.
6. *Migration*— provide a smooth transition while evolving.
7. *Multiplexed support*— provide service to low-capacity personal subscribers as well as to large companies.

27-4-5 ISDN Architecture

Figure 27-17 shows a block diagram of the architecture for ISDN functions. The ISDN network is designed to support an entirely new physical connector for the customer, a digital subscriber loop, and a variety of transmission services.

A common physical connection is defined to provide a standard interface connection. A single interface will be used for telephones, computer terminals, and video equipment. Therefore, various protocols are provided that allow the exchange of control information between the customer's device and the ISDN network. There are three basic types of ISDN channels:

1. B channel: 64 kbps
2. D channel: 16 kbps or 64 kbps
3. H channel: 384 kbps (H_0), 1536 kbps (H_{11}), and 1920 kbps (H_{12})

27-4-6 ISDN Channel Interfaces

There are two ISDN channel offerings: basic rate interface and primary rate interface.

27-4-6-1 Basic rate interface. ISDN standards specify that residential users of the network (i.e., home subscribers) be provided a *basic access* consisting of three full-duplex, time-division multiplexed digital channels, two operating at 64 kbps (designated the B channels, for *bearer*) and one at 16 kbps (designated the D channel, for *data*). The B and D bit rates were selected to be compatible with existing digital T carrier systems. The B channels are sometimes designated 64-kbps clear, which means they carry user data only, which includes any type of digital information, such as digitized voice or video and digital data. Digital signals from several sources can be multiplexed (combined) and transmitted over a single B channel as long as the combined transmission rate does not exceed 64 kbps. The D channel is used for carrying signaling information and for exchanging network control information. Although the name D implies data, the primary function of a D channel is to carry control-signaling information for B channels.

Since the bit rate of a D channel is typically greater than necessary for signaling alone, packet mode data may also be transported over a D channel. Most data transmission protocols use in-channel signaling, where control information (such as call establishment, ringing, call interrupt, or synchronization) is carried by the same channel that carries the message data. With ISDN a special control channel carries the control signaling for all B channels in a given path using a method called *common-channel signaling* (which is sometimes inappropriately called *in-band signaling*). Subscribers use D channels to connect to the network and secure a B channel connection. The B channel is used to send the actual data to the destination. All subscribers connected to a single subscriber loop must use the same D channel for signaling, but each subscriber transmits data over a separate B channel.

The 2B + D service is called the *basic rate interface* (BRI). BRI systems require bandwidths that can accommodate two 64-kbps B channels and one 16-kbps D channel plus an additional 48-kbps maintenance channel that is used for framing, synchronization, and other overhead bits. Therefore, the bit rate for a basic rate interface is

$$64 \text{ kbps} + 64 \text{ kbps} + 16 \text{ kbps} + 48 \text{ kbps} = 192 \text{ kbps}$$

A basic rate interface is the lowest-level ISDN standard. A BRI provides four logical, two-way digital circuits over one physical digital subscriber line connecting a subscriber's location to the central office. One B channel is used for digitally encoded voice telephone service (i.e., PCM-encoded digitized voice) and the other for applications such as digital data transmission (computer service) and videotext.

Figure 27-18 shows how a subscriber is connected to a central office through a basic rate interface. One B channel is connected to a personal computer, and the other B channel is connected to a digital telephone.

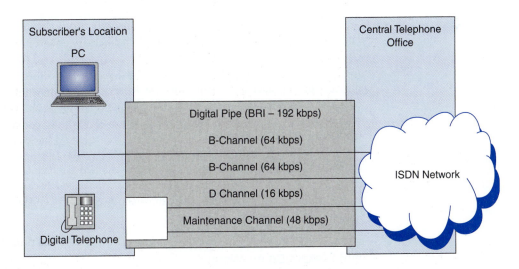

FIGURE 27-18 Subscriber connected to central office through an ISDN basic rate interface

Table 27-5 ISDN Services

Service	Transmission Rate	Channel
Telephone	64 kbps	BC
System alarms	100 kbps	D
Utility company metering	100 kbps	D
Energy management	100 kbps	D
Video	2.4–64 kbps	BP
E-mail	4.8–64 kbps	BP
Facsimile	4.8–64 kbps	BC
Slow-scan television	64 kbps	BC

There are three types of connections that can be set up on a B channel: circuit switched, packet switched, and semipermanent.

Circuit switched. This type of connection is equivalent to a standard switched digital service where the user places a call and a circuit is established to another network user through a switch. However, with ISDN, call setup and termination is accomplished over a common signaling channel, not the B channel itself.

Packet switched. With packet-switched connections, users are connected to a node of a packet-switched network, and data is exchanged using X.25.

Semipermanent. This type of connection is set up in advance and, therefore, requires no setup time to establish a connection.

27-4-6-2 Primary rate interface. There is another ISDN service called the *primary service, primary access,* or *primary rate interface* (PRI) that is capable of carrying 24 digital channels. A PRI is intended for high-volume subscribers to the network. In the United States, Canada, Japan, and Korea, the primary rate interface consists of 23 64-kbps B channels, one 64-kbps D channel, and one 8-kbps framing bit for a combined bit rate of

$$(23 \times 64 \text{ kbps}) + 64 \text{ kbps} + 8 \text{ kbps} = 1.544 \text{ Mbps}$$

The 1.544-Mbps transmission rate is compatible with existing T1 transmission systems. The combination of 23 B channels and one D channel is called 23B + D service. In Europe, the primary rate interface uses 30 64-kbps B channels, one 64-kbps D channel, and one 64-kbps framing channel for a combined bit rate of 2.048 Mbps, which is compatible with the existing E1 transmission bit rate.

It is intended that ISDN provide a circuit-switched B channel with the existing telephone system; however, packet-switched B channels for data transmission at nonstandard rates would have to be created.

The subscriber's loop, as with the twisted-pair cable used with a standard telephone, provides the physical signal path from the subscriber's equipment to the ISDN central office. The subscriber loop must be capable of supporting full-duplex digital transmission for both basic and primary data rates. Ideally, as the network grows, optical fiber cables will replace the metallic cables.

Table 27-5 lists the services provided to ISDN subscribers. BC designates a circuit-switched B channel, BP designates a packet-switched B channel, and D designates a D channel. Figure 27-19 shows how a subscriber is connected to a central office through a primary rate interface. Note in the figure that each B channel is connected to a different digital information source.

27-4-7 ISDN Functional Groupings and System Connections

With ISDN, the devices that enable subscribers to access the services of basic and primary rate interfaces are described by their functions and logical locations within the network.

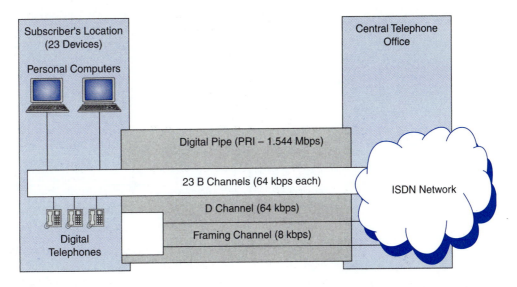

FIGURE 27-19 Subscriber connected to central office through an ISDN primary rate interface

Functional groupings define particular arrangements of physical equipment and combinations of equipment, whereas reference points are abstract points used to separate groups of functions. Network subscribers select the devices best suited to their requirements from the functional groupings. ISDN defines only the functional behavior of a group that can be implemented using devices chosen by the subscriber. Functional groupings include network terminations, terminal equipment, and terminal adapters. Reference points are simply the interface points between the functional equipment groupings.

27-4-7-1 ISDN terminal equipment and terminal adapters.   Figure 27-20 shows how users may be connected to an ISDN. As the figure shows, subscribers must access the network through one of two types of entry devices: *terminal equipment type 1* (TE1) or *terminal equipment type 2* (TE2) and one *terminal adapter* (TA).

TE1 equipment is any equipment that supports standard ISDN interfaces (16 kbps, 64 kbps, or 1.544 Mbps) and, therefore, requires no protocol translation. Data enters the network and is immediately configured into the ISDN protocol format. Examples of TE1 equipment include digital telephones, integrated voice/data terminals, and digital facsimile. TE2 equipment is classified as non-ISDN (bit rates other than standard ISDN rates). TE2 equipment is not immediately compatible with an ISDN network. Therefore, computer terminals are connected to the system through physical interfaces such as the RS-232 and host computers with X.25. Examples of TE2 equipment include computer terminals, workstations, host computers, standard analog telephones, and voice-band data modems.

Translation between non-ISDN data protocols and ISDN is performed in a device called a *terminal adapter* (TA). TAs convert the user's information received in non-ISDN format from a TE2 device into a format compatible with standard ISDN formats, which includes the 64-kbps ISDN channel B and the 16-kbps channel D format. Terminal adapters can also convert X.25 packets into ISDN packet format. If any additional signaling is required, it is added in the terminal adapter. The terminal adapters can also support traditional analog telephones and facsimile signals by using a 3.1-kHz audio service channel. The analog signals are sampled and converted to PCM codes before being put into ISDN format before entering the network.

27-4-7-2 ISDN reference points and network terminations.   ISDN *reference points* refer to categorizations that label and identify individual interfaces between two

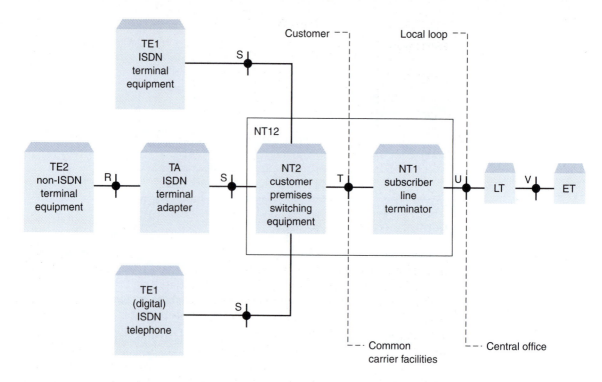

FIGURE 27-20   ISDN connections and reference points

elements of an ISDN installation. Reference points define the functions of the connections between the different types of equipment.

User data at points designated as *S eference points* (system) define a connection between a TE1 or a TA and an NT1 or a TE1 and an NT2. S reference points are presently in ISDN format and operate at the 2B + D data rate of 192 kbps. These reference points separate user terminal equipment from network-related system functions.

T reference points (terminal) define the interface between an NT2 and an NT1. T reference point locations correspond to a minimal ISDN network termination at the user's location. T reference points separate the network provider's equipment from the user's equipment.

R reference points (rate) define a non-ISDN connection between a TE2 and a TA. R reference points provide an interface between non-ISDN-compatible equipment and the terminal adapters. Subscribers can use any of the existing EIA standards, such as RS-232, RS-530, and so on, or any of the X or V series standards, such as X.21.

U reference points define the interface between an NT1 and the ISDN central office. For the U interface connected to either a digital subscriber line or a local loop, the ITU-T specifies a single twisted pair of wire in each direction. U reference points uses a scheme called *two-binary, one-quaternary (2B1Q) encoding*. 2B1Q encoding uses four voltage levels (quaternary) instead of two (binary). Hence, each signal level represents two bits (a dibit) rather than one bit. This reduces the baud rate and required bandwidth by a factor of 2. The four voltage levels for the four binary sequences possible with two bits are shown in Figure 27-21.

An ISDN device called a *network termination 1* (NT1) provides the functions associated with controlling the physical and electrical termination at the user's premise and connects the user's internal system to the ISDN subscriber loop. These functions correspond to OSI layer 1. The letter T designates a network termination. Therefore, an NT1 is a boundary to the network and may be controlled by the ISDN provider. The NT1 performs line

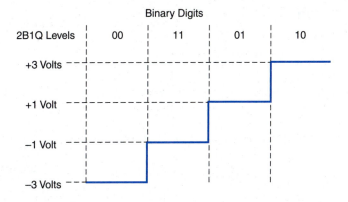

FIGURE 27-21 2B1Q encoding

maintenance functions and supports multiple channels at the physical level (e.g., 2B + D). Data from these channels is time-division multiplexed together. An NT1 organizes data streams transmitted from subscribers into frames that can be transported over the ISDN digital pipeline. NT1s also translate frames received from the ISDN network into a format more appropriate for the subscriber's equipment. NT1 terminators can convert subscriber data from one or more TA, TE1, or NT2 device and combine them into an ISDN frame format.

Network terminal 2 devices perform functions at the physical (multiplexing), data-link (flow control), and network (packetizing) layers of the OSI hierarchy. NT2s devices are intelligent and can perform *concentration,* switching functions (functionally up through OSI level 3), and intermediate signal processing between the data-generating devices, such as NT1 devices. NT2s are used primarily to interface between a multiuser system and an NT1 in a primary rate interface. NT2 devices can be implemented with a variety of equipment, including private branch exchanges and LANs. NT2s coordinate transmissions from a number of incoming links and then multiplex them, which make them more transmittable by an NT1.

NT2 terminations can also be used to terminate several S point connections and provide local switching functions and two-wire-to-four-wire conversions. *U reference points* refer to interfaces between the common carrier subscriber loop and the central office switch. A *U loop* is the media interface point between an NT1 and the central office.

Network termination 1,2 (NT12) constitutes one piece of equipment that combines the functions of NT1 and NT2 terminations. U loops are terminated at the central office by a *line termination* (LT) unit, which provides physical-layer interface functions between the central office and the loop lines. The LT unit is connected to an *exchange termination* (ET) at a *V reference point*. ETs route data to an outgoing channel or central office user.

27-4-8 Additional ISDN Channels

There are several types of transmission channels in addition to the B and D types described in the previous sections, including the following:

Primary rate interface H_0 channel. This interface supports multiple 384-kbps H_0 channels. These structures are $3H_0 + D$ and $4 H_0 + D$ for the 1.544-Mbps interface and $5 H_0 + D$ for the 2.048-Mbps interface.

Primary rate interface H_{11} channel. This interface consists of one 1.536-Mbps H_{11} channel, which is comprised of 24 64-kbps B channels with no framing bit.

Primary rate interface H_{12} channel. This European version of H_{11} uses 30 64-kbps B channels for a combined data rate of 1.92 Mbps plus one 64-kbps D channel with no framing or signaling channels.

E channel 0. This uses packet switching using 64 kbps (similar to standard D channels).

27-4-9 Broadband ISDN

Broadband ISDN (BISDN) is defined by the ITU-T as a service that provides transmission channels capable of supporting transmission rates greater than the primary data rate (i.e., greater than 1.544 Mbps). With BISDN, services requiring data rates of a magnitude beyond those provided by ISDN, such as those required for transporting digitized video information, will become available. With the advent of BISDN, the original concept of ISDN is being referred to as *narrowband ISDN*.

In 1988, the ITU-T first recommended as part of its I-series recommendations relating to BISDN:I. 113, *Vocabulary of Terms for Broadband Aspects of ISDN*, and I.121, *Broadband Aspects of ISDN*. These two documents are a consensus concerning the aspects of the future of BISDN. They outline preliminary descriptions of future standards and development work.

The new BISDN standards are based on the concept of an asynchronous transfer mode (ATM), which incorporates optical fiber cable as the transmission medium for data transmission. The BISDN specifications set a maximum length of 1 kilometer per cable length but are making provisions for repeated interface extensions. The expected data rates on optical fiber cables will be either 11 Mbps, 155 Mbps, or 600 Mbps, depending on the specific application and the location of the fiber cable within the network.

ITU-T classifies the services that could be provided by BISDN as interactive and distributed services. *Interactive services* include those in which there is a two-way exchange of information (excluding control signaling) between two subscribers or between a subscriber and a service provider. *Distribution services* are those in which information transfer is primarily from service provider to subscriber. On the other hand, *conversational services* will provide a means for bidirectional end-to-end data transmission, in real time, between two subscribers or between a subscriber and a service provider.

The authors of BISDN composed specifications that require the new services meet both existing ISDN interface specifications and the new BISDN needs. A standard ISDN terminal and a *broadband terminal interface* (BTI) will be serviced by the *subscriber's premise network* (SPN), which will multiplex incoming data and transfer them to the *broadband node*. The broadband node is called *a broadband network termination* (BNT), which codes the data information into smaller packets used by the BISDN network. Data transmission within the BISDN network can be asymmetric (i.e., access onto and off the network may be accomplished at different transmission rates, depending on system requirements).

27-4-9-1 BISDN configuration.   Figure 27-22 shows how access to the BISDN network is accomplished. Each peripheral device is interfaced to the *access node* of a BISDN network through a broadband distant terminal (BDT). The BDT is responsible for the electrical-to-optical conversion, multiplexing of peripherals, and maintenance of the subscriber's local system. Access nodes concentrate several BDTs into high-speed optical fiber lines directed through a *feeder point* into a *service node*. The service node manages most of the control functions for system access, such as call processing, administrative functions, and switching and maintenance functions. The functional modules are interconnected in a star configuration and include switching, administrative, gateway, and maintenance modules. The interconnection of the function modules is shown in Figure 27-23. The central control hub acts as the end-user interface for control signaling and data traffic maintenance. In essence, it oversees the operation of the modules.

Subscriber terminals near the central office may bypass the access nodes entirely and be directly connected to the BISDN network through a service node. BISDN networks that use optical fiber cables can utilize much wider bandwidths and, consequently, have higher transmission rates and offer more channel-handling capacity than narrowband ISDN systems.

27-4-9-2 BISDN channel rates.   The ITU-T has published preliminary definitions of new broadband channel rates that will be added to the existing ISDN narrowband rates:

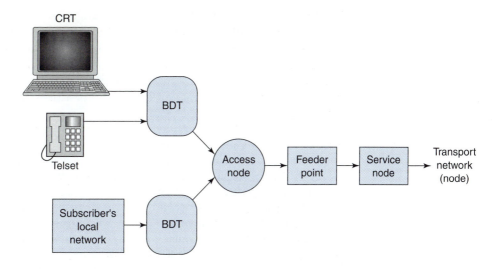

FIGURE 27-22 BISDN access

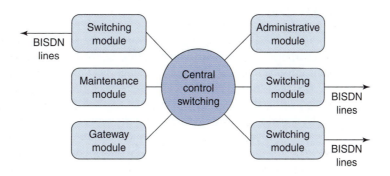

FIGURE 27-23 BISDN functional module interconnections

1. H21: 32.768 Mbps
2. H22: 43 Mbps to 45 Mbps
3. H4: 132 Mbps to 138.23 Mbps

The H32 and H22 data rates are intended for full-motion video transmission for videoconferencing, video telephone, and video messaging. The H4 data rate is intended for bulk data transfer of text, facsimile, and enhanced video information. The H21 data rate is equivalent to 512 64-kbps channels. The H22 and H4 data rates must be multiples of the basic 64-kbps transmission rate.

27-4-10 IEEE 802.9 Integrated Voice, Video, and Data Services

The IEEE 802 committee has devoted a section to integrated services. The IEEE 802.9 specification outlines how LANs and ISDN are interfaced, enabling them to transport voice, data, and video information.

802.9 specifies an *access unit* (AU) as the interface between a subscriber and the integrated network. The network can handle two types of data: packet data and isochronous data. Isochronous data includes voice, image, video, facsimile, and text.

Figure 27-24 shows the format for the 802.9 MAC frame. As shown in the figure, the frame closely resembles an HDLC frame. The frame begins with a start flag sequence,

Start Flag 01111110 (7E hex)	Service Identifier (SID)	Field Control	Destination Address	Source Address	Packet Data Unit (PDU)	CRC-32 Frame Check Sequence	Ending Flag 01111110 (7E hex)

FIGURE 27-24 IEEE 802.9 MAC frame format

which consists of a 0 followed by six 1s and then another 0 (01111110 binary, 7E hex)— exactly the same sequence as an HDLC flag. A MAC header, which is comprised of four fields, follows the start flag. The first field in the header is a *service identifier* (SID), which indicates the type of service provided and the type of data-link frame used. The second field in the header is the *frame control* (FC) field, which specifies the type of MAC frame (information or control) and the frame's priority. The last two fields of the header are for the *destination* and *source addresses*. The *packet data unit* (PDU) holds the higher-layer packet payload. The error-detection mechanism is a CRC-32 frame check sequence, which is followed by the ending flag.

27-5 DIGITAL SUBSCRIBER LINE

The *digital subscriber line* (DSL) is an integrated services *broadband access technology* that incorporates multiplexing, demultiplexing, modulation, and demodulation to transport digital information over existing telephone networks. Integrated services data network were designed to provide affordable high-speed data communications capable of transporting multimedia data between virtually any two computers connected to the network.

A digital subscriber line is more than simply a transmission line. DSL also describes a system or technique of transporting data over conventional analog transmission facilities. In essence, a digital subscriber line is a broadband data link that connects a subscriber to a network over a twisted-pair wire transmission line. *Broadband* is a relative term and has no meaning unless the reference is clearly defined. In general, access technologies are defined as narrowband, wideband, or broadband. The word *band* is short for *bandwidth,* and the wider the bandwidth, the higher the transmission rates possible. Narrowband access technologies accommodate transmission rates between 300 bps and 64 kbps, such as a standard voice-band telephone circuit. Wideband systems typically operate between 64 kbps and 1.5 Mbps, which includes the ISDN primary rate interface. Broadband systems are intended to operate at transmission rates between 1.5 Mbps and well over 1 Gbps. Broadband services include high-speed data services, such as live multicast video on demand (VoD), high-quality Internet gaming, videoconferencing, video telephone, and high-resolution imaging.

DSL is a *copper-loop access technology* that is actually a family of technologies collectively called XDSL. XDSL allows service providers, such as local telephone companies, to provide high-speed multimedia data services over existing twisted-pair copper cables, which is certainly more economical than installing new cable systems. Copper-loop access technologies are sometimes called "*last mile*" technologies because they generally provide service over the last mile or so to the subscriber's location.

XDSL began in the early 1980s with the ISDN (integrated services digital network). ISDN provides access to an integrated services digital network, which is an all-digital data network from subscriber to subscriber. The telecommunications industry (namely, AT&T) developed DSL to provide data service over the millions of miles of existing copper-loop infrastructure. With literally billions of potential users worldwide, the possibility of installing new copper cables for each new customer was indeed an intimidating prospect. Consequently, networks designers explored new ways of utilizing the already existing base of the omnipresent twisted-pair wire that links virtually all residences and businesses to local telephone networks.

XDLS supports both *asymmetrical* and *symmetrical* services. Asymmetrical service provides higher bandwidths (i.e., higher bit rates) in one direction. The higher data rate is typically on the *downstream* channel (from the telephone central office to the customer's site). The *upstream* channel goes from the subscriber's location to the telephone central office. Symmetrical service provides the same bandwidths (bit rates) for upstream and downstream channels. Asymmetrical service is well suited to *video on demand* (VoD) and related services. In addition, subscribers to the Internet generally wish to download high-volume files quickly from remote Web sites while sending much smaller files, such as e-mail. This imbalance in load distribution led to the development of ADSL (*asymmetric digital subscriber line*).

There are currently a variety of DSL technologies collectively known as XDSL. XDSL enables telephone companies with existing twisted-pair copper cables to offer single and/or multiple video channels, high-speed data (symmetrical and asymmetrical), and basic telephone service over the same network using the same access medium. All XDSL technologies share the following characteristics:

1. They are copper-loop access technologies known as the "*last mile*" technology.
2. They do not provide an end-to-end connection through the public switched telephone network as is provided by plain old telephone service (POTS) or ISDN.
3. They provide each customer with a dedicated copper pair (or pairs) to the nearest central telephone office.
4. All XDSL technologies require identical technologies at both ends of the copper loop.

The XDSL technologies listed here either have recently been standardized or are presently undergoing standardization:

1. ADSL—asymmetric digital subscriber line
2. ADSL "lite"—G.Lite
3. HDSL— high-bit-rate digital subscriber line
4. HDSL2—high-bit-rate digital subscriber line version 2
5. DSL—integrated services digital network digital subscriber line (ISDN-DSL)
6. RADSL—rate-adaptive digital subscriber line
7. SDSL—symmetric digital subscriber line
8. UDSL—universal digital subscriber line
9. VDSL—very high-bit-rate digital subscriber line

27-5-1   Asymmetric Digital Subscriber Line

The telecommunications industry developed the *asymmetric digital subscriber line* (ADSL) in the early 1980s in response to the cable television industry's interest in providing video on demand. By the mid-1990s, ADSL was recognized as a practical means of enabling subscribers to access high-speed data services, such as the Internet, over unshielded twisted-pair cable referred to as the *local loop* or the *local subscriber loop*.

ADSL provides asymmetric transmission rates up to 9 Mbps downstream and from 16 kbps to 640 kbps upstream, depending on the length of the local loop. Obviously, the shorter the local loop, the higher the bit rate possible. Standard ADSL uses the entire available bandwidth and, therefore, is sometimes called *full-rate ADSL*. ADSL was designed specifically to coexist on the same line (local loop) at the same time as plain old telephone service (POTS) using frequency-division multiplexing (FDM), as shown in Figure 27-25a. With FDM, POTS signals use a relatively narrow band of low frequencies, and ADSL signals use a relatively wide band of high frequencies that are located above the POTS signals.

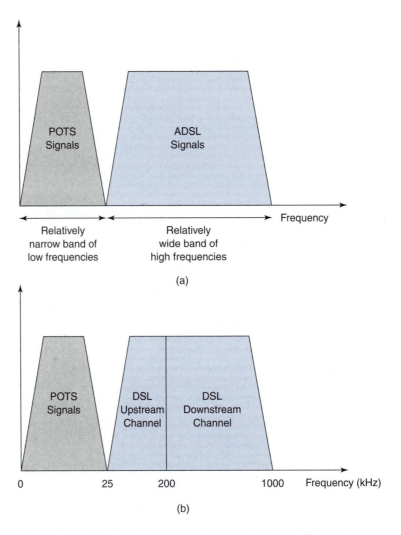

FIGURE 27-25 ADSL: (a) frequency-division multiplexing; (b) typical frequency bands

ADSL divides the usable bandwidth of the twisted-pair transmission facility into three bands—one for POTS voice telephone service, one for upstream DSL data service, and one for downstream DSL data service, as shown in Figure 27-25b. The POTS band is generally between 300 Hz and 25 kHz and is used only for standard voice-band telephone services, such as analog telephone conversation and data modems operating at bit rates of 56 kbps or less. These applications are allocated only a 4-kHz bandwidth; therefore, the remaining 21 kHz is used for a guard band. The middle band (the upstream DSL data channel) is generally between 25 kHz and 200 kHz and the upper band (the downstream DSL data channel) generally extends from 200 kHz to 1 MHz or higher.

With ADSL, the subscriber's equipment is connected to telephone company equipment and wiring at a *demarcation* ("*demarc*") point, as shown in Figure 27-26. At the demarcation point, a device called a *network interface device* (NID) isolates subscriber equipment from telephone company equipment. A *signal splitter* is installed on the subscriber side of the NID (or possibly in the NID itself). Splitters are comprised of two filters that separate the POTS (voice) frequencies from the ADSL modem frequencies. There are three pairs of wires connected to the splitter. One pair of wire is connected to the NID, and two additional pairs are connected to subscriber equipment. One subscriber pair is connected to

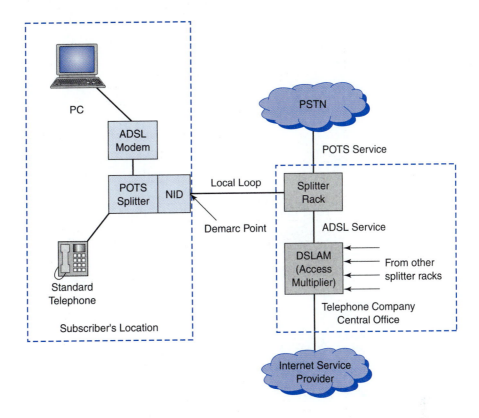

FIGURE 27-26 Basic ADSL network architecture

the telephone through internal wiring to provide voice service. The second pair is connected to the ADSL modem to provide ADSL service. The ADSL modem at the subscriber's location is called an *ADSL termination unit, remote* (ATU-R).

Splitters are also installed on the central office side of the local loop. With ADSL, each wire pair is then connected to a central office (CO) side splitter (actually a bank of splitters, one for each loop that has ADSL service). As with the customer premise splitters, two pairs of wire leave the CO splitter. The first pair is connected to the public switched telephone network through a voice switch, such as Lucent #5 ESS or a Nortel DMS 100, to provide traditional POTS service. The second pair connects to the central office counterpart of the DSL modem called an *ADSL termination unit, central* (ATU-C).

Figure 27-27 shows a functional block diagram of an ADSL connection. In essence, the ATU-R and ATU-C are the DSL modems at either end of the ADSL loop. The splitter at the subscriber is called an *R splitter* and the splitter in the central office is called a *C splitter*. For efficiency reasons, a bank of ATU-Cs is combined in a *DSL access multiplexer* (DSLAM) in the central office before being connected to the service provider of a data network, such as the Internet.

The system components shown in Figure 27-27 are summarized here:

1. Splitter C—interface between telephone company equipment and local loop (CO side of local loop)
2. Splitter R—interface between subscriber equipment and local loop (remote side of local loop)
3. U-C—interface (CO side)
4. U-C2—interface (CO side from splitter to ATU-C)
5. U-R—interface (remote side)

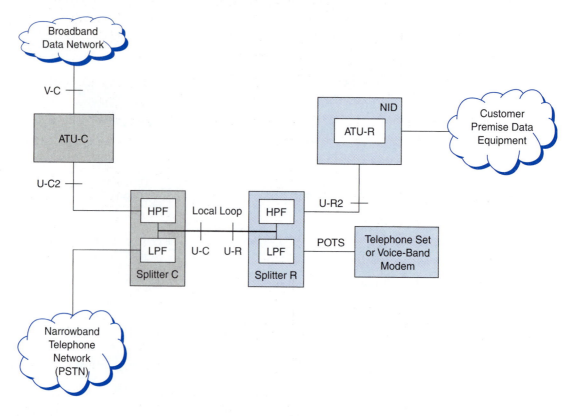

FIGURE 27-27 System reference model for full-rate ADSL

6. U-R2—interface (remote side from splitter to ATU-R)
7. V-C—interface (CO side from access node to network interface)

27-5-2 ADSL Modulation Schemes

The modulation scheme originally used by ADSL was *carrierless amplitude modulation* (CAP). The American National Standards Institute (ANSI) later standardized a second modulation scheme called *discrete multitone* (DMT). CAP and DMT are efficient line-encoding techniques designed to take advantage of the relatively wide bandwidth available above the POTS signals. CAP and DMT are so significantly different that they are incompatible.

27-5-2-1 Carrierless amplitude modulation. CAP is a variation of quadrature amplitude modulation (QAM) except that the carrier is removed. With CAP, the phase and amplitude of two sine waves whose frequencies are within the passband of the subscriber line are varied in accordance with a defined signal constellation. The data rate is divided in half and modulated onto the two sine waves before being filtered, combined, and fed to a digital-to-analog converter and being transmitted.

Like QAM, CAP uses combines *multilevel amplitude modulation* (i.e., multilevel voltage levels per pulse) with phase modulation. The primary difference between CAP and standard QAM is that QAM (like CAP) combines two signals in the analog domain. However, since with CAP the carrier does not carry any information, it is suppressed (making it carrierless). Modulation is performed digitally using two filters with the same amplitude characteristics but different phase responses. With QAM the constellation is fixed; however, with CAP the constellation is free to rotate since there is no carrier reference. To compensate for this, CAP receivers must include a rotation function to detect the relative position of the constellation.

CAP has several advantages over QAM:

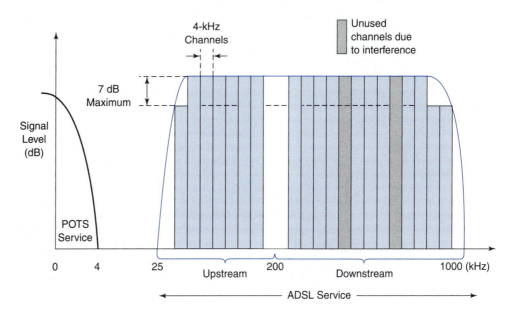

4-kHz
Channels

Unused
channels due
to interference

7 dB
Maximum

Signal
Level
(dB)

POTS
Service

0   4   25   200   1000 (kHz)
         Upstream   Downstream

◄─────────────── ADSL Service ───────────────►

FIGURE 27-28   Discrete multitone (DMT)

1. CAP has a mature technology using V.34 modems, making it easier to implement than DMT.
2. With CAP, it is easier to perform rate adaptation by changing the constellation size (i.e., 4-CAP, 64-CAP, 512-CAP, and so on) or by changing the range of the frequency spectrum utilized.

27-5-2-2 Discrete multitone.   Discrete multitone (DMT) differs from CAP in that it uses a combination of frequency-division multiplexing (FDM) and QAM. DMT is sometimes thought of as combining the outputs from numerous minimodems operating simultaneously, as shown in Figure 27-28. With DMT, the entire usable bandwidth is separated into 256 frequency bands (subchannels), of which each band is 4 kHz wide and carries a fraction of the total information. The lower six subchannels are used for voice, and the remaining subchannels are used for upstream and downstream data.

The ANSI standard for DMT specifies up to a 60-kbps transmission rate for each 4-kHz ADSL channel. With DMT, the data in each subchannel modulates a carrier frequency corresponding to the center frequency of the 4-kHz subchannel. Each subchannel can carry a theoretical maximum of 15 bits per symbol per hertz, which yields a baud rate equal to 1/15th the bit rate (60 kbps/15 bits per baud = 4000 baud). The upstream channel is generally 60 kHz wide (15 4-kHz channels), which theoretically supports a transmission rate of $25 \times 60 = 1.5$ Mbps. However, more realistic rates range between 64 kbps and 1 Mbps. The downstream channel is generally 800 kHz wide (200 4-kHz channels), which theoretically supports a transmission rate of 200×60 kbps $= 12$ Mbps. However, more realistic rates range between 500 kbps and 8 Mbps. However, some subchannels may not be used because of external interference, such as radio signals from AM broadcast band radio.

In contrast to CAP, a DMT signal does not use the entire available bandwidth (except for the POTS channel). DMT uses QAM in each subchannel, where CAP evenly distributes energy across the entire range of frequencies.

27-5-3   ADSL Frame Format
Figure 27-29 shows the format for an ADSL data frame used to carry data between and ATU-C and ATU-R. The format closely resembles an HDLC frame. The ADSL frame carries

Beginning Flag 01111110	PPP Address Code — FF03	PPP Protocol ID	Protocol Data Unit (PDU) — Payload	Frame Check Sequence (CRC-16)	Ending Flag 01111110
1 byte	2 bytes	2 bytes	Variable Length	2 bytes	1 byte

FIGURE 27-29   ADSL frame format

point-to-point protocol (PPP) data packets. PPP is a low-level protocol used to carry higher-level networking protocols, such as IP, in its protocol data unit (PDU). PPP is a protocol that is often used to provide Internet access to subscriber locations that use a dial-up connection.

The frame begins with an HDLC flag (01111110) immediately followed by the PPP address code FF03 hex. The next two bytes contain the protocol ID, which identifies the payload protocol carried in the PDU. The next two bytes are the frame check sequence, which uses CRC-16. The frame ends with another HDLC flag.

27-5-4   ADSL Applications

The inherent asymmetry of ADSL makes it well suited to almost any application that requires high downstream bandwidth but lower upstream bandwidth. Several ADSL applications are listed here:

1. Video over data (VoD).
2. Telecommuting, such as working from home.
3. Video streaming or real-time information, which enables delivery of real-time, bandwidth-intensive applications, such as news, stocks, and weather, where viewers can take advantage of the "*always on*" connection.
4. Distance learning with guaranteed quality of service (QoS). For example, ADSL can support an MPEG-2 video stream, which allows training centers to simulcast training videos to multiple sites and communicate with trainees individually.
5. Telemedicine so that doctors can diagnose X rays and other video images sent from distant locations.
6. Videoconferencing is possible with ADSL, as the entire available downstream bandwidth is not required.

27-5-5   ADSL Lite

In 1998, a simpler form of ADSL emerged called *ADSL lite* (also known as G.lite, splitterless ADSL, or "*consumer* ADSL"). Full-rate ADSL requires splitters at both ends of the local subscriber loop to separate and isolate POTS signals from ADSL signals. ADSL lite eliminates (or at least makes optional) the subscriber's splitter. Eliminating the splitter removes the filters, thus allowing the possibility of data and voice to interfering with one another if they are used at the same time. The downstream bandwidths (bit rates) for ADSL lite are less than those possible with full-rate ADSL (i.e., in the 1.5-Mbps range rather than the 9-Mbps range).

Figure 27-30 shows the layout for an ADSL circuit. The only difference between the layout for ADSL lite and full rate ADSL are the optional splitter at the customer premise and replacing the full-rate ADSL modem with an ADSL lite modem.

ADSL lite provides procedures that allow modems to retrain quickly. Retraining is necessary for the modems to resynchronize and adapt to changes in transmission characteristics that change line conditions. Fast retraining is an advantage when POTS and ADSL services operate concurrently without a splitter. Retraining causes a temporary loss of service; therefore, it is important that the retraining procedures be completed as quickly as possible.

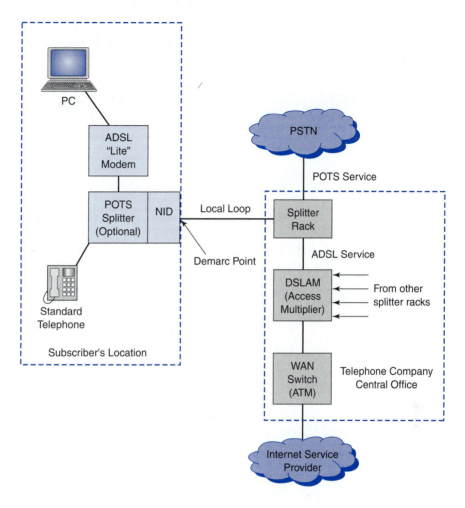

FIGURE 27-30 Basic network architecture for ADSL lite

27-5-6 High-Bit-Rate Digital Subscriber Line

High-bit-rate digital subscriber line (HDSL) and HDSL2 (currently under development) are symmetric baseband transmission systems with equal upstream and downstream data rates. HDSL was developed by Bellcore and was originally used to provide enhanced telephone company services, such as PBX connectivity.

To eliminate placing repeaters every 6000 feet (as is necessary with standard T1 transmission), HDSL incorporates a more advanced and efficient coding technique that allows transmission rates up to 1.544 Mbps over copper lines for distances up to 12,000 feet. T1 transmission uses BPRZ-AMI, which has a line efficiency of one bit per baud; therefore, 1.544 Mbps requires a 1.544-MHz bandwidth. HDSL adopted 2B1Q encoding from ISDN, which has a line efficiency of two bits per baud, allowing a 1.544-Mbps transmission rate over 774 kHz of bandwidth.

HDSL was originally designed to provide a better way of transmitting T1 (1.544 Mbps) and E1 (2.049 Mbps) data rates over copper wires, using less bandwidth and no repeaters (i.e., *repeaterless* T1). However, HDSL requires two pairs of copper wires for T1 compatibility and three pairs of copper wire for E1 compatibility, as shown in Figure 27-31.

Figure 27-32 shows the basic network architecture for HDSL using T1 transmission rates, which requires two pairs of wire where each pair carries 784 kbps (half rate). The 784 kbps includes 768 usable data bits plus 16 kbps of overhead. For full-rate (two pair), each

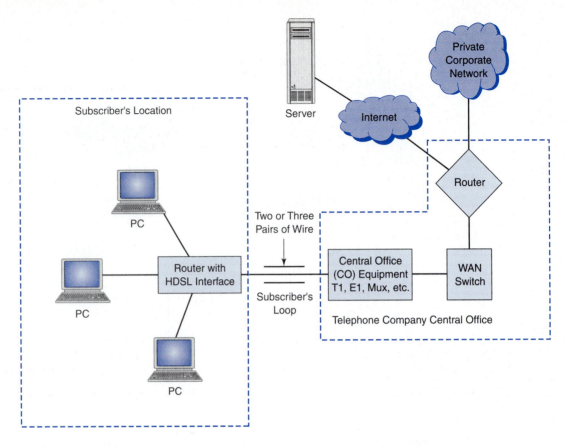

FIGURE 27-31 HDSL architecture

pair carries 784 kbps for a combined equivalent transmission rate of 2 x 784 kbps = 1.568 Mbps. The 1.568 Mbps includes 1.536 Mbps of usable data plus 32 kbps of overhead.

Figure 27-33 shows the basic network architecture for early implementations of HDSL using E1 transmission rates, which requires three pairs of wire where each pair carries 800 kbps (one-third rate). The 800 kbps includes 680 kbps of usable data plus 120 kbps of overhead. For full rate, each pair carries 800 kbps for a combined equivalent transmission rate of 3 x 800 kbps = 2.4 Mbps. The 2.4 Mbps includes 2.04 Mbps of usable data plus 360 kbps of overhead.

Later implementations of HDSL using E1 transmission rates required only two pairs of wire where each pair carries 1.168 Mbps (half rate). The 1.168 Mbps includes 1.02 Mbps of usable data plus 148 kbps of overhead. For full rate, each pair carries 1.168 Mbps for a combined equivalent transmission rate of 2 x 1.168 Mbps = 2.336 Mbps. The 2.336 Mbps includes 2.04 Mbps of usable data plus 296 kbps of overhead.

HDSL version 2 (HDSL2) requires only one pair of copper wire to support T1 data rates (1.544 Mbps), and MDSL (medium-speed digital subscriber line) increases the loop length to nearly 22,000 feet but at the expense of reducing the line capacity to 784 kbps.

Symmetric applications that support HDSL and related technologies include the following:

1. Private T1 line:
    a. Frame relay or ATM line from a business site to a data switch
    b. Frame relay or X.25 link between business sites
    c. Videoconference line to a business site
    d. Voice trunk to wireless cell site (cellular base station)

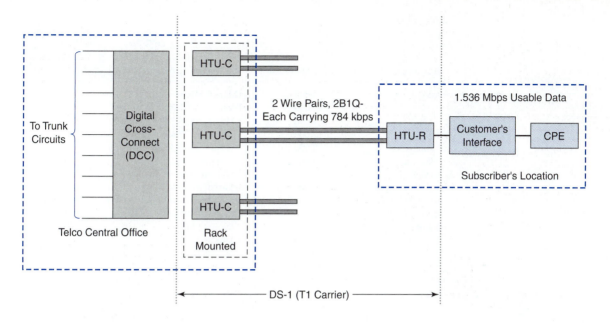

FIGURE 27-32   Basic network architecture for HDSL for T1

HTU-C/R [HDSL *Termination Unit* (C-central office, R-residence)], which consists of a transceiver, a device to map the T1 bits into an HDSL frame structure and vice-versa, and an interface module to accept a standard T1 interface. Another term used for the HTU-R is *line termination unit (LTU)* and another name for the HTU-C is *network termination unit (NTU)*.

CPE (*customer provided equipment*), which is typically a router with a serial interface that connects directly into a port on the HUT-R.

FIGURE 27-32   Basic network architecture for HDSL for T1

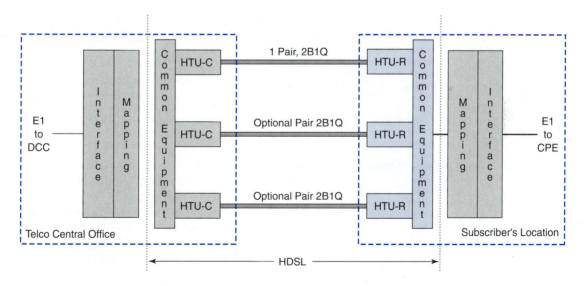

FIGURE 27-33   Basic network architecture for HDSL for E1

Table 27-6 XDSL Family

XDSL Technology	Maximum Distance (ft)	Maximum Upstream/Downstream Transmission Rate (bps)	Modulation Type
ADSL	3000	1Mbps/9 Mbps	DMT
	5000	1 Mbps/8.448 Mbps	
	9000	1 Mbps/7 Mbps	
	12,000	640 kbps/6.312 Mbps	
	18,000	16 kbps–1.544 Mbps	
ADSI lite	18,000	128 kbps/1 Mbps	DMT
HDSL	5000	1.544 Mbps/1.544 Mbps	2B1Q/CAP
	12,000	1.544 Mbps/1.544 Mbps	
HDSL2	12,000	1.544 Mbps/1.544 Mbps	2B1Q/CAP
IDSL	18,000	144 kbps/144 kbps	2B1Q
RADSL	3000	1 Mbps/12 Mbps	CAP
	9000	1 Mbps/9 Mbps	
	12,000	1 Mbps/6 Mbps	
	18,000	1 Mbps/128 kbps	
SDSL	10,000	1.1 Mbps/1.1 Mbps	2B1Q/CAP
UDSL	0–15,000	2 Mbps/2 Mbps	DMT
	15,000–18,000	1 Mbps/1 Mbps	
VDSL	1000	2.3 Mbps/51.84 Mbps	QAM/DMT
	3000	2.3 Mbps/25.83 Mbps	
	4500	1.6 Mbps/12.96 Mbps	

2. Integrated services to businesses (voice and data)
3. Web hosting to provide Web storefront home page
4. Videoconferencing, which easily supports the 384 kbps data rates required for acceptable video transmission
5. LAN-to-LAN interconnection for the purpose of file sharing
6. Telecommuting for bulk file transfers
7. Shared Internet access (multiple PCs on a LAN at a business can share the bidirectional high-speed bandwidth)
8. Reverse "ADSL" option when an ADSL upstream bandwidth is greater than its downstream bandwidth (reverse ADSL results in a significant degradation of service because of crosstalk induced by opposing systems)

Table 27-6 summarizes the technologies included in the XDSL family.

27-6 ASYNCHRONOUS TRANSFER MODE

Asynchronous transfer mode (ATM) is a relatively new data communications technology that uses a high-speed form of packet switching network for the transmission media. ATM was developed in 1988 by the ITU-T as part of the BISDN. ATM is one means by which data can enter and exit the BISDN network in an asynchronous (time-independent) fashion. ATM is intended to be a carrier service that provides an integrated, high-speed communications network for corporate private networks. ATM can handle all kinds of communications traffic, including voice, data, image, video, high-quality music, and multimedia. In addition, ATM can be used in both LAN and WAN network environments, providing seamless internetworking between the two. Some experts claim that ATM may eventually replace both private leased T1 digital carrier systems and on-premise switching equipment.

Most data communications implementations are based on packet switching technologies using packet switching networks. Transporting data in packet form requires adding lengthy headers to the information. The headers contain control information that includes overhead such as routing, flow control, and error-control information. Depending on the protocol, the size and complexity of the packet varies; therefore, the amount of overhead also varies. Adding more overhead means larger headers and a less efficient protocol.

Cell networking can solve (or at least alleviate) many of the problems inherent to packet internetworking. ATM is a *cell-relay protocol*. Cell networks use the cell as the basic unit of data transfer. User information is loaded into identical cells that are transported uniformly and predictably through the network. Packets of data are divided into equal-length data units (cells), and then the cells are multiplexed with other cells and transported through the cell network. Multiplexing is easily accomplished because all cells are the same length.

Conventional electronic switching (ESS) machines currently utilize a central processor to establish switching paths and route traffic through a network. ATM switches, in contrast, include self-routing procedures where individual cells (short, fixed-length data packets) containing subscriber data will route their own way through the ATM switching network in real time using their own address instead of relying on an external process to establish the switching path.

ATM uses *virtual channels* (VCs) and *virtual paths* (VPs) to route cells through a network. In essence, a virtual channel is merely a connection between a source and a destination, which may entail establishing several ATM links between local switching centers. With ATM, all communications occur on the virtual channel, which preserves cell sequence. On the other hand, a virtual path is a group of virtual channels connected between two points that could compromise several ATM links.

ATM incorporates *labeled channels* that are transferable at fixed data rates anywhere from 16 kbps up to the maximum rate of the carrier system. Once data has entered the network, it is transferred into fixed time slots called *cells*. An ATM cell contains all the network information needed to relay individual cells from node to node over a preestablished ATM connection. Figure 27-34 shows the ATM cell structure, which is a fixed-length data packet (cell) that is only 53 bytes long, which includes a five-byte header and a 48-byte information field that is used to encapsulate upper-layer protocols. Small cell sizes reduce *latency,* which is the time it takes to transport a packet through a network.

Fixed-length cells provide the following advantages:

1. A uniform transmission time per cell ensures a more uniform transit-time characteristic for the network as a whole.
2. A short cell requires a shorter time to assemble and, thus, shorter delay characteristics for digitized voice.
3. Shorter cells are easier to transfer over fixed-width processor busses, it is easier to buffer the data in link queues, and they require less processor logic.

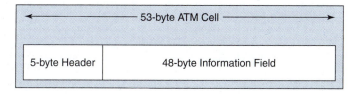

FIGURE 27-34 ATM cell structure

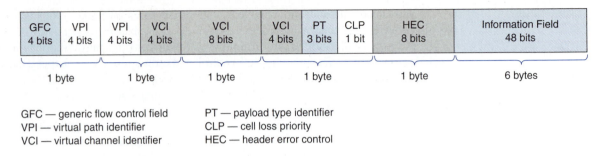

GFC 4 bits	VPI 4 bits	VPI 4 bits	VCI 4 bits	VCI 8 bits	VCI 4 bits	PT 3 bits	CLP 1 bit	HEC 8 bits	Information Field 48 bits
1 byte		1 byte		1 byte	1 byte			1 byte	6 bytes

GFC — generic flow control field   PT — payload type identifier
VPI — virtual path identifier   CLP — cell loss priority
VCI — virtual channel identifier   HEC — header error control

FIGURE 27-35   ATM five-byte header field structure

27-6-1   ATM Header

Figure 27-35 shows the five-byte ATM header, which includes the following fields: generic flow control, virtual path identifier, virtual channel identifier, payload type identifier, cell loss priority, and header error control.

Generic flow control field (GFC). The GFC field uses the first four bits of the first byte of the header. The GFC controls the flow of traffic across the user network interface (UNI) and into the network. The GFC identifies the level of congestion control and cell priority, which helps determine if a cell is discarded when the network is experiencing congestion.

Virtual path identifier (VPI) and virtual channel identifier (VCI). The 24 bits immediately following the GFC are used for the ATM address. VPIs can be either four or eight bits long and are used to identify a group of VCIs. VPIs also provide a logical link between networks or nodes within a network where the VCIs are located. VCIs are used to identify a single virtual channel across an ATM network.

Payload type identifier (PTI). The first three bits of the second half of byte 4 specify the type of message (payload) in the cell. With three bits, there are eight different types of payloads possible. At the present time, however, types 0 to 3 are used for identifying the type of user data; types 4, 5, and 6 indicate management information; and type 7 is reserved for future use. The payload type identifier codes are

000	EFCI = 0 and IND = 0
001	EFCI = 0 and IND = 1
010	EFCI = 1 and IND = 0
011	EFCI = 1 and IND = 1
100	OAM F5 segment
101	OAM F5 end-to-end
110	Resource management
111	Reserved for future use

where   EFCI = explicit forward congestion indication
        IND = last cell indicator (AAL5)
        OAM = operations, administration, and maintenance

Cell loss priority (CLP). The last bit of byte 4 is used to indicate whether a cell is eligible to be discarded by the network during congested traffic periods. The CLP bit is set by the user or cleared by the user. When set, the network may discard the cell during times of heavy utilization.

Header error control (HEC). The last byte of the header is for error control and is used to detect the correct single-bit errors that occur in the header only; the HEC does not serve as an entire cell check character. The value placed in the HEC is computed from the four previous bytes of the header. The HEC provides some protection against the delivery of cells to the wrong destination address.

27-6-2 ATM Information Field

The 48-byte information field is reserved for user data. Insertion of data into the information field of a cell is a function of the upper half of layer 2 of the OSI protocol hierarchy. This layer is specifically called the *ATM adaptation layer* (AAL). The AAL gives ATM the versatility necessary to facilitate, in a single format, a wide variety of different types of services ranging from continuous process signals, such as voice transmission, to messages carrying highly fragmented bursts of data, such as those produced from LANs. Because most user data occupy more than 48 bytes, the AAL divides information into 48-byte segments and places them into a series of segments. The four primary types of AALs are the following:

Constant bit rate (CBR). CBR information fields are designed to accommodate PCM-TDM traffic, which allows the ATM network to emulate voice or DSN services.

Variable bit rate (VBR) timing-sensitive services. This type of AAL is currently undefined; however, it is reserved for future data services requiring transfer of timing information between terminal points as well as data (i.e., packet video).

Connection-oriented VBR data transfer. Type 3 information fields transfer VBR data, such as impulsive data generated at irregular intervals, between two subscribers over a preestablished data link. The data link is established by network signaling procedures that are similar to those used by the public switched telephone network. This type of service is intended for large, long-duration data transfers, such as file transfers or file backups.

Connectionless VBR data transfer. This AAL type provides for transmission of VBR data that does not have a preestablished connection. Type 4 information fields are intended to be used for short, highly bursty types of transmissions, such as those generated from LANs.

27-6-3 ATM Network Components

Figure 27-36 shows a typical ATM network, which is comprised of three primary components: ATM endpoints, ATM switches, and transmission paths.

27-6-3-1 ATM endpoints   ATM endpoints are shown in Figure 27-37. As shown in the figure, endpoints are the source and destination of subscriber data, and therefore they are sometimes called *end systems.* Endpoints can be connected directly to either a public or a private ATM switch. An ATM endpoint can be as simple as an ordinary personal computer equipped with an ATM network interface card. An ATM endpoint could also be a special-purpose network component that services several ordinary personal computers, such as an Ethernet LAN.

27-6-3-2 ATM switches.   The primary function of an *ATM switch* is to route information from a source endpoint to a destination endpoint. ATM switches are sometimes called *intermediate systems,* as they are located between two endpoints. ATM switches fall into two general categories: public and private.

Public ATM switches. A public ATM switch is simply a portion of a public service provider's switching system where the service provider could be a local telephone company or a long-distance carrier, such as AT&T. An ATM switch is sometimes called a *network node.*

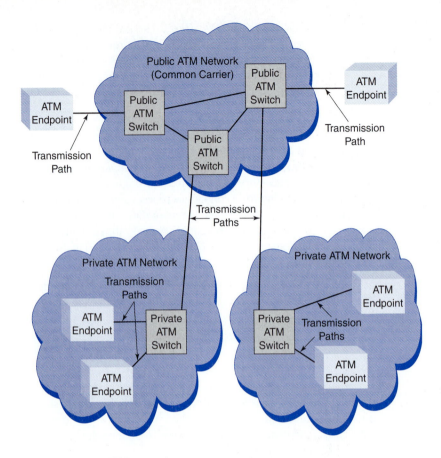

FIGURE 27-36 ATM network components

Private ATM Switches. Private ATM switches are owned and maintained by a private company and sometimes called *customer premise nodes*. Private ATM switches are sold to ATM customers by many of the same computer networking infrastructure vendors who provide ATM customers with the network interface cards and connectivity devices, such as repeaters, hubs, bridges, switches, and routers.

27-6-3-3 ATM switching paths.   ATM switches and ATM endpoints are interconnected with physical communications paths called *ATM transmission paths*. An ATM transmission path can be any of the common transmission media, such as twisted-pair cable or optical fiber cable.

27-6-4   ATM Protocol Layers

Figure 27-38 shows the ATM protocol stack, which compares somewhat to the lower three layers of the OSI seven-layer model: physical, data link, and network. The lowest layer is the ATM physical layer, which is further subdivided into two sublayers: the transmission convergence (TC) sublayer and the physical medium dependent (PMD) sublayer. The primary responsibilities of the convergence layer are the following:

1. Regulating the cell rate
2. Establishing the header error control sequence and method
3. Generating and recovering cells
4. Converting between bit stream rates and ATM cell rates

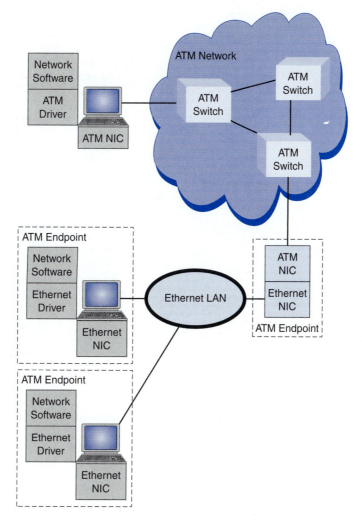

FIGURE 27-37   ATM endpoint implementations

5. Calculating the HEC
6. Decoupling the cell rate, which deals with inserting idle or unassigned cells into the ATM data stream when a cell is expected but has not yet arrived

The ATM physical-medium sublayer does not directly specify parameters. Instead, it relies on existing physical-layer protocols, such as those listed in Table 27-7. The primary responsibility of the physical-medium layer is establishing a means of accessing the transmission medium and to establish bit timing (clocking) required by the physical medium to receive ATM cell transmissions.

The second layer of the ATM hierarchy is the cell layer. The primary responsibilities of the cell layer are the following:

1. Controlling flow
2. Producing and extracting the cell header
3. Translating VPI and VCI addresses
4. Multiplexing and demultiplexing ATM cells
5. Processing cell loss priority
6. Controlling explicit forward congestion

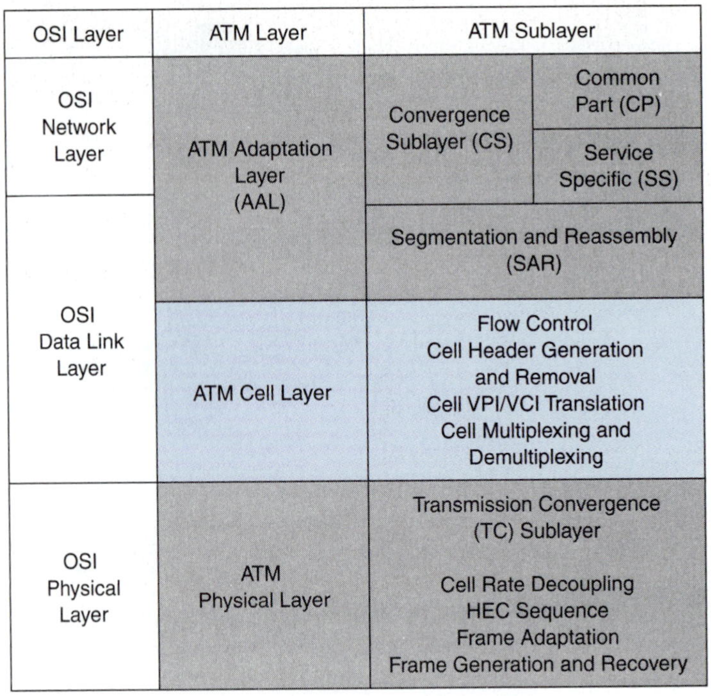

OSI Layer	ATM Layer	ATM Sublayer	
OSI Network Layer	ATM Adaptation Layer (AAL)	Convergence Sublayer (CS)	Common Part (CP)
			Service Specific (SS)
		Segmentation and Reassembly (SAR)	
OSI Data Link Layer	ATM Cell Layer	Flow Control Cell Header Generation and Removal Cell VPI/VCI Translation Cell Multiplexing and Demultiplexing	
OSI Physical Layer	ATM Physical Layer	Transmission Convergence (TC) Sublayer Cell Rate Decoupling HEC Sequence Frame Adaptation Frame Generation and Recovery	

FIGURE 27-38 ATM protocol stack

Table 27-7 Services Supported by ATM

Service	Data Rate
ITU DS-1, ISDN PRI	1.544 Mbps
ITU E, ISDN PRI	2.048 Mbps
ITU DS-2	6.312 Mbps
ITU E3	34.368 Mbps
ITU DS-3, ANSI DQDB	44.736 Mbps
SONET STS-1	51.84–84 Mbps
FDDI	100 Mbps
ITU E4	139.264 Mbps
SONET STS-3c	155.5 Mbps
ITU STM-1	155.5 Mbps
Fiber channel	155.5 Mbps
SONET STS-12c	622.08 Mbps
STS-12c	622.08 Mbps
ITU STM-4	622.08 Mbps

Cell loss priority and explicit forward congestion control are techniques used to reduce traffic load during periods of heavy network use to reduce congestion.

The ATM adaptation layer performs functions similar to the OSI network layer; however, some of the functions are generally performed in the OSI data-link layer. The ATM adaptation layer (AAL) specifies the classes of service for the ATM network. The adaptation layer is divided into two sublayers: the convergence layer and the segmentation and reassembly (SAR) layer. The convergence layer is responsible for placing data into a common format and mapping cells onto the transmission medium. The SAR sublayer is responsible for segmenting upper-layer protocol data units (PDUs) into ATM cells and desegmenting them after they arrive at their destination.

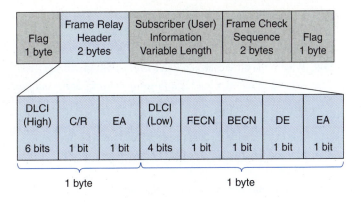

Flag 1 byte	Frame Relay Header 2 bytes	Subscriber (User) Information Variable Length	Frame Check Sequence 2 bytes	Flag 1 byte

DLCI (High) 6 bits	C/R 1 bit	EA 1 bit	DLCI (Low) 4 bits	FECN 1 bit	BECN 1 bit	DE 1 bit	EA 1 bit

1 byte   1 byte

FIGURE 27-39 Frame relay format

27-7 FRAME RELAY

In 1988, the ITU-T introduced recommendation I.122, which defined a new form of packet transmission called *frame-mode bearer service* or simply *frame relay*. Frame relay, unlike ATM, segments its payload into frames that do not necessarily have the same length. Frame relay uses packet switching and routing tables to transport frames through a multitude of networks.

Frame relay is divided into *operating planes: control* and *user*. The control plane establishes (sets up) and terminates (disconnects) logical connections and translates user protocols to network protocols and vice versa. The control plane also provides flow and error control.

The user plane controls end-to-end functions, such as delimiting frames, aligning packets, and multiplexing and demultiplexing frames. The user plane also provides *size control,* which ensures that a frame is the proper length for a specific system.

Figure 27-39 shows the frame format used with frame relay, which is comprised of five fields: two flag fields (beginning and ending), a frame relay header field, a subscriber information field, and a frame check sequence field. The frame relay header contains the following subfields:

Data-link connection identifier (DLCI). A 10-bit address field that can be extended an additional two bytes.

Command/response bit. An application-specific bit that indicates the flow direction.

Extended address bit. Indicates whether the address field is extended beyond the default length of two bytes.

Forward (FECN) and backward (BECN) explicit congestion notification. Defines whether a frame is discarded during times of network congestion.

Discard eligibility (DE). Similar to the loss priority bit used with ATM (when DE is set, it indicates that the frame is eligible to be discarded during times of network congestion).

QUESTIONS

27-1. Define *integrated services.*

27-2. Define the purpose of *integrated services data networks.*

27-3. List five integrated services data networks.

27-4. Define a *public switched data network.*

27-5. Define *switching nodes* and *transmission links.*

27-6. Describe a *value-added network.*

27-7. List and describe three types of *packet switching techniques.*

27-8. Explain *blocking.*

27-9. What is the difference between a *transparent switch* and a *transactional switch?*

27-10. What is meant by a *hold-and-forward network? Store-and-forward network?*

27-11. Describe *datagrams* and *virtual circuits.*

27-12. What is *X.25?*

27-13. List and describe the three types of *X.25 network devices.*

27-14. List and describe the three *X.25 protocol layers.*

27-15. Describe *LAPB.*

27-16. Describe the two types of *virtual circuits.*

27-17. Describe an *X.25 datagram.*

27-18. Describe the fields that comprise an *X.25 call request packet.*

27-19. Describe the fields that comprise an *X.25 data transfer packet.*

27-20. What is the purpose of the *integrated services digital network?*

27-21. Explain the basic principles of *ISDN.*

27-22. Describe the evolution of ISDN.

27-23. Describe the conceptual view of ISDN.

27-24. Describe the *ISDN architecture.*

27-25. List and describe the two types of *ISDN interfaces.*

27-26. List and describe the *functional groupings* used with ISDN.

27-27. List and describe the *terminal equipment* and *terminal adaptors* used with ISDN.

27-28. List and describe the *reference points* used with ISDN.

27-29. List and describe several additional *ISDN channels.*

27-30. Briefly describe the differences between *ISDN* and *broadband ISDN.*

27-31. Describe how access to a *BISDN network* is achieved.

27-32. What are the *BISDN channel rates?*

27-33. Explain the *IEEE 802.9 integrated voice, video,* and *data services.*

27-34. Describe a *digital subscriber line.*

27-35. What does the term *broadband* mean?

27-36. List the *XDSL technologies.*

27-37. What is the difference between *asymmetric* and *symmetric* DSL?

27-38. List and describe the *circuit components* for ADSL.

27-39. List and describe the two predominant *modulation schemes* used with ADSL.

27-40. Describe the fields that comprise an *ADSL frame.*

27-41. Explain the differences between *full-rate ADSL* and *ADSL lite.*

27-42. Briefly describe *HDSL.*

27-43. What is the purpose of *asynchronous transfer mode?*

27-44. How is *ATM* different from *ISDN?*

27-45. What is meant by a *cell-relay protocol?*

27-46. Describe the fields that comprise an *ATM header.*

27-47. List and describe the four primary types of *AALs.*

27-48. List the *ATM network components.*

27-49. What is the difference between *public* and *private* ATM switches?

27-50. Describe *ATM switching paths.*

27-51. Describe *ATM protocol layers* and compare them to the *OSI protocol hierarchy.*

27-52. What is the primary difference between *ATM* and *frame relay?*

27-53. List and describe the fields that comprise a frame relay *frame* and *header.*

Answers to Selected Problems

2-9. 50,000,000 or 77 dB

2-11. a. 8

   b. 32

   c. 128

   d. 4096

2-13. a. bandwidth = 200 Hz

   baud = 200

   BW% = 1 bps/Hz

   b. bandwidth = 1 kHz

   baud = 1000

   BW% = 1 bps/Hz

   c. bandwidth = 10 kHz

   baud = 10000

   BW% = 1 bps/Hz

2-15. BPSK

   a. bandwidth = 2400 Hz

   baud = 2400

   BW% = 1 bps/Hz

   b. bandwidth = 4800 Hz

   baud = 4800

   BW% = 1 bps/Hz

   c. bandwidth = 9600 Hz

   baud = 9600

   BW% = 1 bps/Hz

   QPSK

   a. bandwidth = 1200 Hz

   baud = 1200

   BW% = 2 bps/Hz

   b. bandwidth = 2400 Hz

   baud = 2400

   BW% = 2 bps/Hz

   c. bandwidth = 4800 Hz

   baud = 4800

   BW% = 2 bps/Hz

   8-PSK

   a. bandwidth = 800 Hz

   baud = 800

   BW% = 3 bps/Hz

   b. bandwidth = 1600 Hz

   baud = 1600

   BW% = 3 bps/Hz

   16-PSK

   a. bandwidth = 600 Hz

   baud = 600

   BW% = 4 bps/Hz

   b. bandwidth = 1200 Hz

   baud = 1200

   BW% = 4 bps/Hz

c. bandwidth = 2400 Hz

    baud = 2400

    BW% = 1 bps/Hz

2-17. one bit error for every 1 million bits transmitted

CHAPTER 3

3-1. 1 kHz = 300 km

    100 kHz = 3 km

    1 MHz = 300 m

    1 GHz = 0.3 m

3-3. 261 ohms

3-5. 111.8 ohms

CHAPTER 4

4-1. a. 869 nm, 8690 Å

    b. 828 nm, 8280 Å

    c. 935 nm, 9350 Å

4-3. 38.57°

4-5. 56°

CHAPTER 5

5-1. a. 8 kHz

    b. 20 kHz

    c. 14 kHz

5-3. 7 kHz

5-5. 15 bits (1 sign and 14 magnitude)

5-7. a. + 49 or 100000110001

    b. 10101000

    c. 100000110001

    d. +1.47 V

    e. 1.36%

5-9. a. 10001000

    b. 10010000

    c. 01110000

    d. 10001001

    e. 00100000

    f. 01110000

CHAPTER 6

6-1. 1.521 Mbps

6-4. a. 5 kHz

    b. 1.41 Mbps

6-5. a. 6 kHz

    b. 1.62 Mbps

CHAPTER 7

7-1. 0.2 $\mu W/m^2$

7-3. power density is reduced by a factor of 4

7-5. 20 dB

7-7. 0.125

CHAPTER 8

No problems

CHAPTER 9

9-1. a. 22-gauge wire with 44 mH inductance every 3000 feet

b. 19-gauge wire with 88 mH inductance every 6000 feet

c. 24-gauge wire with 44 mH inductance every 3000 feet

d. 16-gauge wire with 135 mH inductance every 3000 feet

9-3. a. -3 dBmO

b. 18 dBrnc

c. 40 dBrncO

d. 47 dB

9-5. a. 51 dBrncO

b. -100 dBm

c. -1 dBm

d. 36 dB

9-7. 2300 μs

CHAPTER 10

No problems

CHAPTER 11

11-1. 112 channels per cluster, 1344 total channel capacity

11-3. 8

11-5. a. 112

b. 448

c. 1792

CHAPTER 12

No problems

CHAPTER 13

13-1. Baudot   00101

ASCII   P1001000

EBCDIC   11001000

ARQ   1010010

13-3. 0111011   1000100

13-5. 23 bits in error

13-7. 04 hex

13-9. AEA1

13-11. 10100000 binary or A0 hex

13-13. 1001010100000 or 12A0 hex

13-15. five Hamming bits

CHAPTER 14

14-1. 01010101 binary or 55 hex

14-3. 11000011 binary or C3 hex

14-5. 11101100 binary or EC hex

14-7. bit time $= 1/600 = 1.67$ ms

receive clock rate $= 19.2$ kHz

maximum detection error $= 52$ μs

14-9. 14 V and 9 V

CHAPTER 15

No problems

CHAPTER 16

16-1. SPA = G

DA = " "

16-3. SSA = 0

DA = period

16-5. 93 hex

16-7. four inserted zeros

CHAPTER 17

No problems

CHAPTER 18

18-1. **a.** 06 hex

b. 42 hex

c. BC hex

d. F4

18-3. b, multicast

18-5. b, multicast source addresses are not allowed

18-7. c, unicast address

CHAPTER 19

19-1. **a.** 53.202.55.7

b. 142.46.204.14

c. 220.174.15.93

d. 240.177.29.142

19-3. a. and c

19-5. c

19-7. i

19-9. d, e, and h

19-11. none

19-13. b, f, and i

19-15. c and g

19-17. d

19-19. a

19-21. a. 255.0.0.0

     b. 255.255.0.0

     c. 255.255.255.0

     d. 255.255.255.0

     e. 255.0.0.0

CHAPTER 20

20-1. c

20-3. 255.255.224.0

20-5. 255.255.255.240

20-7. 49,140

20-9. a. 40.144.0.0

     b. 40.144.0.1

     c. 40.151.255.254

     d. 40.151.255.255

20-11. 130.40.63.255

20-13. a. 80.0.0.0/12

     b. 168.22.0.0/18

     c. 220.10.1.0/27

20-15. a. subnet address: 204.238.7.32

     subnet mask: 255.25.255.224

     b. subnet address: 204.238.7.64

     subnet mask: 255.25.255.224

     c. subnet address: 204.238.7.96

     subnet mask: 255.25.255.224

     d. subnet address: 204.238.7.128

     subnet mask: 255.25.255.224

     e. subnet address: 204.238.7.160

     subnet mask: 255.25.255.224

     f. subnet address: 204.238.7.192

     subnet mask: 255.25.255.224

20-17. a. 255.255.255.0

     b. 254

     c. 255.255.255.224

     d. 209.100.78.32

     209.100.78.64

     209.100.78.96

     209.100.78.128

     209.100.78.160

     209.100.78.192

     e. 30

     f. 180

     g. 209.100.78.33–209.100.78.62

     209.100.787.63

     209.100.78.65–209.100.78.95

     209.100.787.96

     209.100.78.97–209.100.78.126

     209.100.787.127

209.100.78.129–209.100.78.158

209.100.787.159

209.100.78.161–209.100.78.190

209.100.787.191

209.100.78.193–209.100.78.222

209.100.787.223

    h. 0 and 224

20-19. a and e

20-21. a. 256

    b. 8

    c. 8

    d. 4

    e. 16

    f. 64

20-23. a. 7

    b. 78

CHAPTER 21

21-1. 47 hex (01000111 binary)

21-2. 4F hex (01001111 binary)

21-5. a. 14, category 1, IETF

    b. 38, category 1, IETF

    c. 15, category 2, local

    d. 27, category 2, local

    e. 21, category 3, temporary

    f. 45, category 3, temporary

21-7. a. ARCNET

    b. IP

    c. six bytes

    d. 4

    e. reply

    f. 00 00 BD 41 2A 1C

    g. 00 00 BD A3 B1 CC

    h. 192.168.160.20

    i. 192.168.160.16

21-9. a. Z

    b. X

    c. Y

    d. 1000

21-11. a. 241 fragments

    b. offset value, binary, hex, decimal

        1st = 0 0000 0000 0000, 0, 0

        2nd = 0 0000 0010 0010, 22, 34

        3rd = 0 0000 01000100, 44, 68

        4th = 0 0000 01100110, 66, 02

        240th = 1 1111 1011 1110, 1FBE, 8126

        241st = 1 1111 1110 0000, 1FE0, 8160

        last fragment contains 256 bytes

CHAPTER 22

22-1. **a.** router advertisement

b. address mask request

c. destination unreachable, port unreachable

d. time exceeded, fragments did not all arrive

e. parameter problem, required field missing

f. destination unreachable, network unreachable

g. redirect, host specific route

h. source quench

22-3. **a.** 0

b. 10

c. 9

d. 1

e. 3

f. 2

g. 4

22-5. **a.** query

b. echo reply

c. 0

d. B2 16

e. yes

22-7. **a.** 2

b. 70.16.74.1 and 70.16.74.10

c. 00 00 00 00

d. default router

e. yes

f. should never be selected as the default router

22-9. **a.** redirect for host specific route

b. 32.4.16.8

c. yes

d. immediate

e. TCP

CHAPTER 23

23-1. **a.** well known

b. registered

c. registered

d. well known

e. dynamic

f. well known

23-3. 40 bytes

23-5. **a.** ACK/FIN

b. ACK

c. ACK/PSH

d. RST

e. SYN

f. ACK/PSH/FIN

23-7. **a.** 51

b. well known

c. server

d. 53410

e. dynamic

f. client

g. DNS

h. 20 bytes

i. SYN/ACK

j. 4096

k. 04 A2 B1 23 hex

l. 00 1A 46 A2

CHAPTER 24

24-1. **a.** reserved

b. reserved

c. site local address

d. NSAP

e. reserved

f. reserved

g. multicast

h. multicast

i. site local address

24-7. **a.** no specific traffic

b. non–congestion controlled

c. interactive traffic

d. invalid

24-9. fragmentation

CHAPTER 25

25-1. 300 bytes for all packets

25-3. eng

25-7. 127.0.0.0

25-9. **a.** PQDNs

b. PQDNs

c. FQDNs

d. FQDNs

25-11. no

CHAPTER 26

26-1. No problems

CHAPTER 27

27-1. No problems

Index

Horizontal redundancy checking (HRC), 391

Host file, 808–9

Host table, 827

Howler tones, 236

Hubs, 15, 531, **531, 532,** 533–35, **534, 535**

Huygen's principle, 199

Hybrid coil, 228

Hybrid network, 228

Hybrid set, 271, 273, **273**

Hybrid topology, **29,** 30

Hypertext transfer protocol (HTTP), 886–95
  features, 886–87
  headers, **891,** 891–95, **892, 893, 894**
  transactions, **887,** 887–91, **888, 889, 890, 891**

IBM (International Business Machines), 4, 8

ICMPv6, 784–803, **794,** 794–803, 795, **795**
  ICMP checksum, 802–3
  message processing rules, 794–96

ICMPv6 checksum, 802–3

ICMPv6 header, 796, **797**
  error-reporting messages, 796–99, **797, 798, 799, 800**
  query (informational) messages, **800,** 800–802, **801, 802**

Ideal Nyquist bandwidth, 61

Idle line ones, 399

Idle time, 398

IEEE 802.9 integrated voice, video, and data services, 923–24, **924**

Ignored parity, 388

Imperfect surface finish, 125, **126**

Implied noise margin, 426

Improved Mobile Telephone Services (IMTS), 308, 310

Impulse noise, 55, **265,** 265–66

In-band signaling, 233

Incident waves, 80

Inclined orbits, 212

Index profile of optical fiber, 115, **116**

Industrial noise, 52

Information, 2, 367

Information capacity, 58–59

Information-carrying capacity, 101

Information density, 73

Information Superhighway, 9

Information theory, 59

Infrared, 102

Infrared absorption, 121

Initial voice channel designation messages, 336

Injection laser diode (ILD), 105, 127, 128

INMARSAT, 361

Institute of Electrical and Electronics Engineers (IEEE), 9–10
  Project 802, 554

Integrated services data networks, 898

Integrated services digital network (ISDN), 9, 358, 913–24
  additional channels, 921

architecture, 916, **916**

broadband, 922–23, **923**

channel interfaces, 916–18, **917, 918, 919**

conceptual view of, 914, **915**

evolution of, 914

functional groupings and system connections, 918–21, **920**

IEEE 802.9 integrated voice, video, and data services, 923–24, **924**

objectives of, 916

principles of, 913–14

Intelligent Network, 337

Intelligent terminal, 406–7

Intelligent time-division multiple, 178

Interdigit time, 235

Interdomain routing, 650
  classless, 650, 652

Interexchange carrier, 303

Interface parameters, 262–63

Interference avoidance scheme, 341

Interferences, 199, **200,** 201, **201,** 315–18, **316, 317**
  adjacent-channel, **317,** 317–18
  cochannel, **316,** 316–17

Interlinked nodes, 898

Intermediate links, 298

Intermediate trunks, 289

Intermodulation distortion, 54, **54,** 55, 261, **262**

International Morse Code, 369, **369**

International Standards Organization (ISO), 8

International Telecommunications Union-Telecommunications Sector (ITU-T), 8–9

International Telecommunications Union-Telephone Sector (ITU-T), 463

International Telecommunications Union-Telephony (ITU-T), 913

Internet, 4, 10, 31

Internet Architecture Board, 10

Internet assigned number authority port number ranges, 737–38, **739**

Internet Assigned Numbers Authority (IANA), 832–33

Internet Control Message Protocol (ICMP), **705,** 705–31
  categories, types of service, types, and codes, 707
  checksum, 727–28, **728,** 802–3
  encapsulation, 706, **706**
  message format, 706, **707**
  packet Internet groper, 728–31
  query (diagnostic) messages, 707–16, **708, 709, 710, 711, 712, 713, 715, 716**
  variation (error-reporting) messages, 717–27, **718, 719, 720, 721, 723, 724, 725, 726, 727**

Internet Corporation for Assigned Names and Numbers (ICANN), 831

Internet Engineering Task Force (IETF), 10, 594, 650

Internet group management protocol (IGMP), 801–2

Internet Protocol (IP), 590, 666–67, **667, 668,** 735–36
  shortcomings of, 705

Internet Protocol (IP) address, 622–24

Internet Protocol (IP) address classes, 602–6, **604**

Internet Protocol (IP) addressing, 596, 666
  classful, **597,** 597–99, **598**
  classless, 639–50, **640, 641, 642, 643, 644, 645, 647, 648, 649, 650, 651**
  hardware addressing versus, 668–69, **669**

Internet Protocol (IP) address notation, 596–97

Internet Protocol (IP) address summary, 601–2, **602**

Internet Protocol (IP) datagram, 669–87, **671**
  address fields, 687
  differentiated services field, 671, 675–77, **676**
  fragmentation and reassembly, 679–85, **680, 681, 682, 683**
  fragmentation field, **678,** 678–79, **679**
  header checksum field, 686–87
  ID number field, 678
  P header length field, 671
  protocol field, 686, **686**
  service type field, 673–75, **675**
  time-to-live field, 685–86
  total length field, 677
  version field, 671

Internet Protocol (IP) next generation (IPng), 596

Internet Protocol (IP) options, 687–700
  code field, 687–88
  data fields, 688
  header, 687, **688**
  length fields, 688
  types, **688,** 688–700, **689, 690, 691, 692, 693, 694, 696, 697, 698, 699, 700**

Internet Protocol switches, 545

Internet Protocol version 6 (IPv6), 770
  address allocation, **773,** 773–74, **774**
  addressing format, 771, **772,** 773–81, **774, 775, 776, 777, 778, 780, 781, 782**
  address types, 773
  advantages of, 770–71
  extensions and extension headers, 785–93, **786, 787, 788, 789, 790, 791, 792, 793**
  header, 781–85, **783, 784**
  unicast addresses, **774,** 774–80, **775, 776, 777, 778, 779**

Internet Research Task Force (IRTF), 11

Internet standards process, 595–96

Internetworking address, 657

Internetworking devices, 526, 542

Metallic transmission, 79
Metallic transmission line equivalent circuit, 92–94, **93**
Metallic transmission line losses, 96–98, **97**
Metallic transmission lines, 79
  types of, 83–91
  wave propagation on, 94–96, **95**
Metropolitan area networks (MANs), 32–33, **34,** 280
Microbending, 123
Microcells, 311, 319
Microcellular systems, 340
Microphone, 228
Microwave communications systems, **206,** 206–10, **207, 208, 209, 210**
Microwave radio communications, advantages of, 207–8
Microwave radio link, 208–9, **209**
Microwave radio relay systems, 207
Microwave radio repeaters, 210, **210**
Minimum bandwidth, 61–62
Minimum Nyquist bandwidth, 61
Minimum Nyquist frequency, 61
Minimum sampling rate, 137
Missing segments, 763, **764**
Mixed topologies, **29,** 30
Mobile-assisted handoff (MAHO), 347
Mobile identification number (MIN), 333, 336
Mobile Reported Interferences (MRI), 341
Mobile Satellite Systems (MSS), 360
Mobile station control message, 336
Mobile switching center (MSC), 360
Mobile telephone service, 308–9
Mobile Telephone Switching Office (MTSO), 322
Mobile telephone units, 325
Modal dispersion, **123,** 123–24, **124, 125**
Modem control, AT command set, 469–72, **471, 472**
Modems, 449
  asymmetric digital subscriber line, 472
  asynchronous voice-band, 454–58, **455, 456, 458**
  Bell System 103-compatible, 455–57, **456**
  Bell System 202-compatible, 457–58, **458**
  Bell System-compatible voice-band, 452–53
  broadband, 451
  cable, 451, 472
  data, 407
  data communications, 451
  56K, 468–69
  synchronous voice-band, 458–60, **459**
  telephone-loop modems, 451
  voice-band, 451, **453,** 453–54
  voice-band data communications, 451–52, **452**
Modem synchronization, 398, 460–62, **461, 462, 463**
Modulated wave, 56
Modulating signal carrier, 41
Modulation, 41, 56

Morse, Samuel F. B., 3, 368–69
Morse Code, 3, 368–69, **369**
Most-significant bit (MSB), 371
Muldem (multiplexers/demultiplexer), 167
Multicast addresses, 773, **780,** 780–81
Multicasting, 5, 604
Multifrequency (MF) tones, 233–34, **234**
Multifrequency signaling (MF), 232
Multimedia data communications, 553
Multimode, 114
Multimode graded-index optical fiber, 118, **118,** 119
Multimode step-index optical fiber, **117,** 117–18
  advantages of, 118–19
  disadvantages of, 119
Multiple access, 555
Multiple access with collision detect (CSMA/CD), 517
Multiple-bit error, 384
Multiple-byte options, 691–92, **692, 693, 694,** 694–95, **696, 697,** 697–700, **699, 700**
Multiple destination, 218
Multiple-strand cable, 106, **107**
Multiplexer, 3
Multiplexing, 159, 513
  European time-division, 176–77, **177**
  frequency-division, 181–85, **182, 183**
  statistical time-division, 178–80, **179**
  T carriers and, 158–89
  time-division, **159,** 159–61, **160**
  wavelength-division, 185–89, **187, 188**
Multipoint configuration, 20
Multipoint topology, 28
Multiport repeaters, 531–32

Name servers, 809
Name space, 829
N-AMPS, 340–41
National Association of Food Chains, 379
National Bureau of Standards, U.S., 4
National Electric Code (NEC), 90
National Science Foundation (NSF), 4
Natural sampling, 137, **137**
Negative acknowledgment, 395
Neighbor solicitation and neighbor advertisement, 801, **802**
Network, 2
  components, **22,** 22–23
  features, 22–23
  functions, 22–23
Network address hierarchy, **616,** 616–18, **617, 618**
Network address translator (NAT), 601
Network and host identification, **599,** 599–600, **600**
Network architectures, 5
Network classifications, 30–36, **31, 32**
Network interface cards (NICs), 24, **24,** 334, 807
Network interface layer, 590

Network layer, **13,** 15–16
Network layer protocols, 591
Network models, 25–28
Network operating system, 24–25, **25**
Network Switching Subsystem (NSS), 360
Network topologies, 28, **29,** 30
Network topologies and connectivity devices, 512–49
  broadcast domains, 524–25, **526**
  collision domains, 520–23, **521, 522, 524**
  connectivity devices, 525–29, **527, 528, 529, 530,** 531, **531, 532, 533,** 533–47, **534, 535, 536, 538, 540, 541, 542, 543, 544, 546, 547, 548**
  LAN topologies, **515,** 515–20, **517, 518, 519**
  standard logic symbols, 547, **549**
  transmission formats, 513–14, **514**
Network virtual terminal (NVT), 844, 845, 857
Nodes, 2, 19
Noise, 248
  background, 248
  C-message, 263–65, **264, 265**
  C-notched, 264
  common-mode, 442
  correlated, **54,** 54–55
  cross-modulation, 261
  electrical, 52–55, **53**
  fluctuation, 261
  impulse, 55, 265–66. *265*
  industrial, 52
  man-made, 52
  quantization, 139
  random, 263–64
  thermal, 52–54, 263
  white, 52
Noise interference, 83
Noise margin (NM), 426
Noise power density, 53
Nonlinear amplification, 54–55
Nonlinear crosstalk, 275
Nonlinear distortion, 260, 261
Nonlinear encoding, 145
Nonlinear mixing, 54
Nonreturn-to-zero inverted (NRZI), **506,** 506–7
Nonreturn to zero (NRZ), 170
Nonroutable protocol, 593
Nonroutine protocol, 593
Nonsynchronous satellites, 212
Nonuniform encoding, 145
No parity, 388
North American Digital Cellular (NADC), 346
North American digital multiplexing hierarchy, 167–69, **168, 169**
North American telephone numbering plan areas, 291–93, **294**
North American telephone switching hierarchy, **296,** 296–99, **298**
Numbering plan areas (NPAs), 291

Protocol data unit (PDU), 11, 562–63, **563**
Protocol stack, 6, 475
Provider-based unicast addresses, **774,** 774–76, **775**
Proxy ARP, 660–61, **661**
Proxy server, 601
Psophometric noise weighting, 254
P-type-intrinsic-n-type (PIN) diodes, 128
Public data networks (PDN), 19
Public land mobile radio, 309
Public switched data network (PSDN), 898–902, **899, 901, 902, 903**
Public switched telephone network (PSTN), 223, 280, 308
Public telephone networks (PTNs), 2, 19, 223, 278–305
Pulse, 41
Pulse code modulation (PCM), 134
    delta modulation, 156, **156**
    differential, 156
    linear versus nonlinear codes, 145–46, **146**
    line speed, 155–56
    sampling, 136–37
Pulse spreading, 123, 124
Pulse width dispersion, 124

Quadbits, 69, 465
Quadrature amplitude modulation (QAM), 62, 71, **72,** 454, 458–59
Quad shielding, 90
Quantization, 138–41, **139**
Quantization error, 139
Quantization interval or quantum, 139
Quantization noise, 139
Quantization range equal, 139
Quaternary digital signal, 41
Quaternary phase-shift keying (QPSK), 68–69, **71**
Question records, 834
Quick protocols, 22
Quintbits, 465

Radiant flux, 108
Radiation losses, 97–98, 98, 123
Radiation of laser, 130
Radio-frequency interference (RFI), 85
Radio frequency (RF) propagation, 192
Radio horizon, 203
Radiometric terms, 108
Radio transceivers, 325
Radio waves, optical properties of, 196–201, **197, 198, 200, 201**
Random access, 554–59, **556, 558**
Random access channel (RACH), 344
Random noise, 263–64
Rayleigh scattering loss, 121
Rays, **193,** 193–94
Rebinding process, 815
Rebinding timer, 823
Receive data available, 423
Receive data available flag, 416

Receive data available reset (RDAR), 416
Receive parity error (RPE), 423
Receiver, 19
Receiver diversity, 325
Receiver off-hook tones, 236
Receiver on/off hook, 236
Receiver overrun (ROR), 423
Receive sync strobe (RSS) signal, 422
Redirect message, 722–24, **724, 725**
Redundancy
    character, 385
    in error detection, 385
    message, 385
Redundancy checking, 387–93, **388**
Reeves, Alex H., 133, 134
Reflected waves, 80
Reflection, **198,** 198–99
Refraction, 109, **110,** 197, 197–98, **198**
Refractive index, 109, **110**
Regenerative repeaters, 173
Registered jacks, 226
Registration requests, 336
Registry domain sponsors, 831
Relay agent process, 821
Relay agent software, 815
Remington Rand, 4
Remote terminal protocol, 844
Repeaters, 136, 526–28
    microwave radio, 210, **210**
    regenerative, 173
Repetition rate, 47
Request for comments, 593–96
Request for service, 236
Reserved field, 750–51, **751**
Reserved IP addresses, **600,** 600–601, **601**
Residue checksum, 390
Resolution, 139
Resource records, 834
Retransmission in error correction, 394–95
Return to zero (RZ), 170
Reverse address resolution protocol (RARP), 657, 666, **666**
Reverse control channel (RCC), 326, 335
Reverse digital traffic channel (RDTC), 347
Reverse links, 331–32
Reverse voice channel, 326
Right-hand characters, 380
Ring-back signals, 6, 236
Ringdown system, 284
Ringer circuit, 227
Ringing, 6
Ringing signal, 235
Ring topology, **29,** 30, 518–20, **519**
RJ-11 connector, 226, **226**
Rn, 252
Roaming, 322–24
Robbed-digit framing, 180
Root server, 834
Rotary dial pulsing, 234
Routable protocol, 592–93
Route, 296

Router discovery messages, 714–15, **715, 716**
Router interface, 544
Routers, 543–44, **544,** 618
Router solicitation and advertisement messages, 800–801, **801**
Routing Information Protocol (RIP), 593
Routing Internet Protocol (RIPv2), 603
RS-232 asynchronous data transmission, 434–35, **436, 437**
RS-232 electrical equivalent circuit, 428, **429**
RS-232 functional description, 428, **429, 430,** 430–34
RS-232 serial interface standards, **425,** 425–27, **426, 427**
RS-449 serial interface standards, 439, **441,** 441–42, **443,** 444, **444**
RS-530 serial interface standards, 444–45, **445**
RS-232 signals, 434, **435**

Sample-and-hold circuit, 136, 137
Sampling error, 418
Sampling rate, 137–38
Satellite antenna radiation patterns, 217–18, **219**
Satellite communications systems, 210–20, **211, 213, 214, 215, 217, 218, 219, 220**
Satellite elevation categories, 211–12
Satellite look angles, 216–17, **217**
Satellite multiple-accessing arrangements, 218, **220**
Satellite orbits and orbital patterns, 212–14, **213, 214**
Satellites, geosynchronous, 214–16
Screen mode, 339
Secondary control channels, 343
Secondary data, 430
Second-generation cellular telephone systems, 340
Second harmonic, 47
Sectoring, 319–20, **320**
Segmentation, 320–21
Segments received out of order, 763–64
Seizure, 236
Selective repeat, 395–98, **396, 397**
Selective retransmission, 394
Semantics, 7
Semiconductor lasers, 130
Sender window size field, 751–52
Sendmail, 877
Sequential (serial) comparisons, 387
Serial by character, 20, **20**
Serial data
    asynchronous, 398–99, **399, 400**
    synchronous, 401–2
Serial interfaces, 424–36, **425, 426, 427, 428, 429, 430, 435, 436, 437, 438,** 438–39, **439, 440, 441,** 441–42, **443, 444,** 444–45, **445,** 446

Wheatstone, Charles, 3
White light, 109
White noise, 52, 264
Wide-area networks (WANs), 31, 33–34, **35,** 280
Wide Area Telephone Service (WATS), 301
Wireless communications systems, 191–220
  electromagnetic polarization, 192–93
  electromagnetic radiation, 194
  inverse square law, 195
  microwave, **206,** 206–10, **207, 208, 209, 210**
  optical properties of radio waves, 196–201, **197, 200, 201**
  rays and wavefronts, **193,** 193–94
  satellite, 210–20, **211, 213, 214, 215, 217, 218, 219, 220**
  skip distance, 204–5, **205**
  spherical wavefront, **194,** 194–95
  terrestrial propagation of electromagnetic waves, 201–4, **202**
  wave attenuation and absorption, 195–96
Workstation, 407
World Wide Web (WWW), 4, 5

X.25 user-to-network interface protocol, 903–13, **904**
  layers, 903–9, **905, 906, 907, 908**
  packet formats, 909–13, **910, 911, 912, 913**
XDSL, 924–25
XMODEM, **482,** 482–83

YMODEM, 483

Zero-bit insertion, 505
Zero suppression, 379
ZIP code, 382
ZMODEM, 483
Zonal beans, 218, **219**
Zuis, Konrad, 3